AS YOU EXAMINE HOLT
ALGEBRA 2
WITH TRIGONOMETRY

Please Notice   HOLT ALGEBRA 2 WITH TRIGONOMETRY provides a complete, balanced course designed to be mastered by all second-year algebra students. All of the basic concepts usually studied in a second-year algebra course, along with the fundamental concepts of trigonometry, are found in HOLT ALGEBRA 2 WITH TRIGONOMETRY.

Format   HOLT ALGEBRA 2 WITH TRIGONOMETRY's proven four-part lesson format facilitates student learning:
- Lesson objectives appear on the pupil page
- Worked-out examples demonstrate solution process
- Hints and suggestions on the pupil page make learning difficult concepts easier
- Graded exercises provide for continuous skill development

Contents

1. The course begins with a review of fundametal algebraic concepts of linear equations and inequalities (see pp. 1—2, 4—5).

2. Worked-out examples focus on teaching the solution process without complicated explanations (see pp. 115—116).

3. Suggestions, hints, and comments encourage students to attempt more difficult skills and concepts (see pp. 248, 266).

4. Oral exercises appear before written exercise sets so that the teacher can quickly test understanding and readiness (see pp. 252).

5. Exercises are plentiful and are graded according to three levels of difficulty. Also, exercises within each level are graded from easiest to most challenging. Examples presented in the lesson match every type of exercise in levels Ⓐ and Ⓑ (see pp. 118, 127, 287).

6. Reading in Algebra helps students analyze and better understand the technical language of algebra.

7. Each chapter ends with a Chapter Review and Chapter Test. The Review is diagnostic and keys items to lesson sections in which skills and concepts necessary to solve the item are taught.

8. Special topics pages vary the pace and enrich the course. They include practical applications of algebra in the real world, as well as an introduction to careers that use mathematics (see pp. 83, 138, 181, 275).

9. Calculator and computer activities familiarize students with accepted methods of logical thought processes (see pp. 133, 184).

10. Answers to selected exercises are provided in the back of the pupil edition (see pp. 601—622). Worked-out solutions to every problem in the text are available in the separate Solutions Manual.

11. A teacher's edition with a commentary for each lesson, a complete set of answers, and daily assignments with three levels of difficulty designed to reinforce the current lesson while maintaining the previous two lessons makes teaching HOLT ALGEBRA 2 WITH TRIGONOMETRY easy.

Supplementary Materials on Duplicating Masters

- **HOLT ALGEBRA 2 Skillmasters** provide two parallel sections of additional exercises for every two lessons and cumulative reviews.
- **HOLT ALGEBRA 2 Testmasters** provide an alternate form of the Chapter Tests and Cumulative and Final Tests.
- **HOLT ALGEBRA 2 Teacher's Resource Package** contains Enrichment masters, Testmasters, and Skillmasters in blackline form.

HOLT ALGEBRA 2 WITH TRIGONOMETRY

Teacher's Edition

Eugene D. Nichols
Mervine L. Edwards
E. Henry Garland
Sylvia A. Hoffman
Albert Mamary
William F. Palmer

HOLT, RINEHART AND WINSTON, PUBLISHERS
New York · Toronto · Mexico City · London · Sydney · Tokyo

Staff Credits

Editorial Development	Everett T. Draper, Earl D. Paisley, Antoine Y. Maksoud
Product Manager	Daniel M. Loch
Field Advisory Board	Douglas A. Nash, Sam Sherwood, Robert Wolff, Wendell Anthony, Gary Crump, Jack M. Custer, Roy Eliason, Kenneth C. Scupp, Dennis Spurgeon, Jeffra Ann Nicholson
Marketing Research	Erica S. Felman, Linda A. DeLora
Editorial Processing	Margaret M. Byrne, Holly L. Massey
Art and Design	Carol Steinberg, Mary Ciuffitelli
Production	Joan McNeil, Heidi Henney
Photo Resources	Linda Sykes, Rita Longabucco

Copyright © 1986 by Holt, Rinehart and Winston, Publishers
All rights reserved
Printed in the United States of America

ISBN: 0-03-002174-X

5678901234   032   98765432

BENEFITS

HOLT ALGEBRA 2 is designed to help students obtain the skills necessary for success in subsequent mathematics courses. The book gives students a strong foundation in algebra, including an introduction to probability theory and trigonometry.

FOUR-PART LESSON FORMAT	A consistent four-part lesson format enhances the gradual development of skills and concepts.
PROBLEM-SOLVING STRAND	A clear, logical problem-solving strategy, special problem-solving application lessons, and non-routine problems provide a meaningful approach to problem-solving.
ENRICHMENT	Calculator Activities and special topics such as Statistics provide a wide selection of experiences for students, depending on each student's readiness and learning aptitude.
COMPUTER STRAND	Every chapter contains a Computer Application providing hands-on exercises in BASIC. References are made to further computer applications of algebraic concepts.
TESTING AND REVIEW	Chapter Reviews, Chapter Tests, and Cumulative Reviews geared to the material taught in each chapter keep students and teachers aware of progress.
TEACHER'S ANNOTATED EDITION	Teacher's Edition includes full-size annotated pupil's pages, chapter objectives, Cumulative Reviews, Activities, additional chalkboard work, and approaches to common student errors, as well as detailed teaching suggestions.
MOTIVATIONAL FEATURES	College Prep Tests acquaint students with College Board test-taking strategies. In addition to an expanded use of calculators and computers, historical puzzles and information regarding related careers also furnish motivation. Special photographic sections pique students' interest and accentuate the relationship of algebra to our technological world.
TEACHER'S RESOURCE PACKAGE	Contains Blackline Masters, for Enrichment, Geometry Reviews, Testmasters, Skillmasters, and Competency Tests.
SUPPLEMENTARY MATERIALS	Skillmasters (duplicating masters) for review and Extra Practice. Each Skillmaster, with two parallel forms (A and B) can be used as a pre-test or a post-test, as a diagnostic tool, as extra practice for skill mastery, or as a regular assignment.
	Testmasters (duplicating masters) for review, Extra Practice, diagnosing areas of difficulty, and supplementary testing, as well as a complete testing program containing chapter tests, cumulative tests, and final tests.

FOUR-PART LESSON FORMAT

HOLT ALGEBRA 2 uses a carefully structured four-part lesson format to actively involve students in the learning process.

Objectives tell the students what they will learn in each lesson.

Hints, Comments, and Suggestions guide the student through the solution process.

Examples develop the mathematical concepts being taught in a series of carefully structured steps.

LINEAR EQUATIONS 1.1

Objective **To solve a linear equation in one variable**

Each of the equations below is a **linear equation** in one variable.

$6x + 32 - 10x = 7 + 8x - 39 \quad \frac{3}{5}y - 3 = 8 - \frac{1}{5}y \quad 3.8n + 2.3 = 7.58 - 0.6n$

You can solve a linear equation in one variable by
(1) combining the like terms, if any, on each side of the equation, and
(2) using one or more of the following equation properties.

Equation properties

For all numbers a, b, and c,
- Addition: if $a = b$, then $a + c = b + c$.
- Subtraction: if $a = b$, then $a - c = b - c$.
- Multiplication: if $a = b$, then $a \cdot c = b \cdot c$.
- Division: if $a = b$, then $\frac{a}{c} = \frac{b}{c}$; $c \neq 0$.

Example 1 Find the root of $\frac{2}{5}n + 3 = 7 - \frac{1}{5}n$.

$\frac{2}{5}n + 3 = 7 - \frac{1}{5}n$
$\frac{3}{5}n + 3 = 7$ ◂ Add $\frac{1}{5}n$ to each side.
$\frac{3}{5}n = 4$ ◂ Subtract 3 from each side.
$\frac{5}{3} \cdot \frac{3}{5}n = \frac{5}{3} \cdot 4$ ◂ Multiply each side by $\frac{5}{3}$,
$1n = 6\frac{2}{3}$ the reciprocal of $\frac{3}{5}$.

Thus, the root is $6\frac{2}{3}$.

Sometimes the variable appears on the right side of the equation but not on the left side. In that case, it may be more convenient to get the variable alone on the right side of the equation as shown below.

Example 2 Solve $78 = 22 - 8t$. Check the solution by substitution.

$78 = 22 - 8t$ Check: $\begin{array}{c|c} 78 & 22 - 8t \\ 78 & 22 - 8(-7) \\ & 22 + 56 \\ & 78 \end{array}$
$56 = -8t$
$\frac{56}{-8} = \frac{-8t}{-8}$
$-7 = t$ True: $78 = 78$

Thus, the solution is -7.

Linear Equations   1

T-4

Oral Exercises

Simplify each expression.
1. $3(5x - 8)$ **15x − 24**
2. $-4(3y + 11)$ **−12y − 44**
3. $-(5n - 7)$ **−5n + 7**
4. $-(8 + 6t)$ **−8 − 6t**
5. $\frac{1}{2}(6a + 20)$ **3a + 10**
6. $\frac{1}{3}(12x - 27)$ **4x − 9**
7. $0.4(0.3t - 11)$ **0.12t − 4.4**
8. $1.32(10y + 0.1)$ **13.2y + 0.132**

Written Exercises

Solve each equation.

Ⓐ
1. $10 + \frac{2}{3}t = 6$ **−6**
2. $7 = \frac{3}{4}n - 24$ **$41\frac{1}{3}$**
3. $-27 = 12w + 27$ **−4.5**
4. $126 - a = -9a$ **−15.75**
5. $3(2n + 4) + 5(3 - 2n) = 7 - 2n$ **10**
6. $2(7x - 4) - 4(2x - 6) = 3x + 31$ **5**
7. $5y - (2 - 3y) = 54$ **7**
8. $6(3 + 2c) - (7c - 2) = 3c - 4$ **−12**
9. $\frac{1}{4}(12x - 24) - 4x = \frac{1}{3}(18 + 15x) + 24$ **−6**
10. $\frac{1}{5}(30n + 25) - \frac{1}{8}(32n + 16) = 14n - 33$ **3**
11. $0.4t + 3(1.2t + 0.9) = 21.1$ **4.6**
12. $0.7(5a - 1.2) = 2a - 0.39$ **0.3**
13. $6(3.2y - 2.3) = 7.2y - 13.32$ **0.04**
14. $2.6(4x + 2.3) = 8.2x + 49.98$ **20**

Ⓑ
15. $\frac{2}{7}x + 18 = 8 - \frac{3}{7}x$ **−14**
16. $\frac{4}{5}y - 22 = \frac{1}{5}y + 14$ **60**
17. $\frac{2}{9}n - 31 = 16 + \frac{7}{9}n$ **−84.6**
18. $18 - \frac{2}{11}a = \frac{5}{11}a - 19$ **$58\frac{1}{7}$**
19. $3.2n - 0.05 = n + 0.06$ **0.05**
20. $x - 0.01 = 12.2 - 2.3x$ **3.7**
21. $x - (9x - 5) = -(3x + 7)$ **2.4**
22. $-4(2y - 5) - (7 - 3y) = 4(y + 3)$ **$\frac{1}{9}$**
23. $\frac{3}{5}(15n + 20) + 3n = \frac{2}{7}(14n - 28) - 12$ **−4**
24. $21 + \frac{5}{6}(30n + 18) = 19x - \frac{2}{5}(35x - 15)$ **−1.5**
25. $2.5(1.2x + 6) = 1.92 + 8(4.3x + 2.42)$ **−0.2**
26. $7(3.2n + 9.4) = 4.4(3.5n + 4.5) - 0.2$ **−6.6**

Ⓒ
27. $5(2n + 4) - 3(4n - 6) = 2(20 - n)$ **no solution**
28. $4(x - 3) - (x + 6) = 3(x - 6)$ **all numbers**
29. $5[3(5y - 4) - 2(7y + 6)] + 3y - 8 = 0$ **16**
30. $10 - 4[6(2a + 5) - (20 + 10a)] - 2a = 0$ **−3**

NON-ROUTINE PROBLEMS

1. Solve the equation $x - y + z - w = 6$ for integers x, y, z, and w with the following ratios: $x:y = 3:2$, $y:z = 2:5$, and $x:w = 5:8$. **x:15, y:10, z:25, w:24**
2. Mrs. Jones invested $12,000 in three different parts with the ratio $1:2:3$. The first part was invested at 11% and the second part at 9%. At what rate did she invest the third part if the yearly income was $1,000? **7%**

Linear Equations 3

A variety of features guide students through activities designed to stimulate thinking.

Reading in Algebra helps students develop the skills needed to read, interpret, and use the language of algebra effectively.

Chapter Openers provide opportunities to practice problem formulation, estimation, and application. Related careers and historical puzzles also furnish motivation.

Reading in Algebra

Find the solution set.
1. $x < -2$ or $x < 3$ $\{x | x < 3\}$
2. $x < -2$ and $x < 3$ $\{x | x < -2\}$
3. $x < -2$ or $x > 3$ $\{x | x < -2 \text{ or } x > 3\}$
4. $x < -2$ and $x > 3$ $\emptyset$
5. $x > -2$ or $x > 3$ $\{x | x > -2\}$
6. $x > -2$ and $x > 3$ $\{x | x > 3\}$
7. $x > -2$ or $x < 3$ (all numbers)
8. $x > -2$ and $x < 3$ $\{x | -2 < x < 3\}$
9. $x \leq -2$ or $x > -2$ (all numbers)
10. $-6 < 3x < 12$ $\{x | -2 < x < 4\}$
11. $-3 < x - 1 < 5$ $\{x | -2 < x < 6\}$
12. $-6 < x + 3 < 10$ $\{x | -9 < x < 7\}$

3 FACTORING AND SPECIAL PRODUCTS

Formulating a Problem-Solving Situation

Exploration: Special-Interest Sections — On Land, In Space, and Under Land and Sea — spark the imagination and stress the interrelation of progress and mathematics.

ENRICHMENT MATERIALS

HOLT ALGEBRA 2 provides many resources for enriching the curriculum, depending on the students' readiness and learning aptitude.

Optional special topics, such as Statistics, provide a broadened view of algebraic applications.

Many application lessons feature proof and problem solving to develop reasoning and to challenge the student.

Calculator exercises expand students' practical proficiency in algebraic computation and estimation.

T-7

COMPUTER STRAND

Every chapter has a computer lesson that includes computer applications of algebraic content. Hands-on programming exercises in BASIC encourage the student with computer expertise to extend the algebra concepts further.

Computer Section

WRITING AND RUNNING A PROGRAM

OBJECTIVE: To test solutions of linear equations

For you to communicate with another human being, both of you must understand the same language. Similarly, to communicate with a computer, both you and the computer must understand the same language. Computers are designed to understand a variety of programming languages. In this text, we will use a language used by many microcomputers—BASIC (Beginner's All-purpose Symbolic Instruction Code).

The computer understands BASIC because the BASIC language has been programmed into its circuitry. In this section of the book, you will learn to use BASIC in programming your computer to solve various mathematical problems.

COMPUTER ACTIVITIES

Using String Variables

OBJECTIVE: To check factors of quadratic polynomials by using character strings

The letters A, B, C, etc., are *variables* used to represent numbers. Letters can be used to represent words or algebraic expressions by attaching a dollar sign to them: A$, B$, C$, etc. They are then called *string variables*. Consider the assignment statements: LET A$ = "DOG" and LET B$ = "2X + 3". "DOG" and "2X + 3" are called *character strings*. Character strings may be linked together by means of the + sign. PRINT A$ + B$ for A$ and B$ above would make the computer print DOG2X + 3.

Enter and run this program.

```
110  A = INT ( RND (1) * 12) + 1
120  B = INT ( RND (1) * 12) + 1
130  A$ = "(X+" + STR$ (A) + ") (X+" + STR$ (B) + ")"
140  B$ = "(X+" + STR$ (B) + ")(X+" + STR$ (A) + ")"
150  PRINT "FACTOR: ";
160  PRINT "X^2 + "A + B"X +"A * B
170  INPUT C$
180  IF C$ = A$ THEN 220
190  IF C$ = B$ THEN 220
200  PRINT "THOSE FACTORS ARE NOT CORRECT."
210  GOTO 150
220  PRINT "CORRECT"
230  GOTO 110
```

See the Computer Section beginning on page 574 for more information.

Lines 110 and 120 ask the computer to randomly pick a number from 1 to 12. The STR$ function in lines 130 and 140 converts the numbers represented by A and B into character strings, which are then incorporated into the larger strings assigned to A$ and B$.

Notice the way this program determines a polynomial's factored solutions in lines 110–140 before forming the polynomial itself in line 160.

EXERCISES

1. Make the changes necessary in the program above so that the factored solutions take the form (X + A) (X − B).
2. Write a program involving factoring out a common constant factor so that the results take the form A(X + B).

For Ex. 1–2, see TE Answer Section.

86   Chapter Three

The back of the book contains a 14-page Computer Section designed to give students experiences in programming that relate to exploring ideas in algebra. Looping commands, as well as special computer functions, are employed.

PROBLEM-SOLVING STRAND

Problem-Solving Applications in real-world careers and in mathematics usage offer abundant motivation to practice problem-solving strategies.

APPLICATIONS

A certain investment plan requires regular payments of $30 a month with an expected earnings rate of 2% per month.
The expected value in the plan at the end of the 1st month is the payment plus the interest the payment earned for the month.

$$30 + 30(0.02) = 30(1 + 0.02)$$
or $30(1.02)$

The expected value in the plan at the end of the 2nd month is the payment and the 1st month's value, plus the 2nd month's interest on both.

$$[30 + 30(1.02)] + [30 + 30(1.02)](0.02)$$
$$= [30 + 30(1.02)](1 + 0.02)$$
or $30(1.02) + 30(1.02)^2$

...the end of the 3rd month.
...+ 3rd month's earnings on both

$$[30 + 30(1.02) + 30(1.02)^2](0.02)$$

... Let $x = 1 + 0.01r$.

...ed monthly to the investment plan
...d of n months.

$\cdots + px^{n-1} + px^n$

...ents of $40 a month. The plan has
...h.
...ths if she doubles the payments
...05.66
...plan after 6 months if she makes
...but only receives half the monthly

83

PROBLEM SOLVING: MIXTURE PROBLEMS 2.5

Objective To solve a word problem involving a collection or mixture of items

To solve certain problems that involve a *collection* or *mixture* of items, it is helpful to convert the values of the items to a common unit. For example, a collection of x nickels, $4x$ dimes, and $x + 6$ quarters is worth

$$5 \cdot x + 10 \cdot 4x + 25(x + 6) \text{ cents.}$$

You will use this technique in solving the problem below.

Example 1

READ ▶ The number of dimes in Martha's coin collection is 4 less than the number of nickels and the number of quarters is 3 times the number of dimes. If the value of her collection is $7.40, how many nickels, dimes, and quarters does she have?

PLAN ▶
Represent the data.

Let n = number of nickels.
$n - 4$ = number of dimes
$3(n - 4)$ = number of quarters

Write an equation.

$$\begin{pmatrix}\text{Value of}\\\text{nickels in ¢}\end{pmatrix} + \begin{pmatrix}\text{Value of}\\\text{dimes in ¢}\end{pmatrix} + \begin{pmatrix}\text{Value of}\\\text{quarters in ¢}\end{pmatrix} = \begin{pmatrix}\text{Value of}\\\text{collection in ¢}\end{pmatrix}$$

$$5n + 10(n - 4) + 25 \cdot 3(n - 4) = 100(7.40)$$

SOLVE ▶
Solve the equation.

$$5n + 10n - 40 + 75n - 300 = 740$$
$$90n - 340 = 740$$
$$90n = 1,080$$
$$n = 12$$

$n = 12, n - 4 = 8, 3(n - 4) = 24$

INTERPRET ▶
Check the solution.

12 nickels + 8 dimes + 24 quarters	$7.40
$12 \cdot 5¢ + 8 \cdot 10¢ + 24 \cdot 25¢$	$7.40 \cdot 100¢$
$60¢ + 80¢ + 600¢$	$740¢$
$740¢$	

Answer the question.

Thus, there are 12 nickels, 8 dimes, and 24 quarters.

Example 2 on page 44 is an example of a *dry mixture* problem. In such situations, a mixture is obtained by combining two types of items. The intent is to gain the same amount of money for the mixture as would be realized if the items to be mixed were sold individually. In other words, (Value of 1st Item) + (Value of 2nd Item) = (Value of Mixture).

Problem Solving: Mixture Problems   43

A four-part problem-solving strategy, continually highlighted, encourages students to approach problem solving in a logical manner.

REINFORCING SKILLS

Summary provides a quick review of the key concepts and procedures taught in the lesson.

Non-Routine Problems provide students with the opportunity to solve unusual problems using non-standard aproaches or strategies.

Summary

To solve a quadratic equation by factoring:
- **Step 1** Write the equation in standard form.
- **Step 2** Factor.
- **Step 3** Set each factor equal to 0.
- **Step 4** Solve the linear equations.

Sometimes it is more convenient to use the standard form, $0 = ax^2 + bx + c$, so that the coefficient of x^2 is a positive number.

Solve $25 = 9n^2$.

$$0 = 9n^2 - 25$$
$$0 = (3n + 5)(3n - 5)$$
$$3n + 5 = 0 \quad \text{or} \quad 3n - 5 = 0$$
$$3n = -5 \qquad\qquad 3n = 5$$
$$n = -\frac{5}{3} \qquad\qquad n = \frac{5}{3}$$

Thus, the solutions are $-\frac{5}{3}$ and $\frac{5}{3}$.

extended to any number of factors. In the to solve the *fourth degree equation* ree polynomial is factored completely, set equal to 0.

C 23. A rectangular lawn measures 26 m by 36 m and is surrounded by a uniform sidewalk. The outer edge of the sidewalk is a rectangle with an area of 1,200 m². Find the width of the sidewalk. **2 m**

24. A rectangular picture measures 32 cm by 50 cm. It is surrounded by a uniform frame whose outer edge is a rectangle with an area of 2,320 cm². Find the width of the frame. **4 cm**

25. A rectangular piece of cardboard is 5 cm longer than it is wide. A 3-centimeter square (3 cm by 3 cm) is cut out of each corner. Then the four flaps are turned up to form an open box with a volume of 450 cm³. Find the length and the width of the original piece of cardboard. **16 cm, 21 cm**

26. A 4-decimeter square is cut from each corner of a rectangular piece of plastic whose length is 2 dm less than twice its width. Then the four flaps are folded to form an open carton with a volume of 1,040 dm³. Fin

CALCULATOR ACTIVITY

Find the area of each rectangle described below, in terms of the giv
1. 296 in. by 225 in., in square feet 2. 17.5 ft by 38.75
   462.5 ft²
3. 1.24 m by 0.35 m, in square centimeters 4. 4,225 cm by 2,
   4,340 cm²
5. 222 ft by 363 ft, in square yards 6. 497 yd by 268 y
   8,954 yd²

The lengths of the three sides of a triangle are given. Determine wh a right triangle by using the Pythagorean relation.
7. 98 cm, 336 cm, 350 cm **yes**   8. 0.75 yd, 0.4 yd
9. 110 in., 226 in., 286 in. **no**   10. 0.65 m, 4.2 m,

NON-ROUTINE PROBLEM

Each edge of a cube is 3 in. long. The cube is painted on each of its six faces, and then it is cut up to form 27 one-inch cubes.
1. What is the least number of straight cuts needed to form the 27 one-inch cubes? **6**

Find the number of one-inch cubes that are painted on the given number of faces.
2. three faces **8**   3. exactly two faces **12**
4. exactly one face **6**   5. zero faces **1**

82

COLLEGE PREP TEST

DIRECTIONS: Choose the *one* best answer to each question or problem.

1. Which investment earns the most simple interest? **E**
   (A) $4,444 at $8\frac{1}{2}$% per year for 2yr 3mo
   (B) $4,444 at $4\frac{1}{4}$% per year for 4 yr 6 mo
   (C) $2,222 at $8\frac{1}{2}$% per year for 4 yr 6 mo
   (D) $1,111 at $8\frac{1}{2}$% per year for 9 yr
   (E) They all earn the same amount of interest.

2. If 2 more than $\frac{2}{3}$ of a certain number is 8, then the number is **C**
   (A) 6   (B) 8   (C) 9
   (D) 15   (E) 18

3. In the equation $4x + 6x = 3(8 + tx)$, for what value of t is x equal to 6? **A**
   (A) 2   (B) 4   (C) 6
   (D) 8   (E) 10

4. If $ax < b$, where a, x, and b are integers, then **E**
   (A) $x < \frac{b}{a}$ (B) $x < b - a$
   (C) $x > \frac{b}{a}$ (D) $x > b - a$
   (E) None of these

5. Which fraction is greater than $\frac{1}{5}$ but less than $\frac{1}{4}$? **B**
   (A) $\frac{9}{20}$ (B) $\frac{2}{9}$ (C) $\frac{1}{9}$ (D) $\frac{2}{7}$ (E) $\frac{1}{6}$

6. The figure above contains two rectangular regions and has the dimensions shown. The area of the figure in square units is **D**
   (A) $2xyz$ (B) $xy - yz - 2x + 2z$
   (C) $2x + yz$ (D) $2x + yz - 2z$
   (E) None of these

7. If $0 < x < 1$, then which one of the following is false? **C**
   (A) $x^2 < x$ (B) $2x < 2$
   (C) $\frac{2}{x} < 2$ (D) $\frac{x}{2} < \frac{1}{2}$
   (E) None of these

8. If $5(3x - 4) - 6x = 4(2x + 6) + x$, then **D**
   (A) $x = -20$
   (B) $x = 24$
   (C) $x = -20$ or $x = 24$
   (D) There is no solution to the equation.
   (E) Every number is a solution to the equation.

9. The hypotenuse of a certain right triangle is 5 cm long and one leg is 3 cm long. Find the area of the triangle. **D**
   (A) 3 cm²   (B) 4 cm²   (C) 5 cm²
   (D) 6 cm²   (E) 12 cm²

College Prep Test   29

College Prep Tests give students practice and experience in strategies they will use in taking multiple-choice tests that they will encounter in the future.

T-10

TESTING AND REVIEW

Tests and Reviews assist the teacher in measuring the students' understanding of the major skills and concepts taught.

Cumulative Reviews test the students' understanding of algebraic concepts and corresponding problem-solving skills that have been previously taught.

Chapter Tests are designed to measure the students' mastery of the chapter objectives. The difficulty level is indicated.

Chapter Reviews contain exercise sets that are lesson-coded for easy reference to corresponding lessons when review is needed. Exercises are also coded to difficulty level: A, B, or C.

CHAPTER ONE TEST

Solve each equation.
1. $8 + 5c - 6 = 9c - 22 - 12c$ **−3**
2. $6 + \frac{3}{5}n = 14 - \frac{1}{5}n$ **10**
3. $3 + 5(2y - 1) = 2y - 4(y + 1)$ **−$\frac{1}{6}$**
4. $2.5(6y - 1.2) = 22.8 - 2.2y$ **1.5**

17. Sides a and b of a triangle have the same length. Side c is 12 cm less than 3 times the length of b. Find the length of each side if the perimeter is 23 cm.
    a: 7 cm, b: 7 cm, c: 9 cm
18. Find p in dollars if $A = \$5,620$, $r = 9\%$, Use the formula

R = 4.8 ohms and the formula
12 ohms

It 10 years before ears after building C/ In ined ages of buildings A es that of building B. at age of each building?
A: 15, B: 5, C: 20

CHAPTER ONE REVIEW

Vocabulary
absolute value [1.4]
conjunction [1.3]
consecutive integers [1.5]
disjunction [1.3]
linear equation [1.1]
linear inequality [1.2]
perimeter [1.6]

Solve each problem.
21. Twice the sum of a number and 3 is the same as 24 less than 7 times the number. What is the number? [1.5] **6**
22. The sum of three consecutive even integers is 20 more than twice the second integer. Find the three integers. [1.5] **18, 20, 22**

Solve each equation.
1. $5a + 8 - 3a = 4a - 5 - 6a$ **−3.25** [1.
2. $12 + \frac{3}{5}x = \frac{2}{5}x - 3$ **−75**
3. $7.2c + 2.3 = 1.4c - 0.6$ **−0.5**
4. $0.05 + 2.7(3n - 0.3) = 4.3n$ **0.2**
5. $-2x - (4 - 5x) = x - 3$ **$\frac{1}{3}$**
6. $4c - 2(4c - 3) = 3(2c - 5)$ **$2\frac{1}{10}$**
7. $\frac{2}{3}(6x - 9) + \frac{3}{4}(8x + 4) = 28 - 4x$ **$2\frac{3}{14}$**
★ 8. $12 - 7[6(2x + 5) - 3(7x + 8)] = 0$ **$\frac{10}{21}$**
9. $|3x - 15| = 6$ **3, 7** [1.4]
10. $|4y + 3| = 9$ **−3, $1\frac{1}{2}$**
★11. $|n + 4| + |3 - 2n| = 16$ **−$5\frac{2}{3}$, 5**
13. $\{x | x < -2$ or x

Find the solution set of each inequality. Gra the solutions on a number line.
12. $-5n - 3 \leq 17$ $\{n | n \geq -4\}$ [1.2]
13. $3x < x - 4$ or $5x - 3 > 7$ [1.3]
14. $|y + 4| < 6$ $\{y | -10 < y < 2\}$ [1.4]
15. $|5z - 15| \geq 30$ $\{z | z \leq -3$ or $z \geq 9\}$
★16. $|x - 4| < 6$ or $|x - 4| > 12$
    $\{x | x < -8$ or $-2 < x < 10$ or x

Find the solution set of each inequality.
17. $2(3a + 1) > 3(4a - 5) - 2a$ $\{a | a < 4\frac{1}{4}$
★18. $\frac{4(y - 6)}{-3} > 12 - 2y$ $\{y | y > 6\}$
19. $-2x < 6$ and $3x > -15$ $\{x | x > -3\}$ [1.

Represent each of the three numbers in term one variable. [1.5]
20. A second number is 8 times a first num A third number is 5 more than 3 times second number. **1st: x, 2nd: $8x$, 3rd: 24**

Chapter One Review

CUMULATIVE REVIEW
(Chapters 1–4)

DIRECTIONS: Choose the *one* best answer to each question or problem. (Exercises 1–11)

1. Solve $\frac{5}{9}x - 17 = 25 - \frac{2}{9}x$. **C**
   (A) $10\frac{2}{7}$ (B) 24
   (C) 54 (D) 126

2. Solve $|7 - 4x| = 21$. **D**
   (A) 7 (B) −7
   (C) −7, $3\frac{1}{2}$ (D) 7, −$3\frac{1}{2}$

3. Find two numbers such that the first number is 9 more than 4 times the second number, and 4 more than the second number is twice the first number. **B**
   (A) 7, −$\frac{1}{2}$ (B) 1, −2
   (C) −1, 2 (D) −7, $\frac{1}{2}$

4. Solve $5^{3x-2} = 625$. **A**
   (A) 2 (B) 4 (C) 125 (D) 209

5. Simplify $-3xy^3(-2x^2yz^3)^4$. **D**
   (A) $6x^9y^7z^{12}$ (B) $16x^8y^4z^{12}$
   (C) $-48x^7y^7z^7$ (D) $-48x^9y^7z^{12}$

6. Some powdered milk worth $3.50/kg and some cocoa worth $4.25/kg form a 10-kg mixture that is worth $39.20. How much cocoa is in the mixture? **C**
   (A) 2.5 kg (B) 4.4 kg
   (C) 5.6 kg (D) 5.8 kg

7. Factor $-4c^3 + 20c^2 + 24c$ completely. **B**
   (A) $-4c(c - 3)(c - 2)$
   (B) $-4c(c - 6)(c + 1)$
   (C) $-c(c - 6)(4c + 4)$
   (D) $4(6 - c)(c + 1)$

8. Solve $7x^2 - 8 = 8 - 2x^2$. **D**
   (A) 0 (B) $\frac{4}{3}$
   (C) $-\frac{3}{4}, \frac{3}{4}$ (D) $-\frac{4}{3}, \frac{4}{3}$

9. Choose the graph of the solutions of $x^3 - 9x \leq 0$. **B**
   (A) ⟵●———●———●⟶
       −3   0   3
   (B) ⟵●———●———●⟶
       −3   0   3
   (C) ⟵●———●———●⟶
       −3   0   3
   (D) ⟵●———●———●⟶
       −3   0   3

10. If $x = 7$ and $y = -2$, find the value of $\frac{5x + 4y}{x^2 - y^2}$. **C**
    (A) Does not exist
    (B) $\frac{43}{53}$ (C) $\frac{3}{5}$ (D) 0

11. Simplify $\frac{5a - 10}{6a + 12} \cdot \frac{2a^2 - 8}{a^2 - 2a - 8}$. **C**
    (A) $\frac{5a - 20}{12a + 24}$ (B) $\frac{12a + 24}{5a - 20}$
    (C) $\frac{5(a - 2)^2}{3(a + 2)(a - 4)}$ (D) $\frac{5}{12}$

112   Cumulative Review (Chapters 1–4)

T-11

TEACHER'S EDITION

Clarifies process and techniques for students.

Highlights vocabulary.

The Teacher's Edition of **HOLT ALGEBRA 2** provides an annotated full-size reproduction of each pupil page and includes detailed chapter and lesson commentaries.

Enables the teacher to anticipate difficulties that are typical for students.

Focuses on the key concepts to be taught.

2.1 Properties of Exponents (page 31)
Vocabulary
base, exponent, power, exponential equation

Background
Five properties of positive integer exponents are reviewed. These properties will be extended to include all integer exponents in the next lesson. Simple exponential equations are introduced. Other types of exponential equations are introduced in Chapters 6 and 15.

The exponential equations in this lesson are solved by using the following property.

$$\text{If } y^a = y^b, \text{ then } a = b.$$

For Additional Chalkboard Work
Simplify each expression.
1. $5a^2 \cdot 2a^4$ $10a^6$
2. $5a^2(2a)^4$ $80a^6$
3. $-5(c^2)^3$ $-5c^6$
4. $(-5c^2)^3$ $-125c^6$
5. $\dfrac{8a^2bc^6}{12a^bbc^3}$ $\dfrac{2c^6}{3a^4}$
6. $\left(\dfrac{-2c}{5w^4}\right)^3$ $\dfrac{-8c^3}{125w^{12}}$
7. Solve $3^{2x+4} = 243$. $\tfrac{1}{2}$

Errors Students Make
1. They confuse $x^4 \cdot x^3$ with $(x^4)^3$; wrong answers: $x^{4 \cdot 3}$, or x^{12}, for $x^4 \cdot x^3[=x^7]$; and x^{4+3}, or x^7, for $(x^4)^3[=x^{12}]$.
2. They treat $5(a^2)^3$ as $(5a^2)^3$; wrong answer: $125a^6$ for $5(a^2)^3[=5a^6]$.
3. They treat $(5a^2)^3$ as $5(a^2)^3$; wrong answer: $5a^6$ for $(5a^2)^3[=125a^6]$.
4. They divide exponents in a quotient of powers; wrong answer: $x^{\frac{12}{4}}$, or x^3, for $\dfrac{x^{12}}{x^4}[=x^8]$.

Cumulative Review
Find the solution set of each inequality.
1. $8x - \tfrac{2}{3}(9x + 15) < 12x$ $\{x|x > -2\}$
2. $|2x - 5| > 11$ $\{x|x < -3 \text{ or } x > 8\}$
3. $|x + 7| < 2$ $\{x|-9 < x < -5\}$

2.2 Zero and Negative Integral Exponents (page 34)
Vocabulary
scientific notation, ordinary notation

Background
The zero exponent and negative integer exponents are defined by extending the properties of positive integer exponents as...

Suggestions
Students should interpret exponents loosely as numerator (denominator) exponent in the denominator...
Some students confuse ax^n with expressions... this contrast is shown...

For Additional ...
Simplify and write e... exponents.
1. $\dfrac{2c^4}{5}$
2. $\dfrac{5}{(2c)^{-4}}$
4. $(3a^{-4}b^5)^{-2}$ 5. ...

Find the value of each... notation.
6. $\dfrac{9.6 \times 10^2}{8 \times 10^{-3}}$ 7. ...
1. $\dfrac{2}{5c^4}$
2. $80c^4$
5. $\dfrac{a^2b^5}{c^7}$
6. $120...$

Errors Students ...
1. They might confu...
$(10a)^{-3}\left[=\dfrac{1}{1000...}\right]$
2. They might confu...
$\dfrac{3}{(2c)^{-5}}[=96c^5]$

Cumulative Re...
Alex is 15 years yo... from now she will b... How old is Alex no...

1.8 Problem Solving: Using Formulas (page 22)
Vocabulary
principal, simple interest, rate of interest, total amount, initial velocity

Background
Formulas involving simple interest and total amount, resistances in an electrical circuit, and the height of an object after launch are used to show applications of algebra to problem solving. (Compound interest is delayed until pages 406–408.)

Suggestions
The formula $A = p + prt$ can lead to an expression like $p + 0.16p$. If a student cannot combine these like terms, suggest that it be rewritten as $1p + 0.16p$, or as $1.00p + 0.16p$.
In Exercises 31 and 32, notice that $1,000 and $700, respectively, must be added to the principal in order to answer the questions.
A hand-held claculator can be used to advantage in all of the exercises.

Applications (page 26)
The eleven properties of a field are investigated. These properties will be used in proving theorems and solving problems in this textbook.
This is the first of three applications that form a strand of proofs:
1. Properties of a Field, p. 26
2. Proofs with Rational Numbers, pp. 106–107
3. Proofs with Complex Numbers, pp. 201–202

For Additional Chalkboard Work
1. Find p if $A = \$3,600$, $r = 12\tfrac{1}{2}\%$, $t = 4$ yr, and $A = p + prt$. $\$2,400$
2. Find r_2 if $R = 5$ ohms, $r_1 = 7.5$ ohms, and $R(r_1 + r_2) = r_1 \cdot r_2$. 15 ohms
3. Find v if $h = 175$m, $t = 5$s, and $h = vt - 5t^2$. 60 m/s

Cumulative Review
1. Solve $|3x + 9| = 6$. $-5, -1$
2. Find the solution set of $|3x + 9| < 6$.
3. Find the solution set of $|3x + 9| > 6$.
2. $\{x|-5 < x < -1\}$ 3. $\{x|x < -5 \text{ or } x > -1\}$

T-20

Chapter 2 and Lesson Commentaries

Chapter 2 Polynomials and Problem Solving

Chapter Objectives
To simplify expressions containing integral exponents
To solve an exponential equation
To evaluate a polynomial
To add and subtract polynomials
To multiply two or more polynomials
To solve a word problem involving a collection or mixture of items
To solve problems involving two objects in motion

Chapter Overview
Properties of positive integer exponents are reviewed and then extended to include zero and negative integer exponents.
Synthetic substitution is introduced as a method of evaluating a polynomial in one variable. This powerful tool will be used in Chapter 5 to determine whether a binomial is a factor of an integral polynomial $P(x)$ and again in Chapter 15 to factor $P(x)$ into first-degree factors, find the zeros of $P(x)$, draw the graph of $y = P(x)$, and solve the polynomial equation $P(x) = 0$.
Problem solving is extended to include mixture problems and motion problems involving two objects.

Special Features
Plan How to Solve a Problem   p. 30
Reading in Algebra   pp. 33, 36, 39
Cumulative Review   pp. 40, 46
Non-Routine Problems   pp. 42, 46
Chapter Review   p. 53
Chapter Test   p. 54
College Prep Test   p. 55

Keeps previously learned skills current.

Reinforces the lesson skills.

References all special feature pages.

Indicates mathematical content.

T-12

SUPPLEMENTARY MATERIALS

HOLT ALGEBRA 2's supplementary materials include Testmasters, Skillmasters, and a Teacher's Resource Package with Blackline Masters so that the teacher can instruct according to each student's needs and skill levels.

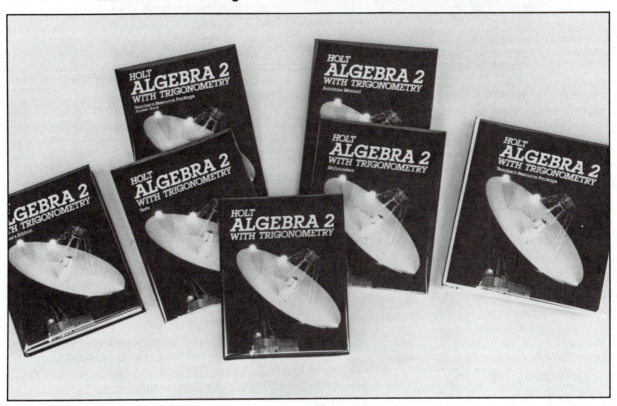

Skillmasters	contain two parallel sections for every two lessons which can be used as pre- or posttests for diagnosis, extra practice, quizzes, and assignments. There is also a Cumulative Review for every few lessons.
Testmasters	supply Chapter Tests, Cumulative Tests, and Final Tests.
Teacher's Resource Package	contains Skillmasters and Testmasters in blackline form plus Enrichment masters, Competency Tests, and Geometry Reviews. The Enrichment Pages provide additional material that can be used to challenge all students. The Competency Tests contain the type of material found on standard competency tests used in many states. The Geometry Reviews are a set of extra materials that can be used in reviewing previously learned geometric concepts.
Solutions Manual	provides worked-out solutions to every problem and exercise in the text.

ABOUT HOLT ALGEBRA II

Success Pattern HOLT ALGEBRA 2 guides students through the process of developing their mathematical abilities by helping them experience a pattern of successful learning that builds self-confidence. Students review fundamental ideas of basic algebra as they gradually deepen their understanding of concepts and skills. Constant attention to problem solving and applications develops a meaningful focus on the subject.

Organization of the Course HOLT ALGEBRA 2 contains all of the major algebra concepts necessary to meet the instructional demands in your school. In addition to feedback materials, which include tests for each unit with items keyed to lesson objectives, HOLT ALGEBRA 2 provides materials in logic, geometry, and probability and statistics to supplement basic coursework. It also provides a computer strand incorporated into every chapter, to correlate algebraic concepts with relevant and motivating BASIC-language programming skills. These pages are referenced to on-hand materials in the Computer Section at the back of the book. Lessons using calculators and practice for standardized college admission tests offer further supplementation.

Instructional Advantage The instructional advantages of Holt's approach are clear and well documented. First, frequent evaluation — Chapter Reviews, tests at the end of chapters, and Cumulative Reviews every four chapters — provides continuous feedback for both you and the student. Second, a varied instructional technique provides students with the opportunity to participate actively in learning and to find an instructional method that best suits their needs. Students who learn quickly and can work independently need the challenge provided by the supplements and extensions to the basic course. Others benefit from selected problems that challenge them at their own levels of competence. HOLT ALGEBRA 2 addresses the needs of both types of students.

Students with Handicapping Conditions Instructional materials are also available for helping students who need to overcome learning problems that may be due to physical disabilities (orthopedic or visual impairments), communicative disabilities (hearing loss, dyslexia), mild mental retardation, or disabling emotional disturbance. Teachers should be on the alert to such possibilities so that students with undiagnosed disabilities can be treated professionally.

Whenever possible, students with handicapping conditions should participate either partially or totally in regular school programs and classes, as part of the requirement of Federal Law PL94-142, which states that handicapped children must be educated in the least restrictive environment possible.

By using the chapter and lesson objectives in the book, the teacher can select appropriate lessons goals for each student and communicate them easily to parents or special-education teachers who may be assisting the handicapped. The Skillmasters and Testmasters in the Teacher's Resource Package are valuable tools for assessing each individual student's level of comprehension and skill in a given lesson or topic.

Success of HOLT ALGEBRA 2 HOLT ALGEBRA 2 has been used successfully with a variety of students. It has been found to be stimulating, highly teachable, and motivational, with a good balance of skill, theory, and application. This new edition of HOLT ALGEBRA 2, with its expanded and new features, is based on suggestions from users of the previous editions.

Reference Chart for Algebra 2 Skillmasters

LESSON PAGES	SKILLMASTER NUMBER	LESSON PAGES	SKILLMASTER NUMBER
1, 4, 7, 10	1	291, 294	29
13, 18	2	303, 306	30
20, 22	3	310, 314	31
31, 34, 37	4	319, 324, 327	32
41, 43, 47	5	338, 341	33
57, 60, 62	6	346, 349	34
65, 68, 71	7	357, 361, 364	35
74, 78	8	368, 372, 375	36
89, 91	9	379, 382, 384	37
94, 96	10	390, 392, 394	38
99, 102	11	397, 400	39
115, 119	12	403, 406	40
122, 125	13	409, 412, 416	41
128, 130	14	424, 428, 431	42
139, 142, 145	15	433, 435	43
148, 152, 155, 157	16	438, 441, 445	44
165, 167, 169, 172	17	455, 458	45
175, 178	18	462, 466	46
187, 190, 193	19	470, 473, 475	47
195, 198	20	477, 481, 483	48
209, 212	21	486, 489	49
216, 220, 223	22	497, 500, 503	50
226, 228, 231	23	506, 509, 512	51
240, 243, 247	24	515, 518, 521	52
250, 253, 257	25	523, 526, 529, 533	53
259, 263, 267	26	539, 542, 544	54
276, 279, 282	27	546, 549, 552	55
285, 288	28	556, 560	56
		563, 566	57

Acknowledgements

Holt is grateful to thousands of teachers and to hundreds of thousands of students who have used our algebra texts in the past and who have contributed their comments and constructive criticism to help make the teachable textbooks of the past even more teachable and enjoyable.

We are also grateful to teachers, supervisors, and administrators across the country who have participated in systematic surveys so that Holt can give America's high school students the kind of materials that best facilitate their learning.

Chapter and Lesson Commentaries

Chapter 1 Linear Equations and Inequalities

Chapter Objectives
To solve a linear equation in one variable
To find the solution set of a linear inequality
To graph the solutions of a linear inequality
To find the solution set of a compound inequality
To graph the solutions of a compound inequality
To solve an equation involving absolute value
To find the solution set of an inequality involving absolute value
To graph the solutions of an inequality involving absolute value
To write a word phrase in algebraic terms
To solve word problems

Chapter Overview
Conjunctions and disjunctions of inequalities are introduced and serve as a tool for solving absolute value inequalities in this chapter and quadratic inequalities in Chapter 3. Problem solving is introduced with a series of four types of word problems: numbers, perimeters, ages, and formulas.

Special Features
Non-Routine Problems   pp. x, 3
Four Steps in Problem Solving   pp. x, 15
Reading in Algebra   pp. 6, 9, 12, 15, 20
Cumulative Review   p. 12
Applications: Properties of a Field   p. 26
Chapter Review   p. 27
Chapter Test   p. 28
College Prep Test   p. 29

1.1 Linear Equations (page 1)

Vocabulary
distributive property, linear equation

Suggestions
Basic skills are maintained in this lesson and throughout the book by including expressions that contain decimals or common fractions. Some students may need to review the rule for placing the decimal point in a product of two decimal numbers.

For Additional Chalkboard Work
Solve each equation.
1. $6x + 32 - 10x = 7 + 8x - 39$ $5\frac{1}{3}$
2. $\frac{3}{5}y - 3 = 8 - \frac{1}{5}y$ $13\frac{3}{4}$
3. $3.8n + 2.3 = 0.2(37.9 - 3n)$ 1.2
4. $9x - (4x - 6) = 17 + 2(3x + 8)$ -27
5. $21 = \frac{3}{4}(12 - 8t)$ -2

Enrichment
Some linear equations, often called identities, have every real number as a solution, and some linear equations have no solution.
1. Solve $8x + 3(4x - 5) = 15 - 5(6 - 4x)$.
$$8x + 12x - 15 = 15 - 30 + 20x$$
$$20x - 15 = 20x - 15$$
1. Every number is a solution:
   $20x - 15 = 20x - 15$, for each x.
2. Solve $6(2x + 5) = 3x + 3(9 + 3x)$.
$$12x + 30 = 3x + 27 + 9x$$
$$12x + 30 = 12x + 27$$
2. There is no solution:
   $12x + 30 > 12x + 27$, for each x.

Cumulative Review
1. Compute $-18 + 12 - (-15) - 19$. -10
2. Compute $8 - 2 \cdot 5 - (-4)(-3)$. -14
3. Simplify $-14x + y - x - 15y$. $-15x - 14y$

1.2 Linear Inequalities (page 4)

Vocabulary
linear inequality, order of an inequality: $<$, $>$

Background
Lessons 1.1 and 1.2 illustrate that, in this textbook, the language and symbolism associated with sets is used only when there is a definite advantage in doing so. For example, (1) in Lesson 1.1 we say
  "The *solution* of $78 = 22 - 8t$ is -7" rather than
  "The *solution set* of $78 = 22 - 8t$ is $\{-7\}$."
However, (2) in Lesson 1.2 we say that
  "The *solution set* of $8 - 5x \leq 28$ is $\{x | x \geq -4\}$" rather than
  "The *solutions* of $8 - 5x \leq 28$ are -4 and all numbers greater than -4."

Suggestions
Many students believe erroneously that the graph of "$a > x$" is a ray to the *right* of a because they see the symbol $>$. These students should rewrite "$a > x$" as "$x < a$" (rewrite "$b < x$" as "$x > b$") so that the variable is on the left side. This is shown in Example 2.

For Additional Chalkboard Work
Solve.
1. $7 - 2x > 4x + 49$ $\{x | x < -7\}$
2. $x - 8 \leq 5x + 12$ $\{x | x \geq -5\}$

Errors Students Make
1. They draw the graph of $a \leq x$ (e.g., $7 \leq x$) as a ray to the left of a because of the symbol "$<$."
2. They solve $-ax < b$ as $x < -\frac{b}{a}$ where $-a < 0$.

Cumulative Review
1. Compute: $\dfrac{-16 - 4 \cdot 3 - 22}{9 - 7 \cdot 7}$.
2. Solve: $1 - \frac{3}{5}x = 7$.
3. Simplify: $-4(2x - 3) - (7 - 12x)$.

1. $\frac{5}{4}$ 2. -10 3. $4x + 5$

1.3 Compound Linear Inequalities (page 7)

Vocabulary
conjunction, disjunction, between: $a < x < b$, compound inequality

Background
[$x > 5$ *or* $x > 9$] is a *disjunction* of two inequalities because two inequalities are separated by the connective *or*. The numbers 7 and 11 are two of the infinite number of solutions of this *compound sentence*. The number 3 is *not* a solution.
[$7 > 5$ or $7 > 9$] is true because one clause is true.
[$11 > 5$ or $11 > 9$] is true because both clauses are true.
[$3 > 5$ or $3 > 9$] is false because both clauses are false.
A disjunction [*p or q*] is true under each of three conditions: (1) p is true, q is false; (2) p is false, q is true; (3) p is true, q is true.
  The compound sentence [$x > 5$ *and* $x > 9$] is a *conjunction* of two inequalities separated by *and*. The number 11 is one of its many solutions, while neither 3 nor 7 is a solution.
[$11 > 5$ and $11 > 9$] is true because both clauses are true.
[$3 > 5$ and $3 > 9$] is false because both clauses are false.
[$7 > 5$ and $7 > 9$] is false because one clause is false.
A conjunction [*p and q*] is true only under the condition that both p and q are true.

For Additional Chalkboard Work
Solve.
1. $x - 2 < 5$ or $x + 3 < 5$
2. $x - 2 < 5$ and $x + 3 < 5$
3. $-4 \leq 3x + 5 < 17$
4. $3x + 5 \geq 17$ or $6 > x + 8$
5. $3x + 5 \leq 17$ and $6 < x + 8$

1. $\{x | x < 7\}$ 2. $\{x | x < 2\}$ 3. $\{x | -3 \leq x < 4\}$
4. $\{x | x < -2$ or $x \geq 4\}$ 5. $\{x | -2 < x \leq 4\}$

Cumulative Review
1. Solve $3x + 4(6 - 5x) = -4 - (x + 4)$. 2
2. Solve $3x + 4(6 - 5x) < -4 - (x + 4)$.
   $\{x | x > 2\}$

1.4 Absolute Value (page 10)

Vocabulary
absolute value, $|x|$

Suggestions
Some students have trouble with the part of the definition of absolute value that states

$$|x| = -x, \text{ if } x < 0.$$

The "$-$" in "$-x$" should not be thought of as "negative"; $-x$ can be read (thought of) as:
1. the opposite of x,
2. the additive inverse of x, or
3. -1 (negative one) times x.

The Reading in Algebra Exercises 7–9 serve as a hint for the Written Exercises 37–40.

For Additional Chalkboard Work
Display the following on the chalkboard to help the students see the relationships of the three properties in this lesson.

A. $|x| = k$; $x = -k$ or $x = k$

B. $|x| < k$; $-k < x < k$

C. $|x| > k$; $x < -k$ or $x > k$

Solve.
1. $|5n + 11| = 14$ $-5, \frac{3}{5}$
2. $|2y - 3| > 7$ $\{y | y < -2 \text{ or } y > 5\}$
3. $|4t + 6| \leq 14$ $\{t | -5 \leq t \leq 2\}$

Errors Students Make
Writing $5 < x < 2$ or $2 > x > 5$ for $2 < x < 5$
1. $a < x < b$ means $[a < x \text{ and } x < b]$.
2. $a > x > b$ means $b < x < a$.

Cumulative Review (Needed for Lesson 1.8)
1. Write $12\frac{1}{2}\%$ as a decimal. 0.125
2. Write 0.0975 as a percent. $9\frac{3}{4}\%$
3. Simplify $p + p(0.06)(10)$. $1.6p$
4. Solve $116 = 1600r$. $r = 0.0725$

1.5 Problem Solving: Number Problems (page 13)

Vocabulary
less than, decreased by, more than, sum, increased by, consecutive integers, even integers, odd integers, consecutive multiples

Background
Problem solving is a central theme in this book. This lesson and the next two provide a simple introduction to solving word problems by using algebra. The exercises lend themselves readily to the five steps in solving a word problem:

Step 1 Represent the data in terms of one variable.
Step 2 Write an equation.
Step 3 Solve the equation.
Step 4 Check in the problem.
Step 5 Answer the question.

Step 5 is necessary, as shown in Example 3 in which the solution (7) of the equation is not the solution (7, 21 and 23) of the problem.

Suggestions
Stress the translation of "less than" where "a less than b" is "$b - a$" rather than "$a - b$." This should not be confused with "a decreased by b," which translates to "$a - b$".

For Additional Chalkboard Work
1. The second of three numbers is 5 times the first. The third number is 4 less than the second. Ten more than the second is the same as 7 less than twice the third. Find the three numbers.
2. Find three consecutive multiples of 4 such that the sum of the first number and twice the third number is 60 more than the second number.
1. 5, 25, 21   2. 24, 28, 32

Cumulative Review (Needed for Lesson 1.6)
1. Represent the perimeter of a rectangle whose length is $5x - 6$ and whose width is $9 - x$.
2. Represent the perimeter of an isosceles right triangle where one leg is $6x$ feet long and the hypotenuse is $(2x + 40)$ feet long.
1. $8x + 6$ 2. $14x + 40$ ft

1.6 Problem Solving: Perimeter Problems (page 18)

Vocabulary
perimeter, pentagon, hexagon, rhombus

Background
Step 1 in problem solving is to *represent the data*. Lessons 1.5, 1.6, and 1.7 show three different ways to organize the data represented in terms of a variable.
1. In Lesson 1.5, the data in Example 3 was organized in *paragraph* form: Let x = the first number. Then $3x$ = the second number and $3x + 2$ = the third number.
2. In this lesson (1.6), the data is organized by sketching and labeling a *geometric figure*.
3. In Lesson 1.7, the data will be organized in the form of a *table*.

Suggestions
Encourage the students to represent the data by sketching and labeling a geometric figure as shown in Examples 1 and 2.

This is an appropriate lesson in which to review the units of length and area in the metric system.

In Exercise 14, students will need to recall that the ratio of $2:3:5:6$ can be represented by $2x:3x:5x:6x$.

For Additional Chalkboard Work
1. The length of a rectangle is 7 m less than 3 times the width. The perimeter is 18 m. Find the area of the rectangle. 20m^2
2. Side a of a triangle is 5 cm shorter than side b, and side c is twice as long as side a. Find the length of each side if the perimeter is 45 cm.
   10cm, 15cm, 20cm

Errors Students Make
They represent the perimeter of a rectangle, x units by $3x$ units, as $x + 3x$ instead of $2x + 2 \cdot 3x$.

Cumulative Review
1. Solve $4t + 10 = 3[(t + 6) + (5t - 12)]$. 2
2. Find the solution set of
   $3x - (9x - 5) \leq 15 - 4(7 - 3x)$.
   $\{x \mid x \geq 1\}$

1.7 Problem Solving: Age Problems (page 20)

Background
In this lesson, the data is organized in the form of a table. This method of organizing data will be used with many other types of word problems.

Suggestions
Continue to encourage students to follow the five steps in solving a word problem. A simpler opening example is given below.

"Carol is 6 times as old as Howard and Betty is 22 years younger than Carol. One year ago the sum of their ages was 27. How old is each now?"

	Howard	Carol	Betty
Age Now →	x	$6x$	$6x - 22$
Age 1 yr ago →	$x - 1$	$6x - 1$	$6x - 22 - 1$

$(x - 1) + (6x - 1) + (6x - 23) = 27$
$13x - 25 = 27$
$13x = 52$
$x = 4$

Howard is 4, Carol is 24, and Betty is 2.

For Additional Chalkboard Work
Bill is 5 times as old as Alice. Cindy is 2 years older than Bill. Three years ago, Cindy was 12 times as old as Alice. Find their ages now.
Alice: 5, Bill: 25, Cindy: 27

Errors Students Make
They multiply the wrong age. For example, they may write $3(x + 4) = x - 12 + 4$ instead of $x + 4 = 3(x - 12 + 4)$ as the equation to be solved, in "*she* is x years old, *he* is 12 years younger, and in 4 years *she* will be 3 times as old as *he* will be."

Cumulative Review (Needed for Lesson 1.8)
Evaluate each expression for the given values of the variable.
1. $445xy$ if $x = 0.08$ and $y = 2.5$ 89
2. $x + xyz$ if $x = 800$, $y = 0.15$, and $z = 3.25$ 1,190
3. Change $9\frac{3}{4}\%$ to a decimal. 0.0975
4. Change 0.0825 to a percent. $8\frac{1}{4}\%$

1.8 Problem Solving: Using Formulas (page 22)

Vocabulary
principal, simple interest, rate of interest, total amount, initial velocity

Background
Formulas involving simple interest and total amount, resistances in an electrical circuit, and the height of an object after launch are used to show applications of algebra to problem solving. (Compound interest is delayed until pages 406–408.)

Suggestions
The formula $A = p + prt$ can lead to an expression like $p + 0.16p$. If a student cannot combine these like terms, suggest that it be rewritten as $1p + 0.16p$, or as $1.00p + 0.16p$.

In Exercises 31 and 32, notice that $1,000 and $700, respectively, must be added to the principal in order to answer the questions.

A hand-held claculator can be used to advantage in all of the exercises.

Applications (page 26)
The eleven properties of a field are investigated. These properties will be used in proving theorems and solving problems in this textbook.

This is the first of three applications that form a strand of proofs:
1. Properties of a Field, p. 26
2. Proofs with Rational Numbers, pp. 106–107
3. Proofs with Complex Numbers, pp. 201–202

For Additional Chalkboard Work
1. Find p if $A = \$3,600$, $r = 12\frac{1}{2}\%$, $t = 4$ yr, and $A = p + prt$.   **$2,400**
2. Find r_2 if $R = 5$ ohms, $r_1 = 7.5$ ohms, and $R(r_1 + r_2) = r_1 \cdot r_2$.   **15 ohms**
3. Find v if $h = 175$m, $t = 5$s, and $h = vt - 5t^2$.   **60 m/s**

Cumulative Review
1. Solve $|3x + 9| = 6$.   **−5, −1**
2. Find the solution set of $|3x + 9| < 6$.
3. Find the solution set of $|3x + 9| > 6$.

2. $\{x | -5 < x < -1\}$ 3. $\{x | x < -5 \text{ or } x > -1\}$

Chapter 2
Chapter and Lesson Commentaries

Chapter 2   Polynomials and Problem Solving

Chapter Objectives
To simplify expressions containing integral exponents
To solve an exponential equation
To evaluate a polynomial
To add and subtract polynomials
To multiply two or more polynomials
To solve a word problem involving a collection or mixture of items
To solve problems involving two objects in motion

Chapter Overview
Properties of positive integer exponents are reviewed and then extended to include zero and negative integer exponents.

Synthetic substitution is introduced as a method of evaluating a polynomial in one variable. This powerful tool will be used in Chapter 5 to determine whether a binomial is a factor of an integral polynomial $P(x)$ and again in Chapter 15 to factor $P(x)$ into first-degree factors, find the zeros of $P(x)$, draw the graph of $y = P(x)$, and solve the polynomial equation $P(x) = 0$.

Problem solving is extended to include mixture problems and motion problems involving two objects.

Special Features
Plan How to Solve a Problem   p. 30
Reading in Algebra   pp. 33, 36, 39
Cumulative Review   pp. 40, 46
Non-Routine Problems   pp. 42, 46
Chapter Review   p. 53
Chapter Test   p. 54
College Prep Test   p. 55

2.1 Properties of Exponents (page 31)

Vocabulary
base, exponent, power, exponential equation

Background
Five properties of positive integer exponents are reviewed. These properties will be extended to include all integer exponents in the next lesson. Simple exponential equations are introduced. Other types of exponential equations are introduced in Chapters 6 and 15.

The exponential equations in this lesson are solved by using the following property.

$$\text{If } y^a = y^b, \text{ then } a = b.$$

For Additional Chalkboard Work
Simplify each expression.
1. $5a^2 \cdot 2a^4$ $\quad 10a^6$
2. $5a^2(2a)^4$ $\quad 80a^6$
3. $-5(c^2)^3$ $\quad -5c^6$
4. $(-5c^2)^3$ $\quad -125c^6$
5. $\dfrac{8a^2bc^9}{12a^6bc^3}$ $\quad \dfrac{2c^6}{3a^4}$
6. $\left(\dfrac{-2t}{5w^4}\right)^3$ $\quad \dfrac{-8t^3}{125w^{12}}$
7. Solve $3^{2x+4} = 243$. $\quad \frac{1}{2}$

Errors Students Make
1. They confuse $x^4 \cdot x^3$ with $(x^4)^3$; wrong answers: $x^{4 \cdot 3}$, or x^{12}, for $x^4 \cdot x^3 [=x^7]$; and x^{4+3}, or x^7, for $(x^4)^3 [=x^{12}]$.
2. They treat $5(a^2)^3$ as $(5a^2)^3$; wrong answer: $125a^6$ for $5(a^2)^3 [=5a^6]$.
3. They treat $(5a^2)^3$ as $5(a^2)^3$; wrong answer: $5a^6$ for $(5a^2)^3 [=125a^6]$.
4. They divide exponents in a quotient of powers; wrong answer: $x^{\frac{12}{4}}$, or x^3, for $\dfrac{x^{12}}{x^4} [=x^8]$.

Cumulative Review
Find the solution set of each inequality.
1. $8x - \frac{2}{3}(9x + 15) < 12x$ $\quad \{x \mid x > -1\}$
2. $|2x - 5| > 11$ $\quad \{x \mid x < -3 \text{ or } x > 8\}$
3. $|x + 7| < 2$ $\quad \{x \mid -9 < x < -5\}$

2.2 Zero and Negative Integral Exponents (page 34)

Vocabulary
scientific notation, ordinary notation

Background
The zero exponent and negative integer exponents are defined by extending two properties of positive integer exponents as shown on page 34.

Suggestions
Students should interpret the definition of negative exponents loosely as: A negative exponent in the numerator (denominator) becomes a positive exponent in the denominator (numerator).

Some students confuse expressions of the form ax^n with expressions of the form $(ax)^n$. Notice that this contrast is shown in the exercises.

For Additional Chalkboard Work
Simplify and write each expression with positive exponents.
1. $\dfrac{2c^{-4}}{5}$
2. $\dfrac{5}{(2c)^{-4}}$
3. $-3x^{-6} \cdot 4x^0 \cdot 2x^3$
4. $(3a^{-4}b^5)^{-2}$
5. $\dfrac{a^{-3}b^3c^{-2}}{a^{-5}b^{-2}c^5}$

Find the value of each expression in ordinary notation.
6. $\dfrac{9.6 \times 10^2}{8 \times 10^{-3}}$
7. $\dfrac{2.4 \times 10^{-5}}{7.2 \times 10^{-2}}$

1. $\dfrac{2}{5c^4}$ 2. $80c^4$ 3. $\dfrac{-24}{x^3}$ 4. $\dfrac{a^8}{9b^{10}}$
5. $\dfrac{a^2b^5}{c^7}$ 6. $120{,}000$ 7. 0.0003

Errors Students Make
1. They might confuse $10a^{-3} \left[= \dfrac{10}{a^3} \right]$ with $(10a)^{-3} \left[= \dfrac{1}{1000a^3} \right]$.
2. They might confuse $\dfrac{3}{2c^{-5}} \left[= \dfrac{3c^5}{2} \right]$ with $\dfrac{3}{(2c)^{-5}} [=96c^5]$.

Cumulative Review
Alex is 15 years younger than Carol and 2 years from now she will be 4 times as old as he will be. How old is Alex now? 3 years old

2.3 Polynomials and Synthetic Substitution (page 37)

Vocabulary
polynomial, term, monomial, binomial, trinomial, degree of a polynomial, linear, quadratic, synthetic substitution, descending order of exponents

Suggestions
List several polynomials on the chalkboard. Identify the degree of each, those that are linear, and those that are quadratic. For example:

$x^5 + 6x^3 - 7x + 4$ 5th degree
$5x^2 - 6$ 2nd degree: quadratic
$-x + 3$ 1st degree: linear
$7x^2$ 2nd degree: quadratic
$8m^2n - 3mn^2 + mn - 5$ 3rd degree
$5m^2 + 6mn - 8n^2$ 2nd degree: quadratic
$4mn$ 2nd degree: quadratic
$8m + 4n - 3$ 1st degree: linear
$-2m$ 1st degree: linear.

Students should use the words "linear" and "quadratic" fluently. This vocabulary is tested by the Reading in Algebra Exercises 7–12. Notice the following.
1. Every monomial is a polynomial.
2. Every real number is both a monomial and a polynomial.

For students who wonder as to the degree of a real number such as 7:

7 is of degree 0 since $7 = 7 \cdot 1 = 7x^0$.

For Additional Chalkboard Work
Use synthetic substitution to evaluate

$3x^4 - 2x^2 - 6x + 10$ for $x = 2$ and for $x = -2$.
38; 62

Cumulative Review
1. Multiply $-2a^2(a^3 - 4a + 2)$.
2. Simplify $4m^2 \cdot 5m^3n(-3mn^5)$.
3. Simplify $6t(4t^3v^4)^2$. $96t^7v^8$
1. $-2a^5 + 8a^3 - 4a^2$ 2. $-60m^6n^6$

2.4 Multiplying Polynomials (page 41)

Vocabulary
FOIL method

Background
The FOIL method for multiplying two linear bionomials in one variable is reviewed and then extended to:
1. binomials in more than one variable
2. binomials of degree greater than one, and
3. the product of a monomial and a pair of binomials.

Suggestions
It should be emphasized that the FOIL method can only be applied to finding the product of two binomials. The product of a binomial and a trinomial can be simplified using the following scheme.

$(3x + 2) \quad (5x^2 - x - 4)$

$15x^3 - 3x^2 - 12x + 10x^2 - 2x - 8$
$15x^3 + 7x^2 - 14x - 8$

For Additional Chalkboard Work
1. $(5x - 3y)(7x + 4y)$ $35x^2 - xy - 12y^2$
2. $4a(2a^2 - 5)(a^2 + 2)$ $8a^5 - 4a^3 - 40a$
3. $(2x + 3)(5x^2 - 6x - 4)$
$10x^3 + 3x^2 - 26x - 12$

Cumulative Review
Find three consecutive multiples of 4 for which the second number, decreased by 3 times the third number, is 56 less than the first number.
12, 16, 20

2.5 Problem Solving: Mixture Problems (page 43)

Suggestions
In Example 1, the data may also be represented in the form of a table as follows.

	Number	Value in ¢
nickels	n	$5n$
dimes	$n - 4$	$10(n - 4)$
quarters	$3(n - 4)$	$25 \cdot 3(n - 4)$

In Example 2, emphasize that: x kg and $(20 - x)$ kg (not $x - 20$) make 20 kg of the mixture.

The Chapter 2 opener, page 30, provides a good introduction to wet-mixture problems and may be studied in connection with Example 3 on page 44.

In Example 4, the equation is based on the fact that

$$\begin{array}{c} \text{Iodine} \\ \text{before} \end{array} + \begin{array}{c} \text{Iodine} \\ \text{added} \end{array} = \begin{array}{c} \text{Iodine} \\ \text{after} \end{array}$$

$$0.16(20) + 0.80x = .30(20 + x)$$

For Additional Chalkboard Work
1. Brand A, worth $3.10/kg, is mixed with brand B, worth $2.20/kg, to make 30 kg of a mixture worth $2.56/kg. How many kilograms of each brand are in the mixture?   A: 12 kg, B: 18 kg
2. How many liters of water must be added to 40 L of an 18% sugar solution to make it a 15% sugar solution?   8 L
3. How many liters of a 60% salt solution must be added to 12 L of a 40% salt solution to obtain a 48% salt solution?   8 L

Errors Students Make
They use $x - 20$, instead of $20 - x$, to represent the number of kilograms of a brand B that must be added to x kilograms of a brand A in order to form a 20-kg mixture of brands A and B.

Cumulative Review
1. Solve $|3 - 2x| = 21$.   2. Find the solution set of $-5(x + 2) > 4 - 7x$.
1. -9 and 12 2. $\{x | x > 7\}$

2.6 Problem Solving: Motion Problems (page 47)

Background
This lesson covers a first type of motion problem: two objects moving in either the same or opposite directions. A second type of motion problem, a round trip, is introduced in Chapter 5.

Suggestions
"Vector diagrams," as used in Examples 1 and 2, can assist the student in deciding whether to add two distances or to set them equal.

This lesson (2.6) and the previous lesson (2.5) can occupy 3 class periods by assigning previously unassigned exercises from the two lessons for the third class period.

For Additional Chalkboard Work
1. Bus A left the city northbound at 60 km/h at the same time that bus B left the city southbound at 70 km/h. In how many hours were they 325 km apart?   2.5 h
2. A bus left the city at 8:00 A.M. at 50 km/h. At 10:00 A.M., an auto left the city at 90 km/h to overtake the bus. At what time did the auto reach the bus?   12:30 P.M.

Cumulative Review
1. Solve $\frac{1}{5}x - 18 = 14 - \frac{3}{5}x$.   40
2. Solve $x + 1.2(0.5x - 3) = 0.4(x + 3)$.   4

2.7 Introduction to Vectors (optional) (page 50)

Vocabulary
displacement, directed line segment, vectors, initial point, terminal point, force, equivalent vectors, direction, magnitude

Background
This introduction to *vectors* is placed early in this book at the request of students and teachers of high school physics. *Displacement, force,* and *velocity* are used as examples of vector quantities. The properties of *magnitude* and *direction* are attached to each vector. The *sum* of several vectors is drawn after studying *equivalent* vectors. The *difference* of two vectors is drawn with the aid of *opposite* vectors. A treatment of a *scalar multiple* of a vector is included after a study of *collinear* vectors. The words *tail* and *head* of a vector are incorporated in the text for *initial point* and *terminal point* of a vector.

Suggestions
Emphasize that a vector is a mathematical quantity that has not only size or magnitude, but also direction. Some physical interpretations of vectors are forces, velocities, accelerations, or displacements.

A physical example of a vector is wind that has both direction and magnitude. Wind out of the west at 10 mph is an example.

Another example might be that of a certain force having a strength of 75 lb acting at an angle of 20° with the horizontal. Once again, notice both the size and direction.

In general, a vector can be represented by a line segment drawn on a Cartesian coordinate system with its initial point at the origin that makes an angle with the x-axis. Note that the line segment has size and that the angle gives the direction.

To give students a good understanding of adding vectors, use a physical model of adding 2 vectors. Start at Point A and walk 3 steps east to Point B. From Point B walk 4 steps north to Point C. Point out that to go from Point A to Point C directly, you must walk 5 steps in a northeasterly direction.

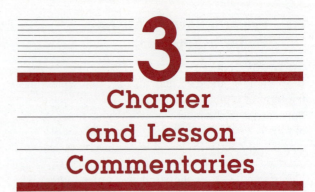

3 Chapter and Lesson Commentaries

Chapter 3 Factoring and Special Products

Chapter Objectives
To factor a polynomial completely
To multiply a pair of related polynomials
To factor a difference of two squares
To factor a perfect square trinomial
To factor a sum or difference of two cubes
To solve a quadratic equation by factoring
To find the solution set of a quadratic inequality
To graph the solutions of a quadratic inequality
To solve a word problem that leads to a quadratic equation

Chapter Overview
Factoring is extended to the special products $a^2 - b^2$, $a^2 \pm 2ab + b^2$, $a^3 \pm b^3$, and to *grouping pairs* of terms.

Quadratic inequalities are solved using the logic of conjunctions and disjunctions introduced in Lesson 1.3.

Word problems include those with two ordered pairs or triples of numbers as answers.

Special Features
Formulating A Problem-Solving Situation   p. 56
Non-Routine Problems   pp. 77, 82
Reading in Algebra   pp. 64, 67, 70, 73
Cumulative Review   pp. 59, 70, 77
Calculator Activities   p. 82
Applications   p. 83
Chapter Review   p. 84
Chapter Test   p. 85
Computer Activities   p. 86
College Prep Test   p. 87

3.1 Factoring Trinomials (page 57)

Vocabulary
quadratic trinomial

Background
This chapter assumes familiarity with the meaning of *factor, prime factor, greatest common factor,* etc. The teacher may wish to introduce the chapter by reviewing the following vocabulary and skills.
1. Factor 15 as the product of two prime numbers.
2. Write the prime factorization of 72.
3. Factor 400 completely.
4. Find the greatest common factor (GCF) of 18, 24, and 36.
5. Find the GCF of x^4, x^2, and x.
6. Find the GCF of x^3y^2, x^2y^3, and xy^4.

Trinomials of the form $ax^2 + bx + c$ in which $a = 1$ are found in the Oral Exercises.

Fourth-degree polynomials that can be factored into three or four binomials are delayed until Lesson 3.4, which involves factoring completely.

For Additional Chalkboard Work
Factor.
1. $2x^2 - 3x - 5$ $(2x - 5)(x + 1)$
2. $11x^2 + 4x - 7$ $(11x - 7)(x + 1)$
3. $8y^2 - 6y - 9$ $(4y + 3)(2y - 3)$
4. $24y^2 + 5y - 14$ $(8y + 7)(3y - 2)$
5. $15x^2 + 26xy + 8y^2$ $(3x + 4y)(5x + 2y)$
6. $20a^2 - 9ab - 18b^2$ $(5a - 6b)(4a + 3b)$
7. $n^4 + 7n^2 - 18$ $(n^2 + 9)(n^2 - 2)$
8. $3p^4 + 5p^2 - 2$ $(3p^2 - 1)(p^2 + 2)$

Cumulative Review
Multiply.
1. $(3x^2 - 4y)(3x^2 + 4y)$
2. $(8a + 5)(8a + 5)$
3. $(n - 5)(n^2 + 5n + 25)$
1. $9x^4 - 16y^2$ 2. $64a^2 + 80a + 25$
3. $n^3 - 125$

3.2 Special Products (page 60)

Vocabulary
product of sum and difference of two terms, square of a binomial, perfect square trinomial, sum of two cubes, difference of two cubes

Background
Three special products are introduced.
1. $(a + b)(a - b) = a^2 - b^2$
2. $(a + b)^2 = a^2 + 2ab + b^2$ and $(a - b)^2 = a^2 - 2ab + b^2$
3. $(a + b)(a^2 - ab + b^2) = a^3 + b^3$ and $(a - b)(a^2 + ab + b^2) = a^3 - b^3$

These three formulas will be reversed in Lesson 3.3 and called special factors.

Suggestions
The teacher may want the Level A students to omit this lesson and continue to use the FOIL method for $(a + b)(a - b)$ and $(a \pm b)^2$. In such a case, emphasize that $(a + b)^2$ should be written as $(a + b)(a + b)$ as a first step.

For Additional Chalkboard Work
Simplify.
1. $(7x - 3y)(7x + 3y)$ $49x^2 - 9y^2$
2. $(5x^2 - 4)^2$ $25x^4 - 40x^2 + 16$
3. $(2x + 5y)(4x^2 - 10xy + 25y^2)$ $8x^3 + 125y^2$

Cumulative Review
Some 50-cent paper tablets are mixed with some 30-cent tablets to make a package worth $14.40. If the number of 30-cent tablets is 4 more than twice the number of 50-cent tablets, how many tablets of each kind are in the package?
12 50-cent tablets, 28 30-cent tablets

3.3 Special Factors (page 62)

Vocabulary
difference of two cubes, difference of two squares, perfect square trinomial, sum of two cubes

Suggestions
Additional practice to accompany Example 5.

1. $x^2 - y^2 - 8y - 16$
   $x^2 - (y^2 + 8y + 16)$
   $x^2 - (y + 4)^2$
   $[x + (y + 4)][x - (y + 4)]$
   $(x + y + 4)(x - y - 4)$

2. $a^2 - 2am - 10a + m^2 + 10m + 25$
   $a^2 - 2a(m + 5) + (m^2 + 10m + 25)$
   $a^2 - 2a(m + 5) + (m + 5)^2$
   $x^2 - 2xy + y^2$, where $x = a$, $y = m + 5$
   $(x - y)^2 = [a - (m + 5)]^2$
   $= (a - m - 5)^2$

For Additional Chalkboard Work
Factor.
1. $81c^2 - 25d^2$ $(9c + 5d)(9c - 5d)$
2. $25m^2 - 70mn + 49n^2$ $(5m - 7n)^2$
3. $8x^3 - 125$ $(2x - 5)(4x^2 + 10x + 25)$
4. $4x^2 + 4xy + y^2 - 25$
   $(2x + y + 5)(2x + y - 5)$

Errors Students Make
They may not recognize perfect cubes: 1, 8, 27, 64, 125, 1000, x^6, x^9, x^{12}, ...
   They do not rewrite $8x^3 + 125$ as $(2x)^3 + 5^3$ before using the formula for factoring $a^3 + b^3$.

Cumulative Review
1. Factor 180 into prime factors.
2. Simplify $\frac{-10x^{-6}y^{-4}z^3}{25x^2y^{-6}z^{-5}}$.
3. Find the GCF of 18, 24, and 36. 6

1. $2 \cdot 2 \cdot 3 \cdot 3 \cdot 5$ 2. $\frac{-2y^2z^8}{5x^8}$

3.4 Combined Types of Factoring (page 65)

Vocabulary
common monomial factor, common binomial factor, greatest common factor (GCF), factor completely

Suggestions
Emphasize throughout the remainder of the chapter that the first step in factoring is to seek a common factor of the terms.
   After factoring out the greatest common monomial factor, if any, the students may use the following hints according to the number of terms in the remaining polynomial.
1. Two terms: Look for difference of squares, sum or difference of cubes.
2. Three terms: Look for perfect square trinomial.
3. Four or more terms: Look for ways to group the terms to fit one of the patterns.

(Additional Practice for Example 7)
1. $ax + bx - ay - by$
   $(a + b)x - (a + b)y$
   $(a + b)(x - y)$
2. $3x^2 + 9xy + 2x + 6y$
   $3x(x + 3y) + 2(x + 3y)$
   $(3x + 2)(x + 3y)$

For Additional Chalkboard Work
Factor.
1. $6x^2y^2 - 15xy^3 - 3xy^2$ $3xy^2(2x - 5y - 1)$
2. $12x^2 - 4x - 40$ $4(3x + 5)(x - 2)$
3. $16x^4 - 100x^2y^2$ $4x^2(2x + 5y)(2x - 5y)$
4. $12x^4 - 23x^2 - 9$ $(2x + 3)(2x - 3)(3x^2 + 1)$
5. $-50x^2 + 60x - 18$ $-2(5x - 3)^2$
6. $2x^3 - 54$ $2(x - 3)(x^2 + 3x + 9)$
7. $3x^3 + 5x^2 - 6x - 10$ $(x^2 - 2)(3x + 5)$

Cumulative Review
Solve.
1. $5^{2x+3} = 625$ $\frac{1}{2}$
2. $|2x + 15| = 37$ 11 and -26
3. $x - 3.4 = 0.2x$ 4.25

3.5 Quadratic Equations (page 68)

Vocabulary
quadratic equation, standard form of a quadratic equation, fourth-degree equation, common numerical factor

Suggestions
Call attention to the fact that both $0 = ax^2 + bx + c$ and $ax^2 + bx + c = 0$ are in standard form.
(Additional Practice for Example 4)
1. $y^4 = 20 - y^2$
   $y^4 + y^2 - 20 = 0$
   $(y^2 - 4)(y^2 + 5) = 0$
   $(y + 2)(y - 2)(y^2 + 5) = 0$
   $y = -2$ or $y = 2$ or $y^2 + 5 = 0$
   -2 and 2 $[y^2 + 5 > 0$ for each $y]$
2. $y^4 - 10y^2 + 9 = 0$
   $(y^2 - 1)(y^2 - 9) = 0$
   $(y + 1)(y - 1)(y + 3)(y - 3) = 0$
   $-1, 1, -3, 3$

For Additional Chalkboard Work
1. $5c^2 + 7c - 6 = 0$ $\frac{3}{5}, -2$
2. $7t = 20 - 3t^2$ $\frac{5}{3}, -4$
3. $36 = 25x^2$ $\frac{6}{5}, -\frac{6}{5}$
4. $-2x^2 = 5x$ $0, -\frac{5}{2}$
5. $7n^2 + 14n - 56 = 0$ $-4, 2$
6. $y^4 - 5y^2 + 4 = 0$ $-1, 1, -2, 2$
7. $y^4 = 12 + y^2$ $-2, 2$

Cumulative Review
Find the solution set of each inequality.
1. $4 - 5x \leq 2(x + 9)$ $\{x | x \geq -2\}$
2. $|x - 5| < 7$ $\{x | -2 < x < 12\}$
3. $|x + 3| > 8$ $\{x | x < -11$ or $x > 5\}$

3.6 Quadratic Inequalities (page 71)

Vocabulary
quadratic inequality, conjunction, polynomial inequality

Suggestions
In Lessons 1.3 and 1.4, compound inequalities were introduced and solved. The teacher can introduce this lesson with a review similar to the following.
1. If $[x < 5$ and $x < 8]$, then $x < 5$.
2. If $[x < -6$ and $x < -3]$, then $x < -6$.
3. If $[x < 2$ and $x > 7]$, then there is no solution.
4. If $[x > 2$ and $x < 7]$, then $2 < x < 7$.
5. If $(m)(n) > 0$, then $[m > 0$ and $n > 0]$
   or $[m < 0$ and $n < 0]$.
6. If $(m)(n) < 0$, then $[m > 0$ and $n < 0]$
   or $[m < 0$ and $n > 0]$.

The procedures shown in Examples 1 and 2 on page 71 are used with quadratic (second-degree) inequalities. For polynomial inequalities of degree greater than two, the graphic technique shown in Examples 3 and 4 should be used.

For Additional Chalkboard Work
1. $x^2 + 3x - 10 < 0$ $\{x | -5 < x < 2\}$
2. $-x^2 + x + 12 \leq 0$ $\{x | x \leq -3$ or $x \geq 4\}$
3. $x^3 + x^2 - 20x < 0$ $\{x | x < -5$ or $0 < x < 4\}$
4. $x^4 - 8x^2 - 9 > 0$ $\{x | x < -3$ or $x > 3\}$

The factor $x^2 + 1$ is positive for each value of x.

Errors Students Make
Because $(x + 2)(x - 3) = 0$ implies $x + 2 = 0$ or $x - 3 = 0$, they believe that $(x + 2)(x - 3) < 0$ implies $x + 2 < 0$ or $x - 3 < 0$.

Cumulative Review
1. Find three consecutive odd integers such that the second integer, decreased by twice the third, is the same as 20 less than the first.
2. Each type A box contains a dozen oranges and each type B box holds 18 oranges. A truck is loaded so that the number of type A boxes is 50 less than the number of type B boxes. The truck contains 7,200 oranges. How many boxes of each type are in the truck?

1. 7, 9, 11   2. A: 210 boxes, B: 260 boxes

3.7 Problem Solving: Number Problems (page 74)

Vocabulary
consecutive integers, consecutive multiples of a number

Background
Many word problems lead to quadratic equations that can be solved by factoring. Since a quadratic equation may have two distinct solutions, these solutions need to be interpreted in the problem.
1. In Example 1, $x = -5$ or $x = 10$ and there are two sets of answers: $(-5, 0, 5)$ and $(10, 15, 20)$.
2. In Example 2, $x = -6$ or $x = 10$ where x is the number of boxes/carton. In this case, -6 cannot be the number of boxes in a carton so there is only one answer.

For Additional Chalkboard Work
1. Find three consecutive multiples of 4 so that the sum of the squares of the first and third numbers is 320.   8,12,16 or $-16,-12,-8$
2. Forty-eight fuses are placed in rows so that the number of fuses in each row is 2 more than the number of rows. Find the number of fuses in one row.   8

Cumulative Review
1. The length of a rectangle is 8 units more than the width. Represent the area of the rectangle in algebraic terms.
2. The lengths of the sides of a right triangle are x, $x + 1$, and $x + 2$ units. Represent the area of the triangle in terms of x.

1. $x(x + 8)$ square units
2. $\frac{1}{2}x(x + 1)$ square units

3.8 Problem Solving: Geometric Problems (page 78)

Vocabulary
Pythagorean relation, area, square units

Suggestions
This lesson contains enough material for two full class periods and may be covered in the following way.

Period 1   Examples 1–3   Selected Exercises from 1–10 and 15–18
Period 2   Example 4   Selected Exercises from 11–14 and 19–22

In Example 3, each side is multiplied by 2 *before* expanding the product, base times height. This avoids introducing the fractional equation

$$\tfrac{3}{2}b^2 - 3b = 36.$$

In Example 4, stress the need to begin by sketching and labeling the two figures whose areas are to be added or subtracted.

For Additional Chalkboard Work
After Example 1, use Exercise 2, 4, or 16.
After Example 2, use Exercise 6.
After Example 3, use Exercise 8 or 10.
After Example 4, use Exercise 14.

Cumulative Review
Simplify.
1. $-5x^{-2}y^4z^0 \cdot 3xy^{-1}z^{-2}$
2. $(-2x^{-2}y^4)^3$
3. $\dfrac{16x^{-5}yz^4}{12xy^{-3}z^{-2}}$

1. $\dfrac{-15y^3}{xz^2}$ 2. $\dfrac{-8y^{12}}{x^6}$ 3. $\dfrac{4y^4z^6}{3x^6}$

Computer Activities (page 86)
The idea of character strings is not too difficult, and most students comprehend it easily. However, be sure that they understand the difference between A = 6 and A$ = 6. Most important in this difference is that 3 * A will produce a value of 18, but 3 * A$ not only does not produce 18 but has no meaning at all.

Chapter and Lesson Commentaries

Chapter 4 Rational Expressions

Chapter Objectives
To show that a given number is a rational number
To find the values of the variables for which a rational expression is undefined
To simplify a rational expression
To multiply and divide rational expressions
To convert the units of a measurement to other units by using dimensional analysis
To add and subtract rational expressions
To simplify a complex rational expression

Chapter Overview
Rational numbers are identified as decimals that either terminate or repeat. Rational expressions, along with their products, quotients, sums, and differences, are simplified.

Dimensional analysis is introduced as an important application of multiplying rational expressions.

Special Features
Is 2 Equal to 1?   p. 88
Calculator Activities   p. 90
Reading in Algebra   p. 93
Non-Routine Problems   p. 101
Applications: Proofs with Rational Numbers   pp. 106–107
Computer Activities   p. 110
Chapter Review   p. 108
Chapter Test   p. 109
College Prep Test   p. 111
Cumulative Review (Chapters 1–4)   pp. 112–113

4.1 Rational Numbers and Decimals (page 89)

Vocabulary
rational number, terminating decimal, repeating decimal

Suggestions
Emphasize that the student is *already familiar* with many types of *rational numbers:*
1. all *integers:* e.g., $-3 = \frac{-3}{1}$
2. all *common fractions:* e.g., $\frac{7}{4}$
3. all *mixed numbers:* e.g., $5\frac{2}{3} = \frac{17}{3}$
4. all *terminating decimals:* e.g., $6.39 = \frac{639}{100}$
5. *some repeating decimals:* e.g., $0.333\ldots = \frac{1}{3}$

Students may wonder if there are decimals that are nonterminating and nonrepeating. These are the irrational numbers and they are introduced in Chapter 6. An example of a nonterminating nonrepeating decimal is $2.6161161116\ldots$.

Emphasize the importance of the restriction $b \neq 0$ in the definition of a rational number. For example: $\frac{8}{2} = 4$ because $4 \cdot 2 = 8$, but $\frac{8}{0}$ is undefined since no number times 0 is equal to 8. This concept is important in Lesson 4.2.

For Additional Chalkboard Work
Show that each is a rational number $\frac{a}{b}$ where a and b are integers.
1. $-3\frac{4}{5}$ 2. 23.67 3. $7.32\overline{2}$ 4. $6.23\overline{23}$
1. $\frac{-19}{5}$ 2. $\frac{2,367}{100}$ 3. $\frac{659}{90}$ 4. $\frac{617}{99}$

Cumulative Review
1. A spruce tree is 1 year older than twice the age of a pine tree, and an oak tree is 1 year older than twice the age of the spruce tree. In 3 years, the oak tree will be 2 years older than the sum of the ages of the pine and the spruce. How old is the oak tree now?   **15 years old**
2. Three brands of canned soup are placed in 12 boxes so that there are twice as many cans of brand A as brand B and 6 dozen more cans of brand C than brand A. If brands A, B, and C are worth 60¢, 75¢, and 80¢ per can, respectively, and if the 12 boxes of soup are worth $185.40, how many cans of each brand are in the boxes?
2. **A: 72 cans, B: 36 cans, C: 144 cans**

4.2 Simplifying Products and Quotients (page 91)

Vocabulary
rational expression

Background
A rational expression is the quotient $\frac{P}{Q}$ of two polynomials P and Q, $Q \neq 0$. Thus, the following expressions *are not* rational expressions.

1. $\frac{\sqrt{x+3}}{2x}$, since $\sqrt{x+3}$ is not a polynomial.
2. $\frac{8x+4}{0}$, since the denominator is zero.

Suggestions
Students should be aware that a rational expression cannot be simplified if either the numerator or the denominator cannot be factored.

Emphasize that it is common *factors*, not *terms*, that are divided out (canceled) when simplifying a rational expression.

For Additional Chalkboard Work
The expression is undefined for what values of x?

1. $\frac{x-2}{x^2-36}$ **6, −6**
2. $\frac{x^2-4}{x^2-2x-15}$ **5, −3**

Simplify.

3. $\frac{3n^2+6n-9}{6n+18}$ $\frac{n-1}{2}$
4. $\frac{2x^2-50}{3x+21} \cdot \frac{x^2+5x-14}{10-2x}$ $\frac{(x+5)(x-2)}{-3}$
5. $\frac{x^2-6x}{2x^2-11x-6} \div \frac{5x^3}{4x^2-1}$ $\frac{2x-1}{5x^2}$

Errors Students Make
They "cancel" a term or a factor of a term in the numerator and denominator of a rational expression.

1. $\frac{\cancel{5}x+15y}{2\cancel{10}} = \frac{x+15y}{2^2}$
2. $\frac{\cancel{3}x+9y}{\cancel{3}x+10} = \frac{9y}{10}$

Cumulative Review
1. At 7 A.M. a car left city A and headed toward city B at 70 km/h and at 8 A.M. another car left city B at 60 km/h and headed toward city A. If the two cities are 265 km apart, at what time did the cars pass each other?   **9:30 A.M.**
2. Factor $x^3 + 8y^3$. $(x+2y)(x^2-2xy+4y^2)$

4.3 Dimensional Analysis (page 94)

Vocabulary
conversion factor, dimensional analysis, volume, cubic units

Background
Some students know facts such as 1 ft = 12 in., 1 yd = 3 ft, 1 m = 100 cm, etc., but do not know whether to multiply or to divide in order to *convert* measurements like 72 in., 24 ft, 12 yd, 2,000 cm, 3 m, etc., to other units. *Dimensional analysis* provides a method for making the correct decision. This technique is required in the study of chemistry and physics.

Suggestions
Review the following facts before introducing the lesson: 1 mi = 5,280 ft, 1 min = 60 s, 1 qt = 2 pt, 1 gal = 4 qt, 1 lb = 16 oz, 1 T (ton) = 2,000 lb, 1 g = 1,000 mg, 1 kg = 1,000 g, 1 L = 1,000 mL, 1 m = 100 cm = 1,000 mm. Then consider areas and volumes like the following.

$\frac{1 \text{ yd}}{3 \text{ ft}} = 1$ $\left(\frac{1 \text{ yd}}{3 \text{ ft}}\right)^2 = \frac{1 \text{ yd}^2}{9 \text{ ft}^2} = 1$

$\left(\frac{1 \text{ yd}}{3 \text{ ft}}\right)^3 = \frac{1 \text{ yd}^3}{27 \text{ ft}^3} = 1$

$\frac{1 \text{ m}}{100 \text{ cm}} = 1$ $\left(\frac{1 \text{ m}}{100 \text{ cm}}\right)^2 = \frac{1 \text{ m}^2}{10,000 \text{ cm}^2} = 1$

For Additional Chalkboard Work
Use dimensional analysis to convert each of the following:

1. 25 m/s to km/h   **90 km/h**
2. 80 lb/min to T/h   **2.4 T/h**
3. 18 g/cm² to kg/m²   **180 kg/m²**
4. 24 oz/in.³ to lb/ft³   **2,592 lb/ft³**

Errors Students Make
1. 1 ft² = 12 in.², 1 ft³ = 12 in.³
2. 1 yd² = 3 ft², 1 yd³ = 3 ft³

Cumulative Review
1. Solve $2^{3-2x} = 16$.
2. Simplify $(3x^2y^{-3})^{-2}$.

1. $-\frac{1}{2}$ 2. $\frac{y^6}{9x^4}$

4.4 Sums and Differences (page 96)

Vocabulary
common denominator, least common denominator (LCD)

Suggestions
In Example 2, emphasize the need for parentheses where $-\dfrac{2x + 5}{6x + 12} = \dfrac{-(2x + 5)}{6x + 12} = \dfrac{-2x - 5}{6x + 12}$.
A common *error* with a binomial numerator is to write $-\dfrac{2x + 5}{6x + 12}$ as $\dfrac{-2x + 5}{6x + 12}$.

Additional practice for Example 2:
$$\dfrac{5a}{2a + 6} - \dfrac{a - 12}{2a + 6} = \dfrac{5a}{2a + 6} + \dfrac{-(a - 12)}{2a + 6}$$
$$= \dfrac{5a}{2a + 6} + \dfrac{-a + 12}{2a + 6}$$
$$= \dfrac{4a + 12}{2a + 6} = \dfrac{4(a + 3)}{2(a + 3)} = 2$$

Additional Practice for Example 4:
$$\dfrac{4}{x + 7} - \dfrac{3}{7 - x} + \dfrac{x - 2}{49 - x^2}$$
$$= \dfrac{4}{x + 7} + \dfrac{3}{-(7 - x)} + \dfrac{-(x - 2)}{-(49 - x^2)}$$
$$= \dfrac{4}{x + 7} + \dfrac{3}{x - 7} + \dfrac{-x + 2}{x^2 - 49} = \dfrac{6x - 5}{(x + 7)(x - 7)}$$

For Additional Chalkboard Work

1. $\dfrac{5x}{9} - \dfrac{x + 3}{6}$ $\dfrac{7x - 9}{18}$
2. $\dfrac{5}{2x + 8} - \dfrac{2x}{3x + 12}$ $\dfrac{-4x + 15}{6(x + 4)}$
3. $\dfrac{5}{2n - 6} - \dfrac{2 - n}{n^2 - 9}$ $\dfrac{7n + 11}{2(n + 3)(n - 3)}$
4. $\dfrac{x}{n - 5} - \dfrac{y}{5 - n} + \dfrac{z}{5 - n}$ $\dfrac{x + y - z}{n - 5}$
5. $\dfrac{2}{x + 6} - \dfrac{3x}{36 - x^2} + \dfrac{4}{6 - x}$ $\dfrac{x - 36}{(x - 6)(x + 6)}$

Errors Students Make
They may write $\dfrac{a}{5} - \dfrac{3a - 7}{4}$ as $\dfrac{a}{5} + \dfrac{-3a - 7}{4}$ instead of $\dfrac{a}{5} + \dfrac{-(3a - 7)}{4}$, or $\dfrac{a}{5} + \dfrac{-3a + 7}{4}$.

Cumulative Review
A second number is 7 less than a first number. If the second number is multiplied by 3 times the first number, the product is -30. Find both numbers. $(5, -2)$ and $(2, -5)$

4.5 Complex Rational Expressions (page 99)

Vocabulary
complex fraction, complex rational expression

Background
In Example 1, notice that $5x^2y$ is the *LCM* of the four denominators x, xy, x^2, and 5 and that $5x^2y$ is also the *LCD* of the four fractions $\dfrac{2}{x}, \dfrac{3}{xy}, \dfrac{3}{x^2}$, and $\dfrac{1}{5}$.
The technique used in this lesson is based on the fact that $\dfrac{x}{y} = \dfrac{k(x)}{k(y)}$ where k is the LCD of all the subfractions found in the complex fraction.

Suggestions
For Exercises 25–27, students should recall that $x^{-n} = \dfrac{1}{x^n}$ in order to write the complex rational expression to be simplified.
For Non-Routine Problems 1 and 2, it is helpful to know that $\dfrac{1}{\dfrac{a}{b}} = \dfrac{b}{a}$.

For Additional Chalkboard Work

1. $\dfrac{5 + \dfrac{2}{3}}{4 - \dfrac{3}{4}}$
2. $\dfrac{\dfrac{4x}{x^2 - 9} - \dfrac{2}{x + 3}}{\dfrac{5}{x - 3} + \dfrac{x + 1}{x^2 - 9}}$
3. $\dfrac{1 - \dfrac{4}{a} - \dfrac{12}{a^2}}{1 - \dfrac{3}{a} - \dfrac{18}{a^2}}$

1. $\dfrac{68}{39}$ 2. $\dfrac{x + 3}{3x + 8}$ 3. $\dfrac{a + 2}{a + 3}$

Cumulative Review
Simplify.
1. $\left(\dfrac{-2x^{-2}y}{3a^3b^{-3}}\right)^4$ $\dfrac{16b^{12}y^4}{81a^{12}x^8}$
2. $-4x^2(-2x^3yz^4)^3$ $32x^{11}y^3z^{12}$

Factor completely.
3. $8x^3 - 125$ $(2x - 5)(4x^2 + 10x + 25)$
4. $6x^3 + 15x^2y - 9xy^2$ $3x(x + 3y)(2x - y)$

4.6 Statistics: Averages and Variability (optional) (page 102)

Vocabulary
mean, median, mode, range, variance, standard deviation

Background
Using averages alone can be misleading. Usually large measures can skew the mean and reduce its importance. Measures such as range, variance, and standard deviation when used in conjunction with the mean may give a more valuable measure.

Suggestions
You may wish to show students how misleading a mean can be by choosing a set of data that contains one or two larger measures.

Applications (page 106)
This is the second in a strand of three applications concerned with proof in algebra.
1. Properties of a Field,   page 26
2. Proofs with Rational Numbers,   pages 106–107
3. Proofs with Complex Numbers,   pages 201–202

The proof of Theorem 7 (closure under addition) can serve as a model for the exercise (closure under multiplication).

Computer Activities (page 110)
The program on this page is a good illustration of one program serving two purposes. The first time through, the computer simply determines if a particular fraction converts to a terminating or repeating decimal. The second time through, at the option of the operator, it will print out the decimal. You may point out line 150, which controls the output. The first time through the program, A$ has not yet been defined. The test is not met, and line 160 is bypassed. Lines 220 and 240 also test for the value of A$.

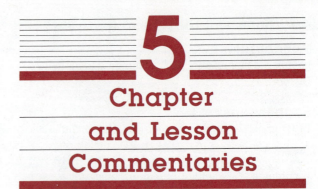

5 Chapter and Lesson Commentaries

Chapter 5 Using Rational Expressions

Chapter Objectives
To solve a fractional equation
To solve a work problem that leads to a fractional equation
To solve a motion problem that leads to a fractional equation
To solve a literal equation for one of its variables
To divide a polynomial by a binomial
To divide a polynomial by a binomial, using synthetic division
To determine if a binomial is a factor of a given polynomial
To evaluate a polynomial for a value of its variable, using synthetic substitution

Chapter Overview
Fractional equations and literal equations are solved. Work problems, round-trip problems, and upstream-downstream problems lead to fractional equations to be solved. Synthetic substitution, introduced in Lesson 2.3, is used to evaluate a polynomial and determine whether a binomial of the form $x - a$ is a factor of the polynomial.

Special Features
Non-Routine Problem Solving   p. 114
Non-Routine Problems   p. 118
Reading in Algebra   pp. 118, 123, 132
Cumulative Review   p. 124
Calculator Activities   pp. 127, 133
Computer Activities   p. 136
College Prep Test   p. 137
Chapter Review   p. 134
Chapter Test   p. 135

5.1 Fractional Equations (page 115)

Vocabulary
fractional equation, derived equation, extraneous solution, LCD proportion, extremes and means of a proportion

Background
An extraneous solution may be introduced if each side of an equation is multiplied by the same expression containing a variable. This is shown below.

$$(1) \quad x - 3 = 4$$
$$(2) \quad (x - 3)(x - 2) = 4(x - 2)$$
$$x^2 - 5x + 6 = 4x - 8$$
$$x^2 - 9x + 14 = 0$$
$$(x - 7)(x - 2) = 0$$
$$x = 7, \text{ or } x = 2$$

The solution of Equation (1) is 7. If each side of (1) is multiplied by $x - 2$, the derived Equation (2) has two solutions: 7 and 2.

NOTE: For all numbers a, b, and c, $a = b$ if and only if $ac = bc$ and $c \neq 0$. In the example above, $c = x - 2$ and $c = 0$ if $x - 2 = 0$, or $x = 2$.

Suggestions
In checking for extraneous solutions to a fractional equation, it is sufficient to substitute the possible solutions in only the *denominators*. If the value of a denominator is zero, then the possible solution is extraneous. See Example 3 on page 116.

For Additional Chalkboard Work
1. $\dfrac{2}{3x} + \dfrac{1}{4x} = 2$ 2. $\dfrac{1}{6}x^2 + \dfrac{1}{3}x = 4$
3. $\dfrac{n-1}{n-5} + \dfrac{n}{3} = \dfrac{4}{n-5}$ 4. $\dfrac{3}{x-2} = \dfrac{x}{5}$
5. $\dfrac{9}{x+3} - \dfrac{2x-8}{x^2-9} = \dfrac{2}{x-3}$

1. $\dfrac{11}{24}$ 2. $-6, 4$ 3. -3 (5 is extran.)
4. $-3, 5$ 5. 5

Cumulative Review
Factor completely.
1. $24x^2 + 44x - 140$ 2. $75x^2y - 48y$
3. $9x^4 + 30x^3 + 25x^2$ 4. $2n^3 + 16$
1. $4(3x - 5)(2x + 7)$ 2. $3y(5x + 4)(5x - 4)$
3. $x^2(3x + 5)^2$ 4. $2(n + 2)(n^2 - 2n + 4)$

5.2 Problem Solving: Work Problems (page 119)

Background
This lesson (5.2) and the next (5.3) are problem solving lessons in which three types of word problems lead to fractional equations.

Suggestions
Encourage students to represent the data in a table as shown in Examples 1 and 2 where the first row shows "the part done in one unit of time" and the second row shows "the part done in ☐ units of time" where ☐ is the time needed to complete the job.

Example 3 illustrates the type of "work" problem where *one* person or machine works alone for a given time and then *two or more* persons or machines work together until the *entire* job is done.

For Additional Chalkboard Work
1. Machine A can do a certain job in 4 h. The job can be done in 6 h by machine B. How long will the job take if machines A and B operate at the same time? 2.4 h
2. Machines A and B can do a certain job in 8 h if they operate at the same time. The job would take 12 h if A operated alone. How long will the job take if B operates alone? 24 h
3. Machine A can do a certain job in 6 h, and machine B can do the job in 10 h. If B operates alone for 3 h and then A and B operate together, find the *total* time needed to do the complete job.

Solution: B, alone, does $\dfrac{3}{10}$ of the job in the first 3 h. Then A and B together do $\dfrac{x}{6} + \dfrac{x}{10}$ of the job in the remaining x hours.
$\dfrac{3}{10} + \left(\dfrac{x}{6} + \dfrac{x}{10}\right) = 1$, $9 + 5x + 3x = 30$,
$x = 2\dfrac{5}{8}$. Thus, the total time, $3 + x$,
is $3 + 2\dfrac{5}{8}$, or $5\dfrac{5}{8}$ h.

Cumulative Review
1. Solve $3^{2-3x} = 243$ for x. -1
2. Factor $9x^2 - 12xy + 4y^2 - 25$.
2. $(3x - 2y + 5)(3x - 2y - 5)$

5.3 Problem Solving: Round Trips (page 122)

Vocabulary
headwind, tailwind, upstream, downstream

Suggestions
The students will be familiar with the formula $d = rt$. Have them solve this formula for $t \left(t = \frac{d}{r} \right)$, since the equations needed to solve the problems will be based on either

1. $t_1 + t_2 = t$ round trip, or
$$\frac{d_1}{r_1} + \frac{d_2}{r_2} = t \text{ round trip, or}$$
2. $t_1 = t_2$, or
$$\frac{d_1}{r_1} = \frac{d_2}{r_2}.$$

Some of the exercises will involve an airplane flying with a tailwind or against a headwind. This is analogous to Example 2 where a boat travels downstream (with the current) or upstream (against the current). The following example may be used after Example 2 to illustrate travel with or against a wind.

A certain plane can fly at 400 km/h in still air. It can fly 126 km with a certain tailwind in the same time it could fly 114 km in the opposite direction (against the headwind). Find the rate of the tailwind.

Let w = the rate of the tailwind.
$$\frac{126}{400 + w} = \frac{114}{400 - w}$$
$$126(400 - w) = 114(400 + w)$$
$$50{,}400 - 126w = 45{,}600 + 114w$$
$$4800 = 240w$$
$$20 = w$$

Thus, the rate of the tailwind is 20 km/h.

For Additional Chalkboard Work
Arnold drove from his home to Grass Lake at 75 km/h. He fished for 2 h and then returned home at 50 km/h. If he was away from home for 5 h, find the distance from his home to the lake.
$\frac{d}{75} + \frac{d}{50} = 5 - 2$ and $d = 90$ km

Cumulative Review
1. $3c^2 = 5c + 28$ 2. $4x^2 = 8x$
3. $4n^4 - 17n^2 + 4 = 0$
1. $-2\frac{1}{3}, 4$ 2. $0, 2$ 3. $-\frac{1}{2}, \frac{1}{2}, 2, -2$

5.4 Literal Equations and Formulas (page 125)

Suggestions
After Example 1, help the students to understand that solving a literal equation for one variable in terms of the other variables does not provide a *unique* expression but does provide an expression that is equivalent to other correct responses. A helpful type of exercise is to have the class create other expressions that are equivalent to a given rational expression, such as
$$\frac{x - a}{-y - b}.$$
Some appropriate answers are:
$$-\frac{x - a}{y + b}, \frac{-x + a}{y + b}, \frac{a - x}{y + b}.$$

Example 4 and Exercises 16–21 provide a type of equation that will occur in Chapter 9 when the point-slope form of the equation of a line is developed.

In Exercises 22–25 and 36–41, the student should solve for the specified variable before finding the value of the variable. These exercises can be facilitated with the use of a calculator.

For Additional Chalkboard Work
Solve for x. (Ex. 1–3)

1. $ax - 7 = 3x + b$ $x = \frac{b + 7}{a - 3}$

2. $a(bx + 5) = cx - d$ $x = \frac{-5a - d}{ab - c} = \frac{5a + d}{c - ab}$

3. $\frac{2}{A} = \frac{3}{x} - \frac{4}{y}$ $x = \frac{3Ay}{2y + 4A}$

4. Solve $\frac{y - 4}{z + 3} = \frac{t}{5}$ for y. Then find the value of y if $t = 6$ and $z = 2.5$. $y = \frac{tz + 3t + 20}{5}$
$= 10.6$

Cumulative Review
Simplify.
1. $-3c(-5a^{-2}bc^2)^3$ $\frac{375b^3c^7}{a^6}$

2. $\left(\frac{-2x^{-2}y^3}{3mn^{-3}} \right)^4$ $\frac{16y^{12}n^{12}}{81x^8m^4}$

3. $\frac{5}{2x - 8} - \frac{3x + 2}{x^2 - 7x + 12}$ $\frac{-x - 19}{2(x - 3)(x - 4)}$

4. $\dfrac{\frac{1}{x} - \frac{2}{y}}{x - \frac{1}{y}}$ $\dfrac{y - 2x}{x^2y - x}$

5.5 Dividing Polynomials (page 128)

Vocabulary
divisor, dividend, quotient, remainder

Suggestions
Review the meaning of *divisor, dividend, quotient,* and *remainder*.

The division work can be checked as follows: (divisor) × (polynomial quotient) + remainder = dividend. This is analogous to checking division with whole numbers.

Example 1 may be followed by the example below where the dividend is of a degree greater than 2.

$(15x^3 - 2x^2 - 7x - 3) \div (3x - 1)$.

$$
\begin{array}{r}
5x^2 + x - 2 \\
3x - 1 \overline{) 15x^3 - 2x^2 - 7x - 3} \\
\underline{15x^3 - 5x^2} \\
3x^2 - 7x \\
\underline{3x^2 - 1x} \\
-6x - 3 \\
\underline{-6x + 2} \\
-5
\end{array}
$$

Thus, the polynomial quotient is $5x^2 + x - 2$, the remainder is -5, and $3x - 1$ is not a factor of the dividend.

In Example 2, the dividend needs to be arranged in descending order of exponents.

Example 3 illustrates the need to examine the dividend for any missing terms. In this example, the missing term may be written as either 0 or $0x^2$.

For Additional Chalkboard Work
Find the polynomial quotient, the remainder, and whether the divisor is a factor of the remainder.
1. $(4x^4 - 14x^2 - 8) \div (2x + 4)$
   $Q = 2x^3 - 4x^2 + x - 2, R = 0, 2x + 4$ is a factor.
2. $(5x^2 + 195 + 3x^3 - 10x) \div (x + 5)$
   $Q = 3x^2 - 10x + 40, R = -5, x + 5$ is not a factor.

Cumulative Review
1. Simplify $\dfrac{16x^{-4}y^6(3x^2 - 108)}{10x^{-1}y^{-2}(2x^2 - 14x + 12)}$.
2. Factor $16x^3 - 2y^3$ completely.
3. Find three consecutive even integers such that the product of the second and the third integers is 168. (10, 12, 14) and (−16, −14, −12)

1. $\dfrac{12y^8(x + 6)}{5x^3(x - 1)}$ 2. $2(2x - y)(4x^2 + 2xy + y^2)$

5.6 Remainder Theorem (page 130)

Suggestions
Some students are confused about the sign of a in the divisor: $\underline{a}\rfloor$. The following summary may be helpful.

To *divide* a polynomial, P, by $x - a$, use: $\underline{a}\rfloor$.
To divide by $x + a$, use: $\underline{-a}\rfloor$.
To *evaluate* a polynomial, P, if $x = a$, use $\underline{a}\rfloor$.

Explanation of Calculator Activity Procedure
$5x^4 - 17x^3 + 23x^2 - 49x + 90$, if $x = 7$,
$= x(5x^3 - 17x^2 - 23x + 49) + 90$
$= x[x(5x^2 - 17x - 23) + 49] + 90$
$= x\{x[x(5 \cdot x - 17) - 23] + 49\} + 90$
$= \{[(5 \cdot x - 17) \cdot x - 23] \cdot x + 49\} \cdot x + 90$
$= 5 \otimes 7 \ominus 17 \otimes 7 \ominus 23 \otimes 7 \oplus 49 \otimes$
   $7 \oplus 90 \ominus 5,480$

The sequence of calculator keys shown is for a simple (nonscientific) calculator. For a scientific calculator, press $\ominus$ following the addition or subtraction of a number.

$5 \otimes 7 \ominus 17 \ominus \otimes 7 \ominus 23 \ominus \otimes 7 \oplus 49 \ominus \otimes$
$7 \oplus 90 \ominus 5,480$

For Additional Chalkboard Work
1. Find the polynomial quotient and remainder.
   $(2x^4 - 10x^2 - 12) \div (x - 3)$
   $Q = 2x^3 + 6x^2 + 8x + 24, R = 60$
2. Evaluate $4x^4 - 8x^3 - 33x^2 + 2x + 8$ using each of $-1, -2, 2,$ and 4 for x. Name the binomial factors found, if any.
   $-15, 0, -120, 0; x + 2$ and $x - 4$

Cumulative Review
One leg of a right triangle is 3 cm longer than the other leg and 3 cm shorter than the hypotenuse. Find the lengths of the three sides.
9 cm, 12 cm, 15 cm

Computer Activities (page 136)
The important new concept on this page is the subscripted variables. Explain that there is no basic difference between using A, B, C for variables or using C(1), C(2), C(3). The advantage to the latter is that it facilitates the use of the loop and saves lines in the program.

T-35

Chapter 6 Radicals and Rational Number Exponents

Chapter Objectives
To solve a quadratic equation by using the definition of square root
To approximate a square root to the nearest hundredth by using a table of square roots
To determine whether a given real number is rational or irrational
To simplify a radical, radical expressions involving sums, differences, products, or quotients
To write a radical expression in exponential form
To write an exponential expression in radical form
To evaluate an expression containing rational number exponents
To simplify an expression containing rational number exponents
To solve an exponential equation

Chapter Overview
Radicals are introduced as solutions of equations like $x^2 = 8$, $x^3 = 16$, and $x^4 = 32$. Properties of exponents are extended to include rational number exponents. Irrational numbers are introduced to complete the system of real numbers.

Special Features
Careers: Chemical Science   p. 138
Non-Routine Problems   pp. 144, 151
Reading in Algebra   pp. 141, 143, 150
Cumulative Review   pp. 141, 151
Chapter Review   p. 160   Chapter Test   p.161

Computer Activities   p. 162
College Prep Test   p. 163

6.1 Square Roots (page 139)

Vocabulary
approximately equal to ($\doteq$), irrational number, real number, square root ($\sqrt{k}$), radical, radical sign, radicand, rational number

Suggestions
Review the process of finding the square root of a number as the *inverse* of squaring the square root, as shown in the following examples.

$$4^2 = 16 \text{ and } \sqrt{16} = \sqrt{4^2} = 4$$
$$7^2 = 49 \text{ and } \sqrt{49} = \sqrt{7^2} = 7$$
In general, $\sqrt{x^2} = x$, for $x \geq 0$.

Some students confuse these two situations:
1. The equation $x^2 = 7$ has *two* solutions, $\sqrt{7}$ and $-\sqrt{7}$.
2. The radical $\sqrt{9}$ has but *one* value, 3.

Emphasize that an *equation* like $x^2 = 16$ has two solutions. The solutions of $x^2 = 16$ are $\sqrt{16}$ and $-\sqrt{16}$, or 4 and -4, where $\sqrt{16} = 4$ and $-\sqrt{16} = -4$.

Stress the terms *radical*, *radicand*, and *radical sign*. The word *radicand* will be used widely in this chapter.

For Additional Chalkboard Work
Solve each equation.
1. $6y^2 - 20 = 8 - y^2$ $-2, 2$
2. $3n^2 + 9 = 7n^2 - 35$ $-\sqrt{11}, \sqrt{11}$

Determine whether each number is rational or irrational.
3. 2.34234234 $\cdots$
4. 1.51151115 $\cdots$
5. 3.5416   6. $\sqrt{90}$ 7. $-\sqrt{900}$
8. $\sqrt{0.9}$ 9. $\sqrt{0.09}$

rational: 3,5,7,9; irrational: 4,6,8

Cumulative Review
Solve each problem.
1. A triangle's height is 5 cm less than twice the length of its base. Find the height if the area of the triangle is 44 cm^2.
2. The length of a rectangle is 3 m more than 3 times its width. Find the length if the area of the rectangle is 60 m^2.

1. 11 cm   2. 15 m

6.2 Simplifying Radicals (page 142)

Background
The first condition for *simplest radical form* is introduced. More conditions are added in the next few lessons. These conditions are listed below for $\sqrt[n]{k}$.

1. No radicand contains a perfect nth power.
2. No denominator contains a radical.
3. No radicand contains a fraction.

Suggestions
Emphasize that when computations with approximate data are involved, no rounding should occur until all other computations have been done. In Example 1, if 4.583 was rounded to 4.58 and then multiplied by 3, the result would be 13.74 which is in error by 0.01.

A classroom set of the table of square roots on page 590 can be duplicated (dry-copied) for student use on tests and quizzes.

For Additional Chalkboard Work
Simplify.
1. $-10\sqrt{128}$ 2. $\sqrt{c^2}$ 3. $\sqrt{c^4}$ 4. $\sqrt{c^{10}}$
5. $\sqrt{c^{12}}$ 6. $\sqrt{c^3}, c \geq 0$ 7. $\sqrt{c^9}, c \geq 0$
8. $\sqrt{c^6 d^8}$ 9. $\sqrt{12x^5 y^6}, x \geq 0$
10. $xy\sqrt{x^6 y^7}, y \geq 0$

1. $-80\sqrt{2}$ 2. $|c|$ 3. c^2 4. $|c^5|$
5. c^6 6. $c\sqrt{c}$
7. $c^4\sqrt{c}$ 8. $|c^3|d^4$ 9. $2x^2|y^3|\sqrt{3x}$
10. $x|x^3|y^4\sqrt{y}$

Cumulative Review
Factor completely.
1. $x^2 y - y^2 + 5x^2 - 5y$
2. $36 - 4x^2 + 4xy - y^2$
3. $5x^3 - 40y^3$

1. $(x^2 - y)(y + 5)$ 2. $(6 + 2x - y)(6 - 2x + y)$
3. $5(x - 2y)(x^2 + 2xy + 4y^2)$

6.3 Sums, Differences, and Products of Radicals (page 145)

Vocabulary
like radicals

Background
For the remaining lessons of Chapter 6, all variables are quantified so that absolute values are not needed in the answers as they often were in Lesson 6.2.

Suggestions
Stress the importance of the quantifier: If $x \geq 0$ and $y \geq 0$, in the property of the product of square roots, $\sqrt{x} \cdot \sqrt{y} = \sqrt{x \cdot y}$. If $x = -4$ and $y = -9$, notice that $\sqrt{-4}$ and $\sqrt{-9}$ are not real numbers

and $\sqrt{-4} \cdot \sqrt{-9} \neq \sqrt{-4(-9)}$
since $\sqrt{-4(-9)} = \sqrt{36}$, or 6,
while $\sqrt{-4} \cdot \sqrt{-9} = 2i \cdot 3i = 6i^2$, or -6.

Level C Exercises 51–58 require students to factor polynomials with integer coefficients over the set of real numbers. Previously, students factored polynomials with integer coefficients over the set of integers. Exercises 51–58 introduce a skill that is required in a first course in calculus.

For Additional Chalkboard Work
Simplify.
1. $3\sqrt{5} + 4\sqrt{3} + \sqrt{20} - 2\sqrt{27}$
   $5\sqrt{5} - 2\sqrt{3}$
2. $\sqrt{25c^2 d^3} + 2cd\sqrt{d}; c \geq 0, d \geq 0$ $7cd\sqrt{d}$
3. $2\sqrt{5}(3\sqrt{10} + \sqrt{5} - 2\sqrt{3})$
   $30\sqrt{2} + 10 - 4\sqrt{15}$
4. $\sqrt{3ab^5} \cdot \sqrt{6a^7 b^2}; a \geq 0, b \geq 0$
   $3a^4 b^3 \sqrt{2b}$
5. $(2\sqrt{7} - 3\sqrt{2})^2$ $46 - 12\sqrt{14}$

Cumulative Review
Machine A can do a certain job in 12 h. If Machine B works at the same time, the job is done in 8 h. How long would the job take Machine B working alone?
24 h

6.4 Quotients of Radicals (page 148)

Vocabulary
conjugates, rationalizing a denominator

Suggestions
In simplifying $\dfrac{\sqrt{5}}{3\sqrt{2}}$, notice that you can multiply by $\dfrac{\sqrt{2}}{\sqrt{2}}$ rather than $\dfrac{3\sqrt{2}}{3\sqrt{2}}$. This saves one step, as shown below.

$\dfrac{\sqrt{5}}{3\sqrt{2}} \cdot \dfrac{\sqrt{2}}{\sqrt{2}} = \dfrac{\sqrt{10}}{6}$ compared to $\dfrac{\sqrt{5}}{3\sqrt{2}} \cdot \dfrac{3\sqrt{2}}{3\sqrt{2}}$
$= \dfrac{3\sqrt{10}}{18} = \dfrac{\sqrt{10}}{6}$.

Example 3 shows that it is better to simplify each radical *before* rationalizing the denominator.

Compare the methods of simplifying $\dfrac{5a}{\sqrt{8x^6y^5}}$:

Method 1	*Method 2*
$\dfrac{5a}{\sqrt{8x^6y^5}}$	$\dfrac{5a}{\sqrt{8x^6y^5}} \cdot \dfrac{\sqrt{8x^6y^5}}{\sqrt{8x^6y^5}}$
$= \dfrac{5a}{2x^3y^2\sqrt{2y}} \cdot \dfrac{\sqrt{2y}}{\sqrt{2y}}$	$= \dfrac{5a \cdot 2x^3y^2\sqrt{2y}}{8x^6y^5}$
$= \dfrac{5a\sqrt{2y}}{4x^3y^3}$	$= \dfrac{5a\sqrt{2y}}{4x^3y^3}$

For Additional Chalkboard Work
Simplify.
1. $\dfrac{\sqrt{30c^{10}}}{\sqrt{3c^3}}$, $c > 0$ $c^3\sqrt{10c}$
2. $\dfrac{10\sqrt{3}}{\sqrt{50}}$ $\sqrt{6}$
3. $\dfrac{4\sqrt{2y}}{\sqrt{12x^6y^5}}$, $x > 0$, $y > 0$ $\dfrac{2\sqrt{6}}{3x^3y^2}$
4. $\dfrac{2\sqrt{3}}{3\sqrt{2} - \sqrt{6}}$ $\dfrac{\sqrt{6} + \sqrt{2}}{2}$
5. $\sqrt{\dfrac{7c}{18d^5}}$; $c > 0$, $d > 0$ $\dfrac{\sqrt{14cd}}{6d^3}$

Cumulative Review
Solve.
1. $2^{3x} = 64$
2. $\dfrac{y+1}{y+2} - \dfrac{y-3}{3} = \dfrac{y-1}{y+2}$

1. 2   2. 4, −3

6.5 Other Radicals (page 152)

Vocabulary
index for a radical, *n*th root of k ($\sqrt[n]{k}$)

Background
Properties for square roots are extended to cube roots, fourth roots, fifth roots, and *n*th roots in general.

Suggestions
The students should memorize third, fourth, and fifth powers of some bases, as shown in the table on page 152.

This lesson deals with *odd* numbered indexes for $\sqrt[n]{k}$, where k is *any* real number, and *even*-numbered indexes for $\sqrt[n]{k}$, where k is a *positive* number or zero. Some students may ask how to deal with $\sqrt[n]{k}$ where n is even and k is a negative number, for example, $\sqrt{-9}$ and $\sqrt{-25}$. Explain that expressions such as these will be defined and used in Chapter 8, where $\sqrt{-1}$ is defined so that $\sqrt{-1} = i$ and $i^2 = -1$.

For Additional Chalkboard Work
Simplify.
1. $\sqrt[3]{-64}$
2. $-2\sqrt[4]{81}$
3. $-3\sqrt[5]{32}$
4. $\sqrt[3]{24c^6}$
5. $\sqrt[4]{32c^8d^5}$, $d \geq 0$
6. $5\sqrt[4]{2} + \sqrt[4]{32} - \sqrt[4]{162}$
7. $\sqrt[3]{4t^4} \cdot \sqrt[3]{6t}$
8. $\dfrac{5}{\sqrt[4]{27c^2}}$, $c \geq 0$
9. $\sqrt[3]{\dfrac{2}{25c}}$, $c \neq 0$
10. $\dfrac{6}{\sqrt[3]{5} - \sqrt[3]{2}}$

1. −4   2. −6   3. −6   4. $2c^2\sqrt[3]{3}$
5. $2c^2d\sqrt[4]{2d}$ 6. $4\sqrt[4]{2}$ 7. $2t\sqrt[3]{3t^2}$
8. $\dfrac{5\sqrt[4]{3c^2}}{3c}$ 9. $\dfrac{\sqrt[3]{10c^2}}{5c}$
10. $2(\sqrt[3]{25} + \sqrt[3]{10} + \sqrt[3]{4})$

Cumulative Review
Brands *A*, *B*, and *C* of peanuts are worth $1.20, $1.50, and $1.80 per kilogram, respectively. If 3 kg of brand *A* are added to twice as much brand *C* as brand *B* to form a mixture that is worth $24.00, how much brand *C* is in the mixture?   8 kg

6.6 Rational Number Exponents (page 155)

Vocabulary
rational number exponent ($x^{\frac{m}{n}}$), radical form, exponential form

Suggestions
The student should associate the *denominator* n in $x^{\frac{m}{n}}$ with the *index* n in $(\sqrt[n]{x})^m$ and the *numerator* m in $x^{\frac{m}{n}}$ with the *exponent* m in $(\sqrt[n]{x})^m$. The student should note also that the base x can be raised to the power either *before* or *after* the root is taken, since $x^{\frac{m}{n}} = \sqrt[n]{x^m}$ and $x^{\frac{m}{n}} = (\sqrt[n]{x})^m$.

For Additional Chalkboard Work
Rewrite in exponential form.
1. $5\sqrt{x}, x \geq 0$
2. $\sqrt[3]{5x}$
3. $2\sqrt[4]{n^7}, n \geq 0$
4. $(\sqrt[5]{3c})^2$

Rewrite in radical form.
5. $14^{\frac{1}{2}}$
6. $7c^{\frac{1}{4}}, c \geq 0$
7. $2^{\frac{3}{5}}$
8. $(5n)^{\frac{2}{3}}$

Find the value.
9. $36^{\frac{1}{2}}$
10. $8^{-\frac{1}{3}}$
11. $10 \cdot 16^{\frac{5}{4}}$
12. $-8 \cdot 32^{-\frac{2}{5}}$

1. $5x^{\frac{1}{2}}$
2. $(5x)^{\frac{1}{3}}$
3. $2n^{\frac{7}{4}}$
4. $(3c)^{\frac{2}{5}}$
5. $\sqrt{14}$
6. $7\sqrt[4]{c}$
7. $\sqrt[5]{8}$ or $(\sqrt[5]{2})^3$
8. $\sqrt[3]{25n^2}$ or $(\sqrt[3]{5n})^2$
9. 6
10. $\frac{1}{2}$
11. 320
12. -2

You may wish to assign these Level C exercises. Write in simplest radical form.
49. $(x^2y^3)^{\frac{5}{n}}$
50. $(x^ay^b)^{\frac{c}{n}}$
51. $(x^ny^{3n})^{\frac{1}{n}}$
52. $(x^{5n}y^n)^{\frac{a}{n}}$
53. $(x^{n+1}y^{2n+2})^{\frac{1}{n}}$
54. $(x^{4n}y^{4n+3})^{\frac{k}{n}}$
49. $\sqrt[n]{x^{10}y^{15}}$
50. $\sqrt[n]{x^{ac}y^{bc}}$
51. xy^3
52. $x^{5a}y^a$
53. $xy^2\sqrt[n]{xy^2}$
54. $x^{4k}y^{4k}\sqrt[n]{y^{3k}}$

Cumulative Review
Simplify.
1. $\frac{2a^2 - 50}{15 - 3a}$
2. $\frac{6 - \frac{3}{x}}{6 + \frac{3}{y}}$

1. $-\frac{2(a+5)}{3}$
2. $\frac{2xy - y}{2xy + x}$

6.7 Expressions With Rational Number Exponents (page 157)

Suggestions
Examples 1–4 illustrate the extension of all of the properties for integral exponents to expressions containing rational number exponents. Emphasize that an expression containing rational number exponents is *not* in simplest form:
1. if an exponent is a *negative* number
2. if the expression's *denominator* contains a fractional exponent.

For example, neither $5x^{-\frac{1}{2}}$ nor $\frac{5}{x^{\frac{1}{2}}} = \left(\frac{5}{\sqrt{x}}\right)$ is in simplest form.

Examples 1 and 3 will need additional practice.

Example 1: Simplify $5x^{-2}y^{\frac{2}{3}} \cdot 2x^{\frac{1}{3}}y^{-\frac{1}{4}}; x > 0, y > 0$.
$$\frac{5 \cdot 2x^{\frac{1}{3}}y^{\frac{2}{3}}}{x^2y^{\frac{1}{4}}} = \frac{10x^{\frac{1}{3}}y^{\frac{8}{12}}}{x^{\frac{6}{3}}y^{\frac{3}{12}}} = \frac{10y^{\frac{5}{12}}}{x^{\frac{5}{3}}} \cdot \frac{x^{\frac{1}{3}}}{x^{\frac{1}{3}}} = \frac{10x^{\frac{1}{3}}y^{\frac{5}{12}}}{x^2}$$

Example 2: Simplify $\frac{a^{\frac{1}{2}}b^{-\frac{2}{5}}}{a^{\frac{3}{5}}b^{-\frac{6}{5}}}; a > 0, b > 0$.
$$\frac{a^{\frac{5}{10}}b^{\frac{6}{5}}}{a^{\frac{6}{10}}b^{\frac{2}{5}}} = \frac{b^{\frac{4}{5}}}{a^{\frac{1}{10}}} \cdot \frac{a^{\frac{9}{10}}}{a^{\frac{9}{10}}} = \frac{a^{\frac{9}{10}}b^{\frac{4}{5}}}{a}$$

For Additional Chalkboard Work
Simplify. Values of the variables are positive numbers.
1. $\frac{a^{\frac{8}{7}}}{a^{\frac{2}{7}}}$
2. $\frac{a^{\frac{2}{7}}}{a^{\frac{8}{7}}}$
3. $x^{-\frac{1}{5}} \cdot x^{\frac{2}{5}} \cdot x^{-\frac{3}{5}}$
4. $\frac{x^{\frac{1}{2}}y^{-\frac{1}{5}}z^{-\frac{1}{2}}}{x^2y^{-\frac{3}{5}}z^{\frac{1}{3}}}$
5. $(8m^6n^{12})^{\frac{1}{3}}$
6. $(3m^4n^{-6})^{-2}$
7. $\left(\frac{27m^9n^{-6}}{a^3b^{-12}}\right)^{\frac{2}{3}}$

Solve for x.
8. $25^x = 125$
9. $8^{2x-1} = 32$
10. $2^{3x-2} = \frac{1}{16}$

1. $a^{\frac{6}{7}}$
2. $\frac{1}{a^{\frac{6}{7}}}$
3. $\frac{x^{\frac{3}{5}}}{x}$
4. $\frac{x^{\frac{1}{2}}y^{\frac{2}{5}}z^{\frac{1}{6}}}{x^2z}$
5. $2m^2n^4$
6. $\frac{n^{12}}{9m^8}$
7. $\frac{9m^6b^8}{a^2n^4}$
8. $\frac{3}{2}$
9. $\frac{4}{3}$
10. $-\frac{2}{3}$

Computer Activities (page 162)
Be sure your students understand the test ABS ( ) <.05 is a test for equality, made necessary by the fact that the composite function S ∧ (1/x) compounds the possibility of values like 4 turning up as 3.9999999 or 4.000001.

T-39

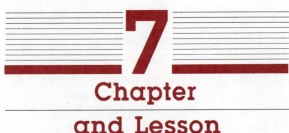

Chapter and Lesson Commentaries

Chapter 7 Quadratic Formula

Chapter Objectives
To solve a quadratic equation by completing the square
To find the maximum value of a quadratic polynomial by completing the square
To solve a word problem involving a maximum value
To solve a quadratic equation by using the quadratic formula
To solve a word problem with irrational solutions by using the quadratic formula
To solve a radical equation

Chapter Overview
The technique of *completing the square* is developed. It is used to derive the quadratic formula and to find the maximum value of $ax^2 + bx + c$, $a < 0$. This skill will be used again in Chapter 12 to write equations in the standard forms for conic sections.

Word problems with irrational answers are solved.

Special Features
Non-Routine Problem Solving   p. 164
Non-Routine Problems   p. 177
Reading in Algebra   p. 171
Applications   p. 181
Calculator Activities   pp. 174, 177
Cumulative Review   p. 180
Computer Activities   p. 184
College Prep Test   p. 185
Chapter Review   p. 182
Chapter Test   p. 183

7.1 Completing the Square (page 165)

Vocabulary
completing the square

Background
An equation of the form $ax^2 + bx + c = 0$, $a \neq 0$, is called a *quadratic equation* in standard form. In Chapter 3, quadratic equations, such as $3x^2 + 10x - 8 = 0$, $5x^2 - 15x = 0$, and $4x^2 - 9 = 0$, were solved by factoring the quadratic polynomial. In this lesson quadratic equations are solved by completing the square. The method is generalized in Lesson 7.3 and results in the quadratic formula. The skill of completing the square has immediate application to maximum value in Lesson 7.2 and extensive application in Chapter 12, where equations of conic sections are rewritten in standard form.

Suggestions
Leave a space for the third term when completing the square, as shown in Examples 1–3. The symbol "$\pm$" is introduced to represent the abbreviated form of the two square roots of a positive number.

For Additional Chalkboard Work
Solve each equation by completing the square.
1. $y^2 - 12y = -7$ **2.** $n^2 + 5n = 24$
3. $16x^2 - 24x + 7 = 0$

1. $6 \pm \sqrt{29}$ **2.** $3, -8$ **3.** $\dfrac{3 \pm \sqrt{2}}{4}$

Cumulative Review
1. Simplify $\dfrac{3}{x-9} + \dfrac{7}{9-x} - \dfrac{2}{x-9} \cdot \dfrac{-6}{x-9}$
2. Solve $\dfrac{3x+5}{2x-3} = \dfrac{2x+7}{x-1}$ $2, -8$
3. Evaluate by using synthetic substitution.
$2x^4 + 10x^3 - 11x^2 - 40$ if $x = -6$ -4

7.2 Maximum Value (page 167)

Vocabulary
maximum value

Suggestions
The steps shown in Example 4 are needed by most students in order to write two binomial factors with one variable. Use the following example after Example 4.

A company can sell 600 games per month if each game sells for $4.00. It can sell 50 more games per month for each 10-cent reduction in the price. (1) What price per game will give the maximum income? (2) What is the maximum monthly income? (3) How many games will be sold in a month to produce the maximum income?

No. of 10-cent drops	No. of games	Price per game
0	600	4
1	600 + 50	4 − 0.10
2	600 + 50(2)	4 − 0.10(2)
3	600 + 50(3)	4 − 0.10(3)
x	600 + 50x	4 − 0.10x

$(600 + 50x)(4 - 0.1x)$
$= -5x^2 + 140x + 2{,}400$
$= -5(x - 14)^2 + 3{,}380$

(1) $4 - 0.10x = 4 - 0.10(14)$, or 2.60/game
(2) $3,380
(3) $600 + 50x = 600 + 50(14)$, or 1,300 games/mo

For Additional Chalkboard Work
Find the maximum value and the corresponding value of x.
1. $16x - 2x^2$ 2. $3200 + 240x - 20x^2$
3. $(40 + 20x)(8 - 2x)$
1. 32; 4   2. 3,920; 6   3. 360; 1

Cumulative Review
Solve.
1. $\frac{a - 26}{a^2 - 16} + \frac{3}{a - 4} = \frac{6}{a + 4}$ 5
2. $x^2 + 7x = 0$ 0, −7
3. $\frac{3}{4}x - 2 = 7$ 12

7.3 The Quadratic Formula (page 169)

Vocabulary
quadratic formula

Suggestions
Encourage the B- and C-level students to study the derivation of the quadratic formula on page 169 in preparation for a future quiz.

Point out to students that a pair of irrational solutions can be written in either of two ways. For example, either $\frac{1 \pm 2\sqrt{3}}{2}$ or $\frac{1 + 2\sqrt{3}}{2}$, $\frac{1 - 2\sqrt{3}}{2}$ is an acceptable answer in Example 2. However, a pair of rational solutions, such as $\frac{9 \pm 9}{10}$, must be listed separately as $\frac{9}{5}$, 0, as shown in Example 3.

For Additional Chalkboard Work
You may wish to use this as a B- and C-level quiz. Solve $mx^2 + nx + p = 0$, $m \neq 0$, by completing the square. Show each step.
$$x = \frac{-n \pm \sqrt{n^2 - 4mp}}{2m}$$
Solve. Use the quadratic formula.
1. $4x^2 - 7x + 2 = 0$ 2. $9x^2 = 12x - 1$
3. $6x^2 + 5x = 0$ 4. $72 - x^2 = 0$

1. $\frac{7 \pm \sqrt{17}}{8}$
2. $\frac{12 \pm \sqrt{108}}{18} = \frac{12 \pm 6\sqrt{3}}{18} = \frac{2 \pm \sqrt{3}}{3}$
3. $0, -\frac{5}{6}$ 4. $\pm 6\sqrt{2}$

Cumulative Review
1. Factor $x^3 - 125$.
2. Factor $ax^2 - nx - amx + mn$.
3. A second number is 1 more than twice a first number. The product of the second number and 3 less than the first number is 30. Find both numbers. $-3, -5; \frac{11}{2}, 12$
1. $(x - 5)(x^2 + 5x + 25)$
2. $(x - m)(ax - n)$

7.4 Problem Solving: The Quadratic Formula (page 172)

Background
In Chapter 3 the students solved word problems that led to quadratic equations with rational solutions. The equations were solved by factoring. In this lesson and in the next lesson, the students will solve problems that lead to quadratic equations with irrational solutions.

Suggestions
In some of the exercises, answers are to be given to the nearest tenth (0.1). Emphasize that rounding to the nearest tenth should be delayed until all other operations have been done. The reason for this is shown below.

Correct Way	Wrong Way
$\dfrac{3 + 6\sqrt{17}}{2}$	$\dfrac{3 + 6\sqrt{17}}{2}$
$\dfrac{3 + 6(4.123)}{2}$	$\dfrac{3 + 6(4.1)}{2}$
$\dfrac{3 + 24.738}{2}$	$\dfrac{3 + 24.6}{2}$
13.869	13.8
13.9	

Notice that the answers differ by 0.1.

For Additional Chalkboard Work
1. One number is 6 more than another number, and their product is 2. Find two pairs of such numbers. $(-3 + \sqrt{11}, 3 + \sqrt{11})$, $(-3 - \sqrt{11}, 3 - \sqrt{11})$
2. Use $h = vt - 5t^2$ to find t to the nearest tenth of a second if $h = 45$m and $v = 40$m/s. **1.4s, 6.6s**

Cumulative Review
1. Simplify $\dfrac{-16x^{-3}y^2z^{-4}}{24x^6y^{10}z^{-12}} \cdot \dfrac{-2z^8}{3x^9y^8}$
2. Simplify $3x^2(-5xy^4)^3$. $-375x^5y^{12}$
3. Solve $3^{4-x} = 243$. -1
4. Find the value of 8.76×10^{-3}.
4. 0.00876

7.5 Problem Solving: Irrational Measurements (page 175)

Suggestions
The most difficult part of this lesson is representing the data in exercises such as Example 2. Use the chalkboard and complete the diagram in Example 2 step by step.

Represent the data on diagrams for Exercises 3 and 7 during the class discussion. Then, assign the parallel Exercises 4 and 8. The diagrams for Exercises 3 and 7 are shown below.

3.

7.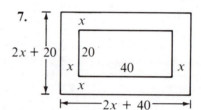

Cumulative Review
Find the solution set. Graph the solutions on a number line.
1. $x^2 - 3x - 10 < 0$ 2. $x^4 - 10x^2 + 9 > 0$
3. $|3x - 6| > 12$
1. $\{x \mid -2 < x < 5\}$
2. $\{x \mid x < -3 \text{ or } -1 < x < 1 \text{ or } x > 3\}$
3. $\{x \mid x < -2 \text{ or } x > 6\}$

7.6 Radical Equations (page 178)

Vocabulary
extraneous solution, radical equations

Background
A radical equation is solved by raising each side of the equation to the same power. This step can introduce an extraneous solution, as shown below:
$$x + 1 = 3 \text{ (the solution is 2)}$$
$$(x + 1)^2 = 3^2$$
$$x^2 + 2x + 1 = 9$$
$$x^2 + 2x - 8 = 0$$
$$(x - 2)(x + 4) = 0$$
$$x = 2 \text{ or } x = -4$$

Suggestions
Stress, as shown in all of the examples, that the "apparent" solutions of a radical equation must be checked in the original equation to identify any extraneous solutions.

Notice that Exercises 28 and 30 have no solution. They can be solved as the other radical equations and an extraneous solution will appear. Some students should notice that an equation such as $\sqrt{2t + 5} = -3$ has no solution since $\sqrt{2t + 5} \geq 0$ for each t, but $-3 < 0$.

For Additional Chalkboard Work
Solve. Check for extraneous solutions.
1. $\sqrt[3]{2x - 5} = -3$ -11 2. $2\sqrt{2y + 1} = 5\sqrt{y - 8}$ 12
3. $3 + \sqrt{2x + 2} = x$ 7 4. $\sqrt[3]{y^{-2}} = 4$ $\frac{1}{8}, -\frac{1}{8}$
5. $\sqrt{4x + 1} - \sqrt{x - 2} = 3$ $6, 2$

Computer Activities (page 184)
The solutions to the quadratic equations in this program do not appear in the most desirable form. Fractions are not always reduced to lowest terms, and radicals are not always simplified. The answers are correct but not in best form. Simplification can be done on the computer, but that was beyond the scope of this particular lesson and would have lengthened the program beyond the size of the page. You may want to challenge your students to add the extra refinements. Both fraction reduction and simplification of radicals are covered elsewhere in these computer pages.

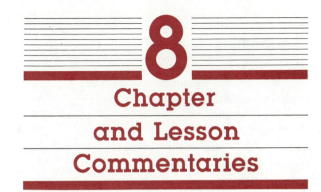

8 Chapter and Lesson Commentaries

Chapter 8   Complex Numbers

Chapter Objectives
To solve an equation whose solutions are imaginary numbers
To add, subtract, multiply, and divide complex numbers
To find the three cube roots of a nonzero integer
To determine the nature of the solutions of a quadratic equation by examining its discriminant
To find the sum and the product of the solutions of a quadratic equation
To write a quadratic equation, given its solutions

Chapter Overview
Imaginary numbers are introduced as the real number system is extended to the complex number system. Operations with complex numbers are practiced. The three cube roots of a nonzero integer that is a perfect cube are found as one real number and a pair of conjugate imaginary numbers. The coefficients of a quadratic equation are related to the sum, the product, and the nature of the solutions of the equation.

Special Features
Sonya Kovalevsky   p. 186
Reading in Algebra   pp. 188, 192, 197
Non-Routine Problems   p. 197
Cumulative Review (Chapters 1–8)   pp. 206–207
Applications: Proofs with Complex Numbers
  pp. 201–202
Chapter Review   p. 202
Chapter Test   p. 203
College Prep Test   p. 205

8.1 Sums and Differences of Complex Numbers (page 187)

Vocabulary
complex number, imaginary number, pure imaginary number

Suggestions
In Example 1, point out that we write $i\sqrt{7}$ rather than $\sqrt{7}\,i$ to avoid confusion with $\sqrt{7i}$.

Have students complete the following table after they study pages 187–188.

Number	Complex	Imaginary	Pure Imag.	Real	Rational	Irrat.
-8	*			*	*	
$3i$	*	*	*			
$5 - 2i$	*	*				
0	*			*	*	
$3 - 2\sqrt{5}$	*			*		*
i^2	*			*	*	

For Additional Chalkboard Work
Simplify.
1. $-\sqrt{-36}$ 2. $2\sqrt{-28}$
3. $(-4 + 3i) - (7 - 5i)$
4. $(8 - 2\sqrt{-2}) + (-2 + \sqrt{-50})$

Solve.
5. $x^2 = -49$ 6. $3x^2 + 18 = 0$
7. $5x^2 + 60 = 0$

1. $-6i$ 2. $4i\sqrt{7}$ 3. $-11 + 8i$
4. $6 + 3i\sqrt{2}$
5. $-7i, 7i$ 6. $-i\sqrt{6}, i\sqrt{6}$ 7. $-2i\sqrt{3}, 2i\sqrt{3}$

Cumulative Review
Samuel is 6 years older than Tina and Renée is 4 years younger than Samuel. Three years ago, 3 times the sum of the two younger ages was 4 times the oldest age. Find their present ages.
Tina: 12, Renée: 14, Samuel: 18

8.2 Products and Quotients of Complex Numbers (page 190)

Vocabulary
conjugate complex numbers

Suggestions
In Example 1, stress that $\sqrt{-6} \cdot \sqrt{-3}$ is *not equal* to $\sqrt{-6(-3)}$. Introduce Examples 5 and 6 by showing the analogy to dividing by an irrational number in the real number system.

An expression such as $\dfrac{3}{7i} = \dfrac{3}{7\sqrt{-1}}$ is not in simplest radical form. $\dfrac{3}{7i}$ is simplified to $\dfrac{3i}{-7}$ to remove the radical in the denominator.

Some standardized tests require the student to simplify large integral powers of i such as i^{98} and i^{100}. Present similar examples that can be simplified as integral powers of i^2 as follows:

$$i^{98} = (i^2)^{49} = (-1)^{49} = -1 \text{ and}$$
$$i^{100} = (i^2)^{50} = (-1)^{50} = 1.$$

Point out that consecutive powers of i take on the values $i, -1, -i, 1,$ and repeat in cycles of four:

$$i^1 = i, i^2 = -1, i^3 = -i, i^4 = 1;$$
$$i^5 = i, i^6 = -1, i^7 = -i, i^8 = 1; \text{ and so on.}$$

For Additional Chalkboard Work
Simplify.
1. $-3i \cdot 4i$ 2. $2\sqrt{-8} \cdot 3\sqrt{-3}$
3. $(5 - 4i)(2 + 3i)$
4. $-4 \div (5i)$ 5. $\dfrac{7}{-3 + \sqrt{-2}}$ 6. $(-2i)^9$

1. 12 2. $-12\sqrt{6}$ 3. $22 + 7i$ 4. $\dfrac{4i}{5}$
5. $\dfrac{-21 - 7i\sqrt{2}}{11}$ 6. $-512i$

Cumulative Review
Machines A, B, and C can each complete a certain job in 45 min, 1 h, and 90 min, respectively. How long will the job take if the machines work at the same time? 20 minutes

8.3 Equations with Imaginary Number Solutions (page 193)

Suggestions
Before undertaking Example 3, review the factoring of the sum or difference of cubes.
$a^3 + b^3 = (a + b)(a^2 - ab + b^2)$
$a^3 - b^3 = (a - b)(a^2 + ab + b^2)$

For Additional Chalkboard Work
Solve. (Ex. 1–2)
1. $2x^2 + 7 = 6x$
2. $27 - 6y^2 = y^4$
3. Find the three cube roots of 27.

1. $\dfrac{3 \pm i\sqrt{5}}{2}$

2. $0 = y^4 + 6y^2 - 27$
$0 = (y^2 + 9)(y^2 - 3)$
$y^2 + 9 = 0$ or $y^2 - 3 = 0$
$y^2 = -9$ or $y^2 = 3$
$\pm 3i; \pm\sqrt{3}$

3. Let $x^3 = 27$.
$x^3 - 27 = 0$
$x^3 - 3^3 = 0$
$(x - 3)(x^2 + 3x + 9) = 0$
$x - 3 = 0$ or $x^2 + 3x + 9 = 0$
$3; \dfrac{-3 \pm 3i\sqrt{3}}{2}$

Cumulative Review
Compute.
1. $16^{-\frac{3}{2}}$ $\dfrac{1}{64}$
2. 9.72×10^{-2} 0.0972
3. $\dfrac{4.73 \times 10^2}{1.1 \times 10^{-3}}$ $430{,}000$
4. $\dfrac{6^{-2}}{5^{-3}}$ $\dfrac{125}{36}$

8.4 The Discriminant (page 195)

Vocabulary
discriminant, nature of solutions

Suggestions
Introduce the lesson by writing four types of numerals that can appear when using the quadratic formula.

1. $\dfrac{2 \pm \sqrt{9}}{5}$ 2. $\dfrac{-1 \pm \sqrt{7}}{2}$
3. $\dfrac{3 \pm \sqrt{0}}{4}$ 4. $\dfrac{-2 \pm \sqrt{-5}}{3}$

For each of the four types above, answer these questions:
(a) What is the radicand?
(b) How many numbers are represented, one or two?
(c) Are the numbers real or imaginary?
(d) If the numbers are real, are they rational or irrational?
(e) How did the radicand help in answering questions (b), (c), and (d)?

Answers: 1. (a) 9, (b) two, (c) real, (d) rational
2. (a) 7, (b) two, (c) real, (d) irrational
3. (a) 0, (b) one, (c) real, (d) rational
4. (a) −5, (b) two, (c) imaginary

(e) If the radicand is a positive perfect square, there are two real, rational numbers. If the radicand is positive and not a perfect square, there are two real, irrational numbers. If the radicand is zero, there is one real, rational number. If the radicand is a negative number, there are two imaginary numbers.

With this introduction, the student will be more likely to understand the boxed statements (page 195) that involve variables.

For Additional Chalkboard Work
Find the discriminant and determine the nature of the solutions. (Ex. 1–4)
1. $4x^2 + 9 = 12x$ 0; one, real, rational
2. $5x + 3 = 2x^2$ 49; two, real, rational
3. $x^2 + 7 = 5x$ −3; two, imaginary
4. $x^2 = 5x + 3$ 37; two, real, irrational
5. For what values of k will $y^2 + (k + 3)y + 16 = 0$ have exactly one solution?   −11, 5

Cumulative Review
1. Change 45 mph to ft/s.   66 ft/s
2. Change 45 g/cm² to kg/m².   450 kg/m²

T-45

8.5 Sum and Product of Solutions
(page 198)

Suggestions
In Example 2, point out that *in standard form* mean *integral coefficients*. Thus $12x^2 + x - 6 = 0$, *not* $x^2 + \frac{1}{12}x - \frac{1}{2} = 0$, is the correct equation.

After introducing Examples 1 and 2, show that the two lesson objectives are inverses of each other. This can be done as follows:

Find $r + s$ and $r \cdot s$ for $3x^2 - 2x - 1 = 0$.
$$3x^2 - 2x - 1 = 0$$
$$x^2 - \tfrac{2}{3}x - \tfrac{1}{3} = 0$$
$$r + s = \tfrac{2}{3} \text{ and } r \cdot s = \tfrac{-1}{3}.$$

Write an equation for which $r = -\tfrac{1}{3}$ and $s = 1$.
$$r + s = \tfrac{2}{3} \text{ and } r \cdot s = -\tfrac{1}{3}$$
$$x^2 - \tfrac{2}{3}x - \tfrac{1}{3} = 0$$
$$3x^2 - 2x - 1 = 0$$

For Additional Chalkboard Work
1. Find the sum and the product of the solutions of $3x^2 - 5x - 2 = 0$. **sum, ⅔; product, −⅔**
2. Write a quadratic equation whose solutions are -5 and 3. **$x^2 + 2x - 15 = 0$**
3. Write a quadratic equation whose only solution is 3. **$r + s = 6$, $rs = 9$; $x^2 - 6x + 9 = 0$**
4. Write a quadratic equation whose solutions are $3 \pm \sqrt{2}$. **$r + s = 6$, $rs = 7$; $x^2 - 6x + 7 = 0$**

Cumulative Review
Simplify.

1. $4m^2n\sqrt{75m^8n^7p^3}$
2. $\dfrac{5t}{\sqrt{18t^3}}$
3. $\dfrac{x^3 + 125}{2x^2 - 10x + 50} \cdot \dfrac{14x}{3x + 15}$

1. $20m^6n^4p\sqrt{3np}$ 2. $\dfrac{5\sqrt{2t}}{6t}$ 3. $\dfrac{7x}{3}$

Applications, page 201
This is the third in a strand of three applications concerned with proof in algebra.
1. Properties of a Field, page 26
2. Proofs with Rational Numbers, pages 106–107
3. Proofs with Complex Numbers, pages 201–202

Chapter and Lesson Commentaries

Chapter 9 Coordinate Geometry

Chapter Objectives
To find the slope of a line
To graph a line; to write an equation of a line
To determine whether a point lies on a line
To determine the distance between any two points in a plane
To determine whether two lines are parallel or perpendicular, given their slopes
To find the slope of a line parallel to or perpendicular to a given line
To write an equation of a line parallel to or perpendicular to a given line
To apply the slope, distance, and midpoint formulas
To operate on vectors

Chapter Overview
Finding the slope, graphing, and writing an equation of a line are introduced in this chapter. Writing an equation of a line parallel to or perpendicular to a line and determining whether two lines are parallel or perpendicular are also introduced. Determining distance between two points and finding the coordinates of the midpoint or endpoints of a line segment are included. The chapter concludes with applications.

Special Features
Careers: Computer Programming   p. 208
Applications   p. 234; Computer   p. 235
Reading in Algebra   pp. 211, 215, 219
Cumulative Review   pp. 215, 225, 230
Non-Routine Problems   p. 222
College Prep Test   p. 238
Chapter Review   p. 236
Chapter Test   p. 237

9.1 The Coordinate Plane (page 209)

Vocabulary
Cartesian coordinate plane, quadrant, abscissa, ordered pairs, ordinate

Background
The word *Cartesian* is derived from the last name of the brilliant mathematician René Des*cartes* (1596–1650). It was Descartes who determined that every point in a plane can be determined by a pair of numbers (scalars) and, conversely, every point in a plane determines a pair of numbers.

Suggestions
The idea of locating points in a quadrant and graphing an equation is not new to the students and will be used throughout the text. When other methods fail, graphing by using a table of values will often yield results.

It is important to use correct terminology in referring to graphs. One does not graph a line or graph a point. A line or a point is a *graph*. An ordered pair of numbers or an equation *is graphed*.

Students should not always call the horizontal axis the *x*-axis and the vertical axis the *y*-axis. You may wish to have students graph ordered pairs (s, t) or other such pairs.

For Additional Chalkboard Work
Draw the graph of the line described by each equation.
1. $s = -2t + 3$
2. $s = 3t - 1$
3. $4t - 3s = 9$
4. $2t + 4s = -5$

Errors Students Make
Students frequently do not choose values for the abscissa judiciously when formulating a table of values. In Example 2 we chose 0, 2 and 4 for *x* in order to reduce unnecessary calculations.

Cumulative Review
1. Simplify. $\dfrac{\frac{3}{4} + \frac{1}{5}}{\frac{4}{5} + \frac{1}{2}}$. $\dfrac{19}{26}$
2. Simplify. $2(3 - 2a) + 3(4a - 2)$. $8a$
3. Find the solution set of $|3x - 7| \leq 2$. $\{x | \frac{5}{3} \leq x \leq 3\}$

9.2 Slope of a Line (page 212)

Vocabulary
slope, 0-slope, positive slope, undefined slope, negative slope

Background
The concept of a slope is one of the most important concepts in mathematics, especially in differential calculus. In differential calculus two major types of problems involving slope emerge. They are problems involving rate, speed, velocity and acceleration, and those involving finding the slope of a tangent line at any given point on a curve.

Suggestions
Students can be guided to discover that lines that slant up to the right have a positive slope and those that slant down to the right have a negative slope. Emphasize that a vertical line has no slope. It is incorrect to say that a vertical line has a slope of infinity.

The concept of the slope of a line can also be introduced through a discussion of the ratio of vertical change to horizontal change $\left(\dfrac{\text{rise}}{\text{run}}\right)$, or $\dfrac{\Delta y}{\Delta x}$. Either approach leads to the same definition of slope, $m = \dfrac{y_2 - y_1}{x_2 - x_1}$.

Students should understand that when finding the slope of a line, the change in *y*-coordinates and change in *x*-coordinates can be accomplished in either direction, i.e., either of the points can have coordinates of (x_1, y_1).

For Additional Chalkboard Work
Are points *P*, *Q* and *R* on the same straight line? Verify.
1. $P(6, -3)$, $Q(0, 0)$, $R(-2, 1)$ Yes
2. $P(-1, 2)$, $Q(7, 4)$, $R(11, 5)$ Yes
3. $P(-3, -4)$, $Q(2, 5)$, $R(6, 12)$ No

Errors Students Make
A puzzling point to students is that a vertical line has an undefined slope not an infinite slope. Students sometimes will subtract the *y*'s in one direction and the *x*'s in another when using
$$m = \dfrac{y_2 - y_1}{x_2 - x_1}.$$

9.3 Equation of a Line (page 216)

Vocabulary
point-slope, slope-intercept, y-intercept, standard form, x-intercept

Background
Descartes stated that every point on a coordinate plane can be represented by an ordered pair. Since a line is a set of points, the obvious question is, can a line be described by an equation? An equation of a line does precisely that.

It is not necessary to use $y - y_1 = m(x - x_1)$ to write an equation of a line. The slope-intercept form $y = mx + b$ is sufficient to write an equation, given any of the examples and exercises in this lesson. But, it is often easier to write equations using the point-slope form of an equation.

Suggestions
When finding the equation of a line in which the coordinates of two points are given, the coordinates of either point can be used to substitute in the point-slope form. In Example 4, let $x_1 = 5$ and $y_1 = -4$.
$y + 4 = -\frac{5}{8}(x - 5); 8y + 32 = -5x + 25;$
$5x + 8y = -7$. This yields the same results.

The equation of a horizontal line is of the form $ax + by = c$, where $a = 0$ and $b \neq 0$. Thus, $by = c$ is an equation of a horizontal line. The equation of a vertical line is of the form $ax + by = c$, where $a \neq 0$ and $b = 0$. Thus, $ax = c$ is an equation of a vertical line.

For Additional Chalkboard Work
Write an equation, in standard form, of each line. First use $y - y_1 = m(x - x_1)$ and then $y = mx + b$.

1. $P(-4, -3), Q(5, 6)$ $x - y = -1$
2. $P\left(-\frac{2}{3}, \frac{5}{2}\right), Q\left(-4, +\frac{2}{3}\right)$ $33x - 60y = -172$
3. $P\left(-3, \frac{1}{2}\right), m = \frac{2}{3}$ $4x - 6y = -15$
4. $P(6, -1), Q(0, 3)$ $2x + 3y = 9$

Cumulative Review
Find the solution.
1. $-6 < 3x + 3 < 12$ 2. $|x| \leq 4$
3. $|3n| > 12$ $n > 4$, or $n < -4$
1. $-3 < x < 3$ 2. $-4 \leq x \leq 4$

9.4 Slope-Intercept Method of Drawing Graphs (page 220)

Background
To graph any linear equation it is sufficient to use only the slope-intercept method of drawing graphs. However, it is not always convenient or wise to do so. Therefore, other methods of graphing linear equations should be discussed. Other approaches such as plotting points and using tables should continue to be encouraged. Another method of graphing involves finding both intercepts of a line.

Suggestions
The approach used to draw graphs of equations of lines in this lesson is to find the slope and y-intercept of the line. Students should first graph the coordinates of the y-intercept and then use the numerator and denominator of the slope to locate another point on the line.

You may also wish to have students use a calculator to determine whether a given point is on a line described by a given equation.

Enrichment
Find n so that $\dfrac{3^{n+1} - 3^n + 6}{3^n + 7}$ equals an integer.

$\dfrac{3^{0+1} - 3^0 + 6}{3^0 + 7} = \dfrac{3 - 1 + 6}{1 + 7} = \dfrac{8}{8}$, or 1.

For Additional Chalkboard Work
Have students discuss the graph of the equation $y = mx + b$ for different values of x, y, m and b. For example, if $x = 0$, $y = b$. If $b < 0$, the line crosses the y-axis below the origin. Consider positions of the line when $b = 0$ and m assumes different values.

Errors Students Make
Students sometimes have difficulty determining which part of the fraction representing the slope determines the horizontal distance and direction and which determines the vertical distance and direction. The expression $\dfrac{\text{rise}}{\text{run}}$ may help to clarify. The sign of the slope can go with either the numerator or the denominator.

9.5 The Distance Between Two Points (page 223)

Vocabulary
Pythagorean relation, converse of the Pythagorean relation

Suggestions
Review for the students the definition of length of a horizontal and vertical line segment. These two definitions are important for the understanding of the distance formula.

The distance formula can be used for any two points even if these points happen to form a horizontal or vertical line segment. However, the methods developed earlier for finding the length of a horizontal and vertical line segment would probably be easier to use.

In deciding which of the three given lengths is the length of the hypotenuse, students may need to use the table of square roots in the back of the book. You also may encourage them to use a calculator.

For Additional Chalkboard Work
Find the coordinates of vertex C, the vertex of the right angle of $\triangle ABC$ where A and B are given.   **1.** (6, 5)   **2.** (9, 1)
1. $A(-3, 5)$, $B(6, 8)$ **2.** $A(3, 1)$, $B(9, 4)$
3. $A(2, 8)$, $B(5, 3)$ **4.** $A(-3, 9)$, $B(6, -4)$
3. (5, 8)   **4.** (6, 9)

Errors Students Make
Students sometimes have difficulty finding the coordinates of point C, the intersection of the perpendicular segments from A and B. You may wish to use some numerical examples to help remove this difficulty.

Cumulative Review
1. Factor $5x^3 - 14x^2 - 3x$.
2. Factor $8a^2 - 50$.
3. Solve $x^2 + 2x - 15 = 0$.
1. $x(5x + 1)(x - 3)$ 2. $2(2a + 5)(2a - 5)$
3. $-5, 3$

9.6 The Midpoint of a Segment (page 226)

Vocabulary
midpoint of a segment, endpoint of a segment

Suggestions
You may wish to develop the formula for finding the midpoint of a line segment by determining the midpoints of horizontal and vertical line segments. Draw right triangle PQR with T, S, and M as midpoints.

Midpoint $\overline{PR}$ is $T(5, 2)$, or $\left(\dfrac{9+1}{2}, 2\right)$.

Midpoint $\overline{QR}$ is $S(9, 4)$, or $\left(9, \dfrac{6+2}{2}\right)$.

Thus midpoint $\overline{PQ}$ is $(5, 4)$.

In general, M has coordinates $\left(\dfrac{x_1 + x_2}{2}, \dfrac{y_1 + y_2}{2}\right)$. Exercise 42 leads to the generalization that the segments joining, in order, the midpoints of any quadrilateral form a parallelogram.

For Additional Chalkboard Work
Determine the coordinates (x, y) of M, the midpoint of the segment joining P and Q.
1. $P(2, 1)$, $Q(8, 7)$ (5, 4)
2. $P(-6, -2)$, $Q(4, -6)$ $(-1, -4)$
3. $P(4, -1)$, $Q(-6, 5)$ $(-1, 2)$
4. $P(3, -3)$, $Q(6, 4)$ $\left(\dfrac{9}{2}, \dfrac{1}{2}\right)$

Errors Students Make
Students sometimes subtract coordinates when finding the midpoint. Show some examples on the chalkboard to avoid this difficulty.

Cumulative Review
Simplify.

1. $\dfrac{\dfrac{3a}{a^2 - 4} + \dfrac{2}{a - 2}}{\dfrac{4}{a + 2} + \dfrac{5}{a - 2}}$ $\dfrac{5a + 4}{9a + 2}$

2. $\dfrac{1 - \dfrac{9}{a^2}}{\dfrac{1}{a} - \dfrac{3}{a^2}}$ $a + 3$

9.7 Parallel and Perpendicular Lines (page 228)

Background
Two lines may be perpendicular to each other with one of the lines having undefined slope. For example, a vertical line and a horizontal line are always perpendicular, yet the slope of the vertical line is undefined.

Another way to show that two lines are perpendicular is to show that the product of their slopes is -1 (no vertical lines) and the converse is true. If the slope of one line is $-\frac{3}{2}$ then the slope of a line perpendicular to it is $\frac{2}{3}$. Hence, $\left(-\frac{3}{2}\right)\left(\frac{2}{3}\right) = -1$.

Suggestions
In this lesson the point-slope form of an equation is used to write an equation of a line. The slope-intercept form can also be used. In Example 1, $m = \frac{3}{2}$ and the line passes through the point $P(4, 2)$. Use $y = mx + b$ and find b.

$$2 = \tfrac{3}{2}(4) + b; \; b = 2 - 6 = -4$$

Thus, $y = \frac{3}{2}x - 4$, or $3x - 2y = 8$.

This form is a good alternative of the development in the lesson.

Write an equation in standard form of the line that passes through the point $A(-3, -2)$ and is perpendicular to the line whose equation is $-4x - 3y = 10$.

$m = -\frac{4}{3}$; the slope of the line perpendicular to it is $\frac{3}{4}$. Use $y = mx + b$;

$$-2 = \tfrac{3}{4}(-3) + b; \; -8 = -9 + 4b;$$
$$4b = 1; \; b = \tfrac{1}{4}.$$

Thus, $y = \frac{3}{4}x + \frac{1}{4}$ is an equation of the line. In standard form, $3x - 4y = -1$ is the equation.

Cumulative Review
Solve each equation.
1. $-4a - 3 + 5a = 7 - 2a + 2$ 4
2. $-3x - (6 + 2x) = -2(x - 3)$ -4
3. $|2x - 3| = 5$ $-1, 4$

9.8 Using Coordinate Geometry (page 231)

Vocabulary
collinear, trapezoid, parallelogram, median, quadrilateral, rectangle, isosceles triangle, rhombus, equilateral triangle, diagonals

Suggestions
Many of the concepts developed in this chapter are applied in this lesson. It might be appropriate to review some geometric definitions and properties.

To show a triangle is a right triangle, students can show either that two sides have negative reciprocal slopes or that the converse of the Pythagorean relation is true. That is, if $c^2 = a^2 + b^2$, then $\triangle ABC$ is a right triangle.

For Additional Chalkboard Work
Show that the lines joining the points $A(-3, -4)$, $B(2, 5)$ and $C(-5, 3)$ form a right triangle. Use the converse of the Pythagorean relation.

$$d(\overline{AB}) = \sqrt{(-3 - 2)^2 + (-4 - 5)^2}$$
$$= \sqrt{25 + 81} = \sqrt{106}$$
$$d(\overline{BC}) = \sqrt{(5 - 3)^2 + (2 + 5)^2}$$
$$= \sqrt{4 + 49} = \sqrt{53}$$
$$d(\overline{AC}) = \sqrt{(-3 + 5)^2 + (-4 - 3)^2}$$
$$= \sqrt{4 + 49} = \sqrt{53}$$
$$(\sqrt{106})^2 = (\sqrt{53})^2 + (\sqrt{53})^2$$
$$106 = 53 + 53$$

Cumulative Review
1. Three times the sum of a number and 2 is 3 more than twice the product of the number and 2. What is the number?
2. Find p in dollars if $A = \$9{,}350$, $r = 10\frac{1}{2}\%$, and $t = 3$ years 3 months. Use the formula $A = p + prt$.

1. 3   2. $6971.11

Computer Activities (page 235)
This is a rather straightforward program, except for the need to test (lines 130 and 140) for vertical and horizontal lines.

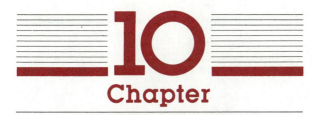

Chapter 10 and Lesson Commentaries

Chapter 10 Linear Systems, Matrices, and Determinants

Chapter Objectives
To solve a system of two linear equations graphically and algebraically
To solve a system of three linear equations algebraically
To find the value of a 2 by 2 and 3 by 3 determinant
To solve systems of linear equations using determinants (Cramer's rule) and minors
To solve a system of linear equations using the inverse of a matrix and matrix transformations
To find sums, products, and inverses of matrices
To apply linear systems

Chapter Overview
Many methods for solving systems of linear equations are introduced. Matrix algebra, which is introduced, is an important topic in the fields of mathematics, science, computer science, economics, and engineering.

Special Features
Non-Routine Problem Solving   p. 239
Reading in Algebra   pp. 241, 249
Cumulative Review   p. 263
Calculator Activities   p. 242
College Prep Test   p. 274
Chapter Review   p. 272
Chapter Test   p. 273

10.1 Solving Linear Systems Graphically (page 240)

Vocabulary
consistent system, dependent system, inconsistent system, independent system, linear system

Background
Since the intersection of the lines represented by a system of linear equations is the solution of the system, students should have experience in graphing equations to understand algebraic solutions.

Suggestions
You may wish to give an overview of the lesson by reviewing that, given two lines, they may: be parallel and have no points in common, intersect and have exactly one point in common, or coincide and have all their points in common. You may also want your students to use slopes and y-intercepts to determine, without graphing, whether the lines for a given system of equations: are parallel and therefore form an inconsistent system, coincide and therefore form a dependent system, or intersect in exactly one point and therefore form an independent system.

The greater the number of squares used to represent one unit, the easier it is to estimate solutions to the nearest tenth of a unit.

Errors Students Make
Students sometimes have difficulty in reading the solutions from their graphs. Emphasizing careful drawing of graphs will help to remove this difficulty. Some students have difficulty with classifying solutions, but finding them is now more important than classifying them.

Cumulative Review
Simplify. Use positive exponents.

1. $\dfrac{-2a^4 b^4}{a^2 b^4}$ 2. $\dfrac{-12a^{-4} b^6 c^{-3}}{6a^{-2} b^{-4} c^5}$ 3. $\left(\dfrac{2x^{-2} y^3}{6xy^{-5}}\right)^4$

1. $-2a^2$ 2. $\dfrac{-2b^{10}}{a^2 c^8}$ 3. $\dfrac{y^{32}}{81 x^{12}}$

10.2 Solving Linear Systems Algebraically (page 243)

Suggestions
In solving systems of equations by substitution, students should be reminded that they can solve for either variable in either equation. However, students should be encouraged to solve for the variable that is easier to isolate. Emphasize that after the value of one variable is found, the other variable's value can be found by substituting in either of the original equations, but to check the solution both values must be substituted in both of the original equations.

You may wish to note that the addition method is based on the *Addition Property of Equality*.

Another method for solving linear systems of equations involves solving both equations for the same variable, then using the *Transitive Property of Equality* to write an equation in one variable. This process is often easier to use than either the substitution or addition method. You may wish to apply this to the system in Example 1:

$y = 2x - 5$ and $y = x - 3$; then $2x - 5 = x - 3$; $x - 5 = -3$ $x = 2; y = -1$.

For Additional Chalkboard Work
You may wish to point out the advantage of solving an equation for the variable that is easier to isolate when solving a system of equations such as
$2x - y = 4$
$3x + 2y = 1$ by substitution.

Errors Students Make
Remind students that a solution for one equation may not be a solution for both equations.

They must check in the original equations, since errors may cause false solutions to satisfy the derived equations.

Cumulative Review
Solve each equation.
1. $5^{x-2} = 125$ **2.** $81 = 3^{3x-2}$ **3.** $2^{4x-2} = 64$
1. 5 2. 2 3. 2

10.3 Solving Systems of Three Linear Equations (page 247)

Vocabulary
ordered triple

Background
Two planes may: coincide, be parallel, or intersect in a line. Three planes may: be parallel, intersect, or intersect in a point. For example, two walls and a ceiling intersect in a point at a corner of a room. If this were put on a coordinate system, the point of intersection would be an ordered triple, as shown.

Suggestions
A system of three linear equations in three variables is: an independent system if its solution is an ordered triple of real numbers, an inconsistent system if it has no solution, or a dependent system if it has more than one solution. Be sure that students understand that $(2, -1, 3) \neq (3, 2, -1)$

For Additional Chalkboard Work
Solve the system: $2x - y + z = 5$ (1)
$x + 2y - z = 0$ (2)
$4x + 2y + 5z = 6$ (3).
$(2, -1, 0)$.
In the second equation, $z = x + 2y$. Substitute in equations (1) and (3): $2x - y + (x + 2y) = 5$
$4x + 2y + 5(x + 2y) = 6$

Errors Students Make
Students forget which equations they were working with and which variable was eliminated. Numbering the equations and eliminating variables in some specific order, for example, first the third variable, then the second, will help.

Cumulative Review
Find the solution set. Graph the solution set.
1. $-3n + 5 \leq -13$ **2.** $3a - 12 > 5a$
3. $|x + 5| < 8$
1. $\{n | \geq 6\}$ 2. $\{a | a < -6\}$ 3. $\{x | -13 < x < 3\}$

10.4 Two by Two Determinants (page 250)

Vocabulary
determinant, matrix, square matrix, 2 by 2 determinant

Background
The method for solving a system of linear equations by determinants is known as Cramer's rule.

Suggestions
Distinguish between a matrix and a determinant. A matrix is any array of elements or numbers arranged in rows and columns. A determinant is a real number assigned to the matrix. For example, $\begin{bmatrix} 1 & 2 \\ -3 & -5 \end{bmatrix}$ is a matrix and $\begin{vmatrix} 1 & 2 \\ -3 & -5 \end{vmatrix}$ is the determinant assigned to the matrix. Thus, the number assigned to the matrix is $(-5) - (-6)$, or 1. Emphasize that all equations must be put in standard form before determinants can be used to solve them.

For Additional Chalkboard Work
1. Solve for x. $\begin{vmatrix} -3 & 2 \\ x & 4 \end{vmatrix} = 6$
2. For what values of x is $\begin{vmatrix} x & 3 \\ 27 & x \end{vmatrix} > 0$?

1. $-12 - 2x = 6$
$-2x = 18$
$x = -9$

2. $x^2 - 81 > 0$
$x^2 > 81$
$x < -9$ or $x > 9$

Errors Students Make
Students are sometimes careless with signs. Using parentheses freely can help avoid such errors. For example, $\begin{vmatrix} 2 & -2 \\ -3 & -4 \end{vmatrix} = 2(-4) - (-3)(-2)$.

Cumulative Review
Evaluate each polynomial for the given values of the variables.
1. $-2xy^2 + 3x^3y^3 - 4$; $x = -2$ and $y = 3$
2. $4x^3y^2 - 2xy^3$; $x = 4$ and $y = 5$
3. $1.3a^2 - 4.2b$; $a = 0.2$ and $b = 1.3$

1. -616 2. $5,400$ 3. -5.408

10.5 Three by Three Determinants (page 253)

Background
Some of the many properties associated with determinants are listed below. You may wish to have some students prove these properties.

1. Corresponding elements of rows and columns of a determinant can be interchanged and the value does not change.

$$\begin{vmatrix} a_1 & a_2 & a_3 \\ b_1 & b_2 & b_3 \\ c_1 & c_2 & c_3 \end{vmatrix} = \begin{vmatrix} a_1 & b_1 & c_1 \\ a_2 & b_2 & c_2 \\ a_3 & b_3 & c_3 \end{vmatrix}$$

2. If each element of a row or column is 0, the value of the determinant is 0.

$$\begin{vmatrix} a_1 & a_2 & a_3 \\ 0 & 0 & 0 \\ c_1 & c_2 & c_3 \end{vmatrix} = \begin{vmatrix} 0 & a_2 & a_3 \\ 0 & b_2 & b_3 \\ 0 & c_2 & c_3 \end{vmatrix} = 0$$

3. If any two rows or columns are reversed, the sign of the determinant is reversed.

$$\begin{vmatrix} a_1 & a_2 & a_3 \\ b_1 & b_2 & b_3 \\ c_1 & c_2 & c_3 \end{vmatrix} = -\begin{vmatrix} c_1 & c_2 & c_3 \\ b_1 & b_2 & b_3 \\ a_1 & a_2 & a_3 \end{vmatrix}$$

4. If any two rows or columns of a determinant are identical, the value of the determinant is 0.

$$\begin{vmatrix} a_1 & a_2 & a_3 \\ b_1 & b_2 & b_3 \\ a_1 & a_2 & a_3 \end{vmatrix} = 0$$

Suggestions
This alternative method for finding the value of the determinant in Example 1 may save time.

$$\begin{vmatrix} 2 & 3 & 5 \\ 4 & 2 & 1 \\ -1 & -3 & 2 \end{vmatrix} =$$

$(2)(2)(2) + (3)(1)(-1) + (5)(4)(-3) - (-1)(2)(5) - (-3)(1)(2) - (2)(4)(3) = 8 - 3 - 60 + 10 + 6 - 24 = -63.$

However, you may not want to present both methods.

10.6 Matrix Addition (page 257)

Vocabulary
additive identity matrix, dimension of a matrix, equal matrices, matrix addition

Background
Matrices form a *commutative (Abelian) group* under addition. More specifically, if A, B, and C are any three matrices, then: $A + B$ is a matrix, Closure Property; $A + B = B + A$, Commutative Property; $(A + B) + C = A + (B + C)$, Associative Property; $A + I = I + A = A$, Identity Property; and $A + (-A) = I$, Inverse Property.

Suggestions
You may wish to discuss the difference of two matrices. If $A = \begin{bmatrix} a_{11} & a_{12} \\ a_{21} & a_{22} \end{bmatrix}$ and $B = \begin{bmatrix} b_{11} & b_{12} \\ b_{21} & b_{22} \end{bmatrix}$,

then $A - B = \begin{bmatrix} a_{11} - b_{11} & a_{12} - b_{12} \\ a_{21} - b_{21} & a_{22} - b_{22} \end{bmatrix}$; thus,

if $A = \begin{bmatrix} -3 & 5 \\ 6 & -2 \end{bmatrix}$ and $B = \begin{bmatrix} 8 & -3 \\ -2 & 7 \end{bmatrix}$, then

$A - B = \begin{bmatrix} -3 - (8) & 5 - (-3) \\ 6 - (-2) & -2 - (7) \end{bmatrix} = \begin{bmatrix} -11 & 8 \\ 8 & -9 \end{bmatrix}$.

In general, each element of a matrix can be written a_{ij}, where i is the row and j the column.

If each element of a matrix is a real number, the matrix is called a *real matrix*.

Errors Students Make
Students may have to be reminded that two matrices can be added only if the two matrices have the same dimension.

Enrichment
If A, B, and C are 3 by 3 matrices, prove that A, B, and C form a commutative group under addition.

Cumulative Review
1. Add $(4ab + 2ab^2 - 3a^2b) + (-3ab^2 + 6a^2b - ab)$.
2. Subtract $(10xy^2 - 3x^2y) - (2xy - 2xy^2)$.
3. Multiply $(2x - 5)(4x + 3)$.

1. $3ab - ab^2 + 3a^2b$ 2. $12xy^2 - 3x^2y - 2xy$
3. $8x^2 - 14x - 15$

10.7 Matrix Multiplication (page 260)

Vocabulary
matrix multiplication, multiplicative identity matrix, multiplicative inverse matrix, scalar

Background
Matrices do not form a group under multiplication. The commutative property does not hold nor does every square matrix have a multiplicative inverse. The set of square matrices is closed under multiplication and the associative property holds. Thus, the set of square matrices forms a semigroup under multiplication.

Suggestions
Emphasize that a product of two matrices is defined if the second dimension of the first matrix is the same as the first dimension of the second matrix. Matrices do not necessarily have to be square matrices for multiplication.

The following example can be used to show that not every square matrix has an inverse. Given $\begin{bmatrix} -3 & 4 \\ 0 & 0 \end{bmatrix}$, let $\begin{bmatrix} a_{11} & a_{12} \\ a_{21} & a_{22} \end{bmatrix}$ be its multiplicative inverse. Then,

$\begin{bmatrix} -3 & 4 \\ 0 & 0 \end{bmatrix} \cdot \begin{bmatrix} a_{11} & a_{12} \\ a_{21} & a_{22} \end{bmatrix} = \begin{bmatrix} 1 & 0 \\ 0 & 1 \end{bmatrix}$;

$\begin{bmatrix} -3a_{11} + 4a_{21} = 1 & 0a_{11} + 0a_{21} = 0 \\ -3a_{12} + 4a_{22} = 0 & 0a_{12} + 0a_{22} = 1 \end{bmatrix}$;

but $0a_{12} + 0a_{22} = 0$, *not* 1. Thus, no multiplicative inverse exists. A necessary and sufficient condition that A^{-1} exists for square matrix A is that the determinant associated with A is not 0.

Cumulative Review
1. How many milliliters of a 30% iodine solution should be added to 70 mL of a 55% iodine solution to obtain a 40% iodine solution?
2. The number of dimes in a coin holder is 12 more than the number of quarters and the number of nickels is twice the number of dimes. Find the number of each kind of coin if their total value is $9.15.

1. 105 mL   2. *n*, 54; *d*, 27; *q*, 15

10.8 Solving Linear Systems Using Matrices (page 264)

Vocabulary
inverse of a matrix, matrix transformations, triangular form

Background
An alternate method for finding the multiplicative inverse of a 2 by 2 matrix follows: Interchange the first and fourth elements, find the additive inverses of the second and third elements, and multiply this matrix by $\frac{1}{\text{determinant}}$.

Suggestions
To provide an overview of this lesson: Show that linear systems can be written as a product of matrices involving coefficients (called *the matrix of detached coefficients*) as one matrix, the variables as a second matrix, and the constants as a third matrix. Show that the first transformation is one that transforms three equations in standard form to one in triangular form. Show that it is sufficient to represent coefficients and constants as one matrix (called the *augmented matrix*).

You may not want to include both methods for solving linear systems.

Errors Students Make
The major difficulty occurs in the elementary transformation of the augmented matrix. Encourage students to establish a procedure. For example, get 1 in a_{11}, then 0 in a_{21}, then 0 in a_{31}, then 1 in a_{22}, then 0 in a_{32}, etc.

Enrichment
Solve using inverses.
$$\begin{aligned} x + 2y - 3z &= 10 \\ -4x + y - z &= -10 \\ 3x - 7y + 2z &= 5 \end{aligned} \quad \left(\frac{43}{14}, -\frac{1}{14}, -\frac{33}{14}\right)$$

Cumulative Review
Solve. Express answers to the nearest tenth.
1. $5x^2 - 8x - 3 = 0$ 2. $-3x^2 + 7x = -2$
3. $x^2 - 3x - 8 = 0$ $-1.7, 4.7$
1. $-0.3, 1.9$ 2. $-0.3, 2.6$

10.9 Problem Solving Using Linear Systems (page 268)

Suggestions
You may wish to have students use various methods to solve the system of equations involved in the applied problems.

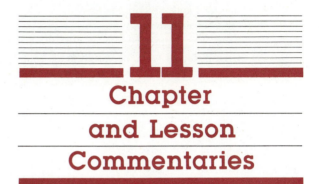

Chapter and Lesson Commentaries

Chapter 11 Functions

Chapter Objectives
To determine the domain and the range of a relation and of a function

To determine whether a relation is a function

To find the value and zeros of a function and of a composition of functions

To determine whether a given function is a linear function or a constant function, given either an equation of the function or a graph of the function

To draw the graphs of some special functions

To find the inverse of a relation

To draw the graph of a function and its inverse

To determine whether the inverse of a function is a function

To determine whether a relation expresses a direct variation, an inverse variation, a joint variation, or a combined variation

Chapter Overview
A function is defined as a special kind of relation. The distinction between a relation and a function and the distinction between an inverse of a function and inverse function are made. Two different checks, *the definitions* and *the vertical line test*, are used to determine whether a relation is a function and whether the inverse of a function is a function. Evaluating functions and the composition of functions precede identifying and drawing graphs of special functions. The chapter concludes with a discussion of the different types of variations.

Special Features
Careers: Pharmacy   p. 275
Non-Routine Problems   p. 287
Reading in Algebra   pp. 284, 287
Cumulative Review   pp. 284, 296
Applications   p. 297
Computer Activity   p. 300
College Prep Test   p. 301
Chapter Review   p. 298
Chapter Test   p. 299

11.1 Relations and Functions
(page 276)

Vocabulary
domain, function, range, relation

Background
A relation is defined as a set of ordered pairs of real numbers. This definition is preferable to one that defines a relation in terms of a rule or a correspondence. However, a rule or a correspondence is usually associated with a set of ordered pairs of real numbers. Frequently the complete set of ordered pairs of real numbers is generated by a rule.

Suggestions
In this lesson, the two major points are: determining the domain and the range of a relation and determining whether a relation is a function. Students should be able to use both the definition of a function and the vertical-line test to determine whether a relation is a function.

Errors Students Make
Students sometimes confuse the domain and the range of a function, but the real problem often occurs with the definition of a function. To help remove this difficulty, you may wish to use this alternate definition: A *function* is a set of ordered pairs of real numbers such that for each first element there exists *one and only* one second element. In other words, no first element can have more than one second element.

Cumulative Review
Write each expression in simplest radical form.
1. $3a^{\frac{4}{3}}$ **2.** $(2x)^{\frac{4}{5}}$ **3.** $-2^{\frac{3}{2}}$

1. $3a\sqrt[3]{a}$ 2. $\sqrt[5]{16x^4}$ 3. $-2\sqrt{2}$

11.2 Values and Composition of Functions (page 279)

Vocabulary
rational function, zeros of a rational function

Background
Composition of functions is an important topic in mathematics. Almost all derivatives involve composition of functions. Exercises such as Exercises 49–52 are used in defining a derivative in the study of calculus.

Suggestions
When a function, denoted by f, g, or any other letter, is described by a rule, it may be necessary to remind students that a function is a set of ordered pairs of real numbers. Letters such as f and g refer to sets of ordered pairs, but $f(x)$ and $g(x)$ refer to function values; that is, $f(x)$ and $g(x)$ refer to the range, or the elements of the set of second coordinates, of f and g, respectively. Thus, every ordered pair of f and g has the form $(x, f(x))$ and $(g, g(x))$, respectively. Point out that finding a function value means finding the value of the function for a given element in the domain.

For a composition such as $f(g(x))$, point out that this is an ordered pair $(g(x), f(g(x)))$ which was derived from the ordered pair $(x, g(x))$. Note that $g(x)$ is an element from the range which then becomes an element in the domain of the function f.

In Example 5, emphasize that 0 and 2 are nonpermissible elements of the domain because the function would not be defined for these values.

For Additional Chalkboard Work
Let $g(x) = \sqrt{x - 1}$ and $h(x) = x^2 + 1$. Find each of the following.
1. $g(1) - h(3)$ 2. $h(2) + g(5)$ 3. $h(g(2))$
4. $g(h(3))$
1. -10 2. 7   3. 2   4. 3

Cumulative Review
Simplify.
1. $\dfrac{\sqrt{2}}{3 - \sqrt{3}}$ 2. $\dfrac{\sqrt{3}}{\sqrt{2} - \sqrt{5}}$ 3. $\dfrac{\sqrt{27a^5 b^6}}{\sqrt{3ab^3}}$
1. $\dfrac{3\sqrt{2} + \sqrt{6}}{6}$ 2. $\dfrac{\sqrt{6} + \sqrt{15}}{-3}$ 3. $3a^2 b\sqrt{b}$

11.3 Linear and Constant Functions (page 282)

Vocabulary
constant function, linear function

Background
Because the mathematical ideas of slope, rate of change, maximum, and minimum are often defined in terms of intervals and increasing and decreasing functions, you may wish to discuss special classes of functions, such as: (1) A function f is an *increasing function* in some interval *if and only if* for all $x_1 < x_2$ in the interval, $f(x_1) \le f(x_2)$. (2) A function f is a *decreasing function* in some interval *if and only if* for all $x_1 < x_2$ in the interval, $f(x_1) \ge f(x_2)$. (3) A function h is called an *even function if and only if* $h(-x) = h(x)$ for all x in the domain. (4) A function g is called an *odd function if and only if* $g(x) = -g(-x)$ for all x in the domain. (5) A function g is called a *periodic function if and only if* there is some real number p such that $g(x + p) = g(x)$ for all x in the domain.

Suggestions
The special functions discussed in this lesson include: functions whose graphs are not continuous, functions that have a restricted domain, and functions that are defined in two or more different domains. The graphs in Example 4 and Exercises 25, 27–32 illustrate *step functions*.

Errors Students Make
Distinguishing between a linear and a constant function and the fact that a vertical line does not represent any function may be the most confusing part of this lesson.

Enrichment
$f = \{(x, y) | f(x) = x^2\}$ in the interval $-4 \le x \le 4$. Where is f decreasing? Where is f increasing? Prove your answer by using the definitions above.

Cumulative Review
Simplify. Write each with positive exponents.
1. $\dfrac{27x^4 y^{-7} z^2}{-3x^2 yz^{-3}}$ 2. $\dfrac{-25a^4 bc^{-3}}{5a^2 b^3 c^4}$ 3. $\dfrac{81mn^4 p^{-2}}{27m^2 n^{-2} p^{-4}}$
1. $\dfrac{-9x^2 z^5}{y^8}$ 2. $\dfrac{-5a^2}{b^2 c^7}$ 3. $\dfrac{3n^6 p^2}{m}$

11.4 Inverse Relations and Functions (page 285)

Vocabulary
inverse function, inverse relation, inverse of a function, inverse of a relation

Background
Three different ideas emerge: Finding the inverse of a relation, or a function, by reversing the order of the elements in the set; finding the inverse of a function, given the formula that describes the function; and graphing the inverse of a function by finding the set of ordered pairs of the function, then reversing the order of the function, and finally reversing the elements of each ordered pair.

Suggestions
Careful terminology must be used to distinguish between an inverse of a function and an inverse function. When the inverse of a function is a function, it is called an *inverse function*.

Exercise 28 states that the graph of a function and its inverse are reflections through the line described by $y = x$. This line is the axis of symmetry. The exercise implies that the graph of an inverse can be formed by graphing the function on a sheet of paper, rotating the paper 90° counterclockwise, flipping the paper over bottom to top, and then tracing the graph.

Errors Students Make
Solving for y after the domain and the range are interchanged may cause problems.

Enrichment
1. Show that the inverse of a linear function is a linear function. Start with $y = mx + b$.
2. Show that if $f(x) = x + 3$ and $f^{-1}(x)$ is its inverse, then $f(f^{-1}(x)) = f^{-1}(f(x))$.

1. $y = mx + b$; $x = my + b$; $y = \frac{1}{m}x - \frac{b}{m}$; $y = m_1 x - b_1$ 2. $f(f^{-1}(x)) = (x - 3) + 3 = x$; (note $f^{-1}(x) = x - 3$); $f^{-1}(f(x)) = (x + 3) - 3 = x$

Cumulative Review
Simplify.
1. $\sqrt{24a^8 b^9}$ 2. $xy^2\sqrt{36x^4 y^5 z^{13}}$ 3. $xy^2\sqrt{162 x^{20} y^{31}}$
1. $2a^4 b^4 \sqrt{6b}$ 2. $6x^3 y^4 z^6 \sqrt{yz}$ 3. $9x^{11} y^{17}\sqrt{2y}$

11.5 Direct Variation (page 288)

Vocabulary
constant of proportionality, direct variation

Suggestions
The graph of a direct variation is a line that passes through the origin. Therefore, you may wish to have students plot the ordered pairs of a relation to see whether its graph passes through the origin. Students may observe that the slope of the line is the constant of variation (proportionality). Since the constant of variation depends on the order in which the variables are compared, the answers to Exercises 1–13 can be the reciprocals of those given.

Emphasize that a direct variation is *always* a linear function, but a linear function *does not necessarily* yield a direct variation.

Example 4 and the narrative preceding it present an alternate approach, one that does not involve the constant of proportionality, to teach direct variation. Both methods are useful.

For Additional Chalkboard Work
1. The number of liters of gasoline used by an automobile traveling at a constant speed is directly proportional to the distance traveled. At 60 km/h, a certain automobile uses 8 L of gasoline. How many liters of gasoline will the automobile use if it travels 560 km?
   9.3 L, to the nearest tenth of a liter
2. The amount of money a person earns varies directly as the number of hours worked or the number of units completed. If a person earns $280 in a 40-h wk, how many weeks would the person have to work in order to earn enough money to buy a used car, if the car costs $3,800? Assume that 60% of the earnings are set aside for the car.
   22.6 wk, to the nearest tenth of a week

Cumulative Review
Solve each equation by completing the square.
1. $x^2 + 8x = 2$ 2. $n^2 + 6n - 6 = 0$
1. $-4 \pm 3\sqrt{2}$ 2. $-3 \pm \sqrt{15}$

11.6 Inverse Variation (page 291)

Vocabulary
constant of variation, inverse variation

Suggestions
In inverse variation, an increase in one variable causes a decrease in the other variable. Moreover, inverse variation is characterized by the concept that when one variable doubles, the other is halved; when one variable triples, the other is divided by three, and so fourth. Contrast this type of variation with direct variation.

Emphasize that an inverse variation is *never* linear and that the graph of an inverse variation is *never* a straight line.

Two different methods for solving inverse variation problems are presented. Both methods are acceptable.

Errors Students Make
Students sometimes confuse the two types of variations. Emphasizing that for an inverse variation the product of the two variables is a constant, while for a direct variation the quotient of the two variables is a constant may help remove this difficulty.

For Additional Chalkboard Work
1. If y varies inversely as x and x is multiplied by c, how is y affected?
2. If y varies inversely as the square of x and x is doubled, how is y affected?
3. The time it takes to travel a given distance varies inversely as the average speed. If it takes 45 min to drive a certain distance at 60 km/h, how long would it take to drive the same distance at 50 km/h?

1. y is divided by c. 2. y is divided by 4.
3. 54 min

Cumulative Review
Solve each equation.
1. $3\sqrt{2x - 1} = 4$ 2. $2\sqrt{x + 4} = 7$
1. $\frac{25}{18}$ 2. $\frac{33}{4}$

11.7 Joint and Combined Variation (page 294)

Vocabulary
combined variation, joint variation

Suggestions
Indicate the similarities between joint variation and direct variation. A joint variation is a function in which one variable varies directly as the product of two other variables. Combined variation is a function in which direct and inverse variation are combined. Joint and combined variations usually yield quadratic equations, but these simplify to linear equations after substitution.

Errors Students Make
Initial substitution for a joint or a combined variation may cause difficulty. Also, some students may become confused if they are required to use both methods for solving variation problems.

For Additional Chalkboard Work
1. The volume, V, of a given mass of a gas varies directly as the absolute temperature, T, and inversely as the pressure, P. If $V = 300$ L when T is $400°$ and P is 16 lb/in.^{2}, what is the volume when T is $468°$ and P is 18 lb/in.^{2}? 312 L
2. The simple interest on a given amount of money varies jointly as the rate and the time. If the interest is \$1,320 when the rate is 11% and the time is 2 yr, find the rate when the interest is \$1,710 and the time is 3 yr. What amount of money was invested? 9.5%; \$6,000

Cumulative Review
Find the value of each expression.
1. $27^{\frac{2}{3}}$ 2. $8^{\frac{4}{3}}$ 3. $25^{\frac{3}{2}}$ 4. $64^{\frac{2}{3}}$
1. 9   2. 16   3. 125   4. 16

Computer Activities (p. 300)
Explain the need for user-defined functions in terms of the fact that every function defined in BASIC uses memory in the computer and the consequent need to restrict the number of them. Later, if and when you get to the trig section where Arctan is the only inverse trig function defined, challenge your better students to define the inverse functions for the other standard trig functions.

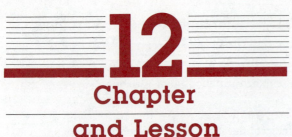

Chapter 12 Chapter and Lesson Commentaries

Chapter 12 Conic Sections

Chapter Objectives
To determine a pair of points that are symmetric with respect to the x-axis, the y-axis, or the origin

To find the center and radius of a circle, given its equation

To find the center and intercepts of an ellipse, given its equation

To draw the graphs of the equations of the conic sections—the parabola, circle, ellipse and hyperbola

To find the coordinates of the vertex and the x-intercepts of a parabola and to determine an equation for its axis of symmetry

To write an equation of a circle, an ellipse, and a hyperbola whose center or vertex is at the origin

To find the equations of the asymptotes of a hyperbola and to determine the hyperbola's intercepts

To use translations in the graphing of an equation of an absolute value function or a conic section whose vertex or center is not at the origin

Chapter Overview
The text does not provide derivations of all the general equations of conic sections. The parabola is simply presented as the graph of a quadratic equation. However, the text *does* explain how to derive the general equation for a circle and the equations for the ellipse and the hyperbola are derived for a specific example and then generalized. This approach will help students understand some

Chapter Overview (cont.)
of the theory without being overwhelmed with tedious algebraic derivations. It is advised however, that the student, does see a formal derivation. Some are included in the Theory of Conics Lesson. The students should be able to draw the graphs of the equations of the conic sections as well as write the equations, given certain conditions. The correspondence between the solutions to the equations and the coordinates of the graph should be stressed throughout the chapter.

An understanding of concepts such as intercepts, vertices, asymptotes and latus rectum will eliminate the need for the use of a table of values for graphing.

The concept of symmetry is presented in Lesson 12.1. Symmetric points on a circle will be used in the study of trigonometry and will be discussed later in this Teacher's Edition. The term *conic section* is not introduced until Lesson 12.6. The parabola, circle, ellipse, and hyperbola are each covered separately before any mention of a cone.

Special Features
Non-Routine Problem Solving   p. 302
Non-Routine Problems   p. 323
Cumulative Review   p. 326
Computer Activity   p. 331
College Prep Test   p. 334
Chapter Review   p. 332
Chapter Test   p. 333
Cumulative Review (Chapters 1–12)   pp. 335–336

12.1 Symmetric Points in a Coordinate Plane (page 303)

Vocabulary
symmetric, line of symmetry, axis of symmetry, point of symmetry

Background
To determine whether a function is even, that is, symmetric with respect to the y-axis, replace x by $-x$ and if no change occurs in the equation, the function is even. Show that $y = x^2$ is an even function. Similarly, to determine whether the graph of a relation is symmetric with respect to the x-axis, replace y by $-y$ and if no change occurs in the equation, the graph is symmetric with respect to the x-axis. To determine whether a function is odd, that is, symmetric with respect to the origin, replace x by $-x$ and y by $-y$ and if no change occurs in the equation, the function is odd. Show that $y = \frac{3}{x}$ is odd.

Suggestions
If the graph of an equation is symmetric with respect to the x-axis, it is sufficient to graph the equation in the first and second quadrants and then reproduce it in the 3rd and 4th quadrants. Similarly, a graph symmetric with respect to the y-axis reflects itself from the first and fourth quadrants into the second and third quadrants. If the graph of an equation is symmetric with respect to the origin, then it is sufficient to graph the equation in two adjacent quadrants and then reproduce it in the diagonally opposite quadrants.

Enrichment
Determine which of the following are even, odd, symmetric with respect to the y-axis, or none of these.
1. $\frac{x^2}{16} + \frac{y^2}{4} = 1$ 2. $xy = -3$ 3. $|y| = x$
4. $y = |x|$ 5. $x^2 - y^2 = 6$ 6. $x^2 + y^2 = 36$
7. $y = |x - 3|$ 5. x, y, O   6. x, y, O   7. N
1. x, y, O   2. O   3. x   4. y

Cumulative Review
Solve each system by substitution.
1. $3x + 5y = 2$ 2. $x + 7y = -9$
   $4x - y = -5$ $-5x + 3y = 7$
1. $(-1, 1)$ 2. $(-2, -1)$

12.2 The Circle (page 306)

Vocabulary
circle

Background
The equation of a circle does not yield a function. Solving $x^2 + y^2 = r^2$ for y, $y = \pm\sqrt{r^2 - x^2}$. Thus, for each value of x there are two values for y. However, a circle has x-axis, y-axis, and origin symmetry. Notice from $y = \pm\sqrt{r^2 - x^2}$ that $y = \pm r$ when $x = 0$. Similarly, $x = \pm r$ when $y = 0$. The extent of the circle is $-r \leq x \leq r$ and $-r \leq y \leq r$.

Suggestions
Students are expected to write an equation of any circle whose center is known, given a point through which the circle passes or given the radius of the circle. Students should also be able to graph the equation of a circle without using a table of values.

In general, if f is a function such that $g(x) = f(x + b)$, where b is any constant, the graph of g has the same shape as that of f but is translated b units to the left if $b > 0$ and $|b|$ units to the right if $b < 0$. Also, if f is any function and $g(x) = f(x) + c$, where c is any constant, the graph of g has the same shape as that of f but is translated c units up if $c > 0$ and $|c|$ units down if $c < 0$.

Emphasize that each of the ordered pairs of coordinates of the points on the circle is a solution of the equation of the circle.

It is advisable to review the methods of finding the distance between two points and completing the square before the start of the lesson.

Errors Students Make
Completing the square when the coefficient of the squared term is not 1 is often a source of difficulty.

Enrichment
Cut a string of desired radius. Attach a tack at one end and a pencil on the other. Tack it at the origin and, holding the string taut, trace a circle. Explain how this illustrates the definition of a circle.

Cumulative Review
Find the value of each determinant.
1. $\begin{vmatrix} 4 & -1 \\ 6 & 3 \end{vmatrix}$ 2. $\begin{vmatrix} 3 & -1 & 2 \\ -1 & 2 & 1 \\ 3 & 2 & -1 \end{vmatrix}$

1. 18   2. -30

12.3 The Ellipse (page 310)

Vocabulary
ellipse, focus (foci) of an ellipse

Background
When $a^2 > b^2$, in the general equation of an ellipse whose center is at the origin, the major axis of the ellipse is on the x-axis. When $b^2 > a^2$, the major axis is on the y-axis. When $a^2 = b^2$, the ellipse is a circle.

The extent of an ellipse can be determined by first solving the equation of the ellipse for y, then for x, and observing permissible values. You get $y = \pm \frac{b}{a}\sqrt{a^2 - x^2}$ and $x = \pm \frac{a}{b}\sqrt{b^2 - y^2}$. To avoid a negative under the radical, $x \leq a$ and $y \leq b$. The extents, or limiting values, of x and y are $-a \leq x \leq a$ and $-b \leq y \leq b$. You may also wish to derive the formula for the length of the latus rectum, $\frac{2b}{a}\sqrt{a^2 - c^2}$.

Suggestions
The length of the latus rectum, which is introduced in the C exercises of Lesson 7, will contribute to a more accurate graph of an ellipse. Lesson 7 also formally derives the equation of an ellipse, whereas this lesson enables students to derive an equation in a specific example, given the coordinates of the foci.

Students may have difficulty in determining the intercepts when the center is not at the origin. Label the new axes x' and y' and count along these axes. The new axes are obtained by translating the x, y axes. The center (h, k) is moved h units horizontally and k units vertically.

Errors Students Make
When the intercepts are fractions, as in Example 3, convert $\frac{6x^2}{27} + \frac{9y^2}{27} = 1$ to $\frac{x^2}{\frac{27}{6}} + \frac{y^2}{\frac{27}{9}} = 1$ to avoid difficulty in determining a and b.

Cumulative Review
Solve each system of equations.
1. $2x - 3y + z = -9$ 2. $2x - 3y = 7$
   $-2x + y - 3z = 7$ $4x + 5z = 13$
   $x - 2y + 3z = -8$ $-2y - 3z = -1$
1. $(-1, 2, -1)$ 2. $(2, -1, 1)$

12.4 The Parabola (page 314)

Vocabulary
parabola, vertex or turning point

Background
The parabola can be flat or steep. For $y = ax^2$, if $a = \frac{1}{100,000}$, then the parabola is quite flat.

Suggestions
In the equation of the parabola $y = ax^2 + bx + c$, if $a > 0$, the graph opens upward, and if $a < 0$, the graph opens downward. Thus, when $a > 0$, the parabola's vertex is a minimum point. When $a < 0$, the vertex is a maximum point.

If the graph crosses the x-axis, the two roots are real and unequal. If the graph of a parabola is tangent to the x-axis, the two roots, or x-intercepts, are real and equal. If the graph does not cross the x-axis, there are no x-intercepts, hence no real roots. In this case the roots are imaginary numbers. Roots, solutions, or x-intercepts are also called zeros of a quadratic function since these values make the function value (y-value) zero.

Encourage students to use completing the square and translations to graph equations of parabolas.

Errors Students Make
Students might have some difficulty choosing appropriate values of x when constructing a table of values. Determining the axis of symmetry first is one way to alleviate this problem. Larger-square graph paper will help students better estimate solutions to the nearest tenth.

Enrichment
Have students determine why transmitting reflectors and receivers, and car headlights are shaped in the form of a parabola.

Cumulative Review
Solve each system by addition.
1. $3x - 4y = 2$ 2. $4x - 2y = -12$
   $-5x + 6y = -2$ $-3x + 5y = 23$
1. $(-2, -2)$ 2. $(-1, 4)$

12.5 The Hyperbola (page 319)

Vocabulary
hyperbola, foci of a hyperbola, asymptotes, rectangular hyperbola

Background
The asymptotes of the hyperbola whose equation is $\dfrac{x^2}{a^2} - \dfrac{y^2}{b^2} = 1$ are $y = \pm \dfrac{b}{a}x$.

If a is very large, the slope $\dfrac{b}{a}$ approaches 0, hence the arms of the hyperbola get very close together and, conversely, the arms spread out when a is very small. This illustrates that a limiting position of a hyperbola is either an axis or a straight line.

The extent of a hyperbola can be determined by solving the equation of an hyperbola for y.

$$y = \pm \frac{b}{a}\sqrt{x^2 - a^2}, \text{ thus } x \geq a \text{ or } x \leq -a$$

For Additional Chalkboard Work
Additional work may be necessary to overcome the difficulty of determining the equations of the asymptotes. Tell orally the equations of the asymptotes.

1. $\dfrac{x^2}{25} - \dfrac{y^2}{9} = 1$ 2. $\dfrac{x^2}{36} - \dfrac{y^2}{100} = 1$

1. $y = \dfrac{3}{5}x, y = -\dfrac{3}{5}x$ 2. $y = \dfrac{5}{3}x, y = -\dfrac{5}{3}x$

Errors Students Make
Additional chalkboard work may be necessary to overcome the difficulty of determining the equations of the asymptotes.

To overcome the difficulty in Example 7 of drawing the graph after finding the new center, construct the new axes x' and y' through the new center and then graph the basic function on the new axes.

Cumulative Review
Find the sum, if possible.

1. $\begin{bmatrix} 3 & -1 \\ -4 & 6 \end{bmatrix} + \begin{bmatrix} 6 & -2 \\ 3 & 9 \end{bmatrix}$ $\begin{bmatrix} 9 & -3 \\ -1 & 15 \end{bmatrix}$

2. $\begin{bmatrix} 3 & 4 & 3 \\ 2 & -1 & 1 \\ -1 & 2 & 2 \end{bmatrix} + \begin{bmatrix} 2 & 5 \\ 4 & 2 \\ -1 & -2 \end{bmatrix}$ not possible

12.6 Identifying Conic Sections (page 324)

Vocabulary
cone, conic section

Background
One way to identify a conic section, given its equation, is to examine the general formula $Ax^2 + Bxy + Cy^2 + Dx + Ey + F = 0$. If $A = C = D = E = 0$, then $Bxy = -F$. This is an equation of a rectangular hyperbola. If $B = 0$, then $Ax^2 + Dx + Cy^2 + Ey = -F$ is an equation of a conic section whose center is not at the origin. If $B = D = E = 0$, then $Ax^2 + Cy^2 = -F$ is an equation of a conic section whose origin is at the center. Note, $-F$ may or may not be negative. Let $G = -F \neq 0$, neither A nor C being 0. So, $Ax^2 + Cy^2 = G$. If $A = C$ and A, C, and G have the same sign, the equation represents a circle. If $A \neq C$ and A, C and G have the same sign, the equation represents an ellipse. If A and C have different signs, the equation represents a hyperbola. Finally, if Bxy is present with the other terms, the graph of the conic is rotated.

Suggestions
The standard form of the equations of the conics presented are for an ellipse and circle whose center is at the origin and for a hyperbola whose foci are on one of the axes. While there are other ways to identify a conic, in this lesson students are asked to put the equation in standard form before deciding which conic is represented by the equation. Some exercises in Part B do not have the center at the origin. This is covered more extensively in 12.7.

For Additional Chalkboard Work
Identify the conic section $x^2 + 8x + y^2 - 4y = 26$. By completing the square, you get $x^2 + 8x + 16 + y^2 - 4y + 4 = 26 + 16 + 4$. $(x + 4)^2 + (y - 2)^2 = 46$. This is an equation of a circle whose center is at $(-4, 2)$ and whose radius is $\sqrt{46}$.

Cumulative Review
Let $g(x) = -x^2 + 4x - 1$. Find each of the following function values.
1. $g(0)$ 2. $g(-1)$ 3. $g(a)$
1. -1 2. -6 3. $-a^2 + 4a - 1$

12.7 Equations and Properties of Conic Sections (page 327)

Background

As stated earlier, a useful aid in sketching graphs of the equations of conic sections is to determine the length of the latus rectum, the perpendicular segment through a focus terminating on the graph.

For the parabola $y^2 = 4px$, let $x = p$ (the x-coordinate of a focus). Then $y^2 = 4p^2$. Thus the length of the latus rectum is $2y = |4p|$. For an ellipse, Exercise 31 asks the student to find the length of the latus rectum. Let $x = c$ with $(\pm c, 0)$ as the coordinates of the foci, then $\frac{c^2}{a^2} + \frac{y^2}{b^2} = 1$ or $\frac{y^2}{b^2} = 1 - \frac{c^2}{a^2}$, $y = \pm \frac{b}{a}\sqrt{a^2 - c^2}$. But it was learned that $a^2 - c^2 = b^2$. Thus, $y = \frac{b}{a} \cdot b$ or $y = \frac{b^2}{a}$, $2y = \frac{2b^2}{a}$. Note that $b^2 = a^2 - c^2$ is an equation that connects the intercepts and foci.

Another useful aid in graphing conics is to determine the eccentricity of the conic. The eccentricity, e, of an ellipse is $e = \frac{c}{a}$ where $c = \sqrt{a^2 - b^2}$. Hence, $e = \frac{\sqrt{a^2 - b^2}}{a}$. When $a \to b$, $e \to 0$ and the ellipse approximates a circle when $b \to 0$, $e \to 1$. When b is very small the hyperbola approaches a horizontal line; when b is very large the hyperbola approaches a vertical line.

Suggestions

Derive the equation $\frac{x^2}{a^2} + \frac{y^2}{b^2} = 1$ to clarify the relationship between the intercepts and the foci. You may also wish to formally derive the equation of a hyperbola.

Computer Activities (page 331)

The program here will draw a good parabola on the monitor but not on a printer. To do it on a printer, the TAB statements have to be replaced by SPC, which counts spaces between successive PRINTS. In this particular program, only the very best students should be encouraged to try it.

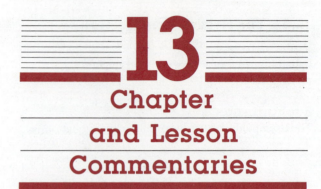

Chapter and Lesson Commentaries

Chapter 13 Linear Quadratic Systems

Chapter Objectives

To find the real-number solutions of a system composed of one quadratic equation and one linear equation through graphing

To solve a system made up of two quadratic equations both graphically and algebraically

To find the real-number solutions of a system of two linear inequalities, two quadratic inequalities and one linear and one quadratic inequality graphically

To draw the graph of a linear inequality in two variables

To draw the graph of a quadratic inequality

Chapter Overview

In this chapter students are expected to:
1. solve systems of linear-quadratic equations
2. solve quadratic equations both algebraically and graphically.

Some examples should be done both algebraically and graphically to show that both methods do arrive at the same solutions. Use the graphing method first, having students estimate the answers, and then solve algebraically. This will give students confidence in estimating, yet will also reinforce the fact that at times graphing gives only an estimation of the solution.

Chapter Overview (cont.)

The chapter goes on to cover linear inequalities. Simple linear inequalities are introduced first, followed by systems of different combinations of inequalities. The systems include two linear inequalities, a linear-quadratic inequality and two quadratic inequalities. Students are expected to solve these systems graphically. Emphasize that each pair of coordinates of the infinite number of points in the cross-hatched region is a solution to both inequalities. Encourage students to utilize all the aids in graphing the conic sections—asymptotes, extents, latus rectum, and so forth, to enable them to draw smooth, accurate curves. This is important if you wish them to find points of intersection to estimate solutions.

Special Features
Non-Routine Problem Solving   p. 337
Non-Routine Problems   p. 340
Reading in Algebra   p. 348
Cumulative Review   pp. 340, 351
Applications   pp. 344–345
Computer Activity   p. 354
College Prep Test   p. 355
Chapter Review   p. 352
Chapter Test   p. 353

13.1 Solving Linear-Quadratic Systems of Equations (page 338)

Vocabulary
linear-quadratic systems

Background
From symbolic logic, the word *conjunction* is the connective "and" that is used to join two simple sentences into one compound sentence. For the compound sentence to be true, both simple sentences must be true. When it is stated that students should solve a system of equations like $\frac{x^2}{16} + \frac{y^2}{9} = 1$, $2x - 3y = 6$, it is implied that students should find values for x and y such that $\frac{x^2}{16} + \frac{y^2}{9} = 1$ and $2x - 3y = 6$. Thus, the solution would be $\left\{ (x,y) \mid \frac{x^2}{16} + \frac{y^2}{9} = 1 \wedge 2x - 3y = 6 \right\}$.

Suggestions
Review the standard forms of the equations of the conic sections. Also remind the students that the coordinates of every point on the graph is a solution of the equation. By substituting the coordinates of each point in the equation, the equation checks. In this lesson the system is first solved graphically. This approach does not give the accuracy that one would get algebraically, but it is believed that students will get a better understanding by this visual approach.

Remind students that the purpose of this lesson is to find coordinates of points that solve both the quadratic and linear equations.

Students should substitute into the linear equation to find the second coordinate after finding the first coordinate. There is a possibility of introducing extraneous roots if they don't. For example, solving $x^2 + y^2 = 10$ and $x - 2y = 1$ algebraically yields an extraneous root when you substitute into the quadratic equation instead of into the linear equation.

Cumulative Review
Solve each of the following.
1. $-3x + 6 \leq -9$ **2.** $|x - 4| \leq 14$
3. $|x + 1| > 3$ 3. $\{x \mid x > 2 \text{ or } x < -4\}$
1. $\{x \mid x \geq 5\}$ 2. $\{x \mid -10 \leq x \leq 18\}$

T-65

13.2 Solving Quadratic Systems of Equations (page 341)

Vocabulary
quadratic systems

Suggestions
This lesson illustrates the intersection of the graphs of two quadratic equations. It is shown that there may be no, one, two, three, or four common points of intersection. Therefore, there may be no, one, two, three, or four solutions.

The ordered pairs obtained by graphing or estimating are approximate solutions and may not check exactly. You may wish to have students use a calculator in checking solutions.

Errors Students Make
Example 3 shows a system solved algebraically where there are four solutions. Students should be careful to include all of them.

For Additional Chalkboard Work
Solve the system $x^2 + y^2 = 10$ algebraically.
$$xy = 3$$
$y = \frac{3}{x}; x^2 + \left(\frac{3}{x}\right)^2 = 10; x^2 + \frac{9}{x^2} = 10; x^4 + 9 = 10x^2; x^4 - 10x^2 + 9 = 0$
Factor.
$(x^2 - 9)(x^2 - 1) = 0; x^2 = 9; x = \pm 3; x^2 = 1; x = \pm 1$
So $x = 3, -3, 1, -1$, and $y = 1, -1, 3, -3$; thus, $(3, 1), (-3, -1), (1, 3)$ and $(-1, -3)$ are solutions. They check.

Enrichment
Solve the following systems algebraically.
1. $x^2 + xy + y^2 = 19$ 2. $3y^2 + 2xy = 7$
   $xy = 6$ $-2y^2 + 3xy = 4$
1. $(3, 2), (-3, -2), (2, 3), (-2, -3)$
2. $(2, 1), (-2, -1)$

Cumulative Review
Solve.
Fred mixed some brand A cereal worth $1.30/lb and some brand B cereal worth $1.94/lb to make a 10-lb package that was worth $15.54. Find the amount of each brand used.
6.03 lbs of brand A and 3.97 lbs of brand B

13.3 Solving Linear Inequality Systems (page 346)

Vocabulary
linear inequality systems

Background
While it is sometimes tempting to describe the solution set as the double-shaded region, it is not correct to do so. The double-shaded region is the graph of the solution set and not the solution set itself. The solution set is the set of all ordered pairs for the points in the double-shaded region. Distinguish between finding the solution set and graphing the solution set. Graphing the solution set yields the double-shaded region.

Suggestions
Testing with a checkpoint is not so much used as a check but as a method to determine which half-plane is to be shaded. To check after shading is only a verification check.

Whenever the equation of a line is graphed, the plane is separated into three regions; the region above the line, the region below the line, and the line itself. Discuss the use of dashed lines and solid lines when graphing inequalities.

Some students may wish to use colored pencils to show the double-shaded region. Others may wish to draw horizontal lines for one space and vertical lines for another. The graph of the solution set would be the intersection of the vertical and horizontal lines or the cross-hatched space.

Errors Students Make
Difficulty might arise when students try to represent the region graphing the solution set. Colored pencils will help avoid problems.

Whether to include or exclude the line may also cause difficulty. Emphasize that a solid line is generated by $\leq$, $\geq$ or $=$ and a dashed line is used with $<$ or $>$.

Cumulative Review
Solve.
How many liters of 32 percent alcohol solution must be added to 24 L of a 58 percent alcohol solution to obtain a 48 percent alcohol solution?
15 L

13.4 Solving Linear-Quadratic Inequality Systems (page 349)

Vocabulary
linear-quadratic inequality

Suggestions
Review graphs of the equations of the conic sections. Whenever the equation of a conic is graphed, the plane is divided into three regions: the region within the conic, outside the conic, and on the conic. Students should graph the inequalities as though they were equations and then use two checkpoints, one point within, and one point outside the graph of the conic. The inequalities using $\leq$ or $\geq$ will include all points on the curve. Thus the curve is solid. The inequalities with $<$ or $>$ will have a dashed curve.

For Additional Chalkboard Work
Solve each system of inequalities graphically.

1. $\dfrac{x^2}{16} + \dfrac{y^2}{9} \leq 1$

   $\dfrac{x^2}{9} + \dfrac{y^2}{4} \geq 1$

2. $\dfrac{x^2}{25} - \dfrac{y^2}{4} \leq 1$

   $x^2 + y^2 \geq 4$

3. $\dfrac{x^2}{36} + \dfrac{y^2}{25} \leq 1$

   $\dfrac{x^2}{9} - \dfrac{y^2}{4} \geq 1$

4. $\dfrac{x^2}{4} + \dfrac{y^2}{9} \geq 1$

   $x^2 + y^2 < 16$

5. $\dfrac{(x-3)^2}{9} + \dfrac{(y+2)^2}{4} \leq 1$

   $(x-3)^2 + (y+2) \geq 4$

Cumulative Review
Solve.

One leg of a right triangle is 7 m longer than the other leg and 1 m shorter than the hypotenuse. Find the area of the triangle. 30 m²

Computer Activities (page 354)
An extension of this program, again for more capable students, would be to do the same for circles whose center is not at the origin.

14 Chapter and Lesson Commentaries

Chapter 14 Progressions, Series, and Binomial Expansions

Chapter Objectives
To write consecutive terms of an arithmetic progression (A.P.) and a geometric progression (G.P.)

To find a specified term of a given A.P. and G.P.

To find arithmetic means and geometric means between two given numbers

To find arithmetic mean and the geometric mean of two numbers

To find the sum of the first n terms of an arithmetic series and a geometric series

To write an arithmetic series using Σ-notation

To find the sum, if it exists, of an infinite geometric series

To write a repeating decimal as an infinite geometric series and an infinite indicated sum

To write a repeating decimal in the form $\dfrac{x}{y}$, where x and y are integers

To expand a positive integral power of a binomial

To find a specified term of a binomial expansion

To apply a special infinite series

Chapter Overview

Arithmetic and geometric progressions, or sequences, are introduced with emphasis on finding the nth term for a given counting number n. Arithmetic and geometric *series* are related to their corresponding A.P. and G.P. with emphasis on finding the sum of the first n terms of a series. Word problems are included in each lesson.

  Infinite series of several types are introduced. Infinite *geometric* series and their sums, if they exist, are introduced first. Then special infinite series for sin x, cos x, e, and e^x are presented; they are evaluated for a given value of x to several decimal places.

Special Features

Blaise Pascal   p. 356
Reading in Algebra   pp. 359, 370, 377
Calculator Activities   pp. 360, 371, 378
Cumulative Review   pp. 363, 371
Chapter Review   p. 386
Chapter Test   p. 387
College Prep Test   p. 388

14.1   Arithmetic Progressions (page 357)

Vocabulary
arithmetic progression (or sequence), common difference

Background
An arithmetic progression with its common difference d and nth term l provides the background for the study of arithmetic means and arithmetic series.

Suggestions
You may wish to introduce and use the term *sequence* as a synonym for *progression,* since some students will meet it in their continuing mathematical studies.

  Before Example 4, you can use a simpler example:
Find the 21st term of $8, 3, -2, -7, \ldots$.
$$a = 8 \quad n = 21 \quad d = -5$$
$l = a + (n - 1)d = 8 + (21 - 1)(-5) = 8 + 20(-5) = -92$

For Additional Chalkboard Work

Find the common difference, if it exists. (Ex. 1–3)
1. $-4.3, -2.8, -1.3, 0.2, \ldots$ 1.5
2. $6.2, 4.4, 2.6, 0.8, \ldots$ -1.8
3. $5, 10, 20, 40, \ldots$ Not an A.P.
4. Write the next 3 terms: $\frac{1}{8}, \frac{1}{2}, \frac{7}{8}, \frac{5}{4}, \ldots$
5. Write the first four terms of the A.P. with $a = 7.5$ and $d = -3$.
6. Find the 31st term of $7, 10, 13, 16, \ldots$.
7. Find the 36th term of $14, 10, 6, 2, \ldots$.

4. $\frac{13}{8}, 2, \frac{19}{8}$
5. $7.5, 4.5, 1.5, -1.5$ 6. 97 7. -126

Cumulative Review
1. Find the slope of $\overrightarrow{AB}$ for $A(-2, 4)$, $B(3, 3)$.
2. Solve $x^2 + 3x - 1 = 0$.
3. Simplify $\dfrac{x^5}{x^{-3}}$.

1. $\dfrac{-1}{5}$ 2. $\dfrac{-3 \pm \sqrt{13}}{2}$ 3. x^8

14.2 The Arithmetic Mean (page 361)

Vocabulary
arithmetic means, *the* arithmetic mean

Suggestions
Encourage students to draw the appropriate number of blanks between the two given numbers when finding two or more means, as shown in Example 1. When finding exactly one mean between two numbers x and y, students should use the formula $\frac{x+y}{2}$. Example 2 is used only to introduce "the arithmetic mean of x and y."

Example 1 can follow a simpler example: Find the two arithmetic means between 29 and 8.

$$29, \underline{}, \underline{}, 8$$
$$a = 29 \quad n = 4 \quad l = 8$$
$$l = a + (n-1)d$$
$$8 = 29 + (4-1)d$$
$$d = -7$$

The A.P. is 29, 22, 15, 8, and the means are 22 and 15.

For Additional Chalkboard Work
1. Find the five arithmetic means between 30 and 21. **28.5, 27, 25.5, 24, 22.5**
2. Find the arithmetic mean of -8 and 22. **7**
3. The arithmetic mean of two numbers is 14. One number is 5 less than twice the other number. Find both numbers. **11, 17**

Cumulative Review
1. Find the slope and the y-intercept of the line with equation $6x + 2y = 8$.
2. Multiply $(3 + 2i)(2 + i)$.

1. $-3; 4$ 2. $4 + 7i$

14.3 Arithmetic Series and Σ-Notation (page 364)

Vocabulary
arithmetic series, summation, index of summation, sigma, sigma notation

Suggestions
Emphasize the differences and the similarities between an arithmetic progression and an arithmetic series. For each progression: $x_1, x_2, x_3, \ldots$, there is a series, or indicated sum: $x_1 + x_2 + x_3 + \ldots$. The series is arithmetic if the progression is arithmetic. An arithmetic series and an arithmetic progression have the same first term, the same common difference, and the same nth term.

Students should know that there are two sum formulas and also understand when to use each of them. Use $S = \frac{n}{2}(a + l)$ if the values of n, a, and l are given and use $S = \frac{n}{2}[2a + (n-1)d]$ if the values of n, a, and d are given.

For Additional Chalkboard Work
Find the sum of each arithmetic series for the given data. (Ex. 1–3)
1. $n = 20$, $a = -4$, $l = 35$ **310**
2. $n = 40$; $5 + 4.5 + 4 + 3.5 + \ldots$ **-190**
3. $\sum_{k=3}^{7}(4k - 3)$ **85**

Write each series using Σ-notation. (Ex. 4–5)
4. $12 + 15 + 18 + 21 + 24$
5. $7 + 12 + 17 + 22 + \ldots + 42$

4. $\sum_{j=4}^{8} 3k$ 5. $\sum_{k=1}^{8}(5k + 2)$

Cumulative Review
Given $f(x) = 5x - 4$, find the following.
1. $f(6)$ 2. $f(-3)$ 3. $f(f(2))$
4. $f(c) - f(h)$ 5. $\frac{f(c+h) - f(c)}{h}$

1. 26   2. -19 3. 26
4. $5c - 5h$ 5. 5

14.4 Geometric Progressions (page 368)

Vocabulary
geometric progression (or sequence), common ratio

Suggestions
Have each student use a calculator until the following expressions are computed correctly.
1. 2^7 2. $(-2)^9$ 3. $12.5(3^6)$ 4. $14(-3)^7$
1. 128   2. -512 3. 9,112.5   4. $-30,618$

Impress upon the students that decimals for $\frac{1}{3}$ and $\frac{2}{3}$ cannot be used on the calculator if exact answers are required for ar^{n-1} with $r = \frac{1}{3}$ or $r = \frac{2}{3}$.

To find $\left(\frac{1}{2}\right)^{10}$, use $2^{10} = 1,024$ and $\left(\frac{1}{2}\right)^{10} = \frac{1}{1,024}$.

To find 2^{10} using a nonscientific calculator, depress ② once, ⊗ once, and then ⊜ $10 - 1$, or 9, times.

Before Example 4, you can use a simpler example. Find the 10th term of the G.P.

$$\frac{1}{4}, \frac{1}{2}, 1, \ldots.$$

$r = \frac{1}{2} \div \frac{1}{4} = 2$ $a = \frac{1}{4}$ $n = 10$

$l = a \cdot r^{n-1} = \frac{1}{4} \cdot 2^{10-1} = \frac{2^9}{4} = \frac{512}{4} = 128$

For Additional Chalkboard Work
Find the common ratio, if it exists. (Ex. 1–4)
1. $-8, 4, -2, 1, \ldots$ 2. $-2, -6, -18, -54, \ldots$
3. $2, 4, 6, 8, \ldots$ 4. $3, 0.6, 0.12, 0.024, \ldots$
5. Write the next 3 terms for $-\frac{1}{27}, \frac{1}{9}, -\frac{1}{3}, 1, \ldots.$
6. Write the first four terms of the G.P. with $a = 0.04$ and $r = -10$.
7. Find the 7th term of the G.P. with $a = \frac{1}{8}$ and $r = -2$.   8
8. Find the 10th term of the G.P. 64, -32, 16, $-8, \ldots.$

1. $-\frac{1}{2}$ 2. 3   3. not a G.P.   4. 0.2
5. $-3, 9, -27$ 6. $0.04, -0.4, 4, -40$ 8. $-\frac{1}{8}$

Cumulative Review
1. Find the distance between $P(1, -1)$ and $Q(-2, 3)$.   2. Simplify $6 - 5\sqrt{-12}$.
3. Find the solution set of $|x| < 7$.
1. 5   2. $6 - 10i\sqrt{3}$ 3. $\{x | -7 < x < 7\}$

14.5 The Geometric Mean (page 372)

Vocabulary
geometric means, *the* geometric mean, mean proportional

Suggestions
Introduce the lesson with the following review of skills needed in the lesson.
Find the real number solutions (Ex. 1–7).
1. $r^2 = 5$ 2. $r^3 = -8$ 3. $r^3 = -\frac{64}{125}$
4. $\sqrt{x + 3} = 2$ 5. $\sqrt{x^2 + 6x} = 4$
6. $7\sqrt{7} = r^3$ 7. $-5\sqrt{5} = r^3$
8. Solve the system $y = x + 5$ and $\sqrt{xy} = 6$.

1. $-\sqrt{5}, \sqrt{5}$ 2. -2 3. $-\frac{4}{5}$
4. 1   5. $-8, 2$
6. $\sqrt{7}$ 7. $-\sqrt{5}$ 8. $(-9, -4)$ and $(4, 9)$

Encourage students to draw the appropriate number of blanks between the two given numbers when finding two or more means, as shown in Examples 1 and 2. When finding exactly one mean between two numbers x and y, students should use the formula: $\sqrt{xy}$ or $-\sqrt{xy}$.

Emphasize that there are two sets of n real geometric means between two numbers if n is an odd number.

For Additional Chalkboard Work
1. Find the two real geometric means between $-\frac{3}{8}$ and 24.
2. Find three geometric means between 32 and 2.
3. Find the positive geometric mean of 4 and 8.
4. The positive geometric mean of two numbers is 6, and one number is 5 more than the other number. Find both numbers.

1. $\frac{3}{2}, -6$ 2. 16, 8, 4 or $-16, 8, -4$
3. $4\sqrt{2}$ 4. 4 and 9 or -9 and -4

Cumulative Review
1. Write $\sqrt[3]{5x}$ in exponential form.
2. Write $5\sqrt[3]{x}$ in exponential form.
3. Write $2y^{\frac{3}{4}}$ in radical form.
4. Write $(2y)^{\frac{3}{4}}$ in radical form.
1. $(5x)^{\frac{1}{3}}$ 2. $5x^{\frac{1}{3}}$ 3. $2\sqrt[4]{y^3}$
4. $\sqrt[4]{(2y)^3}$, or $\sqrt[4]{8y^3}$

14.6 Geometric Series (page 375)

Vocabulary
geometric series

Suggestions
Before starting the lesson, have the students perform the following typical "messy" computations so that they may later focus on the lesson's concepts.

1. $\dfrac{3-(-5)(375)}{1-(-5)}$ 313

2. $\dfrac{3-5(234,375)}{1-5}$ 292,968

3. $\dfrac{8-\left(-\frac{1}{2}\right)\left(-\frac{1}{16}\right)}{1-\left(-\frac{1}{2}\right)}$ $5\dfrac{5}{16}$

4. $\dfrac{81\left[1-\left(\frac{1}{3}\right)^6\right]}{1-\frac{1}{3}}$ $121\dfrac{1}{3}$

Students should know that there are two sum formulas and also understand when to use each of them. Use $S = \dfrac{a-rl}{1-r}$ if a, r, and l are given and use $S = \dfrac{a(1-r^n)}{1-r}$ or $\dfrac{a-a\cdot r^n}{1-r}$ if a, r, and n are given.

Give special attention to Exercises 41 and 42. In Exercise 41, there are 8 descents (128 m, 64 m, 32 m, . . .) and 7 upward flights (64 m, 32 m, 16 m, . . .), which can be summed separately and then combined.

You may also wish to show how the concept of a geometric series applies to the study of the half-life of substances.

For Additional Chalkboard Work
Find the sum of each geometric series for the given data.

1. $a=4$, $r=-6$, $l=-31{,}104$ $-26{,}660$
2. $27-9+3-\cdots+\dfrac{1}{27}$ $20\dfrac{7}{27}$
3. $n=10$; $-5-10-20-40-\cdots$ $-5{,}115$
4. $a=40$, $r=0.1$, $n=7$ 44.44444
5. $\displaystyle\sum_{k=1}^{4} 3(-5)^{k-1}$ -312

Cumulative Review
Solve the system $\begin{array}{l} 3x+2y=8 \\ 5x-3y=7 \end{array}$ for (x,y). $(2,1)$

14.7 Infinite Geometric Series (page 379)

Vocabulary
limit

Suggestions
Notice that Examples 2 and 3 provide a new approach to the problem of writing a repeating decimal in the form $\dfrac{x}{y}$ where x and y are integers.

Stress that an infinite geometric series like $3+6+12+24+\ldots$ does not have a sum, and that $r=2$ (r is not between -1 and 1) in this case.

For Additional Chalkboard Work
Find the sum, if it exists, of each infinite geometric series. (Ex. 1–3)

1. $3+1+\dfrac{1}{3}+\dfrac{1}{9}+\cdots$ $4\dfrac{1}{2}$
2. $18-12+8-\dfrac{16}{3}+\cdots$ $10\dfrac{4}{5}$
3. $\dfrac{1}{12}+\dfrac{1}{6}+\dfrac{1}{3}+\dfrac{2}{3}+\cdots$ does not exist ($r=2$)
4. Write $27.2727\cdots$ as an infinite geometric series and then in the form $\dfrac{x}{y}$, where x and y are integers.
5. Write $4.3222\cdots$ as an infinite indicated sum and then in the form $\dfrac{x}{y}$, where x and y are integers.

4. $27+0.27+0.0027+\cdots$; $r=0.01$; $a=27$; $\dfrac{x}{y}=\dfrac{2{,}700}{99}$, or $\dfrac{300}{11}$

5. $4.3+(0.02+0.002+0.0002+\cdots)$; $\dfrac{389}{90}$

Cumulative Review
Rationalize the denominator. Simplify.

1. $\dfrac{\sqrt{35}}{\sqrt{7}}$ 2. $\dfrac{9\sqrt{2}}{4\sqrt{5}}$ 3. $\dfrac{4}{\sqrt[3]{9x}}$

4. $\dfrac{3}{2\sqrt{5}-\sqrt{6}}$ 5. $\dfrac{5}{6+3i}$

1. $\sqrt{5}$ 2. $\dfrac{9\sqrt{10}}{20}$ 3. $\dfrac{4\sqrt[3]{3x^2}}{3x}$

4. $\dfrac{6\sqrt{5}+3\sqrt{6}}{14}$ 5. $\dfrac{2-i}{3}$

14.8 Binomial Expansions (page 382)

Suggestions
After studying the seven patterns on page 382, have the students complete the following **Reading in Algebra** exercise.

Supply the missing expressions in the following paragraph.

In the ___ of $(a + b)^5$, there are ___ terms, among which the first term is ___ and the ___ is b^5. The exponents of a ___ from ___ to 0, and the ___ of b increase from 0 to ___. The coefficients of the terms form the progression $1, \frac{5}{1}, \frac{5 \cdot 4}{1 \cdot 2},$ ___, ___, $\frac{5 \cdot 4 \cdot 3 \cdot 2 \cdot 1}{1 \cdot 2 \cdot 3 \cdot 4 \cdot 5}$.

expansion, six, a^5, last term, decrease, 5, exponents, 5, $\frac{5 \cdot 4 \cdot 3}{1 \cdot 2 \cdot 3}, \frac{5 \cdot 4 \cdot 3 \cdot 2}{1 \cdot 2 \cdot 3 \cdot 4}$

Errors Students Make
A typical student error in Example 1 is to omit the parentheses in $(3x)$ or $(-y^2)$.

For Additional Chalkboard Work
Expand each power of a binomial.
Simplify each term. (Ex. 1–3)
1. $(a + b)^6$ 2. $(a - b)^7$ 3. $(x^3 - 3y)^4$
4. Find the 5th term in the expansion of $(4x - y^2)^7$.
5. Find the first four terms in the expansion of $(x - y^2)^{12}$. Simplify each term.

1. $a^6 + 6a^5b + 15a^4b^2 + 20a^3b^3 + 15a^2b^4 + 6ab^5 + b^6$
2. $a^7 - 7a^6b + 21a^5b^2 - 35a^4b^3 + 35a^3b^4 - 21a^2b^5 + 7ab^6 - b^7$
3. $x^{12} - 12x^9y + 54x^6y^2 - 108x^3y^3 + 81y^4$
4. $\frac{7 \cdot 6 \cdot 5 \cdot 4}{1 \cdot 2 \cdot 3 \cdot 4}(4x)^3(-y^2)^4 = 2{,}240x^3y^8$
5. $x^{12} - 12x^{11}y^2 + 66x^{10}y^4 - 220x^9y^6 + \cdots$

Cumulative Review
Solve for x.
1. $\sqrt{x + 2} = 2\sqrt{x - 1}$ 2. $x - \sqrt{4x - 7} = 1$
3. $10^{2x-1} = 1000$ 4. $10^x = \frac{1}{100}$

1. 2   2. 2, 4   3. 2   4. -2

14.9 Special Infinite Series (optional) (page 384)

Background
Sine, cosine, e, and powers of e are each defined in terms of an infinite series. Values of sin x are approximated for real number values of x, that is, radian measures for x.

Suggestions
1. Students should divide by $n!$ when evaluating sin x or cos x and add values of $\frac{1}{n!}$ when evaluating e.
2. To compute the value of a series to n decimal places, compute each term until the first $n + 1$ decimal places are zeros.

The use of a calculator is essential to working the exercises in this lesson.

For Additional Chalkboard Work
1. After Example 1, assign Exercise 1 or 6.
2. After Example 2, assign Exercise 3.

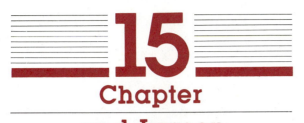

Chapter 15 and Lesson Commentaries

Chapter 15   Exponential, Logarithmic, and Polynomial Functions

Chapter Objectives
To draw the graph of an exponential function and a logarithmic function
To interpret the graph of an exponential function and a logarithmic function
To find the $\log_b$ of a positive number
To solve a logarithmic equation
To write an equation for the inverse of an exponential function and a logarithmic function
To expand the $\log_b$ of a product, quotient, power, and radical
To simplify a sum and a difference of logarithms
To simplify a multiple of a logarithm
To find $\log_{10}$ of a positive rational number
To find antilog_{10} of a rational number
To solve an exponential equation using base-10 logs
To find $\log_b$ of a positive number using a table of base-$_{10}$ logs
To solve a word problem by using an exponential equation
To find the zeros of an integral polynomial of degree greater than two
To factor an integral polynomial into first-degree factors
To solve an integral polynomial equation of degree greater than two
To find an upper bound and a lower bound for the real zeros of an integral polynomial
To sketch the graph of an integral polynomial function of degree n with n distinct real zeros

Chapter Overview
The availability of calculators has shifted the study of logarithms from computational objectives to a study of logarithmic functions and equations and their inverses, exponential functions, and equations. The graphs of exponential and logarithmic functions are drawn to reveal the properties of these functions.

The technique of synthetic substitution, the binomial factor theorem, and the remainder theorem (introduced in Chapter 5) are applied to integral polynomial functions and equations. For an integral polynomial $P(x)$ of degree greater than two,
1. the zeros of $P(x)$ and the solutions of the equation $P(x) = 0$ are found,
2. $P(x)$ is factored into its first-degree factors,
3. an upper bound and a lower bound for the real zeros of $P(x)$ are found, and
4. the graph of $y = P(x)$ is sketched.

Special Features
Non-Routine Problem Solving   p. 389
Non-Routine Problems   p. 405
Reading in Algebra   pp. 396, 405, 408, 411, 417
Cumulative Review   pp. 393, 405, 415, 419
Applications: Descartes' Rule of Signs   pp. 418–419
Chapter Review   p. 420
Chapter Test   p. 421
College Prep Test   p. 422

T-73

15.1 Graphing Exponential Functions (page 390)

Vocabulary
base-b exponential function, constantly increasing, constantly decreasing

Suggestions
1. Before the example (page 391), review the fact that $\left(\frac{1}{n}\right)^x = \frac{1}{n^x} = n^{-x}$. It is easier for students to find pairs (x, y) for $y = n^{-x}$ than it is for $y = \left(\frac{1}{n}\right)^x$ when $x < 0$ and n is a positive integer.
2. Discuss the choosing of convenient, but different, scales for the axes after a table of pairs is constructed.

For Additional Chalkboard Work
1. After presenting page 390, use or assign Exercise 2 or 9.
2. After presenting page 391, use or assign Exercise 4 or 10.

Cumulative Review
1. Write $\sqrt{x}$ in exponential form.
2. Write $x^{\frac{1}{3}}$ in radical form.
3. Write $\sqrt[3]{x^2}$ in exponential form.
4. Write $x^{\frac{3}{4}}$ in radical form.

1. $x^{\frac{1}{2}}$ 2. $\sqrt[3]{x}$ 3. $x^{\frac{2}{3}}$ 4. $\sqrt[4]{x^3}$

15.2 Logarithmic Functions (page 392)

Vocabulary
base-b logarithmic function

Background
Recall that the *inverse* of a relation is found by interchanging the variables x and y. For example,
   A: $3x + 2y = 12$ and A^{-1}: $3y + 2x = 12$;
   B: $y = 4x^2 - 2$ and B^{-1}: $x = 4y^2 - 2$.
The inverse of the base-b exponential function
   E: $y = b^x$ is E^{-1}: $x = b^y$
and E^{-1} is the base-b logarithmic function L, where
   E^{-1}: $x = b^y$ is L: $y = \log_b x$.

Suggestions
Have the students find the base-3 logs of some powers of 3 by using the definition on page 392.
1. $\log_3 9$ **2**   2. $\log_3 81$ **4**   3. $\log_3 1$ **0**
4. $\log_3 \frac{1}{3}$ **−1**   5. $\log_3 \sqrt{3}$ **½**   6. $\log_3 \frac{1}{27}$ **−3**

Have the students practice *rewriting* logarithmic equations in exponential form and the reverse procedure.
7. $a = \log_b c$ $b^a = c$ 8. $t = b^v$ $\log_b t = v$
9. $4 = \log_2 16$ $2^4 = 16$
10. $\frac{9}{16} = \left(\frac{3}{4}\right)^2$ $\log_{\frac{3}{4}} \frac{9}{16} = 2$
11. $5 = \log_b 70$ $b^5 = 70$
12. $60 = 4^q$ $\log_4 60 = q$

For Additional Chalkboard Work
Find each logarithm.
1. $\log_5 625$ **4**   2. $\log_7 1$ **0**   3. $\log_7 7$ **1**
4. $\log_3 \frac{1}{9}$ **−2**   5. $\log_5 \sqrt{5}$ **$\frac{1}{2}$**
6. $\log_3 \sqrt[5]{81}$ **$\frac{4}{5}$**   7. $\log_{\frac{1}{3}} \frac{1}{9}$ **2**   8. $\log_{\frac{1}{3}} 9$ **−2**

Solve each equation.   **9.** 10   **10.** 243
9. $\log_b 10{,}000 = 4$ 10. $\log_3 y = 5$

Cumulative Review
Specify the domain and the range of each relation. Is the relation a function?
1. $A = \{(3, 4), (4, 6), (5, 8), (6, 10)\}$
2. $B = \{(5, 1), (6, -4), (3, 2), (6, -2)\}$
1. $\{3, 4, 5, 6\}$; $\{4, 6, 8, 10\}$; yes
2. $\{5, 6, 3\}$; $\{1, -4, 2, -2\}$; no

15.3 Graphing Logarithmic Functions
(page 394)

Suggestions
1. To find ordered pairs (x, y) for $y = \log_4 x$, use the exponential equation $x = 4^y$ and use integers for y to find x.
2. To find ordered pairs (x, y) for $y = \log_{\frac{1}{4}} x$, use the exponential equation $x = \left(\frac{1}{4}\right)^y$ and rewrite it as $x = 4^{-y}$. Use integers for y to find x.

The four pairs of graphs below summarize and contrast the graphs of the basic exponential and logarithmic functions.

Base-a Exponential Functions

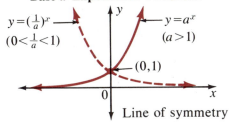

Inverse Functions ($a > 1$)

Base-a Logarithmic Functions

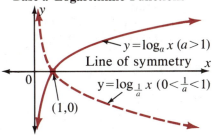

Inverse Functions ($0 < \frac{1}{a} < 1$)

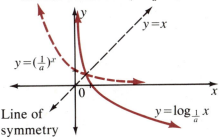

15.4 Logs of Products and Quotients
(page 397)

Background
Lessons 15.1–15.3 provided the graphs of logarithmic and exponential functions. Lessons 15.4–15.7 will furnish a means for solving logarithmic and exponential equations.

Suggestions
With some students, it may be necessary to review the definition of extraneous solutions. Emphasize that all apparent solutions of a logarithmic equation should be checked in the original equation. An apparent solution should be rejected if there is any term $\log_b x$ where $x \leq 0$, since $\log_b x$ is defined only for $x > 0$.

For Additional Chalkboard Work
1. Write $\log_b \frac{7m}{3n}$ in expanded form.
2. Given $T = \frac{8rc}{d}$, write $\log_3 c$ in expanded form.
3. Write $\log_2 9 - \log_2 3c + \log_2 c - \log_2 5c$ as the $\log_2$ of one expression.

Solve each equation. Check.

4. $\log_b 8t - \log_b 2 = \log_b 48$
5. $\log_5 x = \log_5 12 - \log_5 (x - 4)$

1. $\log_b 7 + \log_b m - \log_b 3 - \log_b n$
2. $\log_3 T + \log_3 d - \log_3 8 - \log_3 r$
3. $\log_2 \frac{3}{5c}$ 4. 12   5. 6

Cumulative Review
1. Find the arithmetic mean of -7 and 11.
2. Find the positive geometric mean of 4 and 12.
3. Find the sum of the first 40 terms of $3 + 5 + 7 + 9 + \ldots$.
4. Find the 10th term of the G.P. whose first term is $\frac{1}{4}$ and whose common ratio is 2.

1. 2   2. $4\sqrt{3}$ 3. 1,680   4. 128

15.5 Logs of Powers and Radicals (page 400)

Suggestions
Begin this lesson with a review like the following.

1. $\sqrt[3]{14} = 14^{\frac{1}{3}}$ **2.** $(x-3)^{\frac{1}{4}} = \sqrt[4]{x-3}$

This review will give the background needed for applying $\log_b \sqrt[n]{x} = \log_b x^{\frac{1}{n}} = \frac{1}{n}\log_b x$ and $\frac{1}{n}\log_b x = \log_b x^{\frac{1}{n}} = \log_b \sqrt[n]{x}$.

In Example 4, $\log_5 \sqrt[3]{\frac{tv^4}{w}}$ is an acceptable answer, even though $\sqrt[3]{\frac{tv^4}{w}}$ is not in simplest radical form, because it is the log of one expression.

For Additional Chalkboard Work
Write each expression in expanded form.

1. $\log_2\left(\frac{5c^2}{d}\right)^3$ **2.** $\log_5 \sqrt[3]{4x^2}$

3. Given $y = \frac{3}{5}tx^2$, write $\log_b x$ in expanded form.

Write each expression as the $\log_2$ of one expression.

4. $3\log_2 t + \frac{1}{2}\log_2 5 - 4\log_2 v$

5. $\frac{1}{4}(\log_2 3t + \log_2 5v)$

Solve each equation. Check.

6. $2\log_b x = \log_b 3 + \log_b(x+6)$

7. $\frac{1}{2}\log_4 x + \frac{1}{2}\log_4(x-8) = \log_4 3$

1. $3(\log_2 5 + 2\log_2 c - \log_2 d)$
2. $\frac{1}{3}(\log_5 4 + 2\log_5 x)$
3. $\frac{1}{2}(\log_b 5 + \log_b y - \log_b 3 - \log_b t)$
4. $\log_2 \frac{t^3\sqrt{5}}{v^4}$ 5. $\log_2 \sqrt[4]{15tv}$ 6. 6 7. 9

Cumulative Review
Find the slope of $\overrightarrow{AB}$ for the given coordinates. Also, describe the slant of $\overrightarrow{AB}$.

1. $A(2,-4), B(-3,-11)$ **2.** $A(15,4), B(12,4)$
3. $A(2,-4), B(4,-10)$ **4.** $A(4,15), B(4,12)$

1. $\frac{7}{5}$; up to the right 2. 0; horizontal
3. -3; down to the right 4. undefined; vertical

15.6 Common Logarithms and Antilogarithms (page 403)

Vocabulary
common logarithm, antilogarithm, characteristic, mantissa, scientific notation

Background
Common logarithms and antilogarithms are introduced in this lesson and applied, in Lesson 15.7, to solving exponential equations, approximating the decimal value of $\log_b x$ for given values of b and x, and solving problems in compound interest and growth of bacteria.

Teachers who want to present a lesson on computation of products, quotients, powers, and radicals will find such a lesson on pages 588–589.

Suggestions
Encourage students to use scientific notation as shown in the examples.

Students will wonder why log 0.0546 may be written as $8.7372 - 10$, rather than $-2 + 0.7372$. You can explain this by discussing the table of logs of the trigonometric functions on page 590, where 10 must be subtracted from each entry. See the footnote on page 590.

Errors Students Make
Students may write a log such as $-2 + 0.7372$ as -2.7372. Point out that $-2.7372 = -2 + (-0.7372)$, which is not the same as $-2 + (0.7372)$. Also, note that $-2 + 0.7372 = -1.2628$, which is the same as $-1 + (-0.2628)$ and means that the mantissa is -0.2628. The mantissas in the table are positive, *not* negative.

For Additional Chalkboard Work
Find each logarithm or antilogarithm.
1. log 1,000,000 6 **2.** antilog -4
3. log 3,280   **4.** antilog 2.5198   331
5. log 0.003280   3. 3.5159
6. antilog $7.5198 - 10$ 0.00331
2. 0.0001 5. $7.5159 - 10$ or $-3 + .5159$

Cumulative Review
Solve for x.
1. $5^{3x-2} = 125$ **2.** $8^{x+4} = 32$ **3.** $6^{2x-1} = \frac{1}{36}$
1. $\frac{5}{3}$ **2.** $-\frac{7}{3}$ **3.** $-\frac{1}{2}$

15.7 Using Exponential Equations (page 406)

Vocabulary
exponential equation, compound interest, annual, semiannual, quarterly, fission

Suggestions
In this lesson, base-10 logarithms are used to solve exponential equations such as $3^x = 17$ and $4^{2x-1} = 7.65$. It would be very advantageous to permit the use of a pocket calculator because the division can become tedious.

For Additional Chalkboard Work
1. Solve $6^{3x-2} = 59.7$ for x to three significant digits.
2. Find $\log_6 9.56$ to three significant digits.

1.
$$6^{3x-2} = 59.7$$
$$\log 6^{3x-2} = \log 59.7$$
$$(3x - 2) \log 6 = \log 59.7$$
$$3x - 2 = \frac{\log 59.7}{\log 6}$$
$$3x - 2 = \frac{1.7760}{0.7782}$$
$$3x - 2 = 2.282$$
$$3x = 4.282$$
$$x = 1.427 = 1.43$$

2. Let $\log_6 9.56 = x$
$$6^x = 9.56$$
$$\log 6^x = \log 9.56$$
$$x \log 6 = \log 9.56$$
$$x = \frac{\log 9.56}{\log 6}$$
$$x = \frac{0.9805}{0.7782}$$
$$x = 1.260 = 1.26.$$

Cumulative Review
Factor completely.
1. $3x^3 - 18x^2 + 27x$ 2. $3x^3 - 4x^2 + 6x - 8$
3. $2x^3 - 16$ 4. $x^2 + 6x + 9 - 4y^2$
1. $3x(x - 3)^2$ 2. $(x^2 + 2)(3x - 4)$
3. $2(x - 2)(x^2 + 2x + 4)$
4. $(x + 3 + 2y)(x + 3 - 2y)$

15.8 Higher-Degree Polynomial Equations (page 409)

Vocabulary
integral polynomial, first-degree (binomial) factor, zero of a polynomial, constant term, multiplicity of a zero

Suggestions
Begin this lesson with a review of synthetic substitution, the factor theorem, and the remainder theorem from Lessons 5.5 and 5.6.
1. *Factor theorem:*
   A binomial B is a factor of a polynomial P if $P \div B = Q$ or $P = B \cdot Q + R$, where Q is a polynomial and the remainder $R = 0$.
2. *Remainder theorem:*
   For polynomials P, Q, and $x - a$, if $P = (x - a) \cdot Q + R$ for some number R, then R is the value of P when $x = a$ [that is, $P(a) = R$].
3. Complete the synthetic substitutions below.

	1	−3	3	−5	0	12
2⌋						
	1	−1	1	−3	−6	0
−1⌋						
	1	−4	7	−12	12	0
1⌋						
	1	−2	1	−4	−4	8

4. Given $P(x) = x^5 - 3x^4 + 3x^3 - 5x^2 + 12$ and item 3 above, what are the values of $P(x)$ for $x = 2, -1,$ and 1? Give two first-degree factors of $P(x)$.
   0, 0, 8; $(x - 2), (x + 1)$

For Additional Chalkboard Work
Factor $P(x)$ into first-degree factors. Find the zeros of $P(x)$, or the solutions of $P(x) = 0$.
1. $x^4 - x^3 - 5x^2 + 3x + 6$
2. $2x^3 - 7x^2 + 2x + 3$
3. $x^3 - 2x^2 + 9x - 18$
1. $(x + 1)(x - 2)(x - \sqrt{3})(x + \sqrt{3})$; $-1, 2, \pm\sqrt{3}$
2. $(x - 1)(x - 3)(2x + 1)$; $1, 3, -\frac{1}{2}$
3. $(x - 2)(x - 3i)(x + 3i)$; $2, 3i, -3i$

Cumulative Review
A triangle's height is 2 cm less than twice the length of its base. Find the height if the area of the triangle is 20 cm². **8 cm**

15.9 Rational Zero Theorem (page 412)

Vocabulary
lower bound, upper bound

Suggestions
One purpose for finding an upper bound and a lower bound (for the real zeros of a polynomial) is to eliminate some of the "potential" zeros given by the integral zero theorem and the rational zero theorem. For example, $x^2 - 36$ has 18 potential integral zeros: $\pm(1, 2, 3, 4, 6, 9, 12, 18, 36)$. However, 7 is an upper bound and -7 is a lower bound, as shown below. This eliminates $\pm(9, 12, 18, 36)$.

$$
\begin{array}{r|rrr}
 & 1 & 0 & -36 \\
7 & & 7 & 13 \\
\hline
 & 1 & 7 & 13
\end{array}
\qquad
\begin{array}{r|rrr}
 & 1 & 0 & -36 \\
-7 & & -7 & 13 \\
\hline
 & 1 & -7 & 13
\end{array}
$$

In Exercises 4 and 18, the first coefficient is negative (-1), and all the numbers will be nonpositive when an upper bound is found.

For Additional Chalkboard Work
1. Find an upper bound and a lower bound for the real zeros of $6x^4 - 7x^3 + 3x^2 - 7x - 3$.
2. Find the zeros of $6x^4 - 7x^3 + 3x^2 - 7x - 3$.

1. 2 and -1 2. $-\frac{1}{3}, \frac{3}{2}, i, -i$

Cumulative Review
1. Give the x-intercepts and the y-intercepts of the ellipse with equation $\frac{x^2}{25} + \frac{y^2}{16} = 1$.
2. Solve $a - 2x = bx - cx$ for x.
3. Simplify $(2a^{-3}b^4)^3$ and write the answer with positive exponents.

1. $\pm 5; \pm 4$ 2. $x = \frac{a}{b - c + 2}$ 3. $\frac{8b^{12}}{a^9}$

15.10 Graphing Polynomial Functions (page 416)

Background
This lesson covers the case of graphs of nth degree integral polynomial functions with n distinct *real* zeros. Hence, each graph will *cross* the x-axis at n different points.

Suggestions
Use synthetic substitution and *all* of the *integral* values of x between an upper bound and a lower bound (of the real zeros) to find ordered pairs (x, y).

For Additional Chalkboard Work
Sketch the graph of each function.
1. $y = 4$ 2. $y = 3x - 2$
3. $y = x^2 - 2x - 3$
4. $y = x^3 - 3x^2 - 2x + 6$
5. $y = x^4 - 10x^2 + 9$

1. Line through (1, 4) and (3, 4)
2. Line through (0, -2) and (1, 1)
3. Parabola: (-2, 5), (-1, 0), (0, -3), (1, -4), (2, -3), (3, 0), (4, 5)
4. S shape: (-2, -10), (-1, 4), (0, 6), (1, 2), (2, -2), (3, 0), (4, 14)
5. W shape: (± 4, 105), (± 3, 0), (± 2, -15), (± 1, 0), (0, 9)

Cumulative Review
1. Simplify $\frac{-16x^{-3}y^2z^{-4}}{24x^6y^{10}z^{-12}} \cdot \frac{-2z^8}{3x^9y^8}$
2. Simplify $3x^2(-5xy^4)^3$. $-375x^5y^{12}$
3. Solve $3^{4-x} = 243$. -1
4. Find the value of 8.76×10^{-3}.   0.00876

Applications (page 418)
Descartes' rule of signs and the conjugate zero theorem are introduced. Some teachers and some curricula require one or both of these topics.

This lesson is designed so that the teacher can include both of the topics or one of the topics and not the other.

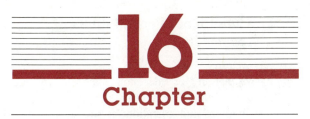

Chapter 16 and Lesson Commentaries

Chapter 16 Permutations, Combinations, and Probability

Chapter Objectives
To find the number of possible arrangements of objects by using the Fundamental Counting Principle

To find the number of permutations of objects when conditions are attached to the arrangement

To find the quotient of numbers given in factorial notation

To find the number of distinguishable permutations when some of the objects in an arrangement are alike

To find the number of possible permutations of objects in a circle

To find the number of combinations of n things taken r at a time

To find the value of $\binom{n}{r}$ for nonnegative integers n and r where $n \geq r$

To determine the probability of a simple event

To determine the probability of a compound event using a sample space

To determine the probability of a compound event using the Fundamental Counting Principle, permutations, and combinations

Chapter Overview
The Fundamental Counting Principle is used to find the number of possible arrangements, or permutations, of
1. n distinct objects, taken r at a time, on a straight line
2. n objects when some of the objects are alike
3. n distinct objects on a circle.

The formula $\binom{n}{r} = \dfrac{n!}{r!\,(n-r)!}$ is used to find the number of combinations of n distinct objects taken r at a time.

The lessons on probability cover the probability of simple events, independent events, dependent events, inclusive events, and mutually exclusive events.

Special Features
Non-Routine Problem Solving   p. 423
Non-Routine Problems   pp. 430, 432, 434, 437
Reading in Algebra   pp. 433, 436
Chapter Review   p. 448
Chapter Test   p. 449
Computer Activities   p. 450
College Prep Test   p. 451
Cumulative Review (Chapters 1–16)   pp. 452–453

16.1 Fundamental Counting Principle (page 424)

Vocabulary
choice, decision, number of choices, number of decisions, permutation, $n!$

Suggestions
For the truck driver's problem at the top of page 424, the number of decisions is 2. "Choose a letter" is one decision. "Choose a number" is the other decision. For the first decision, the number of choices is 4. Each of A, C, G, and T is a choice. For the second decision, the number of choices is 3. Each of 1, 7, and 9 is a choice.

The strategy for problem solving in this lesson is as follows:
1. Determine the number of decisions
2. Draw a blank (_____) for each decision
3. Determine the number of choices for each decision
4. Write the number of choices in the blank for that decision
5. Use the Fundamental Counting Principle

For Additional Chalkboard Work
1. How many four-digit numbers can be formed from the digits 1, 2, 4, 5, 7, 8, 9 if no digit may be repeated in a number? If a digit may be repeated in a number?
2. A park has nine gates—three on the west side, four on the north side, and two on the east side. In how many different ways can you enter the park from the west side and later leave from the east side? Enter from the north and later exit from the north? Enter the park and later leave the park?

1. $7 \times 6 \times 5 \times 4 = 840$; $7 \times 7 \times 7 \times 7 = 2,401$
2. $3 \times 2 = 6$; $4 \times 4 = 16$; $9 \times 9 = 81$

Cumulative Review
1. Expand $(x^2 - y)^4$. Simplify each term.
2. Find the sum of the infinite geometric series:
$8 + 2 + \frac{1}{2} + \frac{1}{8} + \cdots$.

1. $x^8 - 4x^6y + 6x^4y^2 - 4x^2y^3 + y^4$
2. $\frac{32}{3}$, or $10\frac{2}{3}$

16.2 Conditional Permutations (page 428)

Suggestions
Emphasize that an "or" decision involves addition, as shown in Example 2, and that an "and" decision involves multiplication, as shown in Example 3.

Exercise 15 involves both "and" and "or" decisions:
(novels left *and* short stories right) *or* (short stories left *and* novels right), which leads to $(5! \times 6!) + (6! \times 5!)$.

For Additional Chalkboard Work
1. How many permutations of all the letters in *PATRON* begin with *NO*? Begin with either *N* or *O*?
2. How many *odd* numbers containing *one or more* digits can be formed from 1, 2, 3, 4 if no digit may be repeated in a number?
3. How many permutations of 1, *A*, 2, *B*, 3, *C*, 4 have all the letters together and to the right of the digits?

1. $1 \times 1 \times 4 \times 3 \times 2 \times 1 = 24$; $2 \times 5 \times 4 \times 3 \times 2 \times 1 = 240$
2. $2 + 3 \times 2 + 3 \times 2 \times 2 + 3 \times 2 \times 1 \times 2 = 2 + 6 + 12 + 12 = 20$
3. $(4 \times 3 \times 2 \times 1) \times (3 \times 2 \times 1) = 144$

Cumulative Review
1. Add. $\frac{3}{4x^2} + \frac{5}{6x}$
2. Find the positive geometric mean of 3 and 12.
3. Find the solution set of $|2x| > 10$.

1. $\frac{10x + 9}{12x^2}$ 2. 6
3. $\{x \mid x < -5 \text{ or } x > 5\}$

16.3 Distinguishable Permutations (page 431)

Background
Finding the value of an expression like $\frac{8!}{4! \times 3!}$ in this lesson will occur again in Lesson 16.5 where $\binom{10}{4} = \frac{10!}{4! \times 6!}$.

Suggestions
The Short Method shown in Example 1 requires the skill of rewriting factorial notation like 8! as $8 \times 7 \times 6 \times 5 \times (4!)$.

For Additional Chalkboard Work
1. $\frac{6!}{4! \times 2!} = \frac{6 \times 5 \times (4!)}{(4!) \times 2 \times 1} = \frac{6 \times 5}{2 \times 1} = 15$
2. $\frac{12!}{3! \times 9!} = \frac{12 \times 11 \times 10 \times (9!)}{3 \times 2 \times 1 \times (9!)}$
$= \frac{12 \times 11 \times 10}{3 \times 2 \times 1} = 220$
3. How many distinguishable seven-digit numbers can be formed from the digits of 6446346?
4. How many distinguishable permutations can be formed using all the letters in *SUCCESS*?
5. In how many distinguishable ways can 3 dimes and 5 quarters be distributed to 8 children if each child is to receive one coin?

3. $\frac{7!}{3!\,3!} = \frac{7 \cdot 6 \cdot 5 \cdot 4 \cdot 3!}{3 \cdot 2 \cdot 1 \cdot 3!} = 7 \cdot 5 \cdot 4 = 140$
4. $\frac{7!}{2!\,3!} = \frac{7 \cdot 6 \cdot 5 \cdot 4 \cdot 3!}{2 \cdot 1 \cdot 3!} = 7 \cdot 6 \cdot 5 \cdot 2 = 420$
5. $\frac{8!}{3!\,5!} = \frac{8 \cdot 7 \cdot 6 \cdot 5!}{3 \cdot 2 \cdot 1 \cdot 5!} = 8 \cdot 7 = 56$

Cumulative Review
1. Identify the conic section whose equation is $\frac{x^2}{16} + \frac{y^2}{36} = 1$.
2. Solve $x^2 + 2x = 8$.
3. Rationalize the denominator of $\frac{3}{2+i}$.

1. ellipse   2. $-4, 2$ 3. $\frac{6 - 3i}{5}$

16.4 Circular Permutations (page 433)

Suggestions
If Exercises 7 and 8 are to be assigned, then extend the Example on page 433 to create the following second example.

In how many different ways can the 6 guests be seated around the circular table if men and women must be seated alternately?
Solution: Draw and interpret the following diagrams.

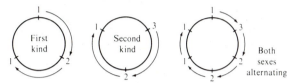

First kind   Second kind   Both sexes alternating

For Additional Chalkboard Work
1. In how many different ways can a teacher and four students be seated around a circular table?
2. In how many different ways can six men and six women be seated around a circular table if no two men may be seated next to each other?

After Example 2, use the following exercise.
3. In how many ways can five keys be arranged on a key ring?

1. $(5 - 1)! = 4! = 24$
2. $1 \cdot 6 \cdot 5 \cdot 5 \cdot 4 \cdot 4 \cdot 3 \cdot 3 \cdot 2 \cdot 2 \cdot 1 \cdot 1 = 86{,}400$
3. $\frac{(5-1)!}{2} = \frac{4 \cdot 3 \cdot 2 \cdot 1}{2} = 12$

Cumulative Review
Given $f(x) = 5x + 4$ and $g(x) = x^2 - 3$, find the following.
1. $f(g(2))$ 2. $g(f(1))$
3. $g(f(a))$ 4. $f(g(c))$

1. 9   2. 78   3. $25a^2 + 40a + 13$ 4. $5c^2 - 11$

16.5 Combinations (page 435)

Vocabulary
combination, zero factorial (0!)

Suggestions
Emphasize the difference between a combination and a permutation:
Combination: a *selection* of objects *without* regard to order
Permutation: an arrangement of objects in a *definite order*
Use the Reading in Algebra exercises to discuss this difference.

For Additional Chalkboard Work
1. List one 3-letter name for each triangle determined by the points A, B, C, and D located on a circle.

Find the value of each expression. (Ex. 2–5)

2. $\binom{7}{4}$ 3. $\binom{7}{7}$ 4. $\binom{7}{0}$
5. $\binom{3}{0} + \binom{3}{1} + \binom{3}{2} + \binom{3}{3}$

6. How many different 4-member Brain Bowl teams can be formed if 20 students are eligible for membership?
7. Six points, A through F, are on a circle. How many lines are determined by the 6 points?

1. ABC, ABD, ACD, BCD   2. 35   3. 1   4. 1
5. 1 + 3 + 3 + 1 = 8   6. 4,845   7. 15

Cumulative Review
1. Find the coordinates of M, the midpoint of $\overline{PQ}$ for $P(-3,4)$, $Q(7,6)$.
2. Write an equation of the circle with radius 4 and center at the origin.

1. $M(2,5)$ 2. $x^2 + y^2 = 16$

16.6 Probability: Simple Events (page 438)

Vocabulary
favorable outcomes, possible outcomes, simple event, probability, odds

Background
If an event can occur successfully in s different ways and the total number of equally likely events is t, then the probability of success of an event happening is $P(s) = \frac{s}{t}$.

If an event can occur unsuccessfully in f different ways and the total number of equally likely events is t, then the probability of an event being unsuccessful is $P(f) = \frac{f}{t}$.

However, $P(s) + P(f)$, of the same event is $\frac{s}{t} + \frac{f}{t}$ or $\frac{s}{t} + \frac{t-s}{t} = \frac{t}{t} = 1$. Note: $s + f = t$ or $f = t - s$.

This discussion suggests that in order to find the probability of an event *not* happening, the probability of the event happening must be found and subtracted from 1.

Suggestions
Only ideal situations are considered in this study of elementary probability. It is assumed that a die is not heavier at one corner and that each face has an equally likely chance to come up as any other face. It is assumed that a coin is fair, that it is equally balanced, and a head is just as likely to occur on a flip as a tail. It is assumed that, in a roll of a die or a flip of a coin, the person rolling the die or flipping the coin will not introduce any bias.

Errors Students Make
Students might make errors in determining the number of favorable outcomes. This type of error might be caused by the students' lack of knowledge of the event itself. Some students may not know that there are four Jacks in a deck of cards, for example. You may wish to review such details.

Cumulative Review
Simplify each expression.
1. $-3a^4b^3c^{-2} \cdot 5a^{-6}b^2c$
2. $-8xy^{-3}z^5 \cdot -2x^{-2}y^{-3}z$

1. $\frac{-15b^5}{a^2c}$ 2. $\frac{16z^6}{xy^6}$

16.7 Probability: Compound Events (page 441)

Vocabulary
sample space, inclusive events, mutually exclusive events, independent events, dependent events

Background
Compound sentences in mathematics are formed with the connectives *or* and *and*. The connective *or* is the inclusive *or*. Two simple sentences connected with *or* form a compound sentence that is true if one or both parts are true. For a compound sentence formed with the connective *and* to be true, both parts must be true.

The $P(A \text{ or } B) = P(A) + P(B) - P(A \text{ and } B)$ if the two events are inclusive, since the sets A and B overlap and are counted twice. To remove that space counted twice, subtract $P(A \text{ and } B)$. If the events are mutually exclusive events, then there are no points in common. Thus, $P(A \text{ and } B) = 0$.

Suggestions
Encourage students to use both the sample-space approach and the formulas and to use one as a check of the other.

Errors Students Make
This lesson has much new vocabulary and care must be taken not to confuse terms. Sometimes independent events are confused with mutually exclusive events. Events that are mutually exclusive occur if not more than one of them can happen in a single trial. That is, mutually exclusive events have nonoverlapping sets. Independent events A and B mean that the probability of event A does not depend on event B. Independent events must have equally likely outcomes.

Cumulative Review
A line has the given slope or y-intercept or contains the indicated point or points. Write, in standard form, an equation of each line.
1. $P(-3, -4)$, $m = -\frac{2}{3}$ 2. $m = \frac{3}{5}$, $b = -2$
3. $P(-6, 1)$, $Q(-3, -1)$ 4. $P(1, -4)$, $Q(-2, -6)$
1. $2x + 3y = -18$ 2. $3x - 5y = 10$
3. $2x + 3y = -9$ 4. $2x - 3y = 14$

16.8 Probability and Arrangements (page 445)

Suggestions
Review the Fundamental Counting Principle: If one choice can be made in a ways and a second choice can be made in b ways, then the choices in order can be made in $a \times b$ different ways.

Also review that a combination is a selection of objects without regard to order and that the number of combinations $\binom{n}{r} = \dfrac{n!}{r!(n-r)!}$.

This lesson combines probability and the notions of the Fundamental Counting Principle and Combinations. In this lesson objects are selected without regard to order.

Errors Students Make
Unless students know how to reduce factorial division, computational errors may occur.

$$\binom{52}{5} = \frac{52!}{5!47!} = \frac{52 \cdot 51 \cdot 50 \cdot 49 \cdot 48 \cdot 47!}{5! \, 47!}$$
$$= \frac{52 \cdot 51 \cdot 50 \cdot 49 \cdot 48}{5 \cdot 4 \cdot 3 \cdot 2}$$

In general,

$$\binom{n}{r} = \frac{n!}{r!(n-r)!} = \frac{n(n-1)(n-2) \cdots (n-r)!}{r!(n-r)!}.$$

Cumulative Review
Factor each polynomial completely.
1. $x^2 + 3x - 10$ 2. $4x^2 - 25$
3. $x^3 - 2x^2 - 3x$ 4. $x^3 - 16x$
5. $x^3 - 8$ 6. $x^3 + 27$
1. $(x + 5)(x - 2)$ 2. $(2x + 5)(2x - 5)$
3. $x(x + 1)(x - 3)$ 4. $x(x + 4)(x - 4)$
5. $(x - 2)(x^2 + 2x + 4)$ 6. $(x + 3)(x^2 - 3x + 9)$

Computer Activities (p. 450)
Once the assignment $P = 1$ has been made in the first program on page 450, the loop could read I = 2 TO N. The choice 1 TO N was made to emphasize the definition of *factorial*. It will be interesting to see if any of your students will, on their own, suggest a change in line 40 of the second program. It could be rewritten as FOR I = U − L + 1 TO U. This means you can drop the "step" portion of the command. If they do not see it on their own, give a hint.

Chapter and Lesson Commentaries

Chapter 17 Trigonometric Functions

Chapter Objectives
To sketch an angle of rotation associated with a directed degree measure and to determine its corresponding reference angle and reference triangle

To find lengths of sides in a 30°–60° or 45°–45° right triangle and to find the coordinates of the points that their terminal sides make with a circle

To define the six trigonometric functions

To find the trigonometric functions of special degree measure like 30°, 45° and 90°

To determine the quadrants where the trig functions are positive and the quadrants where they are negative

To solve linear trigonometric equations

To find the trig functions of angles of negative measure

To find the other five trigonometric functions of an angle of rotation given one function

To use trigonometric tables

To interpolate with trigonometry

To define and use trigonometry of a right triangle

Chapter Overview
Trigonometry is initially approached through the study of the six trigonometric functions of angles of rotation. Special angle measure like 0°, 30°, 45°, and 60° are first considered and then the trigonometric functions of any positive or negative angle degree measure are presented. After explaining how to use trig tables, interpolation, an art that is fast becoming extinct in this age of pocket calculators, is also discussed.

Chapter Overview (cont.)
Once the student has an understanding of the six trigonometric functions, he or she utilizes them in the lesson on solving linear trigonometric equations. Finally, the trigonometry of a right triangle, which is considered as a special case of the trigonometry of an angle of rotation, is presented along with applications.

For students who have taken geometry, the names of the trigonometric functions should sound familiar. Mention that secant and tangent are defined differently in trigonometry than in geometry. It might interest students to find out where the strange names of *sine, secant* and *tangent* originated. Have some students look up the roots of the words and their formal definitions in a dictionary.

You might have some students give a short presentation on the life of Leonhard Euler. He led a very interesting life and some of his many mathematics discoveries dealt with the sine and cosine functions. This type of independent project helps to "humanize" the subject matter. Another project might be a report on how the conventional angle of rotation is different when used in navigation, where the initial side of an angle starts on the positive *y*-axis and is rotated in a clockwise direction. The measure associated with this type of rotation is called *positive measure*.

Some mnemonics are presented in the lesson commentaries. There are quite a few details that the students must remember in this chapter. After they understand the theory, memory aids will be helpful.

Special Features
Karl Weierstrass   p. 454
Reading in Algebra   pp. 457, 465
Calculator Activities   p. 482
Cumulative Review   pp. 465, 472, 474, 476, 480, 485
Computer Activities   p. 494
Chapter Review   p. 492
Chapter Test   p. 493
College Prep Test   p. 495
Non-Routine Problems   p. 472

17.1 Angles of Rotation (page 455)

Vocabulary
angle of rotation, initial ray, terminal ray, clockwise rotation, counterclockwise rotation, quadrantal angles, coterminal angles, reference triangle, reference angle

Background
An angle of rotation with a measure of 90° has a corresponding revolution measure of $\frac{1}{4}$ revolution, which is $\frac{1}{4}$ the circumference of $\frac{1}{4}(2\pi r) = \frac{\pi r}{2}$. Similarly, an angle of rotation with a measure of 180° has a corresponding measure of πr.

Suggestions
The idea of rotation is usually thought of as it occurs in the physical world, as in the rotation of a wheel or a fan blade.

A distinction should be made between an angle of rotation and an angle. An angle of rotation is the angle associated with a rotating ray while an angle is defined as a set of points.

A distinction should also be made between an angle of rotation and angle measure. An angle of rotation can be measured by degrees, number of revolutions, or radians.

Errors Students Make
Errors may be made by the students in the formation of the reference triangles. Stress that triangles are always formed by drawing a perpendicular to the *x*-axis and that the reference angle has the *x*-axis as one of its rays and is acute.

Cumulative Review
Determine whether *P* and *Q* are symmetric with respect to the *x*-axis, the origin, or none of these.
1. $P(3, -6), Q(-3, -6)$ **2.** $P(-5, 3), Q(5, -3)$
1. *y*-axis   2. origin

17.2 Special Right Triangles (page 458)

Vocabulary
30°–60° right triangle, 45°–45° right triangle

Background
It might be a good idea to review rationalizing a denominator:

Suggestions
It is important that students memorize the properties of the 30°–60° right triangle and those of the isosceles right triangle since they will be used throughout the study of trigonometry.

For Additional Chalkboard Work
Determine the point (x, y) and the radius *r* for each of the following.
$(-5\sqrt{2}, -5\sqrt{2}), 10$ $(-2, -2\sqrt{3}), 4$

1.   2.

3.   4.

$(-3\sqrt{3}, 3), 6$ $(2, -2), 2\sqrt{2}$

Errors Students Make
In Examples 5 and 6 students must know in what quadrant *x* and *y* lie: If $|x| = 3$, and *x* is in quadrant II, then $|x| = -x = 3$ or $x = -3$.

Cumulative Review
For each point, find the coordinates of the corresponding point symmetric with respect to the *x*-axis, the *y*-axis, and the origin.
1. $A(-4, -9)$ **2.** $B(6, -3)$ **3.** $C(-5, 9)$
1. $(-4, 9)(4, -9)(4, 9)$ 2. $(6, 3)(-6, -3)$ $(-6, 3)$ 3. $(-5, -9)(5, 9)(5, -9)$

17.3 Sine and Cosine Functions (page 462)

Vocabulary
sine, cosine, unit circle

Background
The sine and cosine functions are defined in terms of an angle of rotation. The sine is defined as the ratio of the ordinate to the radius and the cosine is defined as the ratio of the abscissa to the radius.

Suggestions
Let the radius of a circle have length 1 as shown. The circle intersects the axes at $(1, 0)$, $(0, 1)$, $(-1, 0)$ and $(0, -1)$. Thus, since the radius is 1, sin function is y and the cos is x.
$\sin 0° = 0$, $\sin 90° = 1$,
$\sin 180° = 0$, $\sin 270° = -1$,
$\cos 0° = 1$, $\cos 90° = 0$,
$\cos 180° = -1$, $\cos 270° = 0$

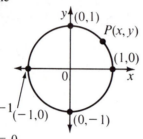

Errors Students Make
A systematic approach to finding the sine or cosine of the measure of an angle will help to reduce errors.
1. Draw the angle of rotation.
2. Draw the reference angle and triangle.
3. Determine x, y and r.
4. Find sin and cos.

Enrichment
From the unit circle above find each of these.
1. sin 360° **0** 2. cos 360° **1** 3. sin (−90°)
4. cos (−90°) **0** 5. sin (−180°) **0** 6. cos (−180°)
7. sin (−270°) **1** 8. cos (−270°) **0** 9. sin (−360°)
10. cos (−360°) 11. cos 720° 3. **−1** 6. **−1**
9. **0**   10. **1**   11. **1**

Cumulative Review
Solve each system of equations.
1. $-3x - 5y = -7$
   $-4x + 2y = 8$
2. $-3(x - 2) + (y - 1) = 5$
   $-(x - 3y) - (2 - x) = 7$
1. $(-1, 2)$ 2. $(1, 3)$

17.4 Tangent And Cotangent Functions (page 466)

Vocabulary
tangent, cotangent

Suggestions
The tangent function is defined as the ratio of the ordinate to the abscissa. Thus, the tangent is the ratio of the sine to the cosine. Similarly, the cotangent function is defined as the ratio of the abscissa to the ordinate. Thus, it is the ratio of the cosine to the sine. Students should make use of both definitions of tan and cot.

In Example 1, $\tan 45° = \dfrac{y}{x} = \dfrac{1}{1} = 1$ or

$\tan 45° = \dfrac{\sin 45°}{\cos 45°} = \dfrac{\frac{\sqrt{2}}{2}}{\frac{\sqrt{2}}{2}} = 1$

Errors Students Make
Follow the same systematic approach suggested in the previous TE lesson.
1. Draw the angle of rotation.
2. Draw the reference angle and triangle.
3. Determine x, y and r. Be careful to get the correct sign.
4. Find tan and cot.

Enrichment
From the unit circle, find each of the following.
1. tan 360° **0**   2. cot 360° **undef.**
3. tan (−90°) **undef.**   4. cot (−90°) **0**
5. tan (−180°) **0**   6. cot (−180°) **undef.**
7. tan (−270°) **undef.**   8. cot (−270°) **0**
9. tan (−360°) **0**   10. cot (−360°) **undef.**
11. tan (720°) **0**   12. cot (720°) **undef.**

Cumulative Review
Solve each system of equations.
1. $-2x + 3y - z = 2$
   $-x - 2y + 2z = -5$
   $3x + 2y + z = 4$
2. $3x - 2y + 2z = 10$
   $2x + 3y - 2z = -1$
   $-3x - 3y - z = -4$
1. $(1, 1, -1)$ 2. $(2, -1, 1)$

17.5 Secant and Cosecant Functions (page 470)

Vocabulary
secant, cosecant

Suggestions
The secant function is defined as the ratio of the radius to the abscissa. Thus, the secant is $\frac{1}{\cos}$. The cosecant function is defined as the ratio of the radius to the ordinate, or $\frac{1}{\sin}$.

Show students that if $\sec u = \frac{1}{\cos u}$, then $\cos u = \frac{1}{\sec u}$. Also, if $\csc u = \frac{1}{\sin u}$, then $\sin u = \frac{1}{\csc u}$. Thus, secant is positive in the same quadrants that the cosine is positive and the cosecant is positive in the same quadrants that the sine is positive.

Errors Students Make
Sketching the angle of rotation can help determine the reference angle and the signs of x and y.

Enrichment
From the unit circle find the following.
1. sec 360° **1**
2. csc 360° **undef.**
3. sec (−90°) **undef.**
4. csc (−90°) **−1**
5. sec (−270°) **undef.**
6. csc (−270°) **1**
7. sec (−180°) **−1**
8. csc (−180°) **undef.**

Cumulative Review
Graph the following equations.
1. $y = |x - 1|$
2. $y = |x| - 2$
3. $y = |x + 1| + 3$

17.6 Functions of Negative Angles (page 473)

Vocabulary
negative measure

Background
Refer to the unit circle: then $(-u)$ determines point $Q(a, -b)$ and u determines $P(a, b)$, so, $\sin u = b$, $\sin(-u) = -b$, and $\sin(-u) = -\sin u$, $\cos u = a$, $\cos(-u) = a$, and $\cos(-u) = \cos u$, $\tan u = \frac{b}{a}$, $\tan(-u) = -\frac{b}{a}$, and $\tan(-a) = -\tan u$.

Suggestions
There are basically two ways to solve problems involving functions of negative angles. Students can use the formulas for negative angles or they can sketch the angle of rotation, determine the reference angle and triangle, find the sign of the trig function, and then determine the function. Have students make up a saying for ASTC: A Smart Trig Class is an old reliable. Remind students that a function has the same sign as its reciprocal function.

Errors Students Make
Sometimes, students make errors with a double negative.

For example, $\sin(-300°) = -\sin 300° = -[-\sin 60°] = +\frac{\sqrt{3}}{2}$.

This can be avoided by sketching the angle of rotation.

Cumulative Review
Graph each equation.
1. $y = (x + 1)^2$
2. $y = x^2 - 1$
3. $y = (x - 1)^2 + 2$

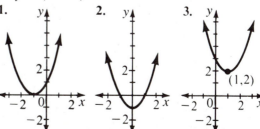

T-87

17.7 Values of Trigonometric Functions (page 475)

Suggestions
Students should memorize lengths of sides of special right triangles, such as 3-4-5, 5-12-13, and 8-15-17. It will save them much time. Students should also understand that multiples of these sides also yield right triangles. For example, if a triangle has sides with lengths $6(2 \cdot 3)$, $8(2 \cdot 4)$ and $10(2 \cdot 5)$, the triangle is a right triangle. Also, if a right triangle has two sides with lengths $15(3 \cdot 5)$ and $36(3 \cdot 12)$, then the third side has length $39 (3 \cdot 13)$.

Review the quadrants in which each trig function is positive and in which each trig function is negative.

After the students understand why the functions have different signs in different quadrants, the coordinate systems below may be useful.

Errors Students Make
In Example 1, $|x| = 3$. Since the terminal side is in quadrant 2, x is negative. Hence, $x = -3$. Students sometimes have difficulty with this idea.

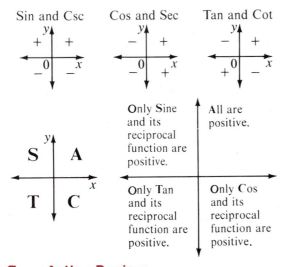

Cumulative Review
Graph each equation.
1. $(x - 1)^2 + (y + 2)^2 = 4$
2. $(x + 2)^2 + (y - 1)^2 = 16$

1. a circle whose center is at the point $(1, -2)$ and whose radius is 2
2. a circle whose center is at the point $(-2, 1)$ and whose radius is 4

17.8 Using Trigonometric Tables (page 477)

Vocabulary
trigonometric table, complementary or cofunction

Suggestions
Students are expected to read a trig table for any degree measure even though the trig tables only go from 0° to 45° in the left-hand column and from 45° to 90° in the right-hand column. Show students that corresponding measure in the left-hand/right-hand columns add to 90°. Also show that corresponding functions at the top of the table and at the bottom of the table are cofunctions.

Students are also expected to find the corresponding angular measure if its trig function value is given. Finally, the use of trigonometric tables to solve trigonometric equations is illustrated.

Errors Students Make
Difficulty occurs when students must read tables from bottom to top using the right-hand column. Some oral exercises may avoid such difficulty.

For Additional Chalkboard Work
Use the table to find each of the following.
1. sin 38°   2. cos 73°   3. tan 18°
4. cot 43°20′   5. sec 50°40′   6. csc 74°30′
7. sin 52°50′   8. cos 10°20′   9. cot 80°10′

Use the table to find u, $u < 90°$.
10. sin u = 0.1965   11. cos u = 0.3719
12. tan u = 0.6577   13. cot u = 0.5280
14. sec u = 1.2779   15. csc u = 1.1471

Cumulative Review
Graph the following equations.
1. $\dfrac{(x + 1)^2}{9} + \dfrac{(y - 2)^2}{4} = 1$
2. $\dfrac{(x - 1)^2}{4} + \dfrac{(y + 2)^2}{16} = 1$

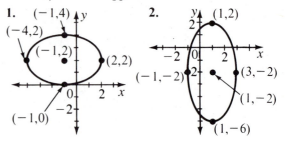

17.9 Linear Trigonometric Equations (page 481)

Vocabulary
linear trigonometric equations

Suggestions
Solving linear trigonometric equations is like solving algebraic equations with the additional steps of determining the reference measure and then the angle. Students should again be encouraged to make diagrams showing the reference angle in the indicated quadrant. Also remind students that they have not necessarily solved the equation after finding the reference angle but must still find the measure of the angle of rotation. Students should understand that, unlike finding one solution for linear algebraic equations, it is frequently necessary to find at least two solutions for linear trigonometric equations.

Errors Students Make
Remind students that both sin and cos must have values between or including ±1. Also, students should be cautioned to observe the restrictions placed on θ.

Cumulative Review
Graph the following equations.
1. $\dfrac{(x + 1)^2}{16} - \dfrac{(y - 2)^2}{4} = 1$
2. $\dfrac{(x - 2)^2}{25} - \dfrac{(y + 1)^2}{9} = 1$

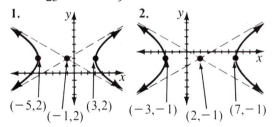

1. y; $(-5,2)$, $(-1,2)$, $(3,2)$
2. y; $(-3,-1)$, $(2,-1)$, $(7,-1)$

17.10 Interpolation with Trigonometry (page 483)

Vocabulary
trigonometric interpolation

Background
An alternate approach to interpolation that can be used with some more advanced students is to explain the theory of linear interpolation so that it applies to interpolation that is not linear. Trigonometric interpolation is not linear. Consider the idea of a slope in the study of linear interpolation. To find the sin 33°37′, for example, assume that the graph of the sin function is linear between the points P (33° 30′, sin 33° 30′) and Q (33° 40′, sin 33° 40′). Find the value of sin 33° 37′ or (33° 37′, y).

Since the slope of a line is the same between any points on the line, you get

$$\frac{y - \sin 33°30'}{33°37' - 33°30'} = \frac{\sin 33°40' - \sin 33°30'}{33°40' - 33°30'}.$$

Solve for y and get $y = \sin 33°37'$.

Suggestions
As hand-held calculators become more common, the need to interpolate becomes less. However, it is the feeling of the authors that students should understand the process and have some practice with interpolation.

Errors Students Make
To avoid errors in interpolation, have students set each problem up in the same manner. Students must then be careful with the signs. When they subtract the sin, tan, or sec of a measure, the sign is positive. When they subtract cos, cot or csc of a measure, the sign is negative. Examples 1 and 2 illustrate this.

Cumulative Review
Let $g(x) = -x^2 - 3x + 2$. Find each of the following function values.
1. $g(-3)$ 2. $g(0)$ 3. $g(a + b)$
1. 2   2. 2
3. $-a^2 - 2ab - b^2 - 3a - 3b + 2$

17.11 Trigonometry of a Right Angle (page 486)

Vocabulary
opposite side, adjacent side

Suggestions
Although some courses in trigonometry begin with the study of right triangles, right-triangle trigonometry is just a special case of trigonometry of an angle of rotation. Some of the work encountered here may be a review for some students.

Show students the relationship between the trigonometry of angles of rotation and the trigonometry of the right triangle. The reference triangles formed by an angle of rotation are right triangles.

Only acute angles are used in the trig of right triangles; one of the angles is a right angle and the other two are acute.

You may wish to include the secant and cosecant in this lesson.

$\sec A = \dfrac{\text{hypotenuse}}{\text{adjacent}} = \dfrac{c}{b}$ $\csc A = \dfrac{\text{hypotenuse}}{\text{opposite}} = \dfrac{c}{a}$

Example: For a $30° - 60°$ right triangle find $\sec 30°$, $\csc 30°$, $\sec 60°$, and $\csc 60°$ (see Example 1).

$\sec 30° = \dfrac{2}{\sqrt{3}} = \dfrac{2\sqrt{3}}{3}$; $\csc 30° = \dfrac{2}{1} = 2$;

$\sec 60° = \dfrac{2}{1}$; $\csc 60° = \dfrac{2}{\sqrt{3}} = \dfrac{2\sqrt{3}}{3}$

For Additional Chalkboard Work
Evaluate and simplify. Rationalize all denominators.

1. $\dfrac{\sec 60° - \csc 60°}{\sec 45°}$ 2. $\dfrac{\csc 30° - \sec 45°}{\csc 60°}$

3. $\dfrac{\sec 45° - \sec 30°}{\csc 30°}$ 4. $\dfrac{\csc 45° - \csc 30°}{\sec 60°}$

1. $\dfrac{3\sqrt{2} - \sqrt{6}}{3}$ 2. $\dfrac{2\sqrt{3} - \sqrt{6}}{2}$

3. $\dfrac{3 - \sqrt{3}}{2}$ 4. 0

Cumulative Review
If $f(x) = x^2 - 2$ and $g(x) = 2x - 1$, then find the following function values.
1. $f(g(2))$ 2. $g(f(-1))$ 3. $g(f(-a))$
1. 7 2. -3 3. $2a^2 - 5$

17.12 Using Trigonometry of a Right Triangle (page 489)

Vocabulary
angle of depression, angle of elevation

Suggestions
For those students who have had some geometry, remind them that the measure of the angle of depression is equal to the measure of the angle of elevation because if two parallel lines are cut by a transversal, the measures of the alternate interior angles are equal. Help students set up problems to avoid dividing by a 4-place decimal. If students have calculators, then it makes no difference whether a student multiplies or divides by decimals.

Exercises 18 and 35 are useful exercises but not realistic. It is inconceivable that anyone would want to find the distance from a point directly below a plane to a ship to the nearest meter. Such exercises are used only to provide for right-triangular trigonometric calculation.

Errors Students Make
The most difficult task for students seems to be deciding which function to use. Have students draw a diagram and then list the measures that are given and those to be found.

You might also introduce the old mnemonic SOHCAHTOA to help students remember the functions:

Sine $= \dfrac{\text{opposite}}{\text{hypotenuse}}$ is represented by SOH, and so on.

Cumulative Review
Draw the graph of each equation.
1. $3x - 2y = 8$ 2. $\frac{1}{2}x - \frac{3}{2}y = 6$
1. a straight line with a y-intercept of -4 and a slope of $\frac{3}{2}$
2. a straight line with a y-intercept of -4 and a slope of $\frac{1}{3}$

Computer Activities (page 494)
From a mathematical viewpoint, you should discuss with the class why it is necessary for at least one of the two known facts about the right triangle to be a side. (Is AAA a congruence theorem?)

Chapter and Lesson Commentaries

Chapter 18 Trigonometric Graphs and Identities

Chapter Objectives
To graph $y = \sin x$ and $y = \cos x$
To determine the amplitude and the period of the sine, cosine, tangent, and cotangent functions
To sketch graphs of $y = a \sin bx$, $y = a \cos bx$, $y = \tan x$, and $y = \cot x$, and to solve a system of trigonometric equations graphically
To prove trigonometric identities
To graph the inverses of the sin, cos, tan, and cot functions and to apply inverse function notation

Chapter Overview
This chapter presents the graphs of the sine, cosine, tangent, and cotangent as functions of degree measure and shows how the coefficients a and b in the function $y = a \sin bx$ control the amplitude and period of its graph. Trigonometric equations are solved both graphically and algebraically. The basic identities are presented and examples show how these can be used to prove more complex identities.

Special Features
Careers: Biological Sciences p. 496
Reading in Algebra pp. 508, 532
Cumulative Review pp. 499, 505, 508
Non-Routine Problems pp. 499, 505, 511
Calculator Activities p. 517
College Prep Test p. 537
Chapter Review p. 535
Chapter Test p. 536

18.1 Graphing $y = \sin x$ and $y = \cos x$ (page 497)

Vocabulary
amplitude, periodic function

Suggestions
Understanding the graphs of $y = \sin x$ and $y = \cos x$ is important because many of the properties of these functions can be observed. These properties can also be observed by using a unit circle.

By observing the changes in the length of the vertical line segment, notice that the maximum value is 1, the minimum value is -1, and the period is 360°. Also, notice:

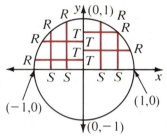

Change in Angle	Change in sin (RS)
0° to 90°	increase from 0 to 1
90° to 180°	decrease from 1 to 0
180° to 270°	decrease from 0 to -1
270° to 360°	increase from -1 to 0

Similarly, by observing the changes in the length of RT, properties of the cosine can be observed.

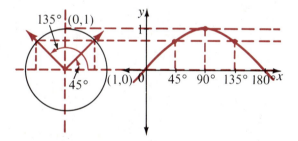

Errors Students Make
Labeling the x-axis sometimes causes difficulty. Have students label 90°, 180°, 270°, 360°, and the corresponding negative angles, then bisect the spaces to get the multiples of 45°.

Cumulative Review
Solve.

1. $35 + \dfrac{5x}{6} = \dfrac{1x}{6} + 29$;

2. $\dfrac{1}{2}(14n - 8) - (2 - 3n) = 4$ 1. -9;   2. 1

18.2 Determining Amplitude
(page 500)

Background
The amplitude of a periodic function is often defined as *one half* of the *absolute* value of the *difference* of the maximum and minimum function values, if such values exist. This definition is essential if the graph of the function is translated so that the absolute value of the maximum is *not* the same as the absolute value of the minimum.

Suggestions
In this lesson, students will learn how a change in amplitude affects the graph of $y = \sin x$, or $y = \cos x$. When $\sin x$ or $\cos x$ is multiplied by a nonzero constant, that is, when $y = a \sin x$, or $y = a \cos x$, there will be either a *vertical* "stretching" or "shrinking" of the function $y = \sin x$, or $y = \cos x$, for each $a \neq 1$.
Specifically:
1. When $a > 1$, there is a vertical "stretching" of the function $y = \sin x$, or $y = \cos x$.
2. When $0 < a < 1$, there is a vertical "shrinking" of the function $y = \sin x$, or $y = \cos x$.
3. When $a < 0$, there is a reflection of the function $y = \sin x$, or $y = \cos x$, across the x-axis.

Having students verbalize these properties will save much time and effort when they sketch the graphs, as assigned in the exercise set.

Errors Students Make
Students who always begin by constructing a table of values to graph a sine, or cosine function in which the amplitude has been altered will have difficulty. Students should have a general understanding of how the change in amplitude will affect the appearance of the graph prior to graphing the function. To avoid problems, provide drill, using the above discussion.

Cumulative Review
Solve each equation.
1. $3.2(2x - 0.3) = 1.3x + 9.24$;
2. $|3x + 7| = 10$
1. 2;   2. $1, -\frac{17}{3}$

18.3 Determining Period
(page 503)

Suggestions
In this lesson, students will learn how a change in period affects the graph of $y = \sin x$, or $y = \cos x$. When the abscissa, x, is multiplied by a nonzero constant, that is, when $y = \sin bx$, or $y = \cos bx$, there is a *horizontal* "stretching" or "shrinking" of the function $y = \sin x$, or $y = \cos x$.
Specifically:
1. When $b > 1$, there is a horizontal "shrinking" of the function $y = \sin x$, or $y = \cos x$.
2. When $0 < b < 1$, there is a horizontal "stretching" of the function $y = \sin x$, or $y = \cos x$.
3. When $b < 0$, there is a reflection across the x-axis for the function $y = \sin x$ and there is a reflection across the y-axis for the function $y = \cos x$.

Errors Students Make
Students sometimes do not have a general understanding of how the change in period will affect the appearance of the graph prior to graphing the function. To avoid problems, provide drill, using the above discussion.

Enrichment
Suppose $h(x) = f(ax)$. Describe the graph as: shrinking, stretching, a reflection, a combination of these, for the given values of a.
1. $\frac{1}{3}$ 2. -2 3. 3 4. $\frac{-1}{2}$ 5. -3
1. stretching   2. a reflection and shrinking
3. shrinking   4. a reflection and stretching
5. a reflection and shrinking

Cumulative Review
Solve each equation.
1. $2^{2x-4} = 64$ 2. $6x^2 + 7x = 3$
1. 5   2. $\frac{1}{3}, \frac{-3}{2}$

18.4 Change of Period and Amplitude
(page 506)

Background
The concepts of Lessons 18.2 and 18.3 are combined in this lesson. As a result of multiplying both the amplitude and the period by nonzero constants, a vertical and a horizontal "stretching," or "shrinking," can occur simultaneously.

Suggestions
Students should be able to discuss how the graphs of $y = a \sin bx$ and $y = a \cos bx$ change for different values of a and b. As preparation for this, you may wish to compare the graphs of the trigonometric functions with the graphs of algebraic equations. A good topic of discussion is, "How does a change in a for a family of parabolas of the form $y = ax^2$ affect the shape of the graph of the basic function $y = x^2$?"

Graphing $y = a \sin bx$ and $y = a \cos bx$ by first constructing a table of values is very tedious. Encourage more advanced students to sketch the graphs by determining key values mentally from the graph of the basic function, as shown below.

Key Values	$y = \cos x$	$y = \frac{1}{2}\cos 2x$
x-intercepts at	90°, 270°	45°, 135°
maximum at	0°, 360°	0°, 180°
minimum at	180°	90°
maximum	1	$\frac{1}{2}$
minimum	-1	$\frac{-1}{2}$

Errors Students Make
Labeling the axes and determining where the graph will cross the axes causes difficulty for some students. If students do not set up a table of values to lessen this difficulty, they should determine maximum points, minimum points, x-intercepts, and y-intercepts before sketching the graph.

Cumulative Review
Solve. Graph the solutions on a number line.
1. $3(2x - 1) \leq 3x + 6$ 2. $|n - 5| \leq 7$
1. $x \leq 3$ 2. $-2 \leq n \leq 12$

18.5 Graphing $y = \tan x$ and $y = \cot x$
(page 509)

Suggestions
Students should realize that the functions *tangent* and *cotangent* are reciprocals of each other, that is, $\tan x = \frac{1}{\cot x}$ and $\cot x = \frac{1}{\tan x}$. Therefore,
1. whenever the cotangent is positive, the tangent is positive; and
2. whenever the cotangent increases, the tangent decreases.

Use the graphs of $y = \tan x$ and $y = \cot x$ on pages 509 and 510, respectively, to emphasize that:
1. the graph of $y = \tan x$ is asymptotic to the vertical lines through $-90°$ and $90°$; the graph of $y = \cot x$ is asymptotic to the vertical lines through $0°$ and $180°$;
2. the graph of $y = \tan x$ passes through $(0, 0)$ and the graph of $y = \cot x$ passes through $(90, 0)$;
3. the period for both graphs is 180° and the graphs of $y = \tan x$ and $y = \cot x$ repeat this cycle every 180°;
4. neither function has a maximum point and therefore neither has an amplitude; and
5. the tangent function always increases while the cotangent function always decreases.

Encourage students to sketch graphs of the tangent and cotangent by plotting key points. These points can be found by substituting 0°, 45°, 90°, $-45°$, and $-90°$ for x in the definitions $\tan x = \frac{\sin x}{\cos x}$ and $\cot x = \frac{\cos x}{\sin x}$. For example, if $x = 45°$, then $\tan 45° = \frac{\sin 45°}{\cos 45°} = 1$ and $\cot 45° = 1$.

Errors Students Make
Errors will be avoided if students lightly draw the vertical lines that will serve as asymptotes and find the x-intercepts before they sketch the graphs of the tangent and cotangent functions.

Cumulative Review
Find p in dollars if $A = \$5,350$, $r = 12.5\%$, and $t = 7$ yr. Use the formula $A = p + prt$.
$2,853

18.6 Finding Logarithms of Trigonometric Functions (page 512)

Background
If $0° < x < 90°$, then $0 < \sin x < 1$ and $0 < \cos x < 1$
Therefore, for $0° < x < 90°$, $\log \sin x$ and $\log \cos x$ will have negative characteristics. Similarly, for $0° < x < 45°$, $\log \tan x$ will have a negative characteristic and for $45° < x < 90°$, $\log \cot x$ will have a negative characteristic.

Suggestions
Review the basic notions of logarithms, as stated on page 512. Point out that the tables of logarithms of trigonometric functions were formed by combining the tables of trigonometric functions and the tables of common logarithms.

Errors Students Make
Students sometimes have difficulty when interpolating is involved. Encouraging them to stay with a single format will help to avoid confusion. Students also have difficulty with signs. To avoid confusion, encourage students to set up the entire problem before they use the tables.

For Additional Chalkboard Work
Use logarithms to find each value.

1. $\dfrac{(32.6)^3 \csc 48°30'}{\sec 18°40'}$ 2. $\dfrac{\sqrt{48.3 \sec 50°42'}}{(\csc 37°58')^3}$

1. $\dfrac{(32.6)^3 \dfrac{1}{\sin 48°30'}}{\dfrac{1}{\cos 18°40'}}$; 43,800

2. $\dfrac{\sqrt{\dfrac{48.3}{\cos 50°42'}}}{\dfrac{1}{(\sin 37°58')^3}}$; 2.03

Cumulative Review
Simplify.
1. $-6a^3b^5c^2 \cdot 5a^2b^2c^5$ 2. $-10(3x^5y^7)^4$
3. $\dfrac{5n - 15}{2n^2 - 12n + 18}$

1. $-30a^5b^7c^7$ 2. $-810x^{20}y^{28}$ 3. $\dfrac{5}{2(n-3)}$

18.7 Quadratic Trigonometric Equations (page 515)

Vocabulary
quadratic trigonometric equations

Background
The method of completing the square can be used with quadratic trigonometric equations.

$2 \sin^2 \theta - 6 \sin \theta - 1 = 0$
$2\left(\sin^2 \theta - 3 \sin \theta + \left(\dfrac{3}{2}\right)^2\right) = 1 + \dfrac{9}{2}$
$\left(\sin \theta - \dfrac{3}{2}\right)^2 = \dfrac{11}{4}$
$\sin \theta - \dfrac{3}{2} = \pm\dfrac{1}{2}\sqrt{11}$
$\sin \theta = \dfrac{3}{2} \pm \dfrac{1}{2}\sqrt{11}$, or $\dfrac{3 \pm \sqrt{11}}{2}$

Since $-1 \leq \sin \theta \leq 1$, reject $\dfrac{3 + \sqrt{11}}{2}$.

Suggestions
You may wish to review the methods for solving quadratic algebraic equations, such as those below, which are similar to the trigonometric equations in this lesson.
1. $x^2 = 16$ 2. $x^2 - 4 = 0$ 3. $x^2 - 3x = 0$
4. $x^2 - 2x + 1 = 0$ 5. $x^2 + 5x - 3 = 0$
1. $\pm\sqrt{16}$; ± 4 2. $(x-2)(x+2)$; $2, -2$
3. $x(x-3)$; $0, 3$ 4. $(x-1)^2$; 1
5. $\dfrac{-5 \pm \sqrt{25 + 12}}{2}$; $\dfrac{-5 \pm \sqrt{37}}{2}$

Emphasize that while a quadratic algebraic equation usually yields two solutions, a quadratic trigonometric equation may yield four solutions if θ is between $0°$ and $360°$.

Errors Students Make
Students sometimes find the reference angle and then stop. Remind them that they must determine all possible quadrants in which the reference angle may be and include all measures as solutions.

Cumulative Review
Multiply.
1. $(3a^2 - 2)(4a^2 + 1)$ 2. $2(2x - 3)(3x + 2)$
3. $(8x^2 - 1)(8x^2 + 1)$
1. $12a^4 - 5a^2 - 2$
2. $12x^2 - 10x - 12$ 3. $64x^4 - 1$

18.8 Basic Trigonometric Identities (page 518)

Background
Reciprocal identities, quotient identities, and Pythagorean identities are verified for the general case, then for particular degree measures.

Suggestions
You can use a unit circle to show how the identities got their names. For example,

1. $\sin x = \dfrac{1}{\csc x}$ is called a *reciprocal identity* because $\sin x, \dfrac{y}{r}$, and $\csc x, \dfrac{r}{y}$ are reciprocals;

2. $\tan x = \dfrac{\sin x}{\cos x}$ is called a *quotient identity* because $\dfrac{\sin x}{\cos x} = \dfrac{y}{r} \div \dfrac{x}{r} = \dfrac{y}{x} = \tan x$; and

3. $\cos^2 x + \sin^2 x = 1$ is called a *Pythagorean identity* because $x^2 + y^2 = 1$.

Students should memorize the identities. A line representation of the six trigonometric functions is useful for this purpose.

$\sin^2 \theta + \cos^2 \theta = 1$
$\tan^2 \theta + 1 = \sec^2 \theta$
$\cot^2 \theta + 1 = \csc^2 \theta$
$\tan \theta = \dfrac{\sin \theta}{\cos \theta}$
$\cot \theta = \dfrac{\cos \theta}{\sin \theta}$
$\sec \theta = \dfrac{1}{\cos \theta}$
$\csc \theta = \dfrac{1}{\sin \theta}$

Errors Students Make
One source of error occurs when students do not understand that $\sin^2 \theta = (\sin \theta)^2$. Simplifying $(\sin \theta)^2 + (\cos \theta)^2$ for $\theta = 150°$ will help. Note that $\left(\dfrac{1}{2}\right)^2 + \left(-\dfrac{\sqrt{3}}{2}\right)^2 = \dfrac{1}{4} + \dfrac{3}{4} = 1$, or $\sin^2 \theta + \cos^2 \theta = 1$.

Cumulative Review
Simplify.
1. $(3a + 4b)^2$
2. $\dfrac{2a + 10}{4a - 12} \cdot \dfrac{a^2 - 9}{a^2 + 7a + 10}$

1. $9a^2 + 24ab + 16b^2$
2. $\dfrac{a + 3}{2(a + 2)}$

18.9 Proving Trigonometric Identities (page 521)

Background
A common misconception in proving trigonometric identities, or identities in general, is to write a series of identities until both members are the same. In doing so, restrictions may be introduced, but not recorded. For example, when $\dfrac{\sin x}{\cos x}$ replaces $\tan x$, the restriction $\cos x \neq 0$ is introduced. When one member simplifies into the other, all restrictions must be considered.

Manipulating one member to arrive at the second does not prove an identity. To prove that the given expressions are really identical, one should start with the conclusion and reverse the process until arriving at the given expressions. Thus, one would proceed as follows in Example 1:

$$\tan x = \dfrac{\sin x}{\cos x} = \dfrac{\dfrac{1}{\csc x}}{\dfrac{1}{\sec x}} = \dfrac{\sec x}{\csc x}.$$

Suggestions
In mathematics it is often important to know when two expressions are identical. Therefore, in this lesson students will be given opportunities to determine if certain trigonometric expressions are identical.

Point out that it is often helpful to use known identities for one member to arrive at the other member in verifying two trigonometric expressions. Caution students that they must work on each side separately because they have not yet proved that the equality holds.

Errors Students Make
There is no well-defined method that will eliminate all errors. Whenever students have a problem with proving a trigonometric identity, they should try to express each side of the identity in terms of sine and cosine.

Cumulative Review
Factor.
1. $8a^2 + 22a + 5$
2. $6x^4 + 7x^2 - 3$
3. $25a^2 - 9b^2$

1. $(4a + 1)(2a + 5)$ 2. $(2x^2 + 3)(3x^2 - 1)$
3. $(5a + 3b)(5a - 3b)$

18.10 Radian Measure (page 523)

Vocabulary
radian

Background
The radian measure of an angle of rotation, θ, is the ratio of the distance, s, on the circumference of a circle to the length of a radius, r, of the circle. Thus, $\theta = \frac{s}{r}$ and when $s = r$, $\theta = 1$ radian. Note that $\frac{s}{r}$ is dimensionless because s and r are both measured in the same unit. It is not necessary, then, to use the word *radian* to express angular rotations in radians.

This simple conversion proportion can be used to change from one system of angular measure to the other: $\frac{\text{radian measure}}{\pi} = \frac{\text{degree measure}}{180°}$

Suggestions
Explain that radian measure is a real number and that this type of angular measure will be used almost exclusively in advanced mathematics. Point out that students will use the new system to review many concepts which they already learned for degree measures. To save time and to avoid errors, students should memorize some basic conversions, such as the following.

0°	30°	45°	60°	90°	180°	270°	360°
↕	↕	↕	↕	↕	↕	↕	↕
0	$\frac{\pi}{6}$	$\frac{\pi}{4}$	$\frac{\pi}{3}$	$\frac{\pi}{2}$	π	$\frac{3\pi}{2}$	2π

For Additional Chalkboard Work
Evaluate each of the following.

1. $\dfrac{\cos \frac{\pi}{6} - \sin \frac{\pi}{6}}{\tan \frac{\pi}{4}}$ 2. $\dfrac{\sin \frac{\pi}{4} + \cos \frac{\pi}{4}}{\cot \frac{\pi}{3}}$

1. $\dfrac{\sqrt{3} - 1}{2}$ 2. $\dfrac{\sqrt{6}}{3}$

Cumulative Review
Factor completely.
1. $6y^2 + 21y - 45$ 2. $6a^3 - 7a^2 - 5a$
1. $3(y + 5)(2y - 3)$
2. $a(2a + 1)(3a - 5)$

18.11 Applying Radian Measure (page 526)

Background
This lesson serves as a good review of many of the concepts developed earlier. Using radian measure, students will get additional practice in sketching graphs of trigonometric functions whose periods and amplitudes have been altered and in solving trigonometric equations.

Suggestions
Although this lesson may serve as a review, you should emphasize how important the lesson really is. Most advanced mathematics courses do not introduce degree measure until late because those courses are defined for real numbers, and radian measure, not degree measure.

As with degree measure, you may wish to have some students graph a circular function by referring to a unit circle. This method is shown below for $\sin \theta$. Note that $\sin \theta = y$ and $\cos \theta = x$; therefore, vertical segments represent the values of $\sin \theta$.

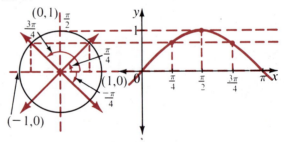

Errors Students Make
Problems may occur when students label the x-axis, because most students must translate from radian measure to degree measure. To avoid this, you may wish to label the x-axis in degrees, then have students write the corresponding radian measure beneath each degree measure.

Cumulative Review
Solve.
The greater of two numbers is 11 less than 3 times the smaller number. Twenty one less than 3 times the greater number is equal to 6 times the smaller number. Find both numbers.
18 and 43

18.12 Inverses of Trigonometric Functions (page 529)

Vocabulary
inverse of a trigonometric function, inverse trigonometric function, principal values

Background
Recall that if a function is strictly increasing or strictly decreasing in an interval, then the inverse of the function is an inverse function in that interval. The sine function is strictly increasing in the interval $\left[-\frac{\pi}{2}, \frac{\pi}{2}\right]$; therefore, its inverse is an inverse function in that interval. Similarly, the inverses of the cosine and tangent functions are inverse functions in the intervals $[0, \pi]$ and $\left[-\frac{\pi}{2}, \frac{\pi}{2}\right]$, respectively. Note that each of the three functions has more than one interval where it is strictly increasing, or strictly decreasing, but the interval closest to the origin is used.

Suggestions
Emphasize that the inverses of the trigonometric functions are *not* functions. However, it is possible to make the inverses functions by restricting their range to values of the range called *principal values*. Using only part of the graph of an inverse of a trigonometric function is sufficient because to define a function, it is necessary to restrict only one choice for y for each x in the domain. Use the graph to help students visualize this discussion.

Cumulative Review
Side a of a triangle is $1\frac{1}{2}$ times as long as side b. Side c is 5 cm longer than side a. The perimeter of the triangle is 53 cm. Find a, b, and c.
18 cm, 12 cm, 23 cm

18.13 Complex Numbers and DeMoivre's Theorem (page 533)

Background
$r = \sqrt{a^2 + b^2}$ is often referred to as the *modulus* of the complex number. Cauchy (1821) was the mathematician given credit for this terminology.

The symbol $|a + bi|$ was used first by Weierstrass to represent absolute value. He also used the terminology of absolute value for $\sqrt{a^2 + b^2}$. Gauss called $\sqrt{a^2 + b^2}$ the *norm* of $a + bi$.

Suggestions
Indicate that the absolute value, or modulus, of $a + bi$, $|a + bi|$, is the nonnegative real number $\sqrt{a^2 + b^2}$.

For Additional Chalkboard Work
Write in polar form:
1. $1 - i$ 2. $4 - 3i$ 3. $-1 + i$ 4. $4 + 3i$
5. $\frac{1}{\sqrt{2}} - \frac{i}{\sqrt{2}}$ 6. $-1 - i$

1. $\sqrt{2}\left(\cos \frac{7\pi}{4} + i \sin \frac{7\pi}{4}\right)$
2. $5\left(\cos \frac{323\pi}{180} + i \sin \frac{323\pi}{180}\right)$
3. $\sqrt{2}\left(\cos \frac{3\pi}{4} + i \sin \frac{3\pi}{4}\right)$
4. $5\left(\cos \frac{37\pi}{180} + i \sin \frac{37\pi}{180}\right)$
5. $1\left(\cos \frac{7\pi}{4} + i \sin \frac{7\pi}{4}\right)$
6. $\sqrt{2}\left(\cos \frac{5\pi}{4} + i \sin \frac{5\pi}{4}\right)$

Enrichment
Prove:
1. If $\bar{z} = a - bi$ when $z = a + bi$, then $|\bar{z}| = |z|$.
2. $|z_1 \cdot z_2| = |z_1| \cdot |z_2|$
3. $\left|\frac{z_1}{z_2}\right| = \frac{|z_1|}{|z_2|}$
4. $|z|^2 = |\bar{z} \cdot z|$

Cumulative Review
Factor:
1. $16m^2 - 81$ 2. $2x^2 - x - 15$ 3. $8a^3 - 27$
1. $(4m + 9)(4m - 9)$ 2. $(2x + 5)(x - 3)$
3. $(2a - 3)(4a^2 + 6a + 9)$

Chapter and Lesson Commentaries

Chapter 19 Trigonometric Laws and Formulas

Chapter Objectives
To use the Law of Cosines to find the length of a side or the degree measure of an angle of a triangle

To use the formula $K = \frac{1}{2} ab \sin C$ to find the area of a triangle or, given the area, find an unknown in the formula

To apply the Law of Sines to find the length of a side or the degree measure of an angle of a triangle

To investigate the "Ambiguous Case" in determining a triangle

To derive and apply the formulas for
sin $(A + B)$ and sin $2A$;
cos $(A + B)$ and cos $2A$; and
tan $(A + B)$ and tan $2A$

To derive and apply the formulas for sine, cosine and tangent of half-angles

Chapter Overview
The first half of this chapter deals with the applications of the trigonometric relations in triangles. It attempts to give students practice in manipulating identities and formulas to gain information about a triangle.

It may be a good idea to present a particular problem to the students, have them consider it, and then show how it can be solved. This may be more motivating than deriving the formula and plugging in given values. Ask the students if it is possible to find the area of a triangle given only the measures of two sides and the included angle.

Chapter Overview (cont.)
Give them a specific example and some time to think about it. Some may come to the conclusion that it is impossible while others may be on their way to deriving the formula themselves. Ask the students, before doing Lesson 19.1, to find the third side of a triangle given two sides and the included angle, or to find the measure of an angle given three sides of a triangle. Again, give a specific example and ample time for consideration. Then, derive the formula for the Law of Cosines and show how it can be used to solve the problems.

If two angles and a side of a triangle are known, can the third angle be found? That is easy, but can the lengths of the other two sides also be found? This problem will lead into the formula for the Law of Sines.

Students may also find the Ambiguous Case more interesting if the problem is presented first. Have students try to construct an "unconstructable triangle" as a lead-in to the lesson on the Amgibuous Case.

The second half of the chapter derives and applies the formulas for the sine, cosine, and tangent of the sum of two angles, double-angles, and half-angles. The last section gives students an opportunity to draw on their knowledge of trigonometric relations and formulas to solve identities. Proving identities is an excellent way for students to exercise independent thinking based on logic and knowledge of algebra.

Special Features
Non-Routine Problem Solving   p. 538
Cumulative Review   pp. 541, 559, 567
Non-Routine Problems   p. 565
Calculator Activities   p. 548
Computer Activity   p. 570
College Prep Test   p. 571
Chapter Review   p. 568
Chapter Test   p. 569
Cumulative Review (Chapters 17–19) pp. 572–573

19.1 Law of Cosines (page 539)

Vocabulary
Law of Cosines

Suggestions
The Law of Cosines is used to find the measure of a side when the measures of two sides and the included angle are known. The Law of Cosines is a generalization of the Pythagorean theorem or the Pythagorean theorem is a special case of the Law of Cosines.

Let $m\angle A = 90°$, then for $a^2 = b^2 + c^2 - 2bc \cdot \cos A$, $a^2 = b^2 + c^2 - 2bc \cos 90°$ or $a^2 = b^2 + c^2$ since $\cos 90° = 0$.

Background
The following is a coordinate proof of the Law of Cosines and serves as a good alternative to the one found on page 539.

$c = PQ =$
$\sqrt{(a\cos\theta - b\cos\alpha)^2 + (a\sin\theta - b\sin\alpha)^2}$
$C^2 = a^2\cos^2\theta - 2ab\cos\theta\cos\alpha + b^2\cos^2\alpha$
$+ a^2\sin^2\theta - 2ab\sin\theta\sin\alpha + b^2\sin^2\alpha$
$= a^2(\cos^2\theta + \sin^2\theta) + b^2(\cos^2\alpha + \sin^2\alpha) - 2ab(\cos\theta\cos\alpha + \sin\theta\sin\alpha); c^2 = a^2 + b^2 - 2ab(\cos\alpha - \theta)$, but $\alpha - \theta$ is angle C
$c^2 = a^2 + b^2 - 2ab\cos C$.

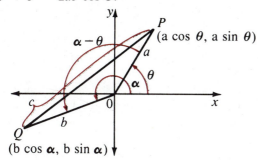

Errors Students Make
Students might make errors in computing the square root. Let them use a table of values, a calculator, or review how to find square roots.

Cumulative Review
Solve each equation for x.

1. $a(2x + b) = c(x + b)$
2. $\dfrac{4}{a} = \dfrac{3}{b} + \dfrac{2}{x}$

1. $\dfrac{bc - ab}{2a - c}$
2. $\dfrac{2ab}{4b - 3a}$

19.2 Using the Law of Cosines to Find Angle Measures (page 542)

Suggestions
This short lesson deals with finding the measure of an angle using the Law of Cosines. To do so, first solve for the cosine of the angle, then use trigonometric tables to find the measure of the angle. Remind students that the longest side of a triangle is opposite the greatest angle, and the smallest side is opposite the smallest angle.

For Additional Chalkboard Work
Have students *set up* the formula for finding the measure of $\angle A$, $\angle B$ and $\angle C$. Do not have them solve.

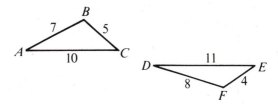

Do the same for $\angle D$, $\angle E$ and $\angle F$.

$(10)^2 = 7^2 + 5^2 - 2(7)(5) \cos B$
$7^2 = (10)^2 + 5^2 - 2(10)(5) \cos C$
$5^2 = (10)^2 + 7^2 - 2(10)(7) \cos A$
$(11)^2 = 8^2 + 4^2 - 2(8)(4) \cos F$
$8^2 = (11)^2 + 4^2 - 2(11)(4) \cos E$
$4^2 = (11)^2 + 8^2 - 2(11)(8) \cos D$

Errors Students Make
Determining which formula to use from the given data may cause some problems. To lessen the chance of error, provide some chalkboard drill.

Cumulative Review
Simplify each expression.

1. $(-2a^{-3}b^2)^{-4}$
2. $\dfrac{-28x^{-5}y^2}{7x^2y^{-5}}$
3. $-2(a^4b^{-6})^{-4}$

1. $\dfrac{a^{12}}{16b^8}$
2. $\dfrac{-4y^7}{x^7}$
3. $\dfrac{-2b^{24}}{a^{16}}$

19.3 Area of a Triangle (page 544)

Background
A formula for finding the area of a triangle given the measures of three sides of a triangle is given by $A = \sqrt{s(s-a)(s-b)(s-c)}$ where $s = \frac{1}{2}(a + b + c)$. This formula is known as *Hero's Formula*. Find the area of triangle ABC if $a = 40$, $b = 60$, and $c = 50$.

$$A = \sqrt{75(75-40)(75-60)(75-50)}$$
$$= \sqrt{75(35)(15)(25)} \quad s = \frac{1}{2}(40 + 60 + 50)$$
$$= 992.2 \quad\quad\quad\quad\quad = 75$$

Suggestions
A formula for finding the area of a triangle given the measures of two sides and the included angle is developed. The formulas presented in this lesson will be used to derive the Law of Sines as well as to derive the formulas for $\sin(A + B)$.

An acute triangle was used to derive the formulas for the area of a triangle. Consider now an obtuse triangle.

$\sin(180° - C) = \frac{h}{a}$
$h = a \sin(180° - C)$
but $\sin(180° - C) = \sin C$
So, $h = a \sin C$
but area of $\triangle ABC = \frac{1}{2} bh$
So, area of $\triangle ABC = \frac{1}{2} ba \sin C$ or $\frac{1}{2} ab \sin C$.

For Additional Chalkboard Work
Find the area of each of the given triangles.
1. $a = 46$, $b = 53$, $c = 58$ 1,542.6
2. $a = 9.6$, $b = 10.3$, $c = 15.3$ 48.6
3. $a = 70$, $b = 80$, $c = 90$ 2,683.3
4. $a = 123.5$, $b = 79.6$, $c = 100.5$ 3,988.9

Errors Students Make
Not knowing which formula to use may cause some difficulty. Some oral drill could reduce error.

Cumulative Review
Evaluate the following if $a = -1$, $b = 2$.
1. $\dfrac{3a^2(2a)^2}{(a^3 + 4)^2}$ 2. $((ab)^2 - ab^2 + a^2b)^{-2}$

1. $\dfrac{4}{3}$ 2. $\dfrac{1}{100}$

19.4 Law of Sines (page 546)

Vocabulary
Law of Sines

Background
Place $\triangle ABC$ so that vertex A is at the origin and $\overline{BC} \parallel$ to the x-axis as shown.
Since $\overline{BC} \parallel$ x-axis, $\angle 1 \cong \angle 2$,
and $b \sin(A + B) = c \sin B$
$A + B + C = 180°$
$A + B = 180° - C$
$b \sin(180° - C) = c \sin B$
$\dfrac{c}{\sin C} = \dfrac{b}{\sin B}$ or $\dfrac{\sin C}{c} = \dfrac{\sin B}{b}$.

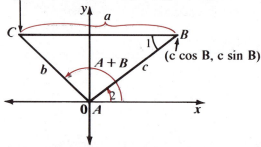

Suggestions
The Law of Sines is a useful proportion for solving triangles when the measures of two angles and a side opposite one of them are given, or when two sides and an angle opposite one of them are given.

While it is possible to solve oblique triangles by drawing altitudes and forming right triangles, it is easier to solve such triangles using the Law of Sines.

It is possible to solve triangles for either sides or angles using the Law of Sines.

Logarithms can be used to solve the derived proportion.

Errors Students Make
At times the computation becomes tedious. Logarithms and calculators will help reduce error.

Cumulative Review
Evaluate each of the following.
1. $2n^2 - 12n + 7$ if $n = 2$
2. $3n^3 - 6n^2 + 5n - 8$ if $n = -2$

1. -9 2. -66

19.5 The Ambiguous Case
(page 549)

Vocabulary
Ambiguous Case

Suggestions
Students may wish to create a physical model that demonstrates the Ambiguous Case. Although students can memorize all the rules of the Ambiguous Case, it seems more appropriate to have them understand the concepts and determine the number of constructable triangles directly from the given conditions.

Solving triangles given two sides and the angle opposite one of them is known as the Ambiguous Case. In studying the Ambiguous Case, students should realize that all cases can be tested by inspection except the case in which the angle is acute and the opposite side is less than the other given side. This is the only time the opposite side must be compared with the length of the altitude, h, that is drawn to the third side.

Errors Students Make
When students can't determine the number of triangles that can be constructed, they should sketch figures and label the parts.

Cumulative Review
Simplify each expression.
1. $-3a^2b\sqrt{32a^5b^7c^4}$
2. $\sqrt{4a^3b^3} + a\sqrt{9ab^3} - ab\sqrt{ab}$

1. $-12a^4b^4c^2\sqrt{2ab}$
2. $4ab\sqrt{ab}$

19.6 Sin $(A \pm B)$ and Sin $2A$
(page 552)

Vocabulary
sum of two angles, double-angle, difference of two angles

Background
You may wish to derive the formula for $\cos(A - B) = \cos A \cos B + \sin A \sin B$ first, and then use it to derive the formula for $\sin(A + B)$.

$$\sin(A + B) = \cos\left[\frac{\pi}{2} - (A + B)\right]$$
$$= \cos\left[\left(\frac{\pi}{2} - A\right) - B\right]$$
$$= \cos\left(\frac{\pi}{2} - A\right)\cos B +$$
$$\sin\left(\frac{\pi}{2} - A\right)\sin B$$
$$\sin(A + B) = \sin A \cos B + \cos A \sin B$$

Suggestions
The $\sin(A + B)$ is derived from using the three formulas for finding the area of a triangle. However, the derivation shown for $\sin(A + B)$ is valid only if $x + y < 180°$. Nevertheless, a rigorous proof of the more general case can be developed using the appropriate reduction formulas. The formula for $\sin(A + B)$ is basic to the development of the other sum and difference formulas. The formula for $\sin 2A$ is derived for $\sin(A + B)$, where $A = B$.

To derive $\sin(A - B)$, rewrite $A - B = A + (-B)$ and use $\sin[A + (-B)]$.

To derive $\sin 2A$, rewrite $2A = A + A$ and use $\sin(A + A)$.

For Additional Chalkboard Work
Show $\sin(x + 2\pi)$ is periodic with a period of 2π. That is, show $\sin(x + 2\pi) = \sin x$.
$\sin(x + 2\pi) = \sin x \cos 2\pi + \cos x \sin 2\pi$
$= \sin x (1) + \cos x (0)$.
So, $\sin(x + 2\pi) = \sin x$.

Cumulative Review
Write an equation of each circle whose center is at the origin with the given radius.
1. 12   2. 15   3. 20

1. $x^2 + y^2 = 144$ 2. $x^2 + y^2 = 225$
3. $x^2 + y^2 = 400$

T-101

19.7 cos (A ± B) and cos 2A
(page 556)

Vocabulary
sum of two angles, double-angle

Background
Start with identity $\cos 2A = 1 - 2\sin^2 A$ and solve for $\sin^2 A$.

$$2\sin^2 A = 1 - \cos 2A \text{ or}$$
$$\sin^2 A = \frac{1 - \cos 2A}{2}$$

Similarly, $\cos 2A = 2\cos^2 A - 1$
$$2\cos^2 A = 1 + \cos 2A$$
$$\cos^2 A = \frac{1 + \cos 2A}{2}.$$

The identities for $\sin^2 A$ and $\cos^2 A$ are defined in terms of $\cos 2A$. These forms are very useful and important in calculus since they allow us to integrate easily the $\sin^2 A$ and $\cos^2 A$ functions.

Suggestions
The $\cos (A + B)$ and $\cos 2A$ formulas are easily derived from the results of the previous lesson. Students should be encouraged to derive them.

To derive $\cos (A + B)$, use $\sin [90° - (A + B)]$ and expand. Remind students that $\cos \theta = \sin (90° - \theta)$.

To derive $\cos (A - B)$, rewrite $A - B = A + (-B)$ and use $\cos (A + B)$.

To derive $\cos 2A$, rewrite $2A$ as $A + A$ and use $\cos (A + A)$.

Students are expected to be familiar with all three cases for $\cos 2A$.

Show $\cos (1 + 2\pi)$ is periodic with a period of 2π. That is, show $\cos (x + 2\pi) = \cos x$.
$$\cos(x + 2\pi) = \cos x \cos 2\pi - \sin x \sin 2\pi$$
$$= \cos x (1) \quad - \sin x (0)$$
$$= \cos x$$
So, $\cos (x + 2\pi) = \cos x$.

Cumulative Review
Identify each conic section whose equation is given.
1. $\frac{x^2}{25} - \frac{y^2}{9} = 1$ 2. $\frac{x^2}{15} + \frac{y^2}{15} = 1$
3. $5x^2 + 3y^2 = 36$ 4. $xy = -3$
1. hyperbola   2. circle   3. ellipse
4. hyperbola

19.8 Tan (A ± B) and Tan 2A
(page 560)

Suggestions
If time permits, develop $\cot (A + B)$ and $\cot 2A$.

$$\cot (A + B) = \frac{\cos (A + B)}{\sin (A + B)}$$
$$= \frac{\cos A \cos B - \sin A \sin B}{\sin A \cos B + \cos A \sin B}$$
$$= \frac{\frac{\cos A \cos B}{\sin A \sin B} - \frac{\sin A \sin B}{\sin A \sin B}}{\frac{\sin A \cos B}{\sin A \sin B} + \frac{\cos A \sin B}{\sin A \sin B}}$$
$$= \frac{\cot A \cot B - 1}{\cot B + \cot A}$$

Thus, $\cot (2A) = \frac{\cot A \cot A - 1}{\cot A + \cot A} = \frac{\cot^2 A - 1}{2 \cot A}$.

Replace B by $(-B)$ in derivation of $\cot (A + B)$ above to find

$$\cot (A - B) = \frac{\cot A \cot B + 1}{\cot B - \cot A}.$$

For Additional Chalkboard Work
1. Show $\tan (x + \pi)$ is periodic with a period of π.

$$\tan (x + \pi) = \frac{\tan x + \tan \pi}{1 - \tan x \tan \pi} = \frac{\tan x + 0}{1 - \tan x(0)}$$
$$= \tan x$$

Find the value of each. Use special angles and the formula for $\cot (A + B)$ or $\cot (A - B)$.
2. cot 105°   3. cot 15°   4. cot 135° -1
5. cot 120°   6. cot 375° -11.4301
Find $\tan (A + B)$ for each of the following. $\frac{39}{2}$, or $19\frac{1}{2}$
7. $\cot A = \frac{4}{3}$ and $\cot B = \frac{5}{6}$
8. $\cot A = -\sqrt{3}$ and $\cot B = \frac{\sqrt{2}}{2}$ $4\sqrt{2} - 3\sqrt{3}$
2. $\frac{\sqrt{3} - 3}{\sqrt{3} + 3}$ 3. $\frac{\sqrt{3} + 1}{\sqrt{3} - 1}$ 5. $-\frac{\sqrt{3}}{3}$

Errors Students Make
Students may need to review methods of rationalizing the denominator.

Cumulative Review
Find the coordinates of the center of each circle and the radius.
1. $x^2 + 6x + y^2 + 4y = -12$
2. $x^2 + y^2 - 4x + 8y = -11$
1. $(-3, -2)$, $r = 1$ 2. $(2, -4)$, $r = 3$

19.9 Half-Angle Formulas (page 563)

Vocabulary
sin of half an angle, cos of half an angle, tan of half an angle

Suggestions
Half-angle formulas are used to find trigonometric values of certain angular measures such as tan 15° or sin 75°.

To derive $\sin \frac{A}{2}$, start with the identity $\cos 2A = 1 - 2\sin^2 A$, replace A by $\frac{x}{2}$ and solve for $\sin \frac{x}{2}$.

To derive $\cos \frac{A}{2}$, start with the identity $\cos 2A = 2\cos^2 A - 1$, replace A by $\frac{x}{2}$ and solve for $\cos \frac{x}{2}$.

To derive $\tan \frac{x}{2}$, start with the identity $\tan \frac{x}{2} = \dfrac{\sin \frac{x}{2}}{\cos \frac{x}{2}}$ and substitute for $\sin \frac{x}{2}$ and $\cos \frac{x}{2}$.

If time permits derive the formula for $\cot \frac{x}{2}$. Start with

$$\cot \frac{x}{2} = \frac{\cos \frac{x}{2}}{\sin \frac{x}{2}}$$

$$= \frac{\pm\sqrt{\dfrac{1+\cos x}{2}}}{\pm\sqrt{\dfrac{1-\cos x}{2}}} = \pm\sqrt{\dfrac{1+\cos x}{2} \cdot \dfrac{2}{1-\cos x}}$$

This is asked in Exercise 33.

Errors Students Make
Determining the quadrant for $\frac{x}{2}$, given the quadrant of x, may cause some difficulty, since the sign of the trig function of an angle may not be the same sign of the trig function of a half-angle.

For example, $\tan x > 0$ if $180° \leq x \leq 270°$, but $\tan \frac{x}{2} < 0$ since $90° \leq \frac{x}{2} \leq 135°$.

Cumulative Review
Solve algebraically.
1. $y = x^2 + 3x - 13$
   $2x - 3y = 13$
2. $x^2 + y^2 = 10$
   $5x + 2y = -13$

1. $(2, -3)$ 2. $(-3, 1)$

19.10 Double- and Half-Angle Identities (page 566)

Vocabulary
double-angle identities, half-angle identities

Suggestions
Double-angle and half-angle formulas are used to prove identities. While it is not always necessary, it is often useful to express functions of double-angle and half-angle formulas as functions of single angles. Remind students that

$$\csc 2x = \frac{1}{\sin 2x} \text{ and } \sec 2x = \frac{1}{\cos 2x}. \text{ Similarly,}$$

$$\csc \frac{x}{2} = \frac{1}{\sin \frac{x}{2}} \text{ and } \sec \frac{x}{2} = \frac{1}{\cos \frac{x}{2}}.$$

Have students put one side of the identity in one column and the other side in another column. Then, have them try to write each side in terms of sin and cos. Watch from one column to another to see if a formula from the chapter occurs. The last step is to have the final line in each column identical.

In Example 1 in the text, one might have replaced $\cos 2\theta$ on the left with $1 - 2\sin^2 \theta$. If so, then the final line on the right would also be $1 - 2\sin^2 \theta$.

Students may go about proving an identity in a variety of ways.

Errors Students Make
When students have difficulty, they should express functions of half-angle and double-angle formulas as functions of a single angle. They should also express functions in terms of sine and cosine.

Cumulative Review
1. Find the length of $\overline{AB}$, $A(-3, -5)$, $B(7, -2)$.
2. Find the slope of $\overrightarrow{PQ}$, $P(6, -8)$, $Q(-3, -9)$.

1. $\sqrt{109}$ 2. $m = \dfrac{-1}{-9}$ or $\dfrac{1}{9}$

Computer Activities (page 570)
There are many more variations for this program than the one given. Here students can go off on their own and rewrite the program in terms of two sides and the included angle. Let them be creative. Also, let them find out for themselves that some combinations do not work.

T-103

Computer Section (page 574)

Every effort has been made to write the programs in this section in a generic BASIC that is common to all computers, especially Apple, Radio Shack, and Commodore. For this reason, some of the programs may not be as elegant as they could be. For example, the screen may sometimes seem cluttered. The reason for this is that the computers mentioned have different ways of clearing the screen for the next problem. Graphs that are drawn do not make use of the best graphic modes because they vary too much from model to model. These and other types of refinements are left to your students and you.

Pages 574–575 are a general introduction to BASIC as a language. Only one programming word, PRINT, is used. You might want to point out that with just this one word the computer becomes a useful tool. Here it is used to test the solutions to linear equations by plugging the solutions into both sides of the equation and comparing results.

Page 576 expands vocabulary to include LET. Although X = 5 without LET accomplishes the same thing, students should be taught to use LET since it emphasizes that X = 5 is not a statement of equality but an assignment of value.

Typing in the equations a second time because of errors has not been too pleasant. On page 577 students learn that it is possible to type in the equation just once and change the value of the variable by means of the INPUT or READ....DATA. GOTO is also added to eliminate the use of multiple RUNs.

The PRINT statement returns on page 578 and is used with quotation marks to include words and symbols in the printout. From this point on, students should be encouraged to make every printout a complete statement rather than just a number.

IF . . . THEN is used on page 579. Even though some versions of BASIC permit a THEN to be followed by LET, PRINT, and other statements, here we use only line numbers, which is common to all forms. The flag tactic is used in line 20. Be sure to emphasize that the flag should be a number that very likely will not show up as a possible answer.

The FOR . . . NEXT loop is explained on pages 580–581. INT is also introduced and used in two ways. One is to get an integral answer to a division (line 140), and the other is to round a number (line 240). To understand the first usage, students should be aware that each answer in a long division is actually a rounded integer. To understand why line 240 makes the INT function round a number, ask the students to compare INT(2.3 + .5), which rounds to 2, and INT(2.75 + .5), which rounds to 3.

SQR and ABS are the topics on pages 582–583. They are used in a program to simplify radicals. In Exercise 3 on page 583, students may have difficulty with rounding to the nearest hundredth. Explain that by multiplying by 100 they can then use INT (including the +.5) and then divide by 100 to get the decimal point back to where it came from.

ON...GOTO is a simple extension of the GOTO command. It does away with the necessity of several IF...THENs. There is, however, the restriction that the testing values must be natural numbers and must begin with 1. On page 584 the conversion of values of powers of i is possible since the values used are 0 to 3, which can be easily converted to 1 to 4 by adding 1.

The random number function, RND, is explained and used on page 585. Four random numbers are generated in the program, which then asks the student to evaluate a 2 by 2 determinant.

Concerning the comment about rounding errors on page 582, the same thing is true about raising to a power, especially 3 or higher. Therefore, in the exercises given on page 577 students might want to use x*x*x instead of x^3.

Assignments for Reinforcement and Maintenance

Following is a suggested schedule of assignments for 170 school days. It is designed to provide a program of reinforcement and maintenance of the major skills and concepts of Algebra 2 with Trigonometry.

Each day's assignments provide a double dimension.

In a spiral view, there is provision for:

1. REINFORCEMENT of the current lesson (major part),
2. MAINTENANCE of the previous lesson,
3. MAINTENANCE of the lesson taught two days ago.

In a lateral view, there is a provision for:

A BASIC COURSE,
B AVERAGE COURSE,
C ENRICHED COURSE.

The reinforcement and maintenance program is provided as a guide for you to adopt and use in the way that you discern to be best for your class.

Included in the text are optional materials that are ancillary to the program. You may use these materials in a variety of ways:

1. Additional examples for teaching the lesson,
2. Practice exercises for seat work or chalkboard work,
3. Items for quizzes,
4. Extra testing,
5. Enrichment Exercises.

OPTIONAL MATERIALS		
Practice	Testing	Enrichment
Cumulative Reviews in PE*	Cumulative Review Tests (1–4, 1–8, 1–12, 1–16, and 17–19) in PE	Calculator Activities in PE Non-Routine Problems in PE Computer Activities in PE
Chalkboard Exercises in TE**	College Prep Tests in PE Cumulative Review in TE	Chapter Opener pages in PE Special Applications pages in PE Enrichment sections in TE

*PE: Pupil's Edition   **TE: Teacher's Edition

Assignment Chart

Day	Page	Level A	Page	Level B	Page	Level C
1	3	Orals All Ex. Odd 1–21	3 6	Orals All Ex. Odd 7–25, 29 Reading All Ex. 3, 6, 9, 12, 15 Ex. Odd 17–27	3 6	Ex. Even 16–26 Reading All Ex. Even 16–28
2	6 3	Reading All Ex. 1, 2, 4, 5, 7, 8, 10, 11, 13, 14, 16, 17, 19, 20 Ex. 6, 8, 12	9 3 6	Reading All Ex. Odd 1–7, 11–21 Ex. 12, 22, 30 Ex. 20, 22, 24	9 3 6	Reading All Ex. 16–24, 29 Ex. 25, 27, 29 Ex. 29, 30, 32
3	9 6 3	Reading All Ex. Odd 1–13 Ex. 3, 6, 12 Ex. 10, 14	12 9 6	Reading All Orals All Ex. Odd 5, 7, 13–31 Ex. 14, 16, 22 Ex. 8	12 9 3 6	Ex. Odd 17–33 Ex. 34, 35, 37, 39, 41, 43 Ex. 25, 27 Ex. 28, 30 Ex. 31, 33
4	12 9 6	Orals All Ex. 1–6, 25, 27 Ex. Odd 9–21 Ex. 4, 6, 8 Ex. 9, 15	15–17 12 9	Reading All Orals All Ex. Odd 5, 13–27 Ex. 24, 26, 28 Ex. 20, 26	17 12 9	Ex. Odd 19–29 Ex. 36, 38, 42 Ex. 28, 30
5	15–17 12 9	Reading All Orals All Ex. 1–8, Odd 9–17 Ex. 7, 14, 18 Ex. 2, 12	19 16–17 12	Ex. Odd 3–15 Ex. 22, 24, 26 Ex. 30, 32	19 17 12	Ex. 7–11 Ex. 30, 31, 33 Ex. 40, 44
6	19 16 12	Ex. Odd 1–9 Ex. 10, 12, 14 Ex. 8, 16	20–21 19 17	Reading All Orals All Ex. Odd 5–13 Ex. 8, 10, 12 Ex. 20, 28	20–21 19 17	Reading All Orals All Ex. 6–9 Ex. 12, 13, 15 Ex. 32, 34
7	20–21 19 16–17	Reading All Orals Ex. 1–6 Ex. Odd 1–9 Ex. 2, 4, 8 Ex. 16, 19	24–25 21	Ex. Odd 1–7, 11–27 Ex. 16 Ex. 8, 10, 14	25 21 19	Ex. 22, 23, 25, 29–31, 33–35 Ex. 10, 11, 13 Ex. 14, 16
8	24–25 21 19	Ex. Odd 1–15, 25 Ex. Even 8–16, 17 Ex. 2, 4, 6 Ex. 6, 10	26 24–25 21 19	Applications All Ex. 20, 24, 26 Ex. 6 Ex. 16	26 25 21	Applications All Ex. 32, 36, 38 Ex. 12, 14

Day	Page	Level A	Page	Level B	Page	Level C
9	27	Chapter Review	27	Chapter Review	27	Chapter Review
	28	Chapter Test	28	Chapter Test	28	Chapter Test
10	33	Ex. 1–3, 5–7, 9–11, 13–15, 17–19, 21–23, 25–27	33	Reading All Ex. Even 2–8, 14–20 Ex. 25–27, 29–31, 33–35, 37–39, 41–43	33	Reading All Ex. 27–30, 35–39, 41–43
					36	Orals All Ex. 11, 12, 17–23 49–51, 59–66, 69
11	36	Ex. Odd 1–51	36	Reading All Orals All Ex. 1–4, 7–9, 11–13, 16–26, 29–31, 35, 36, 42–45, 49, 50, 53–55, 57–59, 61–63, 65, 67	39–40	Reading Ex. 1–12, 19–21; Oral Ex. 1–8 Ex. 6, 20, 22–24, 28, 34
	33	Ex. 8, 16, 24			42	Ex. Even 26–38 Ex. 39–43
					33	Ex. 25, 40, 44
			33	Ex. 21, 28, 36	36	Ex. 52, 68, 70
12	36	Ex. Even 2–52	39–40	Reading All Orals All Ex. Odd 3–31	45–46	Ex. 4, 7–11, 14, 16
	33	Ex. 12, 20, 28			40	Ex. 25, 35
					42	Ex. 37, 44
			36	Ex. 51, 60, 64	36	Ex. 71
			33	Ex. 22, 32		
13	39–40	Reading All Orals All Ex. Odd 1–21	42	Orals All Ex. Odd 13–37	49	Ex. 7–11
					46	Ex. 12, 15, 17
			40	Ex. 24, 26, 30	40	Ex. 33
	36	Ex. 53, 57, 61	36	Ex. 52, 68	42	Ex. 45
14	42	Orals All Ex. Odd 1–23	45–46	Ex. 1–4, 6, 8, 10, 14	52	Ex. Odd 1–23 Ex. Even 12–24
	40	Ex. 6, 10, 14	42	Ex. 24, 28, 36	49	Ex. 12, 13
	36	Ex. 54, 62	40	Ex. 4, 32	46	Ex. 13, 18
15	45–46	Ex. 1–4, 6, 8, 10	49	Ex. Odd 3–11	53	Chapter Review
	42	Ex. 2, 16, 20	45–46	Ex. 5, 11, 16	54	Chapter Test
	40	Ex. 4, 16	42	Ex. 30, 38		
16	49	Ex. 1, 3, 5, 9	52	Ex. 1–12, 18, 29	59	Ex. Even 40–58 Ex. 61–68
	45–46	Ex. 5, 7, 11	49	Ex. 4, 8, 10		
	42	Ex. 18, 22	46	Ex. 13, 15	61	Ex. 18–21, 25–27, 29–31
17	53	Chapter Review	52	Ex. 13–17, 19–20	64	Reading All Ex. Even 14–34 Ex. 35–37, 41, 43
	54	Chapter Test	49	Ex. 6, 12		
					59	Ex. 69, 71, 73
					61	Ex. 24, 28, 32

T–107

Day	Page	Level A	Page	Level B	Page	Level C
18	59	Orals All Ex. 1–18 Ex. Odd 19–35	53 54	Chapter Review Chapter Test	67 64 59	Reading All Orals All Ex. Even 20–44 Ex. 45–47, 51–53 Ex. 25, 38, 42 Ex. 70, 72, 74
19	61 59	Ex. 1–16 Ex. 28, 30, 34	59	Orals All Ex. Odd 25–61	70 67 64	Ex. Even 38–56 Ex. 58–60, 64 Ex. 41, 48, 49 Ex. 39, 44
20	64 61 59	Ex. Odd 1–21 Ex. 6, 8, 10, 22 Ex. 17, 20, 21 Ex. 32, 36	61 64 59	Ex. 5–8, 11–13, 21–23, 28–30 Reading All Orals All Ex. 3–35 Ex. 46, 48, 62	73 70 67	Ex. Even 20–32, 36, 38 Ex. 40–42 Ex. 61, 62, 65 Ex. 43, 50
21	67 64 61	Reading All Orals All Ex. 1–11 Ex. Odd 13–19 Ex. 12, 14, 16 Ex. 18, 22	67 61 64	Reading All Orals All Ex. Odd 7–19, 23–31, 35–43 Ex. 19, 20 Ex. 16, 26	76–77 73 70	Orals All Ex. 16–18, 20, 21 Ex. 34, 39, 44 Ex. 63, 66
22	70 67 64	Reading All Ex. Odd 1–35 Ex. 12, 14, 16 Ex. 4, 26	70 67 61 64	Reading All Ex. 3, 9, 27, 29, 33, 35, 37, 41, 43, 47, 49, 53 Ex. 38, 40 Ex. 27 Ex. 28, 36	81–82 77 73	Ex. 14, 16, 18, 20, 21, 23 Ex. 22, 23, 25 Ex. 43, 45
23	73 70 67	Reading All Orals All Ex. 1–7, 9, 13, 15, 25 Ex. 4, 14, 16 Ex. 18, 20	73 70 67	Reading All Orals All Ex. 3, 5, 11, 13, 15, 23, 25, 27, 31, 37 Ex. 39, 51, 55 Ex. 14, 42	83 82 77	Applications All Ex. 24–26 Ex. 24, 26
24	76 73 70	Orals All Ex. Odd 1–13 Ex. 8, 14, 26 Ex. 6, 20	76–77 73 70	Orals All Ex. Odd 1–5, 9, 13–21 Ex. 16, 26, 39 Ex. 42, 46	84 85	Chapter Review Chapter Test

Day	Page	Level A	Page	Level B	Page	Level C
25	80–82	Ex. Odd 1–15 Non-Routine Prob.	81	Ex. Odd 7, 9, 13–21 Ex. 22	90	Ex. 1–4, 8, 11, 15, 16, 19, 20, 25
	76	Ex. 2, 8, 12	76–77	Ex. 7, 18, 20	92–93	Reading All
	73	Ex. 10, 16	73	Ex. 14, 30		Ex. Even 4–8, 12, 20–30
						Ex. 31–34, 38
26	84	Chapter Review	84	Chapter Review	95	Ex. 7–17, 19
	85	Chapter Test	85	Chapter Test	90	Ex. 22, 27
					93	Ex. 35, 36, 39
27	90	Ex. 1–8, 10	90	Ex. 1–7, 13, 14, 17, 18,	98	Ex. Even 18–32
	93	Reading Ex. 6–13		21, 22, 25		Ex. 35–37
		Ex. 1–3, 5	92–93	Reading Ex. 1–16	95	Ex. 18, 20, 21
		Ex. Odd 7–23		Ex. Odd 1–31	90	Ex. 28
					93	Ex. 37, 40
28	95	Ex. 1–9, 11, 14–16	95	Ex. 1–11	101	Ex. 1, 3, 14, 16, 18,
	90	Ex. 9, 11		Ex. Odd 13–19		19, 22, 25
	93	Ex. 4, 8, 18	90	Ex. 11, 12, 21	98	Ex. 38–40
			93	Ex. 26, 28	95	Ex. 22
29	98	Orals All	98	Orals All	105	Ex. Even 2–22
		Ex. Odd 1–23		Ex. 7, 9, 11, 15, 17,	101	Ex. 21, 23, 26
	95	Ex. 10, 12, 17		21, 23, 27, 29, 33	98	Ex. 41, 42
	90	Ex. 12	95	Ex. 14, 18, 20		
	93	Ex. 6, 20	90	Ex. 23, 26		
30	101	Ex. Odd 1–15	101	Ex. 1, 5, 11, 15, 17	106–107	Applications All
	98	Ex. 10, 14, 20		Ex. Odd 21–27	105	Ex. 9, 13, 19
	95	Ex. 13, 18	98	Ex. 20, 34, 35	101	Ex. 24, 27
			95	Ex. 12, 16		
31	108	Chapter Review	105	Ex. Even 2–22	108	Chapter Review
	109	Chapter Test	101	Ex. 18, 20, 26	109	Chapter Test
			98	Ex. 14, 18		
32	118	Reading All	108	Chapter Review	118	Reading All
		Ex. Odd 1–23	109	Chapter Test		Ex. 9–11, 15, 21, 24, 27, 29–31, 34
33	121	Orals All	118	Reading All	121	Orals All
		Ex. Odd 1–9		Ex. 1, 11, 13, 19		Ex. 5–9
	118	Ex. 14, 18, 22		Ex. Odd 23–37	118	Ex. 12, 32, 38
34	123–124	Reading All	121	Orals All	123–124	Reading All
		Ex. 1, 3, 5, 6		Ex. Odd 1–11		Ex. 2, 4, 5–8
	121	Ex. 2, 6, 10	118	Ex. 15, 26, 28, 34	121	Ex. 10, 11
	118	Ex. 20, 24			118	Ex. 36, 39

Day	Page	Level A	Page	Level B	Page	Level C
35	126–127	Orals All Ex. Odd 1–25	123–124	Reading All Ex. Odd 1–11	126–127	Ex. 8, 14, 20, 24, 25, 28, 32, 34, 38–43
	124	Ex. 2, 4	121	Ex. 2, 4, 10	124	Ex. 9, 11
	121	Ex. 4, 8	118	Ex. 14, 30	121	Ex. 12
36	129	Orals All Ex. Odd 1–11	126–127	Orals All Ex. 3, 9 Ex. Odd 11–15, 23–39	129	Orals All Ex. Even 8–16 Ex. 17–21
	126–127	Ex. 4, 8, 14, 22			127	Ex. 37, 44, 46
	124	Ex. 8	124	Ex. 2, 4, 12	124	Ex. 10, 12
			121	Ex. 6, 12		
37	132–133	Orals All Ex. Odd 1–19, 29, 33, 35	129	Orals All Ex. Odd 5–19	132–133	Reading All Orals All Ex. Even 4–8, 14–18, 26–32 Ex. 33–36
			127	Ex. 22, 30, 36		
	129	Ex. 10, 12	124	Ex. 6, 8		
	126–127	Ex. 6, 24				
					129	Ex. 22, 23
					127	Ex. 45, 47
38	134	Chapter Review	132–133	Ex. 1, 5, 7, 13 Ex. Odd 19–25, 29–35	134	Chapter Review
	135	Chapter Test			135	Chapter Test
			129	Ex. 10, 14, 16		
			127	Ex. 24, 38		
39	141	Reading Ex. 1–12 Ex. 1–3, 5–7, 9–11, 13–15, 17–19, 21–23, 25–27, 29–31, 33–35	134	Chapter Review	141	Reading All Ex. 6, 8, 14, 16–36, 38–41
			135	Chapter Test		
40	143–144	Reading Ex. 1–10 Ex. 1–15	141	Reading All Ex. 3, 7, 15, 17–34, 37–39, 43	143–144	Reading All Ex. 6, 8, 10, 12–16 Ex. Even 18–30 Ex. 33–36
	141	Ex. 4, 8, 20, 24				
					141	Ex. 42, 43, 45
41	147	Orals All Ex. Odd 1–27	143–144	Reading All Ex. Odd 1–31	147	Ex. Even 26–40 Ex. 41–49, 51–54
	144	Reading Ex. 11–12 Ex. 16–18	141	Ex. 35, 40, 41	144	Ex. 32, 37, 39
	141	Ex. 12, 16			141	Ex. 44, 46
42	150–151	Reading All Orals All Ex. Odd 1–27	147	Orals All Ex. 3, 5 Ex. Odd 9–13, 17–43	150–151	Reading All Orals All Ex. 4, 8, 12, 16, 20, 42 Ex. Even 24–38
	147	Ex. 4, 14, 24	144	Ex. 14, 16, 22	147	Ex. 50, 55, 57
	144	Ex. 19, 20	141	Ex. 36, 42	144	Ex. 38, 40

Day	Page	Level A	Page	Level B	Page	Level C
43	154	Ex. Odd 1–35	150–151	Reading All Orals All Ex. Odd 1–7, 15–37	154	Ex. 26, 28, 32 Ex. Even 36–54 Ex. 55, 56, 59–61, 63
	151	Ex. 18, 22, 26	147	Ex. 32, 36, 44	151	Ex. 27, 40, 41
	147	Ex. 6, 28	144	Ex. 20, 24	147	Ex. 56, 58
44	156	Ex. Odd 1–47	154	Ex. Odd 3, 7–11, 15, 19–23, 27, 31, 35–57	156	Ex. 4–6, 11, 12, 22–24, 28–31, 34–38, 41–43, 47, 48
	154	Ex. 10, 30, 32				
	151	Ex. 20, 24	151	Ex. 18, 30, 34	154	Ex. 53, 57, 64
			147	Ex. 34, 38	151	Ex. 29, 44
45	158	Ex. Odd 1–27	156	Ex. 4, 5, 9–12, 16–23, 26–35, 37–39, 42–46	158–159	Ex. 25–29 Ex. Even 30–44 Ex. Odd 45–67
	156	Ex. 16, 28, 40, 46				
	154	Ex. 28, 36				
			154	Ex. 44, 50, 52, 56	156	Ex. 33, 39, 40, 44–46
			151	Ex. 28, 36	154	Ex. 58, 65
46	160	Chapter Review	158	Ex. Odd 13–19 Ex. 21–28 Ex. Odd 29–43	160	Chapter Review
	161	Chapter Test			161	Chapter Test
			156	Ex. 36, 41, 47, 48		
			154	Ex. 54, 57		
47	166	Orals All Ex. Odd 1–13	160	Chapter Review	166	Orals All Ex. 3, 6, 9, 12, 18, 24, 25, 27, 29
			161	Chapter Test		
48	168	Ex. 1, 3, 5, 7–9	166	Orals All Ex. 4, 5, 7, 11, 17, 20, 23, 26	168	Ex. 2, 4–9
	166	Ex. 2, 4, 8			166	Ex. 21, 23, 26
49	171	Reading All Orals All Ex. 1–6 Ex. Odd 7–19	168	Ex. 1–4, 7–9, 11	171	Reading All Ex. 1–6, 12, 15, 18, 27, 30, 33, 36 Ex. Odd 37–45
			166	Ex. 6, 8, 16		
	168	Ex. 2, 4, 10			168	Ex. 10, 11
	166	Ex. 6, 14			166	Ex. 22, 28
50	173	Ex. Odd 1–7	171	Ex. 1–7, 10, 13, 16, 19, 22, 25, 28, 31, 34, 35	173–174	Ex. 4, 6, 7, 9, 10, 12–14
	171	Ex. 8, 10, 16				
	168	Ex. 6			171	Ex. 32, 38
	182	Ex. 9	168	Ex. 5, 10, 12	168	Ex. 12
			166	Ex. 19		
51	176–177	Ex. 1, 3, 5	173–174	Ex. 3, 5, 7, 9–11, 13	176–177	Ex. Odd 3–11
	173	Ex. 2, 6	171	Ex. 17, 32, 36	174	Ex. 8, 11, 15
	171	Ex. 12, 18	168	Ex. 6	171	Ex. 34, 40
			182	Ex. 9		

Day	Page	Level A	Page	Level B	Page	Level C
52	180	Orals All Ex. Odd 1–23	176–177	Ex. Odd 1, 3, 7–11	180	Orals All Ex. Even 18–22, 26, 28
	176–177	Ex. 2, 4, 6	174	Ex. 8, 12	177	Ex. 4, 8, 10
	173–174	Ex. 4, 8	171	Ex. 18, 33	174	Ex. 16
53	182	Chapter Review	180	Orals All Ex. Odd 9–13, 21–25 Ex. 29, 31, 35, 37, 41, 43, 47	181	Applications All
	183	Chapter Test			180	Ex. 12, 42, 51
					177	Ex. 12
			177	Ex. 4, 8, 10		
			173	Ex. 4, 6		
54	189	Ex. Odd 1–39	181	Applications All	182	Chapter Review
			180	Ex. 20, 33, 38, 40	183	Chapter Test
			176–177	Ex. 2, 12		
55	192	Ex. Odd 1–35	182	Chapter Review	188–189	Reading All Ex. 13–15, 19–21, 27–30 Ex. Even 32–48 Ex. 49–53, 55
	189	Ex. Even 12–18 Ex. 28, 38	183	Chapter Test		
56	194	Ex. Odd 1–23	188–189	Orals All Ex. Odd 1–47 Ex. 49–54	192	Reading All Ex. Even 2, 4, 8, 12–24, 28–52 Ex. 53–55
	192	Ex. 6, 20, 22, 26, 34				
	189	Ex. 20, 26			189	Ex. 35, 41, 54, 56
57	197	Reading All Ex. Odd 1–17	192	Reading All Ex. Odd 1–9, 13–23, 27–37, 41– 45, 49, 51	194	Ex. Even 10–16 Ex. 22, 23 Ex. Even 26–40
	194	Ex. 4, 16, 20			192	Ex. 37, 43, 56
	192	Ex. 24, 32, 36	189	Ex. 34, 40, 44	189	Ex. 39, 47
58	200	Ex. Odd 1–23	194	Ex. Odd 3, 7–13, 17–33	197	Reading All Ex. Even 4–8, 14–32
	197	Ex. 2, 6, 14, 16				
	194	Ex. 18, 22	192	Ex. 34, 42, 46	194	Ex. 18, 24, 35, 41
			198	Ex. 38, 48	192	Ex. 39, 43
59	203	Chapter Review	197	Ex. 2; Odd 3–7, 13–17, 21–27	200	Ex. Even 2–10, 14–18, 24–30, 34, 36
	204	Chapter Test				
			194	Ex. 22, 26, 28		Ex. 17, 29, 31
			192	Ex. 40, 44	194	Ex. 39, 41
60	211	Reading Ex. 1–7 Ex. 1–13, 14, 16, 17, 20, 23, 24, 26, 28, 29	200	Ex. Odd 1–9, 17, 21–29, 33, 35 Ex. 10, 14	201–202	Applications All
					200	Ex. 25, 35, 37, 39
			197	Ex. 16, 22, 24	197	Ex. 15, 25
	201	Ex. 9–11	194	Ex. 16, 24		
	200	Ex. 4, 8				

Day	Page	Level A	Page	Level B	Page	Level C
61	215	Reading Ex. 1–8 Ex. 1–26	203 204	Chapter Review Chapter Test	203 204	Chapter Review Chapter Test
	211	Ex. 15, 18, 25				
	201	Ex. 12				
62	219	Reading Ex. 1–5 Ex. Even 2–26, 28–36	211	Reading Ex. 1–7 Ex. 6–13, 15, 18, 21, 22, 24, 30–33, 38, 39, 42, 43	211	Reading Ex. 1–7 Ex. 6–13, 15, 18, 21, 22, 24, 30–33, 38, 39, 42, 43
	215	Ex. 27, 28, 29			201	Ex. 6, 10, 22
	211	Ex. 19, 27			197	Ex. 10, 13
63	221–222	Orals All Ex. Even 2–20, 21–26	215	Reading Ex. 1–8 Ex. Even 2–22 Ex. 25–38	215	Reading Ex. 1–8 Ex. Odd 1–23 Ex. 27, 28, 32–44
	219	Ex. 3, 7, 15	211	Ex. 14, 16, 17	211	Ex. 14, 20, 40
	215	Ex. 30, 31			200	Ex. 10, 26
64	225	Ex. 5–7, 10, 13, 16, 19, 23–37, 43–47	219	Reading Ex. 1–5 Ex. 1–13, 28, 31, 34, 37–43	219	Reading Ex. 1–5 Ex. 14–27, 30, 33, 36, 41, 43–52
	221–222	Ex. 7, 15, 17	215	Ex. 5, 9, 23	215	Ex. 8, 14, 26
	219	Ex. 5, 27	211	Ex. 19, 25	211	Ex. 16, 36
65	227	Orals All Ex. 1–19, 26–31	221–222	Orals All Ex. Odd 1–9 Ex. 25–32, 34, 35, 37, 38	221–222	Orals All Ex. Odd 11–21 Ex. 22, 29, 32, 33, 36, 39–42
	225	Ex. 9, 15, 21				
	221–222	Ex. 3, 19				
			219	Ex. 14, 19, 35	219	Ex. 8, 28, 42
			215	Ex. 7, 17	215	Ex. 6, 20
66	230	Ex. 1–21	225	Ex. 5–7, 10, 13, 16, 19, 23–37, 43–47	225	Ex. 5, 6, 9, 12, 15, 18, 21, 25, 28, 31, 34, 37, 43–55, 57
	227	Ex. 20, 25	222	Ex. 8, 15, 20		
	225	Ex. 11, 20	219	Ex. 16, 29	221–222	Ex. 10, 24, 35
					219	Ex. 34, 40
67	236	Chapter Review Ex. Even 2–40	227	Orals All Ex. 3, 6, 9, 12, 15, 17–37	227	Orals All Ex. 2, 5, 8, 11, 14, 18, 21, 22, 25, 28, 30, 34, 36–42
	237	Chapter Test				
			225	Ex. 8, 15, 20	225	Ex. 10, 16, 26
			222	Ex. 10, 16	221–222	Ex. 14, 38
68	241–242	Reading Ex. 1–4 Ex. 1–9	230	Ex. Even 2–18 Ex. 19–30	230	Ex. 4–6, 10–12, 16–18, 22–24, 28–34
	230	Ex. 22, 23, 25	227	Ex. 8, 10, 14	227	Ex. 16, 24, 32
	221	Ex. 21, 24	225	Ex. 12, 17	225	Ex. 14, 30

Day	Page	Level A	Page	Level B	Page	Level C
69	246	Ex. 1–16	233	Ex. 2, 4, 5–10	233	Ex. 4–10; Even 12–18
	241–242	Ex. 10–12	234	Applications All	234	Applications 2, 4, 6
	230	Ex. 24, 26	230	Ex. 5, 7, 11	230	Ex. 8, 14, 25
			227	Ex. 7, 11	227	Ex. 20, 26
70	249	Reading Ex. 1–4	236	Chapter Review	236	Chapter Review
		Ex. 1–16	237	Chapter Test	237	Chapter Test
	246	Ex. 17, 18				
	241–242	Ex. 13, 14				
71	252	Orals All	241–242	Reading Ex. 1–4	241–242	Reading Ex. 1–4
		Ex. 1–23		Ex. 1–9; Odd 11–13		Ex. Even 2–14, 16–22
	249	Ex. 7, 8, 10	233	Ex. 1, 12, 13	233	Ex. 13, 15, 19
	246	Ex. 19, 20	230	Ex. 9, 15	230	Ex. 27
72	256	Ex. 1–11	246	Ex. Odd 1–19	246	Ex. Even 2–20
	252	Ex. 24, 25	242	Ex. 10, 12, 14	241–242	Ex. 3, 11, 15
	249	Ex. 9, 11	233	Ex. 3, 11	234	Ex. 7
73	271	Ex. 1–6	249	Reading Ex. 1–4	249	Reading Ex. 1–4
	255	Ex. 12–14		Ex. Odd 1–15		Ex. Even 2–16
	252	Ex. 26, 27	246	Ex. 2, 6, 12	246	Ex. 3, 9, 17
					241–242	Ex. 5, 13
74	272	Chapter Review	252	Ex. Odd 1–31	252	Ex. Even 2–28
		Ex. Even 2–28	249	Ex. 4, 8, 12		Ex. 30–33
	273	Chapter Test	246	Ex. 4, 16	249	Ex. 5, 9, 13
					246	Ex. 15, 19
75	278	Orals All	256	Ex. Odd 1–17	256	Ex. 5–12
		Ex. 1–16	252	Ex. 2, 10, 14		Ex. Even 16–20
	271	Ex. 7, 8	249	Ex. 2, 6	252	Ex. 13, 17, 25
	255	Ex. 15, 16			249	Ex. 11, 15
76	281	Ex. 1–18; Odd 19–33	259	Ex. 1–7; Even 10–16	259	Ex. 1–4, 6–8, 10–15
	278	Ex. 17, 18	256	Ex. 4, 8, 10	256	Ex. 13, 15, 17
	271	Ex. 9, 10	252	Ex. 4, 12	252	Ex. 23, 27
77	284	Reading Ex. 1–4	263	Ex. Odd 1–11	263	Ex. Even 2–14; Odd
		Ex. Odd 1–27		Ex. 12–21		17–21; Ex. 22–27
	281	Ex. 11–13	259	Ex. 8, 13, 15	259	Ex. 5, 9
	278	Ex. 19	256	Ex. 2, 12	256	Ex. 4, 19
78	287	Reading Ex. 1–5	267	Ex. Odd 1–9	267	Ex. Even 2–26
		Ex. 1–18		Ex. 10–15	263	Ex. 15, 16, 18
	284	Ex. 4, 8, 12	263	Ex. 4, 8, 10	259	Ex. 16
	281	Ex. 15, 17	259	Ex. 9, 11		

Day	Page	Level A	Page	Level B	Page	Level C
79	290	Orals All Ex. Even 2–30	271	Ex. Even 2–12	271	Ex. Odd 1–13
	287	Ex. 19, 20, 24	267	Ex. 2, 6, 16	267	Ex. 7, 11, 19
	284	Ex. 2, 18	263	Ex. 2, 6	263	Ex. 11, 20
80	293	Orals All Ex. Odd 1–25	272	Chapter Review	272	Chapter Review
	290	Ex. 5, 7, 17	273	Chapter Test	273	Chapter Test
	287	Ex. 21, 25				
81	296	Ex. 1–17	278	Orals All	278	Orals All
	293	Ex. 4, 6, 16	271	Ex. 13, 15, 16	271	Ex. 12
	290	Ex. 9, 19	267	Ex. 4, 8	267	Ex. 21
82	298	Chapter Review Ex. Even 2–40	281	Ex. Odd 1–51	281	Ex. Even 2–52
			278	Ex. 13, 15, 16		Ex. 53–56
			271	Ex. 7, 9	278	Ex. 11, 15, 17
					271	Ex. 10
83	299	Chapter Test	284	Reading Ex. 1–4 Ex. Odd 1–27	284	Reading Ex. 1–4 Ex. Even 2–24
			281	Ex. 4, 10, 20		Ex. 25–33
			278	Ex. 12, 14	281	Ex. 39, 43, 47
					278	Ex. 9, 13
84	305	Orals All Ex. 1–27	287	Reading Ex. 1–5 Ex. Odd 1–17, 19–27 Non-Routine Prob.	287	Reading Ex. 1–5 Ex. Even 2–26, 28–38 Non-Routine Prob.
	296	Ex. 17, 19	284	Ex. 6, 10, 20	284	Ex. 11, 17, 25
	293	Ex. 10, 18	281	Ex. 6, 16	281	Ex. 37, 45
85	309	Orals All Ex. Odd 1–43	290	Orals All	290	Orals All Ex. Even 2–22
	305	Ex. 28–30	287	Ex. 4, 8, 12		Ex. 26–33
	296	Ex. 18, 21	284	Ex. 8, 26	287	Ex. 13, 17, 21
					284	Ex. 15, 23
86	313	Ex. 1–18	293	Orals All Ex. Even 2–26	293	Orals All Ex. Even 2–18, 20–26
	309	Ex. 4, 8, 12	290	Ex. 4, 8, 22	290	Ex. 5, 7, 19
	305	Ex. 31, 33	287	Ex. 6, 10	287	Ex. 15, 27
87	318	Ex. 1–15	296	Ex. Odd 1–21	296	Ex. Even 2–20
	313	Ex. 19, 23	297	Applications All		Ex. 22–25
	309	Ex. 6, 16	293	Ex. 3, 7, 17	293	Ex. 11, 17, 19
			290	Ex. 6, 16	290	Ex. 13, 21
88	323	Ex. Even 2–36	298	Chapter Review	297	Applications Even
	318	Ex. 16, 17, 19	299	Chapter Test	296	Ex. 15, 17, 21
	313	Ex. 21, 25			293	Ex. 9, 15

Day	Page	Level A	Page	Level B	Page	Level C
89	326	Ex. 1–27	305	Orals All	298	Chapter Review
	323	Ex. 5, 19, 25		Ex. 1–21, 29–37	299	Chapter Test
	318	Ex. 18, 20	296	Ex. 4, 12, 16		
			293	Ex. 9, 19		
90	330	Ex. Even 10–16	309	Orals All	305	Orals All
		Ex. 17–20, 23, 24		Ex. Odd 1–53		Ex. Even 2–32
		Ex. 1–8	305	Ex. 22–24	297	Ex. 1, 3
	326	Ex. 2, 4, 20	296	Ex. 6, 14	296	Ex. 5, 7
	323	Ex. 7, 27				
91	332	Chapter Review	313	Ex. Odd 1–27	309	Orals All
		Ex. Even 2–50	309	Ex. 4, 12, 26		Ex. Even 2–38
			305	Ex. 25, 27		Ex. 39–42
					305	Ex. 5, 13, 21
					297	Ex. 5, 7
92	333	Chapter Test	318	Ex. Even 2–12	313	Ex. Even 2–28
				Ex. 13–25		Ex. 31–36, 39–41
			313	Ex. 4, 12, 26	309	Ex. 7, 15, 25
			309	Ex. 10, 18	305	Ex. 7, 9
93	339–340	Orals All	323	Ex. Odd 1–39	318	Ex. Odd 1–11, 19–43
		Ex. 1, 2, 5, 6, 9, 10,	318	Ex. 5, 9, 11		Ex. Even 14–18
		13, 14, 17, 18, 21, 22	313	Ex. 2, 16	313	Ex. 9, 19
	330	Ex. 3, 5, 11			309	Ex. 17
	326	Ex. 6, 18				
94	343	Orals All	326	Ex. Odd 1–43	323	Ex. Even 2–48
		Ex. 1, 2, 5, 6, 9, 10,	323	Ex. 6, 26, 38		Ex. 49–53, 56–62
		13, 14	318	Ex. 3, 7		Non-Routine Prob.
	339–340	Ex. 3, 7, 12			318	Ex. 10, 15, 17
	330	Ex. 7, 15			313	Ex. 23, 25
95	345	Ex. 1–6	330	Ex. Odd 1–29	326	Ex. Even 2–58
	343	Ex. 3, 7, 10	326	Ex. 4, 18, 28	323	Ex. 11, 13, 37
	339–340	Ex. 4, 11	323	Ex. 18, 28	318	Ex. 8, 13, 44
96	348	Reading Ex. 1–8	332	Chapter Review	330	Ex. Even 2–30
		Ex. 1, 2, 5, 6, 9, 10,	333	Chapter Test		Ex. 31, 32
		13, 14, 17, 18, 21,			326	Ex. 5, 27, 43
		22, 25, 26			313	Ex. 31, 55
	345	Ex. 7, 8				
	343	Ex. 4, 8				
97	351	Ex. 1, 2, 5, 6, 9, 10,	339–340	Ex. 3, 4, 7, 8, 11,	332	Chapter Review
		13, 14, 17, 18, 21,		12, 15, 16, 19, 20,	333	Chapter Test
		22, 25, 26, 29, 30		23, 24		
	348	Ex. 3, 12, 24	330	Ex. 6, 12, 14		
	345	Ex. 9, 10	326	Ex. 6, 10		

Day	Page	Level A	Page	Level B	Page	Level C
98	352	Chapter Review Ex. Even 2–32	343 340 330	Orals All Ex. 3, 4, 7, 8, 11, 12, 15–17 Ex. 5, 9, 17 Ex. 4, 16	339–340 330 326	Orals All Ex. 5–8, 13–16, 21–28 Non-Routine Prob. Ex. 7, 11, 19 Ex. 4, 11, 23
99	353	Chapter Test	345 343 340	Ex. Even 2–12 Ex. 2, 6, 10 Ex. 10, 18	343 339–340 330	Orals All Ex. Even 2–26 Ex. 17–19 Ex. 9, 17, 20 Ex. 5, 17
100	359–360 351 348	Reading Ex. 1–4 Orals All Ex. Odd 1–15 Ex. 33, 35 Ex. 4, 8, 16 Ex. 4, 15	348 345 343	Reading Ex. 1–8 Ex. 3, 4, 7, 8, 11, 12, 15, 16, 19, 20, 23, 24, 27, 28, 31–34 Ex. 1, 3, 7 Ex. 1, 13	345 343 339–340	Ex. Even 1–12 Ex. 3, 7, 15 Ex. 10, 19
101	362–363 359–360 351	Orals All Ex. Odd 1–13 Ex. 2, 6, 12, 32 Ex. 7, 15	351 348 345	Orals All Ex. 3, 4, 7, 8, 11, 12, 15, 16, 19, 20, 23, 24, 27, 28, 31–35 Ex. 2, 6, 18 Ex. 5, 11	348 345 343	Reading Ex. 1–8 Ex. 9–16, 21–24, 29–36 Ex. 3, 5, 9 Ex. 5, 9
102	366–367 362–363 359–360	Oral Ex. 1–3, 9–11 Ex. Odd 1–9, 27, 31 Ex. 2, 10, 14 Ex. 8, 14, 34	352 353	Chapter Review Chapter Test	351 348 345	Orals All Ex. 3, 4, 7, 8, 11, 12, 15, 16, 19, 20, 23, 24, 27, 28, 31, 32, 36–40 Ex. 17, 18, 25 Ex. 7, 11
103	367 362–363	Oral Ex. 4–8, 12–14 Ex. 11–14, 17, 19 Ex. 4, 12, 15	359 351 348	Reading All, Orals All Ex. 1, 2, 7–12, 16, 18, 20, 21, 25, 26, 28, 29, 32–34, 36, 37 Ex. 2, 10, 18 Ex. 10, 14	352 353	Chapter Review Chapter Test
104	370–371 367	Reading Ex. 1–4 Orals All Ex. Odd 1–25, 43 Ex. 15, 16, 18, 32	362–363 359–360 351	Orals All Ex. 5–7, 9–11, 25, 27 Ex. Odd 13–21 Ex. 3, 13, 17, 35 Ex. 1, 13	359–360 351 348	Reading All Orals All Ex. 2, 6, 8 Ex. Even 12–34, 38–48 Ex. 5, 9, 29 Ex. 19, 26

T–117

Day	Page	Level A	Page	Level B	Page	Level C
105	374	Ex. Odd 1–25	366–367	Orals All	362–363	Orals All
	370–371	Ex. 2, 8, 20, 44		Ex. 3, 5, 7, 8, 13–17, 19, 22, 23, 25, 27, 28, 29, 31		Ex. Even 6–18, 22–32
			362–363	Ex. 12, 14, 16	359–360	Ex. 21, 33, 35, 43
			359	Ex. 5, 14		
106	377–378	Reading All Orals All	370–371	Reading All, Orals All	366–367	Orals All Ex. 5–8, 13–15, 19–23, 26–31, 33, 34, 36
		Ex. Odd 1–25, 43		Ex. 1, 3, 5, 7, 10, 13, 14, 20–23, 27–29, 31, 37, 39, 43, 44, 46		
	374	Ex. 2, 4, 24			363	Ex. 15, 17, 29
	370–371	Ex. 10, 22			360	Ex. 49
			367	Ex. 24, 30, 32		
			362	Ex. 8		
107	381	Ex. Odd 1–17	374	Ex. Odd 1, 11–31	370–371	Ex. 3, 10–12, 23–26 Ex. Even 30–42
	377–378	Ex. 4, 6, 12, 24	370–371	Ex. 11, 24, 45		
	374	Ex. 8, 16, 26	367	Ex. 10, 21	367	Ex. 16, 24, 32
					363	Ex. 13, 25
108	383	Ex. 4–7, 10–13, 22, 23	377–378	Reading All, Orals All	374	Ex. Even 12–36 Ex. Odd 37–45
	381	Ex. 2, 6, 16		Ex. Odd 5–15	371	Ex. 44, 46, 51
	378	Ex. 8, 14, 26		Ex. 25–28, 31, 35, 37, 41, 43	367	Ex. 35, 37
			374	Ex. 24, 33, 35		
			370	Ex. 26, 30		
109	386	Chapter Review	381	Ex. Odd 1–9 Ex. Even 14–30	377–378	Ex. 2, 8, 21–23, 28–35, 41, 43–45
			378	Ex. 14, 29, 42	374	Ex. 25, 27, 38
			374	Ex. 26, 30	370–371	Ex. 22, 53
110	387	Chapter Test	383	Ex. 3, 6, 11, 13, 15–17, 20, 22, 23	381	Ex. 1–3, 31, 32, 34 Ex. Even 16–26
			381	Ex. 19, 23, 29	378	Ex. 27, 38, 42
			378	Ex. 30, 44	374	Ex. 23, 46
111	391	Ex. 1, 4, 7, 10	385	Ex. Odd 1–13; Ex. 4	383	Ex. Even 6–28
			383	Ex. 12, 19, 24	381	Ex. 19, 33, 35
			381	Ex. 6, 17	378	Ex. 14, 46
112	393	Ex. 1–8, 12–17, 21, 22	386	Chapter Review	385	Ex. Even 2–10 Ex. 13–16
	391	Ex. 2, 5, 11	387	Chapter Test	383	Ex. 15, 23, 29
113	396	Ex. 1, 3, 5–7, 10	391	Ex. 2, 5, 8, 11, 13	386	Chapter Review
	393	Ex. 9, 10, 18, 23			387	Chapter Test
	391	Ex. 3, 6				

Day	Page	Level A	Page	Level B	Page	Level C
114	399	Orals All Ex. Odd 1–19	393	Ex. Odd 1–11 Ex. 13–18, 21, 22, 25, 26	391	Ex. 3, 6, 8, 10, 16
	396	Ex. 2, 9				
	393	Ex. 11, 24	391	Ex. 9, 12, 14		
115	402	Ex. 1–4, 7, 8, 10–13, 16, 18	396	Reading All Ex. 1–3, 5, 6, 11, 14, 17, 19	393	Ex. 2–5, 8–11, 14–19, 23–27, 29–31, 33
	399	Ex. 6, 8, 12, 18			391	Ex. 9, 15, 17
	396	Ex. 4, 8	393	Ex. 2, 19, 23		
			391	Ex. 7		
116	405	Reading Ex. 10–26 Ex. 1–6	399	Ex. 6, 9, 12–15 Ex. Odd 17–29	396	Reading All Ex. 1–3, 7, 10, 13, 14, 17, 19, 21
	402	Ex. 5, 9, 14, 19	396	Ex. 4, 12, 18		
	399	Ex. 2, 14, 16	393	Ex. 20, 24	393	Ex. 20, 28, 34
					391	Ex. 14
117	408	Reading Ex. 1–3 Ex. 1–3, 6, 7, 12, 15	402	Ex. 3, 6, 7, 9, 11–14, 16, 18, 21, 23, 26, 28, 31–33	399	Orals All Ex. Odd 9, 13–31
	405	Ex. 7–9			396	Ex. 4, 11, 16, 18
	402	Ex. 6, 15, 17	399	Ex. 22, 28, 31	393	Ex. 32, 35
			396	Ex. 15, 20		
118	589	Reading All Ex. 1, 4, 14 Ex. Odd 5–11	405	Reading Ex. 1–3, 5, 10–26; Ex. 1–4, 7–10	402	Ex. 8–14, 40 Ex. Even 18–36
			402	Ex. 20, 22, 27	399	Ex. 22, 26, 28
	408	Ex. 5, 8, 14, 16	399	Ex. 20, 30	396	Ex. 20, 22
119	411	Reading Ex. 1–3, 5, 10 Ex. 1, 3, 4, 7, 11, 13	408	Reading All Ex. Odd 3–9, 15–19	405	Reading All Ex. 1–12
	589	Ex. 3, 6, 8, 15		Ex. 12	408	Reading Ex. 3, 4 Ex. Even 8–20; Ex. 21
	408	Ex. 4, 17	405	Ex. 5, 6, 11		
			402	Ex. 24, 35	402	Ex. 21, 31
					399	Ex. 32
120	415	Ex. Odd 1–9 Ex. 10	589	Reading All Ex. 5, 8, 12, 15, 16, 19, 22	589	Reading All Ex. 5, 12, 15, 19, 21–23
	411	Ex. 2, 6, 12			405	Ex. 13–15
	589	Ex. 16	408	Ex. 10, 14, 16	408	Ex. 9, 17, 24, 25
			405	Ex. 12	402	Ex. 23, 37, 41
121	417	Reading All Ex. 1–6	411	Reading All Ex. 1, 3, 6, 7, 11, 13	411	Reading All Ex. Even 2, 6–14
	415	Ex. 2, 6, 8	589	Ex. 14, 23		Ex. 17
	411	Ex. 5, 8	408	Ex. 8, 18	408	Ex. 15
122	420	Chapter Review	415	Ex. 3, 4, 6, 7, 9, 13, 15, 17, 18	415	Ex. 2, 4, 6, 7, 10, 11, 14, 16, 17, 19, 21, 25
			411	Ex. 4, 8, 14	411	Ex. 15, 16, 19
					408	Ex. 15

T–119

Day	Page	Level A	Page	Level B	Page	Level C
123	421	Chapter Test	417	Reading All Ex. 1, 2, 5, 7, 9, 10	417	Reading All Ex. 1, 2, 7–10
			415	Ex. 8, 16, 19	415	Ex. 18, 23, 24
					411	Ex. 18, 20
124	427	Orals All Ex. 1–5, 7, 9–12, 15	419	Applications Ex. 2, 5, 6, 9–11	419	Ex. 4–11
			417	Ex. 3, 8, 12	417	Ex. 6, 11, 12
					415	Ex. 20, 26
125	429–430	Ex. Odd 1–13	420	Chapter Review	420	Chapter Review
	427	Ex. 6, 8, 13	421	Chapter Test	421	Chapter Test
126	432	Ex. Odd 1–17	427	Orals All Ex. 1–5, 7, 9–12, 15, 17–20	427	Ex. 1–15
	429	Ex. Even 2–8				
	427	Ex. 14, 16				
127	433	Reading All	429–430	Ex. Odd 1–13 Ex. 6, 14, 17, 21	429–430	Ex. 1–3, 5–7 Ex. Odd 9–21
	434	Orals All Ex. 1, 3, 5	427	Ex. 6, 8, 13	427	Ex. 16, 17, 19
	432	Ex. Even 4–12				
	430	Ex. 10, 12, 14				
128	436–437	Reading All Ex. 1–7, 9, 11, 13	432	Ex. 1–7, 12–15, 18	432	Ex. 3–5, 8, 9, 14–18
	434	Ex. 2, 4	429–430	Ex. 2, 4, 8, 12	433–434	Orals All, Reading All Ex. Odd 1–9
	432	Ex. 14, 16	427	Ex. 14, 16	430	Ex. 16, 18
					427	Ex. 20
129	440	Ex. 1–8	433–434	Orals All Reading All Ex. 1–5, 7, 9	436–437	Reading All Ex. 3–9, 11, 13–15, 17, 19, 21
	437	Ex. 14, 15	432	Ex. 8, 16, 19	432	Ex. 13, 19
	434	Ex. 6, 7	430	Ex. 15, 18	434	Ex. 8, 10
					430	Ex. 20, 22
130	447	Ex. 1–4	436–437	Reading All Ex. 1–15	440	Ex. Even 2–16
	440	Ex. 9–11	434	Ex. 8, 10	437	Ex. 10, 16, 18
	437	Ex. 16, 17	432	Ex. 9, 17	434	Ex. 6
131	448	Chapter Review Ex. Even 2–22	440	Ex. Even 2–8 Ex. 9–12	444	Ex. Odd 1–61
			437	Ex. 16, 17	440	Ex. 1, 5, 9
			434	Ex. 6	433–434	Ex. 15, 16
132	449	Chapter Test	444	Ex. Even 2–30	447	Ex. Even 2–28
			440	Ex. 3, 7, 13	444	Ex. 10, 22, 44
			437	Ex. 18	440	Ex. 7, 11

Day	Page	Level A	Page	Level B	Page	Level C
133	457	Reading Ex. 1–5 Orals All; Ex. 1–4	447	Ex. 2, 4–12, 15, 16, 19–21	448 449	Chapter Review Chapter Test
	447	Ex. 5, 6	444	Ex. 5, 13, 42		
	440	Ex. 12, 13	440	Ex. 1, 7		
134	457	Ex. Odd 5–41	448	Chapter Review	457	Reading Ex. 1–5 Ex. Even 2–42
	447	Ex. 7, 8			447	Ex. 3, 7, 13
					444	Ex. 14, 26, 48
135	460–461	Ex. 1–6	449	Chapter Test	460–461	Ex. 1–16, 22–23
	457	Ex. 8, 10			457	Ex. 7, 19, 31
	447	Ex. 9, 10			447	Ex. 5, 11, 19
136	461	Ex. Even 8–16 Ex. 7, 13	457	Reading Ex. 1–5 Orals All Ex. Odd 1–15 Ex. 25–30, 37–45	465	Reading Ex. 1–5 Ex. Even 2–48 Ex. 49–54
	457	Ex. 12, 14			460–461	Ex. 17, 19, 21
			447	Ex. 3, 13, 18	457	Ex. 17, 29
			444	Ex. 1, 27		
137	465	Reading Ex. 1–5 Ex. 1–6	460–461	Orals All Ex. Odd 1–21	469	Ex. 1–6, 19–24, 31–36 Ex. Even 38–50 Ex. 52–55
	461	Ex. 9, 11, 15	457	Ex. 16, 23, 36		
	457	Ex. 26, 28	447	Ex. 5, 17	465	Ex. 15, 27, 39
					460–461	Ex. 18, 20
138	465	Ex. 7–35	465	Reading Ex. 1–5 Ex. Odd 1–43 Ex. 45–50	472	Orals All Ex. Even 2–44 Ex. 45–48
	457	Ex. 16, 18, 20				
			461	Ex. 2, 4, 12	469	Ex. 7, 25, 39
			457	Ex. 18, 24	465	Ex. 17, 43
139	469	Ex. Odd 1–29	469	Ex. Odd 1–43 Ex. 44–51	474	Ex. Even 2–42 Ex. 44–47
	465	Ex. 36, 38, 41				
	457	Ex. 22, 24	465	Ex. 2, 16, 40	472	Ex. 9, 23, 37
			461	Ex. 6, 14	469	Ex. 17, 41
140	469	Ex. 4, 8, 22 Ex. Odd 31–41 Ex. 44–46	472	Ex. Odd 1–35 Ex. 37–44 Non-Routine Prob.	476	Orals All Ex. Even 2–20 Ex. 21–30
	465	Ex. 37, 39	469	Ex. 4, 8, 28	474	Ex. 11, 19, 27
			465	Ex. 6, 22	472	Ex. 7, 35
141	472	Orals All Ex. Odd 1–19	474	Ex. Odd 1–29	480	Ex. 5, 10, 15, 20, 25, 28, 31, 34, 37, 45–47, 49, 52, 55, 67–89 Ex. Even 58–66
			472	Ex. 6, 16, 24		
	469	Ex. 6, 14, 32	469	Ex. 10, 26		
	465	Ex. 40, 44				
					476	Ex. 3, 13, 19
					474	Ex. 3, 35

T–121

Day	Page	Level A	Page	Level B	Page	Level C
142	472	Ex. 20–36, 38	476	Orals All	482	Ex. Even 2–26, 28–42
		Ex. 4, 8, 16		Ex. Even 2–12	480	Ex. 9, 33, 53
	469	Ex. 12, 16, 38		Ex. 13–24	476	Ex. 7, 17
			474	Ex. 2, 18, 26		
			472	Ex. 4, 14		
143	474	Ex. Odd 1–25, 29	480	Ex. Odd 1–47	485	Ex. 4, 8, 12, 15, 18,
	472	Ex. 10, 18, 37		Ex. 48–66		21, 25–27, 34–39,
	469	Ex. 18, 24, 34	476	Ex. 5, 7, 13		43–52
			474	Ex. 4, 22	482	Ex. 9, 23, 27
					490	Ex. 53, 57
144	474	Ex. Even 2–30, 40	482	Ex. Odd 1–9	488	Orals All
		Ex. 3, 5, 7		Ex. 11–30		Ex. Even 2–26
	472	Ex. 2, 6	480	Ex. 6, 14, 46		Ex. 27–32
			476	Ex. 9, 17	485	Ex. 7, 16, 30
					482	Ex. 17, 19
145	476	Orals All	485	Ex. 8–12, 18–21,	490–491	Orals All
		Ex. Odd 1–13		25–45		Ex. Even 2–14
	474	Ex. 9, 11, 13	482	Ex. 2, 4, 6		Ex. 16–19, 26–35
			480	Ex. 12, 40	488	Ex. 7, 15, 21
					485	Ex. 11, 22
146	476	Ex. Even 2–14	488	Orals All	492	Chapter Review
		Ex. 15–17, 19		Ex. 1–15, 21–26	493	Chapter Test
	474	Ex. 15, 17	490–491	Orals All		
				Ex. Odd 1–15		
				Ex. 16–28		
			485	Ex. 4, 13, 22		
			482	Ex. 8, 10		
147	480	Ex. Even 2–40	492	Chapter Review	499	Ex. 1–6, 21–30
		Ex. 49, 57				Ex. Even 8–18
	476	Ex. 18, 20, 22			490–491	Ex. 7, 11, 21
					488	Ex. 9, 23
148	480	Ex. Odd 1–43	493	Chapter Test	502	Orals All
		Ex. 48, 60, 61				Ex. 3, 6, 8, 10, 16,
		Ex. 42, 46, 50				18–30
	476	Ex. 21, 23			499	Ex. 7, 15, 17
					490	Ex. 9, 25
149	482	Ex. Odd 1–15	499	Ex. 1–14	505	Orals All
		Ex. 20	488	Ex. 17–19		Ex. 4, 9–12, 17–27
	480	Ex. 44, 45, 47	485	Ex. 7, 14		Non-Routine Prob.
					502	Ex. 5, 11, 17
					499	Ex. 11, 19

Day	Page	Level A	Page	Level B	Page	Level C
150	482	Ex. Even 2–16 Ex. 17–19, 21	502 505 499 491	Ex. 2, 4, 6, 10–20, 25, 26 Ex. Even 2–12 Ex. 13–22 Ex. 15, 17, 19 Ex. 10, 12	508 505 502	Reading Ex. 1–3 Ex. 6–8, 14–16, 21–34 Ex. 3, 5, 15 Ex. 7, 13
151	485 482	Ex. Odd 1–17 Ex. 22, 26 Ex. 22–24	508 502 499	Reading Ex. 1–3 Ex. Even 2–16 Ex. 17–26 Ex. 3, 7, 8 Ex. 16, 18	511 508 505	Ex. 1–7, 15–18, 20, 24–26, 28, 31, 33, 35–38 Ex. 5, 11, 17 Ex. 1, 16
152	485	Ex. Even 2–18 Ex. 19–21, 23, 27	511 508 505	Ex. 1–7, 27–34 Ex. Even 8–26 Ex. 1, 7, 15 Ex. 1, 11	514 511 508	Ex. 4, 8, 11, 12, 19–31 Ex. 9, 19, 27 Ex. 9, 18
153	488 485	Orals All Ex. Odd 1–13, 23 Ex. 24, 25, 28	514 511 508	Ex. 1–4, 8–10, 16–20, 22–27 Ex. 9, 11, 25 Ex. 3, 9	517 514 511	Orals All Ex. 10–15, 22–33 Ex. 1, 10, 13 Ex. 10, 21
154	488 485	Ex. Even 2–14, 24 Ex. 29–31	517 514 511	Orals All Ex. Odd 1–15, 16–24 Ex. 5, 7, 21 Ex. 13, 21	520 517 514	Ex. Even 2–28 Ex. 29–32 Ex. 7, 9, 16 Ex. 9, 15
155	490–491 488	Orals All Ex. Odd 1–17 Ex. 15–17	520 517 514	Ex. 1–4, 21–28, 30 Ex. Even 6–20 Ex. 4, 12, 14 Ex. 6, 11	522 520 517	Ex. 4–6, 15–18 Ex. Even 8–14 Ex. 3, 9, 13 Ex. 2, 17
156	491 488	Ex. Even 2–18 Ex. 23 Ex. 18–20	522 520 517	Ex. Even 2–14 Ex. 5, 11, 19 Ex. 6, 10	515 522 520	Ex. 1–28 Ex. Even 30–42 Ex. 3, 9, 13 Ex. 1, 15
157	492	Chapter Review Ex. Even 2–66	525 522 520	Reading Ex. 1–5 Ex. Even 1–36 Ex. 1, 7, 9 Ex. 7, 17	528 525 522	Ex. 1–10, 23–26, 41–56 Ex. Odd 29–39 Ex. 31, 33, 37 Ex. 5, 11
158	493	Chapter Test	528 525 522	Ex. 1–6, 27, 28 Ex. Even 8–26, 30–40 Ex. 3, 15, 21 Ex. 3, 11	532 534 528 525	Reading Ex. 1–6 Ex. 1–17, 19, 21–23, 30–35 Ex. Even 2–8 Ex. 11, 15, 27 Ex. 29, 35

T–123

Day	Page	Level A	Page	Level B	Page	Level C
159	520	Ex. Even 2–20	532	Reading Ex. 1–6	535	Chapter Review
	491	Ex. 19–21		Ex. 5–29	536	Chapter Test
	488	Ex. 21, 22	534	Ex. Odd 1–7		
			528	Ex. 9, 15, 25		
			525	Ex. 5, 19		
160	525	Reading Ex. 1–5	535	Chapter Review	541	Orals All
		Ex. 1–29, 31	536	Chapter Test		Ex. Even 2–18
	520	Ex. 3, 5, 7				Ex. 19–28
	491	Ex. 22			532	Ex. 18, 20, 24
					528	Ex. 12, 32
161	202	Chapter Review	541	Orals All	543	Ex. Even 2–14
		Ex. Odd 1–29		Ex. Odd 1–9, 11–22		Ex. 15, 16, 18, 20, 21
			532	Ex. 1, 3, 4	541	Ex. 3, 11, 19
			528	Ex. 3, 13	534	Ex. 1, 7
162	236	Chapter Review	543	Ex. Odd 1–13	545	Orals All
		Ex. Odd 1–39		Ex. 15–19		Ex. Even 2–14
			541	Ex. 4, 8, 10		Ex. 15–18
			534	Ex. 2, 4	543	Ex. 7, 13, 17
					544	Ex. 7, 15
163	272	Chapter Review	545	Orals All	548	Orals All
		Ex. Odd 1–27		Ex. 2, 3, 5, 9–16		Ex. Even 2–12
			543	Ex. 4, 14		Ex. 13–16
			541	Ex. 2, 6	545	Ex. 5, 7, 13
					543	Ex. 5, 11
164	298	Chapter Review	548	Orals All	551	Ex. Odd 1–9
		Ex. Odd 1–39		Ex. Even 2–8, 10–14	548	Ex. 5, 7, 9
			551	Ex. Even 2–10	545	Ex. 3, 11
			545	Ex. 4, 6, 8		
			543	Ex. 2, 12		
165	332	Chapter Review	554–555	Orals All	554–555	Orals All
		Ex. Odd 1–51		Ex. 1–10, 26–36		Ex. Even 2–34
				Ex. Odd 11–25		Ex. 35–46
			548	Ex. 1, 3, 7	551	Ex. 4, 6, 8
			545	Ex. 1, 7	548	Ex. 3, 11
166	352	Chapter Review	559	Ex. 1–10, 24–28, 32	559	Ex. 2, 7, 9, 10, 14–
		Ex. Odd 1–31		Ex. Odd 11–23		16, 18, 20, 22, 26–40
			555	Ex. 12, 14, 18	554–555	Ex. 5, 13, 19
			548	Ex. 5, 9	551	Ex. 2, 10
167	386	Chapter Review	562	Ex. Odd 1–31	562	Ex. Even 2–32
		Ex. Odd 1–39		Ex. 34, 35		Ex. 33–38
			559	Ex. 12, 14, 16	559	Ex. 3, 11, 19
			555	Ex. 16, 20	554–555	Ex. 3, 23

Day	Page	Level A	Page	Level B	Page	Level C
168	420	Chapter Review Ex. Odd 1–41	565 562 559	Ex. Odd 1–21 Ex. 23–29, 33, 34 Ex. 4, 12, 22 Ex. 18, 20	565 562 559	Ex. Even 2–28 Ex. 29–36 Non-Routine Prob. Ex. 3, 11, 19 Ex. 5, 21
169	448	Chapter Review Ex. Odd 1–67	567 565 562	Ex. 1–12 Ex. 4, 8, 20 Ex. 14, 24	567 565 562	Ex. 1–10 Ex. Even 12–18 Ex. 7, 15, 19 Ex. 5, 13
170	493	Chapter Review Ex. Odd 1–67	568 569	Chapter Review Chapter Test	568 569	Chapter Review Chapter Test

See pages 601–622 in the back of this book for answers too long for student pages.

Answers for Computer Section

Page 575
1.
```
10 PRINT 7
20 PRINT 3 / 4 * (124 / 3) - 24
30 PRINT 126 - ( - 15.75)
40 PRINT ( - 9) * ( - 15.75)
50 PRINT 2 * (7 * 5 - 4) - 4 * (2 * 5 - 6)
60 PRINT 3 * 5 + 31
70 PRINT 6 * (3 + 2 * ( - 12)) - (7 * ( - 12) - 2)
80 PRINT 3 * ( - 12) - 4
```

Page 576
1.
```
10 LET W = - 4.5
20 PRINT - 27: PRINT 12 * W + 27
25 LET N = 10
30 PRINT 3 * (2 * N + 4) + 5 * (3 - 2 * N): PRINT 7 - 2 * N
40 LET Y = 7
50 PRINT 5 * Y - (2 - 3 * Y): PRINT 54
```
2.
```
10 A = - 3
20 PRINT 10 - 4 * (6 * (2 * A + 5) - (20 + 10 * A)) - 2 * A
30 PRINT 0
```
3.
```
10 LET X = 3
20 PRINT 4 * (X - 3) - (X + 6)
30 PRINT 3 * (X - 6)
```

Page 577
1.
```
10 INPUT X
20 PRINT 9 * X ^ 4 + 8 * X ^ 3 + 7 * X ^ 2 + 6 * X + 5
30 PRINT 4 * X ^ 5 + 4 * X ^ 3 + 4 * X
40 PRINT - X ^ 5 - 2 * X ^ 4 - 3 * X ^ 3 - 4 * X ^ 2 - 5 * X - 6
50 PRINT - 7 * X ^ 6 - 7 * X ^ 4 - 7 * X ^ 2 - 7
60 GOTO 10
```
2.
```
10 READ X
20 PRINT 9 * X ^ 4 + 8 * X ^ 3 + 7 * X ^ 2 + 6 * X + 5
30 PRINT 4 * X ^ 5 + 4 * X ^ 3 + 4 * X
40 PRINT - X ^ 5 - 2 * X ^ 4 - 3 * X ^ 3 - 4 * X ^ 2 - 5 * X - 6
50 PRINT - 7 * X ^ 6 - 7 * X ^ 4 - 7 * X ^ 2 - 7
60 GOTO 10
70 DATA 1,2,3,4,5,6,7,8,9,10
```

Page 578
1.
```
10 INPUT A,B
20 INPUT C,D
30 PRINT A * C;"X^2 ";A * D + C * B;"X +";B * D
40 GOTO 10
```
2.
```
10 INPUT A,B
20 INPUT C,D,E
30 PRINT A * C;"X^3 +";A * D + B * C;"X^2 +";B * D + A * E;"X +";B * E
40 GOTO 10
```
3.
```
10 INPUT A,B
20 INPUT C,D
30 PRINT A * C;"X^2 ";
40 IF A * D + C * B < 0 THEN 60
50 PRINT "+";
60 PRINT A * D + B * C;"X ";
70 IF B * D < 0 THEN 90
80 PRINT "+";
90 PRINT B * D
95 GOTO 10
```

Page 579
1.
```
10 READ X
20 IF X = - 999 THEN 80
30 IF X * X * X * X - 26 * X * X + 25 = 0 THEN 60
40 PRINT "THAT SOLUTION WAS NOT CORRECT."
50 GOTO 10
60 PRINT "THAT WAS A CORRECT SOLUTION."
70 GOTO 10
80 STOP
90 DATA 5,-5,1,-1,-999
```

For Exercise 34, change lines 30 and 90 in the above program as follows:

30 IF $(X - 7) * (X - 7) = 29 - X * X$
90 DATA 5, 2, −999

For Exercise 37, change lines 30 and 90 as follows:

30 IF $2 * B * B - 7 * B + 6 = 0$ THEN 60
90 DATA 2, 1.5, −999

2. 10 INPUT X
   20 IF X = - 999 THEN 80
   30 IF X * (X + 5) * (X - 4) = X * X * X THEN 60
   40 PRINT "THAT SOLUTION WAS NOT CORRECT."
   50 GOTO 10
   60 PRINT "THAT WAS A CORRECT SOLUTION."
   70 GOTO 10
   80 STOP

For Exercise 56, change line 30 in the above program as follows:

30 IF 3 * Y * (Y − 2) * (Y + 4) =
    3 * Y * Y * Y THEN 60

For Exercise 57, change line 30 in the above program as follows:

30 IF (2 * X * X + 3) * (2 * X * X + 3)
    = 4 * X * (X * X * X + 6) THEN 60

Page 581

1. 110 PRINT "NUMERATOR";: INPUT N
   120 PRINT "DENOMINATOR";: INPUT D
   130 PRINT "HOW MANY DECIMAL PLACES?";: INPUT P
   140 A = INT (N / D)
   150 R = N − A * D
   160 IF R = 0 THEN 270
   170 FOR I = 1 TO P − 1
   180 N = R * 10
   190 A = INT (N / D)
   200 R = N − A * D
   210 PRINT A;
   220 IF R = 0 THEN 270
   230 NEXT I
   240 N = R * 10
   250 A = INT (N / D + .5)
   260 PRINT A
   270 PRINT : GOTO 110

2. 10 FOR I = 1 TO 25
   20 PRINT I,I * I,I * I * I
   30 NEXT I

3. 10 PRINT "WHAT IS YOUR NAME?";: INPUT N$
   20 PRINT "HOW OLD ARE YOU?";: INPUT A
   30 FOR I = 1 TO A
   40 PRINT I;" HAPPY BIRTHDAY, ";N$
   50 NEXT I

Page 583

1. 100 PRINT "WHAT IS THE RADICAND TO BE SIMPLIFIED?";: INPUT R
   110 C = 1:S = R
   120 IF ABS (S ^ (1 / 3) − INT (S ^ (1 / 3))) < .05 THEN 190
   130 FOR I = 2 TO SQR (S)
   140 IF ABS (S / I / I / I − INT (S / I / I / I)) > .005 THEN 160
   150 C = C * I:S = S / I / I / I: GOTO 140
   160 NEXT I
   170 PRINT "3RD ROOT(";R;")=";C;" * 3RD ROOT(";S;")"
   180 GOTO 100
   190 XX
   200 GOTO 100

2. 100 PRINT "WHAT IS THE RADICAND TO BE SIMPLIFIED?";: INPUT R
   110 C = 1:S = R
   120 IF ABS (S ^ .25 − INT (S ^ .25)) < .05 THEN 190
   130 FOR I = 2 TO S ^ .25
   140 IF ABS (S / I / I / I / I − INT (S / I / I / I / I)) > .005 THEN 160
   150 C = C * I:S = S / I / I / I / I: GOTO 140
   160 NEXT I
   170 PRINT "4TH RT (";R;") = ";C;" * 4TH RT(";S;")"
   180 GOTO 100
   190 PRINT "4TH RT(";S;") = "; INT (S ^ .25)
   200 GOTO 100

4. 100 PRINT "WHAT IS THE RADICAND TO BE SIMPLIFIED?";: INPUT R
   110 C = 1:S = R
   120 IF ABS (ABS (S) ^ (1 / 3) − INT (ABS (S)) ^ (1 / 3))) < .05 THEN 190
   130 FOR I = 2 TO SQR (ABS (S))
   140 IF ABS (S / I / I / I − INT (S / I / I / I)) > .005 THEN 160
   150 C = C * I:S = S / I / I / I: GOTO 140
   160 NEXT I
   170 PRINT "3RD ROOT(";R;")=";C;" * 3RD ROOT(";S;")"
   180 GOTO 100
   190 PRINT "3RD RT (";S;") = ";
   200 IF S > 0 THEN 220
   210 PRINT "−";
   220 PRINT INT (ABS (S) ^ (1 / 3))
   230 GOTO 100

5. 10 FOR I = 1 TO 10
   20 PRINT I; TAB(3);I * I; TAB(7);I * I * I; TAB(15); SQR (I); TAB(28);I ^ (1 / 3)
   40 NEXT I

6. 10 PRINT "WHAT ARE A,B, AND C?";: INPUT A,B,C
   15 IF B ^ 2 - 4 * A * C < 0 THEN 80
   20 X = (- B + SQR (B ^ 2 - 4 * A * C)) / (2 * A)
   30 Y = (- B - SQR (B ^ 2 - 4 * A * C)) / (2 * A)
   40 X = INT (X * 100 + .5) / 100
   50 Y = INT (Y * 100 + .5) / 100
   60 PRINT "THE SOLUTIONS OF ";A;" X^2+ ";B;"X+"; C;"=0 ARE ";X;" AND ";Y
   70 GOTO 10
   80 PRINT "THERE ARE NO REAL SOLUTIONS.": GOTO 10

Page 584
1. 10 PRINT "THIS PROGRAM WILL MULTIPLY TWO COMPLEX"
   20 PRINT "NUMBERS OF THE FORM (A+Bi)(C+Di)"
   30 PRINT "WHAT ARE A,B?";: INPUT A,B
   40 PRINT "WHAT ARE C,D?";: INPUT C,D
   50 PRINT "(";A;"+";B;"i")(";C;"+";D;"i") = ";A * C - B * D;"+";B * C + A * D;"i"
   60 PRINT : GOTO 10

Page 585
1. 110 PRINT "NUMERATOR";: INPUT N
   120 A(I) = INT (RND (1) * 19) - 9: NEXT I
   130 PRINT "HOW MANY DECIMAL PLACES";: INPUT P
   140 A = INT (N / D
   150 R = N - A * D
   160 PRINT N"/"D" = "A".";
   170 FOR I = 1 TO P - 1
   180 N = R * 10
   190 A = INT (N / D)
   200 R = N - A * D
   210 PRINT A;
   220 NEXT I
   230 N = R * 10
   240 A = INT (N / D + .5)
   250 PRINT A
   260 GOTO 110

2. 10 FOR I = 1 TO 9
   20 A(I) = INT (RND (1) * 12) - 3: NEXT I
   30 PRINT "WHAT IS THE VALUE OF THE DETERMINANT?"
   40 FOR I = 1 TO 3: PRINT TAB(8);A(I); TAB(13);A(I + 3); TAB(18);A(I + 6): NEXT I
   50 P = A(1) * A(5) * A(9) + A(4) * A(8) * A(3) + A(7) * A(2) * A(6)
   60 N = A(7) * A(5) * A(3) + A(8) * A(6) * A(1) + A(9) * A(4) * A(2)
   70 INPUT V
   80 IF V = P - N THEN 95
   90 PRINT "NO, THE VALUE IS ";P - N: GOTO 10
   95 PRINT "YES": GOTO 10

Page 586
1. 10 PRINT "THIS PROGRAM WILL SOLVE EQUATIONS OF"
   20 PRINT "THE FORM A^X=B."
   30 PRINT "WHAT IS A?";: INPUT A
   40 PRINT "WHAT IS B?";: INPUT B
   50 PRINT "X = "; LOG (B) / LOG (A)
   60 PRINT : GOTO 10

2. 100 PRINT "THIS PROGRAM WILL APPROXIMATE ONE"
   110 PRINT "IRRATIONAL ROOT OF A QUADRATIC"
   120 PRINT "EQUATION.": PRINT
   130 PRINT "WHAT ARE THE VALUES OF A,B,C";: INPUT A,B,C
   140 IF B * B - 4 * A * C < 0 THEN 270
   150 PRINT "HOW MANY DECIMAL PLACES? (LIMIT 8)";: INPUT N
   160 DEF FN A(X) = A * X * X + B * X + C
   170 L = - 10:U = 10:C = 1:D = FN A(L - C)
   180 FOR M = 1 TO N + 1
   190 FOR I = L TO U STEP C
   200 E = FN A(I): IF D = 0 THEN 260
   210 IF D * E < 0 THEN 260
   220 D = E: NEXT I
   230 PRINT I - C;" ";I
   240 L = I - C:U = I:C = C / 10: NEXT M
   250 GOTO 100
   260 PRINT "X = "I - C: GOTO 100
   270 PRINT "THERE ARE NO REAL SOLUTIONS.": GOTO 100

HOLT
ALGEBRA 2
WITH TRIGONOMETRY

The three photographic special interest sections in this book are designed to spark the imagination of students and stress the interrelationship of progress and mathematics.

Land Exploration provides examples of these relationships ranging from ancient tools and charts to modern computer-aided compilers and plotters.

Air-Space Exploration pictures the state of the art in spacecraft research as well as early inventions in communication.

Undersea and Underground Exploration portrays acoustic sensors for investigating the deep sea and illustrates inventions and techniques used in mining the earth.

Any one of the photographs may serve as a starting point for further investigation about the subject and the mathematics that supports it. You may find it useful to ask for volunteers, either alone or in small groups, to pursue topics illustrated, formulate problems, and look for solutions. Papers can be written and displays made which are then presented to the class or put on bulletin boards.

About the Authors

Eugene D. Nichols
Robert O. Lawton Distinguished Professor
of Mathematics Education
Florida State University
Tallahassee, Florida

Mervine L. Edwards
Chairman of Mathematics Department
Shore Regional High School
West Long Branch, New Jersey

E. Henry Garland
Head of the Mathematics Department
Developmental Research School
DRS Professor
Florida State University
Tallahassee, Florida

Sylvia A. Hoffman
Resource Consultant in Mathematics
Illinois State Board of Education
State of Illinois

Albert Mamary
Superintendent of Schools for Instruction
Johnson City Central School District
Johnson City, New York

William F. Palmer
Professor of Education and Director
Center for Mathematics and Science Education
Catawba College
Salisbury, North Carolina

Computer Consultant

Michael Murphy
Mathematics Department Chairman
Bonny Eagle High School
West Buxton, Maine

HOLT ALGEBRA 2 WITH TRIGONOMETRY

Eugene D. Nichols
Mervine L. Edwards
E. Henry Garland
Sylvia A. Hoffman
Albert Mamary
William F. Palmer

HOLT, RINEHART AND WINSTON, PUBLISHERS
New York · Toronto · Mexico City · London · Sydney · Tokyo

Acknowledgments for Photographs

Cover: Radio telescope from Very Large Array, Dennis Brack/Black Star.

Page x, Michael Melford/Peter Arnold; *30*, Tom Tracy/Medichrome/Stock Shop; *56*, HRW photo by Richard Haynes; *88*, Geoffrey Gove/Image Bank; *114*, HRW photo by Richard Haynes; *138*, S.L. Craig, Jr./Bruce Coleman; *164*, John Iacono/*Sports Illustrated;* *186*, Bettmann; *208*, Nancy J. Pierce/Photo Researchers; *234*, Georg Gerster/Photo Researchers; *239*, Harvey Olsen; *275*, Larry Mulvehill/Photo Researchers; *297*, Erwin & Peggy Bauer/Bruce Coleman; *302*, Allan Seiden/Image Bank; *337*, Heinz Kluetmeier/*Sports Illustrated;* *345*, Judi Buie/Bruce Coleman; *356*, Bettmann; *389*, John Moss/Medichrome/Stock Shop; *423*, Focus on Sports; *454*, Bettmann; *496*, *FDA Consumer;* *538*, William James Warren/West Light.

Insert 1:
Page 1, Harvey Olsen; *2*, *tl*, Robert Harding Picture Library, London; *tr*, Courtesy of the Wellcome Trustees/Robert Harding Picture Library; *b*, Harald Sund/Image Bank; *3*, *tl*, Robert Farber/Image Bank; *r*, British Museum of Ethnography/NYPL Picture Collection; *bl*, Raoul Hackel/Stock, Boston; *4*, *tl*, Granger; *tr*, Granger; *br*, Courtesy of the Trustees, National Gallery, London; *5*, *tc*, Granger; *br*, Earth Satellite Corporation; *bl*, Brown Brothers; *tl*, Harvey Olsen; *r*, Chuck O'Rear/Woodfin Camp & Assoc.; *bl*, Harvey Olsen; *7*, *tl*, Landsat/Rainbow; *tr*, Harvey Olsen; *b*, John Running/After Image; *8*, *t*, James A. Cook/Stock Broker; *b*, Bill Gallery/Stock, Boston.

Insert 2:
Page 1, Harvey Olsen; *2*, *t*, Granger; *br*, Robert Harding Picture Library, London; *bl*, illustration, *London News*/Robert Harding Picture Library; *3*, *tl*, Robert Harding Picture Library; *tr*, Mark Sherman/Bruce Coleman; *br*, Cameron Davidson/Bruce Coleman; *4*, *tl*, Granger; *r*, Granger; *bl*, Chuck O'Rear/West Light; *5*, *tl*, Mark Tuschman/Discover Magazine; *tr*, Jehangir Gazdar/Woodfin Camp & Assoc.; *b*, Nicholas Foster/Image Bank; *6*, *tl*, Terry A. Renna; *tr*, Keith Brofsky/Stock Broker; *bl*,Granger; *7*, *tc*, Werner H. Muller/Peter Arnold; *br*, Harvey Olsen; *bl*, NASA/Grant Heilman; *8*, *t*, Dr. E.P. Miles/Miles Color Art; *b*, E.R. Degginger/Bruce Coleman.

Insert 3:
Page 1, Ron Church/Photo Researchers; *2*, *l*, Bob Evans/Peter Arnold; *tr*, Granger; *br*, Culver; *3*, *t*, Jeff Simon/Bruce Coleman; *br*, Culver; *bl*, Jack Fields/Photo Researchers; *4*, *l*, Culver; *tr*, Granger; *br*, NASA/Photo Researchers; *5*, *tl*, Granger; *r*, Culver; *bl*, Granger; *6*, *l*, Jeff Simon/Bruce Coleman; *br*, Bettmann; *7*, *tl*, Keith Bradley/Woods Hole Oceanographic Institute; *r*, Soames Summerhays/Photo Researchers; *bl*, Shelly Lauzon/Woods Hole Oceanographic Institute; *8*, Granger.

Copyright © 1986 by Holt, Rinehart and Winston, Publishers
All rights reserved
Printed in the United States of America

ISBN: 0-03-002173-1

5678901234   032   98765432

Contents

1 Linear Equations and Inequalities   x

- 1   Linear Equations   1.1
- 4   Linear Inequalities   1.2
- 7   Compound Linear Inequalities   1.3
- 10 Absolute Value   1.4
- 13 Problem Solving: Number Problems   1.5
- 18 Problem Solving: Perimeter Problems   1.6
- 20 Problem Solving: Age Problems   1.7
- 22 Problem Solving: Using Formulas   1.8
- 26 Applications: Properties of a Field
- 27 Chapter One Review
- 28 Chapter One Test
- 29 College Prep Test

2 Polynomials and Problem Solving   30

- 31 Properties of Exponents   2.1
- 34 Zero and Negative Integral Exponents   2.2
- 37 Polynomials and Synthetic Substitution   2.3
- 41 Multiplying Polynomials   2.4
- 43 Problem Solving: Mixture Problems   2.5
- 47 Problem Solving: Motion Problems   2.6
- 50 Introduction to Vectors   2.7
- 53 Chapter Two Review
- 54 Chapter Two Test
- 55 College Prep Test

3 Factoring and Special Products   56

- 57 Factoring Trinomials   3.1
- 60 Special Products   3.2
- 62 Special Factors   3.3
- 65 Combined Types of Factoring   3.4
- 68 Quadratic Equations   3.5
- 71 Quadratic Inequalities   3.6
- 74 Problem Solving: Number Problems   3.7
- 78 Problem Solving: Geometric Problems   3.8
- 83 Applications
- 84 Chapter Three Review
- 85 Chapter Three Test
- 86 Computer Activities
- 87 College Prep Test

4 Rational Expressions   88

- 89   Rational Numbers and Decimals   4.1
- 91   Simplifying Products and Quotients   4.2
- 94   Dimensional Analysis   4.3
- 96   Sums and Differences   4.4
- 99   Complex Rational Expressions   4.5
- 102 Statistics: Averages and Variability   4.6
- 106 Applications: Proofs with Rational Numbers
- 108 Chapter Four Review
- 109 Chapter Four Test
- 110 Computer Activities
- 111 College Prep Test
- 112 Cumulative Review (Chapters 1–4)

5 Using Rational Expressions 114

- 115 Fractional Equations 5.1
- 119 Problem Solving: Work Problems 5.2
- 122 Problem Solving: Round Trips 5.3
- 125 Literal Equations and Formulas 5.4
- 128 Dividing Polynomials 5.5
- 130 Remainder Theorem 5.6
- 134 Chapter Five Review
- 135 Chapter Five Test
- 136 Computer Activities
- 137 College Prep Test

6 Radicals and Rational Number Exponents 138

- 139 Square Roots 6.1
- 142 Simplifying Radicals 6.2
- 145 Sums, Differences, and Products of Radicals 6.3
- 148 Quotients of Radicals 6.4
- 152 Other Radicals 6.5
- 155 Rational Number Exponents 6.6
- 157 Expressions with Rational Number Exponents 6.7
- 160 Chapter Six Review
- 161 Chapter Six Test
- 162 Computer Activities
- 163 College Prep Test

7 Quadratic Formula 164

- 165 Completing the Square 7.1
- 167 Maximum Value 7.2
- 169 The Quadratic Formula 7.3
- 172 Problem Solving: The Quadratic Formula 7.4
- 175 Problem Solving: Irrational Measurements 7.5
- 178 Radical Equations 7.6
- 181 Applications
- 182 Chapter Seven Review
- 183 Chapter Seven Test
- 184 Computer Activities
- 185 College Prep Test

8 Complex Numbers 186

- 187 Sums and Differences of Complex Numbers 8.1
- 190 Products and Quotients of Complex Numbers 8.2
- 193 Equations with Imaginary Number Solutions 8.3
- 195 The Discriminant 8.4
- 198 Sum and Product of Solutions 8.5
- 201 Applications: Proofs with Complex Numbers
- 203 Chapter Eight Review
- 204 Chapter Eight Test
- 205 College Prep Test
- 206 Cumulative Review (Chapters 1–8)

9 Coordinate Geometry 208

- 209 The Coordinate Plane 9.1
- 212 Slope of a Line 9.2
- 216 Equation of a Line 9.3
- 220 Slope-Intercept Method of Drawing Graphs 9.4
- 223 The Distance Between Two Points 9.5
- 226 The Midpoint of a Segment 9.6
- 228 Parallel and Perpendicular Lines 9.7
- 231 Using Coordinate Geometry 9.8
- 234 Applications
- 235 Computer Activities
- 236 Chapter Nine Review
- 237 Chapter Nine Test
- 238 College Prep Test

10 Linear Systems, Matrices, and Determinants 239

- 240 Solving Linear Systems Graphically 10.1
- 243 Solving Linear Systems Algebraically 10.2
- 247 Solving Systems of Three Linear Equations 10.3
- 250 Two-by-Two Determinants 10.4
- 253 Three-by-Three Determinants 10.5
- 257 Matrix Addition 10.6
- 260 Matrix Multiplication 10.7
- 264 Solving Linear Systems Using Matrices 10.8
- 268 Problem Solving Using Linear Systems 10.9
- 272 Chapter Ten Review
- 273 Chapter Ten Test
- 274 College Prep Test

11 Functions 275

- 276 Relations and Functions 11.1
- 279 Values and Composition of Functions 11.2
- 282 Linear and Constant Functions 11.3
- 285 Inverse Relations and Functions 11.4
- 288 Direct Variation 11.5
- 291 Inverse Variation 11.6
- 294 Joint and Combined Variation 11.7
- 297 Applications
- 298 Chapter Eleven Review
- 299 Chapter Eleven Test
- 300 Computer Activities
- 301 College Prep Test

12 Conic Sections 302

- 303 Symmetric Points in a Coordinate Plane 12.1
- 306 The Circle 12.2
- 310 The Ellipse 12.3
- 314 The Parabola 12.4
- 319 The Hyperbola 12.5
- 324 Identifying Conic Sections 12.6
- 327 Equations and Properties of Conic Sections 12.7
- 331 Computer Activities
- 332 Chapter Twelve Review
- 333 Chapter Twelve Test
- 334 College Prep Test
- 335 Cumulative Review (Chapters 1–12)

13 Linear-Quadratic Systems 337

- 338 Solving Linear-Quadratic Systems of Equations 13.1
- 341 Solving Quadratic Systems of Equations 13.2
- 344 Applications
- 346 Solving Linear Inequality Systems 13.3
- 349 Solving Linear-Quadratic Inequality Systems 13.4
- 352 Chapter Thirteen Review
- 353 Chapter Thirteen Test
- 354 Computer Activities
- 355 College Prep Test

14 Progressions, Series, and Binomial Expansions 356

- 357 Arithmetic Progressions 14.1
- 361 The Arithmetic Mean 14.2
- 364 Arithmetic Series and Σ Notation 14.3
- 368 Geometric Progressions 14.4
- 372 The Geometric Mean 14.5
- 375 Geometric Series 14.6
- 379 Infinite Geometric Series 14.7
- 382 Binomial Expansions 14.8
- 384 Special Infinite Series 14.9
- 386 Chapter Fourteen Review
- 387 Chapter Fourteen Test
- 388 College Prep Test

15 Exponential, Logarithmic, and Polynomial Functions 389

- 390 Graphing Exponential Functions 15.1
- 392 Logarithmic Functions 15.2
- 394 Graphing Logarithmic Functions 15.3
- 397 Logs of Products and Quotients 15.4
- 400 Logs of Powers and Radicals 15.5
- 403 Common Logarithms and Antilogarithms 15.6
- 406 Using Exponential Equations 15.7
- 409 Higher-Degree Polynomial Equations 15.8
- 412 Rational Zero Theorem 15.9
- 416 Graphing Polynomial Functions 15.10
- 418 Applications: Descartes' Rule of Signs
- 420 Chapter Fifteen Review
- 421 Chapter Fifteen Test
- 422 College Prep Test

16 Permutations, Combinations, and Probability 423

- 424 Fundamental Counting Principle 16.1
- 428 Conditional Permutations 16.2
- 431 Distinguishable Permutations 16.3
- 433 Circular Permutations 16.4
- 435 Combinations 16.5
- 438 Probability: Simple Events 16.6
- 441 Probability: Compound Events 16.7
- 445 Probability and Arrangements 16.8
- 448 Chapter Sixteen Review
- 449 Chapter Sixteen Test
- 450 Computer Activities
- 451 College Prep Test
- 452 Cumulative Review (Chapters 1–16)

17 Trigonometric Functions 454

- 455 Angles of Rotation 17.1
- 458 Special Right Triangles 17.2
- 462 Sine and Cosine Functions 17.3
- 466 Tangent and Cotangent Functions 17.4
- 470 Secant and Cosecant Functions 17.5
- 473 Functions of Negative Angles 17.6
- 475 Values of Trigonometric Functions 17.7
- 477 Using Trigonometric Tables 17.8
- 481 Linear Trigonometric Equations 17.9
- 483 Interpolation with Trigonometry 17.10
- 486 Trigonometry of a Right Triangle 17.11
- 489 Using Trigonometry of a Right Triangle 17.12
- 492 Chapter Seventeen Review
- 493 Chapter Seventeen Test
- 494 Computer Activities
- 495 College Prep Test

18 Trigonometric Graphs and Identities 496

- 497 Graphing $y = \sin x$ and $y = \cos x$ 18.1
- 500 Determining Amplitude 18.2
- 503 Determining Period 18.3
- 506 Change of Period and Amplitude 18.4
- 509 Graphing $y = \tan x$ and $y = \cot x$ 18.5
- 512 Finding Logarithms of Trigonometric Functions 18.6
- 515 Quadratic Trigonometric Equations 18.7
- 518 Basic Trigonometric Identities 18.8
- 521 Proving Trigonometric Identities 18.9
- 523 Radian Measure 18.10
- 526 Applying Radian Measure 18.11
- 529 Inverses of Trigonometric Functions 18.12
- 533 Complex Numbers and De Moivre's Theorem 18.13
- 535 Chapter Eighteen Review
- 536 Chapter Eighteen Test
- 537 College Prep Test

19 Trigonometric Laws and Formulas 538

- 539 Law of Cosines 19.1
- 542 Using the Law of Cosines to Find Angle Measures 19.2
- 544 Area of a Triangle 19.3
- 546 Law of Sines 19.4
- 549 The Ambiguous Case 19.5
- 552 Sin $(A \pm B)$ and Sin $2A$ 19.6
- 556 Cos $(A \pm B)$ and Cos $2A$ 19.7
- 560 Tan $(A \pm B)$ and Tan $2A$ 19.8
- 563 Half-Angle Formulas 19.9
- 566 Double and Half-Angle Identities 19.10
- 568 Chapter Nineteen Review
- 569 Chapter Nineteen Test
- 570 Computer Activities
- 571 College Prep Test
- 572 Cumulative Review (Chapters 17–19)

- 574 **Computer Basic Section**
- 588 **Computations Using Logarithms**
- 590 **Tables**
- 597 **Glossary and Symbol List**
- 601 **Answers to Selected Exercises**
- 623 **Index**

1 LINEAR EQUATIONS AND INEQUALITIES

Problem Solving

Problem　　Solve the equation $x + y + z = 74$ for integers x, y, and z with the following ratios: $x:y = 2:3$ and $y:z = 5:4$.

Four Steps in Problem Solving

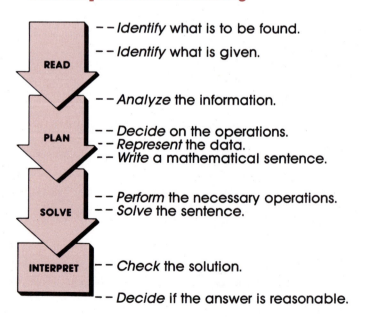

-- *Identify* what is to be found.
-- *Identify* what is given.

-- *Analyze* the information.

-- *Decide* on the operations.
-- *Represent* the data.
-- *Write* a mathematical sentence.

-- *Perform* the necessary operations.
-- *Solve* the sentence.

-- *Check* the solution.

-- *Decide* if the answer is reasonable.

You are to find three numbers such that

(1) they are integers, (2) their sum is 74, and (3) the two ratios are true.

$x:y = 2:3 = 4:6 = 6:9 = 8:12 = 10:15$
$y:z = 5:4 = 10:8 = 15:12$

$x:y:z = 10:15:12 = 10n:15n:12n$
$\qquad 10n + 15n + 12n = 74$

$\qquad\qquad\qquad 37n = 74$
$\qquad\qquad\qquad\ \ n = 2$
$x = 10n = 20,\ y = 15n = 30,\ z = 12n = 24$

$\qquad\quad 20 + 30 + 24 = 74$
$\quad 20:30 = 2:3$ and $30:24 = 5:4$

LINEAR EQUATIONS 1.1

Objective **To solve a linear equation in one variable**

Each of the equations below is a **linear equation** in one variable.
$$6x + 32 - 10x = 7 + 8x - 39 \quad \frac{3}{5}y - 3 = 8 - \frac{1}{5}y \quad 3.8n + 2.3 = 7.58 - 0.6n$$

You can solve a linear equation in one variable by
(1) combining the like terms, if any, on each side of the equation, and
(2) using one or more of the following equation properties.

Equation properties

For all numbers a, b, and c,
Addition: if $a = b$, then $a + c = b + c$.
Subtraction: if $a = b$, then $a - c = b - c$.
Multiplication: if $a = b$, then $a \cdot c = b \cdot c$.
Division: if $a = b$, then $\frac{a}{c} = \frac{b}{c}$; $c \neq 0$.

Example 1 **Find the root of $\frac{2}{5}n + 3 = 7 - \frac{1}{5}n$.**

$\frac{2}{5}n + 3 = 7 - \frac{1}{5}n$

$\frac{3}{5}n + 3 = 7$ ◀ Add $\frac{1}{5}n$ to each side.

$\frac{3}{5}n = 4$ ◀ Subtract 3 from each side.

$\frac{5}{3} \cdot \frac{3}{5}n = \frac{5}{3} \cdot 4$ ◀ Multiply each side by $\frac{5}{3}$,

$1n = 6\frac{2}{3}$ the reciprocal of $\frac{3}{5}$.

Thus, the root is $6\frac{2}{3}$.

Sometimes the variable appears on the right side of the equation but not on the left side. In that case, it may be more convenient to get the variable alone on the right side of the equation as shown below.

Example 2 **Solve $78 = 22 - 8t$. Check the solution by substitution.**

$78 = 22 - 8t$ Check: $\begin{array}{c|c} 78 & 22 - 8t \\ 78 & 22 - 8(-7) \\ & 22 + 56 \\ & 78 \end{array}$

$56 = -8t$

$\frac{56}{-8} = \frac{-8t}{-8}$

$-7 = t$ True: $78 = 78$

Thus, the solution is -7.

Linear Equations

A linear equation may contain parentheses. You can remove parentheses by using the **Distributive Property**.

Distributive property	For all numbers a, b, and c, $a(b + c) = ab + ac$ and $(b + c)a = ba + ca$.

Even though the Distributive Property is stated for multiplication over addition, the property remains true for multiplication over subtraction as well.

Example 3   **Solve $2(3x + 8) - 3(x - 2) = 9x - (4x - 6)$.**

$$2(3x + 8) + (-3)(x - 2) = 9x + (-1)(4x - 6)$$
$$6x + 16 - 3x + 6 = 9x - 4x + 6$$
$$3x + 22 = 5x + 6$$
$$-2x = -16$$
$$x = 8$$

◄ $-(4x - 6) = -1(4x - 6)$
◄ *Use the Distributive Property three times.*

Thus, the solution is 8.

Example 4   **Solve $\frac{1}{3}(6y - 15) - \frac{3}{4}(12y + 16) = 11$.**

$$\frac{1}{3}(6y) + \frac{1}{3}(-15) - \frac{3}{4}(12y) - \frac{3}{4}(16) = 11$$
$$2y - 5 - 9y - 12 = 11$$
$$-7y - 17 = 11$$
$$-7y = 28$$
$$y = -4$$

Thus, the solution is -4.

Example 5   **Solve $0.1 + 1.2(4n - 0.3) = 5(0.3n + 0.74)$.**

$$0.1 + 1.2(4n) + 1.2(-0.3) = 5(0.3n) + 5(0.74)$$
$$0.1 + 4.8n - 0.36 = 1.5n + 3.70$$
$$3.3n = 3.96$$
$$n = 1.2$$

Thus, the solution is 1.2.

Oral Exercises

Simplify each expression.
1. $3(5x - 8)$ **15x − 24**
2. $-4(3y + 11)$ **−12y − 44**
3. $-(5n - 7)$ **−5n + 7**
4. $-(8 + 6t)$ **−8 − 6t**
5. $\frac{1}{2}(6a + 20)$ **3a + 10**
6. $\frac{1}{3}(12x - 27)$ **4x − 9**
7. $0.4(0.3t - 11)$ **0.12t − 4.4**
8. $1.32(10y + 0.1)$ **13.2y + 0.132**

Written Exercises

Solve each equation.

Ⓐ
1. $10 + \frac{2}{3}t = 6$ **−6**
2. $7 = \frac{3}{4}n - 24$ **$41\frac{1}{3}$**
3. $-27 = 12w + 27$ **−4.5**
4. $126 - a = -9a$ **−15.75**
5. $3(2n + 4) + 5(3 - 2n) = 7 - 2n$ **10**
6. $2(7x - 4) - 4(2x - 6) = 3x + 31$ **5**
7. $5y - (2 - 3y) = 54$ **7**
8. $6(3 + 2c) - (7c - 2) = 3c - 4$ **−12**
9. $\frac{1}{4}(12x - 24) - 4x = \frac{1}{3}(18 + 15x) + 24$ **−6**
10. $\frac{1}{5}(30n + 25) - \frac{1}{8}(32n + 16) = 14n - 33$ **3**
11. $0.4t + 3(1.2t + 0.9) = 21.1$ **4.6**
12. $0.7(5a - 1.2) = 2a - 0.39$ **0.3**
13. $6(3.2y - 2.3) = 7.2y - 13.32$ **0.04**
14. $2.6(4x + 2.3) = 8.2x + 49.98$ **20**

Ⓑ
15. $\frac{2}{7}x + 18 = 8 - \frac{3}{7}x$ **−14**
16. $\frac{4}{5}y - 22 = \frac{1}{5}y + 14$ **60**
17. $\frac{2}{9}n - 31 = 16 + \frac{7}{9}n$ **−84.6**
18. $18 - \frac{2}{11}a = \frac{5}{11}a - 19$ **$58\frac{1}{7}$**
19. $3.2n - 0.05 = n + 0.06$ **0.05**
20. $x - 0.01 = 12.2 - 2.3x$ **3.7**
21. $x - (9x - 5) = -(3x + 7)$ **2.4**
22. $-4(2y - 5) - (7 - 3y) = 4(y + 3)$ **$\frac{1}{9}$**
23. $\frac{3}{5}(15n + 20) + 3n = \frac{2}{7}(14n - 28) - 12$ **−4**
24. $21 + \frac{5}{6}(30x + 18) = 19x - \frac{2}{5}(35x - 15)$ **−1.5**
25. $2.5(1.2x + 6) = 1.92 + 8(4.3x + 2.42)$ **−0.2**
26. $7(3.2n + 9.4) = 4.4(3.5n + 4.5) - 0.2$ **−6.6**

Ⓒ
27. $5(2n + 4) - 3(4n - 6) = 2(20 - n)$ **no solution**
28. $4(x - 3) - (x + 6) = 3(x - 6)$ **all numbers**
29. $5[3(5y - 4) - 2(7y + 6)] + 3y - 8 = 0$ **16**
30. $10 - 4[6(2a + 5) - (20 + 10a)] - 2a = 0$ **−3**

NON-ROUTINE PROBLEMS

1. Solve the equation $x - y + z - w = 6$ for integers x, y, z, and w with the following ratios: $x:y = 3:2$, $y:z = 2:5$, and $x:w = 5:8$. **x:15, y:10, z:25, w:24**

2. Mrs. Jones invested $12,000 in three different parts with the ratio 1:2:3. The first part was invested at 11% and the second part at 9%. At what rate did she invest the third part if the yearly income was $1,000? **7%**

Linear Equations

LINEAR INEQUALITIES 1.2

Objectives
To find the solution set of a linear inequality in one variable
To graph the solutions of a linear inequality in one variable

A mathematical sentence such as $5 + 8x > 16$ or $7 - 5x < 27$ is called a **linear inequality** in one variable. The procedures used to solve a linear inequality are similar to those you have been using to solve a linear equation, with one major exception. This exception occurs when you need to multiply or divide by a negative number.

$$-8 < 6$$
$$-8(-2) \mid 6(-2)$$
$$16 > -12$$

$$-8 < 6$$
$$\frac{-8}{-2} \mid \frac{6}{-2}$$
$$4 > -3$$

$$-8 < 6$$
$$-8(2) \mid 6(2)$$
$$-16 < 12$$

$$-8 < 6$$
$$\frac{-8}{2} \mid \frac{6}{2}$$
$$-4 < 3$$

Notice that the *order* changes from $<$ to $>$ when each side of $-8 < 6$ is multiplied or divided by -2. Multiplying or dividing each side by 2 does not reverse the order. Also, adding the same number to each side of an inequality or subtracting the same number from each side does not reverse the order.

Inequality properties

For all numbers a, b, and c,

Addition: if $a < b$, then $a + c < b + c$.

Subtraction: if $a < b$, then $a - c < b - c$.

Multiplication: if $a < b$, $c > 0$, then $a \cdot c < b \cdot c$.
if $a < b$, $c < 0$, then $a \cdot c > b \cdot c$.

Division: if $a < b$, $c > 0$, then $\frac{a}{c} < \frac{b}{c}$.
if $a < b$, $c < 0$, then $\frac{a}{c} > \frac{b}{c}$.

Example 1 Find the solution set of $8 - 3x < 2x + 28$.
Graph the solutions on a number line.

$$8 - 3x < 2x + 28$$
$$8 - 5x < 28$$
$$-5x < 20$$
$$\frac{-5x}{-5} \mid \frac{20}{-5}$$ ◀ Divide each side by -5.
$$x > -4$$ ◀ Reverse the order from $<$ to $>$.

Thus, $\{x \mid x > -4\}$ is the solution set. ◀ Read: "The set of all x such that x is greater than -4."
To graph the solutions, plot all numbers greater than -4.

You will find it easier to read and graph an inequality like $5 > x$ if you write the variable first. Thus, $5 > x$ can be replaced by $x < 5$, and read as *x is less than 5*.

Even though the inequality properties on page 5 are stated for $<$, they are true for $>$ as well. In Example 2, the properties are applied to the $>$ relation.

Example 2   **Find the solution set of $3x - 2 > 5x + 7$.**

$$3x - 2 > 5x + 7$$
$$-2 > 2x + 7 \quad \blacktriangleleft \text{ Subtract } 3x \text{ from each side.}$$
$$-9 > 2x \quad \blacktriangleleft \text{ Subtract 7 from each side.}$$
$$-4\tfrac{1}{2} > x \quad \blacktriangleleft \text{ Divide each side by 2.}$$
$$x < -4\tfrac{1}{2} \quad \blacktriangleleft \text{ Write the variable first.}$$

Thus, the solution set is $\left\{x \mid x < -4\tfrac{1}{2}\right\}$.

Some linear inequalities contain the relation $\geq$, *is greater than or equal to*. You can solve such an inequality by using both equation and inequality properties. This is shown in Example 3.

Example 3   **Find the solution set of $3(4 - 2x) \geq 4x - 23$. Graph the solutions on a number line.**

$$3(4 - 2x) \geq 4x - 23$$
$$12 - 6x \geq 4x - 23 \quad \text{Combined division properties:}$$
$$-10x \geq -35 \qquad \text{If } a = b, \text{ then } \dfrac{a}{c} = \dfrac{b}{c}, c \neq 0.$$
$$x \leq 3\tfrac{1}{2} \qquad \text{If } a > b, c < 0, \text{ then } \dfrac{a}{c} < \dfrac{b}{c}.$$

Thus, the solution set is $\left\{x \mid x \leq 3\tfrac{1}{2}\right\}$.

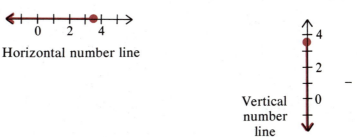

Horizontal number line

Vertical number line

Linear Inequalities

Reading in Algebra

Match each inequality to exactly one graph.
1. $5 < x$ **C**
2. $5 \leq x$ **A**
3. $x < 5$ **D**
4. $x \leq 5$ **B**
5. $x < 2$ **E**
6. $x > 2$ **F**

A: number line with closed circle at 5, arrow right past 6
B: number line with closed circle at 5, arrow left past 4
C: number line with open circle at 5, arrow right past 6
D: number line with open circle at 5, arrow left past 4
E: vertical number line, open circle at 2, arrow down past 1
F: vertical number line, closed circle at 2, arrow up past 3

Oral Exercises

True or false?
1. $-7 < 3$ **T**
2. $-4 \leq -6$ **F**
3. $5 > -9$ **T**
4. $8 \geq 8$ **T**
5. If $5a + 2 < 12$, then $5a > 10$. **F**
6. If $8 \leq y$, then $y \geq 8$. **T**
7. If $-a < -8$, then $a < 8$. **F**
8. If $-\frac{1}{2}x < 6$, then $x > -12$. **T**

Written Exercises

Find the solution set of each inequality. Graph the solutions on a number line.

Ⓐ
1. $-5y \leq 15$ $\{y|y \geq -3\}$
2. $-7a > -28$ $\{a|a < 4\}$
3. $32 < 8x$ $\{x|x > 4\}$
4. $3 > x - 2$ $\{x|x < 5\}$
5. $4 \leq n + 6$ $\{n|n \geq -2\}$
6. $-4 \geq y - 6$ $\{y|y \leq 2\}$

Find the solution set of each inequality.   10. $\{n|n > 7\}$ 11. $\{x|x \leq -3\}$ 12. $\{y|y > -3\}$
7. $6 - 4x < 30$ $\{x|x > -6\}$
8. $3y - 6 > 9y$ $\{y|y < -1\}$
9. $5n + 27 \leq -4n$ $\{n|n \leq -3\}$
10. $8n - 2 > 6n + 12$
11. $4 - 2x \geq 3x + 19$
12. $7y - 6 > 3(y - 6)$
13. $3(2y + 6) < 8y + 11$
14. $4(2n - 3) > 5(n + 4)$
15. $-2(3x - 2) \geq 7(2 - x)$
$\{y|y > 3\frac{1}{2}\}$ $\{n|n > 10\frac{2}{3}\}$ $\{x|x \geq 10\}$

Find the solution set of each inequality. Graph the solutions on a number line.

Ⓑ
16. $-x < 0$ $\{x|x > 0\}$
17. $22 - 3a > 29$ $\{a|a < -2\frac{1}{3}\}$
18. $\frac{1}{2}y \leq 4$ $\{y|y \leq 8\}$
19. $18 - 5y \geq 7 - 3y$
20. $\frac{2}{3}n < 12$ $\{n|n < 18\}$
21. $8x - 7 - 10x \leq 7 + 2x$
$\{y|y \leq 5\frac{1}{2}\}$ $\{x|x \geq -3\frac{1}{2}\}$

Find the solution set of each inequality.   22. $\{x|x > 1\frac{1}{2}\}$ 23. $\{n|n \geq 1\frac{1}{3}\}$ 24. $x|x < 2.75$
22. $8 - 2(2x + 1) < 2x - 3$
23. $2n + 3(n - 2) \geq 2(3 - 2n)$
24. $5(x - 4) - (2 - 3x) < 0$
25. $\frac{1}{4}(8y - 12) > \frac{1}{3}(9 - 12y)$
26. $\frac{3}{5}(10a + 25) < \frac{2}{3}(6a - 9)$
27. $8n - \frac{3}{4}(12n + 28) \leq 0$
$\{y|y > 1\}$ $\{a|a < -10.5\}$ $\{n|n \geq -21\}$

Ⓒ
28. $\dfrac{2x - 1}{-3} < 5$ $\{x|x > -7\}$
29. $\dfrac{2(2y + 3)}{-5} > 6 - 2y$
30. $\dfrac{3(5a + 6)}{-4} < \dfrac{8 - 5a}{-4}$
$\{y|y > 6\}$ $\{a|a > -0.5\}$

31. Use algebraic properties to prove: If $x < y$, then $-y < -x$.
$x < y; \; -1 \cdot x > -1 \cdot y; \; -x > -y; \; -y < -x$

Show that each statement is false by giving a numerical counterexample.

32. If $b \neq 0$ and $a < b$, then $a^2 < b^2$.

Let $a = -3, b = -2$. (Answers may vary.)

33. If $a \neq 0$, $b \neq 0$, and $a < b$, then $\dfrac{1}{a} > \dfrac{1}{b}$.

Let $a = -2, b = 4$. (Answers may vary.)

COMPOUND LINEAR INEQUALITIES 1.3

Objectives **To find the solution set of a disjunction or a conjunction of two linear inequalities in one variable**
To graph the solutions of a disjunction or a conjunction of two linear inequalities

A first linear inequality $x < 3$ and a second linear inequality $x < 5$ may be joined by the connective *or* or the connective *and* to form a compound inequality.
1. $[x < 3 \text{ or } x < 5]$ is a *disjunction* of two inequalities. One solution is 4 since at least one part of the disjunction $[4 < 3 \text{ or } 4 < 5]$ is true.
2. $[x < 3 \text{ and } x < 5]$ is a *conjunction* of two inequalities. Four is *not* a solution since the two parts of the conjunction $[4 < 3 \text{ and } 4 < 5]$ are not both true. Two is a solution because both parts of the conjunction $[2 < 3 \text{ and } 2 < 5]$ are true.

You can find the solution set of a compound inequality and graph its solutions on a number line as shown below.

First, draw the graphs of the parts. Then,
1. For a disjunction (*or*), draw the graph of *both parts* (their union) on one number line.
2. For a conjunction (*and*), draw the graph of the *overlap* (intersection) of both parts on one number line.

Finally, use this graph of the solutions to describe the solution set.

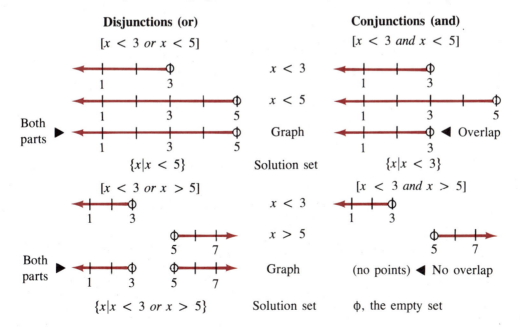

Compound Linear Inequalities

Disjunction (or)

[$x > 3$ or $x < 5$]

Conjunction (and)

[$x > 3$ and $x < 5$]

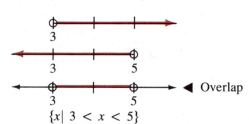

	$x > 3$	
	$x < 5$	
Both parts ▶	Graph	◀ Overlap
{all numbers}	Solution set	{x | $3 < x < 5$}

In the last conjunction, $3 < x < 5$ means [$3 < x$ and $x < 5$]. You can read $3 < x < 5$ as "x is *between* 3 and 5."

> **Definition:** $a < x < b$ means [$a < x$ and $x < b$]. The solutions of $a < x < b$ are the
> $a < x < b$ numbers *between* a and b if $a < b$.

Example 1

Find the solution set of [$2x + 9 \leq 3$ or $3x - 7 > 5$].
Graph the solutions on a number line.

$$\begin{aligned} 2x + 9 &\leq 3 & \text{or} && 3x - 7 &> 5 \\ 2x &\leq -6 & \text{or} && 3x &> 12 \\ x &\leq -3 & \text{or} && x &> 4 \end{aligned}$$

Thus, the solution set is {x \| $x \leq -3$ or $x > 4$}.

Example 2

Find the solution set of $-3 < 2y - 7 \leq 1$.
Graph the solutions on a number line.

Method 1

$$\begin{aligned} -3 < 2y - 7 & \text{ and } & 2y - 7 &\leq 1 \\ 4 < 2y & \text{ and } & 2y &\leq 8 \\ 2 < y & \text{ and } & y &\leq 4 \end{aligned}$$

◀ Add 7. ▶
◀ Divide by 2. ▶

Method 2

$$\begin{aligned} -3 < 2y - 7 &\leq 1 \\ 4 < 2y &\leq 8 \\ 2 < y &\leq 4 \end{aligned}$$

Thus, the solution set is {y \| $2 < y \leq 4$}.

Reading in Algebra

Find the solution set.
1. $x < -2$ or $x < 3$ **{x | x < 3}**
2. $x < -2$ and $x < 3$ **{x | x < -2}**
3. $x < -2$ or $x > 3$ **{x | x < -2 or x > 3}**
4. $x < -2$ and $x > 3$ **∅**
5. $x > -2$ or $x > 3$ **{x | x > -2}**
6. $x > -2$ and $x > 3$ **{x | x > 3}**
7. $x > -2$ or $x < 3$ **{all numbers}**
8. $x > -2$ and $x < 3$ **{x | -2 < x < 3}**
9. $x \leq -2$ or $x > -2$ **{all numbers}**
10. $-6 < 3x < 12$ **{x | -2 < x < 4}**
11. $-3 < x - 1 < 5$ **{x | -2 < x < 6}**
12. $-6 < x + 3 < 10$ **{x | -9 < x < 7}**

Written Exercises

Find the solution set of each compound inequality. Graph the solutions on a number line.

Ⓐ
1. $x + 7 < 3$ or $x - 2 > 2$ **{x | x < -4 or x > 4}**
2. $x + 5 \leq 7$ or $x - 3 \leq 2$ **{x | x ≤ 5}**
3. $3x \geq -6$ or $5x \geq 15$ **{x | x ≥ -2}**
4. $4x > -8$ or $3x < 12$ **{all numbers}**
5. $x + 3 < 9$ and $2x > 6$ **{x | 3 < x < 6}**
6. $5x < -10$ and $x + 4 > 7$ **∅**
7. $-2 < x - 3 < 2$ **{x | 1 < x < 5}**
8. $4 \leq x + 6 \leq 8$ **{x | -2 ≤ x ≤ 2}**
9. $-9 < 3x < 12$ **{x | -3 < x < 4}**
10. $8 \leq 4x < 16$ **{x | 2 ≤ x < 4}**

Ⓑ
11. $-11 \leq 2x - 7 < 11$ **{x | -2 ≤ x < 9}**
12. $6 < 3x + 12 \leq 21$ **{x | -2 < x ≤ 3}**
13. $4x + 9 < 1$ or $4x - 9 < 3$ **{x | x < 3}**
14. $2x - 5 < -3$ and $2x + 5 < 9$ **{x | x < 1}**
15. $5 < 4x + 5 < 25$ **{x | 0 < x < 5}**
16. $-12 \leq 5x - 2 \leq -2$ **{x | -2 ≤ x ≤ 0}**
17. $3 > 5x - 2$ or $4 < 3x - 2$ **{x | x < 1 or x > 2}**
18. $-2 > 3x - 5$ and $7 < 4x - 5$ **∅**
19. $4 - 3x \leq 19$ or $2 - 5x \leq 17$ **{x | x ≥ -5}**
20. $15 > 7 - 2x$ and $11 > 4x + 3$ **{x | -4 < x < 2}**
21. $8 - 6x < 23$ and $15 < 4x + 1$ **{x | x > 3.5}**
22. $-2 - 7x < -16$ or $-5 > 3 - 2x$ **{x | x > 2}**

Ⓒ
23. $-6 < -3x < 9$ **{x | -3 < x < 2}**
24. $-4 \leq 1 - 5x \leq 11$ **{x | -2 ≤ x ≤ 1}**
25. $3x - 7 \leq 5x + 6 \leq 3x + 9$ **{x | -6.5 ≤ x ≤ 1.5}**
26. $2 - 3x < -4$ or $2 - 3x > 14$ **{x | x < -4 or x > 2}**
27. $-4 < 3x - 1 < 5$ or $3x - 1 > 11$ **{x | -1 < x < 2 or x > 4}**
28. $-7 < 2x + 3 < -1$ or $1 < 2x + 3 < 7$ **{x | -5 < x < -2 or -1 < x < 2}**

29. Let p and q represent any pair of statements such that each can be assigned the truth-value T for true or F for false. Complete the following truth table for the disjunction [p or q] and the conjunction [p and q] by writing either T or F in each empty cell.

p	q	p or q	p and q
T	T	T	T
T	F	T	F
F	T	T	F
F	F	**F**	F

Compound Linear Inequalities

ABSOLUTE VALUE   1.4

Objectives **To solve an equation involving absolute value
To find the solution set of an inequality involving absolute value
To graph the solutions of an inequality involving absolute value**

The **absolute value** of any number x is the distance between x and the origin, 0, on a number line. Therefore, the absolute value of -5 is 5 and the absolute value of 5 is 5.

The symbol $|x|$ means the absolute value of x, so $|-5| = |5| = 5$.

Definition: Absolute value	$|x| = -x$, if x is a negative number. $|x| = x$, if x is a positive number or 0.

From the definition of absolute value, the solutions of the equation $|x| = 12$ are -12 and 12 since $|-12| = 12$ and $|12| = 12$. This leads to the following equation property.

Equation property for absolute value	For each number x and each number $k \geq 0$, if $|x| = k$, then $x = -k$ or $x = k$.

Example 1 **Solve $|3n - 5| = 7$.**

$$|3n - 5| = 7$$
$$3n - 5 = -7 \quad \text{or} \quad 3n - 5 = 7 \qquad \blacktriangleleft \text{ If } |x| = k, \text{ then } x = -k \text{ or } x = k.$$
$$3n = -2 \qquad\qquad\qquad 3n = 12$$
$$n = -\frac{2}{3} \qquad\qquad\qquad n = 4$$

Check: $n = -\frac{2}{3}$

$|3n - 5|$	7
$\left\| 3 \cdot \frac{-2}{3} - 5 \right\|$	7
$|-2 - 5|$	
$|-7|$	
7	

True: $7 = 7$

Check: $n = 4$

$|3n - 5|$	7
$|3 \cdot 4 - 5|$	7
$|12 - 5|$	
$|7|$	
7	

True: $7 = 7$

Thus, the solutions are $-\frac{2}{3}$ and 4.

The graph for $|x| < 4$ is shown below on the left. The graph for $|x| > 4$ is shown below on the right.

All numbers between -4 and 4 are solutions of the inequality. In symbols, $-4 < x < 4$. The compound inequality $-4 < x < 4$ means $-4 < x$ and $x < 4$.

All numbers less than -4 and all numbers greater than 4 are solutions of the inequality. In symbols, $x < -4$ or $x > 4$.

The graphs suggest the following inequality properties:

Inequality properties for absolute value	For all numbers x and k, if $k > 0$ and $|x| < k$, then $-k < x < k$. if $k \geq 0$ and $|x| > k$, then $x < -k$ or $x > k$.

The inequality properties for absolute value are applied in the following examples. Notice that Example 3 involves the $\geq$ relation, so combined equation and inequality properties are used.

Example 2 **Find the solution set of $|4y + 2| < 10$. Graph the solutions on a number line.**

$$\begin{aligned}
|4y + 2| &< 10 \\
-10 < 4y + 2 &< 10 \quad &\blacktriangleleft \text{ If } |x| < k, \text{ then } -k < x < k. \\
-12 < 4y &< 8 \quad &\blacktriangleleft \text{ Subtract 2 from each side.} \\
-3 < y &< 2 \quad &\blacktriangleleft \text{ Divide each side by 4.}
\end{aligned}$$

The solution set is $\{y | -3 < y < 2\}$.

Example 3 **Find the solution set of $|2n - 1| \geq 5$. Graph the solutions on a number line.**

$$|2n - 1| \geq 5$$

$$\begin{aligned}
2n - 1 &\leq -5 \quad \text{or} \quad & 2n - 1 &\geq 5 \\
2n &\leq -4 & 2n &\geq 6 \\
n &\leq -2 & n &\geq 3
\end{aligned}$$

Combined properties:
If $|x| = k$, then $x = -k$, or $x = k$.
If $|x| > k$, then $x < -k$, or $x > k$.

The solution set is $\{n | n \leq -2 \text{ or } n \geq 3\}$.

Absolute Value

Reading in Algebra

True or false? If a statement is false, give a numerical counterexample which shows that it is false.
1. $|-x| = x$, if $x > 0$. **T**
2. $|-x| = -x$, if $x < 0$. **T**
3. $|x - 2| = x - 2$, if $x > 0$.
4. $|x - 2| = 2 - x$, if $x \leq 0$. **T**
5. $|4 - x| = 4 - x$, if $x \geq 0$.
6. $|-x - 6| = x + 6$, if $x < 0$.
7. $|5 - x| = |x - 5|$, for each x. **T**
8. $|-x - 8| = |x + 8|$, for each x. **T**
9. $|x| = |-x|$, for each x. **T**

3. F: Let x = 1.   5. F: Let x = 5.   6. F: Let x = −8.

Oral Exercises

Find the value of each expression.
1. $|-9 + 4|$ **5**
2. $|-5 - 8|$ **13**
3. $|-6 + 9|$ **3**
4. $|7 - 7|$ **0**
5. $|-3 \cdot 5|$ **15**
6. $|-2(-4)|$ **8**
7. $|2(-5) - 10|$ **20**
8. $|-5(-4) - 2|$ **18**

Written Exercises

Solve each equation.

Ⓐ
1. $|x - 3| = 7$ **−4, 10**
2. $|a + 6| = 2$ **−8, −4**
3. $|4 - y| = 6$ **−2, 10**
4. $|2 - x| = 3$ **5, −1**
5. $|2x + 15| = 5$ **−10, −5**
6. $|3y - 9| = 18$ **−3, 9**
7. $|4n + 6| = 2$ **−2, −1**
8. $|5 - 2a| = 9$ **−2, 7**

9. $\{x | -3 < x < 3\}$ 10. $\{n | n \leq -6 \text{ or } n \geq 6\}$ 11. $\{y | y < -8 \text{ or } y > 8\}$ 12. $\{a | -4 \leq a \leq 4\}$

Find the solution set of each inequality. Graph the solutions on a number line.
9. $|x| < 3$
10. $|2n| \geq 12$
11. $|y| > 8$
12. $|5a| \leq 20$
13. $|x - 3| \geq 1$
14. $|y + 4| < 7$
15. $|a - 2| \leq 6$
16. $|n + 5| > 3$

$\{x | x \leq 2 \text{ or } x \geq 4\}$ $\{y | -11 < y < 3\}$ $\{a | -4 \leq a \leq 8\}$ $\{n | n < -8 \text{ or } n > -2\}$

Solve each equation. 17. $-2, \frac{1}{2}$ 18. $-1\frac{2}{3}, 5$ 19. $-2.8, -2$ 20. $4, \frac{2}{3}$

Ⓑ
17. $|4y + 3| = 5$
18. $|3x - 5| = 10$
19. $|5n + 12| - 2 = 0$
20. $|14 - 6z| = 10$
21. $|2x - 3.5| = 10.5$ **−3.5, 7**
22. $|4y + 6.8| = 9.2$ **−4, 0.6**
23. $2.7 - |3z + 9.3| = 0$ **−4, −2.2**
24. $|14.2 - 5n| = 0.8$ **3, 2.68**

Find the solution set of each inequality. Graph the solutions on a number line.
25. $|2x - 5| < 17$
26. $|3y + 6| > 21$
27. $|3z - 15| \geq 6$
28. $|2n + 17| \leq 1$
29. $|4y + 6| \geq 2$
30. $10 \geq |5a - 15|$
31. $|5x + 10| - 35 < 0$
32. $20 < |4z - 4|$

25. $\{x | -6 < x < 11\}$ 26. $\{y | y < -9 \text{ or } y > 5\}$ 27. $\{z | z \leq 3 \text{ or } z \geq 7\}$

Ⓒ
33. $|18 - 3x| < 6$ $\{x | 4 < x < 8\}$
34. $|75 - 2y| > 5$ $\{y | y < 35 \text{ or } y > 40\}$
35. $|x - 2| < 6$ or $|x - 2| > 10$
36. $|y + 5| \leq 2$ or $|y - 5| \leq 2$

$\{x | -4 < x < 8 \text{ or } x < -8 \text{ or } x > 12\}$ $\{y | -7 \leq y \leq -3 \text{ or } 3 \leq y \leq 7\}$

Solve each equation.
37. $|x - 2| + |2 - x| = 10$ **7, −3**
38. $|2y + 4| + |6 - 2y| = 22$ **6, −5**
39. $|n + 2| + |-n - 6| = 18$ **−13, 5**
40. $|1 - 2x| + |-x - 7| = 21$ **−9, 5**

Find the solution set of each inequality.
41. $|x - 4| < 0.1$ $\{x | 3.9 < x < 4.1\}$
42. $|y + 6| < 0.001$ $\{y | -6.001 < y < -5.999\}$
43. $|x - 2.5| < 0.01$ $\{x | 2.49 < x < 2.51\}$
44. $|x + 3.4| < 0.0001$ $\{x | -3.4001 < x < -3.3999\}$

28. $\{n | -9 \leq n \leq -8\}$
29. $\{y | y \leq -2 \text{ or } y \geq -1\}$
30. $\{a | 1 \leq a \leq 5\}$
31. $\{x | -9 < x < 5\}$
32. $\{z | z < -4 \text{ or } z > 6\}$

CUMULATIVE REVIEW

1. Solve $\frac{2}{3}(12y - 15) - \frac{3}{5}(15y + 20) = 3y - 2$ **−5**
2. Solve $0.9n + 2.2(0.5n - 3) = 0.4(10n - 1.5)$. **−3**

PROBLEM SOLVING: NUMBER PROBLEMS   1.5

Objectives **To write a word phrase in algebraic terms**
To solve a word problem about one or more numbers

Many mathematical problems are presented using a word phrase or several word phrases. To solve such problems, you begin by writing the word phrases in algebraic terms. Some typical word phrases and their algebraic translations are listed below.

Word Phrase	Written in Algebraic Terms
Ten decreased by a number	$10 - n$
Twice a number, increased by 12	$2x + 12$
Six less than a number	$y - 6$
Seven more than a number	$n + 7$ or $7 + n$
Eight decreased by 5 times the sum of a number and 14	$8 - 5(x + 14)$
Is, or is the same as	$=$

Notice that two translations are given for the word phrase "Seven more than a number." The two translations are equivalent since addition is commutative. However, there is only one correct translation for the word phrase "Six less than a number." Subtraction is not commutative. Remember, for *4 less than x* write $x - 4$, not $4 - x$.

You can solve a word problem about one number if you are given enough information about the number. Example 1 illustrates this situation.

Example 1 **Seven less than twice a number is 20 more than 5 times the number. What is the number?**

Represent the data. Let x be the number.
7 less than twice x is   20 more than $5x$.

Write an equation. $2x - 7 = 5x + 20$
                         $-27 = 3x$
Solve the equation. $-9 = x$

Check in the problem.

Seven less than twice x	20 more than $5x$
$2(-9) - 7$	$5(-9) + 20$
$-18 - 7$	$-45 + 20$
-25	-25

Answer the question. **Thus,** the number is -9.

Problem Solving: Number Problems

In some word problems you will be asked to find more than one number. To do this, you begin by representing each of the numbers in terms of one variable, as shown in Example 2.

Example 2 **The greater of two numbers is 8 less than twice the sum of the smaller number and 5. Represent each number in terms of one variable.**

The *greater* number is *described in terms of* the *smaller* number. Represent the *smaller* number by a variable.

Let s = smaller number.

Twice the sum of s and 5   8 less than

$$2(s + 5) \; - \; 8$$

Thus, s = smaller number and $2(s + 5) - 8$ = greater number.

The four steps you should use in solving any word problem are given at the beginning of this chapter. They are illustrated again in Example 3 below.

Example 3 **The second of three numbers is 3 times the first. The third number is 2 more than the second number. Seven less than twice the second is the same as 12 more than the third. What are the three numbers?**

READ ▶   You are asked to find three numbers. The third number is described *in terms of* the second number. The second number is described *in terms of* the first number.

PLAN ▶   Let f = 1st number
               $3f$ = 2nd number
               $3f + 2$ = 3rd number

       7 less than twice the 2nd   is   12 more than the 3rd.
              $2(3f) - 7$ = $(3f + 2) + 12$

SOLVE ▶ $6f - 7$ = $3f + 14$
                       $3f$ = 21
                        f = 7
   $f = 7$ $3f = 21$ $3f + 2 = 23$

INTERPRET ▶   7 less than twice the 2nd | 12 more than the 3rd
                   $2(21) - 7$ | $23 + 12$
                    $42 - 7$ | 35
                       35

Thus, the three numbers are 7, 21, and 23.

In Example 3, you saw the four major steps in solving a word problem.

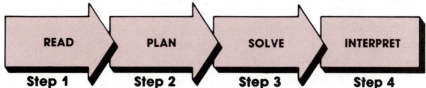

Step 2 may require some information that is not stated directly in the problem. For example, some problems require a knowledge of *consecutive integers*. Four such typical word phrases and their algebraic translations are shown in the table below, along with a numerical example of each.

Word Phrase	Example	Written in Algebraic Terms	
Consecutive integers	−1, 0, 1, 2, 3	$x, x + 1, x + 2, \ldots$	(x is an integer.)
Consecutive odd integers	5, 7, 9, 11, 13	$y, y + 2, y + 4, \ldots$	(y is an odd integer.)
Consecutive even integers	−2, 0, 2, 4, 6	$n, n + 2, n + 4, \ldots$	(n is an even integer.)
Consecutive multiples of 4	12, 16, 20, 24	$a, a + 4, a + 8, \ldots$	(a is a multiple of 4.)

A word problem about consecutive odd integers is solved in Example 4.

Example 4

The sum of three consecutive odd integers is 69 more than twice the third odd integer. What are the three odd integers?

Let y = 1st odd integer, $y + 2$ = 2nd odd integer, and $y + 4$ = 3rd odd integer.

Their sum is 69 more than twice the third.
$$y + (y + 2) + (y + 4) = 2(y + 4) + 69$$
$$3y + 6 = 2y + 77$$
$$y = 71$$

$y = 71 \qquad y + 2 = 73 \qquad y + 4 = 75$

Check:

The sum of the 3 integers	69 more than twice the 3rd
71 + 73 + 75	2 · 75 + 69
219	150 + 69
	219

Thus, the three consecutive odd integers are 71, 73, and 75.

Reading in Algebra

State each word phrase in algebraic terms.
1. Twelve less than a number **$n - 12$**
2. The sum of a number and 14 **$x + 14$**
3. Twenty increased by a number **$20 + y$**
4. A number decreased by 15 **$n - 15$**
5. Eight more than twice a number **$2x + 8$**
6. Five less than 3 times a number **$3y - 5$**
7. Four times the sum of 8 and n **$4(8 + n)$**
8. Ten times the difference of n and 2 **$10(n - 2)$**

Problem Solving: Number Problems

Oral Exercises

1. State 4 consecutive even integers, beginning with 6. **6, 8, 10, 12**
2. State 4 consecutive multiples of 10, beginning with 30. **30, 40, 50, 60**
3. State 4 consecutive integers, beginning with -10. **$-10, -9, -8, -7$**
4. State 4 consecutive odd integers, beginning with -3. **$-3, -1, 1, 3$**

Written Exercises

Solve each problem.

Ⓐ
1. Nine less than 5 times a number is 1 less than the number. What is the number? **2**
2. Sixteen more than a number is 20 more than twice the number. Find the number. **-4**
3. Five times the sum of 6 and a number is 2 more than the number. Find the number. **-7**
4. Fourteen more than twice a number is 6 times the difference of the number and 7. What is the number? **14**

Represent each number in terms of one variable.

5. The greater of two numbers is 7 more than 5 times the smaller number. **smaller: x, greater: $5x+7$**
6. The second of two numbers is 6 less than 9 times the first number. **first: n, second: $9n-6$**
7. The second of three numbers is 7 times the first number. The third number is 6 less than the second number. **1st: x, 2nd: $7x$, 3rd: $7x-6$**
8. A second number is 5 times a first number. A third number is 8 more than twice the second number. **1st: n, 2nd: $5n$, 3rd: $10n+8$**

Solve each problem.

9. The greater of two numbers is 3 less than twice the smaller number. The sum of the two numbers is 18. What are the two numbers? **7, 11**
10. The smaller of two numbers is 10 less than 4 times the greater number. The greater number is 19 more than the smaller. Find the two numbers. **$-3, -22$**
11. The second of three numbers is 5 times the first number. The third number is 12 less than the first. Find the three numbers if their sum is 51. **9, 45, -3**
12. The second of three numbers is 6 more than the first and the third is 7 times the second. If the third number is 12 more than 10 times the first number, what are the three numbers? **10, 16, 112**
13. The sum of three consecutive integers is 12 less than 4 times the first integer. What are the three integers? **15, 16, 17**
14. There are three consecutive even integers such that the sum of the first two is 22 more than the third. Find the three even integers. **24, 26, 28**
15. The population of city A is 4 times the population of city B. City C has 670 fewer people than city A. What is the population of each city if their combined population is 8,330? **A: 4,000, B: 1,000, C: 3,330**
16. A company's total sales were $150,000 for a 3-year period. Sales for the second year were $20,000 more than the first year's sales. The third year's sales were 3 times those of the second year. Find the sales amount for each year. **1st yr: $14,000, 2nd yr: $34,000, 3rd yr: $102,000**

B 17. There are four consecutive multiples of 3 such that the sum of the first three multiples is twice the fourth. Find the four multiples of 3. **9, 12, 15, 18**

18. The sum of three consecutive multiples of 6 is 30 more than the fifth consecutive multiple of 6. Find the fourth of these five consecutive multiples of 6. **36**

19. Twice the sum of three consecutive multiples of 15 is 270. What are the three multiples of 15? **30, 45, 60**

20. One-half of the sum of three consecutive multiples of 10 is 75. Find the three multiples of 10. **40, 50, 60**

21. The third of three numbers is 7 times the first and the second is 8 more than the first. Six more than the third, decreased by 3 times the second, is 9 less than the first. Find the three numbers. **3, 11, 21**

22. The second of three numbers is twice the first and the third is 4 less than the second. If 9 less than the first is decreased by the third, the result is -12. What are the three numbers? **7, 14, 10**

23. A size A box holds twice as many jars as a size B box and a size C box holds one dozen fewer jars than a size A box. Ten size A boxes, 5 size B boxes, and 3 size C boxes will hold a total of 522 jars. Find the number of jars that each type of box will hold. **A: 36, B: 18, C: 24**

24. A company's sales for February increased $400 over its January sales figure. The March sales were twice the February figure and the sales for April doubled the figure for March. If the sales total was $18,000 for the four months, find the sales figure for each month.
Jan.: $1,900, Feb.: $2,300, Mar.: $4,600 Apr.: $9,200

25. Find three consecutive integers so that 8 times the first, decreased by the third, is 488. **70, 71, 72**

26. Find three consecutive odd integers so that 6 times the second, decreased by the third, is 363. **71, 73, 75**

27. Find three consecutive even integers so that 7 times the third is equal to 20 less than the first. **$-8, -6, -4$**

28. Find three consecutive integers so that 25 more than the second is equal to 5 times the third. **4, 5, 6**

C 29. Find three consecutive integers so that 3 less than twice the third integer is the sum of the first and second integers.
any three consecutive integers

30. Find three consecutive multiples of 5 for which the sum of the first and third multiples is twice the second multiple of 5.
any three consecutive multiples of 5

31. Six decreased by 3 times a number is less than 12. Find the set of all such numbers. **$\{x|x > -2\}$**

32. Seven more than 4 times a number is greater than 3. Find the set of all these numbers. **$\{x|x > -1\}$**

33. Eight less than 5 times a number is greater than or equal to 8 more than the number. Find the set of all these numbers.
$\{x|x \geq 4\}$

34. Twenty increased by twice a number is less than or equal to 6 times the number. Find the set of all such numbers.
$\{x|x \geq 5\}$

Problem Solving: Number Problems

PROBLEM SOLVING: PERIMETER PROBLEMS

1.6

Objective   **To solve a word problem involving the perimeter of a geometric figure**

The **perimeter** of a geometric figure is the sum of the lengths of its sides. To solve a problem that involves perimeter, you may find it helpful to draw the geometric figure and label the lengths of the sides. If the measures of any of the sides are unknown, represent these lengths in terms of one variable.

Example 1   **The length of a rectangle is 4 meters (m) more than twice the width. The perimeter is 26 m. Find the length and the width. Find the area of the rectangle in square meters (m^2).**

READS ▶

PLAN ▶   Let w = width of rectangle and $2w + 4$ = length of rectangle. Draw the figure and label each side.

$$\begin{array}{c} 2w + 4 \\ w \;\fbox{}\; w \\ 2w + 4 \end{array}$$

SOLVE ▶ $w + (2w + 4) + w + (2w + 4) = 26$ or $2(w) + 2(2w + 4) = 26$
$6w + 8 = 26$
$6w = 18$
$w = 3$ $2w + 4 = 10$

INTERPRET ▶   Check: $\begin{array}{c|c} 2(w) + 2(2w + 4) & 26 \\ \hline 2(3) + 2(10) & 26 \\ 26 & \end{array}$

The area, length $\times$ width, is $10 \cdot 3$, or 30.
Thus, the length is 10 m, the width is 3 m, and the area is 30 m^2.

Example 2   **Side a of a triangle is 3 times as long as side b. Side c is 2 cm longer than side a. Find the length of each side of the triangle if its perimeter is 72 cm.**

Let x = length of side b, $3x$ = length of side a, and $3x + 2$ = length of side c.

$\begin{array}{c} 3x \quad\quad x \\ \overline{} \\ 3x + 2 \end{array}$ $\begin{aligned} x + 3x + (3x + 2) &= 72 \\ 7x &= 70 \\ x &= 10 \end{aligned}$ $3x = 30$ $3x + 2 = 32$

Check: $\begin{array}{c|c} x + 3x + (3x + 2) & 72 \\ \hline 10 + 30 + 32 & 72 \\ 72 & \end{array}$

Thus, the lengths of the three sides are 10 cm, 30 cm, and 32 cm.

Written Exercises

Solve each problem.

(A) 1. The length of a rectangle is 12 cm more than the width. The perimeter is 44 cm. Find the length and the width. Find the area of the rectangle in cm².
width: 5 cm, length: 17 cm, area: 85 cm²

2. The width of a rectangle is 4 m less than the length and the perimeter is 40 m. Find the width, the length, and the area of the rectangle.
width: 8 m, length: 12 m, area: 96 m²

3. Side a of a triangle is 4 cm longer than side b. Side c is twice as long as side b. What is the length of each side of the triangle if its perimeter is 28 cm?
a: 10 cm, b: 6 cm, c: 12 cm

4. Side c of a triangle is 6 m longer than side a. Side b is 8 m shorter than side a. The perimeter is 52 m. Find the length of each of the three sides.
a: 18 m, b: 10 m, c: 24 m

5. The length of a rectangle is 5 dm more than 3 times the width. Find the length, the width, and the area of the rectangle if its perimeter is 42 dm.
width: 4 dm, length: 17 dm, area: 68 dm²

6. The length of side b of a triangle is 6 m less than twice the length of side a. Sides b and c have the same length. If the perimeter is 28 m, find the lengths of the three sides. **a: 8 m, b: 10 m, c: 10 m**

7. The two longest sides of a *pentagon* (5-sided figure) are each 3 times as long as the shortest side. The other sides are each 8 m longer than the shortest side. Find the length of each side if the perimeter is 79 m.
7 m, 21 m, 21 m, 15 m, 15 m

8. Each of the three longest sides of a *hexagon* (6-sided figure) is 3 cm less than 4 times the length of each of the three shortest sides. If the perimeter is 111 cm, find the length of each side.
8 cm, 8 cm, 8 cm, 29 cm, 29 cm, 29 cm

(B) 9. A rectangle and a square have the same width. The rectangle is 6 cm longer than the square. One perimeter is twice the other. Find the dimensions of the rectangle.
width: 3 cm, length: 9 cm

10. One side of an equilateral triangle is 7 dm shorter than one side of a square. The sum of the perimeters of the two figures is 49 dm. Find the perimeter of each figure.
square: 40 dm, triangle: 9 dm

11. The hypotenuse of a right triangle is 10 m long and one leg is 2 m longer than the other leg. If the perimeter is 24 m, find the area of the right triangle in square meters. **24 m²**

12. One leg of a right triangle is 2 cm longer than twice the length of the other leg. The hypotenuse is 1 cm longer than the longer leg. Find the area if the perimeter is 30 cm. **30 cm²**

13. A longer side of a parallelogram is 4 dm shorter than 5 times the length of a shorter side. One side of a *rhombus* (equilateral quadrilateral) and a longer side of the parallelogram are equal in length. If the difference in the two perimeters is 48 dm, find the perimeter of the rhombus. **124 dm**

14. The lengths of the sides of a quadrilateral are in the ratio of 2 to 3 to 5 to 6. What is the length of each side if the perimeter is 104 m?
13 m, $19\frac{1}{2}$ m, $32\frac{1}{2}$ m, 39 m

(C) 15. The length of a rectangle is 4 cm more than 5 times the width. The perimeter is less than 44 cm. Find the set of all such possible widths.
{w|0 cm < w < 3 cm}

16. The length of a rectangle is 7 dm less than 3 times the width. Find the set of all possible lengths if the perimeter must be greater than 18 dm.
{l|l > 5 dm}

Problem Solving: Perimeter Problems

PROBLEM SOLVING: AGE PROBLEMS 1.7

Objective **To solve a word problem about ages**

Word problems about people's ages may include information about their present ages, past ages, and future ages. If you are x years old now, then 3 years ago you were $x - 3$ years old. Ten years from now, you will be $x + 10$ years old.

To solve a word problem about ages, it may be helpful to use a table to organize the information.

Example **Carol is 6 times as old as her nephew Howard. Betty is 22 years younger than her aunt Carol. In 4 years, Carol's age will be twice the sum of Howard's and Betty's ages. How old is each person now?**

In a table, represent their ages *now* and their ages *in 4 years*. Let $h =$ Howard's age now.

	Howard	Carol	Betty
Ages now	h	$6h$	$6h - 22$
Ages in 4 yr	$h + 4$	$6h + 4$	$(6h - 22) + 4$, or $6h - 18$

In 4 years, Carol's age will be 2 times the sum of Howard's and Betty's ages.

$$6h + 4 = 2[(h + 4) + (6h - 18)]$$
$$6h + 4 = 2(7h - 14)$$
$$6h + 4 = 14h - 28$$
$$32 = 8h$$
$$4 = h$$
$$h = 4 \qquad 6h = 24 \qquad 6h - 22 = 2$$

Check:

Carol's age in 4 years	twice the sum of Howard's and Betty's ages in 4 years
$24 + 4$	$2[(4 + 4) + (2 + 4)]$
28	$2(14)$
	28

Thus, Howard is now 4 years old, Carol is 24, and Betty is 2.

Reading in Algebra

Select the correct equation for each situation.

1. He is x years old and she is y years old. She is 4 times as old as he is. **(a)**
   (a) $y = 4x$ (b) $4y = x$ (c) $y = x + 4$ (d) $x + 4 = y + 4$
2. She is $x - 10$ years old and he is $x - 2$ years old. He is 3 times as old as she is. **(c)**
   (a) $3(x - 2) = x - 10$ (b) $x - 2 = (x - 10) + 3$ (c) $x - 2 = 3(x - 10)$ (d) $x = 3(x - 10)$

Oral Exercises

1. Jane: x, Karen: x + 7
2. Frank: x, Ed: x − 5
3. Robert: x, Eloise: 5x
4. Richard: x, Sarah: $\frac{1}{2}$x

Represent each age now in terms of one variable.

1. Karen is 7 years older than Jane.
2. Ed is 5 years younger than Frank.
3. Eloise is 5 times as old as Robert.
4. Sarah is half as old as Richard.
5. A horse is 10 years older than twice the age of a colt. **colt: x, horse: 2x + 10**
6. A redwood tree is 80 years older than 4 times the age of an oak tree. **oak: x, redwood: 4x + 80**
7. A second city was chartered 20 years before a first city. A third city is 4 times as old as the first city. **1st city: x, 2nd city: x + 20, 3rd city: 4x**
8. An eagle is 3 times as old as a hawk. A robin was hatched 15 years after the eagle was hatched. **hawk: x, eagle: 3x, robin: 3x − 15**

Written Exercises

Solve each problem.

Ⓐ
1. Fred is 4 times as old as his niece, Selma. Ten years from now, he will be twice as old as she will be. How old is each now? **Selma: 5, Fred: 20**
2. Raymond is 12 years younger than Susan. Four years ago, she was 4 times as old as he was. Find their present ages. **Sue: 20, Ray: 8**
3. Byron is 2 years younger than Cindy. Eight years ago, the sum of their ages was 14. Find their present ages. **Cindy: 16, Byron: 14**
4. A fir tree is 5 times as old as a pine tree. Seven years from now, the sum of their ages will be 32. How old is the fir tree now? **15**
5. Brenda is 4 years older than Walter and Carol is twice as old as Brenda. Three years ago, the sum of their ages was 35. How old is each now? **Walt: 8, Brenda: 12, Carol: 24**
6. Denise is 3 times as old as Conrad and Billy is 8 years younger than Denise. Five years from now, the sum of their ages will be 49. How old is Billy? **10**

Ⓑ
7. An eagle is 4 times as old as a falcon. Three years ago, the eagle was 7 times as old as the falcon. Find the present age of each bird. **falcon: 6, eagle, 24**
8. An oak tree is 20 years older than a pine tree. In eight years, the oak will be 3 times as old as the pine will be. How old is each tree now? **pine: 2, oak: 22**
9. Adam is 3 times as old as Cynthia and Fred is 16 years younger than Adam. One year ago, Adam's age was twice the sum of Cynthia's and Fred's ages. Find their present ages. **Cynthia: 7, Adam: 21, Fred: 5**
10. Building A was built 20 years before building B and 30 years after building C. In 10 years, building C will be 20 years younger than the combined ages of buildings A and B. What is the present age of each building? **A: 60, B: 40, C: 90**
11. Phyllis is 6 years older than Keith and Manuel is 10 years older than twice Keith's age. If Manuel's age is added to 5 years less than twice Phyllis' age, the result is the same as 7 years more than 5 times Keith's age. How old is each now? **Keith: 10, Phyllis: 16, Manuel: 30**
12. The present ages of three portraits are in the ratio of 1 to 3 to 4. In 3 years, the sum of their ages will be 49 years. Find the present ages of the portraits. **5, 15, 20**

Ⓒ
13. A person is x years old now and 10 years ago was t years old. Represent the person's age f years from now in terms of t and f. **t + 10 + f**
14. In x years, a person will be y years old. Represent the person's age z years ago in terms of x, y, and z. **y − x − z**

Problem Solving: Age Problems

PROBLEM SOLVING: USING FORMULAS 1.8

Objective **To solve a word problem by using a given formula**

Many applications of mathematics involve the use of **formulas.** One such application is the formula for finding *simple interest* on money borrowed or loaned.

If you borrow $2,500 at $8\frac{1}{2}$% per year for 3 years (yr) 6 months (mo), you will pay $743.75 in simple interest, since $2,500 × 0.085 × 3.5 = $743.75. At the end of 3 yr 6 mo, you will pay $3,243.75, since $2,500 + $743.75 = $3,243.75.

In general, a principal p borrowed at a rate of r% per year for t years is charged a simple interest i, where $i = prt$.
The simple interest i, or prt, added to the principal p gives the total amount A that is owed, or $A = p + prt$.

To find the value of one variable in a formula, you substitute given values for the other variables and then solve the resulting equation. This procedure is shown in the following examples, using the formulas $i = prt$ and $A = p + prt$.

Example 1 **Jack borrowed $4,000 and paid $585 in simple interest after 1 yr 6 mo. What was the annual rate of interest?**

Use the formula $i = prt$. Find r.
$p = \$4,000$ $i = \$585$ $t = 1$ yr 6 mo, or 1.5 yr

$$i = p \cdot r \cdot t$$

Substitute for p, i, and t. ▶ $585 = 4,000 \cdot r \cdot 1.5$
Solve the equation for r. ▶ $585 = 6,000r$
$0.0975 = r$ $0.0975 = 9.75\%$, or $9\frac{3}{4}\%$

Thus, the rate of interest was $9\frac{3}{4}$% per year.

Example 2 **At the end of 2 yr, Maria paid $3,780 to cover a loan for her junior college tuition. If she was charged 4% in simple interest per year, how much money did Maria borrow?**

Use the formula $A = p + prt$. Find p.
$A = \$3,780$ $r = 4\%$, or 0.04 $t = 2$ yr

$$A = p + prt$$
$$3,780 = p + p(0.04)(2)$$
$$3,780 = 1.08\,p$$
$$3,500 = p$$

Thus, Maria borrowed $3,500.

A parallel electric circuit with two resistances is shown at the right. The bulbs have resistances of r_1 and r_2 (r-sub-one and r-sub-two) ohms. The circuit has a total resistance of R ohms, where
$$R(r_1 + r_2) = r_1 \cdot r_2.$$

Source

Example 3   Find r_2 in ohms if $R = 4$ ohms and $r_1 = 6$ ohms. Use the formula $R(r_1 + r_2) = r_1 \cdot r_2$.

$R = 4$ ohms $r_1 = 6$ ohms

$R(r_1 + r_2) = r_1 \cdot r_2$
$4(6 + r_2) = 6 \cdot r_2$ ◀ *Substitute for R and r_1.*
$24 + 4r_2 = 6r_2$ ◀ *Solve the equation for r_2.*
$24 = 2r_2$
$12 = r_2$

Thus, r_2 is 12 ohms.

An object is launched into the air with an initial velocity of 50 meters per second (m/s). The figure at the right shows the height, in meters, of the object at the end of t seconds. The height h is determined by the formula
$$h = 50t - 5t^2.$$
Use this formula to verify the heights for the times given in the figure.

$t=5$ s — 125 m — $t=5$ s
$t=2$ s — 80 m — $t=8$ s
$t=1$ s — 45 m — $t=9$ s
$t=0$ s — 0 meters — $t=10$ s
Earth's Surface

If an object is shot upward from the surface of the earth with an initial velocity of v meters per second, then at the end of t seconds its height h in meters is given by the formula below.

$$h = vt - 5t^2$$

Example 4   Find v in meters per second if $h = 120$ m and $t = 3$ s. Use the formula $h = vt - 5t^2$.

$h = vt - 5t^2$
$120 = v \cdot 3 - 5 \cdot 3^2$
$165 = 3v$
$55 = v$

Thus, v is 55 m/s.

Problem Solving: Using Formulas

Written Exercises

Solve each problem. Use the formula $i = prt$.

(A)
1. For a loan of $6,500 for 8 yr, Jose was charged $3,640 in simple interest. What was the annual interest rate? **7%**

2. For a loan of $2,500 for 5 yr 9 mo, Muna was charged $1,150 in simple interest. What was the annual interest rate? **8%**

3. When would a borrower owe $6,528 in simple interest on a loan of $6,400 at $8\frac{1}{2}\%$ per year? **12 yr**

4. When would a borrower owe $1,152 in simple interest on a loan of $3,200 at 8% per year? **4 yr 6 mo**

5. Ms. Brown borrowed some money on February 1, 1972, at 6% per year. How much did she borrow if Ms. Brown paid off the debt, including $3,000 in simple interest, on February 1, 1982? **$5,000**

6. Mr. Adamshick borrowed some money on July 1, 1974, at $7\frac{1}{2}\%$ per year. How much did he borrow if Mr. Adamshick paid off the debt, including $1,485 in simple interest, on October 1, 1982? **$2,400**

Solve each problem. Use the formula $A = p + prt$.

7. Find p in dollars if $A = \$9,780$, $r = 9\%$ and $t = 7$ yr. **$6,000**

8. Find p in dollars if $r = 8\frac{1}{2}\%$, $t = 6$ yr, and $A = \$5,285$. **$3,500**

Solve each problem. Use the formula $R(r_1 + r_2) = r_1 \cdot r_2$.

9. Find R in ohms if $r_1 = 50$ ohms and $r_2 = 30$ ohms. **18.75 ohms**

10. Find r_2 in ohms if $R = 19.2$ ohms and $r_1 = 32$ ohms. **48 ohms**

11. Find r_2 in ohms if $R = 24$ ohms and $r_1 = 60$ ohms. **40 ohms**

12. Find r_1 in ohms if $R = 21$ ohms and $r_2 = 70$ ohms. **30 ohms**

Use the formula $h = vt - 5t^2$ for Exercises 13–20.

13. Find v in meters per second if $h = 150$ m and $t = 5$ s. **55 m/s**

14. Find h in meters if $v = 60$ m/s and $t = 4$ s. **160 m**

15. At the end of 3 seconds, what will be the height of an arrow if an archer shoots it upward with an initial velocity of 120 meters per second? **315 m**

16. A certain baseball pitcher can throw a ball with an initial velocity of 35 m/s. If the pitcher throws the ball straight up, what will its height be at the end of 2 seconds and at the end of 5 seconds? **50 m, 50 m**

17. Explain how the two answers in Exercise 16 can be equal. **The first is on the way up and the second is on the way down.**

(B)
18. If an object is sent upward, what initial velocity is required to reach a height of 132 m at the end of 4.4 s? **52 m/s**

19. Find the initial velocity needed for a missile, launched upward, to achieve an altitude of 3,000 m at the end of 12 s. **310 m/s**

20. Find v in meters per second if $h = 26.25$ m and $t = 3.5$ s. **25 m/s**

21. If $h = vt - 5t^2$, the maximum value of h is $\frac{v^2}{20}$. What initial velocity is required to launch an object to a maximum height of 8,000 meters? **400 m/s**

22. Find the maximum height achieved by a missile launched with an initial velocity of 250 m/s. Use the maximum value of h as $\frac{v^2}{20}$ from Ex. 21. **3,125 m**

23. Find r_2 in ohms if $R = 3.2$ ohms and $r_1 = 4.8$ ohms. **9.6 ohms**

24. Find r_1 in ohms if $R = 4.2$ ohms and $r_2 = 8.7$ ohms. **8.12 ohms**

Solve each problem.

25. Find p in dollars if $A = \$8,295$, $r = 8\frac{1}{2}\%$, and $t = 4$ yr 6 mo. **$6,000**

26. Find p in dollars if $A = \$4,650$, $r = 7\frac{1}{4}\%$, and $t = 6$ yr 3 mo. **$3,200**

27. Find t in years and months if $A = \$12,160$, $p = \$8,000$, and $r = 8\%$. **6 yr 6 mo**

28. Find the annual interest rate r if $A = \$697$, $p = \$400$, and $t = 9$ yr. **$8\frac{1}{4}\%$**

29. When will an investment of $1,000 at 8% per year in simple interest be doubled in total value? **12 yr 6 mo**

30. When will an investment of p dollars at 12.5% per year in simple interest be doubled in total value? **8 yr**

31. You buy a truck and pay $1,000 down. The finance charge is 8% per year on the original unpaid balance. If you must pay the finance company $9,900 in 48 equal monthly payments, what is the original price of the truck? **$8,500**

32. Susan earned some money last summer, spent $700 of it, and put the remainder in an account that paid 6% per year. At the end of 6 mo, she had $2,575 in her account. Find the total amount she earned last summer. **$3,200**

For Ex. 34, see TE Answer Section.

Ⓒ 33. When will an investment of p dollars at $r\%$ per year in simple interest be doubled in total value? **$\frac{1}{r}$ yr**

34. Prove: If $R(r_1 + r_2) = r_1 \cdot r_2$, then $\frac{1}{R} = \frac{1}{r_1} + \frac{1}{r_2}$.

The diagram shows an object launched from a platform 80 m above the earth's surface. If an object is sent upward from d meters above the surface of the earth with an initial velocity of v meters per second, then its height h in meters at the end of t seconds is determined by the formula
$$h = d + vt - 5t^2.$$

Use the formula $h = d + vt - 5t^2$ for Exercises 35–39.

35. Find h if $d = 120$ m, $v = 90$ m/s, and $t = 5$ s. **445 m**

36. Find v if $h = 195$ m, $d = 75$ m, and $t = 3$ s. **55 m/s**

37. Find d if $h = 765$ m, $v = 150$ m/s, and $t = 6$ s. **45 m**

38. Find v for the diagram. **30 m/s**

39. At the end of 3 s, what is the height of an object shot upward with an initial velocity of 75 m/s from the *bottom* of a shaft that extends 60 m below the earth's surface? **120 m**

Problem Solving: Using Formulas

APPLICATIONS: PROPERTIES OF A FIELD

A set F is a *field* for the operations of addition and multiplication if the eleven properties below exist for all numbers x, y, and z in F.

1–A.	*Closure* for addition	$x + y$ is in F.
1–M.	*Closure* for multiplication	$x \cdot y$ is in F.
2–A.	*Commutative* for addition	$x + y = y + x$
2–M.	*Commutative* for multiplication	$x \cdot y = y \cdot x$
3–A.	*Associative* for addition	$(x + y) + z = x + (y + z)$
3–M.	*Associative* for multiplication	$(x \cdot y) \cdot z = x \cdot (y \cdot z)$
4–A.	*Zero* (additive identity)	There exists a unique number 0 in F such that $x + 0 = x$.
4–M.	*One* (multiplicative identity)	There exists a unique number 1 in F such that $x \cdot 1 = x$.
5–A.	*Opposite* (additive inverse)	For each x, there exists a unique y such that $x + y = 0$. y is the additive inverse of x, or $y = -x$.
5–M.	*Reciprocal* (multiplicative inverse)	For each $x \neq 0$, there exists a unique y such that $x \cdot y = 1$. y is the multiplicative inverse of x, or $y = \frac{1}{x}$.
6.	*Distributive* for multiplication over addition	$x \cdot (y + z) = x \cdot y + x \cdot z$

Written Exercises

Each set of numbers below is *not* a field for addition and multiplication. Identify the first property in the list above that does not exist for the set. (Ex. 1–4)

1. {all whole numbers: 0, 1, 2, 3, 4, . . .}   **5–A**
2. {all odd integers: . . . , −5, −3, −1, 1, 3, 5, . . .}   **1–A**
3. {all even integers: . . . , −4, −2, 0, 2, 4, . . .}   **4–M**
4. {all integers: . . . , −3, −2, −1, 0, 1, 2, 3, . . .}   **5–M**

5. Does the set of all rational numbers $\frac{a}{b}$ where a and b are integers, $b \neq 0$, form a field for addition and multiplication?   **yes**

6. Does {0, 1, 2}, along with the addition and multiplication tables at the right, form a field?   **yes**

+	0	1	2
0	0	1	2
1	1	2	0
2	2	0	1

·	0	1	2
0	0	0	0
1	0	1	2
2	0	2	1

7. Construct an addition (+) table and a multiplication (·) table for {0, 1, 2, 3} using a "four-hour" clock. Does {0, 1, 2, 3} along with the operations + and · form a field?   **No; for tables, see TE Answer Section.**

> The following exercises are types that appear in the (A) section of the lessons' exercises: 1, 3, 4, 9, 12, 14, 19–24, 27. The starred exercises are types that appear in the (C) section. All the remaining exercises appear in the (B) section.

CHAPTER ONE REVIEW

Vocabulary
absolute value [1.4]
conjunction [1.3]
consecutive integers [1.5]
disjunction [1.3]
linear equation [1.1]
linear inequality [1.2]
perimeter [1.6]

Solve each equation.
1. $5a + 8 - 3a = 4a - 5 - 6a$ **−3.25** [1.1]
2. $12 + \frac{3}{5}x = \frac{2}{5}x - 3$ **−75**
3. $7.2c + 2.3 = 1.4c - 0.6$ **−0.5**
4. $0.05 + 2.7(3n - 0.3) = 4.3n$ **0.2**
5. $-2x - (4 - 5x) = x - 3$ **$\frac{1}{2}$**
6. $4c - 2(4c - 3) = 3(2c - 5)$ **$2\frac{1}{10}$**
7. $\frac{2}{3}(6x - 9) + \frac{3}{4}(8x + 4) = 28 - 4x$ **$2\frac{3}{14}$**
★ 8. $12 - 7[6(2x + 5) - 3(7x + 8)] = 0$ **$\frac{10}{21}$**
9. $|3x - 15| = 6$ **3, 7** [1.4]
10. $|4y + 3| = 9$ **−3, $1\frac{1}{2}$**
★ 11. $|n + 4| + |3 - 2n| = 16$ **$-5\frac{2}{3}$, 5**

Find the solution set of each inequality. Graph the solutions on a number line.
12. $-5n - 3 \leq 17$ **$\{n|n \geq -4\}$** [1.2]
13. $3x < x - 4$ or $5x - 3 > 7$ [1.3] **$\{x|x < -2$ or $x > 2\}$**
14. $|y + 4| < 6$ **$\{y|-10 < y < 2\}$** [1.4]
15. $|5z - 15| \geq 30$ **$\{z|z \leq -3$ or $z \geq 9\}$**
★ 16. $|x - 4| < 6$ or $|x - 4| > 12$
$\{x|x < -8$ or $-2 < x < 10$ or $x > 16\}$

Find the solution set of each inequality.
17. $2(3a + 1) > 3(4a - 5) - 2a$ **$\{a|a < 4\frac{1}{4}\}$**
★ 18. $\frac{4(y - 6)}{-3} > 12 - 2y$ **$\{y|y > 6\}$**
19. $-2x < 6$ and $3x > -15$ **$\{x|x > -3\}$** [1.3]

Represent each of the three numbers in terms of one variable. [1.5]
20. A second number is 8 times a first number. A third number is 5 more than 3 times the second number. **1st: x, 2nd: 8x, 3rd: 24x + 5**

Solve each problem.
21. Twice the sum of a number and 3 is the same as 24 less than 7 times the number. What is the number? [1.5] **6**

22. The sum of three consecutive even integers is 20 more than twice the second integer. Find the three integers. [1.5] **18, 20, 22**

23. Side a of a triangle is 3 times as long as side b and side c is 2 cm shorter than side a. Find the length of each side if the perimeter is 26 cm. [1.6]
a: 12 cm, b: 4 cm, c: 10 cm

24. Paul is 7 years younger than Carol. Three years from now, she will be twice as old as he will be. How old is each now? [1.7]
Carol: 11, Paul: 4

25. An oak tree is 6 times as old as a pine tree and a spruce tree is 20 years younger than the oak tree. Three years ago, the age of the oak was 3 times the sum of the ages of the pine and the spruce. Find the present ages of the trees. [1.7]
pine: 5, oak: 30, spruce: 10

26. Find p in dollars if $A = \$7,623$, $r = 8\frac{1}{2}\%$, and $t = 4$ yr 3 mo. Use the formula $A = p + prt$. [1.8] **$5,600**

27. Find r_2 in ohms if $R = 8.4$ ohms and $r_1 = 12$ ohms. Use the formula $R(r_1 + r_2) = r_1 \cdot r_2$. [1.8] **28 ohms**

28. Find the initial velocity needed for a missile, launched upward, to reach a height of 4,875 m at the end of 15 s. Use the formula $h = vt - 5t^2$. [1.8] **400 m/s**

★ 29. The length of a rectangle is 6 cm more than the width. Find the set of all possible lengths if the perimeter must be less than 54 cm. [1.6] **$\{l|6$ cm $< l < 16.5$ cm$\}$**

★ 30. In $x + 4$ years, a tree will be y years old. Represent the age of the tree z years ago in terms of x, y, and z. [1.7] **$y - x - 4 - z$**

Chapter One Review

CHAPTER ONE TEST

The following exercises are types that appear in the Ⓐ section of the lessons' exercises: 1, 4, 6–8, 13, 14, 16, 17, 19. The starred exercises are types that appear in the Ⓒ section. All the remaining exercises appear in the Ⓑ section.

Solve each equation.
1. $8 + 5c - 6 = 9c - 22 - 12c$ **−3**
2. $6 + \frac{3}{5}n = 14 - \frac{1}{5}n$ **10**
3. $3 + 5(2y - 1) = 2y - 4(y + 1)$ **$-\frac{1}{6}$**
4. $2.5(6y - 1.2) = 22.8 - 2.2y$ **1.5**
5. $\frac{1}{2}(16x + 24) + \frac{2}{3}(6x - 9) = 42$ **3**
6. $|2a - 7| = 15$ **−4, 11**

Find the solution set of each inequality. Graph the solutions on a number line.
7. $3 - 4n \geq 19$ **{n|n ≤ −4}**
8. $|y + 2| < 5$ **{y|−7 < y < 3}**

Find the solution set of each inequality.
9. $4(y - 2) + 1 \leq 3(2y + 1)$ **{y|y ≥ −5}**
10. $|2a + 3| > 11$ **{a|a < −7 or a > 4}**
11. $5x > x - 12$ or $3x > x + 12$ **{x|x > −3}**
12. $-8 \leq 3x - 2 \leq 16$ **{x|−2 ≤ x ≤ 6}**

Represent each number in terms of one variable.
13. A third number is 7 less than a first number. A second number is 5 times the first number. **1st: x, 2nd: 5x, 3rd: x − 7**

Solve each problem.
14. The sum of three consecutive integers is 40 less than 4 times the third integer. Find the three integers. **35, 36, 37**

15. Room A has twice as many chairs as room B. Room C has 20 more chairs than room A. If the number of chairs in C is 10 less than the total number of chairs in A and B, find the number of chairs in each room. **A: 60, B: 30, C: 80**

16. Gloria is 4 times as old as Theotis. Three years ago, she was 7 times as old as he was. How old is each now? **Theotis: 6, Gloria: 24**

17. Sides a and b of a triangle have the same length. Side c is 12 cm less than 3 times the length of b. Find the length of each side if the perimeter is 23 cm. **a: 7 cm, b: 7 cm, c: 9 cm**

18. Find p in dollars if $A = \$5{,}620$, $r = 9\%$, and $t = 4$ yr 6 mo. Use the formula $A = p + prt$. **$4,000**

19. Find r_2 in ohms if $R = 4.8$ ohms and $r_1 = 8$ ohms. Use the formula $R(r_1 + r_2) = r_1 \cdot r_2$. **12 ohms**

20. Building A was built 10 years before building B and 5 years after building C. In 20 years, the combined ages of buildings A and C will be 3 times that of building B. What is the present age of each building? **A: 15, B: 5, C: 20**

21. If an object is sent upward, what initial velocity is required to reach a height of 200 m at the end of 3.2 s? Use the formula $h = vt - 5t^2$. **78.5 m/s**

★ 22. Solve. $5[4(6x - 3) - 7(5x - 2)] = 12(1 - 5x)$ **$\frac{2}{5}$**

★ 23. Find and graph the solution set of $|x + 7| \leq 3$ or $|x - 7| \leq 3$. **{x|−10 ≤ x ≤ −4 or 4 ≤ x ≤ 10}**

★ 24. The length of a rectangle is 10 m less than 3 times the width. Find the set of all possible widths if the perimeter must be less than 36 m. **{w|$3\frac{1}{3}$ m < w < 7 m}**

★ 25. Solve $|n - 5| + |-n - 3| = 12$. **−5, 7**

COLLEGE PREP TEST

DIRECTIONS: Choose the *one* best answer to each question or problem.

1. Which investment earns the most simple interest? **E**

   (A) $4,444 at $8\frac{1}{2}\%$ per year for 2yr 3mo
   (B) $4,444 at $4\frac{1}{4}\%$ per year for 4 yr 6 mo
   (C) $2,222 at $8\frac{1}{2}\%$ per year for 4 yr 6 mo
   (D) $1,111 at $8\frac{1}{2}\%$ per year for 9 yr
   (E) They all earn the same amount of interest.

2. If 2 more than $\frac{2}{3}$ of a certain number is 8, then the number is **C**

   (A) 6 (B) 8 (C) 9
   (D) 15 (E) 18

3. In the equation $4x + 6x = 3(8 + tx)$, for what value of t is x equal to 6? **A**

   (A) 2 (B) 4 (C) 6
   (D) 8 (E) 10

4. If $ax < b$, where a, x, and b are integers, then **E**

   (A) $x < \frac{b}{a}$ (B) $x < b - a$
   (C) $x > \frac{b}{a}$ (D) $x > b - a$
   (E) None of these

5. Which fraction is greater than $\frac{1}{5}$ but less than $\frac{1}{4}$? **B**

   (A) $\frac{9}{20}$ (B) $\frac{2}{9}$ (C) $\frac{1}{9}$ (D) $\frac{2}{7}$ (E) $\frac{1}{6}$

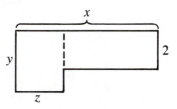

6. The figure above contains two rectangular regions and has the dimensions shown. The area of the figure in square units is **D**

   (A) $2xyz$ (B) $xy - yz - 2x + 2z$
   (C) $2x + yz$ (D) $2x + yz - 2z$
   (E) None of these

7. If $0 < x < 1$, then which one of the following is false? **C**

   (A) $x^2 < x$ (B) $2x < 2$
   (C) $\frac{2}{x} < 2$ (D) $\frac{x}{2} < \frac{1}{2}$
   (E) None of these

8. If $5(3x - 4) - 6x = 4(2x + 6) + x$, then **D**

   (A) $x = -20$
   (B) $x = 24$
   (C) $x = -20$ or $x = 24$
   (D) There is no solution to the equation.
   (E) Every number is a solution to the equation.

9. The hypotenuse of a certain right triangle is 5 cm long and one leg is 3 cm long. Find the area of the triangle. **D**

   (A) 3 cm^2 (B) 4 cm^2 (C) 5 cm^2
   (D) 6 cm^2 (E) 12 cm^2

College Prep Test

2 POLYNOMIALS AND PROBLEM SOLVING

Plan How to Solve a Problem

The second step in problem solving is to *plan* how you will solve the problem. Consider this problem and the analysis that must be done before an equation is written to solve it.

How much water must be added to 250 mL of a 15% sugar solution to dilute it to a 12% solution?

Suppose x mL of water is added.

	Total	Sugar	Water
15%	250	0.15(250)	0.85(250)
12%	250 + x	0.12(250 + x)	0.88(250 + x)

To write an equation, you need to understand the following:
(1) A sugar solution consists of (a) sugar and (b) water.
(2) If water alone is added to a sugar solution, the *amount of sugar* in the total solution *remains the same*.

Using this information, an equation can be written for the problem.

Sugar in 15% solution = sugar in 12% solution
$$0.15(250) = 0.12(250 + x) \quad \textbf{62.5 mL}$$

PROPERTIES OF EXPONENTS 2.1

Objectives **To simplify expressions containing positive integral exponents**
To solve an exponential equation

The *third power* of x, or the *cube* of x, is written as x^3. In the expression x^3, x is the **base** and 3 is the **exponent.** Exponents that are positive integers indicate the number of times the base is used as a factor. Thus, x^3 means $x \cdot x \cdot x$. When no exponent is written for a base, the exponent is understood to be 1. For example, the first power of y may be written as either y^1 or y.

You can use the meaning of exponents to discover how to multiply powers of the same base.

$$x^3 \cdot x \cdot x^4 = (xxx)(x)(xxxx) = x^8$$

Notice that the sum of the exponents in $x^3 \cdot x^1 \cdot x^4$ is $3 + 1 + 4$, or 8. This suggests the following property of exponents.

Product of powers For each number x and all positive integers m and n,
$$x^m \cdot x^n = x^{m+n}.$$

Example 1 **Simplify $-5a^4b \cdot 6a^3b^2$.**

$-5 \cdot a^4 \cdot b^1 \cdot 6 \cdot a^3 \cdot b^2$
$(-5 \cdot 6)(a^4 \cdot a^3)(b^1 \cdot b^2)$ ◀ *Group powers with the same base.*
$-30a^7b^3$

You can use the meaning of exponents to discover how to divide powers of the same base.

$$\frac{x^5}{x^2} = \frac{xx \cdot xxx}{xx} = x^3 \qquad \frac{y^3}{y^6} = \frac{yyy}{yyy \cdot yyy} = \frac{1}{y^3} \qquad \frac{z^3}{z^3} = \frac{zzz}{zzz} = 1$$

Notice that $\frac{x^5}{x^2} = x^{5-2}$ and that $\frac{y^3}{y^6} = \frac{1}{y^{6-3}}$. This leads to the following *Quotient of Powers* properties.

Quotient of powers For each number $x \neq 0$ and all positive integers m and n,
$$\frac{x^m}{x^n} = x^{m-n} \text{ if } m > n, \quad \frac{x^m}{x^n} = \frac{1}{x^{n-m}} \text{ if } m < n, \text{ and}$$
$$\frac{x^m}{x^n} = 1 \text{ if } m = n.$$

Example 2 **Simplify $\frac{-9a^7b^5c}{12a^4b^5c^6}$.**

$$\frac{-9}{12} \cdot \frac{a^7}{a^4} \cdot \frac{b^5}{b^5} \cdot \frac{c^1}{c^6} = \frac{-3}{4} \cdot a^{7-4} \cdot 1 \cdot \frac{1}{c^{6-1}} = \frac{-3a^3}{4c^5}$$

Properties of Exponents

The Product of Powers and the Quotient of Powers properties on page 31 apply only when the bases are the same. Neither property can be used with $a^7 \cdot b^3$ in Example 1 or with $\dfrac{a^3}{c^5}$ in Example 2.

An expression such as $(x^4)^3$ is called a *power of a power*. For the exponent 3, the base is x^4. From the meaning of exponents and the Product of Powers property, it follows that
$$(x^4)^3 = x^4 \cdot x^4 \cdot x^4 = x^{4+4+4} = x^{12}.$$

Notice that the product of the two exponents in $(x^4)^3$ is $4 \cdot 3$, or 12. This leads to the following property.

Power of a power

For each number x and all positive integers m and n,
$$(x^m)^n = x^{m \cdot n}.$$

Example 3

Simplify $6(c^4)^5$.

$$6(c^4)^5 = 6 \cdot c^{4 \cdot 5} = 6c^{20}$$

An expression such as $(2a)^3$ or $\left(\dfrac{a}{b}\right)^4$ can be simplified using the following Power of a Product or Power of a Quotient property, respectively.

Power of a product
Power of a quotient

For all numbers x and y and each positive integer n,
(1) $(xy)^n = x^n \cdot y^n$; and
(2) $\left(\dfrac{x}{y}\right)^n = \dfrac{x^n}{y^n}$, $(y \neq 0)$.

Using these properties, $(2a)^3 = 2^3 \cdot a^3$, or $8a^3$ and $\left(\dfrac{a}{b}\right)^4 = \dfrac{a^4}{b^4}$.

Sometimes you must use more than one property of exponents to simplify an expression. This is shown in Examples 4 and 5.

Example 4

Simplify $4a^2(5a^4b^2c)^3$.

$$\begin{aligned}
4a^2(5a^4b^2c)^3 &= 4a^2 \cdot 5^3(a^4)^3(b^2)^3(c^1)^3 &&\blacktriangleleft \text{ Power of a Product} \\
&= 4a^2 \cdot 125a^{12}b^6c^3 &&\blacktriangleleft \text{ Power of a Power} \\
&= 500a^{14}b^6c^3 &&\blacktriangleleft \text{ Product of Powers}
\end{aligned}$$

Example 5

Simplify $\left(\dfrac{-5a^3b}{2c^2}\right)^4$.

$$\left(\dfrac{-5a^3b}{2c^2}\right)^4 = \dfrac{(-5a^3b)^4}{(2c^2)^4} = \dfrac{(-5)^4(a^3)^4 b^4}{2^4(c^2)^4} = \dfrac{625a^{12}b^4}{16c^8}$$

Equations such as $2^x = 16$, $4^{5x} = 64$, and $5^{4x+1} = 125$ are called **exponential equations** since the variable appears in the exponent. To solve an exponential equation, you must be familiar with powers of numbers like 2^4, 3^4, 4^3, and 5^3.

Example 6

Solve $5^{4x+1} = 125$.

$$5^{4x+1} = 125$$
$$5^{4x+1} = 5^3 \quad \blacktriangleleft \ 125 = 5^3$$
$$4x + 1 = 3 \quad \blacktriangleleft \ \text{If } 5^a = 5^b, \text{ then } a = b.$$
$$4x = 2$$
$$x = \frac{1}{2}$$

Check:
$$5^{4 \cdot \frac{1}{2}+1} = 5^3 = 125$$

Thus, the value of x is $\frac{1}{2}$.

Reading in Algebra

Supply the missing words and expressions in the following paragraph.

In the expression $9y^4$, the exponent is _____ and its base is _____, but in the expression $(9y)^4$, the _____ is 4 and its _____ is _____. Notice that in the expression _____, the base is $7c^3$, the _____ is 2, and the second power of $7c^3$ is multiplied by $5c$.

    4   y   exponent   base   9y $5c(7c^3)^2$ exponent

Written Exercises

Simplify each expression.

(A)
1. $5a^2b^3 \cdot 4ab^5$ **$20a^3b^8$**
2. $-12mn^7 \cdot 6m^3n^2$ **$-72m^4n^9$**
3. $-4x^5y^4(-9x^4y)$ **$36x^9y^5$**
4. $7c^3d^2 \cdot 4c^2d \cdot 10c^4d^6$ **$280c^9d^9$**
5. $\dfrac{10a^4b^7}{-5a^9b^2} \ \ \mathbf{-\dfrac{2b^5}{a^5}}$
6. $\dfrac{8x^9y^3z^2}{12x^3y^3z^8} \ \ \mathbf{\dfrac{2x^6}{3z^6}}$
7. $\dfrac{-4m^3np^5}{6m^{12}np^{10}q^2} \ \ \mathbf{\dfrac{-2}{3m^9p^5q^2}}$
8. $\dfrac{-9a^9bc^2d^7}{-6ac^{12}d^7} \ \ \mathbf{\dfrac{3a^8b}{2c^{10}}}$
9. $(y^6)^5$ **y^{30}**
10. $(n^4)^7$ **n^{28}**
11. $(5x)^3$ **$125x^3$**
12. $(-3c)^4$ **$81c^4$**
13. $7(10n^5)^3$ **$7,000n^{15}$**
14. $3(-3x^2)^4$ **$243x^8$**
15. $4y(11y)^2$ **$484y^3$**
16. $2a(5a^3)^4$ **$1,250a^{13}$**
17. $\left(\dfrac{-5a}{4}\right)^3 \ \ \mathbf{\dfrac{-125a^3}{64}}$
18. $\left(\dfrac{x^2}{y^6}\right)^4 \ \ \mathbf{\dfrac{x^8}{y^{24}}}$
19. $\left(\dfrac{-7c^5}{10d^3}\right)^2 \ \ \mathbf{\dfrac{49c^{10}}{100d^6}}$
20. $\left(\dfrac{2ab^2}{3c^3d}\right)^5 \ \ \mathbf{\dfrac{32a^5b^{10}}{243c^{15}d^5}}$
29. $-36a^9b^{11}c^4$
30. $70m^8n^6p^9$
31. $-1,250x^6y^9$
32. $1,875m^{12}n^8p^{16}$

Solve each equation.

21. $3^{x-2} = 81$ **6**
22. $10^{n+1} = 1,000$ **2**
23. $2^{3y} = 64$ **2**
24. $6^{-x} = 36$ **-2**

(B)
25. $5^{3x+1} = 625$ **1**
26. $4^{6-y} = 256$ **2**
27. $3^{1-2n} = 243$ **-2**
28. $2^{8x+6} = 1,024$ **$\frac{1}{2}$**

33. $384a^7b^9$
34. $-405x^{17}y^{12}$
35. $-4,000c^6d^{18}$
36. $162a^6b^{21}$

Simplify each expression.
41. $x^{4a}y^{2b}$
42. $x^{3a+2b}y^{a-b}$
43. $x^{2c+3}y^{3d-4}$
44. $x^{4n+6}y^{4n-2}$

29. $-4a^6b^4c^3 \cdot 9a^3b^7c$
30. $5m^7n^2p^6 \cdot 14mn^4p^3$
31. $10(-5x^2y^3)^3$
32. $3(5m^3n^2p^4)^4$
33. $6a(4a^2b^3)^3$
34. $-5x(3x^4y^3)^4$
35. $4d^3(-10c^2d^5)^3$
36. $2a^2b(-3ab^5)^4$

(C)
37. $x^ay^{3b} \cdot x^{2c}y^{4d}$
38. $x^{2a+3}y^{2b} \cdot x^{5a}y^{3b-1}$
39. $x^a(x^by^c)^d$
40. $x^{2m}y^n(x^{m+3}y^n)^3$
41. $\dfrac{x^{6a}y^{3b}}{x^{2a}y^b}$
42. $\dfrac{x^{4a+3b}y^{2a-4b}}{x^{a+b}y^{a-3b}}$
43. $\dfrac{x^{3c+2}y^{5d-1}}{x^{c-1}y^{2d+3}}$
44. $\dfrac{x^{6n-2}y^{3n+2}}{x^{2n-8}y^{4-n}}$

37. $x^{a+2c}y^{3b+4d}$
38. $x^{7a+3}y^{5b-1}$
39. $x^{a+bd}y^{cd}$
40. $x^{5m+9}y^{4n}$

Properties of Exponents

ZERO AND NEGATIVE INTEGRAL EXPONENTS   2.2

Objectives **To evaluate expressions containing zero and negative integral exponents**
To simplify expressions containing zero and negative integral exponents

In general, the Quotient of Powers properties indicate that to divide powers of the same base, you should subtract the exponents of the powers. Observe below what happens when this is applied to the quotient $\frac{5^3}{5^3}$.

$$\frac{5^3}{5^3} = 5^{3-3} = 5^0$$

But you know that $\frac{5^3}{5^3}$ is 1: $\frac{5^3}{5^3} = \frac{125}{125} = 1$

This implies that $5^0 = 1$, which leads to the following definition.

Definition: Zero exponent
For each number $x \neq 0$, $x^0 = 1$.

You can define 5^{-3} to mean $\frac{1}{5^3}$ and define x^{-n} to mean $\frac{1}{x^n}$, for each $x \neq 0$, by using the definition of zero exponent and extending the Quotient of Powers properties to include negative exponents.

$$\frac{5^0}{5^3} = 5^{0-3} = 5^{-3} \text{ and } \frac{5^0}{5^3} = \frac{1}{5^3}. \text{ Thus, } 5^{-3} = \frac{1}{5^3}.$$

This view of zero and negative exponents may be seen in the following pattern where any term is divided by 5 to obtain the next term. Notice that $5^0 = 1$ and $5^{-3} = \frac{1}{5^3}$.

... 5^3	5^2	5^1	5^0	5^{-1}	5^{-2}	5^{-3} ...
... 125	25	5	1	$\frac{1}{5}$	$\frac{1}{25}$, or $\frac{1}{5^2}$	$\frac{1}{125}$, or $\frac{1}{5^3}$ ...

Sometimes a negative exponent appears in a denominator, as in $\frac{1}{2^{-4}}$. Study the steps below which show that $\frac{1}{2^{-4}} = 2^4$.

$$\frac{1}{2^{-4}} = 1 \div (2^{-4}) = 1 \div \frac{1}{2^4} = 1 \div \frac{1}{16} = 1 \times 16 = 16 = 2^4$$

Definition: Negative exponent
For each number $x \neq 0$ and each positive integer n,
$$x^{-n} = \frac{1}{x^n} \text{ and } \frac{1}{x^{-n}} = x^n.$$

Chapter Two

In expressions like $6c^{-2}$ and $(6c)^{-2}$, you should be careful in determining the base for the exponent -2. The base for the exponent -2 in $6c^{-2}$ is c and the base in $(6c)^{-2}$ is $(6c)$.

Example 1 Find the value of $2 \cdot 5^{-3}$ and $\dfrac{7}{(2 \cdot 5)^{-3}}$.

$$2 \cdot 5^{-3} = \frac{2}{1} \cdot \frac{1}{5^3} = \frac{2}{125} \qquad \frac{7}{(2 \cdot 5)^{-3}} = 7(2 \cdot 5)^3 = 7 \cdot 10^3 = 7000$$

A number is written in **scientific notation** when it is in the form

$$a \times 10^c, \text{ where } 1 \le a < 10 \text{ and } c \text{ is an } integer.$$

Notice below that the integral exponent c indicates the number of places and the direction to shift the decimal point in a to obtain **ordinary notation**.

$$8.64 \times 10^3 = 8640.0 \quad \text{and} \quad 8.64 \times 10^{-3} = 0.00864$$

Example 2 Find the value of $\dfrac{6.6 \times 10^{-1}}{8.8 \times 10^2}$ in ordinary notation.

$$\frac{6.6 \times 10^{-1}}{8.8 \times 10^2} = \frac{6.6}{8.8} \times 10^{-1} \times 10^{-2} = 0.75 \times 10^{-3} = 0.00075$$

Each of the five properties of exponents can now be extended to, and used with, expressions containing zero and negative integral exponents.

Example 3 Simplify and write each expression with positive exponents.

$$-5x^4 \cdot 3y^{-2} \cdot x^{-7} \cdot y^5 \qquad\qquad \frac{-8a^8b^{-3}c^{-4}}{12a^{-2}b^{-5}c^5}$$

$$\begin{aligned}
&-5 \cdot x^4 \cdot 3 \cdot y^{-2} \cdot x^{-7} \cdot y^5 \\
&(-5 \cdot 3)(x^4 \cdot x^{-7})(y^{-2} \cdot y^5) \\
&-15x^{4+(-7)}y^{-2+5} \\
&-15x^{-3}y^3 \\
&\frac{-15y^3}{x^3} \blacktriangleleft \; x^{-3} = \frac{1}{x^3}
\end{aligned} \qquad \begin{aligned}
&\frac{-8}{12} \cdot \frac{a^8}{a^{-2}} \cdot \frac{b^{-3}}{b^{-5}} \cdot \frac{c^{-4}}{c^5} \\
&\frac{-2}{3} \cdot \frac{a^8 \cdot a^2}{1} \cdot \frac{b^5}{b^3} \cdot \frac{1}{c^5 \cdot c^4} \blacktriangleleft \; \frac{b^{-m}}{b^{-n}} = \frac{b^n}{b^m} \\
&\frac{-2a^{10}b^2}{3c^9}
\end{aligned}$$

Example 4 Simplify $(5a^{-3}b^4)^{-2}$ and $\left(\dfrac{-4x^{-2}}{y^4z^{-3}}\right)^3$.

$$\begin{aligned}
(5a^{-3}b^4)^{-2} &= 5^{-2}(a^{-3})^{-2}(b^4)^{-2} \\
&= 5^{-2}a^6b^{-8} \\
&= \frac{a^6}{25b^8}
\end{aligned} \qquad \begin{aligned}
\left(\frac{-4x^{-2}}{y^4z^{-3}}\right)^3 &= \frac{(-4x^{-2})^3}{(y^4z^{-3})^3} = \frac{(-4)^3(x^{-2})^3}{(y^4)^3(z^{-3})^3} = \frac{-64x^{-6}}{y^{12}z^{-9}} \\
&= \frac{-64z^9}{y^{12}x^6}
\end{aligned}$$

Zero and Negative Integral Exponents

Reading in Algebra

For each phrase at the left, find its one correct match at the right.
1. The base of -3 in $5 \cdot 2^{-3}$ **D**
2. The base of -3 in $(5 \cdot 2)^{-3}$ **E**
3. An exponent in $5 \cdot 2^{-3}$ **B**
4. An exponent in $(5 \cdot 2)^{-3}$ **B**
5. The value of $5 \cdot 2^{-3}$ **H**
6. The value of $(5 \cdot 2)^{-3}$ **G**

A 3   B -3 C 5
D 2   E $5 \cdot 2$ F 1,000
G $\dfrac{1}{1{,}000}$ H $\dfrac{5}{8}$

Oral Exercises

Evaluate each expression if $x = 2$.
1. x^0 **1**
2. $4x^0$ **4**
3. $(4x)^0$ **1**
4. x^{-1} **$\tfrac{1}{2}$**
5. $5x^{-1}$ **$\tfrac{5}{2}$**
6. $(5x)^{-1}$ **$\tfrac{1}{10}$**

Written Exercises

Find the value of each expression.

Ⓐ
1. 8^0 **1**
2. $6 \cdot 4^0$ **6**
3. $(6 \cdot 4)^0$ **1**
4. 4^{-3} **$\tfrac{1}{64}$**
5. 2^{-4} **$\tfrac{1}{16}$**
6. $6^2 \cdot 6^0$ **36**
7. $4 \cdot 3^{-2}$ **$\tfrac{4}{9}$**
8. $(4 \cdot 3)^{-2}$ **$\tfrac{1}{144}$**
9. $\dfrac{2}{5^{-3}}$ **250**
10. $\dfrac{10}{3^{-4}}$ **810**
11. $\dfrac{2}{3 \cdot 4^{-2}}$ **$10\tfrac{2}{3}$**
12. $\dfrac{2}{(3 \cdot 4)^{-2}}$ **288**
13. $\dfrac{3^2}{2^{-3}}$ **72**
14. $\dfrac{3^{-2}}{2^3}$ **$\tfrac{1}{72}$**
15. $\dfrac{3^{-2}}{2^{-3}}$ **$\tfrac{8}{9}$**
16. $\dfrac{5^{-4}}{2^{-4}}$ **$\tfrac{16}{625}$**
17. $7^4 \cdot 7^{-6}$ **$\tfrac{1}{49}$**
18. $7^{-4} \cdot 7^6$ **49**
19. 2.7×10^{-2} **0.027**
20. $\dfrac{6.5}{10^{-3}}$ **6,500**
21. 1.23×10^{-4} **0.000123**
22. $\dfrac{3.57}{10^{-2}}$ **357**
23. 46.8×10^{-3} **0.0468**
24. $\dfrac{0.097}{10^0}$ **0.097**

Simplify and write each expression with positive exponents.
25. $9x^{-3}$ **$\tfrac{9}{x^3}$**
26. $(3y)^{-4}$ **$\tfrac{1}{81y^4}$**
27. $-10a^{-3}$ **$\tfrac{-10}{a^3}$**
28. $(-2c)^{-5}$ **$\tfrac{1}{-32c^5}$**
29. $x^{-4} \cdot x^7$ **x^3**
30. $4y^3 \cdot y^{-7}$ **$\tfrac{4}{y^4}$**
31. $n^{-5} \cdot 7n^{-3}$ **$\tfrac{7}{n^8}$**
32. $6a^{-4} \cdot 2a^0$ **$\tfrac{12}{a^4}$**
33. $\dfrac{1}{10x^{-3}}$ **$\tfrac{x^3}{10}$**
34. $\dfrac{1}{(10x)^{-3}}$ **$1{,}000x^3$**
35. $\dfrac{5}{6a^{-2}}$ **$\tfrac{5a^2}{6}$**
36. $\dfrac{5}{(6a)^{-2}}$ **$180a^2$**
37. $\dfrac{x^6}{x^{-3}}$ **x^9**
38. $\dfrac{a^{-4}}{a^2}$ **$\tfrac{1}{a^6}$**
39. $\dfrac{10n^{-5}}{15n^{-8}}$ **$\tfrac{2n^3}{3}$**
40. $\dfrac{-7c^{-6}}{14c^{-4}}$ **$\tfrac{-1}{2c^2}$**
41. $(x^{-2})^5$ **$\tfrac{1}{x^{10}}$**
42. $(4a^3)^{-2}$ **$\tfrac{1}{16a^6}$**
43. $(2n^{-4})^{-5}$ **$\tfrac{n^{20}}{32}$**
44. $5(-5y^{-2})^4$ **$\tfrac{3{,}125}{y^8}$**
45. $\left(\dfrac{y^{-4}}{-5}\right)^3$ **$\tfrac{1}{-125y^{12}}$**
46. $\left(\dfrac{x^3}{y^{-2}}\right)^5$ **$x^{15}y^{10}$**
47. $\left(\dfrac{c^{-3}}{d^5}\right)^4$ **$\tfrac{1}{c^{12}d^{20}}$**
48. $\left(\dfrac{a^{-3}}{b^{-2}}\right)^6$ **$\tfrac{b^{12}}{a^{18}}$**

Find the value of each expression in ordinary notation.

Ⓑ
49. $\dfrac{8.4 \times 10^2}{4.2 \times 10^{-2}}$ **20,000**
50. $\dfrac{6.3 \times 10^{-7}}{0.9 \times 10^{-9}}$ **700**
51. $\dfrac{9.63 \times 10^{-1}}{3 \times 10^0}$ **0.321**
52. $\dfrac{6.6 \times 10^{-20}}{8.8 \times 10^{-18}}$ **0.0075**

61. $\tfrac{10}{x^5 y^2}$ 62. $\tfrac{30a^4}{b^6}$ 63. $-\tfrac{c^9}{27d^{12}}$ 64. $\tfrac{16c^8}{d^{24}}$

Simplify and write each expression with positive exponents.
53. $\dfrac{a^5 b^{-4}}{c^{-3} d^{-6}}$ **$\tfrac{a^5 d^6}{b^4 c^3}$**
54. $\dfrac{x^3 y^{-4}}{z^{-5}}$ **$\tfrac{x^3 z^5}{y^4}$**
55. $\dfrac{4a^{-2} b^5}{9c^3 d^{-4}}$ **$\tfrac{4b^5 d^4}{9a^2 c^3}$**
56. $\dfrac{6x^{-5}}{8y^{-2} z^{-1}}$ **$\tfrac{3y^2 z}{4x^5}$**
57. $\dfrac{x^8 y^{-14}}{x^{-2} y^{12}}$ **$\tfrac{x^{10}}{y^{26}}$**
58. $\dfrac{a^6 b^{-6}}{a^{10} b^{-8}}$ **$\tfrac{b^2}{a^4}$**
59. $\dfrac{6c^{-6} d^8}{9c^{-3} d^{-2}}$ **$\tfrac{2d^{10}}{3c^3}$**
60. $\dfrac{25m^{-12} n^{-10}}{45m^{-15} n^5}$ **$\tfrac{5m^3}{9n^{15}}$**
61. $2x^{-8} \cdot 5y^{-2} \cdot x^3$
62. $10a^7 \cdot b^{-6} \cdot 3a^{-3}$
63. $(-3c^{-3} d^4)^{-3}$
64. $(-2c^2 d^{-6})^4$
65. $\left(\dfrac{12x^{-3}}{y^3 z^{-5}}\right)^2$ **$\tfrac{144 z^{10}}{x^6 y^6}$**
66. $\left(\dfrac{5a^{-2} b^3}{3c^{-1}}\right)^4$ **$\tfrac{625 b^{12} c^4}{81 a^8}$**
67. $\left(\dfrac{-2x^{-4} y}{3z^3 w^{-2}}\right)^3$ **$\tfrac{-8y^3 w^6}{27x^{12} z^9}$**
68. $\left(\dfrac{ab^{-2} c^3}{d^{-3} ef^2}\right)^4$ **$\tfrac{a^4 c^{12} d^{12}}{b^8 e^4 f^8}$**

Ⓒ
69. x^{a-b} if $0 < a < b$ **$\tfrac{1}{x^{b-a}}$**
70. y^{c-d} if $c < d < 0$ **$\tfrac{1}{y^{d-c}}$**
71. z^{3m-3n} if $m < 0 < n$ **$\tfrac{1}{z^{3n-3m}}$**

POLYNOMIALS AND SYNTHETIC SUBSTITUTION 2.3

Objectives **To evaluate a polynomial for given values of its variables**
To add polynomials
To subtract polynomials

Five examples of algebraic **terms** are shown below.
$$-7 \quad z \quad 3t \quad x^2y \quad 2.3mn^2$$
Each such term is a numeral, a variable, or a product of a numeral and one or more variables.

An algebraic expression consisting of either one term or a sum of several terms is called a **polynomial**. Two examples of polynomials are shown below.
$$-3a^2b \quad 5x^2y - 6x + 4y - 20$$
The first polynomial consists of one term, and the second is the sum of four terms. Expressions like $\frac{3}{x}$ (dividing by a variable) and $\sqrt{x}$ (square root of a variable) are not classified as polynomials.

Some polynomials are classified by the number of terms they contain. For example, $5ab^3$ is a **monomial** (one term), $9n^4 - 1$ is a **binomial** (two terms), and $x^2 + xy - 5$ is a **trinomial** (three terms).

The **degree of a term** is the sum of the exponents of its variables. For example, $5a^2b^4c$ is of degree 7. The sum of the exponents is $2 + 4 + 1$, or 7. The degree of the number 5 is 0, since $5 = 5 \cdot 1 = 5x^0$.

The **degree of a polynomial** is the same as the degree of its term that has the greatest degree. For example, $4xy^2 + 7x^2y^3 - 3x$ is of degree 5. Its terms are of degrees 3, 5, and 1, respectively, and 5 is the greatest of these degrees.

Example 1 **Give the degree of each polynomial.**

Polynomial	Degree	Polynomial	Degree
$5x + 7$	1	$3x + 2y - 6$	1
$3x^2 - 2x - 1$	2	$x^2 - 4xy + 4y^2$	2
$x^3 + 9x^2 - 6x + 4$	3	$a^2 + 3ab^3 + a^3bc$	5

First-degree polynomials, such as $5x + 7$ and $3x + 2y - 6$, are called **linear polynomials**. Second-degree polynomials, such as $3x^2 - 2x - 1$ and $x^2 - 4xy + 4y^2$, are called **quadratic polynomials**.

You can evaluate a polynomial for given values of its variables by **direct substitution** as shown below.

If $P = 5x^3y^2 - xy + 14$ and $x = -2$ and $y = 3$,
then, $P = 5(-2)^3 \cdot 3^2 - (-2) \cdot 3 + 14 = 5 \cdot -8 \cdot 9 - (-6) + 14 = -340$.

Given a polynomial $P = 3x^3 + 10x^2 - 5x - 4$ *with one variable,* you can evaluate P for $x = 2$ by **synthetic substitution**. First, notice that:

$P = 3x^3 + 10x^2 - 5x - 4 = (3x^2 + 10x - 5) \cdot x - 4$

$$ = ([3 \cdot x + 10] \cdot x - 5) \cdot x - 4$$
$$= ([3 \cdot 2 + 10] \cdot 2 - 5) \cdot 2 - 4 = \boxed{50} \text{ if } x = 2.$$

Write the coefficients of P.
Repeat the cycle:
Multiply by 2 and add.

```
       3    10   -5   -4
             6   32   54
    2│ 3    16   27   (50)
```

By **synthetic substitution,** $P = 50$ if $x = 2$.

There is a **shorter form** of synthetic substitution.

```
      3  10  -5  -4      ◄ Do not write the
   2│ 3  16  27  (50)       multiples of 2.
```

Example 2

Evaluate $2x^4 - x^3 + 5x + 3$ if $x = -2$ and again if $x = 3$. Use the shorter form of synthetic substitution.

```
         2  -1   0    5     3    ◄ Write 0 for the coefficient of
   -2│   2  -5   10  -15  (33)     the missing x² term.
    3│   2   5   15   50  (153)
```

Thus, the values are 33 if $x = -2$ and 153 if $x = 3$.

You can add and subtract polynomials by combining the like (similar) terms. For example, $3a^2b$ and $4a^2b$ are like terms, but $5a^2b^3$ and $7a^3b^2$ are not like terms. To simplify the **difference** of two polynomials, use the rule: $x - y = x + (-y)$.

Example 3

Simplify $(7a^2b - 3ab + b^2) - (8ab - 9a^2b + 2b^2)$.

$(7a^2b - 3ab + b^2) + (-8ab + 9a^2b - 2b^2)$ ◄ Add the opposite of the second polynomial.

$(7a^2b + 9a^2b) + (-3ab - 8ab) + (1b^2 - 2b^2)$ ◄ Group the like terms.
$16a^2b - 11ab - b^2$ ◄ Combine the like terms.

Example 4   Subtract: $-8x^3 - 2x^2 - 4x$
$-8x^3 + 7x^2 - 10$

Multiply the second polynomial by -1. ▶ $\begin{array}{l}-8x^3 - 2x^2 - 4x\\ 8x^3 - 7x^2 + 10\end{array}$
Then add. ▶ $-9x^2 - 4x + 10$

It is often useful to arrange the terms of a polynomial, with one variable, in *descending order* of the exponents. In the polynomial $8x^2 - 5x - 7x^4 + 6 + 2x^5$, the exponents in order from left to right are 2, 1, 4, 0, 5. This polynomial can be rewritten as $2x^5 - 7x^4 + 8x^2 - 5x + 6$, with its terms in descending (or decreasing) order of the exponents.

Reading in Algebra

Determine whether each expression is a polynomial (Yes or No).
1. $5a - 3b + 6$ **Yes**
2. 9 **Yes**
3. $2y + \sqrt{y}$ **No**
4. $\frac{1}{4}a + b^2 + \frac{2}{3}$ **Yes**
5. $7m + n^4 + \frac{4}{p^2}$ **No**
6. $\frac{1}{3}x^2 y$ **Yes**

Determine whether each polynomial is linear (L), quadratic (Q), or neither (N).
7. $8y^2 - 6y - 2$ **Q**
8. $-6x + 12$ **L**
9. $3n^3 + 2n^2 - 1$ **N**
10. $10x + 5y + 15$ **L**
11. $3x^2y + 4xy + 5y$ **N**
12. $xy - x^2 + y^2$ **Q**

Evaluate each polynomial if $x = 2$ and $y = -1$.
13. $y - x$ **-3**
14. $xy - y$ **-1**
15. $-xy$ **2**
16. $7x - 5y$ **19**
17. $-10x - 2y$ **-18**
18. $x^3 - x^2 + y$ **3**

Write the next row to complete each synthetic substitution.
19. $\begin{array}{rrrr}2 & 3 & -1 & 1\end{array}$
$\underline{-1|}\begin{array}{rrrr}2 & 1 & -2 & 3\end{array}$

20. $\begin{array}{rrrr}3 & -1 & 0 & -2\end{array}$
$\underline{2|}\begin{array}{rrrr}3 & 5 & 10 & 18\end{array}$

21. $\begin{array}{rrrr}2 & 0 & 0 & 16\end{array}$
$\underline{-2|}\begin{array}{rrrr}2 & -4 & 8 & 0\end{array}$

Oral Exercises

Classify each polynomial by the number of terms.
1. $y^2 - 9$ **binomial**
2. $4a^2x^2$ **monomial**
3. $4a^2 - 4a + 1$ **trinomial**
4. $8 - c^3$ **binomial**

Give the degree of each polynomial.
5. $5x - 8y + 9$ **1**
6. $4xy$ **2**
7. $a^2 - 5a^2b^3c$ **6**
8. 7 **0**

Arrange the terms of each polynomial in descending order of the exponents.
9. $x^2 - x + x^4 - 5$ **$x^4 + x^2 - x - 5$**
10. $8 + 3y + 4y^2 + y^3$ **$y^3 + 4y^2 + 3y + 8$**
11. $-n^2 + n^6 - n^4$ **$n^6 - n^4 - n^2$**
12. $-8 - x^3$ **$-x^3 - 8$**

Polynomials and Synthetic Substitution

Written Exercises

Evaluate each polynomial for the given values of its variables.

Ⓐ 1. $3x^2y - xy + 5y$ if $x = 3$ and $y = -2$ **−58**
2. $8x^2y + 5xy + 2xy^2$ if $x = -4$ and $y = 5$ **340**

Use synthetic substitution to evaluate each polynomial for $x = 3$ and again for $x = -2$.

3. $x^4 + x^3 - x^2 - x + 1$ **97, 7**
4. $2x^4 - 8x^2 - 5x - 5$ **70, 5**
5. $2x^5 - 8x^3 + 10x - 20$ **280, −40**
6. $3x^5 - 10x^4 + x^3 - 4x^2 + 5x + 5$ **−70, −285**

Add. Arrange the terms of the sum in descending order of the exponents. 7. **−5y + 41** 8. **$16n^2 + n - 3$**

7. $(3y - 12y^2 + 24) + (17 - 8y + 12y^2)$
8. $(16 + n^2 - 12n) + (13n + 15n^2 - 19)$
9. $(-15a^4 + a - 25a^3) + (6a^2 + 25a^3 - 9a)$ **$-15a^4 + 6a^2 - 8a$**
10. $(12x^3 - 6x^2 - 6x) + (5x^2 + 6x + x^5)$ **$x^5 + 12x^3 - x^2$**
11. $6 + 2a + a^2$
    $3 - 2a + a^2$ **$2a^2 + 9$**
12. $8 - 2n - 4n^2$
    $-6n + n^2 - 8n^3$ **$-8n^3 - 3n^2 - 8n + 8$**

Subtract. Arrange the terms of the difference in descending order of the exponents. 14. **$-4n^2 + 10n + 4$**

13. $(5x^2 + 3x - 4) - (2x^2 - 4x - 5)$ **$3x^2 + 7x + 1$**
14. $(2n - 3n^2 + 9) - (5 + n^2 - 8n)$
15. $(12y - 12y^2 + 12) - (-8y^2 + 14y + 15)$ **$-4y^2 - 2y - 3$**
16. $(-x^4 + x^6 - x^2 + 10) - (5x^2 + 6x^6 - x^4)$ **$-5x^6 - 6x^2 + 10$**
17. $5a^2 + 9a - 2$
    $5a^2 - 3a + 7$ **$12a - 9$**
18. $-4n^3 \qquad + 8n - 2$
    $-4n^3 - 5n^2 - n$ **$5n^2 + 9n - 2$**

Evaluate each polynomial for the given values of its variables.

Ⓑ 19. $10xy - 5x^2$ if $x = 0.4$ and $y = 0.5$ **1.2**
20. $a^3 + 100ab$ if $a = 0.2$ and $b = 1.25$ **25.008**
21. $1.4xy + 2.2y^2$ if $x = 2.5$ and $y = 4$ **49.2**
22. $3.6x^2 - 0.4xy$ if $x = 0.5$ and $y = 10$ **−1.1**

Use synthetic substitution to evaluate each polynomial for each of the given values of x.

23. $2x^4 + x^3 - 11x^2 - 4x + 12$; $-1, -3, 2,$ and 3 for x **6, 60, 0, 90**
24. $x^6 - x^4 - 9x^3 + 8x + 8$; $1, -1, 2, -2,$ and 3 for x **7, 9, 0, 112, 437**
25. $3x^4 - 26x^3 + 72x^2 - 54x - 27$; $0, 1, -1, 2,$ and 3 for x **−27, −32, 128, −7, 0**

Subtract. 26. **$14x^2y^2 - 7xy + 12$** 27. **$30x^2 + 27xy$**

26. $(6x^2y^2 - 2xy + 4) - (5xy - 8 - 8x^2y^2)$
27. $(29x^2 - y^2 + 28xy) - (xy - x^2 - y^2)$
28. $12n^3 - n^2 + 6n$
    $8n - 5n^2 - 3n^3$ **$15n^3 + 4n^2 - 2n$**
29. $6n^4 + 5n^2 - 4n - 30$
    $6n^4 + 2n^3 - 9n + 20$ **$-2n^3 + 5n^2 + 5n - 50$**

Evaluate each polynomial for $x = 10$. 32. **−123,456**

30. $9x^4 + 8x^3 + 7x^2 + 6x + 5$ **98,765**
31. $4x^5 + 4x^3 + 4x$ **404,040**
32. $-x^5 - 2x^4 - 3x^3 - 4x^2 - 5x - 6$
33. $-7x^6 - 7x^4 - 7x^2 - 7$ **−7,070,707**

Evaluate each polynomial for the given values of its variables.

Ⓒ 34. $5xy^3 + 5x^2y^2 + 10x^3y$
    if $x = 2a$ and $y = -3a$ **$-330a^4$**
35. $10x^4y^2 - 2x^3y^3 - x^2y^4 + 5xy^5$
    if $x = c^2$ and $y = -2c^2$ **$-120c^{12}$**

CUMULATIVE REVIEW

1. Multiply: $3n^2(5n^2 - n - 4)$. **$15n^4 - 3n^3 - 12n^2$**
2. Simplify $3a \cdot 5a^4b \cdot -4a^3b^3$. **$-60a^8b^4$**
3. Simplify $10c(5c^4d)^2$. **$250c^9d^2$**
4. Solve $|x - 9| = 3$. **6, 12**
5. Find the solution set of $|x - 9| > 3$.
   $\{x \mid x < 6 \text{ or } x > 12\}$
6. Solve $5 + 0.75x = 23$. **24**

MULTIPLYING POLYNOMIALS 2.4

Objective **To multiply two or more polynomials**

To multiply a monomial and a trinomial, you use the Distributive Property.

$3x(2x^2 - 4x + 6) = 3x \cdot 2x^2 + 3x(-4x) + 3x \cdot 6 = 6x^3 - 12x^2 + 18x.$

To simplify $(a + b)(c + d)$, the indicated product of any two binomials, you can use the Distributive Property three times as follows.

$$(a + b)[c + d] = a[c + d] + b[c + d] = ac + ad + bc + bd$$

The result, $ac + ad + bc + bd$, is the sum of the products of the: **F**irst terms (ac), **O**uter terms (ad), **I**nner terms (bc), **L**ast terms (bd). This suggests the **FOIL** method, an efficient method for multiplying any pair of binomials.

Example 1 **Simplify $(3x + 4)(2x - 5)$. Use the FOIL method.**

$$(3x + 4)(2x - 5) = 3x \cdot 2x + 3x(-5) + 4 \cdot 2x + 4(-5)$$
$$= 6x^2 - 15x + 8x - 20$$
$$= 6x^2 - 7x - 20$$

To simplify $3n(2n^2 - 6)(3n^2 - 5)$, it is easier to find the product of the binomial factors first and then multiply the result by the monomial factor.

Example 2 **Simplify $3n(2n^2 - 6)(3n^2 - 5)$.**

$$3n[(2n^2 - 6)(3n^2 - 5)] = 3n[2n^2 \cdot 3n^2 + 2n^2 \cdot -5 - 6 \cdot 3n^2 - 6 \cdot -5]$$
$$= 3n[6n^4 - 10n^2 - 18n^2 + 30]$$
$$= 3n[6n^4 - 28n^2 + 30]$$
$$= 18n^5 - 84n^3 + 90n$$

In Example 3, you are to multiply a binomial and a trinomial. To do this, distribute each term of the binomial to each term of the trinomial.

Example 3 **Simplify $(3x - 2y)(5x + y - 4)$.**

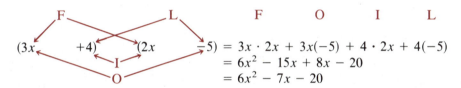

$$3x(5x + y - 4) - 2y(5x + y - 4)$$
$$15x^2 + 3xy - 12x - 10xy - 2y^2 + 8y$$
$$15x^2 - 7xy - 12x - 2y^2 + 8y$$

Multiplying Polynomials 41

Oral Exercises

Simplify.
1. $-2(8a^2 + 4a - 5)$ $-16a^2 - 8a + 10$
2. $5x^2(2x^2 - x + 3)$ $10x^4 - 5x^3 + 15x^2$
3. $-3x(4x - 5y - 1)$ $-12x^2 + 15xy + 3x$

Written Exercises

27. $12a^2 + 7ab + 9a - 10b^2 - 6b$
28. $2x^2 + 3xy + 6x - 5y^2 + 15y$
29. $3n^4 + 4n^3 - 17n^2 - 16n + 20$
30. $8a^4 - 2a^3 + 24a^2 - 3a + 18$
31. $2.24x^2 + 1.24x + 0.12$
32. $2.65n^2 - 3.96n + 0.35$

Simplify.

Ⓐ
1. $(4t + 5)(2t - 4)$ $8t^2 - 6t - 20$
2. $(6n - 3)(2n - 1)$ $12n^2 - 12n + 3$
3. $(2n + 8)(2n - 8)$ $4n^2 - 64$
4. $(3y - 4)(3y + 4)$ $9y^2 - 16$
5. $(3x - 2y)(2x + 4y)$ $6x^2 + 8xy - 8y^2$
6. $(5m + 4n)(3m + 2n)$ $15m^2 + 22mn + 8n^2$
7. $(x^2 + 10)(x^2 + 15)$ $x^4 + 25x^2 + 150$
8. $(n^3 - 8)(n^3 - 9)$ $n^6 - 17n^3 + 72$
9. $(2y^3 - 3)(5y^3 - 7)$ $10y^6 - 29y^3 + 21$
10. $(8x^2 + 5)(3x^2 + 4)$ $24x^4 + 47x^2 + 20$
11. $(7x + 0.4)(3x + 0.8)$ $21x^2 + 6.8x + 0.32$
12. $(5n - 0.6)(4n - 0.7)$ $20n^2 - 5.9n + 0.42$
13. $(0.8a + 6)(0.4a - 9)$ $0.32a^2 - 4.8a - 54$
14. $(0.7x - 3)(0.6x + 5)$ $0.42x^2 + 1.7x - 15$
15. $(5x + 4)(x^2 - 3x + 2)$ $5x^3 - 11x^2 - 2x + 8$
16. $(4y - 3)(2y^2 + 6y - 1)$ $8y^3 + 18y^2 - 22y + 3$
17. $(3y - 2)(3y^2 - y + 4)$ $9y^3 - 9y^2 + 14y - 8$
18. $(2x + 4)(4x^2 + 2x - 3)$ $8x^3 + 20x^2 + 2x - 12$
19. $4(2x - 5)(x + 3)$ $8x^2 + 4x - 60$
20. $3(4a + 5)(2a - 3)$ $24a^2 - 6a - 45$

Ⓑ
21. $-2(3y - 4)(3y - 4)$ $-18y^2 + 48y - 32$
22. $-4(2x + 10)(2x - 10)$ $-16x^2 + 400$
23. $(2x^2 + 3y)(4x^2 - 8y)$ $8x^4 - 4x^2y - 24y^2$
24. $(8a - 5b^2)(3a + 5b^2)$ $24a^2 + 25ab^2 - 25b^4$
25. $(3x^2 + 4y^2)(3x^2 - 5y^2)$ $9x^4 - 3x^2y^2 - 20y^4$
26. $(2c^3 - 3d^2)(2c^3 + 6d^2)$ $4c^6 + 6c^3d^2 - 18d^4$
27. $(3a - 2b)(4a + 5b + 3)$
28. $(2x + 5y)(x - y + 3)$
29. $(n^2 - 4)(3n^2 + 4n - 5)$
30. $(2a^2 + 3)(4a^2 - a + 6)$
31. $(3.2x + 0.4)(0.7x + 0.3)$
32. $(5.3n - 0.5)(0.5n - 0.7)$
33. $(3.5x + 0.9)(3.5x - 0.9)$ $12.25x^2 - 0.81$
34. $(5.4n - 0.7)(5.4n + 0.7)$ $29.16n^2 - 0.49$
35. $4x(3x + 2)(2x + 5)$ $24x^3 + 76x^2 + 40x$
36. $-2y(4y - 3)(5y + 2)$ $-40y^3 + 14y^2 + 12y$
37. $2y(5y^2 - 2)(5y^2 - 2)$ $50y^5 - 40y^3 + 8y$
38. $-3x(10x^3 - 5)(10x^3 + 5)$ $-300x^7 + 75x$

Ⓒ
39. Prove that $(a + b + c)(a + b + c) = a^2 + b^2 + c^2 + 2ab + 2ac + 2bc$. **See TE Answer Section.**

Simplify.
40. $(2x + y + 5)(2x + y + 5)$ $4x^2 + y^2 + 25 + 4xy + 20x + 10y$
41. $(8x - y - 4z)(8x - y - 4z)$ $64x^2 + y^2 + 16z^2 - 16xy - 64xz + 8yz$
42. $(x^n + 3)(x^n - 5)$ $x^{2n} - 2x^n - 15$
43. $(3x^{2a} + 4)(x^{2a} + 1)$ $3x^{4a} + 7x^{2a} + 4$
44. $(x^c + y^{2c})(x^c + 4y^{2c})$ $x^{2c} + 5x^cy^{2c} + 4y^{4c}$
45. $(4x^{2a} - 3y^b)(5x^{2a} + 2y^b)$ $20x^{4a} - 7x^{2a}y^b - 6y^{2b}$

NON-ROUTINE PROBLEMS

1. If $16 \cdot 2^x = 6^{y-8}$ and $y = 8$, then $x =$ **B**
   A. -8.   B. -4.   C. -3.   D. 0.   E. 8.

2. $[7 - 2(5 - 7)^{-1}]^{-1} =$ **B**
   A. 8 B. $\dfrac{1}{8}$ C. 11 D. $\dfrac{1}{11}$ E. -8

PROBLEM SOLVING: MIXTURE PROBLEMS 2.5

Objective **To solve a word problem involving a collection or mixture of items**

To solve certain problems that involve a *collection* or *mixture* of items, it is helpful to convert the values of the items to a common unit. For example, a collection of x nickels, $4x$ dimes, and $x + 6$ quarters is worth

$$5 \cdot x + 10 \cdot 4x + 25(x + 6) \text{ cents.}$$

You will use this technique in solving the problem below.

Example 1

READ ▶ The number of dimes in Martha's coin collection is 4 less than the number of nickels and the number of quarters is 3 times the number of dimes. If the value of her collection is $7.40, how many nickels, dimes, and quarters does she have?

PLAN ▶
Represent the data.

Let n = number of nickels.
$n - 4$ = number of dimes
$3(n - 4)$ = number of quarters

Write an equation.

$$\binom{\text{Value of}}{\text{nickels in } \cent} + \binom{\text{Value of}}{\text{dimes in } \cent} + \binom{\text{Value of}}{\text{quarters in } \cent} = \binom{\text{Value of}}{\text{collection in } \cent}$$

$$5n \quad + \quad 10(n - 4) \quad + \quad 25 \cdot 3(n - 4) \quad = \quad 100(7.40)$$

SOLVE ▶
Solve the equation.

$$5n + 10n - 40 + 75n - 300 = 740$$
$$90n - 340 = 740$$
$$90n = 1{,}080$$
$$n = 12$$

$n = 12$, $n - 4 = 8$, $3(n - 4) = 24$

INTERPRET ▶
Check the solution.

12 nickels + 8 dimes + 24 quarters	$7.40
$12 \cdot 5\cent + 8 \cdot 10\cent + 24 \cdot 25\cent$	$7.40 \cdot 100\cent$
$60\cent \quad + \quad 80\cent \quad + \quad 600\cent$	$740\cent$
$740\cent$	

Answer the question.

Thus, there are 12 nickels, 8 dimes, and 24 quarters.

Example 2 on page 44 is an example of a *dry mixture* problem. In such situations, a mixture is obtained by combining two types of items. The intent is to gain the same amount of money for the mixture as would be realized if the items to be mixed were sold individually. In other words, (Value of 1st Item) + (Value of 2nd Item) = (Value of Mixture).

Example 2

Peanuts worth $2.10/kg are mixed with almonds worth $3.60/kg to make 20 kg of a mixture worth $2.40/kg. How many kilograms of peanuts and how many kilograms of almonds are there in the mixture?

Let x = number of kilograms (kg) of peanuts.

	No. of kg	Value/kg	Value in ¢
Peanuts	x	210¢	$210x$
Almonds	$20 - x$	360¢	$360(20 - x)$
Mixture	20	240¢	$240 \cdot 20$

(Value of peanuts) + (Value of almonds) = (Value of mixture)
$$210x \quad + \quad 360(20 - x) \quad = \quad 240 \cdot 20$$
$$210x + 7{,}200 - 360x = 4{,}800$$
$$-150x = -2{,}400$$
$$x = 16 \qquad 20 - x = 4$$

Check: (16 kg at $2.10/kg) + (4 kg at $3.60/kg) | 20 kg at $2.40/kg
$(16 \cdot \$2.10) \quad + \quad (4 \cdot \$3.60)$ | $20 \cdot \$2.40$
$\$33.60 \quad + \quad \$14.40$ | $\$48.00$
$\$48.00$

Thus, there are 16 kg of peanuts and 4 kg of almonds in the mixture.

If a given liquid is classified as a 25% iodine solution, this means that 25% of the solution is iodine and 75% is water. Thus, 6 liters (L) of a 25% iodine solution is made up of 1.5 L of iodine and 4.5 L of water: $0.25 \cdot 6 \text{ L} = 1.5 \text{ L}$ and $0.75 \cdot 6 \text{ L} = 4.5 \text{ L}$. Example 3 presents a situation in which a solution is obtained that contains a given percentage of a particular ingredient.

Example 3

How many liters of water must be evaporated from 60 L of a 12% vinegar solution to make it a 36% vinegar solution?

Let x = number of liters (L) of water to be evaporated.

	Total L	L of vinegar
12% solution	60	$0.12(60)$
36% solution	$60 - x$	$0.36(60 - x)$

Vinegar in 12% solution = Vinegar in 36% solution
$$0.12(60) = 0.36(60 - x)$$
$$7.20 = 21.60 - 0.36x$$
$$0.36x = 14.40$$
$$x = 40 \qquad \text{(Check in the problem.)}$$

Thus, 40 L of water must be evaporated.

A pharmacist can mix a 90% iodine solution with a 20% iodine solution to obtain a 45% iodine solution. Notice that the

$$\begin{pmatrix} \text{Iodine in the} \\ \text{20\% solution} \end{pmatrix} + \begin{pmatrix} \text{Iodine in the} \\ \text{90\% solution} \end{pmatrix} = \begin{pmatrix} \text{Iodine in the} \\ \text{45\% solution} \end{pmatrix}.$$

This concept is used in Example 4.

Example 4   **How many milliliters (mL) of an 80% iodine solution should be added to 20 mL of a 16% iodine solution to obtain a 30% iodine solution?**

Let x = number of milliliters of 80% solution added.

	Total milliliters	Milliliters of iodine
16% solution	20	0.16(20)
80% solution	x	0.80(x)
Mixture ▶ 30% solution	20 + x	0.30(20 + x)

$$\begin{pmatrix} \text{Iodine in} \\ \text{16\% sol.} \end{pmatrix} + \begin{pmatrix} \text{Iodine in} \\ \text{80\% sol.} \end{pmatrix} = \begin{pmatrix} \text{Iodine in} \\ \text{30\% sol.} \end{pmatrix}$$

$$0.16(20) + 0.80x = 0.30(20 + x)$$
$$3.20 + 0.80x = 6.00 + 0.30x$$
$$0.50x = 2.80$$
$$x = 5.6$$

Check: (16% of 20 mL) + (80% of 5.6 mL) | 30% of (20 + 5.6) mL
            0.16(20 mL) + 0.80(5.6 mL) | 0.30(25.6 mL)
            3.20 mL + 4.48 mL | 7.68 mL
                 7.68 mL

Thus, 5.6 mL of the 80% solution should be added.

Written Exercises

3. 4 kg at $3.50, 8 kg at $2.00

Ⓐ 1. A parking meter takes only dimes and quarters. At the end of one day, the meter held $7.55 and the number of dimes was 8 more than 20 times the number of quarters. What was the number of dimes and quarters? **68 dimes, 3 quarters**

2. Some brand A dog food worth 70¢/kg is to be added to 9 kg of brand B dog food worth 50¢/kg. The mixture will be worth 58¢/kg. How many kilograms of brand A should be added? **6 kg**

3. Some salt tablets worth $3.50/kg were mixed with some cheaper tablets worth $2.00/kg to make 12 kg of a mixture worth $2.50/kg. How many kilograms of each kind were mixed?

4. How many milliliters of water must be evaporated from 160 mL of a 20% iodine solution to make it a 40% iodine solution? **80 mL**

5. How many liters of water must be added to 15 L of a 28% salt solution to dilute it to a 12% solution? **20 L**

6. How many liters of an 80% sulfuric acid solution must be added to 60 L of a 30% sulfuric acid solution to obtain a 50% acid solution? **40 L**

7. How many liters of a 52% citrus solution must be added to 30 L of a 24% citrus solution to make a 40% citrus solution? **40 L**

Problem Solving: Mixture Problems

8. A vending machine contains 12 more dimes than quarters and twice as many nickels as dimes. Find the number of each kind of coin if their total value is $6.00. **8 quarters, 20 dimes, 40 nickels**

15. 8 at 80¢, 16 at 60¢   **16.** cheese: 15, meat: 25

B 9. Gerry paid $11.92 to buy some 10¢, 13¢, and 50¢ stamps. If the number of 13¢ stamps was twice the number of 50¢ stamps, and there were 4 more 10¢ stamps that 13¢ stamps, how many of each kind did she buy? **12 at 50¢, 24 at 13¢, 28 at 10¢**

10. A: 10 kg, B: 6 kg

10. Some brand A bulk tea worth $1.40/kg and some brand B tea worth $3.00/kg are mixed to make 16 kg of a brand C worth $2.00/kg. How many kilograms of each brand are used?

11. Fifty kilograms of oats worth 90¢/kg are added to some corn worth 60¢/kg to make some animal feed worth 75¢/kg. How many kilograms of corn should be used? **50 kg**

12. Skimmed milk contains no butterfat. How many liters of skimmed milk must be added to 600 L of milk that is 3.5% butterfat to obtain milk with 2% butterfat? **450 L**

13. How many liters of a 72% alcohol solution must be added to 15 L of an 18% alcohol solution to obtain a 45% alcohol solution? **15 L**

14. A store sells the Red Dot golf ball for 60¢, the Black Dot ball for 96¢, and the Gold Dot ball for $1.08. A golf pro bought two dozen more Red Dot than Gold Dot balls and three times as many Black Dot as Red Dot balls. If the golf balls and one dollar's worth of tees cost $130.12, how many golf balls of each kind were bought? **10 Gold, 34 Red, 102 Black**

15. A package of paper tablets sells for $16.00 and contains twice as many 60¢ tablets as 80¢ tablets. How many paper tablets of each kind are in the package?

16. Some 8¢ slices of cheese and some 16¢ slices of meat make up a package of 40 slices worth $5.20. How many slices of each kind are in the package?

17. A chemist added 9 mL of a 40% citrus solution and some milliliters of a 60% citrus solution to 210 mL of a 20% citrus solution to make a 30% solution. How many milliliters of the 60% solution were added? **67 mL**

18. Chemicals A, B, and C are worth 60¢, 40¢, and 80¢/g, respectively. They are mixed so that the mass of B is twice that of A and is 3 g less than the mass of C. How many grams of each chemical should be used if the mixture is worth $11.40?

A: 3 g, B: 6 g, C: 9 g

NON-ROUTINE PROBLEMS

A grocer bought some apples at 3 for 26¢. He bought the same number of apples at 5 for 40¢. To "break even" (no profit and no loss), the grocer must sell all of the apples at what rate? **C**

A. 4 for 35¢   **B.** 5 for 42¢   **C.** 6 for 50¢
D. 8 for 66¢   **E.** 11 for 92¢

CUMULATIVE REVIEW

Find the solution set. Graph the solutions on a number line.

1. $7 - 2x > 15$ or $3x - 8 > 7$ $\{x | x < -4 \text{ or } x > 5\}$
2. $|4x - 2| > 14$ $\{x | x < -3 \text{ or } x > 4\}$
3. $9 - 2x < 17$ and $3x - 7 < 8$
   $\{x | -4 < x < 5\}$
4. $|2x + 3| < 9$ $\{x | -6 < x < 3\}$

PROBLEM SOLVING: MOTION PROBLEMS   2.6

Objective **To solve a word problem involving two objects in motion**

An automobile that is driven at an average rate of 90 kilometers per hour (km/h) for 3 hours travels 90 · 3, or 270 km. An Olympic sprinter who runs at an average rate of 8 meters per second (m/s) for 50 seconds travels 8 · 50, or 400 m. The relationship between distance, rate, and time of an object in motion is stated below.

> **Distance of object in motion**
>
> The distance (d) traveled by a moving object is equal to its rate of speed (r) times the time in motion (t): $d = rt$.

The formula $d = rt$ is a mathematical *model* for solving problems involving motion.

Example 1

READ ▶ Two trains started toward each other at the same time from stations 732 km apart. One train traveled at 148 km/h and the other train traveled at 96 km/h. In how many hours did they meet?

PLAN ▶ Represent the data in a diagram and in a table.
Let t = number of hours for each train.

	rate (r)	time (t)	distance (d)
Faster train	148	t	$148t$
Slower train	96	t	$96t$

◀ Use $d = rt$.

(Distance of faster train) + (Distance of slower train) = Total distance
$\qquad 148t \qquad + \qquad\qquad 96t \qquad\qquad = \qquad 732$

SOLVE ▶
$$244t = 732$$
$$t = 3$$

INTERPRET ▶ Check:

(Distance of faster train) + (Distance of slower train)	Total distance
148 km/h · 3 h + 96 km/h · 3 h	732 km
444 km + 288 km	
732 km	

Thus, the trains met in 3 hours.

Problem Solving: Motion Problems

Suppose a car leaves a certain city at 2:00 P.M., traveling at 60 km/h. Then, at 4:00 P.M. a second car leaves the same city, traveling the same route at 90 km/h. The second car will overtake the first car as indicated below.

The second car will overtake the first car at 8:00 P.M. Notice that the two cars travel the same distance: 60 km/h · 6 h = 90 km/h · 4 h = 360 km.

Motion problems of this type can also be solved using the formula $d = rt$ along with appropriate equation properties. The procedure is shown in the next example.

Example 2 **An aircraft carrier, traveling at 30 km/h, left a port at 7:00 A.M. At 9:00 A.M., a destroyer left the same port, traveling the same route at 40 km/h. At what time did the destroyer overtake the carrier?**

Let t = number of hours the carrier traveled.

	rate	time	distance
Carrier	30	t	$30t$
Destroyer	40	$t - 2$	$40(t - 2)$

◀ Use $d = rt$.

(Distance traveled by carrier) = (Distance traveled by destroyer)
$$30t = 40(t - 2)$$
$$30t = 40t - 80$$
$$-10t = -80$$
$$t = 8$$

Check: | (Distance traveled by carrier) | (Distance traveled by destroyer) |
|---|---|
| 30 km/h · 8 h | 40 km/h · 6 h |
| 240 km | 240 km |

The time, 8 hours after 7:00 A.M., is 3:00 P.M.
Thus, the destroyer overtook the carrier at 3:00 P.M.

Written Exercises

(A)

1. A passenger train and a freight train start toward each other at the same time from towns 870 km apart. The passenger train travels 80 km/h and the freight train travels 65 km/h. In how many hours will they meet? **6 h**

2. Two hockey players start toward each other at the same time from opposite ends of an ice rink that is 60 m long. In how many seconds will they meet if one player skates at a rate of 9.5 m/s and the other skates at a rate of 10.5 m/s? **3 s**

3. A ship left port at 2:00 P.M. and traveled at 30 km/h. At 6:00 P.M., a helicopter left the same port and flew at 120 km/h in the same direction as the ship. At what time did the helicopter reach the ship? **7:20 P.M.**

4. John began the marathon run at 8:30 A.M. and averaged 12 km/h. Mary began at 9:00 A.M. and averaged 14 km/h. At what time did Mary pass John on the marathon course? **12:00 P.M.**

5. Two cars start from the same point at the same time but travel in opposite directions. One car averages 80 km/h and the other averages 70 km/h. In how many hours will they be 600 km apart? **4 h**

6. From a dock in a river, Cindy swam downstream at 4.4 m/s while George swam upstream at 2.8 m/s. In how many seconds were they 108 m apart? **15 s**

(B)

7. At 7:00 A.M., an armored van left city A and traveled at 70 km/h toward city B. At 11:00 A.M. a helicopter, headed toward the van, left city B and traveled at 110 km/h. At approximately what time did the copter reach the van if the cities are 550 km apart? **12:30 P.M.**

8. One bus left the city at 2:00 P.M. and averaged 75 km/h. Another bus, traveling in the opposite direction, left the city at 5:00 P.M. and averaged 85 km/h. At what time were the buses 705 km apart? **8:00 P.M.**

9. A plane left an airfield and averaged 600 km/h. Two hours later, a pursuit plane left the same field and traveled at 900 km/h in the same direction. How long did it take the pursuit plane to catch the slower plane? **4 h**

10. A thief left a bank by car at 8:45 A.M. and averaged 110 km/h. The police left the scene at 8:48 A.M. and pursued the thief at 140 km/h. At what time was the thief caught? **8:59 A.M.**

11. A carrier and a destroyer left the same port at 8:00 A.M. and sailed in the same direction. The carrier averaged 30 km/h and the destroyer traveled at 75 km/h. At what time will they be 375 km apart? **4:20 P.M.**

12. A jet plane and a cargo plane left the same terminal at the same time and flew in the same direction. The jet flew at 1,000 km/h. After 6 hours, the planes were 2,100 km apart. Find the rate of the cargo plane. **650 km/h**

(C)

13. An Olympic cyclist rode the first 6 hours of a race with a planned strategy. The cyclist pedaled at a constant rate for the first 3 hours and then reduced his speed by one-half for 1 hour. The last 2 hours, the rate was 3 times that of the fourth hour. Find the three different speeds if 78 km were traveled in the first 6 hours. **12 km/h, 6 km/h, 18 km/h**

14. An object moved in one direction for 6.25 seconds at a constant speed. It then reversed direction for 2.50 seconds with its rate doubled, and then reversed direction again for 0.75 seconds at 40% of its initial speed. Find the three speeds if the object was 5.89 m from its start at the end of 9.50 seconds. **3.80 m/s, 7.60 m/s, 1.52 m/s**

Problem Solving: Motion Problems

INTRODUCTION TO VECTORS (optional)   2.7

Objectives
To identify the magnitude and direction of a vector
To add and subtract vectors by drawing
To multiply a scalar and a vector by drawing

The **displacement** (movement) of an object a distance of 4 km to the northeast is shown by the **directed** line segment $\overrightarrow{AB}$ at the left. Such directed line segments are called **vectors**. The vector shown is written as: $\overrightarrow{AB}$ (with a half-arrow: $\rightarrow$) and read as "vector AB."
1. $\overrightarrow{AB}$ has a **magnitude** of 4 km and its **direction** is northeast.
2. The **initial point** of $\overrightarrow{AB}$ is A. The **terminal point** is B.

The vector $\overrightarrow{CD}$ represents the *force* exerted by an 80-g weight (magnitude) suspended on a coiled spring, stretching it downward (direction).

Sometimes, a single letter is used to name a vector. The **velocity** vector $\vec{F}$ shows a magnitude of 45 mi/h in the westward direction.

Four vectors are shown at the left. $\vec{R}$ and $\vec{R}'$ are **equivalent** vectors since they have the same magnitude and the same direction. $\vec{T}$ and $-\vec{T}$ are **opposite** vectors. They have the same magnitude and opposite directions.

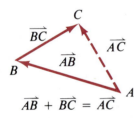

The addition of vectors can be interpreted as follows. If you move from A to B and then from B to C, your *total* displacement is from A to C. Thus, $\overrightarrow{AB} + \overrightarrow{BC} = \overrightarrow{AC}$. This is called the **triangle** method of addition.

Notice that the **tail** of $\overrightarrow{BC}$ is at the **head** of $\overrightarrow{AB}$ and that the sum $\overrightarrow{AC}$ is drawn *from* the tail of $\overrightarrow{AB}$ *to* the head of $\overrightarrow{BC}$.

Given $\vec{A}$ and $\vec{B}$ as shown below, you can use equivalent vectors to draw $\vec{A} + \vec{B}$. Move the tail of $\vec{B}$ to the head of $\vec{A}$. Draw the sum from the tail of $\vec{A}$ to the head of $\vec{B}'$.

You can add three vectors by extending the method above.

If $\vec{F}$ and $\vec{G}$ have the *same* initial point, you can use the **parallelogram** method to draw the sum $\vec{F} + \vec{G}$. Draw $\vec{F'}$ and $\vec{G'}$ to form a parallelogram. Draw the sum $\vec{F} + \vec{G}$ from the tail of $\vec{F}$ to the head of $\vec{G'}$. The sum is a **diagonal** of the parallelogram.

To add two *collinear* vectors $\vec{X}$ and $\vec{Y}$, move the tail of $\vec{Y}$ to the head of $\vec{X}$. Draw $\vec{X} + \vec{Y}$ from the tail of $\vec{X}$ to the head of $\vec{Y'}$. Three cases are shown below.

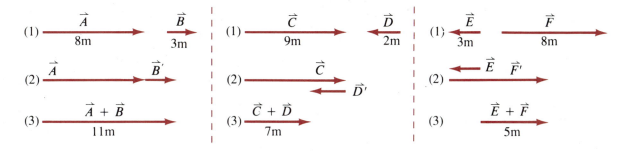

Example 1

Use the figure at the right for the exercises below.

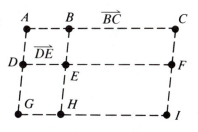

1. Name two vectors equivalent to $\vec{BC}$. **EF, HI**
2. Name three vectors that are opposites of $\vec{BC}$.
3. Name two vectors that are collinear with $\vec{DE}$ and have the same direction as $\vec{DE}$. **DF, EF**
   2. **CB, FE, IH**

Find the following sums.

4. $\vec{GB} + \vec{BF}$ 5. $\vec{AE} + \vec{EG}$ 6. $\vec{AD} + \vec{DG}$ 7. $\vec{AG} + \vec{GD}$
8. $\vec{DE} + \vec{FI}$ 9. $\vec{IE} + \vec{HG}$ 10. $(\vec{DE} + \vec{EB}) + \vec{BC}$ 11. $(\vec{GB} + \vec{EF}) + \vec{CF}$

4. **GF** 5. **AG** 6. **AG** 7. **AD** 8. **DH** 9. **ID** 10. **DC** 11. **GF**

Subtraction with vectors is similar to subtraction with real numbers.

$$a - b = a + (-b) \text{ for all real numbers } a \text{ and } b$$
$$\vec{X} - \vec{Y} = \vec{X} + (-\vec{Y}) \text{ for all vectors } \vec{X} \text{ and } \vec{Y}$$

To draw $\vec{A} - \vec{B}$, you draw $-\vec{B}$ and then draw $\vec{A} + (-\vec{B})$.

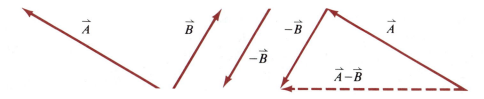

Introduction to Vectors

If a displacement of 2 m to the east occurs each hour for 3 h, the result is a displacement of 6 m to the east.

$\vec{D}$ was multiplied by the real number 3 to give the vector $3 \cdot \vec{D}$. The number 3 is called a **scalar**. The vector $3 \cdot \vec{D}$ is a **scalar multiple** of the vector $\vec{D}$.

Example 2   Draw $\vec{C} - \vec{D}$.   Draw $-2 \cdot \vec{A}$.

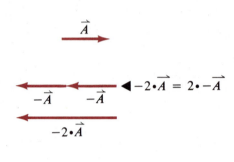

Written Exercises

Draw and label each vector. (Ex. 1–3)

(A) 1. $\vec{A}$, a displacement vector for a move of 10 km to the northwest **10 km ↖ $\vec{A}$**
2. $\vec{B}$, a force vector for the pull exerted by a 20-lb weight suspended by a rope **$\vec{B}$ ↓ 20 lb**
3. $\vec{C}$, a velocity vector for a speed of 70 km/h to the southeast **$\vec{C}$ ↘ 70 km/h**

Use the figure at the right to find the following. (Ex. 4–10)

4. $\overrightarrow{AF} + \overrightarrow{DG}$ **$\overrightarrow{AI}$** 5. $\overrightarrow{HE} + \overrightarrow{FC}$ **$\overrightarrow{HB}$** 6. $\overrightarrow{AG} - \overrightarrow{DG}$ **$\overrightarrow{AD}$**
7. $\overrightarrow{BH} - \overrightarrow{FI}$ **$\overrightarrow{BE}$** 8. $\overrightarrow{CD} + \overrightarrow{DB} + \overrightarrow{BG}$ **$\overrightarrow{CG}$**
9. $\overrightarrow{BF} + \overrightarrow{EH} + \overrightarrow{CB}$ **$\overrightarrow{BH}$** 10. $\overrightarrow{DC} - \overrightarrow{HI} - \overrightarrow{IC}$ **$\overrightarrow{DH}$**

Draw *larger* copies of the vectors below. Use them for Exercises 11–29.

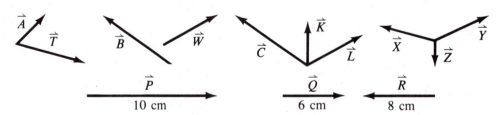

Draw each of the following. Use the vectors shown above. See TE Answer Section.

(B) 11. $\vec{A} + \vec{T}$ 12. $\vec{T} - \vec{A}$ 13. $\vec{B} + \vec{W}$ 14. $\vec{B} - \vec{W}$ 15. $\vec{W} - \vec{B}$
16. $\vec{C} + \vec{K} + \vec{L}$ 17. $\vec{X} + \vec{Y} + \vec{Z}$ 18. $\vec{P} + \vec{R}$ 19. $\vec{Q} + \vec{R}$ 20. $\vec{R} - \vec{Q}$
21. $3 \cdot \vec{Q}$ 22. $\frac{1}{2} \cdot \vec{P}$ 23. $-2 \cdot \vec{R}$ 24. $-1.5 \cdot \vec{P}$ 25. $4(\vec{P} + \vec{R})$
26. $\vec{C} + \vec{K} - \vec{L}$ 27. $\vec{X} - \vec{Y} + \vec{Z}$ 28. $\vec{C} - (\vec{K} + \vec{L})$ 29. $3 \cdot \vec{A} + 2 \cdot \vec{T}$

Ⓐ Exercises are 1, 3, 4, 11, 13, 15, 18, 20, 22, 24–27, 29. Ⓒ Exercises are starred. The rest are Ⓑ Exercises.

CHAPTER TWO REVIEW

Vocabulary
base, exponent [2.1]
exponential equation [2.1]
negative exponent, zero exponent [2.2]
polynomial [2.3]
  binomial   linear   quadratic
  degree of   monomial trinomial
scientific notation [2.2]
synthetic substitution [2.3]

Simplify and write each expression with positive exponents. 1. $-40a^6b^7c^6$ 2. $128m^4n^{15}$
1. $-5a^4bc^5 \cdot 8a^2b^6c$ 2. $2m(4mn^5)^3$ [2.1]
3. $\dfrac{10x^3y^7z^8}{15x^7y^7z^7}$ $\dfrac{2z}{3x^4}$ 4. $\left(\dfrac{-5c^4}{2d^3}\right)^3$ $\dfrac{-125c^{12}}{8d^9}$
★ 5. $x^m y^n (x^{2m} y^{n+3})^2$ $x^{5m}y^{3n+6}$
6. $7x^{-3} \cdot 2y^0 \cdot x^{-4} \cdot 4y^2$ $\dfrac{56y^2}{x^7}$
7. $(2a^{-2}b^3)^{-4}$ $\dfrac{a^8}{16b^{12}}$
8. $\dfrac{6c^{-5}d^{-3}}{8c^4d^{-7}}$ $\dfrac{3d^4}{4c^9}$ 9. $\left(\dfrac{-3xy^{-2}}{2z^3w^{-3}}\right)^4$ $\dfrac{81x^4w^{12}}{16y^8z^{12}}$ [2.2]
★ 10. x^{2c-2d} if $c < d < 0$ $\dfrac{1}{x^{2d-2c}}$

Find the value of each expression. [2.2]
11. $\dfrac{3.45}{10^{-3}}$ **3,450**   12. $\dfrac{6.9 \times 10^{-5}}{0.3 \times 10^{-2}}$ **0.023**

Solve each equation. [2.1]
13. $5^{x-3} = 625$ **7**   14. $3^{4x-3} = 81$ $\dfrac{7}{4}$

Evaluate each polynomial for the given values of its variables. (Ex. 15–17) [2.3]
15. $3xy - 2x^2y^3 - 4$ if $x = 3$ and $y = -2$ **122**
16. $3.2c^2 + 5.4d^2$ if $c = -0.5$ and $d = -5$ **135.8**
★ 17. $10x^3 + 5x^2y - 5xy^2$ if $x = a^2$ and $y = 2a^2$ **0**
18. Use synthetic substitution to evaluate $3x^4 - 2x^3 - 3x - 180$ for $x = 3$ and again for $x = -2$. [2.3]   **0, −110**
19. Subtract: $(9xy^2 - 5x^3y) - (3xy - 5x^3y)$. [2.3] $9xy^2 - 3xy$

Simplify. [2.4]
20. $(5x - 2)(3x - 1)$ $15x^2 - 11x + 2$
21. $(4x^2 + 3y^2)(4x^2 + 3y^2)$ $16x^4 + 24x^2y^2 + 9y^4$
22. $(3a - 1)(2a^2 - a - 1)$ $6a^3 - 5a^2 - 2a + 1$
23. $7c(5c^3 - 2)(4c^3 + 3)$ $140c^7 + 49c^4 - 42c$

Solve each problem.
24. The number of nickels in a coin box is 8 more than the number of quarters and the number of dimes is 3 times the number of nickels. Find the number of each kind of coin if their total value is $7.00. [2.5]
7 quarters, 15 nickels, 45 dimes

25. Some caramels worth $2.60/kg are mixed with some peppermints worth $3.00/kg to make 8 kg of a mixture worth $2.85/kg. How many kilograms of each kind are there in the mixture? [2.5]
caramels: 3 kg, mints: 5 kg

26. How many liters of water must be evaporated from 90 L of a 15% saline solution to make it a 45% saline solution? [2.5] **60 L**

27. How many milliliters of a 25% iodine solution should be added to 90 mL of a 65% iodine solution to obtain a 55% iodine solution? [2.5] **30 mL**

28. At 6:00 A.M., a car left city A and headed for city B at 80 km/h. At 9:00 A.M., another car left city B and headed for city A at 90 km/h. At what time did the cars meet if the two cities are 580 km apart? [2.6]
11:00 A.M.

29. A ship left a port at 25 km/h, and 4 hours later, a helicopter left the same port in the same direction at 100 km/h. How long did it take the helicopter to overtake the ship? [2.6] **1 h 20 min**

30. In problem 29 above, how long did it take the helicopter to reach a distance of 10 km behind the ship? [2.6] **1 h 12 min**

31. Steve left the gymnasium at 7:45 A.M. and jogged in one direction at 10 km/h. At 8:15 A.M., Karla left the gym and jogged in the opposite direction at 12 km/h. At what time were they 10.5 km apart? [2.6] **8:30 A.M.**

Chapter Two Review

Ⓐ Exercises are 1, 3, 5, 7, 11, 13, 15, 17–19, 23–25, 27. Ⓒ Exercises are starred.
The rest are Ⓑ Exercises.

CHAPTER TWO TEST

Simplify and write each expression with positive exponents.

1. $-6x^2yz^4 \cdot 3xy^5z^3$ **$-18x^3y^6z^7$**
2. $8a^{-5} \cdot 5b^0 \cdot a^8 \cdot b^{-3}$ **$\dfrac{40a^3}{b^3}$**
3. $\dfrac{5m^8n^3p^2}{15m^4n^3p^{10}}$ **$\dfrac{m^4}{3p^8}$**
4. $\dfrac{4x^{-6}y^{-2}}{6x^{-4}y^{-4}}$ **$\dfrac{2y^2}{3x^2}$**
5. $(-4c^2)^3$ **$-64c^6$**
6. $10m(2m^3n)^4$ **$160m^{13}n^4$**
7. $\left(\dfrac{3x^2}{-2y}\right)^4$ **$\dfrac{81x^8}{16y^4}$**
8. $(3c^{-4}d^5)^{-2}$ **$\dfrac{c^8}{9d^{10}}$**
9. $4a^2(-5a^3b)^3$ **$-500a^{11}b^3$**
10. $\left(\dfrac{6x^{-4}}{x^4y^{-3}}\right)^2$ **$\dfrac{36y^6}{x^{16}}$**

Find the value of each expression.

11. $\dfrac{7^{-2}}{2^{-4}}$ **$\dfrac{16}{49}$**
12. $\dfrac{4.8 \times 10^{-1}}{1.2 \times 10^{-3}}$ **400**

Solve each equation.

13. $2^{x-3} = 64$ **9**
14. $6^{3x+1} = 36$ **$\dfrac{1}{3}$**

Evaluate each polynomial for the given values of its variables. (Ex. 15–16)

15. $8x^2y - 2xy^2$ if $x = 2$ and $y = -2$ **-80**
16. $3cd^3 - cd$ if $c = -2.4$ and $d = -5$ **888**
17. Use synthetic substitution to evaluate $3x^4 - 3x^3 + 4x - 10$ for $x = 2$ and again for $x = -2$. **22, 54**
18. Subtract: $(5x^5 - x^3 + 2x) - (x^3 - 6x - 8x^5)$. **$13x^5 - 2x^3 + 8x$**

Simplify.

19. $(5x - 3)(2x + 1)$ **$10x^2 - x - 3$**
20. $(7m^3 + 5n)(7m^3 + 5n)$
21. $(3.5a - 0.5)(4a + 6)$ **$14a^2 + 19a - 3$**
22. $4n(3n + 7)(3n - 7)$ **$36n^3 - 196n$**
23. $(2x + 3)(x^2 - 4x - 5)$ **$2x^3 - 5x^2 - 22x - 15$**

20. **$49m^6 + 70m^3n + 25n^2$**

Solve each problem.

24. A collection of dimes and quarters is worth $2.85. If the number of dimes is 3 less than twice the number of quarters, find the number of dimes and quarters. **11 dimes, 7 quarters**

25. How many liters of water must be added to 30 L of a 20% chlorine solution to dilute it to a 15% chlorine solution? **10 L**

26. One train left a depot at 3:00 P.M. and traveled at 80 km/h. Another train left the depot at 6:00 P.M. and traveled in the opposite direction at 150 km/h. At what time were the trains 1,620 km apart? **12:00 A.M.**

27. Some Brand A cleanser worth 80¢/kg is to be mixed with some Brand B cleanser worth 60¢/kg to make 15 kg of a mixture worth 68¢/kg. How many kilograms of each brand should be used? **A: 6 kg, B: 9 kg**

28. John and Everett ran in the local marathon. If John averaged 17 km/h and Everett averaged 14 km/h, how long did it take John to be 4.5 km ahead of Everett? **$1\dfrac{1}{2}$ h**

29. Wheat flour worth 40¢/kg was mixed with some bran flour worth 60¢/kg. The amount of bran flour used in the mixture was 10 kg less than the amount of wheat flour used. If the mixture was worth $14.00, how many kilograms of each kind of flour were in the mixture? **wheat: 20 kg, bran: 10 kg**

★ 30. An automobile dealer purchased some model A, B, and C automobiles worth $3,000, $6,000, and $4,000 each, respectively. The ratio of model A to model B to model C automobiles purchased was 8 to 3 to 4. If their total value was $116,000, how many automobiles of each model were purchased? **A: 16, B: 6, C: 8**

★ 31. Simplify and write with positive exponents.
$\dfrac{x^{6a+4}y^{3b-2}}{x^{2a-2}y^{b+3}}$ **$x^{4a+6}y^{2b-5}$**

★ 32. Evaluate the polynomial for the given values of its variables.
$4x^3y + 10x^2y^2 - 10xy^3$ if $x = a$ and $y = 2a$ **$-32a^4$**

★ 33. Multiply: $(3c + 2d - 5)(2c - d + 4)$.
$6c^2 + cd + 2c - 2d^2 + 13d - 20$

COLLEGE PREP TEST

DIRECTIONS: In each item, you are to compare a quantity in Column 1 with a quantity in Column 2. Write the letter of the correct answer from these choices:

- **A** The quantity in Column 1 is greater than the quantity in Column 2.
- **B** The quantity in Column 2 is greater than the quantity in Column 1.
- **C** The quantity in Column 1 is equal to the quantity in Column 2.
- **D** The relationship cannot be determined from the given information.

Notes: Information centered over both columns refers to one or both of the quantities to be compared.
A symbol that appears in both columns has the same meaning in each column.

SAMPLE ITEMS AND ANSWERS

	Column 1	Column 2
	$x = 7$	
S1.	$x + x$	x^2

The answer is *B*, since $7^2 > 7 + 7$.

	Column 1	Column 2
S2.	$2x$	$3x$

The answer is *D*, since $2x > 3x$ if $x = -1$; $3x > 2x$ if $x = 1$; and $2x = 3x$ if $x = 0$.

	Column 1	Column 2	
1.	$21 \cdot 1{,}234 \cdot 20$	$20 \cdot 1{,}234 \cdot 22$	**B**
	$x = -3$		
2.	$x^3 - 2x$	$x^2 + 10x$	**C**
3.	$5 + 2^4 + 7$	$6 + 4^2 + 5$	**A**
4.	$(y + 2)(y + 3)$	$(y + 4)(y + 5)$	**D**
	$x = 10$		
5.	$2x^4 + 3x^2 + 4$	$9x^3 + 8x^2 + 7x + 6$	**A**

	Column 1	Column 2	
6.	The greater solution of $|x + 5| = 2$	The smaller solution of $|x - 5| = 7$	**B**
	$y = \frac{4}{3}x + 15$		
7.	x, if $y = -45$	y, if $x = -45$	**C**
	$x^y = 16$		
8.	x	y	**D**
9.	Number of liters of iodine in 60 L of a 25% iodine solution	Number of liters of water in 40 L of a 20% salt solution	**B**
10.	Area of a rectangle, 1.5 m by 0.3 m	Area of a rectangle, 9 cm by 5 cm	**A**
11.	$(x + 2)^2$	$(x + 2)(x - 2)$	**D**

12.	Average speed for round trip from A to B to A	40 mph	**B**

College Prep Test

3 FACTORING AND SPECIAL PRODUCTS

Formulating a Problem-Solving Situation

Problem 1. How many cans of liquid floor wax must you buy to cover the floor of your recreation room?

Your first step in solving a problem of this type is to *formulate* the problem-solving situation. That is, *analyze* the situation with particular *focus* on the question to be answered.

To answer the question, you must obtain the following information:
  (a) Dimensions of floor
  (b) Sizes of cans of floor wax available
  (c) Surface area covered by one unit of wax of your choice

Suppose your analysis of the situation reveals these facts:
  (a) The dimensions of the floor are 700 cm by 600 cm.
  (b) The floor wax brand of your choice comes only in 1-L cans.
  (c) One liter of floor wax covers an area of 20 m^2.

Problem 2. How many 1-L cans of liquid floor wax must you buy to cover the floor of your recreation room which measures 700 cm by 600 cm, if one liter of wax covers an area of 20 m^2?   **3 cans**

FACTORING TRINOMIALS   3.1

Objective **To factor a trinomial of degree 2 or greater into two binomials**

You have multiplied two binomials, such as $(3x - 5)(2x + 1)$, by the FOIL method as shown below.

$$(3x - 5)(2x + 1) = 6x^2 - 7x - 5$$

with $6x^2$, -5, $-10x$, $+3x$ indicated.

Notice that the middle term of the trinomial is $3x - 10x$, or $-7x$.

A trinomial of the form $ax^2 + bx + c \, (a \neq 0)$ is called a **quadratic trinomial.** You can *factor* a quadratic trinomial into two binomials by reversing the multiplication process above. Begin by listing the possible pairs of binomial factors. Then check each pair until you find the factorization that gives the correct middle term.

Example 1 **Factor $3x^2 - 8x + 5$ into two binomials.**

$$3x^2 - 8x + 5 = (3x \; \underline{\;?\;})(x \; \underline{\;?\;})$$

with $3x^2$ and $+5$ indicated.

The first terms of the binomial factors are $3x$ and x. The last terms are either $+1$ and $+5$ or -1 and -5. Since the middle term of the trinomial is negative, $(-8x)$, try -1 and -5.

Possible Factors	Middle Term	
$(3x - 1)(x - 5)$	$-15x - x = -16x$	
$(3x - 5)(x - 1)$	$-3x - 5x = -8x$	◀ *Correct middle term: $-8x$*

Thus, $3x^2 - 8x + 5 = (3x - 5)(x - 1)$.

Example 2 **Factor $3n^2 + 26n + 16$ into two binomials.**

$$3n^2 + 26n + 16 = (3n + \underline{\;?\;})(n + \underline{\;?\;})$$

Try 1 and 16, 2 and 8, 4 and 4 for last terms of the factors. ▶

Possible Factors	Middle Term	
$(3n + 1)(n + 16)$	$48n + n = 49n$	
$(3n + 16)(n + 1)$	$3n + 16n = 19n$	
$(3n + 8)(n + 2)$	$6n + 8n = 14n$	
$(3n + 2)(n + 8)$	$24n + 2n = 26n$	◀ *Correct middle term: $26n$*
$(3n + 4)(n + 4)$	$12n + 4n = 16n$	

Thus, $3n^2 + 26n + 16 = (3n + 2)(n + 8)$.

Factoring Trinomials

To factor a polynomial of any type, express it as a product of other polynomials of a lower degree. Using integers only, some polynomials cannot be factored in this manner.

Example 3

Factor $y^2 + y + 9$ into two binomials.

$$y^2 + y + 9 = (y + \underline{\ ?\ })(y + \underline{\ ?\ })$$

Possible Factors	Middle Term
$(y + 3)(y + 3)$	$3y + 3y = 6y$
$(y + 9)(y + 1)$	$y + 9y = 10y$

Neither pair of factors gives the correct middle term, y.
Thus, $y^2 + y + 9$ cannot be factored using integers.

Similar procedures should be followed when factoring quadratic trinomials that contain more than one variable or trinomials of a degree greater than 2. This is shown in Examples 4 and 5, respectively.

Example 4

Factor $2x^2 - 11xy + 5y^2$ into two binomials.

$$2x^2 - 11xy + 5y^2 = (2x - \underline{\ ?\ })(x - \underline{\ ?\ })$$

Possible Factors	Middle Term	
$(2x - 5y)(x - y)$	$-2xy - 5xy = -7xy$	
$(2x - y)(x - 5y)$	$-10xy - xy = -11xy$	◀ Correct middle term: $-11xy$

Thus, $2x^2 - 11xy + 5y^2 = (2x - y)(x - 5y)$.

Example 5

Factor $2x^4 - 9x^2 - 5$ into two binomials.

The first terms of the binomial factors are $2x^2$ and x^2.
The last terms are either -5 and $+1$ or $+5$ and -1.

$$2x^4 - 9x^2 - 5 = (2x^2\ \underline{\ ?\ })(x^2\ \underline{\ ?\ })$$

Possible Factors	Middle Term	
$(2x^2 - 5)(x^2 + 1)$	$2x^2 - 5x^2 = -3x^2$	
$(2x^2 + 1)(x^2 - 5)$	$-10x^2 + x^2 = -9x^2$	◀ Correct middle term: $-9x^2$
$(2x^2 - 1)(x^2 + 5)$	$10x^2 - x^2 = 9x^2$	
$(2x^2 + 5)(x^2 - 1)$	$-2x^2 + 5x^2 = 3x^2$	

Thus, $2x^4 - 9x^2 - 5 = (2x^2 + 1)(x^2 - 5)$.

Oral Exercises

1. (x − 3)(x − 4) 2. (x − 12)(x − 1) 3. (x + 4)(x − 3) 4. (x − 4)(x + 3) 5. (x + 2)(x + 6)
6. (x − 6)(x + 2)

Factor each trinomial into two binomials.

1. $x^2 - 7x + 12$
2. $x^2 - 13x + 12$
3. $x^2 + x - 12$
4. $x^2 - x - 12$
5. $x^2 + 8x + 12$
6. $x^2 - 4x - 12$
7. $x^2 + x + 3$ ← **cannot be factored**
8. $x^2 - 5x + 6$ (x − 3)(x − 2)
9. $x^2 - 5x - 6$ (x − 6)(x + 1)

Written Exercises

16. (5n + 4)(3n − 1)   17. (x + 6y)(x − 2y)   19. (x + 3y)(x − 4y)

For complete solutions to Ex. 1–60, see TE Answer Section.

Factor each trinomial into two binomials.   20. $(4a + b)(2a - b)$ 26. $(5c^2 - 1)(2c^2 + 1)$

Ⓐ
1. $7x^2 - 12x + 5$ (7x − 5)(x − 1)
2. $5y^2 - 36y + 7$ (5y − 1)(y − 7)
3. $5y^2 + 14y - 3$ (5y − 1)(y + 3)
4. $7n^2 + 4n - 11$ (7n + 11)(n − 1)
5. $5a^2 - 54a - 11$ (5a + 1)(a − 11)
6. $3x^2 - 10x - 13$ (3x − 13)(x + 1)
7. $6x^2 - 7x - 2$ **below**
8. $8n^2 + 14n + 3$ (4n + 1)(2n + 3)
9. $2a^2 + a - 10$ (2a + 5)(a − 2)
10. $2x^2 - 11x + 14$ (2x − 7)(x − 2)
11. $5x^2 - 28x + 15$ (5x − 3)(x − 5)
12. $3y^2 + 11y + 6$ (3y + 2)(y + 3)
13. $9a^2 + 18a + 8$ (3a + 4)(3a + 2)
14. $6t^2 - 11t - 10$ (3t + 2)(2t − 5)
15. $4c^2 - 15c + 9$ (4c − 3)(c − 3)
16. $15n^2 + 7n - 4$
17. $x^2 + 4xy - 12y^2$
18. $x^2 + 13xy + 36y^2$ (x + 9y)(x + 4y)
19. $x^2 - xy - 12y^2$
20. $8a^2 - 2ab - b^2$
21. $15c^2 - 8cd + d^2$ (5c − d)(3c − d)
22. $21m^2 + 10mn + n^2$ **below**
23. $x^4 - 9x^2 + 14$ $(x^2 - 2)(x^2 - 7)$
24. $n^4 + 8n^2 + 15$ $(n^2 + 3)(n^2 + 5)$
25. $a^4 - 3a^2 - 18$ $(a^2 + 3)(a^2 - 6)$
26. $10c^4 + 3c^2 - 1$
27. $15a^4 + 16a^2 - 7$ $(5a^2 + 7)(3a^2 - 1)$
28. $21d^4 - 8d^2 - 5$
29. $x^2 - 14x - 72$ (x + 4)(x − 18)
30. $72x^2 + 21x - 1$ (3x + 1)(24x − 1)
31. $64y^2 - 12y - 1$
32. $8a^2 - 5a - 10$ ← **cannot be factored**
33. $y^2 + 30y - 64$ (y − 2)(y + 32)

7. **cannot be factored**   22. (7m + n)(3m + n)

Ⓑ
34. $10x^2 + 23x + 12$
35. $12n^2 - 17n + 6$
36. $12y^2 - 8y - 15$ (2y − 3)(6y + 5)
37. $18a^2 - 35a + 12$
38. $24x^2 - 46x - 35$
39. $16n^2 + 10n - 21$ (8n − 7)(2n + 3)
40. $20c^2 - 48c + 27$
41. $24a^2 - 5a - 36$
42. $6y^2 + yz - 12z^2$
43. $5x^2 + 4xy - 12y^2$
44. $36m^2 - 7mn - 15n^2$
45. $24c^2 + 17cd - 20d^2$
46. $15x^4 - 8x^2 - 12$
47. $12x^4 - 23x^2 + 10$
48. $18y^4 + 9y^2 - 20$ $(3y^2 + 4)(6y^2 - 5)$
49. $6y^4 - 23y^2 - 18$
50. $10n^4 - 31n^2 + 24$
51. $9a^4 + 24a^2 - 20$
52. $2x^4y^4 - 7x^2y^2 + 6$
53. $3a^4b^4 - a^2b^2 - 10$
54. $5m^4n^4 + 4m^2n^2 - 12$
55. $6c^4d^4 + 23c^2d^2 + 20$
56. $5x^4 + 23x^2y^2 + 24y^4$
57. $14c^4 - 9c^2d^2 - 18d^4$
58. $2x^8 + 2x^4 - 1$
59. $x^6 - 7x^3y^2 + 6y^4$
60. $5m^4 - 4m^2n^3 - n^6$

28. $(7d^2 - 5)(3d^2 + 1)$ 31. (16y + 1)(4y − 1)

Ⓒ
61. $x^{6m} - 7x^{3m} + 12$ $(x^{3m} - 3)(x^{3m} - 4)$
62. $6x^{2m} + 19x^m + 15$ $(3x^m + 5)(2x^m + 3)$
63. $9x^{4m} + 24x^{2m} - 20$ $(3x^{2m} + 10)(3x^{2m} - 2)$
64. $8x^{6m} - 2x^{3m} - 21$ $(4x^{3m} - 7)(2x^{3m} + 3)$
65. $5x^{2a} + 23x^a y^b + 12y^{2b}$
66. $8x^{4a} - 26x^{2a} y^{2b} + 15y^{4b}$
67. $12x^{6a} - 8x^{3a} y^{2b} - 15y^{4b}$
68. $x^{2n+4} + 7x^{n+2} + 12$
69. $x^{2n-2} + 3x^{n-1} - 10$
70. $x^{8n+6} - 2x^{4n+3} - 3$
71. $x^{2a+6} + 2x^{a+3} y^{a-2} + y^{2a-4}$
72. $2x^{2a+2} + 5x^{a+1} y^a + 3y^{2a}$
73. $x^2 + (m + n)x + mn$ (x + m)(x + n)
74. $acx^2 + (ad + bc)x + bd$ (ax + b)(cx + d)

65. $(5x^a + 3y^b)(x^a + 4y^b)$ 66. $(4x^{2a} - 3y^{2b})(2x^{2a} - 5y^{2b})$ 67. $(2x^{3a} - 3y^{2b})(6x^{3a} + 5y^{2b})$

CUMULATIVE REVIEW

68. $(x^{n+2} + 3)(x^{n+2} + 4)$ 69. $(x^{n-1} - 2)(x^{n-1} + 5)$ 70. $(x^{4n+3} - 3)(x^{4n+3} + 1)$

Multiply.   71. $(x^{a+3} + y^{a-2})(x^{a+3} + y^{a-2})$ 72. $(2x^{a+1} + 3y^a)(x^{a+1} + y^a)$

1. $(x + 6)(x + 6)$
   $x^2 + 12x + 36$
2. $(x + 10)(2x^2 - 3x - 4)$
   $2x^3 + 17x^2 - 34x - 40$
3. $(3x^2 - 2y)(3x^2 + 2y)$
   $9x^4 - 4y^2$

Factoring Trinomials

SPECIAL PRODUCTS 3.2

Objective **To multiply a pair of related polynomials**

When working with polynomials, one special product that often occurs is the *product of the sum and the difference of two terms*. The general form is $(a + b)(a - b)$. As shown below, the FOIL method can be used to complete the multiplication.

$$(4x + 5)(4x - 5) = 16x^2 - 20x + 20x - 25 = 16x^2 - 25$$

Notice that there is no middle term in the product. Since $16x^2$ and 25 are both squares, you can rewrite $16x^2 - 25$ as $(4x)^2 - (5)^2$.

Product of sum and difference of two terms	The product of the sum and difference of two terms is the difference of the squares of the two terms. $(a + b)(a - b) = a^2 - b^2$, for all numbers a and b.

Example 1 **Multiply: $(5x + 3y)(5x - 3y)$.**   **Multiply: $(6n^2 - 10)(6n^2 + 10)$.**

$$(5x + 3y)(5x - 3y) = (5x)^2 - (3y)^2$$
$$= 25x^2 - 9y^2$$

$$(6n^2 - 10)(6n^2 + 10) = (6n^2)^2 - 10^2$$
$$= 36n^4 - 100$$

The *square of a binomial* is another special product. The general form is $(a + b)^2$ or $(a - b)^2$. [Recall that $a - b$ means $a + (-b)$.] You can simplify $(3n + 5)^2$ by using the FOIL method.

$$(3n + 5)^2 = (3n + 5)(3n + 5) = 9n^2 + 15n + 15n + 25 = 9n^2 + 30n + 25$$

Notice that the product $9n^2 + 30n + 25$ can be written as $(3n)^2 + 2 \cdot 3n \cdot 5 + 5^2$. This leads to the following statement and corresponding formulas.

Square of a binomial	The square of a binomial is the sum of the square of the first term, twice the product of the terms, and the square of the last term. $(a + b)^2 = a^2 + 2ab + b^2$ and $(a - b)^2 = a^2 - 2ab + b^2$, for all numbers a and b.

Example 2 **Multiply: $(6c + 5d)^2$.**   **Multiply: $(4x^2 - 3)^2$.**

$$(a + b)^2 = a^2 + 2ab + b^2$$
$$(6c + 5d)^2 = (6c)^2 + 2(6c \cdot 5d) + (5d)^2$$
$$= 36c^2 + 60cd + 25d^2$$

$$(a - b)^2 = a^2 - 2ab + b^2$$
$$(4x^2 - 3)^2 = (4x^2)^2 - 2(4x^2 \cdot 3) + 3^2$$
$$= 16x^4 - 24x^2 + 9$$

The trinomials, $36c^2 + 60cd + 25d^2$ and $16x^4 - 24x^2 + 9$, are *perfect square trinomials*. A perfect square trinomial is the square of a binomial.

Expressions such as $(x + 3)(x^2 - 3x + 9)$ and $(x - 5)(x^2 + 5x + 25)$ represent two other special products. You can use the distributive property to expand such products, as shown below.

$$(x + 3)(x^2 - 3x + 9)$$
$$= x(x^2 - 3x + 9) + 3(x^2 - 3x + 9)$$
$$= x^3 - 3x^2 + 9x + 3x^2 - 9x + 27$$
$$= x^3 + 27, \text{ or } x^3 + 3^3$$

$$(x - 5)(x^2 + 5x + 25)$$
$$= x(x^2 + 5x + 25) - 5(x^2 + 5x + 25)$$
$$= x^3 + 5x^2 + 25x - 5x^2 - 25x - 125$$
$$= x^3 - 125, \text{ or } x^3 - 5^3$$

Notice that each product is either a *sum* or a *difference of two cubes*. This suggests the special product formulas stated below.

Two special products	$(a + b)(a^2 - ab + b^2) = a^3 + b^3$ and $(a - b)(a^2 + ab + b^2) = a^3 - b^3$, for all numbers a and b.

Example 3 **Multiply:** $(4x + 5)(16x^2 - 20x + 25)$.

Use $(a + b)[a^2 - ab + b^2] = a^3 + b^3$.

$$(4x + 5)(16x^2 - 20x + 25) = (4x + 5)[(4x)^2 - 4x \cdot 5 + 5^2] = (4x)^3 + 5^3$$
$$= 64x^3 + 125$$

Example 4 **Multiply:** $(2x - 3y)(4x^2 + 6xy + 9y^2)$.

Use $(a - b)[a^2 + ab + b^2] = a^3 - b^3$.

$$(2x - 3y)(4x^2 + 6xy + 9y^2) = (2x - 3y)[(2x)^2 + 2x \cdot 3y + (3y)^2] = (2x)^3 - (3y)^3$$
$$= 8x^3 - 27y^3$$

Written Exercises

2. $100y^2 - 49$ 6. $64y^2 + 160y + 100$ 7. $100 - 36a^2$
8. $144x^2 - 121$ 9. $625 - 9y^2$ 11. $49c^2 + 140c + 100$ 12. $225 - 60x + 4x^2$

Multiply each pair of polynomials

(A)
1. $(n + 12)(n - 12)$ $n^2 - 144$
2. $(10y - 7)(10y + 7)$ **above**
3. $(9x + 6)(9x - 6)$ $81x^2 - 36$
4. $(2x + 5)^2$ $4x^2 + 20x + 25$
5. $(1 - 6n)^2$ $1 - 12n + 36n^2$
6. $(8y + 10)^2$ **above**
7. $(10 - 6a)(10 + 6a)$ **above**
8. $(12x + 11)(12x - 11)$ **above**
9. $(25 - 3y)(25 + 3y)$ **above**
10. $(1 - 9n)^2$ $1 - 18n + 81n^2$
11. $(7c + 10)^2$ **above**
12. $(15 - 2x)^2$ **above**
13. $(x + 2)(x^2 - 2x + 4)$ $x^3 + 8$
14. $(3y - 1)(9y^2 + 3y + 1)$ $27y^3 - 1$
15. $(4 - 5a)(16 + 20a + 25a^2)$ $64 - 125a^3$
16. $(4c + 10)(16c^2 - 40c + 100)$ $64c^3 + 1,000$

17. $25x^2 - 64y^2$ 18. $a^2 - 36b^2$ 19. $16n^4 - 1$ 21. $9x^2 - 60xy + 100y^2$

(B)
17. $(5x + 8y)(5x - 8y)$ **above**
18. $(a - 6b)(a + 6b)$ **above**
19. $(4n^2 + 1)(4n^2 - 1)$ **above**
20. $(a + 7b)^2$ $a^2 + 14ab + 49b^2$
21. $(3x - 10y)^2$ **above**
22. $(c^2 + 5d)^2$
23. $(7y^2 - 4)(7y^2 + 4)$ $49y^4 - 16$
24. $(12c + 3d^2)(12c - 3d^2)$
25. $(10y^2 - 5z^2)(10y^2 + 5z^2)$
26. $(5x - 4y^2)^2$
27. $(4m^2 + 3n^2)^2$
28. $(10x^3 - 2y^3)^2$
29. $(8x + y)(64x^2 - 8xy + y^2)$ $512x^3 + y^3$
30. $(5c - 6d)(25c^2 + 30cd + 36d^2)$ $125c^3 - 216d^3$
31. $(m^2 - 3n)(m^4 + 3m^2n + 9n^2)$ $m^6 - 27n^3$
32. $(3x + 4y^2)(9x^2 - 12xy^2 + 16y^4)$ $27x^3 + 64y^6$

22. $c^4 + 10c^2d + 25d^2$ 24. $144c^2 - 9d^4$ 25. $100y^4 - 25z^4$ 26. $25x^2 - 40xy^2 + 16y^4$
27. $16m^4 + 24m^2n^2 + 9n^4$ 28. $100x^6 - 40x^3y^3 + 4y^6$

Special Products

SPECIAL FACTORS 3.3

Objectives
To factor a difference of two squares
To factor a perfect square trinomial
To factor a sum or a difference of two cubes

Many polynomials have the form of one of the following special products that were discussed in the previous lesson.

$$a^2 + 2ab + b^2 \qquad a^2 - b^2 \qquad a^3 + b^3 \qquad a^3 - b^3$$

You can factor polynomials of these types by reversing the appropriate special product formulas. For example, since $(y + 6)(y - 6) = y^2 - 6^2 = y^2 - 36$, you can factor $y^2 - 36$ as follows:

$$y^2 - 36 = y^2 - 6^2 = (y + 6)(y - 6).$$

The expression $y^2 - 36$ is called a **difference of two squares.**

> **Factoring a difference of two squares**
>
> $a^2 - b^2 = (a + b)(a - b)$, for all numbers a and b.

Example 1

Factor $49c^2 - 16$.

$$\begin{aligned}49c^2 - 16 &= (7c)^2 - 4^2 \\ &= (7c + 4)(7c - 4)\end{aligned}$$

Factor $9x^4 - 25y^2$.

$$\begin{aligned}9x^4 - 25y^2 &= (3x^2)^2 - (5y)^2 \\ &= (3x^2 + 5y)(3x^2 - 5y)\end{aligned}$$

Trinomials of the form $a^2 + 2ab + b^2$ and $a^2 - 2ab + b^2$ are **perfect square trinomials.** When a given trinomial is to be factored, you should first check to see if it is a perfect square trinomial. For example, $x^2 + 10x + 25$ is a perfect square trinomial since it can be written in the form $a^2 + 2ab + b^2$ as $x^2 + 2 \cdot x \cdot 5 + 5^2$.

You can factor $x^2 + 10x + 25$ by reversing one of the special product formulas.

$$(x + 5)^2 = x^2 + 2 \cdot x \cdot 5 + 5^2 = x^2 + 10x + 25$$
$$x^2 + 10x + 25 = x^2 + 2 \cdot x \cdot 5 + 5^2 = (x + 5)^2$$

> **Factoring a perfect square trinomial**
>
> $a^2 + 2ab + b^2 = (a + b)^2$ and
> $a^2 - 2ab + b^2 = (a - b)^2$, for all numbers a and b.

Example 2

Factor $16n^2 + 40n + 25$.

$$\begin{aligned}&16n^2 + 40n + 25 \\ &= (4n)^2 + 2 \cdot 4n \cdot 5 + 5^2 \\ &= (4n + 5)^2\end{aligned}$$

Factor $36c^4 - 12c^2d + d^2$.

$$\begin{aligned}&36c^4 - 12c^2d + d^2 \\ &= (6c^2)^2 - 2 \cdot 6c^2 \cdot d + d^2 \\ &= (6c^2 - d)^2\end{aligned}$$

The factorization of $16n^2 + 40n + 25$ in Example 2 can be checked by squaring $4n + 5$: $(4n + 5)^2 = (4n)^2 + 2 \cdot 4n \cdot 5 + 5^2 = 16n^2 + 40n + 25$.

You can factor a **sum** or a **difference of two cubes** by using the reverse of the two special product formulas below.

$$(a + b)(a^2 - ab + b^2) = a^3 + b^3$$
$$(a - b)(a^2 + ab + b^2) = a^3 - b^3$$

> **Factoring a sum or difference of two cubes**
>
> Sum: $a^3 + b^3 = (a + b)(a^2 - ab + b^2)$ and
> Difference: $a^3 - b^3 = (a - b)(a^2 + ab + b^2)$, for all numbers a and b.

Example 3   **Factor $8x^3 + 27$.**

$8x^3 + 27$, or $(2x)^3 + 3^3$, is a sum of two cubes.

$$a^3 + b^3 = (a + b)(a^2 - ab + b^2)$$
$$8x^3 + 27 = (2x)^3 + 3^3 = (2x + 3)[(2x)^2 - 2x \cdot 3 + 3^2]$$
$$= (2x + 3)(4x^2 - 6x + 9)$$

Example 4   **Factor $125x^3 - 64y^3$.**

$125x^3 - 64y^3$, or $(5x)^3 - (4y)^3$, is a difference of two cubes.

$$a^3 - b^3 = (a - b)(a^2 + ab + b^2)$$
$$125x^3 - 64y^3 = (5x)^3 - (4y)^3 = (5x - 4y)[(5x)^2 + 5x \cdot 4y + (4y)^2]$$
$$= (5x - 4y)(25x^2 + 20xy + 16y^2)$$

Sometimes a polynomial can be factored after its terms are grouped appropriately.

$$m^2 - 2mn + n^2 - c^2 - 2cd - d^2 = (m^2 - 2mn + n^2) - (c^2 + 2cd + d^2)$$
$$= (m - n)^2 - (c + d)^2$$
$$= [(m - n) + (c + d)][(m - n) - (c + d)]$$
$$= (m - n + c + d)(m - n - c - d)$$

Example 5   **Factor each polynomial.**

$m^2 + 2mn + n^2 - 6m - 6n + 9$ $9x^2 - c^2 + 8c - 16$

$(m^2 + 2mn + n^2) - 6(m + n) + 9$ $9x^2 - (c^2 - 8c + 16)$
$(m + n)^2 - 2(m + n) \cdot 3 + 3^2$ $(3x)^2 - (c - 4)^2$

$a^2 \quad -2 \cdot a \cdot b \quad + b^2$ $a^2 - b^2$

                                 $(a + b)(a - b)$

              $(a - b)^2$ $(3x + c - 4)(3x - [c - 4])$

           $(m + n - 3)^2$ $(3x + c - 4)(3x - c + 4)$

Special Factors

Reading in Algebra

Determine whether each polynomial is a difference of two squares (DS), a perfect square trinomial (PS), a sum of two cubes (SC), a difference of two cubes (DC), or none of these (N).

1. $x^3 - 8$ **DC**
2. $y^2 + 16$ **N**
3. $x^2 + 6x + 9$ **PS**
4. $9c^2 - 4$ **DS**
5. $n^2 - 10n - 25$ **N**
6. $27x^3 + 1$ **SC**
7. $(a + b)^2 - 4$ **DS**
8. $x^6 - 1$ **DS, DC**
9. $z^2 + 2z - 1$ **N**
10. $b^3 - 15c^3$ **N**
11. $x^2y^2 - 1$ **DS**
12. $49 - 14y + y^2$ **PS**

Oral Exercises

1. $(10n)^2 - 3^2$
2. $(5n)^3 + 1^3$
3. $5^2 - (2y)^2$
4. $10^3 - (3n)^3$

State each binomial as a difference of two squares, a sum of two cubes, or a difference of two cubes.

1. $100n^2 - 9$
2. $125n^3 + 1$
3. $25 - 4y^2$
4. $1{,}000 - 27n^3$
5. $x^6 + 8z^3$ $(x^2)^3 + (2z)^3$
6. $64x^2 - 81y^2$ $(8x)^2 - (9y)^2$
7. $x^9 - 125y^3$ $(x^3)^3 - (5y)^3$
8. $x^4 - 36y^2$ $(x^2)^2 - (6y)^2$

Determine whether each trinomial is a perfect square trinomial.

9. $25x^2 + 50x + 16$ **no**
10. $4y^2 - 20y + 25$ **yes**
11. $9n^2 - 13n + 4$ **no**
12. $9m^4 + 24m^2 + 16$ **yes**

Written Exercises

28. $(10x + 3y)(100x^2 - 30xy + 9y^2)$
32. $(3xy - 5z^2)(9x^2y^2 + 15xyz^2 + 25z^4)$
37. $(x + c + d)^2$
38. $(2x - 3y - 5)^2$

Factor each polynomial.

(A)
1. $4b^2 - 49$ $(2b + 7)(2b - 7)$
2. $121 - 36n^2$ $(11 + 6n)(11 - 6n)$
3. $y^2 + 8y + 16$ $(y + 4)^2$
4. $1 - 20a + 100a^2$ $(1 - 10a)^2$
5. $x^3 + 27$ $(x + 3)(x^2 - 3x + 9)$
6. $y^3 - 8$ $(y - 2)(y^2 + 2y + 4)$
7. $c^4 - 25$ $(c^2 + 5)(c^2 - 5)$
8. $49 - d^4$ $(7 + d^2)(7 - d^2)$
9. $9n^2 - 24n + 16$ $(3n - 4)^2$
10. $25a^2 + 20a + 4$ $(5a + 2)^2$
11. $y^4 + 18y^2 + 81$ $(y^2 + 9)^2$
12. $16d^4 - 40d^2 + 25$ $(4d^2 - 5)^2$
13. $125c^3 - 1$ $(5c - 1)(25c^2 + 5c + 1)$
14. $1{,}000x^3 + 1$ $(10x + 1)(100x^2 - 10x + 1)$
15. $x^2 + 6x + 9 - y^2$ $(x + 3 + y)(x + 3 - y)$
16. $m^2 - 2mn + n^2 - 25$ $(m - n + 5)(m - n - 5)$

(B)
17. $x^2 + 2xy + y^2 + 4x + 4y + 4$ $(x + y + 2)^2$
18. $c^2 - 2cd + d^2 + 10c - 10d + 25$ $(c - d + 5)^2$
19. $x^2 - y^2 - 10y - 25$ $(x + y + 5)(x - y - 5)$
20. $t^2 + 2tv + 12t + v^2 + 12v + 36$ $(t + v + 6)^2$

35. $(m - n + x + y)(m - n - x - y)$
36. $(2x + y + c - 2)(2x + y - c + 2)$

21. $49c^4d^4 - 100$ $(7c^2d^2 + 10)(7c^2d^2 - 10)$
22. $36x^6 - y^2$ $(6x^3 + y)(6x^3 - y)$
23. $25c^2 + 20cd + 4d^2$ $(5c + 2d)^2$
24. $36x^4 - 60x^2y + 25y^2$ $(6x^2 - 5y)^2$
25. $64m^3 + n^3$ $(4m + n)(16m^2 - 4mn + n^2)$
26. $c^3 - 125d^3$ $(c - 5d)(c^2 + 5cd + 25d^2)$
27. $125t^3 - 8v^3$ $(5t - 2v)(25t^2 + 10tv + 4v^2)$
28. $1{,}000x^3 + 27y^3$ **above**
29. $16c^6 - 40c^3d + 25d^2$ $(4c^3 - 5d)^2$
30. $4a^6b^4 + 28a^3b^2 + 49$ $(2a^3b^2 + 7)^2$
31. $25x^6 - 36d^4$ $(5x^3 + 6d^2)(5x^3 - 6d^2)$
32. $27x^3y^3 - 125z^6$ **above**
33. $0.25x^2 - 1.21y^2$ $(0.5x + 1.1y)(0.5x - 1.1y)$
34. $1.44a^2 - 6ab + 6.25b^2$ $(1.2a - 2.5b)^2$

(C)
35. $m^2 - 2mn + n^2 - x^2 - 2xy - y^2$ **above**
36. $4x^2 + 4xy + y^2 - c^2 + 4c - 4$ **above**
37. $x^2 + 2cx + 2dx + c^2 + 2cd + d^2$ **above**
38. $4x^2 - 12xy + 9y^2 - 20x + 30y + 25$ **above**
39. $25x^2 - 30x + 9 - 4a^2 + 4ab - b^2$
   39. $(5x - 3 + 2a - b)(5x - 3 - 2a + b)$
40. $25x^2 + 40xy - 30x + 16y^2 - 24y + 9$
   40. $(5x + 4y - 3)^2$
41. $9x^{4m+6} + 12x^{2m+3}y^n + 4y^{2n}$
42. $x^{4n} - y^{6n+2}$ $(x^{2n} + y^{3n+1})(x^{2n} - y^{3n+1})$
43. $x^{3c} + y^{12d}$ $(x^c + y^{4d})(x^{2c} - x^cy^{4d} + y^{8d})$
44. $8x^{6a} - y^{3b+9}$
   $(2x^{2a} - y^{b+3})(4x^{4a} + 2x^{2a}y^{b+3} + y^{2b+6})$
41. $(3x^{2m+3} + 2y^n)^2$

Chapter Three

COMBINED TYPES OF FACTORING   3.4

Objective   **To factor a polynomial completely**

Some numbers can be factored into more than two factors. For example,

$$30 = 3 \cdot 10 = 3 \cdot 2 \cdot 5 \quad \text{and} \quad 36 = 4 \cdot 9 = 2 \cdot 2 \cdot 3 \cdot 3.$$

Likewise, some polynomials can be factored into more than two factors. To factor a polynomial *completely*, you may have to use several types of factoring. The following steps show you the order in which the factoring should be accomplished.

Factoring a polynomial completely

Step 1 Factor out the *greatest common factor* (GCF), if any.
Step 2 Factor the resulting polynomial, if possible.
Step 3 Factor each polynomial factor where possible.

Example 1   **Factor $12x^4y^3 + 18x^2y^2 - 24x^2y$ completely.**

Step 1 The GCF of 12, 18, and 24 is 6. The GCF of x^4 and x^2 is x^2. The GCF of y^3, y^2, and y is y. Factor out $6x^2y$.

$$12x^4y^3 + 18x^2y^2 - 24x^2y = 6x^2y(2x^2y^2 + 3y - 4)$$

Step 2 The resulting polynomial, $2x^2y^2 + 3y - 4$, cannot be factored.

Thus, $12x^4y^3 + 18x^2y^2 - 24x^2y = 6x^2y(2x^2y^2 + 3y - 4)$.

In Example 1, the monomial $6x^2y$ is called a **common monomial factor**. It is a factor of each term of the original polynomial.

Example 2   **Factor $15x^2 + 10x - 40$ completely.**

$$15x^2 + 10x - 40$$
Step 1 Factor out 5, the GCF. $= 5(3x^2 + 2x - 8)$
Step 2 Factor $3x^2 + 2x - 8$. $= 5(3x - 4)(x + 2)$
Step 3 $3x - 4$ and $x + 2$ cannot be factored.

Thus, $15x^2 + 10x - 40 = 5(3x - 4)(x + 2)$.

Example 3   **Factor $27y^4 - 75y^2$ completely.**

$$27y^4 - 75y^2$$
Step 1 Factor out $3y^2$, the GCF. $= 3y^2(9y^2 - 25)$ ◄ *Difference of*
Step 2 Factor $9y^2 - 25$. $= 3y^2(3y + 5)(3y - 5)$ *two squares*
Step 3 $3y + 5$ and $3y - 5$ cannot be factored.

Thus, $27y^4 - 75y^2 = 3y^2(3y + 5)(3y - 5)$.

A complete factorization may contain more than two binomials, as shown in the next example.

Example 4   **Factor $8x^4 + 10x^2 - 3$ completely.**

Step 1   There is no common monomial factor other than 1.

$$8x^4 + 10x^2 - 3$$

Step 2   Factor $8x^4 + 10x^2 - 3$. $= (2x^2 + 3)(4x^2 - 1)$
Step 3   Factor $4x^2 - 1$. $= (2x^2 + 3)(2x + 1)(2x - 1)$

Thus, $8x^4 + 10x^2 - 3 = (2x^2 + 3)(2x + 1)(2x - 1)$.

To factor a polynomial in which the coefficient of the first term is negative, begin by factoring out -1 as a common monomial factor.

Example 5   **Factor $-9x^2 + 12x - 4$ completely.**

$$-9x^2 + 12x - 4$$

Step 1   Factor out -1. $= -1(9x^2 - 12x + 4)$ ◀ *Perfect square trinomial*
Step 2   Factor $9x^2 - 12x + 4$. $= -1(3x - 2)^2$

Thus, $-9x^2 + 12x - 4 = -1(3x - 2)^2$.

Factoring completely may involve factoring a sum or a difference of two cubes. This is shown in Example 6.

Example 6   **Factor $24x^3 - 81$ completely.**

$$24x^3 - 81$$

Step 1   Factor out 3, the GCF. $= 3(8x^3 - 27)$
Step 2   Factor $8x^3 - 27$. $= 3(2x - 3)(4x^2 + 6x + 9)$

Thus, $24x^3 - 81 = 3(2x - 3)(4x^2 + 6x + 9)$.

Some polynomials with an even number of terms may be factored by "grouping pairs of terms" and factoring out a **common binomial factor.**

Example 7   **Factor $5xy - 15x - 2yz + 6z$ completely.**

$5xy - 15x - 2yz + 6z = (5xy - 15x) + (-2yz + 6z) = 5x(y - 3) - 2z(y - 3)$
$= (5x - 2z)(y - 3)$

Thus, $5xy - 15x - 2yz + 6z = (5x - 2z)(y - 3)$.

Reading in Algebra

Determine whether each expression is factored completely. If not, state why not.
1. $17 \cdot 21 \cdot 59$ **No**
2. $19 \cdot 23 \cdot 37$ **Yes**
3. $8(2x + 6y)$ **No**
4. $4a(5a + 4b)$ **Yes**
5. $-7(x^2 + xy)$ **No**
6. $5(x^2 - 2x + 3)$ **Yes**
7. $2y(y^2 - 7y + 12)$ **No**
8. $-3(9n^2 - 1)$ **No**

Oral Exercises

Find the GCF of each group of expressions.
1. $18x^2$; $36x$; 12 **6**
2. $5x^4$; $-2x^2$; $7x$ **x**
3. x^4; x^2; x^3 **x^2**
4. $12y^2$; $6y$; $24y^3$ **6y**
5. $-10x^2$; $15x^3$ **$5x^2$**
6. $x(5a - 3b)$; $-4(5a - 3b)$ **$5a - 3b$**

Factor -1 out of each polynomial.
7. $-5a + 3b$ **$-1(5a - 3b)$**
8. $-2x^2 - 7y$ **$-1(2x^2 + 7y)$**
9. $-y^2 + y - 12$ **$-1(y^2 - y + 12)$**

48. $x^{3c}(x^c + 3)(x^c - 3)$
49. $x^2(x^{2n} + 3)^2$
50. $y^n(y^n - 1)(y^{2n} + y^n + 1)$

Written Exercises

42. $4(c + d)(c - d)(c + 2d)$
39. $(3a - 7d)(4b - 3c)$

Factor each polynomial completely.

A
1. $12n^2 - 15n - 3$ **$3(4n^2 - 5n - 1)$**
2. $3a^4 - 8a^3 + a^2$ **$a^2(3a^2 - 8a + 1)$**
3. $3x^2 - 21x + 36$ **$3(x - 3)(x - 4)$**
4. $2n^2 + 4n - 30$ **$2(n + 5)(n - 3)$**
5. $16y^2 - 4$ **$4(2y + 1)(2y - 1)$**
6. $18c^2 - 50$ **$2(3c + 5)(3c - 5)$**
7. $4n^2 + 40n + 100$ **$4(n + 5)^2$**
8. $5x^2 - 40x + 80$ **$5(x - 4)^2$**
9. $y^4 - 7y^2 - 18$ **$(y^2 + 2)(y + 3)(y - 3)$**
10. $2a^4 - 9a^2 + 4$ **$(2a^2 - 1)(a + 2)(a - 2)$**
11. $-x^2 + 8x - 16$ **$-1(x - 4)^2$**
12. $-y^2 + 36$ **$-1(y + 6)(y - 6)$**
13. $3y^3 + 81$ **$3(y + 3)(y^2 - 3y + 9)$**
14. $5n^3 - 625$ **$5(n - 5)(n^2 + 5n + 25)$**
15. $8xy + 20x + 6y + 15$ **$(4x + 3)(2y + 5)$**
16. $6cd - 21c - 10d + 35$ **$(3c - 5)(2d - 7)$**
17. $-4n^2 + 4n + 3$ **$-1(2n - 3)(2n + 1)$**
18. $-9t^2 - 30t - 25$ **$-1(3t + 5)^2$**
21. $15x^2y^3(5y^2 - 2xy + 3x^2)$
22. $4x^2y^2z^3(11xz - 25yz^2 - 16xy)$

B
19. $x^3 + 4x^2 + 6x + 24$ **$(x^2 + 6)(x + 4)$**
20. $6a^3 + 20a^2 - 21a - 70$ **$(2a^2 - 7)(3a + 10)$**
21. $75x^2y^5 - 30x^3y^4 + 45x^4y^3$ **above**
22. $44x^3y^2z^4 - 100x^2y^3z^5 - 64x^3y^3z^3$ **above**
23. $8a^3 - 4a^2 - 40a$ **$4a(2a - 5)(a + 2)$**
24. $3a^3 - 75ab^2$ **$3a(a + 5b)(a - 5b)$**
25. $2x^3y - 4x^2y + 2xy$ **$2xy(x - 1)^2$**
26. $36a^3b + 120a^2b^2 + 100ab^3$ **$4ab(3a + 5b)^2$**
27. $9x^4 - 7x^2 - 16$ **$(x^2 + 1)(3x + 4)(3x - 4)$**
28. $4n^4 - 17n^2 + 18$ **$(n^2 - 2)(2n + 3)(2n - 3)$**
29. $n^4 - 13n^2 + 36$ **$(n + 2)(n - 2)(n + 3)(n - 3)$**
30. $25y^4 - 101y^2 + 4$ **$(5y + 1)(5y - 1)(y + 2)(y - 2)$**
31. $-3y^2 + 27$ **$-3(y + 3)(y - 3)$**
32. $-18x^2 + 32$ **$-2(3x + 4)(3x - 4)$**
33. $-4ac^2 - 4ac - a$ **$-a(2c + 1)^2$**
34. $-x^3 + 6x^2y - 9xy^2$ **$-x(x - 3y)^2$**
35. $24x^3 - 375$ **$3(2x - 5)(4x^2 + 10x + 25)$**
36. $-54y^3 - 128$ **$-2(3y + 4)(9y^2 - 12y + 16)$**
37. $y^4 + 64y$ **$y(y + 4)(y^2 - 4y + 16)$**
38. $3x^4 - 3{,}000x$ **$3x(x - 10)(x^2 + 10x + 100)$**
39. $12ab - 9ac - 28bd + 21cd$ **above**
40. $8x^3 - 15y - 20xy + 6x^2$ **$(2x^2 - 5y)(4x + 3)$**
41. $a^3 - a^2b - a^2b^2 + ab^3$ **$a(a - b^2)(a - b)$**
42. $4c^3 + 8c^2d - 4cd^2 - 8d^3$ **above**
43. $4y^8 - 13y^4 + 9$
43. $(2y^2 + 3)(2y^2 - 3)(y^2 + 1)(y + 1)(y - 1)$
44. $36x^8 - 13x^4 + 1$
44. $(3x^2 + 1)(3x^2 - 1)(2x^2 + 1)(2x^2 - 1)$

C
45. $x^{a+6} + x^4$ **$x^4(x^{a+2} + 1)$**
46. $x^{5c} - x^{4c}$ **$x^{4c}(x^c - 1)$**
47. $y^{n+5} + y^{n+4}$ **$y^{n+4}(y + 1)$**
48. $x^{5c} - 9x^{3c}$ **above**
49. $x^{4n+2} + 6x^{2n+2} + 9x^2$ **above**
50. $y^{4n} - y^n$ **above**
51. Factor $x^6 - y^6$ completely as a difference of two squares **below**
52. Factor $x^6 - y^6$ completely as a difference of two cubes. **below**
53. Use factoring to show that the answers in Exercises 51 and 52 are equivalent.
51. $(x + y)(x^2 - xy + y^2)(x - y)(x^2 + xy + y^2)$
52. $(x + y)(x - y)(x^4 + x^2y^2 + y^4)$

Combined Types of Factoring

QUADRATIC EQUATIONS 3.5

Objective **To solve a quadratic equation by factoring**

Equations such as $5x^2 + 17x - 12 = 0$, $9x^2 - 25 = 0$, and $7x^2 + 14x = 0$ are called **quadratic equations**. Each equation contains a polynomial of the second degree

> **Standard form of a quadratic equation**
> The *standard form of a quadratic equation* is $ax^2 + bx + c = 0$ where a, b, and c are numbers, and $a \neq 0$.

You can solve some quadratic equations by writing the equation in standard form, factoring, and then setting each factor equal to 0. The method is based upon the following property.

> **Zero-product property**
> For all numbers m and n, if $mn = 0$, then $m = 0$ or $n = 0$.

Example 1

Find the roots of $3c^2 - 10c - 8 = 0$.

The equation is in standard form. $3c^2 - 10c - 8 = 0$
Factor $3c^2 - 10c - 8$. $(3c + 2)(c - 4) = 0$
Use the Zero-Product Property. $3c + 2 = 0$ or $c - 4 = 0$
Solve both linear equations. $3c = -2$ $c = 4$
$c = -\frac{2}{3}$

Check both solutions:

$$\begin{array}{c|c}
3c^2 - 10c - 8 & 0 \\
\hline
3\left(-\frac{2}{3}\right)^2 - 10\left(-\frac{2}{3}\right) - 8 & 0 \\
\frac{4}{3} + \frac{20}{3} - 8, \text{ or } 0 &
\end{array}$$

$$\begin{array}{c|c}
3c^2 - 10c - 8 & 0 \\
\hline
3 \cdot 4^2 - 10 \cdot 4 - 8 & 0 \\
48 - 40 - 8 & \\
0 &
\end{array}$$

Thus, the roots are $-\frac{2}{3}$ and 4.

Example 2

Solve $5x = 6 - 4x^2$.

Step 1 Write in standard form. $4x^2 + 5x - 6 = 0$
Step 2 Factor. $(4x - 3)(x + 2) = 0$
Step 3 Set each factor equal to 0. $4x - 3 = 0$ or $x + 2 = 0$
Step 4 Solve the linear equations. $4x = 3$ $x = -2$
$x = \frac{3}{4}$

Thus, $\frac{3}{4}$ and -2 are the solutions.

Summary

To solve a quadratic equation by factoring:
- **Step 1** Write the equation in standard form.
- **Step 2** Factor.
- **Step 3** Set each factor equal to 0.
- **Step 4** Solve the linear equations.

Sometimes it is more convenient to use the standard form, $0 = ax^2 + bx + c$, so that the coefficient of x^2 is a positive number.

Example 3

Solve $-7x^2 = 21x$.

$$0 = 7x^2 + 21x$$
$$0 = 7x(x + 3) \quad \blacktriangleleft \text{ The GCF is } 7x.$$
$7x = 0 \quad$ or $\quad x + 3 = 0$
$x = 0 \quad\quad\quad\quad\quad x = -3$

Thus, the solutions are 0 and -3.

Solve $25 = 9n^2$.

$$0 = 9n^2 - 25$$
$$0 = (3n + 5)(3n - 5)$$
$3n + 5 = 0 \quad$ or $\quad 3n - 5 = 0$
$3n = -5 \quad\quad\quad\quad\quad 3n = 5$
$n = -\dfrac{5}{3} \quad\quad\quad\quad\quad n = \dfrac{5}{3}$

Thus, the solutions are $-\dfrac{5}{3}$ and $\dfrac{5}{3}$.

The Zero-Product Property can be extended to any number of factors. In the next example, the property is used to solve the *fourth degree equation* $x^4 - 13x^2 + 36 = 0$. The fourth degree polynomial is factored completely, and then each of the four factors is set equal to 0.

Example 4

Solve $x^4 - 13x^2 + 36 = 0$.

$$x^4 - 13x^2 + 36 = 0$$
$$(x^2 - 4)(x^2 - 9) = 0$$
$$(x + 2)(x - 2)(x + 3)(x - 3) = 0 \quad \blacktriangleleft \text{ Factor completely.}$$
$x + 2 = 0 \quad$ or $\quad x - 2 = 0 \quad$ or $\quad x + 3 = 0 \quad$ or $\quad x - 3 = 0$
$x = -2 \quad\quad\quad\quad x = 2 \quad\quad\quad\quad x = -3 \quad\quad\quad\quad x = 3$

Thus, the solutions are -2, 2, -3, and 3.

In some quadratic equations, the terms contain a common numerical factor.

Example 5

Solve $3n^2 - 15n + 18 = 0$.

Divide each side by 3, the GCF.

$$n^2 - 5n + 6 = 0$$
$$(n - 2)(n - 3) = 0$$
$n - 2 = 0 \quad$ or $\quad n - 3 = 0$
$n = 2 \quad\quad\quad\quad\quad n = 3$

Thus, the solutions are 2 and 3.

Quadratic Equations

Reading in Algebra

Determine whether each statement is true or false.
1. $4x^2 - 8x = 0$ is a quadratic equation. **T**
2. $5x - 15 = 0$ is a quadratic equation. **F**
3. A quadratic equation may have two solutions. **T**
4. A fourth degree equation may have four solutions. **T**
5. If $x(x - 5) = 0$, then $x = 0$ or $x - 5 = 0$. **T**
6. If $(x - 1)(x - 3) = 8$, then $x - 1 = 8$ or $x - 3 = 8$. **F**
7. If $(2n - 5)(3n - 2) = 0$, then $n = 5$ or $n = 2$. **F**
8. If $(y - 5)(y - 5) = 0$, then $y = 5$. **T**

Written Exercises

13. **−5, 5, 1, −1** 14. **−5, 5, −2, 2** 15. **−10, 10, −3, 3**

Solve each equation.

Ⓐ
1. $x^2 - 13x + 40 = 0$ **5, 8**
2. $a^2 - a - 42 = 0$ **−6, 7**
3. $0 = x^2 + 15x + 50$ **−10, −5**
4. $y^2 = 12 - y$ **−4, 3**
5. $6 = b^2 - b$ **−2, 3**
6. $6y = 16 - y^2$ **−8, 2**
7. $2a^2 - 10a = 0$ **0, 5**
8. $3x^2 = -12x$ **0, −4**
9. $15a = a^2$ **0, 15**
10. $z^2 - 64 = 0$ **−8, 8**
11. $0 = c^2 - 36$ **−6, 6**
12. $16 = z^2$ **−4, 4**
13. $x^4 - 26x^2 + 25 = 0$ **above**
14. $a^4 - 29a^2 + 100 = 0$ **above**
15. $900 + x^4 = 109x^2$ **above**
16. $3y^2 - 21y + 30 = 0$ **2, 5**
17. $5c^2 + 20c - 60 = 0$ **−6, 2**
18. $0 = 40z + 100 + 4z^2$ **−5**
19. $2c^2 - 9c + 4 = 0$ **$\frac{1}{2}$, 4**
20. $3z^2 = 8z + 3$ **$-\frac{1}{3}$, 3**
21. $-3c^2 = c - 2$ **$\frac{2}{3}$, −1**
22. $8n^2 + 2n - 1 = 0$ **$\frac{1}{4}, -\frac{1}{2}$**
23. $1 = 7w - 10w^2$ **$\frac{1}{5}, \frac{1}{2}$**
24. $0 = 12t^2 + 7t + 1$ **$-\frac{1}{4}, -\frac{1}{3}$**
25. $9c^2 - 16 = 0$ **$-1\frac{1}{3}, 1\frac{1}{3}$**
26. $16z^2 = 1$ **$-\frac{1}{4}, \frac{1}{4}$**
27. $36 = 25c^2$ **$-1\frac{1}{5}, 1\frac{1}{5}$**
28. $y^4 - y^2 - 12 = 0$ **−2, 2**
29. $b^4 = 24b^2 + 25$ **−5, 5**
30. $4y^2 = y^4 - 45$ **−3, 3**
31. $2b^2 + 12b + 18 = 0$ **−3**
32. $36y^2 = 12 - 6y$ **$-\frac{2}{3}, \frac{1}{2}$**
33. $60x - 180 = 5x^2$ **6**
34. $(x - 7)^2 = 29 - x^2$ **2, 5**
35. $(8 - n)^2 = 2(24 - n^2)$ **$1\frac{1}{3}$, 4**
36. $(3y + 2)^2 = (2y - 5)^2$ **$\frac{3}{5}$, −7**

Ⓑ
37. $2b^2 - 7b + 6 = 0$ **$1\frac{1}{2}$, 2**
38. $3y^2 = 15 - 4y$ **$1\frac{2}{3}$, −3**
39. $0 = 7 + 9n + 2n^2$ **$-3\frac{1}{2}$, −1**
40. $25n^2 - 4n = 0$ **0, $\frac{4}{25}$**
41. $12p^2 = 30p$ **0, $2\frac{1}{2}$**
42. $-22y = 6y^2$ **0, $-3\frac{2}{3}$**
43. $30t^2 - 125t + 120 = 0$ **$2\frac{2}{3}, 1\frac{1}{2}$**
44. $30y + 24y^2 = 75$ **$1\frac{1}{4}, -2\frac{1}{2}$**
45. $70 + 2m = 24m^2$ **$1\frac{3}{4}, -1\frac{2}{3}$**
46. $4y^4 - 37y^2 + 9 = 0$ **below**
47. $25b^4 = 34b^2 - 9$ **below**
48. $52y^2 = 9y^4 + 64$ **below**
49. $4x^4 + 31x^2 - 90 = 0$ **$-1\frac{1}{2}, 1\frac{1}{2}$**
50. $16c^4 + 23c^2 = 18$ **$-\frac{3}{4}, \frac{3}{4}$**
51. $10 - 36t^2 = 16t^4$ **$-\frac{1}{2}, \frac{1}{2}$**
52. $a^3 - 7a^2 + 10a = 0$ **0, 2, 5**
53. $5x^3 = 30x - 25x^2$ **0, −6, 1**
54. $4y^3 + 48y^2 = 4y^4$ **0, −3, 4**
55. $x(x + 5)(x - 4) = x^3$ **0, 20**
56. $3y(y - 2)(y + 4) = 3y^3$ **0, 4**
57. $(2x^2 + 3)^2 = 4x(x^3 + 6)$ **$\frac{1}{2}, 1\frac{1}{2}$**

Solve each equation for x in terms of a. 46. $-\frac{1}{2}, \frac{1}{2}, -3, 3$ 47. $-\frac{3}{5}, \frac{3}{5}, -1, 1$ 48. $-1\frac{1}{3}, 1\frac{1}{3}, -2, 2$

Ⓒ
58. $3x^2 + 9a^2 = 12ax$
59. $0 = 6x^2 + 8ax - 8a^2$
60. $x^4 - 10a^2x^2 + 9a^4 = 0$
61. $42a^2 = 24x^2 - 38ax$
62. $36a^4 = 25a^2x^2 - 4x^4$
63. $a^8 + 100x^4 = 29a^4x^2$

Solve. Each equation has four solutions.
64. $|x^2 - 17| = 8$ **±5, ±3**
65. $|x^2 - 10| = 6$ **±4, ±2**
66. $|20 - x^2| = 16$ **±6, ±2**

58. **$a, 3a$**   59. **$\frac{2a}{3}, -2a$**   60. **$-3a, 3a, -a, a$**   61. **$\frac{-3a}{4}, \frac{7a}{3}$**

62. **$\frac{-3a}{2}, \frac{3a}{2}, -2a, 2a$**   63. **$\frac{-a^2}{5}, \frac{a^2}{5}, \frac{-a^2}{2}, \frac{a^2}{2}$**

CUMULATIVE REVIEW

Find the solution set of each inequality. Graph the solutions on a number line.
1. $-3x < 2x + 35$
   $\{x | x > -7\}$
2. $|x - 2| < 5$
   $\{x | -3 < x < 7\}$
3. $|x + 3| > 7$
   $\{x | x < -10 \text{ or } x > 4\}$

QUADRATIC INEQUALITIES 3.6

Objectives **To find the solution set of a quadratic inequality by factoring**
To graph the solutions of a quadratic inequality

An inequality such as $x^2 - x - 6 > 0$ is called a **quadratic inequality**. The standard form of a quadratic inequality is $ax^2 + bx + c > 0$ (or $ax^2 + bx + c < 0$), where $a \neq 0$. Each quadratic inequality contains a polynomial of the second degree.

To solve the quadratic inequality $x^2 - x - 6 > 0$, you can use the following property for the *positive* product of two factors.

If $m \cdot n > 0$, then ($m < 0$ and $n < 0$) or ($m > 0$ and $n > 0$).

This property will lead to two conjunctions that must be solved.

Example 1 **Find the solution set of $x^2 - x - 6 > 0$. Graph the solutions.**

$x^2 - x - 6 > 0$ ◀ *Factor the polynomial.*
$(x + 2)(x - 3) > 0$
$(x + 2 < 0$ and $x - 3 < 0)$ or $(x + 2 > 0$ and $x - 3 > 0)$
$(x < -2$ and $x < 3)$ or $(x > -2$ and $x > 3)$
$\underbrace{x < -2}$ or $\underbrace{x > 3}$

Thus, the solution set is $\{x | x < -2 \text{ or } x > 3\}$.

The quadratic inequality $x^2 - 16 < 0$ can be solved using the following property for the *negative* product of two factors.

If $m \cdot n < 0$, then ($m < 0$ and $n > 0$) or ($m > 0$ and $n < 0$).

This property produces two conjunctions that must be solved.

To solve a quadratic inequality in which the coefficient of the square term is negative, it may be helpful to multiply the inequality by -1.

Example 2 **Find the solution set of $-x^2 + 16 \geq 0$. Graph the solutions.**

$-x^2 + 16 \geq 0$ ◀ *Multiply each side by -1 and*
$x^2 - 16 \leq 0$ *reverse the order to $\leq$.*
$(x + 4)(x - 4) \leq 0$
$(x + 4 \leq 0$ and $x - 4 \geq 0)$ or $(x + 4 \geq 0$ and $x - 4 \leq 0)$
$(x \leq -4$ and $x \geq 4)$ or $(x \geq -4$ and $x \leq 4)$
$\underbrace{\text{no solution}}$ $\underbrace{-4 \leq x \leq 4}$

Thus, the solution set is $\{x | -4 \leq x \leq 4\}$.

For a *polynomial inequality* of degree 3 or higher, it is more convenient to graph the solutions first and then write the solution set.

Example 3

Graph the solutions of $x^3 - 2x^2 - 15x > 0$. Find the solution set.

$$x^3 - 2x^2 - 15x > 0$$
$$x(x^2 - 2x - 15) > 0$$
$$x(x + 3)(x - 5) > 0$$

The solutions of $x(x + 3)(x - 5) = 0$ are 0, -3, and 5. Plot these numbers on a number line.

The points at -3, 0, and 5 separate the number line into four parts, I–IV. Select the coordinate of a point in each part and evaluate $x(x + 3)(x - 5)$ to find whether the value is positive (>0) or negative (<0).

Part		x	$x(x + 3)(x - 5)$	Value
I.	$x < -3$	-4	$-4(-1)(-9)$	$-36 < 0$
II.	$-3 < x < 0$	-2	$-2(+1)(-7)$	$+14 > 0$
III.	$0 < x < 5$	1	$+1(+4)(-4)$	$-16 < 0$
IV.	$x > 5$	6	$+6(+9)(+1)$	$+54 > 0$

The graph of the solutions will be parts II and IV where the value is positive.

Thus, the solution set is $\{x | -3 < x < 0 \text{ or } x > 5\}$.

Example 4

Graph the solutions of $x^4 - 10x^2 + 9 \leq 0$. Find the solution set.

$$x^4 - 10x^2 + 9 \leq 0$$
$$(x^2 - 9)(x^2 - 1) \leq 0$$
$$(x + 3)(x - 3)(x + 1)(x - 1) \leq 0$$

Plot the solutions of $(x + 3)(x - 3)(x + 1)(x - 1) = 0$ on a number line.

Part		$(x + 3)(x + 1)(x - 1)(x - 3)$	Value	
I.	$x < -3$	(neg.) (neg.) (neg.) (neg.)	pos. > 0	
II.	$-3 < x < -1$	(pos.) (neg.) (neg.) (neg.)	neg. < 0	YES
III.	$-1 < x < 1$	(pos.) (pos.) (neg.) (neg.)	pos. > 0	
IV.	$1 < x < 3$	(pos.) (pos.) (pos.) (neg.)	neg. < 0	YES
V.	$x > 3$	(pos.) (pos.) (pos.) (pos.)	pos. > 0	

Draw parts II and IV and their endpoints on the number line.

Thus, the solution set is $\{x | -3 \leq x \leq -1 \text{ or } 1 \leq x \leq 3\}$.

Reading in Algebra

Match each inequality to exactly one graph.
1. $3 < x < 7$ **D**
2. $x < 3$ or $x > 7$ **C**
3. $x < 3$ or $7 < x < 9$ **A**
4. $3 < x < 7$ or $x > 9$ **B**
5. $x^2 \geq 0$ **F**
6. $x^2 < 0$ **E**

Oral Exercises

Tell whether each sentence is true or false.
1. If $x = 4$, then $(x - 2)(x - 6) < 0$. **T**
2. If $x = -2$, then $(x + 4)(x - 3) > 0$. **F**
3. If $x = 8$, then $(x - 2)(x - 6) < 0$. **F**
4. If $x = -6$, then $(x + 4)(x - 3) > 0$. **T**
5. If $x < -5$, then $(x + 5)(x - 5) > 0$. **T**
6. If $x > 7$, then $(x + 1)(x - 7) < 0$. **F**
7. If $-3 < x < 2$, then $(x + 3)(x - 2) > 0$. **F**
8. If $1 < x < 5$, then $(x - 1)(x - 5) < 0$. **T**

Written Exercises

Find the solution set of each inequality. Graph the solutions.

Ⓐ
1. $n^2 - 8n + 15 > 0$ $\{n|n < 3$ or $n > 5\}$
2. $x^2 - 2x - 8 < 0$ $\{x|-2 < x < 4\}$
3. $c^2 + 8c + 12 \leq 0$ $\{c|-6 \leq c \leq -2\}$
4. $n^2 - 7n + 10 \geq 0$ $\{n|n \leq 2$ or $n \geq 5\}$
5. $x^2 - 9 \geq 0$ $\{x|x \leq -3$ or $x \geq 3\}$
6. $y^2 - 25 \leq 0$ $\{y|-5 \leq y \leq 5\}$
7. $-y^2 + 4 < 0$ $\{y|y < -2$ or $y > 2\}$
8. $-c^2 + 36 \geq 0$ $\{c|-6 \leq c \leq 6\}$
9. $n^2 - 5n > 0$ $\{n|n < 0$ or $n > 5\}$
10. $a^2 + 8a < 0$ $\{a|-8 < a < 0\}$
11. $-4c^2 - 12c \geq 0$ $\{c|-3 \leq c \leq 0\}$
12. $-2x^2 + 10x \leq 0$ $\{x|x \leq 0$ or $x \geq 5\}$
13. $y^3 - y^2 - 6y < 0$ $\{y|y < -2$ or $0 < y < 3\}$
14. $x^3 + 6x^2 - 16x > 0$ $\{x|-8 < x < 0$ or $x > 2\}$
25. $\{y|-3 < y < -2$ or $2 < y < 3\}$
26. $\{x|x < -5$ or $-1 < x < 1$ or $x > 5\}$

Ⓑ
15. $x^3 - 25x \geq 0$ $\{x|-5 \leq x \leq 0$ or $x \geq 5\}$
16. $y^3 - 100y \leq 0$ $\{y|y \leq -10$ or $0 \leq y \leq 10\}$
17. $2x^2 - x - 10 < 0$ $\{x|-2 < x < 2.5\}$
18. $4a^2 + 5a \geq 9$ $\{a|a \leq -2.25$ or $a \geq 1\}$
19. $2y^2 + 7y \geq 0$ $\{y|y \leq -3.5$ or $y \geq 0\}$
20. $3a^2 < 10a$ $\{a|0 < a < 3\frac{1}{3}\}$
21. $-c^2 - c + 12 < 0$ $\{c|c < -4$ or $c > 3\}$
22. $-x^2 \geq 8 - 6x$ $\{x|2 \leq x \leq 4\}$
23. $0 < -n^2 + 4n - 3$ $\{n|1 < n < 3\}$
24. $0 \geq 35 - 2y - y^2$ $\{y|y \leq -7$ or $y \geq 5\}$
25. $y^4 - 13y^2 + 36 < 0$ **above**
26. $x^4 - 26x^2 + 25 > 0$ **above**
27. $a^4 - 16 \geq 0$ $\{a|a \leq -2$ or $a \geq 2\}$
28. $n^4 < 81$ $\{n|-3 < n < 3\}$
29. $n^4 - 14n^2 - 32 > 0$ $\{n|n < -4$ or $n > 4\}$
30. $-y^4 + 4y^2 + 45 < 0$ $\{y|y < -3$ or $y > 3\}$
31. $x^2 - 4x + 4 > 0$ $\{x|x \neq 2\}$
32. $x^2 - 6x + 9 \leq 0$ $\{3\}$
33. $3x^2 + 1 \geq 2x^2$ **{all numbers}**
34. $y^3 - 3y^2 \geq 25y - 75$ $\{y|-5 \leq y \leq 3$ or $y \geq 5\}$
35. $y^3 - 2y^2 < 16y - 32$ $\{y|y < -4$ or $2 < y < 4\}$
36. $y^2 - 5 > 2y^2 + 5$ **0**
37. $a^2 < 4$ or $a^2 > 25$ $\{a|a < -5$ or $-2 < a < 2$ or $a > 5\}$
38. $a^2 \geq 4$ and $a^2 \leq 25$ $\{a|-5 \leq a \leq -2$ or $2 \leq a \leq 5\}$
39. $a^2 < 16$ and $a^3 < 4a$ $\{a|-4 < a < -2$ or $0 < a < 2\}$

Ⓒ
40. $|x^2 - 17| < 8$ $\{x|-5 < x < -3$ or $3 < x < 5\}$
41. $|x^2 - 10| > 6$ $\{x|x < -4$ or $-2 < x < 2$ or $x > 4\}$
42. $|x^2| < 16$ $\{x|-4 < x < 4\}$
43. $|(x - 2)^2 - 5| < 4$ $\{x|-1 < x < 1$ or $3 < x < 5\}$
44. $|(x + 3)^2 - 20| > 16$ $\{x|x < -9$ or $-5 < x < -1$ or $x > 3\}$
45. $|x^2| > 9$ $\{x|x < -3$ or $x > 3\}$

Quadratic Inequalities

PROBLEM SOLVING: NUMBER PROBLEMS 3.7

Objective **To solve a word problem that leads to a quadratic equation**

Many word problems can be solved by using quadratic equations. Since a quadratic equation often has two solutions, a given word problem may have two sets of answers.

In Example 1, a problem about *consecutive integers* is solved. Recall that consecutive multiples of 5 can be represented by $x, x + 5, x + 10, \ldots$, where x is a multiple of 5.

Example 1 **Find three consecutive multiples of 5 so that the square of the third number, decreased by 5 times the second number, is the same as 25 more than twice the product of the first two numbers.**

Represent the data. ▶ Let x, $x + 5$, and $x + 10$ be the three consecutive multiples of 5, where x is a multiple of 5.

Write an equation. ▶

$$(\text{third})^2 - (5 \cdot \text{second}) = (2 \cdot \text{first} \cdot \text{second}) + 25$$
$$(x + 10)^2 - 5(x + 5) = 2x(x + 5) + 25$$
$$x^2 + 20x + 100 - 5x - 25 = 2x^2 + 10x + 25$$

Solve the equation. ▶
$$0 = x^2 - 5x - 50$$
$$0 = (x + 5)(x - 10)$$
$$x + 5 = 0 \quad \text{or} \quad x - 10 = 0$$
$$x = -5 \qquad\qquad x = 10$$

There are two sets of answers. The first multiple, x, may be either -5 or 10. If $x = -5$, then $x + 5 = 0$ and $x + 10 = 5$. If $x = 10$, then $x + 5 = 15$ and $x + 10 = 20$.

Check in the problem. ▶

$(-5, 0, 5)$

$(\text{third})^2 - (5 \cdot \text{second})$	$(2 \cdot \text{first} \cdot \text{second}) + 25$
$5^2 \quad - \quad 5 \cdot 0$	$2(-5 \cdot 0) \quad + 25$
$25 \quad - \quad 0$	$0 \quad + 25$
25	25

$(10, 15, 20)$

$(\text{third})^2 - (5 \cdot \text{second})$	$(2 \cdot \text{first} \cdot \text{second}) + 25$
$20^2 \quad - \quad 5 \cdot 15$	$2 \cdot 10 \cdot 15 \quad + 25$
$400 \quad - \quad 75$	$300 \quad + 25$
325	325

Answer the questions. ▶ **Thus,** the three consecutive multiples are $-5, 0, 5$, or $10, 15, 20$.

When using a quadratic equation to solve a word problem, you should determine if both solutions of the equation provide answers that are "reasonable" in the problem. In Example 1, both solutions check in the equation, and both provide logical answers to the word problem.

Example 2

Some light bulbs are placed in boxes, and the boxes are then packed in cartons. The number of bulbs in each box is 4 less than the number of boxes in each carton. Find the number of bulbs in each box if a full carton contains 60 light bulbs.

Let x = number of boxes in each carton.
$x - 4$ = number of bulbs in each box

$$\begin{pmatrix} \text{Number of boxes} \\ \text{in each carton} \end{pmatrix} \cdot \begin{pmatrix} \text{Number of bulbs} \\ \text{in each box} \end{pmatrix} = \begin{pmatrix} \text{Total number of bulbs} \\ \text{in each carton} \end{pmatrix}$$

$$x(x - 4) = 60$$
$$x^2 - 4x = 60$$
$$x^2 - 4x - 60 = 0$$
$$(x + 6)(x - 10) = 0$$
$$x = -6 \quad \text{or} \quad x = 10$$

The number of boxes in each carton cannot be -6. If $x = 10$, then there are 10 boxes in each carton and $10 - 4$, or 6 bulbs in each box.

Check: (No. of boxes in each carton)(No. of bulbs in each box) | 60
$10 \cdot 6$ | 60
60

Thus, there are 6 bulbs in each box.

To solve a word problem involving two numbers, read carefully to see if the second number is described *in terms of* the first number. If so, you should use a variable to represent the first number. This is shown in Example 3.

Example 3

A second number is 5 less than twice a first number. If the second number is multiplied by 3 more than the first number, the result is 21. Find all such pairs of numbers.

Let x = first number.
$2x - 5$ = second number

Second number times 3 more than first number is 21.
$$(2x - 5)(x + 3) = 21$$
$$2x^2 + x - 15 = 21$$
$$2x^2 + x - 36 = 0$$
$$(2x + 9)(x - 4) = 0$$
$$2x + 9 = 0 \quad \text{or} \quad x - 4 = 0$$
$$2x = -9 \quad\quad\quad x = 4$$
$$x = -4.5$$

There are two sets of answers. The first number may be either -4.5, or 4. If $x = -4.5$, then $2x - 5 = -14$. If $x = 4$, then $2x - 5 = 3$. Check both pairs of numbers in the problem.

Thus, the pairs of first and second numbers are $(-4.5, -14)$ or $(4, 3)$.

Oral Exercises

Beginning with 12, give four integers for each description.
1. consecutive integers **12, 13, 14, 15**
2. consecutive even integers **12, 14, 16, 18**
3. consecutive multiples of 12 **12, 24, 36, 48**
4. consecutive multiples of 3 **12, 15, 18, 21**

Beginning with −9, give four integers for each description.
5. consecutive integers **−9, −8, −7, −6**
6. consecutive odd integers **−9, −7, −5, −3**
7. consecutive multiples of 3 **−9, −6, −3, 0**
8. consecutive multiples of 9 **−9, 0, 9, 18**

Written Exercises

Solve each problem. 1. **−6, −4, −2 or 4, 6, 8**

(A)
1. Find three consecutive even integers so that the first integer times the second integer is 24. [Hint: Use x, $x + 2$, and $x + 4$ to represent the integers.]

2. Find three consecutive integers so that the product of the second and the third integers is 42. **−8, −7, −6 or 5, 6, 7**

3. Find three consecutive integers so that the square of the first, increased by the square of the third, is 100. **−8, −7, −6 or 6, 7, 8**

4. Find three consecutive odd integers so that the sum of the squares of the first two integers is 130. **−9, −7, −5 or 7, 9, 11**

5. Forty chairs are placed in rows so that the number of chairs in each row is 3 less than the number of rows. Find the number of chairs in each row. **5 chairs**

6. Forty-four students are seated in rows in a lecture hall. The number of students in each row is 7 more than the number of rows. Find the number of students seated in each row. **11 students**

7. One-hundred forty peaches were packed in some boxes so that the number of boxes was 6 less than twice the number of peaches in each box. Find the number of boxes used. **14 boxes**

8. An album contains 1,020 stamps. How many pages are in the album if the number of pages is 8 more than 4 times the number of stamps on each page? **68 pages**

9. One number is 4 less than another number. Their product is 21. Find all such pairs of numbers. **−3, −7, or 7, 3**

10. The product of two numbers is 33. One number is 2 more than 3 times the other number. Find both numbers. **$-3\frac{2}{3}$, −9 or 3, 11**

11. The product of two numbers is 9, and one number is 4 times the other number. Find the two numbers. **$-\frac{3}{2}$, −6 or $\frac{3}{2}$, 6**

12. Find two numbers whose product is 25 if one number is 9 times the other number. **$-1\frac{2}{3}$, −15 or $1\frac{2}{3}$, 15**

(B)
13. Find three consecutive multiples of 5 so that the product of the first and the third numbers is 200. **−20, −15, −10 or 10, 15, 20**

14. Find three consecutive multiples of 10 so that 15 times the third number is the same as the product of the first two numbers. **20, 30, 40**

15. For which three consecutive multiples of 4 is 400 equal to the square of the first number, increased by the square of the second number? **−16, −12, −8 or 12, 16, 20**

16. For which three consecutive multiples of 6 will the sum of their squares be 504? **−18, −12, −6 or 6, 12, 18**

Chapter Three

17. An 8-story hotel has the same number of rooms on each of the 8 floors. The number of square meters of carpet in each room is 10 more than twice the number of rooms on each floor. Find the area of the carpet in one room if there are 8,000 m² of carpet in the hotel. **50 m²**

18. Twelve kilograms of flour are placed in some bags and the bags are packed in boxes. The number of kilograms of flour in each bag is 3 less than the number of bags in each box, and the number of boxes is 1 less than the number of bags in each box. Find the number of boxes used. **3 boxes**

19. A second number is 5 more than a first number. If the second number is multiplied by 2 less than the first number, the product is 8. Find the numbers. **−6, −1 or 3, 8**

20. A first number is 3 less than twice a second number. If the first number is multiplied by 3 less than the second number, the result is 5. Find the numbers. **$\frac{1}{2}$, −2**
   4, 5

21. Find four consecutive multiples of 5 such that the product of the second and the third numbers is 50 more than 10 times the fourth number.
   −15, −10, −5, 0 or 10, 15, 20, 25

22. For which five consecutive multiples of 3 is the product of the second and the fourth numbers equal to 90 less than the sum of the squares of the first and the fifth numbers?

(C) 23. Find four consecutive multiples of 0.5 in which the product of the first and the third numbers is 0.25 less than 3 times the second number. **−0.5, 0, 0.5, 1 or 2.5, 3, 3.5, 4**

24. Find four consecutive multiples of π in which the product of the second and the third multiples is $2\pi^2$ more than 3π times the fourth multiple.

25. For which three consecutive integers is the sum of the first and the second integers the same as 3 less than twice the third integer?
   every 3 consecutive integers

22. **−9, −6, −3, 0, 3 or −3, 0, 3, 6, 9**

26. Find three consecutive odd integers so that the square of the second integer is equal to the product of the first and the third integers. **no such integers**

24. **−3π, −2π, −π, 0 or 3π, 4π, 5π, 6π**

NON-ROUTINE PROBLEMS

1. Prove that the following statement is true.

   If a, b, and c are *odd integers* and if x is an *integer*, then $ax^2 + bx + c$ *cannot* be an *even* integer ($ax^2 + bx + c$ must be an *odd* integer).

   Hint: x is either an *odd* integer *or* an *even* integer.
   For the proof, see TE Answer Section.

2. If $4^{2x} = 25$, then $4^{-x} = \underline{\ ?\ }$. **$\frac{1}{5}$**

CUMULATIVE REVIEW

1. The length of a rectangle is 5 units more than its width. Represent the perimeter and the area of the rectangle in algebraic terms.
   $P = 2w + 2(w + 5)$
   $A = w(w + 5)$

2. The base of a triangle is 8 units shorter than 4 times its height. Represent the area of the triangle in algebraic terms. **$A = \frac{1}{2}h(4h - 8)$**

Problem Solving: Number Problems

PROBLEM SOLVING: GEOMETRIC PROBLEMS   3.8

Objectives **To find the length of each side of a right triangle by using the Pythagorean relation**
To solve an area problem that leads to a quadratic equation

For each right triangle ABC with right angle C as shown below, the lengths a, b, and c of the three sides are related by the **Pythagorean relation**: $a^2 + b^2 = c^2$.

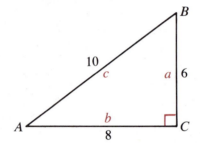

$$a^2 + b^2 = c^2$$
$$6^2 + 8^2 = 10^2$$
$$36 + 64 = 100$$
$$100 = 100$$

Example 1 **One leg of a right triangle is 7 m longer than the other leg. The length of the hypotenuse is 13 m. Find the length of each leg.**

Represent the data. ▶

Let $\quad x$ = length of one leg.
$\quad x + 7$ = length of other leg

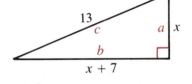

Write an equation. ▶

$$a^2 + b^2 = c^2$$
$$x^2 + (x + 7)^2 = 13^2$$
$$x^2 + (x^2 + 14x + 49) = 169$$
$$2x^2 + 14x - 120 = 0$$

Solve the equation. ▶

$x^2 + 7x - 60 = 0$ ◀ *Divide each side by 2, the GCF.*
$(x + 12)(x - 5) = 0$
$x = -12 \quad$ or $\quad x = 5$

Check in the problem. ▶

The length of a side of a triangle cannot be -12 m.
If $x = 5$, then $x + 7 = 12$. Check to determine if $a^2 + b^2 = c^2$.

$a = 5, b = 12, c = 13$

$a^2 + b^2$	c^2
$5^2 + 12^2$	13^2
$25 + 144$	169
169	

Answer the question. ▶

Thus, the length of one leg is 5 m and the length of the other leg is 12 m.

The surface enclosed by a rectangle is measured in *square units*. This measure is the **area** of the rectangle. The word problem in Example 2 on page 79 involves the area of a rectangle.

Example 2 The length of a rectangle is 3 cm more than twice its width. Find the width and the length if the area is 44 cm².

Let w = width.
$\quad 2w + 3$ = length

width × length = area of rectangle
$$w(2w + 3) = 44$$
$$2w^2 + 3w = 44$$
$$2w^2 + 3w - 44 = 0$$
$$(2w + 11)(w - 4) = 0$$
$$2w + 11 = 0 \quad \text{or} \quad w - 4 = 0$$
$$2w = -11 \qquad\qquad w = 4$$
$$w = -5.5$$

The width of a rectangle cannot be -5.5 cm.
If $w = 4$, then $2w + 3 = 11$.

width × length	area
4 cm × 11 cm	44 cm²
44 cm²	

Thus, the width is 4 cm and the length is 11 cm.

The word problem in Example 3 involves the area of a triangle. The area of a triangle equals $\frac{1}{2}$ times the length of the base times the height of the triangle.

Example 3 The height of a triangle is 6 m less than three times the length of its base. Find the length of the base and the height of the triangle if its area is 36 m².

Let b = length of base.
$\quad 3b - 6$ = height

$\frac{1}{2}$ × base × height = area of triangle

$$\frac{1}{2}b(3b - 6) = 36$$

$2 \cdot \frac{1}{2} \cdot b(3b - 6) = 36 \cdot 2$ ◀ *Multiply each side by 2, the reciprocal of $\frac{1}{2}$.*

$$b(3b - 6) = 72$$
$$3b^2 - 6b - 72 = 0$$
$$b^2 - 2b - 24 = 0 \quad \blacktriangleleft \textit{ Divide each side by 3, the GCF.}$$
$$(b - 6)(b + 4) = 0$$
$b = 6 \quad$ or $\quad b = -4$ ◀ *The length of the base cannot be -4 m.*

If $b = 6$, then $3b - 6 = 12$. Check to see that 6 m and 12 m work in the problem.

Thus, the length of the base is 6 m and the height is 12 m.

Problem Solving: Geometric Problems

Some geometric problems involve addition or subtraction of measures. In the next example, one area is subtracted from another area.

Example 4 The length of a rectangular floor is 4 m shorter than 3 times its width. The width of a rectangular carpet on the floor is 2 m shorter than the floor's width. The carpet is 2 m longer than twice its own width. Find the area of the floor if 44 m² of the floor are not covered by the carpet.

Let x = width of floor.
$3x - 4$ = length of floor

Sketch the two rectangles.

(Area of floor) − (Area of carpet) = (Area not covered)
$$x(3x - 4) - (x - 2)(2x - 2) = 44$$
$$3x^2 - 4x - (2x^2 - 6x + 4) = 44$$
$$x^2 + 2x - 48 = 0$$
$$(x - 6)(x + 8) = 0$$
$$x = 6 \quad \text{or} \quad x = -8 \quad \blacktriangleleft \text{ The width cannot be } -8 \text{ m.}$$

If $x = 6$, then $3x - 4 = 14$, $x - 2 = 4$, and $2(x - 2) + 2 = 10$.

(Area of floor) − (Area of carpet)	(Area not covered)
6 m × 14 m − 4 m × 10 m	44 m²
84 m² − 40 m²	
44 m²	

Thus, the area of the floor is 6 m × 14 m, or 84 m².

Written Exercises

Solve each problem.

Ⓐ 1. One leg of a right triangle is 2 m longer than the other leg. The length of the hypotenuse is 10 m. Find the length of each leg. **6 m, 8 m**

2. One leg of a right triangle is 4 m long. The hypotenuse is 1 m shorter than twice the length of the other leg. Find the length of the other leg and the length of the hypotenuse. **3 m, 5 m**

3. One leg of a right triangle is 7 cm longer than the other leg. The hypotenuse is 1 cm longer than the longer leg. Find the lengths of the three sides. **5 cm, 12 cm, 13 cm**

4. In a right triangle, the hypotenuse is 8 dm longer than one leg and 4 dm longer than the other leg. Find the lengths of the three sides. **12 dm, 16 dm, 20 dm**

5. The length of a rectangle is 3 cm less than twice its width. Find the length and the width if the area is 20 cm². **4 cm, 5 cm**

6. The length of a rectangle is 5 m more than three times its width. Find the length and the width if the area is 42 m². **3 m, 14 m**

7. The height of a triangle is six times the length of its base. Find the length of the base and the height of the triangle if its area is 12 m². **2 m, 12 m**

8. The base of a triangle is 2 m shorter than its height and the area is 24 m². Find the height and the length of the base. **8 m, 6 m**

9. The base of a triangle is 4 dm longer than twice the height. If the area is 63 dm², find the length of the base. **18 dm**

10. The area of a triangle is 54 cm². Find the height of the triangle if the height is 6 cm less than 4 times the length of the base. **18 cm**

11. The width of one square is twice the width of another square. The sum of their areas is 125 m². Find the area of each square. **25 m², 100 m²**

12. One square is 3 times as wide as another square. Find the area of each square if the difference of the two areas is 32 cm². **4 cm², 36 cm²**

13. A square and a rectangle have the same width. The length of the rectangle is 4 times its width. If the sum of the two areas is 45 cm², find the area of each figure. **sq: 9 cm², rect: 36 cm²**

14. The length of a rectangle is 5 m more than twice its width. The width of a square is the same as the length of the rectangle. Find the area of each figure if one area, decreased by the other area, is 88 m². **sq: 121 m², rect: 33 m²**

B 15. A rectangle is 1 m narrower than it is long, and one of its diagonals is 2 m longer than the width of the rectangle. What is the length of the diagonal? **5 m**

16. A diagonal of a rectangle is 5 cm longer than 4 times the width of the rectangle. Find the length of the rectangle if it is 1 cm less than the length of the diagonal. **40 cm**

17. The length of a rectangle is 4 cm more than its width and 4 cm less than the length of a diagonal of the rectangle. What is the area of the rectangle? **192 cm²**

18. The length of a rectangle is 2 m more than twice its width and 1 m less than the length of a diagonal of the rectangle. Find the area of the rectangle. **60 m²**

19. A rectangle is twice as long as it is wide. The height of a triangle is twice the length of its base and the same as the length of the rectangle. Find the area of each figure if the sum of their areas is 108 m². **rect: 72 m², triangle: 36 m²**

20. The base of a triangle is 2 cm shorter than the width of a rectangle. The rectangle's length is 4 times its width and the triangle's height is the same as the rectangle's length. Find the area of each figure if the difference of the two areas is 30 cm². **triangle: 6 cm², rect. 36 cm²**

21. The length of a rectangular wall is 3 m more than twice its height. The width of a picture on the wall is 1 m less than the height of the wall. The picture is 2 m longer than it is wide. Find the area of the wall if 19 m² of the wall are not covered by the picture. **27 m²**

22. The length of a patio floor is 1 m less than twice its width. The floor is extended by building a second patio floor that is 3 m longer and 2 m narrower than the first floor. Find the area of the original floor if the total area of the patio is 48 m² after the extension is built. **28 m²**

Problem Solving: Geometric Problems

Ⓒ 23. A rectangular lawn measures 26 m by 36 m and is surrounded by a uniform sidewalk. The outer edge of the sidewalk is a rectangle with an area of 1,200 m². Find the width of the sidewalk. **2 m**

24. A rectangular picture measures 32 cm by 50 cm. It is surrounded by a uniform frame whose outer edge is a rectangle with an area of 2,320 cm². Find the width of the frame. **4 cm**

25. A rectangular piece of cardboard is 5 cm longer than it is wide. A 3-centimeter square (3 cm by 3 cm) is cut out of each corner. Then the four flaps are turned up to form an open box with a volume of 450 cm³. Find the length and the width of the original piece of cardboard. **16 cm, 21 cm**

26. A 4-decimeter square is cut from each corner of a rectangular piece of plastic whose length is 2 dm less than twice its width. Then the four flaps are folded to form an open carton with a volume of 1,040 dm³. Find the length and the width of the original piece of plastic. **18 dm, 34 dm**

CALCULATOR ACTIVITIES

Find the area of each rectangle described below, in terms of the given unit of measure.

1. 296 in. by 225 in., in square feet
   462.5 ft²
2. 17.5 ft by 38.75 ft, in square inches
   97,650 in.²
3. 1.24 m by 0.35 m, in square centimeters
   4,340 cm²
4. 4,225 cm by 2,416 cm, in square meters
   1,020.76 m²
5. 222 ft by 363 ft, in square yards
   8,954 yd²
6. 497 yd by 268 yd, in square feet
   1,198,764 ft²

The lengths of the three sides of a triangle are given. Determine whether each triangle is a right triangle by using the Pythagorean relation.

7. 98 cm, 336 cm, 350 cm **yes**
8. 0.75 yd, 0.4 yd, 0.85 yd **yes**
9. 110 in., 226 in., 286 in. **no**
10. 0.65 m, 4.2 m, 4.25 m **yes**

NON-ROUTINE PROBLEMS

Each edge of a cube is 3 in. long. The cube is painted on each of its six faces, and then it is cut up to form 27 one-inch cubes.

1. What is the least number of straight cuts needed to form the 27 one-inch cubes?
   6

Find the number of one-inch cubes that are painted on the given number of faces.
2. three faces **8**
3. exactly two faces **12**
4. exactly one face **6**
5. zero faces **1**

Chapter Three

APPLICATIONS

A certain investment plan requires regular payments of $30 a month with an expected earnings rate of 2% per month.

The expected value in the plan at the end of the 1st month is the payment plus the interest the payment earned for the month.

$$30 + 30(0.02) = 30(1 + 0.02)$$

or 30(1.02)

The expected value in the plan at the end of the 2nd month is the payment and the 1st month's value, plus the 2nd month's interest on both.

$$[30 + 30(1.02)] + [30 + 30(1.02)](0.02)$$
$$= [30 + 30(1.02)](1 + 0.02)$$
or $30(1.02) + 30(1.02)^2$

Now find the expected value in the plan at the end of the 3rd month.

Payment + Value already in the plan + 3rd month's earnings on both

$$[30 + 30(1.02) + 30(1.02)^2 + [30 + 30(1.02) + 30(1.02)^2](0.02)$$
$$= [30 + 30(1.02) + 30(1.02)^2](1 + 0.02)$$
or, $30(1.02) + 30(1.02)^2 + 30(1.02)^3$

Generalize the formula for $r\%$ per month. Let $x = 1 + 0.01r$.
$$30x + 30x^2 + 30x^3$$

If p dollars, rather than 30 dollars, is added monthly to the investment plan above, show the expected value at the end of n months.

$$px + px^2 + px^3 + \cdots + px^{n-1} + px^n$$

Solve each problem.
Tina's pension plan requires regular payments of $40 a month. The plan has an expected earnings rate of 3% per month.

1. What is the expected value after 6 months if she doubles the payments beginning with the fourth month? **$405.66**

2. What would the earnings be on Tina's plan after 6 months if she makes double payments from the first month but only receives half the monthly earnings rate? **$25.84**

CHAPTER THREE REVIEW

Ⓐ Exercises are 1, 5–7, 13–15, 20–23, 27, 30, 32, 33, 35, 36, 40, 41, 44, 45.
Ⓒ Exercises are starred. The rest are Ⓑ Exercises.

Vocabulary
area [3.8]
common binomial factor [3.4]
common monomial factor [3.4]
difference of two cubes [3.3]
difference of two squares [3.3]
perfect square trinomial [3.3]
Pythagorean relation [3.8]
quadratic trinomial [3.1]
standard form
 quadratic equation [3.5]
 quadratic inequality [3.6]
sum of two cubes [3.3]

For complete solutions, see TE Answer Section.

Factor each polynomial.
1. $5c^2 - 37c + 14$ **$(5c - 2)(c - 7)$** [3.1]
2. $9y^4 + 18y^2 + 8$ **$(3y^2 + 2)(3y^2 + 4)$**
3. $25a^4b^4 + 5a^2b^2 - 12$
★ 4. $2x^{2n+2} - 11x^{n+1} + 5$
5. $9a^2 - 16$
6. $n^3 + 64$ [3.3]
7. $a^4 - 16a^2 + 64$
8. $100x^4 - 49y^2$
9. $8c^3 - 125d^3$ **$(2c - 5d)(4c^2 + 10cd + 25d^2)$**
10. $25x^2 + 40xy + 16y^2$ **$(5x + 4y)^2$**
11. $25x^2 - y^2 + 6y - 9$
★ 12. $x^{6m} + 10x^{3m}y^n + 25y^{2n}$

7. $(a^2 - 8)^2$ 8. $(10x^2 + 7y)(10x^2 - 7y)$

Multiply each pair of polynomials. [3.2]
13. $(4n + 6)(4n - 6)$ 14. $(3a - 5)^2$
15. $(x + 3)(x^2 - 3x + 9)$ **$x^3 + 27$**
16. $(5x^2 + 4y)^2$ **$25x^4 + 40x^2y + 16y^2$**
17. $(8y^2 - 6z)(8y^2 + 6z)$ **$64y^4 - 36z^2$**
18. $(2m - n)(4m^2 + 2mn + n^2)$ **$8m^3 - n^3$**

Factor each polynomial completely. [3.4]
19. $18x^3 - 15x^2 + 3x$
20. $-2a^2 + 3a + 14$
21. $12y^2 - 27$
22. $-y^2 + 10y - 25$
23. $2n^4 - 15n^2 - 27$
24. $4c^4 - 25c^2 + 36$
25. $2a^3b - 4a^2b - 30ab$
26. $3x^3 - 24y^3$
27. $x^3 + 4x^2 + 5x + 20$
28. $15ac - 6bc - 20ad + 8bd$
★ 29. $2x^{7n} - 18x^{3n}$

Solve each equation. [3.5]
30. $x^2 = x + 12$ **4, −3**
31. $-6y^2 = 14y$ **0, $-2\frac{1}{3}$**
32. $25a^2 = 4$ **$-\frac{2}{5}, \frac{2}{5}$**
33. $n^4 - 29n^2 + 100 = 0$ **−5, 5, −2, 2**
★ 34. $|x^2 - 5| = 4$ **−3, 3, −1, 1**

Find the solution set of each inequality.
Graph the solutions. [3.6]
35. $y^2 + 2y - 8 < 0$ **{y | −4 < y < 2}**
36. $x^3 - 2x^2 - 24x > 0$ **below**
37. $2n^2 + n \geq 15$ **{n | n ≤ −3 or n ≥ 2.5}**
38. $y^4 - 23y^2 - 50 > 0$ **below**
★ 39. $|x^2 - 20| > 16$
 {x | x < −6 or −2 < x < 2 or x > 6}

Solve each problem.
40. Find three consecutive odd integers so that the product of the second integer and the third integer is 63. [3.7]
 −11, −9, −7 or 5, 7, 9
41. Fifty-four cans are packed in some boxes so that the number of boxes is 3 more than the number of cans in each box. Find the number of boxes used. [3.7]
 9 boxes
42. Find three consecutive multiples of 6 so that 36 less than the product of the second and third numbers is equal to the sum of the squares of the first and the second numbers. [3.7] **0, 6, 12 or 6, 12, 18**
43. A second number is 1 less than 3 times a first number. If the second number is multiplied by 2 more than the first number, the product is 10. Find both numbers. [3.7]
 $\frac{4}{3}$, 3 or −3, −10
44. One leg of a right triangle is 3 m longer than the other leg. The hypotenuse is 6 m longer than the shorter leg. Find the lengths of the three sides. [3.8]
 9 m, 12 m, 15 m
45. The height of a triangle is 8 cm less than twice the length of its base. Find the length of the base and the height of the triangle if its area is 12 cm². [3.8] **6 cm, 4 cm**
46. The length of one rectangle is 6 m more than 2 times its width. The length and width of a second rectangle are twice the length and width, respectively, of the first rectangle. Find the area of each rectangle if the difference of the two areas is 324 m². [3.8] **108 m², 432 m²**

36. **{x | −4 < x < 0 or x > 6}**   38. **{y | y < −5 or y > 5}**

(A) Exercises are 1, 3, 6, 8, 9, 13, 15–17, 20. (C) Exercises are starred. The rest are (B) Exercises.

CHAPTER THREE TEST

Factor each polynomial.
1. $3x^2 - 13x - 10$ **$(3x + 2)(x - 5)$**
2. $9a^2 + 30ab + 25b^2$ **$(3a + 5b)^2$**
3. $4c^2 - 81$ **$(2c + 9)(2c - 9)$**
4. $27n^3 - 125$ **$(3n - 5)(9n^2 + 15n + 25)$**
5. $9x^2 + 6xy + y^2 - 16$
   $(3x + y + 4)(3x + y - 4)$

Multiply each pair of polynomials.
6. $(7y + 8)(7y - 8)$ **$49y^2 - 64$**
7. $(4a + 5b)^2$ **$16a^2 + 40ab + 25b^2$**
8. $(2y - 3)(4y^2 + 6y + 9)$ **$8y^3 - 27$**

Factor each polynomial completely.
9. $98 - 32x^2$ **$2(7 - 4x)(7 + 4x)$**
10. $2xy^2 - 2xy - 12x$ **$2x(y - 3)(y + 2)$**
11. $-24x^2 - 6x + 45$ **$-3(4x - 5)(2x + 3)$**
12. $4a^4 - a^2 - 18$ **$(2a + 3)(2a - 3)(a^2 + 2)$**

Solve each equation.
13. $a^2 = 11a$ **0, 11**
14. $2x^2 = 12 - 5x$ **$1\frac{1}{2}$, -4**
15. $0 = a^4 - 37a^2 + 36$ **$-6, 6, -1, 1$**

Find the solution set of each inequality. Graph the solutions.
16. $x^2 + 3x - 18 \leq 0$ **$\{x \mid -6 \leq x \leq 3\}$**
17. $x^3 - 64x \geq 0$ **$\{x \mid -8 \leq x \leq 0 \text{ or } x \geq 8\}$**
18. $a^4 - 29a^2 + 100 < 0$
    $\{a \mid -5 < a < -2 \text{ or } 2 < a < 5\}$

Solve each problem.
19. One number is 6 less than another number. The product of the smaller number and 3 more than the larger number is 10. Find both numbers. **7, 1 or $-4, -10$**

20. The length of the base of a triangle is 10 cm less than 4 times the height. Find the length of the base and the height of the triangle if its area is 25 cm². **5 cm, 10 cm**

21. Find three consecutive even integers such that the square of the second integer, increased by twice the product of the first two integers, is the same as 24 more than twice the square of the first integer.
    $-10, -8, -6$ or 2, 4, 6

22. The length of a rectangle is 2 m more than its width and 2 m shorter than the length of a diagonal. Find the length of the diagonal and the area of the rectangle.
    10 m, 48 m²

23. A rectangular garden is 10 m longer than 4 times its width. The garden is to be enlarged by adding a second rectangular garden, 3 times as wide but half as long as the first garden. Find the area of the original garden if the total area of the enlarged garden will be 665 m². **266 m²**

★ 24. Factor $2x^{2n+2} - x^{n+1}y^{2n} - y^{4n}$.
    $(x^{n+1} - y^{2n})(2x^{n+1} + y^{2n})$

★ 25. Solve $25x^2 = 40ax - 16a^2$ for x in terms of a. **$\frac{4a}{5}$**

★ 26. Solve $|x^2 - 20| = 16$. **$-6, 6, -2, 2$**

Find the solution set. Graph the solutions.
★ 27. $x^2 < 25$ and $x^2 > 9$
    $\{x \mid -5 < x < -3 \text{ or } 3 < x < 5\}$
★ 28. $|x^2 - 5| < 4$
    $\{x \mid -3 < x < -1 \text{ or } 1 < x < 3\}$

COMPUTER ACTIVITIES

Using String Variables

OBJECTIVE: To check factors of quadratic polynomials by using character strings

The letters A, B, C, etc., are *variables* used to represent numbers. Letters can be used to represent words or algebraic expressions by attaching a dollar sign to them: A$, B$, C$, etc. They are then called *string variables*. Consider the assignment statements: LET A$ = "DOG" and LET B$ = "2X + 3". "DOG" and "2X + 3" are called *character strings*. Character strings may be linked together by means of the + sign. PRINT A$ + B$ for A$ and B$ above would make the computer print DOG2X + 3.

Enter and run this program.

```
110  A =  INT ( RND (1) * 12) + 1
120  B =  INT ( RND (1) * 12) + 1
130  A$ = "(X+" +  STR$ (A) + ")(X+" +  STR$ (B) + ")"
140  B$ = "(X+" +  STR$ (B) + ")(X+" +  STR$ (A) + ")"
150    PRINT "FACTOR: ";
160    PRINT "X^2 + "A + B"X +"A * B
170    INPUT C$
180    IF C$ = A$ THEN 220
190    IF C$ = B$ THEN 220
200    PRINT "THOSE FACTORS ARE NOT CORRECT."
210    GOTO 150
220    PRINT "CORRECT"
230    GOTO 110
```

See the Computer Section beginning on page 574 for more information.

Lines 110 and 120 ask the computer to randomly pick a number from 1 to 12. The STR$ function in lines 130 and 140 converts the numbers represented by A and B into character strings, which are then incorporated into the larger strings assigned to A$ and B$.

Notice the way this program determines a polynomial's factored solutions in lines 110–140 before forming the polynomial itself in line 160.

EXERCISES

1. Make the changes necessary in the program above so that the factored solutions take the form (X − A) (X − B).

2. Write a program involving factoring out a common constant factor so that the results take the form A(X + B).

   For Ex. 1–2, see TE Answer Section.

LAND EXPLORATION

Discovery. From ancient sailors mapping the globe to satellite photography enhanced by advanced computers, land exploration has been a challenge throughout history.

Janus display. This computer graphics display uses color to highlight topographical features for sophisticated mapping.

Assyrian astrolabe. Early mariners used this delicate disk to "sight" stars and observe their apparent movement across the sky.

Cross-staff being used by two people. Simpler and more sturdy than the astrolabe, a cross-staff could be aimed at the sun and stars to determine latitude.

Silhouette of First Officer with sextant. Combining the features and functions of such navigational tools as the astrolabe, cross-staff, and quadrant, sextants are used in marine navigation today.

Navigator pictured at work. Once a ship's latitude and longitude have been determined, the navigator is responsible for charting the course.

Micronesian sailing chart of the Marshall Islands. Ribs of coconut palms tied together formed a map indicating winds and sea currents; shells indicated groups of islands.

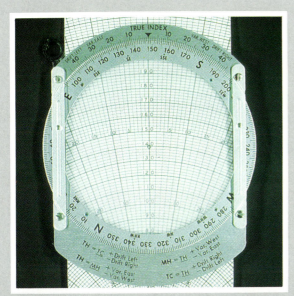

Charting-navigation tools. To chart a course accurately, navigational tools must account for the curvature of the earth on flat maps.

Keere's map of the world. Early maps were based on information derived from the ships' logs of the time and were very inaccurate.

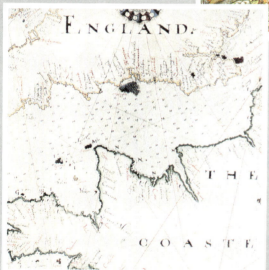

Detail of Thomas Hood chart. The carefully kept charts and logs of ancient navigators enabled cartographers to update their maps with details of coastal outlines and more accurate scales of distance.

"The Cognoscenti"—painting of English explorers. The great explorers, anxious to find less treacherous trade routes to Asia, were responsible for discovering untold riches in the New World.

Fr. Jacques Marquette and Louis Jolliet making a portage during their descent of the Mississippi River in 1673. Transporting boats over land was often necessary to chart the interior of the New World.

Computer-enhanced satellite photo of the Luxor region of the Nile River. High-altitude photography makes extremely accurate mapping possible when details are statistically accentuated by computer calculation.

Matt Henson, a member of Peary's expedition to the North Pole. In 1909, one of the earth's last frontiers was reached when Commander Robert Peary and his party reached the Pole.

Geological cross-section. Stratigraphy is the subdivision of rock strata (layers) into mappable units to determine time relationships with similar sequences at other places.

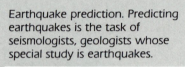

Earthquake prediction. Predicting earthquakes is the task of seismologists, geologists whose special study is earthquakes.

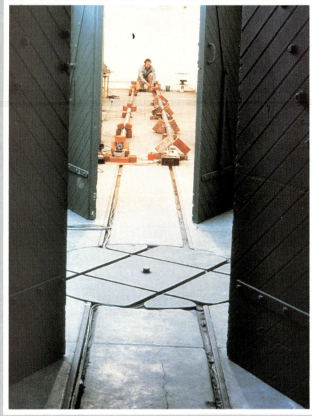

Computer graph. Data can be entered into a computer to reconstruct a topographical image in three dimensions.

Computer-aided information. Compiling data relating to areas is simplified when digitized maps are used, as in this environmental data chart.

Computer graphics. Some computers can be programmed to rotate a reconstructed image so that it can be viewed from different angles.

Lorrie Brow, geologist. A geologist studies not only the materials that make up our planet, but how their synthesis affects the earth's topology.

A cartographer works with the concept of distance. A chief value of maps drawn accurately to scale is that simple measurements can be converted to actual distances.

Stereoscopic plotter. A stereoscopic plotter generates a three-dimensional image on a flat surface.

COLLEGE PREP TEST

DIRECTIONS: Choose the *one* best answer to each question or problem.

1. Which value of x is not a solution of the inequality $(x + 40)(x - 20)(x + 60) \geq 0$? **B**

    (A) -53 (B) -7 (C) 20
    (D) 36   (E) None of these

2. For which equation is the product of its solutions not a negative number? **D**

    (A) $x^2 - 10x - 24 = 0$ (B) $x^2 - 25 = 0$
    (C) $x^2 + 7x - 18 = 0$ (D) $x^2 + 6x = 0$
    (E) None of these

3. Which figure has an area different from the other four figures? **C**
   [1 meter (m) = 100 centimeters (cm)]

    (A) A rectangle, 1.2 m by 0.3 m
    (B) A rectangle, 2 m by 0.18 m
    (C) A rectangle, 9 cm by 4 cm
    (D) A triangle, height of 120 cm and base 60 cm long
    (E) A parallelogram, height of 60 cm and base 60 cm long

4. If x is an integer, for what value(s) of x will $x^2 + 12x + 36$ be a positive perfect square? **D**

    (A) -6 (B) 0   (C) 6
    (D) Every integer except -6
    (E) Every integer except 0

5. If $y = \frac{5}{4}x + 20$, then $y = x$ only if **D**

    (A) $x = 0$ (B) $y = 20$ (C) $x = 100$
    (D) $y = -80$ (E) None of these

6. Howard is n years old now. Ten years ago, he was t years old. Find his age y years from now. **A**

    (A) $t + y + 10$ (B) $t + 10$
    (C) $t + y$ (D) $y + 10$
    (E) $n + t + y$

7. Represent the total value of x nickels, y dimes, and z quarters. **C**

    (A) $x + y + z$ cents   (B) 40 cents
    (C) $0.05x + 0.10y + 0.25z$ dollars
    (D) $5x + 10y + 25z$ dollars
    (E) None of these

8. What is the GCF of $2(6x^2y - 9xy^3) + (15a^3x + 10ay^2)$? **E**

    (A) 2   (B) $6xy$ (C) $10a$
    (D) $30axy$ (E) None of these

9. For which equation is the product of its solutions the greatest number? **E**

    (A) $x^2 - 36 = 0$
    (B) $x^2 + 8x = 0$
    (C) $x^2 + 5x + 6 = 0$
    (D) $x^2 - 2x - 15 = 0$
    (E) $x^2 - 7x + 12 = 0$

10. A car traveled x miles in a hours, then y miles in b hours, and finally z miles in c hours. Find the average speed in miles per hour for the trip of $x + y + z$ miles. **B**

    (A) $\frac{x}{a} + \frac{y}{b} + \frac{z}{c}$ (B) $\frac{x + y + z}{a + b + c}$
    (C) $\left(\frac{x}{a} + \frac{y}{b} + \frac{z}{c}\right) \div 3$
    (D) $(x + y + z)(a + b + c)$
    (E) None of these

College Prep Test

4 RATIONAL EXPRESSIONS

Non-Routine Problem Solving

Problem

Is 2 equal to 1?
Study each step in the following "proof." Can you find the error?

1. Let $a = b$.
2. $a \cdot a = a \cdot b$ — Multiply each side by a.
3. $a^2 = ab$ — Simplify.
4. $a^2 - b^2 = ab - b^2$ — Subtract b^2 from each side.
5. $(a + b)(a - b) = b(a - b)$ — Factor each side.
6. $\dfrac{(a + b)(a - b)}{(a - b)} = \dfrac{b(a - b)}{(a - b)}$ — Divide each side by $(a - b)$.
7. $a + b = b$ — Divide out the common factor, $a - b$.
8. $b + b = b$ — From Step 1, $a = b$. Substitute b for a.
9. $2b = 1b$ — Simplify.
10. $\dfrac{2b}{b} = \dfrac{1b}{b}$ — Divide each side by b.
11. **Thus,** $2 = 1$. — Divide out the common factor, b.

RATIONAL NUMBERS AND DECIMALS 4.1

Objective **To show that a given number is a rational number**

Every integer, mixed number, and terminating decimal can be expressed in the form $\frac{a}{b}$ where a and b are integers. These numbers are examples of **rational numbers.**

> **Definition: Rational number**
> A *rational number* is one that can be written in the form $\frac{a}{b}$, where a and b are integers, $b \neq 0$.

Example 1 **Show that each of 18, $-7\frac{3}{4}$ and 27.89 is a rational number.**

Rewrite each number in the form $\frac{a}{b}$, where a and b are integers.

$$18 = \frac{18}{1} \qquad -7\frac{3}{4} = \frac{-31}{4} \qquad 27.89 = \frac{2{,}789}{100}$$

You can use division to express a rational number as a decimal.

$$\frac{5}{4} = 4\overline{)5.00}^{1.25} \qquad \frac{2}{3} = 3\overline{)2.000\ldots}^{0.666\ldots}$$

The decimal for $\frac{5}{4}$ is a **terminating decimal.** The decimal for $\frac{2}{3}$ is a **repeating decimal** and can be written either as $0.666\ldots$ or as $0.6\overline{6}$ with a bar above the repeating digit.

A repeating decimal may have more than one digit in its repeating block of digits. For example,

$$\frac{26}{11} = 11\overline{)26.00\,00\,00\ldots}^{2.36\,36\,36\ldots}$$

Notice that $2.3\overline{36}$ is a repeating decimal with a repeating block of two digits.

As shown on page 90, the procedure above can be reversed. That is, every repeating decimal can be written in the form $\frac{a}{b}$, where a and b are integers.

Summary Every rational number has an equivalent decimal that either terminates or repeats.
Every terminating decimal is a rational number.
Every repeating decimal is a rational number.

Example 2   **Show that $2.3\overline{36}$ and $1.62\overline{2}$ are rational numbers.**

Each must be rewritten in the form $\frac{a}{b}$, where a and b are integers.

Let $n = 2.363636\ldots$
$$100n = 236.363636\ldots$$
Subtract. $\underline{n = 2.363636\ldots}$
$$99n = 234.0$$
$$n = \frac{234}{99}, \text{ or } \frac{26}{11}$$

Let $n = 1.6222\ldots$
$$10n = 16.2222\ldots$$
Subtract. $\underline{n = 1.6222\ldots}$
$$9n = 14.60$$
$$n = \frac{14.6}{9} \quad \blacktriangleleft \text{ 14.6 is not an integer.}$$
$$n = \frac{146}{90}, \text{ or } \frac{73}{45}$$

Notice in Example 2 that when a two-digit block like $\overline{36}$ repeats, you multiply each side of the equation by 100. When a one-digit block like $\overline{2}$ repeats, you multiply by 10.

Written Exercises

Show that each of the following is a rational number.

Ⓐ   1. $-9 \quad \frac{-9}{1}$ 2. $-2\frac{3}{10} \quad \frac{-23}{10}$ 3. $-34.719 \quad \frac{-34,719}{1000}$ 4. $0 \quad \frac{0}{1}$ **(Answers may vary.)**

5. $0.8\overline{8} \quad \frac{8}{9}$ 6. $0.25\overline{25} \quad \frac{25}{99}$ 7. $0.78\overline{8} \quad \frac{71}{90}$ 8. $0.70\overline{70} \quad \frac{70}{99}$

9. $0.43\overline{43} \quad \frac{43}{99}$ 10. $0.624\overline{4} \quad \frac{281}{450}$ 11. $0.463\overline{63} \quad \frac{51}{110}$ 12. $0.517\overline{7} \quad \frac{233}{450}$

Ⓑ   13. $4.4\overline{4} \quad \frac{40}{9}$ 14. $68.68\overline{68} \quad \frac{6800}{99}$ 15. $5.05\overline{05} \quad \frac{500}{99}$ 16. $22.2\overline{2} \quad \frac{200}{9}$

17. $5.38\overline{8} \quad \frac{97}{18}$ 18. $8.74\overline{141} \quad \frac{4327}{495}$ 19. $3.81\overline{6} \quad \frac{229}{60}$ 20. $47.23\overline{939} \quad \frac{15,589}{330}$

21. $0.571\overline{571} \quad \frac{571}{999}$ 22. $0.743\overline{8438} \quad \frac{2477}{3330}$ 23. $8.31\overline{8318} \quad \frac{2770}{333}$ 24. $1.2347\overline{2347} \quad \frac{12,346}{9,999}$

Ⓒ   25. $0.999\ldots \quad \frac{9}{9}$ 26. $0.4999\ldots \quad \frac{1}{2}$ 27. $3.74999\ldots \quad \frac{15}{4}$ 28. $-4.124999\ldots \quad \frac{-33}{8}$

CALCULATOR ACTIVITIES

Three different ways of expressing the repeating block of digits in the decimal for $\frac{1810}{333}$ are given below.

$1810 \div 333 = 5.4354354$ $5.4\overline{3543}$ $5.43\overline{5435}$ $5.435\overline{4354}$

Find all of the different ways to express the repeating block of digits in the decimal for each rational number.

1. $\frac{80}{33}$ **2.4$\overline{24}$; 2.4$\overline{242}$**

2. $\frac{532}{99}$ **5.3$\overline{737}$; 5.37$\overline{373}$**

3. $\frac{235}{111}$ **2.117$\overline{117}$; 2.1171$\overline{171}$; 2.11711$\overline{711}$**

4. $\frac{7700}{777}$ **9.90$\overline{9909}$; 9.909$\overline{9099}$; 9.9099$\overline{0990}$**

Express each rational number as a decimal and tell whether it terminates or repeats.

5. $\frac{19}{40}$ 6. $\frac{37}{30}$ 7. $\frac{5}{8}$ 8. $\frac{4}{9}$ 9. $\frac{119}{125}$ 10. $\frac{22}{7}$ 11. $\frac{234,500}{9,999}$

0.475; terminates   **1.23$\overline{3}$; repeats**   **0.625; terminates**   **0.4$\overline{4}$; repeats**   **0.952; terminates**   **3.142857$\overline{142847}$; repeats**   **23.4523$\overline{4523}$; repeats**

SIMPLIFYING PRODUCTS AND QUOTIENTS

4.2

Objectives **To determine the values of the variables for which a rational expression is undefined**
To simplify a rational expression, a product of rational expressions, and a quotient of rational expressions

Each expression below is a **quotient of polynomials**.

$$\frac{2x + y}{3x - 2y} \qquad \frac{a^2 - 7a + 12}{3a - 9} \qquad \frac{0}{4n} \qquad \frac{x + 3}{1}, \text{ or } x + 3$$

Such expressions are called **rational expressions**. An expression like $\frac{\sqrt{x + 3}}{2y}$ is *not* a rational expression since $\sqrt{x + 3}$ is not a polynomial.

> **Definition: Rational expression**
>
> A *rational expression* is either a polynomial or a quotient of polynomials.

You can *evaluate* a rational expression, given values of the variables.
If $x = 4$ and $y = -5$, the value of $\frac{2x + y}{3x - 2y}$ is $\frac{2 \cdot 4 - 5}{3 \cdot 4 - 2 \cdot -5}$, or $\frac{3}{22}$.
If $x = 4$ and $y = 6$, the value of $\frac{2x + y}{3x - 2y}$ is **undefined** because $\frac{2 \cdot 4 + 6}{3 \cdot 4 - 2 \cdot 6}$, or $\frac{14}{0}$, is undefined. **Division by 0 is undefined.**

Example 1 For what values of x is the rational expression $\frac{5}{x^2 - 7x + 12}$ undefined?

The rational expression is undefined when $x^2 - 7x + 12 = 0$.
If $x^2 - 7x + 12 = 0$, then $(x - 3)(x - 4) = 0$, and $x = 3$ or $x = 4$.
Thus, $\frac{5}{x^2 - 7x + 12}$ is undefined if $x = 3$ or $x = 4$.

To simplify a fraction, you can factor its terms and divide out common factors. Rational expressions are simplified in a similar way in Example 2.

$$\frac{30}{42} = \frac{\overset{1}{\cancel{2}} \cdot 5 \cdot \overset{1}{\cancel{3}}}{\underset{1}{\cancel{2}} \cdot \underset{1}{\cancel{3}} \cdot 7} = \frac{5}{7}$$

Example 2 Simplify $\frac{8x - 12}{12x - 18}$ and $\frac{3n^2 - 15n + 18}{5n^2 - 45}$.

$$\frac{8x - 12}{12x - 18} = \frac{4(2x - 3)}{6(2x - 3)} = \frac{4}{6} = \frac{2}{3}$$

$$\frac{3n^2 - 15n + 18}{5n^2 - 45} = \frac{3(n^2 - 5n + 6)}{5(n^2 - 9)} = \frac{3(n - 2)(n - 3)}{5(n + 3)(n - 3)} = \frac{3(n - 2)}{5(n + 3)}$$

Simplifying Products and Quotients

If $\frac{a}{b}$ and $\frac{c}{d}$ are rational expressions, $b \neq 0$, $d \neq 0$, then $\frac{a}{b} \cdot \frac{c}{d} = \frac{a \cdot c}{b \cdot d}$.

To multiply either rational numbers or rational expressions, it is sometimes convenient to begin by factoring and dividing out common factors as shown.

$$\frac{2}{21} \cdot \frac{35}{11} = \frac{2 \cdot \cancel{7} \cdot 5}{\cancel{7} \cdot 3 \cdot 11} = \frac{10}{33} \qquad \frac{2}{5x + 20} \cdot \frac{3x + 12}{7} = \frac{2 \cdot 3\cancel{(x + 4)}}{5\cancel{(x + 4)} \cdot 7} = \frac{6}{35}$$

Example 3 **Simplify** $\frac{10c + 15}{10x - 50} \cdot \frac{x^2 - 25}{6c + 9}$ and $\frac{3x^8}{5x + 10} \cdot \frac{x^2 - 4}{6x^3}$.

$$\frac{10c + 15}{10x - 50} \cdot \frac{x^2 - 25}{6c + 9} = \frac{5(2c + 3) \cdot (x + 5)(x - 5)}{10(x - 5) \cdot 3(2c + 3)} = \frac{x + 5}{6}$$

$$\frac{3x^8}{5x + 10} \cdot \frac{x^2 - 4}{6x^3} = \frac{3 \cdot x^8(x + 2)(x - 2)}{6 \cdot x^3 \cdot 5(x + 2)} = \frac{x^5(x - 2)}{10}$$

Before identifying the common factors, be sure that each polynomial is factored *completely* as shown in Example 4.

Example 4 **Simplify** $\frac{2x^2 - 18}{24 - 6x} \cdot \frac{2x^2 - 9x + 4}{3x^2 + 24x + 45}$.

$$\frac{2(x^2 - 9)}{-6(x - 4)} \cdot \frac{(2x - 1)(x - 4)}{3(x^2 + 8x + 15)} = \frac{2(x + 3)(x - 3) \cdot (2x - 1)(x - 4)}{-6(x - 4) \cdot 3(x + 5)(x + 3)} = \frac{(x - 3)(2x - 1)}{-9(x + 5)}$$

The reciprocal of $\frac{2}{3}$ is $\frac{3}{2}$ and $\frac{8}{9} \div \frac{2}{3} = \frac{8}{9} \cdot \frac{3}{2} = \frac{4}{3}$.

To divide a rational number by a nonzero second rational number, you multiply the first number by the *reciprocal* (*multiplicative inverse*) of the second number.

Rational expressions are divided in a similar way.

If $\frac{a}{b}$ and $\frac{c}{d}$ are rational expressions, $b \neq 0$, $c \neq 0$, $d \neq 0$, then $\frac{a}{b} \div \frac{c}{d} = \frac{a}{b} \cdot \frac{d}{c}$.

Example 5 **Simplify** $\frac{4a + 8}{5a - 20} \div \frac{a^2 - 3a - 10}{a^2 - 4a}$.

The reciprocal of $\frac{a^2 - 3a - 10}{a^2 - 4a}$ is $\frac{a^2 - 4a}{a^2 - 3a - 10}$.

$$\frac{4a + 8}{5a - 20} \cdot \frac{a^2 - 4a}{a^2 - 3a - 10} = \frac{4(a + 2) \cdot a(a - 4)}{5(a - 4) \cdot (a - 5)(a + 2)} = \frac{4a}{5(a - 5)}$$

Reading in Algebra

Determine whether each expression is a rational expression. If it is not, then tell why not.

1. $\dfrac{3x^2 - 15x + 18}{4x - 8}$ **yes**
2. $\dfrac{\sqrt{y - 3}}{2y - 6}$ **no**
3. $4a^2 - 25b^2$ **yes**
4. $\dfrac{5n + 10}{10}$ **yes**
5. $\dfrac{0}{x + 6}$ **yes**

Evaluate each rational expression for $x = 4$ and $y = -3$, if possible.

6. $\dfrac{x + y}{3x + 4y}$
7. $\dfrac{x^2 + 2xy + y^2}{x + y}$ **1**
8. $\dfrac{y^2 + y}{x^2 - 16}$
9. $\dfrac{x^2 - y^2}{x^2 + y^2}$ $\dfrac{7}{25}$
10. $\dfrac{x^3}{y^3 - 9y}$

6. **not possible** 8. **not possible** 10. **not possible**

Written Exercises

17. $\dfrac{2(x - 5)}{-3}$
18. $\dfrac{5a}{8(a - 6)}$
21. $\dfrac{y + 6}{6(y + 4)}$
31. $\dfrac{1}{2}$

For what value(s) of the variable is each rational expression undefined?

(A)
1. $\dfrac{5}{3x}$ **0**
2. $\dfrac{n - 3}{n + 4}$ **-4**
3. $\dfrac{7}{3 - y}$ **3**
4. $\dfrac{x + 4}{x^2 - 25}$ **-5, 5**
5. $\dfrac{2c}{c^2 - 8c + 15}$ **3, 5**
6. $\dfrac{a^2 - 2a - 24}{a^2 + 2a - 15}$ **-5, 3**

Simplify.

19. $\dfrac{4(n - 3)}{3(n + 5)}$ 20. $\dfrac{3(a - 2)}{a + 2}$ 24. $\dfrac{5c^9(c + 3)}{2d^4(c - 7)}$ 34. $\dfrac{x^{2n} + 3}{x^n + 5}$ 35. $\dfrac{x^{4n} + x^{2n}y^n + y^{2n}}{x^{2n} + y^n}$

7. $\dfrac{10y - 20}{25y - 75}$ $\dfrac{2(y - 2)}{5(y - 3)}$
8. $\dfrac{3a - 15}{10 - 2a}$ $-\dfrac{3}{2}$
9. $\dfrac{4 - c^2}{5c - 10}$ $\dfrac{-(c + 2)}{5}$

10. $\dfrac{3n^2 + 21n + 36}{2n^2 - 4n - 48}$ $\dfrac{3(n + 3)}{2(n - 6)}$
11. $\dfrac{6x^2 - 30x + 24}{10x^3 - 100x + 240}$ $\dfrac{3(x - 1)}{5(x - 6)}$
12. $\dfrac{5y^2 + 15y - 50}{50 - 2y^2}$ $\dfrac{5(y - 2)}{-2(y - 5)}$

13. $\dfrac{-xy}{6a^6} \cdot \dfrac{4a^2}{-5xy^3z}$ $\dfrac{2}{15a^4y^2z}$
14. $\dfrac{5bc^4}{3} \cdot \dfrac{3}{10ab^4c} \cdot \dfrac{c^3}{2ab^3}$
15. $\dfrac{9c^3}{8x + 32} \cdot \dfrac{2x + 8}{-3c^4}$ $\dfrac{3}{-4c}$

16. $\dfrac{7x + 35}{15n} \div \dfrac{9x + 45}{10n^2}$ $\dfrac{14n}{27}$
17. $\dfrac{-4a + 8}{-2x - 10} \div \dfrac{6 - 3a}{x^2 - 25}$ **above**
18. $\dfrac{5a + 25}{4a - 12} \div \dfrac{2a^2 - 2a - 60}{a^2 - 3a}$ **above**

19. $\dfrac{n^2 - 5n + 6}{3n + 12} \cdot \dfrac{4n + 16}{n^2 + 3n - 10}$
20. $\dfrac{9a + 27}{2a^2 + 8a + 8} \cdot \dfrac{4a^2 - 16}{6a + 18}$ **above**
21. $\dfrac{y^2 + 6y}{3y^2 + 6y - 24} \cdot \dfrac{10 - 5y}{-10y}$ **above**

(B)
22. $\dfrac{y^3 + 64}{y^2 - 4y + 16}$ $y + 4$
23. $\dfrac{5x^2 + 15x + 45}{x^3 - 27}$ $\dfrac{5}{x - 3}$ **above**
24. $\dfrac{15c^2d^{-6}(-5c^2 - 30c - 45)}{-10c^{-7}d^{-2}(3c^2 - 12c - 63)}$

25. $\dfrac{6a}{2n^2 + 3n - 9} \div \dfrac{8a^4}{4n^2 - 9}$ $\dfrac{3(2n + 3)}{4a^3(n + 3)}$
26. $\dfrac{3x^2 - 7x + 2}{4a^5b^2} \div \dfrac{2x^2 - 3x - 2}{8a^4b^2}$ $\dfrac{2(3x - 1)}{a(2x + 1)}$

27. $\dfrac{3c^2 - 5c - 2}{6c^2} \cdot \dfrac{4c^2 - 8c}{c^2 - 4c + 4}$ $\dfrac{2(3c + 1)}{3c}$
28. $\dfrac{2x^2 + 7x + 6}{2x^2 - 6x - 20} \cdot \dfrac{15 - 3x}{6x + 6}$ $\dfrac{-(2x + 3)}{4(x + 1)}$

29. $\dfrac{x^2 - 5xy + 6y^2}{x + y} \div (4x - 12y)$ $\dfrac{x - 2y}{4(x + y)}$
30. $\dfrac{c^2 + 2cd + d^2}{7a^2 - 7b^2} \div \dfrac{8c + 8d}{4b - 4a}$ $\dfrac{-(c + d)}{14(a + b)}$

31. $\dfrac{10x^3 - 4x^2 + 15x - 6}{10x^2 - 24x + 8} \cdot \dfrac{2x^3 - 4x^2 - 3x + 6}{4x^4 - 9}$ **above**
32. $\dfrac{y^3 - 8}{y^2 - 4y + 16} \cdot \dfrac{y^3 + 64}{3y - 6}$ $\dfrac{(y + 4)(y^2 + 2y + 4)}{3}$

(C)
33. $\dfrac{x^{4n+1} + x^{5n}}{x^{6n} - x^{5n+2}}$ $\dfrac{x + x^n}{x^n(x^n - x^2)}$
34. $\dfrac{x^{4n} + 7x^{2n} + 12}{x^{3n} + 5x^{2n} + 4x^n + 20}$
35. $\dfrac{x^{6n} - y^{3n}}{x^{4n} - y^{2n}}$

36. $\dfrac{x^{2a} - y^{2b}}{4x^{2c} - 20x^c + 25} \cdot \dfrac{8x^c - 20}{3x^a + 3y^b}$ $\dfrac{4(x^a - y^b)}{3(2x^c - 5)}$
37. $\dfrac{x^{3a} + x^{2a+1}}{x^{b+c}} \div \dfrac{x^{3a} - x^{a+2}}{x^{a+2b} - x^{2b+1}}$ $\dfrac{x^a x^b}{x^c}$ or x^{a+b-c}

For what values of the variables is each rational expression undefined?

38. $\dfrac{5x}{x^2 - 2xy + y^2}$ $x = y$
39. $\dfrac{x + 1}{x^2 - y^2}$ $x = y, x = -y$
40. $\dfrac{3x + 4}{x^3 + 4x^2y - 5xy^2}$ $x = 0, x = y, x = -5y$

Simplifying Products and Quotients

DIMENSIONAL ANALYSIS 4.3

Objective   **To convert a measurement in terms of one unit to an equivalent measurement in terms of a different unit using dimensional analysis**

In designing a new type of tire, an engineer needed to convert a speed of 90 *miles per hour* to *yards per second*. To do this, the engineer used the following four facts:

$$1 \text{ h} = 60 \text{ min} \qquad 1 \text{ min} = 60 \text{ s} \qquad 1 \text{ mi} = 5{,}280 \text{ ft} \qquad 1 \text{ yd} = 3 \text{ ft}$$

Four **conversion factors** were set up:

$$\frac{1 \text{ h}}{60 \text{ min}} \qquad \frac{1 \text{ min}}{60 \text{ s}} \qquad \frac{5{,}280 \text{ ft}}{1 \text{ mi}} \qquad \frac{1 \text{ yd}}{3 \text{ ft}}$$

These conversion factors were used to convert 90 mi/h to yards per second (yd/s) as shown below.

$$\frac{90 \text{ mi}}{1 \text{ h}} \times \frac{1 \text{ h}}{60 \text{ min}} \times \frac{1 \text{ min}}{60 \text{ s}} \times \frac{5{,}280 \text{ ft}}{1 \text{ mi}} \times \frac{1 \text{ yd}}{3 \text{ ft}}$$

$$\frac{90 \times 5{,}280 \times 1 \text{ yd}}{60 \times 60 \text{ s} \times 3} = \frac{44 \text{ yd}}{1 \text{ s}}, \text{ or } 44 \text{ yd/s}$$

This procedure for converting from one measurement unit to another measurement unit is called **dimensional analysis.** The process is *similar* to the one you have used for *multiplying rational expressions*. You are cautioned to inspect each conversion factor to determine whether it *or* its reciprocal is needed. For example,

to convert 180 ft to inches, use $\frac{12 \text{ in.}}{1 \text{ ft}}$ as shown on the left below;

to convert 180 in. to feet, use $\frac{1 \text{ ft}}{12 \text{ in.}}$ as shown on the right below.

$$\frac{180 \text{ ft}}{1} \times \frac{12 \text{ in.}}{1 \text{ ft}} = 2{,}160 \text{ in.} \qquad \frac{180 \text{ in.}}{1} \times \frac{1 \text{ ft}}{12 \text{ in.}} = 15 \text{ ft}$$

When working with *area* or *volume*, you sometimes have to convert a measurement with *square* units or *cubic* units, respectively. To find the number of square inches in a square foot, or cubic inches in a cubic foot, you can proceed as follows.

$$\frac{12 \text{ in.}}{1 \text{ ft}} = 1 \qquad \left(\frac{12 \text{ in.}}{1 \text{ ft}}\right)^2 = \frac{144 \text{ in.}^2}{1 \text{ ft}^2} \qquad \left(\frac{12 \text{ in.}}{1 \text{ ft}}\right)^3 = \frac{1{,}728 \text{ in.}^3}{1 \text{ ft}^3}$$

Example 1   **Convert 5 square yards (yd²) to square feet (ft²).**

$$\frac{5 \text{ yd}^2}{1} \cdot \left(\frac{3 \text{ ft}}{1 \text{ yd}}\right)^2 = \frac{5 \text{ yd}^2}{1} \cdot \frac{9 \text{ ft}^2}{1 \text{ yd}^2} = 45 \text{ ft}^2$$

Written Exercises

Convert each measure as indicated. Use only one conversion factor in each exercise.

(A)
1. 24 pt to quarts **12 qt**
2. 12 gal to quarts **48 qt**
3. 15 yd to feet **45 ft**
4. 144 in. to yards **4 yd**
5. 42 oz to pounds **$2\frac{5}{8}$ lb**
6. 6 T to pounds **12,000 lb**
7. 20 g to milligrams **20,000 mg**
8. 400 g to kilograms **0.4 kg**
9. 750 mL to liters **0.75 L**
10. 5 m to centimeters **500 cm**
11. 24 ft^2 to square inches **3,456 in.^{2}**
12. 36 ft^2 to square yards **4 yd^2**
13. 135 ft^3 to cubic yards **5 yd^3**
14. 2.5 m^2 to square centimeters **25,000 cm^2**

Use dimensional analysis to convert each measure to the indicated unit.

(B)
15. 30 mi/h to feet per second **44 ft/s**
16. 15 T/h to pounds per minute **500 lb/min**
17. 10 ounces per pint to pounds per gallon **5 lb/gal**
18. 8 ounces per square inch to pounds per square foot **72 lb/ft^2**

Example

For 11 h a test car was driven at an average speed of 45 mi/h. The car averaged 22 mi/gal of gasohol. Determine the dollar cost of the fuel used if gasohol costs 129¢/gal.

$$\frac{11 \text{ h}}{1} \times \frac{45 \text{ mi}}{1 \text{ h}} \times \frac{1 \text{ gal}}{22 \text{ mi}} \times \frac{129¢}{1 \text{ gal}} \times \frac{\$1}{100¢}$$

$$= \frac{11 \times 45 \times 129 \times \$1}{22 \times 100}, \text{ or } \$29.03$$

Use dimensional analysis to solve each problem.

19. How many platforms can be built with 14 kegs of rivets if each keg contains 660 rivets, each platform contains 21 braces, and each brace requires 44 rivets? **10 platforms**

20. A construction crew uses 135 ft^3 of sand per day during a 5-day work week. If sand weighs 1.215 tons per cubic yard (T/yd^3), how many pounds of sand does the crew use in one work week? **60,750 lb**

(C)
21. Each football player must have both ankles taped before each practice session and before each game. One roll of tape is needed for every 3 ankles that are taped. One carton of tape costs $12.75 and contains 6 rolls of tape. What is the cost for all the tape used by a 45-player football team during a season that consists of 60 practice sessions and 10 games? **$4,462.50**

22. A city recreation department enrolled 450 youths in its summer softball program and assigned 15 players to each team. Each team was furnished with 20 bats for the season. They were purchased at $37.80 per carton with a dozen bats in each carton. What was the cost of the bats if the department received a 5% discount for buying in quantity? **$1,795.50**

Dimensional Analysis

SUMS AND DIFFERENCES   4.4

Objective **To add and subtract rational expressions**

Rational expressions with a *common denominator* are added in the same way that rational numbers with a common denominator are added.

$$\frac{3}{10} + \frac{1}{10} = \frac{3+1}{10} = \frac{4}{10} = \frac{2}{5} \qquad \frac{2y}{3x} + \frac{y}{3x} = \frac{2y+y}{3x} = \frac{3y}{3x} = \frac{y}{x}$$

If $\frac{a}{c}$ and $\frac{b}{c}$ are rational expressions, $c \neq 0$, then $\frac{a}{c} + \frac{b}{c} = \frac{a+b}{c}$.

Example 1 Simplify $\dfrac{n^2 - 6n}{n^2 + 8n + 16} + \dfrac{4n - 24}{n^2 + 8n + 16}$.

$$\frac{n^2 - 6n + 4n - 24}{n^2 + 8n + 16} = \frac{n^2 - 2n - 24}{n^2 + 8n + 16} = \frac{(n-6)(n+4)}{(n+4)(n+4)} = \frac{n-6}{n+4}$$

You can subtract rational expressions by using the definition of subtraction: $x - y = x + (-y)$. The **opposite** *(additive inverse)* of a rational number can be written in a convenient form using the property: $-\dfrac{a}{b} = \dfrac{-a}{b}$. For example,

$$-\frac{8}{2} = \frac{-8}{2} \quad \text{and} \quad -\frac{3x-2}{x^2-4} = \frac{-(3x-2)}{x^2-4} = \frac{-3x+2}{x^2-4}.$$

Example 2 Simplify $\dfrac{7x + 15}{6x + 12} - \dfrac{2x + 5}{6x + 12}$.

$$\frac{7x+15}{6x+12} + \frac{-(2x+5)}{6x+12} = \frac{7x+15-2x-5}{6x+12} = \frac{5x+10}{6x+12} = \frac{5(x+2)}{6(x+2)} = \frac{5}{6}$$

To add $\dfrac{7}{51}$ and $\dfrac{5}{34}$, where the denominators are different, you use the property: $\dfrac{x}{y} = \dfrac{x(c)}{y(c)}$, $y \neq 0$, $c \neq 0$.

$$\frac{7}{51} + \frac{5}{34} = \frac{7}{3 \cdot 17} + \frac{5}{2 \cdot 17} = \frac{7(2)}{3 \cdot 17(2)} + \frac{5(3)}{2 \cdot 17(3)} = \frac{14 + 15}{3 \cdot 17 \cdot 2} = \frac{29}{104}$$

To add or subtract rational expressions with different denominators, you can proceed through the following steps:

Step 1 Factor each denominator completely.
Step 2 Find the *least common denominator* (LCD) of the rational expressions. [The LCD of the rational expressions is the least common multiple (LCM) of their denominators.]
Step 3 Express each rational expression in terms of the LCD.
Step 4 Add or subtract the equivalent rational expressions with a common denominator.

For example, $\dfrac{4x+3}{5x+10} - \dfrac{x-7}{3x+6} = \dfrac{4x+3}{5(x+2)} + \dfrac{-x+7}{3(x+2)}$

The LCD is $3 \cdot 5(x+2)$. ▶

$= \dfrac{3(4x+3)}{3 \cdot 5(x+2)} + \dfrac{5(-x+7)}{5 \cdot 3(x+2)} = \dfrac{7x+44}{15(x+2)}$

Example 3 Simplify $\dfrac{3}{4n-12} - \dfrac{5-2n}{n^2+2n-15}$.

$\dfrac{3}{4(n-3)} + \dfrac{-5+2n}{(n-3)(n+5)} = \dfrac{3(n+5)}{4(n-3)(n+5)} + \dfrac{4(-5+2n)}{4(n-3)(n+5)}$ ◀ The LCD is $4(n-3)(n+5)$.

$= \dfrac{3n+15-20+8n}{4(n-3)(n+5)} = \dfrac{11n-5}{4(n-3)(n+5)}$

Another convenient form for the opposite of a rational expression is given by the property: $-\dfrac{a}{b} = \dfrac{a}{-b}$. To simplify the *difference* $\dfrac{n+5}{n-7} - \dfrac{3}{7-n}$, notice that the denominators $n-7$ and $7-n$ are a pair of opposites. That is, $-(7-n) = n-7$.

Thus, $\dfrac{n+5}{n-7} - \dfrac{3}{7-n} = \dfrac{n+5}{n-7} + \dfrac{3}{-(7-n)} = \dfrac{n+5}{n-7} + \dfrac{3}{n-7} = \dfrac{n+8}{n-7}$.

You can use another property, $\dfrac{a}{b} = \dfrac{-a}{-b}$, to simplify the *sum* $\dfrac{6}{m-5} + \dfrac{2}{5-m}$ where the denominators are a pair of opposites.

$\dfrac{6}{m-5} + \dfrac{2}{5-m} = \dfrac{6}{m-5} + \dfrac{-2}{-(5-m)} = \dfrac{6}{m-5} + \dfrac{-2}{m-5} = \dfrac{4}{m-5}$

Example 4 Simplify $\dfrac{5a}{a^2-9} - \dfrac{4}{a+3} + \dfrac{2}{3-a}$.

$\dfrac{5a}{(a+3)(a-3)} - \dfrac{4}{a+3} + \dfrac{-2}{-(3-a)} = \dfrac{5a}{(a+3)(a-3)} + \dfrac{-4}{a+3} + \dfrac{-2}{a-3}$

$= \dfrac{5a}{(a+3)(a-3)} + \dfrac{-4}{a+3} \cdot \dfrac{a-3}{a-3} + \dfrac{-2}{a-3} \cdot \dfrac{a+3}{a+3}$

$= \dfrac{5a-4a+12-2a-6}{(a+3)(a-3)}$, or $\dfrac{-a+6}{(a+3)(a-3)}$

Sums and Differences

Oral Exercises

Determine whether each statement is true or false. 4. T 6. T 7. F 8. T

1. $\dfrac{48}{-3} = -\dfrac{48}{3}$ **T**
2. $-\dfrac{6}{9} = \dfrac{-6}{-9}$ **F**
3. $\dfrac{12}{-4} = \dfrac{-12}{4}$ **T**
4. $\dfrac{x-7}{-(x+3)} = \dfrac{-(x-7)}{x+3}$
5. $8 - x = x - 8$ **F**
6. $-(9 - x^2) = x^2 - 9$
7. $-\dfrac{2y}{-(3-y)} = \dfrac{2y}{y-3}$
8. $\dfrac{2n+3}{4-n^2} = \dfrac{-2n-3}{n^2-4}$

Written Exercises

31. $\dfrac{40am + 10an + 4m^2 - 4mn}{5a(m-n)(m-n)}$
32. $\dfrac{2ac - 5bc - 9ab + 6b^2}{2c(3a+2b)(3a-2b)}$

Ⓐ Simplify.
8. $\dfrac{21n-33}{50}$ 19. $\dfrac{8x-1}{(x+3)(x-3)}$ 20. $\dfrac{-5c-24}{(c+5)(c-5)}$ 21. $\dfrac{3x-7}{(x+2)(x-2)}$ 29. $\dfrac{11y-9}{6(y+3)(y-1)}$

12. $\dfrac{23n+44}{6(n-2)(n+4)}$ 13. $\dfrac{6x-8}{(x+2)(x-2)}$

1. $\dfrac{2x+5}{5x+10} + \dfrac{x+1}{5x+10}$ $\dfrac{3}{5}$
2. $\dfrac{7a-10}{3a-12} - \dfrac{a+8}{3a-12}$ $\dfrac{2(a-3)}{(a-4)}$
3. $\dfrac{6n+5}{2n+6} - \dfrac{2n-7}{2n+6}$ 2
4. $\dfrac{2x^2-55}{x^2+7x} + \dfrac{6-x^2}{x^2+7x}$ $\dfrac{x-7}{x}$
5. $\dfrac{7a-15}{a^2-36} - \dfrac{2a+15}{a^2-36}$ $\dfrac{5}{a+6}$
6. $\dfrac{5c^2-8c}{c^2-9} - \dfrac{4c+9c^2}{c^2-9}$ $\dfrac{-4c}{c-3}$
7. $\dfrac{7x}{12} + \dfrac{3x}{8} - \dfrac{5x}{6}$ $\dfrac{x}{8}$
8. $\dfrac{3n+9}{10} - \dfrac{2n-1}{25} + \dfrac{n-8}{5}$
9. $\dfrac{a+2}{a} + \dfrac{a-6}{3a} + \dfrac{a-9}{5a}$
10. $\dfrac{2x}{3x+12} + \dfrac{5x}{6x+24}$ $\dfrac{3x}{2(x+4)}$
11. $\dfrac{3y-5}{2y-6} - \dfrac{4y-2}{5y-15}$ $\dfrac{7}{10}$
12. $\dfrac{5}{2n-4} + \dfrac{4}{3n+12}$ 9. $\dfrac{23a-27}{15a}$
13. $\dfrac{2x}{x^2-4} + \dfrac{4}{x+2}$ **above**
14. $\dfrac{8}{n+5} - \dfrac{2n-15}{n^2-25}$ **below**
15. $\dfrac{3y-20}{y^2-y-20} + \dfrac{6}{y-5}$ **below**
16. $\dfrac{c-10}{c-4} - \dfrac{c+2}{4-c}$ 2
17. $\dfrac{2a+11}{a+2} - \dfrac{2a-3}{-a-2}$ 4
18. $\dfrac{3x}{x-5} + \dfrac{x}{5-x} - \dfrac{2x-3}{5-x}$ $\dfrac{4x-3}{x-5}$
19. $\dfrac{4}{x+3} + \dfrac{x+2}{x^2-9} + \dfrac{3}{x-3}$
20. $\dfrac{3c-4}{c^2-25} - \dfrac{2}{c+5} + \dfrac{6}{5-c}$
21. $\dfrac{6}{x+2} - \dfrac{8x+5}{x^2-4} - \dfrac{5}{2-x}$
22. $2a + 1 + \dfrac{5a}{a-3}$ $\dfrac{2a^2-3}{a-3}$
23. $3x - 4 - \dfrac{10}{x+4}$ **below**
24. $\dfrac{3y+7}{2y+5} - (5y-1)$ **below**

Ⓑ
25. $\dfrac{y^2-8y}{y^2+10y+16} + \dfrac{3y-14}{y^2+10y+16}$ $\dfrac{y-7}{y+8}$
26. $\dfrac{3n^2+11n}{n^2+16n+64} - \dfrac{n^2+40}{n^2+16n+64}$ $\dfrac{2n-5}{n+8}$
27. $\dfrac{4a^2-16}{a^2-8a+16} - \dfrac{a^2+8a}{a^2-8a+16}$ $\dfrac{3a+4}{a-4}$
28. $\dfrac{3x+7}{2x^2+10x+12} + \dfrac{x+5}{2x^2+10x+12}$ $\dfrac{2}{x+2}$
29. $\dfrac{2y+3}{3y^2+6y-9} + \dfrac{3}{2y+6} - \dfrac{1}{3y-3}$ **above**
30. $\dfrac{7}{6x-30} - \dfrac{9x-6}{2x^2-18x+40} + \dfrac{11}{3x-12}$ **below**
31. $\dfrac{8m+2n}{m^2-2mn+n^2} + \dfrac{4m}{5am-5an}$ **above**
32. $\dfrac{2a-5b}{18a^2-8b^2} - \dfrac{3b}{6ac+4bc}$ **above**
33. $\dfrac{3x-5}{x^2-9} + \dfrac{7-x}{9-x^2} - \dfrac{1}{x+3}$ $\dfrac{3}{x+3}$
34. $\dfrac{5y+2}{2y^2-72} - \dfrac{3y-50}{72-2y^2}$ $\dfrac{4}{y+6}$ 30. $\dfrac{x-60}{3(x-5)(x-4)}$
35. $\dfrac{3}{4y+12} + \dfrac{7}{6-2y} - \dfrac{5y}{9-y^2}$ $\dfrac{9y-51}{4(y+3)(y-3)}$
36. $\dfrac{3x}{4-x^2} - \dfrac{4}{5x+10} + \dfrac{5}{6-3x}$ $\dfrac{-82x-26}{15(x+2)(x-2)}$

Ⓒ
37. $\dfrac{y^2+y+1}{y+3} + \dfrac{y^2-3y+9}{y-1}$ $\dfrac{2y^3+26}{(y+3)(y-1)}$
38. $\dfrac{x^2-2x+4}{x-4} - \dfrac{x^2+4x+16}{x+2}$ $\dfrac{72}{(x-4)(x+2)}$
39. $\dfrac{20-x^c}{x^{2c}+4x^c} + \dfrac{x^c+10}{x^c+4}$ $\dfrac{x^c+5}{x^c}$
40. $\dfrac{y^n}{y^n+6} - \dfrac{6-5y^n}{y^{2n}+6y^n}$ $\dfrac{y^n-1}{y^n}$
41. $\dfrac{1}{x^{3a}y^{5n}} + \dfrac{1}{x^{2a+4}y^{n+3}}$ $\dfrac{x^4y^3+x^ay^{4n}}{x^{3a+4}y^{5n+3}}$
42. $\dfrac{5y^b}{6x^c} - \dfrac{2}{9x^cy^b} + \dfrac{4x^c}{3y^{2b}}$ $\dfrac{15y^{3b}-4y^b+24x^{2c}}{18x^cy^{2b}}$

14. $\dfrac{6n-25}{(n+5)(n-5)}$ 15. $\dfrac{9y+4}{(y-5)(y+4)}$ 23. $\dfrac{3x^2+8x-26}{x+4}$ 24. $\dfrac{-10y^2-20y+12}{2y+5}$

COMPLEX RATIONAL EXPRESSIONS   4.5

Objective **To simplify a complex rational expression**

Each of the two expressions at the right contains at least one fraction or one rational expression in the numerator and the denominator. The first expression is a **complex fraction**, and the second is a **complex rational expression**.

$$\frac{\frac{3}{8}+\frac{7}{12}}{\frac{5}{6}-\frac{1}{4}} \qquad \frac{\frac{2}{x}+\frac{3}{xy}}{\frac{3}{x^2}+\frac{1}{5}}$$

You can simplify a complex fraction or a complex rational expression by multiplication. For example, the LCD of the four fractions in $\dfrac{\frac{3}{8}+\frac{7}{12}}{\frac{5}{6}-\frac{1}{4}}$ is 24.

To simplify, multiply the numerator and denominator by 24.

$$\frac{24\left(\frac{3}{8}+\frac{7}{12}\right)}{24\left(\frac{5}{6}-\frac{1}{4}\right)} = \frac{\overset{3}{\cancel{24}}\cdot\frac{3}{\cancel{8}_1}+\overset{2}{\cancel{24}}\cdot\frac{7}{\cancel{12}_1}}{\overset{4}{\cancel{24}}\cdot\frac{5}{\cancel{6}_1}+\overset{6}{\cancel{24}}\cdot\frac{-1}{\cancel{4}_1}} = \frac{9+14}{20-6} = \frac{23}{14}$$

Example 1 **Simplify** $\dfrac{\frac{2}{x}+\frac{3}{xy}}{\frac{3}{x^2}+\frac{1}{5}}$. | **Simplify** $\dfrac{5a-\frac{1}{2a}}{\frac{1}{3b}+4b}$.

Multiply the numerator and denominator by $5x^2y$, the LCD of the fractions.

$$\frac{5x^2y\left(\frac{2}{x}+\frac{3}{xy}\right)}{5x^2y\left(\frac{3}{x^2}+\frac{1}{5}\right)}$$

$$\frac{5x^2y\cdot\frac{2}{x}+5\cdot x^2y\cdot\frac{3}{xy}}{5x^2y\cdot\frac{3}{x^2}+5\cdot x^2y\cdot\frac{1}{5}}$$

$$\frac{10xy+15x}{15y+x^2y}$$

Multiply the numerator and denominator by $6ab$, the LCD of the fractions.

$$\frac{6ab\left(5a+\frac{-1}{2a}\right)}{6ab\left(\frac{1}{3b}+4b\right)}$$

$$\frac{6ab\cdot 5a+6ab\cdot\frac{-1}{2a}}{6ab\cdot\frac{1}{3b}+6ab\cdot 4b}$$

$$\frac{30a^2b-3b}{2a+24ab^2}$$

You should examine every answer to see if it can be simplified further by factoring the numerator or denominator (or both) and then dividing out any common factors.

Complex Rational Expressions

Example 2 Simplify $\dfrac{1 - \dfrac{6}{x} + \dfrac{5}{x^2}}{1 - \dfrac{3}{x} - \dfrac{10}{x^2}}$.

$$\dfrac{x^2\left(1 + \dfrac{-6}{x} + \dfrac{5}{x^2}\right)}{x^2\left(1 + \dfrac{-3}{x} + \dfrac{-10}{x^2}\right)} = \dfrac{x^2 \cdot 1 + x^2 \cdot \dfrac{-6}{x} + x^2 \cdot \dfrac{5}{x^2}}{x^2 \cdot 1 + x^2 \cdot \dfrac{-3}{x} + x^2 \cdot \dfrac{-10}{x^2}}$$ ◀ The LCD is x^2.

$$= \dfrac{x^2 - 6x + 5}{x^2 - 3x - 10}$$ ◀ Factor numerator and denominator.

$$= \dfrac{(x-1)(x-5)}{(x+2)(x-5)}, \text{ or } \dfrac{x-1}{x+2}$$

For some complex rational expressions, you will need to factor the denominators of the individual fractions to find their LCD. This is shown in Example 3.

Example 3 Simplify $\dfrac{\dfrac{2x}{x^2 - 25} + \dfrac{5}{x - 5}}{\dfrac{4}{x + 5} + \dfrac{3x}{x^2 - 25}}$.

$$\dfrac{\dfrac{2x}{(x+5)(x-5)} + \dfrac{5}{x-5}}{\dfrac{4}{x+5} + \dfrac{3x}{(x+5)(x-5)}}$$

$$\dfrac{(x+5)(x-5) \cdot \dfrac{2x}{(x+5)(x-5)} + (x+5)(x-5) \cdot \dfrac{5}{x-5}}{(x+5)(x-5) \cdot \dfrac{4}{x+5} + (x+5)(x-5) \cdot \dfrac{3x}{(x+5)(x-5)}}$$ ◀ The LCD is $(x+5)(x-5)$.

$$\dfrac{2x + 5(x+5)}{4(x-5) + 3x}$$

$$\dfrac{2x + 5x + 25}{4x - 20 + 3x}$$

$$\dfrac{7x + 25}{7x - 20}$$

Chapter Four

Written Exercises

Simplify each complex expression.

(A)

1. $\dfrac{\frac{3}{5} - \frac{1}{4}}{\frac{5}{2} - \frac{7}{10}}$ $\dfrac{7}{36}$

2. $\dfrac{5 + \frac{1}{4}}{2 + \frac{2}{3}}$ $\dfrac{63}{32}$

3. $\dfrac{3\frac{1}{2} + 4\frac{2}{3}}{5\frac{1}{6} + 2\frac{3}{4}}$ $\dfrac{98}{95}$

4. $\dfrac{\frac{4}{xy} - \frac{6}{y^2}}{\frac{8}{x^2} + \frac{3}{xy}}$ $\dfrac{4xy - 6x^2}{8y^2 + 3xy}$

5. $\dfrac{\frac{4}{3a} + \frac{3}{2b}}{\frac{1}{6a} - \frac{3}{4b}}$ $\dfrac{16b + 18a}{2b - 9a}$

6. $\dfrac{\frac{5}{3x^2} + \frac{7}{2x}}{\frac{1}{6x} + \frac{10}{x^2}}$ $\dfrac{10 + 21x}{x + 60}$

7. $\dfrac{\frac{3}{5x} + 4}{2 - \frac{3}{10y}}$ $\dfrac{6y + 40xy}{20xy - 3x}$

8. $\dfrac{2a - \frac{1}{4c^2}}{\frac{5}{6c} - 3a}$ $\dfrac{24ac^2 - 3}{10c - 36ac^2}$

9. $\dfrac{7m - 3n + \frac{2}{mn}}{\frac{9}{2m} + \frac{4}{3n} - 5}$

9. $\dfrac{42m^2n - 18mn^2 + 12}{27n + 8m - 30mn}$

10. $\dfrac{1 + \frac{8}{x} + \frac{12}{x^2}}{1 + \frac{6}{x} + \frac{8}{x^2}}$ $\dfrac{x + 6}{x + 4}$

11. $\dfrac{1 - \frac{2}{a} - \frac{8}{a^2}}{1 - \frac{7}{a} + \frac{12}{a^2}}$ $\dfrac{a + 2}{a - 3}$

12. $\dfrac{2 + \frac{5}{c} + \frac{3}{c^2}}{\frac{2}{c^2} + \frac{5}{c} + 3}$ $\dfrac{2c + 3}{3c + 2}$

13. $\dfrac{\frac{2y}{y^2 - 4} + \frac{3}{y - 2}}{\frac{5}{y + 2} + \frac{4}{y - 2}}$ $\dfrac{5y + 6}{9y - 2}$

14. $\dfrac{\frac{3a}{a^2 - 3a - 10} + \frac{2}{a + 2}}{\frac{2a}{a^2 - 3a - 10} + \frac{3}{a - 5}}$

14. $\dfrac{5a - 10}{5a + 6}$

15. $\dfrac{\frac{5n}{2n^2 - n - 1} - \frac{4}{n - 1}}{\frac{6}{2n + 1} + \frac{7n}{2n^2 - n - 1}}$ $\dfrac{-3n - 4}{13n - 6}$

(B)

16. $\dfrac{\frac{4}{x - 2} + \frac{3}{x}}{\frac{5}{x} + \frac{3}{x - 2}}$ $\dfrac{7x - 6}{8x - 10}$

17. $\dfrac{\frac{6}{a + 3} - \frac{4}{a - 4}}{\frac{2}{a - 4} + \frac{5}{a + 3}}$ $\dfrac{2a - 36}{7a - 14}$

18. $\dfrac{1 - \frac{9}{x^2}}{\frac{1}{x} - \frac{3}{x^2}}$ $x + 3$

19. $\dfrac{\frac{10a}{a^2 + 6a + 8}}{\frac{7}{a + 4} + \frac{3}{a + 2}}$ $\dfrac{5a}{5a + 13}$

20. $\dfrac{\frac{12}{c + 5} - \frac{4}{c - 5}}{\frac{8c}{c^2 - 25}}$ $\dfrac{c - 10}{c}$

21. $\dfrac{\frac{x - 7}{x^2 - 5x + 6}}{\frac{5}{x - 2} - \frac{4}{x - 3}}$ 1

22. $\dfrac{\frac{x}{y} + 3 - \frac{4y}{x}}{\frac{x}{y} - \frac{y}{x}}$ $\dfrac{x + 4y}{x + y}$

23. $\dfrac{\frac{a}{b} - 1 - \frac{6b}{a}}{\frac{a}{b} + 4 + \frac{4b}{a}}$ $\dfrac{a - 3b}{a + 2b}$

24. $\dfrac{\frac{3}{y} + \frac{3}{x} - \frac{6}{xy}}{\frac{4}{x} + \frac{4}{y} - \frac{8}{xy}}$ $\dfrac{3}{4}$

25. $\dfrac{x^{-2}y^{-2}}{x^{-4} - y^{-4}}$ $\dfrac{x^2y^2}{y^4 - x^4}$

26. $\dfrac{x^{-2} - y^{-2}}{x^{-4} - y^{-4}}$ $\dfrac{x^2y^2}{y^2 + x^2}$

27. $\dfrac{x^{-2} - 1}{x - 1}$ $\dfrac{-x - 1}{x^2}$

NON-ROUTINE PROBLEMS

Simplify.

1. $5 + \dfrac{1}{4 + \dfrac{1}{3 + \dfrac{1}{2 + \dfrac{1}{1 + x}}}}$ $\dfrac{157x + 225}{30x + 43}$

2. $2 - \dfrac{1}{2 - \dfrac{1}{2 - \dfrac{1}{2 - \dfrac{1}{2 - x}}}}$ $\dfrac{6 - 5x}{5 - 4x}$

Complex Rational Expressions

STATISTICS: AVERAGES AND VARIABILITY (optional)   4.6

Objectives
To find the mean, median and mode for a set of data
To find the mean for grouped data
To find the range, variance, and standard deviation for a set of data

One of the most frequently used statistical concepts is that of a **measure of central tendency**. Three measures of central tendency you will study are the mean, median, and mode.

Definition: Mean
The *mean* is the value obtained by adding the scores and dividing the sum by the number of scores.

Example 1
Find the mean, to the nearest hundredth, for the following scores:
88, 92, 48, 68, 71, 83, 88, 74, 64, 58, 59, 98, 96, 90, 88.

Let $\bar{x}$ represent the mean.

Add the scores. ▶ $\bar{x} = \dfrac{88 + 92 + 48 + 68 + 71 + 83 + 88 + 74 + 64 + 58 + 59 + 98 + 96 + 90 + 88}{15}$

Divide by the number of scores. ▶ $\bar{x} = \dfrac{1{,}165}{15} = 77.67$

Thus, the mean, $\bar{x}$, is 77.67.

Definition: Mode
The *mode* is the value that occurs most frequently and is obtained by listing the frequency of each score and choosing the one that occurs most frequently.

Example 2
Find the mode for the following scores:
50, 48, 51, 50, 49, 48, 46, 50, 49, 48, 44, 43, 48.

Group like scores. ▶

51,   50 50 50   49 49   48 48 48 48   46 44 43
once   3 times   twice   4 times   once

The most frequently occurring score is 48. It occurs 4 times.

Thus, the mode is 48.

Sometimes there is more than one mode for a set of data. This occurs when no one score occurs most frequently, but rather two or more scores have the same frequency.

Definition: Median	The *median* is the value that is the middle one after the scores have been arranged in order. To obtain the median, arrange the scores in order and choose the middle one or average the middle two scores.

Example 3 **Find the median for the following scores:**
300, 298, 301, 303, 297, 298, 300, 304, 296, 295, 296.

Arrange the scores in order. Choose the middle score. ▶

▶ 295 296 296 297 298 298 300 300 301 303 304

          5 below   5 above

Thus, the median is 298.

When many scores are given, it is often useful to make a frequency table to represent the data and to find the mean.

Example 4 **Make a frequency table and determine the mean for the following scores:**
95 90 95 75 90 80 95 90 95 85
90 80 75 85 70 85 90 100 90 70
85 85 80 70 75 95 100 85 100 75.

Make a frequency table.

Label the columns. ▶

Score (s)	Tally	Frequency (f)	Sum ($f \cdot s$)
100	|||	3	300
95	⊞	5	475
90	⊞ |	6	540
85	⊞ |	6	510
80	|||	3	240
75	||||	4	300
70	|||	3	210
		30	2575

Count the tallies to get the frequency f.

The sum is the frequency times the score.

$$\text{Mean} = \frac{\text{sum } (f \cdot s)}{\text{frequency}} = \frac{2{,}575}{30} \doteq 85.833$$

Thus, the mean is 85.833.

Statistics: Averages and Variability

Sometimes you may wish to determine whether scores cluster around the mean or are spread out from the mean. This measure is called the *variability* of the scores. The *range, variance,* and *standard deviation* are used to measure variability.

Definition: Range

The *range* is obtained by finding the difference between the highest and lowest scores.

Example 5

Find the range for the following scores: 12, 18, 16, 24, 10, 19, 22, 17, 9, 12, 16, 15.

Subtract the lowest score from the highest score.

$$24 - 9 = 15$$

Thus, the range is 15.

The indicated sum of a set of numbers can be written in a compact form using the capital Greek letter Σ, read "sigma."

Definitions: Variance

Standard deviation

For a set of n measures $x_1, x_2, x_3, \ldots, x_n$ with a mean of $\bar{x}$, the *variance* s^2 of these measures is

$$s^2 = \frac{(x_1 - \bar{x})^2 + (x_2 - \bar{x})^2 + (x_3 - \bar{x})^2 + \ldots + (x_n - \bar{x})^2}{n}.$$

The *standard deviation* of the measures is the square root of the variance, or s. Sometimes the lowercase Greek letter sigma σ is used to represent the standard deviation.

Example 6

Find the variance and standard deviation for the following scores: 8, 10, 12, 9, 8, 12, 11.

x_i	$\bar{x}$	$(x_i - \bar{x})$	$(x_i - \bar{x})^2$
8	10	-2	4
10	10	0	0
12	10	2	4
9	10	-1	1
8	10	-2	4
12	10	2	4
11	10	1	1
Total			18

$$s^2 = \frac{\Sigma(x_i - \bar{x})^2}{n} = \frac{18}{7} \doteq 2.571$$

$$\sigma = s = \sqrt{\frac{\Sigma(x_i - \bar{x})^2}{n}} = \sqrt{2.571} \doteq 1.60$$

Most scores cluster around the mean. Their spread is not large. If the scores had a larger spread, the standard deviation would have been larger. Approximately 68% of the scores will fall between ± one standard deviation of the mean, that is between $\bar{x} \pm 1\sigma$.

Written Exercises

Find the mean.
1. 93, 92, 88, 90, 86, 75, 68, 72, 75, 100, 86, 84, 78, 63, 48, 82 **80**
2. 14, 9, 12, 10, 9, 13, 14, 16, 19, 20, 12, 8, 9, 15, 16, 14, 19, 18, 16, 17 **14**
3. 245, 296, 205, 308, 310, 420, 495, 310, 187, 204 **298**
4. 896, 998, 1,005, 1,496, 400, 1,296, 983, 1,438 **1,064**

Find the mode(s).
5. 7, 8, 7, 6, 6, 6, 5, 7, 8, 7, 9 **7**
6. 48, 47, 47, 46, 48, 45, 43, 42 **47, 48**
7. 98, 97, 98, 96, 100, 96, 95, 96 **96**
8. 101, 100, 99, 100, 99, 100, 100, 98 **100**

Find the median.
9. 100, 98, 86, 99, 100, 93, 84, 93, 79, 76, 93, 98, 100, 99, 98, 96 **97**
10. 12, 11, 13, 10, 10, 9, 8, 7, 9, 12, 9, 15, 16, 10, 9, 8, 7, 9, 8, 10, 14 **10**
11. 98, 97, 96, 98, 100, 101, 83, 96, 101 **98**
12. 53, 54, 56, 58, 93, 92, 73, 68, 93 **68**

For frequency tables, see TE Answer Section.

Make a frequency table and determine the mean for the following scores.
13. 100, 101, 98, 97, 96, 99, 94, 93, 92, 88, 86, 88, 85, 92, 98, 97, 98, 88, 93, 96, 94, 96, 93, 97, 96, 100, 93, 99, 88, 86, 87, 92 **93.75**

14. 10, 12, 11, 10, 8, 12, 15, 16, 14, 12, 11, 8, 7, 4, 5, 9, 10, 18, 17, 16, 9, 8, 14, 16, 18, 12, 13, 14, 20, 19, 18, 16, 14, 6, 9, 17 **$12.4\bar{4}$, or $12\frac{4}{9}$**

Find the range.
15. 93, 98, 100, 98, 96, 94, 93, 92, 87 **13**
16. 10, 15, 16, 19, 21, 18, 17, 9, 28 **19**
17. 100, 103, 195, 201, 93, 200, 99 **108**
18. 48, 37, 36, 34, 33, 19, 16, 18, 19, 10 **38**

Find the variance and standard deviation for the following data.
19. 12, 11, 14, 16, 19, 8, 10, 14, 16, 10 **10.4, 3.22**
20. 100, 98, 93, 92, 96, 93, 88, 76 **48.75, 6.98**
21. 46, 43, 98, 38, 51, 87, 78 **497.71, 22.31**
22. 42, 83, 82, 94, 48, 95 **447.67, 21.16**

Solve each problem.
23. The mean for 10 scores is 94. Nine of the scores are known and are the following: 98, 99, 96, 94, 98, 92, 94, 90, 95. What is the missing score? **84**
24. The mean for 12 scores is 48. Eleven of the scores are known and are the following: 52, 56, 54, 40, 42, 46, 48, 47, 44, 40, 46. What is the missing score? **61**
25. The average of 20 students' test scores is 78. If the 2 lowest and the 2 highest scores are removed, the average of the remaining scores is 74. What is the average of the scores removed? **94**
26. The average of 98 students' test scores is 92. If the 4 lowest and the 4 highest scores are removed, the average of the remaining scores is 94. What is the average of the scores removed? **69.5**

Statistics: Averages and Variability

APPLICATIONS: PROOFS WITH RATIONAL NUMBERS

To prove theorems involving rational numbers, you can use the properties of a field (page 26) and the definitions below. As you study each proof, you will be asked to supply reasons for some statements. Recall the following definitions:

1. $\frac{a}{b}$ is a **rational** number if and only if a and b are integers and $b \neq 0$.

2. $\frac{a}{b} = a \cdot \frac{1}{b}$

Theorem 1 $\quad \frac{a}{c} + \frac{b}{c} = \frac{a+b}{c}$

Proof

1. $\frac{a}{c} + \frac{b}{c} = a \cdot \frac{1}{c} + b \cdot \frac{1}{c}$	1. Definition 2
2. $\phantom{\frac{a}{c} + \frac{b}{c}} = (a+b)\frac{1}{c}$	2. Distributive property
3. $\phantom{\frac{a}{c} + \frac{b}{c}} = \frac{a+b}{c}$	3. Definition 2

The associative property allows you to regroup three factors. Sometimes, it is necessary to regroup four factors. This fact is established in Theorem 2.

Theorem 2 $\quad (xy)(zw) = (xz)(yw)$

Proof

1. $(xy)(zw) = x[y(zw)]$	1. Associative for mult.
2. $ = x[(yz)w]$	2. $\underline{}$ **Associative for mult.**
3. $ = x[(zy)w]$	3. $\underline{}$ **Commutative for mult.**
4. $ = x[z(yw)]$	4. $\underline{}$ **Associative for mult.**
5. $ = (xz)(yw)$	5. $\underline{}$ **Associative for mult.**

To prove $\frac{1}{a} \cdot \frac{1}{b} = \frac{1}{ab}$, use the property of reciprocals: If $x \cdot y = 1$, then $y = \frac{1}{x}$.

Theorem 3 $\quad \frac{1}{a} \cdot \frac{1}{b} = \frac{1}{ab}$

Proof

1. $(ab)\left(\frac{1}{a} \cdot \frac{1}{b}\right) = \left(a \cdot \frac{1}{a}\right)\left(b \cdot \frac{1}{b}\right)$	1. Theorem $\underline{}$ 2
2. $\phantom{(ab)\left(\frac{1}{a} \cdot \frac{1}{b}\right)} = 1 \cdot 1$	2. Prop. of reciprocals: $x \cdot \frac{1}{x} = 1$
3. $\phantom{(ab)\left(\frac{1}{a} \cdot \frac{1}{b}\right)} = 1$	3. Multiplicative identity
4. $(ab)\left(\frac{1}{a} \cdot \frac{1}{b}\right) = 1$	4. Statements 1–3
5. $\frac{1}{a} \cdot \frac{1}{b} = \frac{1}{ab}$	5. Prop. of reciprocals: If $x \cdot y = 1$, then $y = \frac{1}{x}$.

Theorem 4 $\dfrac{a}{b} \cdot \dfrac{c}{d} = \dfrac{ac}{bd}$

Proof

1. $\dfrac{a}{b} \cdot \dfrac{c}{d} = \left(a \cdot \dfrac{1}{b}\right)\left(c \cdot \dfrac{1}{d}\right)$	1. _?_ **Definition 2**
2. $\phantom{\dfrac{a}{b} \cdot \dfrac{c}{d}} = (ac)\left(\dfrac{1}{b} \cdot \dfrac{1}{d}\right)$	2. Theorem _?_ **2**
3. $\phantom{\dfrac{a}{b} \cdot \dfrac{c}{d}} = (ac)\left(\dfrac{1}{bd}\right)$	3. Theorem _?_ **3**
4. $\phantom{\dfrac{a}{b} \cdot \dfrac{c}{d}} = \dfrac{ac}{bd}$	4. _?_ **Definition 2**

Theorem 5 $\dfrac{a}{a} = 1$

Proof

1. $\dfrac{a}{a} = a \cdot \dfrac{1}{a}$	1. _?_ **Definition 2**
2. $\phantom{\dfrac{a}{a}} = 1$	2. _?_ **Prop. of reciprocals**

Theorem 6 $\dfrac{a}{b} + \dfrac{c}{d} = \dfrac{ad + bc}{bd}$

Proof

1. $\dfrac{a}{b} + \dfrac{c}{d} = \dfrac{a}{b} \cdot 1 + \dfrac{c}{d} \cdot 1$	1. _?_ **Mult. identity**
2. $\phantom{\dfrac{a}{b} + \dfrac{c}{d}} = \dfrac{a}{b} \cdot \dfrac{d}{d} + \dfrac{c}{d} \cdot \dfrac{b}{b}$	2. Theorem _?_ **5**
3. $\phantom{\dfrac{a}{b} + \dfrac{c}{d}} = \dfrac{ad}{bd} + \dfrac{cb}{db}$	3. Theorem _?_ **4**
4. $\phantom{\dfrac{a}{b} + \dfrac{c}{d}} = \dfrac{ad}{bd} + \dfrac{bc}{bd}$	4. _?_ **Commutative for mult.**
5. $\phantom{\dfrac{a}{b} + \dfrac{c}{d}} = \dfrac{ad + bc}{bd}$	5. Theorem _?_ **1**

Theorem 7 The set of rational numbers is *closed* under addition.

Prove that $\dfrac{a}{b} + \dfrac{c}{d}$ is a rational number.

Proof

1. $a, b, c,$ and d are integers.	1. Definition _?_ **1**
2. $\dfrac{a}{b} + \dfrac{c}{d} = \dfrac{ad + bc}{bd}$	2. Theorem _?_ **6**
3. $ad, bc,$ and bd are integers.	3. Closure for _?_ of integers **mult.**
4. $ad + bc$ is an integer.	4. Closure for _?_ of _?_ **add., integers**
5. $\dfrac{ad + bc}{bd}$ is a rational number.	5. Definition _?_ **1**
6. $\dfrac{a}{b} + \dfrac{c}{d}$ is a rational number.	6. Statements 2 and 5

Prove that the set of rational numbers is closed under multiplication.

See TE Answer Section

Applications: Proofs with Rational Numbers

Ⓐ Exercises are 1–5, 7–10, 12–14, 20, 21, 24–26, 30, 31.   Ⓒ Exercises are starred.
The rest are Ⓑ Exercises.

CHAPTER FOUR REVIEW

Vocabulary
complex rational expression [4.5]
conversion factor [4.3]
least common denominator (LCD) [4.4]
opposite of rational expression [4.4]
rational expression [4.2]
rational number [4.1]
reciprocal [4.2]
repeating decimal [4.1]
terminating decimal [4.1]

Evaluate each rational expression for $x = 3$ and $y = -2$, if possible. [4.2]

1. $\dfrac{x-y}{4x+6y}$ not possible

2. $\dfrac{y^2+5y+6}{x^2+5x+6}$ 0

For what values of the variable is each rational expression undefined? [4.2]

3. $\dfrac{a-4}{5a+10}$ -2

4. $\dfrac{4c}{c^2-5c+6}$ 2, 3

Simplify each rational expression. [4.2]

5. $\dfrac{9y+18}{12y+48}$ $\dfrac{3(y+2)}{4(y+4)}$

6. $\dfrac{x^3+27}{2x^2-4x-30}$ $\dfrac{x^2-3x+9}{2(x-5)}$

7. $\dfrac{3n-6}{8-4n}$ $-\dfrac{3}{4}$

8. $\dfrac{2a^2-50}{30+9a-3a^2}$ $\dfrac{2(a+5)}{-3(a+2)}$

9. $\dfrac{4a-\dfrac{2}{3a^2}}{\dfrac{5}{6b}+3b^2}$ [4.5] $\dfrac{24a^3b-4b}{5a^2+18a^2b^3}$

10. $\dfrac{\dfrac{4n}{n^2-5n+6}-\dfrac{2}{n-3}}{\dfrac{6}{n-2}+\dfrac{2n}{n^2-5n+6}}$ $\dfrac{n+2}{4n-9}$

★ 11. $\dfrac{1+\dfrac{9}{x^a}+\dfrac{20}{x^{2a}}}{1+\dfrac{7}{x^a}+\dfrac{10}{x^{2a}}}$ $\dfrac{x^a+4}{x^a+2}$

Simplify. [4.2]

12. $\dfrac{10a^2}{3b} \cdot \dfrac{b^4c^2}{-15a^3} \quad -\dfrac{2b^3c^2}{9a}$

13. $\dfrac{x^2-1}{3x+12} \cdot \dfrac{x^2-16}{x^2-5x+4} \quad \dfrac{x+1}{3}$

14. $\dfrac{3c^2+c-2}{6-9c} \cdot \dfrac{9}{8+9c} \quad -\dfrac{3}{8}$

★ 15. $\dfrac{x^{2a}}{y^{2a}} \cdot \dfrac{y^{5a}}{z^{8a}} \cdot \dfrac{z^{3a}}{x^a} \quad \dfrac{x^a y^{3a}}{z^{5a}}$

Simplify. (Ex. 16–18) [4.2]

16. $\dfrac{-9x^7}{2y^5} \div \dfrac{12x^3}{y^3z} \quad \dfrac{-3x^4z}{8y^2}$

17. $\dfrac{5c-10}{3x+3y} \div \dfrac{2c^2-8}{x^2-2xy+y^2} \quad \dfrac{5(x+y)}{6(c+2)}$

18. $\dfrac{n^2-2n-8}{8n^3} \div \dfrac{n^2-6n+8}{6n^2} \quad \dfrac{3(n-4)}{4n(n+4)}$

19. Convert 40 ounces per square inch to pounds per square foot. (1 lb = 16 oz) [4.3] 360 lb/ft²

Simplify. [4.4]

20. $\dfrac{7a+5}{3a+6} - \dfrac{2a-5}{3a+6}$

21. $\dfrac{2x+5}{x^2-9} + \dfrac{3x+10}{x^2-9} \quad \dfrac{5}{x-3}$

22. $\dfrac{2x+7}{x^2-x-12} + \dfrac{2x+5}{x^2-x-12} \quad \dfrac{4}{x-4}$ 20. $1\dfrac{2}{3}$

23. $\dfrac{3n+2}{2n^2+3n-5} - \dfrac{9-4n}{2n^2+3n-5} \quad \dfrac{7}{2n+5}$

24. $\dfrac{2y+1}{5y-10} - \dfrac{5-y}{3y-6} \quad \dfrac{11}{15}$

25. $3x+2+\dfrac{4x}{x+3} \quad \dfrac{3x^2+15x+6}{x+3}$

26. $\dfrac{2y+17}{3y-12} - \dfrac{11-9y}{12-3y} \quad -2\dfrac{1}{3}$

27. $\dfrac{7}{x+5} - \dfrac{x+3}{25-x^2} + \dfrac{1}{5-x} \quad \dfrac{7x-37}{(x+5)(x-5)}$

28. $\dfrac{3n-1}{2n^2+4n-30} + \dfrac{2}{3n-9} - \dfrac{5}{6n+30}$ below

★ 29. $\dfrac{a^2-a+1}{a-3} - \dfrac{a^2+3a+9}{a+1} \quad \dfrac{28}{(a-3)(a+1)}$

Show that each of the following is a rational number. [4.1]

30. $-3\dfrac{4}{5} \quad \dfrac{-19}{5}$ 28. $\dfrac{4(n+4)}{3(n+5)(n-3)}$

31. $0.58\overline{8} \quad \dfrac{53}{90}$

32. $18.63 \quad \dfrac{1863}{100}$

33. $2.73\overline{434} \quad \dfrac{2707}{990}$

108   Chapter Four Review

CHAPTER FOUR TEST

Ⓐ Exercises are 1–6, 9, 10, 12, 15, 17–19. Ⓒ Exercises are starred. The rest are Ⓑ Exercises.

1. Evaluate $\dfrac{4x + 3y}{x - 3y}$ for $x = 5$ and $y = -1$. $2\dfrac{1}{8}$

2. For what values of the variable is the rational expression $\dfrac{n + 3}{n^2 - 6n + 8}$ undefined?   2, 4

Simplify each rational expression.

3. $\dfrac{4y - 12}{21 - 7y}$ $-\dfrac{4}{7}$

4. $\dfrac{3x^2 - 75}{2x^2 - 9x - 5}$ $\dfrac{3(x + 5)}{2x + 1}$

Simplify.

5. $\dfrac{4ac}{b^2} \cdot \dfrac{3b^5}{16a^3c}$ $\dfrac{3b^3}{4a^2}$

6. $\dfrac{6n^2 - 24}{3n^2 - 5n - 2} \cdot \dfrac{9n + 3}{8n + 16}$ $2\dfrac{1}{4}$

Simplify.

7. $\dfrac{n^2 - 4}{2x - 3y} \div \dfrac{5n + 10}{6x - 9y}$ $\dfrac{3(n - 2)}{5}$

8. $\dfrac{3a^2 - 5a - 12}{x^2 - 9y^2} \div \dfrac{4a^2 - 36}{5x + 15y}$ $\dfrac{5(3a + 4)}{4(a + 3)(x - 3y)}$

Simplify.

9. $\dfrac{5c}{12} + \dfrac{14c}{12} - \dfrac{c}{12}$ $\dfrac{3c}{2}$

10. $\dfrac{4a + 21}{3a + 15} + \dfrac{2a - 4}{3a + 15} - \dfrac{a - 8}{3a + 15}$ $1\dfrac{2}{3}$

11. $\dfrac{4y + 3}{y^2 - 3y - 10} - \dfrac{y - 3}{y^2 - 3y - 10}$ $\dfrac{3}{y - 5}$

12. $2x - 3 + \dfrac{6x}{x + 2}$ $\dfrac{2x^2 + 7x - 6}{x + 2}$

13. $\dfrac{3}{4y - 12} - \dfrac{3}{15 - 5y}$ $\dfrac{27}{20(y - 3)}$

Simplify.

14. $\dfrac{2c + 1}{c^2 - 5c + 6} + \dfrac{3}{2c - 6} - \dfrac{1}{c - 2}$ $\dfrac{5c + 2}{2(c - 2)(c - 3)}$

15. $\dfrac{6}{x + 4} - \dfrac{3x}{x^2 - 16} + \dfrac{5}{4 - x}$ $\dfrac{-2x - 44}{(x + 4)(x - 4)}$

16. Convert 5 meters per second to kilometers per hour.   18 km/h

Simplify each complex rational expression.

17. $\dfrac{\dfrac{1}{5a^2} - \dfrac{2}{b}}{\dfrac{7}{10a} + \dfrac{3}{2b^2}}$ $\dfrac{2b^2 - 20a^2b}{7ab^2 + 15a^2}$

18. $\dfrac{\dfrac{2x}{x^2 - 9} + \dfrac{4}{x + 3}}{\dfrac{2}{x - 3} + \dfrac{4x}{x^2 - 9}}$ $\dfrac{x - 2}{x + 1}$

Show that each of the following is a rational number.

19. $-8\dfrac{2}{3}$ $\dfrac{-26}{3}$

20. $2.47\overline{7}$ $\dfrac{223}{90}$

21. $2.6\overline{969}$ $\dfrac{89}{33}$

Simplify.

★ 22. $\dfrac{6x^{a+n}}{x^{2n} + 6x^n} \cdot \dfrac{x^{3n} - 36x^n}{x^{a+n} - 6x^a}$ $6x^n$

★ 23. $\dfrac{c^2 - 2c + 4}{c - 5} + \dfrac{c^2 + 5c + 25}{c + 2}$ $\dfrac{2c^3 - 117}{(c - 5)(c + 2)}$

★ 24. $\dfrac{\dfrac{8y^n}{y^{2n} - 16} - \dfrac{6}{y^n - 4}}{\dfrac{10}{y^n + 4} - \dfrac{4y^n}{y^{2n} - 16}}$ $\dfrac{y^n - 12}{3y^n - 20}$

Chapter Four Test

COMPUTER ACTIVITIES

Decimal Forms of Fractions

OBJECTIVE: To determine whether a given fraction will yield a terminating or repeating decimal, using a loop

The program below uses long division to convert fractions into decimals. A terminating decimal results when the division yields a zero remainder, and a repeating decimal when division yields the same remainder a second time. The program first determines whether a fraction is terminating or repeating. The program will then print out the decimal at the user's request.

```
110   DIM R(40)
120   INPUT "ENTER NUMERATOR.";N
130   INPUT "ENTER DENOMINATOR.";D
140 A = INT (N / D)
150   IF A$ < > "YES" THEN 170          (Can be true only the 2nd time through)
160   PRINT A".";
170 R(1) = N - A * D
180   IF R(1) = 0 THEN 320               (Test for 0 for termination)
190   FOR I = 2 TO D
200 M = R(I - 1) * 10                    (Another test for termination)
210 A = INT (M / D)                      (These lines compare remainders to see if
220   IF A$ < > "YES" THEN 240           the same remainder will show up a second
230   PRINT A;                           time.)
240 R(I) = M - A * D
250   IF R(I) = 0 THEN 320
260   FOR J = 1 TO I - 1
270   IF R(I) = R(J) THEN 300
280   NEXT J
290   NEXT I
300   IF A$ = "YES" THEN END
310   PRINT "THIS IS A REPEATING DECIMAL.": GOTO 340
320   IF A$ = "YES" THEN 370
330   PRINT "THIS IS A TERMINATING DECIMAL."
340   INPUT "DO YOU WANT TO SEE THE DECIMAL?";A$
350   IF A$ < > "YES" THEN 370
360   GOTO 140
370   END
```

See the Computer Section beginning on page 574 for more information.

EXERCISES

Run the program above for the fractions $\frac{1}{2}$, $\frac{3}{10}$, $\frac{8}{2}$, $\frac{3}{8}$, $\frac{5}{16}$, $\frac{1}{3}$, $\frac{5}{6}$, $\frac{3}{7}$, and $\frac{1}{23}$.
0.5, 0.3, 4, 0.3, 0.3, 0.3, 0.83, 0.428571, 0.375, 0.3125

COLLEGE PREP TEST

DIRECTIONS: Choose the *one* best answer to each question or problem.

1. Let $\frac{a}{b}$ represent a rational number where a and b are positive integers. Which statement is always true? **E**

   (A) $3a > 2b$ (B) $b - a < a$
   (C) $\frac{a+b}{a-b} > \frac{a-b}{a+b}$ (D) $\frac{6a}{6b} = 1$
   (E) None of these

2. Let $\frac{a}{b}$ and $\frac{c}{d}$ represent any two *unequal* rational numbers, $b \neq 0$, $d \neq 0$. Which statement is always true? **C**

   (A) $\frac{a}{b} < \frac{c}{d}$ (B) $a \cdot d = b \cdot c$
   (C) If $\frac{a}{b} < \frac{c}{d}$, then $\frac{a}{b} < \frac{a+c}{b+d} < \frac{c}{d}$.
   (D) If $\frac{a}{b} < \frac{c}{d}$, then $a \cdot b < c \cdot d$.
   (E) None of these

3. Let $\frac{a}{b}$ and $\frac{c}{d}$ represent any two nonzero rational expressions. Which statement is true? **A**

   (A) $\frac{a}{b} \div \frac{c}{d} = \frac{d}{c} \div \frac{b}{a}$ (B) $\frac{a}{b} + \frac{c}{d} = \frac{a+c}{b+d}$
   (C) $\frac{a}{b} \cdot \frac{c}{d} = \frac{a \cdot d}{b \cdot c}$ (D) $\frac{a}{b} \div \frac{c}{d} = \frac{b \cdot c}{a \cdot d}$
   (E) None of these

4. There are $x + 6$ males and $y - 3$ females in a choir. Find the ratio of the number of females to the total number of people in the choir. **B**

   (A) $\frac{y - x - 9}{x + y + 3}$ (B) $\frac{y - 3}{x + y + 3}$
   (C) $\frac{y - 3}{x + 6}$ (D) $\frac{x + 6}{y - 3}$
   (E) $\frac{x + y + 3}{y - 3}$

5. If $\frac{x + y}{a - b} = \frac{2}{3}$, find the value of $\frac{9x + 9y}{10a - 10b}$. **B**

   (A) $\frac{2}{3}$ (B) $\frac{3}{5}$ (C) $\frac{9}{10}$ (D) $\frac{20}{27}$
   (E) None of these

6. If $\frac{ax^2 + 2ax + a}{x^2 + 2x + 1} = 6$, find the value of a. **D**

   (A) $x^2 + 2x + 1$ (B) $x + 1$ (C) 0
   (D) 6   (E) $\frac{1}{6}$

7. If $a = \frac{1}{3}$ and $b = \frac{1}{5}$, find the value of $\frac{a+b}{a-b}$. **E**

   (A) $\frac{1}{4}$ (B) $\frac{2}{15}$ (C) $\frac{8}{15}$ (D) $\frac{16}{15}$ (E) 4

8. Find the arithmetic mean (average) of $\frac{x}{2}$, $\frac{x}{3}$, and $\frac{x}{6}$. **C**

   (A) x (B) $\frac{x}{2}$ (C) $\frac{x}{3}$ (D) $\frac{x}{6}$ (E) $\frac{x}{18}$

College Prep Test

CUMULATIVE REVIEW

(Chapters 1–4)

DIRECTIONS: Choose the *one* best answer to each question or problem. (Exercises 1–11)

1. Solve $\frac{5}{9}x - 17 = 25 - \frac{2}{9}x$. C

    (A) $10\frac{2}{7}$ (B) 24
    (C) 54   (D) 126

2. Solve $|7 - 4x| = 21$. D

    (A) 7   (B) -7
    (C) $-7, 3\frac{1}{2}$ (D) $7, -3\frac{1}{2}$

3. Find two numbers such that the first number is 9 more than 4 times the second number, and 4 more than the second number is twice the first number. B

    (A) $7, -\frac{1}{2}$ (B) $1, -2$
    (C) $-1, 2$ (D) $-7, \frac{1}{2}$

4. Solve $5^{3x-2} = 625$. A

    (A) 2   (B) 4   (C) 125   (D) 209

5. Simplify $-3xy^3(-2x^2yz^3)^4$. D

    (A) $6x^9y^7z^{12}$ (B) $16x^8y^4z^{12}$
    (C) $-48x^7y^7z^7$ (D) $-48x^9y^7z^{12}$

6. Some powdered milk worth $3.50/kg and some cocoa worth $4.25/kg form a 10-kg mixture that is worth $39.20. How much cocoa is in the mixture? C

    (A) 2.5 kg   (B) 4.4 kg
    (C) 5.6 kg   (D) 5.8 kg

7. Factor $-4c^3 + 20c^2 + 24c$ completely. B

    (A) $-4c(c - 3)(c - 2)$
    (B) $-4c(c - 6)(c + 1)$
    (C) $-c(c - 6)(4c + 4)$
    (D) $4(6 - c)(c + 1)$

8. Solve $7x^2 - 8 = 8 - 2x^2$. D

    (A) 0   (B) $\frac{4}{3}$
    (C) $-\frac{3}{4}, \frac{3}{4}$ (D) $-\frac{4}{3}, \frac{4}{3}$

9. Choose the graph of the solutions of $x^3 - 9x \leq 0$. B

    (A) number line with closed dots at $-3, 0, 3$
    (B) number line shaded left of -3 and between 0 and 3
    (C) number line shaded between -3 and 0 and right of 3
    (D) number line shaded left of -3 and right of 3

10. If $x = 7$ and $y = -2$, find the value of $\frac{5x + 4y}{x^2 - y^2}$. C

    (A) Does not exist
    (B) $\frac{43}{53}$ (C) $\frac{3}{5}$ (D) 0

11. Simplify: $\frac{5a - 10}{6a + 12} \cdot \frac{2a^2 - 8}{a^2 - 2a - 8}$. C

    (A) $\frac{5a - 20}{12a + 24}$ (B) $\frac{12a + 24}{5a - 20}$
    (C) $\frac{5(a - 2)^2}{3(a + 2)(a - 4)}$ (D) $\frac{5}{12}$

Solve each equation.

12. $23 + \frac{6}{7}x = \frac{1}{7}x + 33$ 14
13. $\frac{2}{3}(18a + 6) - (5a - 12) = 14$ $-\frac{2}{7}$
14. $4.2(3n - 0.8) = 4.9n - 12.6$ -1.2
15. $|3y - 8| = 7$ $\frac{1}{3}, 5$
16. $2^{4x-3} = 32$ $\frac{1}{2}$
17. $2x^2 = 20 - 3x$ $2\frac{1}{2}, -4$

18. Use synthetic substitution to evaluate $5x^4 - 5x^2 + 10$ for $x = -3$.   370

Find the solution set of each inequality. Graph the solutions.

19. $4(7 - x) < 2x - 14$ $\{x|x > 7\}$
20. $|y + 2| \leq 5$ $\{y|-7 \leq y \leq 3\}$
21. $x^2 - 4x - 12 > 0$ $\{x|x < -2 \text{ or } x > 6\}$
22. $6x - 4 < 2x$ and $5 - 3x > -4$ $\{x|x < 1\}$

Simplify each expression.

23. $-8a^3b^4c^{-2} \cdot 4a^{-5}bc^6$ $\frac{-32b^5c^4}{a^2}$
24. $-10x^2y(5x^4y)^3$ $-1{,}250x^{14}y^4$
25. $\frac{9m^{-8}n^7(5n + 25)}{10m^{-2}n^{-2}(3n^2 - 75)}$ $\frac{3n^9}{2m^6(n - 5)}$
26. $\frac{8.4 \times 10^5}{2.4 \times 10^7}$ 0.035

Simplify.

27. $(7m^2 + 5n)(3m^2 - 5n)$ $21m^4 - 20m^2n - 25n^2$
28. $6x(5x - 3)(4x + 5)$ $120x^3 + 78x^2 - 90x$
29. $(9y^2 - 8)(9y^2 + 8)$ $81y^4 - 64$
30. $(5c - 3d)^2$ $25c^2 - 30cd + 9d^2$
31. $\frac{4a - 28}{a^2 - 36} \cdot \frac{a^2 + 2a - 24}{6a - 42}$ $\frac{2(a - 4)}{3(a - 6)}$

32. Convert 90 km/h to m/s.   25 m/s

Factor each polynomial.

33. $12x^2 + 8xy - 15y^2$ $(6x - 5y)(2x + 3y)$
34. $3x^4 - 14x^2 - 5$ $(3x^2 + 1)(x^2 - 5)$
35. $36a^2 - 49b^2$ $(6a - 7b)(6a + 7b)$
36. $8x^3 + 125$ $(2x + 5)(4x^2 - 10x + 25)$
37. $x^2 + 8xy + 16y^2 - 9$ $(x + 4y + 3)(x + 4y - 3)$

Factor each polynomial completely.

38. $6y^2 - 45y - 24$ $3(2y + 1)(y - 8)$
39. $50c^3 + 40c^2 + 8c$ $2c(5c + 2)^2$
40. $10x^4 - 160y^4$ $10(x^2 + 4y^2)(x + 2y)(x - 2y)$
41. $4a^4 - 29a^2 + 25$ below
42. $y^4 - 8y$ $y(y - 2)(y^2 + 2y + 4)$
43. $x^3 + 2x^2 - 6x - 12$ $(x + 2)(x^2 - 6)$
41. $(2a - 5)(2a + 5)(a + 1)(a - 1)$

Solve each problem.

44. The larger of two numbers is 4 less than 7 times the smaller number. Thirty more than the larger number is the same as 6 more than 9 times the smaller number. Find the numbers.   10, 66

45. Side b of a triangle is 4 times as long as side a, and side c is 6 cm shorter than side b. Find the length of each side if the perimeter is 66 cm.
    a: 8 cm; b: 32 cm; c: 26 cm

46. Find p in dollars if $A = \$4{,}832$, $r = 8\frac{1}{2}\%$, and $t = 6$ years. Use the formula $A = p + prt$. $3,200

47. How many liters of water must be added to 20 L of a 40% sugar solution to obtain a 25% solution?   12 L

48. At 8:00 A.M., a car left city A and headed for city B at 75 km/h. At 10:00 A.M., a car left city B at 90 km/h and headed for city A. At what time did the cars meet if the two cities are 645 km apart?   1:00 P.M.

49. Find three consecutive even integers such that the product of the second and the third integers is 80.   6, 8, 10 or $-12, -10, -8$

50. The height of a triangle is 10 cm less than twice the length of its base. Find the length of the base and the height if the area of the triangle is 14 cm^2.   base: 7 cm; height: 4 cm

Cumulative Review (Chapters 1–4)

5 USING RATIONAL EXPRESSIONS

Non-Routine Problem Solving

Problem 1. You must make a choice between three successive discounts of 10%, 10%, and 5% and two successive discounts of 20% and 5% on an article priced at $1,000. How much money, if any, can you save by making the better choice?
$9.50 can be saved with discounts of 20% and 5%.

Problem 2. If a waiter receives a 25% cut in wages, he will return to his original pay with a raise of what percent? **$33\frac{1}{3}$%**

FRACTIONAL EQUATIONS 5.1

Objective **To solve a fractional equation**

The equation $\frac{3}{4} + \frac{1}{6} = \frac{11}{12}$ is a true equation. If you multiply each side by 12, the LCD of all fractions in the equation, the resulting equation is also true, but it contains no fractions.

$$\frac{3}{4} + \frac{1}{6} = \frac{11}{12}: \textit{True}$$
$$12\left(\frac{3}{4} + \frac{1}{6}\right) = 12 \cdot \frac{11}{12}$$
$$12 \cdot \frac{3}{4} + 12 \cdot \frac{1}{6} = 12 \cdot \frac{11}{12}$$
$$9 + 2 = 11: \textit{True}$$

As shown in Examples 1 and 2 below, this multiplication procedure can be used to simplify the work of solving equations that contain rational number coefficients.

Example 1 Solve $\frac{x-2}{6} + \frac{3x}{8} - \frac{3x+8}{24} = 1$.

Multiply each side by 24, the LCD.

$$24\left(\frac{x-2}{6} + \frac{3x}{8} + \frac{-(3x+8)}{24}\right) = 24 \cdot 1$$
$$24 \cdot \frac{x-2}{6} + 24 \cdot \frac{3x}{8} + 24 \cdot \frac{-3x-8}{24} = 24 \cdot 1$$
$$4(x-2) + 3 \cdot 3x + 1(-3x - 8) = 24$$
$$4x - 8 + 9x - 3x - 8 = 24$$
$$10x = 40$$
$$x = 4$$

Thus, the solution is 4.

Example 2 Solve $\frac{1}{3}y^2 + \frac{5}{12}y = \frac{1}{2}$.

Multiply each side by 12, the LCD.

$$12 \cdot \frac{1}{3}y^2 + 12 \cdot \frac{5}{12}y = 12 \cdot \frac{1}{2}$$
$$4y^2 + 5y = 6 \quad \blacktriangleleft \textit{Quadratic equation}$$
$$4y^2 + 5y - 6 = 0 \quad \blacktriangleleft \textit{Standard form}$$
$$(4y - 3)(y + 2) = 0$$
$$4y - 3 = 0 \quad \text{or} \quad y + 2 = 0$$
$$y = \frac{3}{4} \qquad\qquad y = -2$$

Thus, $\frac{3}{4}$ and -2 are the solutions.

Fractional Equations

An equation such as $\frac{2n-9}{n-7} + \frac{n}{2} = \frac{5}{n-7}$, in which a variable appears in a denominator, is called a **fractional equation.** You can use multiplication to help solve a fractional equation. However, multiplying by the LCD may produce an equation that is *not* equivalent to the original equation, as is shown in Example 3.

Example 3   Solve $\frac{2n-9}{n-7} + \frac{n}{2} = \frac{5}{n-7}$.

Multiply each side by $2(n-7)$, the LCD.

$$2(n-7)\left[\frac{2n-9}{n-7} + \frac{n}{2}\right] = 2(n-7) \cdot \frac{5}{n-7}$$
$$2(2n-9) + n(n-7) = 2 \cdot 5$$
$$4n - 18 + n^2 - 7n = 10$$
$$n^2 - 3n - 28 = 0$$
$$(n-7)(n+4) = 0$$
$$n = 7 \qquad n = -4$$

◀ The "possible" solutions are 7 and -4.

Check both numbers in the original equation.

$n = 7$ ▶

$$\begin{array}{c|c} \frac{2n-9}{n-7} + \frac{n}{2} & \frac{5}{n-7} \\ \frac{14-9}{7-7} + \frac{7}{2} & \frac{5}{7-7} \\ \frac{5}{0} + \frac{7}{2} & \frac{5}{0} \end{array}$$

7 is not a solution since $\frac{5}{0}$ is not defined.

$n = -4$ ▶

$$\begin{array}{c|c} \frac{2n-9}{n-7} + \frac{n}{2} & \frac{5}{n-7} \\ \frac{-8-9}{-4-7} + \frac{-4}{2} & \frac{5}{-4-7} \\ \frac{-17}{-11} - 2 & \frac{5}{-11} \\ -\frac{5}{11} & -\frac{5}{11} \end{array}$$

Thus, -4 is the only solution of the original equation.

In Example 3, 7 is a solution of the *derived equation,* $n^2 - 3n - 28 = 0$, but it is not a solution of the original equation. In this case, 7 is called an **extraneous solution.**

Definition: Extraneous solution	An *extraneous solution* of an equation is a solution of a derived equation that is not a solution of the original equation.

Even though multiplying both sides of a fractional equation by a polynomial may produce extraneous solutions, all of the solutions of the original equation are retained. For this reason, all possible solutions must be checked in the original equation. You should notice that the extraneous solutions of a fractional equation are the numbers that make the value of at least one denominator equal to 0. In Example 3, 7 is an extraneous solution and 7 is also the number for which $n-7$ has the value of 0.

Example 4 Solve $\dfrac{5}{2y + 6} - \dfrac{2y - 4}{y^2 - y - 12} = \dfrac{3}{y - 4}$.

$$\dfrac{5}{2(y + 3)} + \dfrac{-(2y - 4)}{(y + 3)(y - 4)} = \dfrac{3}{y - 4} \qquad \blacktriangleleft \text{ The LCD is } 2(y + 3)(y - 4).$$

$$2(y + 3)(y - 4) \cdot \dfrac{5}{2(y + 3)} + 2(y + 3)(y - 4) \cdot \dfrac{-2y + 4}{(y + 3)(y - 4)} = 2(y + 3)(y - 4) \cdot \dfrac{3}{y - 4}$$

$$5(y - 4) + 2(-2y + 4) = 3 \cdot 2(y + 3)$$
$$5y - 20 - 4y + 8 = 6(y + 3)$$
$$y - 12 = 6y + 18$$
$$-5y = 30$$
$$y = -6$$

Check in the original equation. If $y = -6$, none of the denominators has the value of 0.

Thus, -6 is the solution.

A fractional equation like $\dfrac{3}{x + 1} = \dfrac{2}{x - 4}$ is called a **proportion**. A precise definition of *proportion* follows.

Definition: Proportion   A *proportion* is an equation of the form $\dfrac{a}{b} = \dfrac{c}{d}$, where a, b, c, and d are numbers, $b \neq 0$, $d \neq 0$.

For the proportion $\dfrac{a}{b} = \dfrac{c}{d}$, a and d are called the **extremes** and b and c are called the **means**. In the true proportion $\dfrac{2}{3} = \dfrac{10}{15}$, notice that *the product of the extremes equals the product of the means*: $2 \cdot 15 = 3 \cdot 10$. This suggests the Proportion Property below.

Proportion property   For all numbers a, b, c, and d, $b \neq 0$, $d \neq 0$, if $\dfrac{a}{b} = \dfrac{c}{d}$, then $a \cdot d = b \cdot c$.

Example 5 Solve $\dfrac{3}{x + 1} = \dfrac{2}{x - 4}$.

$\dfrac{3}{x + 1} = \dfrac{2}{x - 4}$ $\blacktriangleleft$ $\dfrac{a}{b} = \dfrac{c}{d}$

$3(x - 4) = (x + 1)2$ $\blacktriangleleft$ $a \cdot d = b \cdot c$

$3x - 12 = 2x + 2$

$x = 14$

Thus, the solution is 14.

Check:

$\dfrac{3}{x + 1}$	$\dfrac{2}{x - 4}$
$\dfrac{3}{14 + 1}$	$\dfrac{2}{14 - 4}$
$\dfrac{1}{5}$	$\dfrac{1}{5}$

Fractional Equations

Reading in Algebra

Supply the missing words and expressions in the paragraph below.

$\frac{2x}{15} = \frac{6}{5x}$ is a special type of fractional equation. It is called a __proportion__.
The extremes are $2x$ and __$5x$__, and the __means__ are 15 and __6__. A derived equation is $10x^2 =$ __90__, or $x^2 =$ __9__. So, the solutions are __3__ and __-3__. Neither solution is an __extraneous__ solution, since the value of $5x$ is not equal to __0__ if $x = 3$ or if $x = -3$.

Written Exercises

Solve each equation.

Ⓐ 1. $\frac{x}{10} + \frac{x}{6} + \frac{x}{15} = 1$ **3**   2. $\frac{2n-3}{2} = \frac{3}{4} + \frac{n-4}{8}$ **2**   3. $\frac{a-1}{3} + \frac{a+2}{6} = 2$ **4**

4. $\frac{2}{5} + \frac{2}{y} = 1$ **$\frac{10}{3}$**   5. $\frac{6}{x} + \frac{9}{2x} = 3$ **$\frac{7}{2}$**   6. $\frac{2}{3n^2} = \frac{1}{4n^2} + \frac{5}{6n}$ **$\frac{1}{2}$**

7. $\frac{1}{2}x^2 + x = 12$ **$-6, 4$**   8. $\frac{1}{9}y^2 + \frac{4}{3}y + 3 = 0$ **$-9, -3$**   9. $\frac{1}{10}x^2 - \frac{1}{2}x = 5$ **$-5, 10$**

10. $\frac{3n-7}{n-5} + \frac{n}{2} = \frac{8}{n-5}$ **-6**   11. $\frac{a-4}{a+3} = \frac{3a+2}{a+3} + \frac{a}{4}$ **-8**   12. $\frac{4c-3}{c-4} - \frac{2c}{3} = \frac{2c+5}{c-4}$ **3**

13. $\frac{2}{y-3} + \frac{2}{y} = 1$ **1, 6**   14. $\frac{3}{x} + \frac{2}{x+2} = 2$ **$-\frac{3}{2}, 2$**   15. $\frac{10}{n+4} - \frac{1}{n} = 1$ **1, 4**

16. $\frac{5}{x+4} = \frac{3}{x-2}$ **11**   17. $\frac{4a+3}{3} = \frac{2a+5}{4}$ **$\frac{3}{10}$**   18. $\frac{3}{2y+1} = \frac{2}{3y-2}$ **$\frac{8}{5}$**

19. $\frac{a+1}{8} = \frac{2}{a+1}$ **$-5, 3$**   20. $\frac{x-3}{2} = \frac{1}{x-4}$ **2, 5**   21. $\frac{2z-3}{2z+3} = \frac{z-2}{2z-3}$ **$\frac{5}{2}, 3$**

22. $\frac{5}{x-3} - \frac{6}{x^2-9} = \frac{4}{x+3}$ **-21**   23. $\frac{5}{n+2} = \frac{3}{n-2} - \frac{2n}{n^2-4}$ **4**   24. $\frac{7y-9}{y^2-25} - \frac{3}{y-5} = \frac{2}{y+5}$ **7**

Ⓑ 25. $\frac{3}{4}y^2 + y = 1$ **$-2, \frac{2}{3}$**   26. $\frac{1}{7}x^2 = 1 - \frac{9}{28}x$ **$-4, \frac{7}{4}$**   27. $\frac{7}{4} = \frac{13}{6}y - \frac{2}{3}y^2$ **$\frac{3}{2}, \frac{7}{4}$**

28. $\frac{3}{x+1} + \frac{x-2}{3} = \frac{13}{3x+3}$ **$-2, 3$**   29. $\frac{3}{10}a^2 = \frac{11}{20}a + \frac{1}{2}$ **$-\frac{2}{3}, \frac{5}{2}$**   30. $\frac{n}{n-3} + \frac{n-2}{6} = \frac{5n-1}{4n-12}$

31. $\frac{4y+9}{y+1} = \frac{2y+7}{y-1}$ **$-2, 4$**   32. $\frac{2x+2}{3x+1} = \frac{x-2}{x-1}$ **0, 5**   33. $\frac{a^2+5a-2}{a+3} = \frac{4a-1}{3}$ **1, 3**

34. $\frac{3}{n-2} + \frac{6}{n^2-5n+6} = \frac{4}{n-3}$ **5**   35. $\frac{1}{x-2} = \frac{2x+1}{x^2+2x-8} + \frac{2}{x+4}$ **$\frac{7}{3}$**   **$\frac{3}{2}, 5$**

36. $\frac{x}{2x-6} - \frac{3}{x^2-6x+9} = \frac{x-2}{3x-9}$ **$-6, 5$**   37. $\frac{1}{2a} - \frac{9}{a^2+6a} = \frac{2-a}{2a+12}$ **$-3, 4$**

Ⓒ 38. $\frac{4}{x^2+2x-15} + \frac{5}{x^2-x-6} = \frac{3}{x^2+7x+10}$ **-7**   39. $\frac{6}{y^2-9} + \frac{4}{6+y-y^2} + \frac{2}{y^2+5y+6} = 0$ **$\frac{3}{2}$**

NON-ROUTINE PROBLEMS

Solve $\frac{10}{2x^3 - x^2 - 8x + 4} + \frac{24}{2x^3 - x^2 - 2x + 1} = \frac{40}{x^4 - 5x^2 + 4}$.   **$-\frac{11}{17}, 3$**

PROBLEM SOLVING: WORK PROBLEMS 5.2

Objective **To solve a work problem that leads to a fractional equation**

A certain house can be painted in 6 days. The chart below indicates the part of the house that can be painted in a given number of days.

Number of days	1	2	5	x	6
Part of job done	$\frac{1}{6}$	$\frac{2}{6}$, or $\frac{1}{3}$	$\frac{5}{6}$	$\frac{x}{6}$	$\frac{6}{6}$, or 1

In general, if a job can be done in 6 days, then $\frac{x}{6}$ represents the part of the job completed in x days and $\frac{6}{6}$, or 1, represents the *total job*.

Example 1 **Palmer's crew can do the cement work for a new building in 6 days. Hoffman's crew would need 8 days to complete the same job. How many days will the job take if the two crews work together?**

Let x = the number of days it will take if the crews work together.
Represent the data in a table.

Represent the data. ▶

	Palmer	Hoffman
Part done in 1 day	$\frac{1}{6}$	$\frac{1}{8}$
Part done in x days	$\frac{x}{6}$	$\frac{x}{8}$

Job takes x days. ▶

After x days ▶ (Palmer's part) + (Hoffman's part) = Total job

Write an equation. ▶
$$\frac{x}{6} + \frac{x}{8} = 1$$

The LCD is 24. ▶
$$24 \cdot \frac{x}{6} + 24 \cdot \frac{x}{8} = 24 \cdot 1$$

Solve the equation. ▶
$$4x + 3x = 24$$
$$7x = 24$$
$$x = \frac{24}{7}, \text{ or } 3\frac{3}{7}$$

Check in the problem. ▶ To check, add the parts done in $\frac{24}{7}$ days.

$$\frac{x}{6} + \frac{x}{8} = \frac{1}{6} \cdot \frac{24}{7} + \frac{1}{8} \cdot \frac{24}{7} = \frac{4}{7} + \frac{3}{7} = \frac{7}{7}, \text{ or } 1$$

Answer the question. ▶ Thus, the job will take $3\frac{3}{7}$ days if the two crews work together.

If a job requires x hours to complete, then $\frac{1}{x}$ represents the part of the job that is done in 1 hour and $\frac{n}{x}$ represents the part of the job done in n hours.

Problem Solving: Work Problems

Example 2   Al, Betty, and Carl can harvest a strawberry crop in 12 hours (h) if they work together. If each person worked alone, Al could complete the job in 30 h, and Carl would take twice as long as Betty. How long would it take Carl to do the job alone?

Let x = Betty's time to do the job alone.
        $2x$ = Carl's time to do the job alone

Job takes 12 h if they work together. ▶

	Al	Betty	Carl
Part done in 1 h	$\frac{1}{30}$	$\frac{1}{x}$	$\frac{1}{2x}$
Part done in 12 h	$\frac{12}{30}$	$\frac{12}{x}$	$\frac{12}{2x}$

After 12 h ▶   Al's part + Betty's part + Carl's part = Total job

$$\frac{12}{30} + \frac{12}{x} + \frac{12}{2x} = 1$$

The LCD is 30x. ▶ $30x \cdot \frac{12}{30} + 30x \cdot \frac{12}{x} + 30x \cdot \frac{12}{2x} = 30x \cdot 1$

$$12x + 360 + 180 = 30x$$
$$540 = 18x$$
$$30 = x \qquad 2x = 60$$

To check, add the parts done in 12 h.
$$\frac{12}{30} + \frac{12}{x} + \frac{12}{2x} = \frac{12}{30} + \frac{12}{30} + \frac{12}{60} = \frac{2}{5} + \frac{2}{5} + \frac{1}{5} = \frac{5}{5}, \text{ or } 1$$

Thus, it would take Carl 60 h to do the job alone.

Example 3   Machine A can do a job in 12 h and machine B can do the job in 8 h. If B starts 2 h after A has started, find the total time needed for the two machines to do the complete job.

Let x = the time that A and B work together.
(Part done by A in 2 hours) + (Part done by A and B together in x hours) = Total job

$$\frac{2}{12} + \left(\frac{x}{12} + \frac{x}{8}\right) = 1$$
$$4 + 2x + 3x = 24$$
$$x = 4 \qquad 2 \text{ h} + 4 \text{ h} = 6 \text{ h}$$

Thus, the two machines need 6 h to do the complete job.

Oral Exercises

A farmer takes 15 days to harvest a crop. Represent the part harvested in the given period of time.

1. 1 day $\frac{1}{15}$
2. 4 days $\frac{4}{15}$
3. x days $\frac{x}{15}$
4. 15 days 1

A carpenter can build a cabinet in x hours. Represent the part built in the given period of time.

5. 1 hour $\frac{1}{x}$
6. 3 hours $\frac{3}{x}$
7. y hours $\frac{y}{x}$
8. x hours 1

Written Exercises

Solve each problem.

(A)

1. Mr. Adams can plant a wheat crop in 10 days and his daughter can do it in 15 days. How many days will it take if they work together? **6 days**

2. Mrs. Brown can paint 3 average-sized rooms in 8 hr. Her son would need 12 h to do it. If they work together, how long will it take to paint 6 average-sized rooms? **$9\frac{3}{5}$ h**

3. Paul can put carpet on a floor in 10 h. If Irene helps him, the job is done in 6 h. How long would it take Irene if she worked alone? **15 h**

4. Diane can clean the attic in 6 h. If Sam helps her, the attic can be cleaned in 4 h. How long would it take Sam working alone? **12 h**

5. Machine A can do a job in 15 h. Machines B and C can do the same job in 12 h and 20 h, respectively. How many hours will the job take if the three machines operate at the same time? **5 h**

6. Work crews 1, 2, 3, and 4 can load a freight train in 16 h, 10 h, 12 h, and 15 h, respectively. When will the freight train be loaded if the four crews start at 8:00 A.M. and work together? **11:12 A.M.**

7. Kim can complete a job in 6 weeks (wk) and the same job would take Kevin 10 wk. How long would it take Derek working alone if, working together, all three can complete the job in 2 wk? **$4\frac{2}{7}$ wk**

8. Linda can keypunch 1,000 cards in 150 minutes (min). The same job would take John 3 h. How long would it take Martha to keypunch 1,000 cards if, working together, all three can keypunch 2,000 cards in 2 h? **$3\frac{3}{4}$ h**

(B)

9. Work crew A takes 15 h to do a job and crew B can do it in 10 h. If B starts 3 h after A has begun, what is the total time needed for the two crews to do the job? **$7\frac{4}{5}$ h**

10. Ned can mow a lawn in 75 min and Pedro can do it in 50 min. If Pedro watches Ned mow for 20 min and then helps to finish the job, find the total time for the job. **42 min**

11. Gene, Marie, and Merv can complete a job in 20 h if they work together. Gene can do the job alone in twice the time it would take Marie and half the time that Merv would need. Find the time each person would take to do the job if each worked alone. **Gene: 70 h, Marie: 35 h, Merv: 140 h**

12. Work crews A, B, and C can pave a road in 10 days if they work together. If they work alone, crew A would take 60 days to do the job and crew B would take 3 times as long as crew C to do the job. How long would it take crews B and C if each worked alone? **crew B: 48 days, crew C: 16 days**

Problem Solving: Work Problems

PROBLEM SOLVING: ROUND TRIPS   5.3

Objective **To solve a motion problem that leads to a fractional equation**

An automobile that is driven a distance of 240 km at an average rate of 80 km/h will take $\frac{240 \text{ km}}{80 \text{ km/h}}$, or 3 h, for the trip. You can use the formula below to solve motion problems involving *time*.

$$\text{time}(t) = \frac{\text{distance}(d)}{\text{rate}(r)}$$

Example 1 The Pai family drove from their home into the country at a speed of 80 km/h. They returned over the same road at 70 km/h. If the round trip took 6 hours, how far did the Pais drive into the country?

Let x = the distance of the trip to the country.
Represent the data with a vector diagram.

$$\left(\begin{array}{c}\text{Time for the}\\\text{trip out}\end{array}\right) + \left(\begin{array}{c}\text{Time for the}\\\text{trip back}\end{array}\right) = \left(\begin{array}{c}\text{Time for the}\\\text{round trip}\end{array}\right)$$

Use $t = \frac{d}{r}$. ▶ $\frac{x}{80} + \frac{x}{70} = 6$

The LCD is 560. ▶ $560 \cdot \frac{x}{80} + 560 \cdot \frac{x}{70} = 560 \cdot 6$

$$7x + 8x = 3{,}360$$
$$15x = 3{,}360$$
$$x = 224$$

Check: (Time out) + (Time back) | (Time for round trip)
$\frac{224}{80} + \frac{224}{70}$ | 6
2.8 + 3.2
6

Thus, the Pais drove 224 km into the country.

A certain falcon can fly at a speed of 40 km/h in still air. If a wind is blowing at 12 km/h, the falcon can fly at 40 + 12, or 52 km/h, with the wind and at 40 − 12, or 28 km/h, against the wind. A wind in the direction of an object's motion is called a *tailwind* and a wind in the direction against the object's motion is called a *headwind*.

If a boat can travel at 20 km/h in still water and if a river has a current of 4 km/h, then the boat can travel *downstream,* or with the current, at 24 km/h. The boat can travel *upstream,* or against the current, at 20 − 4, or 16 km/h.

Example 2 **A boat can travel 20 km/h in still water. It can travel 47 km downstream in the same time that it can travel 33 km upstream. What is the rate of the current?**

Let c = the rate of the current. Represent the data in a table.

	distance (d)	rate (r)	time (t)	
Downstream	47	$20 + c$	$\dfrac{47}{20 + c}$	◀ $t = \dfrac{d}{r}$
Upstream	33	$20 - c$	$\dfrac{33}{20 - c}$	◀ $t = \dfrac{d}{r}$

Time downstream = Time upstream
$$\frac{47}{20 + c} = \frac{33}{20 - c}$$
$$47(20 - c) = 33(20 + c)$$
$$940 - 47c = 660 + 33c$$
$$280 = 80c$$
$$3.5 = c$$

Check:

Time downstream	Time upstream
$\dfrac{47}{20 + c}$	$\dfrac{33}{20 - c}$
$\dfrac{47}{20 + 3.5}$	$\dfrac{33}{20 - 3.5}$
$\dfrac{47}{23.5}$	$\dfrac{33}{16.5}$
2	2

Thus, the rate of the current is 3.5 km/h.

Reading in Algebra

A car traveled from P to Q at 50 mi/h for 3 h and then returned from Q to P at 30 mi/h for 5 h. Choose the one correct ending for each statement.

1. The distance of the round trip was
   A 80 mi.
   B 150 mi. **C**
   C 300 mi.

2. The time for the round trip was
   A 8 h.
   B 5 h. **A**
   C 3 h.

3. The average speed for the round trip was
   A 40 mi/h.
   B 37.5 mi/h. **B**
   C 80 mi/h.

4. For the trip from P to Q and for the trip from Q to P, the two
   A times were equal.   B distances were equal.   C rates were equal. **B**
5. At the end of the first 2 h, the car was
   A 50 mi from Q.   B 50 mi from P.   C 60 mi from P. **A**
6. The car was halfway between P and Q at the end of the first
   A $\frac{1}{2}$ h.   B $3\frac{1}{2}$ h.   C $5\frac{1}{2}$ h. **C**

Problem Solving: Round Trips

Written Exercises

Solve each problem.

(A) 1. Alice drove to the ocean at 90 km/h. She returned home on the same road at 60 km/h. If the round trip took 2 h, how far does Alice live from the ocean? **72 km**

2. You left home at 9:00 A.M. and drove to the seaquarium at 75 km/h. After 2 h of sightseeing, you returned home at 50 km/h. If you arrived home at 2:00 P.M., how far do you live from the seaquarium? **90 km**

3. A scout in a canoe can travel 12 km/h in still water. If the scout traveled 30 km downstream in the same time that it took to travel 15 km upstream, what was the rate of the current? **4 km/h**

4. A plane flew 780 km with a tailwind in the same time it would have flown 720 km in the opposite direction. Find the rate of the tailwind if the plane flies at 500 km/h in still air. **20 km/h**

5. You jog along a straight trail at 15 km/h. Then you rest for 1 hour and jog back home on that trail at 18 km/h. If you were gone from home for 3 h, how far away from home did you go? $16\frac{4}{11}$ **km**

6. A certain salmon can swim 3 km/h in still water. In a river, the salmon traveled 8 km upstream in the same time it took to swim 24 km downstream. What was the rate of the current? **1.5 km/h**

(B) 7. A camper rowed 24 km downstream and then returned the same distance upstream in twice the time of the downstream trip. If the camper can row 6 km/h in still water, find the rate of the current. **2 km/h**

8. Harvey hiked along a road at 5 km/h. Sara met him with a truck and they returned on the same road at 40 km/h. If the round trip took Harvey 4 h and 30 min, find the distance that he hiked. **20 km**

9. One plane flew 390 km with the help of a tailwind in twice the time that another plane flew 180 km against the headwind. If each plane can fly 500 km/h in still air, find the rate of the tailwind. **20 km/h**

10. With the help of a tailwind, a plane flew 378 km in $\frac{3}{4}$ of the time that a second plane flew 456 km in the opposite direction. Find the rate of the tailwind if each plane can fly 600 km/h in still air. **30 km/h**

11. A marathon race course is in the shape of a square and each side of the square is called a leg of the race. Find the length of the course if a runner ran each leg at 12 km/h, 15 km/h, 10 km/h, and 16 km/h, respectively, in a total time of 3.75 h. **48 km**

12. A gymnast climbed a vertical rope at 3.2 m/s and then descended at 4.8 m/s. Find the height of the rope if the gymnast was timed at 6.25 s for the round trip. **12 m**

CUMULATIVE REVIEW

Solve each equation.

1. $8(3 - 5x) - (6x + 15) - 4(-7x - 8) = 95$
   −3

2. $10^{3x-1} = 1{,}000$
   $\frac{4}{3}$

3. $|5x + 3| = 18$
   $-4\frac{1}{5}, 3$

LITERAL EQUATIONS AND FORMULAS   5.4

Objective **To solve a formula or a literal equation for one of its variables**

A formula for the perimeter of a rectangle is $p = 2(l + w)$. Since a formula is an equation, $p = 2(l + w)$ can be solved for the length l in terms of the perimeter p and the width w using equation properties, as shown at the right.

$$p = 2(l + w)$$
$$p = 2l + 2w$$
$$p - 2w = 2l$$
$$\frac{p - 2w}{2} = l$$

A formula like $p = 2(l + w)$ and an equation like $ax + b = cx + d$ are called **literal equations**, since they contain more than one letter, or variable. To solve a literal equation for one specific variable, you should rewrite the equation so that all of the terms containing that variable are alone on one side of the equation.

Example 1 **Solve $ax + b = cx - d$ for x.**

$$ax + b = cx - d$$
$$ax - cx = -d - b \quad \blacktriangleleft \text{ All } x\text{-terms are alone on the left side.}$$
$$(a - c)x = -d - b \quad \blacktriangleleft \text{ Factor out } x.$$
$$x = \frac{-d - b}{a - c} \quad \blacktriangleleft \text{ Divide by } a - c.$$

There are several expressions that are equivalent to $\frac{-d - b}{a - c}$ in Example 1.
If $x = \frac{-d - b}{a - c}$, then $x = -\frac{d + b}{a - c} = -\frac{-d - b}{c - a} = \frac{d + b}{c - a}$.

To solve the equation $3(5 - 2y) = 7$, you would simplify the equation by applying the Distributive Property. As shown in Example 2, a similar procedure is followed in solving the literal equation $a(b - cy) = d$ for y.

Example 2 **Solve $a(b - cy) = d$ for y.**

$$a(b - cy) = d$$
$$ab - acy = d \quad \blacktriangleleft \text{ Use the Distributive Property.}$$
$$-acy = d - ab \quad \blacktriangleleft \text{ The } y\text{-term is alone on the left side.}$$
$$y = \frac{d - ab}{-ac}, \text{ or } \frac{ab - d}{ac}$$

The formula $\frac{1}{R} = \frac{1}{x} + \frac{1}{y}$ is a *literal fractional equation*. Recall that a fractional equation can be simplified by multiplying each side by the LCD.

Literal Equations and Formulas   **125**

Example 3   Solve $\frac{1}{R} = \frac{1}{x} + \frac{1}{y}$ for x. Then find the value of x if $R = 3.2$ and $y = 4.8$.

Multiply each side by Rxy, the LCD.

$$Rxy \cdot \frac{1}{R} = Rxy \cdot \frac{1}{x} + Rxy \cdot \frac{1}{y}$$
$$xy = Ry + Rx$$
$$xy - Rx = Ry$$
$$x(y - R) = Ry$$
$$x = \frac{Ry}{y - R}$$

$R = 3.2 \quad y = 4.8$
$$x = \frac{Ry}{y - R}$$
$$= \frac{3.2(4.8)}{4.8 - 3.2}$$
$$= \frac{15.36}{1.6}$$
$$x = 9.6$$

To solve a literal equation like $\frac{y - 3}{x + 4} = -5$ for either x or y, you can first rewrite it as a proportion. This technique is used in Example 4.

Example 4   Solve $\frac{y - 3}{x + 4} = -5$ for y.

$$\frac{y - 3}{x + 4} = \frac{-5}{1}$$ ◀ Rewrite as a proportion.
$$(y - 3)1 = (x + 4)(-5)$$ ◀ If $\frac{a}{b} = \frac{c}{d}$, then $ad = bc$.
$$y - 3 = -5x - 20$$
$$y = -5x - 17$$

Oral Exercises

Solve each equation for x.
1. $a + x + b = c$ **1.** $x = c - a - b$ **2.** $x = -a + b$ **3.** $x = \frac{bc}{a}$ **6.** $x = \frac{b+c}{a}$
2. $-x = a - b$ **3.** $ax = bc$ **4.** $ax = ab$ $x = b$
5. $\frac{x}{a} = b$ $x = ab$ **6.** $ax = b + c$ **7.** $\frac{x}{a} = \frac{b}{c}$ $x = \frac{ab}{c}$ **8.** $1 = \frac{a}{x}$ $x = a$

Written Exercises

1. $x = \frac{b+c}{a}$ **2.** $x = \frac{a-c}{b}$ **3.** $x = \frac{b}{a-c}$ **4.** $x = \frac{b+4}{a-5}$

Solve each equation for x.

5. $x = \frac{b}{4a}$ **6.** $x = \frac{4b+5d}{3a-2c}$ **8.** $x = \frac{-2ab-cd}{a-c}$ **9.** $x = \frac{ab-c}{2a-cd}$

Ⓐ **1.** $ax - b = c$ **2.** $a = bx + c$ **3.** $ax = b + cx$
4. $5x + 4 = ax - b$ **5.** $7ax + b = 3b - ax$ **6.** $3ax - 4b = 2cx + 5d$
7. $a(x - b) = c$ $x = \frac{ab+c}{a}$ **8.** $a(x + 2b) = c(x - d)$ **9.** $a(2x - b) = c(dx - 1)$
10. $\frac{a}{x} = b$ $x = \frac{a}{b}$ **11.** $\frac{a}{b} \cdot x = c + d$ $x = \frac{bc+bd}{a}$ **12.** $\frac{ax}{b} = \frac{c}{d}$ $x = \frac{bc}{ad}$
13. $\frac{a}{b} = \frac{x}{c} + d$ $x = \frac{ac-bcd}{b}$ **14.** $\frac{2}{x} = \frac{3}{a} + \frac{4}{b}$ $x = \frac{2ab}{3b+4a}$ **15.** $\frac{8}{a} = \frac{6}{x} - \frac{4}{b}$ $x = \frac{3ab}{4b+2a}$

Chapter Five

Solve each equation for y.

16. $\dfrac{y-10}{x-8} = \dfrac{1}{2}$ $y = \dfrac{x+12}{2}$

17. $\dfrac{y-2}{x+5} = \dfrac{4}{5}$ $y = \dfrac{4x+30}{5}$

18. $\dfrac{y+3}{x+6} = \dfrac{-2}{3}$ $y = \dfrac{-2x-21}{3}$

19. $\dfrac{y-6}{x+1} = 4$ $y = 4x + 10$

20. $\dfrac{y+8}{x-4} = -3$ $y = -3x + 4$

21. $\dfrac{y-a}{x-b} = m$ $y = mx - bm + a$

Solve each formula for the specified variable. Then find the value of that variable for the given data.

22. Solve $p = 2(l + w)$ for w. $w = \dfrac{p - 2l}{2}$; 6.9
    $p = 23.4$; $l = 4.8$

23. Solve $A = B + BCD$ for B. $B = \dfrac{A}{1 + CD}$; 0.5
    $A = 48.35$; $C = 5.8$; $D = 16.5$

24. Solve $A = \dfrac{1}{2} bh$ for b. $b = \dfrac{2A}{h}$; 17
    $A = 30.6$; $h = 3.6$

25. Solve $\dfrac{1}{x} = \dfrac{1}{a} + \dfrac{1}{b}$ for x. $x = \dfrac{ab}{a+b}$; 4.2
    $a = 8.12$; $b = 8.7$

Solve each equation for x.

Ⓑ 26. $\dfrac{2x}{g} = t^2$ $x = \dfrac{gt^2}{2}$

27. $gt^2 = 2(200t - x)$ $x = \dfrac{400t - gt^2}{2}$

28. $ax + b(x - c) = d(x + e)$ $x = \dfrac{bc + de}{a + b - d}$

29. $3a(2x - b) - a(b - x) = 3ab + 7c(a + b)$

30. $\dfrac{5a}{x+b} = \dfrac{a}{x-b}$ $x = \dfrac{3b}{2}$

31. $\dfrac{x-a}{x-b} = \dfrac{c}{d}$ $x = \dfrac{ad - bc}{d - c}$ $x = \dfrac{ab + ac + bc}{a}$

32. $\dfrac{x-a}{4} + \dfrac{x+b}{12} = \dfrac{c}{3}$ $x = \dfrac{3a - b + 4c}{4}$

33. $\dfrac{x-4}{a} + \dfrac{3}{b} = \dfrac{2x-3}{c}$ $x = \dfrac{4bc - 3ac - 3ab}{bc - 2ab}$

34. $\dfrac{ax-b}{cx+d} = 5$ $x = \dfrac{b + 5d}{a - 5c}$

35. $\dfrac{ax+b}{c} = dx + e$ $x = \dfrac{ce - b}{a - cd}$

Solve each formula for the specified variable. Then find the value of that variable for the given data.

36. Solve $T = mg - mf$ for m. $m = \dfrac{T}{g-f}$; 10
    $T = 912.7$; $g = 93.7$; $f = 2.43$

37. Solve $A = ah + bh + ch$ for h. $h = \dfrac{A}{a+b+c}$; 24
    $A = 252$; $a = 2.8$; $b = 3.5$; $c = 4.2$

38. Solve $\dfrac{1}{R} = \dfrac{1}{a} + \dfrac{1}{b}$ for b. $b = \dfrac{Ra}{a-R}$; 6.4
    $R = 4.8$; $a = 19.2$

39. Solve $V = \dfrac{1}{3} Bh$ for h. $h = \dfrac{3V}{B}$; 1.5×10^3, or 1,500
    $V = 2.4 \times 10^5$; $B = 4.8 \times 10^2$

40. Solve $gt^2 - h = 3(400t - h)$ for h.
    $g = 32$; $t = 9$ $h = \dfrac{1200t - gt^2}{2}$; 4,104

41. Solve $\dfrac{h - 80}{16t} = 30 - t$ for h.
    $t = 4$ $h = -16t^2 + 480t + 80$; 1,744

Solve each equation for x.

Ⓒ 42. $x^4 - 5a^2x^2 + 4a^4 = 0$ $a, -a, 2a, -2a$

43. $16x^4 - 40a^2x^2 + 9a^4 = 0$ $\dfrac{3a}{2}, -\dfrac{3a}{2}, \dfrac{a}{2}, -\dfrac{a}{2}$

44. $x^3 - ax^2 - b^2x + ab^2 = 0$ $a, b, -b$

45. $4x^4 - 36a^2x^2 - b^2x^2 + 9a^2b^2 = 0$ $3a, -3a, \dfrac{b}{2}, -\dfrac{b}{2}$

46. $|ax - b| = c$, if $c \geq 0$ $\dfrac{b+c}{a}, \dfrac{b-c}{a}$

47. $\dfrac{x-a}{x-b} = \dfrac{x-a}{x-c}$ a

CALCULATOR ACTIVITIES

Use the formula $y = \dfrac{m(ab - c) + d}{mc}$ to find the value of y for the given data.

1. $m = 28$; $a = 360$; $b = 212$; $c = 415$; $d = 198,660$ **200**
2. $m = 2.8$; $a = 7.2$; $b = 10.6$; $c = 8.3$; $d = 18.704$ **9**

Literal Equations and Formulas

DIVIDING POLYNOMIALS 5.5

Objectives **To divide a polynomial of two or more terms with one variable by a binomial**
To determine if a binomial is a factor of a polynomial

Two forms for dividing 679 by 32, using the divide-multiply-subtract cycle, are shown below.

ORDINARY FORM

$$\begin{array}{r} 21 \\ 32\overline{)679} \\ \underline{64} \\ 39 \\ \underline{32} \\ 7 \end{array}$$

Divide: $600 \div 30$
Multiply: $20(30 + 2)$
Subtract: $(600 + 70) - (600 + 40)$
Divide: $30 \div 30$
Multiply: $1(30 + 2)$
Subtract: $(30 + 9) - (30 + 2)$

EXPANDED FORM

$$\begin{array}{r} 20 + 1 \\ 30 + 2\overline{)600 + 70 + 9} \\ \underline{600 + 40} \\ 30 + 9 \\ \underline{30 + 2} \\ 7 \end{array}$$

Thus, $679 \div 32 = 21$, with remainder 7. Check: $679 = 32 \times 21 + 7$.

The check shows that:
(1) Dividend = Divisor × Quotient + Remainder.
(2) 32 is *not* a factor of 679 because there is a nonzero remainder.

The *divide-multiply-subtract* cycle shown above in the expanded form can be used to divide a polynomial, with one variable, by a binomial.

Example 1 **Divide: $(4x^2 - 5x - 30) \div (x - 3)$. Is $x - 3$ a factor of $4x^2 - 5x - 30$?**

Step 1

Divide: $4x^2 \div x$.
Multiply: $4x(x - 3)$.
Subtract: $(4x^2 - 5x)$
$-(4x^2 - 12x)$.

$$\begin{array}{r} 4x \\ x - 3\overline{)4x^2 - 5x - 30} \\ \underline{4x^2 - 12x} \\ 7x \end{array}$$

Step 2

Divide: $7x \div x$.
Multiply: $7(x - 3)$.
Subtract: $(7x - 30)$
$-(7x - 21)$.

$$\begin{array}{r} 4x + 7 \\ x - 3\overline{)4x^2 - 5x - 30} \\ \underline{4x^2 - 12x} \\ 7x - 30 \\ \underline{7x - 21} \\ -9 \end{array}$$

Thus, the polynomial quotient is $4x + 7$ and the remainder is -9.

Check: $4x^2 - 5x - 30 = (x - 3)(4x + 7) + (-9)$
So, $x - 3$ is not a factor of $4x^2 - 5x - 30$, since there is a nonzero remainder.

Factor Theorem

A binomial B is a factor of a polynomial P if
$$P \div B = Q, \text{ or } P = B \cdot Q + R,$$
where Q is a polynomial and the remainder R is 0.

Example 2

Divide: $(30 - 2x^2 - 2x + 6x^3) \div (3x + 5)$. Is $3x + 5$ a factor of the dividend?

$$
\begin{array}{r}
2x^2 - 4x + 6 \\
3x + 5 \overline{\smash{\big)}\, 6x^3 - 2x^2 - 2x + 30} \\
\underline{6x^3 + 10x^2} \\
-12x^2 - 2x \\
\underline{-12x^2 - 20x} \\
18x + 30 \\
\underline{18x + 30} \\
0
\end{array}
$$

◀ Arrange the terms of the dividend in descending order of the exponents.
Check: $30 - 2x^2 - 2x + 6x^3$
$= (3x + 5)(2x^2 - 4x + 6) + 0$
Thus, the quotient is $2x^2 - 4x + 6$, the remainder is 0, and $3x + 5$ is a factor of the dividend.

If the dividend has missing powers of the variable, you can use 0 as the coefficient to replace the missing terms.

Example 3

Divide: $(10x^3 - 34x + 4) \div (2x - 4)$. Is the divisor a factor of the dividend?

$$
\begin{array}{r}
5x^2 + 10x + 3 \\
2x - 4 \overline{\smash{\big)}\, 10x^3 + 0x^2 - 34x + 4} \\
\underline{10x^3 - 20x^2} \\
20x^2 - 34x \\
\underline{20x^2 - 40x} \\
6x + 4 \\
\underline{6x - 12} \\
16
\end{array}
$$

◀ Replace the missing x^2-term with $0x^2$.
Check: $10x^3 - 34x + 4$
$= (2x - 4)(5x^2 + 10x + 3) + 16$
Thus, the polynomial quotient is $5x^2 + 10x + 3$, the remainder is 16, and the divisor is not a factor of the dividend.

Oral Exercises

1. $2x^4 - 3x^2$ 2. $5n + 6m$ 3. $6b - 4$ 4. $3m^2 - 5m - 4$
5. $-3d^3 + 5c^4d - 8c^2d^2$ 6. $6x^2y + 4x - 7y$

Simplify each quotient. Assume that no divisor is equal to 0.

1. $(14x^5 - 21x^3) \div (7x)$
2. $(5mn^2 + 6m^2n) \div (mn)$
3. $(18a^2b^2 - 12a^2b) \div (3a^2b)$
4. $\dfrac{15m^4 - 25m^3 - 20m^2}{5m^2}$
5. $\dfrac{-6cd^4 + 10c^5d^2 - 16c^3d^3}{2cd}$
6. $\dfrac{24x^3y^3 + 16x^2y^2 - 28xy^3}{4xy^2}$

Written Exercises

For complete solutions to Ex. 7–8, 10, 12, and 17–22, see TE Answer Section.

Divide to find the polynomial quotient and remainder. Is the divisor a factor of the dividend?

Ⓐ
1. $(2c^2 + c - 2) \div (c + 1)$ **2c − 1, −1, no**
2. $(y^2 + 3y + 1) \div (y + 2)$ **y + 1, −1, no**
3. $(10x^2 - 8x - 24) \div (x - 2)$ **10x + 12, 0, yes**
4. $(2n^2 + 19n + 35) \div (2n + 5)$ **n + 7, 0, yes**
5. $(4y^2 + 7y - 4) \div (y + 3)$ **4y − 5, 11, no**
6. $(6a^2 - 13a + 2) \div (3a - 2)$ **2a − 3, −4, no**
7. $(3x^3 - x^2 - 17x + 9) \div (x - 3)$ **no**
8. $(3n^4 + 13n^3 + 4n^2 - 2n - 8) \div (n + 4)$ **yes**
9. $(15 - 14n + 8n^2) \div (4n - 5)$ **2n − 1, 10, no**
10. $(14x^2 - 3x + 3x^3 + 7) \div (x + 5)$ **no**
11. $(a^3 - 13a - 12) \div (a + 3)$ **a² − 3a − 4, 0, yes**
12. $(9y^4 + 5y^2 - 12) \div (3y - 2)$ **no**

Ⓑ
13. $(6x^3 + 4x^2 - 3x - 8) \div (2x^2 - 1)$ **below**
14. $(3c^4 + 2c^3 - 8c - 48) \div (c^2 - 4)$ **yes**
15. $(a^3 + 1) \div (a + 1)$ **a² − a + 1, 0, yes**
16. $(8y^3 - 125) \div (2y - 5)$ **4y² + 10y + 25, 0, yes**
17. $(2n^4 - 4n^3 + 7n^2 - 12n + 3) \div (n^2 + 3)$ **yes**
18. $(4x^4 + 3x^3 + 13x^2 - 15x + 35) \div (x^2 - 5)$ **no**
19. $(y^6 - y^5 - y^4 + y^2 - y - 1) \div (y^4 + 1)$ **yes**
20. $(9a^6 - 3a^5 + a^2 - 4) \div (3a^3 - 1)$ **no**

Ⓒ
21. $(x^{3m} - 4x^{2m} + 3x^m - 10) \div (x^m - 4)$ **no**
22. $(x^{3c} + 3x^{2c} - x^{2c+1} - 3x^{c+1} + 2) \div (x^c + 3)$ **no**
23. $(24x^3y + 7x^2y^2 - 6x^2y - 6xy^3 - 4xy^2) \div (3x + 2y)$ **8x²y − 3xy² − 2xy, 0, yes**

13. **3x + 2, −6, no**

Dividing Polynomials

REMAINDER THEOREM 5.6

Objectives **To divide a polynomial by a binomial, using synthetic division**
To evaluate a polynomial for a given value of its variable, using synthetic substitution
To determine if a given binomial is a factor of a given polynomial, using synthetic substitution.

The division at the right can be done in a shorter way. The shorter method eliminates repetitious rewriting of the variable and some of the coefficients.
The method is called **synthetic division** and is illustrated below.

$$\begin{array}{r} 3x^2 - 4x + 5 \\ x - 2 \overline{\smash{)}3x^3 - 10x^2 + 13x - 26} \\ \underline{3x^3 - 6x^2} \\ -4x^2 + 13x \\ \underline{-4x^2 + 8x} \\ 5x - 26 \\ \underline{5x - 10} \\ -16 \end{array}$$

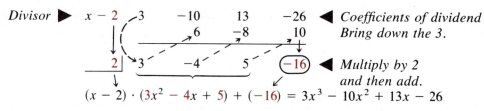

$(x - 2) \cdot (3x^2 - 4x + 5) + (-16) = 3x^3 - 10x^2 + 13x - 26$

Check: Divisor × Quotient + Remainder = Dividend

Notice the following.
1. The remainder is -16.
2. $x - 2$ is *not* a factor of the dividend, since there is a nonzero remainder, -16.
3. If $x = 2$, then $3x^3 - 10x^2 + 13x - 26 = -16$ since
$3x^3 - 10x^2 + 13x - 26 = (x - 2)(3x^2 - 4x + 5) - 16$
$= (2 - 2)(3x^2 - 4x + 5) - 16$
$= 0 - 16$, or -16.

Example 1 Given $P = 4x^4 - 14x^3 - x + 27$, use synthetic division to divide P by $x - 3$. Find the polynomial quotient Q and the remainder R.

$$\begin{array}{c|ccccc} x - 3 & 4 & -14 & 0 & -1 & 27 \\ & & 12 & -6 & -18 & -57 \\ \hline 3 & 4 & -2 & -6 & -19 & \boxed{-30} \end{array}$$

◀ *Replace the missing x^2-term with $0x^2$.*

$P = (x - 3)(4x^3 - 2x^2 - 6x - 19) - 30$

Thus, $Q = 4x^3 - 2x^2 - 6x - 19$ and $R = -30$.

In Example 1, a polynomial P was divided by $x - 3$ to obtain a polynomial quotient Q and a remainder R of -30, so that

$$P = 4x^4 - 14x^3 - x + 27$$
$$= (x - 3)(4x^3 - 2x^2 - 6x - 19) - 30$$
$$= (x - 3) \cdot Q - 30.$$

If $x = 3$, then $P = (3 - 3) \cdot Q - 30 = 0 - 30 = -30$. The value of P is -30, the same as the remainder R, when $x = 3$. This suggests the Remainder Theorem, which is stated below.

Remainder Theorem For the polynomials P, Q, and $x - a$, if $P = (x - a) \cdot Q + R$ for some number R, then R is the value of P when $x = a$.

The synthetic division in Example 1 can be shortened to the method of synthetic substitution studied in Lesson 2.3 (page 37).

Synthetic Division

Synthetic Substitution

The multiples of 3 are not written.

The Remainder Theorem states that the binomial divisor must be in the form $x - a$. A divisor like $x + 5$ can be rewritten as $x - (-5)$ where $a = -5$.

Example 2 Given: $P = 2x^5 + 11x^4 - x^3 - 30x^2 + 6x + 30$. Use synthetic substitution to divide P by $x + 5$. Find (1) the polynomial quotient and the remainder, (2) the value of P if $x = -5$, and (3) whether $x + 5$ is a factor of P.

Rewrite $x + 5$ as $x - (-5)$.

```
          2    11    -1    -30    6    30      Find and add
    -5|         -10   -5    30    0   -30      the multiples
          2     1    -6     0    6    (0)      of -5.
```

$P = (x + 5)(2x^4 + x^3 - 6x^2 + 0x + 6) + 0$

Thus, (1) the quotient is $2x^4 + x^3 - 6x^2 + 6$ and the remainder is 0, (2) the value of P is 0 when $x = -5$, and (3) $x + 5$ is a factor of P.

Example 3 Evaluate $2x^4 + 2x^3 - 13x^2 - x + 6$ using each of -3, -2, 2, and 3 for x. Name the binomial factors found, if any.

```
         2     2    -13    -1     6
   -3|   2    -4     -1     2    (0)
   -2|   2    -2     -9    17   (-28)
    2|   2     6     -1    -3    (0)
    3|   2     8     11    32   (102)
```

Thus, the values of P are 0, -28, 0, and 102 when the values of x are -3, -2, 2, and 3, respectively; $x + 3$ and $x - 2$ are binomial factors of the polynomial.

Remainder Theorem

Reading in Algebra

An example of synthetic substitution is shown at the right. Complete the paragraph below using the information in the example.

$$\begin{array}{r|rrrr} & 2 & 11 & 0 & -10 \\ -5 & & -10 & -5 & 25 \\ \hline & 2 & 1 & -5 & 15 \end{array}$$

The dividend is __$2x^3 + 11x^2 - 10$__, the divisor is __$x + 5$__, 15 is the __remainder__, and the polynomial quotient is __$2x^2 + x - 5$__. The value of $2x^3 + 11x^2 - 10$ is __15__ when the value of x is __-5__. The binomial $x + 5$ is not a __factor__ of $2x^3 + 11x^2 - 10$.

Oral Exercises

Determine whether each statement is true or false.
1. If $P = (x + 2)(x^2 - 3x - 12) + 4$, then $x + 2$ is a factor of P. **F**
2. If $P = (x - 4)(2x^2 + 7x - 3)$, then $x - 4$ is a factor of P. **T**
3. If $P = (x - 3)(x^2 - 7x + 12)$, then $P = 0$ when $x = 3$. **T**
4. If $P = (x + 6)(x^2 + 8x + 16) + 5$, then $P = 5$ when $x = -6$. **T**

Written Exercises

3. $3y^3 - 8y^2 + 10y - 20, 51$ 4. $-4y^2 + 6y - 9, 10$

Find the polynomial quotient and the remainder. Use synthetic division or synthetic substitution.

(A)
1. $(x^3 + 2x^2 - 8x - 3) \div (x - 1)$ $x^2 + 3x - 5, -8$
2. $(x^3 - 13x^2 + 20x + 40) \div (x - 5)$ $x^2 - 8x - 20, -60$
3. $(3y^4 + y^3 - 14y^2 + 10y - 9) \div (y + 3)$ above
4. $(-4y^3 + 2y^2 - 3y + 1) \div (y + 1)$ above
5. $(n^5 - 2n^3 + n - 2) \div (n + 2)$ above $n^4 - 2n^3 + 2n^2 - 4n + 9, -20$
6. $(2n^5 - 6n^4 - 12n^2 - 15) \div (n - 4)$ $2n^4 + 2n^3 + 8n^2 + 20n + 80, 305$

Evaluate each polynomial for the given data. Use synthetic substitution.
7. $6x^5 - 5x^4 - 4x^3 - 3x^2 - 2x + 1$ if $x = 2$ **65**
8. $2x^5 + 9x^4 + 8x^3 - 7x^2 - 6x + 8$ if $x = -3$ **−10**
9. $3y^3 + 22y^2 + 10y + 20$ if $y = -10$ **−880**
10. $5y^3 - 80y^2 + 80y - 5$ if $y = 15$ **70**
11. $n^4 - n^3 + n^2 + 1$ if $n = 4$ **209**
12. $n^4 + n^3 - n^2 + n - 1$ if $n = -5$ **469**
13. $6x^4 + 4x^2 - 2$ if $x = -4$ **1,598**
14. $4x^5 - 6x^3 + 7x$ if $x = 2$ **94**

Determine whether the binomial is a factor of the polynomial.
15. $x - 3;\ 2x^3 - 11x^2 + 18x - 12$ **no**
16. $x + 6;\ 3x^3 + 17x^2 - 8x - 12$ **yes**
17. $y + 2;\ y^4 + 2y^3 - 3y^2 + 2y + 16$ **yes**
18. $y - 5;\ 3y^4 - 12y^3 - 20y^2 + 30y + 5$ **no**
19. $n - 1;\ 4n^5 - 2n^3 + 6n^2 - 9n + 1$ **yes**
20. $n + 3;\ 5n^5 + 10n^4 - 10n^3 - 34n + 33$ **yes**

(B)
21. $y + 8;\ 3y^4 + 21y^3 - 31y^2 - 51y + 40$ **yes**
22. $y - 9;\ 12y^3 - 8y^2 + 100y - 8,100$ **no**
23. $n - 10;\ n^3 - 9,000$ **no**
24. $n + 20;\ n^3 + 8,000$ **yes**
25. $x + 2;\ x^6 + x^4 + x^2 - 84$ **yes**
26. $x - 3;\ x^5 - x^3 - 9$ **no**
27. $y - 2;\ y^6 + 64$ **no**
28. $y + 1;\ y^7 + 1$ **yes**

Evaluate each polynomial for the given data. Use synthetic substitution.
29. $n^6 - 2n^4 - 3n^2 - 4$ if $n = -3$ **536**
30. $4n^7 + 3n^5 + 2n^3 + n$ if $n = 2$ **626**
31. $3x^5 + 7x^4 + 9x^3 + 6x^2 + 4x + 2$ if $x = 10$ **379,642**
32. $7x^6 - 8x^4 + 9x^2$ if $x = -2$ **356**

Evaluate each polynomial using each of $-3, -2, -1, 0, 1, 2,$ and 3 for x.
Use synthetic substitution. Name the binomial factors found, if any.

33. $x^4 - 2x^3 - 6x^2 + 6x + 9$
34. $x^4 - 3x^2 - 4$
35. $2x^4 + 2x^3 - 7x^2 - 3x + 6$
36. $4x^4 - 4x^2 - 3$

For complete solutions to Ex. 33–36, see TE Answer Section.

CALCULATOR ACTIVITIES

You can evaluate a polynomial with one variable for a *positive* number by using the *multiply-and-add* cycle of synthetic substitution. Evaluate $5x^4 - 17x^3 - 23x^2 + 49x + 90$ if $x = 7$.

$5 \otimes 7 \ominus 17 \otimes 7 \ominus 23 \otimes 7 \oplus 49 \otimes 7 \oplus 90 \ominus 5{,}480$,
the value of the polynomial when $x = 7$.

> The sequence shown above is for a simple (nonscientific) calculator. If you use a scientific calculator, press $\ominus$ after each addition or subtraction of a number as shown below.

$5 \otimes 7 \ominus 17 \ominus \otimes 7 \ominus 23 \ominus \otimes 7 \oplus 49 \ominus \otimes 7 \oplus 90 \ominus 5{,}480$

Evaluate each polynomial for $x = 12$.
1. $18x^3 - 47x^2 + 62x - 75$ **25,005**
2. $24x^4 + 25x^3 - 26x^2 - 27x - 28$ **536,768**
3. $8x^5 - 7x^4 - 6x^3 + 5x^2 + 4x - 3$ **1,835,901**

You can evaluate a polynomial for a *negative* number as follows:

Evaluate $5x^4 - 17x^3 - 23x^2 + 49x + 90$ if $x = -6$.

Given: $x = -6$. (1) Let $x = -y$, (2) simplify, and (3) evaluate for $y = 6$.
(1) $5(-y)^4 - 17(-y)^3 - 23(-y)^2 + 49(-y) + 90$ if $-y = -6$.
(2) $5y^4 \quad + 17y^3 \quad - 23y^2 \quad - 49y \quad + 90$ if $y = 6$.
(3) $5 \otimes 6 \oplus 17 \otimes 6 \ominus 23 \otimes 6 \ominus 49 \otimes 6 \oplus 90 \ominus 9{,}120$,
the value of the polynomial when $x = -6$.

Evaluate each polynomial for $x = -15$.
4. $3x^4 - 8x^3 - 7x^2 + 6x + 9$ **177,219**
5. $5x^3 - 4x^2 + 9x + 800$ **−17,110**
6. $-2x^5 - 2x^4 + 3x^3 + 3x^2 - 4x + 4$ **1,408,114**

There are no missing terms in the exercises above. If one or more terms are missing, you can evaluate as follows.

Evaluate $4x^5 - 6x^3 + 7x$ if $x = 2$.
$4x^5 + 0x^4 - 6x^3 + 0x^2 + 7x \quad + 0$

$4 \otimes 2 \underbrace{\otimes 2}_{0x^4} \ominus 6 \otimes 2 \underbrace{\otimes 2}_{0x^2} \oplus 7 \otimes 2 \oplus 0 \ominus 94$

Evaluate each polynomial for $x = 5$ and again for $x = -3$.
7. $7x^6 - 2x^4 + 6x^3 - 8x + 12$ **108,847; 4,815**
8. $8x^6 + 4x^5 - 2x^2 - 10x$ **137,400; 4,872**

Remainder Theorem

CHAPTER FIVE REVIEW

Ⓐ Exercises are 1–3, 5, 6, 8, 10–13, 15, 18, 20, 23–25, 27. Ⓒ Exercises are starred. The rest are Ⓑ Exercises.

Vocabulary
extraneous solution [5.1]
fractional equation [5.1]
literal equation [5.4]
proportion [5.1]
    extremes of [5.1]
    means of [5.1]
synthetic division [5.6]

Solve each equation. [5.1]

1. $\dfrac{n-4}{3} - \dfrac{n-2}{4} = \dfrac{n}{2}$ **−2**

2. $\dfrac{1}{6}y^2 = \dfrac{3}{2}y - 3$ **3, 6**

3. $\dfrac{8}{5n-2} = \dfrac{4}{2n+7}$ **16**

4. $\dfrac{a-1}{3a-7} = \dfrac{a-2}{a+1}$ **$\frac{3}{2}$, 5**

5. $\dfrac{3}{x+2} + \dfrac{2}{x} = 3$ **$-\frac{4}{3}$, 1**

6. $\dfrac{2y-1}{y-3} = \dfrac{y+2}{y-3} - \dfrac{y}{4}$ **−4**

7. $\dfrac{7}{n-2} = \dfrac{n-3}{n^2-7n+10} + \dfrac{1}{n-5}$ **6**

8. $\dfrac{a}{a+5} - \dfrac{5}{a^2-25} = \dfrac{a-6}{2a-10}$ **4**

★9. $\dfrac{3}{x^2+2x-24} + \dfrac{4}{x^2+8x+12} = \dfrac{5}{x^2-2x-8}$ **20**

Solve each problem.

10. Mr. Brown can plant his crops in 30 days. His son would take 40 days and his daughter 60 days. How many days will it take if they work together? [5.2] **$13\frac{1}{3}$ days**

11. Lori can put carpet on a floor in 9 h. If Brent helps her, the job is done in 6 h. How many hours would it take Brent if he works alone? [5.2] **18 h**

12. A train traveled from city A to city B at 40 km/h and then returned at 60 km/h. If the round trip took 8 h, what is the rail distance between cities A and B? [5.3] **192 km**

13. A family can row their boat 5 km/h in still water. They traveled 30 km downstream in the same time that they traveled 20 km upstream. What was the rate of the current? [5.3] **1 km/h**

14. One plane flew 498 km with a tailwind in $\frac{3}{4}$ of the time that another plane flew 616 km against the headwind. If each plane can fly 400 km/h in still air, find the rate of the tailwind. [5.3] **15 km/h**

Solve each equation for x. [5.4]

15. $a(x-b) = c(2x+3)$ $x = \dfrac{3c+ab}{a-2c}$

16. $\dfrac{7a}{x+b} = \dfrac{3c}{x-b}$ $x = \dfrac{3bc+7ab}{7a-3c}$

★17. $x^4 - a^2x^2 - 9b^2x^2 + 9a^2b^2 = 0$ **$a, -a, 3b, -3b$**

Solve each formula for the specified variable. Then find the value of that variable for the given data. [5.4]

18. Solve $\dfrac{1}{a} = \dfrac{2}{b} + \dfrac{3}{c}$ for b. $b = \dfrac{2ac}{c-3a}$; **22.08**
$a = 2.4$; $c = 9.2$

19. Solve $A = \dfrac{1}{2}h(a+b)$ for h. $h = \dfrac{2A}{a+b}$; **15**
$A = 300$; $a = 17$; $b = 23$

        20. $3x^2 + 5x + 2$, R: 3

Divide. Assume that no divisor is equal to 0. [5.5]

20. $(9x^3 - 19x - 7) \div (3x - 5)$ **above**

21. $(4y - 3y^2 + 2y^3 - 6) \div (y^2 + 2)$ **$2y - 3$**

★22. $(x^{2n+2} + 2x^{2n+1} - 5x^{n+2} - 10x^{n+1}) \div (x^n - 5)$ $x^{n+2} + 2x^{n+1}$

23. Divide: $(n^3 + 5n^2 - 4n - 6) \div (n - 2)$. Use synthetic division. [5.6] **below**

Determine whether the binomial is a factor of the polynomial in each case. [5.6]

24. $x + 5$; $3x^3 + 17x^2 + 8x - 10$ **yes**
25. $y - 3$; $4y^3 - 10y^2 - 12$ **no**
26. $c + 5$; $4c^4 - 80c^2 + 500$ **no**

        23. $n^2 + 7n + 10$, R: 14

Evaluate each polynomial for the given data. Use synthetic substitution. Name the binomial factors found, if any. [5.6]

27. $5n^3 - 7n^2 - 10n - 22$ if $n = 3$ **20**
28. $x^6 - 24x^4 - 75x^2 + 250$ if $x = -5$ **−1,000**
29. $2x^4 - 7x^3 - 23x^2 + 28x + 60$ using each of $-2, -1, 2$ and 5 for x.

        0, 18, 0, 0, $x + 2$, $x - 2$, $x - 5$

Ⓐ Exercises are 1–4, 6, 7, 9, 12, 14, 15, 17, 18. Ⓒ Exercises are starred. The rest are Ⓑ Exercises.

CHAPTER FIVE TEST

Solve each equation.

1. $\dfrac{y+8}{4} - \dfrac{2y-3}{6} = \dfrac{y}{3}$ **6**

2. $\dfrac{1}{4}n^2 = \dfrac{5}{2}n - 6$ **4, 6**

3. $\dfrac{9}{2x-12} = \dfrac{3}{x-2}$ **–6**

4. $\dfrac{2y+1}{y-1} + \dfrac{y}{5} = \dfrac{y+2}{y-1}$ **–5**

5. $\dfrac{n-6}{n^2-2n-8} + \dfrac{3}{n-4} = \dfrac{2}{n+2}$ **–4**

Solve each problem.

6. Sandra can decorate a room for a party in 8 h and Edward can do it in 12 h. If Lois helps, all three can do the job in 3 h, working together. How long would it take Lois if she works alone?   **8 h**

7. Samuel jogged along a straight trail at 18 km/h and returned on the same trail at 9 km/h. If the round trip took 2 h, how far did he jog in one direction?   **12 km**

8. With the help of a tailwind, a plane flew 1,230 km in twice the time that a second plane flew 585 km against the headwind. Find the rate of the wind if both planes can fly 800 km/h in still air.   **20 km/h**

Solve each equation for x.

9. $5ax + c = 4c - ax$ $x = \dfrac{c}{2a}$

10. $\dfrac{x+a}{d} = \dfrac{bx+c}{4}$ $x = \dfrac{cd - 4a}{4 - bd}$

11. Solve $A = \dfrac{h}{2}(b+c)$ for c. Then find c if $A = 26$, $h = 4$, and $b = 5$. $c = \dfrac{2A - bh}{h}$; **8**

Divide. Assume that no divisor is equal to 0.

12. $(6x^3 + 4 - 2x - 17x^2) \div (3x + 2)$
    $2x^2 - 7x + 4$, R: –4

13. $(27n^3 - 8) \div (3n - 2)$ **$9n^2 + 6n + 4$**

14. Divide: $(y^5 - 16y^3 + 5y - 20) \div (y - 4)$. Use synthetic division.   **$y^4 + 4y^3 + 5$**

Determine whether the binomial is a factor of the polynomial in each case.

15. $n + 3$; $4n^3 + 8n^2 - 10n + 6$ **yes**

16. $x - 5$; $x^4 - 20x^2 - 25$ **no**

Evaluate each polynomial for the given data. Use synthetic substitution. Name the binomial factors found, if any.

17. $7y^3 - 18y^2 + 25y - 39$ if $y = 2$ **–5**

18. $8n^4 + 50n^3 - 60n + 80$ if $n = -6$ **8**

19. $x^4 - x^3 - 22x^2 - 44x - 24$ using each of –1, –3 and 6 for x.
    0, 18, 0, $x + 1$, $x - 6$

★ 20. Divide: $(3x^{4n} + 4x^{3n} + x^{2n}) \div (x^{2n} + x^n)$.
    $3x^{2n} + x^n$

★ 21. Solve $\dfrac{5}{x^2 - 25} + \dfrac{4}{x^2 - 9x + 20} = \dfrac{3}{x^2 + x - 20}$.   **$-\dfrac{5}{2}$**

COMPUTER ACTIVITIES

Dividing Polynomials

OBJECTIVE: To divide a polynomial by a binomial of the form X + A

The computer program below uses the concept of synthetic division to divide a polynomial by a binomial. The program also contains two loops, one to INPUT the coefficients of the polynomials, and one to carry out synthetic division.

```
110    PRINT "WHAT IS THE DEGREE OF THE POLYNOMIAL";:
       INPUT N
120    DIM C(2 * N)
130    HOME
140    FOR I = N TO 0 STEP  - 1
150    PRINT "WHAT IS THE COEFFICIENT OF THE "I"
       DEGREE TERM?";
160    INPUT C(N + I - 1)
170    NEXT I
180    PRINT "WHAT IS THE CONSTANT IN THE BINOMIAL?";:
       INPUT K
190    PRINT C(1)"X^"N - 1;
200 D(1) = C(1)
210    FOR I = 2 TO N
220 D(I) = C(I) - K * D(I - 1)
230    IF D(I) < 0 THEN 250
240    PRINT "+";
250    PRINT D(I);: IF N - I = 0 THEN 270
260    PRINT "X^"N - I;
270    NEXT I
280 D(I) = C(I) - K * D(I - 1)
290    PRINT " REM "D(I)
300    GOTO 110
310    END
```

See the Computer Section beginning on page 574 for more information.

Note the use of the STEP option in line 140's FOR statement. STEP −1 makes the computer count backward by ones. Using STEP, you can increment the variable following FOR by any value. Another example is FOR I = 2 TO 20 STEP 2. This statement will make the computer count by twos: 2, 4, 6, ..., 20.

EXERCISES

1. Write a computer program that prints the squares of the odd numbers from 15 to 39 inclusive. **See TE Answer Section.**

2. Run the program above to do the division in Exercises 1–6 on page 132.

COLLEGE PREP TEST

DIRECTIONS: Choose the *one* best answer to each question or problem.

1. The expression $\frac{5}{0}$ is undefined. Which expression could be undefined if x and y are two *different positive* numbers? **D**

   (A) $\frac{x+y}{x-y}$ (B) $\frac{x}{y}$ (C) $\frac{2x-y}{x+y}$

   (D) $\frac{x+y}{3x-4y}$ (E) None of these

2. If 15, 22, and 30 are each factors of a certain number, then another factor of the number must be **D**

   (A) 4   (B) 9   (C) 25
   (D) 55   (E) None of these

3. Which numeral does not represent $\frac{11}{4}$? **E**

   (A) $2\frac{3}{4}$ (B) 2.75   (C) 2.75000...
   (D) 2.74999...   (E) None of these

4. $\dfrac{\frac{1}{2}+\frac{1}{4}+\frac{1}{8}}{\frac{1}{16}+\frac{1}{32}+\frac{1}{64}} =$ **C**

   (A) $\frac{1}{24}$ (B) $\frac{1}{8}$ (C) 8   (D) 24
   (E) None of these

5. If $\frac{1}{6}+\frac{1}{1.5}=\frac{1}{x}$, then $x =$ **B**

   (A) $\frac{5}{6}$ (B) 1.2   (C) 7.5   (D) 12
   (E) None of these

6. Juan drove 6 mi to the airport in 12 min and returned in 18 min. Find the average speed for the round trip in miles per hour. **A**

   (A) 24 mi/h   (B) 30 mi/h   (C) 35 mi/h
   (D) 40 mi/h   (E) None of these

7. If $\frac{1}{x+y}=\frac{1}{x-y}$, then **B**

   (A) $x=0$ and $y\neq 0$ (B) $y=0$ and $x\neq 0$
   (C) $x=y$ and $x\neq 0$ (D) $x=-y$ and $x\neq 0$
   (E) None of these

8. Albert can do a certain job in 6 days, and Brenda can do it in x days. Carol works 3 times as fast as Albert. What part of the job will be done at the end of y days if all three work together? **A**

   (A) $\frac{y}{6}+\frac{y}{x}+\frac{y}{2}$ (B) $\frac{1}{6}+\frac{1}{x}+\frac{1}{2}$
   (C) $\frac{y}{6}+\frac{y}{x}+\frac{y}{18}$ (D) $\frac{1}{6}+\frac{1}{x}+\frac{1}{18}$
   (E) $\frac{6}{y}+\frac{x}{y}+\frac{2}{y}$

9. A certain plane can fly 400 mi/h in still air. The plane can go round trip between points M and N in 4 h when the wind is blowing at a rate of 20 mi/h from M toward N. Find the distance between M and N. **B**

   (A) 780 mi   (B) 798 mi
   (C) 800 mi   (D) 802 mi
   (E) 805 mi

College Prep Test

6 RADICALS AND RATIONAL NUMBER EXPONENTS

CAREERS
Chemical Science

If you complete courses in chemistry and mathematics at the secondary or college level, you may qualify for an interesting and well-paid career in chemical science. The chemical industry is among the five leading manufacturing industries in the United States.

Chemical laboratory research has led to the production of textiles of high quality. More than 25% of textiles produced are made from synthetic fibers such as rayon or nylon. Chemical treatments are used to provide us with fabrics that are water-repellent, shrink-proof, and wrinkle-resistant.

Project

A chemist was asked to change 50 mL of 32% iodine solution to 28% iodine solution. He had 40% and 24% iodine solutions available. The amount of the 24% solution he used was 4 times the amount of the 40% solution. How many milliliters of each solution did he add to the 50 mL of the 32% solution?

40%:50 mL, 24%:200 mL

To write the equation, you need to understand the following:
(1) An iodine solution consists of iodine and water.
(2) The amount of iodine in the separate solutions is the same as in the total solution.

SQUARE ROOTS 6.1

Objectives **To solve a quadratic equation by using the definition of square root
To approximate a square root to the nearest hundredth by using a square root table
To determine whether a given number is rational or irrational**

You can solve the quadratic equation $x^2 = 16$ by factoring:

$$x^2 = 16$$
$$x^2 - 16 = 0$$
$$(x + 4)(x - 4) = 0$$

The solutions are -4 and 4.

The solutions to this equation can also be found by using the concept of **square root** as shown below.

$\sqrt{16}$ represents the *positive* square root of 16, which is 4.
$-\sqrt{16}$ represents the *negative* square root of 16, which is -4.
Notice that $(\sqrt{16})^2 = 16$ since $4^2 = 16$ and $(-\sqrt{16})^2 = 16$ since $(-4)^2 = 16$.

Definition: Square root

If $x^2 = k$, then $x = \sqrt{k}$ or $x = -\sqrt{k}$, for each number $k \geq 0$. $\sqrt{k}$ is the *principal*, or *positive*, square root of k and $-\sqrt{k}$ is the *negative square root* of k.

The symbol $\sqrt{k}$ (read "the square root of k") is called a **radical**, where k is the **radicand**, and $\sqrt{}$ is the **radical sign**, or **square root sign**.

The number $\sqrt{25}$ can easily be rewritten without a radical sign since the radicand, 25, is a *perfect square*. That is, 25 is the square of an integer. The number $\sqrt{10}$ cannot be simplified in this way since the radicand, 10, is not a perfect square.

Example 1 **Solve each equation.**

$x^2 + 5 = 15$ | $3y^2 = 75$

Subtract 5 from each side of the equation to obtain the form $x^2 = k$. | Divide each side of the equation by 3 to obtain the form $y^2 = k$.

$$x^2 = 10$$
$$x = \sqrt{10} \quad \text{or} \quad x = -\sqrt{10}$$

| $y^2 = 25$
| $y = \sqrt{25} \quad \text{or} \quad y = -\sqrt{25}$
| $y = \sqrt{5^2} = 5 \quad \text{or} \quad y = -\sqrt{5^2} = -5$

Thus, the solutions are $\sqrt{10}$ and $-\sqrt{10}$. | **Thus, the solutions are 5 and -5.**

Square Roots

In Example 1, $\sqrt{10}$ and $-\sqrt{10}$ cannot be rewritten as integers, but each number can be approximated to the nearest hundredth by using the table of square roots in the back of the book. To the nearest hundredth, $\sqrt{10} \doteq 3.16$ and $-\sqrt{10} \doteq -3.16$. The symbol $\doteq$ means *is approximately equal to*.

Example 2 Solve each equation. Give the solutions to the nearest hundredth.

$5y^2 - 9 = 206$ $\qquad\qquad$ $20n^2 = 4n^2 + 81$

$5y^2 = 215$
$y^2 = 43$
$y = \sqrt{43}$ or $y = -\sqrt{43}$

Use the table of square roots.

The solutions are 6.56 and -6.56.

$16n^2 = 81$
$n^2 = \dfrac{81}{16}$
$n = \sqrt{\dfrac{81}{16}} = \dfrac{9}{4}$ or $n = -\sqrt{\dfrac{81}{16}} = -\dfrac{9}{4}$

The solutions are 2.25 and -2.25.

The numbers $\sqrt{\dfrac{81}{16}}$ and $\sqrt{1.44}$ can be rewritten as rational numbers:

$\sqrt{\dfrac{81}{16}} = \sqrt{\left(\dfrac{9}{4}\right)^2} = \dfrac{9}{4}$ $\qquad$ $\sqrt{1.44} = \sqrt{(1.2)^2} = 1.2$

Notice that $\dfrac{81}{16}$ and 1.44 are *perfect squares* since they are the squares of the rational numbers $\dfrac{9}{4}$ and 1.2, respectively.

The number $\sqrt{35}$ is not a rational number. It cannot be written in the form $\dfrac{a}{b}$ where a and b are integers since 35 is not a perfect square. To 13 decimal places, $\sqrt{35} \doteq 5.9160797830996$. The decimal for $\sqrt{35}$ is *nonterminating and nonrepeating*. Numbers of this type are called **irrational numbers**. In general, for $k \geq 0$,

(1) $\sqrt{k}$ is a rational number if k is a perfect square, and
(2) $\sqrt{k}$ is an irrational number if k is not a perfect square.

Example 3 Determine whether each number is rational or irrational.

Number	Answer	(Reason)
$3.888\ldots$	Rational	A repeating decimal
5.37	Rational	A terminating decimal
$4.3131131113\ldots$	Irrational	A nonterminating, nonrepeating decimal
$\sqrt{\dfrac{2}{9}}$	Irrational	$\dfrac{2}{9}$ is not a perfect square.
$\sqrt{\dfrac{169}{9}}$	Rational	$\dfrac{169}{9} = \left(\dfrac{13}{3}\right)^2$
$\sqrt{99}$	Irrational	99 is not a perfect square.
$-\sqrt{6.25}$	Rational	$6.25 = 2.5^2$

Your work in mathematics often involves both rational and irrational numbers. All the rational numbers together with all the irrational numbers are called the **real numbers**.

Reading in Algebra

Determine whether each statement is true or false.
1. $\sqrt{17}$ is a radical sign. **F**
2. The radical sign in $\sqrt{6}$ is $\sqrt{}$. **T**
3. $\sqrt{15}$ is a radical. **T**
4. The radicand in $\sqrt{52}$ is 52. **T**
5. $\frac{4}{9}$ is a perfect square. **T**
6. $\frac{25}{11}$ is not a perfect square. **T**
7. 0.9 is a perfect square. **F**
8. 0.49 is not a perfect square. **F**
9. $\sqrt{2}$ is a real number. **T**
10. $\sqrt{4}$ is a real number. **T**
11. $\sqrt{17} \doteq 4.12$ to the nearest hundredth. **T**
12. $\sqrt{19} \doteq 4.35$ to the nearest hundredth. **F**

Written Exercises

Solve each equation.

Ⓐ
1. $x^2 = 64$ **−8, 8**
2. $y^2 = 3$ **$-\sqrt{3}, \sqrt{3}$**
3. $4n^2 = 64$ **−4, 4**
4. $c^2 + 6 = 21$ **$-\sqrt{15}, \sqrt{15}$**
5. $a^2 - 7 = 22$ **$-\sqrt{29}, \sqrt{29}$**
6. $900 = x^2$ **−30, 30**
7. $6y^2 + 14 = 434$ **$-\sqrt{70}, \sqrt{70}$**
8. $16n^2 = 7n^2 + 4$ **$-\frac{2}{3}, \frac{2}{3}$**

Solve each equation. Approximate the solutions to the nearest hundredth.
9. $x^2 = 87$ **−9.33, 9.33**
10. $3y^2 = 195$ **below**
11. $4n^2 = 25$ **below**
12. $25c^2 - 7 = 9$ **below**
13. $9a^2 = 2a^2 + 294$ **−6.48, 6.48**
14. $4n^2 = 20n^2 - 9$ **−0.75, 0.75**
15. $3x^2 - 86 = x^2 + 86$ **−9.27, 9.27**
16. $6n^2 + 75 = n^2 + 455$ **−8.72, 8.72**

Determine whether each number is rational or irrational.
17. 5.282282228... **I**
18. 0.123123123... **R**
19. 4.343343334... **I**
20. 7.65765765... **R**
21. $\sqrt{169}$ **R**
22. $\sqrt{47}$ **I**
23. $-\sqrt{400}$ **R**
24. $-\sqrt{75}$ **I**
25. 87.6543 **R**
26. 19.4747 **R**
27. $-\sqrt{3,000}$ **I**
28. $\sqrt{2,025}$ **R**

Ⓑ
29. $\sqrt{\frac{49}{81}}$ **R**
30. $\sqrt{\frac{18}{100}}$ **I**
31. $\sqrt{\frac{18}{8}}$ **R**
32. $\sqrt{\frac{20}{40}}$ **I**
33. $\sqrt{2.5}$ **I**
34. $\sqrt{2.25}$ **R**
35. $\sqrt{0.36}$ **R**
36. π **I**

10. **−8.06, 8.06**   11. **−2.50, 2.50**   12. **−0.80, 0.80**

Solve each equation. Approximate the solutions to the nearest hundredth.
37. $144n^2 = 64$ **−0.67, 0.67**
38. $60x^2 = 9 - 4x^2$ **−0.38, 0.38**
39. $7y^2 - 9 = 16y^2 - 58$ **−2.33, 2.33**
40. $y^2 = 0.25$ **−0.50, 0.50**
41. $n^2 - 0.0009 = 0.004$ **−0.07, 0.07**
42. $5x^2 - 1 = 4x^2 + 0.21$ **−1.10, 1.10**

Ⓒ
43. $x^4 - 30x^2 + 125 = 0$ **−5.00, 5.00, −2.24, 2.24**
44. $6x^4 = 30x^2 - 36$ **−1.73, 1.73, −1.41, 1.41**
45. $\frac{x^2 + 3}{4} + \frac{2x^2 - 5}{3} = \frac{3x^2 + 2}{6}$ **−1.73, 1.73**
46. $\frac{2}{3}(9x^2 + 21) - \frac{3}{4}(12x^2 - 20) + 148 = 0$ **−7.68, 7.68**

CUMULATIVE REVIEW

Simplify each expression.
1. $(-7x^2y^5z^9)^2$ **$49x^4y^{10}z^{18}$**
2. $(-4ab^2c^7)^3$ **$-64a^3b^6c^{21}$**
3. $10m^2n(3m^3np^6)^4$ **$810m^{14}n^5p^{24}$**

Square Roots

141

SIMPLIFYING RADICALS 6.2

Objective **To simplify a radical in which the radicand contains a perfect square factor**

At the right, you see that $\sqrt{100 \cdot 36} = \sqrt{100} \cdot \sqrt{36}$. In general, the square root of the product of two or more real numbers is the product of the square roots of the numbers.

$\sqrt{100 \cdot 36}$	$\sqrt{100} \cdot \sqrt{36}$
$\sqrt{3{,}600}$	$10 \cdot 6$
60	60

Square root of a product

$\sqrt{x \cdot y} = \sqrt{x} \cdot \sqrt{y}$ for all real numbers $x \geq 0$, $y \geq 0$.

Example 1 **Simplify $\sqrt{189}$. Approximate the result to the nearest hundredth.**

$\sqrt{189} = \sqrt{9 \cdot 21}$ ◀ 9 is a perfect square factor of 189.
$\phantom{\sqrt{189}} = \sqrt{9} \cdot \sqrt{21}$ ◀ $\sqrt{x \cdot y} = \sqrt{x} \cdot \sqrt{y}$
$\phantom{\sqrt{189}} = 3 \cdot \sqrt{21}$, or $3\sqrt{21}$
$\phantom{\sqrt{189}} \doteq 3 \cdot 4.583$ ◀ $\sqrt{21} \doteq 4.583$
$\phantom{\sqrt{189}} \doteq 13.749$
$\phantom{\sqrt{189}} \doteq 13.75$ ◀ Round to the nearest hundredth.
Thus, $\sqrt{189} \doteq 13.75$.

In Example 1, the expression $3\sqrt{21}$ is the **simplest radical form** of $\sqrt{189}$. A radical or square root expression is in simplest radical form whenever the radicand is the smallest possible integer. In other words, the radicand must be an integer that does not contain a perfect square factor, other than 1. Unless otherwise indicated, all answers involving radicals should be given in simplest radical form.

Example 2 **Simplify $-5\sqrt{72}$.**

$-5\sqrt{72} = -5\sqrt{36 \cdot 2}$
$\phantom{-5\sqrt{72}} = -5\sqrt{36} \cdot \sqrt{2}$
$\phantom{-5\sqrt{72}} = -5 \cdot 6 \cdot \sqrt{2}$
$\phantom{-5\sqrt{72}} = -30\sqrt{2}$

Thus, $-5\sqrt{72} = -30\sqrt{2}$.

When simplifying a radical, the work is easier if you identify the *greatest perfect square* factor of the radicand. In Example 2, 9 and 4 are also perfect square factors of 72. Notice what happens when these numbers are used to simplify $-5\sqrt{72}$.

$-5\sqrt{72} = -5\sqrt{9 \cdot 4 \cdot 2} = -5\sqrt{9} \cdot \sqrt{4} \cdot \sqrt{2} = -5 \cdot 3 \cdot 2\sqrt{2} = -30\sqrt{2}$

In this case, the Square Root of a Product property is extended to three numbers.

Notice that $\sqrt{5^2} = 5$ but $\sqrt{(-5)^2} \neq -5$ because $\sqrt{(-5)^2} = \sqrt{25} = 5$. This shows that it is *not* true that:

$$\sqrt{x^2} = x \text{ for } each \text{ real number } x.$$

You can simplify radicals like $\sqrt{x^2}$, $\sqrt{x^4}$, and $\sqrt{x^6}$ where the exponents are *even* integers and x is a *negative* number ($x < 0$). Study the statements below.

1. $\sqrt{(-6)^2} \neq -6$ but $\sqrt{(-6)^2} = \sqrt{36} = 6 = |-6|$.
   $\sqrt{x^2} \neq x^1$ but $\sqrt{x^2} = |x^1|$ if $x < 0$.

2. $\sqrt{(-3)^4} = \sqrt{81} = 9 = (-3)^2$ and $\sqrt{x^4} = x^2$ if $x < 0$.

3. $\sqrt{(-2)^6} \neq (-2)^3$ but $\sqrt{(-2)^6} = \sqrt{64} = 8 = |(-2)^3|$.
   $\sqrt{x^6} \neq x^3$ but $\sqrt{x^6} = |x^3|$ if $x < 0$.

Thus, if x is *any* real number, then

$$\sqrt{x^2} = |x|, \quad \sqrt{x^4} = x^2, \quad \sqrt{x^6} = |x^3|, \quad \sqrt{x^8} = x^4, \text{ and so on, } or$$

$$\sqrt{x^{2n}} = |x^n| \text{ if } n \text{ is an } odd \text{ integer and}$$
$$\sqrt{x^{2n}} = x^n \text{ if } n \text{ is an } even \text{ integer.}$$

Radicals like $\sqrt{x^1}$ and $\sqrt{x^3}$, where the exponents are *odd* integers, are *not* real numbers if x is a *negative* number ($x < 0$), as shown below.

$$\sqrt{-2} \text{ and } \sqrt{(-2)^3}, \text{ or } \sqrt{-8}, \text{ are } not \text{ real numbers.}$$

To simplify $\sqrt{x^7}$ and $\sqrt{x^9}$, where the exponents are *odd* integers and x is a *non-negative* number ($x \geq 0$), rewrite the radicands as shown below.

$$\sqrt{x^7} = \sqrt{x^6 \cdot x^1} = \sqrt{x^6}\sqrt{x} = x^3\sqrt{x} \text{ if } x \geq 0.$$
$$\sqrt{x^9} = \sqrt{x^8 \cdot x^1} = \sqrt{x^8}\sqrt{x} = x^4\sqrt{x} \text{ if } x \geq 0.$$

Example 3 **Simplify $7ab\sqrt{24a^{10}b^7}$ if $b \geq 0$.**

$$\begin{aligned}
7ab\sqrt{24a^{10}b^7} &= 7ab\sqrt{4 \cdot 6 \cdot a^{10} \cdot b^6 \cdot b^1} \\
&= 7ab \cdot \sqrt{4}\sqrt{a^{10}}\sqrt{b^6} \cdot \sqrt{6b} \\
&= 7a^1b^1 \cdot 2 \cdot |a^5| \cdot b^3\sqrt{6b} \\
&= 14a|a^5| \cdot b^4\sqrt{6b}
\end{aligned}$$

Reading in Algebra

Determine whether each statement is true or false.
1. The square root of the product of three positive real numbers is the product of their square roots. **T**
2. The greatest perfect square factor of 200 is 25. **F**
3. The greatest perfect square factor of 88 is 4. **T**
4. The simplest radical form of $\sqrt{12}$ is $4\sqrt{3}$. **F**
5. The simplest radical form of $\sqrt{32}$ is $2\sqrt{8}$. **F**
6. There are exactly nine integers between 0 and 101 that are perfect squares. **F**
7. $4x^{14}$ is a perfect square. **T**
8. $9y^9$ is a perfect square. **F**

Simplifying Radicals

Match each expression at the left with its *one* equivalent expression at the right.

9. $\sqrt{a^2b^5}$ if $b \geq 0$ **B**   (A) $ab^2\sqrt{b}$ (B) $|a|b^2\sqrt{b}$ (C) $a|b^2|\sqrt{b}$

10. $a\sqrt{a^2b^5}$ if $b \geq 0$ **C**   (A) $a^2b^2\sqrt{b}$ (B) $|a^2|b^2\sqrt{b}$ (C) $a|a|b^2\sqrt{b}$

11. $ab\sqrt{a^3b^6}$ if $a \geq 0$ **B**   (A) $a^2b^4\sqrt{a}$ (B) $a^2b|b^3|\sqrt{a}$ (C) $a^2|b^4|\sqrt{a}$

12. $ab^2\sqrt{a^3b^6}$ if $a \geq 0$ **B**   (A) $a^2b^5\sqrt{a}$ (B) $a^2b^2|b^3|\sqrt{a}$ (C) $|a^2|b^5\sqrt{a}$

Written Exercises

25. $3a^2b^7|c^9|\sqrt{10b}$ 28. $m^7|n^7|p\sqrt{mp}$ 32. $\dfrac{9yz^5\sqrt{3xyz}}{7}$

Simplify each expression. Approximate the result to the nearest hundredth.

(A) 1. $\sqrt{250}$ **$5\sqrt{10}$, 15.81**
2. $-\sqrt{135}$ **$-3\sqrt{15}$, -11.62**
3. $\sqrt{180}$ **$6\sqrt{5}$, 13.42**
4. $-\sqrt{243}$ **$-9\sqrt{3}$, -15.59**

5. $3\sqrt{120}$ **$6\sqrt{30}$, 32.86**
6. $-10\sqrt{490}$ **$-70\sqrt{10}$, -221.34**
7. $2.5\sqrt{162}$ **$22.5\sqrt{2}$, 31.82**
8. $0.75\sqrt{512}$ **$12\sqrt{2}$, 16.97**

Simplify each expression. Assume that no denominator has the value of 0.

9. $2\sqrt{27}$ **$6\sqrt{3}$**
10. $5\sqrt{32}$ **$20\sqrt{2}$**
11. $-6\sqrt{45}$ **$-18\sqrt{5}$**
12. $-10\sqrt{96}$ **$-40\sqrt{6}$**

13. $5\sqrt{4x^6}$ **$10|x^3|$**
14. $\sqrt{12b^2}$ **$2|b|\sqrt{3}$**
15. $9a\sqrt{16a^{12}}$ **$36a^7$**
16. $6y^3\sqrt{24y^{10}}$ **$12y^3|y^5|\sqrt{6}$**

17. $\sqrt{n^9}$, $n \geq 0$ **$n^4\sqrt{n}$**
18. $8\sqrt{y^{15}}$, $y \geq 0$ **$8y^7\sqrt{y}$**
19. $-2y^4\sqrt{25y^5}$, $y \geq 0$ **$-10y^6\sqrt{y}$**
20. $-12c^2\sqrt{8c^{17}}$, $c \geq 0$ **$-24c^{10}\sqrt{2c}$**

(B) 21. $\sqrt{24c^{10}d^3}$, $d \geq 0$ **$2|c^5|d\sqrt{6d}$**
22. $ab^2\sqrt{a^{11}b^{16}}$, $a \geq 0$ **$a^6b^{10}\sqrt{a}$**
23. $8x^2\sqrt{9x^5y^5}$; $x \geq 0$, $y \geq 0$ **$24x^4y^2\sqrt{xy}$**

24. $5mn\sqrt{8m^{20}n^7}$, $n \geq 0$ **$10m^{11}n^4\sqrt{2n}$**
25. $\sqrt{90a^4b^{15}c^{18}}$, $b \geq 0$ **above**
26. $c^2d\sqrt{180c^8d^9}$, $d \geq 0$ **$6c^6d^5\sqrt{5d}$**

27. $7c^2\sqrt{128c^{12}d}$, $d \geq 0$ **$56c^8\sqrt{2d}$**
28. $m^2n^4\sqrt{m^{11}n^6p^3}$; $m \geq 0$, $p \geq 0$ **above**
29. $x^{10}\sqrt{243x^{30}y^{25}}$, $y \geq 0$ **$9|x^{25}|y^{12}\sqrt{3y}$**

30. $\dfrac{\sqrt{x^7y^8z^9}}{x^2y^3z^4}$; $x > 0$, $z > 0$ **$xy\sqrt{xz}$**
31. $\dfrac{\sqrt{27a^8b^{11}c^{12}}}{6ab^5c^{10}}$, $b > 0$ **$\dfrac{a^3\sqrt{3b}}{2c^4}$**
32. $\dfrac{3z\sqrt{108x^5y^7z^9}}{14x^2y^2}$; $x > 0$, $y > 0$, $z > 0$ **above**

Simplify. Assume that m and n are integers and that x and y are positive real numbers.

(C) 33. $\sqrt{x^{4m}}$ **x^{2m}**
34. $\sqrt{x^{4m+1}}$ **$x^{2m}\sqrt{x}$**
35. $\sqrt{x^{2m+3}}$ **$x^{m+1}\sqrt{x}$**
36. $x^m\sqrt{x^{6m+5}}$ **$x^{4m+2}\sqrt{x}$**

37. $x^my^n\sqrt{x^{6m}y^{8n}}$ **$x^{4m}y^{5n}$**
38. $x^{2m-3}\sqrt{x^{4m+9}}$ **$x^{4m+1}\sqrt{x}$**
39. $xy\sqrt{x^{6m}y^{8n+3}}$ **$x^{3m+1}y^{4n+2}\sqrt{y}$**
40. $x^{2m}y^{3n}\sqrt{x^{2m}y^{3n}}$ **$x^{3m}y^{4n}\sqrt{y^n}$**

NON-ROUTINE PROBLEMS

1. If n is any integer, then $n^2(n^2 - 1)$ is *always* divisible by what positive integers? **1, 2, 3, 4, 6, 12**

2. At a party, Mantia placed 8 saucers around a rectangular table. She put 24 peanuts in the 8 saucers so that there were 9 peanuts along each side of the table.

   Ira said, "I can pick up the 24 peanuts, add 8 more, and place the 32 peanuts in the 8 saucers so there will still be 9 peanuts along each side of the table." How can Ira do this? **Place 1–7–1 peanuts along each side.**

144   Chapter Six

SUMS, DIFFERENCES, AND PRODUCTS OF RADICALS 6.3

Objectives **To simplify an expression containing sums or differences of radicals**
To simplify an expression containing products of radicals

You have simplified algebraic expressions by combining (adding or subtracting) like terms, using the Distributive property. For example, $5x^2y + 3x^2y + 2xy^2 = (5 + 3)x^2y + 2xy^2 = 8x^2y + 2xy^2$. Expressions containing like radicals can be simplified in a similar way.

$$6\sqrt{10} + 8\sqrt{10} + 2\sqrt{7} = (6 + 8)\sqrt{10} + 2\sqrt{7} = 14\sqrt{10} + 2\sqrt{7}$$

The expression $14\sqrt{10} + 2\sqrt{7}$ is in its simplest radical form since $14\sqrt{10}$ and $2\sqrt{7}$ are *not* like radicals and, therefore, cannot be combined. The like radicals in this case are $6\sqrt{10}$ and $8\sqrt{10}$.

Example 1 Simplify $8\sqrt{3} + 4\sqrt{6} - 2\sqrt{3} - 7\sqrt{6} + 9\sqrt{5}$.

$$8\sqrt{3} + 4\sqrt{6} - 2\sqrt{3} - 7\sqrt{6} + 9\sqrt{5}$$
$$= (8\sqrt{3} - 2\sqrt{3}) + (4\sqrt{6} - 7\sqrt{6}) + 9\sqrt{5}$$ ◀ *Group the like radicals together.*
$$= 6\sqrt{3} - 3\sqrt{6} + 9\sqrt{5}$$ ◀ *Combine the like radicals.*

Example 2 **Simplify each expression.**
$$4\sqrt{18} - \sqrt{8} + \sqrt{2} \qquad\qquad \sqrt{36ab^3} - \sqrt{ab^3} + 5b\sqrt{ab}; \; a \geq 0, b \geq 0$$

$$\begin{aligned} & 4\sqrt{9 \cdot 2} - \sqrt{4 \cdot 2} + \sqrt{2} \\ &= 4 \cdot 3\sqrt{2} - 2\sqrt{2} + \sqrt{2} \\ &= 12\sqrt{2} - 2\sqrt{2} + \sqrt{2} \\ &= 11\sqrt{2} \end{aligned} \qquad \begin{aligned} & \sqrt{36 \cdot a \cdot b^2 \cdot b} - \sqrt{a \cdot b^2 \cdot b} + 5b\sqrt{ab} \\ &= 6b\sqrt{ab} - b\sqrt{ab} + 5b\sqrt{ab} \\ &= 10b\sqrt{ab} \end{aligned}$$

You can simplify $\sqrt{50} \cdot \sqrt{12}$ as shown at the right. In general, the product of the square roots of two or more real numbers is the square root of the product of the numbers.

$$\begin{aligned} \sqrt{50} \cdot \sqrt{12} &= \sqrt{50 \cdot 12} \\ &= \sqrt{600} \\ &= \sqrt{100 \cdot 6} \\ &= 10\sqrt{6} \end{aligned}$$

Product of square roots

For all real numbers $x \geq 0$, $y \geq 0$, $\sqrt{x} \cdot \sqrt{y} = \sqrt{x \cdot y}$
and $\sqrt{x} \cdot \sqrt{x} = x$.

Using the first product property above,
$$\sqrt{12} \cdot \sqrt{2} = \sqrt{12 \cdot 2} = \sqrt{24} = \sqrt{4 \cdot 6} = 2\sqrt{6}.$$

Sums, Differences, and Products of Radicals

The second product property on page 145 involves a special case in the multiplication of radicals. Notice that $\sqrt{x} \cdot \sqrt{x} = (\sqrt{x})^2 = x$. Using this property, $\sqrt{17} \cdot \sqrt{17} = 17$ and $\sqrt{m+n} \cdot \sqrt{m+n} = m+n$.

Example 3 **Simplify each expression.**

$-7\sqrt{30} \cdot 6\sqrt{30}$

$ -7 \cdot 6 \cdot \sqrt{30} \cdot \sqrt{30}$
$= -42 \cdot 30 \quad \blacktriangleleft \sqrt{x} \cdot \sqrt{x} = x$
$= -1{,}260$

$\sqrt{6a^3b^8} \cdot \sqrt{8a^5b^3}; \ a \geq 0, \ b \geq 0$

$ \sqrt{6a^3b^8 \cdot 8a^5b^3} \quad \blacktriangleleft \sqrt{x} \cdot \sqrt{y} = \sqrt{x \cdot y}$
$= \sqrt{6 \cdot 8 \cdot a^3 \cdot a^5 \cdot b^8 \cdot b^3}$
$= \sqrt{48a^8b^{11}}$
$= \sqrt{16 \cdot 3 \cdot a^8 \cdot b^{10} \cdot b}$
$= 4a^4b^5\sqrt{3b}$

You can also simplify expressions containing products of radicals in which binomials or trinomials are involved. Two such cases are shown in Example 4.

Example 4 **Simplify each expression.**

$5\sqrt{3}(2\sqrt{6} + 4\sqrt{3} - \sqrt{7})$

Use the Distributive property to multiply.
$5\sqrt{3} \cdot 2\sqrt{6} + 5\sqrt{3} \cdot 4\sqrt{3} - 5\sqrt{3} \cdot \sqrt{7}$
$= 10\sqrt{18} + 20 \cdot 3 - 5\sqrt{21}$
$= 10 \cdot \sqrt{9 \cdot 2} + 60 - 5\sqrt{21}$
$= 30\sqrt{2} + 60 - 5\sqrt{21}$

$(3\sqrt{5} + \sqrt{2})(4\sqrt{5} - 6\sqrt{2})$

Use the FOIL method to multiply.
$3\sqrt{5} \cdot 4\sqrt{5} - 3\sqrt{5} \cdot 6\sqrt{2} + \sqrt{2} \cdot 4\sqrt{5} - \sqrt{2} \cdot 6\sqrt{2}$
$= 12 \cdot 5 - 18\sqrt{10} + 4\sqrt{10} - 6 \cdot 2$
$= 60 - 18\sqrt{10} + 4\sqrt{10} - 12$
$= 48 - 14\sqrt{10}$

The special product formulas, $(a + b)(a - b) = a^2 - b^2$ and $(a - b)^2 = a^2 - 2ab + b^2$, can sometimes be used to simplify expressions containing products of radicals.

Example 5 **Simplify each expression.**

$(2\sqrt{7} + 4\sqrt{3})(2\sqrt{7} - 4\sqrt{3})$

$(a+b)(a-b) = a^2 - b^2 \blacktriangleright (2\sqrt{7} + 4\sqrt{3})(2\sqrt{7} - 4\sqrt{3})$
$= (2\sqrt{7})^2 - (4\sqrt{3})^2$
$= 28 - 48, \text{ or } -20$

$(3\sqrt{c} - 5\sqrt{d})^2; \ c \geq 0, \ d \geq 0$

$ (3\sqrt{c} - 5\sqrt{d})^2$
$= (3\sqrt{c})^2 - 2 \cdot 3\sqrt{c} \cdot 5\sqrt{d} + (5\sqrt{d})^2$
$= 9c - 30\sqrt{cd} + 25d$

In Example 5, notice how the Power of a Product property, $(xy)^n = x^n \cdot y^n$, is used with radical expressions:

$(2\sqrt{7})^2 = (2)^2 \cdot (\sqrt{7})^2 = 4 \cdot 7 = 28$ and
$(5\sqrt{d})^2 = (5)^2 \cdot (\sqrt{d})^2 = 25 \cdot d = 25d.$

Oral Exercises

2. $-3\sqrt{30}$ 3. $10\sqrt{cd}$ 4. $\sqrt{6ab}$

Simplify. Assume that values of the variables are nonnegative numbers.

1. $\sqrt{11} \cdot \sqrt{7}$ $\sqrt{77}$ 2. $-3\sqrt{6} \cdot \sqrt{5}$ 3. $5\sqrt{c} \cdot 2\sqrt{d}$ 4. $\sqrt{2a} \cdot \sqrt{3b}$ 5. $(\sqrt{c})^2$ c
6. $\sqrt{2} \cdot \sqrt{32}$ 8   7. $\sqrt{3} \cdot \sqrt{12}$ 6   8. $\sqrt{5} \cdot \sqrt{2} \cdot \sqrt{10}$ 10   9. $\sqrt{3} + \sqrt{3}$ $2\sqrt{3}$ 10. $5\sqrt{2} - \sqrt{2}$ $4\sqrt{2}$

Written Exercises

11. $60 - 15\sqrt{2} + 6\sqrt{35}$ 2. $3\sqrt{5} - 6\sqrt{6} + 10\sqrt{3}$

Simplify. Assume that values of the variables are nonnegative numbers.

A
1. $3\sqrt{10} + 9\sqrt{7} - 8\sqrt{10} - 3\sqrt{7}$ $-5\sqrt{10} + 6\sqrt{7}$ 2. $4\sqrt{5} - 7\sqrt{6} - \sqrt{5} + 10\sqrt{3} + \sqrt{6}$ above
3. $7\sqrt{45} - \sqrt{20} + \sqrt{5} - \sqrt{80}$ $16\sqrt{5}$ 4. $3\sqrt{50} + 4\sqrt{18} - 2\sqrt{32} - \sqrt{8}$ $17\sqrt{2}$
5. $\sqrt{25n} - 7\sqrt{n} + \sqrt{16n}$ $2\sqrt{n}$ 6. $\sqrt{x} - 6\sqrt{4x} - \sqrt{9x} + 5\sqrt{36x}$ $16\sqrt{x}$
7. $8\sqrt{6} \cdot 5\sqrt{10}$ $80\sqrt{15}$ 8. $-2\sqrt{15} \cdot 4\sqrt{5} \cdot \sqrt{6}$ $-120\sqrt{2}$
9. $\sqrt{6y} \cdot \sqrt{3y^9}$ $3y^5\sqrt{2}$ 10. $4\sqrt{3x^3} \cdot 5\sqrt{15x^8}$ $60x^5\sqrt{5x}$
11. $3\sqrt{5}(4\sqrt{5} - \sqrt{10} + 2\sqrt{7})$ above   12. $-4\sqrt{6}(\sqrt{15} + 8\sqrt{7} - 2\sqrt{6})$ below
13. $(4\sqrt{11} + 6)(2\sqrt{11} - 8)$ $40 - 20\sqrt{11}$ 14. $(\sqrt{10} - 2\sqrt{3})(6\sqrt{10} - \sqrt{3})$ $66 - 13\sqrt{30}$
15. $(2\sqrt{7} - 4\sqrt{2})(5\sqrt{7} + 6\sqrt{2})$ $22 - 8\sqrt{14}$ 16. $(8\sqrt{30} + 4\sqrt{7})(\sqrt{30} + \sqrt{7})$ $268 + 12\sqrt{210}$
17. $(-3\sqrt{11})^2$ 99   18. $(7\sqrt{x})^2$ $49x$
19. $(20\sqrt{17})^2$ 6,800   20. $(-8\sqrt{y} + 2)^2$ $64y + 128$
21. $(5\sqrt{3} + 7)(5\sqrt{3} - 7)$ 26   22. $(6\sqrt{5} + 4\sqrt{6})(6\sqrt{5} - 4\sqrt{6})$ 84
23. $(\sqrt{7} + \sqrt{6})^2$ $13 + 2\sqrt{42}$ 24. $(5\sqrt{2} - 4\sqrt{3})^2$ $98 - 40\sqrt{6}$
25. $(4\sqrt{x} + 2\sqrt{5})(4\sqrt{x} - 2\sqrt{5})$ $16x - 20$ 26. $(3\sqrt{7} - 8\sqrt{y})(3\sqrt{7} + 8\sqrt{y})$ $63 - 64y$
27. $(6\sqrt{n} - 6)^2$ $36n - 72\sqrt{n} + 36$ 28. $(5\sqrt{c} + 4\sqrt{10})^2$ $25c + 40\sqrt{10c} + 160$

12. $-12\sqrt{10} - 32\sqrt{42} + 48$ 30. $15\sqrt{2} - 17\sqrt{5}$ 36. $-160x^5y^6\sqrt{3xy}$

B
29. $\sqrt{8} + \sqrt{12} + \sqrt{18} + \sqrt{27}$ $5\sqrt{2} + 5\sqrt{3}$ 30. $5\sqrt{2} - 4\sqrt{20} - 3\sqrt{45} + 2\sqrt{50}$
31. $2\sqrt{c^2d} + \sqrt{36c^2d} - 3c\sqrt{d}$ $5c\sqrt{d}$ 32. $12b^2\sqrt{ab} - \sqrt{25ab^5} + 10\sqrt{ab^5}$ $17b^2\sqrt{ab}$
33. $5\sqrt{3m^2n^3} + 2mn\sqrt{27n}$ $11mn\sqrt{3n}$ 34. $2x^2\sqrt{20x^3y} - x^2\sqrt{45x^3y} - x^3\sqrt{5xy}$ 0
35. $\sqrt{3c^3d^5} \cdot 8\sqrt{8cd^7}$ $16c^2d^6\sqrt{6}$ 36. $-8x\sqrt{6xy^5} \cdot 5y\sqrt{8x^5y^3} \cdot xy\sqrt{xy}$ above
37. $(4\sqrt{c} + \sqrt{7})(8\sqrt{c} - 3\sqrt{7})$ $32c - 4\sqrt{7c} - 21$ 38. $(5\sqrt{a} - \sqrt{b})(4\sqrt{a} - \sqrt{b})$ $20a - 9\sqrt{ab} + b$
39. $(\sqrt{x} - 3\sqrt{y})(3\sqrt{x} + \sqrt{y})$ $3x - 8\sqrt{xy} - 3y$ 40. $(x\sqrt{y} + y\sqrt{x})(2x\sqrt{y} + 3y\sqrt{x})$ below
41. $(5\sqrt{c} - 4\sqrt{d})(5\sqrt{c} + 4\sqrt{d})$ $25c - 16d$ 42. $(7\sqrt{a} + 2b)(7\sqrt{a} - 2b)$ $49a - 4b^2$
43. $(\sqrt{a} + \sqrt{b})^2$ $a + 2\sqrt{ab} + b$ 44. $(6\sqrt{c} - 5\sqrt{d})^2$ $36c - 60\sqrt{cd} + 25d$

40. $2x^2y + 5xy\sqrt{xy} + 3xy^2$ $x + 39 + 12\sqrt{x + 4}$ $3x - 23$

C
45. $(\sqrt{x + 4} + 5)(\sqrt{x + 4} + 7)$ 46. $(\sqrt{3x - 7} - 4)(\sqrt{3x - 7} + 4)$, $3x - 7 \geq 0$
47. $(\sqrt{x + y} + \sqrt{y})(\sqrt{x + y} - \sqrt{y})$ x   48. $(6 - \sqrt{y - 3})^2$, $y \geq 3$ $y + 33 - 12\sqrt{y - 3}$
49. $(\sqrt{n + 5} - \sqrt{n})^2$ $2n + 5 - 2\sqrt{n^2 + 5n}$ 50. $(3\sqrt{c - 8} + 4\sqrt{c})^2$, $c \geq 8$ $25c - 72 + 24\sqrt{c^2 - 8c}$
51. $(\sqrt{m} - \sqrt{n})(\sqrt{m} + \sqrt{n})$ 52. $(2\sqrt{x} - \sqrt{y})(2\sqrt{x} + \sqrt{y})$ 53. $(\sqrt{c} - 6\sqrt{d})(\sqrt{c} + 6\sqrt{d})$
54. $(\sqrt{5a} - \sqrt{3b})(\sqrt{5a} + \sqrt{3b})$
56. $(3\sqrt{c} - d\sqrt{7})(3\sqrt{c} + d\sqrt{7})$

Factor each binomial as a difference of two squares.
[*Hint:* Rewrite $x - y$ as $(\sqrt{x})^2 - (\sqrt{y})^2$.]

51. $m - n$ 52. $4x - y$ 53. $c - 36d$ 54. $5a - 3b$
55. $x^2 - 2y$ 56. $9c - 7d^2$ 57. $6m^2 - 25n$ 58. $49a^2b - 10cd^2$
$(x - \sqrt{2y})(x + \sqrt{2y})$ 57. $(m\sqrt{6} - 5\sqrt{n})(m\sqrt{6} + 5\sqrt{n})$ $(7a\sqrt{b} - d\sqrt{10c})(7a\sqrt{b} + d\sqrt{10c})$

Sums, Differences, and Products of Radicals

QUOTIENTS OF RADICALS 6.4

Objective **To simplify an expression containing a quotient of radicals**

The quotient of the square roots of two real numbers is the square root of the quotient of the numbers.

> **Quotient of square roots**
>
> For all real numbers $x \geq 0$, $y > 0$, $\dfrac{\sqrt{x}}{\sqrt{y}} = \sqrt{\dfrac{x}{y}}$.

The expression $\dfrac{\sqrt{24}}{\sqrt{3}}$ contains a quotient of two square roots in which the radicand in the numerator is exactly divisible by the radicand in the denominator. Expressions of this type can be simplified by applying the Quotient of Square Roots property, as shown in Example 1.

Example 1 **Simplify each expression.**

$\dfrac{\sqrt{24}}{\sqrt{3}}$ $\qquad\qquad$ $\dfrac{\sqrt{28n^6}}{\sqrt{7n^2}}, n \neq 0$ $\qquad\qquad$ $\dfrac{\sqrt{18c^7}}{\sqrt{6c^2}}, c > 0$

$= \sqrt{\dfrac{24}{3}}$ ◄ $\dfrac{\sqrt{x}}{\sqrt{y}} = \sqrt{\dfrac{x}{y}}$ $\qquad$ $= \sqrt{\dfrac{28n^6}{7n^2}}$ $\qquad\qquad$ $= \sqrt{\dfrac{18c^7}{6c^2}}$

$= \sqrt{8}$ $\qquad\qquad\qquad\qquad\qquad$ $= \sqrt{4n^4}$ $\qquad\qquad\qquad$ $= \sqrt{3c^5}$

$= 2\sqrt{2}$ $\qquad\qquad\qquad\qquad\qquad$ $= 2n^2$ $\qquad\qquad\qquad\quad$ $= c^2\sqrt{3c}$

To simplify an expression like $\dfrac{\sqrt{5}}{3\sqrt{2}}$, in which the radicand in the numerator is *not* exactly divisible by the radicand in the denominator, a procedure called **rationalizing the denominator** is used. The denominator, $3\sqrt{2}$, is an irrational number. Notice what happens when $\dfrac{\sqrt{5}}{3\sqrt{2}}$ is multiplied by $\dfrac{\sqrt{2}}{\sqrt{2}}$, or 1.

$$\dfrac{\sqrt{5}}{3\sqrt{2}} = \dfrac{\sqrt{5}}{3\sqrt{2}} \cdot \dfrac{\sqrt{2}}{\sqrt{2}} = \dfrac{\sqrt{5} \cdot \sqrt{2}}{3\sqrt{2} \cdot \sqrt{2}} = \dfrac{\sqrt{10}}{3 \cdot 2} = \dfrac{\sqrt{10}}{6}$$

The new denominator, 6, is a rational number.

Example 2 **Simplify $\dfrac{6\sqrt{8}}{\sqrt{6}}$.**

$\dfrac{6\sqrt{8}}{\sqrt{6}} = \dfrac{6\sqrt{8}}{\sqrt{6}} \cdot \dfrac{\sqrt{6}}{\sqrt{6}} = \dfrac{6\sqrt{8} \cdot \sqrt{6}}{\sqrt{6} \cdot \sqrt{6}} = \dfrac{6\sqrt{48}}{6} = \sqrt{48} = \sqrt{16 \cdot 3} = 4\sqrt{3}$

Contrast the following two methods of rationalizing the denominator of $\dfrac{1}{\sqrt{12}}$:

Method 1: $\quad \dfrac{1}{\sqrt{12}} = \dfrac{1}{2\sqrt{3}} \cdot \dfrac{\sqrt{3}}{\sqrt{3}} = \dfrac{\sqrt{3}}{6}$

Method 2: $\quad \dfrac{1}{\sqrt{12}} = \dfrac{1}{\sqrt{12}} \cdot \dfrac{\sqrt{12}}{\sqrt{12}} = \dfrac{\sqrt{12}}{12} = \dfrac{2\sqrt{3}}{12} = \dfrac{\sqrt{3}}{6}$

In Method 1, simplifying the radical in the denominator before rationalizing made the work easier. The numerator and denominator of the original fraction were multiplied by a smaller number, $\sqrt{3}$, as opposed to $\sqrt{12}$, which was used in Method 2.

Example 3 Simplify $\dfrac{5\sqrt{7}}{\sqrt{12}}$ and $\dfrac{9\sqrt{5x}}{\sqrt{24x^4y^3}}$; $x > 0, y > 0$.

$$\dfrac{5\sqrt{7}}{\sqrt{12}} = \dfrac{5\sqrt{7}}{2\sqrt{3}} = \dfrac{5\sqrt{7}}{2\sqrt{3}} \cdot \dfrac{\sqrt{3}}{\sqrt{3}}$$
$$= \dfrac{5\sqrt{7} \cdot \sqrt{3}}{2\sqrt{3} \cdot \sqrt{3}}$$
$$= \dfrac{5\sqrt{21}}{6}$$

$$\dfrac{9\sqrt{5x}}{\sqrt{24x^4y^3}} = \dfrac{9\sqrt{5x}}{2x^2y\sqrt{6y}}$$
$$= \dfrac{9\sqrt{5x}}{2x^2y\sqrt{6y}} \cdot \dfrac{\sqrt{6y}}{\sqrt{6y}}$$
$$= \dfrac{9\sqrt{30xy}}{12x^2y^2}, \text{ or } \dfrac{3\sqrt{30xy}}{4x^2y^2}$$

The denominator of $\dfrac{7}{\sqrt{15} - \sqrt{10}}$ contains two terms. To rationalize this denominator, you use the **conjugate** of $\sqrt{15} - \sqrt{10}$, which is $\sqrt{15} + \sqrt{10}$. The product of these two irrational expressions is $(\sqrt{15} - \sqrt{10})(\sqrt{15} + \sqrt{10})$, which is equal to $15 - 10$, or the rational number 5.

| Property of conjugates | The two binomials, $a + b$ and $a - b$, are a pair of *conjugates*. Their product is a difference of two squares: $(a + b)(a - b) = a^2 - b^2$. |

The Property of Conjugates is used in Example 4 to simplify two radical expressions whose denominators are binomials.

Example 4 Simplify $\dfrac{7}{\sqrt{15} - \sqrt{10}}$ and $\dfrac{6\sqrt{y}}{5x + 2\sqrt{y}}$; $y > 0, 5x + 2\sqrt{y} \neq 0$.

$\dfrac{7}{\sqrt{15} - \sqrt{10}} = \dfrac{7}{\sqrt{15} - \sqrt{10}} \cdot \dfrac{\sqrt{15} + \sqrt{10}}{\sqrt{15} + \sqrt{10}}$

$\phantom{\dfrac{7}{\sqrt{15} - \sqrt{10}}} = \dfrac{7(\sqrt{15} + \sqrt{10})}{(\sqrt{15} - \sqrt{10})(\sqrt{15} + \sqrt{10})}$

$(a - b)(a + b)$
$= a^2 - b^2$ ▶ $\phantom{\dfrac{7}{\sqrt{15} - \sqrt{10}}} = \dfrac{7\sqrt{15} + 7\sqrt{10}}{(\sqrt{15})^2 - (\sqrt{10})^2}$

$\phantom{\dfrac{7}{\sqrt{15} - \sqrt{10}}} = \dfrac{7\sqrt{15} + 7\sqrt{10}}{5}$

$\dfrac{6\sqrt{y}}{5x + 2\sqrt{y}} = \dfrac{6\sqrt{y}}{5x + 2\sqrt{y}} \cdot \dfrac{5x - 2\sqrt{y}}{5x - 2\sqrt{y}}$

$\phantom{\dfrac{6\sqrt{y}}{5x + 2\sqrt{y}}} = \dfrac{6\sqrt{y}(5x - 2\sqrt{y})}{(5x + 2\sqrt{y})(5x - 2\sqrt{y})}$

$\phantom{\dfrac{6\sqrt{y}}{5x + 2\sqrt{y}}} = \dfrac{30x\sqrt{y} - 12y}{(5x)^2 - (2\sqrt{y})^2}$

$\phantom{\dfrac{6\sqrt{y}}{5x + 2\sqrt{y}}} = \dfrac{30x\sqrt{y} - 12y}{25x^2 - 4y}$

Quotients of Radicals

To simplify $\sqrt{\frac{7}{10}}$, you use the Quotient of Square Roots property to rewrite it as $\frac{\sqrt{7}}{\sqrt{10}}$ and then proceed to rationalize the denominator.

Example 5 **Simplify $\sqrt{\frac{7}{10}}$.**

$$\sqrt{\frac{7}{10}} = \frac{\sqrt{7}}{\sqrt{10}} = \frac{\sqrt{7}}{\sqrt{10}} \cdot \frac{\sqrt{10}}{\sqrt{10}} = \frac{\sqrt{70}}{10}$$

Example 6 **Simplify $\sqrt{\frac{5c}{12d^3}}$; $c > 0$, $d > 0$.**

$$\sqrt{\frac{5c}{12d^3}} = \frac{\sqrt{5c}}{\sqrt{12d^3}} = \frac{\sqrt{5c}}{2d\sqrt{3d}} = \frac{\sqrt{5c}}{2d\sqrt{3d}} \cdot \frac{\sqrt{3d}}{\sqrt{3d}} = \frac{\sqrt{15cd}}{6d^2}$$

Summary An expression is in simplest radical form when
(1) each radicand contains no factor, other than 1, that is a perfect square,
(2) the denominator contains no radicals, and
(3) each radicand contains no fractions.

Reading in Algebra

Indicate why each expression is not in simplest radical form.

1. $\frac{5x^2}{\sqrt{3x}}$ **The denominator contains a radical.**
2. $\frac{\sqrt{8y}}{5y}$ **The radicand contains a perfect square factor.**
3. $\sqrt{\frac{7m}{3n}}$ **The radicand contains a fraction.**
4. $\frac{4\sqrt{3}}{6 - 5\sqrt{2}}$ **The denominator contains a radical.**

Oral Exercises

Determine the conjugate of each expression.

1. $3x + 2y$ **$3x - 2y$**   2. $2\sqrt{3} - 5\sqrt{2}$ **$2\sqrt{3} + 5\sqrt{2}$**   3. $-2 + \sqrt{6}$ **$-2 - \sqrt{6}$**   4. $-\sqrt{30} - \sqrt{10}$ **$-\sqrt{30} + \sqrt{10}$**

Written Exercises

Simplify. Assume that no denominator has the value of 0 and that values of the variables and radicands are nonnegative numbers.

Ⓐ   1. $\frac{\sqrt{21}}{\sqrt{7}}$ **$\sqrt{3}$**   2. $\frac{\sqrt{90}}{\sqrt{10}}$ **3**   3. $\frac{\sqrt{88}}{\sqrt{11}}$ **$2\sqrt{2}$**   4. $\frac{\sqrt{150}}{\sqrt{2}}$ **$5\sqrt{3}$**

5. $\frac{\sqrt{a^{10}}}{\sqrt{a^6}}$ **a^2**   6. $\frac{\sqrt{x^6}}{\sqrt{x}}$ **$x^2\sqrt{x}$**   7. $\frac{\sqrt{45n^9}}{\sqrt{5n^3}}$ **$3n^3$**   8. $\frac{\sqrt{24x^{12}}}{\sqrt{3x^5}}$ **$2x^3\sqrt{2x}$**

9. $\frac{1}{\sqrt{7}}$ **$\frac{\sqrt{7}}{7}$**   10. $\frac{8}{5\sqrt{3}}$ **$\frac{8\sqrt{3}}{15}$**   11. $\frac{\sqrt{11}}{2\sqrt{6}}$ **$\frac{\sqrt{66}}{12}$**   12. $\frac{-7\sqrt{6}}{6\sqrt{5}}$ **$\frac{-7\sqrt{30}}{30}$**

13. $\dfrac{7}{\sqrt{12}}$ $\dfrac{7\sqrt{3}}{6}$
14. $\dfrac{6}{\sqrt{8}}$ $\dfrac{3\sqrt{2}}{2}$
15. $\dfrac{-5}{\sqrt{32}}$ $\dfrac{-5\sqrt{2}}{8}$
16. $\dfrac{5\sqrt{44}}{\sqrt{18}}$ $\dfrac{5\sqrt{22}}{3}$

17. $\dfrac{5}{\sqrt{x^3}}$ $\dfrac{5\sqrt{x}}{x^2}$
18. $\dfrac{6}{\sqrt{9c^5}}$ $\dfrac{2\sqrt{c}}{c^3}$
19. $\dfrac{-10d}{\sqrt{32d^3}}$ $\dfrac{-5\sqrt{2d}}{4d}$
20. $\dfrac{15x^2}{\sqrt{5x^7}}$ $\dfrac{3\sqrt{5x}}{x^2}$

21. $\dfrac{7}{6-\sqrt{30}}$ $\dfrac{42+7\sqrt{30}}{6}$
22. $\dfrac{-5}{3+\sqrt{2}}$ $\dfrac{-15+5\sqrt{2}}{7}$
23. $\dfrac{\sqrt{2}}{\sqrt{15}-3}$ $\dfrac{\sqrt{30}+3\sqrt{2}}{6}$
24. $\dfrac{-\sqrt{7}}{\sqrt{26}+4}$ $\dfrac{-\sqrt{182}+4\sqrt{7}}{10}$

25. $\sqrt{\dfrac{5}{11}}$ $\dfrac{\sqrt{55}}{11}$
26. $\sqrt{\dfrac{3c}{5d}}$ $\dfrac{\sqrt{15cd}}{5d}$
27. $\sqrt{\dfrac{23x}{8y}}$ $\dfrac{\sqrt{46xy}}{4y}$
28. $\sqrt{\dfrac{27m}{32n}}$ $\dfrac{3\sqrt{6mn}}{8n}$

B 29. $\sqrt{\dfrac{4x^7}{7y^3}}$ $\dfrac{2x^3\sqrt{7xy}}{7y^2}$
30. $\sqrt{\dfrac{7a^3}{8m^3n^5}}$ $\dfrac{a\sqrt{14amn}}{4m^2n^3}$
31. $\sqrt{\dfrac{3c-5}{6}}$ $\dfrac{\sqrt{18c-30}}{6}$
32. $\sqrt{\dfrac{5}{4x+9}}$ $\dfrac{\sqrt{20x+45}}{4x+9}$

33. $\dfrac{10}{\sqrt{22}+\sqrt{11}}$ below
34. $\dfrac{2}{5\sqrt{3}-\sqrt{15}}$ $\dfrac{5\sqrt{3}+\sqrt{15}}{30}$
35. $\dfrac{\sqrt{5}}{2\sqrt{5}+3\sqrt{2}}$ $\dfrac{10-3\sqrt{10}}{2}$
36. $\dfrac{2\sqrt{3}}{4\sqrt{2}-3\sqrt{3}}$ $\dfrac{8\sqrt{6}+18}{5}$

37. $\dfrac{5+\sqrt{10}}{8-\sqrt{10}}$ $\dfrac{50+13\sqrt{10}}{54}$
38. $\dfrac{\sqrt{7}+2}{3\sqrt{7}+9}$ $\dfrac{1-\sqrt{7}}{-6}$
39. $\dfrac{3\sqrt{x}}{5\sqrt{x}-2\sqrt{y}}$ below
40. $\dfrac{\sqrt{cd}}{6\sqrt{c}-3\sqrt{d}}$ below

C 41. $\dfrac{3\sqrt{x+y}}{\sqrt{x+y}+2}$ below
42. $\dfrac{\sqrt{c+d}}{5-\sqrt{c+d}}$ below
43. $\dfrac{4\sqrt{5}}{(\sqrt{5}+\sqrt{3})-2}$ below
44. $\dfrac{\sqrt{2}}{(\sqrt{30}-\sqrt{2})+3}$ below

33. $\dfrac{10\sqrt{22}-10\sqrt{11}}{11}$
39. $\dfrac{15x+6\sqrt{xy}}{25x-4y}$
40. $\dfrac{2c\sqrt{d}+d\sqrt{c}}{12c-3d}$
41. $\dfrac{3x+3y-6\sqrt{x+y}}{x+y-4}$
42. $\dfrac{5\sqrt{c+d}+c+d}{25-c-d}$

NON-ROUTINE PROBLEMS

1. Find the value of $\dfrac{x-1}{x+1}$ in simplest radical form if you replace each x with $\dfrac{n-1}{n+1}$ and replace each n with $\sqrt{2}$. $\dfrac{-\sqrt{2}}{2}$

2. $1+\sqrt{3}+\dfrac{1}{1-\sqrt{3}}+\dfrac{1}{1+\sqrt{3}}=$ **C**
   (A) $2+\sqrt{3}$ (B) $2+2\sqrt{3}$ (C) $\sqrt{3}$ (D) $1+\sqrt{3}$ (E) $1-\sqrt{3}$

Simplify. Answer in simplest radical form.

3. $\dfrac{\sqrt{\sqrt{\sqrt{16x^{16}}}}}{\sqrt{\sqrt{25y^{24}}}}$ $\dfrac{x^2\sqrt{10}}{5y^6}$

4. $\dfrac{\sqrt{\sqrt{\sqrt{\sqrt{256x^{64}}}}}}{\sqrt{\sqrt{\sqrt{81y^{32}}}}}$ $\dfrac{x^4\sqrt{6}}{3y^4}$

CUMULATIVE REVIEW

Brand A: 12 lb, Brand B: 8 lb

1. Tonya mixed some Brand A dog food worth 45¢/lb and some Brand B dog food worth 63¢/lb to make a 20-lb package that was worth $10.44. Find the amount of each brand that was used.

2. One bus left a terminal at 7:00 A.M. and headed north at 80 km/h. A second bus left the same terminal at 8:30 A.M. and traveled south at 90 km/h. At what time were the buses 460 km apart? **10:30 A.M.**

3. Find three consecutive multiples of 4 such that the product of the second and the third numbers is 80 more than 20 times the first number. **−4, 0, 4 or 12, 16, 20**

4. How many milliliters of water must be added to 240 mL of an $87\tfrac{1}{2}$% iodine solution to dilute it to a $62\tfrac{1}{2}$% solution? **96 mL**

43. $\dfrac{-10-6\sqrt{15}+8\sqrt{5}-20\sqrt{3}}{-11}$
44. $\dfrac{74-69\sqrt{2}+38\sqrt{15}-12\sqrt{30}}{289}$

Quotients of Radicals

OTHER RADICALS   6.5

Objective **To simplify expressions containing cube roots, fourth roots, or fifth roots**

You have solved equations like $x^2 = 49$ by using the definition of square root. In a similar way, you can solve equations such as $x^3 = -8$, $x^4 = 16$, and $x^5 = -32$.

If $x^3 = -8$,	If $x^4 = 16$,	If $x^5 = -32$,
then $x = \sqrt[3]{-8} = -2$	then $x = \sqrt[4]{16} = 2$ or $x = -\sqrt[4]{16} = -2$	then $x = \sqrt[5]{-32} = -2$
since $(-2)^3 = -8$.	since $2^4 = 16$ and $(-2)^4 = 16$.	since $(-2)^5 = -32$.

The three illustrations above suggest that:

(1) if $x^3 = k$, then $x = \sqrt[3]{k}$,
(2) if $x^4 = k$, and $k \geq 0$, then $x = \sqrt[4]{k}$ or $x = -\sqrt[4]{k}$, and
(3) if $x^5 = k$, then $x = \sqrt[5]{k}$.

This leads to the following definition of the **nth root** of a real number k.

Definition: nth root

For all odd integers $n > 1$ and all real numbers k, if $x^n = k$, then $x = \sqrt[n]{k}$.

For all even integers $n > 1$ and all real numbers $k \geq 0$, if $x^n = k$, then $x = \sqrt[n]{k}$ or $x = -\sqrt[n]{k}$.

The symbol $\sqrt[n]{k}$ (read "the nth root of k") is called a **radical** where k is the **radicand**, $\sqrt[n]{}$ is the **radical sign**, and n is the **index**. [Note: The index 2 is usually omitted so that $\sqrt{k}$ and $\sqrt[2]{k}$ both mean "the square root of k."]
You may find the following table of powers helpful when you are required to simplify cube roots, fourth roots, or fifth roots.

Base	n	2	3	4	5	10	x^2	x^3	x^4
3rd power	n^3	8	27	64	125	1,000	x^6	x^9	x^{12}
4th power	n^4	16	81	256	625	10,000	x^8	x^{12}	x^{16}
5th power	n^5	32	243	1,024	3,125	100,000	x^{10}	x^{15}	x^{20}

All the properties involving operations with square roots can be extended to other radicals. For example, the Square Root of a Product property can be extended to cube roots as shown below.

$$\sqrt[3]{8x^{12}} = \sqrt[3]{8} \cdot \sqrt[3]{x^{12}} = 2x^4$$

Use the table above to verify that $\sqrt[3]{8} = 2$ and $\sqrt[3]{x^{12}} = x^4$.

---

nth root of a product

For all odd integers $n > 1$ and all real numbers x and y, $\sqrt[n]{x \cdot y} = \sqrt[n]{x} \cdot \sqrt[n]{y}$.

For all even integers $n > 1$ and all real numbers $x \geq 0$, $y \geq 0$, $\sqrt[n]{x \cdot y} = \sqrt[n]{x} \cdot \sqrt[n]{y}$.

You can use the nth Root of a Product property to simplify radicals such as $\sqrt[5]{-64}$. First find the greatest 5th power that is also a factor of -64, as shown in Example 1.

Example 1 Simplify $\sqrt[5]{-64}$, $\sqrt[3]{c^8}$, and $\sqrt[4]{64a^{21}}$, $a \geq 0$.

$$\sqrt[5]{-64} = \sqrt[5]{-32 \cdot 2} = \sqrt[5]{-32} \cdot \sqrt[5]{2} = -2\sqrt[5]{2}$$

$$\sqrt[3]{c^8} = \sqrt[3]{c^6} \cdot \sqrt[3]{c^2} = c^2\sqrt[3]{c^2}$$

$$\sqrt[4]{64a^{21}} = \sqrt[4]{16 \cdot 4 \cdot a^{20} \cdot a} = \sqrt[4]{16} \cdot \sqrt[4]{4} \cdot \sqrt[4]{a^{20}} \cdot \sqrt[4]{a} = 2a^5\sqrt[4]{4a}$$

You can combine (add or subtract) like radicals and simplify products of radicals for nth roots as you have done previously with square roots.

Example 2 Simplify $5\sqrt[3]{3} + \sqrt[3]{24} + \sqrt[3]{81}$.

$$5\sqrt[3]{3} + \sqrt[3]{24} + \sqrt[3]{81} = 5\sqrt[3]{3} + \sqrt[3]{8 \cdot 3} + \sqrt[3]{27 \cdot 3} = 5\sqrt[3]{3} + 2\sqrt[3]{3} + 3\sqrt[3]{3} = 10\sqrt[3]{3}$$

Example 3 Simplify $\sqrt[4]{8d^3} \cdot \sqrt[4]{10d} \cdot \sqrt[4]{d^2}$, $d \geq 0$.

$$\sqrt[4]{8d^3} \cdot \sqrt[4]{10d} \cdot \sqrt[4]{d^2} = \sqrt[4]{8d^3 \cdot 10d \cdot d^2} = \sqrt[4]{80d^6} = \sqrt[4]{16 \cdot 5 \cdot d^4 \cdot d^2} = 2d\sqrt[4]{5d^2}$$

Recall that $\sqrt{5} \cdot \sqrt{5} = (\sqrt{5})^2 = 5$. Similarly, notice that

$$\sqrt[3]{27} \cdot \sqrt[3]{27} \cdot \sqrt[3]{27} = (\sqrt[3]{27})^3 = (3)^3 = 27 \text{ and}$$
$$\sqrt[4]{16} \cdot \sqrt[4]{16} \cdot \sqrt[4]{16} \cdot \sqrt[4]{16} = (\sqrt[4]{16})^4 = (2)^4 = 16.$$

In general, $(\sqrt[n]{k})^n = k$. This property may be used to *rationalize the denominator* of $\dfrac{5}{\sqrt[3]{36}}$.

$$\frac{5}{\sqrt[3]{36}} = \frac{5}{\sqrt[3]{6 \cdot 6}} = \frac{5}{\sqrt[3]{6} \cdot \sqrt[3]{6}} = \frac{5}{\sqrt[3]{6} \cdot \sqrt[3]{6}} \cdot \frac{\sqrt[3]{6}}{\sqrt[3]{6}} = \frac{5 \cdot \sqrt[3]{6}}{(\sqrt[3]{6})^3} = \frac{5\sqrt[3]{6}}{6}$$

> **nth root of a quotient** For all odd integers $n > 1$ and all real numbers x and y, $y \neq 0$, and for all even integers $n > 1$ and all real numbers $x \geq 0$, $y > 0$,
> $$\sqrt[n]{\frac{x}{y}} = \frac{\sqrt[n]{x}}{\sqrt[n]{y}}.$$

Other Radicals

Example 4   Simplify $\sqrt[4]{\dfrac{7}{25a^3}}$, $a > 0$.

$$\sqrt[4]{\dfrac{7}{25a^3}} = \dfrac{\sqrt[4]{7}}{\sqrt[4]{25a^3}} = \dfrac{\sqrt[4]{7}}{\sqrt[4]{5 \cdot 5 \cdot a \cdot a \cdot a}} = \dfrac{\sqrt[4]{7}}{\sqrt[4]{5 \cdot 5 \cdot a \cdot a \cdot a}} \cdot \dfrac{\sqrt[4]{5 \cdot 5 \cdot a}}{\sqrt[4]{5 \cdot 5 \cdot a}} = \dfrac{\sqrt[4]{7 \cdot 25a}}{\sqrt[4]{5^4 a^4}} = \dfrac{\sqrt[4]{175a}}{5a}$$

If a denominator has two terms and at least one term involves a cube root, then the factors of $a^3 \pm b^3$ can be used to rationalize the denominator.

Example 5   Simplify $\dfrac{3}{2\sqrt[3]{x} + \sqrt[3]{5}}$.

$$\dfrac{3}{2\sqrt[3]{x} + \sqrt[3]{5}} \cdot \dfrac{4\sqrt[3]{x^2} - 2\sqrt[3]{5x} + \sqrt[3]{25}}{4\sqrt[3]{x^2} - 2\sqrt[3]{5x} + \sqrt[3]{25}} = \dfrac{3(4\sqrt[3]{x^2} - 2\sqrt[3]{5x} + \sqrt[3]{25})}{(2\sqrt[3]{x})^3 + (\sqrt[3]{5})^3}$$

$$(a + b) \quad (a^2 - ab + b^2) = a^3 + b^3$$

$$= \dfrac{12\sqrt[3]{x^2} - 6\sqrt[3]{5x} + 3\sqrt[3]{25}}{8x + 5}$$

Written Exercises

Simplify each expression. Assume that no denominator has the value of 0.

(A)
1. $\sqrt[3]{-27}$ -3
2. $\sqrt[3]{-125}$ -5
3. $\sqrt[4]{625}$ 5
4. $-\sqrt[4]{81}$ -3
5. $\sqrt[3]{x^9}$ x^3
6. $\sqrt[4]{y^8}$ y^2
7. $\sqrt[4]{y^{20}}$, $y \geq 0$ y^5
8. $\sqrt[3]{x^3 y^6}$ xy^2
9. $\sqrt[4]{32}$ $2\sqrt[4]{2}$
10. $\sqrt[3]{-40}$ $-2\sqrt[3]{5}$
11. $\sqrt[4]{128}$ $4\sqrt[4]{2}$
12. $\sqrt[4]{243}$ $3\sqrt[4]{3}$
13. $\sqrt[3]{a^5}$ $a\sqrt[3]{a^2}$
14. $\sqrt[4]{b^7}$, $b \geq 0$ $b\sqrt[4]{b^3}$
15. $\sqrt[4]{x^{10}}$ $x^3\sqrt[4]{x}$

16. $\sqrt[4]{y^9}$, $y \geq 0$ $y^2\sqrt[4]{y}$
17. $\sqrt[4]{8} + \sqrt[4]{8}$ $2\sqrt[4]{8}$
18. $6\sqrt[3]{4} + \sqrt[3]{4}$ $7\sqrt[3]{4}$
19. $\sqrt[4]{32} + 3\sqrt[4]{2}$ $5\sqrt[4]{2}$
20. $\sqrt[3]{54} + \sqrt[3]{2}$ $4\sqrt[3]{2}$
21. $\sqrt[3]{-64x^6}$ $-4x^2$
22. $\sqrt[4]{81y^{16}}$ $3y^4$
23. $\sqrt[3]{250a^7}$ $5a^2\sqrt[3]{2a}$
24. $\sqrt[4]{162b^{19}}$, $b \geq 0$ $3b^4\sqrt[4]{2b^3}$
25. $\sqrt[4]{7a^2} \cdot \sqrt[4]{3a^2}$, $a \geq 0$
26. $\sqrt[3]{4a} \cdot \sqrt[3]{16a}$ $4\sqrt[3]{a^2}$
27. $\sqrt[4]{8x^3} \cdot \sqrt[4]{20x^6}$, $x \geq 0$
28. $\sqrt[3]{25y^4} \cdot \sqrt[3]{10y}$ $5y\sqrt[3]{2y^2}$
29. $\dfrac{2}{\sqrt[3]{7}}$ $\dfrac{2\sqrt[3]{49}}{7}$ $a\sqrt[4]{21}$
30. $\dfrac{6x}{\sqrt[4]{3}}$ $2x\sqrt[4]{27}$
31. $\dfrac{7\sqrt[3]{32}}{\sqrt[3]{4}}$ 14 $2x^2\sqrt[4]{10x}$
32. $\dfrac{3\sqrt[4]{144}}{5\sqrt[4]{9}}$ $\dfrac{6}{5}$
33. $\sqrt[4]{\dfrac{1}{5}}$ $\dfrac{\sqrt[4]{125}}{5}$
34. $\sqrt[3]{\dfrac{5}{6}}$ $\dfrac{\sqrt[3]{180}}{6}$
35. $\sqrt[4]{\dfrac{11}{8}}$ $\dfrac{\sqrt[4]{22}}{2}$
36. $\sqrt[3]{\dfrac{10}{9}}$ $\dfrac{\sqrt[3]{30}}{3}$

(B)
37. $\sqrt[3]{216}$ 6 $2xy^5\sqrt[4]{125x^3y^2}$
38. $\sqrt[4]{256}$ 4
39. $\sqrt[5]{-243}$ -3
40. $\sqrt[4]{2{,}000x^7y^{22}}$, $x \geq 0$
41. $5ab\sqrt[3]{270a^5b^{10}}$ $15a^2b^4\sqrt[3]{10a^2b}$
42. $\sqrt[5]{64m^{20}n^{14}}$ $2m^4n^2\sqrt[5]{2n^4}$
43. $8\sqrt[3]{4} - \sqrt[3]{32} + \sqrt[3]{108}$ $9\sqrt[3]{4}$
44. $\sqrt[4]{2} + 3\sqrt[4]{32} - \sqrt[4]{162}$ $4\sqrt[4]{2}$
45. $\sqrt[5]{2} - \sqrt[5]{64} + \sqrt[5]{486}$ $2\sqrt[5]{2}$
46. $\sqrt[4]{16c^7} + \sqrt[4]{81c^7}$, $c \geq 0$
47. $\sqrt[3]{8a^5} + \sqrt[3]{27a^5}$ $5a\sqrt[3]{a^2}$
48. $\sqrt[5]{32y^9} - \sqrt[5]{243y^9}$ $-y\sqrt[5]{y^4}$
49. $\sqrt[3]{2x^2} \cdot \sqrt[3]{8x^2} \cdot \sqrt[3]{4x^2}$ $4x^2$
50. $\sqrt[3]{3y^3} \cdot \sqrt[3]{6y^3} \cdot \sqrt[3]{9y^3}$, $y \geq 0$
51. $\sqrt[5]{3n^4} \cdot \sqrt[5]{8n^3} \cdot \sqrt[5]{4n^2}$ $2n\sqrt[5]{3n^4}$
52. $\sqrt[3]{\dfrac{5}{4y^2}}$ $\dfrac{\sqrt[3]{10y}}{2y}$ $5c\sqrt[4]{c^3}$
53. $\dfrac{6a^2}{\sqrt[4]{8a^5}}$, $a > 0$ $\dfrac{3a\sqrt[4]{2a^2}}{3y^2\sqrt[4]{2y}}$
54. $\dfrac{12n}{\sqrt[3]{9n^7}}$ $\dfrac{4\sqrt[3]{3n^2}}{n^2}$
55. $\dfrac{4}{\sqrt[3]{7} - \sqrt[3]{5}}$ below
56. $\dfrac{-2}{\sqrt[3]{a} + 4}$ below
57. $\dfrac{c}{5\sqrt[3]{c} + 2d}$ below
58. $\dfrac{2t}{\sqrt[3]{6t} - 4\sqrt[3]{2v}}$ below

Simplify. Assume that n is an integer, $n > 1$, and that all other variables represent positive real numbers.

55. $2\sqrt[3]{49} + 2\sqrt[3]{35} + 2\sqrt[3]{25}$

(C)
59. $\sqrt[3]{x^{6n}}$ x^{2n}
60. $\sqrt[4]{y^{4n+12}}$ y^{n+3}
61. $\sqrt[3]{x^{2n+1}} \cdot \sqrt[3]{x^{4n+2}}$ x^{2n+1}
62. $y^n\sqrt[4]{y^{3n}} + \sqrt[4]{y^{7n}}$ $2y^n\sqrt[4]{y^{3n}}$
63. $\sqrt[n]{x^{4n}}$ x^4
64. $\sqrt[n]{x^{3n+2}}$ $x^3\sqrt[n]{x^2}$
65. $\sqrt[n]{a^n x^{5n+1}}$ $ax^5\sqrt[n]{x}$
66. $\sqrt[n]{ab^{2n}x^{2n+3}}$ $b^2x^2\sqrt[n]{ax^3}$

56. $\dfrac{-2\sqrt[3]{a^2} + 8\sqrt[3]{a} - 32}{a + 64}$

57. $\dfrac{25c\sqrt[3]{c^2} - 10cd\sqrt[3]{c} + 4cd^2}{125c + 8d^3}$

58. $\dfrac{t\sqrt[3]{36t^2} + 4t\sqrt[3]{12tv} + 16t\sqrt[3]{4v^2}}{3t - 64v}$

RATIONAL NUMBER EXPONENTS   6.6

Objectives
To write a radical expression in exponential form
To write an exponential expression in radical form
To evaluate an expression containing rational number exponents

Each of the properties and definitions for integral exponents can be extended to rational number exponents. Using the Product of Powers property, $3^{\frac{1}{2}} \cdot 3^{\frac{1}{2}} = 3^1 = 3$. But you know that $\sqrt{3} \cdot \sqrt{3} = 3$. Thus, since $3^{\frac{1}{2}}$ and $\sqrt{3}$ are two different symbols for the same number, $3^{\frac{1}{2}} = \sqrt{3}$.
Similarly, $5^{\frac{1}{3}} = \sqrt[3]{5}$ since $5^{\frac{1}{3}} \cdot 5^{\frac{1}{3}} \cdot 5^{\frac{1}{3}} = 5$ and $\sqrt[3]{5} \cdot \sqrt[3]{5} \cdot \sqrt[3]{5} = 5$. Notice that $3^{\frac{1}{2}}$ and $5^{\frac{1}{3}}$ are written in *exponential form* and that $\sqrt{3}$ and $\sqrt[3]{5}$ are written in *radical form*.

Definition:
$x^{\frac{1}{n}}$

For all integers $n > 1$ and all real numbers x,
$$x^{\frac{1}{n}} = \sqrt[n]{x},$$
except when n is even and $x < 0$.

You will use the definition of $x^{\frac{1}{n}}$ to write radical expressions in exponential form, to write exponential expressions in radical form, and to evaluate expressions containing rational number exponents.

Example 1 Write $7\sqrt[4]{a}$ and $\sqrt[4]{7a}$, $a \geq 0$, in exponential form.

$7\sqrt[4]{a} = 7 \cdot \sqrt[4]{a} = 7a^{\frac{1}{4}}$ $\qquad$ $\sqrt[4]{7a} = (7a)^{\frac{1}{4}}$

Example 2 Write $5y^{\frac{1}{3}}$ and $(5y)^{\frac{1}{3}}$ in radical form.

$5y^{\frac{1}{3}} = 5 \cdot y^{\frac{1}{3}} = 5\sqrt[3]{y}$ $\qquad$ $(5y)^{\frac{1}{3}} = \sqrt[3]{5y}$

Example 3 Find the value of $36^{\frac{1}{2}}$ and $10 \cdot 81^{\frac{1}{4}}$.

$36^{\frac{1}{2}} = \sqrt{36} = 6$ $\qquad$ $10 \cdot 81^{\frac{1}{4}} = 10 \cdot \sqrt[4]{81} = 10 \cdot 3 = 30$

Since $\frac{2}{3} = 2 \cdot \frac{1}{3}$, you can rewrite $8^{\frac{2}{3}}$ as $(8^2)^{\frac{1}{3}}$ or $(8^{\frac{1}{3}})^2$, using the Power of a Power property. Then, $(8^2)^{\frac{1}{3}} = \sqrt[3]{8^2}$ and $(8^{\frac{1}{3}})^2 = (\sqrt[3]{8})^2$. Notice that $\sqrt[3]{8^2} = \sqrt[3]{64} = 4$ and $(\sqrt[3]{8})^2 = (2)^2 = 4$. **Thus, $\sqrt[3]{8^2} = (\sqrt[3]{8})^2$.**

Rational Number Exponents

> **Definition:** For all integers m and n, with $n > 1$, and all nonzero real numbers x,
> $$x^{\frac{m}{n}} = (\sqrt[n]{x})^m = \sqrt[n]{x^m},$$
> except when n is even and $x < 0$.

Example 4   Write $\sqrt[5]{x^4}$ and $(\sqrt[4]{15})^3$ in exponential form.

$\sqrt[5]{x^4} = x^{\frac{4}{5}}$ ◀ $\sqrt[n]{x^m} = x^{\frac{m}{n}}$ $(\sqrt[4]{15})^3 = 15^{\frac{3}{4}}$ ◀ $(\sqrt[n]{x})^m = x^{\frac{m}{n}}$

Example 5   Write $5y^{\frac{2}{3}}$ and $(5y)^{\frac{2}{3}}$ in radical form.

$5y^{\frac{2}{3}} = 5 \cdot y^{\frac{2}{3}} = 5\sqrt[3]{y^2}$ $(5y)^{\frac{2}{3}} = \sqrt[3]{(5y)^2} = \sqrt[3]{25y^2}$

Example 6   Find the value of $2 \cdot 81^{\frac{5}{4}}$ and $8^{-\frac{2}{3}}$.

$2 \cdot 81^{\frac{5}{4}} = 2(\sqrt[4]{81})^5 = 2 \cdot 3^5 = 2 \cdot 243 = 486$

$8^{-\frac{2}{3}} = \frac{1}{8^{\frac{2}{3}}} = \frac{1}{(\sqrt[3]{8})^2} = \frac{1}{2^2} = \frac{1}{4}$ ◀ $x^{-n} = \frac{1}{x^n}$

Notice in Example 6 how an expression containing a negative rational number exponent is changed into an equivalent expression containing a positive rational number exponent.

Written Exercises

Write each expression in exponential form.

Ⓐ 1. $\sqrt{5}$ $5^{\frac{1}{2}}$ 2. $\sqrt[3]{30}$ $30^{\frac{1}{3}}$ 3. $\sqrt[4]{21}$ $21^{\frac{1}{4}}$ 4. $7\sqrt[3]{a}$ $7a^{\frac{1}{3}}$ 5. $\sqrt{3b}, b \geq 0$ $(3b)^{\frac{1}{2}}$ 6. $3\sqrt[5]{9c}$ $3(9c)^{\frac{1}{5}}$

Write each expression in radical form.

7. $17^{\frac{1}{2}}$ $\sqrt{17}$ 8. $a^{\frac{1}{5}}$ $\sqrt[5]{a}$ 9. $6c^{\frac{1}{3}}$ $6\sqrt[3]{c}$ 10. $(6c)^{\frac{1}{3}}$ $\sqrt[3]{6c}$ 11. $(7n)^{\frac{1}{4}}, n \geq 0$ $\sqrt[4]{7n}$ 12. $7n^{\frac{1}{4}}, n \geq 0$ $7\sqrt[4]{n}$

Find the value of each expression.

13. $25^{\frac{1}{2}}$ 5   14. $27^{\frac{1}{3}}$ 3   15. $81^{\frac{1}{4}}$ 3   16. $20 \cdot 32^{\frac{1}{5}}$ 17. $-7 \cdot 64^{\frac{1}{2}}$ 18. $-10 \cdot 625^{\frac{1}{4}}$
19. $8^{-\frac{1}{3}}$ $\frac{1}{2}$ 20. $36^{-\frac{1}{2}}$ $\frac{1}{6}$ 21. $16^{-\frac{1}{4}}$ $\frac{1}{2}$ 22. $6 \cdot 9^{-\frac{1}{2}}$ 2   23. $8^{\frac{1}{3}} \cdot 16^{\frac{1}{4}}$ 24. $-8 \cdot 32^{-\frac{1}{5}}$

16. 40   17. −56   18. −50   23. 4   24. −4

Ⓑ 25. $64^{\frac{2}{3}}$ 16   26. $16^{\frac{5}{4}}$ 32   27. $243^{\frac{2}{5}}$ 9   28. $11 \cdot 25^{\frac{5}{2}}$ 29. $5 \cdot 81^{\frac{3}{4}}$ 30. $9 \cdot 100^{\frac{5}{2}}$
31. $625^{-\frac{3}{4}}$ $\frac{1}{125}$ 32. $4^{-\frac{5}{2}}$ $\frac{1}{32}$ 33. $27^{-\frac{5}{3}}$ $\frac{1}{243}$ 34. $12 \cdot 8^{-\frac{4}{3}}$ $\frac{3}{4}$ 35. $10 \cdot 16^{-\frac{5}{4}}$ $\frac{5}{16}$ 36. $25^{-\frac{3}{2}} \cdot 16^{\frac{3}{4}}$ $\frac{8}{125}$

28. 1,375   29. 135   30. 900,000

Write each expression in exponential form.

37. $\sqrt[5]{c^3}$ $c^{\frac{3}{5}}$ 38. $(\sqrt[3]{10})^5$ $10^{\frac{5}{3}}$ 39. $\sqrt[6]{x^5}, x \geq 0$ $x^{\frac{5}{6}}$ 40. $(\sqrt[4]{23})^3$ $23^{\frac{3}{4}}$ 41. $\sqrt{n^7}, n \geq 0$ $n^{\frac{7}{2}}$ 42. $(\sqrt{53})^9$ $53^{\frac{9}{2}}$

Write each expression in simplest radical form.

43. $5^{\frac{3}{2}}$ $5\sqrt{5}$ 44. $2^{\frac{5}{3}}$ $2\sqrt[3]{4}$ 45. $10x^{\frac{2}{3}}$ $10\sqrt[3]{x^2}$ 46. $(10x)^{\frac{2}{3}}$ $\sqrt[3]{100x^2}$ 47. $(6t)^{\frac{3}{4}}, t \geq 0$ $\sqrt[4]{216t^3}$ 48. $6t^{\frac{3}{4}}, t \geq 0$ $6\sqrt[4]{t^3}$

EXPRESSIONS WITH RATIONAL NUMBER EXPONENTS   6.7

Objectives **To simplify an expression containing rational number exponents**
To solve an exponential equation using properties of rational number exponents

To simplify an expression containing rational number exponents, the properties for integral exponents and rationalizing denominators can be used as shown below.

$$8^{-\frac{3}{7}} = \frac{1}{8^{\frac{3}{7}}} = \frac{1}{8^{\frac{3}{7}}} \cdot \frac{8^{\frac{4}{7}}}{8^{\frac{4}{7}}} = \frac{8^{\frac{4}{7}}}{8^{\frac{7}{7}}}, \text{ or } \frac{8^{\frac{4}{7}}}{8}$$

Notice that such an expression is not in *simplest form* until the exponent is a *positive* number and the *denominator* is a rational number.

Example 1 Simplify $5a^{-1} \cdot 2a^{\frac{1}{2}} \cdot 4a^{-\frac{3}{4}}$, $a > 0$.

$$5a^{-1} \cdot 2a^{\frac{1}{2}} \cdot 4a^{-\frac{3}{4}} = \frac{5 \cdot 2 \cdot 4a^{\frac{1}{2}}}{a^1 \, a^{\frac{3}{4}}} = \frac{40a^{\frac{2}{4}}}{a^{\frac{4}{4}} \, a^{\frac{3}{4}}} = \frac{40}{a^{\frac{5}{4}}} \cdot \frac{a^{\frac{3}{4}}}{a^{\frac{3}{4}}} = \frac{40a^{\frac{3}{4}}}{a^2}$$

Example 2 Simplify $(a^{\frac{1}{2}}b^{\frac{2}{3}})^6$ and $(27c^6d^{-3})^{\frac{2}{3}}$; $a \geq 0$, $d \neq 0$.

$(xy)^n = x^n y^n$ ▶ $(a^{\frac{1}{2}}b^{\frac{2}{3}})^6 = (a^{\frac{1}{2}})^6 (b^{\frac{2}{3}})^6$ $(27c^6d^{-3})^{\frac{2}{3}} = 27^{\frac{2}{3}}(c^6)^{\frac{2}{3}}(d^{-3})^{\frac{2}{3}}$

$(x^m)^n = x^{mn}$ ▶ $= a^3 b^4$ $= (\sqrt[3]{27})^2 c^4 d^{-2}$ ◀ $x^{\frac{m}{n}} = (\sqrt[n]{x})^m$

$= \dfrac{9c^4}{d^2}$

The Quotient of Powers and Power of a Quotient properties are used in simplifying expressions in Examples 3 and 4, respectively.

Example 3 Simplify $\dfrac{t^{\frac{1}{2}}v^{-\frac{1}{5}}}{t^{\frac{2}{3}}v^{-\frac{4}{5}}}$; $t > 0$, $v \neq 0$.

$$\frac{t^{\frac{1}{2}}v^{-\frac{1}{5}}}{t^{\frac{2}{3}}v^{-\frac{4}{5}}} = \frac{t^{\frac{3}{6}}v^{\frac{4}{5}}}{t^{\frac{4}{6}}v^{\frac{1}{5}}} = \frac{v^{\frac{3}{5}}}{t^{\frac{1}{6}}} \cdot \frac{t^{\frac{5}{6}}}{t^{\frac{5}{6}}} = \frac{v^{\frac{3}{5}} t^{\frac{5}{6}}}{t}$$

Example 4 Simplify $\left(\dfrac{8a^{12}}{27b^9}\right)^{\frac{2}{3}}$, $b \neq 0$.

$\left(\dfrac{x}{y}\right)^n = \dfrac{x^n}{y^n}$ ▶ $\left(\dfrac{8a^{12}}{27b^9}\right)^{\frac{2}{3}} = \dfrac{(8a^{12})^{\frac{2}{3}}}{(27b^9)^{\frac{2}{3}}} = \dfrac{8^{\frac{2}{3}}(a^{12})^{\frac{2}{3}}}{27^{\frac{2}{3}}(b^9)^{\frac{2}{3}}} = \dfrac{(\sqrt[3]{8})^2 a^8}{(\sqrt[3]{27})^2 b^6} = \dfrac{4a^8}{9b^6}$

Expressions with Rational Number Exponents

You have solved an exponential equation like $3^{2x-1} = 3^7$ by setting the exponents equal to each other and solving the equation $2x - 1 = 7$. This method of solving an exponential equation is based upon the property that if $a^m = a^n$, then $m = n$. As shown in Example 5, some exponential equations that involve rational number exponents can be solved in a similar way.

Example 5 **Solve each equation.**

$8^x = 16$
Write 8 and 16 as powers of 2.
$(8)^x = 16$
$(2^3)^x = 2^4$
$2^{3x} = 2^4$
$3x = 4$ ◀ If $a^m = a^n$, then $m = n$.
$x = \dfrac{4}{3}$

Thus, the solution is $\dfrac{4}{3}$.

$3^{8x} = \dfrac{1}{81}$
Write $\dfrac{1}{81}$ as a power of 3.
$3^{8x} = 3^{-4}$ ◀ $\dfrac{1}{81} = \dfrac{1}{3^4} = 3^{-4}$
$8x = -4$
$x = -\dfrac{1}{2}$

Thus, the solution is $-\dfrac{1}{2}$.

Written Exercises

Simplify each expression. Assume that values of the variables are positive numbers.

(A) 1. $6^{\frac{3}{8}} \cdot 6^{-\frac{1}{8}} \cdot 6^{\frac{5}{8}}$ **$6^{\frac{7}{8}}$**
2. $x^{\frac{4}{3}} \cdot x^{\frac{5}{3}} \cdot x^{-\frac{7}{3}}$ **$x^{\frac{2}{3}}$**
3. $y^3 \cdot y^{-\frac{1}{2}}$ **$y^{\frac{5}{2}}$**
4. $5a^2 \cdot 6a^{-\frac{3}{8}}$ **$30a^{\frac{13}{8}}$**
5. $(x^{\frac{2}{5}})^{10}$ **x^4**
6. $(y^{12})^{\frac{2}{3}}$ **y^8**
7. $(n^{10})^{\frac{3}{5}}$ **n^6**
8. $(b^{-\frac{2}{3}})^{-9}$ **b^6**
9. $(x^{\frac{1}{4}}y^{\frac{3}{4}})^8$ **x^2y^6**
10. $(25a^2b^8)^{\frac{1}{2}}$ **$5ab^4$**
11. $(-4m^{\frac{1}{2}}n^{\frac{3}{2}})^2$ **$16mn^3$**
12. $(16c^{12}d^8)^{\frac{3}{4}}$ **$8c^9d^6$**
13. $\dfrac{a^{\frac{9}{7}}}{a^{\frac{3}{7}}}$ **$a^{\frac{6}{7}}$**
14. $\dfrac{x^{\frac{4}{5}}}{x^{-\frac{2}{5}}}$ **$x^{\frac{6}{5}}$**
15. $\dfrac{n^{-\frac{1}{3}}}{n^{-\frac{5}{3}}}$ **$n^{\frac{4}{3}}$**
16. $\dfrac{b^3}{b^{\frac{3}{4}}}$ **$b^{\frac{9}{4}}$**
17. $\left(\dfrac{x^{\frac{3}{4}}}{y^{\frac{1}{2}}}\right)^8$ **$\dfrac{x^6}{y^4}$**
18. $\left(\dfrac{a^{\frac{2}{3}}}{b^{\frac{3}{5}}}\right)^{15}$ **$\dfrac{a^{10}}{b^9}$**
19. $\left(\dfrac{25m^6}{9n^8}\right)^{\frac{1}{2}}$ **$\dfrac{5m^3}{3n^4}$**
20. $\left(\dfrac{27c^6}{8d^{12}}\right)^{\frac{2}{3}}$ **$\dfrac{9c^4}{4d^8}$**

Solve each equation.

21. $9^x = 27$ **$\dfrac{3}{2}$**
22. $3^{9x} = \dfrac{1}{27}$ **$-\dfrac{1}{3}$**
23. $125^x = 25$ **$\dfrac{2}{3}$**
24. $4^{6x} = \dfrac{1}{64}$ **$-\dfrac{1}{2}$**

(B) 25. $25^{2x+1} = 125$ **$\dfrac{1}{4}$**
26. $2^{2x+2} = \dfrac{1}{16}$ **-3**
27. $32^{3x-2} = 16$ **$\dfrac{14}{15}$**
28. $10^{3x-1} = \dfrac{1}{10{,}000}$ **-1**

Simplify each expression. Assume that values of the variables are positive numbers.

29. $x^{\frac{2}{5}} \cdot x^{\frac{1}{5}} \cdot x^{-\frac{4}{5}}$ **$x^{\frac{4}{5}}$**
30. $b^{-6} \cdot b^{\frac{5}{8}} \cdot \dfrac{b^{\frac{5}{8}}}{b^6}$
31. $7a^{\frac{3}{5}} \cdot 6a^{\frac{1}{2}} \cdot 5a^{\frac{3}{4}}$ **$210a^{\frac{37}{20}}$**
32. $-10n^{-\frac{3}{4}} \cdot 2n^{-\frac{5}{8}}$ **$\dfrac{-20n^{\frac{5}{8}}}{n^2}$**
33. $(x^{\frac{2}{3}})^{-9}$ **$\dfrac{1}{x^6}$**
34. $(x^{\frac{1}{2}}y^{-\frac{3}{4}})^8$ **$\dfrac{x^4}{y^6}$**
35. $(8m^9n^{-6})^{\frac{1}{3}}$ **$\dfrac{2m^3}{n^2}$**
36. $(125a^{-9}b^{12})^{-\frac{2}{3}}$ **$\dfrac{a^6}{25b^8}$**

37. $\dfrac{x^{\frac{1}{3}}y^{\frac{5}{4}}}{x^{\frac{2}{3}}y^{\frac{3}{4}}}$ $\dfrac{x^{\frac{2}{3}}y^{\frac{1}{2}}}{x}$

38. $\dfrac{a^{\frac{2}{5}}b^{\frac{1}{6}}}{a^{\frac{3}{10}}b^{\frac{1}{4}}}$ $\dfrac{a^{\frac{1}{10}}b^{\frac{11}{12}}}{b}$

39. $\dfrac{m^{-\frac{2}{3}}n^{-4}}{m^{-\frac{1}{3}}n^{-5}}$ $\dfrac{m^{\frac{2}{3}}n}{m}$

40. $\dfrac{27^{\frac{3}{4}}x^{-\frac{1}{2}}y^{\frac{1}{4}}}{27^{\frac{1}{12}}x^{\frac{3}{4}}y^{-\frac{1}{6}}}$ $\dfrac{9x^{\frac{3}{4}}y^{\frac{5}{12}}}{x^2}$

41. $\left(\dfrac{x^{-\frac{3}{4}}}{y^{-\frac{1}{4}}}\right)^8$ $\dfrac{y^2}{x^6}$

42. $\left(\dfrac{-2a^{-\frac{1}{3}}}{3b^{-\frac{2}{3}}}\right)^3$ $\dfrac{-8b^2}{27a}$

43. $\left(\dfrac{64m^{-3}}{27n^{-6}}\right)^{\frac{2}{3}}$ $\dfrac{16n^4}{9m^2}$

44. $\left(\dfrac{4c^{-6}}{25d^{-4}}\right)^{\frac{3}{2}}$ $\dfrac{8d^6}{125c^9}$

Simplify and write each expression with the least possible number of radical signs.

Example Simplify $\sqrt[3]{x} \cdot \sqrt[4]{x}$ and $(\sqrt[4]{x^3} + \sqrt[6]{y})(\sqrt[4]{x^3} - \sqrt[6]{y})$.

Two radical signs ▶ $\sqrt[3]{x} \cdot \sqrt[4]{x} = x^{\frac{1}{3}} \cdot x^{\frac{1}{4}}$ ◀ $\sqrt[n]{x} = x^{\frac{1}{n}}$

$\qquad\qquad\qquad\qquad\quad = x^{\frac{7}{12}}$ ◀ $x^m \cdot x^n = x^{m+n}$

One radical sign ▶ $\qquad = \sqrt[12]{x^7}$ ◀ $x^{\frac{m}{n}} = \sqrt[n]{x^m}$

Four radical signs ▶ $(\sqrt[4]{x^3} + \sqrt[6]{y})(\sqrt[4]{x^3} - \sqrt[6]{y})$

$\qquad\qquad\qquad\quad (x^{\frac{3}{4}} + y^{\frac{1}{6}})(x^{\frac{3}{4}} - y^{\frac{1}{6}})$ ◀ $\sqrt[n]{x^m} = x^{\frac{m}{n}}$

$\qquad\qquad\qquad\quad (x^{\frac{3}{4}})^2 - (y^{\frac{1}{6}})^2$ ◀ $(a+b)(a-b) = a^2 - b^2$

$\qquad\qquad\qquad\quad x^{\frac{3}{2}} - y^{\frac{1}{3}}$

$\qquad\qquad\qquad\quad \sqrt{x^3} - \sqrt[3]{y}$

Two radical signs ▶ $\quad x\sqrt{x} - \sqrt[3]{y}$

Ⓒ 45. $\sqrt{x} \cdot \sqrt[5]{x}$ $\sqrt[10]{x^7}$
46. $\sqrt[6]{y} \cdot \sqrt{y}$ $\sqrt[3]{y^2}$
47. $\sqrt[3]{z} \cdot \sqrt[4]{z} \cdot \sqrt[6]{z}$ $\sqrt[4]{z^3}$

48. $\sqrt[3]{x^2} \cdot \sqrt[4]{x^3}$ $x\sqrt[12]{x^5}$
49. $\sqrt{a} \cdot \sqrt[3]{b} \cdot \sqrt[4]{c}$ $\sqrt[12]{a^6b^4c^3}$
50. $\sqrt[6]{m^5} \cdot \sqrt[3]{n^2}$ $\sqrt[6]{m^5n^4}$

51. $(\sqrt[3]{x} + \sqrt[4]{x})(\sqrt[3]{x} - \sqrt[4]{x})$
52. $(\sqrt{x} - \sqrt[5]{y^2})(\sqrt{x} + \sqrt[5]{y^2})$
53. $(\sqrt[4]{x^3} + \sqrt[8]{y^5})(\sqrt[4]{x^3} - \sqrt[8]{y^5})$

54. $(\sqrt{x} + \sqrt[4]{x})^2$
55. $(3\sqrt{x} - 2\sqrt[6]{y})^2$
56. $(\sqrt[4]{x} + \sqrt[4]{y})(\sqrt[4]{x} + \sqrt[4]{y})$

57. $\dfrac{\sqrt{x}}{\sqrt[3]{x}} \cdot \sqrt[6]{x}$
58. $\dfrac{\sqrt[4]{x^3}}{\sqrt[5]{x^2}} \cdot \sqrt[20]{x^7}$
59. $\dfrac{\sqrt[5]{x}}{\sqrt[4]{x}} \cdot \dfrac{\sqrt[20]{x^{19}}}{x}$ $\sqrt{x} + 2\sqrt[4]{xy} + \sqrt{y}$

60. $(\sqrt[3]{x} - \sqrt[3]{2})(\sqrt[3]{x^2} + \sqrt[3]{2x} + \sqrt[3]{4})$ $x - 2$
61. $(\sqrt[3]{x} + \sqrt[3]{y})(\sqrt[3]{x^2} - \sqrt[3]{xy} + \sqrt[3]{y^2})$ $x + y$

51. $\sqrt[3]{x^2} - \sqrt{x}$
52. $x - \sqrt[5]{y^4}$
53. $x\sqrt{x} - y\sqrt[4]{y}$
54. $x + 2\sqrt[4]{x^3} + \sqrt{x}$
55. $9x - 12\sqrt[6]{x^3y} + 4\sqrt[3]{y}$

Use one of the special products, $(a+b)(a^2 - ab + b^2) = a^3 + b^3$ **or**
$(a-b)(a^2 + ab + b^2) = a^3 - b^3$, **to rationalize each denominator.**

62. $\dfrac{14}{5^{\frac{1}{3}} + 2^{\frac{1}{3}}}$ $2(5^{\frac{2}{3}} - 10^{\frac{1}{3}} + 2^{\frac{2}{3}})$

63. $\dfrac{-2}{x^{\frac{1}{3}} - y^{\frac{1}{3}}}$ $\dfrac{-2x^{\frac{2}{3}} - 2(xy)^{\frac{1}{3}} - 2y^{\frac{2}{3}}}{x - y}$

64. $\dfrac{8}{6^{\frac{2}{3}} - 12^{\frac{1}{3}} + 2^{\frac{2}{3}}}$ $6^{\frac{1}{3}} + 2^{\frac{1}{3}}$

65. $\dfrac{1}{2x^{\frac{1}{3}} + 3^{\frac{1}{3}}}$ $\dfrac{4x^{\frac{2}{3}} - 2(3x)^{\frac{1}{3}} + 3^{\frac{2}{3}}}{8x + 3}$

66. $\dfrac{c}{3c^{\frac{1}{3}} - 5d^{\frac{1}{3}}}$ $\dfrac{c[9c^{\frac{2}{3}} + 15(cd)^{\frac{1}{3}} + 25d^{\frac{2}{3}}]}{27c - 125d}$

67. $\dfrac{x - y}{x^{\frac{2}{3}} + x^{\frac{1}{3}}y^{\frac{1}{3}} + y^{\frac{2}{3}}}$ $x^{\frac{1}{3}} - y^{\frac{1}{3}}$

Expressions with Rational Number Exponents

CHAPTER SIX REVIEW

(A) Exercises are 1–5, 7, 8, 10, 11, 13, 14, 16, 17, 20, 24, 29, 33, 34, 38, 40. (C) Exercises are starred. The rest are (B) Exercises.

Vocabulary
conjugates [6.4]
index [6.5]
irrational number [6.1]
like radicals [6.3]
nth root, $\sqrt[n]{k}$ [6.5]
radical [6.1]
radical sign [6.1]
radicand [6.1]
rational number exponent, $x^{\frac{1}{n}}$, $x^{\frac{m}{n}}$ [6.6]
real number [6.1]
simplest radical form [6.4]
square root, $\sqrt{k}$ [6.1]

Solve each equation. Approximate the solutions to the nearest hundredth. [6.1]
1. $5y^2 + 24 = 174$ **−5.48, 5.48**
2. $n^2 - 7 = 18 - 3n^2$ **−2.50, 2.50**

Determine whether each number is rational or irrational. [6.1]
3. $8.43843843\ldots$ **R**
4. $\sqrt{160}$ **I**
5. $-\sqrt{1600}$ **R**
6. $\sqrt{0.16}$ **R**

Simplify. Assume that values of the variables are all real numbers.
7. $\sqrt{20c^2}$ **$2|c|\sqrt{5}$**
8. $3m\sqrt{8m^4}$ **$6m^3\sqrt{2}$** [6.2]

In the remaining exercises, assume that (1) values of the variables are nonnegative numbers and (2) no denominator has the value of 0 (unless stated otherwise).

9. $m^2n\sqrt{32m^7n^3}$ **$4m^5n^2\sqrt{2mn}$** [6.2]
10. $\sqrt{8} + \sqrt{50} - \sqrt{18}$ **$4\sqrt{2}$** [6.3]
11. $(4\sqrt{3} - 2\sqrt{5})^2$ **$68 - 16\sqrt{15}$**
12. $6\sqrt{c^5d} + c\sqrt{16c^3d}$ **$10c^2\sqrt{cd}$**
13. $\dfrac{\sqrt{27a^8}}{\sqrt{3a^3}}$ **$3a^2\sqrt{a}$**
14. $\dfrac{-5\sqrt{2}}{3\sqrt{7}}$ **$\dfrac{-5\sqrt{14}}{21}$** [6.4]
15. $\dfrac{\sqrt{2}}{2\sqrt{7} + \sqrt{10}}$ **$\dfrac{\sqrt{14} - \sqrt{5}}{9}$**
16. $\sqrt[3]{-16x^{12}}$ **$-2x^4\sqrt[3]{2}$**
17. $\sqrt[5]{2c^2} \cdot \sqrt[5]{16c^8}$ **$2c^2$** [6.5]
18. $\sqrt[4]{32x^8y^{11}}$ **$2x^2y^2\sqrt[4]{2y^3}$**
19. $\sqrt[3]{27a^7} + \sqrt[3]{64a^7}$ **$7a^2\sqrt[3]{a}$**
20. $\sqrt[3]{\dfrac{7}{9}}$ **$\dfrac{\sqrt[3]{21}}{3}$**
21. $\dfrac{2c^2}{\sqrt[4]{4c^3}}$ **$c\sqrt[4]{4c}$**
22. $\dfrac{a}{\sqrt[3]{x} + 2\sqrt[3]{y}}$ **$\dfrac{a(\sqrt[3]{x^2} - 2\sqrt[3]{xy} + 4\sqrt[3]{y^2})}{x + 8y}$**

Write each expression in exponential form. [6.6]
23. $(\sqrt[3]{17})^2$ **$17^{\frac{2}{3}}$**
24. $6\sqrt[4]{3a}$ **$6(3a)^{\frac{1}{4}}$**
25. $\sqrt{c^5}$ **$c^{\frac{5}{2}}$**

Write each expression in radical form. [6.6]
26. $2a^{\frac{5}{3}}$ **$2\sqrt[3]{a^5}$, or $2a\sqrt[3]{a^2}$**
27. $4y^{\frac{3}{5}}$ **$4\sqrt[5]{y^3}$**
28. $(4y)^{\frac{3}{5}}$ **$\sqrt[5]{64y^3}$, or $2\sqrt[5]{2y^3}$**

Find the value of each expression. [6.6]
29. $-15 \cdot 32^{\frac{1}{5}}$ **−30**
30. $125^{\frac{4}{3}}$ **625**
31. $20 \cdot 625^{\frac{3}{4}}$ **2,500**
32. $32^{-\frac{2}{5}}$ **$\frac{1}{4}$**

Simplify each expression. [6.7]
33. $-9a^{\frac{3}{4}} \cdot 2a^{-\frac{1}{2}}$ **$-18a^{\frac{1}{4}}$**
34. $(49a^2b^{10})^{\frac{1}{2}}$ **$7ab^5$**
35. $(a^{\frac{1}{3}}b^{-\frac{2}{3}})^{-9}$ **$\dfrac{b^6}{a^3}$**
36. $\dfrac{x^{\frac{1}{4}}y^{-\frac{1}{5}}}{x^{\frac{3}{8}}y^{-\frac{3}{5}}}$ **$\dfrac{x^{\frac{7}{8}}y^{\frac{2}{5}}}{x}$**
37. $\left(\dfrac{8x^9}{27y^{-12}}\right)^{\frac{2}{3}}$ **$\dfrac{4x^6y^8}{9}$**

Solve each equation. (x is a real number.) [6.7]
38. $4^x = 8$ **$\frac{3}{2}$**
39. $27^{2x-1} = 9$ **$\frac{5}{6}$**
40. $3^{10x} = \frac{1}{9}$ **$-\frac{1}{5}$**
41. $5^{4x+1} = \frac{1}{125}$ **−1**

★42. Solve $\dfrac{2x^2 - 3}{4} + \dfrac{x^2 + 5}{6} = \dfrac{5x^2 + 4}{8}$. Approximate the solutions to the nearest hundredth. [6.1] **−3.16, 3.16**

Simplify each expression.
★43. $x^m y^{2n}\sqrt{x^{4m+5}y^{6n}}$ [6.3] **$x^{3m+2}y^{5n}\sqrt{x}$**
★44. $\dfrac{4\sqrt{x} - y}{\sqrt{x - y} + 3}$ [6.4] **$\dfrac{4x - 4y - 12\sqrt{x - y}}{x - y - 9}$**
★45. $\sqrt[n]{x^{2n}y^{3n+5}}$ [6.7] **$x^2y^3\sqrt[n]{y^5}$**

★46. Factor $9x - 4y$ as a difference of two squares. [6.3] **$(3\sqrt{x} - 2\sqrt{y})(3\sqrt{x} + 2\sqrt{y})$**

★47. Write $(x^{2n}y^{n+2})^{\frac{a}{n}}$ in simplest radical form. [6.6] **$x^{2a}y^a\sqrt[n]{y^{2a}}$**

Ⓐ Exercises are 1–9, 15–17, 19. Ⓒ Exercises are starred. The rest are Ⓑ Exercises.

CHAPTER SIX TEST

Solve each equation. Approximate the solutions to the nearest hundredth.

1. $7a^2 + 19 = 236$ **−5.57, 5.57**
2. $3y^2 - 11 = 7y^2 - 60$ **−3.50, 3.50**

Determine whether each number is rational or irrational.

3. $\sqrt{90}$ **I**
4. $-\sqrt{900}$ **R**

Simplify. Assume that values of the variables are *all* real numbers.

5. $\sqrt{27n^8}$ **$3n^4\sqrt{3}$**
6. $5\sqrt{12a^6}$ **$10|a^3|\sqrt{3}$**

In the remaining exercises, assume that (1) values of the variables are nonnegative numbers and (2) no denominator has the value of 0 (unless stated otherwise).

7. $5\sqrt{12x} - 6\sqrt{48x} + \sqrt{75x}$ **$-9\sqrt{3x}$**
8. $(3\sqrt{2} + \sqrt{5})(5\sqrt{2} - \sqrt{5})$ **$25 + 2\sqrt{10}$**
9. $\sqrt[3]{9x^4} \cdot \sqrt[3]{3x^5}$ **$3x^3$**
10. $8\sqrt{16x^5y^9}$ **$32x^2y^4\sqrt{xy}$**
11. $m\sqrt[4]{2n^5} + mn\sqrt[4]{32n}$ **$3mn\sqrt[4]{2n}$**
12. $\dfrac{4}{3\sqrt{5} + \sqrt{30}}$
13. $\dfrac{5}{2\sqrt[3]{x} - \sqrt[3]{y}}$
14. $\dfrac{2x}{\sqrt[3]{5x^2}}$ **$\dfrac{2\sqrt[3]{25x}}{5}$**
15. $\sqrt[4]{\dfrac{3}{8}}$ **$\dfrac{\sqrt[4]{6}}{2}$**
16. $\dfrac{8}{5\sqrt{12c^3}}$ **$\dfrac{4\sqrt{3c}}{15c^2}$**
17. $(27x^6y^{12})^{\frac{2}{3}}$ **$9x^4y^8$**
18. $\dfrac{x^{\frac{5}{7}}y^{-\frac{1}{4}}\, x^{\frac{6}{7}}y^{\frac{1}{4}}}{x^{-\frac{1}{7}}y^{\frac{1}{2}}}$ **$\dfrac{y}{\ }$**
19. $\left(\dfrac{c^{\frac{2}{3}}}{d^{\frac{1}{2}}}\right)^{12}$ **$\dfrac{c^8}{d^6}$**

20. Write $\sqrt[4]{a^3}$ in exponential form. **$a^{\frac{3}{4}}$**

21. Write $(2x)^{\frac{4}{5}}$ in radical form. **$\sqrt[5]{16x^4}$, or $(\sqrt[5]{2x})^4$**

Find the value of each expression.

22. $25^{\frac{3}{2}}$ **125**
23. $6 \cdot 81^{-\frac{3}{4}}$ **$\dfrac{2}{9}$**

Solve each equation. (x is a real number.)

24. $25^{2x+1} = 125$ **$\dfrac{1}{4}$**
25. $2^{4x-1} = \dfrac{1}{32}$ **−1**

★26. Solve $\dfrac{x^2 - 10}{5} + \dfrac{x^2}{10} = \dfrac{x^2 + 10}{4}$.

Approximate the solutions to the nearest hundredth. **−9.49, 9.49**

Simplify each expression.

★27. $x^{2c}y^d\sqrt{x^{2c+3}y^{6d+5}}$ **$x^{3c+1}y^{4d+2}\sqrt{xy}$**

★28. $\dfrac{\sqrt{x+y}}{\sqrt{x+y}+6}$ **$\dfrac{x+y-6\sqrt{x+y}}{x+y-36}$**

★29. Simplify and write $(\sqrt{x} + \sqrt[6]{y})(\sqrt{x} + \sqrt[6]{y})$ with the least possible number of radical signs. **$x + 2\sqrt[6]{x^3y} + \sqrt[3]{y}$**

12. $\dfrac{12\sqrt{5} - 4\sqrt{30}}{15}$
13. $\dfrac{5(4\sqrt[3]{x^2} + 2\sqrt[3]{xy} + \sqrt[3]{y^2})}{8x - y}$

COMPUTER ACTIVITIES

Simplifying Radicals

OBJECTIVE: To simplify radicals of any integral index

On page 582, there is another version of the program below that simplifies square roots. You may wish to scrutinize that program before undertaking the program below.

This program functions very similarly to the program on page 582. This program, however, allows you to simplify radicals for any integral index you specify. The program on page 582 tests only the squares of whole numbers 2 and greater for divisibility into the radicand. This program tests the *n*th power of whole numbers 2 and greater for divisibility into the radicand, for any index *n* you specify for the radical.

Like the program on page 582, this program tests the divisibility of one number by another by checking whether the quotient of the numbers is equal to the greatest INTeger of the quotient. When they are equal, we know that the number is divisible by the other.

```
110   PRINT "WHAT IS THE RADICAND TO BE SIMPLIFIED";:
       INPUT R
120   PRINT "WHAT IS THE INDEX";: INPUT X
130 C = 1:S = R
140   IF ABS (S ^ (1 / X) - INT (S ^ (1 / X))) < .05
       THEN 240
150   FOR I = 2 TO SQR (S)
160 B = 1
170   FOR N = 1 TO X:B = B * I: NEXT N
180   IF ABS (S / B - INT (S / B)) > .0001 THEN 210
190 C = C * I:S = S / B: IF S < B THEN 220
200   GOTO 180
210   NEXT I
220   PRINT "RADICAL("R") = "C" * RAD ("S")"
230   GOTO 110
240   PRINT "RADICAL ("S") = S ^ (1 / X)
250   GOTO 110
```

See the Computer Section on page 574 for more information.

EXERCISES   For Ex. 1–2, see TE Answer Section.

1. Write a program that multiplies two radicals of the form $(a\sqrt{b})\,(c\sqrt{d})$ and then simplifies the product.

2. Expand the program above to multiply two radicals of the form $(a\sqrt[3]{b})\,(c\sqrt[3]{d})$ and to simplify the result.

COLLEGE PREP TEST

Directions: In each item, you are to compare a quantity in Column 1 with a quantity in Column 2. Write the letter of the correct answer from these choices:

A The quantity in Column 1 is greater than the quantity in Column 2.
B The quantity in Column 2 is greater than the quantity in Column 1.
C The quantity in Column 1 is equal to the quantity in Column 2.
D The relationship cannot be determined from the given information.

Notes: Information centered over both columns refers to one or both of the quantities to be compared.
A symbol that appears in both columns has the same meaning in each column.
All variables represent real numbers.

SAMPLE ITEMS AND ANSWERS

	Column 1	Column 2
		$x = 8$ and $y = 9$
S1.	$\sqrt{2x}$	$\sqrt[3]{3y}$

The answer is **A** because $\sqrt{2 \cdot 8} = \sqrt{16} = 4$, $\sqrt[3]{3 \cdot 9} = \sqrt[3]{27} = 3$, and $4 > 3$.

	Column 1	Column 2
		$0 < x < y$
S2.	$\sqrt{2x}$	$\sqrt{y}$

The answer is **D**. If $x = 8$ and $y = 9$, then $\sqrt{2x} > \sqrt{y}$. If $x = 2$ and $y = 4$, then $\sqrt{2x} = \sqrt{y}$.

	Column 1	Column 2	
1.	$\sqrt[3]{65}$	4	**A**
2.	$\dfrac{1}{3}$	$\sqrt[3]{\dfrac{2}{27}}$	**B**
3.	$\sqrt[4]{x+y}$	$\sqrt[4]{x-y}$	**D**
4.	The sum of the two solutions of $x^2 = 10$	The sum of the two solutions of $x^4 = 16$	**C**

	Column 1	Column 2	
	$x = 12$ and $y = 9$		
5.	$\sqrt{xy}$	$\sqrt{6x + 4y}$	**C**
	$x = y = 50$		
6.	$7 + \sqrt{x}$	$\sqrt{x} + \sqrt{y}$	**B**
	$x > 0$ and $y > 0$		
7.	$\sqrt{x^2 + y^2}$	$x + y$	**B**
	$x = -2$ and $y = 2$		
8.	$\sqrt{8x^2 y^3}$	$2xy\sqrt{2y}$	**A**

Items 9 and 10 refer to Rectangles I and II below. The rectangles are not drawn to scale.

$\sqrt{2}$ [I] $\sqrt{50}$ $\sqrt{5}$ [II] $\sqrt{20}$

	Column 1	Column 2	
9.	Area of Rectangle I	Area of Rectangle II	**C**
10.	Perimeter of Rectangle I	Perimeter of Rectangle II	**A**

College Prep Test

7 QUADRATIC FORMULA

Non-Routine Problem Solving

Problem 1. Automobiles A and B had equal amounts of fuel. The two cars began at the same time and were driven at the same speed. Car A ran out of fuel in 4 hours and car B ran out of fuel in 3 hours. In how many hours was A's amount of fuel twice B's amount of fuel? $2\frac{2}{5}$ h

Problem 2. How many hours will it take a car averaging 80 km/h between stops to travel x kilometers if it is to make k stops of n minutes each? Simplify your answer.

$$\frac{3x + 4kn}{240}$$

COMPLETING THE SQUARE 7.1

Objective **To solve a quadratic equation by completing the square**

As shown below, you can solve a *quadratic equation* like $y^2 = 49$ or $(n + 3)^2 = 16$ by using the definition of square root: if $x^2 = k$, then $x = \sqrt{k}$ or $x = -\sqrt{k}$.

$$y^2 = 49$$
$$y = \sqrt{49} \quad \text{or} \quad y = -\sqrt{49}$$
$$y = 7 \qquad\qquad y = -7$$
Solutions: 7, −7

$$(n + 3)^2 = 16$$
$$n + 3 = \sqrt{16} \quad \text{or} \quad n + 3 = -\sqrt{16}$$
$$n + 3 = 4 \qquad \text{or} \quad n + 3 = -4$$
$$n = 1 \qquad\qquad n = -7$$
Solutions: 1, −7

The equation $y^2 + 10y = 3$ can be solved in a similar way.

Example 1 Solve $y^2 + 10y = 3$.

$$y^2 + 10y = 3$$
$$y^2 + 10y + 25 = 3 + 25 \quad \blacktriangleleft \text{ Add 25 to each side.}$$
$$(y + 5)^2 = 28$$
$$y + 5 = \pm\sqrt{28} \quad \blacktriangleleft \text{ If } x^2 = k, \text{ then } x = \pm\sqrt{k}.$$
$$y + 5 = \pm 2\sqrt{7}$$
$$y = -5 \pm 2\sqrt{7} \quad \blacktriangleleft \text{ Add } -5 \text{ to each side.}$$

Thus, the solutions are $-5 + 2\sqrt{7}$ **and** $-5 - 2\sqrt{7}$.

The procedure used to solve the quadratic equation in Example 1 is called **completing the square.** By adding 25 to each side of the equation, the left side becomes a *perfect square trinomial*, $y^2 + 10y + 25$, that can be factored into the square of a binomial, $(y + 5)^2$.

To solve a quadratic equation of the form $x^2 + bx = c$ by completing the square, you divide b by 2, square the result, then add $\left(\dfrac{b}{2}\right)^2$ to each side.

Example 2 Solve $n^2 - 3n = 10$ by completing the square.

$$n^2 - 3n = 10 \quad \blacktriangleleft \, b = -3$$
$$n^2 - 3n + \frac{9}{4} = 10 + \frac{9}{4} \quad \blacktriangleleft \text{ Add } \left(\frac{-3}{2}\right)^2, \text{ or } \frac{9}{4}, \text{ to each side.}$$
$$\left(n - \frac{3}{2}\right)^2 = \frac{49}{4}$$
$$n - \frac{3}{2} = \pm\sqrt{\frac{49}{4}} \quad \blacktriangleleft \text{ If } x^2 = k, \text{ then } x = \pm\sqrt{k}.$$
$$n = \frac{3}{2} \pm \frac{7}{2}, \text{ or } \frac{3 \pm 7}{2}$$

Thus, the solutions are 5 and −2.

Completing the Square

To solve $9y^2 + 12y - 8 = 0$ by completing the square, you must rewrite the equation in the general form of $y^2 + by = c$.

Example 3   Solve $9y^2 + 12y - 8 = 0$ by completing the square.

$$9y^2 + 12y - 8 = 0$$
$$9y^2 + 12y = 8 \quad \blacktriangleleft \text{ Add 8 to each side.}$$
$$y^2 + \frac{12}{9}y = \frac{8}{9} \quad \blacktriangleleft \text{ Divide each side by 9.}$$
$$y^2 + \frac{4}{3}y = \frac{8}{9} \quad \blacktriangleleft \frac{12}{9} = \frac{4}{3}, \text{ so } b = \frac{4}{3}.$$
$$y^2 + \frac{4}{3}y + \frac{4}{9} = \frac{8}{9} + \frac{4}{9} \quad \blacktriangleleft \text{ Add } \left(\frac{1}{2} \cdot \frac{4}{3}\right)^2, \text{ or } \frac{4}{9}, \text{ to each side.}$$
$$\left(y + \frac{2}{3}\right)^2 = \frac{12}{9}$$
$$y + \frac{2}{3} = \pm\sqrt{\frac{12}{9}}$$
$$y = -\frac{2}{3} \pm \frac{2\sqrt{3}}{3} \quad \blacktriangleleft \pm\sqrt{\frac{12}{9}} = \pm\frac{\sqrt{12}}{\sqrt{9}} = \pm\frac{2\sqrt{3}}{3}$$

Thus, the solutions are $\dfrac{-2 + 2\sqrt{3}}{3}$ and $\dfrac{-2 - 2\sqrt{3}}{3}$.

Oral Exercises

Find the number to be added to each expression to complete the square.

1. $a^2 + 14a$ **49**
2. $n^2 - 8n$ **16**
3. $x^2 + 5x$ $\dfrac{25}{4}$
4. $y^2 - y$ $\dfrac{1}{4}$
5. $b^2 + \dfrac{3}{2}b$ $\dfrac{9}{16}$
6. $c^2 - \dfrac{2}{5}c$ $\dfrac{1}{25}$

Written Exercises

Solve each equation by completing the square.

Ⓐ
1. $x^2 + 4x = 21$ **−7, 3**
2. $y^2 - 8y = 33$ **−3, 11**
3. $n^2 + 12n = -20$ **−10, −2**
4. $a^2 - 6a = 5$ $3 \pm \sqrt{14}$
5. $b^2 + 10b = -5$ $-5 \pm 2\sqrt{5}$
6. $c^2 - 2c = 74$ $1 \pm 5\sqrt{3}$
7. $y^2 + 3y - 6 = 0$ $\dfrac{-3 \pm \sqrt{33}}{2}$
8. $t^2 - 5t + 2 = 0$ $\dfrac{5 \pm \sqrt{17}}{2}$
9. $x^2 - x - 7 = 0$ $\dfrac{1 \pm \sqrt{29}}{2}$
10. $5x^2 + 3x - 2 = 0$ $-1, \dfrac{2}{5}$
11. $2a^2 - 7a - 4 = 0$ $-\dfrac{1}{2}, 4$
12. $3y^2 + 10y + 3 = 0$ $-3, -\dfrac{1}{3}$
13. $a^2 - 5a - 5 = 0$ $\dfrac{5 \pm 3\sqrt{5}}{2}$
14. $y^2 + 7y = 0$ **−7, 0**
15. $n^2 = n$ **1, 0**

Ⓑ
16. $2n^2 = 5 - 5n$ $\dfrac{-5 \pm \sqrt{65}}{4}$
17. $7x - 1 = 5x^2$ $\dfrac{7 \pm \sqrt{29}}{10}$
18. $3a^2 + 4a = 3$ $\dfrac{-2 \pm \sqrt{13}}{3}$
19. $\dfrac{1}{3}y^2 - 2y - 1 = 0$ $3 \pm 2\sqrt{3}$
20. $\dfrac{1}{8}n^2 - n = \dfrac{1}{2}$ $4 \pm 2\sqrt{5}$
21. $\dfrac{5}{3}x^2 - \dfrac{5}{2}x - \dfrac{5}{6} = 0$ $\dfrac{3 \pm \sqrt{17}}{4}$
22. $\dfrac{x^2 + 12x}{24} = \dfrac{1}{12}$ $-6 \pm \sqrt{38}$
23. $\dfrac{y^2 + 15}{10} = y$ $5 \pm \sqrt{10}$
24. $2n = \dfrac{3 - n^2}{4}$ $-4 \pm \sqrt{19}$

Ⓒ
25. $x^2 + (2\sqrt{3})x - 24 = 0$ $-4\sqrt{3}, 2\sqrt{3}$
26. $y^2 - 4\sqrt{2} \cdot y + 8 = 0$ $2\sqrt{2}$
27. $n^2 + 6n\sqrt{5} + 45 = 0$ $-3\sqrt{5}$
28. $a^2 - 2a\sqrt{3} + 1 = 0$ $\sqrt{3} \pm \sqrt{2}$
29. $b^2 - 6b\sqrt{2} + 15 = 0$ $3\sqrt{2} \pm \sqrt{3}$
30. $4x^2 + 4x\sqrt{5} + 3 = 0$ $\dfrac{-\sqrt{5} \pm \sqrt{2}}{2}$

MAXIMUM VALUE   7.2

Objectives **To find the maximum value of a quadratic polynomial**
To solve a real-life problem involving a maximum value

The value of $-5t^2 + 40t$ will increase from 0 to some **maximum value** and then decrease to 0, as the value of t increases from 0 to 8.

t	0	1	2	5	6	8
$-5t^2 + 40t$	0	35	60	75	60	0

Example 1 Find the maximum value of $-5t^2 + 40t$ and the corresponding value of t.

$-5t^2 + 40t$
$-5(t^2 - 8t)$ ◀ Factor out -5.
$-5(t^2 - 8t + 16) + 5(16)$ ◀ Complete the square. Add $-5(16) + 5(16)$, or 0.
$-5(t - 4)^2 + 80$

If $t = 4$, then:
$(t - 4)^2 = 0$
$-5(t - 4)^2 = 0$
$-5(t - 4)^2 + 80 = 80.$

If $t \neq 4$, then:
$(t - 4)^2 > 0$
$-5(t - 4)^2 < 0$
$-5(t - 4)^2 + 80 < 80.$

Thus, the maximum value is 80, and this occurs when $t = 4$.

Example 2 Find the maximum value of $-10x^2 + 100x + 6{,}000$ and the corresponding value of x.

$-10x^2 + 100x + 6{,}000$
$-10(x^2 - 10x) + 6{,}000$ ◀ Factor out -10 from $-10x^2 + 100x$.
$-10(x^2 - 10x + 25) + 6{,}000 + 10(25)$ ◀ Complete the square. Add $-10(25) + 10(25)$, or 0.
$-10(x - 5)^2 + 6{,}250$

Thus, the maximum value is 6,250, and this occurs when $x = 5$.

In Chapter 1, you learned a method for finding the maximum height of an object that is shot upward from the earth's surface with a given initial velocity.

If $h = vt - 5t^2$, the maximum value of h is $\frac{v^2}{20}$.

The maximum height can also be found by the method of completing the square.

Example 3 Find the maximum height and the corresponding time of an object shot upward from the earth's surface at an initial velocity of **40 m/s**.

$h = -5t^2 + vt = -5t^2 + 40t = -5(t - 4)^2 + 80$ ◀ See Example 1 for the steps.

Thus, the maximum height is 80 m at the end of 4 s.

Maximum Value 167

Example 4 An orange grove has 20 trees per acre, and the average yield is 300 oranges per tree. For each additional tree per acre, the average yield will be reduced by 10 oranges per tree. How many trees per acre will yield the maximum number of oranges per acre, and what is the maximum number of oranges per acre?

Additional trees per acre	(Trees per acre) × (Oranges per tree)	= (Oranges per acre)
2 ▶	$(20 + 2)(300 - 10 \cdot 2)$	= 6,160
3 ▶	$(20 + 3)(300 - 10 \cdot 3)$	= 6,210
⋮	⋮	⋮
x ▶	$(20 + x)(300 - 10x)$	$= 6{,}000 + 100x - 10x^2$
		$= -10x^2 + 100x + 6{,}000$
	See Example 2 for the steps. ▶	$= -10(x - 5)^2 + 6{,}250$

The maximum value is 6,250, and this occurs when $x = 5$.
Thus, $20 + 5$, or 25 trees per acre, will yield the maximum number of 6,250 oranges per acre.

Written Exercises

Find the maximum value of each polynomial and the corresponding value of x. (Ex. 1–6)

Ⓐ 1. $-2x^2 + 12x$ **18; x = 3**   2. $64x - 16x^2$ **64; x = 2**   3. $10 + 6x - x^2$ **19; x = 3**
  4. $-3x^2 + 6x + 25$ **28; x = 1**   5. $625 + 200x - 20x^2$ 6. $-2x^2 + 5x + 20$ **$23\frac{1}{8}$; $x = \frac{5}{4}$**
  1,125; x = 5

  7. An arrow is shot upward at 60 m/s. Find the maximum height of the arrow and the corresponding time. Use the method of completing the square and the formula $h = -5t^2 + vt$. **180 m, 6 s**

Ⓑ 8. A chartered bus ride costs $30 per ticket if there are 20 passengers. The ticket price is reduced by $1 for each additional passenger beyond 20. How many passengers will produce the maximum income, and what is the maximum income? **25 passengers; $625**

  9. A rectangular chicken pen next to a barn wall is enclosed by 120 m of chicken wire on three sides and the barn wall on the fourth side. Find the maximum area that can be enclosed in this way. Find the dimensions of the pen. **1,800 m²; 30 m by 60 m**

  10. A rectangular field adjacent to the straight bank of a river is to be fenced, but there is to be no fencing along the riverbank. If 180 m of fencing is available, what is the maximum area that can be enclosed? **4,050 m²**

  11. Four hundred people will attend a sidewalk display if tickets cost $1 each. Attendance will decrease by 20 people for each 10-cent increase in ticket price. What ticket price will yield the maximum income, and what is the maximum income from ticket sales?
  $1.50; $450

  12. A publishing company can get 1,000 subscribers for a new magazine if the monthly subscription rate is $5. It will get 100 more subscribers for each 10-cent decrease in the monthly rate. What monthly rate will produce the maximum monthly income, and what will that income be?
  $3; $9,000

THE QUADRATIC FORMULA   7.3

Objective **To solve a quadratic equation by using the quadratic formula**

Every quadratic equation can be written in the standard form of $ax^2 + bx + c = 0$, where $a \neq 0$. You can solve this general equation for x by the method of completing the square. The result is called the **quadratic formula.**

$$ax^2 + bx + c = 0$$
$$ax^2 + bx = -c$$
$$x^2 + \frac{b}{a}x = \frac{-c}{a} \qquad \blacktriangleleft \text{ Divide each side by } a.$$
$$x^2 + \frac{b}{a}x + \left(\frac{b}{2a}\right)^2 = \frac{-c}{a} + \left(\frac{b}{2a}\right)^2 \qquad \blacktriangleleft \text{ Add } \left(\frac{b}{2a}\right)^2 \text{ to each side.}$$
$$x^2 + \frac{b}{a}x + \frac{b^2}{4a^2} = \frac{-c}{a} + \frac{b^2}{4a^2}$$
$$\left(x + \frac{b}{2a}\right)^2 = \frac{-4ac + b^2}{4a^2}$$
$$x + \frac{b}{2a} = \pm\sqrt{\frac{b^2 - 4ac}{4a^2}} \qquad \blacktriangleleft \text{ If } x^2 = k, \text{ then } x = \pm\sqrt{k}.$$
$$x = \frac{-b}{2a} \pm \frac{\sqrt{b^2 - 4ac}}{2a}$$
$$x = \frac{-b \pm \sqrt{b^2 - 4ac}}{2a}$$

The quadratic formula   The solutions of a quadratic equation of the form $ax^2 + bx + c = 0$, $a \neq 0$, are given by the formula $x = \dfrac{-b \pm \sqrt{b^2 - 4ac}}{2a}$.

The quadratic formula can be used to solve any quadratic equation. You must first identify the values of a, b, and c. Then, you substitute these values in the formula, as shown in Example 1 below.

Example 1   Solve $3x^2 + 5x - 4 = 0$ by using the quadratic formula.

$$ax^2 + bx + c = 0$$
$$\downarrow \quad \downarrow \quad \downarrow$$
$$3x^2 + 5x - 4 = 0$$
$$a = 3, b = 5, c = -4$$

$$x = \frac{-b \pm \sqrt{b^2 - 4ac}}{2a}$$
$$= \frac{-(5) \pm \sqrt{5^2 - 4 \cdot 3 \cdot (-4)}}{2 \cdot 3}$$
$$= \frac{-5 \pm \sqrt{25 + 48}}{6}$$
$$= \frac{-5 \pm \sqrt{73}}{6}$$

Thus, the solutions are $\dfrac{-5 + \sqrt{73}}{6}$ and $\dfrac{-5 - \sqrt{73}}{6}$.

If a quadratic equation is not given in standard form, you should rewrite it in standard form in order to identify the values of a, b, and c.

Example 2 Solve $4x^2 = 11 + 4x$ by using the quadratic formula.

$4x^2 - 4x - 11 = 0$ ◀ Write the equation in the standard form of $ax^2 + bx + c = 0$.

$a = 4, b = -4, c = -11$ ◀ Identify the values of a, b, and c.

$$x = \frac{-b \pm \sqrt{b^2 - 4ac}}{2a}$$

$$= \frac{-(-4) \pm \sqrt{(-4)^2 - 4 \cdot 4 \cdot (-11)}}{2 \cdot 4}$$ ◀ Substitute.

$$= \frac{4 \pm \sqrt{192}}{8}$$

$$= \frac{4 \pm 8\sqrt{3}}{8}$$ ◀ $\sqrt{192} = \sqrt{64 \cdot 3}$

$$= \frac{4(1 \pm 2\sqrt{3})}{4 \cdot 2}$$

$$= \frac{1 \pm 2\sqrt{3}}{2}$$ ◀ Divide numerator and denominator by 4.

Thus, the solutions are $\frac{1 + 2\sqrt{3}}{2}$ and $\frac{1 - 2\sqrt{3}}{2}$.

An equation of the form $ax^2 + bx + c = 0$, where $a = 0$, is *not* a quadratic equation since the square term is eliminated: $0 \cdot x^2 = 0$. However, there is no such restriction on the value of b or c. The standard form of the quadratic equation $5x^2 - 9x = 0$ is $5x^2 - 9x + 0 = 0$, where $c = 0$. The standard form of the quadratic equation $y^2 - 150 = 0$ is $y^2 + 0y - 150 = 0$. Equations of either type can be solved by using the quadratic formula.

Example 3 Solve $5x^2 - 9x = 0$ and $y^2 - 150 = 0$ by using the quadratic formula.

$5x^2 - 9x = 0$
$5x^2 - 9x + 0 = 0$
$a = 5, b = -9, c = 0$

$$x = \frac{-(-9) \pm \sqrt{(-9)^2 - 4 \cdot 5 \cdot 0}}{2 \cdot 5}$$

$$= \frac{9 \pm \sqrt{81 - 0}}{10}$$

$$= \frac{9 \pm 9}{10}, \text{ or } \frac{9 + 9}{10} \text{ and } \frac{9 - 9}{10}$$

The solutions are $\frac{9}{5}$ and 0.

$y^2 - 150 = 0$
$1y^2 + 0y - 150 = 0$
$a = 1, b = 0, c = -150$

$$y = \frac{-(0) \pm \sqrt{0^2 - 4 \cdot 1 \cdot (-150)}}{2 \cdot 1}$$

$$= \frac{0 \pm \sqrt{0 + 600}}{2}$$

$$= \frac{\pm 10\sqrt{6}}{2}, \text{ or } \pm 5\sqrt{6}$$

The solutions are $5\sqrt{6}$ and $-5\sqrt{6}$.

Reading in Algebra

For each equation in the left column, identify every correct description in the right column.
1. $9x + 5^2 = 0$ **C**
2. $x^2 - 14 = 0$ **A, D, E**
3. $3x - 5 = x^2$ **B, D**
4. $5x^2 - 9x = 0$ **A, F**
5. $2x^2 - 7x - 6 = 0$ **A**

A   A quadratic equation in standard form
B   A quadratic equation not in standard form
C   Not a quadratic equation
D $a = 1$ **E** $b = 0$
F $c = 0$

6. Identify three different methods for solving the quadratic equation $x^2 - 3x - 10 = 0$.
factoring; completing the square; quadratic formula

Oral Exercises

Identify the values of a, b, and c for each quadratic equation.
1. $6x^2 + 3x - 2 = 0$ 2. $y^2 - 8y + 7 = 0$ 3. $-2n^2 - n = 0$ 4. $z^2 - 8 = 0$
$a = 6, b = 3, c = -2$ $a = 1, b = -8, c = 7$ $a = -2, b = -1, c = 0$ $a = 1, b = 0, c = -8$

Give each quadratic equation in standard form.
5. $x^2 - 6x = 7$ 6. $4 + 3y^2 = -y$ 7. $-3 = t - 4t^2$ 8. $5x = -10x^2$
$x^2 - 6x - 7 = 0$ $3y^2 + y + 4 = 0$ $4t^2 - t - 3 = 0$ $10x^2 + 5x + 0 = 0$

Written Exercises

Simplify each expression.

Ⓐ
1. $\dfrac{-5 \pm \sqrt{8}}{2}$ $\dfrac{-5 \pm 2\sqrt{2}}{2}$
2. $\dfrac{6 \pm \sqrt{20}}{4}$ $\dfrac{3 \pm \sqrt{5}}{2}$
3. $\dfrac{-10 \pm \sqrt{32}}{2}$ $-5 \pm 2\sqrt{2}$
4. $\dfrac{8 \pm \sqrt{12}}{4}$ $\dfrac{4 \pm \sqrt{3}}{2}$
5. $\dfrac{10 \pm \sqrt{36 - 36}}{4}$ $\dfrac{5}{2}$
6. $\dfrac{0 \pm \sqrt{24}}{6}$ $\pm \dfrac{\sqrt{6}}{3}$

Solve each equation by using the quadratic formula.
7. $4x^2 + x - 2 = 0$ $\dfrac{-1 \pm \sqrt{33}}{8}$
8. $3y^2 + 7y + 3 = 0$ $\dfrac{-7 \pm \sqrt{13}}{6}$
9. $2n^2 - n - 2 = 0$ $\dfrac{1 \pm \sqrt{17}}{4}$
10. $2y^2 - 4y + 1 = 0$ $\dfrac{2 \pm \sqrt{2}}{2}$
11. $t^2 + 4t - 1 = 0$ $-2 \pm \sqrt{5}$
12. $2x^2 + 6x + 3 = 0$ $\dfrac{-3 \pm \sqrt{3}}{2}$
13. $4n^2 + 7 = 12n$ $\dfrac{3 \pm \sqrt{2}}{2}$
14. $0 = 4x^2 + 8x + 1$ $\dfrac{-2 \pm \sqrt{3}}{2}$
15. $9y^2 = 6y + 7$ $\dfrac{1 \pm 2\sqrt{2}}{3}$
16. $3x^2 + 6x = 2$ $\dfrac{-3 \pm \sqrt{15}}{3}$
17. $2y^2 = 6y - 1$ $\dfrac{3 \pm \sqrt{7}}{2}$
18. $5n^2 = 1 - 2n$ $\dfrac{-1 \pm \sqrt{6}}{5}$
19. $y^2 - 8y + 16 = 0$ **4**
20. $x^2 - 6x = 0$ **0, 6**
21. $t^2 - 11 = 0$ $\pm\sqrt{11}$
22. $t^2 - 20 = 0$ $\pm 2\sqrt{5}$
23. $0 = 9y^2 + 12y + 4$ $-\dfrac{2}{3}$
24. $3x^2 + 2x = 0$ $-\dfrac{2}{3}, 0$
25. $4x^2 = 7x$ $0, \dfrac{7}{4}$
26. $5t^2 - 11 = 0$ $\pm\dfrac{\sqrt{55}}{5}$
27. $4y^2 = 20y - 25$ $\dfrac{5}{2}$

Ⓑ
28. $\dfrac{3}{2}x^2 + \dfrac{1}{2}x - 3 = 0$ $\dfrac{-1 \pm \sqrt{73}}{6}$
29. $\dfrac{1}{2}y^2 - \dfrac{3}{2}y + \dfrac{5}{6} = 0$ $\dfrac{9 \pm \sqrt{21}}{6}$
30. $\dfrac{1}{4}n^2 = \dfrac{1}{2}n + \dfrac{3}{8}$ $\dfrac{2 \pm \sqrt{10}}{2}$
31. $\dfrac{y^2}{15} + \dfrac{5}{3} = \dfrac{2y}{3}$ **5**
32. $\dfrac{n^2}{4} = \dfrac{n}{2} - \dfrac{1}{6}$ $\dfrac{3 \pm \sqrt{3}}{3}$
33. $\dfrac{x^2}{3} - \dfrac{1}{2} = \dfrac{x}{3}$ $\dfrac{1 \pm \sqrt{7}}{2}$
34. $6\left(2x^2 + \dfrac{4}{3}x\right) = 3(3x^2 - 1)$ $\dfrac{-4 \pm \sqrt{7}}{3}$
35. $\dfrac{y+3}{2y-1} = \dfrac{2y+3}{y+5}$ $\dfrac{2 \pm \sqrt{58}}{3}$
36. $\dfrac{n^2 + 1}{n} + \dfrac{n-2}{2n} = \dfrac{1}{3n}$ $\dfrac{-3 \pm \sqrt{57}}{12}$

Ⓒ
37. $x^2 - (2\sqrt{2})x - 6 = 0$ $-\sqrt{2}, 3\sqrt{2}$
38. $x^2\sqrt{6} - 4x - 2\sqrt{6} = 0$ $\dfrac{-\sqrt{6}}{3}, \sqrt{6}$
39. $5x^2 + 2x\sqrt{10} - 1 = 0$ $\dfrac{-\sqrt{10} \pm \sqrt{15}}{5}$
40. $(y + 2)^2 + 7(y + 2) - 3 = 0$ [Hint: Let $x = y + 2$.] $\dfrac{-11 \pm \sqrt{61}}{2}$
41. $(n + \sqrt{5})^2 - 5(n + \sqrt{5}) - 5 = 0$ [Hint: Let $x = n + \sqrt{5}$.] $\dfrac{5 - 5\sqrt{5}}{2}, \dfrac{5 + \sqrt{5}}{2}$
42. $(n^2)^2 - 3n^2\sqrt{2} + 4 = 0$ [Hint: Let $x = n^2$.] $\pm\sqrt[4]{2}, \pm\sqrt[4]{8}$
43. $x^{-2} - 6x^{-1} + 4 = 0$ $\dfrac{3 \pm \sqrt{5}}{4}$
44. $(x^{\frac{1}{2}})^2 - (3\sqrt{3})x^{\frac{1}{2}} + 6 = 0$ **3, 12**
45. $8x^{-4} - 6x^{-2} + 1 = 0$ $-2, 2, -\sqrt{2}, \sqrt{2}$

The Quadratic Formula   **171**

PROBLEM SOLVING: THE QUADRATIC FORMULA   7.4

Objective **To solve a word problem by using the quadratic formula**

Sometimes a statement about the relationships between two numbers is in the form of a quadratic equation. Since a given quadratic equation may have two real number solutions, there may be two values for each of the numbers.

Example 1 **A second number is 6 more than twice a first number. If the second number is multiplied by 3 more than the first number, the product is 9. Find the two numbers.**

Let f = first number.
$2f + 6$ = second number
(second number) · (3 more than first number) = 9
$$(2f + 6)(f + 3) = 9$$
$$2f^2 + 12f + 18 = 9$$
$$2f^2 + 12f + 9 = 0$$

Use the quadratic formula: $a = 2$, $b = 12$, and $c = 9$.

$$f = \frac{-12 \pm \sqrt{12^2 - 4 \cdot 2 \cdot 9}}{2 \cdot 2}$$
$$= \frac{-12 \pm \sqrt{72}}{4}$$
$$= \frac{-12 \pm 6\sqrt{2}}{4}, \text{ or } \frac{-6 \pm 3\sqrt{2}}{2}$$

If $f = \frac{-6 + 3\sqrt{2}}{2}$, then $2f + 6 = 2 \cdot \frac{-6 + 3\sqrt{2}}{2} + 6 = 3\sqrt{2}$.

If $f = \frac{-6 - 3\sqrt{2}}{2}$, then $2f + 6 = 2 \cdot \frac{-6 - 3\sqrt{2}}{2} + 6 = -3\sqrt{2}$.

Each pair of answers checks in the problem.

Thus, the two numbers are $\frac{-6 + 3\sqrt{2}}{2}$, $3\sqrt{2}$ or $\frac{-6 - 3\sqrt{2}}{2}$, $-3\sqrt{2}$.

Some problems involve the use of formulas so that when values are substituted for the variables, a quadratic equation is obtained. The resulting quadratic equation can then be solved using the quadratic formula. One formula of this type is $h = vt - 5t^2$, which you used earlier in this chapter. It is used again in Example 2 on page 173.

172   Chapter Seven

Example 2 **An object is to be shot upward from the earth's surface with an initial velocity of 25 m/s. To the nearest tenth of a second, when will the height of the object be 15 m?**

$$h = vt - 5t^2$$
$$15 = 25t - 5t^2 \quad \blacktriangleleft \text{ Substitute 15 for } h \text{ and 25 for } v.$$
$$3 = 5t - t^2$$
$$t^2 - 5t + 3 = 0$$
$$t = \frac{5 \pm \sqrt{25 - 12}}{2} = \frac{5 \pm \sqrt{13}}{2}$$
$$= \frac{5 \pm 3.606}{2}, \text{ or } 4.303 \text{ and } 0.697$$

Check the irrational solutions, $\frac{5 \pm \sqrt{13}}{2}$, in the original equation.

Check: $\frac{5 + \sqrt{13}}{2}$

15	$25t - 5t^2$
15	$25 \cdot \frac{5 + \sqrt{13}}{2} - 5\left(\frac{5 + \sqrt{13}}{2}\right)^2$
	$\frac{125 + 25\sqrt{13}}{2} - 5\left(\frac{25 + 10\sqrt{13} + 13}{4}\right)$
	$\frac{125 + 25\sqrt{13}}{2} - \frac{95 + 25\sqrt{13}}{2}$
	$\frac{30}{2}$, or 15

Check: $\frac{5 - \sqrt{13}}{2}$

15	$25t - 5t^2$
15	$25 \cdot \frac{5 - \sqrt{13}}{2} - 5\left(\frac{5 - \sqrt{13}}{2}\right)^2$
	$\frac{125 - 25\sqrt{13}}{2} - 5\left(\frac{25 - 10\sqrt{13} + 13}{4}\right)$
	$\frac{125 - 25\sqrt{13}}{2} - \frac{95 - 25\sqrt{13}}{2}$
	$\frac{30}{2}$, or 15

Thus, the height will be 15 m at the end of 0.7 s and again at the end of 4.3 s, to the nearest 0.1 s.

Written Exercises

1. $1 - \sqrt{5}, -1 - \sqrt{5}$ or $1 + \sqrt{5}, -1 + \sqrt{5}$ 2. $-2 - \sqrt{5}, 2 - \sqrt{5}$ or $-2 + \sqrt{5}, 2 + \sqrt{5}$

Solve each problem. Give the answers in simplest radical form.

(A)
1. A second number is 2 less than a first number. Their product is 4. Find the numbers.
2. The product of two numbers is 1. Find the numbers if one of them is 4 more than the other.
3. One number is 2 less than twice another number. Find these numbers if their product is 3. $\frac{1 - \sqrt{7}}{2}, -1 - \sqrt{7}$ or $\frac{1 + \sqrt{7}}{2}, -1 + \sqrt{7}$
4. Find two numbers such that their product is -2 and one number is 6 more than 3 times the other number. $\frac{-3 - \sqrt{3}}{3}, 3 - \sqrt{3}$ or $\frac{-3 + \sqrt{3}}{3}, 3 + \sqrt{3}$

Solve each problem using the formula $h = vt - 5t^2$. Approximate each answer to the nearest 0.1 s, unless directed otherwise.

5. An object is shot upward from the earth's surface with an initial velocity of 50 m/s. When will the height of the object be 100 m? **2.8 s, 7.2 s**
6. If a pellet is launched straight up from the surface of the earth with an initial velocity of 75 m/s, when will the altitude of the pellet be 225 m? **4.1 s, 10.9 s**

(B)
7. A certain baseball pitcher can throw a ball with an initial velocity of 40 m/s. If the ball is thrown straight up, when will its height be 15 m? **0.4 s, 7.6 s**

Problem Solving: The Quadratic Formula

8. An archer shot an arrow upward with an initial velocity of 125 m/s. When will its height be 755 m on its upward flight? When will its height be 755 m on its downward flight? **10.2 s, 14.8 s**

9. An object was shot upward from the earth with an initial velocity of 38 m/s. When did it return to earth? **7.6 s**

10. The *maximum value* of h in the formula $h = vt - 5t^2$ is $\frac{v^2}{20}$. What initial velocity is needed for an object to reach a maximum height of 400 m?. Give the answer to the nearest meter per second. **89 m/s**

Solve each problem. Give the answers in simplest radical form.

11. A second number is 4 more than 3 times a first number. If the second number is multiplied by 2 more than the first number, the product is 2. Find the numbers. $\frac{-5 - \sqrt{7}}{3}, -1 - \sqrt{7}$ **or** $\frac{-5 + \sqrt{7}}{3}, -1 + \sqrt{7}$

12. The smaller of two numbers is 3 less than twice the greater number. The product of the smaller number and 8 more than 4 times the greater number is -16. Find the numbers. $\frac{-1 - \sqrt{17}}{4}, \frac{-7 - \sqrt{17}}{2}$ **or** $\frac{-1 + \sqrt{17}}{4}, \frac{-7 + \sqrt{17}}{2}$

Solve each problem using the formula $h = d + vt - 5t^2$. Approximate each answer to the nearest 0.1 s. (See explanation on page 24, if necessary.)

Ⓒ 13. An object is launched upward from the top of a tower 80 m tall with an initial velocity of 40 m/s. When will the object be 130 m above the earth's surface? **1.6 s, 6.4 s**

14. A pellet was sent up from the top of a tower 60 m tall with an initial velocity of 35 m/s. When will it be 115 m above the earth? **2.4 s, 4.6 s**

15. When will the pellet in Exercise 14 return to earth? **8.4 s**

16. When would the pellet in Exercise 14 be 35 m above the surface if it were launched from 20 m below the surface? **2.4 s, 4.6 s**

CALCULATOR ACTIVITIES

In Example 2, you can verify that the height of the object will be within 0.01 m of 15 m (15 ± 0.01 m) when the time $t = 0.697$ s.

First, rewrite the equation $15 = 25t - 5t^2$ as $t(25 - 5t) = 15$.
Second, show that $0.697 \times (25 - 5 \times 0.697)$ is between 14.99 and 15.01.

Compute ▶ ⊖ 5 ⊗ .697 ⊕ 25 ⊗ .697 ⊜ 14.995955, which is between 14.99 and 15.01.

1. In Example 2, verify that $h = 15 \pm 0.01$ m when $t = 4.303$ s. **h = 14.995955**

2. If $h = 25t - 5t^2$, verify that $h = 22 \pm 0.01$ m if $t = 1.1399$ s or $t = 3.8601$ s. **h = 22.000639**

3. If $h = 82t - 5t^2$, verify that $h = 334.6 \pm 0.01$ m when $t = 8.765$ s. **h = 334.60387**

4. If $h = 647t - 5t^2$, verify that $h \neq 633.7 \pm 0.01$ m when $t = 0.987$ s. **h = 633.71815**

PROBLEM SOLVING: IRRATIONAL MEASUREMENTS   7.5

Objective **To solve a word problem involving a geometric figure with irrational dimensions**

You can use the Pythagorean relation, $a^2 + b^2 = c^2$, and the quadratic formula to solve some problems involving a right triangle. As shown in Example 1, measurements may be given as irrational numbers in simplest radical form.

Example 1 **One leg of a right triangle is twice as long as the other leg. The hypotenuse is 3 cm longer than the longer leg. Find the length of each side of the triangle.**

Let w = length of the shorter leg.
     $2w$ = length of the longer leg
  $2w + 3$ = length of the hypotenuse

$$a^2 + b^2 = c^2$$

$w^2 + (2w)^2 = (2w + 3)^2$
$w^2 + 4w^2 = 4w^2 + 12w + 9$
$w^2 - 12w - 9 = 0$ ◀ *Quadratic equation: $a = 1$, $b = -12$, $c = -9$*

$$w = \frac{-(-12) \pm \sqrt{(-12)^2 - 4 \cdot 1(-9)}}{2 \cdot 1}$$

$$= \frac{12 \pm \sqrt{180}}{2}$$

$$= \frac{12 \pm 6\sqrt{5}}{2}, \text{ or } 6 \pm 3\sqrt{5}$$

One solution of the quadratic equation is $6 - 3\sqrt{5}$, but it is *not* a solution to the problem. The length of a side of a triangle cannot be a negative number: $(6 - 3\sqrt{5}) < 0$ since $6 - 3 \cdot 2.236 < 0$.

$w = 6 + 3\sqrt{5}$ $2w = 12 + 6\sqrt{5}$ $2w + 3 = 15 + 6\sqrt{5}$

Check:

$a^2 + b^2$	c^2
$(6 + 3\sqrt{5})^2 + (12 + 6\sqrt{5})^2$	$(15 + 6\sqrt{5})^2$
$36 + 36\sqrt{5} + 45 + 144 + 144\sqrt{5} + 180$	$225 + 180\sqrt{5} + 180$
$405 + 180\sqrt{5}$	$405 + 180\sqrt{5}$

Thus, the lengths of the two legs and the hypotenuse are $6 + 3\sqrt{5}$ cm, $12 + 6\sqrt{5}$ cm, and $15 + 6\sqrt{5}$ cm, respectively.

Sometimes measurements in the form of irrational numbers may have to be changed into decimal form. This is shown in Example 2 on page 176.

Example 2 A 10-cm square (10 cm by 10 cm) is cut from each corner of a rectangular piece of cardboard that is 4 times as long as it is wide. The flaps are folded up to form an open carton with a volume of 7,000 cm³. Find the dimensions of the original piece of cardboard to the nearest 0.1 cm.

Let x be the width and $4x$ the length of the original piece of cardboard.

Open carton

length × width × height = volume

$(4x - 20)(x - 20) \cdot 10 = 7{,}000$
$(4x - 20)(x - 20) = 700$
$4x^2 - 100x + 400 = 700$
$4x^2 - 100x - 300 = 0$
$x^2 - 25x - 75 = 0$

$x = \dfrac{-(-25) \pm \sqrt{(-25)^2 - 4 \cdot 1(-75)}}{2}$

$= \dfrac{25 \pm \sqrt{925}}{2}$

$= \dfrac{25 \pm 5\sqrt{37}}{2}$

$\doteq \dfrac{25 \pm 5(6.083)}{2}$

$= 27.7075 \text{ or } -2.7075$

The width cannot be negative, so $x = 27.7075$.
The length, $4x = 4 \cdot 27.7075$, or 110.8300.

Check: See the Non-Routine Problems on page 177.

Thus, the width of the original piece of cardboard was 27.7 cm, and the length was 110.8 cm, to the nearest 0.1 cm.

Written Exercises

Solve each problem. Give the answers in simplest radical form, unless directed otherwise.

Ⓐ 1. One leg of a right triangle is twice as long as the other leg and the hypotenuse is 6 m long. Find the length of each leg of the triangle. $\dfrac{6\sqrt{5}}{5}$ m; $\dfrac{12\sqrt{5}}{5}$ m

2. In a certain right triangle, one leg is 1 cm shorter than the other leg and the hypotenuse is 3 cm longer than the shorter leg. Find the length of each side of the triangle.
$(3 + 2\sqrt{3})$ cm; $(2 + 2\sqrt{3})$ cm; $(5 + 2\sqrt{3})$ cm

3. A 2-cm square is cut from each corner of a rectangular piece of cardboard that is 20 cm longer than it is wide. The four flaps are turned up to form an open box with a volume of 50 cm³. Find the dimensions of the original piece of cardboard to the nearest 0.1 cm. **5.2 cm by 25.2 cm**

4. A rectangular piece of plastic is 3 times as long as it is wide. After a 6-cm square is cut from each corner, the flaps are pressed upward to form an open pan with a volume of 720 cm³. To the nearest 0.1 cm, find the length and the width of the original piece of plastic. **15.5 cm; 46.5 cm**

5. A diagonal of a square is 8 cm long. Find the width of the square. **4√2 cm**

6. Find the width of a square if one diagonal is 3√22 mm long. **3√11 mm**

B 7. A rectangular lawn measures 20 m by 40 m and is surrounded by a sidewalk of uniform width. The outer edge of the sidewalk is a rectangle with an area of 1,000 m². Find the width of a strip of the sidewalk to the nearest 0.1 m. **1.6 m**

8. A rectangular piece of mirror glass measures 40 cm by 60 cm. It is surrounded by a uniform frame whose outer edge is a rectangle with an area of 3,200 cm². To the nearest 0.1 cm, find the width of a strip of the frame. **3.7 cm**

9. A rectangle is 2 dm longer than it is wide. A diagonal of the rectangle is 1 dm longer than the rectangle's length. Find the length of the diagonal and the area of the rectangle. **(4 + √6) dm; (9 + 4√6) dm²**

10. A rectangle's length is 1 mm more than 3 times its width. The length of one of its diagonals is 1 mm more than the rectangles' length. Find the length, the width, and the area of the rectangle. **(10 + 6√3) mm; (3 + 2√3) mm; (66 + 38√3) mm²**

11. A diagonal of a square is 2 m longer than a side of the square. Find the length of the diagonal and the area of the square. **(4 + 2√2) m; (12 + 8√2) m²**

12. One side of a square is 3 dm shorter than one diagonal of the square. Find the length of the side and the area of the square. **(3 + 3√2) dm; (27 + 18√2) dm²**

CALCULATOR ACTIVITIES

1. 11,950.653, which is within 11,950 ± 1   **2. 2,238.4375, which is not within 2,250 ± 1**

In Example 2 on page 176, the open carton measures $4x - 20$ cm by $x - 20$ cm by 10 cm and the volume is 7,000 cm³. Determine whether the volume is within 7,000 ± 1 cm³ if $x = 27.7075$ cm.

If $x = 27.7075$, then $4x - 20 = 90.8300$ and $x - 20 = 7.7075$.

$90.83 \otimes 7.7075 \otimes 10 \ominus 7000.7222$, which is within 7000 ± 1.

A certain box measures $x - 10$ cm by $3x - 10$ cm by 20 cm. Determine whether the volume of the box is within 1 cm³ of the given volume for the given value of x.

1. 11,950 cm³ if $x = 21.168$ cm
2. 2,250 cm³ if $x = 13.625$ cm
3. 39,240 cm³ if $x = 32.456$ cm
4. 1,270 cm³ if $x = 12.347$ cm

3. 39,238.716, which is not within 39,240 ± 1   **4. 1,269.3045, which is within 1,270 ± 1**

NON-ROUTINE PROBLEMS

Show that $10(x - 20)(4x - 20) = 7,000$ if $x = \dfrac{25 + 5\sqrt{37}}{2}$.

See TE Answer Section.

Problem Solving: Irrational Measurements

RADICAL EQUATIONS 7.6

Objective **To solve a radical equation**

Equations like $\sqrt[3]{6x + 10} = -2$ and $3\sqrt{2y + 2} = 2\sqrt{5y - 1}$ contain a variable in a radicand. They are called **radical equations.**

Example 1 **Solve $\sqrt[3]{6x + 10} = -2$.**

$$\sqrt[3]{6x + 10} = -2$$
$$(\sqrt[3]{6x + 10})^3 = (-2)^3 \quad \blacktriangleleft \text{ If } a = b, \text{ then } a^3 = b^3.$$
$$6x + 10 = -8$$
$$6x = -18$$
$$x = -3$$

Check in the original equation.

$\sqrt[3]{6x + 10}$	-2
$\sqrt[3]{6(-3) + 10}$	-2
$\sqrt[3]{-18 + 10}$	
$\sqrt[3]{-8}$	
-2	

Thus, the solution is -3.

Example 2 **Solve $3\sqrt{2y + 2} = 2\sqrt{5y - 1}$.**

$$3\sqrt{2y + 2} = 2\sqrt{5y - 1}$$
$$(3\sqrt{2y + 2})^2 = (2\sqrt{5y - 1})^2 \quad \blacktriangleleft \text{ If } a = b, \text{ then } (a)^2 = (b)^2.$$
$$9(2y + 2) = 4(5y - 1)$$
$$18y + 18 = 20y - 4$$
$$22 = 2y$$
$$11 = y$$

Check in the original equation.

$3\sqrt{2y + 2}$	$2\sqrt{5y - 1}$
$3\sqrt{2 \cdot 11 + 2}$	$2\sqrt{5 \cdot 11 - 1}$
$3\sqrt{24}$	$2\sqrt{54}$
$3 \cdot 2\sqrt{6}$	$2 \cdot 3\sqrt{6}$
$6\sqrt{6}$	$6\sqrt{6}$

Thus, the solution is 11.

It is necessary to check the apparent solutions of a radical equation since raising each side of an equation to the same power may introduce *extraneous solutions.*

Example 3 **Solve $3 + \sqrt{3x + 1} = x$.**

"Isolate" the radical. Rewrite the equation so that the term containing the radical is alone on one side.

$$\sqrt{3x + 1} = x - 3$$
$$(\sqrt{3x + 1})^2 = (x - 3)^2$$
$$3x + 1 = x^2 - 6x + 9$$
Quadratic equation ▶ $$0 = x^2 - 9x + 8$$
$$0 = (x - 1)(x - 8)$$
$$x = 1 \text{ or } x = 8$$

Check: $x = 1$

$3 + \sqrt{3x + 1}$	x
$3 + \sqrt{3 \cdot 1 + 1}$	1
$3 + \sqrt{4}$	
$3 + 2$, or 5	

1 does *not* check. ▶

Check: $x = 8$

$3 + \sqrt{3x + 1}$	x
$3 + \sqrt{3 \cdot 8 + 1}$	8
$3 + \sqrt{25}$	
$3 + 5$, or 8	

Thus, the solution is 8.

In some radical equations, the radicand contains an exponent.

Example 4

Solve $\sqrt[3]{2y^{-1}} = -4$.

$(\sqrt[3]{2y^{-1}})^3 = (-4)^3$
$2y^{-1} = -64$
$y^{-1} = -32$
$\dfrac{1}{y} = -32$
$y = \dfrac{1}{-32}, \text{ or } -\dfrac{1}{32}$

Check:
$\sqrt[3]{2y^{-1}} \;\Big|\; -4$
$\sqrt[3]{2 \cdot \left(\dfrac{1}{-32}\right)^{-1}} \;\Big|\; -4$
$\sqrt[3]{2(-32)}$
$\sqrt[3]{-64}, \text{ or } -4$

Thus, the solution is $-\dfrac{1}{32}$.

Example 5

Solve $\sqrt[4]{x^2 - 8} = 2$.

$\left(\sqrt[4]{x^2 - 8}\right)^4 = 2^4$
$x^2 - 8 = 16$
$x^2 = 24$
$x = \pm 2\sqrt{6}$

Check:
$\sqrt[4]{x^2 - 8} \;\Big|\; 2$
$\sqrt[4]{(\pm 2\sqrt{6})^2 - 8} \;\Big|\; 2$
$\sqrt[4]{24 - 8}$
$\sqrt[4]{16}, \text{ or } 2$

Thus, the solutions are $2\sqrt{6}$ and $-2\sqrt{6}$.

An equation such as $\sqrt{5x + 1} - \sqrt{3x - 5} = 2$ in Example 6 contains two radicals and a third term. To solve this equation, you need to isolate the radicals, one at a time.

Example 6

Solve $\sqrt{5x + 1} - \sqrt{3x - 5} = 2$.

Isolate one of the two radicals.
$\sqrt{5x + 1} = 2 + \sqrt{3x - 5}$
$(\sqrt{5x + 1})^2 = (2 + \sqrt{3x - 5})^2$
$5x + 1 = 4 + 4\sqrt{3x - 5} + 3x - 5$

Isolate the other radical.
$2x + 2 = 4\sqrt{3x - 5}$
$x + 1 = 2\sqrt{3x - 5}$
$(x + 1)^2 = (2\sqrt{3x - 5})^2$
$x^2 + 2x + 1 = 4(3x - 5)$
$x^2 + 2x + 1 = 12x - 20$
$x^2 - 10x + 21 = 0$
$(x - 3)(x - 7) = 0$
$x = 3 \quad \text{or} \quad x = 7$

Check: $x = 3$
$\sqrt{5x + 1} - \sqrt{3x - 5} \;\Big|\; 2$
$\sqrt{5 \cdot 3 + 1} - \sqrt{3 \cdot 3 - 5} \;\Big|\; 2$
$\sqrt{16} - \sqrt{4}$
$4 - 2, \text{ or } 2$

Check: $x = 7$
$\sqrt{5x + 1} - \sqrt{3x - 5} \;\Big|\; 2$
$\sqrt{5 \cdot 7 + 1} - \sqrt{3 \cdot 7 - 5} \;\Big|\; 2$
$\sqrt{36} - \sqrt{16}$
$6 - 4, \text{ or } 2$

Thus, the solutions are 3 and 7.

Radical Equations

Oral Exercises

For each equation, the only apparent solutions are 3 and 5. Check 3 and 5 in each equation to find the solution, or solutions, of the equation.

1. $\sqrt{x-1} = 2$ **5**
2. $\sqrt{x+1} = x - 1$ **3**
3. $\sqrt{x+1} = -2$ **neither**
4. $\sqrt{x-1} = x - 3$ **5**
5. $\sqrt{2x-6} = x - 3$ **3, 5**

Written Exercises

Solve each equation.

(A)
1. $\sqrt{x-2} = 7$ **51**
2. $\sqrt{2y-6} = 6$ **21**
3. $\sqrt{3c} = \sqrt{c+6}$ **3**
4. $\sqrt[3]{2a} = 4$ **32**
5. $\sqrt[3]{7x-1} = \sqrt[3]{5x+7}$ **4**
6. $\sqrt[4]{5x+1} - \sqrt[4]{6x-2} = 0$ **3**
7. $2\sqrt{x} = \sqrt{3x+5}$ **5**
8. $\sqrt{12y+3} - 3\sqrt{2y} = 0$ **$\frac{1}{2}$**
9. $3\sqrt{2a-4} = 2\sqrt{a-2}$ **2**
10. $2\sqrt[3]{4y} = -1$ **$-\frac{1}{32}$**
11. $3\sqrt[4]{2x} = 2$ **$\frac{8}{81}$**
12. $2\sqrt[3]{3n} = 3\sqrt[3]{n-1}$ **9**
13. $x - 4 = \sqrt{2x}$ **8**
14. $y - 5 = \sqrt{y-3}$ **7**
15. $\sqrt{2w-3} = w - 3$ **6**
16. $\sqrt[3]{y^{-1}} = 5$ **$\frac{1}{125}$**
17. $\sqrt[3]{c^2} = 2$ **$\pm 2\sqrt{2}$**
18. $\sqrt[3]{x^2+9} = 3$ **$\pm 3\sqrt{2}$**
19. $\sqrt[4]{x^3} = 2$ **$2\sqrt[3]{2}$**
20. $\sqrt[4]{y^{-2}} = 3$ **$\pm \frac{1}{9}$**
21. $\sqrt[4]{a^2+16} = 4$ **$\pm 4\sqrt{15}$**
22. $\sqrt{8y} = \sqrt{2y} + 2$ **2**
23. $\sqrt{x+7} = \sqrt{x} + 1$ **9**
24. $\sqrt{3n-2} = \sqrt{n-2}$ **3**

(B)
25. $\sqrt[3]{3y^{-1}} = -6$ **$-\frac{1}{72}$**
26. $\sqrt[4]{2y^{-2}} = \frac{1}{3}$ **$\pm 9\sqrt{2}$**
27. $\sqrt{(x-3)^{-2}} = \frac{1}{4}$ **$-1, 7$**
28. $3 + \sqrt{2t+5} = 0$ **no solution**
29. $-2 + \sqrt{4x+3} = 0$ **$\frac{1}{4}$**
30. $-5 - \sqrt{7y-3} = 0$ **no solution**
31. $6 + \sqrt{3x} = x$ **12**
32. $y - \sqrt{4y-3} = 2$ **7**
33. $2 + \sqrt{5x+6} = 3x$ **2**
34. $4 - \sqrt{10-3a} = a$ **2, 3**
35. $2\sqrt{3x-2} = x + 2$ **2, 6**
36. $3\sqrt{2n+3} = 2n + 5$ **$-1, \frac{1}{2}$**
37. $\sqrt{x^2+x-3} = 3$ **$-4, 3$**
38. $\sqrt[3]{y^2-12y} = 4$ **$-4, 16$**
39. $\sqrt[4]{5x^2+3} = 2\sqrt[4]{x}$ **$\frac{1}{5}, 3$**
40. $\sqrt{2y+3} = 1 + \sqrt{y+1}$ **$-1, 3$**
41. $\sqrt{4x-3} = 2 + \sqrt{2x-5}$ **3, 7**
42. $\sqrt{3a+1} - \sqrt{a-4} = 3$ **5, 8**
43. $\sqrt{\frac{x-3}{x+2}} = \frac{2}{3}$ **7**
44. $\sqrt{\frac{2}{y+3}} = \sqrt{\frac{3}{2y+2}}$ **5**
45. $\frac{\sqrt{x-2}}{x-2} = \frac{x-5}{\sqrt{x-2}}$ **6**
46. $\sqrt{y+15} - \sqrt{2y+7} = 1$ **1**
47. $\sqrt{3x-5} + \sqrt{x-1} = 2$ **2**
48. $\sqrt{3x+9} - \sqrt{2x+7} = 1$ **9**

(C)
49. The time T for one period of a simple pendulum is determined by the formula $T = 2\pi\sqrt{\frac{l}{g}}$. Solve for l, the length of the pendulum's string. **$l = \frac{T^2 g}{4\pi^2}$**

50. The time t on an upward launch is found from the formula $t = \frac{v \pm \sqrt{v^2 + 20(d-h)}}{10}$. Solve for h, the height at the time t.

51. Solve $\sqrt{x-1} = \sqrt[4]{8x+1}$ for x. **10**

52. Solve $\sqrt[3]{x+6} = \sqrt[6]{32x}$ for x. **2, 18**

53. The radius r of a sphere is given by the formula $r = \sqrt[3]{\frac{3V}{4\pi}}$. Solve for V, the volume of the sphere. **$V = \frac{4\pi r^3}{3}$**

54. One side s of an equilateral triangle is given by the formula $s = \frac{2\sqrt{A}}{\sqrt[4]{3}}$. Solve for A, the area of the triangle. **$A = \frac{s^2}{4}\sqrt{3}$**

55. Solve $x = \frac{N \pm \sqrt{N^2 - 4M(T-D)}}{2M}$ for T. **$T = -Mx^2 + Nx + D$**

56. Solve $2Lx + R = \sqrt{R^2 - \frac{4L}{C}}$ for $\frac{-1}{C}$. **$\frac{-1}{C} = Lx^2 + Rx$**

50. $h = -5t^2 + tv + d$

CUMULATIVE REVIEW

Simplify each expression.

1. $5\sqrt{72}$ **$30\sqrt{2}$**
2. $3x^2y\sqrt{24x^7y^8}$ **$6x^5y^5\sqrt{6x}$**
3. $5\sqrt{2} + 2\sqrt{18} + 3\sqrt{2}$ **$14\sqrt{2}$**
4. $(2\sqrt{7} - \sqrt{10})(2\sqrt{7} + \sqrt{10})$ **18**

APPLICATIONS

If a weight is suspended on a string, the result is a simple pendulum. Ideally, the motion of a pendulum is **periodic**. If the weight on the pendulum is pulled to one side and released, it will swing away and back. One such back-and-forth motion is called a **vibration**.

The action of some clocks is based on the periodic motion of a pendulum. Of course, some mechanical contrivance is usually employed to keep the periodic motion stable. The time, T, for one period of a simple pendulum is given by the radical equation

$$T = 2\pi \sqrt{\frac{l}{g}}$$

where l is the length of the string, and g is the acceleration due to gravity. Notice that the period, T, depends on the length, l, of the string and not upon the amount of weight that is used.

The formula above can be solved for l.

$$\left(\frac{T}{2\pi}\right)^2 = \left(\sqrt{\frac{l}{g}}\right)^2 \qquad \frac{l}{g} = \frac{T^2}{4\pi^2} \qquad \text{So, } l = \frac{gT^2}{4\pi^2}.$$

Solve each problem. Use the more appropriate of the two formulas above. Use 10 m/s² for g and 3.14 for π.

1. Find the length of the string for a simple pendulum if the time for one period is 1.57 s. **0.625 m**
2. Find the time for one period of a simple pendulum if the length of the string is 0.225 m. **0.942 s**
3. How should the length of the string be changed in order to double the period of a given simple pendulum? [*Hint*: Replace T by $2T$ in the second formula.] **4 times as long**

For a sphere of radius r and volume V, $r = \sqrt[3]{\frac{3V}{4\pi}}$.

4. Solve the formula for V. $V = \frac{4\pi r^3}{3}$
5. Find V if $r = 30$ cm. Use 3.14 for π. **113,040 cm³**
6. Find the change in the volume of a sphere if the radius of a given sphere is tripled. **27 times larger**

For a cylinder of radius r and volume V, $r = \sqrt{\frac{V}{\pi h}}$.

7. Solve the formula for V. $V = \pi r^2 h$
8. Find V if $r = 4.5$ cm and $h = 8$ cm. Use 3.14 for π. **508.68 cm³**
9. What is the change in the volume of a cylinder if the radius and the height of a given cylinder are each doubled? **8 times larger**

CHAPTER SEVEN REVIEW

(A) Exercises are 1, 4, 5, 8, 9, 11, 12, 15, 16.
The rest are (B) Exercises.
(C) Exercises are starred.

Vocabulary
completing the square [7.1]
quadratic formula [7.3]
radical equation [7.6]

Solve each equation by completing the square. [7.1]

1. $x^2 - 6x = 3$ $\quad 3 \pm 2\sqrt{3}$

2. $2n^2 = 1 - 5n$ $\quad \dfrac{-5 \pm \sqrt{33}}{4}$

★ 3. $x^2 + 2x\sqrt{2} - 48 = 0$ $\quad -6\sqrt{2}, 4\sqrt{2}$

Solve each equation by using the quadratic formula. [7.3]

4. $3y^2 + 2y = 2$ $\quad \dfrac{-1 \pm \sqrt{7}}{3}$

5. $5c^2 = 7$ $\quad \dfrac{\pm\sqrt{35}}{5}$

6. $\dfrac{x^2}{8} = \dfrac{x}{2} - \dfrac{1}{4}$ $\quad 2 \pm \sqrt{2}$

★ 7. $x^{-2} - 8x^{-1} + 3 = 0$ $\quad \dfrac{4 \pm \sqrt{13}}{3}$

8. Find the maximum value of $-3x^2 + 24x + 152$ and the corresponding value of x. [7.2] **200, $x = 4$**

9. An object is sent upward at 80 m/s. Find the maximum height it reaches and the corresponding time. Use the formula $h = vt - 5t^2$. [7.2] **320 m, $t = 8$ s**

10. A second number is 2 more than 4 times a first number. Find the numbers if their product is 5. [7.4]
    $\dfrac{-1 - \sqrt{21}}{4}, 1 - \sqrt{21}$ or $\dfrac{-1 + \sqrt{21}}{4}, 1 + \sqrt{21}$

11. An object is shot upward from the earth's surface with an initial velocity of 60 m/s. To the nearest 0.1 s, when will its height be 150 m? Use the formula $h = vt - 5t^2$. [7.4] **3.6 s, 8.4 s**

12. A 4-cm square is cut from each corner of a rectangular piece of sheet metal that is twice as long as it is wide. Find the dimensions of the original rectangle if the flaps are turned up to form a pan with a volume of 800 cm³. Express each dimension in simplest radical form. [7.5]
    $(6 + 2\sqrt{26})$ cm, $(12 + 4\sqrt{26})$ cm

13. A rectangle is 1 m longer than it is wide. The length of one of its diagonals is 3 m more than the rectangle's width. Find the length of the diagonal and the area of the rectangle. Give the answers in simplest radical form. [7.5]
    $(5 + 2\sqrt{3})$ m, $(18 + 10\sqrt{3})$ m²

★ 14. A pellet is sent up from the top of a tower 50 m tall with an initial velocity of 45 m/s. To the nearest 0.1 s, when will its height be 125 m? Use the formula $h = d + vt - 5t^2$. [7.4] **2.2 s, 6.8 s**

Solve each equation. [7.6]

15. $3\sqrt{2x + 4} = 12$ **6**

16. $x - 1 = \sqrt{x + 5}$ **4**

17. $\sqrt[4]{5y^{-1}} = \dfrac{1}{2}$ **80**

18. $\sqrt[3]{n^2 - 6n} = 3$ **−3, 9**

19. $\sqrt{3x - 5} - \sqrt{x - 2} = 1$ **2, 3**

★ 20. $\sqrt[3]{4x + 6} = 2\sqrt[6]{2x}$ $\dfrac{1}{2}, \dfrac{9}{2}$

Ⓐ Exercises are 1, 3, 4, 6, 9–12.
The rest are Ⓑ Exercises.

Ⓒ Exercises are starred.

CHAPTER SEVEN TEST

Solve each equation by completing the square.

1. $x^2 - 8x - 4 = 0$ $4 \pm 2\sqrt{5}$

2. $3c^2 = -7c - 1$ $\dfrac{-7 \pm \sqrt{37}}{6}$

Solve each equation by using the quadratic formula.

3. $2y^2 + 3y = 3$ $\dfrac{-3 \pm \sqrt{33}}{4}$

4. $5x^2 = 11x$ $0, \dfrac{11}{5}$

5. $\dfrac{x^2}{4} = \dfrac{x}{2} - \dfrac{1}{6}$ $\dfrac{3 \pm \sqrt{3}}{3}$

6. One leg of a right triangle is 3 times as long as the other leg and 2 cm shorter than the hypotenuse. Find the length of the hypotenuse. Give the answer in simplest radical form. **$(20 + 6\sqrt{10})$ cm**

7. A second number is 2 less than 4 times a first number. If the second number is multiplied by 1 less than the first number, the product is 1. Find the numbers.
$\dfrac{3-\sqrt{5}}{4}, 1 - \sqrt{5}$ or $\dfrac{3+\sqrt{5}}{4}, 1 + \sqrt{5}$

8. A rectangular piece of land is to be fenced on three sides with 240 m of fencing. Find the maximum area of the piece of land. **7,200 m²**

9. To the nearest 0.1 s, when will an object reach a height of 15 m if it is sent up from the earth with an initial velocity of 25 m/s? Use the formula $h = vt - 5t^2$. **0.7 s, 4.3 s**

10. A 3-cm square is cut from each corner of a rectangular piece of plastic that is 3 times as long as it is wide. Find the dimensions of the original rectangle if the four flaps are turned up to form an open box with a volume of 270 cm³. Express each dimension in simplest radical form.
$(4 + \sqrt{34})$ cm, $(12 + 3\sqrt{34})$ cm

11. Find the maximum value of $-2x^2 + 24x + 228$ and the corresponding value of x. **300, $x = 6$**

Solve each equation.

12. $\sqrt[4]{7x + 4} = 3$ **11**

13. $y - \sqrt{2y - 1} = 2$ **5**

14. $\sqrt[3]{n^2 - 2n} = 2$ **−2, 4**

15. $\sqrt{3x + 1} - \sqrt{x - 1} = 2$ **1, 5**

★ 16. Solve $x^2 + 4x\sqrt{5} + 15 = 0$ by completing the square. **$-3\sqrt{5}, -\sqrt{5}$**

★ 17. Solve $\sqrt{x - 1} = \sqrt[4]{3x + 1}$. **5**

COMPUTER ACTIVITIES

Solutions of Quadratic Equations

OBJECTIVE: To find solutions of quadratic equations in radical form

On page 583, you find the solutions of a quadratic equation by inserting coefficient values into the quadratic formula. Unfortunately, the computer only outputs answers in decimal form. The program below will output the solutions in radical form as an alternative to decimal solutions.

In this program, line 130 tests whether or not the discriminant is a perfect square. The loop from lines 140 to 170 simplifies the radical. Lines 180 through 230 reduce the numerator and denominator to simplest terms.

Notice the use of the symbol < > in lines 150, 190, 200, and 210. This combination of the greater than and less than is the BASIC symbol for "is not equal to." Notice also the use of the ABS function in line 180.

```
110   PRINT "GIVE THE VALUES OF A,B, AND C.";: INPUT
      A,B,C
120   D = - B:E = 1:F = B * B - 4 * A * C:G = 2 * A
125   IF F < 0 THEN 270
130   IF SQR (F) = INT ( SQR (F)) THEN 280
140   FOR I = 2 TO SQR (F)
150   IF F / I ^ 2 < > INT (F / I ^ 2) THEN 170
160   E = E * I:F = F / I ^ 2
170   NEXT I
180   FOR I = 2 TO ABS (D)
190   IF D / I < > INT (D / I) THEN 230
200   IF E / I < > INT (E / I) THEN 230
210   IF G / I < > INT (G / I) THEN 230
220   D = D / I:E = E / I:G = G / I: GOTO 190
230   NEXT I
240   PRINT "THE SOLUTIONS ARE ";
250   PRINT "("D"/"G" + "E" SQR ("F" ))/"G"  AND "
260   PRINT "("D"/"G" - "E"SQR ("F"))/"G: GOTO 110
270   PRINT "NO REAL SOLUTIONS": GOTO 110
280   PRINT "THE SOLUTIONS ARE "D +  SQR (F)"/"G" AND";
290   PRINT D -  SQR (F)"/"G: GOTO 110
```

See the Computer Section beginning on page 574 for more information.

EXERCISES   For Ex. 2, see TE Answer Section.

1. Run the program above by inputting the values from the illustrative exercises on pages 165 and 166.
2. Write a program that solves problems similar to the ones in Exercise 2 on page 167.

COLLEGE PREP TEST

DIRECTIONS: Choose the *one* best answer to each question or problem.

1. One diagonal of a rectangle is 10 cm long. The rectangle's length is 8 cm. Find the area of the rectangle. **C**

   (A) 24 cm²   (B) 36 cm²   (C) 48 cm²
   (D) 80 cm²   (E) None of these

2. Which equation has two different positive integers as its solutions? **D**

   (A) $(x - 5)^2 = 0$ (B) $(x - 1)^2 = 4$
   (C) $(x + 4)^2 = 9$ (D) $(x - 6)^2 = 16$
   (E) None of these

3. For which pair of numbers is the product of *2 more than the first* and *3 less than the second* greatest? **D**

   (A) 9 and 7   (B) 8 and 8
   (C) 7 and 9   (D) −12 and −3
   (E) All products are the same.

4. If $x = -2$, then **D**

   (A) $\sqrt{x^2} = x$ (B) $\sqrt[3]{x^3} = -x$
   (C) $\sqrt[4]{x^4} = x$ (D) $\sqrt[3]{x^3} = x$
   (E) All of these

5. For which equation is the sum of its two solutions greater than their product? **B**

   (A) $x^2 - 7x + 12 = 0$
   (B) $x^2 + 9x - 10 = 0$
   (C) $x^2 + 8x = 0$
   (D) $x^2 + 4 = 0$
   (E) None of these

6. Find a solution of
   $\sqrt{4x} + \sqrt{2x} = \sqrt{4x - 12} + \sqrt{2x + 12}$. **C**

   (A) 2   (B) 4   (C) 6   (D) 8
   (E) None of these

7. A rectangular piece of cardboard is 70 cm by 50 cm. A 10-cm square is cut from each corner and the four flaps are folded up to form an open box. Find the volume of the box. **B**

   (A) 1,500 cm³   (B) 15,000 cm³
   (C) 24,000 cm³   (D) 35,000 cm³
   (E) None of these

8. Choose the number that is not equal to the other three? **E**

   (A) $4 + 2\sqrt{3}$ (B) $\dfrac{8 + \sqrt{48}}{2}$
   (C) $6 + \sqrt{48} - 2 - \sqrt{12}$
   (D) $\sqrt{2}(\sqrt{8} + \sqrt{6})$
   (E) They are all equal.

9. The formula, $h = 5t(20 - t)$, gives the height h of an object at t seconds. The greatest height is achieved at which time? **C**

   (A) 0 s   (B) 5 s   (C) 10 s   (D) 20 s
   (E) All heights are the same.

10. Find an equation with integral coefficients having $3 - 2\sqrt{3}$ as a solution. **B**

    (A) $x^2 = 9x - 12$ (B) $x^2 - 6x = 3$
    (C) $x^2 - 6x + 15 = 0$ (D) $x^2 = 21$
    (E) None of these

College Prep Test

8 COMPLEX NUMBERS

HISTORY

Sonya Kovalevsky 1850–1891

Sonya Kovalevsky was born in Moscow, Russia, on January 15, 1850. She is best known for her work on partial differential equations and elliptic functions.

At age 18, Sonya left Russia and moved to Germany. She overcame many obstacles to study mathematics. She was unable to enter the University of Berlin because she was a woman. However, after studying with Karl Weierstrass she was awarded a Ph.D. degree in absentia from the University of Gottingham. Sonya was one of the few women of her time to receive the Prix Bordin award from the Paris Academy for outstanding work in mathematics.

In 1884, she became a professor of mathematics at Stockholm University in Sweden, and in 1889 the university named her a professor for life.

Project Write a report on Sonya Kovalevsky, emphasizing her work in elliptic functions.

SUMS AND DIFFERENCES OF COMPLEX NUMBERS 8.1

Objectives **To simplify a square root expression where the radicand is a negative real number**
To solve a quadratic equation whose solutions are imaginary numbers
To simplify an expression containing a sum or difference of complex numbers

The equation $x^2 = -1$ does not have a real number solution since there is no real number whose square is negative. In order to solve equations of this type, mathematicians invented the numbers i and $-i$.
If $x^2 = -1$, then x would have to be $\sqrt{-1}$ or $-\sqrt{-1}$, where $(\sqrt{-1})^2 = -1$ and $(-\sqrt{-1})^2 = -1$. The number $\sqrt{-1}$ is represented by i and the number $-\sqrt{-1}$ is represented by $-i$.

Definition: i and $-i$

$i = \sqrt{-1}$ and $i^2 = -1$
$-i = -\sqrt{-1}$ and $(-i)^2 = -1$

Notice that i and $-i$ are *not* variables; each of them names one specific number. You can use the definition of i to simplify a square root expression where the radicand is a negative real number.

Example 1   **Simplify $\sqrt{-16}$, $\sqrt{-7}$, and $\sqrt{-12}$.**

$$\sqrt{-16} = \sqrt{-1(16)}$$
$$= \sqrt{-1} \cdot \sqrt{16}$$
$$= 4i$$

$$\sqrt{-7} = \sqrt{-1(7)}$$
$$= \sqrt{-1} \cdot \sqrt{7}$$
$$= i\sqrt{7}$$

$$\sqrt{-12} = \sqrt{-1(12)}$$
$$= \sqrt{-1} \cdot \sqrt{12}$$
$$= i \cdot 2\sqrt{3}$$
$$= 2i\sqrt{3}$$

In Example 1, the property $\sqrt{xy} = \sqrt{x} \cdot \sqrt{y}$ for $x \geq 0, y \geq 0$, was extended to include $\sqrt{-1 \cdot y} = \sqrt{-1} \cdot \sqrt{y}$ for $y \geq 0$. In general, $\sqrt{-y} = i\sqrt{y}$ for $y \geq 0$.

Example 2   **Simplify each expression.**

$-3 + \sqrt{-4}$ $-5\sqrt{-8}$ $2 - \sqrt{-10}$

$-3 + i\sqrt{4}$ $-5 \cdot i\sqrt{8}$ $2 - i\sqrt{10}$
$-3 + i \cdot 2$ $-5 \cdot i \cdot 2\sqrt{2}$
$-3 + 2i$ $-10i\sqrt{2}$

The numbers $-3 + 2i$, $-10i\sqrt{2}$ and $2 - i\sqrt{10}$ are examples of **complex numbers.** A complex number is one that can be written in the form $a + bi$, where a and b are real numbers. Note that $-10i\sqrt{2} = 0 + (-10\sqrt{2})i$, where $a = 0$, $b = -10\sqrt{2}$, and $2 - i\sqrt{10} = 2 + (-\sqrt{10})i$, where $a = 2$, $b = -\sqrt{10}$.

Sums and Differences of Complex Numbers

> **Definition: Complex number**
> A *complex number* is a number that can be written in the form $a + bi$ where a and b are real numbers and $i = \sqrt{-1}$.

Complex numbers like $-3 + 2i$ and $-10i\sqrt{2}$, where $b \neq 0$, are called **imaginary numbers**. A complex number like $-10i\sqrt{2}$, where $a = 0$, is called a **pure imaginary number**. Notice that every *real number* is also a complex number. For example, the real number -5 can be written as $-5 + 0i$, where $b = 0$.

You can solve equations like $x^2 = -25$ and $2y^2 + 16 = 0$ if imaginary numbers are permitted as solutions.

Example 3   **Solve $x^2 = -25$.**   **Solve $2y^2 + 16 = 0$.**

$$x^2 = -25$$
$$x = \pm\sqrt{-25}$$
$$= \pm 5i$$

The solutions are $5i$ and $-5i$.

$$2y^2 = -16$$
$$y^2 = -8$$
$$y = \pm\sqrt{-8} = \pm 2i\sqrt{2}$$

The solutions are $2i\sqrt{2}$ and $-2i\sqrt{2}$.

An expression containing a sum of two complex numbers can be simplified by combining the like terms as you would with a pair of binomials. For example,

$$(3 + 5i) + (2 - 8i) = (3 + 2) + (5i - 8i) = (3 + 2) + (5 - 8)i = 5 - 3i.$$

This leads to the following property:

> **Sum of complex numbers**
> For all real numbers a, b, c, and d,
> $(a + bi) + (c + di) = (a + c) + (b + d)i.$

To simplify an expression containing a difference of two complex numbers, $(a + bi) - (c + di)$, add the additive inverse (opposite) of $c + di$, which is $-c - di$.

Example 4   **Simplify each expression.**

$$(8 - 9i) - (6 - 4i) \qquad\qquad (-9 + \sqrt{-2}) + (6 - \sqrt{-18})$$

$$(8 - 9i) + (-6 + 4i) \qquad\qquad (-9 + i\sqrt{2}) + (6 - 3i\sqrt{2})$$
$$(8 - 6) + (-9 + 4)i \qquad\qquad (-9 + 6) + (\sqrt{2} - 3\sqrt{2})i$$
$$2 - 5i \qquad\qquad\qquad\qquad -3 - 2i\sqrt{2}$$

Reading in Algebra

Which of the words at the right apply to each complex number?

1. $5i$ **A, E**
2. $8 - 4i$ **A**
3. -25 **B, D**
4. i **A, E**
5. i^2 **B, D**
6. $5 + 6\sqrt{2}$ **B, C**

A imaginary   B real   C irrational
D rational   E pure imaginary

Oral Exercises

Show that each number is a complex number $a + bi$ by identifying the real numbers a and b.

1. $6i$ $a = 0, b = 6$
2. 7 $a = 7, b = 0$
3. i $a = 0, b = 1$
4. $-\sqrt{10}$ $a = -\sqrt{10}, b = 0$
5. $-9i$ $a = 0, b = -9$
6. $i\sqrt{3}$ $a = 0, b = \sqrt{3}$
7. -2.5 $a = -2.5, b = 0$
8. $-i$ $a = 0, b = -1$
9. $-i\sqrt{5}$ $a = 0, b = -\sqrt{5}$
10. $7 - 12i$ $a = 7, b = -12$

Written Exercises

Simplify each expression.

Ⓐ
1. $\sqrt{-36}$ $6i$
2. $-\sqrt{-81}$ $-9i$
3. $\sqrt{-64}$ $8i$
4. $\sqrt{-15}$ $i\sqrt{15}$
5. $\sqrt{-22}$ $i\sqrt{22}$
6. $-\sqrt{-7}$ $-i\sqrt{7}$
7. $\sqrt{-40}$ $2i\sqrt{10}$
8. $-\sqrt{-27}$ $-3i\sqrt{3}$
9. $2\sqrt{-45}$ $6i\sqrt{5}$
10. $-5\sqrt{-4}$ $-10i$
11. $-6\sqrt{-20}$ $-12i\sqrt{5}$
12. $10\sqrt{-90}$ $30i\sqrt{10}$
13. $4 + \sqrt{-9}$ $4 + 3i$
14. $-20 - \sqrt{-30}$ $-20 - i\sqrt{30}$
15. $8 - \sqrt{-50}$ $8 - 5i\sqrt{2}$

Solve each equation.

16. $x^2 = -49$ $\pm 7i$
17. $y^2 = -3$ $\pm i\sqrt{3}$
18. $z^2 = -48$ $\pm 4i\sqrt{3}$
19. $3y^2 = -30$ $\pm i\sqrt{10}$
20. $x^2 = -100$ $\pm 10i$
21. $4a^2 = -200$ $\pm 5i\sqrt{2}$

Simplify each expression.

22. $(3 + 6i) + (4 + 3i)$ $7 + 9i$
23. $(7 + 8i) - (3 + 6i)$ $4 + 2i$
24. $(-2 - 5i) + (8 + 12i)$ $6 + 7i$
25. $(6 + 12i) - (-3 + 8i)$ $9 + 4i$
26. $(5 + i) + (-6 - 3i)$ $-1 - 2i$
27. $(-7 + i) - (5 - i)$ $-12 + 2i$
28. $(7 - i) + (-10 + 10i)$ $-3 + 9i$
29. $(-14 + 6i) - (-6 - 6i)$ $-8 + 12i$
30. $(-8 + 4i) + (-8 - 4i)$ -16

Ⓑ
31. $7 + 8\sqrt{-16}$ $7 + 32i$
32. $-5 - 5\sqrt{-25}$ $-5 - 25i$
33. $\sqrt{60} - 4\sqrt{-15}$ $2\sqrt{15} - 4i\sqrt{15}$
34. $20 - 10\sqrt{-12}$ $20 - 20i\sqrt{3}$
35. $-1 + 4\sqrt{-80}$ $-1 + 16i\sqrt{5}$
36. $-\sqrt{8} - 2\sqrt{-32}$ $-2\sqrt{2} - 8i\sqrt{2}$

Solve each equation.

37. $x^2 + 81 = 0$ $\pm 9i$
38. $y^2 + 45 = 0$ $\pm 3i\sqrt{5}$
39. $0 = n^2 + 65$ $\pm i\sqrt{65}$
40. $5t^2 + 21 = t^2 - 43$ $\pm 4i$
41. $4x^2 + 21 = 3 - 2x^2$ $\pm i\sqrt{3}$
42. $8y^2 + 12 = 8 - y^2$ $\pm \frac{2}{3}i$

Simplify each expression.

43. $(3 + \sqrt{-36}) + (-8 - \sqrt{-16})$ $-5 + 2i$
44. $(-9 + 2\sqrt{-25}) - (7 - 3\sqrt{-49})$ $-16 + 31i$
45. $(6 - 5\sqrt{-3}) - (1 - 2\sqrt{-3})$ $5 - 3i\sqrt{3}$
46. $(-10 + 6\sqrt{-20}) + (-4 - 4\sqrt{-5})$ **below**
47. $(12 + \sqrt{-18}) + (-4 + 4\sqrt{-2})$ $8 + 7i\sqrt{2}$
48. $(7 - 3\sqrt{-8}) - (17 - 2\sqrt{-50})$ $-10 + 4i\sqrt{2}$

46. $-14 + 8i\sqrt{5}$

The **absolute value** of a *complex number* $a + bi$ is defined as $|a + bi| = \sqrt{a^2 + b^2}$, where a and b are real numbers.

Simplify. Use the definition of $|a + bi|$.

Ⓒ
49. $|3 - 4i|$ 5
50. $|2 + 6i|$ $2\sqrt{10}$
51. $|-3|$ 3
52. $|5i|$ 5
53. $|6 - 8i|$ and $|6 + 8i|$ $10, 10$
54. $|3 + 4i|$ and $|-3 - 4i|$ $5, 5$

Prove each theorem.
55. $|a + bi| = \sqrt{a^2 + b^2}$ and $|a - bi| = \sqrt{a^2 + (-b)^2} = \sqrt{a^2 + b^2}$
55. $|a + bi| = |a - bi|$
56. $|a + bi| = |-a - bi|$
   $|a + bi| = \sqrt{a^2 + b^2}$ and $|-a - bi| = \sqrt{(-a)^2 + (-b)^2} = \sqrt{a^2 + b^2}$

Sums and Differences of Complex Numbers

PRODUCTS AND QUOTIENTS OF COMPLEX NUMBERS 8.2

Objectives **To simplify an expression containing a product of complex numbers**
To simplify an expression containing a quotient of complex numbers

The fact that $i^2 = -1$ permits you to simplify products that involve imaginary numbers. Before multiplying two pure imaginary numbers, each number should be written as the product of i and a real number.

Example 1 **Simplify each product.**

$6i \cdot 4i$ $\qquad\qquad -\sqrt{-6} \cdot 5\sqrt{-3}$

$\quad 6i \cdot 4i$
$= 24i^2$
$i^2 = -1 \blacktriangleright = 24(-1)$
$= -24$

$-\sqrt{-6} \cdot 5\sqrt{-3} = -i\sqrt{6} \cdot 5i\sqrt{3}$ ◀ $\sqrt{-y} = i\sqrt{y}$
$\qquad\qquad\qquad = -5i^2\sqrt{18}$
$\qquad\qquad\qquad = -5(-1) \cdot 3\sqrt{2}$
$\qquad\qquad\qquad = 15\sqrt{2}$

In Example 1, notice that $\sqrt{-6} \cdot \sqrt{-3}$ is *not* equal to $\sqrt{-6(-3)}$, or $\sqrt{18}$. The correct way to simplify this product is shown below.

$$\sqrt{-6} \cdot \sqrt{-3} = i\sqrt{6} \cdot i\sqrt{3} = i^2\sqrt{18} = (-1) \cdot 3\sqrt{2} = -3\sqrt{2}$$

The Product of Square Roots property on page 145, $\sqrt{x} \cdot \sqrt{y} = \sqrt{x \cdot y}$, applies only when each radicand is a positive real number or 0. The following property is used to multiply two pure imaginary numbers.

Product of pure imaginary numbers

$$\sqrt{-x} \cdot \sqrt{-y} = i\sqrt{x} \cdot i\sqrt{y} = i^2\sqrt{x \cdot y} = -\sqrt{xy},$$
where $-x$ and $-y$ are negative real numbers.

The multiplication in Example 2 involves both real and imaginary numbers.

Example 2 **Simplify each product.**

$5\sqrt{10} \cdot 6\sqrt{-2}$ $\qquad\qquad 2\sqrt{5}(3\sqrt{10} - 4i\sqrt{15})$

$\quad 5\sqrt{10} \cdot 6\sqrt{-2}$
$= 5\sqrt{10} \cdot 6i\sqrt{2}$
$= 30i\sqrt{20}$
$= 30i \cdot 2\sqrt{5}$
$= 60i\sqrt{5}$

$\quad 2\sqrt{5}(3\sqrt{10} - 4i\sqrt{15})$ ◀ *Use the Distributive Property.*
$= 6\sqrt{50} - 8i\sqrt{75}$
$= 6 \cdot 5\sqrt{2} - 8i \cdot 5\sqrt{3}$
$= 30\sqrt{2} - 40i\sqrt{3}$

You should notice in Example 1 that the product of two pure imaginary numbers is a real number, whereas in Example 2 the product of a real number and an imaginary number is an imaginary number.

To simplify a product that involves complex numbers that are imaginary numbers, but not pure imaginary, the FOIL method of multiplying two binomials can be used.

Example 3 **Simplify $(3 + 2i)(5 + 4i)$.**

$$\begin{array}{c} \text{F} \text{O}\text{I}\text{L} \\ (3 + 2i)(5 + 4i) = 3 \cdot 5 + 3 \cdot 4i + 2i \cdot 5 + 2i \cdot 4i \\ = 15 + 12i + 10i + 8i^2 \\ = 7 + 22i \quad \blacktriangleleft \quad 8i^2 = 8(-1) = -8 \end{array}$$

In Example 3, notice that the product of the two imaginary numbers is an imaginary number. This is not always the case, however. Two numbers of the form $a + bi$ and $a - bi$ are imaginary, but $(a + bi)(a - bi)$ is a real number.

> **Definition: Conjugate complex numbers**
>
> Two complex numbers of the form $a + bi$ and $a - bi$ are *conjugates*. Their product, $(a + bi)(a - bi)$, is a real number, $a^2 + b^2$.

Example 4 **Simplify $(-2 + 3i)(-2 - 3i)$.**

$$(-2 + 3i)(-2 - 3i) = (-2)^2 + (3)^2 = 4 + 9 = 13 \quad \blacktriangleleft \quad (a + bi)(a - bi) = a^2 + b^2$$

Recall that $\dfrac{2}{3\sqrt{5}} = \dfrac{2}{3\sqrt{5}} \cdot \dfrac{\sqrt{5}}{\sqrt{5}} = \dfrac{2\sqrt{5}}{15}$ and

$\dfrac{7}{4 - \sqrt{6}} = \dfrac{7}{4 - \sqrt{6}} \cdot \dfrac{4 + \sqrt{6}}{4 + \sqrt{6}} = \dfrac{7(4 + \sqrt{6})}{16 - 6} = \dfrac{28 + 7\sqrt{6}}{10}.$

You can simplify quotients of complex numbers by rationalizing the denominators in a similar way.

Example 5 **Simplify $-5 \div (6i)$ by rationalizing the denominator.**

Write the quotient as a fraction.

$$\frac{-5}{6i} = \frac{-5}{6i} \cdot \frac{i}{i} = \frac{-5i}{6i^2} = \frac{-5i}{6(-1)} = \frac{-5i}{-6} = \frac{5i}{6}$$

Example 6 **Simplify $\dfrac{2 - 4i}{5 + 3i}$ by rationalizing the denominator.**

$$\frac{2 - 4i}{5 + 3i} = \frac{2 - 4i}{5 + 3i} \cdot \frac{5 - 3i}{5 - 3i} = \frac{10 - 6i - 20i + 12i^2}{25 + 9} = \frac{-2 - 26i}{34} = \frac{-1 - 13i}{17}$$

Powers of i can be simplified by using i^2 as a factor as many times as needed. For example,

$$i^9 = i^2 \cdot i^2 \cdot i^2 \cdot i^2 \cdot i = (-1)(-1)(-1)(-1)i = i.$$

Products and Quotients of Complex Numbers

Example 7   Simplify $(2i)^7$.

$(2i)^7 = 2^7 \cdot i^7 = 2^7 \cdot i^4 \cdot i^2 \cdot i = 128 \cdot 1(-1) \cdot i = -128i$ ◀ $i^4 = i^2 \cdot i^2 = 1$

Reading in Algebra

Determine whether each statement is always true (A), sometimes true (S), or never true (N).

1. The product of two pure imaginary numbers is an imaginary number. **N**
2. The product of two imaginary numbers is a real number. **S**
3. The product of a real number and an imaginary number is an imaginary number. **A**
4. The quotient of two imaginary numbers is an imaginary number. **S**

Written Exercises

Simplify each product.

Ⓐ
1. $6i \cdot 7i$ **−42**
2. $-i\sqrt{3} \cdot i\sqrt{5}$ **$\sqrt{15}$**
3. $i\sqrt{3} \cdot i\sqrt{2} \cdot i\sqrt{6}$ **−6i**
4. $\sqrt{-5} \cdot \sqrt{-6}$ **$-\sqrt{30}$**
5. $-3\sqrt{-2} \cdot 4\sqrt{-11}$ **$12\sqrt{22}$**
6. $2\sqrt{3} \cdot 3\sqrt{-2}$ **$6i\sqrt{6}$**
7. $2\sqrt{-6} \cdot 4\sqrt{-2}$ **$-16\sqrt{3}$**
8. $-4\sqrt{-15} \cdot 2\sqrt{6}$ **$-24i\sqrt{10}$**
9. $\sqrt{-5} \cdot \sqrt{-10} \cdot \sqrt{-2}$ **−10i**
10. $\sqrt{3}(2\sqrt{6} + i\sqrt{15})$
11. $3\sqrt{6}(\sqrt{2} - \sqrt{-10})$
12. $\sqrt{-6}(\sqrt{11} + \sqrt{-3})$
13. $(4 + 3i)(5 + 6i)$ **$2 + 39i$**
14. $(3 - 4i)(2 + 2i)$ **$14 - 2i$**
15. $(6 - 5i)(-3 + i)$ **$-13 + 21i$**
16. $(9 + 11i)(9 - 11i)$ **202**
17. $(-7 - 6i)(-7 + 6i)$ **85**
18. $(-3 + 4i)(-3 - 4i)$ **25**
19. $(4 + 5i)^2$ **$-9 + 40i$**
20. $(6 - 2i)^2$ **$32 - 24i$**
21. $(-5 - i)^2$ **$24 + 10i$**
22. $(-4i)^3$ **64i**
23. $(i\sqrt{3})^4$ **9**
24. $(-2i)^5$ **−32i**

10. $6\sqrt{2} + 3i\sqrt{5}$
11. $6\sqrt{3} - 6i\sqrt{15}$
12. $i\sqrt{66} - 3\sqrt{2}$

Simplify each quotient by rationalizing the denominator.

25. $7 \div (4i)$ $\frac{7i}{-4}$
26. $-3 \div (5i)$ $\frac{3i}{5}$
27. $8 \div (-6i)$ $\frac{4i}{3}$
28. $-10 \div (-10i)$ **−i**
29. $\frac{-2}{5i}$ $\frac{2i}{5}$
30. $\frac{1}{-9i}$ $\frac{i}{9}$
31. $\frac{7}{i}$ **−7i**
32. $\frac{-8}{12i}$ $\frac{2i}{3}$
33. $\frac{5}{3 + 2i}$ $\frac{15 - 10i}{13}$
34. $-3 \div (4 - 2i)$ $\frac{-6 - 3i}{10}$
35. $\frac{i}{-1 + i}$ $\frac{1 - i}{2}$
36. $5i \div (-3 - i)$ $\frac{-1 - 3i}{2}$

Ⓑ
37. $\frac{2 + 3i}{3 - 4i}$ $\frac{-6 + 17i}{25}$
38. $\frac{5 - 3i}{1 - i}$ $4 + i$
39. $\frac{1 - 3i}{-2 - 4i}$ $\frac{1 + i}{2}$
40. $\frac{50}{3\sqrt{5} + \sqrt{-5}}$ $3\sqrt{5} - i\sqrt{5}$

Simplify each product. (Ex. 41–52)

41. $(10i)^6$ **−1,000,000**
42. $(-2i)^8$ **256**
43. i^{11} **−i**
44. $(2i)^{10}$ **−1,024**
45. $(4 + 2i\sqrt{3})^2$
46. $(6 - 2\sqrt{-6})^2$
47. $(2i\sqrt{7} + 3i\sqrt{2})^2$
48. $(\sqrt{-17} - \sqrt{-10})^2$
49. $(3i\sqrt{7})^4$ **3,969**
50. $(-2\sqrt{-2})^{10}$ **−32,768**
51. $(1 - i\sqrt{3})^3$ **−8**
52. $(2 + 2i\sqrt{3})^3$ **−64**

45. $4 + 16i\sqrt{3}$
46. $12 - 24i\sqrt{6}$
47. $-46 - 12\sqrt{14}$
48. $-27 + 2\sqrt{170}$

Simplify. Use the definition of $|a + bi|$. (See page 189.)

Ⓒ
53. $|4 + 3i| \cdot |2 - 2i|$ and $|(4 + 3i)(2 - 2i)|$
    $10\sqrt{2}, 10\sqrt{2}$
54. Prove that $|a + bi| \cdot |c + di| \stackrel{?}{=} |(a + bi)(c + di)|$. **See TE Answer Section.**
55. Prove that the product of two conjugate complex numbers is a real number.
    $(a + bi)(a - bi) = a^2 - abi + abi - b^2i^2 = a^2 + b^2$
56. Prove that the sum of two conjugate complex numbers is a real number.
    $(a + bi) + (a - bi) = 2a$

EQUATIONS WITH IMAGINARY NUMBER SOLUTIONS

8.3

Objectives **To solve an equation whose solutions are imaginary numbers**
To find the three cube roots of a nonzero integer that is a perfect cube

You have used the quadratic formula, $x = \dfrac{-b \pm \sqrt{b^2 - 4ac}}{2a}$, to find the real number solutions of a given quadratic equation. As shown in Example 1, the formula can also be used when the solutions are imaginary numbers.

Example 1 Solve $3x^2 + 2 = 4x$.

Write the equation in standard form and use the quadratic formula.
$$3x^2 - 4x + 2 = 0$$

$a = 3, b = -4, c = 2$ ▶ $x = \dfrac{-(-4) \pm \sqrt{(-4)^2 - 4 \cdot 3 \cdot 2}}{2 \cdot 3} = \dfrac{4 \pm \sqrt{-8}}{6} = \dfrac{4 \pm 2i\sqrt{2}}{6} = \dfrac{2 \pm i\sqrt{2}}{3}$

Thus, the solutions are $\dfrac{2 + i\sqrt{2}}{3}$ and $\dfrac{2 - i\sqrt{2}}{3}$.

A fourth-degree equation may have imaginary number solutions. One such equation is solved by factoring in Example 2. Notice that two of the solutions of the equation are imaginary numbers and two are real numbers.

Example 2 Solve $2x^4 + 3x^2 - 20 = 0$.

$$2x^4 + 3x^2 - 20 = 0$$
$$(x^2 + 4)(2x^2 - 5) = 0$$

$x^2 + 4 = 0$ or $2x^2 - 5 = 0$
$x^2 = -4$ $2x^2 = 5$
$x = \pm\sqrt{-4}$ $x^2 = \dfrac{5}{2}$
$x = \pm 2i$ $x = \pm\sqrt{\dfrac{5}{2}} = \pm\dfrac{\sqrt{5}}{\sqrt{2}} \cdot \dfrac{\sqrt{2}}{\sqrt{2}} = \pm\dfrac{\sqrt{10}}{2}$

Thus, the four solutions are $2i$, $-2i$, $\dfrac{\sqrt{10}}{2}$, and $-\dfrac{\sqrt{10}}{2}$.

Every real number, except 0, has two different square roots and three different cube roots. However, some of the roots may be imaginary numbers. You can use a third-degree equation to find the three cube roots of a nonzero integer that is a perfect cube, such as -8.

Equations with Imaginary Number Solutions

Example 3   **Find the three cube roots of −8.**

The cube roots of −8 are the solutions of the equation $x^3 = -8$.
$$x^3 = -8$$
$$x^3 + 8 = 0 \quad \blacktriangleleft \; x^3 + 8 \text{ is a sum of two cubes.}$$
$$x^3 + 2^3 = 0$$
$$(x + 2)(x^2 - 2x + 4) = 0 \quad \blacktriangleleft \; m^3 + n^3 = (m + n)(m^2 - mn + n^2)$$
$$x + 2 = 0 \quad \text{or} \quad x^2 - 2x + 4 = 0$$
$$x = -2 \qquad x = \frac{2 \pm \sqrt{4 - 16}}{2} = \frac{2 \pm \sqrt{-12}}{2} = \frac{2 \pm 2i\sqrt{3}}{2} = 1 \pm i\sqrt{3}$$

Thus, the three cube roots of −8 are -2, $1 + i\sqrt{3}$, and $1 - i\sqrt{3}$.

In Example 3, the three cube roots of −8 were found as solutions of the equation $x^3 = -8$. You can check these roots by showing that the third power of each number is −8.

-2: $(-2)^3 = (-2)(-2)(-2) = -8$
$1 + i\sqrt{3}$: $(1 + i\sqrt{3})^3 = (1 + i\sqrt{3})(1 + i\sqrt{3})^2 = (1 + i\sqrt{3})(1 + 2i\sqrt{3} + 3i^2)$
$\qquad\qquad\qquad = (1 + i\sqrt{3})(-2 + 2i\sqrt{3}) = -2 + 6i^2, \text{ or } -8$
$1 - i\sqrt{3}$:   See exercise 51 on page 192.

Written Exercises

Solve each equation.

(A)
1. $x^2 - 3x + 3 = 0$ $\frac{3 \pm i\sqrt{3}}{2}$
2. $2y^2 + 3y + 3 = 0$ $\frac{-3 \pm i\sqrt{15}}{4}$
3. $3n^2 - 3n + 1 = 0$ $\frac{3 \pm i\sqrt{3}}{6}$
4. $2y^2 + 4 = 5y$ $\frac{5 \pm i\sqrt{7}}{4}$
5. $c^2 + 4 = 3c$ $\frac{3 \pm i\sqrt{7}}{2}$
6. $3x^2 = x - 2$ $\frac{1 \pm i\sqrt{23}}{6}$
7. $x^2 - 4x + 5 = 0$ $2 \pm i$
8. $n^2 + 2n + 2 = 0$ $-1 \pm i$
9. $y^2 - 8y + 20 = 0$ $4 \pm 2i$
10. $2y^2 - 2y + 13 = 0$ $\frac{1 \pm 5i}{2}$
11. $3t^2 + 2 = 2t$ $\frac{1 \pm i\sqrt{5}}{3}$
12. $4x^2 = 2x - 1$ $\frac{1 \pm i\sqrt{3}}{4}$
13. $x^4 + 7x^2 - 18 = 0$
14. $x^4 - 16 = 0$ $\pm 2i, \pm 2$
15. $y^4 = 7y^2 + 8$ $\pm i, \pm 2\sqrt{2}$
16. $y^4 + 2y^2 - 15 = 0$
17. $c^4 = 10c^2 + 24$
18. $x^4 = 24 - 5x^2$
13. $\pm 3i, \pm \sqrt{2}$ 16. $\pm i\sqrt{5}, \pm\sqrt{3}$ 17. $\pm i\sqrt{2}, \pm 2\sqrt{3}$ 18. $\pm 2i\sqrt{2}, \pm\sqrt{3}$

Find the three cube roots of each number.

19. -1 $-1, \frac{1 + i\sqrt{3}}{2}, \frac{1 - i\sqrt{3}}{2}$
20. 8 $2, -1 + i\sqrt{3}, -1 - i\sqrt{3}$
21. -27 $-3, \frac{3 + 3i\sqrt{3}}{2}, \frac{3 - 3i\sqrt{3}}{2}$

(B) 22. 64 $4, -2 + 2i\sqrt{3}, -2 - 2i\sqrt{3}$
23. -125 $-5, \frac{5 + 5i\sqrt{3}}{2}, \frac{5 - 5i\sqrt{3}}{2}$
24. $1,000$ $10, -5 + 5i\sqrt{3}, -5 - 5i\sqrt{3}$

Solve each equation.
      28. $\pm 2i, \pm 3i$ 29. $\pm i\sqrt{2}, \pm 2i\sqrt{2}$

25. $\frac{1}{2}x^2 = x - \frac{3}{4}$ $\frac{2 \pm i\sqrt{2}}{2}$
26. $\frac{1}{3}x^2 + \frac{3}{2} = \frac{1}{3} - x$ $\frac{-3 \pm i\sqrt{5}}{2}$
27. $\frac{5x}{12} - \frac{x^2}{6} = \frac{1}{2}$ $\frac{5 \pm i\sqrt{23}}{4}$
28. $y^4 + 13y^2 + 36 = 0$
29. $n^4 + 10n^2 + 16 = 0$
30. $c^4 + 16c^2 + 64 = 0$ $\pm 2i\sqrt{2}$
31. $4x^4 + 19x^2 + 12 = 0$
32. $16x^4 + 16x^2 + 3 = 0$
33. $2x^4 + 11x^2 + 9 = 0$
      31. $\pm \frac{i\sqrt{3}}{2}, \pm 2i$ 32. $\pm \frac{i}{2}, \pm \frac{i\sqrt{3}}{2}$ 33. $\pm \frac{3i\sqrt{2}}{2}, \pm i$

(C) 34. $x^2 + 2ix + 3 = 0$ $-3i, i$
35. $x^2 - 6ix + 8 = 0$ $3i \pm i\sqrt{17}$
36. $2ix^2 + 5x - 2i = 0$ $\frac{i}{2}, 2i$
37. $\sqrt{x} = 3i$ -9
38. $\sqrt{x - 1} = 2i$ -3
39. $\sqrt{x + 8} = 3 - i$ $-6i$

Find the three cube roots of each number. Then raise each cube root to the third power.

40. $\frac{1}{8}$ $\frac{1}{2}, \frac{-1 + i\sqrt{3}}{4}, \frac{-1 - i\sqrt{3}}{4}$
41. $-\frac{8}{27}$ $-\frac{2}{3}, \frac{1 + i\sqrt{3}}{3}, \frac{1 - i\sqrt{3}}{3}$
42. $\frac{125}{64}$ $\frac{5}{4}, \frac{-5 + 5i\sqrt{3}}{8}, \frac{-5 - 5i\sqrt{3}}{8}$

THE DISCRIMINANT 8.4

Objectives
To determine the nature of the solutions of a quadratic equation by examining its discriminant
To find the coefficients of a quadratic equation having exactly one solution by using the value of its discriminant

The quadratic formula, $x = \dfrac{-b \pm \sqrt{b^2 - 4ac}}{2a}$, can be used to find the solutions of any quadratic equation. However, it is not necessary to solve a given quadratic equation in order to tell the *number* and *type* of solutions that it will have. All of this information about the nature of the solutions can be obtained from the value of $b^2 - 4ac$, the radicand in the quadratic formula.

Equation	Solutions	$b^2 - 4ac$	Nature of Solutions
$6x^2 - x - 2 = 0$	$\dfrac{1 \pm \sqrt{49}}{12} = \dfrac{1 \pm 7}{12}$	49: positive perfect square	Two real number solutions: rational
$x^2 + 3x - 5 = 0$	$\dfrac{-3 \pm \sqrt{29}}{2}$	29: positive	Two real number solutions: irrational
$\dfrac{3}{2}x^2 - 2x + \dfrac{2}{3} = 0$	$\dfrac{2 \pm \sqrt{0}}{3} = \dfrac{2}{3}$	0: perfect square	One real number solution: rational
$2x^2 + 3x + 3 = 0$	$\dfrac{-3 \pm \sqrt{-15}}{4} = \dfrac{-3 \pm i\sqrt{15}}{4}$	-15: negative	Two imaginary number solutions

The expression $b^2 - 4ac$ is called the **discriminant** because its value determines the nature of the solutions of a quadratic equation.

Nature of solutions of quadratic equation

If $ax^2 + bx + c = 0$, where a, b, and c are real numbers ($a \neq 0$), then $x = \dfrac{-b \pm \sqrt{b^2 - 4ac}}{2a}$, and $b^2 - 4ac$ is the discriminant of the equation.

The nature of the solutions of the equation is determined as follows:
(1) if $b^2 - 4ac > 0$, there are two real number solutions.
(2) if $b^2 - 4ac = 0$, there is exactly one real number solution.
(3) if $b^2 - 4ac < 0$, there are two imaginary number solutions.

If a, b, and c are rational numbers and $b^2 - 4ac \geq 0$, the solutions are *rational* when $b^2 - 4ac$ is a perfect square, and the solutions are *irrational* when $b^2 - 4ac$ is not a perfect square.

The Discriminant

Example 1 **Determine the nature of the solutions of $x^2 + 5x - 3 = 0$ without solving the equation.**

Identify a, b, and c and compute the value of $b^2 - 4ac$.
$$a = 1, b = 5, c = -3$$
$$b^2 - 4ac = 5^2 - 4(1)(-3) = 25 + 12 = 37$$

$b^2 - 4ac = 37$; positive but not a perfect square.
Thus, $x^2 + 5x - 3 = 0$ has two real number solutions that are irrational.

Example 2 **Determine the nature of the solutions of $3y^2 = 4y - 2$ without solving the equation.**

$$3y^2 = 4y - 2$$
$$3y^2 - 4y + 2 = 0 \quad \blacktriangleleft \text{ Rewrite the equation in standard form.}$$
$$a = 3, b = -4, c = 2 \quad \blacktriangleleft \text{ Identify } a, b, \text{ and } c.$$
$$b^2 - 4ac = (-4)^2 - 4(3)(2) = 16 - 24 = -8$$

$b^2 - 4ac = -8$; negative
Thus, $3y^2 = 4y - 2$ has two imaginary number solutions.

Example 3 **Determine the nature of the solutions of $\frac{1}{9}n^2 + \frac{2}{3}n + 1 = 0$ without solving the equation.**

$$\frac{1}{9}n^2 + \frac{2}{3}n + 1 = 0$$
$$a = \frac{1}{9}, b = \frac{2}{3}, c = 1$$
$$b^2 - 4ac = \left(\frac{2}{3}\right)^2 - 4\left(\frac{1}{9}\right)(1) = \frac{4}{9} - \frac{4}{9} = 0$$

$b^2 - 4ac = 0$; a perfect square
Thus, $\frac{1}{9}n^2 + \frac{2}{3}n + 1 = 0$ has exactly one real number solution that is rational.

Example 4 **For what values of k will $2x^2 + (k + 2)x + 8 = 0$ have exactly one solution?**

If $b^2 - 4ac = 0$, there is exactly one solution to the equation.
$$b^2 - 4ac = 0 \qquad a = 2, b = k + 2, c = 8$$
$$(k + 2)^2 - 4(2)(8) = 0$$
$$k^2 + 4k + 4 - 64 = 0$$
$$k^2 + 4k - 60 = 0$$
$$(k - 6)(k + 10) = 0$$
$$k = 6 \quad \text{or} \quad k = -10$$

Thus, the equation will have exactly one solution if $k = 6$ or $k = -10$.

Reading in Algebra

You are given a quadratic equation $ax^2 + bx + c = 0$, where a, b, and c are real numbers ($a \neq 0$). Match each expression or statement at the left to exactly one phrase at the right.

1. $b^2 - 4ac < 0$ **C**
2. $b^2 - 4ac > 0$ **D**
3. $b^2 - 4ac = 0$ **B**
4. $b^2 - 4ac$ **A**
5. $\sqrt{b^2 - 4ac}$ **F**
6. $b^2 - 4ac$ is a positive number but not a perfect square. **E**

A the discriminant
B exactly one solution
C two imaginary solutions
D two real solutions
E two irrational solutions
F the radical in the quadratic formula

(In Ex. 1–12 and 19–30 below, R = real, Im = imaginary, Rt = rational, and Ir = irrational.)

Written Exercises
1. two R:Rt 2. two R:Ir 3. two Im 4. one R:Rt 5. one R:Rt
6. two Im 7. two R:Rt 8. two R:Ir 9. two Im 10. two R:Rt 11. two R:Ir 12. one R:Rt

Determine the nature of the solutions without solving the equation.

Ⓐ
1. $x^2 + 6x - 16 = 0$
2. $y^2 - 8y + 2 = 0$
3. $n^2 + 2n + 3 = 0$
4. $z^2 + 16 = 8z$
5. $4y^2 + 12y + 9 = 0$
6. $3n^2 - 4n + 3 = 0$
7. $6x^2 = 4 - 5x$
8. $15d^2 - 25d + 9 = 0$
9. $9x^2 + 0x + 4 = 0$
10. $4y^2 - 5y = 0$
11. $n^2 = 15$
12. $25t^2 = 0$

For what value(s) of k will each quadratic equation have exactly one solution?

Ⓑ
13. $x^2 - 20x + k^2 = 0$ **−10, 10**
14. $25y^2 + 10ky + 16 = 0$ **−4, 4**
15. $9ky^2 + 1 = 2ky$ **9**
16. $x^2 - 16x + (k-1)^2 = 0$ **−7, 9**
17. $4y^2 - (k-3)y + 9 = 0$ **−9, 15**
18. $(2k+1)x^2 = 12x - k$ **$-\frac{9}{2}$, 4**

Determine the nature of the solutions without solving the equation.

19. $\frac{1}{3}x^2 - \frac{5}{6}x - 2 = 0$ **two R:Rt**
20. $\frac{5}{2}y^2 + 2y + \frac{2}{5} = 0$ **one R:Rt**
21. $\frac{2}{9}n^2 = \frac{4}{3}n - 2$ **one R:Rt**
22. $(2y-3)(3y+4) = -10$
23. $(5x+1)(2x+1) = 2$
24. $(t-5)(t-3) + 3 = 0$ **two Im**
25. $-5n^2 + 8n - 3 = 0$
26. $-3y^2 - 3y - 108 = 0$
27. $-2x(3x+4) + 4 = 0$

22. two R:Rt 23. two R:Ir 25. two R:Rt 26. two Im 27. two R:Ir

Determine the nature of the solutions without solving the equation. The coefficient of x in each equation is an irrational number.

Ⓒ
28. $x^2 + 2\sqrt{2}x - 6 = 0$ **two R:Ir**
29. $x^2 - 2\sqrt{3}x + 3 = 0$ **one R:Ir**
30. $x^2 + 4\sqrt{5}x + 4 = 0$ **two R:Ir**

31. For what values of k will $2x^2 + kx + 8 = 0$ have two real solutions? **$k < -8$ or $k > 8$**
32. For what values of k will $2x^2 + (k+1)x + 18 = 0$ have two imaginary solutions? **$-13 < k < 11$**

NON-ROUTINE PROBLEMS

Given a pair of quadratic equations,
(1) $px^2 + qx + r = 0$ and (2) $ptx^2 + qtx + rt = 0$,
prove that the discriminant of equation (2) is t^2 times the discriminant of equation (1). **discriminant of (1) $q^2 - 4pr$; discriminant of (2) $(qt)^2 - 4(pt)(rt)$, or $q^2t^2 - 4prt^2$, or $t^2(q^2 - 4pr)$**

The Discriminant

SUM AND PRODUCT OF SOLUTIONS 8.5

Objectives **To find the sum and the product of the solutions of a quadratic equation**
To write a quadratic equation, given its solutions

The solutions of the quadratic equation $3x^2 - 4x - 20 = 0$ are -2 and $\frac{10}{3}$.

The equation is in the standard form of $ax^2 + bx + c = 0$, where $a = 3$, $b = -4$, and $c = -20$. Let r and s represent the solutions of $3x^2 - 4x - 20 = 0$. Then,

$$(1)\quad r + s = -2 + \frac{10}{3} = \frac{4}{3} = \frac{-b}{a} \quad \text{and}$$

$$(2)\quad r \cdot s = -2 \cdot \frac{10}{3} = \frac{-20}{3} = \frac{c}{a}.$$

The *sum*, $r + s$, and the *product*, $r \cdot s$, of the solutions of $3x^2 - 4x - 20 = 0$ are related to the value of the coefficients a, b, and c.

$$3x^2 - 4x - 20 = 0 \qquad\qquad ax^2 + bx + c = 0$$
$$x^2 - \frac{4}{3}x - \frac{20}{3} = 0 \qquad\qquad x^2 + \frac{b}{a}x + \frac{c}{a} = 0$$

Notice that $\quad r + s = \frac{4}{3} = -\frac{b}{a} \quad$ or $\quad \frac{b}{a} = -(r + s)$, and

$$r \cdot s = -\frac{20}{3} = \frac{c}{a}.$$

These relationships remain true for any quadratic equation.

Sum and product of solutions of a quadratic equation

If r and s are the solutions of $ax^2 + bx + c = 0$ ($a \neq 0$), then
$$r + s = -\frac{b}{a} \quad \text{and} \quad r \cdot s = \frac{c}{a}.$$

Example 1 Find the sum and the product of the solutions of $4x^2 + 7x + 3 = 0$.

$$4x^2 + 7x + 3 = 0$$
$$x^2 + \frac{7}{4}x + \frac{3}{4} = 0 \quad \blacktriangleleft \text{ Divide each side by } a, \text{ or } 4.$$

$\frac{b}{a} = \frac{7}{4}$, or $-\frac{b}{a} = -\frac{7}{4}$, and $\frac{c}{a} = \frac{3}{4}$.

Thus, the sum of the solutions is $-\frac{7}{4}$ and the product is $\frac{3}{4}$.

If you are given two numbers, either real or imaginary, you can write a quadratic equation that has the two numbers as its solutions.

Example 2   Write a quadratic equation, in standard form, that has $-\frac{3}{4}$ and $\frac{2}{3}$ as its solutions.

Let $r = -\frac{3}{4}$ and $s = \frac{2}{3}$. Find $\frac{b}{a}$ and $\frac{c}{a}$.

$$\frac{b}{a} = -(r + s) = -\left(-\frac{3}{4} + \frac{2}{3}\right) \qquad \frac{c}{a} = r \cdot s$$
$$= -\left(-\frac{9}{12} + \frac{8}{12}\right) \qquad\qquad = -\frac{3}{4} \cdot \frac{2}{3}$$
$$= \frac{1}{12} \qquad\qquad\qquad\qquad = -\frac{1}{2}$$

$$x^2 + \frac{b}{a}x + \frac{c}{a} = 0$$

Substitute. ▶ $x^2 + \left(\frac{1}{12}\right)x + \left(-\frac{1}{2}\right) = 0$

Write in standard form. ▶ $12x^2 + x - 6 = 0$

Thus, $12x^2 + x - 6 = 0$ is a quadratic equation, in standard form, whose solutions are $-\frac{3}{4}$ and $\frac{2}{3}$.

A similar procedure is used to write a quadratic equation that has exactly one given real number solution. You begin by letting $r = s$.

Example 3   Write a quadratic equation, in standard form, that has $\frac{3}{7}$ as its only solution.

Let $r = s = \frac{3}{7}$.

$$\frac{b}{a} = -(r + s) = -\left(\frac{3}{7} + \frac{3}{7}\right) = -\frac{6}{7} \qquad \frac{c}{a} = r \cdot s = \frac{3}{7} \cdot \frac{3}{7} = \frac{9}{49}$$

$$x^2 + \frac{b}{a}x + \frac{c}{a} = 0$$

Substitute. ▶ $x^2 + \left(-\frac{6}{7}\right)x + \left(\frac{9}{49}\right) = 0$

Write in standard form. ▶ $49x^2 - 42x + 9 = 0$

Thus, the required equation is $49x^2 - 42x + 9 = 0$.

The two numbers represented by $\frac{4 \pm 2\sqrt{3}}{3}$ are a pair of conjugate irrational numbers, $\frac{4 + 2\sqrt{3}}{3}$ and $\frac{4 - 2\sqrt{3}}{3}$. As shown in Example 4 on page 200, a quadratic equation can be written that has these two numbers as its solutions.

Sum and Product of Solutions

Example 4   Write a quadratic equation, in standard form, that has $\dfrac{4 \pm 2\sqrt{3}}{3}$ as its solutions.

Let $r = \dfrac{4 + 2\sqrt{3}}{3}$ and $s = \dfrac{4 - 2\sqrt{3}}{3}$.

$\dfrac{b}{a} = -\left(\dfrac{4 + 2\sqrt{3}}{3} + \dfrac{4 - 2\sqrt{3}}{3}\right) = -\dfrac{4 + 2\sqrt{3} + 4 - 2\sqrt{3}}{3} = -\dfrac{8}{3}$

$\dfrac{c}{a} = \dfrac{4 + 2\sqrt{3}}{3} \cdot \dfrac{4 - 2\sqrt{3}}{3} = \dfrac{4^2 - (2\sqrt{3})^2}{3 \cdot 3} = \dfrac{16 - 12}{9} = \dfrac{4}{9}$

$x^2 - \dfrac{8}{3}x + \dfrac{4}{9} = 0$

Thus, $9x^2 - 24x + 4 = 0$ is the required equation.

9. $x^2 - 8x + 15 = 0$ 10. $x^2 + 4x - 12 = 0$ 12. $x^2 + 9x = 0$ 13. $8x^2 - 6x + 1 = 0$

Written Exercises   For complete solutions, see TE Answer Section.

1. sum: $-\frac{9}{2}$, prod.: -3 2. sum: $-\frac{4}{3}$, prod.: $-\frac{1}{3}$ 3. sum: 3, prod.: $-\frac{1}{2}$ 4. sum: -9, prod.: 14

Without solving the equation, find the sum and the product of its solutions.

(A)   1. $2x^2 + 9x - 6 = 0$ 2. $3y^2 + 4y - 1 = 0$ 3. $4n^2 - 12n - 2 = 0$ 4. $x^2 + 9x + 14 = 0$

5. $6t^2 + 5t = 0$ 6. $3x^2 - 9 = 0$ 7. $y^2 = 6y - 9$ 8. $12 = 6n + 3n^2$

sum: $-\frac{5}{6}$, prod.: 0   sum: 0, prod.: -3 sum: 6, prod.: 9   sum: -2, prod.: -4

Write a quadratic equation, in standard form, that has the given solution(s).

9. 3 and 5 **above**   10. -6 and 2 **above**   11. 0 and 4 $x^2 - 4x = 0$ 12. -9 and 0 **above**

13. $\frac{1}{4}$ and $\frac{1}{2}$ **above**   14. $-\frac{3}{4}$ and $-\frac{5}{8}$ **below**   15. -6 and $\frac{2}{3}$ **below**   16. $\frac{9}{10}$ and 2 **below**

17. 7 $x^2 - 14x + 49 = 0$ 18. -8 $x^2 + 16x + 64 = 0$ 19. $-\frac{1}{2}$ $4x^2 + 4x + 1 = 0$ 20. $\frac{10}{7}$ $49x^2 - 140x + 100 = 0$

$x^2 - 7 = 0$

(B)   21. $-\sqrt{7}$ and $\sqrt{7}$ 22. $-4\sqrt{3}$ and $4\sqrt{3}$ 23. $-i\sqrt{10}$ and $i\sqrt{10}$ 24. $-2i\sqrt{6}$ and $2i\sqrt{6}$

25. $4 \pm \sqrt{5}$ 26. $-3 \pm 2\sqrt{10}$ 27. $5 \pm 2i\sqrt{7}$ 28. $-6 \pm 3i\sqrt{5}$ **below**

29. $\dfrac{2 \pm \sqrt{3}}{2}$ 30. $\dfrac{-2 \pm 3\sqrt{2}}{3}$ 31. $\dfrac{-3 \pm 2i\sqrt{3}}{3}$ 32. $\dfrac{4 \pm 2i\sqrt{5}}{3}$

$4x^2 - 8x + 1 = 0$ $9x^2 + 12x - 14 = 0$ $3x^2 + 6x + 7 = 0$ $3x^2 - 8x + 12 = 0$

Without solving the equation, find the sum and the product of its solutions.

33. $4x^2 + \frac{1}{3}x - \frac{1}{2} = 0$ sum: $-\frac{1}{12}$, prod.: $-\frac{1}{8}$ 34. $\frac{1}{3}y^2 - 3y + 1 = 0$ sum: 9, prod.: 3

35. $2.5x^2 + 5x - 10 = 0$ sum: -2, prod.: -4 36. $0.2y^2 - 4y + 0.6 = 0$ sum: 20, prod.: 3

(C)   37. Prove that the sum and the product of $\dfrac{-b + \sqrt{b^2 - 4ac}}{2a}$ and $\dfrac{-b - \sqrt{b^2 - 4ac}}{2a}$, $a \neq 0$, are $-\dfrac{b}{a}$ and $\dfrac{c}{a}$, respectively.

38. For what values of k will the *sum* of the solutions of $x^2 - (k^2 - 2k)x + 12 = 0$ be 8? $-2, 4$

39. For what values of k will the *product* of the solutions of $2x^2 + x + (4k^2 - 4k - 3) = 0$ be 0? $-\frac{1}{2}, \frac{3}{2}$

40. Prove: If a, b, and c are odd integers, then $ax^2 + bx + c = 0$ has no integral solutions. [Hint: First prove that there are no *even* integral solutions and then that there are no *odd* integral solutions.]   14. $32x^2 + 44x + 15 = 0$ 15. $3x^2 + 16x - 12 = 0$ 16. $10x^2 - 29x + 18 = 0$ 22. $x^2 - 48 = 0$ 23. $x^2 + 10 = 0$ 24. $x^2 + 24 = 0$ 25. $x^2 - 8x + 11 = 0$ 26. $x^2 + 6x - 31 = 0$ 27. $x^2 - 10x + 53 = 0$ 28. $x^2 + 12x + 81 = 0$

APPLICATIONS: PROOFS WITH COMPLEX NUMBERS

You can use the properties of a field (page 26) for the *real* numbers and some definitions to prove theorems involving complex numbers. As you study each proof, you will be asked to supply a property, definition, or theorem as a reason for a statement in the proof.

Definitions:
1. $a + bi$ is a complex number if and only if a and b are real numbers.
2. $a + bi = c + di$ if and only if $a = c$ and $b = d$.
3. $(a + bi) + (c + di) = (a + c) + (b + d)i$
4. $x - y = x + (-y)$ for all real numbers x and y.
5. $(a + bi) - (c + di) = (a + bi) + [-(c + di)]$
6. $(a + bi)(c + di) = (ac - bd) + (ad + bc)i$

Theorem 1   The set of complex numbers is *closed* under addition.
Prove that $(a + bi) + (c + di)$ is a complex number.

Proof

1. $(a + bi) + (c + di) = (a + c) + (b + d)i$	1. Definition $\underline{?}$ **3**
2. $a + c$ and $b + d$ are real numbers.	2. Closure for addition of real numbers
3. $(a + c) + (b + d)i$ is a complex number.	3. Definition $\underline{?}$ **1**
4. $(a + bi) + (c + di)$ is a complex number.	4. Statement 1 (Substitution)

Theorem 2   Addition of complex numbers is *commutative*.
Prove that $(a + bi) + (c + di) = (c + di) + (a + bi)$.

Proof

1. $(a + bi) + (c + di) = (a + c) + (b + d)i$	1. $\underline{?}$ **Definition 3**
2. $ = (c + a) + (d + b)i$	2. $\underline{?}$ **Commutative of add. for real numbers**
3. $(c + di) + (a + bi) = (c + a) + (d + b)i$	3. $\underline{?}$ **Definition 3**
4. $(a + bi) + (c + di) = (c + di) + (a + bi)$	4. Statements 2 and 3 (Transitive prop. for =)

Theorem 3 $0 + 0i$ is *the* additive identity.
Solve $(a + bi) + (x + yi) = a + bi$ for $x + yi$, the additive identity.

Proof

1. $(a + bi) + (x + yi) = a + bi$	1. Given
2. $(a + x) + (b + y)i = a + bi$	2. $\underline{?}$ **Definition 3**
3. $a + x = a$ and $b + y = b$	3. Definition $\underline{?}$ **2**
4. $x = 0$ and $y = 0$	4. Additive identity for real numbers
5. $x + yi = 0 + 0i$	5. Substitution

Applications: Proofs with Complex Numbers

Theorem 4 $-a + (-bi)$ is the additive inverse of $a + bi$.
Solve $(a + bi) + (x + yi) = 0 + 0i$ for $x + yi$, or $-(a + bi)$.

Proof
1. $(a + bi) + (x + yi) = 0 + 0i$
2. $(a + x) + (b + y)i = 0 + 0i$
3. $a + x = 0$ and $b + y = 0$
4. $x = -a$ and $y = -b$
5. $x + yi = -a + (-bi)$

1. $\underline{}$ **Given**
2. $\underline{}$ **Definition 3**
3. $\underline{}$ **Definition 2**
4. $\underline{}$ **Additive inverse for real numbers**
5. $\underline{}$ **Substitution**

Theorem 5 $(a + bi) - (c + di) = (a - c) + (b - d)i$

Proof
1. $(a + bi) - (c + di) = (a + bi) + [-(c + di)]$
2. $ = (a + bi) + [-c + (-di)]$
3. $ = [a + (-c)] + [b + (-d)]i$
4. $ = (a - c) + (b - d)i$

1. Definition $\underline{}$ 5
2. Theorem $\underline{}$ 4
3. $\underline{}$ Definition 3
4. Definition $\underline{}$ 4

Theorem 6 $1 + 0i$ is the multiplicative identity.
Solve $(a + bi)(x + yi) = a + bi$ for $x + yi$, the multiplicative identity.

Proof
1. $(a + bi)(x + yi) = a + bi; a + bi \neq 0 + 0i$
2. $(ax - by) + (ay + bx)i = a + bi$
3. $ax - by = a$ and $ay + bx = b$

$x = \dfrac{by + a}{a}$ and $x = \dfrac{b - ay}{b}$

$\dfrac{by + a}{a} = \dfrac{b - ay}{b}$

$b^2y + ab = ab - a^2y$

1. Given
2. Definition $\underline{}$ 6
3. $\underline{}$ Definition 2

$a^2y + b^2y = 0$
$(a^2 + b^2)y = 0$
$y = 0 \quad [a^2 + b^2 \neq 0]$
$x = \dfrac{b - ay}{b} = \dfrac{b}{b} = 1$

Thus, $x + yi = 1 + 0i$.
$[(0 + 0i)(1 + 0i) = 0 + 0i]$

Prove each theorem. **For proofs, see TE Answer Section.**

Theorem 7 The set of complex numbers is closed under multiplication. (Review the proof of Theorem 1.)

Theorem 8 Multiplication of complex numbers is commutative. (Review the proof of Theorem 2.)

Theorem 9 Addition of complex numbers is associative.
Prove: $[(a + bi) + (c + di)] + (e + fi) = (a + bi) + [(c + di) + (e + fi)]$.

Theorem 10 $\dfrac{a}{a^2 + b^2} + \dfrac{-b}{a^2 + b^2} \cdot i$ is the multiplicative inverse of $a + bi$.
Proof: Solve $(a + bi)(x + yi) = 1 + 0i$ for $x + yi$ where $a + bi \neq 0 + 0i$. (Review the proof of Theorem 6.)

Ⓐ Exercises are 1, 2, 4, 6, 8, 9, 11, 13, 14, 17–19, 23, 25, 26.
The rest are Ⓑ Exercises. Ⓒ Exercises are starred.

CHAPTER EIGHT REVIEW

Vocabulary
complex number [8.1]
conjugate complex numbers [8.2]
discriminant [8.4]
imaginary number [8.1]
pure imaginary number [8.1]

Simplify each expression.
1. $5\sqrt{-12}$ [8.1] **$10i\sqrt{3}$**
2. $(-3 + 4i) - (5 - i)$ **$-8 + 5i$**
3. $(9 - 4\sqrt{-18}) + (-2 + 3\sqrt{-50})$ **$7 + 3i\sqrt{2}$**
4. $2\sqrt{-6} \cdot 3\sqrt{-15}$ [8.2] **$-18\sqrt{10}$**
5. $(2i)^{11}$ **$-2{,}048i$**
6. $(7 - 2i)(-6 + 3i)$ **$-36 + 33i$**
7. $(4 - \sqrt{-10})^2$ **$6 - 8i\sqrt{10}$**

Simplify each quotient by rationalizing the denominator. [8.2]
8. $-3 \div (4i)$ **$\dfrac{3i}{4}$**
9. $4i \div (5 - 3i)$ **$\dfrac{-6 + 10i}{17}$**
10. $\dfrac{3 + 2i}{-2 + i}$ **$\dfrac{-4 - 7i}{5}$**

Solve each equation. (Exercises 11–16)
11. $x^2 = -60$ [8.1] **$\pm 2i\sqrt{15}$**
12. $4y^2 + 100 = 0$ **$\pm 5i$**
13. $x^2 + 4x + 6 = 0$ [8.3] **$-2 \pm i\sqrt{2}$**
14. $2x^2 + 13 = 2x$ **$\dfrac{1 \pm 5i}{2}$**
15. $y^4 + 12y^2 + 32 = 0$ **$\pm 2i, \pm 2i\sqrt{2}$**
★ 16. $x^2 - 2ix + 15 = 0$ **$-3i, 5i$**

17. Find the three cube roots of 27. [8.3] **$3, \dfrac{-3 + 3i\sqrt{3}}{2}, \dfrac{-3 - 3i\sqrt{3}}{2}$**

Determine the nature of the solutions without solving the equation. [8.4]
18. $2x^2 + 5x + 2 = 0$ **two R:Rt**
19. $4y^2 + 25 = 0$ **two Im**
20. $\dfrac{1}{2}n^2 - n = \dfrac{2}{3}$ **two R:Ir**
★ 21. $x^2 - 6x\sqrt{2} - 14 = 0$ **two R:Ir**

22. For what values of k will $2kx^2 + 12x + (4k + 1) = 0$ have exactly one solution? [8.4] **$-\dfrac{9}{4}, 2$**

Without solving the equation, find the sum and the product of its solutions. [8.5]
23. $4x^2 + 3x - 2 = 0$ **sum: $-\dfrac{3}{4}$, prod.: $-\dfrac{1}{2}$**
24. $\dfrac{1}{6}y^2 - \dfrac{4}{3}y + 2 = 0$ **sum: 8, prod.: 12**

Write a quadratic equation, in standard form, that has the given pair of numbers as its solutions. [8.5]
25. -8 and 5 **$x^2 + 3x - 40 = 0$**
26. $\dfrac{1}{6}$ and $\dfrac{3}{4}$ **$24x^2 - 22x + 3 = 0$**
27. $-5 \pm 2\sqrt{7}$ **$x^2 + 10x - 3 = 0$**
28. $\dfrac{1 \pm 2i\sqrt{5}}{2}$ **$4x^2 - 4x + 21 = 0$**

★ 29. For what values of k will the sum of the solutions of $x^2 - (k^2 - 3k)x + 24 = 0$ be 10? [8.5] **$-2, 5$**

Chapter Eight Review

**Ⓐ Exercises are 1–4, 6, 7, 9, 10, 13, 14, 17, 18.
The rest are Ⓑ Exercises.**   **Ⓒ Exercises are starred.**

CHAPTER EIGHT TEST

Simplify each expression.

1. $-8 + 4\sqrt{-20}$ **$-8 + 8i\sqrt{5}$**

2. $(12 - 10i) - (-3 + 4i)$ **$15 - 14i$**

3. $5\sqrt{-3} \cdot 2\sqrt{-6}$ **$-30\sqrt{2}$**

4. $(6 - 5i)(6 + 5i)$ **61**

5. $(3 + 2i\sqrt{5})^2$ **$-11 + 12i\sqrt{5}$**

Simplify each quotient by rationalizing the denominator.

6. $-7 \div (10i)$ **$\frac{7i}{10}$**

7. $\dfrac{3i}{4 + 3i}$ **$\dfrac{9 + 12i}{25}$**

Solve each equation. (Ex. 8–11)

8. $2x^2 + 24 = 0$ **$\pm 2i\sqrt{3}$**

9. $y^2 - 6y + 10 = 0$ **$3 \pm i$**

10. $3x^2 + 5 = 2x$ **$\dfrac{1 \pm i\sqrt{14}}{3}$**

11. $t^4 + 7t^2 + 12 = 0$ **$\pm 2i, \pm i\sqrt{3}$**

12. Find the three cube roots of 125.
   $5, \dfrac{-5 + 5i\sqrt{3}}{2}, \dfrac{-5 - 5i\sqrt{3}}{2}$

Determine the nature of the solutions without solving the equation. (Ex. 13–15)

13. $3x^2 + 4x = 4$ **two R:Rt**

14. $2y^2 - 3y + 4 = 0$ **two Im**

15. $\dfrac{5}{3}x^2 + 2x + \dfrac{3}{5} = 0$ **one R:Rt**

16. For what values of k will $y^2 + (k + 2)y + 16 = 0$ have exactly one solution?   **−10, 6**

17. Without solving the equation, find the sum and the product of the solutions of $12x^2 + 2 = 3x$.   **sum: $\frac{1}{4}$, prod.: $\frac{1}{6}$**

Write a quadratic equation, in standard form, that has the given pair of numbers as its solutions. (Ex. 18–19)

18. $-\dfrac{1}{3}$ and $\dfrac{2}{3}$ **$9x^2 - 3x - 2 = 0$**

19. $-2 \pm i\sqrt{6}$ **$x^2 + 4x + 10 = 0$**

★ 20. Compute the value of the discriminant for $x^2 - 2x\sqrt{3} - 24 = 0$ and determine the nature of the solutions.   **108; two R:Ir**

★ 21. For what values of k will $5x^2 - kx + 5 = 0$ have two real number solutions?
   $k < -10$ or $k > 10$

★ 22. Factor $x^2 + 9$ as a difference of two squares. [Hint: Use the fact that $i^2 = -1$.]   **$(x + 3i)(x - 3i)$**

Use the definition $|a + bi| = \sqrt{a^2 + b^2}$ in Ex. 23–24.

★ 23. Simplify $|5 + 5i|$ and $|-3 - 3i|$.
   $5\sqrt{2}, 3\sqrt{2}$

★ 24. Prove that $|a + bi| = |a|\sqrt{2}$ if $a = b$.
   $|a + bi| = |a + ai| = \sqrt{a^2 + a^2} = \sqrt{2a^2} = |a|\sqrt{2}$

COLLEGE PREP TEST

DIRECTIONS: Choose the *one* best answer to each question or problem.

1. Let x and y represent two imaginary numbers. Which statement(s) could be true? **E**

   I xy is a real number.
   II xy is an imaginary number.
   III $x + y$ is a real number.

   (A) I only   (B) II only
   (C) I or II only   (D) II or III only
   (E) I, II, or III

2. If $i = \sqrt{-1}$ and $i^2 = -1$, which statement is *false*? **E**

   (A) $i^4 = 1$ (B) $i^6 = i^2$ (C) $i^3 = -i$
   (D) $i^5 = i$ (E) None of these

3. For what value(s) of k will
   $$kx^2 + 2x + k = 0$$
   have exactly one solution? **A**

   (A) -1, 0, and 1 (B) 0 and 1 only
   (C) -1 and 1 only   (D) 1 only
   (E) None of these

4. If $n \div 7$ gives a remainder of 5, then $(n + 3) \div 7$ gives what remainder? **A**

   (A) 1   (B) 2   (C) 3   (D) 4   (E) 5

5. For which equation(s) is the sum of the solutions greater than the product of the solutions? **C**

   (A) $7x^2 - 28x + 28 = 0$
   (B) $8x^2 + 12x - 8 = 0$
   (C) $9x^2 - 3x - 2 = 0$
   (D) All of these
   (E) None of these

6. The conjugate of $-(-2i + 3)$ is **B**

   (A) $3 - 2i$ (B) $-3 - 2i$ (C) $3 + 2i$
   (D) $-3 + 2i$ (E) None of these

7. For every complex number x, $x\star$ is defined to be x^2. Find the value of $[(-2i)\star]\star$. **B**

   (A) $16i$ (B) 16 (C) $-16i$
   (D) -16 (E) None of these

8. Which equation does *not* have two imaginary number solutions? **C**

   (A) $x^2 + 4 = 0$ (B) $\dfrac{x}{2} + \dfrac{2}{x} = 0$
   (C) $(x + 2)^2 = 0$ (D) $\dfrac{(-x)^2}{3} + \sqrt{\dfrac{16}{9}} = 0$
   (E) None of these

9. If $-2i \cdot 3i \cdot 4i \cdot N = 3i \cdot 4i$, then $N =$ **C**

   (A) $-\dfrac{i}{2}$ (B) $-\dfrac{1}{2}$ (C) $\dfrac{i}{2}$
   (D) $\dfrac{1}{2}$ (E) None of these

10. For what value(s) of k will
    $$x^2 + kx + 1 = 0$$
    have two imaginary number solutions? **B**

    (A) $k < -2$ or $k > 2$ (B) $-2 < k < 2$
    (C) $-4 < k < 4$ (D) 0 only
    (E) None of these

11. If $x = a + bi$ and $y = a - bi$, where a and b are real numbers and $i = \sqrt{-1}$, which statement(s) is (are) true? **E**

    I xy is always an imaginary number.
    II xy is always a real number.
    III $x + y$ is always a real number.

    (A) I only   (B) II only   (C) III only
    (D) I and III only   (E) II and III only

College Prep Test

CUMULATIVE REVIEW

(Chapters 1–8)

DIRECTIONS: Choose the *one* best answer to each question or problem. (Exercises 1–12)

1. Factor $3x^3 - 4x^2 + 15x - 20$ completely by grouping pairs of terms. **D**

   (A) $(3x^2 + 5)(x - 4)$ (B) $(x^2 - 4)(3x + 5)$
   (C) $(3x^2 - 4)(x + 5)$ (D) $(x^2 + 5)(3x - 4)$

2. If $x^2 - 5x - 14 > 0$, then **B**

   (A) $-2 < x < 7$ (B) $x < -2$ or $x > 7$
   (C) $-7 < x < 2$ (D) $x < -7$ or $x > 2$

3. Multiply: $\dfrac{8x^2 - 72y^2}{10a^{-4}b^{-3}c^2} \cdot \dfrac{-6a^{-2}b^3c^{-6}}{6x^2 + 3xy - 45y^2}$. **A**

   (A) $-\dfrac{8a^2b^6(x - 3y)}{5c^8(2x - 5y)}$ (B) $\dfrac{5(x + 3y)}{16a^2bc^3(2x - 5y)}$
   (C) $\dfrac{-8a^2b^6(x - 3y)}{5c^8(2x + 5y)}$ (D) $\dfrac{8(x + 3y)}{-5a^2c^4(2x + 5y)}$

4. Machine X can do a certain job in 4 h and machine Y can do the same job in 6 h. If machines X, Y, and Z can do the job in 2 h working together, how long would it take Z to do the job working alone? **A**

   (A) 12 h   (B) 10 h   (C) 8 h   (D) 6 h

5. Solve $\dfrac{x + 1}{3x - 1} = \dfrac{x + 4}{2x + 6}$. **D**

   (A) $\dfrac{3}{5}, \dfrac{1}{4}$ (B) 2, 5
   (C) $-2, 5$ (D) $-5, 2$

6. Solve $32^{2x+1} = 8$ for x. **C**

   (A) 1   (B) $\dfrac{7}{2}$ (C) $-\dfrac{1}{5}$ (D) $\dfrac{1}{4}$

7. Solve $a(bx - c) = x - 2(x + c)$ for x. **C**

   (A) $\dfrac{ac - 2c - x}{ab}$ (B) $\dfrac{ac - 2c}{ab}$
   (C) $\dfrac{ac - 2c}{ab + 1}$ (D) $\dfrac{2c - ac}{ab + 1}$

8. Use synthetic substitution to find the value of $3x^4 + 20x^3 - 2x + 20$ if $x = -6$. **A**

   (A) -400 (B) 32
   (C) 104   (D) 8,216

9. Find the value of $\dfrac{81^{-\frac{5}{4}}}{9^{-\frac{3}{2}}}$. **C**

   (A) 9   (B) $\dfrac{1}{4}$ (C) $\dfrac{1}{9}$ (D) $\dfrac{5}{2}$

10. Solve $n - 1 = \sqrt{3n + 7}$. **B**

    (A) -1 (B) 6
    (C) $-1, 6$ (D) None of these

11. Factor $4x^2 - 12x + 9 - 25y^2$ as a difference of squares. **B**

    (A) $(2x - 3 - 5y)^2$
    (B) $(2x + 5y - 3)(2x - 5y - 3)$
    (C) $(2x - 3 - 5y)(2x - 3 - 5y)$
    (D) $(2x - 5y + 3)(2x - 5y - 3)$

12. Find the three cube roots of 64. **A**

    (A) $4, -2 + 2i\sqrt{3}, -2 - 2i\sqrt{3}$
    (B) $4, -2 + 4i\sqrt{3}, -2 - 4i\sqrt{3}$
    (C) $-4, 2 + 2i\sqrt{3}, 2 - 2i\sqrt{3}$
    (D) $4, 2 + 4i\sqrt{3}, 2 - 4i\sqrt{3}$

Solve each equation.
13. $|5x - 4| = 26$ $-\frac{22}{5}, 6$
14. $\dfrac{5}{x + 4} = \dfrac{3x - 4}{x^2 + 2x - 8} + \dfrac{1}{2x - 4}$ $\dfrac{16}{3}$
15. $\dfrac{y + 5}{2y + 2} = \dfrac{y - 4}{y - 3}$ $1, 7$
16. $\dfrac{9}{n - 2} - \dfrac{6}{n} = 3$ $-1, 4$
17. $8^{3x - 1} = 4$ $\dfrac{5}{9}$
18. $5^{2x - 3} = \dfrac{1}{25}$ $\dfrac{1}{2}$
19. $n + 1 = \sqrt{n + 7}$ 2
20. $\sqrt[3]{y^2 - 3y + 10} = 4$ $-6, 9$
21. $n^4 + 16n^2 + 48 = 0$ $\pm 2i, \pm 2i\sqrt{3}$
22. $x^2 + 13 = 6x$ $3 \pm 2i$

Solve each equation for x.
23. $a(3x + t) = b + c(4x - v)$ $x = \dfrac{b - cv - at}{3a - 4c}$
24. $\dfrac{6}{m} = \dfrac{4}{n} + \dfrac{2}{x}$ $x = \dfrac{mn}{3n - 2m}$

Factor each polynomial completely.
25. $2n^3 - 54$ $2(n - 3)(n^2 + 3n + 9)$
26. $6x^3 - 3x^2 + 10x - 5$ $(3x^2 + 5)(2x - 1)$

Find the solution set of each inequality. Graph the solutions on a number line.
27. $8(2x - 3) - (14x - 19) \leq 3x - 2$ $\{x | x \geq -3\}$
28. $y^2 - 2y - 24 > 0$ $\{y | y < -4 \text{ or } y > 6\}$
29. $n^4 - 26n^2 + 25 < 0$
$\{n | -5 < n < -1 \text{ or } 1 < n < 5\}$

Simplify each expression. Assume that no divisor has the value of 0.
30. $(-3a^{-5}b^2)^{-4}$ $\dfrac{a^{20}}{81b^8}$
31. $\dfrac{-14x^{-2}y^3}{28x^6y^{-6}}$ $\dfrac{-y^9}{2x^8}$
32. $\dfrac{40mn^2 - 90m}{30n^2 + 75n + 45}$ $\dfrac{2m(2n - 3)}{3(n + 1)}$
33. $\dfrac{6 - 2a}{3a^2 + 12a} \div \dfrac{a^2 - 9}{a^3 + 3a^2 - 4a}$ $\dfrac{-2(a - 1)}{3(a + 3)}$
34. $\dfrac{4}{x + 4} - \dfrac{x + 1}{16 - x^2} + \dfrac{2}{12 - 3x}$ $\dfrac{13x - 53}{3(x + 4)(x - 4)}$
35. $\dfrac{\frac{4}{x} + \frac{6}{y}}{\frac{4}{x} + \frac{6}{y}}$ $\dfrac{4x + 6y}{4x + 6y}$

35. $\dfrac{\frac{4}{x} + \frac{6}{y}}{}$ $\dfrac{4x + 6y}{}$ $\dfrac{xy(2x + 3y)}{2y + 3x}$
36. $-5a^2b\sqrt{80a^7b^9c^{11}}$; $a \geq 0, b \geq 0, c \geq 0$ $-20a^5b^5c^5\sqrt{5abc}$
37. $\sqrt{8c^3d} + c\sqrt{50cd} - 4\sqrt{18c^3d}$; $c \geq 0, d \geq 0$ $-5c\sqrt{2cd}$
38. $\sqrt{\dfrac{25a^5}{12b^3}}$; $a \geq 0, b > 0$ $\dfrac{5a^2\sqrt{3ab}}{6b^2}$
39. $\dfrac{6c^2}{\sqrt[4]{8c^2}}$, $c > 0$ $3c\sqrt[4]{2c^2}$
40. $\sqrt[3]{-8x^7} + \sqrt[3]{125x^7}$ $3x^2\sqrt[3]{x}$
41. $(-5 + \sqrt{-27}) - (4 - 2\sqrt{-3})$ $-9 + 5i\sqrt{3}$
42. $\dfrac{6i}{1 + 7i}$ $\dfrac{21 + 3i}{25}$

Use synthetic substitution in Exercises 43–45.
43. $(5x^3 + 8 - 77x) \div (x + 4)$
  $5x^2 - 20x + 3, R: -4$
44. Is $x + 4$ a factor of $5x^3 - 77x + 8$? (See Exercise 43.) No
45. Evaluate $2n^3 - 15n^2 - 10n + 33$ if $n = 8$. 17

Determine whether each number is rational or irrational.
46. $\sqrt{40}$ I
47. $83.43434\ldots$ R
48. $-\sqrt{400}$ R
49. $9.29929992\ldots$ I

Find the value of each expression.
50. $-10 \cdot 32^{\frac{3}{5}} \cdot 36^{\frac{1}{2}}$ -480
51. $\dfrac{5^{-3}}{8^{-\frac{2}{3}}}$ $\dfrac{4}{125}$
52. Write $4\sqrt[3]{x^2}$ in exponential form. $4x^{\frac{2}{3}}$
53. Write $(5y)^{\frac{3}{4}}$ in radical form. $\sqrt[4]{125y^3}$

54. How many liters of a 20% salt solution must be added to 8 L of a 60% salt solution to obtain a 50% salt solution? $2\frac{2}{3}$ L

55. Machines A, B, and C can complete a certain job in 30 min, 40 min, and 1 h, respectively. How long will the job take if the machines work together? $13\frac{1}{3}$ min

56. One plane flew 390 km with a tailwind in the same time that another plane flew 360 km against the wind. Find the rate of the wind if each plane can fly at 500 km/h in still air. 20 km/h

57. The length of a rectangle is twice its width and 5 m less than the length of one of its diagonals. Find the length of the diagonal and the area of the rectangle.
$(25 + 10\sqrt{5})$ m, $(450 + 200\sqrt{5})$ m^2

58. Without solving $5y^2 + 3y - 10 = 0$, find the sum and the product of its solutions.
sum: $-\dfrac{3}{5}$, prod.: -2

Cumulative Review (Chapters 1–8)

9 COORDINATE GEOMETRY

CAREERS
Computer Programming

If you are organized in your work, consider alternative ways to solve difficult problems, pay attention to details, look for shortcuts, and enjoy mathematics and problem solving, you may want to consider a career in the field of programming.

The person who writes a computer program is called a computer programmer. In addition to the high school mathematics program, a computer programmer must receive further education. This additional training may be taken at a college that offers courses in computer science or at a vocational school that specializes in computer programming.

Project

Write a program to determine if a triangle is a right triangle, given the measures of the three sides.

To write the program, you need to know the following:
(1) If the square of the hypotenuse is equal to the sum of the squares of the two legs, then the triangle is a right triangle.
(2) The dimensions of the three sides.

THE COORDINATE PLANE 9.1

Objectives **To graph an ordered pair of real numbers in a coordinate plane**
To give the coordinates of a point in a coordinate plane
To draw the graph of a linear equation in two variables

When a horizontal number line is perpendicular to a vertical number line, a **coordinate plane** is formed. The horizontal line is called the **x-axis** and the vertical line is called the **y-axis**. The two axes intersect in a point called the **origin**, and they separate the plane into four sections called **quadrants**. The four quadrants are numbered in a counterclockwise direction, as shown in the figure.

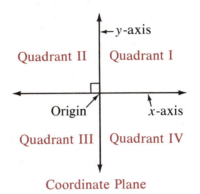

Each point in a coordinate plane is assigned a unique **ordered pair** of real numbers, (x, y). The two numbers paired with a given point are called the **coordinates** of that point. In the second figure, the coordinates of point A are written as the ordered pair of numbers, (6, 4). [The order is important; (6, 4) and (4, 6) are two different ordered pairs.] For each ordered pair of real numbers, there is a unique point in the coordinate plane. The point is called the **graph** of the ordered pair. Thus, the point A is the graph of (6, 4).

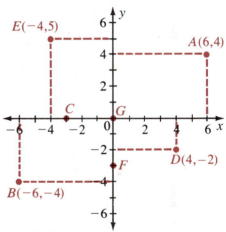

You can "graph the ordered pair of numbers" (6, 4) or "plot the point" $A(6, 4)$ in the following way. Starting at the origin, move to the right 6 units and then up 4 units. Place a dot in the coordinate plane to represent the graph. In general, the x-coordinate, or **abscissa**, indicates the direction and distance to move horizontally from the origin along the x-axis. The y-coordinate, or **ordinate**, indicates the direction and distance to move vertically, or parallel to the y-axis.

$A(6, 4)$	$B(-6, -4)$	$D(4, -2)$	$E(-4, 5)$
right 6, up 4	left 6, down 4	right 4, down 2	left 4, up 5

In the second figure, notice that the points C, F, and G are not located in a quadrant. Each point falls on one of the axes. Point C is the graph of $(-3, 0)$ and point F is the graph of $(0, -3)$. Point G is at the origin, so it is the graph of $(0, 0)$.

The Coordinate Plane

An equation of the form $ax + by = c$, where a, b, and c are real numbers with a and b both not equal to 0, is called a **linear equation in two variables** x and y. One solution of a linear equation in two variables is an ordered pair of real numbers that satisfies the equation. For example, (0, 4) is a solution of $-2x + y = 4$ since $-2 \cdot 0 + 4 = 0 + 4 = 4$.

The graph of a given linear equation in two variables is the set of all the points whose coordinates (ordered pairs) satisfy the equation. As shown in the examples, the graph of any linear equation in two variables is a line.

Example 1

Draw the graph of $-2x + y = 4$.

First, solve the equation for y in terms of x.

$$-2x + y = 4$$
$$y = 2x + 4$$

Second, make a table by choosing values for x and finding corresponding values for y.

x	$2x + 4$	y	ordered pair
0	$2 \cdot 0 + 4$	4	(0, 4)
1	$2 \cdot 1 + 4$	6	(1, 6)
-2	$2 \cdot -2 + 4$	0	$(-2, 0)$

Third, plot the points and draw the graph.

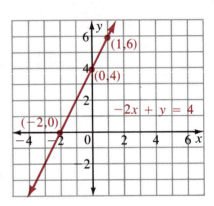

Example 2

Draw the graph of $3x + 2y = 5$.

$$3x + 2y = 5$$
$$2y = -3x + 5$$
$$y = -\frac{3}{2}x + \frac{5}{2}$$

x	$-\frac{3}{2}x + \frac{5}{2}$	y	ordered pair
0	$-\frac{3}{2} \cdot 0 + \frac{5}{2}$	$\frac{5}{2}$	$\left(0, \frac{5}{2}\right)$
2	$-\frac{3}{2} \cdot 2 + \frac{5}{2}$	$-\frac{1}{2}$	$\left(2, -\frac{1}{2}\right)$
4	$-\frac{3}{2} \cdot 4 + \frac{5}{2}$	$-\frac{7}{2}$	$\left(4, -\frac{7}{2}\right)$

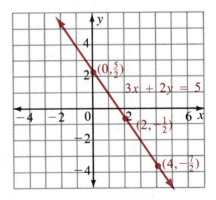

You could have used different values than those used for x in the tables. Although only two points are needed to determine the graph of a linear equation, it is wise to plot a third point as a check.

Reading in Algebra

Let $P(x, y)$ represent any point in a coordinate plane. Match each term at the left to exactly one description at the right.

1. Quadrant I **G**
2. Quadrant II **C**
3. Quadrant III **E**
4. Quadrant IV **A**
5. x-axis **F**
6. y-axis **B**
7. origin **D**

A $x > 0$ and $y < 0$
B $x = 0$ and y is a real number
C $x < 0$ and $y > 0$
D $(0, 0)$
E $x < 0$ and $y < 0$
F $y = 0$ and x is a real number
G $x > 0$ and $y > 0$

Written Exercises

On the same coordinate plane, graph each ordered pair. **Check student graphs.**

A
1. $(-3, 2)$
2. $(4, -1)$
3. $(0, 1)$
4. $(-1, -3)$
5. $(4, 3)$
6. $(0, 0)$
7. $(-1, 4)$
8. $(5, 0)$
9. $(-3, 0)$
10. $(0, -5)$

Use the coordinate system at the right for Exercises 11–13.

11. Give the coordinates of each of the labeled points.

12. Which of the labeled points do not lie in a quadrant? **A and E**

13. In which quadrant does each of the other labeled points lie?
I: D; II: B, F; III: C, G; IV: H, I

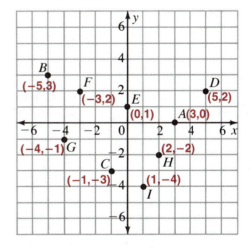

Draw the graph of each equation. **Check student graphs. See TE Answer Section.**

14. $y = -3x - 1$
15. $y = -2x + 3$
16. $y = 3x + 2$
17. $y - 4x = -2$
18. $y = 4x + 3$
19. $y = 2x + 4$
20. $y = -2x - 3$
21. $x - y = 5$
22. $3x + 3y = 9$
23. $4x - 2y = 10$
24. $-4x - 2y = 10$
25. $10x - 5y = 15$
26. $x - y = 0$
27. $6x + 3y = 9$
28. $6x - 3y = 15$
29. $4x + 2y = 8$

B
30. $2x - 4y = 12$
31. $3x + 2y = 12$
32. $-3x - y = 5$
33. $5x + 3y = 9$
34. $x - 2y + 5 = 0$
35. $-x + 2y - 3 = 0$
36. $-4x - 3y = 9$
37. $4x - 3y = 8$
38. $\frac{2}{3}x + \frac{1}{6}y = 4$
39. $-\frac{1}{5}x + \frac{3}{10}y = 2$
40. $\frac{1}{5}x + \frac{2}{3}y = -1$
41. $\frac{1}{3}x + 3y = 8$
42. $-1.2x + 3.6y = 2.4$
43. $0.2x + 0.3y = 1.2$
44. $0.12x - 0.02y = 0.06$

C
45. $\sqrt{2}x + 3y = 9$
46. $x - \sqrt{3}y = 10$
47. $\sqrt{3}x + \sqrt{2}y = -2$

The Coordinate Plane

SLOPE OF A LINE   9.2

Objectives To find the length of a vertical or a horizontal segment, given the coordinates of its endpoints
To find the slope of a line, given the coordinates of any two points on the line

If you are given the coordinates of the endpoints of a vertical or horizontal segment in a coordinate plane, you can find the *length* of the segment.

Notice that $\overline{AB}$ (read: segment AB) in the figure at the right is vertical and that the points A and B have the same x-coordinate. You can find AB (read: the length of $\overline{AB}$) by subtracting the y-coordinates of the two points A and B and then taking the absolute value of the difference.

$AB = |y\text{-coord. of } A - y\text{-coord. of } B| = |3 - (-2)| = 5$

It is necessary to use absolute value since a length cannot be negative.

Length of a vertical segment	For a vertical segment $\overline{AB}$ with $A(x_1, y_1)$ and $B(x_1, y_2)$, $AB = |y_2 - y_1|$, or $|y_1 - y_2|$.	

In the figure at the right, $\overline{CD}$ is horizontal and the points C and D have the same y-coordinate. You can find CD by subtraction of the x-coordinates, as shown below.

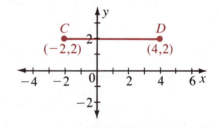

$CD = |x\text{-coord. of } D - x\text{-coord. of } C| = |4 - (-2)| = 6$

Length of a horizontal segment	For a horizontal segment $\overline{CD}$ with $C(x_1, y_1)$ and $D(x_2, y_1)$, $CD = |x_2 - x_1|$, or $|x_1 - x_2|$.	

When variables are used for coordinates in this chapter, you should assume that the variables represent real numbers.

Example 1

Find the length of $\overline{BC}$ and $\overline{AC}$ in the figure. Then find the ratio $\dfrac{BC}{AC}$.

Find the coordinates of C.

$\overline{BC}$ is a vertical segment. So, C has the same x-coordinate as B(3, 4).

$\overline{AC}$ is a horizontal segment. So, C has the same y-coordinate as A(−2, 1).

$$BC = |y_2 - y_1| \qquad AC = |x_2 - x_1|$$
$$= |4 - 1|, \text{ or } 3 \qquad = |3 - (-2)|, \text{ or } 5$$

Thus, $BC = 3$ and $AC = 5$ and the ratio $\dfrac{BC}{AC} = \dfrac{3}{5}$.

Notice that for $\overleftrightarrow{AB}$ (read: line AB) in the figure above, the ratio $\frac{3}{5}$ shows how the change in the y-coordinates compares to the change in the x-coordinates as you move from point A to point B on the line:

$$\frac{\text{change in } y\text{-coordinates}}{\text{change in corresponding } x\text{-coordinates}} = \frac{3}{5}.$$

This ratio is called the **slope** of $\overleftrightarrow{AB}$. The letter m is used to represent slope.

Definition: Slope of a line

The *slope, m, of a line* that contains the points $A(x_1, y_1)$ and $B(x_2, y_2)$ is given by the following formula:

$$m = \frac{\text{change in } y\text{-coordinates}}{\text{change in corresponding } x\text{-coordinates}}, \text{ or } m = \frac{y_2 - y_1}{x_2 - x_1}.$$

Example 2

Find the slope of each line.

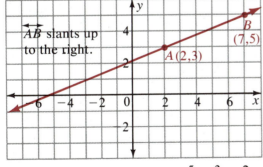

$m(\overleftrightarrow{AB}) = \dfrac{y_2 - y_1}{x_2 - x_1} = \dfrac{5 - 3}{7 - 2} = \dfrac{2}{5}$

The slope of $\overleftrightarrow{AB}$ is $\frac{2}{5}$.

$m(\overleftrightarrow{PQ}) = \dfrac{y_2 - y_1}{x_2 - x_1} = \dfrac{-2 - 2}{4 - (-1)} = -\dfrac{4}{5}$

The slope of $\overleftrightarrow{PQ}$ is $-\frac{4}{5}$.

Slope of a Line

The slope of a line is a measure of its "steepness." In Example 2, you have seen that if a line slants up to the right it has a positive slope and that if a line slants down to the right it has a negative slope. You will now see what happens when different pairs of points on the same line are used to compute its slope.

Example 3

Find the slope of the line in the figure, using points A and B, B and C, C and D.

$$m = \frac{y_2 - y_1}{x_2 - x_1}$$

$$m(\overrightarrow{AB}) = \frac{0 - (-1)}{-3 - (-6)} = \frac{1}{3}$$

$$m(\overrightarrow{BC}) = \frac{2 - 0}{3 - (-3)} = \frac{2}{6}, \text{ or } \frac{1}{3}$$

$$m(\overrightarrow{CD}) = \frac{3 - 2}{6 - 3} = \frac{1}{3}$$

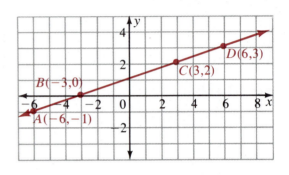

Thus, the slope of the line is $\frac{1}{3}$ using all three pairs of points.

Example 3 shows you that the slope of a given line is uniform. That is, for any two points on the line, the slope is the same.

Example 4

Find the slope of each line.

Vertical line

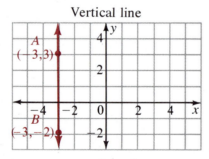

$$m = \frac{y_2 - y_1}{x_2 - x_1}$$

$$m(\overrightarrow{AB}) = \frac{3 - (-2)}{-3 - (-3)} = \frac{5}{0}$$

The slope is undefined.

Horizontal line

$$m = \frac{y_2 - y_1}{x_2 - x_1}$$

$$m(\overrightarrow{PQ}) = \frac{3 - 3}{3 - (-5)} = \frac{0}{8}$$

The slope is 0.

Since all points of a vertical line have the same x-coordinates, the slope of any vertical line is *undefined*. Similarly, since all points of a horizontal line have the same y-coordinates, the slope of any horizontal line is *0*.

Reading in Algebra

Match each sentence to exactly one of the statements at the right.
1. Points A, B, C, and D are on the same line. **C**
2. The coordinates of P and Q are $P(3, 1)$ and $Q(5, 1)$. **E**
3. The length of segment PQ is 2. **D**
4. The slope of line PQ is 2. **A**
5. A line is vertical. **H**
6. The coordinates of four points are $A(1, 5)$, $B(6, 5)$, $C(1, 1)$ and $D(4, 1)$. **B**
7. A line slants up to the right. **G**
8. A line slants down to the right. **F**

A $m(\overrightarrow{PQ}) = 2$
B $m(\overrightarrow{AB}) = m(\overrightarrow{CD})$
C $m(\overrightarrow{AB}) = m(\overrightarrow{BC}) = m(\overrightarrow{CD})$
D $PQ = 2$
E $m = 0$
F $m < 0$
G $m > 0$
H m is undefined

Written Exercises

1. **5**   2. **3**   3. **5**   4. **2**   5. **4**   6. **2**   7. **10**   8. **12**

Find the length of $\overline{AB}$ for the given coordinates.

(A)
1. $A(3, 2)$, $B(3, -3)$ 2. $A(1, -4)$, $B(-2, -4)$ 3. $A(2, 3)$, $B(-3, 3)$ 4. $A(-2, 1)$, $B(-2, 3)$
5. $A(2, -4)$, $B(-2, -4)$ 6. $A(5, -1)$, $B(5, 1)$ 7. $A(-2, 5)$, $B(-2, -5)$ 8. $A(6, -4)$, $B(-6, -4)$

UR = up to the right, DR = down to the right, H = horizontal, V = vertical, UN = undefined

Find the slope of $\overrightarrow{AB}$ for the given coordinates. Tell whether the line slants up to the right, slants down to the right, is horizontal, or is vertical.

9. $A(-6, 1)$, $B(4, 4)$ $\frac{3}{10}$, **UR**   10. $A(-1, -4)$, $B(-3, 2)$ **−3, DR**   11. $A(-2, 9)$, $B(1, -3)$ **−4, DR**
12. $A(3, 8)$, $B(-3, 8)$ **0, H**   13. $A(2, -3)$, $B(2, 3)$ **UN, V**   14. $A(6, 5)$, $B(-4, -5)$ **1, UR**
15. $A(7, 8)$, $B(1, 11)$ $-\frac{1}{2}$, **DR**   16. $A(-4, -1)$, $B(-2, 4)$ $\frac{5}{2}$, **UR**   17. $A(-4, -7)$, $B(3, -7)$ **0, H**
18. $A(-3, -1)$, $B(-6, -4)$ **1, UR**   19. $A(-9, -7)$, $B(10, 8)$ $\frac{15}{19}$, **UR**   20. $A(6, 2)$, $B(-1, -4)$ $\frac{6}{7}$, **UR**
21. $A(-4, -2)$, $B(-1, 7)$ **3, UR**   22. $A(4, -2)$, $B(8, -6)$ **−1, DR**   23. $A(-3, -5)$, $B(-5, -5)$ **0, H**
24. $A(6, 0)$, $B(0, -8)$ $\frac{4}{3}$, **UR**

Show that $m(\overrightarrow{AB}) = m(\overrightarrow{BC}) = m(\overrightarrow{AC})$ for the given coordinates.

25. $A(-2, -4)$, $B(0, 0)$, $C(4, 8)$ $m = 2$ 26. $A(-6, -4)$, $B(-4, -3)$, $C(-2, -2)$ $m = \frac{1}{2}$
27. $A(3, 6)$, $B(0, 4)$, $C(-3, 2)$ $m = \frac{2}{3}$ 28. $A(5, -1)$, $B(0, -1)$, $C(-6, -1)$ $m = 0$

Find the slope of $\overrightarrow{PQ}$ for the given coordinates.

(B)
29. $P(-2, -3)$, $Q(-\frac{1}{2}, -\frac{1}{3})$ $\frac{16}{9}$ 30. $P(-\frac{1}{3}, \frac{1}{2})$, $Q(\frac{5}{6}, 4)$ **3**   31. $P(\frac{1}{5}, \frac{1}{4})$, $Q(\frac{3}{5}, \frac{3}{4})$ $\frac{5}{4}$
32. $P(-\frac{1}{6}, \frac{1}{4})$, $Q(\frac{1}{2}, \frac{1}{2})$ $\frac{3}{8}$ 33. $P(\frac{2}{3}, \frac{3}{4})$, $Q(\frac{1}{6}, -\frac{1}{2})$ $\frac{5}{2}$ 34. $P(\frac{2}{5}, -\frac{1}{3})$, $Q(-\frac{1}{3}, \frac{1}{3})$ $-\frac{10}{11}$

35. $P(a, -b)$, $Q(-a, b)$ $-\frac{b}{a}$ 36. $P(4x, 5y)$, $Q(-2x, -7y)$ $\frac{2y}{x}$
37. $P(6a, -2b)$, $Q(7a, 5b)$ $\frac{7b}{a}$ 38. $P(-10a, 12b)$, $Q(-2a, -7b)$ $-\frac{19b}{8a}$

(C)
39. $P(x_1, y_1)$, $Q(x_1 + t, y_1 + s)$ $\frac{s}{t}$ 40. $P(2m - 5n, t + v)$, $Q(-n, 3v - t)$ $\frac{v-t}{2n-m}$

Find the missing coordinate for the given data.

41. $A(4, y)$, $B(5, 0)$, $m(\overrightarrow{AB}) = -2$ **2**   42. $M(-3, 2)$, $N(x, -4)$, $m(\overrightarrow{MN}) = -\frac{1}{2}$ **9**
43. $P(-\frac{1}{3}, -\frac{1}{2})$, $Q(\frac{4}{3}, y)$, $m(\overrightarrow{PQ}) = \frac{1}{5}$ $-\frac{1}{6}$ 44. $F(x, -\frac{1}{5})$, $G(\frac{2}{5}, -\frac{4}{5})$, $m(\overrightarrow{FG}) = -\frac{3}{4}$ $-\frac{2}{5}$

CUMULATIVE REVIEW

Solve each equation.

1. $3x - 2(1 - x) = 13$ **3**
2. $\dfrac{x-5}{2} = \dfrac{x-3}{5}$ $\dfrac{19}{3}$
3. $\dfrac{3}{x^2 - 7x + 12} - \dfrac{x}{x-3} = \dfrac{2x-1}{x-4}$ $0, \dfrac{11}{3}$

Slope of a Line

EQUATION OF A LINE 9.3

Objectives **To write an equation of a line, given its slope and the coordinates of a point on the line**
To write an equation of a line, given its slope and y-intercept
To find the slope and the y-intercept of a line, given an equation of the line
To write an equation of a line, given the coordinates of two points on the line

The line in the figure passes through the points $P(x_1, y_1)$ and $Q(x_2, y_2)$. Let $R(x, y)$ be any other point on $\overrightarrow{PQ}$. You can find an equation of this line by applying the definition of slope. Consider the points P and R.

| Point-slope form of an equation of a line | $y - y_1 = m(x - x_1)$ is an equation of a nonvertical line whose slope is a real number, m, and passes through the point $P(x_1, y_1)$. |

The point-slope form enables you to write an equation of a nonvertical line when its slope and the coordinates of one point of the line are known. The process is shown in Example 1.

Example 1 **Write an equation of the line that passes through the point $P(3, -4)$ and whose slope is $-\frac{2}{3}$.**

$$y - y_1 = m(x - x_1) \quad \blacktriangleleft \text{ Use the point-slope form.}$$
$$y - (-4) = -\frac{2}{3}(x - 3)$$
$$y + 4 = -\frac{2}{3}(x - 3)$$
$$3y + 12 = -2x + 6 \quad \blacktriangleleft \text{ Multiply each side by 3.}$$
$$2x + 3y = -6$$

Thus, an equation of the line is $2x + 3y = -6$.

An equation of a line is a *linear equation in two variables*. The **standard form** is $ax + by = c$, where a, b, and c are integers, and a and b both not equal to 0. Notice that the equation in Example 1 is written in this standard form.

The **y-intercept** of a line is the y-coordinate of the point where the line intersects the y-axis. Generally, the letter b is used to represent the y-intercept of a line.

In the figure at the right, the point $P(0, b)$ is the point where the given line intersects the y-axis, with b the y-intercept. Let $R(x, y)$ be any other point on the line. You can find an equation of the line by using the point-slope form of an equation.

$P(0, b)$ $R(x, y)$ $y - y_1 = m(x - x_1)$
↑ ↑   ↑ ↑ $y - b = m(x - 0)$
x_1 y_1 x y $y - b = mx$
                                 $y = mx + b$

Slope-intercept form of an equation of a line	$y = mx + b$ is an equation of a nonvertical line whose slope is a real number, m, and whose y-intercept is b.

Example 2

Write an equation, in standard form, of the line whose slope is $-\frac{5}{2}$ and whose y-intercept is -3.

$y = mx + b$ ◀ Use the slope-intercept form.
$y = -\frac{5}{2}x + (-3)$ ◀ $m = -\frac{5}{2}, b = -3$
$2y = -5x - 6$ ◀ Multiply each side by 2.
$5x + 2y = -6$

Thus, an equation of the line is $5x + 2y = -6$.

Given an equation of a line, you can find the slope and the y-intercept of the line.

Example 3

Find the slope and the y-intercept of the line whose equation is $3x + 2y = 9$.

Rewrite the equation in the form $y = mx + b$.
$3x + 2y = 9$
$2y = -3x + 9$
$y = -\frac{3}{2}x + \frac{9}{2}$ ◀ Divide each side by 2.

Thus, the slope is $-\frac{3}{2}$ and the y-intercept is $\frac{9}{2}$.

Example 4 on page 218 shows you how to find an equation of a line when the coordinates of two points of the line are known.

Equation of a Line

Example 4   Write an equation, in standard form, of the line that passes through the points $P(-3, 1)$ and $Q(5, -4)$.

Find the slope of $\overrightarrow{PQ}$. ▸ $m = \dfrac{y_2 - y_1}{x_2 - x_1} = \dfrac{-4 - 1}{5 - (-3)} = -\dfrac{5}{8}$

Use the point-slope form. ▸ $y - y_1 = m(x - x_1)$

$y - 1 = -\dfrac{5}{8}[x - (-3)]$

$y - 1 = -\dfrac{5}{8}(x + 3)$

$8y - 8 = -5x - 15$

$5x + 8y = -7$

Thus, an equation of the line is $5x + 8y = -7$.

In the figure at the right, $\overleftrightarrow{AB}$ is a horizontal line and $\overleftrightarrow{PQ}$ is a vertical line. You can write an equation of $\overleftrightarrow{AB}$ and an equation of $\overleftrightarrow{PQ}$ as shown below. Recall that the slope of a horizontal line is 0 and the slope of a vertical line is undefined.

The y-intercept of $(\overleftrightarrow{AB})$ is 4 and $m(\overleftrightarrow{AB})$ is 0.
$y = mx + b$
$y = 0 \cdot x + 4$
$y = 4$

For each point C on $\overleftrightarrow{AB}$, the y-coordinate is 4 and the x-coordinate is a real number. Therefore, $y = 4$ is an equation of $\overleftrightarrow{AB}$.

$\overleftrightarrow{PQ}$ does not intersect the y-axis, so there is no y-intercept; $m(\overleftrightarrow{PQ})$ is undefined. Let $R(x, y)$ be any point on $\overleftrightarrow{PQ}$. Then $m(\overleftrightarrow{PQ}) = m(\overleftrightarrow{PR}) = \dfrac{y - 3}{x - 2}$, which is undefined only for $x = 2$. For each point R on $\overleftrightarrow{PQ}$, the x-coordinate is 2 and the y-coordinate is a real number. Therefore, $x = 2$ is an equation of $\overleftrightarrow{PQ}$.

Example 5   For the given data, write an equation of each line.

Data	Equation
Line passes through $A(7, 3)$ and $B(7, -2)$	$x = 7$
Line passes through $C(-5, -4)$ and $D(6, -4)$	$y = -4$
Line passes through $T(-10, 8)$ and m is undefined	$x = -10$
Line with $m = 0$ and $b = 6$	$y = 6$

The linear equations $y = c$ and $x = c$, where c is a constant, represent the *standard form* of the equation of a horizontal line and a vertical line, respectively.

Reading in Algebra

A description of each of five lines is given at the left. Match each description to the *one* best form to use to find an equation of each line.

1. $P(-1, -5)$ is on the line and its slope is $-\frac{3}{5}$. **D**
2. The line is vertical and $P(-2, 3)$ is on the line. **B**
3. The y-intercept of the line is -4 and its slope is 3. **A**
4. The line is horizontal and $P(-2, 3)$ is on the line. **C**
5. $P(-3, 5)$ and $Q(2, -1)$ are on the line. **D**

A. $y = mx + b$
B. $x = c$
C. $y = c$
D. $y - y_1 = m(x - x_1)$

Written Exercises

A line has the given slope, y-intercept, or contains the indicated point(s). Write an equation, in standard form, of each line.   4. $x + y = -5$ 14. $3x - 5y = 15$ 18. $9x - 2y = -19$

(A)
1. $P(2, -5)$, $m = \frac{2}{3}$ $2x - 3y = 19$
2. $Q(2, -2)$, $m = -\frac{1}{2}$ $x + 2y = -2$
3. $R(-1, 6)$, $m = \frac{5}{6}$ $5x - 6y = -41$
4. $A(-2, -3)$, $m = -1$ above
5. $M(-2, 4)$, $m = -4$ $4x + y = -4$
6. $R(3, -2)$, $m = -3$ $3x + y = 7$
7. $m = -\frac{4}{3}$, $b = 2$ $4x + 3y = 6$
8. $m = \frac{3}{5}$, $b = -10$ $3x - 5y = 50$
9. $m = -\frac{5}{8}$, $b = -\frac{3}{8}$ $5x + 8y = -3$
10. $m = 2$, $b = 3$ $2x - y = -3$
11. $m = -4$, $b = -4$ $4x + y = -4$
12. $m = -3$, $b = \frac{7}{3}$ $9x + 3y = 7$
13. $A(-1, 3)$, $B(0, 4)$ $x - y = -4$
14. $A(0, -3)$, $B(-5, -6)$ above
15. $A(2, -3)$, $B(0, 4)$ $7x + 2y = 8$
16. $M(-1, 6)$, $N(-4, 2)$ below
17. $M(7, 3)$, $N(-3, 6)$ below
18. $M(-1, 5)$, $N(-3, -4)$ above
19. $P(7, 5)$, $Q(-3, 5)$ $y = 5$
20. $P(5, -4)$, $Q(-2, -4)$ $y = -4$
21. $P(8, 4)$, $Q(8, 9)$ $x = 8$
22. $A(0, 7)$, $B(4, 0)$ $7x + 4y = 28$
23. $M(-5, 0)$, $N(0, 9)$ below
24. $R(0, -6)$, $Q(10, 0)$ $3x - 5y = 30$
25. $R(8, 12)$, $m = 0$ $y = 12$
26. $S(-6, -4)$, m is undefined.
27. $T(7, -3)$, line is horizontal.
16. $4x - 3y = -22$
23. $9x - 5y = -45$ $x = -6$
17. $3x + 10y = 51$ $y = -3$

Find the slope and the y-intercept of the line described by the given equation.

28. $2x + 3y = 9$ $-\frac{2}{3}, 3$
29. $3x - 5y = 10$ $\frac{3}{5}, -2$
30. $7x + 3y = -21$ $-\frac{7}{3}, -7$
31. $5x - 3y - 15 = 0$ $\frac{5}{3}, -5$
32. $5y - x + 20 = 0$ $\frac{1}{5}, -4$
33. $3x + 4y - 40 = 0$ $-\frac{3}{4}, 10$
34. $7x - 3(4 - y) = 0$ $-\frac{7}{3}, 4$
35. $15 - 4(-2x - 3y) = 39$ $-\frac{2}{3}, 2$
36. $3y - (18 + 6x) = 0$ $2, 6$
  $-8, -15$ $\frac{3}{2}, -2$ 0, 2

(B)
37. $y - (4x + 5) = -4(3x + 5)$
38. $3x + 5(x - y) - (y - x) = 12$
39. $2(x + 3y) - (2x - y) = 14$
40. $-2(3y + 4x) - 5(2x - 3y) = 15$ $2, \frac{15}{9}$
41. $7y + 5(-3y - 4x) = -(7 - 3x) + 2y$ $-\frac{23}{10}, \frac{7}{10}$
42. $\frac{4}{5}(10y - 20x) - \frac{2}{3}(6x - 15y) = 0$ $\frac{10}{9}, 0$
43. $-\frac{1}{2}(6x - 8y) + \frac{4}{5}(10x + 15y) = 6 - y$ $-\frac{5}{17}, \frac{6}{17}$

(C)
44. $4x - 5[2(6x - 3) - 4(6y - 3)] = 6$ $\frac{7}{15}, \frac{3}{10}$
45. $-8 - 3[-(4 - 3y) - 2(5x + 2y)] = x - y$
46. $0.3x - 4.3[-(x - 2) - 2.1(4 - y)] = 1.8$
47. $1.6y - 3.1[(y + 5) - 1.6(4 - x)] = 7.3$ $-\frac{248}{75}, -\frac{148}{75}$
46. $\frac{460}{903}, \frac{2,572}{903}$
45. $-\frac{29}{4}, -1$

Write an equation, in standard form, of each line.

48. The line having an x-intercept of -2 and a y-intercept of 3 $3x - 2y = -6$
49. The line having an x-intercept of 4 and a y-intercept of -4 $x - y = 4$
50. The line having an x-intercept of $\frac{1}{A}$ and a y-intercept of $\frac{1}{B}$ $Ax + By = 1$
51. The line having an x-intercept of $\frac{C}{A}$ and a y-intercept of $\frac{C}{B}$ $Ax + By = C$

52. If $P(a, 0)$ and $Q(0, b)$ are the x- and y-intercepts, respectively, of $\overleftrightarrow{PQ}$ ($a \ne 0, b \ne 0$), prove that $\frac{x}{a} + \frac{y}{b} = 1$ is an equation of $\overleftrightarrow{PQ}$. **See TE Answer Section.**

Equation of a Line

SLOPE-INTERCEPT METHOD OF DRAWING GRAPHS   9.4

Objectives **To draw the graph of a linear equation in two variables, using the slope-intercept method**
To determine whether a point is on a line, given the coordinates of the point and an equation of the line

To draw the graph of $y = \frac{3}{5}x + 1$ using its slope and y-intercept, proceed as follows:

First, determine the slope and the y-intercept of the line.

$$y = \underset{m}{\tfrac{3}{5}}x + \underset{b}{1}$$

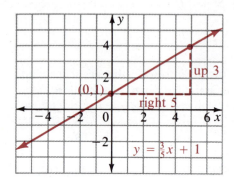

Then, plot the point whose y-intercept is 1. From this point, move to the right 5 units and up 3 units to locate a second point.

Draw the line that contains the two points. The line is the graph of $y = \frac{3}{5}x + 1$.

Example 1   Draw the graph of $2x + 3y = -9$.

Solve for y in terms of x.
$2x + 3y = -9$
$3y = -2x - 9$
$y = -\frac{2}{3}x - 3$

$m = \frac{-2}{3}$ or $\frac{2}{-3}$, and $b = -3$.

Plot the point whose y-intercept is -3. From this point, move to the right 3 units and down 2 units *or* move to the left 3 units and up 2 units.

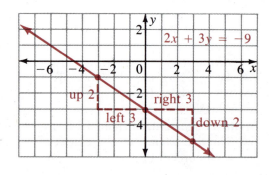

Recall that $m = \frac{y_2 - y_1}{x_2 - x_1}$. Thus, whenever the slope, m, is written as a fraction, the denominator indicates the change in the x-coordinates (distance and direction to move horizontally), and the numerator indicates the change in the y-coordinates (distance and direction to move vertically) as you move from one point to another point. Even though two points determine a line, when drawing graphs it is wise to locate a third point as a check.

Example 2   **Draw the graph of $y = 4$ and the graph of $x = -3$.**

$y = c$, where c is a constant, determines a horizontal line.

$x = c$, where c is a constant, determines a vertical line.

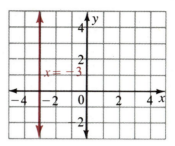

Given the coordinates of a point and an equation of a line, you can determine whether the point is on the line. The process involves substituting the x- and y- coordinate values of the point into the equation and checking to see if a true equation results.

Example 3   **Determine whether each point having the given coordinates is on the line described by the given equation.**

$P(-4, 7)$, $5x + 3y - 1 = 0$ $\quad\quad$ $Q(8, -2)$, $4y - x = 34$

Substitute -4 for x and 7 for y. ▶

$5x + 3y - 1$	0
$5(-4) + 3(7) - 1$	0
$-20 + 21 - 1$	
0	

Thus, $(-4, 7)$ satisfies the equation and so P is a point on the line.

$4y - x$	34
$4(-2) - (8)$	34
$-8 - 8$	
-16	

◀ *Substitute 8 for x and -2 for y.*

Thus, $(8, -2)$ does not satisfy the equation and so Q is not a point on the line.

> **Point on a line**
>
> Each point $P(x, y)$ whose coordinates satisfy an equation of a given line is a point on the line.
>
> Each point $P(x, y)$ on a given line has coordinates that satisfy an equation of the line.

Oral Exercises

Match each equation to exactly one group of ordered pairs.

1. $3x - 8y = 24$ **B**
2. $6x + 4y = -24$ **C**
3. $x - 24y = 24$ **A**
4. $12x + 2y = -24$ **D**

A $(0, -1)$	**B** $(8, 0)$	**C** $(-2, -3)$	**D** $(-1, -6)$
$(24, 0)$	$(-8, -6)$	$(0, -6)$	$(-2, 0)$
$(-24, -2)$	$(0, -3)$	$(-4, 0)$	$(0, -12)$

Slope-Intercept Method of Drawing Graphs

Tell whether each equation describes a horizontal line, a vertical line, or neither.
5. $y = 5$ **H**   6. $x = -4$ **V**   7. $2y = 7$ **H**   8. $y = 0$ **H**   9. $-x = -6$ **V**   10. $y = 7 + x$ **N**

Written Exercises

Draw the graph of each equation. **Check student's graphs. See TE Answer Section.**

Ⓐ
1. $y = \frac{2}{3}x + 3$
2. $y = -\frac{2}{3}x - 2$
3. $y = \frac{1}{3}x + 1$
4. $y = -\frac{1}{2}x + 4$
5. $y = -2x - 1$
6. $y = -2x + 1$
7. $y = -3x - 2$
8. $y = 5x + 3$
9. $y = -2x + 4$
10. $y = -3x - 1$
11. $y = 4x - 3$
12. $y = -x + 5$
13. $3x + 2y = 4$
14. $x - 3y - 9 = 0$
15. $3y = 15$
16. $3x - 4y = 12$
17. $x + y - 4 = 0$
18. $x = 4$
19. $5x + 20 = 0$
20. $2x + 5y + 15 = 0$

Determine whether a point having the given coordinates is on the line described by the given equation.
21. $(1, 2)$, $3x + 2y = 7$ **yes**
22. $(2, 3)$, $x - 3y = -7$ **yes**
23. $(-3, -2)$, $3x - 2y = -6$ **no**
24. $(-2, 1)$, $2x - y = -5$ **yes**
25. $(-1, -2)$, $-2x + y + 4 = 0$ **no**
26. $(-2, 1)$, $-2x + y - 5 = 0$ **yes**

Ⓑ
27. $(2, 3)$, $\frac{2}{3}y - \frac{1}{2}x = 1$ **yes**
28. $(-6, -4)$, $-\frac{1}{3}x + \frac{3}{4}y = -1$
29. $(-2, 2)$, $\frac{3}{4}x - \frac{3}{4}y = -3$ **yes**
30. $(-5, -6)$, $\frac{3}{5}x + \frac{1}{3}y = 5$ **no**
31. $(8, 4)$, $\frac{3}{4}y - \frac{7}{8}x = -4$ **yes**
32. $(10, -6)$, $\frac{2}{5}x + \frac{5}{3}y = 6$ **no**

28. **yes**

Draw the graph of each equation. **For Ex 33–38, see TE Answer Section.**
33. $2(x - 3) - 4(3 - 2y) = 8$
34. $6x - 2(5 - 3y) = 12$
35. $-4x - 3(2y - 1) = 6$
36. $2y - (5 - 2x) = -2x - 3y$
37. $-4(3 - 2x) - (8 - 2y) = x - y$
38. $-(x - 3y) - 2(x + y) = -10$

Find the coordinates of a point on the line that satisfy the given conditions. Do not draw a graph.

Ⓒ
39. An equation of the line is $x + 4y = 18$ and the ordinate of the point is twice the abscissa of the point. **(2, 4)**
40. An equation of the line is $3x - 2y = 12$ and the ordinate of the point is one-half the abscissa of the point. **(6, 3)**

Find each equation that satisfies the given description. Draw the graph of each equation.
41. A set of points equidistant from both axes
$x - y = 0$ or $x + y = 0$
42. The set of all points whose ordinate is two more than three times its abscissa
$-3x + y = 2$

NON-ROUTINE PROBLEMS

What number does the sequence of numbers below seem to approach?
$\left(1 - \frac{1}{2^2}\right), \left(1 - \frac{1}{3^2}\right), \left(1 - \frac{1}{4^2}\right), \left(1 - \frac{1}{5^2}\right), \ldots$ **1**

THE DISTANCE BETWEEN TWO POINTS   9.5

Objectives **To find the length of each side of a right triangle in a coordinate plane
To find the distance between any two points in a coordinate plane
To determine if a triangle with given lengths for its sides is a right triangle**

Recall that the square of the length of the hypotenuse of a right triangle is equal to the sum of the squares of the lengths of the two legs: That is, for right triangle ABC with right angle C, $c^2 = a^2 + b^2$, where c is the length of the hypotenuse and a and b are the lengths of the legs. This relationship, known as the *Pythagorean relation*, is illustrated below.

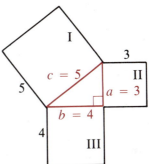

$$\underbrace{\text{Area of Square I}}_{25} = \underbrace{\text{Area of Square II}}_{9} + \underbrace{\text{Area of Square III}}_{16}$$

$$c^2 = a^2 + b^2$$

The Pythagorean relation can be used to find the length of the hypotenuse of a right triangle where the coordinates of the vertices of the triangle are given.

Example 1 **The vertices of right triangle PRS are $P(-1, 2)$, $R(6, 5)$, and $S(6, 2)$. Find the length of $\overline{PR}$.**

Let $c = PR$, $a = RS$, and $b = PS$.

$$c^2 = a^2 + b^2$$
$$(PR)^2 = (RS)^2 + (PS)^2$$

$\overline{RS}$ is vertical.
$\overline{PS}$ is horizontal. ▶

$$= |5 - 2|^2 + |6 - (-1)|^2$$
$$= 3^2 + 7^2$$
$$= 9 + 49$$
$$(PR)^2 = 58$$
$$PR = \pm\sqrt{58}$$ ◀ *A length cannot be negative.*

Thus, the length of $\overline{PR}$ is $\sqrt{58}$.

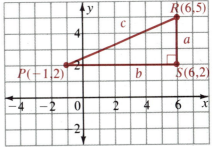

In Example 1, notice that for $P(x_1, y_1)$, $R(x_2, y_2)$, and $S(x_2, y_1)$, $(PS)^2 = (x_2 - x_1)^2$ and $(RS)^2 = (y_2 - y_1)^2$. Therefore, $(PR)^2 = (x_2 - x_1)^2 + (y_2 - y_1)^2$, or $PR = \sqrt{(x_2 - x_1)^2 + (y_2 - y_1)^2}$. On page 224, it is shown that this relationship remains true for any two points in a coordinate plane.

The Distance Between Two Points   223

In the figure at the right, points $A(x_1, y_1)$ and $B(x_2, y_2)$ determine $\overline{AB}$. A horizontal ray from A and a vertical ray from B intersect at a point C. Then, C has the same x-coordinate as B and the same y-coordinate as A. Since ABC is a right triangle,

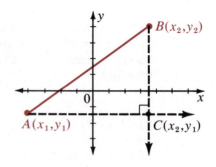

$(AB)^2 = (AC)^2 + (BC)^2$ by the Pythagorean relation.
$\quad\quad\quad = |x_2 - x_1|^2 + |y_2 - y_1|^2$ by the definition of length of a horizontal and a vertical segment.
$\quad\quad\quad = (x_2 - x_1)^2 + (y_2 - y_1)^2$ since $|a - b|^2 = (a - b)^2$.

So, $AB = \sqrt{(x_2 - x_1)^2 + (y_2 - y_1)^2}$ by the definition of square root.

Distance formula

The distance d between any two points $A(x_1, y_1)$ and $B(x_2, y_2)$ in a coordinate plane is given by the formula
$$d = \sqrt{(x_2 - x_1)^2 + (y_2 - y_1)^2}.$$

Notice that the distance between any two points A and B is the length of the segment connecting the two points. The symbol d or AB may be used to indicate this distance.

Example 2 Find the distance between $A(2, -4)$ and $B(4, 6)$.

Let $A(2, -4) = A(x_1, y_1)$ and $B(4, 6) = B(x_2, y_2)$.
$$d = \sqrt{(x_2 - x_1)^2 + (y_2 - y_1)^2}$$
$$= \sqrt{(4 - 2)^2 + [6 - (-4)]^2}$$
$$= \sqrt{4 + 100} = \sqrt{104}, \text{ or } 2\sqrt{26}$$

Thus, the distance between A and B is $2\sqrt{26}$.

The converse of the Pythagorean relation is also true. If $c^2 = a^2 + b^2$ where a, b, and c are the lengths of the sides of a triangle, then the triangle is a right triangle.

Example 3 A triangle has sides of lengths 5, $\sqrt{7}$, and $4\sqrt{2}$. Is the triangle a right triangle?

The hypotenuse is the longest side, so $c = 4\sqrt{2}$. ▶

c^2	$a^2 + b^2$
$(4\sqrt{2})^2$	$(5)^2 + (\sqrt{7})^2$
$16 \cdot 2$	$25 + 7$
32	32

Thus, since $c^2 = a^2 + b^2$, the triangle is a right triangle.

Written Exercises

For complete solutions, see TE Answer Section.

1. $AB = \sqrt{82}$, $BC = 1$, $AC = 9$ 2. $AB = 2\sqrt{26}$, $BC = 10$, $AC = 2$ 3. $AB = \sqrt{85}$, $BC = 7$, $AC = 6$

Find the length of each side of right triangle *ABC* with the given coordinates for its vertices. *C* is the vertex of the right angle.

A
1. $A(-3, 4)$, $B(6, 5)$, $C(6, 4)$
2. $A(2, -4)$, $B(4, 6)$, $C(4, -4)$
3. $A(4, 1)$, $B(10, 8)$, $C(10, 1)$
4. $A(-5, -1)$, $B(-2, 5)$, $C(-2, -1)$
5. $A(-6, -4)$, $B(-2, -1)$, $C(-2, -4)$
6. $A(8, 10)$, $B(10, 15)$, $C(10, 10)$

4. $AB = 3\sqrt{5}$, $BC = 6$, $AC = 3$ 5. $AB = 5$, $BC = 3$, $AC = 4$ 6. $AB = \sqrt{29}$, $BC = 5$, $AC = 2$

Find the distance between the given points. Give each answer in simplest radical form.

7. $A(2, 4)$, $B(6, 6)$
8. $R(-2, -3)$, $S(-7, 5)$
9. $P(-2, 4)$, $Q(7, 8)$
10. $M(7, 8)$, $N(-2, -4)$
11. $A(6, -2)$, $B(-3, 4)$
12. $A(5, 1)$, $B(-2, 4)$
13. $M(8, 4)$, $N(-2, -3)$
14. $A(0, 3)$, $B(4, 0)$
15. $M(0, -4)$, $N(5, 0)$
16. $M(9, 1)$, $N(-2, -1)$
17. $A(5, 1)$, $B(7, -3)$
18. $P(10, 2)$, $Q(-1, -3)$
19. $R(-5, -4)$, $S(-2, -3)$
20. $P(-1, 4)$, $Q(5, -3)$
21. $R(-2, -2)$, $S(-4, -4)$
22. $R(-10, 12)$, $S(2, -5)$

7. $2\sqrt{5}$ 8. $\sqrt{89}$ 9. $\sqrt{97}$

B
23. $P(1, -3)$, $Q(\frac{2}{3}, -\frac{1}{2})$
24. $A(\frac{1}{2}, \frac{1}{4})$, $B(-3, 2)$
25. $A(1, 2)$, $B(\frac{1}{6}, -\frac{1}{3})$
26. $A(\frac{1}{3}, \frac{1}{4})$, $B(-\frac{2}{3}, \frac{3}{4})$
27. $P(\frac{1}{5}, -\frac{1}{4})$, $Q(\frac{3}{5}, \frac{1}{2})$
28. $A(\frac{2}{3}, \frac{1}{3})$, $B(\frac{1}{3}, \frac{2}{3})$
29. $M(a, b)$, $N(3a, 2b)$
30. $M(2x, 3y)$, $N(-3, y)$
31. $M(s, p)$, $N(-3s, -2p)$
32. $P(0, 0)$, $Q(a, 0)$
33. $A(b, c)$, $B(a + b, c)$
34. $P(\frac{b}{2}, \frac{c}{2})$, $Q(\frac{2+a}{2}, \frac{c}{2})$
35. $A(\frac{a}{2}, \frac{b}{3})$, $B(\frac{a}{6}, \frac{b}{4})$
36. $M(\frac{c}{5}, \frac{d}{3})$, $N(\frac{c}{3}, -\frac{d}{2})$
37. $M(\frac{a}{3}, -\frac{b}{2})$, $N(-\frac{a}{6}, \frac{b}{8})$

The lengths of the sides of a triangle are given. Is the triangle a right triangle?
38. 6, 8, 11 **no**
39. $\sqrt{5}$, 1, 2 **yes**
40. 18, $3\sqrt{5}$, 19
41. 15, 12, 9 **yes**
42. $2\sqrt{3}$, $4\sqrt{3}$, 6
43. 4, 6, 9 **no**
44. 6, $2\sqrt{3}$, $2\sqrt{6}$
45. 12, 20, 16 **yes**
46. 10, 24, 26 **yes**
47. $\sqrt{19}$, $3\sqrt{5}$, 8

40. **no**   42. **yes**   44. **yes**   47. **yes**

For each of the following, find the values of *x*, if any, that make the distance from *A* to *B* equal to the given distance.

C
48. $A(x, 4)$, $B(-2, 3)$; $d = 4$ **below**
49. $A(3, x)$, $B(6, 2)$; $d = 5$ **6, −2**
50. $A(3, -1)$, $B(-x, 4)$; $d = 3$ **no values**
51. $A(-2, 1)$, $B(-4, x)$; $d = 6$ **$1 + 4\sqrt{2}, 1 - 4\sqrt{2}$**

48. $-2 + \sqrt{15}, -2 - \sqrt{15}$

Find the perimeter of each polygon with the given coordinates for its vertices. Give each answer in simplest radical form. **For complete solutions, see TE Answer Section.**

52. $P(3, -1)$, $Q(6, -5)$, $R(-4, -5)$ **$15 + \sqrt{65}$**
53. $P(-b, 0)$, $Q(c, 0)$, $R(c - d, e)$
54. $P(-a, 0)$, $Q(a, 0)$, $R(0, a)$ **$2a + 2a\sqrt{2}$**
55. $P(-\frac{2}{3}, -\frac{1}{2})$, $Q(2, -\frac{4}{5})$, $R(\frac{5}{6}, -5)$
56. $P\left(\frac{a+c}{2}, \frac{b+d}{2}\right)$, $Q\left(\frac{c+e}{2}, \frac{f+d}{2}\right)$, $R\left(\frac{g+e}{2}, \frac{h+f}{2}\right)$, $S\left(\frac{a+g}{2}, \frac{h+b}{2}\right)$
57. Prove that the midpoint of the hypotenuse of a right triangle is equidistant from the vertices of the triangle.

CUMULATIVE REVIEW

1. One number is 8 more than another number. Their product is 20. Find the numbers. **−10, −2 or 2, 10**
2. Find three consecutive integers so that 5 times the second, decreased by the third, is 15. **3, 4, 5**

The Distance Between Two Points

THE MIDPOINT OF A SEGMENT 9.6

Objectives **To determine the coordinates of the midpoint of a segment**
To determine the coordinates of an endpoint of a segment, given the coordinates of the other endpoint and the midpoint

In the figure at the right, M is the **midpoint** of $\overline{AB}$ since $AM = MB = \frac{1}{2}(AB)$.

$AM = \sqrt{(7-5)^2 + (12-8)^2} = \sqrt{4+16} = \sqrt{20}$, or $2\sqrt{5}$
$MB = \sqrt{(5-3)^2 + (8-4)^2} = \sqrt{4+16} = \sqrt{20}$, or $2\sqrt{5}$
$AB = \sqrt{(7-3)^2 + (12-4)^2} = \sqrt{16+64} = \sqrt{80}$, or $4\sqrt{5}$

Notice that $M(5, 8) = M\left(\frac{7+3}{2}, \frac{12+4}{2}\right)$. In other words, the x-coordinate of M is the average (arithmetic mean) of the x-coordinates of A and B, and the y-coordinate of M is the average of the y-coordinates of A and B.

Midpoint formula Given $P(x_1, y_1)$ and $Q(x_2, y_2)$, the midpoint of $\overline{PQ}$ is
$$M\left(\frac{x_1 + x_2}{2}, \frac{y_1 + y_2}{2}\right).$$

Example 1 **Determine the coordinates of M, the midpoint of $\overline{PQ}$, given $P(6, -3)$ and $Q(-14, 9)$.**

$$M\left(\frac{x_1 + x_2}{2}, \frac{y_1 + y_2}{2}\right) = M\left(\frac{6 + (-14)}{2}, \frac{-3 + 9}{2}\right) = M(-4, 3)$$

Thus, $(-4, 3)$ are the coordinates of M.

If the coordinates of one endpoint of a segment and the coordinates of its midpoint are known, you can find the coordinates of the other endpoint.

Example 2 **Determine the coordinates of Q, the other endpoint of $\overline{PQ}$, given $P(2, 3)$ and its midpoint $M(4, -7)$.**

$P(2, 3), M(4, -7), Q(x_2, y_2)$

x-coordinate of the midpoint. ▶

$x_m = \dfrac{x_1 + x_2}{2}$

$4 = \dfrac{2 + x_2}{2}$

$8 = 2 + x_2$

$x_2 = 6$

$y_m = \dfrac{y_1 + y_2}{2}$

$-7 = \dfrac{3 + y_2}{2}$

$-14 = 3 + y_2$

$y_2 = -17$

◀ *y-coordinate of the midpoint.*

Thus, the coordinates of the other endpoint are $(6, -17)$.

Chapter Nine

Oral Exercises

Determine the coordinates (x, y) of M, the midpoint of $\overline{AB}$.

1. **(4, 3)** 2. **(2, 1)** 3. **(0, −2)** 4. **(−4, 1½)**

Written Exercises

Determine the coordinates (x, y) of M, the midpoint of the segment joining P and Q.

Ⓐ
1. $P(7, 4)$, $Q(3, 2)$ **(5, 3)**
2. $P(-2, 1)$, $Q(8, 7)$ **(3, 4)**
3. $P(-6, 1)$, $Q(-2, -2)$ **(−4, −½)**
4. $P(-6, 12)$, $Q(-2, -2)$ **(−4, 5)**
5. $P(0, 4)$, $Q(6, 0)$ **(3, 2)**
6. $P(9, -2)$, $Q(1, 6)$ **(5, 2)**
7. $P(-3, 5)$, $Q(7, -3)$ **(2, 1)**
8. $P(19, 0)$, $Q(-4, 9)$ **($\frac{15}{2}$, $\frac{9}{2}$)**
9. $P(5, 4)$, $Q(-1, -2)$ **(2, 1)**
10. $P(8, 2)$, $Q(4, 6)$ **(6, 4)**
11. $P(0, -6)$, $Q(-8, 0)$ **(−4, −3)**
12. $P(-5, 6)$, $Q(-1, -8)$ **(−3, −1)**
13. $P(10, 9)$, $Q(-3, 7)$ **($\frac{7}{2}$, 8)**
14. $P(-9, -3)$, $Q(-1, -7)$ **(−5, −5)**
15. $P(-11, -5)$, $Q(-2, -8)$
16. $P(3, -4)$, $Q(5, 7)$ **(4, $\frac{3}{2}$)**

15. **($-\frac{13}{2}$, $-\frac{13}{2}$)**   20. **($-\frac{7}{20}$, $-\frac{5}{16}$)**   21. **($-\frac{2}{15}$, $-\frac{1}{15}$)**   22. **($-\frac{11}{36}$, $-\frac{5}{12}$)**

Ⓑ
17. $P(\frac{2}{3}, -\frac{1}{2})$, $Q(\frac{1}{6}, \frac{1}{5})$ **($\frac{5}{12}$, $-\frac{3}{20}$)**
18. $P(-\frac{1}{3}, \frac{1}{3})$, $Q(-\frac{1}{2}, -\frac{1}{6})$ **($-\frac{5}{12}$, $\frac{1}{12}$)**
19. $P(\frac{2}{5}, \frac{3}{4})$, $Q(-\frac{2}{3}, \frac{1}{3})$ **($-\frac{2}{15}$, $\frac{13}{24}$)**
20. $P(-\frac{1}{5}, -\frac{3}{4})$, $Q(-\frac{1}{2}, \frac{1}{8})$ **above**
21. $P((\frac{2}{5}, \frac{1}{5}))$, $Q(-\frac{2}{3}, -\frac{1}{3})$ **above**
22. $P(-\frac{1}{9}, -\frac{1}{6})$, $Q(-\frac{1}{2}, -\frac{2}{3})$ **above**
23. $P(3a, 2b)$, $Q(5a, 4b)$ **(4a, 3b)**
24. $P(m, n)$, $Q(n, m)$ **($\frac{m+n}{2}$, $\frac{n+m}{2}$)**
25. $P(m, n)$, $Q(m + 4, n - 6)$ **(m + 2, n − 3)**

Determine the coordinates of Q, the other endpoint of $\overline{PQ}$, given P and the midpoint, M.

26. $P(4, 3)$, $M(6, 5)$ **(8, 7)**
27. $P(-1, -2)$, $M(4, 3)$ **(9, 8)**
28. $P(8, -1)$, $M(4, 6)$ **(0, 13)**
29. $P(9, 2)$, $M(4, 0)$ **(−1, −2)**
30. $P(-3, -2)$, $M(1, 2)$ **(5, 6)**
31. $P(-8, -6)$, $M(-2, -4)$ **(4, −2)**
32. $P(8, -2)$, $M(-2, -1)$ **(−12, 0)**
33. $P(4, 3)$, $M(6, 2)$ **(8, 1)**
34. $P(3, 4)$, $M(4, 8)$ **(5, 12)**
35. $P(\frac{1}{3}, \frac{2}{3})$, $M(-\frac{2}{3}, -\frac{5}{4})$
36. $P(-\frac{7}{5}, \frac{1}{4})$, $M(3, \frac{5}{6})$ **($\frac{37}{5}$, $\frac{17}{12}$)**
37. $P(-5, -\frac{9}{2})$, $M(\frac{4}{5}, -\frac{1}{2})$ **($\frac{33}{5}$, $\frac{7}{2}$)**

35. **($-\frac{5}{3}$, $-\frac{19}{6}$)**

Ⓒ
38. Given triangle ABC with $A(-3, -1)$, $B(3, 5)$, and $C(-5, 13)$, find the length of the median from B. [A median of a triangle is a segment drawn from a vertex to the midpoint of the opposite side.] **5√2**

39. Given triangle ABC with $A(5, -3)$, $B(-1, -1)$, and $C(-2, 6)$, find the length of the median from C. **4√5**

40. In Exercise 38, if K, L, and M are the midpoints of $\overline{AB}$, $\overline{BC}$, and $\overline{AC}$, respectively, find the perimeter of triangle KLM. **12√2**

41. Given $A(-5, 5)$, $B(5, 10)$, $C(3, -1)$, and $D(-7, -6)$, prove that the diagonals of $ABCD$ bisect each other. **See TE Answer Section.**

42. Given quadrilateral $PQRS$ with $P(-1, 2)$, $Q(1, -1)$, $R(5, 1)$, and $S(1, 3)$, and points M, N, T, V, the midpoints of $\overline{PQ}$, $\overline{QR}$, $\overline{RS}$, and $\overline{SP}$, respectively, draw the quadrilateral on a coordinate plane and find the coordinates of M, N, T, and V. Prove that $MN = VT$ and $MV = NT$. **M(0, ½), N(3, 0), T(3, 2), V(0, 5/2), MN = VT = ½√37, MV = NT = 2**

The Midpoint of a Segment

PARALLEL AND PERPENDICULAR LINES 9.7

Objectives To determine whether two lines are parallel, perpendicular, or neither, using their slopes
To write an equation of a line that passes through a point whose coordinates are given and is parallel, or perpendicular, to a line whose equation is given

Recall that **parallel lines** are lines that lie in the same plane but do not intersect. In the figure at the right, $\overleftrightarrow{MN} \parallel \overleftrightarrow{PQ} \parallel \overleftrightarrow{RS}$.

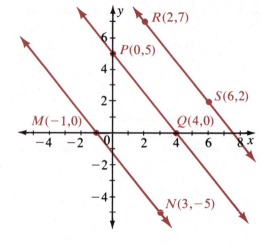

$$m(\overleftrightarrow{MN}) = \frac{-5 - 0}{3 - (-1)} = -\frac{5}{4}$$

$$m(\overleftrightarrow{PQ}) = \frac{0 - 5}{4 - 0} = -\frac{5}{4}$$

$$m(\overleftrightarrow{RS}) = \frac{7 - 2}{2 - 6} = -\frac{5}{4}$$

Notice that each line has the same slope, $-\frac{5}{4}$.

> **Slope of parallel lines**
> In a plane, if two or more different nonvertical lines are parallel, then the lines have the same slope.
> If two or more different nonvertical lines have the same slope, then the lines are parallel.

Notice that the statements above about the slope of parallel lines refer only to *nonvertical* lines. The slope of any vertical line is undefined, but all vertical lines in the same plane are parallel.

You can use slope to find an equation of a line that is parallel to a second line.

Example 1 Write an equation, in standard form, of the line that passes through the point $P(4, 2)$ and is parallel to a line whose equation is $2y - 3x = 6$.

First, find the slope.
$$2y - 3x = 6$$
$$2y = 3x + 6$$
$$m = \frac{3}{2} \blacktriangleright \quad y = \frac{3}{2}x + 3$$

The slope of each line is $\frac{3}{2}$ since the lines are parallel.

Then, use the point-slope form.
$$y - y_1 = m(x - x_1)$$
$$y - 2 = \frac{3}{2}(x - 4) \quad \blacktriangleleft (x_1, y_1) = (4, 2)$$
$$2y - 4 = 3x - 12$$
$$3x - 2y = 8 \quad \blacktriangleleft \text{Standard form}$$

Thus, an equation of the line is $3x - 2y = 8$.

228 Chapter Nine

Recall that two lines are **perpendicular** if they intersect to form a right angle. In the figure, $\overrightarrow{PQ} \perp \overrightarrow{RS}$.

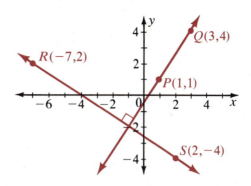

$$m(\overrightarrow{PQ}) = \frac{4-1}{3-1} = \frac{3}{2}$$

$$m(\overrightarrow{RS}) = \frac{-4-2}{2-(-7)} = \frac{-6}{9}, \text{ or } -\frac{2}{3}$$

Notice that the product of their slopes is $\frac{3}{2} \cdot -\frac{2}{3}$, or -1. Hence, their slopes are *negative reciprocals* of each other.

Slope of perpendicular lines	If two lines are perpendicular, then the slopes of the lines are negative reciprocals. If the slopes of two lines are negative reciprocals, then the lines are perpendicular.

The statements above about the slope of perpendicular lines do not refer to horizontal or vertical lines. A vertical line and a horizontal line in the same plane are always perpendicular to each other.

Example 2 Write an equation, in standard form, of the line that passes through the point $Q(4, 2)$ and is perpendicular to a line whose equation is $2y - 3x = 6$.

First, find the slope.

$$2y - 3x = 6$$
$$y = \frac{3}{2}x + 3$$

$m = \frac{3}{2}$ ▶

The slope of the given line is $\frac{3}{2}$ and so the slope of the line perpendicular to it is $-\frac{2}{3}$.

Then, use the point-slope form.

$$y - y_1 = m(x - x_1)$$
$$y - 2 = -\frac{2}{3}(x - 4)$$
$$3y - 6 = -2x + 8$$
$$2x + 3y = 14 \quad \blacktriangleleft \text{ Standard form}$$

Thus, $2x + 3y = 14$ is an equation of the line.

Example 3 Determine whether the lines with equations of $3x = -4y + 7$ and $6x + 8y = 11$ are parallel, perpendicular, or neither.

Rewrite in slope-intercept form. ▶

$$3x = -4y + 7$$
$$4y = -3x + 7$$
$$y = -\frac{3}{4}x + \frac{7}{4}$$
$$m = -\frac{3}{4}$$

$$6x + 8y = 11$$
$$8y = -6x + 11$$
$$y = -\frac{6}{8}x + \frac{11}{8}$$
$$m = -\frac{6}{8}, \text{ or } -\frac{3}{4}$$

Thus, the lines are parallel since their slopes are the same.

Parallel and Perpendicular Lines

Example 4 Write an equation, in standard form, of the line that passes through the point $B(-3, 4)$ and is perpendicular to a line whose equation is $y = 2$.

$y = 2$ is the equation of a horizontal line.
A line perpendicular to this line is vertical. A vertical line has an equation of the form $x = c$, where c is a constant.

Thus, $x = -3$ is an equation of the line.

Written Exercises

Write an equation, in standard form, of the line that passes through the given point and is parallel to the line described by the given equation.

(A)
1. $A(5, 7), x + 4y = 5$
2. $B(6, 3), 5x + 4y = 16$
3. $C(-3, -4), 2x - 5y = 15$
4. $D(3, -4), y = 6$ $y = -4$
5. $E(-3, -8), x = -5$ $x = -3$
6. $F(6, -10), y = x$ $x - y = 16$

1. $x + 4y = 33$
2. $5x + 4y = 42$
3. $2x - 5y = 14$

Determine whether the lines whose equations are given are parallel, perpendicular, or neither.

7. $y = 2x + 3; 2y = 4x + 7$
8. $6x - 4y = 3; 8y - 12x = -7$
9. $x = 5; y = 4$ **perp.**
10. $4x - 5y + 3 = 0; x = 2$
11. $7x + 3y = 7; 8x + 4y = 6$
12. $y = 2; x = 0$ **perp.**

7. par.   10. neither   8. par.   11. neither

Write an equation, in standard form, of the line that passes through the given point and is perpendicular to the line described by the given equation.

13. $G(6, -2), x = 5$ $y = -2$
14. $H(3, -4), y = -4$ $x = 3$
15. $I(6, -10), y = -x$ $x - y = 16$
16. $J(-3, 2), 2x - 5y = 4$
17. $K(-5, -8), 3x + 2y + 7 = 0$
18. $L(8, 0), 3y + 5 = 0$ $x = 8$

$5x + 2y = -11$ $2x - 3y = 14$

Write an equation, in standard form, of the line that is perpendicular to $\overline{AB}$ at point B.

(B)
19. $A(-3, 5), B(6, -2)$
20. $A(-2, -3), B(-1, 4)$
21. $A(6, -1), B(8, 2)$
22. $A(-6, -4), B(10, -3)$
23. $A(8, -7), B(6, -4)$
24. $A(-3, -5), B(0, 0)$

19. $9x - 7y = 68$ $16x + y = 157$ 20. $x + 7y = 27$ $2x - 3y = 24$ 21. $2x + 3y = 22$ $3x + 5y = 0$

Write an equation, in standard form, of the line that is the perpendicular bisector of the segment joining P and Q.

25. $P(-3, -4), Q(6, 5)$ $x + y = 2$
26. $P(7, -3), Q(-8, 2)$
27. $P(-7, -3), Q(4, 8)$ $x + y = 1$
28. $P(-2, 3), Q(4, 7)$
29. $P(1, 5), Q(7, -3)$
30. $P(6, -3), Q(4, 7)$ $x - 5y = -5$

$3x + 2y = 13$ 26. $3x - y = -1$ $3x - 4y = 8$

(C)
31. Given $P(a, 2b), Q(2a, b)$ and $R(a, b)$, write an equation, in standard form, of the line through R that is parallel to $\overrightarrow{PQ}$ if $a \ne 0$.
    $bx + ay = 2ab$
32. Given $P(a, b), Q(3a, 3b), a \ne 0$ and $b \ne 0$, write an equation, in standard form, of the line through P that is perpendicular to $\overrightarrow{PQ}$. $ax + by = a^2 + b^2$
33. Given $P(2a, 2b), Q(2c, 2d), a \ne c$, and $b \ne d$, write an equation, in standard form, of the line that is the perpendicular bisector of $\overline{PQ}$.
    $(a - c)x + (b - d)y = a^2 - c^2 - d^2 + b^2$
34. Given triangle PQR with $P(2a, 2b), Q(2a + 2c, 2b)$, and $R(2a + c, 2d)$, write an equation of the line that contains the median from R. $x = 2a + c$

CUMULATIVE REVIEW

Solve each equation for x.

1. $ax + b = c$ $\dfrac{c - b}{a}$
2. $ax - bx = c$ $\dfrac{c}{a - b}$
3. $\dfrac{1}{a} - \dfrac{1}{b} = \dfrac{1}{x}$ $\dfrac{ab}{b - a}$

USING COORDINATE GEOMETRY 9.8

Objective **To prove that a given property of a geometric figure is true, by using the slope, distance, or midpoint formula**

Coordinate geometry can often be used to prove that a given property of a geometric figure is true. The geometric figure is drawn on a coordinate plane, usually with coordinates assigned to each of the vertices of the figure. The slope, distance, or midpoint formula can then be applied.

Example 1 **Prove that the segments joining the points $A(-3, -4)$, $B(2, 5)$ and $C(-5, 3)$ form a right triangle.**

Draw a diagram and find the slopes.

$$m(\overline{AB}) = \frac{-4 - 5}{-3 - 2} = \frac{9}{5}$$

$$m(\overline{AC}) = \frac{-4 - 3}{-3 - (-5)} = -\frac{7}{2}$$

$$m(\overline{BC}) = \frac{5 - 3}{2 - (-5)} = \frac{2}{7}$$

Since the slopes of $\overline{AC}$ and $\overline{BC}$ are negative reciprocals, $\overline{AC} \perp \overline{BC}$ and angle C is a right angle.

Thus, $\triangle ABC$ is a right triangle.

Points that lie on the same line are said to be *collinear*. You can use the slope formula to prove that certain points are collinear. Recall that the coordinates of any two points of a line can be used to find its slope, and that the slope should be the same regardless of which points are used.

Example 2 **Prove that the points $A(-3, -2)$, $B(0, 0)$, and $C(6, 4)$ are collinear.**

$$m(\overleftrightarrow{AB}) = \frac{0 - (-2)}{0 - (-3)} = \frac{2}{3}$$

$$m(\overleftrightarrow{BC}) = \frac{4 - 0}{6 - 0} = \frac{4}{6}, \text{ or } \frac{2}{3}$$

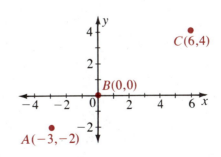

Thus, $m(\overleftrightarrow{AB}) = m(\overleftrightarrow{BC})$ and so the points A, B, and C are collinear.

Using Coordinate Geometry

If you studied geometry, you may have proved that the line connecting the midpoints of two sides of a triangle is parallel to the third side of the triangle. This can be proved algebraically using coordinate geometry.

Example 3

In triangle *ABC*, *D* is the midpoint of $\overline{AB}$ and *E* is the midpoint of $\overline{BC}$. Prove that $\overleftrightarrow{DE} \parallel \overline{AC}$.

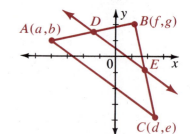

First, find the coordinates of the midpoints.

$$D(x, y) = D\left(\frac{a+f}{2}, \frac{b+g}{2}\right) \qquad E(x, y) = E\left(\frac{f+d}{2}, \frac{g+e}{2}\right)$$

Next, find the slope of $\overleftrightarrow{DE}$ and $\overline{AC}$.

$$m = \frac{y_1 - y_2}{x_1 - y_2}$$

$$m(\overleftrightarrow{DE}) = \frac{\frac{g+e}{2} - \frac{b+g}{2}}{\frac{f+d}{2} - \frac{a+f}{2}} = \frac{\frac{g+e-b-g}{2}}{\frac{f+d-a-f}{2}} = \frac{e-b}{d-a}$$

$$m(\overline{AC}) = \frac{e-b}{d-a}$$

Thus, $m(\overleftrightarrow{DE}) = m(\overline{AC})$, and so $\overleftrightarrow{DE} \parallel \overline{AC}$.

In the next example, you will use the slope formula to prove that a given quadrilateral is a parallelogram. Recall that if both pairs of opposite sides of a quadrilateral are parallel, then the quadrilateral is a parallelogram.

Example 4

The vertices of a quadrilateral are $A(-1, 3)$, $B(0, 0)$, $C(6, 1)$, and $D(5, 4)$. Prove that *ABCD* is a parallelogram.

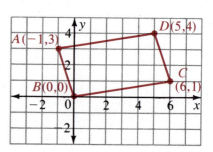

$$m(\overline{AB}) = \frac{0-3}{0-(-1)} = -3; \; m(\overline{DC}) = \frac{1-4}{6-5} = -3$$

$$m(\overline{BC}) = \frac{1-0}{6-0} = \frac{1}{6}; \; m(\overline{AD}) = \frac{4-3}{5-(-1)} = \frac{1}{6}$$

So, $\overline{AB} \parallel \overline{DC}$ and $\overline{BC} \parallel \overline{AD}$.

Thus, *ABCD* is a parallelogram.

Chapter Nine

Written Exercises

For Ex. 1–19, see TE Answer Section.

For each of the following, prove that the segments joining the points A, B, and C form a right triangle.

Ⓐ 1. $A(1, 2)$, $B(2, 5)$, $C(-1, 6)$

2. $A(-1, -2)$, $B(5, 1)$, $C(3, 5)$

3. $A(5, -5)$, $B(7, 3)$, $C(-1, 5)$

4. $A(2, 3)$, $B(4, -1)$, $C(8, 1)$

5. Prove that the points $A(1, 2)$, $B(-1, -8)$, and $C(0, -3)$ are collinear.

6. The vertices of a quadrilateral are $A(0, 0)$, $B(1, 2)$, $C(6, 5)$, and $D(5, 3)$. Prove that $ABCD$ is a parallelogram.

7. The vertices of triangle ABC are $A(2, 1)$, $B(8, 3)$, and $C(4, 5)$. Prove that the segments joining the midpoints of $\overline{AB}$ and $\overline{BC}$ is one-half the length of $\overline{AC}$.

8. The vertices of a triangle are $A(2, 1)$, $B(7, 4)$, and $C(2, 7)$. Prove that triangle ABC is isosceles.

Ⓑ 9. The vertices of a quadrilateral are $A(0, 0)$, $B(a, 0)$, $C(a, b)$, and $D(0, b)$. Prove that $ABCD$ is a rectangle by proving that $\overline{AB} \perp \overline{BC}$, $\overline{AB} \perp \overline{AD}$, $\overline{CD} \perp \overline{AD}$, and $\overline{CD} \perp \overline{BC}$.

10. The vertices of a rectangle are $A(0, 0)$, $B(a, 0)$, $C(a, b)$, and $D(0, b)$. Prove that the segments joining the midpoints of $\overline{AB}$, $\overline{BC}$, $\overline{CD}$, and $\overline{DA}$ form a rhombus.

11. Prove that the midpoints of the sides of trapezoid $ABCD$ determine a parallelogram. The coordinates of the vertices are $A(0, 0)$, $B(a, 0)$, $C(b + d, c)$, and $D(b, c)$.

12. Prove that the diagonals of rhombus $STUV$ are perpendicular to each other. The coordinates of the vertices are $S(0, 0)$, $T(a, 0)$, $U(a + \sqrt{a^2 - b^2}, b)$, and $V(\sqrt{a^2 - b^2}, b)$.

13. Prove that the diagonals of a rectangle bisect each other.

Ⓒ 14. Given $A(0, 0)$, $B(2a, 0)$, and $C(2b, 2c)$, prove that the three medians of triangle ABC intersect at $T\left(\dfrac{2a + 2b}{3}, \dfrac{2c}{3}\right)$. [*Hint:* Use the point-slope form to write an equation of each median and then solve each equation for y.]

15. Prove that the diagonals of a square are perpendicular.

16. Prove that the diagonals of a rectangle are equal in length.

17. Prove that the segments successively joining the midpoints of a quadrilateral form a parallelogram.

18. Prove that the medians of an equilateral triangle are equal in length.

19. Prove that if two lines in the same plane are perpendicular to a given line, then the two lines are parallel.

Using Coordinate Geometry

APPLICATIONS

1. Use the slope-intercept form $y = mx + b$ and the two points given in the figure at the right to show that $C = \frac{5}{9}(F - 32)$, the relationship between Fahrenheit and Celsius. Let $y = C$, $m = \frac{5}{9}$, $x = F$, $b = -\frac{160}{9}$.

2. Show $F = \frac{9}{5}C + 32$.

3. At what temperature do the Celsius and Fahrenheit scales register the same? Show your derivation. **−40°**

4. Use $y = mx + b$ and the two points given in the figure at the right to show that $C = k - 273.15°$, the relationship between Kelvin and Celsius. Let $y = C$, $m = 1$, $x = k$, $b = -273.15$.

5. Thermometers sell for $3.50 each. Let n represent the number of thermometers and C the cost of n thermometers. Write an equation representing C in terms of n. Draw the graph showing this relationship. **$C = 3.50 n$**

6. A car traveled at an average speed of 70 km/h. How far did it travel in 3 h? in 5 h? Draw the graph showing this relationship. **210 km, 350 km**

7. Use the graph at the right and write an equation describing the relationship among distance, rate, and time. **$d = 50t$** How many units would you travel in 3 h? in 10 h? **150 units, 500 units**

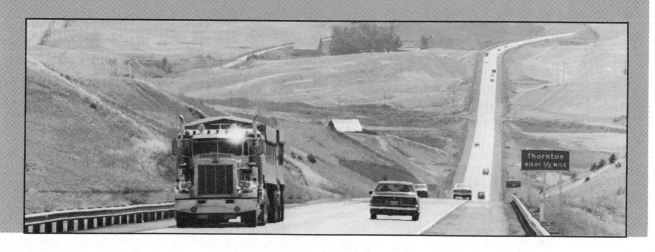

234 Applications

COMPUTER ACTIVITIES

Linear Equations

OBJECTIVE: To determine the equation of a line in which the only INPUTs are two points on the line

To write an equation of a line through two points is not very difficult. Using the coordinates of the two points, you determine the slope and write the equation. The following three-line program INPUTs two points and outputs the resulting equation.

```
100    INPUT A,B
102    INPUT C,D
103    PRINT B - D"X + "C - A"Y = "B * (C - A) - A *
       (D - B)
```

The above program appears to work well enough. However, what if the line is either vertical or horizontal? Then the equation is not totally valid. In fact, when the line is vertical, the computer will output an error message. Therefore, the program must include two additional tests. One tests for lines parallel to the x-axis, and the other tests for lines parallel to the y-axis.

In addition to the two tests, it is a good idea to include a PRINT statement that identifies the output as an equation through two particular points. Add the following lines to the program above, and the new program will do it all.

```
110    PRINT "THE EQUATION THROUGH ("A","B") AND (";
120    PRINT C","D") IS "
130    IF A = C THEN 160
140    IF B = D THEN 180
150    GOTO 100
160    PRINT "X = "A
170    GOTO 100
180    PRINT "Y = "B
190    INPUT "MORE POINTS? (YES OR NO);K$
200    IF K$ = "YES" THEN 100
210    END
```

See the Computer Section beginning on page 574 for more information.

EXERCISES   For Ex. 1–3, see TE Answer Section.

1. Write a program that calculates the slope of a line through two given points.

2. Write a program that calculates the length and determines the midpoint of a segment, given the coordinates of its endpoints.

3. Write a program that finds the missing endpoints of a segment, from the INPUT of the midpoint and the known endpoint.

CHAPTER NINE REVIEW

Ⓐ Exercises are 1–11, 14–17, 21–25, 27, 29, 30, 33, 36–39. Ⓒ Exercises are starred.
The rest are Ⓑ Exercises.

Vocabulary
abscissa [9.1]
coordinate plane [9.1]
coordinates [9.1]
linear equation [9.1]
  point-slope form of [9.3]
  slope-intercept form of [9.3]
ordered pair [9.1]
ordinate [9.1]
origin [9.1]
quadrant [9.1]
slope [9.2]
y-intercept [9.3]

For Ex. 1–6 and 12–13, see TE Answer Section.

On the same coordinate plane, graph each ordered pair. [9.1]

1. $(-2, -3)$ 2. $(4, -6)$ 3. $(-7, 0)$
4. $(0, -3)$ 5. $(5, -5)$ 6. $(0, 2)$

Give the coordinates of each point. [9.1]

7. A **(−2, −3)**
8. B **(4, −1)**
9. C **(1, 3)**
10. D **(−3, 2)**
11. E **(1, 0)**

Draw the graph of each equation.
12. $2x + 4y = 12$ [9.1]
13. $-(x - y) - 3(x - y) = 6$ [9.4]

Find the length of $\overline{AB}$. [9.2]
14. $A(2, -3), B(-6, -3)$ **8**
15. $A(-4, 1), B(-4, 5)$ **4**

Find the slope of $\overrightarrow{PQ}$. [9.2]
16. $P(6, 4), Q(-2, 3)$ **$\frac{1}{8}$**
17. $P(-2, -2), Q(-3, -4)$ **2**
18. $P(\frac{1}{2}, \frac{1}{4}), Q(-\frac{3}{4}, \frac{2}{3})$ **$-\frac{1}{3}$**
★ 19. $P(2x, -3y), Q(-3x, 2y)$ **$-\frac{y}{x}$**

Find the missing coordinate for the given data. [9.2]
★ 20. $A(-3, y), B(6, -8), m(\overline{AB}) = -2$ **10**

A line has the given slope, y-intercept, or contains the indicated point(s). Write an equation, in standard form, of each line. [9.3]
21. $P(2, 5), m = \frac{2}{3}$ 22. $m = 3, b = 7$
23. $A(-3, 4), B(-1, 5)$ **x − 2y = −11**
21. **2x − 3y = −11**   22. **3x − y = −7**

Find the slope and the y-intercept of the line described by each given equation. [9.3]
24. $y = -\frac{2}{3}x + 4$ 25. $y - 2(3 - 2x) = 7$
★ 26. $6 - (3 - 2y) - 2[-3(2x - 1)] = 5$ **−6, 4**
24. **$-\frac{2}{3}$, 4**   25. **−4, 13**

Is the given point on the line described by the given equation? [9.4]
27. $P(-1, -3), 2x - 3y = 7$ **yes**
28. $Q(-4, 2), \frac{3}{5}x + \frac{2}{3}y = 1$ **no**

Find the distance between the given points. Answer in simplest radical form. [9.5]
29. $P(-1, 2), Q(7, 4)$ **$2\sqrt{17}$**
30. $M(3, 0), N(10, 8)$ **$\sqrt{113}$**
31. $P(\frac{1}{2}, -\frac{1}{3}), Q(\frac{2}{3}, \frac{5}{6})$ **$\frac{5}{6}\sqrt{2}$**

The lengths of the sides of a triangle are given. Is the triangle a right triangle?
32. $8, 12, 4\sqrt{13}$ [9.5] **yes**

Determine the coordinates (x, y) of M, the midpoint of $\overline{PQ}$. [9.6]
33. $P(4, -3), Q(7, 4)$ **$(\frac{11}{2}, \frac{1}{2})$**
34. $P(-\frac{1}{4}, \frac{1}{2}), Q(\frac{2}{5}, -\frac{1}{3})$ **$(\frac{3}{40}, \frac{1}{12})$**

Determine the coordinates of Q, the other endpoint of $\overline{PQ}$, given P and the midpoint, M. [9.6]
35. $P(-3, 6), M(1, 2)$ **(5, −2)**

Write an equation, in standard form, of the line that passes through the given point and is parallel to the given line. [9.7]
36. $C(1, 2), 3x - 5y = 15$ **3x − 5y = −7**
37. $D(4, -1), -2x - 3y = 10$ **2x + 3y = 5**

Write an equation, in standard form, of the line that passes through the given point and is perpendicular to the given line. [9.7]
38. $A(-1, -3), 2x + 5y = 7$ **5x − 2y = 1**
39. $B(2, -7), 3y + 6x = 8$ **x − 2y = 16**

40. The vertices of a quadrilateral are $A(-5, 3), B(4, 0), C(6, 6),$ and $D(-3, 9)$. Prove that $ABCD$ is a parallelogram. [9.8]
See TE Answer Section.

CHAPTER NINE TEST

Ⓐ Exercises are 1–5, 7, 8, 10–14, 16, 18, 20, 21, 22. Ⓒ Exercises are starred. The rest are Ⓑ Exercises.

Give the coordinates of each point.

1. A **(1, 0)**
2. B **(3, 1)**
3. C **(−3, −1)**
4. D **(−4, 3)**

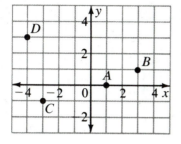

Draw the graph of each equation.
5. $y = -2x + 5$
6. $6x - 2(-3 + 2x) - (3 - y) = 6$

For Ex. 5–6, see TE Answer Section.

Find the length of $\overline{AB}$.
7. $A(3, 4), B(6, -3)$ **√58**

Find the slope of $\overrightarrow{PQ}$.
8. $P(3, -6), Q(5, -2)$ **2**
9. $P(-\frac{1}{2}, \frac{3}{4}), Q(\frac{5}{6}, -\frac{1}{2})$ **$-\frac{15}{16}$**

10. Write an equation, in standard form, of the line that passes through the point $P(-1, 2)$ and whose slope is $\frac{1}{2}$. **$x - 2y = -5$**

11. Write an equation, in standard form, of the line that passes through the points $A(-3, 2)$ and $B(-4, 4)$. **$2x + y = -4$**

Find the slope and the y-intercept of the line described by each given equation.
12. $2x + 4y = 12$ **$-\frac{1}{2}, 3$**
13. $y - (2x - 5) = -8$ **2, −13**

Is the given point on the line described by the given equation?
14. $R(2, -3), 3x - 2y = -12$ **no**
15. $S(-1, 0), \frac{4}{5}x - \frac{1}{3}y = -\frac{4}{5}$ **yes**

16. Find the distance between points $P(-1, -7)$ and $Q(7, -3)$. Give the answer in simplest radical form. **4√5**

17. The lengths of the sides of a triangle are $4\sqrt{3}, 8\sqrt{3},$ and 12. Is the triangle a right triangle? **yes**

18. Determine the coordinates of M, the midpoint of $\overline{PQ}$, given $P(4, -3)$ and $Q(6, -7)$. **(5, −5)**

19. Determine the coordinates of Q, the other endpoint of $\overline{PQ}$, given $P(-2, -3)$ and its midpoint $M(3, -1)$. **(8, 1)**

20. Write an equation, in standard form, of the line the passes through the point $A(-1, -2)$ and is parallel to a line whose equation is $3x + 5y = 15$. **$3x + 5y = -13$**

21. Write an equation, in standard form, of the line that passes through the point $N(-4, 1)$ and is perpendicular to a line whose equation is $5x - 2y = 10$. **$2x + 5y = -3$**

22. The vertices of a quadrilateral are $A(3, 0)$, $B(9, 4), C(6, 9),$ and $D(0, 5)$. Prove that $ABCD$ is a parallelogram.
See TE Answer Section.

★ 23. Find the missing coordinate for the given data: $P(x, -3), Q(9, 6), m(\overline{PQ}) = 3$. **6**

★ 24. Prove that the midpoints of the sides of an isosceles triangle determine an isosceles triangle. **See TE Answer Section.**

Chapter Nine Test

COLLEGE PREP TEST

DIRECTIONS: Choose the *one* best answer to each question or problem.

1. In the figure, the coordinates of M are $(6, 0)$ and the area of triangle MON is 24. What are the coordinates of N? **C**

   (A) $(4, 0)$ (B) $(0, 4)$ (C) $(0, 8)$
   (D) $(8, 0)$ (E) None of these

2. In the figure, triangles PQR and PMN are isosceles right triangles. If the areas of the triangles are the same, what are the coordinates of point M? **B**

   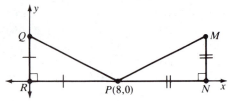

   (A) $(8, 16)$ (B) $(16, 8)$ (C) $(8, 8)$
   (D) $(16, 4)$ (E) $(8, 4)$

3. The distance between $R(-5, 0)$ and S is 4. The coordinates of point S could be any of the following except **D**

   (A) $(-5, 4)$ (B) $(-5, -4)$ (C) $(-9, 0)$
   (D) $(0, -1)$ (E) $(-1, 0)$

4. The point $(a, -b)$ is on the line described by which linear equation in x and y? **D**

   (A) $x + y = a + b$ (B) $x - y = a - b$
   (C) $2x - b = 2a - y$ (D) $x - \dfrac{y + b}{2} = a$
   (E) None of these

5. Which point is the farthest away from the origin? **E**

   (A) $P(7, -5)$ (B) $Q(5, -7)$
   (C) $R(8, -3)$ (D) $S(2, 8)$
   (E) $T(0, -9)$

6. Find the area of right triangle PQR, with right angle R, if the equation of $\overleftrightarrow{PQ}$ is $y = 3x + 9$ and the coordinates of R are $(7, 0)$. **C**

   (A) 70   (B) 100   (C) 150
   (D) 300 (E) None of these

7. Choose the table of ordered pairs (x, y) for which $5ay - 10bx = 15ab$. **B**

   (A) | x | y |
       |---|---|
       | $3b$ | 0 |
       | b | $-a$ |
       | $-b$ | $-2a$ |

   (B) | x | y |
       |---|---|
       | a | $5b$ |
       | $-3a$ | $-3b$ |
       | $-a$ | b |

   (C) | x | y |
       |---|---|
       | 0 | $3b$ |
       | a | b |
       | $-a$ | $-b$ |

   (D) | x | y |
       |---|---|
       | $-2a$ | $-b$ |
       | $-a$ | b |
       | a | $-2b$ |

   (E) None of these

8. Given $P(a, b)$, $Q(3a, 3b)$, $R(-c, -d)$, and $S(c, d)$, where a, b, c, and d are nonzero real numbers, which statement is true? **A**

   (A) $PQ = RS$ if $a = d$ and $b = c$.
   (B) $\overline{PQ} \parallel \overline{RS}$ if $a = b = c$.
   (C) $\overline{PQ} \perp \overline{RS}$ if $ac = bd$.
   (D) $\overline{PQ}$ and $\overline{RS}$ have the same midpoint if $a = b$.
   (E) None of these

10 LINEAR SYSTEMS, MATRICES, AND DETERMINANTS

Non-Routine Problem Solving

Problem 1. A two-digit number is three times the product of the two digits. Find all such numbers. **15 and 24**

Problem 2. Find the number of integers between 1 and 10,000 that can be expressed in the form $13y^2$, where y is a positive integer. **27**

Problem 3. There are only 4 four-digit numbers that are perfect squares with all even digits. Find as many of these 4 numbers as you can. **4,624; 6,084; 6,400; 8,464**

SOLVING LINEAR SYSTEMS GRAPHICALLY   10.1

Objectives **To solve a system of two linear equations in two variables graphically**
To determine whether a system of two linear equations is consistent, inconsistent, dependent, or independent

A pair of linear equations in two variables is called a **system of equations.** A solution of a system of two linear equations is an ordered pair of real numbers that makes both equations true. You can solve a system of linear equations by graphing each equation. You should draw the graph of each equation in the same coordinate plane and then locate the point of intersection of the two graphs, if any. The ordered pair associated with this point is a solution of the system.

Example 1 Solve the system $2x - y = 5$ graphically.
$-x + y = -3$

Write each equation in slope-intercept form.

$$2x - y = 5 \qquad\qquad -x + y = -3$$
$$y = 2x - 5 \qquad\qquad y = x - 3$$

Graph each equation in the same coordinate plane.

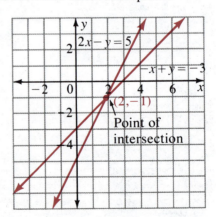

Check that the ordered pair $(2, -1)$ is a solution of both equations.

Thus, $(2, -1)$ is the solution of the system.

In Example 1, the two lines have exactly one point in common, and so the two equations have one common solution. When a system of linear equations has exactly one solution, the system is called an **independent system.** When a system has at least one solution, the system is called a **consistent system.**

Example 2

Solve the system $5x - 3(y + x) = -6$ graphically.
$$2x - 3y = 3$$

$$5x - 3(y + x) = -6$$
$$5x - 3y - 3x = -6$$
$$-3y = -2x - 6$$
$$y = \frac{2}{3}x + 2$$

$$2x - 3y = 3$$
$$-3y = -2x + 3$$
$$y = \frac{2}{3}x - 1$$

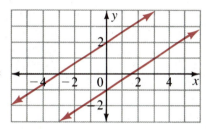

The slopes of the two lines are the same but their y-intercepts are different. The lines are parallel. There is no point of intersection of the graphs.

Thus, there is no solution of the system.

A system of linear equations which has no solution is called an **inconsistent system**.

Example 3

Solve the system $3x + y = 1$ graphically.
$$2y = 2 - 6x$$

$$3x + y = 1$$
$$y = -3x + 1$$

$$2y = 2 - 6x$$
$$y = -3x + 1$$

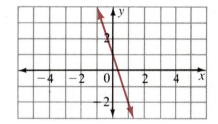

The slopes of the two lines and their y-intercepts are the same. The lines coincide. The two graphs have all of their points in common.

Thus, all (x, y) such that $y = -3x + 1$ are solutions of the system.

In Example 3, the two lines have all of their points in common and so each solution of one equation is also a solution of the other. A system of linear equations that has all solutions in common is called a **dependent system**. A dependent system has infinitely many solutions.

Reading in Algebra

True or false? Give a reason for your answer.
1. A system of two linear equations has at least one solution. **F**
2. A system of two linear equations has at most one solution. **F**
3. An inconsistent system of equations is one that has more than one solution. **F**
4. A dependent system of equations is also a consistent system. **T**

Solving Linear Systems Graphically

Written Exercises

(Note: con = consistent, inc = inconsistent, dep = dependent, ind = independent)

Solve each system of equations graphically. Indicate whether the system is consistent, inconsistent, dependent, or independent.

Ⓐ
1. $2x + y = 3$ **(2, −1);**
   $y - x = -3$ **con, ind**

2. $3x + 2y = 8$
   $3x = 10 - 2y$
   no solution; inc

3. $4x - y + 3 = 0$
   $x - y = 0$
   (−1, −1); con, ind

4. $y = 2 - 3x$
   $x = \frac{2}{3} - \frac{y}{3}$
   all $(x, y) | y = 2 - 3x$; con, dep

5. $\frac{x}{2} = \frac{y}{3}$
   $y - x = 1$
   (2, 3); con, ind

6. $\frac{x + 2y}{2} = x$
   $2y = x$
   all $(x, y) | y = \frac{1}{2}x$; con, dep

7. $3y + 4 = 5x$
   $10x - 3 = 6y$
   no solution; inc

8. $-4x + 2y - 3 = 0$
   $6y - 9 = 12x$
   all $(x, y) | y = 2x + \frac{3}{2}$; con, dep

9. $2x + y = 3$
   $2y - 3x = 6$
   (0, 3); con, ind

Ⓑ
10. $2(x + y) = x + 2$
    $2y + 3(x - 2y + 8) = 0$
    (−4, 3); con, ind

11. $x + 2(y + 5) = 3(x + y + 1)$
    $0 = 5y - 3(x - 10)$
    (5, −3); con, ind

12. $-3(x - y) = 2(y - x) + 8$
    $-(2x - y) = 3(y + 3x - 1)$
    (−1, 7); con, ind

13. $-x - 3y = 2(x - 3y) + 4$
    $y - x - 2(x - y) = -3$
    no solution; inc

14. $0.2x - 0.3y = -0.8$
    $0.3x + 0.5y = 0.7$
    (−1, 2); con, ind

15. $0.3x - 0.2y = 0.7$
    $0.4x - 0.1y = 0.6$
    (1, −2); con, ind

Write a system of two linear equations in two variables that has the given number of solutions. (Answers may vary.)

Ⓒ
16. exactly one solution
    any two equations with $m_1 \neq m_2$

17. no solution
    any two equations with $m_1 = m_2$, $b_1 \neq b_2$

18. infinitely many solutions
    any two equations with $m_1 = m_2$, $b_1 = b_2$

Find the value(s) of k, if any, that satisfies the given condition for each system.

19. $\left.\begin{array}{l} y = kx + 8 \\ y = 3x - 1 \end{array}\right\}$ is inconsistent **3**

20. $\left.\begin{array}{l} kx - 3y = 9 \\ 2x + 3y = 12 \end{array}\right\}$ is consistent
    all values except $k = -2$

21. $\left.\begin{array}{l} 4x - 2y = 7 \\ 12x - ky = 21 \end{array}\right\}$ has infinitely many solutions **6**

22. $\left.\begin{array}{l} 2x + ky = 6 \\ 3x - 4y = -8 \end{array}\right\}$ has exactly one solution
    all values except $k = -\frac{8}{3}$

CALCULATOR ACTIVITIES

To check to see if (1.3, 2.5) is a solution of the system
$12.6x + 15.7y = 55.63$,
$-5.3x + 11.8y = 22.61$
substitute 1.3 for x and 2.5 for y in each equation.

$12.6 \otimes 1.3$ $15.7 \otimes 2.5$ $-5.3 \otimes 1.3$ $11.8 \otimes 2.5$
$=$ $=$ $=$ $=$
16.38 $+$ 39.25 -6.89 $+$ 29.50
       55.63 22.61

Check to determine if the given ordered pair is a solution of the system of equations.

1. $4.6x + 3.5y = 29.23$
   $5.1x - 1.7y = 17.34$, (4.3, 2.7) **yes**

2. $15.9x - 13.3y = 21.44$
   $-18.1x + 21.7y = 13.64$, (6.2, 5.8) **yes**

SOLVING LINEAR SYSTEMS ALGEBRAICALLY   10.2

Objectives **To solve a system of two linear equations by the substitution method**
To solve a system of two linear equations by the addition method

A system of two linear equations can be solved algebraically. In this lesson, you will use two algebraic methods, *substitution* and *addition*.

You can solve a system of two linear equations in two variables by substitution using the following procedure:
(1) Solve one of the equations for either variable.
(2) Substitute this value in the other equation.
(3) Solve for the other variable.
(4) Substitute in either original equation to find the value of the second variable.
(5) Check the ordered pair in both original equations.

Example 1 **Solve the system $2x - y = 5$ by substitution.**
$$-x + y = -3$$

Solve one equation for y. ▶
$$2x - y = 5$$
$$y = 2x - 5$$

Substitute in the other equation. ▶
$$-x + y = -3$$
$$-x + (2x - 5) = -3$$
$$-x + 2x - 5 = -3$$

Solve for x. ▶
$$x = 2$$

Substitute in one of the original equations. ▶
Solve for y. ▶
$$2x - y = 5$$
$$2(2) - y = 5$$
$$y = -1$$

Check in both original equations. ▶

$2x - y$	5
$2(2) - (-1)$	5
$4 + 1$	
5	

$-x + y$	-3
$-(2) + (-1)$	-3
$-2 - 1$	
-3	

Thus, $(2, -1)$ is the solution of the system.

Notice that Example 1 above is an algebraic solution to Example 1 of the preceding lesson.

A system of two linear equations can also be solved algebraically by addition, as shown on page 244.

Solving Linear Systems Algebraically

Example 2

Solve the system $2x + 3y = 9$ by addition.
$-2x - 5y = 1$

Add the two equations and solve for y.
$$2x + 3y = 9$$
$$-2x - 5y = 1$$
$$-2y = 10$$
$$y = -5$$

Substitute -5 for y in either equation. Solve for x.
$$2x + 3y = 9$$
$$2x + 3(-5) = 9$$
$$2x - 15 = 9$$
$$2x = 24$$
$$x = 12$$

Check in both original equations. Substitute 12 for x and -5 for y.

$2x + 3y$	9	$-2x - 5y$	1
$2(12) + 3(-5)$	9	$-2(12) - 5(-5)$	1
$24 - 15$		$-24 + 25$	
9		1	

Thus, $(12, -5)$ is the solution of the system.

In Example 2, notice that the coefficients of x in the original equations are additive inverses. Thus, when the equations are added the x-variable is eliminated.

Sometimes it is necessary to multiply one or both of the linear equations by some number or numbers so that the coefficients of one variable are additive inverses. This is shown in Examples 3 and 4 which follow.

Example 3

Solve the system $5x - 2y = 20$ by addition.
$7x + 4y = 11$

Multiply the first equation by 2. ▶
$$5x - 2y = 20$$
$$2(5x - 2y) = 2(20)$$

Add the equations. ▶
$$10x - 4y = 40$$
$$7x + 4y = 11$$
$$17x = 51$$
$$x = 3$$

Substitute 3 for x in one of the original equations. ▶
$$5x - 2y = 20$$
$$5(3) - 2y = 20$$
$$-2y = 5$$

Solve for y. ▶
$$y = -\frac{5}{2}$$

Check $(3, -\frac{5}{2})$ in both original equations.

Thus, $(3, -\frac{5}{2})$ is the solution of the system.

Example 4

Solve the system $3x + 2y = 5$ by addition.
$4x - 5y = 22$

First Way

Eliminate x. Use $12x$ and $-12x$.

$$4(3x + 2y) = 4(5) \qquad 12x + 8y = 20$$
$$-3(4x - 5y) = -3(22) \qquad \underline{-12x + 15y = -66}$$
$$23y = -46$$
$$y = -2$$

Solve for x. $3x + 2y = 5$
$3x + 2(-2) = 5$
$3x = 9$
$x = 3$

Second Way

Eliminate y. Use $10y$ and $-10y$.

$$5(3x + 2y) = 5(5) \qquad 15x + 10y = 25$$
$$2(4x - 5y) = 2(22) \qquad \underline{8x - 10y = 44}$$
$$23x = 69$$
$$x = 3$$

Solve for y. $3x + 2y = 5$
$3(3) + 2y = 5$
$2y = -4$
$y = -2$

Check in both original equations. Substitute 3 for x and -2 for y.

$3x + 2y$	5
$3(3) + 2(-2)$	5
$9 - 4$	
5	

$4x - 5y$	22
$4(3) - 5(-2)$	22
$12 + 10$	
22	

Thus, $(3, -2)$ is the solution of the system.

When you solve a system of two linear equations either by the method of substitution or addition, sometimes both variables are eliminated. If the resulting equation is a false statement, the system is inconsistent and has no solution. If the resulting equation is a true statement, the system is dependent and has infinitely many solutions.

Example 5

Solve each system by addition.

$2x - 4y = 5$
$-x + 2y = 8$

$-3x + 2y = 4$
$9x - 6y = -12$

Multiply the second equation by 2. ▶

$-x + 2y = 8$
$2(-x + 2y) = 2(8)$
$-2x + 4y = 16$

$-3x + 2y = 4$
$3(-3x + 2y) = 3(4)$
$-9x + 6y = 12$

◀ *Multiply the first equation by 3.*

Add the equations. ▶

$2x - 4y = 5$
$\underline{-2x + 4y = 16}$
$0 + 0 = 21$
$0 = 21$ False

$-9x + 6y = 12$
$\underline{9x - 6y = -12}$
$0 + 0 = 0$
$0 = 0$ True

◀ *Add the equations.*

Thus, there is no solution.

Thus, all (x, y) such that $-3x + 2y = 4$ are the solutions.

Solving Linear Systems Algebraically

Example 6

READ ▶ A riverboat travels downstream 16 km in 2 h and returns the same distance upstream in 3 h. Find the rate of the boat in still water and the rate of the current.

PLAN ▶ Let x = rate of boat in still water and y = rate of current

	rate(r)	time(t)	distance(d)
Downstream	$x + y$	2	$2(x + y)$
Upstream	$x - y$	3	$3(x - y)$

◀ Use $d = rt$.

$$\begin{array}{l} 2(x + y) = 16 \\ 3(x - y) = 16 \end{array} \Rightarrow \begin{array}{l} 2x + 2y = 16 \\ 3x - 3y = 16 \end{array} \Rightarrow \begin{array}{l} x + y = 8 \\ 3x - 3y = 16 \end{array}$$

SOLVE ▶ Multiply the first equation by 3. Then add the equations.

$$\begin{array}{r} 3x + 3y = 24 \\ 3x - 3y = 16 \\ \hline 6x = 40 \\ x = 6\tfrac{2}{3} \end{array} \qquad \begin{array}{l} 2(x + y) = 16 \\ 2(6\tfrac{2}{3} + y) = 16 \\ 6\tfrac{2}{3} + y = 8 \\ y = 1\tfrac{1}{3} \end{array}$$

INTERPRET ▶ Check to see that distance downstream, $2(x + y)$, is equal to distance upstream, $3(x - y)$.

Thus, the rate of the boat in still water is $6\tfrac{2}{3}$ km/h and the rate of the current is $1\tfrac{1}{3}$ km/h.

Written Exercises

Solve each system of equations by substitution.

Ⓐ 1. $4x - 3y = 2$
$\ 2x + y = -4$
$\ (-1, -2)$

2. $5x + 6y = 14$
$\ 4x - y = 17$
$\ (4, -1)$

3. $2a - b + 1 = 0$
$\ 3b - 5a + 1 = 0$
$\ (-4, -7)$

4. $2r - 3s = 1.3$
$\ s - r = -0.5$
$\ (0.2, -0.3)$

5. $2x - 4y = -6$
$\ -x + 2y = 3$
all (x, y) such that $x = 2y - 3$

6. $2y - z = 8$
$\ -4y + 6z = -16$
$\ (4, 0)$

7. $2 = x + 4y$
$\ 8y = 7 + x$
$\ (-1, \tfrac{3}{4})$

8. $5x - 2y = 0$
$\ 2x + y = 3$
$\ (\tfrac{2}{3}, \tfrac{5}{3})$

Solve each system of equations by addition.

9. $2x + 2y = 3$
$\ -5x + 4y = 15$
$\ (-1, \tfrac{5}{2})$

10. $3x - 2y = 10$
$\ 5x + 3y = -15$
$\ (0, -5)$

11. $7c + 9d - 3 = 0$
$\ 9c - 7d - 2 = 0$
$\ (\tfrac{3}{10}, \tfrac{1}{10})$

12. $2y + 3x - 8.1 = 0$
$\ 2x - 3y + 3.7 = 0$
$\ (1.3, 2.1)$

13. $4t - 7u = -13$
$\ -3u - 5 = -7t$
$\ (2, 3)$

14. $5x - 7y + 16 = 0$
$\ x + 4y - 13 = 0$
$\ (1, 3)$

15. $9m = 21 - 7n$
$\ 12n = 36 - m$
$\ (0, 3)$

16. $3x - 5y + 10 = 0$
$\ -9x + 15y = -30$
no solution

Solve each problem by using a system of two linear equations in two variables.

Ⓑ 17. An airplane flew 112 km in 21 min with a tailwind and returned the same distance in 24 min against the wind. What was the rate of the tailwind? **20 km/h**

18. A first number is 10 less than 4 times a second number. The first number, decreased by the second number, is -19. Find the numbers. **$-22, -3$**

19. Alicia is 4 years older than Bill. Two years ago, she was $1\tfrac{1}{2}$ times as old as he was. Find their present ages. **Alicia: 14, Bill: 10**

20. How many milliliters of a 25% iodine solution must be added to 90 mL of a 65% iodine solution to obtain a 55% iodine solution? **30 mL**

AIR-SPACE EXPLORATION

Air and Space explorations, which extend our view from the earth's crust outward, require complex and accurate equipment to correlate and classify all the information gathered.

Deep Space Network Control Center. This is the control and data center for all deep-space probes, including Voyager and Pioneer. On the large screen is a computer-enhanced picture of Saturn.

Alexander Graham Bell demonstrating the telephone at Salem, MA, 1877. Teaching speech to the hearing-impaired provided the inspiration for Alexander Graham Bell's invention, which transmits speech by electric waves.

Edison's loud-speaking telephone. Thomas Edison improved Bell's telephone by adding the carbon transmitter, which markedly increased its audibility.

Canada's domestic communications satellite. Fixed-orbit satellites offer the possibility of reliable communications without the need for cumbersome equipment.

Replica of Marconi's first transmitter, Italy, 1895. Guglielmo Marconi established the first wireless communication, and by 1901 trans-Atlantic radio communications were in operation.

Parabolic dish antenna in stand. A parabolic antenna, by virtue of its shape, concentrates minute amounts of energy to provide data from farther out in space than do other types of telescopes.

Microwave tower. Microwave transmissions, on a very tight beam capable of accurate focusing on a distant receiver, are simplifying earthbound communications systems.

Replica of Isaac Newton's reflecting telescope, 1671. Using mirrors rather than prisms, Newton's reflecting telescope eliminated the spectrum of colors that obscured images.

Astronomers in the Istanbul Observatory, late sixteenth century. Adherence to Ptolemy's geocentric view of the universe impeded much advancement in the science of astronomy, even when observed phenomena did not fit in with what were believed to be facts.

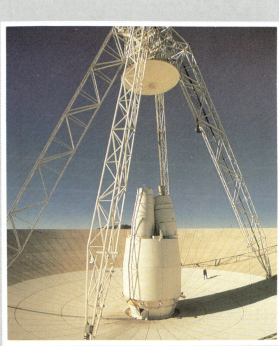

Radio telescope. Radio telescopes are used to pick up signals from and measure distances between other bodies within and beyond our solar system.

Ancient Chinese Observatory. Ancient Chinese astronomers were hindered in their exploration of the mechanics of solar, lunar, and planetary movement because they had not yet acquired a system of geometry.

Eighteenth-century celestial observatory, Jaipur, India. After Galileo's defense of the Copernican theory was verified, astronomical research proceeded with increasingly accurate instruments to plumb the depths of space.

Kitt's Peak, AZ. The National Observatory at Kitt's Peak houses the McMath Solar Tower telescope, the largest of its type; a 158-inch reflecting telescope; and various other telescopes.

Airplane Propeller. With its central hub and radiating blades forming a helical surface, the propeller produces thrust and forward motion when it is rotated.

Guppy. An airplane designed to haul large, bulky objects such as satellite equipment and noncollapsible antennae.

The Pilatre-D'Arlandes hot-air flight, Passy, France, 1783. Pilatre and D'Arlandes were the first humans to experience free flight in a hot air balloon, fabricated from paper and linen by Jacques and Joseph Montgolfier.

ang glider. A source of [in]spiration to the Wright brothers, [O]tto Lilienthal was one of the [ea]rliest inventors of the glider; he [m]ade over 2,000 flights.

Blimp. By 1906, Fernand von Zeppelin had built a rigid airship capable of a speed of 30 miles per hour.

Solar Maximum Mission spacecraft, solar powered. Research into the collection of solar energy by satellite may soon make available tight beam transmissions to earth that can be converted to electrical energy.

Technicolor tornado. The atmospheric variations inherent in a tornado suggest the vast areas yet to be researched within our atmosphere as well as the artistic colorations that are possible in computer-enhanced photography.

SOLVING SYSTEMS OF THREE LINEAR EQUATIONS 10.3

Objective **To solve a system of three linear equations**

You have learned that a system of two linear equations may have no solution, infinitely many solutions, or exactly one solution. A solution of an independent system of two linear equations is always an ordered pair of real numbers. In this lesson, you will learn to solve a system containing three linear equations. The equations in the system contain three variables, and so a solution of an independent system is an **ordered triple** of real numbers.

Example 1 Solve the system: $x + 2y - z = 1$
$2x + y + 3z = 5$
$3x + y + 2z = 8.$

Choose *any two* equations and eliminate one of the variables.

$\begin{aligned} x + 2y - z &= 1 \\ 2x + y + 3z &= 5 \end{aligned} \Rightarrow \begin{aligned} 3x + 6y - 3z &= 3 \\ 2x + y + 3z &= 5 \\ \hline 5x + 7y &= 8 \end{aligned}$

Next, choose a *different* pair of equations and eliminate the same variable.

$\begin{aligned} x + 2y - z &= 1 \\ 3x + y + 2z &= 8 \end{aligned} \Rightarrow \begin{aligned} 2x + 4y - 2z &= 2 \\ 3x + y + 2z &= 8 \\ \hline 5x + 5y &= 10 \end{aligned}$

A system of 2 linear equations in x and y remains. Solve the system.

$\begin{aligned} 5x + 7y &= 8 \\ 5x + 5y &= 10 \end{aligned} \Rightarrow \begin{aligned} 5x + 7y &= 8 \\ -5x - 5y &= -10 \\ \hline 2y &= -2 \\ \boxed{y = -1} \end{aligned}$

Substitute for y in either equation to find x.

$5x + 5y = 10$
$5x + 5(-1) = 10$
$5x = 15$
$\boxed{x = 3}$

Substitute for x and y in one of the original equations to find z.

$x + 2y - z = 1$
$3 + 2(-1) - z = 1$
$1 - z = 1$
$-z = 0$
$\boxed{z = 0}$

Check by substituting 3 for x, -1 for y, and 0 for z in the original equations.

$x + 2y - z$	1	$2x + y + 3z$	5	$3x + y + 2z$	8
$3 + 2(-1) - 0$	1	$2(3) + (-1) + 3(0)$	5	$3(3) + (-1) + 2(0)$	8
$3 - 2$		$6 - 1$		$9 - 1$	
1		5		8	

Thus, $(3, -1, 0)$ is the solution of the system.

Solving Systems of Three Linear Equations 247

Example 2

Solve the system: $x + 2y = -6$
$y + 2z = 11$
$2x + z = 16.$

Use the addition method with the second and third original equations. Eliminate z.

▶ $\begin{array}{l} y + 2z = 11 \\ 2x + z = 16 \end{array}$ ⇒ $\begin{array}{l} y + 2z = 11 \\ -4x - 2z = -32 \\ \hline -4x + y = -21 \end{array}$

The equation $-4x + y = -21$ and the first original equation, $x + 2y = -6$, form a linear system in two variables. Eliminate x and solve for y.

▶ $\begin{array}{l} x + 2y = -6 \\ -4x + y = -21 \end{array}$ ⇒ $\begin{array}{l} 4x + 8y = -24 \\ -4x + y = -21 \\ \hline 9y = -45 \\ \boxed{y = -5} \end{array}$

Substitute -5 for y in the first original equation.

▶ $\begin{array}{l} x + 2y = -6 \\ x + 2(-5) = -6 \\ \boxed{x = 4} \end{array}$

Substitute 4 for x in the third original equation.

▶ $\begin{array}{l} 2x + z = 16 \\ 2(4) + z = 16 \\ \boxed{z = 8} \end{array}$

Check in all three original equations.

Thus, $(4, -5, 8)$ is the solution of the system.

A system of linear equations in three variables may be inconsistent.

Example 3

Solve the system: $2x + 3y - 2z = 4$
$3x - 3y + 2z = 16$
$6x - 12y + 8z = 5.$

Use the addition method with the first and third original equations. Two variables, y and z, are eliminated.

▶ $\begin{array}{l} 2x + 3y - 2z = 4 \\ 6x - 12y + 8z = 5 \end{array}$ ⇒ $\begin{array}{l} 8x + 12y - 8z = 16 \\ 6x - 12y + 8z = 5 \\ \hline 14x = 21 \\ x = \tfrac{3}{2} \end{array}$

Use the addition method with the first and second original equations. Two variables, y and z, are eliminated.

▶ $\begin{array}{l} 2x + 3y - 2z = 4 \\ 3x - 3y + 2z = 16 \\ \hline 5x = 20 \\ x = 4 \end{array}$

Two different values are obtained for x, which is a contradiction. The system is inconsistent.

Thus, there is no solution of the system.

Reading in Algebra

True or false? Give a reason for your answer.
1. A system of three linear equations in three variables has at least one solution. **F**
2. A solution of a system of three linear equations in three variables is an ordered triple of real numbers. **T**
3. A system of three linear equations in three variables may be independent, inconsistent, or dependent. **T**
4. If an ordered triple of real numbers is a solution of one of the linear equations in a system, it is a solution of all equations in the system. **F**

Written Exercises

Solve each system of equations.

Ⓐ
1. $-x + 3y + z = -10$
   $3x + 2y - 2z = 3$
   $2x - y - 4z = -7$
   (7, −3, 6)

2. $a + b - 3c = 8$
   $3a + 4b - 2c = 20$
   $2a - 3b + c = -6$
   (2, 3, −1)

3. $m + n + p = 1$
   $m + 3n + 7p = 13$
   $m + 2n + 3p = 4$
   (1, −3, 3)

4. $x + 2y + 3z = 9$
   $-3x + 5y - 4z = -7$
   $3x - y + 2z = -1$
   (−3, 0, 4)

5. $a + b + c = 2$
   $2a + b + 2c = 3$
   $3a - b + c = 4$
   (2, 1, −1)

6. $2m + 2n + 6p = 9$
   $m - 3n + 2p = 5$
   $-m - 5n - 4p = 4$
   no solution

Ⓑ
7. $x - 2y = 14$
   $y + 2z = 11$
   $2x + z = 16$
   (4, −5, 8)

8. $y + z = -3$
   $-x - 2z = 5$
   $3x + 2y = -5$
   (−1, −1, −2)

9. $2x + 3y = -5$
   $4y - 5z = -32$
   $3x + 2z = 14$
   (2, −3, 4)

Solve each problem by using a system of three linear equations in three variables.

10. Find three numbers in decreasing order such that their sum is 3, the difference of the first two numbers is 4, and the sum of the smallest number and the greatest number is 2. **5, 1, −3**

11. Find three positive numbers in increasing order such that the difference of the first two numbers is 2, the difference of the first and the last numbers is 4, and the quotient of the last two numbers is $1\frac{1}{5}$. **8, 10, 12**

12. A coin box contains pennies, nickels, and dimes. The pennies and nickels are worth 35¢, the nickels and dimes are worth 80¢, and the value of the dimes and the pennies is 75¢. Find the number of each kind of coin in the box.
    pennies: 15, nickels: 4, dimes: 6

13. Find the number of nickels, dimes, and quarters in a collection of 80 such coins if the nickels and the quarters are worth $4.50 and the value of the quarters and the dimes is $5.50.
    nickels: 40, dimes: 30, quarters: 10

Solve each system of equations.

Ⓒ
14. $\dfrac{m}{4} - \dfrac{3n}{2} + \dfrac{p}{2} = -6$
    $\dfrac{m}{6} - \dfrac{n}{4} - \dfrac{p}{3} = 1$
    $\dfrac{m}{3} + \dfrac{n}{2} - p = 7$ **(6, 4, −3)**

15. $\dfrac{1}{a} + \dfrac{1}{b} - \dfrac{2}{c} = 9$
    $\dfrac{3}{a} - \dfrac{2}{b} + \dfrac{1}{c} = -1$
    $\dfrac{2}{a} - \dfrac{1}{b} + \dfrac{3}{c} = 7$ $(\tfrac{14}{65}, \tfrac{2}{17}, \tfrac{14}{29})$

16. $\dfrac{3}{x} - \dfrac{1}{y} + \dfrac{4}{z} = -3$
    $\dfrac{2}{x} + \dfrac{3}{y} - \dfrac{1}{z} = 6$
    $-\dfrac{1}{x} + \dfrac{2}{y} - \dfrac{3}{z} = 2$ **no solution**

Solving Systems of Three Linear Equations

TWO BY TWO DETERMINANTS 10.4

Objectives
**To find the value of a 2 by 2 determinant
To solve a system of two linear equations in two variables using determinants**

A **matrix** is an array of numbers or other elements arranged in rows and columns. A matrix that has the same number of rows as columns is called a **square matrix.** Each square matrix has a corresponding real number assigned to it called a **determinant.**

The square matrix $\begin{bmatrix} 3 & 7 \\ 2 & 8 \end{bmatrix}$ contains four elements, which are arranged in 2 rows and 2 columns. It is called a 2×2 (read: two by two) matrix. The determinant of the matrix is written in the same form as the matrix but with vertical bars instead of brackets. The value of the 2 by 2 determinant $\begin{vmatrix} 3 & 7 \\ 2 & 8 \end{vmatrix}$ is found in the following way.

$$\begin{vmatrix} 3 & 7 \\ 2 & 8 \end{vmatrix} = 3 \cdot 8 - 2 \cdot 7 = 10$$

**Definition:
2 by 2 determinant**

The determinant of $\begin{bmatrix} a & b \\ c & d \end{bmatrix}$ is $\begin{vmatrix} a & b \\ c & d \end{vmatrix}$. Its value is $ad - cb$.

Example 1 Find the value of each determinant.

$\begin{vmatrix} 2 & 1 \\ -3 & -5 \end{vmatrix}$ $\begin{vmatrix} m+n & m-n \\ -3 & 5 \end{vmatrix}$

$\begin{vmatrix} 2 & 1 \\ -3 & -5 \end{vmatrix} = 2(-5) - (-3)(1)$
$= -10 + 3$
$= -7$

$\begin{vmatrix} m+n & m-n \\ -3 & 5 \end{vmatrix} = (m+n)(5) - (-3)(m-n)$
$= 5m + 5n + 3m - 3n$
$= 8m + 2n$

Thus, the value is -7.

Thus, the value is $8m + 2n$.

Using determinants to solve a system of linear equations is known as **Cramer's rule,** named for the Swiss mathematician Gabriel Cramer (1704–1752). Cramer's rule is used to solve a system such as $5x + 2y = 4$
$2x - 3y = 13.$

$$5x + 2y = 4$$
$$2x + -3y = 13$$

$$x = \frac{\begin{vmatrix} 4 & 2 \\ 13 & -3 \end{vmatrix}}{\begin{vmatrix} 5 & 2 \\ 2 & -3 \end{vmatrix}} = \frac{\begin{vmatrix} 4 & 2 \\ 13 & -3 \end{vmatrix}}{\begin{vmatrix} 5 & 2 \\ 2 & -3 \end{vmatrix}} = \frac{4(-3) - 13(2)}{5(-3) - 2(2)} = \frac{-12 - 26}{-15 - 4} = \frac{-38}{-19} = 2$$

$$y = \frac{\begin{vmatrix} 5 & 4 \\ 2 & 13 \end{vmatrix}}{\begin{vmatrix} 5 & 2 \\ 2 & -3 \end{vmatrix}} = \frac{\begin{vmatrix} 5 & 4 \\ 2 & 13 \end{vmatrix}}{\begin{vmatrix} 5 & 2 \\ 2 & -3 \end{vmatrix}} = \frac{5(13) - 2(4)}{5(-3) - 2(2)} = \frac{65 - 8}{-15 - 4} = \frac{57}{-19} = -3$$

So, $x = 2$ and $y = -3$. The solution of the system is $(2, -3)$.

The solutions of a system of linear equations of the form $a_1x + b_1y = c_1$
$a_2x + b_2y = c_2$,
where all coefficients are real numbers, are found as follows:

$$x = \frac{\begin{vmatrix} c_1 & b_1 \\ c_2 & b_2 \end{vmatrix}}{\begin{vmatrix} a_1 & b_1 \\ a_2 & b_2 \end{vmatrix}} = \frac{c_1b_2 - c_2b_1}{a_1b_2 - a_2b_1} \qquad y = \frac{\begin{vmatrix} a_1 & c_1 \\ a_2 & c_2 \end{vmatrix}}{\begin{vmatrix} a_1 & b_1 \\ a_2 & b_2 \end{vmatrix}} = \frac{a_1c_2 - a_2c_1}{a_1b_2 - a_2b_1} \qquad [a_1b_2 - a_2b_1 \neq 0]$$

Each solution is an ordered pair (x, y) of real numbers.

Example 2

Solve the system $2x + 3y = 3$ using determinants.
$6x = y - 11$

Write each equation in the general form.
$2x + 3y = 3 \qquad\qquad 6x + (-1)y = -11$
↑   ↑   ↑   ↑   ↑   ↑
a_1 b_1 c_1 a_2 b_2 c_2

$$x = \frac{\begin{vmatrix} c_1 & b_1 \\ c_2 & b_2 \end{vmatrix}}{\begin{vmatrix} a_1 & b_1 \\ a_2 & b_2 \end{vmatrix}} \qquad\qquad y = \frac{\begin{vmatrix} a_1 & c_1 \\ a_2 & c_2 \end{vmatrix}}{\begin{vmatrix} a_1 & b_1 \\ a_2 & b_2 \end{vmatrix}}$$

$$= \frac{\begin{vmatrix} 3 & 3 \\ -11 & -1 \end{vmatrix}}{\begin{vmatrix} 2 & 3 \\ 6 & -1 \end{vmatrix}} \qquad\qquad = \frac{\begin{vmatrix} 2 & 3 \\ 6 & -11 \end{vmatrix}}{\begin{vmatrix} 2 & 3 \\ 6 & -1 \end{vmatrix}}$$

$$= \frac{3(-1) - (-11)(3)}{2(-1) - 6(3)} \qquad\qquad = \frac{2(-11) - 6(3)}{2(-1) - 6(3)}$$

$$= \frac{-3 + 33}{-2 - 18} \qquad\qquad = \frac{-22 - 18}{-2 - 18}$$

$$= \frac{30}{-20}, \text{ or } -\frac{3}{2} \qquad\qquad = \frac{-40}{-20}, \text{ or } 2$$

Check in the original equations.

Thus, $\left(-\frac{3}{2}, 2\right)$ is the solution of the system.

Two by Two Determinants

Oral Exercises

30. $\left(\dfrac{11}{3m+2n}, \dfrac{4m-n}{3m+2n}\right)$ 31. $\left(\dfrac{nr+ms}{n+2m}, \dfrac{s-2r}{n+2m}\right)$

To solve the system $2x - 3y = -4$ using determinants, state the determinant that
$\qquad -4x + 5y = 7$
you would use for each of the following. Do not find the value.
1. the denominator for x **below**
2. the numerator for x **below**
3. the numerator for y **below**
4. the denominator for y **below**

Written Exercises

1. $\begin{vmatrix} 2 & -3 \\ -4 & 5 \end{vmatrix}$ 2. $\begin{vmatrix} -4 & -3 \\ 7 & 5 \end{vmatrix}$ 3. $\begin{vmatrix} 2 & -4 \\ -4 & 7 \end{vmatrix}$ 4. $\begin{vmatrix} 2 & -3 \\ -4 & 5 \end{vmatrix}$

Find the value of each determinant.

Ⓐ 1. $\begin{vmatrix} 2 & 3 \\ -2 & 5 \end{vmatrix}$ **16**   2. $\begin{vmatrix} 6 & -1 \\ 4 & 8 \end{vmatrix}$ **52**   3. $\begin{vmatrix} -3 & -2 \\ 5 & 3 \end{vmatrix}$ **1**   4. $\begin{vmatrix} -8 & 6 \\ 4 & -2 \end{vmatrix}$ **−8**

5. $\begin{vmatrix} -10 & 3 \\ -4 & 5 \end{vmatrix}$ **−38**   6. $\begin{vmatrix} 1 & 0 \\ 0 & 1 \end{vmatrix}$ **1**   7. $\begin{vmatrix} -3 & 5 \\ -9 & -2 \end{vmatrix}$ **51**   8. $\begin{vmatrix} -2 & 0 \\ 4 & 2 \end{vmatrix}$ **−4**

9. $\begin{vmatrix} 3\sqrt{2} & 5 \\ -2 & 2\sqrt{2} \end{vmatrix}$ **22**   10. $\begin{vmatrix} 0 & 7 \\ 0 & 9 \end{vmatrix}$ **0**   11. $\begin{vmatrix} \frac{1}{4} & \frac{5}{6} \\ \frac{1}{2} & \frac{2}{3} \end{vmatrix}$ **−$\frac{1}{4}$**   12. $\begin{vmatrix} \frac{1}{3} & \frac{1}{6} \\ -\frac{1}{5} & \frac{3}{10} \end{vmatrix}$ **$\frac{2}{15}$**

13. $\begin{vmatrix} m+n & 3 \\ m-n & 5 \end{vmatrix}$ **$2m + 8n$**   14. $\begin{vmatrix} c-d & c+d \\ 3 & 4 \end{vmatrix}$ **$c - 7d$**   15. $\begin{vmatrix} -r & p+r \\ 6 & 10 \end{vmatrix}$ **$-6p - 16r$**

Solve each system of linear equations using determinants.

16. $2x + 3y = 7$
    $x + 2y = 3$ **(5, −1)**

17. $5x - 12y = 4$
    $4x - 7y = -2$ **(−4, −2)**

18. $3x - 4y = 7$
    $4x + 6y = 15$ **(3, ½)**

19. $2x + 3y = 4$
    $4x + y = -2$ **(−1, 2)**

20. $3x + 2y = 1$
    $3x - 2y = -5$ **(−⅔, 3/2)**

21. $-x + y = 1$
    $3x - 4y = -3$ **(−1, 0)**

22. $4x + y = 0$
    $6x - y = 5$ **(½, −2)**

23. $2x + 3y = -2$
    $x + 5y = 3$ **(−19/7, 8/7)**

Ⓑ 24. $2x - 3y = 2.5$
    $-x - 2y = 0.5$ **(0.5, −0.5)**

25. $5x - y = 0.4$
    $x - y = 0.2$ **(0.05, −0.15)**

26. $0.6x - 0.1y = -1$
    $-4x - 0.5y = 2$ **(−1, 4)**

Find the value of each determinant.

27. $\begin{vmatrix} p+r & -r \\ r & p-r \end{vmatrix}$ **p^2**   28. $\begin{vmatrix} m+n & m+n \\ m+n & m-n \end{vmatrix}$ **$-2n^2 - 2mn$**   29. $\begin{vmatrix} m^2 - n^2 & 1 \\ m+n & \dfrac{1}{m+n} \end{vmatrix}$ **$-2n$**

Solve each system of linear equations for (x, y) using determinants. Assume that all variables represent real numbers.

Ⓒ 30. $mx - 2y = 1$ **above**
    $nx + 3y = 4$

31. $x - my = r$ **above**
    $2x + ny = s$

32. $a_1x + b_1y = 1$
    $a_2x + b_2y = 1$

33. $a_1x + b_1y = c_1$
    $a_2x + b_2y = c_2$

32. $\left(\dfrac{b_2 - b_1}{a_1b_2 - a_2b_1}, \dfrac{a_1 - a_2}{a_1b_2 - a_2b_1}\right)$ 33. $\left(\dfrac{c_1b_2 - c_2b_1}{a_1b_2 - a_2b_1}, \dfrac{a_1c_2 - a_2c_1}{a_1b_2 - a_2b_1}\right)$

THREE BY THREE DETERMINANTS   10.5

Objectives **To find the value of a 3 by 3 determinant**
To solve a system of three linear equations in three variables using determinants and minors

The 3 by 3 square matrix $\begin{bmatrix} a_1 & b_1 & c_1 \\ a_2 & b_2 & c_2 \\ a_3 & b_3 & c_3 \end{bmatrix}$ contains 3 rows and 3 columns.
You can find the value of its corresponding determinant as shown below.

(1) Repeat columns 1 and 2 as columns 4 and 5.

(2) Multiply on the *down* diagonals.

(3) Multiply on the *up* diagonals.

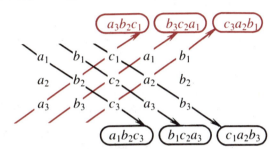

(4) Subtract the sum of the up diagonal products from the sum of the down diagonal products.

$[a_1b_2c_3 + b_1c_2a_3 + c_1a_2b_3] -$
$[a_3b_2c_1 + b_3c_2a_1 + c_3a_2b_1]$

Definition:
3 by 3
determinant

The determinant of $\begin{bmatrix} a_1 & b_1 & c_1 \\ a_2 & b_2 & c_2 \\ a_3 & b_3 & c_3 \end{bmatrix}$ is $\begin{vmatrix} a_1 & b_1 & c_1 \\ a_2 & b_2 & c_2 \\ a_3 & b_3 & c_3 \end{vmatrix}$. Its value is

$[a_1b_2c_3 + b_1c_2a_3 + c_1a_2b_3] - [a_3b_2c_1 + b_3c_2a_1 + c_3a_2b_1]$.

Example 1 Find the value of $\begin{vmatrix} 2 & 3 & 5 \\ 4 & 2 & 1 \\ -1 & -3 & 2 \end{vmatrix}$.

Repeat columns 1 and 2 and follow the steps above. ▶

$[a_1b_2c_3 + b_1c_2a_3 + c_1a_2b_3] - [a_3b_2c_1 + b_3c_2a_1 + c_3a_2b_1]$
$[2(2)(2) + 3(1)(-1) + 5(4)(-3)] - [-1(2)(5) + (-3)(1)(2) + 2(4)(3)]$

$[8 + (-3) + (-60)] - [-10 + (-6) + 24] = (-55) - (8) = -63$
Thus, the value of the determinant is -63.

Three by Three Determinants

You can also use Cramer's rule to solve a system of three linear equations of the form

$$a_1x + b_1y + c_1z = d_1$$
$$a_2x + b_2y + c_2z = d_2$$
$$a_3x + b_3y + c_3z = d_3$$

All coefficients are real numbers, as shown below.

$$x = \frac{\begin{vmatrix} d_1 & b_1 & c_1 \\ d_2 & b_2 & c_2 \\ d_3 & b_3 & c_3 \end{vmatrix}}{D} \qquad y = \frac{\begin{vmatrix} a_1 & d_1 & c_1 \\ a_2 & d_2 & c_2 \\ a_3 & d_3 & c_3 \end{vmatrix}}{D} \qquad z = \frac{\begin{vmatrix} a_1 & b_1 & d_1 \\ a_2 & b_2 & d_2 \\ a_3 & b_3 & d_3 \end{vmatrix}}{D},$$

$$\text{where } D = \begin{vmatrix} a_1 & b_1 & c_1 \\ a_2 & b_2 & c_2 \\ a_3 & b_3 & c_3 \end{vmatrix}$$

Each solution is an ordered triple (x, y, z) of real numbers. The value of the determinant in the denominator must not be equal to 0.

Example 2

Solve the system $2x + 3y + 4z = 4$ using determinants.
$5x + 7y + 8z = 9$
$3x - 2y - 6z = 7$

$$x = \frac{\begin{vmatrix} 4 & 3 & 4 \\ 9 & 7 & 8 \\ 7 & -2 & -6 \end{vmatrix}}{D} \qquad y = \frac{\begin{vmatrix} 2 & 4 & 4 \\ 5 & 9 & 8 \\ 3 & 7 & -6 \end{vmatrix}}{D} \qquad z = \frac{\begin{vmatrix} 2 & 3 & 4 \\ 5 & 7 & 9 \\ 3 & -2 & 7 \end{vmatrix}}{D}$$

$$D = \begin{vmatrix} 2 & 3 & 4 \\ 5 & 7 & 8 \\ 3 & -2 & -6 \end{vmatrix} \begin{matrix} 2 & 3 \\ 5 & 7 \\ 3 & -2 \end{matrix} = [-84 + 72 + (-40)] - [84 + (-32) + (-90)]$$
$$= -14$$

$$x = \frac{[-168 + 168 + (-72)] - [196 + (-64) + (-162)]}{-14} = \frac{-42}{-14}, \text{ or } 3$$

$$y = \frac{[-108 + 96 + 140] - [108 + 112 + (-120)]}{-14} = \frac{28}{-14}, \text{ or } -2$$

$$z = \frac{[98 + 81 + (-40)] - [84 + (-36) + 105]}{-14} = \frac{-14}{-14}, \text{ or } 1$$

Check in the original equations.

Thus, $(3, -2, 1)$ is the solution of the system.

Minors can also be used to find the determinant of a matrix and for solving systems of equations. The **minor** of an element of a determinant is the determinant found by deleting the column and row in which the element lies.

The value of the determinant $\begin{vmatrix} a_1 & b_1 & c_1 \\ a_2 & b_2 & c_2 \\ a_3 & b_3 & c_3 \end{vmatrix}$ can be found by finding the minors of all the elements of any row or column. Using Column 1, the expansion by minors is shown below.

$$D = +a_1 \begin{vmatrix} b_2 & c_2 \\ b_3 & c_3 \end{vmatrix} - a_2 \begin{vmatrix} b_1 & c_1 \\ b_3 & c_3 \end{vmatrix} + a_3 \begin{vmatrix} b_1 & c_1 \\ b_2 & c_2 \end{vmatrix}$$
$$= a_1[b_2c_3 - b_3c_2] - a_2[b_1c_3 - b_3c_1] + a_3[b_1c_2 - b_2c_1]$$
$$= a_1b_2c_3 - a_1b_3c_2 - a_2b_1c_3 + a_2b_3c_1 + a_3b_1c_2 - a_3b_2c_1$$

The product of an element and its minor is multiplied by -1 only if the sum of the number of the row and the number of the column is an odd integer. The value of the determinant is the same value shown in the definition of a 3 by 3 determinant.

Example 3 Using minors, show that the value of $\begin{vmatrix} 2 & 3 & 5 \\ 4 & 2 & 1 \\ -1 & -3 & 2 \end{vmatrix}$ is -63.

See Example 1. ▶ $+2\begin{vmatrix} 2 & 1 \\ -3 & 2 \end{vmatrix} - 4\begin{vmatrix} 3 & 5 \\ -3 & 2 \end{vmatrix} + (-1)\begin{vmatrix} 3 & 5 \\ 2 & 1 \end{vmatrix}$ ◀ Use Column 1.

$= 2[4 + 3] - 4[6 + 15] + (-1)[3 - 10] = 14 - 84 + 7 = -63$

Thus, the value of the determinant is -63.

Example 4 Using minors, show that $(3, -2, 1)$ is the solution of the system $2x + 3y + 4z = 4$
$5x + 7y + 8z = 9$
$3x - 2y - 6z = 7.$

See Example 2. ▶ $x = \dfrac{4\begin{vmatrix} 7 & 8 \\ -2 & -6 \end{vmatrix} - 9\begin{vmatrix} 3 & 4 \\ -2 & -6 \end{vmatrix} + 7\begin{vmatrix} 3 & 4 \\ 7 & 8 \end{vmatrix}}{D}$

$= \dfrac{4[-42 + 16] - 9[-18 + 8] + 7[24 - 28]}{D} = \dfrac{-42}{D}$

$y = \dfrac{+2\begin{vmatrix} 9 & 8 \\ 7 & -6 \end{vmatrix} - 5\begin{vmatrix} 4 & 4 \\ 7 & -6 \end{vmatrix} + 3\begin{vmatrix} 4 & 4 \\ 9 & 8 \end{vmatrix}}{D}$

$= \dfrac{2[-54 - 56] - 5[-24 - 28] + 3[32 - 36]}{D} = \dfrac{28}{D}$

Three by Three Determinants

$$z = \frac{2\begin{vmatrix} 7 & 9 \\ 2 & 7 \end{vmatrix} - 5\begin{vmatrix} 3 & 4 \\ -2 & 7 \end{vmatrix} + 3\begin{vmatrix} 3 & 4 \\ 7 & 9 \end{vmatrix}}{D}$$

$$= \frac{2[49 + 18] - 5[21 + 8] + 3[27 - 28]}{D} = \frac{-14}{D}$$

$$D = 2\begin{vmatrix} 7 & 8 \\ -2 & -6 \end{vmatrix} - 5\begin{vmatrix} 3 & 4 \\ -2 & -6 \end{vmatrix} + 3\begin{vmatrix} 3 & 4 \\ 7 & 8 \end{vmatrix}$$

$$= 2[-42 + 16] - 5[-18 + 8] + 3[24 - 28] = -14$$

$$x = \frac{-42}{-14}, \text{ or } 3; \quad y = \frac{28}{-14}, \text{ or } -2; \quad z = \frac{-14}{-14}, \text{ or } 1$$

Thus, $(3, -2, 1)$ is the solution.

20. $\left(\dfrac{b_2 c_3 d_1 + b_1 c_2 d_3 - b_1 c_3 d_2}{a_1 b_2 c_3 + a_3 b_1 c_2}, \dfrac{a_1 c_3 d_2 + a_3 c_2 d_1 - a_1 c_2 d_3}{a_1 b_2 c_3 + a_3 b_1 c_2}, \dfrac{a_1 b_2 d_3 + a_3 b_1 d_2 - a_3 b_2 d_1}{a_1 b_2 c_3 + a_3 b_1 c_2} \right)$

Written Exercises

Find the value of each determinant.

(A) 1. $\begin{vmatrix} 2 & -4 & -3 \\ 4 & -1 & -2 \\ -3 & 4 & -2 \end{vmatrix}$ -75

2. $\begin{vmatrix} -3 & 2 & -5 \\ 4 & -1 & 2 \\ -1 & -2 & -3 \end{vmatrix}$ 44

3. $\begin{vmatrix} 4 & -1 & -2 \\ -3 & 2 & -1 \\ 2 & -1 & 3 \end{vmatrix}$ 15

4. $\begin{vmatrix} -2 & -3 & 2 \\ -1 & 2 & -2 \\ 3 & -2 & 1 \end{vmatrix}$ 11

Using minors, find the value of each determinant.

5. $\begin{vmatrix} 2 & 3 & 5 \\ 4 & 2 & 1 \\ -1 & -3 & 2 \end{vmatrix}$ -63

6. $\begin{vmatrix} -3 & 1 & 5 \\ -2 & 0 & 2 \\ 6 & 3 & 4 \end{vmatrix}$ 8

7. $\begin{vmatrix} 1 & 3 & -2 \\ 1 & -4 & 5 \\ 1 & 2 & 3 \end{vmatrix}$ -28

8. $\begin{vmatrix} 1 & -1 & 4 \\ 0 & 1 & -7 \\ 0 & 0 & 1 \end{vmatrix}$ 1

Using determinants, solve each system of linear equations.

(B) 9. $2x - 3y + z = 7$
$3x - 2y + 2z = 9$
$-2x - y + 2z = 3$
$(1, -1, 2)$

10. $-x + 2y - 3z = 11$
$2x + y + 2z = -4$
$3x + 3y - z = 5$
$(-1, 2, -2)$

11. $3x - 2y + 4z = 11$
$-2x + 4y + z = -10$
$2x - y + 3z = 7$
$(3, -1, 0)$

12. $-3x + 2y - 2z = -10$
$2x + 3y + z = -1$
$-x - y + 2z = -8$
$(4, -2, -3)$

13. $-3x + 4y + 5z = 3$
$x - y + 3z = 8$
$-2x + 3y - z = -7$
$(1, -1, 2)$

14. $4x - 3y + 2z = 0$
$-2x + y - 3z = 4$
$5x - 6y + 2z = 1$
$(1, 0, -2)$

Using minors, solve each system of linear equations.

15. $a + b + c = -1$
$2a - b + c = 19$
$3a - 2b - 4c = 16$
$(4, -8, 3)$

16. $3x - 4z = 7$
$2y + 5z = 2$
$6x + 5y = 10$
$(5, -4, 2)$

17. $5n + 10d = 70$
$5n + 25q = 270$
$10d + 25q = 300$
$(5, 4, 10)$

(C) 18. $0.6x + 0.4y - 0.6z = 5.8$
$0.8x - 1y - 0.4z = 16$
$1.5x - 1.5y = 19.5$
$(5, -8, -10)$

19. $\dfrac{2x}{5} + \dfrac{3y}{4} - \dfrac{2z}{3} = -\dfrac{9}{10}$
$-\dfrac{x}{3} - \dfrac{2y}{5} + \dfrac{3z}{5} = \dfrac{4}{3}$
$\dfrac{3x}{2} - \dfrac{3y}{5} - \dfrac{z}{5} = -\dfrac{33}{10}$
$(-1, 2, 3)$

20. $a_1 x + b_1 y = d_1$
$b_2 y + c_2 z = d_2$
$a_3 x + c_3 z = d_3$
above

MATRIX ADDITION   10.6

Objectives **To find the sum of two matrices**
To determine the properties of matrix addition

Recall that a matrix (pl. matrices) is an array of real numbers or other elements enclosed by brackets.

$$\begin{bmatrix} a_{11} & a_{12} & a_{13} \\ a_{21} & a_{22} & a_{23} \\ a_{31} & a_{32} & a_{33} \end{bmatrix} \quad \begin{bmatrix} a_{11} & a_{12} \\ a_{21} & a_{22} \\ a_{31} & a_{32} \end{bmatrix} \quad \begin{bmatrix} a_{11} \\ a_{21} \\ a_{31} \end{bmatrix} \quad \begin{bmatrix} a_{11} & a_{12} & a_{13} \end{bmatrix}$$

    3 by 3 square   3 by 2   3 by 1   1 by 3
       matrix   matrix   matrix   matrix
3 rows, 3 columns   3 rows, 2 columns   3 rows, 1 column   1 row, 3 columns

In the matrices above, the first number of the subscript indicates the row and the second number of the subscript indicates the column; a_{32} means the number or element in the third row and second column position.

The number of rows and columns determines the **dimensions** of a matrix. The dimensions of the matrices above, in order, are 3 by 3, 3 by 2, 3 by 1, and 1 by 3.

It is possible for two matrices to be equal. The matrix $\begin{bmatrix} 1 & -2 \\ -1 & 4 \end{bmatrix}$ is equal to

the matrix $\begin{bmatrix} \frac{2}{2} & -\frac{4}{2} \\ -\frac{6}{6} & \frac{12}{3} \end{bmatrix}$ since they have the same dimensions, 2 by 2, and their

corresponding elements are equal.

> **Equality of matrices**
>
> Two matrices are equal if they have the same dimensions and the elements in corresponding positions are equal.

Example 1 Determine which pairs of matrices are equal, if any.

$$\begin{bmatrix} -2 & 1 & 6 \\ 3 & -1 & 2 \end{bmatrix} \quad \begin{bmatrix} -\frac{6}{3} & \frac{7}{7} & \frac{-12}{-2} \\ \frac{12}{4} & -\frac{8}{8} & \frac{4}{2} \end{bmatrix} \quad \bigg| \quad \begin{bmatrix} 1 & -3 & 2 \end{bmatrix} \quad \begin{bmatrix} 1 \\ -3 \\ 2 \end{bmatrix}$$

Dimensions of each matrix are 2 by 3. | Dimensions of one matrix are 1 by 3.
Elements in corresponding positions are equal: | Dimensions of other matrix are 3 by 1.

$-2 = -\frac{6}{3} \quad 1 = \frac{7}{7} \quad 6 = \frac{-12}{-2}$

$3 = \frac{12}{4} \quad -1 = -\frac{8}{8} \quad 2 = \frac{4}{2}$

Thus, the first pair are equal; the second pair are not.

Matrix Addition

You can find the sum of two matrices with the same dimensions by using the following property.

> **Sum of two matrices**
>
> The *sum* of two matrices, A and B, having the same dimensions is a matrix whose elements are the sums of the corresponding elements of A and B.
>
> If $A = \begin{bmatrix} a_{11} & a_{12} \\ a_{21} & a_{22} \end{bmatrix}$ and $B = \begin{bmatrix} b_{11} & b_{12} \\ b_{21} & b_{22} \end{bmatrix}$, then
>
> $A + B = \begin{bmatrix} a_{11} + b_{11} & a_{12} + b_{12} \\ a_{21} + b_{21} & a_{22} + b_{22} \end{bmatrix}.$

Example 2 If $A = \begin{bmatrix} -3 & -2 \\ 5 & 6 \end{bmatrix}$, $B = \begin{bmatrix} 4 & 8 \\ -7 & -3 \end{bmatrix}$, and $C = \begin{bmatrix} 4 \\ -7 \\ 1 \end{bmatrix}$, find $A + B$ and $A + C$, if possible.

A and B have the same dimensions. ▶ $A + B = \begin{bmatrix} -3 + 4 & -2 + 8 \\ 5 + (-7) & 6 + (-3) \end{bmatrix} = \begin{bmatrix} 1 & 6 \\ -2 & 3 \end{bmatrix}$

Since A and C do not have the same dimensions, they cannot be added.

Thus, $A + B = \begin{bmatrix} 1 & 6 \\ -2 & 3 \end{bmatrix}$, but $A + C$ is not defined.

Recall that the *additive identity* is 0 since $a + 0 = a$ for any real number a. Also, the *additive inverse* of any real number a is $-a$ since $a + (-a) = 0$. Similarly, identities and inverses exist for matrix addition.

> **Additive identity matrix**
>
> I is called the *additive identity* (or *zero*) *matrix* for any matrix A if A and I have the same dimensions and $A + I = A$.
>
> **Additive inverse matrix**
>
> $-A$ is called the *additive inverse matrix* of any matrix A if A and $-A$ have the same dimensions and $A + (-A) = I$.

Example 3 If $A = \begin{bmatrix} 2 & -3 \\ 5 & -7 \end{bmatrix}$, $I = \begin{bmatrix} 0 & 0 \\ 0 & 0 \end{bmatrix}$, and $-A = \begin{bmatrix} -2 & 3 \\ -5 & 7 \end{bmatrix}$, show that $A + I = A$ and $A + (-A) = I$.

A and I have the same dimensions. ▶ $A + I = \begin{bmatrix} 2 + 0 & -3 + 0 \\ 5 + 0 & -7 + 0 \end{bmatrix} = \begin{bmatrix} 2 & -3 \\ 5 & -7 \end{bmatrix} = A$

A and −A have the same dimensions. ▶ $A + (-A) = \begin{bmatrix} 2 + -2 & -3 + 3 \\ 5 + -5 & -7 + 7 \end{bmatrix} = \begin{bmatrix} 0 & 0 \\ 0 & 0 \end{bmatrix} = I$

Thus, $A + I = A$ and $A + (-A) = I$.

Matrix addition is *associative* and *commutative*.

> **Associative and commutative property of addition**
>
> For any matrices A, B, and C having the same dimensions,
> $(A + B) + C = A + (B + C)$ and $A + B = B + A$.

Example 4   If $A = \begin{bmatrix} -3 & 2 \\ -1 & 4 \end{bmatrix}$, $B = \begin{bmatrix} 5 & -3 \\ 3 & -6 \end{bmatrix}$, and $C = \begin{bmatrix} -1 & 8 \\ 4 & -2 \end{bmatrix}$, show that $(A + B) + C = A + (B + C)$ and that $A + B = B + A$.

$$(A + B) + C = \left(\begin{bmatrix} -3+5 & 2+-3 \\ -1+3 & 4+-6 \end{bmatrix}\right) + \begin{bmatrix} -1 & 8 \\ 4 & -2 \end{bmatrix} = \begin{bmatrix} 2+-1 & -1+8 \\ 2+4 & -2+-2 \end{bmatrix} = \begin{bmatrix} 1 & 7 \\ 6 & -4 \end{bmatrix}$$

$$A + (B + C) = \begin{bmatrix} -3 & 2 \\ -1 & 4 \end{bmatrix} + \begin{bmatrix} 5+-1 & -3+8 \\ 3+4 & -6+-2 \end{bmatrix} = \begin{bmatrix} -3+4 & 2+5 \\ -1+7 & 4+-8 \end{bmatrix} = \begin{bmatrix} 1 & 7 \\ 6 & -4 \end{bmatrix}$$

$$A + B = \begin{bmatrix} -3+5 & 2+-3 \\ -1+3 & 4+-6 \end{bmatrix} = \begin{bmatrix} 2 & -1 \\ 2 & -2 \end{bmatrix}$$

$$B + A = \begin{bmatrix} 5+-3 & -3+2 \\ 3+-1 & -6+4 \end{bmatrix} = \begin{bmatrix} 2 & -1 \\ 2 & -2 \end{bmatrix}$$

5. $\begin{bmatrix} 7 & 0 & 10 \\ 4 & -2 & -1 \\ 6 & -6 & -3 \end{bmatrix}$

Written Exercises

Determine if the pairs of matrices are equal.

A   1. $\begin{bmatrix} -1 & 0 \\ 3 & -2 \end{bmatrix}$, $\begin{bmatrix} -\frac{2}{2} & 5-5 \\ \frac{-6}{-2} & 2-4 \end{bmatrix}$ 2. $\begin{bmatrix} -2 & 3 \\ 1 & -2 \\ -3 & 4 \end{bmatrix}$, $\begin{bmatrix} -\frac{4}{2} & \frac{6}{2} \\ \frac{3}{3} & -\frac{8}{4} \\ \frac{9}{-3} & 2^2 \end{bmatrix}$ yes   3. $\begin{bmatrix} \frac{1}{2} \\ \frac{1}{3} \\ \frac{1}{4} \end{bmatrix} \begin{bmatrix} -\frac{4}{8} & -\frac{3}{9} & -\frac{5}{20} \end{bmatrix}$, no

yes

Find each sum, if possible.

4. $\begin{bmatrix} -2 & -3 \\ 4 & -6 \\ -5 & 1 \end{bmatrix} + \begin{bmatrix} 4 & 6 \\ -1 & 7 \\ 8 & -2 \end{bmatrix} \begin{bmatrix} 2 & 3 \\ 3 & 1 \\ 3 & -1 \end{bmatrix}$ 5. $\begin{bmatrix} -1 & 2 & 3 \\ -2 & 3 & -4 \\ 7 & -6 & 5 \end{bmatrix} + \begin{bmatrix} 8 & -2 & 7 \\ 6 & -5 & 3 \\ -1 & 0 & -8 \end{bmatrix}$ above

6. $\begin{bmatrix} 4 & -2 & 1 \\ 3 & -6 & -4 \end{bmatrix} + \begin{bmatrix} 3 & 8 \\ 1 & 7 \\ 5 & -2 \end{bmatrix}$ not possible   7. $\begin{bmatrix} 7 & -3 & 6 \\ -1 & 4 & -8 \end{bmatrix} + \begin{bmatrix} -7 & 3 & -6 \\ 1 & -4 & 8 \end{bmatrix}$ $\begin{bmatrix} 0 & 0 & 0 \\ 0 & 0 & 0 \end{bmatrix}$

8. $[9 \ 8 \ -7] + \begin{bmatrix} 6 \\ -3 \\ -2 \end{bmatrix}$ not possible   9. $\begin{bmatrix} 0 & -8 \\ 9 & -1 \end{bmatrix} + \begin{bmatrix} -9 & -7 \\ -6 & -5 \end{bmatrix} \begin{bmatrix} -9 & -15 \\ 3 & -6 \end{bmatrix}$

Let $A = \begin{bmatrix} 3 & 5 \\ -2 & -6 \end{bmatrix}$, $B = \begin{bmatrix} 8 & -3 \\ -7 & -6 \end{bmatrix}$, and $C = \begin{bmatrix} 1 & -2 \\ 7 & -9 \end{bmatrix}$.   10. $\begin{bmatrix} -3 & -5 \\ 2 & 6 \end{bmatrix}, \begin{bmatrix} -8 & 3 \\ 7 & 6 \end{bmatrix}, \begin{bmatrix} -1 & 2 \\ -7 & 9 \end{bmatrix}$

10. Find $-A$, $-B$, and $-C$. **above**
11. Show that $B + I = B$.

B   12. Show that $B + C = C + B$.
13. Show that $(B + C) + A = B + (C + A)$.
14. Show that $C + A = A + C$.
15. Show that $(C + A) + B = C + (A + B)$.
16. Show that $(A + C) + B = A + (C + B)$.

For Ex. 11–16, see TE Answer Section.

Matrix Addition

MATRIX MULTIPLICATION   10.7

Objectives
To find the product of a scalar and a matrix
To find the product of two matrices
To determine the properties of matrix multiplication

When working with matrices, any real number is called a **scalar**. You will now learn how to find the product of a scalar and a matrix.

Product of scalar and matrix

The product of a scalar k and a matrix A is the matrix kA obtained by multiplying each element of A by k.

$$\text{If } A = \begin{bmatrix} a_{11} & a_{12} & a_{13} \\ a_{21} & a_{22} & a_{23} \\ a_{31} & a_{32} & a_{33} \end{bmatrix}, \text{ then } kA = \begin{bmatrix} ka_{11} & ka_{12} & ka_{13} \\ ka_{21} & ka_{22} & ka_{23} \\ ka_{31} & ka_{32} & ka_{33} \end{bmatrix}.$$

Example 1 Find kA if $k = -3$ and $A = \begin{bmatrix} -1 & 2 & -3 \\ 13 & -5 & 4 \end{bmatrix}$.

$$kA = \begin{bmatrix} -3(-1) & -3(2) & -3(-3) \\ -3(13) & -3(-5) & -3(4) \end{bmatrix} = \begin{bmatrix} 3 & -6 & 9 \\ -39 & 15 & -12 \end{bmatrix}$$

Thus, $kA = \begin{bmatrix} 3 & -6 & 9 \\ -39 & 15 & -12 \end{bmatrix}$.

Two matrices can be multiplied if the number of columns of one matrix equals the number of rows of the other.

Product of two matrices

The *product* of two matrices, A and B, where the number of columns of A equals the number of rows of B, is a matrix whose elements are obtained by multiplying the row elements of A by the column elements of B in the following manner.

If $A = \begin{bmatrix} a_{11} & a_{12} & a_{13} \\ a_{21} & a_{22} & a_{23} \end{bmatrix}$ and $B = \begin{bmatrix} b_{11} & b_{12} \\ b_{21} & b_{22} \\ b_{31} & b_{32} \end{bmatrix}$, then

$$A \cdot B = \begin{bmatrix} a_{11}b_{11} + a_{12}b_{21} + a_{13}b_{31} & a_{11}b_{12} + a_{12}b_{22} + a_{13}b_{32} \\ a_{21}b_{11} + a_{22}b_{21} + a_{23}b_{31} & a_{21}b_{12} + a_{22}b_{22} + a_{23}b_{32} \end{bmatrix}.$$

In general, $A_{m \times n} \cdot B_{n \times r} = AB_{m \times r}$ where $m \times n$, $n \times r$, and $m \times r$ are the dimensions of the matrices.

Example 2   If $A = \begin{bmatrix} -1 & 2 & -6 \\ 3 & -1 & 4 \end{bmatrix}$ and $B = \begin{bmatrix} -6 & 4 \\ -2 & 3 \\ 1 & -4 \end{bmatrix}$, find AB, if possible.

$$AB = \begin{bmatrix} (-1)(-6) + 2(-2) + (-6)(1) & (-1)(4) + 2(3) + (-6)(-4) \\ 3(-6) + (-1)(-2) + 4(1) & 3(4) + (-1)(3) + 4(-4) \end{bmatrix} = \begin{bmatrix} -4 & 26 \\ -12 & -7 \end{bmatrix}$$

Thus, $AB = \begin{bmatrix} -4 & 26 \\ -12 & -7 \end{bmatrix}$.

Recall that the *multiplicative identity* is 1 since $a \cdot 1 = a$ for any real number a. Also, the *multiplicative inverse* of any nonzero real number a is $\frac{1}{a}$ since $a \cdot \frac{1}{a} = 1$. Similarly, identities and inverses exist for matrix multiplication.

Multiplicative identity matrix	I is called the *multiplicative identity matrix* for any square matrix A if $A \cdot I = A$.
Multiplicative inverse matrix	A^{-1} is called the *multiplicative inverse matrix* of any square matrix A if $A \cdot A^{-1} = I$.

For a 3 by 3 square matrix, the multiplicative identity matrix is $\begin{bmatrix} 1 & 0 & 0 \\ 0 & 1 & 0 \\ 0 & 0 & 1 \end{bmatrix}$.

Example 3   If $A = \begin{bmatrix} -1 & 2 & 3 \\ -2 & 4 & -1 \\ 5 & -3 & 2 \end{bmatrix}$ and $I = \begin{bmatrix} 1 & 0 & 0 \\ 0 & 1 & 0 \\ 0 & 0 & 1 \end{bmatrix}$, show that $AI = A$.

$$AI = \begin{bmatrix} -1(1) + 2(0) + 3(0) & -1(0) + 2(1) + 3(0) & -1(0) + 2(0) + 3(1) \\ -2(1) + 4(0) + (-1)(0) & -2(0) + 4(1) + (-1)(0) & -2(0) + 4(0) + (-1)(1) \\ 5(1) + (-3)(0) + 2(0) & 5(0) + (-3)(1) + 2(0) & 5(0) + (-3)(0) + 2(1) \end{bmatrix}$$

$$= \begin{bmatrix} -1 & 2 & 3 \\ -2 & 4 & -1 \\ 5 & -3 & 2 \end{bmatrix} = A$$

Thus, $AI = A$.

Example 4   If $A = \begin{bmatrix} 1 & 4 \\ 2 & 9 \end{bmatrix}$ and $B = \begin{bmatrix} 9 & -4 \\ -2 & 1 \end{bmatrix}$, show that B is the multiplicative inverse of A by showing that $A \cdot B = I$.

$$A \cdot B = \begin{bmatrix} 1(9) + 4(-2) & 1(-4) + 4(1) \\ 2(9) + 9(-2) & 2(-4) + 9(1) \end{bmatrix} = \begin{bmatrix} 1 & 0 \\ 0 & 1 \end{bmatrix} = I.$$

Thus, B is the multiplicative inverse of A, or $B = A^{-1}$.

The next example shows you how to find the multiplicative inverse matrix for a given square matrix.

Example 5   If $A = \begin{bmatrix} 3 & -2 \\ -2 & 1 \end{bmatrix}$, find A^{-1}.

Let $A^{-1} = \begin{bmatrix} a_1 & b_1 \\ a_2 & b_2 \end{bmatrix}$.

$\begin{bmatrix} 3 & -2 \\ -2 & 1 \end{bmatrix} \cdot \begin{bmatrix} a_1 & b_1 \\ a_2 & b_2 \end{bmatrix} = \begin{bmatrix} 1 & 0 \\ 0 & 1 \end{bmatrix}$ ◀ $A \cdot A^{-1} = I$

$\begin{bmatrix} 3a_1 - 2a_2 & 3b_1 - 2b_2 \\ -2a_1 + a_2 & -2b_1 + b_2 \end{bmatrix} = \begin{bmatrix} 1 & 0 \\ 0 & 1 \end{bmatrix}$

Since the two matrices are equal, the corresponding elements are equal.

$3a_1 - 2a_2 = 1$ $3b_1 - 2b_2 = 0$
$-2a_1 + a_2 = 0$ $-2b_1 + b_2 = 1$

Solve each system of equations.

$\begin{array}{r} 3a_1 - 2a_2 = 1 \\ -4a_1 + 2a_2 = 0 \\ \hline -a_1 = 1 \end{array}$ $\begin{array}{r} 3b_1 - 2b_2 = 0 \\ -4b_1 + 2b_2 = 2 \\ \hline -b_1 = 2 \end{array}$

$\boxed{a_1 = -1}$ $\boxed{b_1 = -2}$

$-2a_1 + a_2 = 0$ $3b_1 - 2b_2 = 0$
$-2(-1) + a_2 = 0$ $3(-2) - 2b_2 = 0$

$\boxed{a_2 = -2}$ $\boxed{b_2 = -3}$

Thus, $A^{-1} = \begin{bmatrix} -1 & -2 \\ -2 & -3 \end{bmatrix}$.

Matrix multiplication is *associative*.

Associative property of multiplication	For any matrices A, B, and C, if the products exist, $(A \cdot B) \cdot C = A \cdot (B \cdot C)$.

As shown in Example 6, matrix multiplication is *not commutative*.

Example 6   If $A = \begin{bmatrix} 2 & 1 \\ -1 & 3 \end{bmatrix}$ and $B = \begin{bmatrix} -1 & 5 \\ 4 & -2 \end{bmatrix}$, show that $A \cdot B \neq B \cdot A$.

$A \cdot B = \begin{bmatrix} 2(-1) + 1(4) & 2(5) + 1(-2) \\ -1(-1) + 3(4) & -1(5) + 3(-2) \end{bmatrix} = \begin{bmatrix} 2 & 8 \\ 13 & -11 \end{bmatrix}$

$B \cdot A = \begin{bmatrix} -1(2) + 5(-1) & -1(1) + 5(3) \\ 4(2) + -2(-1) & 4(1) + -2(3) \end{bmatrix} = \begin{bmatrix} -7 & 14 \\ 10 & -2 \end{bmatrix}$

Thus, $A \cdot B \neq B \cdot A$.

Written Exercises

Which of the following pairs of matrices can be multiplied? Why or why not?

A 1. $[3 \quad 2 \quad -1], \begin{bmatrix} 2 \\ -1 \\ 5 \end{bmatrix}$ **yes**

2. $\begin{bmatrix} 2 & -1 & 3 \\ 4 & 6 & -1 \end{bmatrix}, \begin{bmatrix} 1 & -3 \\ -5 & 4 \\ 6 & -2 \end{bmatrix}$ **yes**

3. $\begin{bmatrix} 4 & 5 & -6 \\ 3 & -1 & 7 \end{bmatrix}, \begin{bmatrix} -1 & 4 & -6 \\ 5 & 9 & -2 \end{bmatrix}$ **no**

4. $\begin{bmatrix} -1 & -5 \\ 6 & -3 \\ -4 & 7 \end{bmatrix}, \begin{bmatrix} 3 & -1 & 5 \\ -7 & 6 & -7 \end{bmatrix}$ **yes**

For Ex. 5–21, see TE Answer Section.

Find each product, if possible.

5. $-3 \begin{bmatrix} -2 & 3 & 5 \\ -6 & 10 & 6 \\ 8 & 12 & -4 \end{bmatrix}$

6. $9 \begin{bmatrix} -8 & -2 \\ 5 & -1 \\ 9 & 3 \end{bmatrix}$

7. $-1 \begin{bmatrix} 7 & 6 & 9 \\ 8 & -1 & -5 \end{bmatrix}$

8. $\begin{bmatrix} -2 & 1 \\ 4 & -3 \end{bmatrix} \cdot \begin{bmatrix} 4 & -5 \\ -1 & 2 \end{bmatrix}$

9. $\begin{bmatrix} -2 & 3 & 8 \\ 4 & -1 & -3 \end{bmatrix} \cdot \begin{bmatrix} 4 & 6 & 1 \\ -1 & 7 & -2 \\ 2 & -1 & 2 \end{bmatrix}$

10. $\begin{bmatrix} 1 & -3 & 2 \\ -1 & 2 & -4 \\ -2 & 5 & -3 \end{bmatrix} \cdot \begin{bmatrix} 1 & 4 & -1 \\ -2 & 5 & -4 \end{bmatrix}$

11. $\begin{bmatrix} 1 & -8 & -9 \\ 7 & 6 & -3 \end{bmatrix} \cdot \begin{bmatrix} -1 & 0 & 1 \\ 0 & 1 & -1 \\ -1 & 0 & 1 \end{bmatrix}$

Find the multiplicative identity and the multiplicative inverse for each matrix, if possible.

B 12. $\begin{bmatrix} 6 & -7 \\ -4 & 5 \end{bmatrix}$

13. $\begin{bmatrix} -8 & 2 \\ -6 & -1 \end{bmatrix}$

14. $\begin{bmatrix} -1 & 2 & -3 \\ -2 & 4 & 2 \end{bmatrix}$

15. $\begin{bmatrix} 7 & -3 & 2 \\ -1 & -2 & 5 \\ 6 & -4 & 8 \end{bmatrix}$

$A = \begin{bmatrix} -3 & 1 \\ -2 & -4 \end{bmatrix}, B = \begin{bmatrix} 1 & -6 \\ 3 & -2 \end{bmatrix}$ and $C = \begin{bmatrix} 7 & -3 \\ 2 & 5 \end{bmatrix}$. Prove that each of the following statements is true.

16. $(A \cdot B) \cdot C = A \cdot (B \cdot C)$
17. $B \cdot C \neq C \cdot B$
18. $A \cdot B \neq B \cdot A$
19. $A \cdot (B + C) = A \cdot B + A \cdot C$
20. $A \cdot I = I \cdot A$
21. $B \cdot B^{-1} = I$

$A = \begin{bmatrix} a_{11} & a_{12} \\ a_{21} & a_{22} \end{bmatrix}, B = \begin{bmatrix} b_{11} & b_{12} \\ b_{21} & b_{22} \end{bmatrix}$, and $C = \begin{bmatrix} c_{11} & c_{12} \\ c_{21} & c_{22} \end{bmatrix}$. All elements represent real numbers.

Determine which are true. Justify your answer by performing the computations.

C 22. $A \cdot B = B \cdot A$ **False**
23. $I \cdot A = A \cdot I = A$ **True**
24. $(A \cdot B)^{-1} = B^{-1} \cdot A^{-1}$ **True**
25. $A \cdot (B + C) = A \cdot B + A \cdot C$ **True**
26. $(A \cdot B)^2 = A^2 \cdot B^2$ **False**
27. $A \cdot A^{-1} = A^{-1} \cdot A = I$ **True**

CUMULATIVE REVIEW

1. Solve $\sqrt{x - 8} + \sqrt{x} = 4$. **9**

2. Find the distance from $A(3, 5)$ to $B(-6, 4)$. $\sqrt{82}$

Matrix Multiplication

SOLVING LINEAR SYSTEMS USING MATRICES 10.8

Objectives
To solve a system of two linear equations in two variables using the inverse of a matrix
To solve a system of three linear equations in three variables using matrix transformations

The general system of two linear equations $\begin{array}{c} a_1x + b_1y = c_1 \\ a_2x + b_2y = c_2 \end{array}$ can be written in matrix form as

$$\begin{bmatrix} a_1 & b_1 \\ a_2 & b_2 \end{bmatrix} \begin{bmatrix} x \\ y \end{bmatrix} = \begin{bmatrix} c_1 \\ c_2 \end{bmatrix}.$$

This can be verified by multiplying the two matrices on the left side. Using the definition of matrix equality, you get

$$\begin{bmatrix} a_1x + b_1y \\ a_2x + b_2y \end{bmatrix} = \begin{bmatrix} c_1 \\ c_2 \end{bmatrix}, \text{ or } \begin{array}{c} a_1x + b_1y = c_1 \\ a_2x + b_2y = c_2. \end{array}$$

Let $A = \begin{bmatrix} a_1 & b_1 \\ a_2 & b_2 \end{bmatrix}$, $X = \begin{bmatrix} x \\ y \end{bmatrix}$, and $B = \begin{bmatrix} c_1 \\ c_2 \end{bmatrix}$, then $AX = B$.

$A^{-1}(AX) = A^{-1}B$ ◀ Multiply both sides by A^{-1}.
$(A^{-1}A)X = A^{-1}B$ ◀ $A^{-1}A = I$
$IX = A^{-1}B$ ◀ $IX = X$
$X = A^{-1}B$

Substituting, $\begin{bmatrix} x \\ y \end{bmatrix} = \begin{bmatrix} a_1 & b_1 \\ a_2 & b_2 \end{bmatrix}^{-1} \begin{bmatrix} c_1 \\ c_2 \end{bmatrix}.$

So, the solution can be found by multiplying the inverse of A by B.

Example 1 **Using the inverse of a matrix, solve the system** $\begin{array}{c} -2x + 3y = -8 \\ x + 2y = -3. \end{array}$

Write in matrix form.

$$\begin{bmatrix} -2 & 3 \\ 1 & 2 \end{bmatrix} \begin{bmatrix} x \\ y \end{bmatrix} = \begin{bmatrix} -8 \\ -3 \end{bmatrix}$$

Use $X = A^{-1}B$. ▶ $\begin{bmatrix} x \\ y \end{bmatrix} = \begin{bmatrix} -2 & 3 \\ 1 & 2 \end{bmatrix}^{-1} \begin{bmatrix} -8 \\ -3 \end{bmatrix}$

Let $A^{-1} = \begin{bmatrix} a_1 & b_1 \\ a_2 & b_2 \end{bmatrix}$. ▶ Find A^{-1}.

$$\begin{bmatrix} -2 & 3 \\ 1 & 2 \end{bmatrix} \begin{bmatrix} a_1 & b_1 \\ a_2 & b_2 \end{bmatrix} = \begin{bmatrix} 1 & 0 \\ 0 & 1 \end{bmatrix}$$

Multiply. ▶ $\begin{bmatrix} -2a_1 + 3a_2 & -2b_1 + 3b_2 \\ a_1 + 2a_2 & b_1 + 2b_2 \end{bmatrix} = \begin{bmatrix} 1 & 0 \\ 0 & 1 \end{bmatrix}$

$\begin{array}{cc} -2a_1 + 3a_2 = 1 & -2b_1 + 3b_2 = 0 \\ a_1 + 2a_2 = 0 & b_1 + 2b_2 = 1 \end{array}$

Chapter Ten

Solve each system of equations.

$$a_1 = \frac{\begin{vmatrix} 1 & 3 \\ 0 & 2 \end{vmatrix}}{D}, \quad a_2 = \frac{\begin{vmatrix} -2 & 1 \\ 1 & 0 \end{vmatrix}}{D}, \quad b_1 = \frac{\begin{vmatrix} 0 & 3 \\ 1 & 2 \end{vmatrix}}{D}, \quad b_2 = \frac{\begin{vmatrix} -2 & 0 \\ 1 & 1 \end{vmatrix}}{D}$$

$$D = \begin{bmatrix} -2 & 3 \\ 1 & 2 \end{bmatrix} = [-4 - 3], \text{ or } -7$$

$$a_1 = \frac{2 - 0}{-7}, \text{ or } -\frac{2}{7} \qquad a_2 = \frac{0 - 1}{-7}, \text{ or } \frac{1}{7}$$

$$b_1 = \frac{0 - 3}{-7}, \text{ or } \frac{3}{7} \qquad b_2 = \frac{-2 - 0}{-7}, \text{ or } \frac{2}{7}$$

So, $A^{-1} = \begin{bmatrix} -\frac{2}{7} & \frac{3}{7} \\ \frac{1}{7} & \frac{2}{7} \end{bmatrix}$.

$X = A^{-1}B \blacktriangleright \begin{bmatrix} x \\ y \end{bmatrix} = \begin{bmatrix} -\frac{2}{7} & \frac{3}{7} \\ \frac{1}{7} & \frac{2}{7} \end{bmatrix} \begin{bmatrix} -8 \\ -3 \end{bmatrix}$ or $\begin{bmatrix} \frac{16}{7} - \frac{9}{7} \\ -\frac{8}{7} - \frac{6}{7} \end{bmatrix}$ or $\begin{bmatrix} 1 \\ -2 \end{bmatrix}$

Thus, $(1, -2)$ is the solution of the system.
By applying a series of steps, called **transformations,** the coefficients and constants of a system of three linear equations can be changed until the system is in triangular form like the system of linear equations shown on the right.

$$\begin{aligned} 2x - 3y + 2z &= 3 \\ 2y - 3z &= -6 \\ z &= 4 \end{aligned}$$

The solution of a system is easily obtained from triangular form. $z = 4$; so, $2y - 3(4) = -6$, or $y = 3$; and $2x - 3 \cdot 3 + 2 \cdot 4 = 3$, or $x = 2$; and the solution is $(2, 3, 4)$.

The matrix form of the system of equations shown on the right is

$$\begin{aligned} 2x - 3y - 3z &= -1 \\ -x + 2y + 4z &= 5 \\ 3x - 4y - 5z &= -3 \end{aligned}$$

$$\begin{bmatrix} 2 & -3 & -3 \\ -1 & 2 & 4 \\ 3 & -4 & -5 \end{bmatrix} \begin{bmatrix} x \\ y \\ z \end{bmatrix} = \begin{bmatrix} -1 \\ 5 \\ -3 \end{bmatrix}.$$

The matrix of coefficients is sometimes called the **matrix of detached coefficients.**

In working with matrix transformations, it is convenient to write the matrix of detached coefficients and the matrix of constants in one combined matrix, called an **augmented matrix.** The augmented matrix for

$$\begin{bmatrix} 2 & -3 & -3 \\ -1 & 2 & 4 \\ 3 & -4 & -5 \end{bmatrix} \begin{bmatrix} x \\ y \\ z \end{bmatrix} = \begin{bmatrix} -1 \\ 5 \\ -3 \end{bmatrix} \text{ is } \begin{bmatrix} 2 & -3 & -3 & -1 \\ -1 & 2 & 4 & 5 \\ 3 & -4 & -5 & -3 \end{bmatrix}.$$

There are three elementary transformations that can be used to change an augmented matrix into an equivalent augmented matrix.

Solving Linear Systems Using Matrices

> **Elementary matrix transformations**
>
> Two augmented matrices are equivalent if one is obtained from the other by
> (1) interchanging any two rows;
> (2) multiplying each element of any row by a nonzero number; or
> (3) adding the elements of any row to the corresponding elements of another row.

You can solve a system of three linear equations by using elementary matrix transformations. Begin by writing an augmented matrix. Then, transform it to an equivalent augmented matrix in triangular form with ones in the main diagonal.

$$\begin{bmatrix} 1 & b_1 & c_1 & d_1 \\ 0 & 1 & c_2 & d_2 \\ 0 & 0 & 1 & d_3 \end{bmatrix}$$
Augmented matrix in triangular form

Example 2

Solve the system by matrix transformations:
$$5x - 2y + 3z = -7$$
$$-2x + 4y + z = 4$$
$$3x - 3y + 4z = 2.$$

Augmented matrix → $\begin{bmatrix} 5 & -2 & 3 & -7 \\ -2 & 4 & 1 & 4 \\ 3 & -3 & 4 & 2 \end{bmatrix}$ $\begin{bmatrix} 1 & 6 & 5 & 1 \\ 0 & -5 & 0 & 5 \\ 0 & 21 & 11 & 1 \end{bmatrix}$ ◀ Add row 3 to row 2.
◀ Multiply row 3 by -1.

Add 2 times row 2 to row 1. → $\begin{bmatrix} 1 & 6 & 5 & 1 \\ -2 & 4 & 1 & 4 \\ 3 & -3 & 4 & 2 \end{bmatrix}$ $\begin{bmatrix} 1 & 6 & 5 & 1 \\ 0 & 1 & 0 & -1 \\ 0 & 21 & 11 & 1 \end{bmatrix}$ ◀ Multiply row 2 by $-\frac{1}{5}$.

Add 2 times row 1 to row 2. → $\begin{bmatrix} 1 & 6 & 5 & 1 \\ 0 & 16 & 11 & 6 \\ 3 & -3 & 4 & 2 \end{bmatrix}$ $\begin{bmatrix} 1 & 6 & 5 & 1 \\ 0 & 1 & 0 & -1 \\ 0 & 0 & 11 & 22 \end{bmatrix}$

◀ Add -21 times row 2 to row 3.

Add -3 times row 1 to row 3. → $\begin{bmatrix} 1 & 6 & 5 & 1 \\ 0 & 16 & 11 & 6 \\ 0 & -21 & -11 & -1 \end{bmatrix}$ $\begin{bmatrix} 1 & 6 & 5 & 1 \\ 0 & 1 & 0 & -1 \\ 0 & 0 & 1 & 2 \end{bmatrix}$ ◀ Multiply row 3 by $\frac{1}{11}$.

Solve the system $1x + 6y + 5z = 1$ for x. $1x + 6(-1) + 5(2) = 1$
$1y = -1$ $x = -3$
$1z = 2$

Thus, $(-3, -1, 2)$ is the solution of the system.

Whenever a matrix can be transformed into an augmented matrix of the form $\begin{bmatrix} 1 & 0 & 0 & d_1 \\ 0 & 1 & 0 & d_2 \\ 0 & 0 & 1 & d_3 \end{bmatrix}$, it is even easier to solve.

If you continue to transform the augmented matrix in Example 2 above, you get $\begin{bmatrix} 1 & 0 & 0 & -3 \\ 0 & 1 & 0 & -1 \\ 0 & 0 & 1 & 2 \end{bmatrix}$, which is equivalent to $\begin{bmatrix} 1 & 0 & 0 \\ 0 & 1 & 0 \\ 0 & 0 & 1 \end{bmatrix} \begin{bmatrix} x \\ y \\ z \end{bmatrix} = \begin{bmatrix} -3 \\ -1 \\ 2 \end{bmatrix}$.

So, $x = -3$, $y = -1$, and $z = 2$.

Written Exercises

Solve by using the inverse of a matrix.

(A)

1. $2x - 3y = 5$
    $-x + 2y = -3$
    (1, −1)

2. $-2x + y = -6$
    $3x - 3y = 12$
    (2, −2)

3. $x - 4y = 3$
    $-3x + 5y = -2$
    (−1, −1)

4. $4x - y = -6$
    $-2x - 3y = 10$
    (−2, −2)

5. $-x + 3y = -6$
    $-2x + y = -7$
    (3, −1)

6. $3x - 2y = -1$
    $2x + 3y = 8$
    (1, 2)

Solve by using matrix transformations.

(B)

7. $-x + 2y - 3z = -8$
    $2x + 3y + z = -3$
    $-2x - y + 2z = 2$
    (1, −2, 1)

8. $3x - 2y + 2z = -1$
    $-2x + 3y - 3z = -1$
    $-x - y + 4z = 8$
    (−1, 1, 2)

9. $-2x + 3y - 2z = -1$
    $-3x + 2y - z = -5$
    $4x - 3y + 2z = 5$
    (2, −1, −3)

10. $4x - 2y + z = 2$
    $-x - 3y + 2z = -5$
    $2x + y + 3z = -4$
    (1, 0, −2)

11. $-3x + 4y + z = 5$
    $2x + 5y + 3z = 0$
    $-4x - 2y + 2z = 16$
    (−2, −1, 3)

12. $2x - 3y + z = 10$
    $-x + 2y + 3z = 1$
    $-3x + y - 2z = -9$
    (1, −2, 2)

13. $-3x + 2y + 3z = 7$
    $2x + 3y - z = -7$
    $-x - y + z = 4$
    (−1, −1, 2)

14. $2x - 3y + 3z = -12$
    $3x - 2y - 2z = 4$
    $-3x + 2y - z = 5$
    (0, 1, −3)

15. $-4x + 5y + 2z = 2$
    $3x + 2y + 3z = -4$
    $-3x - 5y + 2z = 20$
    (−2, −2, 2)

16. $6x - 3y + 4z = -4$
    $-2x + 5y + 3z = 2$
    $3x - 2y - 2z = -9$
    (−3, −2, 2)

17. $5x + 2y + 3z = 12$
    $3x - 5y + 2z = 13$
    $-2x + 3y + 4z = -19$
    (4, −1, −2)

18. $-3x - 4y + 5z = 4$
    $2x + 5y + 3z = -1$
    $-4x - y + 2z = -8$
    (3, −2, 1)

(C)

19. $-2a + 3b - c + 2d = -11$
    $3a - 2b + 3c + d = 9$
    $-4a + 5b - 3c - 3d = -9$
    $4a - 3b + 4c + 4d = 7$ **(1, −1, 2, −2)**

20. $5a + 2b + 3c - 2d = 4$
    $-3a + 3b - 2c + 4d = 9$
    $-2a - 3b + 4c - 2d = -14$
    $3a + 4b - 5c + 3d = 19$ **(1, 2, −1, 1)**

21. $3r + 2s - 2t + 4v = 4$
    $-2r - 3s + 4t - 3v = 4$
    $4r + 2s - 3t + v = 1$
    $-3r - 4s + 3t + 5v = 9$ **(2, −1, 2, 1)**

22. $-2r + 6s - 3t + 3v = -5$
    $4r - 3s + 2t - v = -3$
    $-3r - 4s - 3t + 4v = -15$
    $5r + 2s - 2t + 3v = -20$ **(−2, 1, 3, −2)**

23. If $A = \begin{bmatrix} a_1 & b_1 \\ a_2 & b_2 \end{bmatrix}$, then prove that $A^{-1} = \dfrac{1}{\text{Determinant } A} \begin{bmatrix} b_2 & -b_1 \\ -a_2 & a_1 \end{bmatrix}$. **See TE Answer Section.**

Find A^{-1} by using the formula in Exercise 23.

24. $A = \begin{bmatrix} 3 & 5 \\ -2 & 4 \end{bmatrix}$ $\dfrac{1}{22}\begin{bmatrix} 4 & -5 \\ 2 & 3 \end{bmatrix}$

25. $A = \begin{bmatrix} -1 & -2 \\ 3 & -4 \end{bmatrix}$ $\dfrac{1}{10}\begin{bmatrix} -4 & 2 \\ -3 & -1 \end{bmatrix}$

26. $A = \begin{bmatrix} -1 & 4 \\ -2 & 3 \end{bmatrix}$ $\dfrac{1}{5}\begin{bmatrix} 3 & -4 \\ 2 & -1 \end{bmatrix}$

27. $A = \begin{bmatrix} c & -d \\ e & f \end{bmatrix}$ $\dfrac{1}{cf + de}\begin{bmatrix} f & d \\ -e & c \end{bmatrix}$

Solving Linear Systems Using Matrices

PROBLEM SOLVING USING LINEAR SYSTEMS

10.9

Objectives **To solve a digit problem using a system of linear equations**
To solve various types of word problems using systems of linear equations

A two-digit number like 76 can be written as follows:
$$76 = 70 + 6$$
$$= 10(7) + 1(6),$$
where 7 is the *tens digit* and 6 is the *units digit*.

To *reverse the digits* of a two-digit number, interchange the positions of the tens and units digits as follows: $76 = 10(7) + 6$; its reverse, $67 = 10(6) + 7$.

The *sum of the digits* of a two-digit number like 76 is $7 + 6$, or 13.

Two-digit number	Tens digit t	Units digit u	Original number $10t + u$	Sum of the digits $t + u$	Number with its digits reversed $10u + t$

Systems of linear equations can often be used to solve problems involving two-digit numbers.

Example 1

READ ▶

In a certain two-digit number, the units digit is 24 less than 3 times the sum of the digits. If the digits are reversed, the new number is 18 more than the original number. Find the two-digit number.

PLAN ▶

Let t = tens digit and u = units digit.
The units digit is 24 less than 3 times the sum of the digits.
$$u = 3(t + u) - 24 \quad \text{or} \quad 2u + 3t = 24$$

When the digits are reversed, the new number is 18 more than the original.
$$10u + t = 10t + u + 18 \quad \text{or} \quad u - t = 2$$

SOLVE ▶

Solve the system $2u + 3t = 24$.
$u - t = 2$

$\begin{aligned} 2u + 3t &= 24 \\ 3u - 3t &= 6 \\ \hline 5u &= 30 \\ \boxed{u} &= \boxed{6} \end{aligned}$

$\begin{aligned} u - t &= 2 \\ 6 - t &= 2 \\ \boxed{t = 4} \end{aligned}$

INTERPRET ▶

The two-digit number is $10t + u = 10 \cdot 4 + 6$, or 46. (Check the answer in the problem.)

Thus, the two-digit number is 46.

A three-digit number like 658 can be written as follows:
$$658 = 600 + 50 + 8$$
$$= 100(6) + 10(5) + 1(8),$$ where 6 is the hundreds digit, 5 is the tens digit, and 8 is the units digit.

When the digits are reversed, you get
$$856 = 100(8) + 10(5) + 6.$$

The sum of the digits of the number is $6 + 5 + 8$, or 19.

Three-digit number	Hundreds digit	Tens digit	Units digit	Original number	Sum of the digits	Number with its digits reversed
	h	t	u	$100h + 10t + u$	$h + t + u$	$100u + 10t + h$

Example 2

In a certain three-digit number, if the sum of the digits is doubled, the result is 16 more than 3 times the tens digit. One more than twice the hundreds digit is the units digit. The hundreds digit is 11 decreased by twice the tens digit. Find the three-digit number.

Represent the data. Let h = hundreds digit, t = tens digit, and u = units digit.

Write a system of equations.

(1) $2(h + t + u) = 3t + 16$
$2h + 2t + 2u = 3t + 16$
$2h - t + 2u = 16$

(2) $2h + 1 = u$
$2h - u = -1$

(3) $h = 11 - 2t$
$h + 2t = 11$

$$2h - t + 2u = 16$$
$$2h - u = -1$$
$$h + 2t = 11$$

Solve the system. Solve the system for h using determinants.

$$h = \frac{\begin{vmatrix} 16 & -1 & 2 \\ -1 & 0 & -1 \\ 11 & 2 & 0 \end{vmatrix} \begin{matrix} 16 & -1 \\ -1 & 0 \\ 11 & 2 \end{matrix}}{\begin{vmatrix} 2 & -1 & 2 \\ 2 & 0 & -1 \\ 1 & 2 & 0 \end{vmatrix} \begin{matrix} 2 & -1 \\ 2 & 0 \\ 1 & 2 \end{matrix}} = \frac{[0 + 11 + (-4)] - [0 + (-32) + 0]}{[0 + 1 + 8] - [0 + (-4) + 0]} = \frac{39}{13}, \text{ or } 3$$

$2h + 1 = u$ \qquad $h + 2t = 11$
$2 \cdot 3 + 1 = u$ \qquad $3 + 2t = 11$
$\boxed{h = 3}$ \qquad $\boxed{7 = u}$ \qquad $\boxed{t = 4}$

Check the answer. The three-digit number is $100h + 10t + u = 100 \cdot 3 + 10 \cdot 4 + 7$, or 347. (Check the answer in the problem.)

Thus, the three-digit number is 347.

Problem Solving Using Linear Systems

In Chapter 2, you learned how to solve various types of mixture problems by writing and then solving one equation in one variable. Many of these same problems can be solved by using a system of two equations in two variables.

Example 3 Some dried fruit worth $3.45/kg is to be mixed with some raisins worth $2.10/kg to make a mixture worth $2.64/kg. How many kilograms of dried fruit and how many kilograms of raisins should be used to make 10.5 kg of the mixture?

Represent the data. Represent the data in a table. Use two variables.

	No. of kg	Value/kg	Value in ¢
Dried fruit	x	345	$345x$
Raisins	y	210	$210y$
Mixture	10.5	264	$264(10.5)$

Write a system of equations.

$$(\text{kg of fruit}) + (\text{kg of raisins}) = (\text{kg of mixture})$$
$$x \quad + \quad y \quad = \quad 10.5$$

$$(\text{Value of fruit}) + (\text{Value of raisins}) = (\text{Value of mixture})$$
$$345x \quad + \quad 210y \quad = \quad 264(10.5)$$

Solve the system. Solve the system $x + y = 10.5$ by substitution.
$345x + 210y = 2772$

$$\begin{aligned} x + y &= 10.5 \\ y &= 10.5 - x \end{aligned} \qquad \begin{aligned} 345x + 210y &= 2772 \\ 345x + 210(10.5 - x) &= 2772 \\ 345x + 2205 - 210x &= 2772 \\ 135x &= 567 \\ \boxed{x = 4.2} \end{aligned} \qquad \begin{aligned} y &= 10.5 - x \\ y &= 10.5 - 4.2 \\ \boxed{y = 6.3} \end{aligned}$$

Check the answer.

(4.2 kg at $3.45/kg) + (6.3 kg at $2.10/kg)	(10.5 kg at $2.64/kg)
$14.49 + $13.23	$27.72
$27.72	

Thus, 4.2 kg of dried fruit and 6.3 kg of raisins should be used.

In the exercises on page 271, you will be asked to solve a variety of word problems using a system of two (three) linear equations in two (three) variables.

Written Exercises

Use a system of two (three) linear equations in two (three) variables to solve each problem.

A

1. The tens digit of a two-digit number is 3 more than twice the units digit. If the digits are reversed, the resulting number is 45 less than the original number. Find the original number. **72**

2. A two-digit number is 9 less than 4 times the sum of its digits. If the digits are reversed, the new number is 54 more than the original number. Find the original number. **39**

3. A two-digit number is 27 more than the number obtained by reversing the digits. The number is also 38 more than twice the sum of the digits. Find the two-digit number. **52**

4. The sum of the digits of a two-digit number is 8. If the digits are reversed, the new number is 18 more than the original number. Find the two-digit number. **35**

5. Brand A coffee worth $5.60/kg is to be mixed with Brand B coffee worth $6.10/kg to make 15 kg of a mixture worth $5.90/kg. How many kilograms of each brand should be mixed? **brand A: 6 kg, brand B: 9 kg**

6. Powdered milk worth $3.60/kg and cocoa worth $5.20/kg are to be mixed so that their combination is worth $82.40. How many kilograms of milk should be used if the weight of the milk is 10 times that of the cocoa? **20 kg**

B

7. The tens digit of a three-digit number is 7 less than 3 times the units digit. Twice the hundreds digit is 10 less than 3 times the tens digit. Three times the sum of the three digits is 4 more than 8 times the hundreds digit. Find the three-digit number. **785**

8. A three-digit number is 198 less than the number obtained by reversing the digits. Twice the sum of the digits is 5 more than 7 times the tens digit. The tens digit is 5 less than twice the hundreds digit. Find the original number. **436**

9. Ralph is 8 years younger than Lisa but 6 years ago, she was three times as old as he was. How old is each now? **Ralph: 10, Lisa: 18**

10. A certain boat can travel 24 km downstream in 48 minutes and 21 km upstream in 72 minutes. Find the rate of the boat in still water and the rate of the current in kilometers per hour. **boat: 23.75 km/h, current: 6.25 km/h**

11. Mrs. Brown invested a total of $6,400, one part at 9% per year and the rest at 8% per year. At the end of 2 years 6 months, she had earned a total of $1,380 in simple interest. How much did she invest at each rate? **$4,000 at 9%, $2,400 at 8%**

12. Sunflower seeds worth $2.90/kg, peanuts worth $2.50/kg, and almonds worth $3.30/kg are to be mixed to make 40 kg worth $3.00/kg. If the combined weight of the peanuts and the almonds is triple that of the seeds, how many kilograms of each should be used? **sunflower: 10 kg, peanut: 10 kg, almond: 20 kg**

C

13. A chemist mixed iodine solutions of 25%, 40%, and $37\frac{1}{2}$% to obtain 15 L of a $33\frac{1}{3}$% iodine solution. Find the volume of each solution if the volume of the 25% solution was $1\frac{1}{2}$ times the volume of the $37\frac{1}{2}$% solution. **25%: 6 L, 40%: 5 L, $37\frac{1}{2}$%: 4 L**

Problem Solving Using Linear Systems

Ⓐ Exercises are 1, 2, 5–9, 11, 13, 14, 17–19, 21–23, 27. Ⓒ Exercises are starred. The rest are Ⓑ Exercises.

CHAPTER TEN REVIEW

Vocabulary
determinant [10.4]
  2 by 2 [10.4]   3 by 3 [10.5]
matrix [10.4]
  additive identity [10.6]
  additive inverse [10.6]
  augmented [10.8]
  dimensions of [10.6]
  multiplicative identity [10.7]
  multiplicative inverse [10.7]
  square [10.4]
ordered triple [10.3]   scalar [10.7]
system of equations [10.1]
  consistent [10.1]   dependent [10.1]
  inconsistent [10.1]   independent [10.1]

Solve each system of equations graphically. Classify each system as consistent, inconsistent, dependent, or independent. [10.1]

1. $3x + 2y - 14 = 0$
   $2y - 3x + 10 = 0$

2. $5y - 2x = 15$
   $2x - 5y = 15$ **below**

3. $3(x + 2y) = 2y + 8$ **all (x, y) such that**
   $4y = 8 - 3x$ $y = -\frac{3}{4}x + 2$; con, dep

4. $-2(y - 3x) = (y + 5x) + 9$ **(−3, −4); con, ind**
   $2(3x - 2y) = -(5x - 7y) + 11$

Solve each system of equations by substitution. [10.2]

5. $5x + 2y = 7$
   $4x - y = 3$ **(1, 1)**

6. $4x + 5y = 12$ **(3, 0)**
   $6x - y = 18$

Solve each system of equations by addition. [10.2]

7. $4x - 3y - 10 = 0$
   $5x + 6y + 7 = 0$ **(1, −2)**

8. $5x + 4y + 8 = 0$
   $7y = 3x + 33$ **(−4, 3)**

9. $4x - 6y = 11$
   $-2x + 3y = 9$

★ 10. $c_1 x + d_1 y = e_1$
     $c_2 x + d_2 y = e_2$

Solve the system of equations. [10.3]

11. $x + 3y + 2z = 11$
    $4x + 2y + 5z = 15$
    $-2x + 3y - z = 5$ **(−1, 2, 3)**

1. **(4, 1) con, ind**
2. **no solution; inc**

Solve by using Cramer's rule. [10.4]

12. $-3x + 4y = -6$
    $5x - 3z = -22$
    $3y + 2z = -1$ **(−2, −3, 4)**

9. **no solution**

Find the value of each determinant. [10.4]

13. $\begin{vmatrix} 3 & -2 \\ 4 & 8 \end{vmatrix}$ **32**

14. $\begin{vmatrix} a - b & -b \\ -3 & 2 \end{vmatrix}$ **below**

15. $\begin{vmatrix} m + n & m \\ m & m - n \end{vmatrix}$

16. $\begin{vmatrix} a - b & a + b \\ a - b & a - b \end{vmatrix}$ **below**

Using minors, find the value of each determinant. [10.5]

17. $\begin{vmatrix} -2 & 1 & 2 \\ 3 & -1 & 4 \\ 5 & 2 & 3 \end{vmatrix}$ **55**

18. $\begin{vmatrix} a & 0 & b \\ 0 & 1 & 0 \\ c & 0 & d \end{vmatrix}$ **ad − bc**

Solve each system of equations by using determinants. [10.4], [10.5]

19. $-2x + 5y = 19$
    $4x - 7y = -29$
    (−2, 3)

20. $3x + 2y - 5z = 5$
    $2x - y + z = 6$
    $x + 4y - 6z = -4$
    (2, −3, −1)

Solve by using the inverse of a matrix. [10.8]

21. $2x - y = 0.7$
    $3x - 4y = -0.2$ **(0.6, 0.5)**

For Ex. 10, 22–25, see TE Answer Section.

22. Find the sum, if possible: [10.6]

$\begin{bmatrix} -1 & 2 & 3 \\ 2 & 3 & -1 \\ 4 & -2 & 5 \end{bmatrix} + \begin{bmatrix} 6 & -1 & 2 \\ -3 & 4 & 5 \\ 2 & -1 & -3 \end{bmatrix}$

23. Find the product, if possible: [10.7]

$\begin{bmatrix} -2 & 3 & 1 \\ 4 & -1 & 2 \end{bmatrix} \cdot \begin{bmatrix} 1 & -1 & 2 \\ -3 & 4 & 1 \\ 2 & -2 & 3 \end{bmatrix}$

Find the multiplicative identity, additive inverse, and the multiplicative inverse for each, if possible. [10.6], [10.7]

24. $\begin{bmatrix} 7 & -3 \\ 2 & 1 \end{bmatrix}$

25. $\begin{bmatrix} 3 & -1 & 2 \\ 2 & 1 & 2 \\ -3 & 2 & -1 \end{bmatrix}$

Solve by using matrix transformations. [10.8]

26. $3x - 2y + 2z = 10$
    $-2x + y - 3z = -8$
    $x + 3y - 2z = -3$ **(2, −1, 1)**

Use a system of two equations in two variables to solve the problem. [10.9]

27. A two-digit number is 45 less than the number obtained by reversing the digits. Find the original number if twice the sum of the digits is 1 more than 7 times the tens digit. **38**

14. $2a - 5b$ 15. $-n^2$ 16. $2b^2 - 2ab$

CHAPTER TEN TEST

Ⓐ Exercises are 1–5, 7, 9, 10, 12, 13, 16, 18, 19. Ⓒ Exercises are starred. The rest are Ⓑ Exercises.

Solve each system graphically. Classify each system as consistent, inconsistent, dependent, or independent.

1. $4y - 3x - 8 = 0$
   $x + 2y + 6 = 0$
   (−4, −1); con, ind

2. $4y - 5x = 8$
   $5x - 4y = 12$
   no solution; inc

3. Solve the system by substitution.
   $4x - 3y = 3$
   $x - 2y = -8$ **(6, 7)**

4. Solve the system by addition.
   $2x + 3y = 10$
   $3x + 5y = 19$ **(−7, 8)**

Solve each system of equations.

5. $2x - 3y + 5z = 13$
   $-3x + 3y - 2z = -8$
   $5x - 2y + 4z = 2$ **(−2, −4, 1)**

6. $2x + 5y = -2$
   $5x - 2z = 14$
   $3y + 4z = 6$ **(4, −2, 3)**

Find the value of each determinant.

7. $\begin{vmatrix} -3 & 4 \\ 6 & -2 \end{vmatrix}$ **−18**

8. $\begin{vmatrix} a+b & 2a+b \\ b & a+b \end{vmatrix}$ **a^2**

Using minors, find the value of the determinant.

9. $\begin{vmatrix} 3 & -1 & 2 \\ -2 & 3 & 4 \\ 2 & 1 & -2 \end{vmatrix}$ **−50**

Solve the system of equations by using determinants.

10. $3x + 5y = 1$
    $-2x + 3y = 12$ **(−3, 2)**

Solve by using Cramer's rule.

11. $3x + 2y + 2z = -3$
    $2x + 3y + 3z = -2$
    $-3x - 5y + z = -9$ **(−1, 2, −2)**

12. Find the sum, if possible:
    $\begin{bmatrix} 3 & -1 & 2 \\ 4 & 5 & 3 \\ -2 & -6 & -1 \end{bmatrix} + \begin{bmatrix} -2 & 3 & 4 \\ -3 & -2 & 1 \\ 4 & 6 & -5 \end{bmatrix}$ **below**

13. Find the product, if possible:
    $\begin{bmatrix} 2 & 1 & 3 \\ -3 & 2 & -4 \end{bmatrix} \cdot \begin{bmatrix} -1 & 4 & 6 \\ 2 & -1 & -3 \\ 3 & -2 & 1 \end{bmatrix}$

13. $\begin{bmatrix} 9 & 1 & 12 \\ -5 & -6 & -28 \end{bmatrix}$
14. $\begin{bmatrix} -3 & -2 \\ 1 & -2 \end{bmatrix}, \begin{bmatrix} 1 & 0 \\ 0 & 1 \end{bmatrix}$

Find the additive inverse and the multiplicative identity for each, if possible.

14. $\begin{bmatrix} 3 & 2 \\ -1 & 2 \end{bmatrix}$

15. $\begin{bmatrix} 4 & -1 & 6 \\ -3 & 2 & 1 \\ 2 & -4 & -3 \end{bmatrix}$ **below**

Solve by using the inverse of a matrix.

16. $2x - 3y = 5$
    $-3x + y = -4$ **(1, −1)**

17. Solve by using matrix transformations.
    $4x - 3y + z = -9$
    $-3x + 2y - 5z = -10$
    $2x - 4y + 3z = 8$ **(−4, −1, 4)**

Use a system of two linear equations in two variables to solve each problem.

18. If 3 more than a first number is divided by a second number and if 5 less than the second number is divided by 1 more than the first number, both quotients are $\frac{3}{4}$. Find the numbers. **1st: 3; 2nd: 8**

19. A certain two-digit number is 27 less than the number obtained by reversing the digits. Three times the sum of the digits is 3 more than 7 times the tens digit. Find the two-digit number. **69**

★20. Solve this system using matrix transformations:
    $3a + 4b - 2c + 5d = -5$
    $-2a - 3b + c + 2d = 14$
    $4a + 2b + 3c - 3d = -8$
    $-5a - 2b + 4c + 3d = 23$ **(−1, −2, 2, 2)**

12. $\begin{bmatrix} 1 & 2 & 6 \\ 1 & 3 & 4 \\ 2 & 0 & -6 \end{bmatrix}$

15. $\begin{bmatrix} -4 & 1 & -6 \\ 3 & -2 & -1 \\ -2 & 4 & 3 \end{bmatrix}, \begin{bmatrix} 1 & 0 & 0 \\ 0 & 1 & 0 \\ 0 & 0 & 1 \end{bmatrix}$

COLLEGE PREP TEST

DIRECTIONS: Choose the *one* best answer to each question or problem.

1. Which ordered pair (x, y) is a solution of
   $ax + by = 7ab$? **B**
   $3ax - 2by = 6ab$

   (A) $(3b, 4a)$ (B) $(4b, 3a)$
   (C) $(2b, 0)$ (D) all of these
   (E) None of these

2. The equation of the graph shown is **D**

   (A) $y = -\frac{1}{2}x + \frac{1}{2}$
   (B) $y = -\frac{2}{3}x + \frac{1}{2}$
   (C) $y = \frac{1}{2}x - 1$
   (D) $y = -\frac{3}{2}x + \frac{1}{2}$
   (E) $y = \frac{1}{2}x - \frac{3}{2}$

3. $\dfrac{\begin{vmatrix} a & b \\ c & d \end{vmatrix}}{\begin{vmatrix} 2b & -a \\ 2d & -c \end{vmatrix}} =$ **C**

   (A) $-\frac{1}{2}$ (B) $-\frac{1}{4}$ (C) $\frac{1}{2}$
   (D) $\frac{1}{4}$ (E) None of these

4. If the value of $\begin{vmatrix} 2x & -4 \\ x & x \end{vmatrix}$ is 6, then x equals **E**

   (A) 1 (B) -1 (C) -1 or 3
   (D) -3 (E) 1 or -3

5. The solution of the system
   $\frac{x}{2} + y + z = 6$
   $x + \frac{y}{3} - z = 6$
   $x + y + 2z = 5$ is **E**

   (A) $(4, 1, 1)$ (B) $(6, 3, 1)$ (C) $(4, 3, -1)$
   (D) $(1, 1, 2)$ (E) None of these

6. In a collection of pennies, nickels, and dimes, the pennies and nickels are worth 35¢, the nickels and dimes are worth 80¢, and the pennies and dimes are worth 75¢. What is the value of the collection? **D**

   (A) $1.90 (B) $1.55 (C) $1.15
   (D) 95¢ (E) None of these

7. Which of the following matrices are equivalent? **A**

   $R = \begin{bmatrix} 1 & 5 & 6 \\ 2 & 3 & -1 \end{bmatrix}$ $S = \begin{bmatrix} 2 & 3 & -1 \\ 1 & 5 & 6 \end{bmatrix}$

   $T = \begin{bmatrix} 3 & 8 & -5 \\ 2 & 3 & -1 \end{bmatrix}$ $U = \begin{bmatrix} 2 & 6 & 7 \\ 2 & 3 & -1 \end{bmatrix}$

   (A) R and S (B) R and T
   (C) T and U (D) R, S, and T
   (E) R, S, T, and U

8. Consider square matrices A, B, C, and I, each having the same dimensions and with I the multiplicative identity element. Which of the following statements is false? **B**

   (A) $A + B = B + A$
   (B) $AB = BA$
   (C) $AI = IA = A$
   (D) $A + (B + C) = (A + B) + C$
   (E) $(A \cdot B) \cdot C = A \cdot (B \cdot C)$

11 FUNCTIONS

CAREERS
Pharmacy

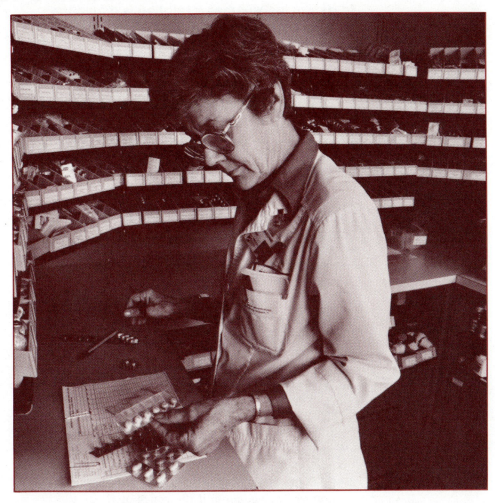

The primary responsibility of a pharmacist is to dispense prescribed medicines. Although the pharmacist may be called on to mix ingredients to obtain a certain form of a given medicine, most medicines prescribed today are produced in consumable form by the drug manufacturers. When situations require the pharmacist to mix ingredients, he or she must possess a clear understanding of the mathematics that is needed to obtain the precise results.

All states require pharmacists to obtain a license to practice. Five years of study and a degree from an accredited college or university are usually the minimum requirements. Most pharmacists study sciences, such as chemistry, biology, and social sciences, as well as mathematics.

Project

A pharmacist has to measure exactly 22 oz of water. How can she do it if only 5-oz beakers and 8-oz beakers are available for measuring liquids? **8 + 8 + (8 − 5) + (8 − 5) = 22**

RELATIONS AND FUNCTIONS 11.1

Objectives **To determine the domain and the range of a relation**
To determine whether a relation is a function

The points shown on the coordinate plane at the right are associated with the following set of ordered pairs of real numbers:

$$A = \{(-1, -2), (-1, 1), (2, -1), (3, 2)\}$$

A set of ordered pairs is called a **relation**. The set of all first coordinates of the ordered pairs is called the **domain** of the relation, and the set of all second coordinates is called the **range** of the relation.

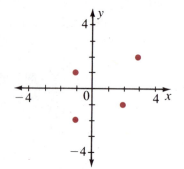

The domain of relation A above is $\{-1, 2, 3\}$ and the range is $\{-2, 1, -1, 2\}$. Notice that even though -1 is the first coordinate of two different points, it is listed only once in the domain.

Definition:	
Relation	A *relation* is a set of ordered pairs of real numbers.
Domain	The *domain* of a relation is the set of all first coordinates of the ordered pairs.
Range	The *range* of a relation is the set of all second coordinates of the ordered pairs.

Example 1 **Write the relation G whose graph is given. Determine the domain and the range of G.**

$G = \{(-3, -2), (-1, 3), (3, 2), (4, 1)\}$

The domain of G is $\{-3, -1, 3, 4\}$.
The range of G is $\{-2, 3, 2, 1\}$.

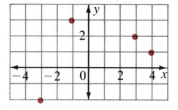

For relation G in Example 1, notice that each ordered pair has a different first coordinate. However, this is not true for all relations. When it is true, the relation is called a **function**.

Definition:	
Function	A *function* is a relation in which different ordered pairs have different first coordinates

276 Chapter Eleven

Example 2

Determine which of the following relations is a function.
$A = \{(-1, 2), (2, 2), (3, -1)\}$ $B = \{(2, 3), (3, 4), (2, 6)\}$

All ordered pairs have different first coordinates.

Thus, relation A is a function.

Two ordered pairs have a first coordinate of 2.

Thus, relation B is not a function.

A given relation may contain infinitely many ordered pairs (x, y) of real numbers. A graph of one such relation is shown in Example 3.

Example 3

Determine the domain and the range of the relation C whose graph is given. Is the relation a function?

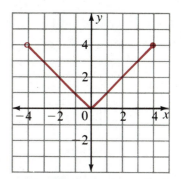

The domain is $\{x \mid -4 < x \leq 4\}$.
The range is $\{y \mid 0 \leq y \leq 4\}$.

For each value of x, there is exactly one value for y. All ordered pairs (x, y) have different first coordinates.

Thus, relation C is a function.

The open circle on the segment in the figure above indicates that this point is not a point of the graph. Notice that -4 is not included in the domain of the relation.

You can tell whether a given relation is a function by examining its graph. If any vertical line intersects the graph of a relation in more than one point, the relation is not a function.

Example 4

Determine which graphs are the graphs of functions.
Draw vertical lines. Check to see if any of the lines intersect the graph in more than one point.

Not a function

A function

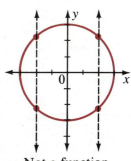
Not a function

Relations and Functions

Oral Exercises

State the domain and the range of each relation.

1. {(2, −1), (3, 2), (2, −5)}
   D = {2, 3}, R = {−5, −1, 2}
2. {(−2, −2), (6, −4), (5, 2)}
   D = {−2, 5, 6}, R = {−4, −2, 2}
3. {(−2, 3), (2, −3), (3, −2), (−3, 2)}
   D = {−3, −2, 2, 3}, R = {−3, −2, 2, 3}

Written Exercises

Determine the domain and the range of each relation. Is the relation a function?

(A)
1. {(3, 1), (2, −3), (−1, 5), (−2, −2), (0, 2)}
2. {(8, 7), (7, 8), (−7, 8), (−8, 7), (5, −2)}
3. {(0, 0), (8, 7), (−3, 6), (4, −1), (5, −1)}
4. {(3, −3), (4, −4), (5, −5), (−3, 3), (−4, 4)}
5. {(3, 2), (4, 2), (−4, −2), (−3, 2), (2, −3)}
6. {(0, 1), (3, 1), (0, 4), (1, −3), (2, −3)}
7. {(1, −2), (−1, 0), (2, −4), (−2, 3), (3, 0)}
8. {(9, 8), (7, 9), (8, 9), (−9, 8), (−8, 7)}
9. {(4, −1), (−1, 4), (−1, 3), (4, −3), (−3, 0)}
10. {(6, −3), (−3, 3), (−6, −3), (6, 5), (−3, 6)}

Write the relation G whose graph is given. Determine the domain and the range of G.

11.
12.
13.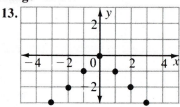

Determine the domain and the range of each relation whose graph is given. Is the relation a function?

(B)
14.
15.
16.

17.
18.
19.

(C)
20.
21.
22.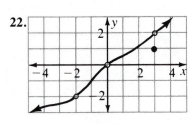

Determine the domain and the range of each relation. Is the relation a function?

23. {..., (−4, −2), (−2, −1), (0, 0), (2, 1), ...}
24. {(2, 4), (3, 9), (4, 16), (5, 25), ...}
25. {(4, 2), (9, 3), (16, 4), (25, 5), (36, 6), ...}
26. {..., (−2, −1), (−1, 1), (0, 3), (1, 5), (2, 7), ...}

For all Written Exercises on this page, see TE Answer Section.

VALUES AND COMPOSITION OF FUNCTIONS   11.2

Objectives
To find the value of a function, given an element in the domain of the function
To find the range and domain of a function
To find a value of a function that is composed of two other functions
To find the zeros of a rational function

The graph of the function, $f = \{(-3, -2), (-2, 3), (3, 2), (4, 1)\}$, is shown at the right. A lowercase letter is usually used to name a function. Notice that each element of the domain is paired with exactly one element of the range.

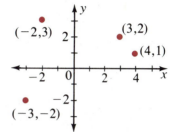

Domain of f: $\{-3, -2, 3, 4\}$
$\updownarrow \; \updownarrow \; \updownarrow \; \updownarrow$
Range of f: $\{-2, \; 3, 2, 1\}$

Each element of the range is called a **value of the function.** The expression $f(-3) = -2$ is read "the value of f at negative 3 equals negative two," or "f at negative three equals negative two," or "f of negative three equals negative two." In general, the notation $f(x)$ means the value of f at x.

Definition: Value of a function	For any ordered pair, (x, y) of a function, f, $f(x) = y$. The *value of f* at x equals y.

Example 1 Let $g = \{(-1, -2), (0, 3), (2, 3), (3, -1)\}$. Find $g(-1), g(0), g(2)$ and $g(3)$.

$$(-1, -2) \quad\quad (0, 3) \quad\quad (2, 3) \quad\quad (3, -1)$$
$$\;\;\;\nearrow \quad\quad\quad\quad \nearrow \quad\quad\quad \nearrow \quad\quad\quad\quad \nearrow$$
$$g(-1) \quad\quad\;\; g(0) \quad\quad\;\; g(2) \quad\quad\;\; g(3)$$

Thus, $g(-1) = -2$, $g(0) = 3$, $g(2) = 3$, and $g(3) = -1$.

A function f is often defined by giving an equation or formula for its range.

Example 2 If $f(x) = 2x - 3$, find $f(1), f(-2),$ and $f(-a)$.

Substitute for x in $f(x) = 2x - 3$.

$f(x) = 2x - 3$ $f(x) = 2x - 3$ $f(x) = 2x - 3$
$f(1) = 2(1) - 3$ $f(-2) = 2(-2) - 3$ $f(-a) = 2(-a) - 3$
$\;\;\;\;\;\; = -1$ $= -7$ $= -2a - 3$

Values and Composition of Functions

Example 3   If $f(x) = 3x + 2$, find $f(2a - 5)$, $f(a + b)$, and $f(a + h) - f(a)$.

$f(x) = 3x + 2$
$f(2a - 5) = 3(2a - 5) + 2$
$\quad = 6a - 15 + 2$
$\quad = 6a - 13$

$f(x) = 3x + 2$
$f(a + b) = 3(a + b) + 2$
$\quad = 3a + 3b + 2$

$f(x) = 3x + 2$
$f(a + h) - f(a) = 3(a + h) + 2 - [3(a) + 2]$
$\quad = 3a + 3h + 2 - 3a - 2$
$\quad = 3h$

Example 4   If $h(x) = -x^2 + 3$, find the range of h for D (domain) $= \{-2, 0, 1\}$.

$h(x) = -x^2 + 3$
$h(-2) = -(-2)^2 + 3$
$\quad = -4 + 3$
$\quad = -1$

$h(x) = -x^2 + 3$
$h(0) = -(0)^2 + 3$
$\quad = 0 + 3$
$\quad = 3$

$h(x) = -x^2 + 3$
$h(1) = -(1)^2 + 3$
$\quad = -1 + 3$
$\quad = 2$

Thus, R (range) $= \{-1, 3, 2\}$.

> **Definition:** **Rational function**
> A *rational function* is a quotient of two polynomial functions of the form $f(x) = \dfrac{p(x)}{q(x)}$, where $q(x) \neq 0$.

The **domain** of a rational function is the set of all real numbers except those numbers that make the denominator equal to 0.

Example 5   If $g(x) = \dfrac{x^2 - 5x + 6}{x^2 - 2x}$, determine the domain of g.

Factor: $g(x) = \dfrac{(x - 3)(x - 2)}{x(x - 2)}$.

If $x(x - 2) = 0$, then $x = 0$ or $x = 2$. So, the domain must exclude 0 and 2.
Thus, the domain is {all real numbers except 0 and 2}.

> **Definition:** **Zeros of a rational function**
> The number r is called a *zero* of a rational function if and only if for $f(x) = \dfrac{p(x)}{q(x)}$, $p(r) = 0$ and $q(r) \neq 0$.

Example 6   If $f(x) = \dfrac{x^2 + x - 2}{x(x - 3)}$, determine the zeros of f.

$f(x) = \dfrac{p(x)}{q(x)}$ ▶

Factor: $f(x) = \dfrac{(x - 1)(x + 2)}{x(x - 3)}$, where $p(x) = (x - 1)(x + 2)$ and $q(x) = x(x - 3)$.

If $p(x) = (x - 1)(x + 2) = 0$, then $x = 1$ or $x = -2$. So, $p(1) = 0$, $p(-2) = 0$, $q(1) \neq 0$, and $q(-2) \neq 0$.
Thus, the zeros are -2 and 1.

Example 7
If $f(x) = x + 2$ and $g(x) = 2x^2 - 3$, find $f(g(-4))$ and $g(f(-4))$.

$g(x) = 2x^2 - 3$ $f(g(-4)) = f(29)$ $f(x) = x + 2$ $g(f(-4)) = g(-2)$
$g(-4) = 2(-4)^2 - 3$ $f(29) = 29 + 2$ $f(-4) = -4 + 2$ $g(-2) = 2(-2)^2 - 3$
$= 2 \cdot 16 - 3$ $= 31$ $= -2$ $= 5$
$= 29$

Thus, $f(g(-4)) = 31$ and $g(f(-4)) = 5$.

Written Exercises

Let $g = \{(2, -3), (-1, 5), (-4, -2), (0, 5), (8, -1), (-7, 9)\}$. Find each of the following function values.

A   1. $g(2)$ **−3**   2. $g(-1)$ **5**   3. $g(-4)$ **−2**   4. $g(0)$ **5**   5. $g(8)$ **−1**   6. $g(-7)$ **9**

Let $f(x) = -2x + 5$. Find each of the following function values.
7. $f(2)$ **1**   8. $f(-1)$ **7**   9. $f(0)$ **5**   10. $f(-2)$ **9**   11. $f(5)$ **−5**   12. $f(-a)$ **2a + 5**   13. $f(b)$ **−2b + 5**

Let $t(x) = x^2 + 3x - 1$. Find each of the following function values.
14. $t(0)$ **−1**   15. $t(-1)$ **−3**   16. $t(-5)$ **9**   17. $t(a + b)$ **$a^2 + 2ab + b^2 + 3a + 3b - 1$**   18. $t(2a + 3)$ **$4a^2 + 18a + 17$**

Find the range of each function for the given domain.
19. $h(x) = x + 5; D = \{-1, 0, 2\}$ **{4, 5, 7}**   20. $f(x) = -3x + 2; D = \{-2, 2, 3\}$ **{8, −4, −7}**
21. $t(x) = 2x - 3; D = \{0, 1, 3\}$ **{−3, −1, 3}**   22. $g(x) = 4x + 3; D = \{-3, 2, 0\}$ **{−9, 11, 3}**
23. $g(x) = x^2 - 3; D = \{-2, 0, 2\}$ **{1, −3}**   24. $t(x) = (x - 2)^2; D = \{-1, 0, 2\}$ **{9, 4, 0}**
25. $f(x) = |x|; D = \{-3, -2, 0, 1, 2\}$ **{3, 2, 1, 0}**   26. $h(x) = \sqrt{x}; D = \{0, 1, 4, 9\}$ **{0, 1, 2, 3}**

Let $f(x) = x - 1$ and $g(x) = 2x$. Find each of the following.
27. $f(g(2))$ **3**   28. $g(f(2))$ **2**   29. $g(f(-3))$ **−8**   30. $f(g(-3))$ **−7**

Find the domain and zeros of each function. **See TE Answer Section.**
31. $h(x) = \dfrac{1}{x}$ 32. $t(x) = \dfrac{x - 1}{x(x + 2)}$ 33. $f(x) = \dfrac{3x - 6}{(x - 3)(x - 4)}$

B   34. $g(x) = -\dfrac{4x^2 - 3x}{(x + 1)(x + 3)}$ 35. $f(x) = \dfrac{2x - 4}{x^2 + 4x - 12}$ 36. $t(x) = \dfrac{-3x^2}{x^2 - 5x - 14}$

Let $f(x) = 2x - 3$ and $g(x) = x^2 - 2$. Find each of the following.
Assume that no denominator has the value of 0. (Ex. 37–52)
37. $f(2) - g(3)$ **−6**   38. $g(-1) - f(-2)$ **6**   39. $f(-3) - g(0)$ **−7**   40. $f(a) + g(b)$
41. $f(a + h) - f(a)$ **2h**   42. $g(a + h) - g(a)$ 43. $f(a) - f(t)$ **2a − 2t**   44. $g(a) - f(b)$
45. $f(g(a))$ **$2a^2 - 7$**   46. $g(f(a))$ 47. $g(f(-3a))$ 48. $f(g(-3a))$ **$18a^2 - 7$**

C   49. $\dfrac{f(a + h) - f(a)}{h}$ **2**   50. $\dfrac{g(a) - g(b)}{a - b}$ **a + b**   51. $\dfrac{f(b) - f(a)}{b - a}$ **2**   52. $\dfrac{g(c + h) - g(c)}{h}$ **2c + h**

40. $2a + b^2 - 5$ 42. $2ah + h^2$ 44. $a^2 - 2b + 1$ 46. $4a^2 - 12a + 7$ 47. $36a^2 + 36a + 7$

53. If $s(x) = -x + 2$ and $t(x) = -x - 2$, find $t(s(x))$ and $s(t(x))$. Does $t(s(x)) = s(t(x))$?

54. If $f(x) = 3x - 4$ and $g(x) = \dfrac{x + 4}{3}$, find $f(g(x))$ and $g(f(x))$. Does $f(g(x)) = g(f(x))$? **x, x, yes**

55. If $f(x) = x^2 - 3$, $g(x) = 5x + 2$, and $h(x) = 4 - x$, find $f(g(h(3)))$ and $h(g(f(-2)))$. **46, −3**   53. **x − 4, x + 4, no**

56. If $t(x) = x^2 - 3x + 2$ and $s(x) = -x - 3$, and $v(x) = 5 - x$, find $t(s(v(-a)))$ and $s(v(t(a + 3)))$. **$a^2 + 19a + 90, a^2 + 3a - 6$**

Values and Composition of Functions

LINEAR AND CONSTANT FUNCTIONS   11.3

Objectives **To determine whether a given function is a linear function or a constant function, given either an equation of the function or a graph of the function**
To draw the graphs of some special functions

As shown in the figures below, the graph of a **linear function** may be a line, a ray, or a segment, depending upon the domain of the function.

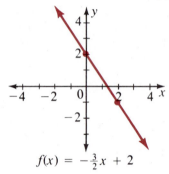
$f(x) = -\frac{3}{2}x + 2$

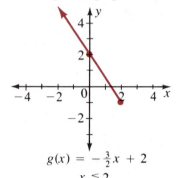
$g(x) = -\frac{3}{2}x + 2$
$x \leq 2$

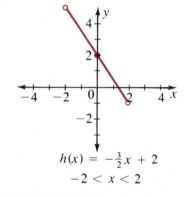
$h(x) = -\frac{3}{2}x + 2$
$-2 < x < 2$

> **Definition:**
> **Linear function**
>
> A *linear function* is a function whose graph is a nonvertical line, ray, or segment.
> The equation $y = mx + b$, where m and b are real numbers, describes a linear function.

Example 1 **Determine which graphs are the graphs of linear functions.**

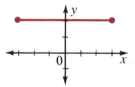

Horizontal segment
A linear function

Ray
A linear function

Vertical line
Not a function
Not a linear function

The graph of a linear function described by the equation $y = mx + b$, where $m = 0$, is a horizontal line, ray, or segment, depending upon the domain of the function. A function of this type is called a **constant function.**

> **Definition:**
> **Constant function**
>
> A constant function is a linear function whose graph is a horizontal line, ray, or segment.
> The equation $y = mx + b$, where $m = 0$ and b is a real number, or $y = b$, describes a constant function.

Example 2
Determine which graphs are the graphs of constant functions.

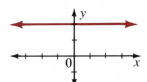

Horizontal line
A linear function
A constant function

Line
A linear function
Not a constant function

Horizontal ray
A linear function
A constant function

Example 3
Determine whether each equation describes a linear function, a constant function, or neither of these.

$3x + 5y - 15 = 0$ $4(x^2 - x + 2) + y = (2x - 1)^2$ $xy = 12$

Solve for y. ▶

$5y = -3x + 15$

$y = -\frac{3}{5}x + 3$

A linear function

$4x^2 - 4x + 8 + y = 4x^2 - 4x + 1$

$y = -7$

Both linear and constant

$y = \frac{12}{x}$

Neither of these

The graphs of two special types of functions are shown in the examples that follow. The function in Example 5 is the absolute value function.

Example 4
Graph the function $f(x) = \begin{cases} 1, & \text{if } x \geq 0 \\ -1, & \text{if } x < 0 \end{cases}$

x	f(x)
0	1
1	1
2	1
-1	-1
-2	-1
-3	-1

Example 5
Graph the function $f(x) = \begin{cases} x, & \text{if } x > 0 \\ 0, & \text{if } x = 0 \\ -x, & \text{if } x < 0 \end{cases}$

x	f(x)
0	0
1	1
2	2
3	3
-1	-(-1), or 1
-2	-(-2), or 2
-3	-(-3), or 3

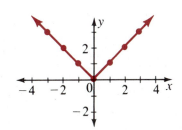

Linear and Constant Functions

Reading in Algebra

True or false?
1. A constant function is a linear function. **True**
2. A vertical line is the graph of a function that is not a linear function. **False**
3. The equation $f(x) = b$, where b is a real number, describes a constant function. **True**
4. The equation $x = 0$ describes a linear function. **False**

Written Exercises

Determine whether each of the following describes a linear function, a constant function, or neither of these.

(A)
1. $y = 2x - 5$ **L**
2. $y = -2x$ **L**
3. $y = -5$ **C, L**
4. $f(x) = 3x + 1$ **L**
5. $h(x) = x^2$ **N**
6. $t(x) = \frac{x}{2}$ **L**
7. $x = -7$ **N**
8. $g(x) = 3$ **C, L**
9. $y = x$ **L**
10. $x = y + 2$ **L**
11. $xy = 9$ **N**
12. $h(x) = 6, -2 \le x \le 2$ **C, L**
13. $5(7x - 2y) - 15x = 20x - 4$
14. $(2x + 3)^2 + 5 = (2x - 3)^2 + 4$ **N**
15. $h(x) = 3x - 2, x \ge 1$ **L**

(B)
16. $(x + 5)^2 - y = 10x$ **N**
17. $(x - 3)^2 + y = (x + 1)^2$ **L**
18. $2x - 3y - 6 = 3(-2y + x)$

13. **C, L** 18. **L**

19. $\{(5, 3), (7, 3), (9, 3), \ldots\}$ **C, L**
20. $\{(1, 7), (2, 9), (3, 11), (4, 13), \ldots\}$ **L**
21. $\{(3, 9), (4, 16), (5, 25)\}$ **N**

22.

23.

24.

N   **C, L**   **L**

Graph each of the following functions. For Ex. 25–31, 33 see TE Answer Section.

25. $f(x) = \begin{cases} 0, & \text{if } x \le 0 \\ -1, & \text{if } x > 0 \end{cases}$

26. $f(x) = \begin{cases} -x, & \text{if } x > 0 \\ 0, & \text{if } x = 0 \\ x, & \text{if } x < 0 \end{cases}$

27. $f(x) = \begin{cases} 2, & \text{if } x \ge 2 \\ 1, & \text{if } 0 \le x < 2 \\ -1, & \text{if } x < 0 \end{cases}$

(C)
28. The **greatest integer function** is the function that is described by the equation $y = [x]$ or $f(x) = [x]$, where $[x]$ is the greatest integer less than or equal to any real number x. Draw the graph of $y = [x]$ for $-3 \le x \le 3$.
29. Graph $y = -[x]$ for $-3 \le x \le 3$.
30. Graph $y = 2 + [x]$ for $-4 \le x \le 4$.
31. Graph $y = [x] - 2$ for $-4 \le x \le 4$.
32. If $f(x) = [x]$, find $f(0), f(4.5), f(-\frac{5}{2})$, and $f(\sqrt{6})$. **0, 4, −3, 2**
33. Graph $h(x) = \begin{cases} 0, & \text{if } x \text{ is rational} \\ 1, & \text{if } x \text{ is irrational.} \end{cases}$

CUMULATIVE REVIEW

Solve each proportion.

1. $\dfrac{5}{n + 4} = \dfrac{3}{n - 2}$ **11**
2. $\dfrac{4a + 3}{3} = \dfrac{2a + 5}{4}$ **$\frac{3}{10}$**
3. $\dfrac{5}{n - 7} = \dfrac{5}{7 - n}$ **no solution**

INVERSE RELATIONS AND FUNCTIONS   11.4

Objectives **To find the inverse of a relation**
To draw the graph of a function and its inverse
To determine whether the inverse of a function is a function

If you reverse the order of the elements of each ordered pair in a relation A, you will obtain the **inverse** of relation A, or A^{-1}.
For example, if $A = \{(1, 2), (2, -3), (5, 2)\}$,

then $A^{-1} = \{(2, 1), (-3, 2), (2, 5)\}$.

Notice that the domain of A is the range of A^{-1}, and the range of A is the domain of A^{-1}. Also, A is a function, but A^{-1} is not a function.

> **Definition:** The inverse of a relation A is the relation A^{-1} obtained by reversing the order of the elements of each ordered pair in A.
> **Inverse relations**
> A and A^{-1} are called *inverse relations*.

When both a relation and its inverse are functions, they are called **inverse functions**.

Example 1 **Let $A = \{(-3, -2), (-1, 2), (3, -5)\}$. Find A^{-1}. Is A^{-1} a function?**

Interchange the elements of each ordered pair in A.
$$A^{-1} = \{(-2, -3), (2, -1), (-5, 3)\}$$
All ordered pairs in A^{-1} have different first coordinates.

Thus, A^{-1} is a function.

When the inverse of a given function f is a function, the symbol f^{-1} is used to denote it. If a function is defined by an equation, the equation of the inverse is obtained by interchanging x and y in the original equation.

Example 2 **A function f is defined by $y = -\frac{2}{3}x + 4$. Find an equation of f^{-1}.**

$$y = -\tfrac{2}{3}x + 4$$
$$x = -\tfrac{2}{3}y + 4 \quad \blacktriangleleft \;\textit{Interchange x and y.}$$
$$3x = -2y + 12$$
$$2y = -3x + 12$$
$$y = -\tfrac{3}{2}x + 6$$

Thus, an equation of f^{-1} is $y = -\frac{3}{2}x + 6$.

Inverse Relations and Functions

In the next two examples, you will learn how to graph the inverse of a function, and to determine whether the inverse is a function by using the vertical line test.

Example 3

Graph the function defined by $y = |x|$ and its inverse. Determine if the inverse is a function.

Make a table of ordered pairs of the function and its inverse.

x	$|x|$	function	inverse
0	0	(0, 0)	(0, 0)
1	1	(1, 1)	(1, 1)
2	2	(2, 2)	(2, 2)
−1	1	(−1, 1)	(1, −1)
−2	2	(−2, 2)	(2, −2)

function

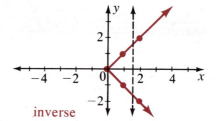
inverse

Since a vertical line intersects the graph of the inverse in two points, the inverse is not a function.

Example 4

Graph the function defined by $4x - 2y = 8$ and its inverse in the same coordinate plane. Determine if the inverse is a function.

$4x - 2y = 8$
$-2y = -4x + 8$
$y = 2x - 4$

x	$2x - 4$	function	inverse
0	−4	(0, −4)	(−4, 0)
1	−2	(1, −2)	(−2, 1)
2	0	(2, 0)	(0, 2)

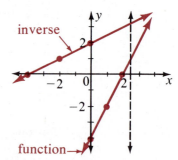

Since no vertical line intersects the graph of the inverse in more than one point, the inverse is a function.

An equation of the inverse function in Example 4 is $x = 2y - 4$, or $y = \frac{1}{2}x + 2$. If a function is defined by an equation of the form $y = mx + b$, $m \neq 0$, then its inverse can be written in the same form, and the inverse is a function.

Reading in Algebra

True or false? Give a reason for your answer.
1. Each relation A has an inverse relation, A^{-1}. **True**
2. Each function f has an inverse function, f^{-1}. **False**
3. If $G = \{(3, 7), (5, 5), (7, 3)\}$, then $G = G^{-1}$. **True**
4. For any relations A and B, if $B = A^{-1}$, then $A = B^{-1}$. **True**
5. For any relation A, $(A^{-1})^{-1} = A$. **True**

(F = function, N = not a function)

Written Exercises
1. $\{(3, -2), (3, -1), (-1, 3)\}$, **N**   2. $\{(6, 0), (-3, 1), (4, 1)\}$, **F**   3. $\{(-1, 4), (-2, 5), (-3, 6)\}$, **F**

Find the inverse of each relation. Determine if the inverse is a function.

(A)
1. $\{(-2, 3), (-1, 3), (3, -1)\}$
2. $\{(0, 6), (1, -3), (1, 4)\}$
3. $\{(4, -1), (5, -2), (6, -3)\}$
4. $\{(4, 9), (5, -1), (6, 9)\}$
5. $\{(7, 11), (8, 10), (9, 9)\}$
6. $\{(5, -2), (-6, -6), (-2, 5)\}$
   $\{(9, 4), (-1, 5), (9, 6)\}$, **N** $\{(11, 7), (10, 8), (9, 9)\}$, **F** $\{(-2, 5), (-6, -6), (5, -2)\}$, **F**

Graph each function and its inverse in the same coordinate plane. Determine if the inverse is a function.

7. $x + y = 6$ **F**
8. $x - y = 6$ **F**
9. $5x - 2y = 10$ **F**
10. $2y - 3x = 4$ **F**
11. $3x + 4y = -8$ **F**
12. $4y - x - 10 = 0$ **F**
13. $y = 6$ **N**
14. $x = y$ **F**
15. $y = 0.5x - 10$ **F**
16. $y = |x| - 2$ **N**
17. $y = 2|x| - 1$ **N**
18. $y = |x| + 3$ **N**
19. $y = \frac{9}{10}x + 15$; **F**
20. $y = -\frac{5}{3}x + 16$; **F**
21. $y = \frac{2}{21}x + \frac{8}{7}$; **F**

A relation is defined by each given equation. Find an equation of the inverse of each relation. Determine if the inverse is a function.

(B)
19. $\frac{2}{3}x - \frac{3}{5}y = 10$ **above**
20. $\frac{5}{4}y + \frac{3}{4}x = 12$ **above**
21. $7x - \frac{2}{3}y - 8 = 0$ **above**
22. $x^2 + y^2 = 1$ **below**
23. $x^2 - y^2 = 4$ **below**
24. $y = x^2$ $y = \pm\sqrt{x}$; **N**
25. $y = x^2 + 5$ **below**
26. $x = y^2$ $y = x^2$; **F**
27. $xy = 9$ $y = \frac{9}{x}$; **F**
22. $y = \pm\sqrt{1 - x^2}$; **N**   23. $y = \pm\sqrt{x^2 + 4}$; **N**   25. $y = \pm\sqrt{x - 5}$; **N**

(C)
28. Show that the graph of $y = mx + b$ and the graph of its inverse are symmetric with respect to the line determined by the equation $y = x$.
[Hint: Select points $P(x, y)$ and $Q(y, x)$ on the graphs. Show that the line determined by $y = x$ is the perpendicular bisector of $\overline{PQ}$.]
For Ex. 28, see TE Answer Section.

If $f(x) = 3x - 5$, find each of the following function values.
29. $f^{-1}(1)$ **2**
30. $f^{-1}(4)$ **3**
31. $f^{-1}(-5)$ **0**
32. $f^{-1}(16)$ **7**
33. $f^{-1}(2.5)$ **2.5**
34. $f^{-1}(3a - 5)$ **a**
35. $f(f^{-1}(10))$ **10**
36. $f^{-1}(f(30))$ **30**
37. $f(f^{-1}(a))$ **a**
38. $f^{-1}(f(a))$ **a**

NON-ROUTINE PROBLEMS

If f is any linear function, is the following statement true or false? Prove your answer.
For each real number x, $f(x + 2) - f(x + 1) = f(x + 1) - f(x)$.
See TE Answer Section.

Inverse Relations and Functions

DIRECT VARIATION   11.5

Objectives **To determine whether a relation expresses direct variation**
To solve a word problem using direct variation

Frequently in mathematics, one variable is related to another variable in such a way that a change in the value of one produces a corresponding change in the other. For example, if you drive an automobile at a constant speed of 60 km/h, the distance traveled is directly related to the time traveled. When the time increases, the distance traveled increases. In the table, notice that the ratio $\frac{d}{t}$ is 60 in each case.

Time in hours (t)	1	2	3	4	5
Distance in kilometers (d)	60	120	180	240	300

The table determines a set of ordered pairs with coordinates (t, d) that expresses **direct variation** between the variables t and d.

Definition: Direct variation

A linear function defined by an equation of the form $y = kx$, or $\frac{y}{x} = k$, where k is a nonzero constant, is called a *direct variation*.
The constant k is called the *constant of variation*.

The equation $\frac{y}{x} = k$, or $y = kx$, is read, "y varies directly as x," or "y varies directly with x."

Example 1 Determine if the table expresses a direct variation between the variables y and x. If so, find the constant of variation and an equation that defines the function.

x	y	ordered pair
1	-6	$(1, -6)$
-2	12	$(-2, 12)$
3	-18	$(3, -18)$

$\frac{y}{x} = \frac{-6}{1} = \frac{12}{-2} = \frac{-18}{3} = -6$

$\frac{y}{x} = -6$, or $y = -6x$

The ratio $\frac{y}{x} = -6$ for each ordered pair in the function.

Thus, y varies directly as x, the constant of variation is -6, and an equation of the function is $y = -6x$.

Example 2 Determine if the equation expresses a direct variation between the variables y and x, in each case.

$y = 2x$ $\qquad\qquad\qquad\qquad\qquad$ $y = 3x + 4$

The equation is of the form $\qquad$ The equation is not of the form
$y = kx$. $\qquad\qquad\qquad\qquad\qquad\qquad$ $y = kx$.

Thus, $y = 2x$ expresses a direct variation; $y = 3x + 4$ does not.

Notice that an equation of a function that expresses direct variation is an equation of the form $y = mx + b$, where m is a constant and $b = 0$.

Example 3 The voltage V of a given electrical circuit varies directly as the current I. If V is 120 volts when I is 9 amperes, find the voltage V when the current I is 15 amperes.

(1) Write a formula and find the value of k.
$V = kI$
$120 = k \cdot 9$
$k = \dfrac{120}{9} = \dfrac{40}{3}$

(2) Substitute for k and I. Find the value of V.
$V = kI$
$V = \dfrac{40}{3} \cdot 15 = 200$

Thus, the voltage is 200 volts when the current is 15 amperes.

A direct variation problem may be solved by using a proportion. If (x_1, y_1) and (x_2, y_2) are any two ordered pairs of a direct variation and neither is $(0, 0)$, then $\dfrac{y_1}{x_1} = k$ and $\dfrac{y_2}{x_2} = k$ for some constant $k \neq 0$. Thus, $\dfrac{y_1}{x_1} = \dfrac{y_2}{x_2}$.

The constant k is called the **constant of proportionality,** and y is said to be "directly proportional to" x.

Example 4 If y varies directly as the cube of x, and $y = 54$ when $x = 3$, find y when $x = 5$.

y varies directly as the cube of x means that y is directly proportional to the cube of x. Write a proportion.

$\dfrac{54}{3^3} = \dfrac{y}{5^3}$

$\dfrac{54}{27} = \dfrac{y}{125}$

$\dfrac{2}{1} = \dfrac{y}{125}$ $\qquad$ ◀ In a proportion, product of means equals product of extremes.

$1 \cdot y = 2 \cdot 125$
$y = 250$

Thus, $y = 250$ when $x = 5$.

Direct Variation

Oral Exercises

If c is a constant, determine whether each equation expresses a direct variation.

1. $\frac{a}{b} = c$ **Yes**
2. $m = nc$ **Yes**
3. $mn = c$ **No**
4. $x = cy$ **Yes**
5. $xy = c$ **No**
6. $p = cq$ **Yes**
7. $\frac{1}{p} = \frac{c}{q}$ **Yes**
8. $\frac{c}{1} = \frac{q}{p}$ **Yes**
9. $x = \frac{c}{y}$ **No**
10. $\frac{y}{x} - c = 0$ **Yes**

Written Exercises

Determine if the table expresses a direct variation between the variables. If so, find the constant of variation and an equation that defines the function.

1.
x	y
-3	1
-9	3
3	-1
6	-2

Yes; $-\frac{1}{3}$; $y = -\frac{1}{3}x$

2.
c	d
7	3
14	6
-21	-9
28	-12

No

3.
A	r
-15	-3
10	2
-20	-4
25	5

Yes, 5; $A = 5r$

4.
R	S
2	1
3	2
4	3
5	4

No

5.
M	N
1	1
2	2
3	3
-4	-4

Yes; 1; $M = 1N$

Determine if the equation expresses a direct variation between the variables. If so, find the constant of variation. 6. Yes; -1 7. Yes; 3 8. No 9. Yes; 2 10. Yes, 2π

6. $y = -x$
7. $y = 3x$
8. $y = -2x + 5$
9. $c = 2d$
10. $c = 2\pi r$
11. $A = 3r + 6$ **No**
12. $A = \frac{4}{l}$ **No**
13. $\frac{y}{x} - 6 = 0$ **Yes; 6**
14. $y = \frac{4}{x}$ **No**
15. $3y = 2x$ **Yes, $\frac{2}{3}$**

In Exercises 16–23, y varies directly as x. Find the value as indicated.

16. If $y = 12$ when $x = 4$, find y when $x = 12$. **36**
17. If $y = -81$ when $x = 9$, find y when $x = 7$. **-63**
18. If $y = 16$ when $x = -4$, find y when $x = 8$. **-32**
19. If $y = -3$ when $x = -15$, find x when $y = 2$. **10**
20. If $y = -12$ when $x = -3$, find y when $x = -8$. **-32**
21. If $y = 5$ when $x = 25$, find y when $x = 30$. **6**
22. If $y = 3$ when $x = 10$, find x when $y = 1.2$. **4**
23. If $y = 3$ when $x = 10$, find x when $y = \frac{1}{2}$. **$\frac{5}{3}$**

24. If y varies directly as the square of x, and $y = 25$ when $x = 3$, find y when $x = 6$. **100**
25. If y is directly proportional to $\sqrt{x}$, and $y = 3.5$ when $x = 4$, find x when $y = 8.75$.
26. If y is directly proportional to $\sqrt[3]{x}$, and $x = 27$ when $y = 7.2$, find x when $y = 12$.
27. If y varies directly as the cube of x, and $x = 0.2$ when $y = 16$, find y when $x = 0.5$.
28. If you earn $22.75 for 7 hours work, how long must you work to earn $39.65? **12.2 h**
29. If a batter averages 81 hits for each 216 times at bat, approximately how many hits will he get in 400 times at bat? **150**
30. The scale on a certain blueprint is $3'' = 8'$. Find the dimensions of a floor that is $5\frac{1}{2}''$ by $7\frac{3}{4}''$ on the blueprint. **14' 8'' by 20' 8''**
31. The stretch S in a spring balance varies directly as the applied weight w. If $S = 5$ in. when $w = 15$ lb, find the weight needed for a stretch of $7\frac{1}{2}$ in. **22.5 lb**

32. In a statement of direct variation in which y varies directly as x, if the value of y remains constant, $(y > 0)$, what happens to the value of k when x increases? when x decreases? **k decreases; k increases**
33. If A varies directly as B, and if A is multiplied by c, how is the corresponding value of B affected? **B is multiplied by c.**

25. 25 26. 125 27. 250

INVERSE VARIATION 11.6

Objectives **To determine whether a relation expresses inverse variation**
To solve a word problem using inverse variation

In some mathematical situations, when the value of one variable increases, the value of a second variable decreases. For example, if you drive an automobile a distance of 180 km, the time required for the trip decreases as the speed increases. In the table, notice that the product rt is 180 in each case.

Speed in kilometers per hour (r)	90	72	60	45	40	36	30
Time in hours (t)	2	2.5	3	4	4.5	5	6

The table determines a set of ordered pairs with coordinates (r, t) that expresses **inverse variation** between the variables r and t.

Definition: Inverse variation

A function defined by an equation of the form $xy = k$, or $y = \dfrac{k}{x}$, where k is a nonzero constant, is called an *inverse variation*.
The constant k is called the *constant of variation*.

The equation $xy = k$, or $y = \dfrac{k}{x}$, is read, "y varies inversely as x," or "y varies inversely with x."

Example 1 Determine if the table expresses an inverse variation between the variables y and x. If so, find the constant of variation and an equation that defines the function.

x	4	8	1	-1	-2	$-\frac{1}{2}$
y	-2	-1	-8	8	4	16

$$\begin{array}{cccccc} 4 \cdot (-2) & 8 \cdot (-1) & 1 \cdot (-8) & -1 \cdot 8 & -2 \cdot 4 & -\frac{1}{2} \cdot 16 \\ -8 & -8 & -8 & -8 & -8 & -8 \end{array}$$

$x \cdot y = -8$ for each ordered pair (x, y) in the function.

Thus, y varies inversely as x, the constant of variation is -8, and an equation of the function is $xy = -8$.

Inverse Variation

Example 2 Determine if the equation expresses an inverse variation between the variables y and x, in each case.
$y = 5x - 3$ $xy = -12$

The equation is not of the form $xy = k$. | The equation is of the form $xy = k$.

Thus, $y = 5x - 3$ does not express an inverse variation; $xy = -12$ does.

Example 3 The height h of a triangle varies inversely as the length of the base b. If $h = 10$ cm when $b = 4.8$ cm, find h when $b = 9$ cm.

(1) Write a formula and find the value of k.
$$bh = k$$
$$4.8 \cdot 10 = k$$
$$48 = k$$

(2) Substitute for k and b. Find the value of h.
$$bh = k$$
$$9 \cdot h = 48$$
$$h = 5\tfrac{1}{3}, \text{ or } 5.3$$

Thus, $h = 5.3$ cm when $b = 9$ cm.

If (x_1, y_1) and (x_2, y_2) are any two ordered pairs of an inverse variation and neither is $(0, 0)$, then $x_1 y_1 = k$ and $x_2 y_2 = k$. Thus, $x_1 y_1 = x_2 y_2$. This equation is often written as a proportion of the form $\frac{x_1}{x_2} = \frac{y_2}{y_1}$; y is said to be "inversely proportional to" x.

Example 4 If y varies inversely as the square of x, and $y = 10$ when $x = 4$, find x when $y = 5$.

y varies inversely as the square of x means that y is inversely proportional to the square of x. Write a proportion.

$$\frac{4^2}{x^2} = \frac{5}{10}$$
$$\frac{16}{x^2} = \frac{1}{2}$$
$$x^2 \cdot 1 = 16 \cdot 2 \quad \blacktriangleleft \text{ In a proportion, product of means}$$
$$x^2 = 32 \quad \quad \quad \text{equals product of extremes.}$$
$$x = \pm\sqrt{32}, \text{ or } \pm 4\sqrt{2}$$

Thus, $x = \pm 4\sqrt{2}$ when $y = 5$.

Example 5 If y is inversely proportional to $\sqrt[3]{x}$ and $y = 2.7$ when $x = 8$, find y when $x = 27$.

$$\sqrt[3]{8} \cdot 2.7 = \sqrt[3]{27} \cdot y \quad \blacktriangleleft \text{ Write an equation in the}$$
$$2 \cdot 2.7 = 3 \cdot y \quad \quad \text{product form.}$$
$$5.4 = 3y$$
$$1.8 = y$$

Thus, $y = 1.8$ when $x = 27$.

Oral Exercises

If c is a constant, determine whether each equation expresses an inverse variation.

1. $\dfrac{x}{y} = c$ **No**
2. $xc = y$ **No**
3. $x = \dfrac{c}{y}$ **Yes**
4. $x = \dfrac{y}{c}$ **No**
5. $y = cx + x$ **No**
6. $\dfrac{1}{c} = \dfrac{x}{y}$ **No**
7. $\dfrac{c}{x} = \dfrac{1}{y}$ **No**
8. $1 = \dfrac{c}{xy}$ **Yes**
9. $xc = \dfrac{c}{y}$ **Yes**
10. $\dfrac{c}{x} = \dfrac{y}{c}$ **Yes**

Written Exercises

Determine if the table expresses an inverse variation between the variables. If so, find the constant of variation and an equation that defines the function.

1.
s	t
12	3
16	4
20	5
24	6

No

2.
x	y
−3	−4
−2	−6
4	3
12	1

Yes; 12; $xy = 12$

3.
c	d
−4	6
8	−3
−12	−2
−24	−1

No

4.
x	y
1.0	0.32
0.4	0.80
0.2	1.60
0.1	3.20

Yes; 0.32; $xy = 0.32$

5.
a	b
$\frac{1}{2}$	$\frac{1}{6}$
$\frac{1}{3}$	$\frac{1}{4}$
$-\frac{1}{6}$	$-\frac{1}{2}$
$-\frac{1}{12}$	-1

Yes; $\frac{1}{12}$; $ab = \frac{1}{12}$

Determine if the equation expresses an inverse variation between the variables. If so, find the constant of variation.

6. $x = 10y$ **No**
7. $xy = 12$ **Yes; 12**
8. $5st = 12$ **Yes; $\frac{12}{5}$**
9. $c = 5r$ **No**
10. $a \cdot b = \pi$ **Yes; π**
11. $A = 6.28r$ **No**
12. $\dfrac{r}{4} = t$ **No**
13. $6m = \dfrac{4}{n}$ **Yes; $\frac{2}{3}$**
14. $x = \dfrac{10}{y}$ **Yes; 10**
15. $y = \dfrac{10}{x}$ **Yes; 10**

In Exercises 16–19, y varies inversely as x. Find the value as indicated.

16. If $y = 27$ when $x = 3$, find y when $x = 9$. **9**
17. If $y = 15$ when $x = -2$, find y when $x = -5$. **6**
18. If $y = 9$ when $x = 4$, find y when $x = 6$. **6**
19. If $y = 81$ when $x = -9$, find x when $y = 27$. **−27**

20. The illumination i from a light varies inversely as the square of its distance d from an object. If $i = 8$ foot-candles when $d = 3$ ft, find i when $d = 4$ ft. **$4\frac{1}{2}$ foot-candles**

21. The pressure P of a gas at a constant temperature varies inversely as the volume V. If $V = 450$ in.^{3} when $P = 30$ lb/in.^{2}, find P when $V = 750$ in.^{3}. **18 lb/in.^{2}**

B

22. If y varies inversely as the square of x, and $y = 3$ when $x = 4$, find y when $x = 2$. **12**

23. If y is inversely proportional to $\sqrt{x}$, and $y = 6$ when $x = 9$, find y when $x = 16$. **$4\frac{1}{2}$**

24. If x is inversely proportional to $\sqrt[3]{y}$, and $x = 1.8$ when $y = 64$, find x when $y = 27$. **2.4**

25. If x varies inversely as the cube of y, and $x = 3$ when $y = 2$, find x when $y = 5$. **$\frac{24}{125}$**

26. The frequency of a radio wave is inversely proportional to its wave length. If a radio wave, 30 m long, has a frequency of 1200 kilocycles per second, what is the length of a wave with a frequency of 900 kilocycles per second? **40 m**

Inverse Variation

JOINT AND COMBINED VARIATION 11.7

Objectives **To determine whether a relation expresses joint variation or combined variation**
To solve a word problem using joint variation or combined variation

In some mathematical situations, one variable is a constant multiple of the product of two or more variables. For example, simple interest, i, received on an investment of a specific amount of money is related to both the rate of interest, r, and the time, t, that the money was invested. Specifically, as the value of $r \cdot t$ increases, the value of i increases. The variation of the variables is called a **joint variation**.

Definition:
Joint variation

A function defined by an equation of the form $y = kxz$, or $\frac{y}{xz} = k$, where k is a nonzero constant, is called a *joint variation*. The constant k is called the *constant of variation*.

The equation $\frac{y}{xz} = k$, or $y = kxz$, is read, "y varies jointly as x and z," or "y varies jointly with x and z." The variation in this case is a direct variation between y and the product xz.

Example 1 Determine if the equation expresses a joint variation between the variables, in each case.
$A = \frac{1}{2}bh$ $\qquad\qquad\qquad\qquad m = \frac{3}{4}n + \frac{1}{3}$

The equation is of the form $y = kxz$. $\qquad$ The equation is not of the form $y = kxz$.

Thus, $A = \frac{1}{2}bh$ expresses a joint variation; $m = \frac{3}{4}n + \frac{1}{3}$ does not.

Example 2 If y varies jointly as x and z and if $y = -24$ when $x = 4$ and $z = 3$, find y when $x = -6$ and $z = 2$.

First Method: $\qquad$ Second Method:

Use the equation $y = kxz$.

$-24 = k \cdot 4 \cdot 3$ $\qquad$ $\frac{-24}{4 \cdot 3} = \frac{y}{-6 \cdot 2}$ $\qquad$ Use a proportion of the form $\frac{y_1}{x_1 z_1} = \frac{y_2}{x_2 z_2}$.
$-24 = 12k$
$-2 = k$ $\qquad\qquad\qquad$ $\frac{-2}{1} = \frac{y}{-12}$

$y = -2(-6)(2)$ $\qquad\qquad$ $1 \cdot y = -2(-12)$
$y = 24$ $\qquad\qquad\qquad\qquad$ $y = 24$

Thus, $y = 24$ when $x = -6$ and $z = 2$.

Example 3   The area A of a triangle varies jointly as the length of a base b and the length of a corresponding altitude h. If $A = 15$ cm when $b = 10$ cm and $h = 3$ cm, find A when $b = 25$ cm and $h = 6$ cm.

$A = kbh$
$15 = k \cdot 10 \cdot 3$ ◀ Write a formula and solve for k.
$15 = 30k$
$\frac{1}{2} = k$

$A = kbh$
$A = \frac{1}{2} \cdot 25 \cdot 6$ ◀ Substitute for k, b, and h. Find the value of A.
$A = 75$

Thus, A is 75 cm² when $b = 25$ cm and $h = 6$ cm.

In some mathematical applications, both direct variation and inverse variation occur at the same time. This type of variation is called **combined variation.**

> **Definition: Combined variation**
> A function defined by an equation of the form $y = \frac{kxz}{w}$, or $\frac{yw}{xz} = k$, where k is a nonzero constant, is called a *combined variation*. The constant k is called the *constant of variation*.

In the equation $y = \frac{kxz}{w}$, or $\frac{yw}{xz} = k$, the variable y varies jointly with the variables x and z and inversely with the variable w.

Example 4   If y varies jointly as x and z and inversely as $\sqrt{w}$, and $y = 12$ when $x = 2$, $z = 6$, and $w = 9$, find y when $x = 5$, $z = 7$, and $w = 25$.

Use an equation of the form $y = \frac{kxz}{\sqrt{w}}$.

First Method:

$12 = \frac{k \cdot 2 \cdot 6}{\sqrt{9}}$ $y = \frac{3 \cdot 5 \cdot 7}{\sqrt{25}}$

$12 = \frac{12k}{3}$ $y = \frac{3 \cdot 5 \cdot 7}{5}$

$12 = 4k$ $y = 21$

$3 = k$

Second Method:

$\frac{12 \cdot \sqrt{9}}{2 \cdot 6} = \frac{y \cdot \sqrt{25}}{5 \cdot 7}$

$\frac{12 \cdot 3}{2 \cdot 6} = \frac{y \cdot 5}{5 \cdot 7}$

$\frac{3}{1} = \frac{y}{7}$

$1 \cdot y = 3 \cdot 7$

$y = 21$

Use a proportion of the form $\frac{y_1 \sqrt{w_1}}{x_1 z_1} = \frac{y_2 \sqrt{w_2}}{x_2 z_2}$.

Thus, $y = 21$ when $x = 5$, $z = 7$, and $w = 25$.

Joint and Combined Variation

Written Exercises

(Note: J = joint variation, C = combined variation, N = neither)
Determine whether each equation expresses a joint variation, a combined variation, or neither. If either joint or combined variation exists, find the constant of variation.

(A)
1. $x = \dfrac{-2}{y}$ **N**
2. $mn = 5$ **N**
3. $\dfrac{r}{s} = \dfrac{6t}{v}$ **C; 6**
4. $-6x = \dfrac{y}{z}$ **J; −6**
5. $w = 4u$ **N**
6. $a \cdot b \cdot c = 15$ **N**
7. $m = \dfrac{5n^2}{2p}$ **J; $\dfrac{5}{2}$**
8. $y = \dfrac{8x}{\sqrt{t}}$ **J; 8**

In Exercises 9–15, y varies jointly as x and z. Find the value as indicated.

9. If $y = 12$ when $x = 4$ and $z = 3$, find y when $x = 9$ and $z = 8$. **72**
10. If $y = 72$ when $x = 3$ and $z = 8$, find y when $x = -2$ and $z = -3$. **18**
11. If $y = 24$ when $x = 2$ and $z = 3$, find y when $x = 4$ and $z = 7$. **112**
12. If $y = 18$ when $x = 1$ and $z = 6$, find y when $x = 4$ and $z = 10$. **120**
13. If $y = 30$ when $x = 3$ and $z = 15$, find y when $x = -5$ and $z = -12$. **40**

14. If y varies directly as the square of x and inversely as z, and if $y = 12$ when $x = 2$ and $z = 7$, find y when $x = 3$ and $z = 9$. **21**

15. If y varies directly as $\sqrt[3]{x}$ and inversely as the square of z, and if $y = 3$ when $x = 8$ and $z = 4$, find y when $x = 27$ and $z = 6$. **2**

16. The area A, of a parallelogram varies jointly as the length of a base, b, and the length of a corresponding altitude, h. If $A = 16$ when $b = 2$ and $h = 8$, find A when $b = 8$ and $h = 16$. **128**

17. The distance, D, traveled at a uniform rate varies jointly as the rate, r, and the time, t. If $D = 120$ when $r = 60$ and $t = 2$, find D when $r = 80$ and $t = 3$. **240**

(B)
18. If y varies jointly as x and $\sqrt{z}$ and inversely as the cube of w, and if $y = 6$ when $x = 3$, $z = 9$, and $w = 3$, find y when $x = 4$, $z = 36$, and $w = 4$. **$\dfrac{27}{4}$**

19. If y varies jointly as x and z and inversely as the square of w, and if $y = 6$ when $x = 3$, $z = 4$, and $w = 49$, find y when $x = 6$, $z = 8$, and $w = 4$. **$\dfrac{7{,}203}{2}$**

20. Interest i varies jointly as the principal p, the rate r, and the time t. If the value of r is halved, what happens to the value of i? **The value of i is halved.**

21. The volume V of a cube varies directly as the cube of its edge e. If the value of e is doubled, what happens to the value of V? **The value of V is multiplied by 8.**

(C)
22. If r varies directly as the square of s and inversely as t, what happens to the value of r if the value of s is doubled and the value of t is tripled? **The value of r is multiplied by $\dfrac{4}{3}$.**

23. If t varies directly as s and inversely as the square of r, how is the value of t affected if the value of r is increased by 25%? **The value of t is multiplied by $\dfrac{16}{25}$.**

24. If 6 people take 14 days to assemble 18 computers, how many days will it take 10 people to assemble 30 computers? **14 days**

25. If y varies directly as x and inversely as z, how is the value of y affected if the value of z is decreased by 25%? **The value of z is multiplied by $\dfrac{4}{3}$.**

CUMULATIVE REVIEW

Solve each equation.
1. $\sqrt{x - 3} + \sqrt{x} = 3$ **4**
2. $\sqrt{x} + \sqrt{x - 5} = 5$ **9**
3. $\sqrt{2x} - \sqrt{\dfrac{x}{2}} = 2$ **8**

APPLICATIONS

1. The frequency of a string under constant tension is inversely proportional to its length. If a string, 40 cm long, vibrates 680 times per second, what length must the string be to vibrate 850 times per second under the same tension? **32 cm**

2. The frequency of a radio wave is inversely proportional to its wavelength. If a radio wave, 30 m long, has a frequency of 1,200 kilocycles per second, what is the length of a wave with a frequency of 900 kilocycles per second? **40 m**

3. Boyle's Law states that the volume of a given mass of gas varies inversely as the absolute pressure, provided the temperature remains constant. If the volume of a gas is 40 in.^{3} when the absolute pressure is 50 lb/in.^{2}, what will be the volume when the pressure is 70 lb/in.^{2}? **$28\frac{4}{7}$ in.^{3}**

4. The kinetic energy, E, of a body varies jointly as its weight, w, and the square of its velocity, V. If a 10-lb body moving at 6 ft/s has 3 ft-lb of kinetic energy, what will the kinetic energy be of a 4-T truck traveling at 45 mi/h? **290,400 ft-lb**

5. The load, l, that a beam of fixed length can support varies jointly as its width, w, and the square of its depth, d. A beam with $w = 4$ and $d = 2$ can support a load of 1,760 kg. Find l when $w = 4$ and $d = 8$. **28,160 kg**

6. The pressure, P, of a gas varies directly as the temperature, T, and inversely as the volume, V. If $P = 50$ when $T = 25$ and $V = 2$, find P when $T = 40$ and $V = 4$. **40**

7. The speed, V, at which a motorcycle must travel on the inside cylinder in order not to fall varies directly as the square root of the radius of the cylinder. If the motorcycle must travel at least 45 km/h to stay on the side of the cylinder with a radius of 12 m, how fast must it travel on a cylinder with radius of 30 m? Answer to the nearest km/h. **71 km/h**

8. A conservationist can approximate the number of deer in a forest. Suppose 650 deer are caught, tagged and released. Suppose later 216 deer are caught and 54 of them were tagged. How many deer are in the forest? Assume that the number of tagged deer caught varies directly as the number of deer caught later. **2,600 deer**

Applications

Ⓐ Exercises are 1, 2, 4, 6–14, 17–19, 22–24, 26–39.   Ⓒ Exercises are starred. The rest are Ⓑ Exercises.

CHAPTER ELEVEN REVIEW

Vocabulary
combined variation [11.7]
constant function [11.3]
direct variation [11.5]
domain [11.1]   function [11.1]
inverse function [11.4]   inverse relation [11.4]
inverse variation [11.6]
joint variation [11.7]   range [11.1]
rational function [11.2]   relation [11.1]
zeros of a rational function [11.2]

Determine the domain and the range of each relation. Is the relation a function? [11.1]
1. $\{(-2, 3), (0, -7), (8, -3), (4, 3)\}$ **F**
2. $\{(0, 6), (1, -3), (4, -6), (0, 0)\}$ **N**
★ 3. $\{\ldots(-1, 1), (0, 0), (1, 1), (2, 4), (3, 9), \ldots\}$ **F**
4.

F

5.

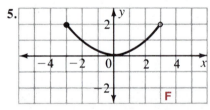

F

Let $f = \{(2, -3), (1, 0), (-3, 5), (-1, -2)\}$. Find each of these function values. [11.2]
6. $f(1)$ **0**   7. $f(-1)$ **−2**   8. $f(-3)$ **5**   9. $f(2)$ **−3**

Let $h(x) = x^2 - 2$. Find each of the following function values. [11.2] $a^2 - 2a - 1$
10. $h(0)$ **−2**   11. $h(-4)$ **14**   12. $h(a-1)$

Find the range of each function for the given domain. [11.2]   15. {all real numbers except 1}; 2
13. $h(x) = 2x + 3$, $D = \{0, 1, -2\}$ **{3, 5, −1}**
14. $g(x) = x^2 + 1$, $D = \{-1, 0, 4\}$ **{2, 1, 17}**
16. {all real numbers except −2 and 5}; −3

Find the domain and zeros of each function. [11.2]
15. $h(x) = \dfrac{x-2}{x-1}$ 16. $t(x) = \dfrac{-4x-12}{x^2 - 3x - 10}$
   above   **above**

Let $f(x) = 3x - 5$ and $g(x) = x^2 + 6$. Find each of the following. [11.2]   **18** ⟵
17. $f(-7)$ **−26**   18. $g(2\sqrt{3})$ 19. $g(f(\tfrac{2}{3}))$ **15**
★ 20. $\dfrac{f(a) - f(b)}{a - b}$; $a \neq b$ ★ 21. $\dfrac{g(c+h) - g(c)}{h}$; $h \neq 0$
   3   **2c + h**

Determine whether each equation describes a linear function, a constant function, or neither of these. [11.3]
22. $2x - 3y = 6$ **L**   23. $y = 2$ 24. $x = -3$ **N**
25. $-3(2x - 3y) - 8x = 3(2x - 1)$ **L**   23. **C, L**

Find the inverse of each relation. Determine if the inverse is a function. [11.4]
26. $\{(-1, -2), (6, -2), (1, 2)\}$
27. $\{(3, 1), (3, 6), (3, 9)\}$ **{(1, 3), (6, 3), (9, 3)}, F**

Graph each function and its inverse in the same coordinate plane. Determine if the inverse is a function. [11.4]   26. $\{(-2, 1), (-2, 6), (2, 1)\}$; **N**
28. $3x - 2y = 8$ **F**   29. $y = |x| - 1$ **N**
30. $xy = 4$ **F**   31. $x^2 = y - 1$ **N**

Determine if the table expresses a direct variation or an inverse variation. Find the constant of variation. [11.5–11.6]

32.
x	y
8	3
12	2
−6	−4

33.
s	t
−6	2
−3	1
6	−2

34.
m	n
12	1
6	2
−4	−3

inverse; 24   **direct; −3**   **inverse; 12**

In Exercises 35–36, y varies directly as x. [11.5]
35. If $y = 18$ when $x = 3$, find y when $x = 5$. **30**
36. If $y = -8$ when $x = 2$, find y when $x = -4$.
   16

In Exercises 37–38, y varies inversely as x. [11.6]
37. If $y = 12$ when $x = 2$, find y when $x = 12$. **2**
38. If $y = -3$ when $x = 5$, find y when $x = -3$.
39. If y varies jointly as x and z, and $y = -12$ when $x = 3$ and $z = -1$, find y when $x = 5$ and $z = 6$. [11.7] **120**   **38. 5**
40. If y varies directly as the square of x and inversely as the positive square root of z, and if $y = 9$ when $x = 3$ and $z = 16$, find y when $x = -1$ and $z = 4$. [11.7] **2**

Ⓐ Exercises are 1–11, 14–16, 18–20, 24–32. Ⓒ Exercises are starred. The rest are Ⓑ Exercises.

CHAPTER ELEVEN TEST

Determine the domain and the range of each relation. Is the relation a function?
1. $\{(-2, -1), (3, -2), (4, -3), (2, 1)\}$ **F**
2. $\{(2, 5), (3, 0), (1, 3), (-2, 0)\}$ **F**

1. D = {4, 3, 2, −2}; R = {1, −1, −2, −3}; function
2. D = {3, 2, 1, −2}; R = {5, 3, 0}; function

Let $h = \{(-1, 4), (0, -3), (1, 4), (2, -6)\}$. Find each of these function values.
3. $h(-1)$ **4**   4. $h(0)$ **−3**   5. $h(2)$ **−6**   6. $h(1)$ **4**

Let $f(x) = x^2 + 3$. Find each of the following function values.
7. $f(-1)$ **4**   8. $f(0)$ **3**   9. $f(a + 2)$ **$a^2 + 4a + 7$**

Find the range of each function for the given domain.
10. $h(x) = x^2 + 2$, $D = \{-1, 0, 4\}$ **{3, 2, 18}**
11. $g(x) = 3x - 5$, $D = \{0, -1, 2\}$ **{−5, −8, 1}**

Find the domain and zeros of each function.
12. $g(x) = \dfrac{x - 3}{x - 2}$
13. $t(x) = \dfrac{4x - 4}{x^2 - 4x - 5}$

12. D = {all real numbers except 2}; 3
13. D = {all real numbers except 5 and −1}; 1

Let $f(x) = 2x + 4$ and $g(x) = x^2 - 1$. Find each of the following.
14. $f(-1)$ **2**   15. $g(3\sqrt{2})$ **17**   16. $f(g(\tfrac{1}{3}))$ **$\tfrac{20}{9}$**

★17. $\dfrac{f(a) - f(b)}{a - b}$; $a \neq b$ **2**

Determine whether each equation describes a linear function, a constant function, or neither of these.
18. $-2x + 3y = 12$ **L**   19. $y = -6$ 20. $x = 0$ **N**
21. $-2(-2y - x) - 3x = 2(3x - 1)$ **L**

19. C, L

Graph each function and its inverse in the same coordinate plane. Determine if the inverse is a function.
22. $y = x^2 + 2$ **not a function**   23. $xy = -2$ **function**

Determine if the table expresses a direct variation or an inverse variation. Find the constant of variation.

24.
x	y
−4	−5
−5	−4
1	20

inverse; 20

25.
r	s
−2	8
16	−1
−32	$\tfrac{1}{2}$

inverse; −16

26.
m	n
2	24
−3	−36
$\tfrac{1}{2}$	6

direct; 12

In Exercises 27–28, y varies directly as x.
27. If $y = 10$ when $x = 2$, find y when $x = 6$. **30**
28. If $y = -10$ when $x = 5$, find y when $x = -6$. **12**

In Exercises 29–30, y varies inversely as x.
29. If $y = 5$ when $x = 6$, find y when $x = 15$. **2**
30. If $y = 8$ when $x = 3$, find y when $x = 12$. **2**

31. If r varies directly as the square of s and if $r = 32$ when $s = 4$, find r when $s = 6$. **72**

32. If x varies jointly as s and t, and $x = 24$ when $s = 4$ and $t = 2$, find x when $s = -2$ and $t = -3$. **18**

33. If x varies directly as the positive square root of y and inversely as the square of z, and if $x = 2$ when $y = 4$ and $z = 5$, find x when $y = 9$ and $z = 6$. **$2\tfrac{1}{12}$**

COMPUTER ACTIVITIES

User Defined Functions

OBJECTIVE: To evaluate composite functions, using computer programs that contain user defined functions

As you have progressed through the computer pages in this text, you have learned about many different functions that are part of the BASIC programming language. These functions are immediately available for you to use. However, you are not limited to only those functions that are part of BASIC. You can use the DEF to DEFine your own functions for use in programs.

To define your own functions using the DEF command, you write DEF, the function name FN followed by a letter of your choice, the variable of the function within parentheses, an equal sign, and the function itself. For example, to DEFine a function $A(X) = 2X + 3$, you would write the statement DEF FN A(X) = 2 * X + 3. Once you have DEFined a function in a program, you can use that defined function in the program. The following program DEFines the function $F(X) = X^2 + 2X - 5$ and allows you to insert values into it.

```
10   DEF   FN A(X) = X ^ 2 + 2 * X - 5
20   INPUT B
30   PRINT "A("B") = " FN A(B)
40   END
```

If you have DEFined two or more functions in a program, then in addition to the functions themselves, you can use composites of those functions in the program. The following program illustrates how composite functions can be used.

```
10   DEF   FN F(X) = X + 3
20   DEF   FN G(X) = 2 * X - 5
30   INPUT "FOR WHAT VALUE OF X DO YOU WANT
         COMPOSITES? ";X
40   PRINT "F(G("X")) = " FN F( FN G(X))
50   PRINT "G(F("X")) = " FN G( FN F(X))
60   PRINT "F(F("X")) = " FN F( FN F(X))
70   PRINT "G(G("X")) = " FN G( FN G(X)): PRINT
80   PRINT : GOTO 10
```

See the Computer Section beginning on page 574 for more information.

For Ex. 1, see TE Answer Section.

1. Use the new DEF function to evaluate the polynomials in Exercises 30–33 on page 40. Use one program that INPUTs values one at a time, to evaluate all four Exercises.

COLLEGE PREP TEST

DIRECTIONS: Choose the *one* best answer to each question or problem.

1. A bag of rabbit food will feed 24 rabbits for 72 days. How many days will it feed 16 rabbits? **A**

   (A) 108   (B) 48   (C) 71
   (D) 49   (E) 96

2. Which of the following equations gives the relationship between x and y in the table? **E**

x	1	2	3	4	5	6
y	3	7	11	15	19	23

   (A) $y = 3x$ (B) $y = x^2 + 2$
   (C) $y = x^2 + 3$ (D) $y = 3x + 5$
   (E) $y = 4x - 1$

3. In making egg salad, a recipe calls for 4 cups of diced eggs to $\frac{1}{4}$ cup of chopped onions. How much onion should be used for 16 cups of egg salad? **E**

   (A) $\frac{1}{4}$ (B) $\frac{1}{8}$ (C) $\frac{3}{4}$
   (D) 4   (E) 1

4. A person sews m boxes of shirts in n days. If there are p shirts in a box, how many shirts does the person sew in a day? **D**

   (A) $\frac{m}{pn}$ (B) $\frac{n}{pm}$ (C) $\frac{pn}{m}$
   (D) $\frac{pm}{n}$ (E) $\frac{mn}{p}$

5. A bell chimes every c minutes and a horn blasts every b minutes. Assume c and b are prime numbers and the bell and horn start simultaneously. In how many minutes will they sound simultaneously? **C**

   (A) $\frac{b}{c}$ (B) $\frac{c}{b}$ (C) bc
   (D) $\frac{cb}{2}$ (E) $2bc$

6. If $2y - 6$ varies directly as the square of $x + 2$ and inversely as $z - 3$, and if $y = 1$ when $x = 2$ and $z = 11$, find y when $x = 4$ and $z = 9$. **C**

   (A) -213 (B) $\frac{17}{6}$ (C) -3
   (D) $\frac{18 - \sqrt{6}}{6}$ (E) None of these

7. A machine can produce 140 valves in 20 seconds. At this rate, how many valves will it produce in 30 minutes? **B**

   (A) 210   (B) 12,600   (C) 114
   (D) 93   (E) None of these

8. If the tax on a $125 bike is $10, at the same rate, what is the tax on a $300 bike? **D**

   (A) $30   (B) $21   (C) $27
   (D) $24

College Prep Test

12 CONIC SECTIONS

Non-Routine Problem Solving

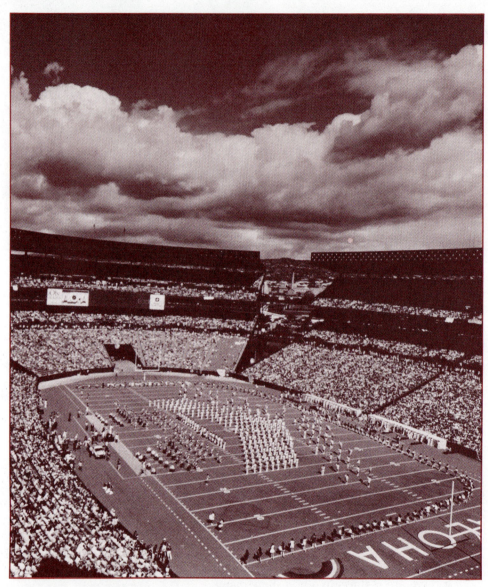

Problem 1. How many positive integers less than or equal to 3,300 are divisible by neither 3 nor 5 nor 11? **1,600**

Problem 2. For the real numbers A, B, and C, in that order, subtraction is associative. What are the possible values for A, B, and C?

A and B are any real numbers; $C = 0$.

Problem 3. Let $s = ax + by$, $t = bx - ay$, and $a^2 + b^2 = 1$. Express $x^2 + y^2$ in terms of s and t. (*Hint:* Square both equations involving s and t.)
$x^2 + y^2 = s^2 + t^2$

SYMMETRIC POINTS IN A COORDINATE PLANE   12.1

Objective **To determine whether two given points are symmetric with respect to the *x*-axis, the *y*-axis, or the origin**

Two points in a coordinate plane are **symmetric** with respect to a given line, l, in the plane if l is the perpendicular bisector of the segment connecting the two points. Line l is called the **line of symmetry** or the **axis of symmetry** for the two points.

In the figure at the right, the two pairs of points $P(-5, 4)$ and $Q(-5, -4)$, $R(4, 3)$ and $S(4, -3)$ are symmetric with respect to the *x*-axis. The *x*-axis, whose equation is $y = 0$, is the axis of symmetry. Notice that the points $T(1, 2)$ and $U(2, -1)$ are not symmetric with respect to the *x*-axis.

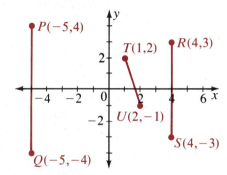

Symmetric points with respect to the *x*-axis

Two points, $P(x, y)$ and $Q(x, -y)$, in a coordinate plane are *symmetric with respect to the x-axis*. Their *x*-coordinates are the same and their *y*-coordinates are additive inverses.

Example 1 **Determine whether the two points are symmetric with respect to the *x*-axis.**
$C(2, -3)$ $D(2, 3)$ $E(5, -4)$ $F(-5, -4)$

Points C and D have the same *x*-coordinates. Their *y*-coordinates are additive inverses.
Thus, C and D are symmetric with respect to the *x*-axis.

Points E and F have different *x*-coordinates.

Thus, E and F are not symmetric with respect to the *x*-axis.

In the figure at the right, the two pairs of points $G(-4, 3)$ and $H(4, 3)$, $I(-5, -2)$ and $J(5, -2)$ are symmetric with respect to the *y*-axis. The *y*-axis, whose equation is $x = 0$, is the axis of symmetry. Notice that the points $K(-1, 2)$ and $L(2, 1)$ are not symmetric with respect to the *y*-axis.

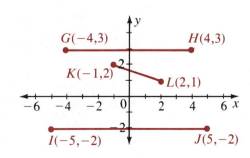

Symmetric Points in a Coordinate Plane   **303**

Symmetric points with respect to the y-axis	Two points, $P(x, y)$ and $Q(-x, y)$, in a coordinate plane are *symmetric with respect to the y-axis*. Their x-coordinates are additive inverses and their y-coordinates are the same.

Example 2 **Determine whether the two points are symmetric with respect to the y-axis.**
  $A(5, -6)$ $B(-5, -6)$ $M(-4, 6)$ $N(-4, -6)$

The x-coordinates of points A and B are additive inverses. Their y-coordinates are the same.	The x-coordinates of points M and N are not additive inverses.
Thus, A and B are symmetric with respect to the y-axis.	**Thus, M and N are not symmetric with respect to the y-axis.**

It is possible for two given points in a coordinate plane to be symmetric with respect to the origin.
In the figure at the right, the two pairs of points, the endpoints of the diameters of the circle, $P(-3, 2)$ and $Q(3, -2)$, $R(3, 2)$ and $S(-3, -2)$ are symmetric with respect to the origin, whose coordinates are $(0, 0)$. The origin is a **point of symmetry**.

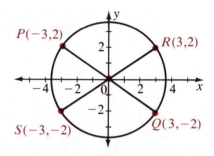

Symmetric points with respect to the origin	Two points, $P(x, y)$ and $Q(-x, -y)$, in a coordinate plane are *symmetric with respect to the origin*. Their x-coordinates are additive inverses and their y-coordinates are additive inverses.

Example 3 **Determine whether the two points are symmetric with respect to the origin.**
  $W(5, -9)$ $X(-5, 9)$ $Y(2, \sqrt{5})$ $Z(-2, \sqrt{5})$

The x-coordinates of points W and X are additive inverse. Their y-coordinates are additive inverses.	The y-coordinates of points Y and Z are not additive inverses.
Thus, W and X are symmetric with respect to the origin.	**Thus, Y and Z are not symmetric with respect to the origin.**

Example 4 **For $P(3, -5)$, find the coordinates of the corresponding point symmetric with respect to the x-axis, the y-axis, and the origin.**

x-axis	y-axis	origin
(3, 5)	(−3, −5)	(−3, 5)

Oral Exercises

Determine which point, A, B, C, or D, is symmetric to each given point with respect to the x-axis, y-axis, and the origin.

1. $P(7, -4)$: $A(-7, -4)$ $B(0, 0)$ $C(-7, 4)$ $D(7, 4)$ **x-axis: D, y-axis: A, origin: C**
2. $Q(-2, 3)$: $A(-2, 0)$ $B(-2, -3)$ $C(2, -3)$ $D(2, 3)$ **x-axis: B, y-axis: D, origin: C**
3. $R(6, 1)$: $A(-6, -1)$ $B(0, 1)$ $C(-6, 1)$ $D(6, -1)$ **x-axis: D, y-axis: C, origin: A**
4. $S(-1, -5)$: $A(1, -5)$ $B(1, 5)$ $C(-1, 5)$ $D(5, 0)$ **x-axis: C, y-axis: A, origin: B**

10. $(6, 4), (-6, -4), (-6, 4)$ 11. $(-4, 5), (4, -5), (4, 5)$ 12. $(6, 6), (-6, -6), (-6, 6)$

25. $(\frac{5}{6}, \frac{1}{3}), (-\frac{5}{6}, -\frac{1}{3}), (-\frac{5}{6}, \frac{1}{3})$

Written Exercises

26. $(\frac{2}{3}, -\frac{5}{7}), (-\frac{2}{3}, \frac{5}{7}), (-\frac{2}{3}, -\frac{5}{7})$ 27. $(-1, \frac{2}{3}), (1, -\frac{2}{3}), (1, \frac{2}{3})$

13. $(-3, -5), (3, 5), (3, -5)$ 14. $(-1, 3), (1, -3), (1, 3)$ 15. $(-2, -5), (2, 5), (2, -5)$

Determine whether P and Q are symmetric with respect to the x-axis, the y-axis, the origin, or none of these. 16. $(-7, 2), (7, -2), (7, 2)$ 17. $(5, -4), (-5, 4), (-5, -4)$ 18. $(7, 2), (-7, -2), (-7, 2)$

(A)
1. $P(-3, 2); Q(3, -2)$ **origin**   2. $P(-5, -4); Q(5, 4)$ **origin**   3. $P(-4, -3); Q(4, -3)$ **y-axis**
4. $P(7, 3); Q(7, -3)$ **x-axis**   5. $P(6, 8); Q(-6, 8)$ **y-axis**   6. $P(-4, -5); Q(4, 5)$ **origin**
7. $P(0, 5); Q(0, -5)$ **x-axis and origin**   8. $P(8, 0); Q(-8, 0)$ **y-axis and origin**   9. $P(-7, 6); Q(6, -7)$ **none of these**

For each point, find the coordinates of the corresponding point symmetric with respect to the x-axis, the y-axis, and the origin. 19. $(-3, 4), (3, -4), (3, 4)$ 20. $(2, 2), (-2, -2), (-2, 2)$ 21. $(-8, 8), (8, -8), (8, 8)$
10. $A(6, -4)$ 11. $B(-4, -5)$ 12. $C(6, -6)$ 13. $D(-3, 5)$ 14. $E(-1, -3)$ 15. $F(-2, 5)$
16. $G(-7, -2)$ 17. $H(5, 4)$ 18. $I(7, -2)$ 19. $J(-3, -4)$ 20. $K(2, -2)$ 21. $L(-8, -8)$

22. $(\frac{1}{3}, \frac{1}{4}), (-\frac{1}{3}, -\frac{1}{4}), (-\frac{1}{3}, \frac{1}{4})$ 23. $(-\frac{2}{3}, -\frac{4}{5}), (\frac{2}{3}, \frac{4}{5}), (\frac{2}{3}, -\frac{4}{5})$ 24. $(-\frac{1}{2}, \frac{1}{3}), (\frac{1}{2}, -\frac{1}{3}), (\frac{1}{2}, \frac{1}{3})$

(B) 22. $M(\frac{1}{3}, -\frac{1}{4})$ 23. $N(-\frac{2}{3}, \frac{4}{5})$ 24. $O(-\frac{1}{2}, -\frac{1}{3})$ 25. $P(\frac{5}{6}, -\frac{1}{3})$ 26. $Q(\frac{2}{3}, \frac{5}{7})$ 27. $R(-1, -\frac{2}{3})$
28. $S(a, -b)$ 29. $T(-a, -b)$ 30. $U(-a, b)$ 31. $V(x, -y)$ 32. $W(-x, y)$ 33. $Z(-x, -y)$

28. $(a, b), (-a, -b), (-a, b)$ 29. $(-a, b), (a, -b), (a, b)$ 30. $(-a, -b), (a, b), (a, -b)$ 31. $(x, y), (-x, -y), (-x, y)$

For the given point on each circle, find the coordinates of the corresponding point symmetric with respect to the x-axis, the y-axis, and the origin. 32. $(-x, -y), (x, y), (x, -y)$ 33. $(-x, y), (x, -y), (x, y)$

34.   35.   36.   37.

$(3, -2), (-3, 2), (-3, -2)$ $(-5, 2), (5, -2), (5, 2)$ $(-4, -\sqrt{5}), (4, \sqrt{5}), (4, -\sqrt{5})$ $(3.3, 2.3), (-3.3, -2.3), (-3.3, 2.3)$

Determine if the graph of each equation is symmetric with respect to the x-axis, y-axis, or the origin. [Hint: For all real numbers a and b, if (a, b) and $(-a, -b)$ satisfy the equation, the graph is symmetric with respect to the origin.] 37. $(3.3, 2.3), (-3.3, -2.3), (-3.3, 2.3)$

(C) 38. $x^2 + y^2 = 7$ 39. $y = x^2$ 40. $x^2 + (-y)^2 = 7$ 41. $-y = x^2$ 42. $-y = (-x)^2$
x-axis, y-axis, and origin   **y-axis**   **x-axis, y-axis, and origin**   **y-axis**   **y-axis**

A function is said to be *even* if its graph has y-axis symmetry and *odd* if it has origin symmetry. Which of the following are odd? even?

43. $y = x^2$ **even**   44. $y = \frac{1}{x}$ **odd**   45. $y = |x|$ **even**   46. $xy = 4$ **odd**   47. $y = x^3$ **odd**

Symmetric Points in a Coordinate Plane   **305**

THE CIRCLE 12.2

Objectives To write an equation of a circle given its center and radius or given its center and a point on the circle
To find the center and radius of a circle given an equation of a circle
To draw a graph of the equation of a circle

| Definition: Circle | A *circle* is the set of all points in a coordinate plane that are a given distance from a given point in the plane. The given distance is the *radius* of the circle and the given point is the *center* of the circle. |

For simplicity, the term "radius" is used to mean "length of the radius." That is, radius 6 means a radius has a length of 6 units.

You can write an equation of a circle with radius r and center at the origin. Let $P(x, y)$ be any point on a circle.

Use the distance formula to find the distance from P to the center of the circle.

$d = \sqrt{(x_2 - x_1)^2 + (y_2 - y_1)^2}$
$r = \sqrt{(x - 0)^2 + (y - 0)^2}$

Squaring both sides and simplifying, you get $x^2 + y^2 = r^2$.

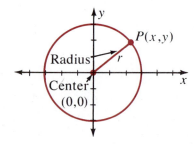

| Standard form of equation of circle | The standard form of the equation of a circle with center at the origin $(0, 0)$ and radius of length r is $x^2 + y^2 = r^2$. |

Example 1 Write an equation of a circle whose center is at the origin and whose radius is 4.

Use the standard form of the equation of a circle.

$x^2 + y^2 = r^2$
Substitute 4 for r. ▶ $x^2 + y^2 = 4^2$
$x^2 + y^2 = 16$

Thus, $x^2 + y^2 = 16$ is an equation of the circle.

You can draw a graph of an equation of a circle after finding the center and the radius of the circle.

Example 2   Find the radius of each circle. Draw the graph of each equation.

Use $x^2 + y^2 = r^2$. ▶

$x^2 + y^2 = 9$

$x^2 + y^2 = 9$
$r^2 = 9$
$r = 3$

Plot some points 3 units from the origin.

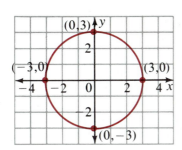

$x^2 + y^2 = 21$

$x^2 + y^2 = 21$
$r^2 = 21$
$r = \sqrt{21} \doteq 4.6$

Plot some points 4.6 units from the origin.

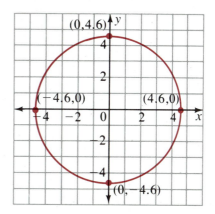

You can also write an equation of a circle with center at $C(h, k)$ and radius of length r, $r > 0$.

Let $P(x, y)$ be any point on the circle. Use the distance formula to find the distance from P to the center of the circle.

$d = \sqrt{(x_2 - x_1)^2 + (y_2 - y_1)^2}$
$r = \sqrt{(x - h)^2 + (y - k)^2}$

Squaring both sides and simplifying, you get

$(x - h)^2 + (y - k)^2 = r^2$.

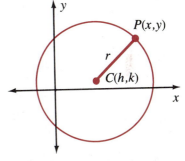

Standard form of equation of circle with center (h, k)

The standard form of the equation of a circle with center at $C(h, k)$ and radius of length r, $r > 0$, is
$$(x - h)^2 + (y - k)^2 = r^2.$$

Example 3   Write an equation of a circle whose center is at $C(7, -6)$ and whose radius is 8.

Use $(x - h)^2 + (y - k)^2 = r^2$.
$(x - 7)^2 + [y - (-6)]^2 = 8^2$ ◀ Substitute 7 for h, -6 for k, and 8 for r.

Thus, $(x - 7)^2 + (y + 6)^2 = 64$ is an equation of the circle.

The Circle

Example 4 Write an equation of a circle whose center is at $C(-3, 5)$ and which passes through the point $P(7, -3)$.

Let $P(x, y) = P(7, -3)$ and $C(h, k) = C(-3, 5)$. Find r^2.
$$(x - h)^2 + (y - k)^2 = r^2$$
$$[7 - (-3)]^2 + (-3 - 5)^2 = r^2$$
So, $r^2 = 100 + 64$, or 164

Thus, $(x + 3)^2 + (y - 5)^2 = 164$ is an equation of the circle.

You can rewrite equations like $x^2 + y^2 - 6x + 12y - 5 = 0$ in standard form for a circle by completing the square. Recall that to solve a quadratic equation of the form $x^2 + bx = c$ by completing the square, divide b by 2, square the result, then add $\left(\frac{b}{2}\right)^2$ to each side.

Example 5 Rewrite $x^2 + y^2 - 6x + 12y - 5 = 0$ in the standard form $(x - h)^2 + (y - k)^2 = r^2$. Find the center $C(h, k)$ and the radius r of the circle. Complete the square.

Group the terms with the same variable. ▶ $(x^2 - 6x + \underline{\ ?\ }) + (y^2 + 12y + \underline{\ ?\ }) = 5$

$\left(-\frac{6}{2}\right)^2 = 9; \left(\frac{12}{2}\right)^2 = 36$ ▶ $x^2 - 6x + 9 + y^2 + 12y + 36 = 5 + 9 + 36$

$(x - 3)^2 + (y + 6)^2 = 50$
$(x - h)^2 + (y - k)^2 = r^2$
So, $h = 3$, $k = -6$, and $r = \sqrt{50}$, or $5\sqrt{2}$

Thus, $(x - 3)^2 + (y + 6)^2 = 50$ is the equation in standard form for the circle with center $C(3, -6)$, and radius $r = 5\sqrt{2}$.

Example 6 Draw the graph of the equation $x^2 + y^2 + 2x - 4y + 1 = 0$.

Rewrite in the standard form $(x - h)^2 + (y - k)^2 = r^2$.

Complete the ▶ $x^2 + 2x + \underline{\ ?\ } + y^2 - 4y + \underline{\ ?\ } = -1$
square.
$x^2 + 2x + 1 + y^2 - 4y + 4 = -1 + 1 + 4$
$(x + 1)^2 + (y - 2)^2 = 4$

Thus, $C(h, k) = C(-1, 2)$ and $r = \sqrt{4}$, or 2.

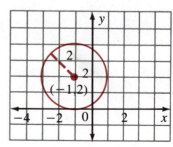

Oral Exercises

1. $(2, 8), 2$ 2. $(-3, -2), \sqrt{21}$ 3. $(1, -2), 2\sqrt{5}$ 4. $(-5, 3), 3\sqrt{2}$

Find the center and the radius for each circle.
1. $(x - 2)^2 + (y - 8)^2 = 4$
2. $(x + 3)^2 + (y + 2)^2 = 21$
3. $(x - 1)^2 + (y + 2)^2 = 20$
4. $(x + 5)^2 + (y - 3)^2 = 18$
5. $x^2 + (y - 3)^2 = 16$
6. $(x + 6)^2 + y^2 = 15$

5. $(0, 3), 4$ 6. $(-6, 0), \sqrt{15}$

Written Exercises

Answers for Ex. 1–10 are of the form $x^2 + y^2 = r^2$. The value of r^2 is given for each exercise.
Write an equation, in standard form, for each circle with its center at the origin and with the given radius.

A
1. 1   1
2. 12   144
3. 3   9
4. 15   225
5. 2   4
6. 13   169
7. $\sqrt{2}$ 2
8. $\sqrt{5}$ 5
9. $\sqrt{3}$ 3
10. $\sqrt{7}$ 7

Graphs for Ex. 11–17 are circles with centers at the origin.
Find each radius. Draw the graph of each equation.
11. $x^2 + y^2 = 25$ 5
12. $x^2 + y^2 = 16$ 4
13. $x^2 + y^2 = 81$ 9
14. $x^2 + y^2 = 100$ 10
15. $x^2 + y^2 = 36$ 6
16. $x^2 + y^2 = 3$ $\sqrt{3}$
17. $x^2 + y^2 = 12$ $2\sqrt{3}$
18. $x^2 + y^2 = 6$ $\sqrt{6}$

19. $(x - 5)^2 + (y - 6)^2 = 49$ 20. $(x + 2)^2 + (y - 8)^2 = 81$ 21. $(x - 11)^2 + (y + 7)^2 = \frac{49}{4}$

Write an equation, in standard form, for each circle given its center C and radius r.

B
19. $C(5, 6); r = 7$
20. $C(-2, 8); r = 9$
21. $C(11, -7); r = 3\frac{1}{2}$
22. $C(-3, -6); r = 2$
23. $C(0, 0); r = \sqrt{5}$
24. $C(0, 0); r = \sqrt{14}$

22. $(x + 3)^2 + (y + 6)^2 = 4$ 23. $x^2 + y^2 = 5$ 24. $x^2 + y^2 = 14$

Write an equation, in standard form, for each circle given its center C and a point P on the circle.
25. $C(2, -3); P(4, 6)$
26. $C(7, 0); P(5, 2)$
27. $C(-6, -3); P(4, -1)$
28. $C(-4, -3); P(-3, -4)$
29. $C(0, 0); P(8, 0)$
30. $C(0, 0); P(0, 4)$

25. $(x - 2)^2 + (y + 3)^2 = 85$ 26. $(x - 7)^2 + y^2 = 8$ 27. $(x + 6)^2 + (y + 3)^2 = 104$

Write each equation of a circle in standard form. Find the center $C(h, k)$ and the radius r of the circle.
31. $x^2 + y^2 - 6y - 7 = 0$ $x^2 + (y - 3)^2 = 16$, (0, 3), 4
32. $x^2 + y^2 + 8x - 20 = 0$ $(x + 4)^2 + y^2 = 36$, (−4, 0), 6
33. $x^2 + y^2 + 8x + 4y - 5 = 0$ $(x + 4)^2 + (y + 2)^2 = 25$, (−4, −2), 5
34. $x^2 + y^2 - 6x - 10y - 2 = 0$ $(x - 3)^2 + (y - 5)^2 = 36$, (3, 5), 6

Draw the graph of each equation. The center and radius are given in Ex. 35–38.
35. $x^2 + y^2 + 4y - 12 = 0$ (0, −2), 4
36. $x^2 + y^2 - 6x - 3 = 0$ (3, 0), $2\sqrt{3}$
37. $x^2 + y^2 + 10x - 6y + 25 = 0$ (−5, 3), 3
38. $x^2 + y^2 - 4x + 8y + 6 = 0$ (2, −4), $\sqrt{14}$

28. $(x + 4)^2 + (y + 3)^2 = 2$ 29. $x^2 + y^2 = 64$ 30. $x^2 + y^2 = 16$ 40. $(x - 6)^2 + (y - 6)^2 = 36$

Write an equation for each circle described. (Ex. 39–43)

C
39. $P(5, -3)$ and $Q(-3, 11)$ are the endpoints of a diameter of the circle. $(x - 1)^2 + (y - 4)^2 = 65$
40. The circle is tangent to the x-axis at $A(6, 0)$ and tangent to the y-axis at $B(0, 6)$.
41. The circle intersects the axes at $Q(0, 3)$, $R(-1, 0)$, $S(0, -3)$, and $T(9, 0)$. $(x - 4)^2 + y^2 = 25$
42. $\overline{AB}$ is a diameter of the circle through $A(-2, 6)$ and $B(8, -4)$. $(x - 3)^2 + (y - 1)^2 = 50$
43. Write an equation of the circle determined by $T(6, -2)$, $V(5, -9)$, and $W(-1, -1)$.
   $(x - 2)^2 + (y + 5)^2 = 25$

Determine the center $C(h, k)$ and the radius r for each circle.
44. $x^2 + y^2 + 2ax + 2by + c = 0$
   $(-a, -b), \sqrt{a^2 + b^2 - c}$
45. $x^2 + y^2 + cx + dy + e = 0$
   $(-\frac{c}{2}, -\frac{d}{2}), \frac{1}{2}\sqrt{c^2 + d^2 - 4e}$

The Circle

THE ELLIPSE   12.3

Objectives **To write an equation of an ellipse given its center and its *x*- and *y*-intercepts**
To find the center and *x*- and *y*-intercepts of an ellipse, given an equation of an ellipse
To draw a graph of the equation of an ellipse

The orbit of the earth around the sun is in the shape of an **ellipse.** An ellipse can be defined as a set of points as follows.

Definition: Ellipse   An *ellipse* is the set of all points in a coordinate plane such that for each point of the set, the sum of its distances from two fixed points is constant. Each of the fixed points is called a *focus* (plural: foci).

You can write an equation of an ellipse whose center is at the origin, foci are $F_1(-3, 0)$ and $F_2(3, 0)$, and where the sum of the distances from any point on the ellipse to the foci is 10.

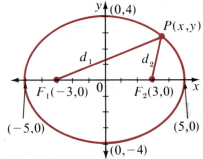

Let $P(x, y)$ be any point on the ellipse. Then find the distance d_1 from $P(x, y)$ to $F_1(-3, 0)$ and d_2 from $P(x, y)$ to $F_2(3, 0)$. Use the distance formula.

$$d_1 = \sqrt{[x - (-3)]^2 + (y - 0)^2}$$
$$d_2 = \sqrt{(x - 3)^2 + (y - 0)^2}$$

The sum of the distances is 10. Use this fact to write one radical equation. Then simplify the radical equation by squaring each side.

$$\sqrt{(x + 3)^2 + y^2} + \sqrt{(x - 3)^2 + y^2} = 10$$
$$\sqrt{x^2 + 6x + 9 + y^2} = 10 - \sqrt{x^2 - 6x + 9 + y^2}$$
$$(\sqrt{x^2 + 6x + 9 + y^2})^2 = (10 - \sqrt{x^2 - 6x + 9 + y^2})^2$$
$$x^2 + 6x + 9 + y^2 = 100 - 20\sqrt{x^2 - 6x + 9 + y^2} + x^2 - 6x + 9 + y^2$$
$$12x - 100 = -20\sqrt{x^2 - 6x + 9 + y^2}$$
$$3x - 25 = -5\sqrt{x^2 - 6x + 9 + y^2}$$
$$(3x - 25)^2 = (-5\sqrt{x^2 - 6x + 9 + y^2})^2$$
$$9x^2 - 150x + 625 = 25x^2 - 150x + 225 + 25y^2$$
$$400 = 16x^2 + 25y^2$$

Divide each side by 400 to get the following equation of the ellipse.

$$\frac{x^2}{25} + \frac{y^2}{16} = 1$$

Notice that the *x*-intercepts of the ellipse in the figure above are $\pm\sqrt{25}$, or ± 5, and the *y*-intercepts are $\pm\sqrt{16}$, or ± 4.

310   Chapter Twelve

> **Standard form of equation of ellipse with center (0, 0)**
>
> The standard form of the equation of an ellipse with center at the origin (0, 0) and whose x- and y-intercepts are $\pm a$ and $\pm b$, respectively, is
> $$\frac{x^2}{a^2} + \frac{y^2}{b^2} = 1; a \neq 0, b \neq 0.$$

Example 1 Write an equation, in standard form, of an ellipse whose center is at the origin and whose x-intercepts are ± 9 and whose y-intercepts are ± 7.

Use standard form of the equation of an ellipse.
$$\frac{x^2}{a^2} + \frac{y^2}{b^2} = 1$$

Substitute ± 9 for a and ± 7 for b. ▶ $\frac{x^2}{(\pm 9)^2} + \frac{y^2}{(\pm 7)^2} = 1$

Thus, $\frac{x^2}{81} + \frac{y^2}{49} = 1$ is an equation of the ellipse.

You can also write an equation of an ellipse whose center is at $C(h, k)$ and whose x'- and y'-intercepts are $\pm a$ and $\pm b$. The x', y' axes are obtained by *translating* the x, y axes. Then center (h, k) is moved h units horizontally and k units vertically.

> **Standard form of equation of ellipse with center (h, k)**
>
> The standard form of the equation of an ellipse with center at $C(h, k)$ and whose x'- and y'-intercepts are $\pm a$ and $\pm b$, respectively, is
> $$\frac{(x - h)^2}{a^2} + \frac{(y - k)^2}{b^2} = 1.$$

Example 2 Sketch the graph of an ellipse whose center is $C(2, -1)$, whose x'-intercepts are ± 4, and whose y'-intercepts are ± 6. Then, write an equation for the ellipse.

Use the standard form
$$\frac{(x - h)^2}{a^2} + \frac{(y - k)^2}{b^2} = 1.$$

$C(h, k) = C(2, -1)$
Substitute ± 4 for a and ± 6 for b. ▶ $\frac{(x - 2)^2}{(\pm 4)^2} + \frac{(y + 1)^2}{(\pm 6)^2} = 1$

Thus, $\frac{(x - 2)^2}{16} + \frac{(y + 1)^2}{36} = 1$
is an equation of the ellipse.

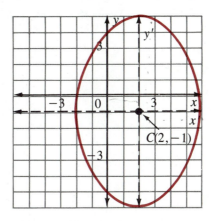

You can also rewrite equations like $4x^2 + 9y^2 + 32x - 90y + 253 = 0$ in standard form for an ellipse by completing the square.

The Ellipse

Example 3

Rewrite $4x^2 + 9y^2 + 32x - 90y + 253 = 0$ in standard form for an ellipse. Find the center $C(h, k)$ and x'- and y'-intercepts. Draw the graph.

Group the terms with the same variable. ▶ $(4x^2 + 32x + \underline{?}) + (9y^2 - 90y + \underline{?}) = -253$

Factor out the coefficients of $4x^2$ and $9y^2$. ▶ $4(x^2 + 8x + \underline{?}) + 9(y^2 - 10y + \underline{?}) = -253$

Complete the squares in x and y. ▶ $4(x^2 + 8x + 16) + 9(y^2 - 10y + 25) = -253 + 4(16) + 9(25)$
$4(x + 4)^2 + 9(y - 5)^2 = 36$

Divide each term by 36. ▶ $\dfrac{4(x + 4)^2}{36} + \dfrac{9(y - 5)^2}{36} = \dfrac{36}{36}$

$\dfrac{(x - h)^2}{a^2} + \dfrac{(y - k)^2}{b^2} = 1$

Thus, (1) $\dfrac{(x + 4)^2}{9} + \dfrac{(y - 5)^2}{4} = 1$ is the equation in standard form.

(2) $h = -4$ and $k = 5$, so the center is at $C(-4, 5)$.

(3) $a = \pm 3$ and $b = \pm 2$, so the x'-intercepts are ± 3 and the y'-intercepts are ± 2. The horizontal and vertical diameters are $2 \cdot 3$ and $2 \cdot 2$ units long, respectively.

(4) Draw the ellipse through the endpoints of the diameters.

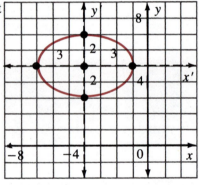

Example 4

Draw the graph of $6x^2 + 9y^2 = 27$. Find the x- and y-intercepts.

Write $6x^2 + 9y^2 = 27$ in standard form.

Divide by 27. ▶ $\dfrac{6x^2}{27} + \dfrac{9y^2}{27} = \dfrac{27}{27}$

Use $\dfrac{x^2}{a^2} + \dfrac{y^2}{b^2} = 1$. ▶ $\dfrac{x^2}{\frac{27}{6}} + \dfrac{y^2}{3} = 1$

$\dfrac{x^2}{\frac{9}{2}} + \dfrac{y^2}{3} = 1$

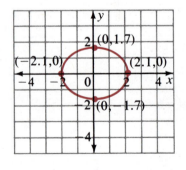

Find a and b. ▶ $a = \pm \dfrac{3\sqrt{2}}{2} \doteq \pm 2.1$ $\quad b = \pm\sqrt{3} \doteq \pm 1.7$

The x-intercepts are ± 2.1 and the y-intercepts are ± 1.7.

Written Exercises

7. $a = \pm 4, b = \pm 3$ **8.** $a = \pm 6, b = \pm 5$ **9.** $a = \pm 5, b = \pm 2$ **10.** $a = \pm 9, b = \pm 3$

Write an equation, in standard form, of an ellipse whose center is at the orgin and whose intercepts are given.

(A)
1. x-intercepts ± 3; y-intercepts ± 2 $\frac{x^2}{9} + \frac{y^2}{4} = 1$ 2. x-intercepts $\pm\sqrt{8}$; y-intercepts ± 3 $\frac{x^2}{8} + \frac{y^2}{9} = 1$
3. x-intercepts ± 2; y-intercepts $\pm\sqrt{3}$ $\frac{x^2}{4} + \frac{y^2}{3} = 1$ 4. x-intercepts $\pm\sqrt{3}$; y-intercepts $\pm\sqrt{5}$ $\frac{x^2}{3} + \frac{y^2}{5} = 1$
5. x-intercepts ± 8; y-intercepts ± 6 $\frac{x^2}{64} + \frac{y^2}{36} = 1$ 6. x-intercepts ± 7; y-intercepts $\pm\sqrt{3}$ $\frac{x^2}{49} + \frac{y^2}{3} = 1$

11. $a = \pm 2, b = \pm 4$ **12.** $a = \pm 10, b = \pm 5$ **13.** $a = \pm 3, b = \pm 10$ **14.** $a = \pm 2, b = \pm 5$

Find the x- and y-intercepts of the ellipse whose equation is given. Then draw the graph.

7. $\frac{x^2}{16} + \frac{y^2}{9} = 1$ above   8. $\frac{x^2}{36} + \frac{y^2}{25} = 1$ above   9. $\frac{x^2}{25} + \frac{y^2}{4} = 1$ above   10. $\frac{x^2}{81} + \frac{y^2}{9} = 1$ above
11. $\frac{x^2}{4} + \frac{y^2}{16} = 1$ above   12. $\frac{x^2}{100} + \frac{y^2}{25} = 1$ above   13. $\frac{x^2}{9} + \frac{y^2}{100} = 1$ above   14. $\frac{x^2}{4} + \frac{y^2}{25} = 1$ above
15. $4x^2 + y^2 = 100$ 16. $3x^2 + y^2 = 75$ 17. $x^2 + 9y^2 = 36$ 18. $4x^2 + 25y^2 = 100$
    $a = \pm 5, b = \pm 10$ $a = \pm 5, b = \pm 5\sqrt{3}$ $a = \pm 6, b = \pm 2$ $a = \pm 5, b = \pm 2$

Write an equation, in standard form, for each ellipse given its center C and its x'- and y'-intercepts.

19. $C(-1, 3), a = \pm 4, b = \pm 5$ below   20. $C(2, -1), a = \pm 3, b = \pm 6$ below
21. $C(-2, -3), a = \pm 2, b = \pm 4$ below   22. $C(4, 3), a = \pm 4, b = \pm 7$ below
23. $C(-5, 2), a = \pm\sqrt{7}, b = \pm 6$ below   24. $C(-1, -3), a = \pm\sqrt{9}, b = \pm\sqrt{5}$ below

For Ex. 25–28, answers are of the form center, horizontal diameter, vertical diameter. Determine the center of the ellipse whose equation is given. Find the lengths of the horizontal and vertical diameters. Draw the ellipse.

(B)
25. $4x^2 + 25y^2 + 16x - 150y + 141 = 0$ 26. $16x^2 + 9y^2 - 96x + 72y + 144 = 0$
27. $4x^2 + y^2 - 32x - 4y + 52 = 0$ 28. $9x^2 + 25y^2 + 54x + 200y + 256 = 0$

25. $(-2, 3), 10, 4$ 26. $(3, -4), 6, 8$ 27. $(4, 2), 4, 8$ 28. $(-3, -4), 10, 6$

Write an equation, in standard form, of each ellipse with center at the origin, F_1 and F_2 the foci, and where c is the *sum of the distances* from any point on the ellipse to the foci.

(C)
29. $F_1(-3, 0), F_2(3, 0); c = 8$ $\frac{x^2}{16} + \frac{y^2}{7} = 1$ 30. $F_1(-2, 0), F_2(2, 0); c = 12$ $\frac{x^2}{36} + \frac{y^2}{32} = 1$
31. $F_1(-4, 0), F_2(4, 0); c = 10$ $\frac{x^2}{25} + \frac{y^2}{9} = 1$ 32. $F_1(-5, 0), F_2(5, 0); c = 14$ $\frac{x^2}{49} + \frac{y^2}{24} = 1$

19. $\frac{(x+1)^2}{16} + \frac{(y-3)^2}{25} = 1$ 20. $\frac{(x-2)^2}{9} + \frac{(y+1)^2}{36} = 1$ 21. $\frac{(x+2)^2}{4} + \frac{(y+3)^2}{16} = 1$

Find the domain and range of the function defined by the given equation by solving the equation for each variable in terms of the other variable. [The *extent* of the graph of the equation is the set of all possible values of x and the set of all possible values of y. This is the region of the coordinate plane that the graph occupies.]

33. $\frac{x^2}{9} + \frac{y^2}{16} = 1$ 34. $\frac{x^2}{36} + \frac{y^2}{25} = 1$ 35. $\frac{x^2}{4} + \frac{y^2}{81} = 1$
$\{x | -3 \le x \le 3\}, \{y | -4 \le y \le 4\}$ $\{x | -6 \le x \le 6\}, \{y | -5 \le y \le 5\}$ $\{x | -2 \le x \le 2\}, \{y | -9 \le y \le 9\}$

Draw the graph of each equation. Determine the extent of the graph.

36. $y = \pm\frac{1}{3}\sqrt{81 - x^2}$ 37. $y = \pm\frac{1}{2}\sqrt{16 - x^2}$ 38. $y = \pm\frac{3}{2}\sqrt{4 - x^2}$
$\{x|-9 \le x \le 9 \text{ and } y|-3 \le y \le 3\}$ $\{x|-4 \le x \le 4 \text{ and } y|-2 \le y \le 2\}$ $\{x|-2 \le x \le 2 \text{ and } y|-3 \le y \le 3\}$

$(x - h)^2 + (y - k)^2 = r^2$ defines a circle whose area is πr^2.
$\frac{(x - h)^2}{a^2} + \frac{(y - k)^2}{b^2} = 1$ defines an ellipse whose area is πab.

22. $\frac{(x-4)^2}{16} + \frac{(y-3)^2}{49} = 1$ 23. $\frac{(x+5)^2}{7} + \frac{(y-2)^2}{36} = 1$ 24. $\frac{(x+1)^2}{9} + \frac{(y+3)^2}{5} = 1$

Find the area of each figure defined.

39. $x^2 + y^2 + 8x + 4y - 5 = 0$ 25π 40. $x^2 + y^2 - 6x - 10y - 2 = 0$ 36π
41. $4x^2 + 25y^2 + 16x - 150y + 141 = 0$ 10π 42. $16x^2 + 9y^2 - 96x + 72y + 144 = 0$ 12π

The Ellipse

THE PARABOLA 12.4

Objectives   **To draw the graph of the equation of a parabola whose vertex is not at the origin**
To find the coordinates of the vertex and write the equation of the axis of symmetry of a parabola
To estimate graphically the roots of a quadratic equation

The graph of a quadratic function is called a **parabola**. In general, any equation of the form $y = ax^2 + bx + c$, where a, b, and c, are real numbers with $a \neq 0$, describes a parabola.

You can graph a parabola determined by the equation $y = x^2$ or $y = -x^2$ as follows. First, make a table by choosing values for x and finding corresponding values for y. Then, plot as many points as needed to determine the parabola.

x	x^2	y
-2	$(-2)^2$	4
-1	$(-1)^2$	1
0	0^2	0
1	1^2	1
2	2^2	4

x	$-x^2$	y
-2	$-(-2)^2$	-4
-1	$-(-1)^2$	-1
0	0^2	0
1	$-(1)^2$	-1
2	$-(2)^2$	-4

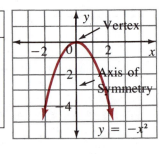

The **vertex**, or **turning point**, of each parabola above is the origin $(0,0)$. The **axis of symmetry** passes through the vertex and is parallel to—or in this case, is—the y-axis. You can draw the graph of an equation of the form $y = (x - h)^2$ by using the basic graph of $y = x^2$ for reference.

Example 1   **Graph $y = (x - 2)^2$ and $y = (x + 3)^2$. Compare with the graph of $y = x^2$.**

x	$(x - 2)^2$	y
0	$(0 - 2)^2$	4
1	$(1 - 2)^2$	1
2	$(2 - 2)^2$	0
3	$(3 - 2)^2$	1
4	$(4 - 2)^2$	4

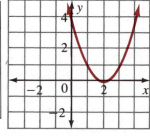

x	$(x + 3)^2$	y
-1	$(-1 + 3)^2$	4
-2	$(-2 + 3)^2$	1
-3	$(-3 + 3)^2$	0
-4	$(-4 + 3)^2$	1
-5	$(-5 + 3)^2$	4

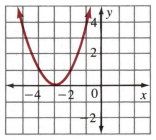

Move the graph of $y = x^2$ two units to the right.

Move the graph of $y = x^2$ three units to the left.

Using the basic graph of $y = x^2$ for reference, you can draw the graph of an equation of the form $y = x^2 + k$.

Example 2

Graph $y = x^2 + 2$ and $y = x^2 - 1$. Compare with the graph of $y = x^2$.

x	$x^2 + 2$	y
-1	$1 + 2$	3
0	$0 + 2$	2
1	$1 + 2$	3

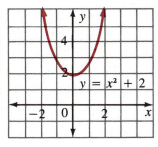

x	$x^2 - 1$	y
-2	$4 - 1$	3
-1	$1 - 1$	0
0	$0 - 1$	-1
1	$1 - 1$	0
2	$4 - 1$	3

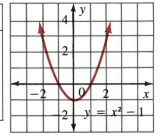

Move the graph of $y = x^2$ two units up.   Move the graph of $y = x^2$ one unit down.

Standard form of equation of parabola symmetric to y-axis

The standard form of the equation of a parabola with vertex at (h, k) and axis of symmetry parallel to the y-axis is
$$y = (x - h)^2 + k, \text{ or } y - k = (x - h)^2.$$

Example 3

Graph $y = x^2 + 8x + 6$. Estimate graphically, to the nearest tenth, the roots of the quadratic equation.

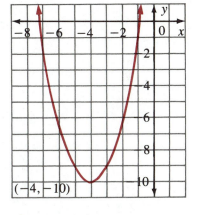

Rewrite in the standard form $y = (x - h)^2 + k$.

Complete the square. ▶

$y = x^2 + 8x + \underline{} + 6$
$y = x^2 + 8x + 16 - 16 + 6$
$y = (x + 4)^2 - 10$
$y = [x - (-4)]^2 - 10$

The vertex has coordinates $(-4, -10)$. Move the graph of $y = x^2$ four units to the left and ten units down.

Thus, the roots (x-intercepts) of the quadratic equation are approximately -7.2 and -0.8.

In Example 3, the graph intersects the x-axis in two points with coordinates of $(-7.2, 0)$ and $(-0.8, 0)$. Notice that the *equation of the axis of symmetry*, $x = -4$, is the *average of the roots.* That is, $x = \dfrac{-7.2 + (-0.8)}{2} = -4$. In general, if r and s are the solutions of the quadratic equation, then an equation of the axis of symmetry is $x = \dfrac{r + s}{2}$. Since the sum $(r + s)$ of the roots of a quadratic equation is $-\dfrac{b}{a}$ and since $x = \dfrac{r + s}{2}$, then $x = \dfrac{-\dfrac{b}{a}}{2}$, or $x = -\dfrac{b}{2a}$.

The Parabola

Equation of axis of symmetry	For a parabola with equation $y = ax^2 + bx + c$, $a \neq 0$, an equation of the axis of symmetry is $x = -\frac{b}{2a}$.

Example 4 **Write an equation of the axis of symmetry of the parabola with equation $y = 4x + x^2 - 3$.**

Write the equation in the form $y = ax^2 + bx + c$.
$y = 1x^2 + 4x - 3$
   ↑   ↑
   a   b   ◀ *Determine a and b.*

The equation for the axis of symmetry is $x = -\frac{b}{2a}$.

Substitute 1 for a ▶ $x = -\frac{4}{2}$, or -2
and 4 for b.

Thus, an equation of the axis of symmetry is $x = -2$.

Example 5 **Graph $y = x^2 - 7x + 5$. Estimate graphically, to the nearest tenth, the roots of the quadratic equation. Determine the coordinates of the vertex, write the equation of the axis of symmetry, and find the x-intercepts.**

Rewrite in the standard form $y = (x - h)^2 + k$.

Complete the square. ▶

$y = x^2 - 7x + \underline{\ ?\ } + 5$
$y = x^2 - 7x + \frac{49}{4} - \frac{49}{4} + 5$
$y = \left(x - \frac{7}{2}\right)^2 - \frac{29}{4}$

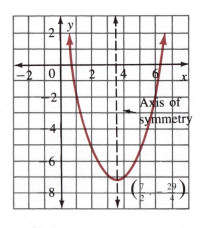

The vertex has coordinates $\left(\frac{7}{2}, -\frac{29}{4}\right)$. Move the graph of $y = x^2$ to the right 3.5 units and down 7.25 units.

The equation of the axis of symmetry is
$x = -\frac{b}{2a}$. $x = -\frac{-7}{2}$, or $x = \frac{7}{2}$, or 3.5.

The x-intercepts are the points of intersection of the graph and the x-axis. They are approximately 0.8 and 6.2.

Thus, the roots of the quadratic equation are approximately 0.8 and 6.2, the coordinates of the vertex are $\left(\frac{7}{2}, -\frac{29}{4}\right)$, and the equation of the axis of symmetry is $x = \frac{7}{2}$.

Example 6 **Draw the graphs of $y = \frac{1}{2}x^2$, $y = x^2$, and $y = 2x^2$ on the same coordinate axes. Write an equation of the axis of symmetry and the coordinates of the vertex of each parabola.**

x	x^2	$\frac{1}{2}x^2$	$2x^2$
-2	4	2	8
-1	1	.5	2
0	0	0	0
1	1	.5	2
2	4	2	8

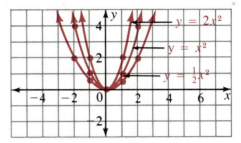

Thus, for each parabola, $x = 0$ is an equation of the axis of symmetry and $(0, 0)$ are the coordinates of the vertex.

In Example 6, notice that each parabola intersects the x-axis at the point whose coordinates are $(0, 0)$. Since there is only one point of intersection, the single x-intercept of each graph is 0. This means also that the corresponding quadratic equations each have exactly one real number solution.

Sometimes a parabola of the form $y = ax^2 + bx + c$ does not intersect the x-axis. In this case, there is no x-intercept and the corresponding quadratic equation has no real number solution. The solutions of such quadratic equations are imaginary numbers.

You can also graph a parabola determined by the equation $x = y^2$ or $x = -y^2$ by making a table of values. To make a table of values for these equations, choose values for y and find corresponding values for x.

y	y^2	x
-2	$(-2)^2$	4
-1	$(-1)^2$	1
0	0^2	0
1	1^2	1
2	2^2	4

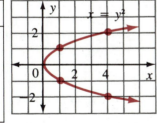

y	$-y^2$	x
-2	$-(-2)^2$	-4
-1	$-(-1)^2$	-1
0	-0^2	0
1	$-(1)^2$	-1
2	$-(2)^2$	-4

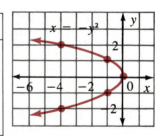

Notice that the graphs of $x = y^2$ and $x = -y^2$ are symmetric with respect to the x-axis. Parabolas of this form are relations but not functions. The standard form of equation for such parabolas is stated below.

> **Standard form of equation of parabola symmetric to x-axis**
>
> The standard form of the equation of a parabola with vertex at (h, k) and axis of symmetry parallel to the x-axis is
>
> $$x = (y - k)^2 + h, \text{ or } x - h = (y - k)^2.$$

The Parabola

Example 7

Graph $x = 2y^2 + 8y + 9$. Determine the coordinates of the vertex and write the equation of the axis of symmetry.

Rewrite $x = 2y^2 + 8y + 9$ in the standard form $x = (y - k)^2 + h$.

Factor. ▶ $x = 2(y^2 + 4y) + 9$
Complete the ▶ $x = 2(y^2 + 4y + 4) - 8 + 9$
square. $x = 2(y + 2)^2 + 1$

$y = -\dfrac{b}{2a} = -\dfrac{8}{4}$,

or $y = -2$.

So, the vertex has coordinates $(1, -2)$. Move the graph of $x = y^2$ one unit to the right and two units down.

▶ The axis of symmetry is $y = -2$.

Thus, the vertex has coordinates $(1, -2)$ and the equation of the axis of symmetry is $y = -2$.

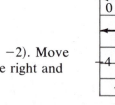

Written Exercises

For Ex. 1–12, see TE Answer Section.

Draw a graph of each of the following equations. Use a basic graph for reference.

(A)
1. $y = (x + 1)^2$
2. $y = (x - 3)^2$
3. $y = (x - 1)^2$
4. $y = (x + 3)^2$
5. $y = x^2 + 1$
6. $y = x^2 - 2$
7. $y = x^2 - 4$
8. $y = x^2 + 3$
9. $y = (x - 2)^2 + 1$
10. $y = (x + 2)^2 - 1$
11. $y = (x - 1)^2 - 2$
12. $y = (x - 1)^2 + 2$

Write an equation of the axis of symmetry of each parabola with the given equation. Determine the coordinates of the vertex. Do not draw the graphs. $x = \frac{3}{2}, (\frac{3}{2}, \frac{5}{4})$ $x = -\frac{3}{2}, (-\frac{3}{2}, -\frac{25}{2})$

13. $y = x^2 - 2x + 3$ **x = 1, (1, 2)** 14. $y = -x^2 + 3x - 1$ 15. $y = 6x + 2x^2 - 8$
16. $y = -5x + 6x^2 - 4$ 17. $y = -4x^2 - x + 7$ 18. $y = -10x^2 + 4x - 6$
    $x = \frac{5}{12}, (\frac{5}{12}, -\frac{121}{24})$ $x = -\frac{1}{8}, (-\frac{1}{8}, \frac{113}{16})$ $x = \frac{1}{5}, (\frac{1}{5}, -\frac{28}{5})$

Graph each quadratic equation. Estimate, to the nearest tenth, the x-intercept(s) of each parabola.

(B)
19. $y = x^2 - 10x + 25$ **5.0**
20. $y = x^2 + 6x + 9$ **-3.0**
21. $x = y^2 - 4y + 4$ **4.0**
22. $y = x^2 + 4x + 2$ **-0.6, -3.4**
23. $y = -x^2 + 4x - 1$ **0.3, 3.7**
24. $x = y^2 + 6y + 7$ **7.0**
25. $y = 2x^2 - 12x + 18$ **3.0**
26. $y = -2x^2 - 8x - 8$ **-2.0**
27. $x = 2y^2 - 16y + 32$ **32.0**
28. $y = 2x^2 + 12x + 13$ **below**
29. $x = 2y^2 - 8y + 3$ **3.0**
30. $y = 3x^2 - 18x + 19$ **4.6, 1.4**
31. $x = -y^2 - 10y - 22$ **-22.0**
32. $y = -3x^2 + 12x - 7$
33. $x = 3y^2 + 18y + 20$ **20.0**
    28. **-1.4, -4.6**   0.7, 3.3

Graph each set of functions on the same coordinate axes. Find an equation of the axis of symmetry and the coordinates of the vertex.

34. $y = -\frac{1}{4}x^2, y = -x^2, y = -4x^2$ **x = 0, (0, 0)**
35. $y = \frac{1}{4}x^2 - 2, y = x^2 - 2, y = 4x^2 - 2$
36. $y = x^2, y = 2x^2, y = 4x^2$ **x = 0, (0, 0)**
37. $y = x^2 + 1, y = 2x^2 + 1, y = 4x^2 + 1$

35. **x = 0, (0, -2)**   37. **x = 0, (0, 1)**

Find the inverse of the function determined by each given equation. Draw the graph of each inverse. Is the inverse an inverse function? Find an equation of the axis of symmetry for the inverse relation.

(C)
38. $y = x^2 - 3x - 4$
39. $y = x^2 + 2x - 6$
40. $y = -x^2 - 4x + 3$
41. $y = -2x^2 - 4x + 5$
42. $y = -3x^2 + 6x + 1$
43. $y = 4x^2 - 5x - 2$

38. $x = y^2 - 3y - 4$; No; $y = \frac{3}{2}$ 39. $x = y^2 + 2y - 6$; No; $y = -1$ 40. $x = -y^2 - 4y + 3$; No; $y = -2$
41. $x = -2y^2 - 4y + 5$; No; $y = -1$ 42. $x = -3y^2 + 6y + 1$; No; $y = 1$ 43. $x = 4y^2 - 5y - 2$; No; $y = \frac{5}{8}$

44. Show that the coordinates of the vertex of the parabola with equation $y = ax^2 + bx + c$ are $\left(-\dfrac{b}{2a}, \dfrac{4ac - b^2}{4a}\right)$. **See TE Answer Section.**

THE HYPERBOLA   12.5

Objectives
To write an equation of a hyperbola
To determine the intercepts of a hyperbola
To write the equations of the asymptotes of a hyperbola
To draw the graph of an equation of a hyperbola

Another curve that can be defined as a set of points is a **hyperbola**.

> **Definition: Hyperbola**
> A *hyperbola* is the set of all points in a coordinate plane such that for each point of the set, the difference of its distances from two fixed points is constant. The fixed points are called the *foci*.

You can write an equation of a hyperbola whose center is at the origin, foci are $F_1(-5, 0)$ and $F_2(5, 0)$, and where the difference of the distances from any point on the hyperbola to the foci is 8.

Let $P(x, y)$ be any point on the hyperbola. Then find the distance d_1 from $P(x, y)$ to $F_1(-5, 0)$ and d_2 from $P(x, y)$ to $F_2(5, 0)$. Use the distance formula.

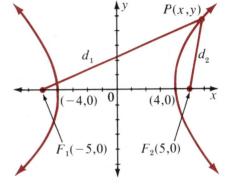

$d_1 = \sqrt{[x - (-5)]^2 + (y - 0)^2}$
$d_2 = \sqrt{(x - 5)^2 + (y - 0)^2}$

The difference of the distance is 8. Use this fact to write an equation for the hyperbola.

$$d_1 - d_2 = 8 \text{ or } d_1 = 8 + d_2$$
$$\sqrt{(x + 5)^2 + y^2} = 8 + \sqrt{(x - 5)^2 + y^2}$$
$$\sqrt{x^2 + 10x + 25 + y^2} = 8 + \sqrt{x^2 - 10x + 25 + y^2}$$
$$(\sqrt{x^2 + 10x + 25 + y^2})^2 = (8 + \sqrt{x^2 - 10x + 25 + y^2})^2$$
$$x^2 + 10x + 25 + y^2 = 64 + 16\sqrt{x^2 - 10x + 25 + y^2} + x^2 - 10x + 25 + y^2$$
$$20x - 64 = 16\sqrt{x^2 - 10x + 25 + y^2}$$
$$5x - 16 = 4\sqrt{x^2 - 10x + 25 + y^2}$$
$$(5x - 16)^2 = (4\sqrt{x^2 - 10x + 25 + y^2})^2$$
$$25x^2 - 160x + 256 = 16x^2 - 160x + 400 + 16y^2$$
$$9x^2 - 16y^2 = 144$$

An equation of the hyperbola is $\dfrac{x^2}{16} - \dfrac{y^2}{9} = 1$.

> **Standard form of equation of hyperbola with x-intercepts**
> The standard form of the equation of a hyperbola with center at the origin $(0, 0)$ and whose x-intercepts are $\pm a$ is $\dfrac{x^2}{a^2} - \dfrac{y^2}{b^2} = 1$; $a \neq 0, b \neq 0$.

The Hyperbola

Example 1 Determine the intercepts of the hyperbola whose equation is $\frac{x^2}{16} - \frac{y^2}{9} = 1$.

If $y = 0$, then $\frac{x^2}{16} = 1$ and $x = \pm 4$.

If $x = 0$, then $-\frac{y^2}{9} = 1$ and $y^2 = -9$ or no real number.

Thus, the x-intercepts are ± 4 and the y-intercepts do not exist.

Example 2 Draw the graph of $\frac{x^2}{25} - \frac{y^2}{9} = 1$.

Use $\frac{x^2}{a^2} - \frac{y^2}{b^2} = 1$. ▶ $\frac{x^2}{25} - \frac{y^2}{9} = 1$
                              ↑   ↑
                             a^2 b^2

Find a and b. ▶ $a = \pm 5$ $b = \pm 3$

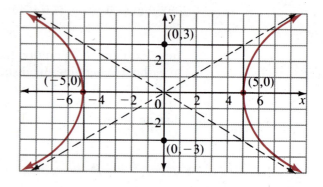

Draw a rectangle using $(5, 0)$, $(-5, 0)$, $(0, 3)$, and $(0, -3)$ as the coordinates of the midpoints of the sides.

Use the extended diagonals of the rectangle and the x-intercepts to draw a smooth curve.

The extended diagonal lines in Example 2 are called the **asymptotes** of the hyperbola. A hyperbola approaches but never intersects its asymptotes.

The graph of $\frac{x^2}{a^2} - \frac{y^2}{b^2} = 1$ is shown at the right. The y-intercept of each of the asymptotes is 0 and the slopes are $\frac{b}{a}$ and $-\frac{b}{a}$, respectively. Therefore, $y = \frac{b}{a}x$ and $y = -\frac{b}{a}x$ are the equations of the asymptotes, where a and b are positive real numbers.

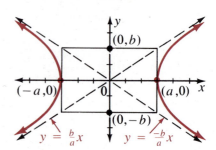

Example 3 Write the equations of the asymptotes of the hyperbola whose equation is $\frac{x^2}{25} - \frac{y^2}{9} = 1$.

$a^2 = 25$ $b^2 = 9$
$a = \pm 5$ $b = \pm 3$

Use $y = \frac{b}{a}x$ and $y = -\frac{b}{a}x$. ▶ Thus, $y = \frac{3}{5}x$ and $y = -\frac{3}{5}x$ are the equations of the asymptotes.

You can also graph a hyperbola that intersects the y-axis. Its equation is the inverse of the standard form $\frac{x^2}{a^2} - \frac{y^2}{b^2} = 1$, which is $\frac{y^2}{a^2} - \frac{x^2}{b^2} = 1$. The graph of the inverse is a hyperbola whose center is at the origin and whose y-intercepts are $\pm a$.

Standard form of equation of hyperbola with y-intercepts

The standard form of the equation of a hyperbola with center at the origin $(0, 0)$ and whose y-intercepts are $\pm a$ is $\frac{y^2}{a^2} - \frac{x^2}{b^2} = 1$; $a \neq 0$, $b \neq 0$.

The equations of the asymptotes for a hyperbola of this type are $y = \frac{a}{b}x$ and $y = -\frac{a}{b}x$.

Example 4

Draw the graph of $4y^2 - x^2 = 16$. Write the equations of the asymptotes.

Write $4y^2 - x^2 = 16$ in standard form.

$$\frac{4y^2}{16} - \frac{x^2}{16} = \frac{16}{16}, \text{ or}$$

$$\frac{y^2}{4} - \frac{x^2}{16} = 1$$

Use $\frac{y^2}{a^2} - \frac{x^2}{b^2} = 1$. ▶ $a^2 = 4 \quad b^2 = 16$
$a = \pm 2 \quad b = \pm 4$

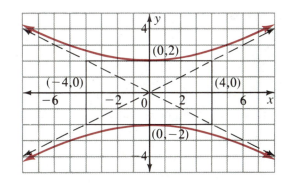

Draw a rectangle using $(0, 2)$, $(0, -2)$, $(4, 0)$, and $(-4, 0)$ as the coordinates of the midpoints of the sides.

Use $y = \frac{a}{b}x$ and $y = -\frac{a}{b}x$. ▶ $y = \frac{2}{4}x \quad y = -\frac{2}{4}x$

Thus, the equations of the asymptotes are $y = \frac{1}{2}x$ and $y = -\frac{1}{2}x$.

The graph of an equation of the form $xy = k$ ($k \neq 0$) is a hyperbola that does not intersect the x- or y-axis. A hyperbola of this type is called a **rectangular hyperbola**. The asymptotes of the hyperbola are the x- and y-axes. If $k > 0$, the branches of the hyperbola lie in Quadrants I and III. If $k < 0$, the branches lie in Quadrants II and IV.

Standard form of equation of rectangular hyperbola

The standard form of the equation of a hyperbola with center at the origin $(0, 0)$ and which does not intersect the x-axis or y-axis is $xy = k$, where k is a nonzero real number. The hyperbola is called a *rectangular hyperbola*.

The Hyperbola

Example 5 **Draw the graph of $xy = 4$. Write the equations of the asymptotes.**

Solve for y in terms of x and make a table.

$xy = 4$, or $y = \dfrac{4}{x}$

x	$\dfrac{4}{x}$	y
1	$\dfrac{4}{1}$	4
2	$\dfrac{4}{2}$	2
4	$\dfrac{4}{4}$	1
-1	$\dfrac{4}{-1}$	-4
-2	$\dfrac{4}{-2}$	-2
-4	$\dfrac{4}{-4}$	-1

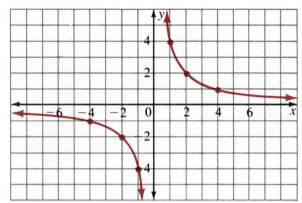

Thus, the equations of the asymptotes are $x = 0$ and $y = 0$.
You can also graph an equation of a hyperbola whose center is not at the origin.

Example 6 **Graph $\dfrac{(x + 1)^2}{9} - \dfrac{(y + 2)^2}{4} = 1$. Use the basic graph of $\dfrac{x^2}{9} - \dfrac{y^2}{4} = 1$.**

The hyperbola defined by $\dfrac{x^2}{9} - \dfrac{y^2}{4} = 1$ has center at the origin and x-intercepts of ± 3.
The hyperbola defined by $\dfrac{(x + 1)^2}{9} - \dfrac{(y + 2)^2}{4} = 1$ has center at $C(-1, -2)$.
Move the basic graph one unit to the left and two units down.

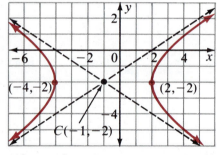

You can also rewrite equations like $25x^2 - 4y^2 - 150x - 16y + 109 = 0$ in the standard form for a hyperbola by completing the square.

Example 7 **Draw the graph of the equation $25x^2 - 4y^2 - 150x - 16y + 109 = 0$.**

Rewrite in the standard form.

Complete the square:
$25(9) = 225$
$-4(4) = -16$

$25x^2 - 150x + \underline{\ ?\ } - 4y^2 - 16y + \underline{\ ?\ } = -109$
$25(x^2 - 6x + \underline{\ ?\ }) - 4(y^2 + 4y + \underline{\ ?\ }) = -109$
$25(x^2 - 6x + 9) - 4(y^2 + 4y + 4) = -109 + 225 - 16$
$25(x - 3)^2 - 4(y + 2)^2 = 100$

Divide by 100.
$\dfrac{(x - 3)^2}{4} - \dfrac{(y + 2)^2}{25} = 1$

$\quad\quad\quad \uparrow \quad\quad\quad \uparrow$
$\quad\quad\quad a^2 \quad\quad\quad b^2$

$a = \pm 2 \quad b = \pm 5$

Move the basic graph three units right and two units down.

The center is at $(3, -2)$.

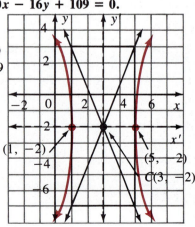

Written Exercises   For complete solutions, see TE Answer Section.

13.–16. Neither x- nor y-intercepts exist.

Draw a graph of each equation. Determine the intercepts.

(A)
1. $\frac{x^2}{16} - \frac{y^2}{4} = 1$ $x = \pm 4$
2. $\frac{x^2}{25} - \frac{y^2}{9} = 1$ $x = \pm 5$
3. $\frac{x^2}{4} - \frac{y^2}{16} = 1$ $x = \pm 2$
4. $\frac{x^2}{36} - \frac{y^2}{49} = 1$ $x = \pm 6$
5. $\frac{x^2}{9} - \frac{y^2}{16} = 1$ $x = \pm 3$
6. $\frac{x^2}{49} - \frac{y^2}{81} = 1$ $x = \pm 7$
7. $\frac{x^2}{49} - \frac{y^2}{4} = 1$ $x = \pm 7$
8. $\frac{x^2}{100} - \frac{y^2}{64} = 1$ $x = \pm 10$
9. $\frac{x^2}{4} - \frac{y^2}{9} = 1$ $x = \pm 2$
10. $\frac{x^2}{81} - \frac{y^2}{64} = 1$ $x = \pm 9$
11. $\frac{x^2}{100} - \frac{y^2}{81} = 1$ $x = \pm 10$
12. $\frac{x^2}{121} - \frac{y^2}{36} = 1$ $x = \pm 11$
13. $xy = 10$
14. $xy = -6$
15. $xy = -8$
16. $4xy = 48$
17. $\frac{y^2}{16} - \frac{x^2}{9} = 1$ $y = \pm 4$
18. $\frac{y^2}{25} - \frac{x^2}{16} = 1$ $y = \pm 5$
19. $\frac{y^2}{81} - \frac{x^2}{25} = 1$ $y = \pm 9$
20. $\frac{y^2}{36} - \frac{x^2}{4} = 1$ $y = \pm 6$
21. $\frac{y^2}{9} - \frac{x^2}{25} = 1$ $y = \pm 3$
22. $\frac{y^2}{36} - \frac{x^2}{16} = 1$ $y = \pm 6$
23. $\frac{y^2}{16} - \frac{x^2}{4} = 1$ $y = \pm 4$
24. $\frac{y^2}{81} - \frac{x^2}{36} = 1$ $y = \pm 9$

Write the equations of the asymptotes. 39. $x = 0, y = 0$ 40. $x = 0, y = 0$

25. $\frac{x^2}{25} - \frac{y^2}{9} = 1$ $y = \pm \frac{3}{5}x$
26. $\frac{x^2}{36} - \frac{y^2}{16} = 1$ $y = \pm \frac{2}{3}x$
27. $\frac{x^2}{4} - \frac{y^2}{25} = 1$ $y = \pm \frac{5}{2}x$
28. $\frac{x^2}{16} - \frac{y^2}{81} = 1$ $y = \pm \frac{9}{4}x$
29. $\frac{y^2}{36} - \frac{x^2}{25} = 1$ $y = \pm \frac{6}{5}x$
30. $\frac{y^2}{81} - \frac{x^2}{25} = 1$ $y = \pm \frac{9}{5}x$
31. $\frac{y^2}{4} - \frac{x^2}{9} = 1$ $y = \pm \frac{2}{3}x$
32. $\frac{y^2}{25} - \frac{x^2}{4} = 1$ $y = \pm \frac{5}{2}x$
33. $xy = 8$ $x = 0, y = 0$
34. $xy = 4$ $x = 0, y = 0$
35. $xy = -16$ $x = 0, y = 0$
36. $xy = 20$ $x = 0, y = 0$

(B)
37. $xy = -6$ $x = 0, y = 0$
38. $xy = -9$ $x = 0, y = 0$
39. $-xy = 12$ above
40. $-3xy = 12$ above
41. $4x^2 - y^2 = 36$
42. $4x^2 - 9y^2 = 36$
43. $16x^2 - 9y^2 = 144$
44. $12x^2 - 3y^2 = 48$
45. $4y^2 - 16x^2 = 64$
46. $5y^2 - 9x^2 = 45$
47. $16y^2 - 4x^2 = 64$
48. $4y^2 - 5x^2 = 100$
  $y = 2x, y = -2x$
  $y = \frac{3\sqrt{5}}{5}x, y = -\frac{3\sqrt{5}}{5}x$
  $y = \frac{1}{2}x, y = -\frac{1}{2}x$
  $y = \frac{\sqrt{5}}{2}x, y = -\frac{\sqrt{5}}{2}x$

Draw the graph of each equation.

49. $\frac{(x-1)^2}{16} - \frac{(y+2)^2}{9} = 1$
50. $\frac{(x+2)^2}{25} - \frac{(y+3)^2}{4} = 1$
51. $\frac{(x-2)^2}{9} - \frac{(y-3)^2}{4} = 1$

41. $y = \pm 2x$ 42. $y = \pm \frac{2}{3}x$ 43. $y = \pm \frac{4}{3}x$ 44. $y = \pm 2x$

Write each equation in the standard form for a hyperbola. Draw the graph of each equation.

52. $16x^2 - 9y^2 - 72y - 288 = 0$ $\frac{x^2}{9} - \frac{(y+4)^2}{16} = 1$
53. $25x^2 - 9y^2 + 200x + 175 = 0$ $\frac{(x+4)^2}{9} - \frac{y^2}{25} = 1$
54. $y^2 - 4x^2 - 6y + 32x - 71 = 0$
    $\frac{(y-3)^2}{16} - \frac{(x-4)^2}{4} = 1$
55. $y^2 - x^2 + 10y + 6x + 12 = 0$
    $\frac{(y+5)^2}{4} - \frac{(x-3)^2}{4} = 1$

Write an equation, in standard form, of each hyperbola with center at the origin, F_1 and F_2 the foci, and where c is the *difference of the distances* from any point on the hyperbola to the foci.

(C)
56. $F_1(-13, 0), F_2(13, 0); c = 24$ $\frac{x^2}{144} - \frac{y^2}{25} = 1$
57. $F_1(-10, 0), F_2(10, 0); c = 16$ $\frac{x^2}{64} - \frac{y^2}{36} = 1$
58. $F_1(-6, 0), F_2(6, 0); c = 10$ $\frac{x^2}{25} - \frac{y^2}{11} = 1$
59. $F_1(-10, 0), F_2(10, 0); c = 18$ $\frac{x^2}{81} - \frac{y^2}{19} = 1$

Write an equation, in standard form, of each hyperbola whose center C is given, and with the given values for a and b. Draw a graph of each equation.

60. $C(-3, 5); a = \pm 3, b = \pm 2$ $\frac{(x+3)^2}{9} - \frac{(y-5)^2}{4} = 1$
61. $C(4, -1); a = \pm 5, b = \pm 3$ $\frac{(x-4)^2}{25} - \frac{(y+1)^2}{9} = 1$
62. $C(1, 7); a = \pm 8, b = \pm 5$ $\frac{(x-1)^2}{64} - \frac{(y-7)^2}{25} = 1$
63. $C(-8, -2); a = \pm 10, b = \pm 8$ $\frac{(x+8)^2}{100} - \frac{(y+2)^2}{64} = 1$

NON-ROUTINE PROBLEMS

Find the average of the abscissas of all the points of $\frac{x^2}{9} + \frac{y^2}{4} = 1$ that lie in the first and second quadrants. **0**

The Hyperbola

IDENTIFYING CONIC SECTIONS 12.6

Objective **To identify an equation of a circle, parabola, ellipse, and hyperbola**

As shown below, a circle, parabola, ellipse, or hyperbola can be formed by passing a plane through a hollow double cone. Therefore, each of these curves is called a **conic section.**

Circle

Parabola

Ellipse

Hyperbola

An equation of any conic section can be written in the form

$$Ax^2 + Bxy + Cy^2 + Dx + Ey + F = 0,$$
where A, B, and C are not all zero.

To identify the conic section described by a given *quadratic* equation of the form above, you should rewrite the equation in one of the standard forms below.

Conic Section	Standard Forms	
Circle	$x^2 + y^2 = r^2$	$(x - h)^2 + (y - k)^2 = r^2$
Parabola	$y = ax^2 + c$ $x = ay^2 + c$	$y = (x - h)^2 + k$ $x = (y - k)^2 + h$
Ellipse	$\dfrac{x^2}{a^2} + \dfrac{y^2}{b^2} = 1$	$\dfrac{(x - h)^2}{a^2} + \dfrac{(y - k)^2}{b^2} = 1$
Hyperbola	$\dfrac{x^2}{a^2} - \dfrac{y^2}{b^2} = 1$ $\dfrac{y^2}{a^2} - \dfrac{x^2}{b^2} = 1$	$\dfrac{(x - h)^2}{a^2} - \dfrac{(y - k)^2}{b^2} = 1$ $\dfrac{(y - k)^2}{a^2} - \dfrac{(x - h)^2}{b^2} = 1$
Rectangular Hyperbola	$xy = k; \ k \neq 0$	

Example 1   Determine whether $9x^2 - 16y^2 = 144$ is an equation of a parabola, a circle, an ellipse, or a hyperbola.

Write the equation in standard form.
$$9x^2 - 16y^2 = 144$$

Divide each term by 144. ▶ $\dfrac{9x^2}{144} - \dfrac{16y^2}{144} = \dfrac{144}{144}$

Standard form: ▶ $\dfrac{x^2}{16} - \dfrac{y^2}{9} = 1$
$\dfrac{x^2}{a^2} - \dfrac{y^2}{b^2} = 1.$

Thus, $9x^2 - 16y^2 = 144$ is an equation of a hyperbola.

Example 2   Determine whether $17 + 3x^2 = y + 5x$ is an equation of a parabola, a circle, an ellipse, or a hyperbola.

There is no y^2 term. ▶ $17 + 3x^2 = y + 5x$ ◀ *Standard form:*
Solve for y. ▶ $3x^2 - 5x + 17 = y$ $y = ax^2 + bx + c.$

Thus, $17 + 3x^2 = y + 5x$ is an equation of a parabola.

When the x-intercepts and y-intercepts of an ellipse are equal, the standard form of the equation of the ellipse becomes $\dfrac{x^2}{a^2} + \dfrac{y^2}{a^2} = 1$. Multiplying by a^2, you get $x^2 + y^2 = a^2$. This is an equation of a circle with radius a. Thus, a circle can be considered to be a special type of ellipse.

Example 3   Identify each conic section whose equation is given.
$$9x^2 = 36 - 4y^2 \qquad\qquad 8x^2 + 8y^2 = 56$$

Standard form: ▶ $\begin{aligned}9x^2 &= 36 - 4y^2 \\ 9x^2 + 4y^2 &= 36 \\ \dfrac{x^2}{4} + \dfrac{y^2}{9} &= 1\end{aligned}$ $\Big|$ $\begin{aligned}8x^2 + 8y^2 &= 56 \\ \dfrac{x^2}{7} + \dfrac{y^2}{7} &= 1 \\ x^2 + y^2 &= 7\end{aligned}$ ◀ *Standard form:*
$\dfrac{x^2}{a^2} + \dfrac{y^2}{b^2} = 1.$ $x^2 + y^2 = r^2.$

Thus, $9x^2 = 36 - 4y^2$ is an equation of an ellipse and $8x^2 + 8y^2 = 56$ is an equation of a circle.

You may have to complete the square in x or y or *both* to rewrite an equation in the standard form for a conic section.

Example 4   Identify the conic section described by the equation $x^2 - 4x + y^2 + 6y = 1$.

Complete the square. ▶ $x^2 - 4x + 4 + y^2 + 6y + 9 = 1 + 4 + 9$
Standard form: $(x - h)^2 + (y - k)^2 = r^2$ ▶ $(x - 2)^2 + (y + 3)^2 = 14$
Thus, $x^2 - 4x + y^2 + 6y = 1$ is an equation of a circle.

Identifying Conic Sections

Example 5   Identify the conic section described by each equation.
$$16x^2 - 9y^2 - 72y - 288 = 0 \qquad\qquad -5xy = 60$$

$$16x^2 - 9(y^2 + 8y) = 288 \qquad\qquad \frac{-5xy}{-5} = \frac{60}{-5}$$
$$16x^2 - 9(y^2 + 8y + 16) = 288 - 9(16) \qquad\qquad xy = -12$$
$$16x^2 - 9(y + 4)^2 = 144$$

Thus, $16x^2 - 9y^2 - 72y - 288 = 0$ and $-5xy = 60$ are equations of hyperbolas.

47. H; x-axis, y-axis, origin   48. E; x-axis, y-axis, origin   49. C; x-axis, y-axis, origin   50. P; none of these

Written Exercises   (In Ex. 1–58, P = parabola, C = circle, E = ellipse, and H = hyperbola.)

Identify each conic section whose equation is given.   16. E   20. P   26. E   29. H   31. H

Ⓐ   1. $\dfrac{x^2}{16} + \dfrac{y^2}{4} = 1$ **E**   2. $\dfrac{x^2}{14} + \dfrac{y^2}{14} = 1$ **C**   3. $\dfrac{x^2}{16} - \dfrac{y^2}{16} = 1$ **H**   4. $\dfrac{x^2}{12} + \dfrac{y^2}{3} = 1$ **E**

5. $\dfrac{y^2}{4} - \dfrac{x^2}{16} = 1$ **H**   6. $x = \dfrac{-3}{y}$ **H**   7. $\dfrac{x^2}{50} + \dfrac{y^2}{50} = 1$ **C**   8. $y = \dfrac{8}{x}$ **H**

9. $6x^2 + 4y^2 = 24$ **E**   10. $x^2 = 4 + y$ **P**   11. $7x^2 - 3x = -3 + y$ **P**   12. $5x^2 - 6y^2 = 30$ **H**
13. $15x^2 + 15y^2 = 255$ **C**   14. $5x^2 - 3y^2 = 15$ **H**   15. $y - 3x^2 = 6 + x$ **P**   16. $30x^2 = -3y^2 + 15$
17. $y + x^2 = 3x + 2$ **P**   18. $x^2 + 2y^2 = 8$ **E**   19. $9x^2 - 4y^2 = 36$ **H**   20. $-y - 3x = x^2 + 5$
21. $3x^2 + 4y^2 = 48$ **E**   22. $7x^2 - 8y^2 = 35$ **H**   23. $4x^2 + 4y^2 = 15$ **C**   24. $2x^2 + 3y^2 = 45$ **E**
25. $16x^2 - 16y^2 = 5$ **H**   26. $12x^2 + 16y^2 = 17$ 27. $19y^2 + 9x^2 = 5$ **E**   28. $6x^2 - 6y^2 = 8$ **H**

Ⓑ   29. $(x + 3y)(x - 3y) = 4$ 30. $y - 3 = (x - 2)^2$ **P**   31. $(3x - 2y)(3x + 2y) = 1$ 32. $\dfrac{x}{9} = \dfrac{1}{y}$ **H**

33. $x^2 - 6x + y^2 + 8y = 24$ **C**   34. $x^2 - 2x + y^2 - 6y = 26$ **C**   35. $x^2 + 10x + y^2 + 12y = 60$ **C**

36. $(2x - 3)^2 = y + 4$ **P**   37. $(3y - 2x)(3y + 2x) = 1$ **H**   38. $x = \dfrac{-4}{y}$ **H**

39. $2y^2 - x + 20y + 50 = 0$ **P**   40. $6x^2 + 6y^2 + 36y + 18 = 24x$ **C**
41. $3x^2 - 12x + y + 7 = 0$ **P**   42. $25x^2 + 200x + 175 = 9y^2$ **H**
43. $x^2 - 24 + 8y = 6x - y^2$ **C**   44. $5y^2 + 50y + 275 = 100x - 2x^2$ **E**

45. H   46. H   51. C; x-axis, y-axis, origin   52. H; x-axis, y-axis, origin   53. H; x-axis, y-axis, origin

Ⓒ   45. $Ax^2 + Bxy + Cy^2 + Dx + Ey + F = 0$ if $A > 0$, $B = 0$, $C < 0$, and D, E, and F are real numbers, $F \neq 0$

46. $Ax^2 + Bxy + Cy^2 + Dx + Ey + F = 0$ if $A = C = D = E = 0$ and B and F are nonzero real numbers

54. E; x-axis, y-axis, origin   56. P; y-axis   58. C; x-axis, y-axis, origin

Identify each conic section whose equation is given. Determine whether the graph is symmetric with respect to the x-axis, the y-axis, the origin, or none of these.

47. $5x^2 - 6y^2 = 30$ 48. $x^2 + 3y^2 = 12$ 49. $4x^2 + 4y^2 = 16$ 50. $y - x^2 - 2x = 8$

51. $\dfrac{x^2}{16} + \dfrac{y^2}{16} = 1$ 52. $\dfrac{x^2}{4} - \dfrac{y^2}{16} = 1$ 53. $\dfrac{y^2}{25} - \dfrac{x^2}{4} = 1$ 54. $\dfrac{x^2}{16} + \dfrac{y^2}{4} = 1$

55. $7y = \dfrac{-16}{x}$ **H; origin**   56. $x^2 - 16 = y$ 57. $\dfrac{16}{y} = \dfrac{-x}{7}$ **H; origin**   58. $\dfrac{x^2}{18} + \dfrac{y^2}{18} = 23$

CUMULATIVE REVIEW

Find the slope of a line perpendicular to each line whose equation is given.

1. $-3x + 5y = 8$ $-\dfrac{5}{3}$ 2. $7x + 2y = 10$ $\dfrac{2}{7}$ 3. $x = -4$ **0**   4. $y = 6$ **undefined**

EQUATIONS AND PROPERTIES OF CONIC SECTIONS 12.7

Objectives **To write an equation of a parabola, using the coordinates of the focus and the equation of the directrix
To draw the graph of an equation of a parabola
To write an equation of an ellipse, using the coordinates of the foci, length of major axis, or length of minor axis**

In the figure at the right, the distance from the point P to the fixed point F is the same as the distance from P to the fixed line l. Notice that the curve traced by all such points P is a parabola.

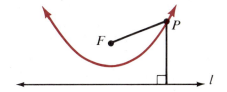

Definition: Parabola

A *parabola* is the set of all points in a coordinate plane that are equidistant from a fixed point and a fixed line in the plane. The fixed point is called the *focus* and the fixed line is called the *directrix*.

Let $P(x, y)$ be any point on a parabola with $F(0, p)$ the focus, and $y = -p$ the equation of the directrix. You can use the distance formula to find an equation of the parabola. Let
d_1 = distance from P to F and
d_2 = distance from P to A.

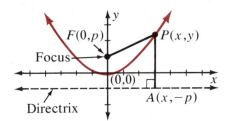

$$d_1 = d_2$$
$$\sqrt{(x-0)^2 + (y-p)^2} = \sqrt{(x-x)^2 + [y-(-p)]^2}$$
$$\sqrt{x^2 + y^2 - 2py + p^2} = \sqrt{y^2 + 2py + p^2}$$
$$x^2 + y^2 - 2py + p^2 = y^2 + 2py + p^2$$
$$x^2 = 4py$$

There are three other types of parabolas with vertex at the origin. Information on all four types is presented in the table.

Coordinates of Focus	$(0, p)$	$(0, -p)$	$(p, 0)$	$(-p, 0)$
Equation of directrix	$y = -p$	$y = p$	$x = -p$	$x = p$
Opening	Up	Down	Right	Left
Equation of the Parabola	$x^2 = 4py$ $p > 0$	$x^2 = -4py$ $p > 0$	$y^2 = 4px$ $p > 0$	$y^2 = -4px$ $p > 0$

Example 1 Write an equation of a parabola with $F(-3, 0)$ the focus and $x = 3$ the equation of the directrix.

The coordinates of the vertex are of the form $(-p, 0)$.
Use $y^2 = -4px$ from the table on page 327.
$$y^2 = -4px$$
Substitute 3 for p. ▶ $y^2 = -4 \cdot 3x$

Thus, $y^2 = -12x$ is an equation of the parabola.

Example 2 Draw a graph of the equation $y^2 = -6x$.

The equation $y^2 = -6x$ is of the form $y^2 = -4px$.

$$y^2 = -6x$$
$$-4p = -6$$
$$p = \tfrac{3}{2}$$

The coordinates of the focus, $(-p, 0)$, are $(-\tfrac{3}{2}, 0)$.
The equation of the directrix, $x = p$, is $x = \tfrac{3}{2}$.
The graph opens to the left.

Recall that an ellipse is the set of all points in a plane the sum of whose distances from two fixed points called the foci is a constant. $\dfrac{x^2}{a^2} + \dfrac{y^2}{b^2} = 1$ is the standard form of the equation of an ellipse with center at the origin and foci on the x-axis. This equation is derived as follows.

Let $P(x, y)$ be any point on an ellipse with $F_1(-c, 0)$ and $F_2(c, 0)$ the foci. Let $d_1 =$ distance from P to F_1 and $d_2 =$ distance from P to F_2.
Then, $d_1 + d_2 = 2a$, a constant.

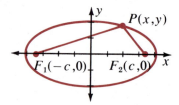

$$\sqrt{[x - (-c)]^2 + (y - 0)^2} + \sqrt{(x - c)^2 + (y - 0)^2} = 2a$$
$$\sqrt{(x + c)^2 + y^2} = 2a - \sqrt{(x - c)^2 + y^2}$$
$$x^2 + 2cx + c^2 + y^2 = 4a^2 - 4a\sqrt{(x - c)^2 + y^2} + x^2 - 2cx + c^2 + y^2$$
$$cx - a^2 = -a\sqrt{(x - c)^2 + y^2}$$
$$c^2x^2 - 2a^2cx + a^4 = a^2x^2 - 2a^2cx + a^2c^2 + a^2y^2$$
$$a^2(a^2 - c^2) = x^2(a^2 - c^2) + a^2y^2$$

$$\frac{x^2\cancel{(a^2 - c^2)}}{a^2\cancel{(a^2 - c^2)}} + \frac{\cancel{a^2}y^2}{\cancel{a^2}(a^2 - c^2)} = \frac{\cancel{a^2}\cancel{(a^2 - c^2)}}{\cancel{a^2}\cancel{(a^2 - c^2)}}$$

Let $a^2 - c^2 = b^2$.
Therefore, $\dfrac{x^2}{a^2} + \dfrac{y^2}{b^2} = 1$ is an equation of the ellipse.

Properties of an ellipse of the form
$$\frac{x^2}{a^2} + \frac{y^2}{b^2} = 1$$

Vertices: $V_1(-a, 0)$ and $V_2(a, 0)$
Major axis: $\overline{V_1 V_2}$ of length $2a$
Minor axis: $\overline{B_1 B_2}$ of length $2b$
x-intercepts: a and $-a$
y-intercepts: b and $-b$
Foci: $F_1(-c, 0)$ and $F_2(c, 0)$
Relationship between intercepts and foci: $b^2 = a^2 - c^2$

Example 3 Write an equation, in standard form, of an ellipse whose center is at the origin, coordinates of vertices are $(\pm 4, 0)$, and coordinates of foci are $(\pm 3, 0)$.

The major axis of the ellipse is on the x-axis. Use $\frac{x^2}{a^2} + \frac{y^2}{b^2} = 1$.

$b^2 = a^2 - c^2$
$b^2 = 4^2 - 3^2$ ◀ Substitute 4 for a and 3 for c.
$b^2 = 16 - 9$
$b^2 = 7$

Thus, $\frac{x^2}{16} + \frac{y^2}{7} = 1$ is an equation of the ellipse.

Example 4 Write an equation, in standard form, of an ellipse whose center is at the origin, coordinates of foci are $(\pm 4, 0)$ and the length of the minor axis is 6.

Use $\frac{x^2}{a^2} + \frac{y^2}{b^2} = 1$. ◀ Major axis is on the x-axis.

Length of minor axis $= 2b = 6$.
Substitute 3 for b.
▶
$b^2 = a^2 - c^2$
$3^2 = a^2 - 4^2$ ◀ The foci has coordinates $(\pm c, 0)$. Substitute 4 for c.
$9 = a^2 - 16$
$25 = a^2$

Thus, $\frac{x^2}{25} + \frac{y^2}{9} = 1$ is an equation of the ellipse.

When the major axis of an ellipse is on the y-axis, the general form of the equation of the ellipse becomes $\frac{x^2}{b^2} + \frac{y^2}{a^2} = 1$.

Properties of an ellipse of the form
$$\frac{x^2}{b^2} + \frac{y^2}{a^2} = 1$$

Vertices: $V_1(0, a)$ and $V_2(0, -a)$
Major Axis: $\overline{V_1 V_2}$ of length $2a$
Minor Axis: $\overline{B_1 B_2}$ of length $2b$
x-intercepts: b and $-b$
y-intercepts: a and $-a$
Foci: $F_1(0, c)$ and $F_2(0, -c)$
Relationship between intercepts and foci: $b^2 = a^2 - c^2$.

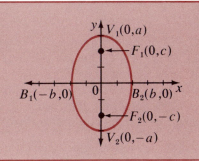

Equations and Properties of Conic Sections

Example 5

Write an equation, in standard form, of an ellipse whose center is at the origin and whose x-intercepts are ± 3 and y-intercepts are ± 4. Find the coordinates of the foci.

The major axis of the ellipse is on the y-axis.

Use $\dfrac{x^2}{b^2} + \dfrac{y^2}{a^2} = 1$.

$b = 3;\ b^2 = 9$
$a = 4;\ a^2 = 16$

$b^2 = a^2 - c^2$
$9 = 16 - c^2$
$c^2 = 7,\ \text{or } c = \sqrt{7}$

Thus, $\dfrac{x^2}{9} + \dfrac{y^2}{16} = 1$ is an equation of the ellipse and $(0, \sqrt{7})$ and $(0, -\sqrt{7})$ are the coordinates of the foci.

Written Exercises

Find an equation of each parabola.

(A) 1. Focus $(2, 0)$, directrix $x = -2$ $y^2 = 8x$
2. Focus $(0, -2)$, directrix $y = 2$ $x^2 = -8y$
3. Focus $(-2, 0)$, directrix $x = 2$ $y^2 = -8x$
4. Focus $(0, 3)$, directrix $y = -3$ $x^2 = 12y$
5. Focus $(0, -1)$, directrix $y = 1$ $x^2 = -4y$
6. Focus $(4, 0)$, directrix $x = -4$ $y^2 = 16x$
7. Focus $(0, 2)$, directrix $y = -2$ $x^2 = 8y$
8. Focus $(-4, 0)$, directrix $x = 4$ $y^2 = -16x$

Draw the graph of each equation. For Ex. 9–16, see TE Answer Section.

9. $x^2 = -4y$
10. $y^2 = -3x$
11. $y^2 = 4x$
12. $x^2 = 10y$
13. $y^2 = 20x$
14. $x^2 = 6y$
15. $-x^2 = 3y$
16. $-y^2 = 2x$

Write an equation, in standard form, of an ellipse with center at the origin and which satisfies the following conditions.

17. Vertices $(\pm 5, 0)$, Foci $(\pm 4, 0)$ $\dfrac{x^2}{25} + \dfrac{y^2}{9} = 1$
18. Vertices $(\pm 7, 0)$, Foci $(\pm 5, 0)$ $\dfrac{x^2}{49} + \dfrac{y^2}{24} = 1$
19. Vertices $(\pm 4, 0)$, Foci $(\pm 2, 0)$ $\dfrac{x^2}{16} + \dfrac{y^2}{12} = 1$
20. Vertices $(\pm \sqrt{7}, 0)$, Foci $(\pm 2, 0)$ $\dfrac{x^2}{7} + \dfrac{y^2}{3} = 1$
21. Vertices $(0, \pm 6)$, Foci $(0, \pm 5)$ $\dfrac{x^2}{11} + \dfrac{y^2}{36} = 1$
22. Vertices $(0, \pm 8)$, Foci $(0, \pm 4)$ $\dfrac{x^2}{48} + \dfrac{y^2}{64} = 1$
23. Foci $(\pm 4, 0)$, length of minor axis 6 $\dfrac{x^2}{25} + \dfrac{y^2}{9} = 1$
24. Foci $(0, \pm 6)$, length of minor axis 8 $\dfrac{x^2}{16} + \dfrac{y^2}{52} = 1$
25. x-intercepts ± 7, y-intercepts ± 9 $\dfrac{x^2}{49} + \dfrac{y^2}{81} = 1$
26. x-intercepts ± 11, y-intercepts ± 10 $\dfrac{x^2}{121} + \dfrac{y^2}{100} = 1$

(B) 27. length of major axis 10 $\dfrac{x^2}{25} + \dfrac{y^2}{9} = 1$
length of minor axis 6 or $\dfrac{x^2}{9} + \dfrac{y^2}{25} = 1$

28. length of major axis 8 $\dfrac{x^2}{16} + \dfrac{y^2}{4} = 1$
length of minor axis 4 or $\dfrac{x^2}{4} + \dfrac{y^2}{16} = 1$

29. length of major axis 6 $\dfrac{x^2}{9} + \dfrac{y^2}{4} = 1$
length of minor axis 4 or $\dfrac{x^2}{4} + \dfrac{y^2}{9} = 1$

30. length of major axis 12 $\dfrac{x^2}{36} + \dfrac{y^2}{25} = 1$
length of minor axis 10 or $\dfrac{x^2}{25} + \dfrac{y^2}{36} = 1$

For Ex. 31–32, see TE Answer Section.

(C) 31. A *latus rectum* of an ellipse is a chord that is perpendicular to the major axis through one of the foci. Show that the length of a latus rectum for the ellipse defined by $\dfrac{x^2}{a^2} + \dfrac{y^2}{b^2} = 1$ is $\dfrac{2b^2}{a}$.

32. The standard form of the equation of a hyperbola with x-intercepts is $\dfrac{x^2}{a^2} - \dfrac{y^2}{b^2} = 1$.

Let $P(x, y)$ be any point on a hyperbola with $F_1(-c, 0)$ and $F_2(c, 0)$ the foci. Show how the standard form of the equation is derived.

COMPUTER ACTIVITIES

Graphs of Parabolas

OBJECTIVE: To draw graphs of vertical parabolas

The program below draws graphs of parabolas of the form $y = (x + a)^2 + b$, where a and b are restricted to values between -10 and 10.

Ordinarily, you would insert the values for x into the equation $y = (x + a)^2 + b$ to find corresponding values for y. However, since the computer must graph from top to bottom, the equation is converted into a form that finds values for x when values for y are plugged into it. The program below uses the equation $x = \pm\sqrt{y - b} - a$, and inserts values for y from 12 to -10 into it.

```
110    PRINT "VALUE OF A";: INPUT A
120    IF  ABS (A) > 10 THEN 110
130    PRINT "VALUE OF B";: INPUT B
140    IF  ABS (B) > 10 THEN 110
150    FOR Y = 12 TO  - 10 STEP  - 1
160    IF Y <  > 0 THEN 180
170    FOR I = 1 TO 39: PRINT "*";: NEXT I: PRINT :
       GOTO 340
180    Y1 = Y - B: IF Y1 < 0 THEN 320
190    A1 =   - A -   INT ( SQR (Y1)):A2 =   - A +   INT
       ( SQR (Y1))
200    IF A1 < 0 THEN 220
210    C = 0:D = A1:E = A2: GOTO 250
220    IF A2 < 0 THEN 240
230    C = A1:D = 0:E = A2: GOTO 250
240    C = A1:D = A2:E = 0
250    IF C <  - 19 THEN 330
260    PRINT  TAB( 20 + C);"*";
270    IF D = C THEN 290
280    PRINT  TAB( 20 + D);"*";
290    IF E > 19 THEN 330
300    IF D = E THEN 330
310    PRINT  TAB( 20 + 3)"*";: GOTO 330
320    PRINT  TAB( 20)"*";
330    PRINT : NEXT Y: GOTO 110
```

See the Computer Section beginning on page 574 for more information.

EXERCISES For Ex. 1–2, see TE Answer Section.
1. 19) $y = (x - 5)^2 + 0$ 20) $y = (x + 3)^2 + 0$ 22) $y = (x + 2)^2 - 2$
Note: Graphs show on monitor, but do not print.

1. Transform equations 19, 20, and 22 on page 318 to the form $Y = (X + A)^2 + B$ and then run the values through the program above.

2. Make all necessary changes in the program above so that it will graph $Y = -(X + A)^2 + B$. Be sure to specify appropriate parameters.

Graphs of Parabolas

Ⓐ Exercises are 1–7, 13–15, 20, 23, 25, 27–31, 34, 38–41, 45–47. Ⓒ Exercises are starred.
The rest are Ⓑ Exercises.

CHAPTER TWELVE REVIEW

Vocabulary

asymptotes [12.5]   axis of symmetry [12.1]
circle [12.2]   conic section [12.6]
ellipse [12.3]   hyperbola [12.5]
parabola [12.4]   rectangular hyperbola
symmetric points [12.1]   [12.5]

For complete solutions, see TE Answer Section.

Determine whether P and Q are symmetric with respect to the x-axis, the y-axis, the origin, or none of these. [12.1]

1. $P(-2, 6)$; $Q(2, -6)$ 2. $P(-3, 5)$; $Q(3, 5)$
3. $P(8, -1)$; $Q(8, 1)$ 4. $P(-5, 3)$; $Q(3, -5)$

For each point find the coordinates of the corresponding point symmetric with respect to the x-axis, the y-axis, and the origin. [12.1]

5. $A(-5, 1)$ 6. $B(7, -2)$ 7. $C(-3, -8)$
8. $D\left(-\frac{1}{7}, -\frac{1}{8}\right)$ 9. $E(m, -n)$ 10. $F(-r, -s)$

Determine whether the following functions are odd, even, or neither. [12.1]

★ 11. $y = x^2$ **even**   ★ 12. $x - 4y^2 = 1$ **neither**

Write an equation of each circle whose center is at the origin and whose radius is given. [12.2]

13. 13   14. 17   15. $\sqrt{8}$

Write an equation of each circle whose center C and radius r are given. [12.2]

16. $C(-2, 3)$, $r = 5$ 17. $C(3, -2)$, $r = 4$

Write an equation of each circle whose center C is given and which passes through point P. [12.2]

18. $C(-6, -5)$, $P(5, -1)$
19. $C(1, -8)$, $P(-1, -6)$

Draw the graph of each equation. [12.2]

20. $x^2 + y^2 = 49$
21. $x^2 + y^2 - 8x + 6y + 11 = 0$
22. $(x - 1)^2 + (y + 2)^2 = 16$

Find the x- and y-intercepts of the ellipse whose equation is given. Then draw the graph. [12.3]

23. $\frac{x^2}{36} + \frac{y^2}{4} = 1$ 24. $\frac{(x - 3)^2}{16} + \frac{(y + 5)^2}{4} = 1$
25. $9x^2 + y^2 = 36$
26. Draw the graph of $3x^2 + 4y^2 + 12x - 8y + 4 = 0$. [12.3]

27. Write an equation, in standard form, of an ellipse whose center is at the origin, whose x-intercepts are ±5, and whose y-intercepts are ±3. [12.3] $\frac{x^2}{25} + \frac{y^2}{9} = 1$

Graph each quadratic function. Draw the axis of symmetry and label the coordinates of the vertex. Write an equation of the axis of symmetry. [12.4]

28. $y = x^2 - 6x + 12$ 29. $y = -\frac{1}{3}x^2$

Graph each equation. [12.4]
30. $y = (x - 2)^2 - 3$ 31. $y = (x + 1)^2 + 4$

Graph each function. Estimate, to the nearest tenth, the x-intercepts of each parabola. [12.4]
32. $y = x^2 + 4x - 3$ 33. $y = -2x^2 + 7x + 1$

Draw the graph of each equation. Find the equations of the asymptotes. [12.5]

34. $\frac{x^2}{9} - \frac{y^2}{4} = 1$ 35. $xy = -6$
    x = 0, y = 0

Draw the graph of each equation. [12.5]
36. $\frac{(x - 2)^2}{9} - \frac{(y + 1)^2}{4} = 1$
37. $9x^2 - 4y^2 - 18x - 16y - 43 = 0$

Identify each conic section. [12.6]
38. $\frac{x^2}{16} - \frac{y^2}{4} = 1$ **H**   39. $\frac{x^2}{8} + \frac{y^2}{8} = 1$ **C**
40. $4x^2 + 5y^2 = 20$ **E**   41. $10x^2 + 20y^2 = 20$ **E**
42. $x^2 - y^2 - 2x - 4y - 4 = 0$ **H**
43. $x^2 + y^2 + 4x - 4y + 7 = 0$ **C**
44. $2x^2 + 3y^2 + 6x - 9y - 15 = 0$ **E**
45. Write an equation of a parabola with $F(-4, 0)$ the focus and $x = 4$ the equation of the directrix. Draw a graph of the equation. [12.7] **y² = −16x**

Write an equation, in standard form, of an ellipse with center at the origin and which satisfies the following conditions. [12.7]
46. Vertices (±6, 0), foci (±5, 0)
47. Foci (0, ±7), length of minor axis 12

A Exercises are 1–6, 9, 11, 12, 14–17, 19, 22–24, 27, 28. **C** Exercises are starred.
The rest are **B** Exercises.

CHAPTER TWELVE TEST

Determine whether P and Q are symmetric with respect to the x-axis, the y-axis, the origin, or none of these.

1. $P(3, -4)$; $Q(3, 4)$ **x-axis**
2. $P(-2, -1)$; $Q(2, 1)$ **origin**

For each point, find the coordinates of the corresponding point symmetric with respect to the x-axis, the y-axis, and the origin.

3. $A(-8, 1)$ **(−8, −1), (8, 1), (8, −1)**
4. $B(-2, -3)$ **(−2, 3), (2, −3), (2, 3)**

Write an equation of each circle whose center is at the origin and whose radius is given.

5. 6 $x^2 + y^2 = 36$
6. $\sqrt{20}$ $x^2 + y^2 = 20$

7. Write an equation of a circle whose center is at $C(1, -4)$ and whose radius r is 3.
   $(x - 1)^2 + (y + 4)^2 = 9$

8. Write an equation of a circle whose center is at $C(-8, -5)$ and which passes through the point $P(3, -2)$.
   $(x + 8)^2 + (y + 5)^2 = 130$

Draw the graph of each equation.

9. $x^2 + y^2 = 4$ **C(0, 0), r = 2**
10. $(x - 3)^2 + (y + 1)^2 = 16$ **C(3, −1), r = 4**

Find the x- and y-intercepts of the ellipse whose equation is given. Then draw the graph.

11. $\frac{x^2}{16} + \frac{y^2}{4} = 1$ **a = ±4, b = ±2**
12. $9x^2 + 4y^2 = 36$ **a = ±2, b = ±3**

13. Draw the graph of $x^2 + 4y^2 + 4x - 8y - 8 = 0$. **See TE Answer Section.**

14. Write an equation, in standard form, of an ellipse whose center is at the origin, whose x-intercepts are ± 3, and whose y-intercepts are ± 2. $\frac{x^2}{9} + \frac{y^2}{4} = 1$

Graph each quadratic function. Draw the axis of symmetry and label the coordinates of the vertex. Write an equation of the axis of symmetry.

15. $y = x^2 + 8x + 3$ **(−4, −13); x = −4**
16. $y = -\frac{2}{3}x^2$ **(0, 0); x = 0**
17. $y = (x + 3)^2 - 2$ **(−3, −2), x = −3**

18. Graph the quadratic function $y = x^2 + 6x + 2$. Estimate, to the nearest tenth, its x-intercepts. **−5.6, −0.4**

Draw the graph of each equation. Write the equations of the asymptotes.

19. $\frac{x^2}{16} - \frac{y^2}{4} = 1$ $y = -\frac{1}{2}x, y = \frac{1}{2}x$
20. $xy = -2$ **x = 0, y = 0**
21. Draw the graph of $\frac{(x - 1)^2}{16} - \frac{(y + 2)^2}{25} = 1$. $y = -\frac{5}{4}x, y = \frac{5}{4}x$

Identify each conic section whose equation is given.

22. $\frac{x^2}{49} + \frac{y^2}{81} = 1$ **E**
23. $xy = -4$ **H**
24. $\frac{x^2}{9} - \frac{y^2}{25} = 1$ **H**
25. $3x^2 + 5y^2 + 6x - 15y + 12 = 0$ **E**
26. $x^2 - 3y^2 - 4x + 9y - 1 = 0$ **H**

27. Write an equation of a parabola with $F(0, 3)$ the focus and $y = -3$ the equation of the directrix. $x^2 = 12y$

28. Write an equation, in standard form, of the ellipse with center at the origin, foci $(0, \pm 10)$, and length of minor axis 8. $\frac{x^2}{16} + \frac{y^2}{116} = 1$

Determine whether the following functions are odd, even, or both.

★ 29. $y = 7x^2$ **even**
★ 30. $x^2 + 16y^2 = 1$ **even and odd**
★ 31. Find the coordinates of the center of a circle and the radius of the circle whose equation is $x^2 - 10x + y^2 + 6y = -33$.   **(5, −3); 1**

Chapter Twelve Test

COLLEGE PREP TEST

DIRECTIONS: Choose the *one* best answer to each question or problem.

1.

   Each circle in the above figure has a diameter of 4. What is the area of the rectangle? **E**

   (A) 64 (B) 32 (C) 80
   (D) 16 (E) 40

2. The circumference of a circle in a coordinate plane is 18π units. The center of the circle is at the origin. Which point is on the circle? **A**

   (A) $P(9, 0)$ (B) $P(-3, -3)$ (C) $P(0, -3)$
   (D) $P(3, 0)$ (E) $P(9, 9)$

3. The area of a circle is 36π square units. The center of the circle is at $C(-2, 3)$. Which point does the circle not pass through? **E**

   (A) $P(-2, -3)$ (B) $P(4, 3)$ (C) $P(-2, 9)$
   (D) $P(-8, 3)$ (E) $P(-2, 3)$

4. $\overline{AB}$ is a diameter of a circle whose center is point O. The diameter is 6, C is a point on the circle, and the measure of $\angle BOC$ is 60°. What is the length of chord $\overline{AC}$? **B**

   (A) 3 (B) $3\sqrt{3}$ (C) 6
   (D) $3\sqrt{2}$ (E) None of these

5. A radius of a given circle measures 10 cm. If the length of the radius is decreased by 2 cm, what is the percent of area decrease? **D**

   (A) 81 (B) 19 (C) 64
   (D) 36 (E) None of these

6. The set of all points in a coordinate plane 10 units from the origin is represented by which of the following equations? **D**

   (A) $x = 10$ (B) $x^2 + y^2 = 10$
   (C) $x^2 + y^2 = \sqrt{10}$ (D) $x^2 + y^2 = 100$
   (E) $y = 10$

7. An equation of a circle with center at $C(-3, 4)$, and radius 5 is **B**

   (A) $(x - 3)^2 + (y + 4)^2 = 25$
   (B) $(x + 3)^2 + (y - 4)^2 = 25$
   (C) $x^2 + y^2 = 25$
   (D) $x^2 + y^2 = 1$ (E) None of these

8. The x- and y-intercepts of the ellipse defined by $\dfrac{x^2}{9} + \dfrac{y^2}{16} = 1$ are **C**

   (A) 3 and 4 (B) 4 and 3
   (C) ± 3 and ± 4 (D) ± 4 and ± 3
   (E) None of these

9. The equations of the asymptotes of a hyperbola defined by $\dfrac{x^2}{9} - \dfrac{y^2}{4} = 1$ are **A**

   (A) $y = \pm \dfrac{2}{3}x$ (B) $y = \pm 2x$
   (C) $y = \pm 3x$ (D) $y = \pm \dfrac{3}{2}x$
   (E) None of these

10. The one equation whose graph is not a conic section is **B**

    (A) $(x + 2)^2 + (y - 6)^2 = 4$
    (B) $\dfrac{x}{4} + \dfrac{y}{9} = 16$ (C) $\dfrac{x^2}{4} + \dfrac{y^2}{16} = 1$
    (D) $\dfrac{x^2}{9} - \dfrac{y^2}{25} = 1$ (E) $xy = 4$

CUMULATIVE REVIEW
(Chapters 1–12)

DIRECTIONS: Choose the *one* best answer to each question or problem.

1. Factor $2x^3 - 5x^2 - 3x$ completely.
   (A) $(x^2 - 3x)(2x + 1)$
   (B) $(2x^2 + x)(x - 3)$
   (C) $x(x - 3)(2x + 1)$
   (D) $x(x + 3)(2x - 1)$ **C**

2. Solve $16^{x-2} = 64$.
   (A) $\frac{7}{2}$ (B) 6 (C) 3 (D) $\frac{9}{4}$ **A**

3. Solve $-3(x - 2c) = 2 + a(bx + c)$ for x.
   (A) $\frac{-(ac + 6c + 2)}{ab + 3}$ (B) $\frac{6c - ac - 2}{ab + 3}$
   (C) $\frac{6c - ac - 2}{3ab}$ (D) $\frac{5ac - 2}{ab + 3}$ **B**

4. Find the value of $2 \cdot 27^{-\frac{4}{3}}$.
   (A) $\frac{2}{81}$ (B) 162 (C) $\frac{1}{6}$ (D) $-\frac{2}{3}$ **A**

5. Solve $n + 2 = \sqrt{2n + 12}$.
   (A) -4 (B) 2
   (C) -4 and 2 (D) 10 **B**

6. Factor $9x^2 + 6x + 1 - 16y^2$ as a difference of two squares.
   (A) $(3x - 1 + 4y)(3x - 1 - 4y)$
   (B) $(3x - 2x + 1 - 4y)^2$
   (C) $(3x + 1 + 4y)(3x + 1 - 4y)$
   (D) $(3x - 4y - 1)(3x - 4y + 1)$ **C**

7. Find the slope of $\overrightarrow{PQ}$ for $P(-3, 5)$ and $Q(6, -2)$.
   (A) $-\frac{9}{7}$ (B) $-\frac{1}{3}$
   (C) $-\frac{7}{9}$ (D) $-\frac{7}{3}$ **C**

8. Write an equation, in standard form, of the line that passes through the point $P(-2, 1)$ and is parallel to the line described by the equation $3x - 2y = 4$.
   (A) $2x + 3y = -1$ (B) $2x - 3y = 1$ **C**
   (C) $3x - 2y = -8$ (D) $3x - 2y = -2$

9. Find the product, if possible.
   $\begin{bmatrix} -3 & 1 & -2 \\ 2 & 5 & -1 \end{bmatrix} \cdot \begin{bmatrix} 1 & -2 & 2 \\ 2 & 3 & 4 \\ -1 & 2 & -2 \end{bmatrix}$
   (A) $\begin{bmatrix} -2 & -1 & 0 \\ 4 & 8 & 0 \\ -1 & 2 & -2 \end{bmatrix}$ (B) $\begin{bmatrix} 1 & 13 \\ 5 & 9 \\ -1 & 11 \end{bmatrix}$
   (C) $\begin{bmatrix} 1 & 5 & -1 \\ 13 & 9 & 11 \end{bmatrix}$ (D) None of these **C**

10. Find the domain of the function defined by $t(x) = \frac{-2x}{x^2 + 2x - 3}$. **D**
    (A) {all real numbers}
    (B) {all real numbers except 0}
    (C) {all real numbers except -3, 0, and 1}
    (D) {all real numbers except -3 and 1}

11. Let $f(x) = 3x - 1$ and $g(x) = x^2 + 1$. Find $f(g(-2))$.
    (A) 14 (B) -10 (C) 50 (D) 5 **A**

12. Write an equation of the inverse of the function defined by $y = 3x - 2$.
    (A) $y = -3x + 2$ (B) $x - 3y = -2$
    (C) $3y = x - 2$ (D) $-3x + y = -2$ **B**

13. Which of the following is an equation of a parabola?
    (A) $y = 2x + 3$ (B) $y = x^2 + 2$
    (C) $y^2 - x^2 = 3$ (D) $y^2 + x^2 = 2$ **B**

14. Which of the following is an equation of a hyperbola? **C**
    (A) $5x^2 + 4y^2 = 20$ (B) $4x^2 + 5y^2 = 20$
    (C) $16x^2 - 16y^2 = 1$ (D) $8x^2 + 8y^2 = 8$

15. Write an equation of a parabola with $F(-2, 0)$ the focus and $x = 2$ the equation of the directrix.
    (A) $x^2 = -8y$ (B) $x^2 = 8y$
    (C) $y^2 = 8x$ (D) $y^2 = -8x$ **D**

Solve each equation.

16. $|3x - 2| = 13$ $\quad 5, -\frac{11}{3}$
17. $\frac{10}{n-1} - \frac{3}{n} = 4$ $\quad 3, -\frac{1}{4}$
18. $3^{x-5} = \frac{1}{9}$ $\quad 3$
19. $n^2 - 9 = -5n$ $\quad \frac{-5 \pm \sqrt{61}}{2}$
20. Solve $\frac{-4}{a} = \frac{3}{b} + \frac{1}{x}$ for x. $\quad \frac{-ab}{3a + 4b}$

Factor each polynomial completely.

21. $36a^2 - 49b^2$ $\quad (6a - 7b)(6a + 7b)$
22. $2x^3 - 16$ $\quad 2(x - 2)(x^2 + 2x + 4)$

Simplify and write each expression with positive exponents.

23. $(-2a^{-3}b^4)^2$ $\quad \frac{4b^8}{a^6}$
24. $\frac{-25x^{-3}y^4}{15x^{-2}y^6}$ $\quad \frac{-5}{3xy^2}$
25. Divide: $\frac{8 - 2n}{4n^2 + 8n} \div \frac{n^2 - 16}{n^3 - n^2 - 6n}$ $\quad \frac{n-3}{-2(n+4)}$
26. Simplify: $ab\sqrt{18ab^3} - 2b\sqrt{8a^3b^3} + a\sqrt{32ab^5}$; $a \geq 0, b \geq 0$. $\quad 3ab^2\sqrt{2ab}$
27. Simplify: $-2 - \sqrt{-8} - (5 - 3\sqrt{-2})$. $\quad -7 + i\sqrt{2}$
28. Some sunflower seeds worth \$3.80/kg are mixed with some raisins worth \$3.20/kg to make 12 kg of a mixture worth \$3.60/kg. How many kg of each kind are there in the mixture? **sunflower: 8 kg, raisins: 4 kg**
29. A second number is 6 more than 3 times a first number. Find the numbers if their difference is 26. **10, 36**

Draw the graph of each equation.

30. $3x - 2y = 8$
31. $\frac{2}{3}y = \frac{1}{2}x + 4$

For Ex. 30–31, see TE Answer Section.

A line has the given slope, y-intercept, or contains the indicated point(s). Write an equation, in standard form, of each line.

32. $A(-1, 2)$, $m = -\frac{3}{5}$ $\quad 3x + 5y = 7$
33. $m = \frac{2}{3}$, $b = -1$ $\quad 2x - 3y = 3$
34. $P(4, -1)$, $Q(-5, 2)$ $\quad x + 3y = 1$

35. The vertices of a quadrilateral are $A(2, -3)$, $B(5, 6)$, $C(3, 2)$ and $D(-2, -1)$. Prove that the segments connecting the midpoints of $ABCD$ form a parallelogram.

For Ex. 35, see TE Answer Section.

Solve each system of equations.

36. $3(x - 2) = 2y - 7$
    $-2(y + 4) = -(x + 5)$ $\quad (-2, -\frac{5}{2})$
37. $2x - 3y + z = 7$
    $-x + 2y - 3z = -9$
    $3x - y + 2z = 8$ $\quad (1, -1, 2)$

38. A two-digit number is 54 less than the number obtained by reversing the digits. Find the original number if 4 times the sum of the digits is 3 more than 5 times the units digit. **39**

Let $g(x) = x^2 - 2x + 3$. Find each of the following function values.

39. $g(-2)$ **11**
40. $g(2n)$ $\quad 4n^2 - 4n + 3$

41. If y varies inversely as x, and if $y = 6$ when $x = 5$, find y when $x = 10$. **3**
42. If y varies directly as the positive square root of x and inversely as the cube of z, and if $y = 6$ when $x = 9$ and $z = 2$, find y when $x = 4$ and $z = 3$. $\quad \frac{32}{27}$
43. For the point whose coordinates are $(-6, 1)$, find the corresponding point symmetric with respect to the x-axis, the y-axis, and the origin. $\quad (-6, -1), (6, 1), (6, -1)$
44. Draw the graph of $y = 2x^2 + 6x - 1$.

For Ex. 44, see TE Answer Section.

45. Find an equation of a circle whose center is at $C(3, -1)$ and which passes through $P(5, 4)$. $\quad (x - 3)^2 + (y + 1)^2 = 29$

Graph each of the following equations.

46. $x^2 - y^2 = 9$
47. $y = (x + 2)^2 - 3$
48. $\frac{(x - 1)^2}{9} + \frac{(y + 2)^2}{4} = 1$
49. $(x + 2)^2 + (y - 2)^2 = 9$

For Ex. 45–49, see TE Answer Section.

Cumulative Review (Chapters 1–12)

13 LINEAR-QUADRATIC SYSTEMS

Non-Routine Problem Solving

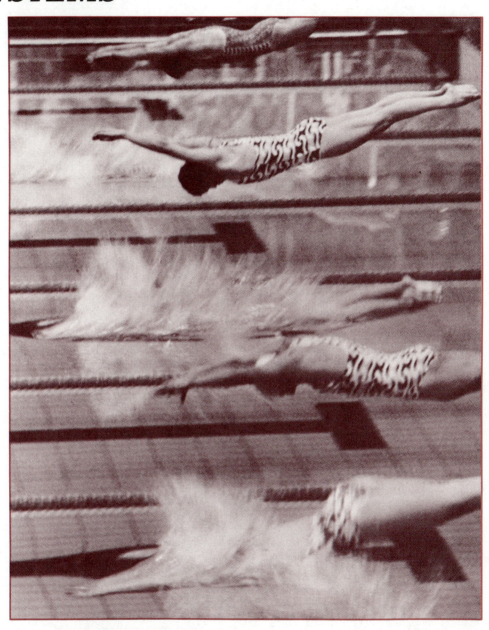

Problem 1. For 3 h a fishing boat traveled at the same rate. For the next hour, the fishing boat traveled at one-half that rate. If the boat traveled a total distance of 105 mi, what was the initial rate? **30 mi/h**

Problem 2. A swimmer can swim at a rate of 80 yd/min in still water. In a river with a current of 30 ft/min, she can swim round-trip, upstream and downstream, between points A and B in 4 min. Find the distance between points A and B. **472.5 ft, or 157.5 yd**

SOLVING LINEAR-QUADRATIC SYSTEMS OF EQUATIONS   13.1

Objective   **To find the real number solutions of a system of one quadratic and one linear equation graphically and algebraically**

The general form of the quadratic equations of some conic sections you have studied are given below.

Parabola: $y = ax^2 + bx + c$ Circle: $x^2 + y^2 = r^2$

Hyperbola: $\frac{x^2}{a^2} - \frac{y^2}{b^2} = 1$ and $xy = k$ Ellipse: $\frac{x^2}{a^2} + \frac{y^2}{b^2} = 1$

You have also learned that the general form of a linear equation in two variables is $ax + by = c$, and that the graph of each such equation is a line.

A system of equations may contain one quadratic equation and one linear equation. Such a system is called a **linear-quadratic system.** A common solution of a linear-quadratic system of equations is an ordered pair of real numbers corresponding to a point of intersection of their graphs. As illustrated by the graphs below, a system of one quadratic and one linear equation may have 0, 1, or 2 solutions.

One solution   No solution   Two solutions

The solutions of a linear-quadratic system may be found graphically.

Example 1   Solve the system $\frac{x^2}{16} + \frac{y^2}{9} = 1$ graphically.
$2x - 3y = 6$
Estimate the solutions to the nearest tenth.

Ellipse ▶ $\frac{x^2}{16} + \frac{y^2}{9} = 1$ 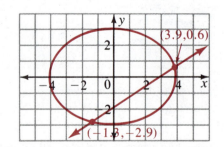 $2x - 3y = 6$ ◀ *Line*
$a^2 = 16 \quad b^2 = 9$ $y = \frac{2}{3}x - 2$
$a = \pm 4, b = \pm 3$ $m = \frac{2}{3}, b = -2$

Thus, $(-1.3, -2.9)$ and $(3.9, 0.6)$ are the solutions of the system.

Example 2

Solve the system $\begin{aligned} x^2 + y^2 &= 13 \\ -2x + y &= -1 \end{aligned}$ algebraically.

Step 1 Solve the linear equation for one of its variables.

$-2x + y = -1$
$y = 2x - 1$

Step 2 Substitute this result in the quadratic equation.

$x^2 + y^2 = 13$
$x^2 + (2x - 1)^2 = 13$
$x^2 + 4x^2 - 4x + 1 = 13$

Step 3 Solve the resulting quadratic equation by factoring, completing the square, or using the quadratic formula.

$5x^2 - 4x - 12 = 0$ ◄ Solve by factoring.
$(5x + 6)(x - 2) = 0$

$\boxed{x = -\tfrac{6}{5}}$ or $\boxed{x = 2}$

Step 4 Find the value of the other variable by substituting in the linear equation.

$-2x + y = -1$ $-2x + y = -1$
$-2\left(-\tfrac{6}{5}\right) + y = -1$ $-2(2) + y = -1$
$\tfrac{12}{5} + y = -1$ $-4 + y = -1$
$\boxed{y = -\tfrac{17}{5}}$ $\boxed{y = 3}$

Step 5 Check by substituting each ordered pair in both equations.

$\begin{array}{c|c}
x^2 + y^2 = 13 \\
\left(-\tfrac{6}{5}\right)^2 + \left(-\tfrac{17}{5}\right)^2 \;\Big|\; 13 \\
\tfrac{36}{25} + \tfrac{289}{25} \\
\tfrac{325}{25} \\
13
\end{array}$
$\begin{array}{c|c}
-2x + y = -1 \\
-2\left(-\tfrac{6}{5}\right) + \left(-\tfrac{17}{5}\right) \;\Big|\; -1 \\
\tfrac{12}{5} - \tfrac{17}{5} \\
-\tfrac{5}{5} \\
-1
\end{array}$
$\begin{array}{c|c}
x^2 + y^2 = 13 \\
2^2 + 3^2 \;\Big|\; 13 \\
4 + 9 \\
13
\end{array}$
$\begin{array}{c|c}
-2x + y = -1 \\
-2(2) + 3 \;\Big|\; -1 \\
-4 + 3 \\
-1
\end{array}$

Thus, $\left(-\tfrac{6}{5}, -\tfrac{17}{5}\right)$ and $(2, 3)$ are the solutions of the system.

Oral Exercises

State the number of solutions of each system of equations whose graphs are given. Give the solution(s) of each system, if any.

1.
2.
3.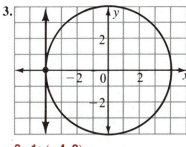

1. 2; (3, −2), (0, 3)

2. 0; no solution

3. 1; (−4, 0)

Solving Linear-Quadratic Systems of Equations

Written Exercises

20. $(\frac{19 + \sqrt{973}}{18}, \frac{-26 + \sqrt{973}}{27})$, $(\frac{19 - \sqrt{973}}{18}, \frac{-26 - \sqrt{973}}{27})$ 21. $(-\frac{11}{19}, \frac{63}{19})$, $(1, -3)$

Solve each system of equations graphically. Estimate the decimal solutions to the nearest tenth.

Ⓐ
1. $\frac{x^2}{16} + \frac{y^2}{9} = 1$
   $2x - 3y = 6$
   (−1.2, −2.8), (3.9, 0.6)

2. $\frac{x^2}{25} - \frac{y^2}{4} = 1$
   $x = 5$ **(5, 0)**

3. $\frac{x^2}{16} + \frac{y^2}{16} = 1$
   $x + y = 2$
   (3.6, −1.6), (−1.7, 3.7)

4. $xy = 4$
   $2x + 3y = 9$
   no solution

5. $y = x^2 - 6x + 8$
   $x - 2y = 2$
   (2, 0), (4.5, 1.3)

6. $x^2 + y^2 = 16$
   $3x - 2y = 4$
   (−1.2, −3.8), (3.1, 2.7)

7. $\frac{x^2}{4} + \frac{y^2}{4} = 1$
   $2x - 6y = 12$
   (1.2, −1.6), (0, −2)

8. $y = x^2 - 9$
   $x + y = 3$
   (−4, 7), (3, 0)

For complete solutions, see TE Answer Section.

Solve each system of equations algebraically. Give the irrational solutions in simplest radical form.

9. $xy = 8$
   $y = x + 2$
   (−4, −2), (2, 4)

10. $y = x^2 - 7x - 3$
    $2x - y = 3$
    (0, −3), (9, 15)

11. $y = 5(x - 3)$
    $2y = x^2 - 6$
    (4, 5), (6, 15)

12. $\frac{x^2}{10} + \frac{y^2}{40} = 1$
    $y = \frac{2}{3}x$
    (−3, −2), (3, 2)

13. $y = 3(x - 3)$
    $2y = x^2 - 10$
    (2, −3), (4, 3)

14. $y = 2x - 3$
    $x = -\frac{y^2}{4} + \frac{21}{4}$
    (−1, −5), (3, 3)

15. $y - 2x = 5$
    $x^2 + y^2 = 25$
    (−4, −3), (0, 5)

16. $\frac{x^2}{1} - \frac{y^2}{2} = 1$
    $y = 2x - 2$
    (1, 0), (3, 4)

Ⓑ
17. $y = -\frac{x}{2} + \frac{1}{2}$
    $x = -3y^2 - 12y + 8$

18. $y = x^2 - 8x + 2$
    $x - 3y = 5$
    $(\frac{25 + \sqrt{493}}{6}, \frac{-5 + \sqrt{493}}{18})$,
    $(\frac{25 - \sqrt{493}}{6}, \frac{-5 - \sqrt{493}}{18})$

19. $3x - 2y = 8$
    $y = 2x^2 - 5x - 3$
    $(\frac{13 + \sqrt{137}}{8}, \frac{-25 + 3\sqrt{137}}{16})$,
    $(\frac{13 - \sqrt{137}}{8}, \frac{-25 - 3\sqrt{137}}{16})$

20. $3y - 2x = -5$
    $y = -3x^2 + 7x + 4$
    above

21. $4x + y = 1$
    $3x^2 + y^2 = 12$

22. $y = 2x + 3$
    $x^2 + y^2 = 25$

23. $x^2 + (y - 5)^2 = 49$
    $y - x = -2$
    (0, −2), (7, 5)

24. $\frac{x^2}{16} + \frac{y^2}{3} = 1$
    $y = x - 1$

22. $(\frac{-6 + 2\sqrt{29}}{5}, \frac{3 + 4\sqrt{29}}{5})$, $(\frac{-6 - 2\sqrt{29}}{5}, \frac{3 - 4\sqrt{29}}{5})$

Ⓒ
25. $(x - 2)^2 + (y + 3)^2 = 96$
    $y = 2x$
    $(\frac{-4 + \sqrt{431}}{5}, \frac{-8 + 2\sqrt{431}}{5})$, $(\frac{-4 - \sqrt{431}}{5}, \frac{-8 - 2\sqrt{431}}{5})$

26. $x^2 + 4x + y^2 - 6y = 12$
    $y = -x + 3$

27. $2x^2 + y^2 = 8x + 4y + 3$
    $2x - 3y = 7$
    $(\frac{31 + 3\sqrt{42}}{11}, \frac{-5 + 2\sqrt{42}}{11})$, $(\frac{31 - 3\sqrt{42}}{11}, \frac{-5 - 2\sqrt{42}}{11})$

28. $3x^2 - 2y^2 - 6x + 5y = -12$
    $3x - 2y = -8$
    (0, 4), (−15, −$\frac{37}{2}$)

NON-ROUTINE PROBLEMS

The number of distinct points in the coordinate plane common to the graphs of $(2x - y + 1)(2x + 3y - 11) = 0$ and $(x + 2y - 7)(3x - 2y + 3) = 0$ is **B**
(A) 0   (B) 1   (C) 2   (D) 3   (E) 4   (F) infinitely many

CUMULATIVE REVIEW

Simplify.

1. $3a^3b(-2a^4b^5)$ **$-6a^7b^6$**

2. $\frac{16x^6yz^2}{-4xy^5z^2} - \frac{4x^5}{y^4}$

3. $(-3a^2b^3c)^3$ **$-27a^6b^9c^3$**

Chapter Thirteen

SOLVING QUADRATIC SYSTEMS OF EQUATIONS   13.2

Objective **To find the real number solutions of a system of two quadratic equations graphically and algebraically**

A system of equations may contain two quadratic equations. A system of this type is called a **quadratic system.** As illustrated by the graphs below, a system of two quadratic equations may have 4, 3, 2, 1, or 0 solutions.

Two solutions

Four solutions

One solution

Three solutions

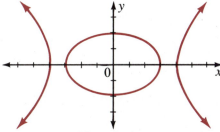
No solution

Example 1 Solve the system $\frac{x^2}{34} + \frac{y^2}{15} = 1$ graphically.
$x^2 + y^2 = 34$
Estimate the solutions to the nearest tenth.

Ellipse ▶ $\frac{x^2}{34} + \frac{y^2}{15} = 1$ $x^2 + y^2 = 34$ ◀ *Circle*

$a^2 = 34 \quad b^2 = 15 \quad r^2 = 34$
$a \doteq \pm 5.8 \quad b \doteq \pm 3.9 \quad r \doteq 5.8$

The graphs intersect at two points with coordinates of $(5.8, 0)$ and $(-5.8, 0)$. Check by substituting $(5.8, 0)$ and $(-5.8, 0)$ in the two equations.

Thus, $(5.8, 0)$ and $(-5.8, 0)$ are the solutions of the system.

Solving Quadratic Systems of Equations

The next examples show you how to find the solutions of a system of two quadratic equations algebraically. Notice that either the addition method or substitution method may be used.

Example 2

Solve the system $\dfrac{x^2}{36} + \dfrac{y^2}{25} = 1$ algebraically.
$x^2 + y^2 = 25$

$\dfrac{x^2}{36} + \dfrac{y^2}{25} = 1$ ◀ Multiply the first equation by 900 to eliminate fractions.

$25x^2 + 36y^2 = 900$
$\underline{-25x^2 - 25y^2 = -625}$ ◀ Multiply the second equation by -25.
$11y^2 = 275$ ◀ Add the equations to eliminate the x^2 term.
$y^2 = 25$

$\boxed{y = 5}$ $\boxed{y = -5}$

Solve for x. Substitute 5 and −5 for y.

$x^2 + y^2 = 25$ $x^2 + y^2 = 25$
$x^2 + (5)^2 = 25$ $x^2 + (-5)^2 = 25$
$x^2 = 0$ $x^2 = 0$
$\boxed{x = 0}$ $\boxed{x = 0}$

Check by substituting (0, 5) and (0, −5) in the original equations.

Thus, (0, 5) and (0, −5) are the solutions of the system.

Example 3

Solve the system $x^2 + y^2 = 49$ algebraically.
$-2y^2 = 30 - 2x^2$
Estimate the solutions to the nearest tenth.

$-2y^2 = 30 - 2x^2$ ◀ Solve the second equation for y^2.
$y^2 = x^2 - 15$

$x^2 + y^2 = 49$ ◀ Substitute in the first equation.
$x^2 + (x^2 - 15) = 49$
$2x^2 - 15 = 49$
$2x^2 = 64$
$x^2 = 32$

$\boxed{x = 4\sqrt{2} \doteq 5.7}$ $\boxed{x = -4\sqrt{2} \doteq -5.7}$

Solve for y. Substitute $4\sqrt{2}$ and $-4\sqrt{2}$ for x.

$x^2 + y^2 = 49$ $x^2 + y^2 = 49$
$(4\sqrt{2})^2 + y^2 = 49$ $(-4\sqrt{2})^2 + y^2 = 49$
$32 + y^2 = 49$ $32 + y^2 = 49$
$y^2 = 17$ $y^2 = 17$
$\boxed{y \doteq \pm 4.1}$ $\boxed{y \doteq \pm 4.1}$

Check by substituting the four ordered pairs in the original equations.

Thus, (5.7, 4.1), (5.7, −4.1), (−5.7, 4.1), and (−5.7, −4.1) are the solutions of the system.

Oral Exercises

12. $(\frac{6\sqrt{430}}{43}, \frac{2\sqrt{129}}{43})$, $(\frac{6\sqrt{430}}{43}, \frac{-2\sqrt{129}}{43})$, $(\frac{-6\sqrt{430}}{43}, \frac{2\sqrt{129}}{43})$, $(\frac{-6\sqrt{430}}{43}, \frac{-2\sqrt{129}}{43})$

State the number of solutions of each system of equations whose graphs are given. Give the solution(s) of each system, if any.

1.
2.
3.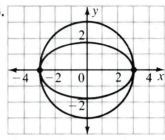

1. 4; (2, 2), (2, −2), (−2, 2), (−2, −2)   2. 2; (1.5, 1.5), (−1.5, 1.5)   3. 2; (3, 0), (−3, 0)

Written Exercises

2. (3.1, 2.4), (3.1, −2.4), (−3.1, 2.4), (−3.1, −2.4)   7. (5, 3.6), (5, −3.6), (−5, 3.6), (−5, −3.6)

Solve each system of equations graphically. Estimate the decimal solutions to the nearest tenth.

Ⓐ 1. $\frac{x^2}{16} + \frac{y^2}{4} = 1$ 2. $\frac{x^2}{25} + \frac{y^2}{9} = 1$ 3. $\frac{x^2}{9} - \frac{y^2}{4} = 1$ 4. $\frac{x^2}{16} + \frac{y^2}{25} = 1$
   $x^2 + y^2 = 16$ $x^2 - y^2 = 4$ $xy = 5$ $x^2 + y^2 = 4$
   (−4, 0), (4, 0)   above   (3.6, 1.4), (−3.6, −1.4)   no solution

Solve each system of equations algebraically. Give the irrational solutions as decimals to the nearest tenth.   9. $(2\sqrt{5}, 2), (2\sqrt{5}, -2), (-2\sqrt{5}, 2), (-2\sqrt{5}, -2)$ 11. $(\frac{\sqrt{3}}{3}, 4), (\frac{\sqrt{3}}{3}, -4), (-\frac{\sqrt{3}}{3}, 4), (-\frac{\sqrt{3}}{3}, -4)$

5. $x^2 + y^2 = 16$ 6. $x^2 + y^2 = 1$ 7. $x^2 + y^2 = 38$ 8. $x^2 + y^2 = 13$
   $x^2 - y^2 = 16$ $y = x^2 + 5$ $x^2 - y^2 = 12$ $xy = -6$
   (−4, 0), (4, 0)   no solution   above   (−3, 2), (3, −2), (−2, 3), (2, −3)

Solve each system of equations algebraically. Give the irrational solutions in simplest radical form.

Ⓑ 9. $\frac{x^2}{36} + \frac{y^2}{9} = 1$ 10. $\frac{x^2}{2} + \frac{y^2}{4} = 1$ 11. $3x^2 + 2y^2 = 33$ 12. $\frac{x^2}{9} + \frac{y^2}{4} = 1$
   $x^2 - y^2 = 16$ $8x^2 + 3y^2 = 12$ $3x^2 + y^2 = 17$ above $3x^2 - 4y^2 = 24$
   above   (0, 2), (0, −2)   14. $(\frac{4\sqrt{5}}{3}, \frac{2}{3}), (\frac{4\sqrt{5}}{3}, -\frac{2}{3}), (\frac{-4\sqrt{5}}{3}, \frac{2}{3}), (\frac{-4\sqrt{5}}{3}, -\frac{2}{3})$

13. $\frac{x^2}{16} + \frac{y^2}{4} = 1$ 14. $\frac{x^2}{10} + \frac{y^2}{4} = 1$ 15. $x^2 + y^2 = 13$ 16. $\frac{x^2}{5} + \frac{y^2}{25} = 1$
    $\frac{x^2}{4} - \frac{y^2}{3} = 1$ $2y^2 = x^2 - 8$ $\frac{x^2}{5} - \frac{y^2}{5} = 1$ below $x^2 + y^2 = 9$

$(\sqrt{7}, \frac{3}{2}), (\sqrt{7}, -\frac{3}{2}), (-\sqrt{7}, \frac{3}{2}), (-\sqrt{7}, -\frac{3}{2})$

$(2, \sqrt{5}), (2, -\sqrt{5}), (-2, \sqrt{5}), (-2, -\sqrt{5})$

Solve each system of equations graphically. Give the solutions as decimals to the nearest tenth.

Ⓒ 17. $\frac{(x-3)^2}{9} + \frac{(y+2)^2}{4} = 1$ 18. $\frac{(x+2)^2}{9} - \frac{(y-3)^2}{4} = 1$ 19. $(x-3)^2 + (y+4)^2 = 25$
    $(x+1)^2 + (y-3)^2 = 16$ $\frac{(x-1)^2}{25} + \frac{(y+2)^2}{16} = 1$ $\frac{(x+2)^2}{9} + \frac{(y-1)^2}{16} = 1$
    (1.0, −0.5)   (1.2, 2.0), (6.0, −2.3)   (1.0, 0.6), (−1.9, −3.0)

15. (3, 2), (3, −2), (−3, 2), (−3, −2)

Solving Quadratic Systems of Equations   343

APPLICATIONS

There are many applications that use a system of equations to determine a solution. The equations can be both linear, both quadratic, or one of each. The following example illustrates such an application using one linear and one quadratic equation.

Example   **Find the length of each side of both squares if the sum of the perimeters is 56 cm and the sum of the areas is 106 cm^2.**

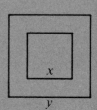

Let x = length of the side of the small square
 y = length of the side of the large square.

The sum of the perimeters is 56 cm, so
$$4x + 4y = 56.$$
The sum of the areas is 106 cm^2, so
$$x^2 + y^2 = 106.$$

Solve the system $4x + 4y = 56.$
$\qquad\qquad\qquad\;\; x^2 + y^2 = 106$

Solve $4x + 4y = 56$ for y. $y = \dfrac{56 - 4x}{4}$, or $y = 14 - x$

Substitute in $x^2 + y^2 = 106$. $\quad x^2 + (14 - x)^2 = 106$
Square. ▶ $\qquad\qquad\qquad\qquad x^2 + 196 - 28x + x^2 = 106$
$\qquad\qquad\qquad\qquad\qquad\quad 2x^2 - 28x + 90 = 0$
$\qquad\qquad\qquad\qquad\qquad\quad x^2 - 14x + 45 = 0$
Factor. ▶ $\qquad\qquad\qquad\quad\; (x - 9)(x - 5) = 0$
$\qquad\qquad\qquad\qquad\qquad\quad x = 9 \text{ or } x = 5$

Find y. ▶
$\qquad\quad 4x + 4y = 56 \qquad\qquad 4x + 4y = 56$
$\qquad\quad 4(9) + 4y = 56 \qquad\qquad 4(5) + 4y = 56$
$\qquad\qquad\quad 4y = 20 \qquad\qquad\qquad\quad 4y = 36$
$\qquad\qquad\quad\; y = 5 \qquad\qquad\qquad\quad\;\; y = 9$

Reject $y = 5$ since y represents the length of the side of the large square. Thus, it appears that the length of each side of both squares is 5 and 9, respectively.
The sum of the perimeters: $(4 \times 5) + (4 \times 9)$ is 56.
The sum of the areas is $5^2 + 9^2$ or $25 + 81$ is 106.
Thus, both measures check in the words of the problem.

Solve.

1. The perimeter of a rectangular lot is 26 cm. The area is 36 cm². Find the lengths of the sides. **4 cm, 9 cm**

2. Find the length of each side of two squares if the sum of their perimeters is 56 cm and the sum of their areas is 100 cm². **6 cm, 8 cm**

3. The area of a right triangle is 25 cm². The length of its hypotenuse is $5\sqrt{5}$ cm. Find the lengths of the legs. **5 cm, 10 cm**

4. The difference of the squares of two numbers is 84. The first number is 2 more than twice the second. What are the numbers? **10, 4**

5. Senior Girl Scout Troop 195 hiked 8 km to the top of High Peak and hiked the same distance down in a total time of 6 hours and 40 minutes. Their rate of ascent was 10 km/h slower than their rate of descent. What was their rate of ascent? **1.3 km/h**

6. Ms. Stanton received $480 interest on an investment for one year. If the interest rate had been $1\frac{1}{2}\%$ higher and the principal $1,000 less, she would have received the same amount of interest. Find the principal and the rate of interest. **$6,178.91, 7.77%**

7. A rectangular sheet of metal has 4 squares with sides 8 cm long cut from the corners of the sheet. The edges are bent up to form an open box with a volume of 12,672 cm³. The area of the sheet is 3,120 cm². Find the dimensions of the box. **8 cm, 36 cm, 44 cm**

8. Find two numbers whose product is 18 and the sum of their reciprocals is $\frac{1}{2}$. **3, 6**

9. The sum of the squares of two numbers is 80 and their product is 32. What are the numbers? **8, 4 or −8, −4**

10. If the numerator of a fraction is increased by 3 and the denominator is decreased by 3, the resulting fraction is the reciprocal of the original fraction. The numerator of the original fraction is 1 more than one half its denominator. What was the original fraction? **$\frac{5}{8}$**

11. The product of a two-digit number and the number obtained by reversing its digits is 2,268. If the difference of the numbers is 27, find the numbers. **63, 36**

Applications

SOLVING LINEAR INEQUALITY SYSTEMS   13.3

Objectives   **To draw the graph of a linear inequality in two variables**
To find the real number solutions of a system of two linear inequalities graphically

The general form of a **linear inequality** in two variables, x and y, is $ax + by > c$ or $ax + by < c$, where a and b are not both equal to 0.

The graph of the linear inequality $3x - 2y < 2$ is a region of the coordinate plane above the line determined by the equation $3x - 2y = 2$. You should use the following steps to draw the graph of $3x - 2y < 2$:

(1) Solve for y. Division by a negative number reverses the order of the inequality.

$$3x - 2y < 2$$
$$-2y < -3x + 2$$
$$y > \tfrac{3}{2}x - 1$$

(2) Draw the graph of the equation $y = \tfrac{3}{2}x - 1$. Use a dashed line with $m = \tfrac{3}{2}$ and $b = -1$.

(3) Shade the region above the line.

(4) Check. Pick a point in the shaded region and one below the line. Test $(-2, 1)$ and $(4, -2)$.

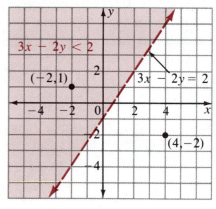

$$\begin{array}{c|c}
3x - 2y < 2 & 3x - 2y < 2 \\
\hline
3(-2) - 2(1) \;\big|\; 2 & 3(4) - 2(-2) \;\big|\; 2 \\
-6 - 2 & 12 + 4 \\
-8 & 16 \\
-8 < 2 & 16 < 2 \\
\text{True} & \text{False}
\end{array}$$

Notice that since $y > \tfrac{3}{2}x - 1$, the region *above* the line is shaded. Also notice that a dashed line is used to indicate that the line is not part of the graph. The graph would include the line if the inequality contained the symbol $\geq$. When the line is part of the graph, a solid line should be used.

Example 1

Draw the graph of $-x - 2y \geq 4$.

Solve for y. ▶ $-x - 2y \geq 4$
$-2y \geq x + 4$
$y \leq -\tfrac{1}{2}x - 2$

Graph the equation. Use a solid line. ▶ $y = -\tfrac{1}{2}x - 2$
$m = -\tfrac{1}{2},\; b = -2$

Shade the region below the line.

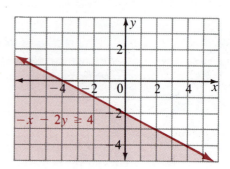

Example 2 **Draw the graph of $y \leq -2$.**

Graph the equation $y = -2$.
Use a solid line.

Shade the region below the line.

Check. Pick a point in the shaded region and one above the line.

The next two examples show you how to solve a system of two linear inequalities graphically.

Example 3 **Solve the system $x + y > -1$ graphically.**
$$3x - 2y > 4$$

Graph $x + y > -1$ or $y > -x - 1$:
(a) Graph $y = -x - 1$. Use a dashed line.
(b) Shade above the line.

Graph $3x - 2y > 4$ or $y < \frac{3}{2}x - 2$:
(a) Graph $y = \frac{3}{2}x - 2$. Use a dashed line.
(b) Shade below the line with a different color.

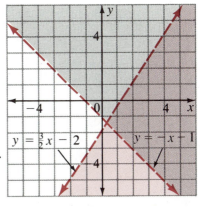

Check. Pick a point in the double-shaded region and one not in the region.

Thus, the solution set is the set of all ordered pairs for the points in the double-shaded region.

Example 4 **Solve the system $2x - y \leq -3$ graphically.**
$$x \leq -2$$

Graph $2x - y \leq -3$ or $y \geq 2x + 3$:
(a) Graph $y = 2x + 3$. Use a solid line.
(b) Shade above the line.

Graph $x \leq -2$:
(a) Graph $x = -2$. Use a solid line.
(b) Shade to the left of the line with a different color.

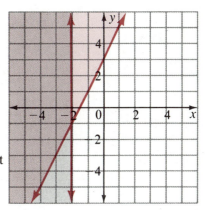

Check. Pick a point in the double-shaded region and one not in the region.

Thus, the solution set is the set of all ordered pairs for the points in the double-shaded region.

Solving Linear Inequality Systems

Reading in Algebra

Examine the inequalities on the left and the expressions describing the graphs of the inequalities on the right. Match each inequality with one or more of the descriptions.

1. $x \leq 3$ **A, F**
2. $y > -4$ **B, C**
3. $y \leq 2x + 3$ **A, D, E**
4. $y > 3x + 5$ **B, C, F**
5. $y \leq -7$ **A, D**
6. $x \geq -8$ **A, E**
7. $y < x + 8$ **B, D, E**
8. $y \geq -2x + 1$ **A, C, E**

A includes the line
B excludes the line
C above the line
D below the line
E right of the line
F left of the line

Written Exercises

For all Written Exercises, see TE Answer Section.
Draw the graph of each of the following inequalities. Check.

Ⓐ
1. $3x - 2y \leq 6$
2. $2y - 3x + 4 \geq 0$
3. $2x + 5y < 10$
4. $5x - 3y > 15$
5. $7x - 3y \geq 21$
6. $2x - 7y > 35$
7. $-y \leq 2x - 8$
8. $-5x - 2y \leq 12$
9. $3x < -y + 8$
10. $-5x > 4y + 8$
11. $-8 + x \geq 2y$
12. $y - 2x < -7$
13. $x \geq 8$
14. $y \geq 4$
15. $y < -4$
16. $x < 5$

Solve each system of inequalities graphically. Check.

Ⓑ
17. $y \leq -2x + 3$
    $y > 4x - 1$
18. $y \leq 3x - 1$
    $y > -2x + 4$
19. $2x + 3y \geq 6$
    $2y - x < -4$
20. $5x + 2y \geq 12$
    $3x + 4y \leq 8$

21. $2x + 3y \geq 9$
    $-3x + 4y \geq -4$
22. $5x + 6y < 30$
    $4x + 3y \geq 9$
23. $-4x + 5y \leq 15$
    $4y \geq 5x + 12$
24. $-9 \leq 2x + 3y$
    $3y + 4x \leq 6$

25. $-2x - 3y \geq 6$
    $-2y + x > -4$
26. $x - 2y > 6$
    $x + 2y \leq 4$
27. $2x - 3y \geq 9$
    $-2y + x > 6$
28. $-3x - y \geq 6$
    $y - 2x > 1$

29. $2x + y > 3$
    $y \geq x$
30. $y \leq 2$
    $x \leq 4$
31. $y \geq -3$
    $x \leq -2$
32. $y \geq 5$
    $x \geq -3$

Solve each system of three inequalities graphically. Check.

Ⓒ
33. $x + y \leq 3$
    $4x - 5y \geq 10$
    $-2x + y > 2$
34. $2x - y \geq -4$
    $x + 3y < 6$
    $x + 2y \geq 4$
35. $-x + 3y < 5$
    $3x - 2y \geq 6$
    $-4x + y < 3$
36. $4x - 2y \leq 2$
    $-4x + y \geq 1$
    $2x - y \leq -3$

348   Chapter Thirteen

SOLVING LINEAR-QUADRATIC INEQUALITY SYSTEMS   13.4

Objectives
**To draw the graph of a quadratic inequality
To find the real number solutions of a system of one linear and one quadratic inequality graphically
To find the real number solutions of a system of two quadratic inequalities graphically**

The graph of the **quadratic inequality** $x^2 + y^2 < 16$ is a region of the coordinate plane in the interior of the circle determined by the equation $x^2 + y^2 = 16$. You should use the following steps to draw the graph of $x^2 + y^2 < 16$:

(1) Draw the graph of the equation $x^2 + y^2 = 16$. The graph is a circle. Use a dashed curve.

(2) Shade the region in the interior of the circle.

(3) Check. Pick a point in the shaded region, one not in the region, and one on the circle.
Test $(2, 3)$, $(4, -3)$, and $(4, 0)$.

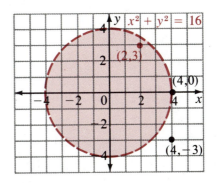

Notice that since $x^2 + y^2 < 16$, the region *in the interior of* the circle is shaded. The dashed curve is used to show that the circle is not part of the graph of the inequality. The graph would include the circle if the inequality contained the symbol $\leq$. The graph of $x^2 + y^2 > 16$ is the region *in the exterior of* the circle.

Example 1 **Draw the graph of $xy \geq 4$.**

Make a table of x and y values for $xy = 4$.

Graph the equation $xy = 4$. Use a solid curve.

Shade the region in the interior of each branch of the hyperbola.

Check. Pick a point in the shaded region, one not in the region, and one on the hyperbola.

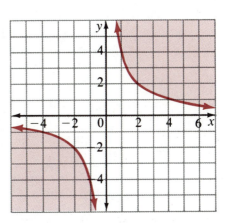

Solving Linear-Quadratic Inequality Systems   **349**

Systems containing linear or quadratic inequalities can be solved graphically.

Example 2

Solve the system $x^2 + y^2 \leq 9$ graphically.
$$y \geq 3x - 2$$

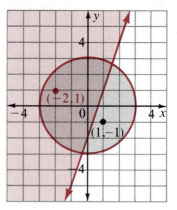

Graph $x^2 + y^2 \leq 9$:
(a) Graph $x^2 + y^2 = 9$. The graph is a circle. Use a solid curve.
(b) Shade the region in the interior of the circle.

Graph $y \geq 3x - 2$:
(a) Graph $y = 3x - 2$. Use a solid line.
(b) Shade the region above the line with a different color.

Check. Pick a point in the double-shaded region and one not in the region. Test $(-2, 1)$ and $(1, -1)$ in both inequalities.

$x^2 + y^2 \leq 9$		$y \geq 3x - 2$		$x^2 + y^2 \leq 9$		$y \geq 3x - 2$	
$(-2)^2 + 1^2$	9	1	$3(-2) - 2$	$1^2 + (-1)^2$	9	-1	$3 \cdot 1 - 2$
5			-8	2			1
$5 \leq 9$		$1 \geq -8$		$2 \leq 9$		$-1 \geq 1$	
True		True		True		False	

Thus, the solution set is the set of all ordered pairs for the points in the double-shaded region, including the points of the boundary of the region.

Example 3

Solve the system $x^2 + y^2 \geq 4$ graphically.
$$\frac{x^2}{9} + \frac{y^2}{4} \leq 1$$

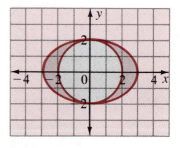

Graph $x^2 + y^2 \geq 4$:
(a) Graph $x^2 + y^2 = 4$. The graph is a circle. Use a solid curve.
(b) Shade the region in the exterior of the circle.

Graph $\frac{x^2}{9} + \frac{y^2}{4} \leq 1$:

(a) Graph $\frac{x^2}{9} + \frac{y^2}{4} = 1$. The graph is an ellipse. Use a solid curve.
(b) Shade the region in the interior of the ellipse with a different color.

Check. Pick a point in the double-shaded region and one not in the region.

Thus, the solution set is the set of all ordered pairs for the points in the double-shaded region, including the points of the boundary of the region.

Oral Exercises

State the inequality that describes the shaded region.

1.
2.
3.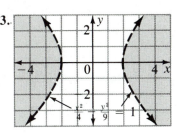

Written Exercises

1. $\dfrac{x^2}{9} + \dfrac{y^2}{4} \le 1$ 2. $y \ge 3x - 2$ 3. $\dfrac{x^2}{4} - \dfrac{y^2}{9} > 1$

For solutions to all Written Exercises, see TE Answer Section.

Draw the graph of each of the following inequalities. Check.

Ⓐ 1. $x^2 + y^2 \le 9$ 2. $x^2 + y^2 \ge 4$ 3. $x^2 + y^2 < 25$ 4. $x^2 + y^2 > 16$
5. $xy \ge 6$ 6. $xy < 8$ 7. $xy > 3$ 8. $xy \le -2$
9. $\dfrac{x^2}{25} + \dfrac{y^2}{4} \le 1$ 10. $\dfrac{x^2}{36} + \dfrac{y^2}{16} \ge 1$ 11. $\dfrac{x^2}{100} + \dfrac{y^2}{9} < 1$ 12. $\dfrac{x^2}{16} + \dfrac{y^2}{1} > 1$
13. $\dfrac{x^2}{9} - \dfrac{y^2}{4} \le 1$ 14. $\dfrac{x^2}{25} - \dfrac{y^2}{8} \ge 1$ 15. $\dfrac{x^2}{4} - \dfrac{y^2}{9} < 1$ 16. $\dfrac{x^2}{25} - \dfrac{y^2}{81} > 1$

Ⓑ 17. $2x^2 + 3y^2 \ge 18$ 18. $3x^2 + 9y^2 \le 27$ 19. $9x^2 - 4y^2 < 36$ 20. $16x^2 - 8y^2 \ge 64$
21. $27x^2 + 9y^2 \ge 81$ 22. $25x^2 - 4y^2 > 100$ 23. $36x^2 - 18y^2 \le 72$ 24. $25x^2 + 4y^2 < 100$

Solve each system of inequalities graphically. Check.

25. $x^2 + y^2 \le 16$
    $2x - 3y \le 6$
26. $x^2 + y^2 > 25$
    $-3x + 4y > 12$
27. $xy \ge 1$
    $4x + 5y \ge 15$
28. $xy < -5$
    $-2x - 3y \le 6$

29. $\dfrac{x^2}{16} + \dfrac{y^2}{4} \ge 1$
    $-2x - 3y \le 2$
30. $\dfrac{x^2}{4} - \dfrac{y^2}{2} \le 2$
    $3x - y > 2$
31. $\dfrac{x^2}{25} + \dfrac{y^2}{81} \le 1$
    $2x + 4y \le -3$
32. $\dfrac{x^2}{9} - \dfrac{y^2}{16} \ge 3$
    $-x - 4y \ge -5$

Ⓒ 33. $x^2 + y^2 \ge 25$
    $xy < -5$
34. $x^2 - y^2 \le 12$
    $x^2 + y^2 \le 36$
35. $25x^2 + 4y^2 \ge 100$
    $x^2 + y^2 \le 36$
36. $3x^2 - 27y^2 \le 81$
    $4x^2 + 25y^2 \ge 100$

37. $4x^2 + 16y^2 \ge 64$
    $3x^2 - 9y^2 \le 27$
38. $-3x^2 + 4y^2 \le 36$
    $x^2 + 5y^2 \ge 25$
39. $4x^2 + 6y^2 \ge 24$
    $-x^2 - y^2 \ge -9$
40. $2x^2 + 4y^2 \le 12$
    $3x^2 - 5y^2 \ge 15$

CUMULATIVE REVIEW

1. Denine has nickels, dimes, and quarters in her coin collection. The number of quarters is 4 less than twice the number of dimes. The number of dimes is 3 times the number of nickels. If the value of her collection is $8.25, how many nickels, dimes, and quarters does she have?
   5 nickels, 15 dimes, 26 quarters

2. Some dry peanuts worth $2.20/kg are mixed with sunflower seeds worth $3.80/kg to make 30 kg of a mixture worth $3.00/kg. How many kilograms of peanuts and how many kilograms of sunflower seeds are there in the mixture?
   peanuts: 15 kg, sunflower seeds: 15 kg

Solving Linear-Quadratic Inequality Systems

CHAPTER THIRTEEN REVIEW

Ⓐ Exercises are 1, 3–5, 10, 11, 13–19, 21. Ⓒ Exercises are starred. The rest are Ⓑ Exercises.

Vocabulary
linear inequality   [13.3]
quadratic inequality   [13.4]
system of equations
 linear-quadratic   [13.1]
 quadratic   [13.2]
system of inequalities
 linear-quadratic   [13.4]
 quadratic   [13.4]

Solve each system of equations graphically. Estimate the decimal solutions to the nearest tenth.

1. $\dfrac{x^2}{25} + \dfrac{y^2}{4} = 1$
   $y = 2x + 1$
   (0.5, 2.0), (−1.5, −1.9)

2. $x^2 - y^2 = 9$ [13.1]
   $x - 3y = 6$
   (3.1, −1.0), (−4.6, −3.5)

3. $4x^2 + 9y^2 = 100$
   $2x - 9y = -10$
   (4, 2), (−5, 0)   [13.1]

4. $\dfrac{x^2}{9} + \dfrac{y^2}{4} = 1$ [13.2]
   $xy = 2$
   (2.8, 0.7), (−2.8, −0.7)
   (1.1, 1.8), (−1.1, −1.8)

5. $\dfrac{x^2}{8} - \dfrac{y^2}{4} = 1$ [13.2]
   $x^2 + 4y^2 = 10$
   (2.9, 0.6), (2.9, −0.6), (−2.9, 0.6), (−2.9, −0.6)

★ 6. $(x - 1)^2 + (y - 3)^2 = 9$ [13.2]
   $\dfrac{(x - 2)^2}{4} + \dfrac{(y + 2)^2}{9} = 1$ **(0.5, 0), (2.9, 0.7)**

Solve each system of equations algebraically. Give the irrational solutions in simplest radical form.

7. $y = x^2 + 7x - 3$
   $x + 2y = -3$

8. $x^2 - y^2 = -5$ [13.1]
   $2x + y = -1$
   (−2, 3), ($\tfrac{2}{3}$, −$\tfrac{7}{3}$)

★ 9. $x^2 + 2x + y^2 - 4y = 8$ [13.1]
   $x - y = 2$ **(2, 0), (1, −1)**

10. $y = 2x^2 - x - 5$ [13.2]
    $y = x^2 + 1$ **(3, 10), (−2, 5)**

7. $\left(\dfrac{-15 - \sqrt{249}}{4}, \dfrac{3 + \sqrt{249}}{8}\right)$, $\left(\dfrac{-15 + \sqrt{249}}{4}, \dfrac{3 - \sqrt{249}}{8}\right)$

11. $x^2 + y^2 = 10$
    $xy = 3$
    (3, 1), (−3, −1),
    (1, 3), (−1, −3)

12. $\dfrac{x^2}{16} + \dfrac{y^2}{16} = 1$ [13.2]
    $\dfrac{x^2}{16} - \dfrac{y^2}{9} = 1$
    (4, 0), (−4, 0)

Draw the graph of each of the following inequalites. Check.   [13.3]

13. $y \geq 2x - 3$

14. $y < -3x + 5$

15. $3x - 2y \leq 8$

16. $y > -3$

17. $x^2 + y^2 \leq 4$

18. $xy > 4$ [13.4]

19. $xy \leq -5$

20. $3x^2 + 6y^2 \geq 18$

21. $\dfrac{x^2}{36} - \dfrac{y^2}{45} \leq 1$

22. $9x^2 + 4y^2 > 36$

Solve each system of inequalities graphically. Check.

23. $y \geq -x + 3$
    $y < 3x - 4$

24. $5x - 2y < 4$ [13.3]
    $x \geq y$

25. $y - 3x < 5$
    $2x + 4y \geq 15$

26. $3x + 3y > 7$
    $5x - 2y \leq 10$

★ 27. $2x - y \leq 2$
    $x + y \geq 4$
    $-2x + 3y \leq 6$ [13.3]

28. $x^2 + y^2 \leq 25$
    $-x + 2y > 6$ [13.4]

29. $xy < 6$
    $y \geq x$

30. $\dfrac{x^2}{36} + \dfrac{y^2}{4} \geq 1$ [13.4]
    $-3x + 2y \leq 8$

31. $4x^2 - 25y^2 < 100$
    $x + 2y \leq 4$

★ 32. $4x^2 + 5y^2 \geq 60$
    $2x^2 - 3y^2 \leq 18$

Check student graphs for Ex. 13–32. See TE Answer Section.

CHAPTER THIRTEEN TEST

(A) Exercises are 1–6, 9–14. (C) Exercises are starred. The rest are (B) Exercises.

For complete solutions, see TE Answer Section.

Solve each system of equations graphically. Estimate the decimal solutions to the nearest tenth.

1. $\dfrac{x^2}{4} - \dfrac{y^2}{9} = 1$
   $-2x + 3y = 6$
   (3.5, 4.3), (−2.0, 0.6)

2. $x^2 + y^2 = 25$
   $-2y = 8 - 4x$
   (3.7, 3.4), (−0.5, −5.0)

3. $x^2 + y^2 = 16$
   $xy = 6$
   (3.6, 1.7), (1.7, 3.6), (−3.6, −1.7), (−1.7, −3.6)

4. $\dfrac{x^2}{25} + \dfrac{y^2}{36} = 1$
   $x^2 - y^2 = 9$
   below

Solve each system of equations algebraically. Give the irrational solutions in simplest radical form.

5. $x^2 - y^2 = 36$
   $y = 6 - x$ **(6, 0)**

6. $y = x^2 - 2x + 5$
   $4x - y = 4$ **(3, 8)**

7. $x^2 + y^2 = 16$
   $\dfrac{x^2}{16} + \dfrac{y^2}{4} = 1$
   (4, 0), (−4, 0)

8. $x^2 + 2y^2 = 24$
   $x^2 - y^2 = 3$
   ($\sqrt{10}$, $\sqrt{7}$), ($\sqrt{10}$, $-\sqrt{7}$), ($-\sqrt{10}$, $\sqrt{7}$), ($-\sqrt{10}$, $-\sqrt{7}$)

Draw the graph of each of the following inequalities. Check.

9. $y \leq 3x + 5$

10. $2x - 3y \geq 9$

11. $x > -1$

12. $x^2 - y^2 \geq 6$

13. $xy > 2$

14. $\dfrac{x^2}{36} + \dfrac{y^2}{4} \leq 1$

Check student graphs for Ex. 9–14.
4. (−4.3, −3.1), (4.3, 3.1), (4.3, −3.1), (−4.3, 3.1)

Solve each system of inequalities graphically. Check.

15. $y < 2x - 1$
    $y \geq -3x + 2$

16. $3x - 4y \leq -8$
    $y - x \leq 2$

17. $2x - 5y \geq 10$
    $3x + 7y < 14$

18. $x^2 - y^2 \geq 4$
    $2x - 3y \leq 6$

19. $\dfrac{x^2}{16} + \dfrac{y^2}{4} \leq 1$
    $4x + 5y \geq 10$

20. $xy \leq 4$
    $-y \leq x$

Check student graphs for Ex. 15–20.

★ 21. Solve the system $(x - 2)^2 + (y + 1)^2 = 16$
       $y = x + 1$
       algebraically. **(2, 3)(−2, −1)**

Solve each system of inequalities graphically. Check.

★ 22. $3x + y \geq 4$
       $x - 2y \leq 6$
       $-2x + 3y < 9$

★ 23. $5x^2 + 25y^2 \leq 125$
       $3x^2 - 9y^2 \geq 27$

Check student graphs for Ex. 22–23.

Chapter Thirteen Test

COMPUTER ACTIVITIES

Equation of a Circle

OBJECTIVE: To determine whether a given point is on, inside, or outside a circle whose equation is given

The program below generates different equations for circles by randomly choosing a value for r^2 (lines 110–130). The program then asks you to enter any point in the coordinate plane (lines 150–180). Then the computer uses the distance formula to calculate the distance from the center, whereupon it describes the location of the point with respect to the circle (lines 190–220).

The empty loop in line 225 works as a delaying device that allows the program output to remain displayed on the screen for a few moments before the HOME statement in line 230 erases the screen. The loop makes the computer count to a certain number before continuing with the rest of the program. To see the difference line 225 makes, run the program without it before doing so with it.

```
110  R =  INT ( RND (1) * 50) + 1
120   PRINT "THE EQUATION OF A CIRCLE IS"
130   PRINT " X^2 + Y^2 = ";R
140   PRINT
150   PRINT "INPUT THE COORDINATES OF A POINT AND I"
160   PRINT "WILL TELL YOU ITS LOCATION WITH"
170   PRINT "RESPECT TO THE CIRCLE."
180   INPUT A,B
190   PRINT "THE POINT (";A;",";B") IS ";
200   IF A ^ 2 + B ^ 2 = R THEN  PRINT "ON THE CIRCLE."
210   IF A ^ 2 + B ^ 2 > R THEN  PRINT "OUTSIDE THE
      CIRCLE."
220   IF A ^ 2 + B ^ 2 < R THEN  PRINT "INSIDE THE
      CIRCLE."
225   FOR P = 1 TO 2500: NEXT P
230   HOME : INPUT "MORE COORDINATES? (ENTER 1 FOR
      YES, 2 FOR NO.)";K
240   IF K = 1 THEN 110
250   END
```

See the Computer Section beginning on page 574 for more information.

EXERCISES
For Ex. 1–2, see TE Answer Section.
Note: Use Control C to stop printing.

1. Alter the program above so that if the point is outside the circle, the computer will tell how far outside.

2. Rewrite the program above so that it generates two concentric circles and describes whether a point lies on either circle, within the smaller, outside the larger, or outside the smaller but inside the larger.

COLLEGE PREP TEST

DIRECTIONS: Choose the *one* best answer to each question or problem.

1. Find the area of the region described by $\frac{x^2}{2} + \frac{y^2}{2} \leq 8$. **C**

   (A) 4π (B) 8π (C) 16π
   (D) 32π (E) None of these

2. The graph of $2(x + y) < y - 3$ does not pass through **A**

   (A) quadrant I (B) quadrant II
   (C) quadrant III (D) quadrant IV
   (E) None of these

3. Find the area of the region determined by the graph of the solutions of the system
   $x^2 + y^2 \geq 4$
   $x^2 + y^2 \leq 25$. **B**

   (A) 4π (B) 21π (C) 25π
   (D) 29π (E) None of these

4. Find a solution of the system
   $3x + 2y < 12$
   $4x + y > 11$. **E**

   (A) (2, 3) (B) (3, 2) (C) (2, 2)
   (D) (3, 3) (E) None of these

5. Find the area of the region determined by the graph of the solutions of the system
   $y \leq x$
   $x \leq 6$
   $y \geq 0$. **B**

   (A) 12 (B) 18 (C) 24
   (D) 30 (E) 36

6. Two distinct parallel lines may intersect another two distinct parallel lines in the same plane in exactly **D**

   (A) one point (B) two points
   (C) three points (D) four points
   (E) None of these

7. The system $x^2 + y^2 \geq a^2$
   $x^2 + y^2 \leq b^2$
   has no solution if **A**

   (A) $a^2 > b^2$ (B) $a^2 < b^2$ (C) $a^2 = b^2$
   (D) All of these (E) None of these

8. Which ordered pair (x, y) is a solution of the system
   $ax + by = 7ab$
   $3ax - 2by = 6ab$? **B**

   (A) $(3b, 4a)$ (B) $(4b, 3a)$
   (C) $(2b, 0)$ (D) All of these
   (E) None of these

9. Find the area of the triangle determined by the lines described by
   $x = -4$
   $y = 0$
   $x + 3y = 8$. **B**

   (A) 12 (B) 24 (C) 32
   (D) 48 (E) None of these

10. Circle I has a radius of 4 with center at $P(-3, 4)$. Circle II has a radius of 2 with center at $Q(2, 4)$. Circles I and II intersect in exactly **B**

    (A) one point (B) two points
    (C) infinitely many points
    (D) no points (E) None of these

College Prep Test

14 PROGRESSIONS, SERIES, AND BINOMIAL EXPANSIONS

HISTORY

Blaise Pascal 1623–1662

Blaise Pascal was born at Clermont-Ferrand in Auvergne on June 19, 1623. Blaise showed phenomenal ability in mathematics at an early age.

Even though his father, Etienne, was an able and noted mathematician, he wanted his son to study languages rather than mathematics. Therefore, he took away from Blaise all books on mathematics. Blaise secretly studied geometry on his own, and he succeeded considerably before his endeavors were realized by his father.

```
            1
          1   1
        1   2   1
      1   3   3   1
    1   4   6   4   1
  1   5  10  10   5   1
1   6  15  20  15   6   1
     PASCAL'S TRIANGLE
```

Blaise Pascal is credited with many mathematical discoveries. At the age of 19, he invented a computing machine that served as a starting point in the development of the modern-day calculator. Blaise wrote extensively on the triangular arrangement of the coefficients of the powers of a binomial, $(a + b)^n$. This arrangement is known today as Pascal's Triangle.

Project

Write a paper on one of Pascal's major works.

ARITHMETIC PROGRESSIONS 14.1

Objectives **To write several consecutive terms of an arithmetic progression**
To find a specified term of a given arithmetic progression

$$12, 17, 22, 27, \ldots \qquad 2, -4, -10, -16, \ldots \qquad -2, -0.5, 1, 2.5$$

An ordered list of numbers formed according to some pattern is called a **progression** or a **sequence**. Three progressions are shown above. The numbers in each progression are called the **terms** of the progression.

In the first progression above, notice that the **common difference** between consecutive terms is 5.

$$17 - 12 = 5 \qquad 22 - 17 = 5 \qquad 27 - 22 = 5$$

A progression formed in this way is called an **arithmetic progression** (abbreviated as A.P.). Note: When "arithmetic" is used as an adjective as in "arithmetic progression," it is pronounced as air-ith-MEH-tik.

> **Definition:**
> **Arithmetic progression**
> An *arithmetic progression* is a progression in which the difference found by subtracting any term from the next term is constant.

In general, if $x_1, x_2, x_3, x_4, \ldots$ is an arithmetic progression (A.P.) with common difference d, then $d = x_2 - x_1 = x_3 - x_2 = x_4 - x_3 = \ldots$.

Example 1 **For each progression that is an A.P., find the common difference d. Give a reason for each answer.**

Progression	Answer	Reason
5, 7, 10, 14, 19, ...	Not an A.P.	$7 - 5 \neq 10 - 7$
$-6, -1, 4, 9, \ldots$	$d = 5$	$-1 - (-6) = 5; 4 - (-1) = 5;$ and so on
$2, -4, -10, -16, \ldots$	$d = -6$	$-4 - 2 = -6; -10 - (-4) = -6;$ and so on
6, 12, 24, 48, ...	Not an A.P.	$12 - 6 \neq 24 - 12$
$-2, -0.5, 1, 2.5, \ldots$	$d = 1.5$	$-0.5 - (-2) = 1.5; 1 - (-0.5) = 1.5;$ and so on
$3, 3 + 2\sqrt{5}, 3 + 4\sqrt{5}, \ldots$	$d = 2\sqrt{5}$	$(3 + 2\sqrt{5}) - 3 = 2\sqrt{5}; (3 + 4\sqrt{5}) - (3 + 2\sqrt{5}) = 2\sqrt{5};$ and so on

You can write several consecutive terms of an arithmetic progression by using the following procedure. First find the common difference d. Then add d to any term to get the next term.

Example 2 **Write the next three terms of the A.P.: 16, 4, -8, -20, ...**

$d = 4 - 16$, or -12. Add -12 to obtain consecutive terms.

Thus, the next three terms are -32, -44, and -56.

Arithmetic Progressions

Example 3 Write the first four terms of the A.P. whose first term a is 5 and common difference d is -3.

$$a = 5 \qquad 5 + (-3) = 2 \qquad 2 + (-3) = -1 \qquad -1 + (-3) = -4$$

Thus, the first four terms are 5, 2, −1, and −4.

The arithmetic progression 12, 17, 22, 27, ... with common difference 5 can be written in a way that will help you find a specified term, such as the 81st term, of the progression.

12	17	22	27	...	412
↓	↓	↓	↓		↑
$12 + 0 \cdot 5$	$12 + 1 \cdot 5$	$12 + 2 \cdot 5$	$12 + 3 \cdot 5$	...	$12 + 80 \cdot 5$
1st	2nd	3rd	4th	...	81st

Notice that the 3rd term is $12 + (3 - 1) \cdot 5$ and that the 4th term is $12 + (4 - 1) \cdot 5$. Thus, the 81st term is $12 + (81 - 1) \cdot 5$, or $12 + 80 \cdot 5$. This leads to the following formula for the nth term of an arithmetic progression.

---

nth term of an arithmetic progression

The nth term of an arithmetic progression is given by the formula
$$l = a + (n - 1)d, \qquad n \geq 1,$$
where l is the nth term, a is the first term, and d is the common difference.

You can use this formula to find a specified term of an arithmetic progression (A.P.).

Example 4 Find the specified term of each A.P.

26th term of 8, 5.4, 2.8, 0.2, ...

$d = 5.4 - 8 = -2.6$
$a = 8$
$n = 26$
$l = a + (n - 1)d$
$ = 8 + (26 - 1)(-2.6)$
$ = 8 + 25(-2.6)$
$ = 8 - 65$
$ = -57$

Thus, the 26th term is −57.

31st term of $3 - \sqrt{2}$, 1, $-1 + \sqrt{2}$, ...

$d = 1 - (3 - \sqrt{2}) = -2 + \sqrt{2}$
$a = 3 - \sqrt{2}$
$n = 31$
$l = a + (n - 1)d$
$ = (3 - \sqrt{2}) + (31 - 1)(-2 + \sqrt{2})$
$ = (3 - \sqrt{2}) + 30(-2 + \sqrt{2})$
$ = 3 - \sqrt{2} - 60 + 30\sqrt{2}$
$ = -57 + 29\sqrt{2}$

Thus, the 31st term is $-57 + 29\sqrt{2}$.

Reading in Algebra

Match each arithmetic progression at the left with exactly one statement at the right.

1. 7, 5, 3, 1, ... **B**
2. 4, 11, 18, 25, ... **A**
3. 5, 6, 7, 8, ... **E**
4. −1, 1, 3, 5, ... **C**
5. 2, 4, 6, 8, ... **D**

A 7 is the common difference of the A.P.
B 7 is the first term of the A.P.
C 7 is the next term of the A.P.
D 7 is the counting number corresponding to the term 14 of the A.P.
E 7 is the third term of the A.P.

Oral Exercises

For each progression that is an A.P., find the common difference d. 6. Not an A.P.

1. 23, 35, 47, 59, ... **12**
2. 36, 23, 10, −3, ... **−13**
3. −10, −8, −6, −4, ... **2**
4. 2, 4, 8, 16, ... **Not an A.P.**
5. 4, 6.6, 9.2, 11.8, ... **2.6**
6. 9.7, 7.4, 6.7, 5, ...
7. 16, $15\frac{1}{3}$, $14\frac{2}{3}$, 14, ... **$-\frac{2}{3}$**
8. $7\frac{1}{8}$, $8\frac{3}{4}$, $10\frac{3}{8}$, 12, ... **$1\frac{5}{8}$**
9. −2, 4, −6, 8, ... **Not an A.P.**

Written Exercises

4. 18, 25, 32, 39   5. 9, 3, −3, −9   6. −7, −10, −13, −16
19. Not an A.P.   20. Not an A.P.   21. $5\sqrt{2}, 6\sqrt{2}, 7\sqrt{2}$

Write the next three terms of each arithmetic progression.

(A)
1. 22, 23.2, 24.4, 25.6, ...
   26.8, 28, 29.2
2. 7, $6\frac{3}{4}$, $6\frac{1}{2}$, $6\frac{1}{4}$, ... **6, $5\frac{3}{4}$, $5\frac{1}{2}$**
3. $\frac{1}{4}$, $\frac{1}{3}$, $\frac{5}{12}$, $\frac{1}{2}$, ... **$\frac{7}{12}$, $\frac{2}{3}$, $\frac{3}{4}$**

Write the first four terms of each A.P. whose first term and common difference are given.

4. $a = 18$ and $d = 7$
5. $a = 9$ and $d = -6$
6. $a = -7$ and $d = -3$
7. $a = \$5.00$ and $d = \$2.40$
8. $a = -6°C$ and $d = 4°C$
9. $a = 15$ kg and $d = -3.4$ kg

$5.00, $7.40, $9.80, $12.20   **−6°C, −2°C, 2°C, 6°C**   **15 kg, 11.6 kg, 8.2 kg, 4.8 kg**

Find the specified term of each arithmetic progression.

10. 21st term of 11, 15, 19, 23, ... **91**
11. 36th term of 10, 4, −2, −8, ... **−200**
12. 31st term of 8.6, 8.3, 8, 7.7, ... **−0.4**
13. 46th term of −9, −7.6, −6.2, −4.8, ... **54**
14. 81st term of 7, $7\frac{1}{2}$, 8, $8\frac{1}{2}$, ... **47**
15. 26th term of $12\frac{4}{5}$, $11\frac{2}{5}$, 10, $8\frac{3}{5}$, ... **$-22\frac{1}{5}$**

16. $3 + 3\sqrt{2}, 3 + 4\sqrt{2}, 3 + 5\sqrt{2}$ 17. $-1 + 3\sqrt{3}, -3 + 4\sqrt{3}, -5 + 5\sqrt{3}$ 18. $7 - 2i, 9 - 3i, 11 - 4i$

For each progression that is an A.P., write the next three terms.

(B)
16. $3, 3 + \sqrt{2}, 3 + 2\sqrt{2}, \ldots$
17. $5, 3 + \sqrt{3}, 1 + 2\sqrt{3} \ldots$
18. $1 + i, 3, 5 - i \ldots$
19. $-2, 4 - \sqrt{2}, 8 - \sqrt{2}, \ldots$
20. $\sqrt{15}, \sqrt{17}, \sqrt{19}, \sqrt{21}, \ldots$
21. $\sqrt{2}, \sqrt{8}, \sqrt{18}, \sqrt{32}, \ldots$
22. $5x + 2, 7x, 9x - 2, \ldots$
23. $8x - y, 5x + 2y, 2x + 5y, \ldots$
24. $x^2 + 1, x^3 + 1, x^4 + 1, \ldots$

11x − 4, 13x − 6, 15x − 8   **−x + 8y, −4x + 11y, −7x + 14y**   **Not an A.P.**

Write the first four terms of each A.P. whose first term and common difference are given.

25. $a = -3$ and $d = 3 + \sqrt{6}$
26. $a = 4 + \sqrt{5}$ and $d = -2 + 3\sqrt{5}$
   $4 + \sqrt{5}, 2 + 4\sqrt{5}, 7\sqrt{5}, -2 + 10\sqrt{5}$
27. $a = x^2 - y$ and $d = x^2 + y$

25. **−3, $\sqrt{6}$, $3 + 2\sqrt{6}$, $6 + 3\sqrt{6}$**   27. **$x^2 - y, 2x^2, 3x^2 + y, 4x^2 + 2y$**

Find the 21st term of each arithmetic progression in Exercises 28–31.

28. $4, 4 + 2\sqrt{3}, 4 + 4\sqrt{3}, \ldots$ **$4 + 40\sqrt{3}$**
29. $5 - \sqrt{2}, 6 - 2\sqrt{2}, 7 - 3\sqrt{2}, \ldots$ **$25 - 21\sqrt{2}$**
30. $4 + 3i, 5, 6 - 3i, \ldots$ **$24 - 57i$**
31. $3x + 2y + 1, 2x + 3y, x + 4y - 1, \ldots$
   $-17x + 22y - 19$

Arithmetic Progressions

32. Some boxes are stacked so that there are 3 in the top row, 5 in the second row, 7 in the third row, and so on. How many boxes are in the fifteenth row? **31**

33. A mechanic's starting salary is $12,500. He is guaranteed a minimum annual increase of $600. What minimum salary may he expect in his twelfth year? **$19,100**

34. Mrs. Prince deposited $200 in a bank on January 15. Then, each month she deposited $35 more than she had deposited the preceding month. What was her deposit on December 15? **$585**

35. A parachutist in free fall travels 5 m in the first second, 15 m during the second, 25 m during the third second. How far would the chutist travel in free fall during the eighth second? **75 m**

Example If 76 is the nth term of $-4, -2, 0, 2, \ldots$, find the value of n.

$$-4, -2, 0, 2, \ldots, 76, \ldots \quad \leftarrow n\text{th term}$$

$$l = a + (n-1)d$$
$$76 = -4 + (n-1) \cdot 2 \quad \blacktriangleleft \; d = -2 - (-4) = 2$$
$$80 = (n-1) \cdot 2$$
$$40 = n - 1 \quad n = 41$$

Thus, 76 is the 41st term of the A.P.

Find the value of n in each exercise.

36. 62 is the nth term of $-4, -1, 2, 5, \ldots$. **23**
37. 129 is the nth term of $5, 9, 13, 17, \ldots$. **32**
38. -54 is the nth term of $11, 6, 1, -4, \ldots$. **14**
39. -106 is the nth term of $-26, -30, -34, \ldots$. **21**
40. 7 is the nth term of $1, 1.2, 1.4, 1.6, \ldots$. **31**
41. 8 is the nth term of $-2, -1.6, -1.2, -0.8, \ldots$. **26**

Ⓒ 42. Find the value of y so that $y + 3, 4y + 1, 8y - 3$ is an A.P. [HINT: $d = x_2 - x_1 = x_3 - x_2$ if x_1, x_2, x_3 is an A.P.] **y = 2**

43. Find two values of c such that $\frac{1}{c}, 1, \frac{10}{c+3}$ is an A.P. **c = 3 or c = $-\frac{1}{2}$**

44. Find the first term of the A.P. whose 16th term is 110 and common difference is 7. **5**

45. Find the 1st term of the A.P. whose 41st term is -124 and common difference is -3. **-4**

46. Find the first three terms of the A.P. whose 9th term is 26 and 20th term is 59. **2, 5, 8**

47. Find the first three terms of the A.P. whose 48th term is -189 and 26th term is -79. **46, 41, 36**

48. Prove that the progression formed by every odd-numbered term of an A.P. is also an A.P.
49. Prove: The progression formed by reversing the order in a finite A.P. is also an A.P.

For Ex. 48–49, see TE Answer Section.

CALCULATOR ACTIVITIES

Find the 93rd term of the A.P. where $a = 62.34$ and $d = -1.25$.

$$l = a + (n-1)d = 62.34 + (93-1)(-1.25) = \ominus 1.25 \otimes 92 \oplus 62.34 \ominus -52.66$$

Find the nth term of each A.P. for the given values of n, a, and d.

1. $n = 119$, $a = -73$, $d = 6.75$ **723.5**
2. $n = 43$, $a = 72.58$, $d = -2.5$ **-32.42**
3. $n = 2{,}000$, $a = -8.6$, $d = -5.4$ **$-10{,}803.2$**

THE ARITHMETIC MEAN 14.2

Objectives: To find arithmetic means between two given terms of an arithmetic progression
To find the arithmetic mean of two numbers

The terms between two given terms of an arithmetic progression are called **arithmetic means** between the given terms. For example, three arithmetic means between 2 and 18 in the progression below are 6, 10, and 14 since 2, 6, 10, 14, 18, ... is an arithmetic progression.

$$2, \underline{6}, \underline{10}, \underline{14}, 18, \ldots$$

As shown in the examples below, you can find any specified number of arithmetic means between two given numbers.

Example 1

Find the four arithmetic means between 24 and 20.

$$24, \underline{}, \underline{}, \underline{}, \underline{}, 20$$

When the four arithmetic means are found, an A.P. with six terms will be formed.
$n = 6 \qquad a = 24 \qquad l = 20$

Use $l = a + (n - 1)d$ to find d.

$20 = 24 + (6 - 1)d$ ◀ *Substitute for l, a, and n.*
$-4 = 5d$
$-0.8 = d$

Add -0.8 to each term.

$$24, \underline{23.2}, \underline{22.4}, \underline{21.6}, \underline{20.8}, 20$$

Thus, the four arithmetic means are 23.2, 22.4, 21.6, and 20.8.

Example 2

Find the one arithmetic mean between 5 and 17.

$$5, \underline{}, 17$$

$l = a + (n - 1)d$
$17 = 5 + (3 - 1)d$ ◀ *Substitute 17 for l, 5 for a, and 3 for n.*
$12 = 2d$
$6 = d$

$$5, \underline{11}, 17$$

Thus, the one arithmetic mean is 11.

In Example 2, notice that the one arithmetic mean is the "average" of 5 and 17, since $(5 + 17) \div 2 = 11$. In this case, 11 is called **the arithmetic mean** of 5 and 17.

> **Arithmetic mean**
>
> The *arithmetic mean* of the real numbers x and y is $\dfrac{x+y}{2}$.

In Exercise 32, you will prove that the statement above is true.

Example 3 **The arithmetic mean of two numbers is 12. If the greater number is decreased by 5 times the smaller number, the result is 6. Find the two numbers.**

Represent the data. Let $x =$ the smaller number and $y =$ the greater number.

Write a system of equations.
(1) $\dfrac{x+y}{2} = 12 \rightarrow x + y = 24$ (2) $y - 5x = 6 \rightarrow y = 5x + 6$

Solve the system.

$x + (5x + 6) = 24$ $y - 5x = 6$
$6x = 18$ $y - 5 \cdot 3 = 6$
$x = 3$ $y = 21$

Check.

Arithmetic mean	12
$\dfrac{3 + 21}{2}$	12
$\dfrac{24}{2}$, or 12	

$y - 5x$	6
$21 - 5 \cdot 3$	6
$21 - 15$	
6	

Answer the question. **Thus,** the two numbers are 3 and 21.

Oral Exercises

For each progression that is an A.P., state the arithmetic means between the first term and the last term.

1. 10, 7, 4 **7**
2. 1, 2, 4, 8 **not an A.P.**
3. 4, 1, −2, −5 **1, −2**
4. −4, −2, 2, 4 **Not an A. P.**
5. −7, −4, −1, 2, 5 **−4, −1, 2**
6. 3, 5, 7, 9, 11, 13 **5, 7, 9, 11**

Written Exercises

Find the indicated number of arithmetic means between each pair of numbers.

1. Two, between 12 and 33 **19, 26**
2. Three, between 30 and 2 **23, 16, 9**
3. Four, between −2 and 18 **2, 6, 10, 14**
4. Five, between 22 and −8 **17, 12, 7, 2, −3**
5. Three, between 6 and 8 **$6\frac{1}{2}, 7, 7\frac{1}{2}$**
6. Four, between 15 and 7 **$13\frac{2}{5}, 11\frac{4}{5}, 10\frac{1}{5}, 8\frac{3}{5}$**
7. Two, between 3.4 and 10.3 **5.7, 8**
8. Five, between 10.6 and −8.6 **7.4, 4.2, 1, −2.2, −5.4**

Find the arithmetic mean of each pair of numbers.

9. 26 and 72 **49**
10. 40 and −12 **14**
11. 2.1 and 3.5 **2.8**
12. 12.8 and 4.2 **8.5**

Solve each problem.

13. The arithmetic mean of two numbers is 15. Two less than the larger number is 3 times the smaller number. Find the two numbers. **7, 23**

14. The arithmetic mean of two numbers is 25. One of the numbers increased by twice the other number is 79. Find the two numbers. **21, 29**

Ⓑ 15. A teacher's annual salary for ten consecutive years was in arithmetic progression from $12,200 to $19,400. Find the salary for each of the ten years. **$12,200, $13,000, $13,800, $14,600, $15,400, $16,200, $17,000, $17,800, $18,600, $19,400**

16. Eight different weights are used in a science laboratory to balance small masses. The weights are in arithmetic progression. Find each weight if the heaviest is 22 g and the lightest is 1 g. **1 g, 4 g, 7 g, 10 g, 13 g, 16 g, 19 g, 22 g**

17. If the arithmetic mean of two numbers is increased by one-fourth of the greater number, the sum is four times the smaller number. Find the two numbers if five times the smaller number is three more than the greater number. **42, 9**

18. The reciprocal of the arithmetic mean of two numbers is $\frac{4}{21}$ and the difference of the two numbers is the reciprocal of $\frac{2}{5}$. Find the two numbers.

18. $4, 6\frac{1}{2}$ **19.** $7\sqrt{7}, 11\sqrt{7}$ **20.** $7 + \sqrt{5}, 10 + 2\sqrt{5}, 13 + 3\sqrt{5}$ **21.** $-2 - 2i, -4 + i, -6 + 4i, -8 + 7i$

Find the indicated number of arithmetic means between each pair of terms.

19. Two, between $3\sqrt{7}$ and $15\sqrt{7}$
20. Three, between 4 and $16 + 4\sqrt{5}$
21. Four, between $-5i$ and $-10 + 10i$
22. Two, between $1 + \sqrt{3}$ and $7 - 5\sqrt{3}$
23. Three, between $-8a + 9b$ and $4a - 3b$
24. Four, between $10x$ and $15y$.

22. $3 - \sqrt{3}, 5 - 3\sqrt{3}$ **23.** $-5a + 6b, -2a + 3b, a$ **24.** $8x + 3y, 6x + 6y, 4x + 9y, 2x + 12y$

For Exercises 25–30, find the arithmetic mean of each pair of terms.

25. $-6x$ and $14x$ **$4x$**
26. $15x - 3y + 2z$ and $9x + 15y - 18z$ **$12x + 6y - 8z$**
27. $3 + \sqrt{2}$ and $7 - 7\sqrt{2}$ **$5 - 3\sqrt{2}$**
28. $9\sqrt{21} - 5\sqrt{10}$ and $9\sqrt{21} + 5\sqrt{10}$ **$9\sqrt{21}$**

Ⓒ 29. The complex number $a + bi$ and its conjugate **a**

30. $(x + y)^2$ and $(x - y)^2$ **$x^2 + y^2$**

31. Find the two arithmetic means between x and y. Write each mean as a rational expression. **$\frac{2x + y}{3}, \frac{x + 2y}{3}$**

32. Prove: The arithmetic mean of x and y is $\frac{x + y}{2}$. [Hint: Use the formula $l = a + (n - 1)d$.]

33. Prove: The arithmetic mean of any two consecutive odd integers is an even integer. [Hint: If x is an integer, then $2x - 1$ is an odd integer and $2x$ is an even integer.] **For Ex. 32–33, see TE Answer Section.**

CUMULATIVE REVIEW

1. Solve the formula $T = \frac{n}{2}(x + y)$ for y. Then find the value of y if $T = 750$, $n = 25$, and $x = 20$. **$y = \frac{2T - nx}{n}$; 40**

2. Solve the formula $2T + nd = 2an + n^2d$ for d. Then find the value of d if $T = 210$, $n = 10$, and $a = 3$. **$d = \frac{2(T - an)}{n(n - 1)}$; 4**

The Arithmetic Mean

ARITHMETIC SERIES AND Σ-NOTATION 14.3

Objectives
To find the sum of the first n terms of an arithmetic series
To write an arithmetic series using Σ-notation

If you take the arithmetic progression 10, 6, 2, −2, −6, ... and write it as the indicated sum $10 + 6 + 2 - 2 - 6 - \ldots$, the result is an **arithmetic series**.

Definition: Arithmetic series	An *arithmetic series* is an indicated sum of the terms of an arithmetic progression.

Example 1 Write each A.P. as an arithmetic series. Find the common difference d of the series.

$$-7, 2, 11, 20, \ldots \qquad\qquad 8, 3, -2, -7, \ldots$$

The arithmetic series is
$-7 + 2 + 11 + 20 + \ldots$
$d = 2 - (-7)$, or 9

The arithmetic series is
$8 + 3 - 2 - 7 - \ldots$
$d = 3 - 8$, or -5

You can find the sum S of the first 20 terms of the arithmetic series $3 + 6 + 9 + 12 + \ldots$ as follows: Write the series in its given order and also in reverse order. Then add the two series.

$$\begin{aligned} S &= 3 + 6 + 9 + \ldots + 54 + 57 + 60 \\ S &= 60 + 57 + 54 + \ldots + 9 + 6 + 3 \\ \hline 2S &= 63 + 63 + 63 + \ldots + 63 + 63 + 63 \end{aligned}$$

$$2S = 20 \cdot 63$$
$$S = 630$$

The steps above suggest a method for obtaining a formula for the sum S of the first n terms of an arithmetic series. Let l represent the nth term.

$$\begin{aligned} S &= a + (a + d) + (a + 2d) + \ldots + (l - 2d) + (l - d) + l \\ S &= l + (l - d) + (l - 2d) + \ldots + (a + 2d) + (a + d) + a \\ \hline 2S &= (a + l) + (a + l) + (a + l) + \ldots + (a + l) + (a + l) + (a + l) \end{aligned}$$

n binomials
$$2S = n(a + l)$$
$$S = \frac{n}{2}(a + l)$$

Sum of n terms of arithmetic series	The sum S of the first n terms of an arithmetic series is given by the formula $$S = \frac{n}{2}(a + l),$$ where a is the first term and l is the nth term.

Example 2 Find the sum of the arithmetic series in which the number of terms $n = 35$, the first term $a = 12$, and the last term $l = 114$.

$$S = \frac{n}{2}(a + l) = \frac{35}{2}(12 + 114)$$
$$= \frac{35}{2} \cdot 126$$
$$= 35 \cdot 63, \text{ or } 2{,}205$$

Thus, the sum of the arithmetic series is 2,205.

You can find the sum of the first 34 terms of the arithmetic series $24.5 + 21.5 + 18.5 + 15.5 + \ldots$ without finding the 34th term, or l. To do this, another sum formula is used. It is obtained by substitution, as shown below:

$$S = \frac{n}{2}[a + l] \text{ and } l = a + (n - 1)d$$
$$S = \frac{n}{2}[a + \underbrace{a + (n - 1)d}_{l}], \text{ or } S = \frac{n}{2}[2a + (n - 1)d]$$

Use this second sum formula when l, the last term, is not known but a, the first term, and d, the common difference, are easily determined for a given arithmetic series.

Example 3 Find the sum of the first 34 terms of the arithmetic series:
$24.5 + 21.5 + 18.5 + 15.5 + \ldots$.

$$n = 34 \qquad a = 24.5 \qquad d = -3$$
$$S = \frac{n}{2}[2a + (n - 1)d] = \frac{34}{2}[2(24.5) + 33(-3)]$$
$$= 17[49 - 99], \text{ or } -850$$

Thus, the sum of the first 34 terms is -850.

The indicated sum, $4 + 8 + 12 + 16 + 20$, can be written in a more compact form as $\sum_{k=1}^{5} 4k$. Recall that the Greek letter Σ (read: sigma) is used as a **summation sign** and k is called the **index of summation**.

$$\sum_{k=1}^{5} 4k = \underbrace{4 \cdot 1 + 4 \cdot 2 + 4 \cdot 3 + 4 \cdot 4 + 4 \cdot 5}_{\text{Expanded Form}} = \underbrace{4 + 8 + 12 + 16 + 20}_{\text{Series Form}} = \underbrace{60}_{\text{Sum}}$$

$\sum_{k=1}^{5} 4k$ is read "the summation from 1 to 5 of $4k$." Notice that the five terms of the series are found by substituting 1, 2, 3, 4, and 5 for k in the general term, $4k$.

Arithmetic Series and Σ-Notation

Example 4 Write the expanded form and the series form of $\sum_{j=3}^{6} (5j - 22)$.
Find the sum of the series.

Substitute 3, 4, 5, and 6 for j in $(5j - 22)$.

$\sum_{j=3}^{6} (5j - 22) = (5 \cdot 3 - 22) + (5 \cdot 4 - 22) + (5 \cdot 5 - 22) + (5 \cdot 6 - 22)$ ◀ *Expanded form*

$= -7 - 2 + 3 + 8$ ◀ *Series form*
$= 2$ ◀ *Sum of the series*

Example 5 Write the expanded form and the series form of $\sum_{c=1}^{35} (90 - 2c)$.
Find the sum of the series.

$\sum_{c=1}^{35} (90 - 2c) = (90 - 2 \cdot 1) + (90 - 2 \cdot 2) + (90 - 2 \cdot 3) + \ldots + (90 - 2 \cdot 35)$ ◀ *Expanded form*
$= 88 + 86 + 84 + \ldots + 20$ ◀ *Series form*
$= \frac{35}{2}(88 + 20)$ ◀ *Use $S = \frac{n}{2}(a + l)$ to find the sum.*
$= 1{,}890$ ◀ *Sum of the series*

You can reverse the procedure above and write certain arithmetic series using Σ-notation. For example,

$-2 - 4 - 6 - 8 - \ldots - 40 = -2 \cdot 1 - 2 \cdot 2 - 2 \cdot 3 - 2 \cdot 4 - \ldots - 2 \cdot 20 = \sum_{n=1}^{20} -2n$

Notice that the common difference d is -2 and that n goes from 1 to 20.

Example 6 Write $10 + 13 + 16 + 19 + \ldots + 52$ using Σ-notation.

The series is arithmetic and the common difference is 3.
Write each term in the form $3k + t$ where

$k = 1, 2, 3, 4, \ldots$ and t is a real number.

Notice that the first term is 10 and that $10 = 3 \cdot 1 + 7$. So, $t = 7$.

$\begin{array}{cccccccccc} 10 & + & 13 & + & 16 & + & 19 & + \ldots + & 52 \\ \downarrow & & \downarrow & & \downarrow & & \downarrow & & \downarrow \\ (3 \cdot 1 + 7) & + & (3 \cdot 2 + 7) & + & (3 \cdot 3 + 7) & + & (3 \cdot 4 + 7) & + \ldots + & (3 \cdot 15 + 7) \end{array}$

$\sum_{k=1}^{15} (3k + 7)$

Oral Exercises

State each A.P. as an arithmetic series.

1. 5, 9, 13, 17, … ←
   5 + 9 + 13 + 17 + …

2. 5, 1, −3, −7, … ←
   5 + 1 − 3 − 7 − …

3. −2, −3, −4, −5, … ←
   −2 − 3 − 4 − 5 − …

4. $5 + 10 + 15 + \ldots + 50$ 5. $-8 - 9 - 10 - 11 - 12$ 6. $3 + 4 + 5 + 6 + 7 + 8$

Read each summation aloud. State the series form.   7. $12 + 11 + 10$ 8. $\frac{1}{5} + \frac{2}{5} + \frac{3}{5} + \frac{4}{5}$

4. $\sum_{k=1}^{10} 5k$ 5. $\sum_{j=8}^{12} -j$ 6. $\sum_{c=1}^{6} (c + 2)$ 7. $\sum_{n=5}^{7} (17 - n)$ 8. $\sum_{p=1}^{4} \frac{p}{5}$

Find the common difference d of each arithmetic series.

9. $12 + 15 + 18 + 21 + \ldots$ $d = 3$ 10. $3 + 1 - 1 - 3 - \ldots$ $d = -2$ 11. $1 + 1.4 + 1.8 + 2.2 + \ldots$ $d = 0.4$

12. $\sum_{j=1}^{4} 5j$ $d = 5$ 13. $\sum_{k=1}^{3} (k + 2)$ $d = 1$ 14. $\sum_{c=1}^{3} -2c$ $d = -2$

21. $\sum_{k=4}^{8} 2k$ 22. $\sum_{k=1}^{9} (4k + 11)$ 23. $\sum_{k=1}^{12} (14 - 5k)$ 24. $\sum_{j=1}^{13} (-4 - 3j)$

Written Exercises

Find the sum of each arithmetic series for the given data.

(A)
1. $n = 20, a = 1, l = 154$ **1,550**
2. $n = 40, a = 2, l = -115$ **−2,260**
3. $n = 25, a = 16, l = 184$ **2,500**
4. $n = 45, a = 3, l = -50$ **−1,057$\frac{1}{2}$**
5. $n = 35, a = -5.2, l = 60$ **959**
6. $n = 30, a = -7, l = -60.2$ **−1,008**
7. $n = 30; 5 + 15 + 25 + 35 + \ldots$ **4,500**
8. $n = 40; 6 + 2 - 2 - 6 - \ldots$ **−2,880**
9. $n = 65; 7 + 7.2 + 7.4 + 7.6 + \ldots$ **871**
10. $n = 45; -2 - 0.9 + 0.2 + 1.3 + \ldots$ **999**

11. $\sum_{k=1}^{6} 3k$ **63**   12. $\sum_{j=1}^{5} -j$ **−15**   13. $\sum_{c=1}^{50} (c + 1)$ **1,325**

14. $\sum_{j=5}^{8} (3j - 4)$ **62**   15. $\sum_{m=40}^{43} (8 - m)$ **−134**   16. $\sum_{k=1}^{25} (15 - 3k)$ **−600**

Write each arithmetic series using Σ-notation.

17. $6 + 12 + 18 + 24$ 17. $\sum_{k=1}^{4} 6k$ 18. $-8 - 16 - 24 - 32 - \ldots - 80$ 18. $\sum_{j=1}^{10} -8j$

19. $7 + 8 + 9 + 10 + \ldots + 26$ 20. $12 + 11 + 10 + 9 + \ldots - 5$

(B)
21. $8 + 10 + 12 + 14 + 16$ 19. $\sum_{k=7}^{26} k$ 22. $15 + 19 + 23 + 27 + \ldots + 47$ 20. $\sum_{j=1}^{18} (13 - j)$

23. $9 + 4 - 1 - 6 - \ldots - 46$ 24. $-7 - 10 - 13 - 16 - \ldots - 43$

Find the sum of each arithmetic series for the given data. 26. $8(-44x + 20y) = -352x + 160y$

25. $n = 41, a = 2\sqrt{3}, l = 198\sqrt{3}$ **4,100$\sqrt{3}$**   26. $n = 16, a = 3x - 10y, l = -47x + 30y$

27. $n = 26; 0.3 + 0.8 + 1.3 + 1.8 + \ldots$ **170.3**   28. $n = 37; -1.3 - 0.1 + 1.1 + 2.3 + \ldots$ **751.1**

29. $\sum_{n=1}^{100} \frac{n}{4}$ **1,262.5**   30. $\sum_{n=1}^{30} \frac{n + 4}{3}$ **195**

31. Cartons are stacked in 20 rows with 2 in the top row, 5 in the 2nd row, 8 in the 3rd row, and so on. How many cartons are stacked? **610 cartons**

32. Gerry saved 1 dime the 1st day, 2 dimes the 2nd day, 3 on the 3rd day, and so on. How much money did she save in 30 days? **$46.50**

(C)
33. Prove: $\sum_{k=1}^{50} 10kx = 10x \cdot \sum_{k=1}^{50} k$.   34. Prove: $\sum_{j=1}^{20} (jx + jy) = \sum_{j=1}^{20} jx + \sum_{j=1}^{20} jy$.

35. Prove: $\sum_{c=1}^{30} (c + t) = 30t + \sum_{c=1}^{30} c$.   36. Solve $\sum_{k=4}^{7} (kx - 3) = 32$ for x.

37. Prove: $\sum_{n=1}^{k} (2n - 1) = k^2$. What does this say about adding odd integers?   38. Prove: $\sum_{n=1}^{k} 2n = k(k + 1)$. What does this say about adding even integers?

For Ex. 33–38, see TE Answer Section.

Arithmetic Series and Σ-Notation

GEOMETRIC PROGRESSIONS 14.4

Objectives **To write several consecutive terms of a geometric progression**
To find a specified term of a given geometric progression

Another type of progression (or sequence) is one such as

$$5, 10, 20, 40, \ldots$$

in which each consecutive term is found by multiplying the preceding term by a given number. Notice that in the progression $5, 10, 20, 40, \ldots$, the **common ratio** of consecutive terms is 2.

$$\frac{10}{5} = 2 \qquad \frac{20}{10} = 2 \qquad \frac{40}{20} = 2$$

A progression formed in this way is called a **geometric progression** (abbreviated as G. P.). Note: Pronounce "geometric" as gee-oh-MEH-trik.

Definition: Geometric progression	A *geometric progression* is a progression in which the ratio found by dividing any term by the preceding term is constant.

In general, if $x_1, x_2, x_3, x_4, \ldots$ is a geometric progression (G. P.) with common ratio $r \neq 0$, then $r = \frac{x_2}{x_1} = \frac{x_3}{x_2} = \frac{x_4}{x_3} = \ldots$.

Example 1 **For each progression that is a G.P., find the common ratio r. Give a reason for each answer.**

Progression	Answer	Reason
$3, 6, 9, 12, \ldots$	Not a G. P.	$6 \div 3 \neq 9 \div 6$
$-2, -8, -32, -128, \ldots$	$r = 4$	$\frac{-8}{-2} = 4; \frac{-32}{-8} = 4;$ and so on
$64, -16, 4, -1, \ldots$	$r = -\frac{1}{4}$	$\frac{-16}{64} = -\frac{1}{4}; \frac{4}{-16} = -\frac{1}{4};$ and so on
$6, 0, 0, 0, \ldots$	Not a G. P.	$0 \div 6 \neq 0 \div 0$ ($0 \div 0$ is undefined)
$7, 0.7, 0.07, 0.007, \ldots$	$r = 0.1$	$0.7 \div 7 = 0.1; 0.07 \div 0.7 = 0.1;$ and so on
$5, 5\sqrt{2}, 10, 10\sqrt{2}, \ldots$	$r = \sqrt{2}$	$\frac{5\sqrt{2}}{5} = \sqrt{2}; \frac{10}{5\sqrt{2}} = \sqrt{2};$ and so on

You can write several consecutive terms of a G.P. as shown below.

Example 2 **Write the next three terms of the G.P.: $128, -64, 32, -16, \ldots$.**

$r = \frac{-64}{128}$, or $-\frac{1}{2}$. Multiply by $-\frac{1}{2}$ to obtain consecutive terms.

Thus, the next three terms are 8, -4, and 2.

Example 3   **Write the first four terms of the G.P. whose first term *a* is -2 and common ratio *r* is -5.**

$$a = -2 \qquad -2(-5) = 10 \qquad 10(-5) = -50 \qquad -50(-5) = 250$$

Thus, the first four terms are -2, 10, -50, and 250.

The geometric progression 5, 10, 20, 40, 80, ... with common ratio 2 can be written in a way that will help you find a specified term, such as the 11th term, of the progression.

5	10	20	40	80	...	5,120
↓	↓	↓	↓	↓		↑
5	$5 \cdot 2^1$	$5 \cdot 2^2$	$5 \cdot 2^3$	$5 \cdot 2^4$	...	$5 \cdot 2^{10}$
1st	2nd	3rd	4th	5th	...	11th

Notice that the 4th term is $5 \cdot 2^{4-1}$, or $5 \cdot 2^3$, and that the 5th term is $5 \cdot 2^{5-1}$, or $5 \cdot 2^4$. Thus, the 11th term is $5 \cdot 2^{11-1}$, or $5 \cdot 2^{10}$. This leads to the following formula for the *n*th term of a geometric progression.

---

nth term of a geometric progression

The *n*th term of a geometric progression is given by the formula

$$l = a \cdot r^{n-1}, \ n \geq 1,$$

where *l* is the *n*th term, *a* is the first term, and *r* is the common ratio.

Example 4   **Find the specified term of each geometric progression, given the first term *a* and the common ratio *r*.**

5th term: $a = 12$ and $r = 0.1$

$$\begin{aligned} l = a \cdot r^{n-1} &= 12(0.1)^{5-1} \\ &= 12(0.1)^4 \\ &= 12(0.0001) \\ &= 0.0012 \end{aligned}$$

Thus, the 5th term is 0.0012.

7th term: $a = -8$ and $r = -\dfrac{1}{2}$

$$l = a \cdot r^{n-1} = -8\left(-\dfrac{1}{2}\right)^{7-1}$$
$$= -8 \cdot \dfrac{1}{64} = -\dfrac{1}{8}$$

Thus, the 7th term is $-\dfrac{1}{8}$.

Example 5   **Find the 10th term of the G.P.: $\dfrac{1}{2}$, -1, 2, -4, ...**

$$r = -1 \div \dfrac{1}{2}, \text{ or } -2 \qquad a = \dfrac{1}{2} \qquad n = 10 \qquad l \text{ is the 10th term.}$$

$$l = a \cdot r^{n-1} = \dfrac{1}{2}(-2)^9 = \dfrac{1}{2}(-512) = -256$$

Thus, the 10th term is -256.

Geometric Progressions

Reading in Algebra

Match each geometric progression at the left with exactly one statement at the right.

1. 0.003, 0.03, 0.3, ...   **C**
2. 5, 15, 45, ...   **A**
3. 3, 1, $\frac{1}{3}$, ...   **B**
4. 48, 24, 12, ...   **E**
5. $x, x^3, x^5, ...$ **D**

A 3 is the common ratio of the G. P.
B 3 is the first term of the G. P.
C 3 is the next term of the G. P.
D 3 is the counting number for the term, x^5.
E 3 is the 5th term of the G. P.

Oral Exercises

For each progression that is a G. P., find the common ratio r.

1. 2, 8, 32, 128, ...   **4**
2. 2, −6, 18, −54, ...   **−3**
3. 16, 8, 4, 2, ...   **$\frac{1}{2}$**
4. 5, 10, 15, 20, ...   **Not a G. P.**
5. −40, −20, −10, −5, ...   **$\frac{1}{2}$**
6. 3, −3, 3, −3, ...   **−1**
7. −9, 3, −1, $\frac{1}{3}$, ...   **−$\frac{1}{3}$**
8. 4, 0.4, 0.04, 0.004, ...   **0.1**
9. $2\sqrt{3}, 6, 6\sqrt{3}, 18, ...$ **$\sqrt{3}$**
10. $\frac{1}{8}, \frac{1}{4}, \frac{1}{2}, 1, ...$ **2**
11. $3, 2, \frac{4}{3}, \frac{8}{9}, ...$ **$\frac{2}{3}$**
12. $5x, 10x^2, 20x^3, 40x^4, ...$ **$2x$**

Written Exercises

Write the next three terms of each geometric progression.

Ⓐ 1. 30, 60, 120, ...   **240, 480, 960**
2. −0.1, −0.2, −0.4, ...   **−0.8, −1.6, −3.2**
3. −$\frac{1}{3}$, 1, −3, ...   **9, −27, 81**

Write the first four terms of each G. P. whose first term and common ratio are given.

4. $a = 20$ and $r = 2$ **20, 40, 80, 160**
5. $a = -\frac{1}{4}$ and $r = 4$
6. $a = 0.25$ and $r = 4$
7. $a = 32$ and $r = \frac{1}{2}$ **32, 16, 8, 4**
8. $a = -8$ and $r = \frac{1}{4}$
9. $a = 6$ and $r = 0.1$
10. $a = \frac{1}{8}$ and $r = -2$
11. $a = 64$ and $r = -\frac{3}{8}$
12. $a = -2$ and $r = -3$

5. $-\frac{1}{4}, -1, -4, -16$ 6. 0.25, 1, 4, 16   8. $-8, -2, -\frac{1}{2}, -\frac{1}{8}$ 9. 6, 0.6, 0.06, 0.006   10. $\frac{1}{8}, -\frac{1}{4}, \frac{1}{2}, -1$

Find the specified term of each geometric progression.   11. $64, -24, 9, -3\frac{3}{8}$ 12. $-2, 6, -18, 54$

13. 5th term: $a = -10, r = 3$
14. 6th term: $a = 20, r = -5$
15. 7th term: $a = 34, r = 10$
16. 7th term: $a = 900, r = 0.1$
17. 6th term: $a = -30, r = 0.2$
18. 5th term: $a = 72, r = -0.1$
19. 10th term: 4, 8, 16, ...
20. 9th term: 10, −20, 40, ...
21. 10th term: $-\frac{1}{4}, -\frac{1}{2}, -1, ...$
22. 11th term: −8, −4, −2, ...
23. 12th term: −32, 16, −8, ...
24. 11th term: 96, 48, 24, ...
25. 8th term: 370, 37, 3.7, ...
26. 9th term: 0.24, 2.4, 24, ...
27. 7th term: 1, 0.2, 0.04, ...

13. **−810** 14. **−62,500** 15. **34,000,000** 16. **0.0009** 17. **−0.0096** 18. **0.0072** 19. **2048** 20. **2560**

Ⓑ 28. 6th term: $a = -9, r = \frac{1}{3}$ **$-\frac{1}{27}$**
29. 4th term: $a = 80, r = -\frac{1}{4}$ **$-\frac{5}{4}$**
30. 5th term: $a = 25, r = \frac{2}{5}$ **$\frac{16}{25}$**
31. 8th term: $a = 6x, r = 2x$ **$768x^8$**
32. 7th term: $a = 4y, r = 2y^2$
33. 8th term: $a = 8xy, r = x^2y^3$
34. 7th term: $a = 10, r = \sqrt{3}$ **270**
35. 6th term: $a = -4, r = \sqrt{5}$
36. 5th term: $a = \sqrt{2}, r = 3\sqrt{2}$

21. **−128** 22. **$-\frac{1}{128}$** 23. **$\frac{1}{64}$** 24. **$\frac{3}{32}$** 25. **0.000037** 26. **24,000,000** 27. **0.000064** 32. **$256y^{13}$** 33. **$8x^{15}y^{22}$**

Write the next three terms of each geometric progression in Exercises 37–42.   35. **$-100\sqrt{5}$**   36. **$324\sqrt{2}$**

37. $2x^2, 8x^5, 32x^8, ...$
38. $1, i, -1, -i, ...$
39. $3 + 2\sqrt{5}, 6 + 4\sqrt{5}, 12 + 8\sqrt{5}, ...$
40. $2 + \sqrt{6}, 6 + 2\sqrt{6}, 12 + 6\sqrt{6}, ...$
41. $-3x^3, 6x^5, -12x^7, ...$
42. $3i, -6, -12i, 24, ...$

37. **$128x^{11}, 512x^{14}, 2048x^{17}$**   38. **$1, i, -1$**   39. **$24 + 16\sqrt{5}, 48 + 32\sqrt{5}, 96 + 64\sqrt{5}$**
40. **$36 + 12\sqrt{6}, 72 + 36\sqrt{6}, 216 + 72\sqrt{6}$**   41. **$24x^9, -48x^{11}, 96x^{13}$**   42. **$48i, -96, -192i$**

43. A tennis ball, dropped from a height of 128 dm, rebounds on each bounce one-half the distance from which it fell. How high does it go on its 9th rebound? **$\frac{1}{4}$ dm, or 0.25 dm**

44. A tank contains 4,000 L of water. Each day, one-half of the water will be removed. How much water will be in the tank at the end of the eighth day? **$15\frac{5}{8}$ L, or 15.625 L**

45. A golf ball, dropped from a height of 81 m, rebounds on each bounce two-thirds of the distance from which it fell. How far does it fall on its 6th descent? **$10\frac{2}{3}$ m**

46. A tank contains 243 L of gasoline. Each time that a valve is opened, one-third of the gasoline is released. How much gas will be in the tank after the valve is operated 6 times? **$21\frac{1}{3}$ L**

47. Find the first term of the geometric progression whose 7th term is 512 and 8th term is 1,024. [Hint: Use the formula $l = ar^{n-1}$.] **8**

$$a, \ldots, 256, 512, 1024, \ldots$$
$$1st \qquad\qquad 7th\ 8th$$

48. Find the first term of the G.P. whose 9th term is $\frac{1}{64}$ and 10th term is $\frac{1}{128}$. **4**

49. What is the first term of a G.P., if its 8th and 9th terms are 6,250 and −31,250, respectively? **$-\frac{2}{25}$**

Ⓒ 50. Find two values of t so that $\frac{1}{3}$, $t - 1$, $4t$ is a geometric sequence.
 $\left[\text{Hint:}\ r = \dfrac{x_2}{x_1} = \dfrac{x_3}{x_2}\ \text{if}\ x_1, x_2, x_3\ \text{is a geometric sequence.}\right]$ **$t = 3$ or $t = \frac{1}{3}$**

51. Find two values of t so that $6t$, $t + 1$, $\frac{3}{4}$ is a G.P. **$t = 2$ or $t = \frac{1}{2}$**

52. Which term of 3, −6, 12, ... is −384? **8th term**

53. Which term of $\frac{1}{8}, \frac{1}{4}, \frac{1}{2}, \ldots$ is 128? **11th term**

CALCULATOR ACTIVITIES

Find the 7th term of the G.P. where $a = 12.5$ and $r = 3$.

$$l = a \cdot r^{n-1} = 12.5 \times 3^6$$
$l = 3\ \otimes\ \ominus\ \ominus\ \ominus\ \ominus\ \ominus\ \otimes\ 12.5\ \ominus\ 9112.5$ ◀ The 7th term

Find the nth term of each G.P. for the given values of n, a, and r.
1. $n = 9, a = -7.2, r = 2$ 2. $n = 6, a = 640, r = 4$ 3. $n = 8, a = 0.2, r = 0.5$
 −1843.2   **655,360**   **0.0015625**

Check student's calculator. Certain calculators take the equal sign.

CUMULATIVE REVIEW

Solve each equation for n.

1. $2^n = 64$ **$n = 6$** 2. $3^n = \dfrac{1}{81}$ **$n = -4$** 3. $5^{3n+1} = 625$ **$n = 1$** 4. $8^n = 128$ **$n = \frac{7}{3}$** 5. $9^n = \dfrac{1}{27}$ **$n = -\frac{3}{2}$**

Geometric Progressions

THE GEOMETRIC MEAN   14.5

Objectives **To find geometric means between two given terms of a geometric progression**
To find the geometric mean (mean proportional) of two numbers

The terms between two given terms of a geometric progression are called **geometric means** between the given terms. For example, four geometric means between 3 and 96 in the progression below are 6, 12, 24, and 48, since 3, 6, 12, 24, 48, 96, ... is a geometric progression.

$$3, \underline{6}, \underline{12}, \underline{24}, \underline{48}, 96, \ldots$$

Example 1 Find three geometric means between -2 and -162.

$$-2, \underline{}, \underline{}, \underline{}, -162$$

When the three geometric means are found, a G.P. with five terms will be formed.
$n = 5 \qquad a = -2 \qquad l = -162$

Use $\quad l = a \cdot r^{n-1}$ to find r.
$\quad -162 = -2 \cdot r^{5-1}$ ◀ *Substitute for l, a, and n.*
$\quad\quad 81 = r^4$
$r = \sqrt[4]{81} \quad$ or $\quad r = -\sqrt[4]{81}$
$r = 3 \quad\quad\quad\quad\quad r = -3 \quad$ ◀ *Two real numbers for r*

Multiply each term by either 3 or -3.
$-2, \underline{-6}, \underline{-18}, \underline{-54}, -162 \quad$ or $\quad -2, \underline{6}, \underline{-18}, \underline{54}, -162$

Thus, the means are $-6, -18, -54 \quad$ or $\quad 6, -18, 54$.

Example 2 Find two real geometric means between 9 and $-\frac{8}{3}$.

$$9, \underline{}, \underline{}, -\frac{8}{3}$$

$l = a \cdot r^{n-1}$
$-\frac{8}{3} = 9 \cdot r^3 \quad$ ◀ *Substitute $-\frac{8}{3}$ for l, 9 for a, and 4 for n.*
$-\frac{8}{27} = r^3$
$r = \sqrt[3]{-\frac{8}{27}} = -\frac{2}{3}$

Multiply each term by $-\frac{2}{3}$: $\quad 9, \underline{-6}, \underline{4}, -\frac{8}{3}$.

Thus, the two real geometric means are -6 and 4.

A single geometric mean between two numbers is called **the geometric mean** or **mean proportional** of the two numbers.

Example 3 **Find the geometric mean (mean proportional) of 5 and 10.**

$$5, \underline{}, 10$$

$l = ar^{n-1}$
$10 = 5 \cdot r^2$ ◀ *Substitute 10 for l, 5 for a, and 3 for n.*
$2 = r^2$
$r = \sqrt{2}$ or $r = -\sqrt{2}$
Multiply by either $\sqrt{2}$ or $-\sqrt{2}$.

$$5, \underline{5\sqrt{2}}, 10 \quad \text{or} \quad 5, \underline{-5\sqrt{2}}, 10$$

Thus, the geometric mean (mean proportional) is $5\sqrt{2}$ or $-5\sqrt{2}$.

In Example 3, the *positive* geometric mean of 5 and 10 is $\sqrt{5 \cdot 10}$, or $5\sqrt{2}$, and the *negative* geometric mean of 5 and 10 is $-\sqrt{5 \cdot 10}$, or $-5\sqrt{2}$. This suggests the following statement about the geometric mean of two nonzero real numbers x and y. You will prove this statement in Exercise 45.

Geometric mean The geometric mean (mean proportional) of the real numbers x and y ($xy > 0$) is $\sqrt{xy}$ or $-\sqrt{xy}$.

Some word problems that involve the geometric mean of two numbers can be solved by using a system of two equations as shown below.

Example 4 **One number is 30 more than another number and the positive geometric mean of the two numbers is 8. Find the two numbers.**

Let x and y represent the two numbers. Write a system of equations.
(1) $y = x + 30$ (2) $\sqrt{xy} = 8$
Solve the system by substitution.

$\sqrt{x(x + 30)} = 8$ ◀ *Substitute for y in equation (2).*
$(\sqrt{x^2 + 30x})^2 = 8^2$
$x^2 + 30x = 64$
$x^2 + 30x - 64 = 0$
$(x - 2)(x + 32) = 0$
$x = 2$ or $x = -32$

If $x = 2$, then $y = x + 30 = 32$. If $x = -32$, then $y = x + 30 = -2$. Check each pair of numbers in both equations, (1) and (2).

Thus, the numbers are 2 and 32 or -32 and -2.

When you are asked to find "n real geometric means" between a pair of given numbers, there are two sets of means if n is odd, as in Examples 1 and 3. If n is even, there is only one set of means, as in Example 2.

The Geometric Mean

Written Exercises

10. −32, −16, −8 or 32, −16, 8   12. 10, 50, 250 or −10, 50, −250

Find the indicated number of real geometric means between each pair of numbers.

A
1. Three, between 1 and 81   **3, 9, 27 or −3, 9, −27**
2. One positive, between 3 and 12   **6**
3. One negative, between 2 and 50   **−10**
4. Two, between 5 and 320   **20, 80**
5. Two, between −3 and 24   **6, −12**
6. Two, between −2 and −128   **−8, −32**
7. One positive, between −4 and −36   **12**
8. One negative, between −5 and −80   **−20**
9. Two, between 16 and −2   **−8, 4**
10. Three, between −64 and −4
11. Three, between −5 and −405
12. Three, between 2 and 1250
13. One positive, between 8 and $\frac{1}{2}$ **2**
14. One negative, between $\frac{2}{5}$ and 10   **−2**
15. Two, between 9 and $\frac{1}{3}$ **3, 1**
16. Two, between $\frac{3}{5}$ and 75   **3, 15**
17. Four, between 11 and 352   **22, 44, 88, 176**
18. Four, between 2 and 200,000
19. One positive, between 3 and 6   **$3\sqrt{2}$**
20. One negative, between 4 and 12   **$−4\sqrt{3}$**
21. One negative, between $5\sqrt{2}$ and $10\sqrt{2}$
22. One positive, between $3\sqrt{5}$ and $15\sqrt{5}$ **15**

11. **15, −45, 135 or −15, −45, −135**   21. **−10**   18. **20, 200, 2,000, 20,000**

23. One number is 4 times another number and the positive geometric mean of the two numbers is 6. Find the two numbers.   **3, 12 or −3, −12**
24. Eight more than the positive geometric mean of two numbers is 20. Find the two numbers if one number is 9 times the other number.   **4, 36 or −4, −36**
25. If one number is 6 more than another number and the positive geometric mean of the two numbers is 4, what are the two numbers?   **2, 8 or −8, −2**
26. Find two numbers whose difference is 5 and whose positive geometric mean is 6.   **9, 4 or −4, −9**

B
27. Seven different weights are used to balance small masses in a laboratory. These weights are in geometric progression. Find each weight if the lightest is 0.5 g and the median (middle) weight is 4 g.   **0.5 g, 1 g, 2 g, 4 g, 8 g, 16 g, 32 g**
28. Crude oil is pumped from a tanker so that the amounts remaining at the end of successive hours are in G.P. At the end of 3 hours, 6,400 kiloliters (kL) remain in the tanker, and at the end of 6 hours, there are 2,700 kL. How much oil remains in the tanker at the end of 4 hours?   **4,800 kL**

Find the indicated number of real geometric means between each pair of terms.

29. One positive, between 0.1 and 0.016   **0.04**
30. Two, between 0.2 and 0.0128   **0.08, 0.032**
31. Two, between 3 and $21\sqrt{7}$ **$3\sqrt{7}$, 21**
32. Two, between $3\sqrt{5}$ and 75   **15, $15\sqrt{5}$**
33. Three, between $5x^3$ and $80x^{11}$
34. Five, between y^2 and $64y^{20}$ **below**
35. One positive, between $2i$ and $−18i$ **6**
36. One negative, between $−3i$ and $15i$ **$−3\sqrt{5}$**

33. **$10x^5, 20x^7, 40x^9$ or $−10x^5, 20x^7, −40x^9$**

Find the geometric mean of each two numbers. (Each exercise has two answers.)

C
37. i and $4i$ **±2i**
38. 3 and −27   **±9i**
39. −4 and 12   **±4$i\sqrt{3}$**
40. $5i$ and $8i$ **±2$i\sqrt{10}$**
41. $1+i$ and $1-i$ **±$\sqrt{2}$**
42. $3-4i$ and $3+4i$ **±5**
43. x^3 and y^3 **±$xy\sqrt{xy}$**
44. $\sqrt{x}$ and $\sqrt{y}$ **±$\sqrt[4]{xy}$**

34. **$2y^5, 4y^8, 8y^{11}, 16y^{14}, 32y^{17}$ or $−2y^5, 4y^8, −8y^{11}, 16y^{14}, −32y^{17}$**

Use the formula $l = a \cdot r^{n-1}$ to prove each statement.

45. The geometric mean of the real numbers x and y ($xy > 0$) is $\sqrt{xy}$ or $-\sqrt{xy}$.
46. The geometric mean of the complex number $a + bi$ and its conjugate is a real number.

For Ex. 45–46, see TE Answer Section.

GEOMETRIC SERIES 14.6

Objective **To find the sum of the first *n* terms of a geometric series**

If you take the geometric progression 2, 10, 50, 250, ... and write it as the indicated sum $2 + 10 + 50 + 250 + \ldots$, the result is a **geometric series**.

> **Definition: Geometric series**
>
> A *geometric series* is an indicated sum of the terms of a geometric progression.

Example 1 **Write each G.P. as a geometric series. Find the common ratio *r* of the series.**

$-2, 8, -32, 128, \ldots$ 　　　　　　$-8, -2, -\dfrac{1}{2}, -\dfrac{1}{8}, \ldots$

The geometric series is 　　　　The geometric series is

$-2 + 8 - 32 + 128 - \ldots$ 　　　$-8 - 2 - \dfrac{1}{2} - \dfrac{1}{8} - \ldots$

$r = \dfrac{8}{-2}$, or -4 　　　　　　$r = \dfrac{-2}{-8}$, or $\dfrac{1}{4}$

You can find the sum S of the terms of the geometric series $5 + 15 + 45 + \ldots + 3{,}645$ with common ratio 3 in the following way:

(1) Write the series in the form $a + ar + ar^2 + ar^3 + \ldots + l$.
(2) Multiply each term by -3 to get a second series.
(3) Then add the two series.

$$S = 5 + 5 \cdot 3 + 5 \cdot 3^2 + 5 \cdot 3^3 + \ldots + 3{,}645$$
$$-3S = - 5 \cdot 3 - 5 \cdot 3^2 - 5 \cdot 3^3 - \ldots - 3{,}645 - 3(3{,}645)$$
$$\overline{-2S = 5 + 0 + 0 + 0 + \ldots + 0 - 3(3{,}645)}$$
$$S = \dfrac{5 - 3(3{,}645)}{-2}, \text{ or } 5{,}465$$

The steps above suggest a method for obtaining a formula for the sum S of the terms of a finite geometric series with first term a, common ratio r, and last term l.

$$S = a + ar + ar^2 + ar^3 + \ldots + l$$
$$-r \cdot S = - ar - ar^2 - ar^3 - \ldots - l - rl$$
$$\overline{S - rS = a + 0 + 0 + 0 + \ldots + 0 - rl}$$

$$S - rS = a - rl$$
$$(1 - r)S = a - rl$$
$$S = \dfrac{a - rl}{1 - r}$$

Geometric Series

> **Sum of the terms of finite geometric series**
>
> The sum S of the terms of a finite geometric series is given by the formula
> $$S = \frac{a - rl}{1 - r}, (r \neq 1)$$
> where a is the first term, r is the common ratio, and l is the last term.

In this book, "the sum of a geometric series" will be used to mean "the sum of the terms of a geometric series."

Example 2 **Find the sum of the geometric series in which the first term $a = 3$, the common ratio $r = -5$, and the last term $l = -375$.**

$$S = \frac{a - rl}{1 - r} = \frac{3 - (-5)(-375)}{1 - (-5)} = \frac{3 - 1875}{6} = -312$$

Thus, the sum of the geometric series is -312.

Example 3 **Find the sum of the geometric series: $8 - 4 + 2 - \ldots - \frac{1}{16}$.**

$$a = 8 \qquad r = \frac{-4}{8}, \text{ or } -\frac{1}{2} \qquad l = -\frac{1}{16}$$

$$S = \frac{8 - \left(-\frac{1}{2}\right)\left(-\frac{1}{16}\right)}{1 - \left(-\frac{1}{2}\right)} = \left(8 - \frac{1}{32}\right) \div \frac{3}{2} \qquad \blacktriangleleft \text{ Use the sum formula. Simplify.}$$

$$= \frac{255}{32} \cdot \frac{2}{3}$$

$$= \frac{85}{16}, \text{ or } 5\frac{5}{16} \qquad \blacktriangleleft \text{ Sum of the series}$$

You can find the sum of the first 9 terms of the geometric series $-\frac{1}{4} + \frac{1}{2} - 1 + 2 - \ldots$ without finding the 9th term, or l. To do this, another sum formula is used. This formula is obtained by substitution.

$$S = \frac{a - r \cdot l}{1 - r} \text{ and } l = a \cdot r^{n-1}$$

$$S = \frac{a - r(a \cdot r^{n-1})}{1 - r} = \frac{a - a \cdot r^1 \cdot r^{n-1}}{1 - r} = \frac{a - a \cdot r^n}{1 - r}, \text{ or } \frac{a(1 - r^n)}{1 - r}$$

Example 4 **Find the sum of the first 9 terms of the geometric series: $-\frac{1}{4} + \frac{1}{2} - 1 + 2 - \ldots$.**

$$S = \frac{a(1 - r^n)}{1 - r} = \frac{-\frac{1}{4}[1 - (-2)^9]}{1 - (-2)} = \frac{-\frac{1}{4}(1 + 512)}{3} = -\frac{1}{4} \cdot \frac{513}{3} = -\frac{1}{4} \cdot 171 = -42\frac{3}{4}$$

$\sum_{k=3}^{6} 2k$ is the arithmetic series $2 \cdot 3 + 2 \cdot 4 + 2 \cdot 5 + 2 \cdot 6$, or $6 + 8 + 10 + 12$, whose common difference is 2. In a similar way, $\sum_{k=3}^{6} 2^k$ is the geometric series $2^3 + 2^4 + 2^5 + 2^6$, or $8 + 16 + 32 + 64$, whose common ratio is 2.

Example 5   **Write the expanded form and the series form of $\sum_{j=1}^{5} 5(-3)^j$. Find the sum of the series.**

Substitute 1, 2, 3, 4, and 5 for j in $5(-3)^j$.

$\sum_{j=1}^{5} = 5(-3)^1 + 5(-3)^2 + 5(-3)^3 + 5(-3)^4 + 5(-3)^5$ ◀ *Expanded form*

$= -15 + 45 - 135 + 405 - 1215$ ◀ *Series form*

$= \dfrac{-15 - (-3)(-1215)}{1 - (-3)}$ ◀ *Use $S = \dfrac{a - rl}{1 - r}$ to find the sum.*

$= -915$ ◀ *Sum of the series*

Reading in Algebra

Determine whether each expression is (A) an arithmetic progression, (B) an arithmetic series, (C) a geometric progression, or (D) a geometric series.

1. $5 + 10 + 15 + 20 + \ldots$ **B**
2. $5, 15, 45, 135, \ldots$ **C**
3. $4 - 12 + 36 - 108 + \ldots$ **D**
4. $4, -1, -6, -11, \ldots$ **A**
5. $\sum_{k=1}^{6} 3k$ **B**
6. $\sum_{k=1}^{6} 3^k$ **D**

Oral Exercises

State each G.P. as a geometric series.

1. $4 + 12 + 36 + 108 + \ldots$
2. $3 - 1 + \frac{1}{3} - \frac{1}{9} + \ldots$
3. $-\frac{2}{5} - 2 - 10 - 50 - \ldots$

1. $4, 12, 36, 108, \ldots$
2. $3, -1, \frac{1}{3}, -\frac{1}{9}, \ldots$
3. $-\frac{2}{5}, -2, -10, -50, \ldots$

Find the common ratio r of each geometric series.

4. $-\frac{1}{9} + \frac{1}{3} - 1 + 3 - \ldots$ **−3**
5. $6 + 0.6 + 0.06 + 0.006 + \ldots$ **0.1**
6. $14 + 0.14 + 0.0014 + \ldots$ **0.01**
7. $25 - 5 + 1 - \frac{1}{5} + \ldots$ **$-\frac{1}{5}$**
8. $\sum_{j=3}^{6} (-2)^j$ **−2**
9. $\sum_{c=1}^{5} 4(5^c)$ **5**

Written Exercises

Find the sum of each geometric series for the given data.

Ⓐ
1. $a = 6, r = 10, l = 60{,}000$ **66,666**
2. $a = -2, r = 3, l = -1{,}458$ **−2,186**
3. $a = -4, r = -2, l = -1{,}024$ **−684**
4. $a = 256, r = -\frac{1}{2}, l = 4$ **172**
5. $16 + 8 + 4 + \ldots + \frac{1}{8}$ **$31\frac{7}{8}$**
6. $\frac{1}{9} + \frac{1}{3} + 1 + \ldots + 243$ **$364\frac{4}{9}$**

Geometric Series

7. $0.3 + 3 + 30 + \ldots + 300{,}000$ **333,333.3**
8. $-1{,}024 - 512 - 256 - \ldots - \frac{1}{4}$ **$-2{,}047\frac{3}{4}$**
9. $0.2 + 0.4 + 0.8 + \ldots + 51.2$ **102.2**
10. $900 - 90 + 9 - \ldots + 0.0009$ **818.1819**
11. $n = 10; 5 + 10 + 20 + 40 + \ldots$ **5,115**
12. $n = 9; -3 - 6 - 12 - 24 - \ldots$ **$-1{,}533$**
13. $n = 8; 4 - 8 + 16 - 32 + \ldots$ **-340**
14. $n = 10; -\frac{1}{8} + \frac{1}{4} - \frac{1}{2} + 1 - \ldots$ **$42\frac{5}{8}$**
15. $n = 7; 0.06 + 0.6 + 6 + 60 + \ldots$ **66,666.66**
16. $n = 8; -1 + 10 - 100 + 1{,}000 - \ldots$ **9,090,909**
17. $a = 12, r = 3, n = 6$ **4,368**
18. $a = -4, r = -3, n = 5$ **-244**
19. $a = 20, r = 0.2, n = 6$ **24.9984**
20. $a = -300, r = 0.3, n = 5$ **-427.53**
21. $a = \frac{1}{5}, r = 10, n = 7$ **222,222.2**
22. $a = \frac{3}{5}, r = -10, n = 8$ **$-5{,}454{,}545.4$**
23. $\sum_{k=1}^{6} 2^k$ **126**
24. $\sum_{j=3}^{6} (-3)^j$ **540**
25. $\sum_{c=1}^{5} \left(\frac{1}{2}\right)^c$ **$\frac{31}{32}$**
26. $\sum_{k=4}^{7} (0.2)^{k-3}$ **.2496**

B 27. $\sum_{j=1}^{4} \frac{1}{2}(4^j)$ **170**
28. $\sum_{k=1}^{4} 6(-2)^{k-1}$ **-30**
29. $\sum_{c=1}^{5} -64\left(\frac{1}{4}\right)^{c-1}$ **$-85\frac{1}{4}$**
30. $\sum_{j=1}^{6} -27\left(-\frac{1}{3}\right)^{j-1}$ **$-20\frac{2}{9}$**
31. $n = 9; -32 + 16 - 8 + 4 - \ldots$ **$-21\frac{3}{8}$**
32. $n = 7; 125 + 25 + 5 + 1 + \ldots$ **$156\frac{31}{125}$**
33. $n = 7; \frac{2}{3} + 2 + 6 + 18 + \ldots$ **$728\frac{2}{3}$**
34. $n = 6; \frac{2}{5} + 2 + 10 + 50 + \ldots$ **$1{,}562\frac{2}{5}$**
35. $a = -243, r = \frac{1}{3}, n = 6$ **-364**
36. $a = 64, r = -\frac{1}{4}, n = 4$ **51**
37. $a = 2{,}500, r = -\frac{1}{5}, n = 5$ **2,084**
38. $a = -128, r = -\frac{1}{4}, n = 6$ **$-102\frac{3}{8}$**
39. $a = 5, r = 4, n = 6$ **6,825**
40. $a = -3, r = 5, n = 6$ **$-11{,}718$**

41. A golf ball, dropped from a height of 128 m, rebounds on each bounce one-half the distance from which it fell. When the ball hits the ground the 8th time, how far has it traveled since it was first dropped? **382 m**

42. A steel ball rebounds on each bounce two-thirds the distance from which it fell. If the ball is dropped from a height of 81 m, how far will it travel before it hits the ground the 6th time? **$362\frac{1}{3}$ m**

43. Smaller and smaller squares are formed consecutively as shown at the right. Find the sum of the perimeters of the first 8 squares so formed, if the width of the first square is 20 cm. **159.375 cm**

44. Find the sum of the areas of the first 5 squares formed, as shown at the right, if the first square is 20 cm wide. **532.8125 cm²**

For Ex. 45–46, see TE Answer Section.

C 45. Prove: $\sum_{k=1}^{4} ax^{k-1} = a + \left(a \cdot \sum_{k=1}^{3} x^k\right)$.
46. Solve $\sum_{k=3}^{5} (x^k + 3) = 9$ for x.

CALCULATOR ACTIVITIES

Find the sum of each geometric series for the given data.
(1) $3 + 15 + 75 + \ldots + 234{,}375$
(2) $n = 7: 2 - 6 + 18 - 54 + \ldots$

(1) $S = \dfrac{a - r \cdot l}{1 - r} = \dfrac{3 - 5 \times 234{,}375}{-4}$: $\ominus 5 \otimes 234375 \oplus 3 \oslash 4$ Answer: 292968

(2) $S = \dfrac{a(1 - r^n)}{1 - r} = \dfrac{2[1 - (-3)^7]}{1 - (-3)} = \dfrac{2(1 + 3^7)}{4} = $ $3 \otimes \ominus \ominus \ominus \ominus \ominus \ominus$
$\oplus 1 \otimes 2 \oslash 4 \ominus 1094$

Check student's calculator. Certain calculators take the equal sign.

INFINITE GEOMETRIC SERIES 14.7

Objectives **To find the sum, if it exists, of an infinite geometric series
To write a repeating decimal as an infinite geometric series or an infinite indicated sum
To write a repeating decimal in the form $\frac{x}{y}$ where x and y are integers**

The geometric series, $2 + 1 + \frac{1}{2} + \frac{1}{4} + \frac{1}{8} + \ldots$, is an **infinite geometric series**. The sum of the first 10 terms of the series is $3\frac{255}{256}$. This sum is found by using the formula $S = \frac{a(1 - r^n)}{1 - r}$, where $S = \frac{2\left[1 - \left(\frac{1}{2}\right)^{10}\right]}{1 - \frac{1}{2}}$. The value of r^n in this sum is $\left(\frac{1}{2}\right)^{10}$, or $\frac{1}{1024}$. Observe how rapidly $\left(\frac{1}{2}\right)^n$ approaches 0 as n increases by 5.

$\left(\frac{1}{2}\right)^0 = 1 \qquad \left(\frac{1}{2}\right)^5 = \frac{1}{32} \qquad \left(\frac{1}{2}\right)^{10} = \frac{1}{1024} \qquad \left(\frac{1}{2}\right)^{15} = \frac{1}{32,768}$

The sum formula $S = \frac{a(1 - r^n)}{1 - r}$ can be written as $S = \frac{a}{1 - r} \cdot (1 - r^n)$. If $-1 < r < 1$ and n increases without limit, then r^n approaches 0 as a limit and $\frac{a}{1 - r} \cdot (1 - r^n)$ approaches $\frac{a}{1 - r} \cdot (1 - 0)$, or $\frac{a}{1 - r}$, as a limit. This limit, $\frac{a}{1 - r}$, is defined to be the sum of the infinite geometric series with first term a and common ratio r.

> **Sum of infinite geometric series**
>
> The sum S of the terms of an infinite geometric series is given by the formula
> $$S = \frac{a}{1 - r}, \qquad [-1 < r < 1, r \neq 0]$$
> where a is the first term and r is the common ratio.

You can see that the infinite geometric series, $2 + 1 + \frac{1}{2} + \frac{1}{4} + \ldots$, appears to **converge** to 4 by finding the sums of the series for $n = 1, 2, 3, \ldots$.

Infinite Geometric Series

Example 1 **Find the sum, if it exists, of each infinite geometric series.**

$2 + 1 + \frac{1}{2} + \frac{1}{4} + \ldots$ $\quad\quad 8 - 2 + \frac{1}{2} - \frac{1}{8} + \ldots$ $\quad\quad \frac{1}{4} + \frac{1}{2} + 1 + 2 + \ldots$

$r = \frac{1}{2}$ $\quad\quad\quad\quad\quad\quad\quad\quad r = -\frac{1}{4}$ $\quad\quad\quad\quad\quad\quad\quad\quad r = 2$

$S = \frac{a}{1-r} = \frac{2}{1-\frac{1}{2}}$ $\quad\quad S = \frac{a}{1-r} = \frac{8}{1-\left(-\frac{1}{4}\right)}$ $\quad$ Since r is not between -1 and 1, the sum of the series does not exist.

$\quad\quad = 2 \div \frac{1}{2}$ $\quad\quad\quad\quad\quad\quad = 8 \div \frac{5}{4}$

$\quad\quad = 4$ $\quad\quad\quad\quad\quad\quad\quad\quad = \frac{32}{5}$, or $6\frac{2}{5}$

The decimal 0.56 is a terminating decimal while 0.434343 . . . and $0.27\overline{7}$ are repeating decimals. Any repeating decimal can be written as an infinite geometric series or as the indicated sum of a terminating decimal and an infinite geometric series.

$0.434343 \ldots = 0.43 + 0.0043 + 0.000043 + \ldots$, an infinite geometric series with $a = 0.43$ and $r = 0.01$

$0.27\overline{7} = 0.2 + (0.07 + 0.007 + 0.0007 + \ldots)$, an infinite indicated sum
$\quad\quad\uparrow\quad\quad\quad\quad\quad\quad\uparrow$
terminating $\quad$ infinite geometric series
decimal $\quad\quad$ with $a = 0.07$ and $r = 0.1$

After a repeating decimal is written in series form, the formula $S = \frac{a}{1-r}$ can be used to write the decimal in the form $\frac{x}{y}$ where x and y are integers.

Example 2 **Write 0.434343 . . . in the form $\frac{x}{y}$ where x and y are integers.**

$0.434343 \ldots = 0.43 + 0.0043 + 0.000043 + \ldots$ ◀ *Infinite geometric series*
$\quad\quad\quad\quad a = 0.43 \quad\quad r = 0.01$
$\quad\quad\quad\quad S = \frac{a}{1-r} = \frac{0.43}{1-0.01} = \frac{0.43}{0.99} = \frac{43}{99}$

Thus, $0.434343 \ldots = \frac{43}{99}$.

Example 3 **Write $6.01\overline{212}$ in the form $\frac{x}{y}$ where x and y are integers.**

$6.01\overline{212} = 6.0 + (0.012 + 0.00012 + 0.0000012 + \ldots)$ ◀ *Infinite indicated sum*
$\quad\quad\quad\quad a = 0.012 \quad\quad r = 0.01$
$\quad\quad S = \frac{a}{1-r} = \frac{0.012}{1-0.01} = \frac{0.012}{0.99} = \frac{2}{165} \quad\quad 6 + \frac{2}{165} = \frac{992}{165}$

Thus, $6.01\overline{212} = \frac{992}{165}$.

Chapter Fourteen

Written Exercises

Find the sum, if it exists, of each infinite geometric series.

Ⓐ
1. $\frac{1}{2} + 1 + 2 + 4 + \ldots$ **does not exist**
2. $8 + 4 + 2 + 1 + \ldots$ **16**
3. $9 - 3 + 1 - \frac{1}{3} + \ldots$ **$6\frac{3}{4}$**
4. $9 + 6 + 4 + \frac{8}{3} + \ldots$ **27**
5. $\frac{1}{2} + \frac{1}{4} + \frac{1}{8} + \frac{1}{16} + \ldots$ **1**
6. $\frac{1}{4} - \frac{1}{8} + \frac{1}{16} - \frac{1}{32} + \ldots$ **$\frac{1}{6}$**
7. $1 - \frac{1}{3} + \frac{1}{9} - \frac{1}{27} + \ldots$ **$\frac{3}{4}$**
8. $2 + 0.2 + 0.02 + 0.002 + \ldots$ **$2\frac{2}{9}$, or $2.2\overline{2}$**
9. $70 + 7 + 0.7 + 0.07 + \ldots$ **$77\frac{7}{9}$, or $77.\overline{7}$**

Write each repeating decimal as an infinite geometric series or as an infinite indicated sum and then in the form $\frac{x}{y}$ where x and y are integers.

10. $0.5\overline{55}\ldots$ **$\frac{5}{9}$**
11. $0.\overline{77}$ **$\frac{7}{9}$**
12. $3.\overline{333}\ldots$ **$\frac{10}{3}$**
13. $22.\overline{22}$ **$\frac{200}{9}$**
14. $0.04\overline{4}$ **$\frac{2}{45}$**
15. $4.\overline{666}\ldots$ **$\frac{14}{3}$**
16. $7.5\overline{44}$ **$\frac{679}{90}$**
17. $6.25\overline{55}\ldots$ **$\frac{563}{90}$**

Solve each problem.

Ⓑ
18. The midpoints of the sides of an equilateral triangle are connected to form a second triangle. This procedure is repeated on each resulting triangle. Find the sum of the perimeters of all triangles so formed, if one side of the first triangle measures 60 cm. **360 cm**

19. The midpoints of the sides of a square are joined to form a second square. This process is repeated on each resulting square. What is the sum of the areas of all squares so formed, if one side of the first square measures 8 m? **128 m²**

20. A golf ball, dropped from a height of 128 m, rebounds on each bounce one-half the distance from which it falls. How far will it travel before coming to rest? **384 m**

21. A steel ball was dropped from a height of 81 m and rebounded, on each bounce, two-thirds of the distance from which it fell. How far did it travel before coming to rest? **405 m**

Find the sum, if it exists, of each infinite geometric series.

22. $\frac{1}{5} + 1 + 5 + 25 + \ldots$ **does not exist**
23. $-16 + 12 - 9 + \frac{27}{4} - \ldots$ **$-9\frac{1}{7}$**
24. $25 - 10 + 4 - \frac{8}{5} + \ldots$ **$17\frac{6}{7}$**

Find the sum of the series for $n = 1, 2, 3, \ldots, 8$. To what value does the series converge?

25. $27 + 9 + 3 + \ldots$ **$40\frac{1}{2}$**
26. $-4 - 2 - 1 - \ldots$ **-8**
27. $0.9 + 0.09 + 0.009 + \ldots$ **1**

Use the formula $S = \frac{a}{1-r}$ to write each repeating decimal in the form $\frac{x}{y}$ where x and y are integers.

28. $53.5\overline{353}$ **$\frac{5300}{99}$**
29. $67.47\overline{474}$ **$\frac{6680}{99}$**
30. $3.02\overline{929}$ **$\frac{2999}{990}$**
31. $17.1\overline{717}$ **$\frac{1700}{99}$**
32. $8.03\overline{232}$ **$\frac{7952}{990}$**
33. $63.53\overline{535}$ **$\frac{6290}{99}$**

For what values of x does each infinite geometric series have a sum? [Hint: $-1 < r < 1, r \ne 0$]

Ⓒ
34. $(x-1)^0 + (x-1)^1 + (x-1)^2 + \ldots$ **$0 < x < 2$ and $x \ne 1$**
35. $\frac{1}{x} + \frac{1}{x^3} + \frac{1}{x^5} + \ldots$ **$x < -1$ or $x > 1$**
36. $\frac{1}{x^2} + \frac{1}{x^3} + \frac{1}{x^4} + \ldots$ **$x < -1$ or $x > 1$**

Infinite Geometric Series

BINOMIAL EXPANSIONS   14.8

Objectives **To expand a positive integral power of a binomial**
To find a specified term of a binomial expansion

The first few positive integral powers of $(a + b)$ reveal patterns that can be used to **expand** any positive integral power of a binomial.

$$(a + b)^0 = 1$$
$$(a + b)^1 = a + b$$
$$(a + b)^2 = a^2 + 2ab + b^2$$
$$(a + b)^3 = a^3 + 3a^2b + 3ab^2 + b^3$$
$$(a + b)^4 = a^4 + 4a^3b + 6a^2b^2 + 4ab^3 + b^4$$
$$(a + b)^5 = a^5 + 5a^4b + 10a^3b^2 + 10a^2b^3 + 5ab^4 + b^5$$

Notice in the expansion of $(a + b)^n$ that:

(1) The number of terms is always $n + 1$.
(2) The first term is a^n and the last term is b^n.
(3) The exponent of a decreases by 1, from one term to the next.
(4) The exponent of b increases by 1, from one term to the next.
(5) For each term, the sum of the exponents of a and b is n.
(6) The coefficients are symmetrical. (They read the same from left to right as right to left.)
(7) There is a pattern for finding the coefficients in an expansion.

Notice that $(a + b)^4 = a^4 + \dfrac{4}{1}a^3b + \dfrac{4 \cdot 3}{1 \cdot 2}a^2b^2 + \dfrac{4 \cdot 3 \cdot 2}{1 \cdot 2 \cdot 3}ab^3 + b^4$
$$= a^4 + 4a^3b + 6a^2b^2 + 4ab^3 + b^4.$$

All of the patterns above can be seen in the statement of the **binomial theorem.**

Binomial theorem   If n is a positive integer, then $(a + b)^n =$
$$a^n + \frac{n}{1}a^{n-1}b + \frac{n(n-1)}{1 \cdot 2}a^{n-2}b^2 + \frac{n(n-1)(n-2)}{1 \cdot 2 \cdot 3}a^{n-3}b^3 + \ldots + b^n.$$

Example 1   **Expand $(3x - y^2)^5$. Simplify each term.**

Use the binomial theorem with $n = 5$, $a = 3x$, and $b = -y^2$: $[3x + (-y^2)]^5$.

$(3x)^5 + \dfrac{5}{1}(3x)^4(-y^2) + \dfrac{5 \cdot 4}{1 \cdot 2}(3x)^3(-y^2)^2 + \dfrac{5 \cdot 4 \cdot 3}{1 \cdot 2 \cdot 3}(3x)^2(-y^2)^3 + \dfrac{5 \cdot 4 \cdot 3 \cdot 2}{1 \cdot 2 \cdot 3 \cdot 4}(3x)(-y^2)^4 + (-y^2)^5$

Thus, $(3x - y^2)^5 = 243x^5 - 405x^4y^2 + 270x^3y^4 - 90x^2y^6 + 15xy^8 - y^{10}$.

Example 1 suggests the following properties for the signs in a binomial expansion.
(1) In the expansion of $(a + b)^n$, the signs will all be $+$.
(2) In the expansion of $(a - b)^n$, the signs will alternate $+$ and $-$.

The notation $k!$ is read as "k factorial" and is defined below as is $0!$.

$$k! = 1 \cdot 2 \cdot 3 \cdot \ldots \cdot k, \text{ where } k \text{ is a positive integer, and } 0! = 1$$

Factorial notation can be used with the binomial theorem to rewrite the denominators of the coefficients, since $1 = 1!$, $1 \cdot 2 = 2!$, $1 \cdot 2 \cdot 3 = 3!$, and so on.

Thus, the binomial theorem can be expressed as $(a + b)^n =$

$$a^n + \frac{n}{1!}a^{n-1}b + \frac{n(n-1)}{2!}a^{n-2}b^2 + \frac{n(n-1)(n-2)}{3!}a^{n-3}b^3 + \ldots + b^n.$$

The 7th term is $\frac{n(n-1)(n-2)\ldots(n-[7-2])}{(7-1)!}a^{n-7+1}b^{7-1}$, or

$$\frac{n(n-1)(n-2)\ldots(n-5)}{6!}a^{n-6}b^6.$$

rth term in a binomial expansion	The rth term in the expansion of $(a + b)^n$ is $$\frac{n(n-1)(n-2)\ldots(n-r+2)}{(r-1)!}a^{n-r+1}b^{r-1}.$$

The formula for the rth term permits you to find a specified term of an expansion without writing the preceding terms.

Example 2 Find the 4th term in the expansion of $(2x - y^3)^7$. Simplify the term.

Use the 4th term of $(a + b)^7$, or $\frac{7 \cdot 6 \cdot 5}{(4-1)!}a^{7-4+1}b^{4-1}$, to find the 4th term of $(2x - y^3)^7$ where $a = 2x$, $b = -y^3$, $n = 7$, and $r = 4$.

$$\frac{7 \cdot 6 \cdot 5}{1 \cdot 2 \cdot 3}(2x)^{7-3}(-y^3)^3 = 35 \cdot 2^4 x^4 (-y^9)$$

Thus, the 4th term is $-560x^4y^9$.

Written Exercises

For complete solutions, see TE Answer Section.
10. $270c^2d^3$ 11. $-20x^6y^3$ 12. $24x^2y^4$
13. $720x^9y^2$ 14. $135c^4d^2$ 15. $160x^9$

Expand each power of a binomial. Simplify.

(A)
1. $(a + b)^8$
2. $(a - b)^9$
3. $(x + 2)^6$
4. $(2c - 3)^5$
5. $(x^2 + y)^8$
6. $(m^2 - 2n)^5$
7. $(2r + 5t)^4$
8. $(y - 0.2)^4$
9. $(2x + 0.1)^6$

Find the specified term of each expansion. Simplify.
10. 4th term of $(c + 3d)^5$
11. 4th term of $(x^2 - y)^6$
12. middle term of $(2x + y^2)^4$

(B) 13. 3rd term of $(2x^3 + 3y)^5$
14. 5th term of $(3c^2 - \sqrt{d})^6$
15. middle term of $(4x^2 + \frac{x}{2})^6$

Expand each power of a binomial. Simplify.
16. $(x^3 - 2y)^4$
17. $(2c^3 + 3d^2)^5$
18. $(x + \sqrt{2})^6$
19. $(2x - \sqrt{y})^4$
20. $(2r^2 + \frac{t}{2})^4$
21. $(\frac{1}{x} + \frac{1}{y})^5$

Find the first four terms of each expansion. Simplify.
22. $(x + y)^{12}$
23. $(x - 2y)^{11}$
24. $(x^2 + 3y^3)^{10}$
25. $(1 + 0.01)^7$
26. $(1 + 0.02)^8$
27. $(1 + 0.03)^6$

The Binomial Theorem is true for $(a + b)^q$, where q is a rational number. For each expansion, write the first four terms in simplest radical form.

(C) 28. $(x + 4)^{\frac{1}{2}}$
29. $(x + 9)^{\frac{2}{3}}$
30. $(x - 9)^{-\frac{1}{3}}$

Binomial Expansions

SPECIAL INFINITE SERIES (optional)   14.9

Objective **To apply an example of a special infinite series**

Some special infinite series involve a product of consecutive counting numbers beginning with 1, such as

$$1 \times 2, \quad 1 \times 2 \times 3, \quad 1 \times 2 \times 3 \times 4, \quad \text{and so on.}$$

These products are abbreviated as follows.

$$1 \times 2 = 2! \qquad 1 \times 2 \times 3 \times 4 = 4!$$
$$1 \times 2 \times 3 = 3! \qquad 1 \times 2 \times 3 \times 4 \times 5 = 5!$$

Thus, $1 \times 2 \times 3 \times \ldots \times (n-1) \times n = n!$

Reciprocals of factorials, $\frac{1}{n!}$, occur in some special series. Shown below is a table of factorials ($n!$) and reciprocals of factorials $\left(\frac{1}{n!}\right)$ for some values of n.

n	$n!$	$\frac{1}{n!}$	n	$n!$	$\frac{1}{n!}$
2	2	0.5	6	720	0.0013888 ...
3	6	0.1666666 ...	7	5,040	0.0001984 ...
4	24	0.0416666 ...	8	40,320	0.0000248 ...
5	120	0.0083333 ...	9	362,880	0.0000027 ...

The trigonometric functions called *sine* (sin) and *cosine* (cos) are defined by the following infinite series, where x is a real number.

$$\sin x = x - \frac{x^3}{3!} + \frac{x^5}{5!} - \frac{x^7}{7!} + \frac{x^9}{9!} - \ldots$$
$$\cos x = 1 - \frac{x^2}{2!} + \frac{x^4}{4!} - \frac{x^6}{6!} + \frac{x^8}{8!} - \ldots$$

Example 1 **Use the infinite series for sin x to compute sin 0.2 to four decimal places.**

Replace x by 0.2 in the series until you find a term whose first five decimal places are each 0.

$$\sin x = x - \frac{x^3}{3!} + \frac{x^5}{5!} - \ldots$$

$$\sin 0.2 = 0.2 - \frac{(0.2)^3}{3!} + \frac{(0.2)^5}{5!} - \ldots$$

$$= 0.2 - \frac{0.008}{6} + \frac{0.00032}{120} - \ldots$$

$$= 0.2 - 0.0013333 \ldots + 0.0000026 \ldots - \ldots$$

$$= 0.19866 \ldots$$

Thus, sin 0.2 = 0.1987, correct to four decimal places.

384   Chapter Fourteen

Another number related to an infinite series is the number *e*, which is used as the base of **natural** logarithms. The number *e* is defined by the infinite series:

$$e = 2 + \frac{1}{2!} + \frac{1}{3!} + \frac{1}{4!} + \ldots + \frac{1}{n!} + \ldots$$

Example 2   Use the infinite series for *e* to compute *e* to three decimal places.

Use the table for $\frac{1}{n!}$ until you find a term whose first four decimal places are each 0.

$e = 2 + .5 + .1666\ldots + .041666\ldots + .008333\ldots$
$\quad + .0013888\ldots + .0001984\ldots + .0000248\ldots + \ldots$

Thus, *e* = 2.718, correct to three decimal places.

Written Exercises

(A) 1. Use the infinite series for sin *x* to compute sin 1 to three decimal places. **0.841**
2. Use the infinite series for cos *x* to compute cos 1 to three decimal places. **0.540**
3. Use the infinite series for *e* to compute *e* to four decimal places. **2.7183**

4. Compute $\sum_{n=2}^{6} n!$. **872**

5. Compute $\sum_{n=2}^{6} \frac{1}{n!}$. $\frac{517}{720} \doteq$ **0.7180555 . . .**

Use the infinite series for sin *x* or for cos *x* to compute each of the following to four decimal places.

(B) 6. sin 0.1   7. sin 0.3   8. sin 0   9. cos 0.1   10. cos 0.2   11. cos 0
   0.0998   **0.2955**   **0.0000**   **0.9950**   **0.9801**   **1.0000**

Use the infinite series $e^x = 1 + x + \frac{x^2}{2!} + \frac{x^3}{3!} + \ldots + \frac{x^{n-1}}{(n-1)!} + \ldots$ to compute each of the following. (Ex. 12–13)

12. e^2, to two decimal places **7.39**
13. $\sqrt{e}$, to four decimal places [*Hint:* Recall that $\sqrt{x} = x^{\frac{1}{2}}$, or $x^{0.5}$.] **1.6487**

(C) 14. Divide 1 by $1 - x$ [$1 - x\overline{)1}$] to show that **14.** $1 - x\overline{)1}$ $\frac{1 + x + x^2 + x^3 + \ldots}{}$

$\frac{1}{1-x} = 1 + x + x^2 + x^3 + \ldots + x^{n-1} + \ldots$, where $|x| < 1$.

A series of the form $1 + x + x^2 + x^3 + \ldots$, where $|x| < 1$, is called an **infinite power series**. This series is an infinite geometric series with a common ratio $r = x$ and $|r| < 1, r \neq 0$.

15. Use the formula $S = \frac{a}{1 - r}$ for an infinite geometric series, $|r| < 1, r \neq 0$, to show that $1 + x + x^2 + x^3 + \ldots = \frac{1}{1-x}$. **15.** $S = \frac{a}{1-r} = \frac{1}{1-x}$

16. Simplify the infinite power series $1 + (y + 1) + (y + 1)^2 + (y + 1)^3 + \ldots$. $\frac{1}{-y}$
17. Simplify the infinite power series $1 + (z - 2) + (z - 2)^2 + (z - 3)^3 + \ldots$. $\frac{1}{3-z}$

Special Infinite Series

CHAPTER FOURTEEN REVIEW

(A) Exercises are 1, 3–6, 8–10, 12, 14–17, 19, 20, 22, 23, 25, 26, 29, 35, 36. The rest are (B) Exercises.

Vocabulary
arithmetic means [14.2]
arithmetic progression [14.1]
arithmetic series [14.3]
binomial expansion [14.8]
geometric means [14.5]
geometric progression [14.4]
geometric series [14.6]
mean proportional [14.5]
progression [14.1]   sequence [14.1]

Write the first four terms of each A.P. [14.1]
1. $a = 14$ and $d = -8$ **14, 6, −2, −10**
2. $a = 3x - 4$ and $d = -x + 2$ **below**
3. Find the 36th term of the A.P.:
   $-7.2, -3.4, 0.4, 4.2, \ldots$ [14.1] **125.8**
4. Find the three arithmetic means between 12 and 22.8. [14.2] **14.7, 17.4, 20.1**
5. Find the arithmetic mean of -11.8 and 19.6. [14.2] **3.9**   **2. $3x - 4, 2x - 2, x, 2$**

Find the sum of each arithmetic series [14.3]
6. $n = 35, a = -15.4, l = 43.8$ **497**
7. $n = 26; 2.5 + 2.9 + 3.3 + 3.7 + \ldots$ **195**
8. $\sum_{k=1}^{14} -5k$ **−525**   9. $\sum_{j=3}^{6} (2j - 5)$ **16**

Write each arithmetic series using Σ-notation.
10. $-10 - 20 - 30 - 40 - 50$ **below** [14.3]
11. $18 + 22 + 26 + 30 + \cdots + 58$

Write the next three terms of each G.P. [14.4]
12. $-\frac{1}{3}, 1, -3, \ldots$ **9, −27, 81** [14.4]
13. $-4x, -8x^2, -16x^3, \ldots$ **$-32x^4, -64x^5, -128x^6$**

Find the specified term of each G.P. [14.4]
14. 10th term: $6, -12, 24, \ldots$ **−3,072**
15. 5th term: $a = -4, r = 3$ **−324**

Find the indicated number of real geometric means between each pair of numbers. [14.5]
16. One positive, between $\frac{1}{5}$ and 20 **2**
17. Three, between -3 and -48
18. Two, between $2\sqrt{6}$ and 72 **12, $12\sqrt{6}$**

Find the sum of each geometric series for the given data. [14.6]
19. $64 + 32 + 16 + \ldots + \frac{1}{4}$ **$127\frac{3}{4}$**
20. $a = 3, r = 5, l = 1,875$ **2,343**
21. $n = 8; 32 - 16 + 8 - 4 + \ldots$ **$21\frac{1}{4}$**
22. $a = -2, r = -4, n = 5$ **−410**
23. $\sum_{k=2}^{6} (-5)^{k-1}$ **−2,605**   24. $\sum_{j=1}^{4} 25\left(\frac{1}{5}\right)^j$ **$6\frac{6}{25}$**

Find the sum, if it exists, of each infinite geometric series. [14.7]
25. $\frac{1}{3} + 1 + 3 + 9 + \ldots$ **does not exist**
26. $64 - 16 + 4 - 1 + \ldots$ **51.2**
27. $16\sqrt{3} + 8\sqrt{3} + 4\sqrt{3} + 2\sqrt{3} + \ldots$ **$32\sqrt{3}$**

28. Use the formula $S = \dfrac{a}{1 - r}$ to write **$\frac{8290}{99}$**
   $83.7\overline{37}$ in the form $\frac{x}{y}$, where x and y are integers. [14.7]

For Ex. 29–31, see TE Answer Section.
Expand each power of a binomial. Simplify. [14.8]
29. $(x + 3y)^4$ 30. $(2x^3 - y)^5$ 31. $\left(\frac{1}{m} + \frac{1}{n}\right)^4$

32. Find the 5th term of $(3x^2 - \sqrt{5})^6$. Simplify. **$3,375x^4$**
33. Find the first four terms of $(x - 2y^2)^{10}$. Simplify. [14.8]
   $x^{10} - 20x^9y^2 + 180x^8y^4 - 960x^7y^6$

Solve each problem.
34. Mr. Edwards began a 30-day training program by jogging 440 m. Each day, he ran 50 m more than he did the preceding day. How far did Mr. Edwards jog on the 30th day? [14.1] **1,890 m**
35. The arithmetic mean of two numbers is 24. Six less than the larger number is 5 times the smaller number. Find the numbers. [14.2] **41, 7**   **12, 3 or −3, −12**
36. If one number is 9 less than another number and their positive geometric mean is 6, what are the two numbers? [14.5]

17. $-6, -12, -24,$ or $6, -12, 24$

10. $\sum_{k=1}^{5} -10k$ 11. $\sum_{k=1}^{11} (4k + 14)$ or $\sum_{k=4}^{14} (4k + 2)$

CHAPTER FOURTEEN TEST

Ⓐ Exercises are 1–4, 6, 9–13, 15, 17–19, 23, 24. Ⓒ Exercises are starred. The rest are Ⓑ Exercises.

1. Write the first four terms of the A.P. whose first term is -2 and common difference is 1.5. **$-2, -0.5, 1, 2.5$**
2. Find the 31st term of the A.P.: 6.0, 6.4, 6.8, 7.2 . . . . **18**
3. Find the three arithmetic means between 18 and 29. **$20\frac{3}{4}, 23\frac{1}{2}, 26\frac{1}{4}$**
4. Find the arithmetic mean of -5.4 and 8.2. **1.4**

Find the sum of each arithmetic series.

5. $n = 26; 7.6 + 7.2 + 6.8 + 6.4 + \ldots$ **67.6**
6. $\sum_{k=1}^{20} (2k + 3)$ **480**
8. $\sum_{k=3}^{7} (3k + 1)$

7. Cartons are stacked in 16 rows with 4 cartons in the top row, 7 in the second row, 10 in the next row, and so on. How many cartons are in the stack? **424**
8. Write the arithmetic series, $10 + 13 + 16 + 19 + 22$, using Σ-notation. **above**
9. Write the next three terms of the G.P.: $\frac{3}{4}, \frac{3}{2}, 3, \ldots$. **6, 12, 24**
10. Find the 10th term of the G.P.: $\frac{1}{4}, \frac{1}{2}, 1, 2, \ldots$. **128**

Find the indicated number of real geometric means between each pair of numbers.

11. Two between -4 and 108 **12, -36**
12. One positive, between $\frac{1}{6}$ and 24 **2**

Find the sum of each geometric series.

13. $a = 5, r = 4, l = 5,120$ **6,825**
14. $n = 6; 100 - 50 + 25 - 12\frac{1}{2} + \ldots$ **$65\frac{5}{8}$**
15. $a = 5, r = -10, n = 4$ **$-4,545$**
16. $\sum_{j=1}^{4} 81\left(\frac{1}{3}\right)^j$ **40**

17. Find the sum of the infinite geometric series: $16 - 4 + 1 - \frac{1}{4} + \ldots$. **$12\frac{4}{5}$**

18. Write 7.2888 . . . as an infinite indicated sum and then in the form $\frac{x}{y}$, where x and y are integers. **$\frac{328}{45}$**
19. Expand $(3x - y^2)^4$. Simplify.
20. Find the 5th term of $(4x - \sqrt{2})^6$. Simplify.
21. Find the first four terms of $(x^2 + \frac{1}{2}y)^{10}$. Simplify. **$x^{20} + 5x^{18}y + 11\frac{1}{4}x^{16}y^2 + 15x^{14}y^3$**
19. **$81x^4 - 108x^3y^2 + 54x^2y^4 - 12xy^6 + y^8$**
20. **$960x^2$**

Solve each problem.

22. An object, shot upward at 200 m/s, travels 195 m during the 1st second, 185 m during the 2nd second, 175 m during the 3rd second, and so on. How far does it travel during the 14th second? **65 m**   **12, 28**
23. The arithmetic mean of two numbers is 20. Four less than one number is twice the other number. Find the two numbers.
24. Two more than the geometric mean of two numbers is 10. Find the two numbers if one number is 12 more than the other number. **$-16, -4$ or 4, 16**

Smaller and smaller squares are formed, one after the other, as shown at the right. The width of the first square is 32 m. Use this information in Ex. 25–26.

25. Find the sum of the perimeters of the first eight squares. **255 m**
26. Find the sum of the areas of all such squares. **$1,365\frac{1}{3}$ m²**

★27. Find two values of c such that $\frac{2}{c}, 4, \frac{15}{c-2}$ is an A.P. **4; $\frac{1}{8}$**
★28. Find two values of x such that $x - 9, 12, 4x$ is a G.P. **12; -3**
★29. Find the geometric mean of $-2i$ and $10i$. **$\pm 2\sqrt{5}$**
★30. For what values of x does the infinite geometric series $\frac{x}{2} - 2x^3 + 8x^5 - 32x^7 + \ldots$ have a sum? **$-\frac{1}{2} < x < \frac{1}{2}$ and $x \neq 0$**

COLLEGE PREP TEST

DIRECTIONS: In each item, you are to compare a quantity in Column 1 with a quantity in Column 2. Write the letter of the correct answer from these choices:

A The quantity in Column 1 is greater than the quantity in Column 2.
B The quantity in Column 2 is greater than the quantity in Column 1.
C The quantity in Column 1 is equal to the quantity in Column 2.
D The relationship cannot be determined from the given information.

Notes: Information centered over both columns refers to one or both of the quantities to be compared.
A symbol that appears in both columns has the same meaning in each column.
All variables represent real numbers.

SAMPLE ITEM AND ANSWER

Column 1	Column 2
$x = 2$ and $y = -2$	
The value of the next term of the A.P.: $x, 2x, 3x, 4x$	The value of the next term of the G.P.: y, y^2, y^3, y^4

The answer is A because $5x = 5 \cdot 2 = 10$, $y^5 = (-2)^5 = -32$, and $10 > -32$.

	Column 1		Column 2
1.	Arithmetic mean of $\frac{1}{2}$ and $\frac{1}{8}$	C	Arithmetic mean of $-\frac{1}{8}$ and $\frac{3}{4}$
2.	The next term in the G.P.: 9, 3, 1	B	The next term in the G.P.: $\frac{1}{16}, \frac{1}{8}, \frac{1}{4}$
3.	Next term in the sequence: 0, 1, 1, 2, 3, 5, 8, 13	A	Next term in the sequence: 75, 64, 53, 42, 31
4.	Sum of the series: $-20 - 15 - 10 - \ldots + 30$	B	Sum of the series: $-25 - 20 - 15 - \ldots + 35$

	Column 1		Column 2
5.	Arithmetic mean of 2 and 8	A	Geometric mean of 2 and 8
6.	$(x + 5)^3$	D	$(x + 5)^4$
7.	$\sum_{k=1}^{5} 2^{k-1}$	C	$\sum_{j=3}^{7} 2^{j-3}$
8.	The value of x if $x + 3, 2x, x - 7$ is an arithmetic progression	A	The value of n if $2n, n - 1, n + 1$ is an arithmetic progression
	$x > 0$ and $n > 0$		
9.	Value of x if 3, $3x$, $12x$ is a G.P.	B	Value of n if $-2, -2n, -10n$ is a G.P.
10.	Coefficient of the 3rd term in the expansion of $(x + 2)^{12}$	C	Coefficient of the 3rd term in the expansion of $(y - 2)^{12}$

15 EXPONENTIAL, LOGARITHMIC, AND POLYNOMIAL FUNCTIONS

Non-Routine Problem Solving

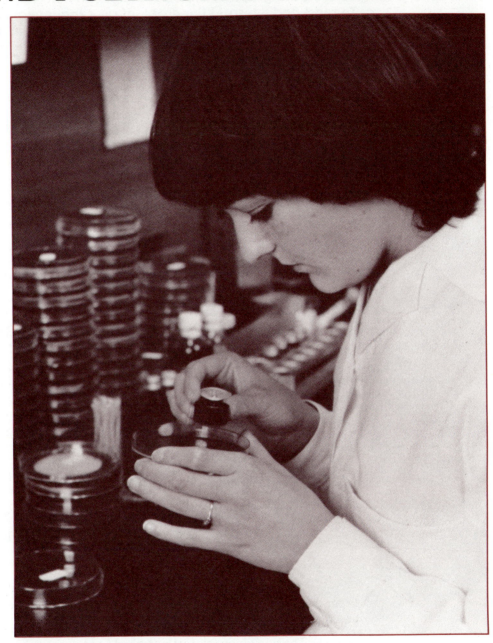

Problem 1. At the end of x hours, the number y of bacteria in a culture is (0, 100), (1, 145), (2, 210), (3, 305). Write an exponential equation of the form $y = a \cdot b^x$. **y = 100(1.45)^x**

Problem 2. Find four integers x, y, z, and w such that $2x + 3y - z - w = 84$ and $x:y = 3:4$, $y:z = 2:3$, and $x:w = 5:6$. **x = 30, y = 40, z = 60, w = 36**

GRAPHING EXPONENTIAL FUNCTIONS 15.1

Objectives **To draw the graph of an exponential function**
To read and interpret the graph of an exponential function

The equation $y = 3 \cdot 2^x$ describes a base-2 **exponential function.** The graph of this function is drawn using a table of ordered pairs (x, y).

x	2^x	$3 \cdot 2^x = y$
-3	$\frac{1}{8}$	$\frac{3}{8}$
-2	$\frac{1}{4}$	$\frac{3}{4}$
-1	$\frac{1}{2}$	$\frac{3}{2}$
0	1	3
1	2	6
2	4	12
3	8	24

The ordered pairs (x, y) in the table are plotted, and a smooth curve is drawn. All rational numbers and all irrational numbers are in the domain of the function above.

To draw the graph for $y = 3 \cdot 2^x$, integers were used as the values of x. You can use a calculator with the y^x function to verify the following:

If $x = 2.5$, then $y = 3 \cdot 2^x = 3 \cdot 2^{2.5} = 16.970562\ldots$, and
If $x = \sqrt{3}$, then $y = 3 \cdot 2^x = 3 \cdot 2^{\sqrt{3}} = 9.9659913$

From the graph above, you can list four properties of the exponential function described by $y = 3 \cdot 2^x$.

(1) The domain is {all real numbers}.

(2) The range is $\{y | y > 0\}$.

(3) The function is *constantly increasing*. That is, if $x_2 > x_1$, then $y_2 > y_1$.

(4) The ordered pair (0, 3) belongs to the function. That is, $3 \cdot 2^0 = 3$.

Example Draw the graphs of the exponential functions $A: y = 4^x$ and $B: y = \left(\frac{1}{4}\right)^x$ in the same coordinate plane with convenient scales for the axes. List four properties of each function.

1. Rewrite $y = \left(\frac{1}{4}\right)^x$ as $y = 4^{-x}$. ◀ $y = \left(\frac{1}{4}\right)^x = \frac{1}{4^x} = 4^{-x}$

2. Construct a table of ordered pairs (x, y) for A and for B.

A			B	
x	$4^x = y$		x	$\left(\frac{1}{4}\right)^x = 4^{-x} = y$
-2	$\frac{1}{16}$		-2	16
-1	$\frac{1}{4}$		-1	4
0	1		0	1
0.5	2		0.5	0.5
1	4		1	$\frac{1}{4}$
$\sqrt{2}$	7.10		$\sqrt{2}$	0.14

3. Plot the ordered pairs (x, y) and draw two smooth curves.

The graphs show the following four properties for $A: y = 4^x$ and $B: y = \left(\frac{1}{4}\right)^x$.

(1) The domain for A and B is {all real numbers}.
(2) The range for A and B is $\{y \mid y > 0\}$.
(3) The function $A: y = 4^x$ is constantly increasing: $y_2 > y_1$ if $x_2 > x_1$.
The function $B: y = \left(\frac{1}{4}\right)^x$ is constantly decreasing: $y_2 < y_1$ if $x_2 > x_1$.
(4) The ordered pair $(0, 1)$ belongs to both functions A and B.

Notice that the graphs of A and B are *symmetrical* with respect to the y-axis.

Written Exercises

See TE Answer Section.

Graph the exponential function described by each equation. Choose convenient scales for the axes. List four properties of the function.

Ⓐ 1. $y = 2^x$ 2. $y = 3^x$ 3. $y = 5^x$
4. $y = \left(\frac{1}{2}\right)^x$ 5. $y = \left(\frac{1}{3}\right)^x$ 6. $y = \left(\frac{1}{5}\right)^x$

Ⓑ 7. $y = 5 \cdot 2^x$ 8. $y = -1 \cdot 3^x$ 9. $y = 3 \cdot 4^x$
10. $y = 3\left(\frac{1}{2}\right)^x$ 11. $y = 2\left(\frac{1}{3}\right)^x$ 12. $y = -\left(\frac{1}{4}\right)^x$
13. $y = 2^x + 4$ 14. $y = 4^x - 8$ 15. $y = 3\left(\frac{1}{2}\right)^x + 6$

Ⓒ 16. $y = 2^{x-3}$ 17. $y = \left(\frac{1}{2}\right)^{x+2}$ 18. $y = 9^{\sqrt{x}}$
     for $1 \leq x \leq 7$ for $-4 \leq x \leq 2$ for $0 \leq x \leq 4$
                                                                                            Use multiples of 0.5 for x.

Graphing Exponential Functions

LOGARITHMIC FUNCTIONS   15.2

Objectives **To find the base-*b* logarithm of a positive number**
To solve a base-*b* logarithmic equation

The base-2 exponential function $A: y = 2^x$ can be viewed as a *mapping* from the exponent (x) to the power (y, or 2^x).

Exponent	$\longrightarrow$	Power
5	$\longrightarrow$	2^5, or 32
3	$\longrightarrow$	2^3, or 8
0	$\longrightarrow$	2^0, or 1
-1	$\longrightarrow$	2^{-1}, or $\frac{1}{2}$

The *inverse* of the base-2 exponential function is a mapping from the power to the exponent.

Power	$\longrightarrow$	Exponent
32, or 2^5	$\longrightarrow$	5
8, or 2^3	$\longrightarrow$	3
1, or 2^0	$\longrightarrow$	0
$\frac{1}{2}$, or 2^{-1}	$\longrightarrow$	-1

The *inverse* of the base-2 *exponential function* is the base-2 **logarithmic function**. The base-2 logarithm of 8 ($\log_2 8$) is found as follows.

$$8 = 2^3$$
$$\log_2 8 = \log_2 2^3 = 3$$

The four sentences below show how some base-5 logarithms ($\log_5$) are found.

$$\log_5 125 = \log_5 5^3 = 3 \qquad 125 = 5 \cdot 5 \cdot 5$$
$$\log_5 1 = \log_5 5^0 = 0 \qquad 1 = 5^0$$
$$\log_5 \frac{1}{25} = \log_5 5^{-2} = -2 \qquad \frac{1}{25} = \frac{1}{5^2} = 5^{-2}$$
$$\log_5 \sqrt{5} = \log_5 5^{\frac{1}{2}} = \frac{1}{2} \qquad \sqrt[n]{x} = x^{\frac{1}{n}}$$

Definition:
Base-*b*
logarithm

For each positive number y and each positive number $b \neq 1$,
if $y = b^x$, then $\log_b y = \log_b b^x = x$.

Example 1 Find $\log_{10} 100$ and $\log_2 \frac{1}{16}$.

$100 = 10^2$; so, $\log_{10} 100 = \log_{10} 10^2 = 2$.

$\frac{1}{16} = \frac{1}{2^4} = 2^{-4}$; so, $\log_2 \frac{1}{16} = \log_2 2^{-4} = -4$.

Example 2 Find $\log_3 \sqrt[4]{3}$ and $\log_2 \sqrt[5]{8}$.

$\sqrt[4]{3} = 3^{\frac{1}{4}}$; so, $\log_3 \sqrt[4]{3} = \log_3 3^{\frac{1}{4}} = \frac{1}{4}$.

$\sqrt[5]{8} = \sqrt[5]{2^3} = 2^{\frac{3}{5}}$; so, $\log_2 \sqrt[5]{8} = \log_2 2^{\frac{3}{5}} = \frac{3}{5}$.

You can solve some logarithmic equations of the form $\log_b y = x$ by rewriting them in the exponential form, $y = b^x$.

Example 3 Solve each logarithmic equation for the variable.

	Logarithmic equation	Exponential equation	Solution	
1.	$\log_b 81 = 4$	$81 = b^4$	3	◀ $3^4 = 81$
2.	$\log_5 625 = x$	$625 = 5^x$	4	◀ $5^4 = 625$
3.	$\log_4 y = 3$	$y = 4^3$	64	
4.	$\log_b \frac{1}{8} = -3$	$\frac{1}{8} = b^{-3}$	2	◀ $2^{-3} = \frac{1}{2^3} = \frac{1}{8}$

Written Exercises

Find each logarithm.

(A)
1. $\log_7 49$ **2**
2. $\log_4 64$ **3**
3. $\log_3 81$ **4**
4. $\log_6 1$ **0**
5. $\log_3 3$ **1**
6. $\log_{12} 1$ **0**
7. $\log_2 \frac{1}{2}$ **−1**
8. $\log_4 \frac{1}{16}$ **−2**
9. $\log_3 \frac{1}{81}$ **−4**
10. $\log_2 \sqrt[3]{2}$ **$\frac{1}{3}$**
11. $\log_3 \sqrt[5]{3}$ **$\frac{1}{5}$**
12. $\log_6 \sqrt{6}$ **$\frac{1}{2}$**

Solve each equation for the variable.

13. $\log_b 1000 = 3$ **10**
14. $\log_2 y = 8$ **256**
15. $\log_b 32 = 5$ **2**
16. $\log_3 243 = x$ **5**

(B)
17. $\log_{\frac{1}{2}} y = 3$ **$\frac{1}{8}$**
18. $\log_b \frac{1}{9} = -2$ **3**
19. $\log_{\frac{1}{3}} 27 = x$ **−3**
20. $\log_b \frac{1}{125} = -3$ **5**

Find each logarithm.

21. $\log_2 \sqrt[5]{16}$ **$\frac{4}{5}$**
22. $\log_3 \sqrt[4]{27}$ **$\frac{3}{4}$**
23. $\log_5 \sqrt{125}$ **$\frac{3}{2}$**
24. $\log_{10} \sqrt[3]{100}$ **$\frac{2}{3}$**
25. $\log_{\frac{1}{2}} \frac{1}{16}$ **4**
26. $\log_{\frac{3}{5}} \frac{27}{125}$ **3**
27. $\log_{\frac{1}{5}} 25$ **−2**
28. $\log_{\frac{3}{4}} \frac{16}{9}$ **−2**

(C)
29. $\log_c c^d$ **d**
30. $\log_b (b^3)^a$ **3a**
31. $\log_{5n} 25n^2$ **2**
32. $\log_{3n} \frac{1}{9n^2}$ **−2**

Find the sum of each series for the given data.

33. $\sum_{k=1}^{5} \log_2 2^{k-2}$ **5**
34. $\sum_{n=2}^{5} \log_{10} \sqrt[n]{10}$ **$\frac{77}{60}$**
35. $\sum_{j=3}^{6} \log_2 4^{5-j}$ **4**

CUMULATIVE REVIEW

A: 6 capsules, B: 24 capsules, C: 18 capsules

1. Box type A holds 18 fewer vitamin capsules than box type B, and box type C contains three times as many capsules as box type A. If 10 type A boxes, 8 type B boxes, and 4 type C boxes hold 27 dozen capsules, find the number of capsules that each box type contains.
2. A rectangular garden is 9 m longer than it is wide. A second rectangular garden is planned so that it will be 6 m wider and twice as long as the first garden. Find the area of the first garden if the sum of the areas of both gardens will be 528 m². **112 m²**

Logarithmic Functions

GRAPHING LOGARITHMIC FUNCTIONS 15.3

Objectives **To write an equation for the inverse of an exponential function or a logarithmic function**
To draw the graph of a logarithmic function
To read and interpret the graph of a logarithmic function

Consider the exponential function $A: y = 2^x$ and its *inverse* function, A^{-1}. You can write the y-form of the equation for the inverse as follows.

$A:$ $\quad y = 2^x$ ◀ To obtain the inverse of A, replace x by y and y by x.

A^{-1}, inverse of $A:$ $\quad x = 2^y$

$\log_2 x = \log_2 2^y = y$ ◀ Change the exponential equation to a logarithmic equation.

$y = \log_2 x$

Thus, the inverse of the base-2 exponential function ($y = 2^x$) is the base-2 logarithmic function ($y = \log_2 x$).

> **Inverse of an exponential function**
>
> The *inverse* of the exponential function
> $$y = b^x \quad [y > 0, b > 0, b \neq 1, x \text{ is a real number}]$$
> is the logarithmic function described by
> $$y = \log_b x \quad [x > 0, b > 0, b \neq 1, y \text{ is a real number}].$$

Example 1 Write the y-form of the equation for the inverse of $A: y = \left(\frac{1}{4}\right)^x$ and the inverse of $B: y = \log_3 x$.

$y = b^x$ and $y = \log_b x$ describe inverse functions.

Thus, the inverse of $A: y = \left(\frac{1}{4}\right)^x$ is described by $y = \log_{\frac{1}{4}} x$.
The inverse of $B: y = \log_3 x$ is described by $y = 3^x$.

You can draw the graph for $B: y = \log_3 x$ by constructing a table of ordered pairs (x, y). First, rewrite $y = \log_3 x$ as $x = 3^y$. Second, choose values of y and use $x = 3^y$ to find the corresponding values of x.

x	$\frac{1}{9}$	$\frac{1}{3}$	1	3	9
$y = \log_3 x$	-2	-1	0	1	2

◀ Second, use $x = 3^y$ to find x.
◀ First, choose values of y.

The inverse of $B: y = \log_3 x$ is the exponential function $B^{-1}: y = 3^x$.
The graphs of these inverse functions are shown on page 395.

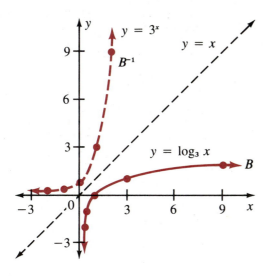

Recall that the graphs of inverse functions are symmetric with respect to the line $y = x$. This can be seen for the graphs of the inverse functions B and B^{-1}.

From the graph of B you can list four properties of the logarithmic function described by $y = \log_3 x$.
(1) The domain is $\{x \mid x > 0\}$.
(2) The range is {all real numbers}.
(3) The function is *constantly increasing*. That is, if $x_2 > x_1$, then $y_2 > y_1$.
(4) The ordered pair $(1, 0)$ belongs to the function. That is, $\log_3 1 = 0$.

Example 2   Draw the graphs of the logarithmic functions $A: y = \log_4 x$ and $B: y = \log_{\frac{1}{4}} x$ in the same coordinate plane with convenient scales for the axes. List four properties of each function.

1. Rewrite $y = \log_4 x$ as $x = 4^y$.

A:

x	$\frac{1}{16}$	$\frac{1}{4}$	1	4	16
y	-2	-1	0	1	2

◀ $x = 4^y$

2. Rewrite $y = \log_{\frac{1}{4}} x$ as $x = \left(\frac{1}{4}\right)^y = 4^{-y}$.

B:

x	16	4	1	$\frac{1}{4}$	$\frac{1}{16}$
y	-2	-1	0	1	2

◀ $x = 4^{-y}$

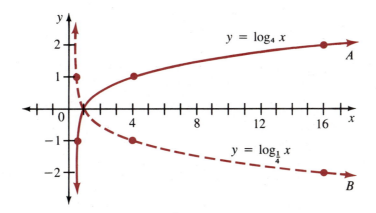

Graphing Logarithmic Functions

The graphs show the following four properties for
$A: y = \log_4 x$ and $B: y = \log_{\frac{1}{4}} x$.

(1) The domain for A and B is $\{x \mid x > 0\}$.
(2) The range for A and B is {all real numbers}.
(3) The function $A: y = \log_4 x$ is constantly increasing.
    The function $B: y = \log_{\frac{1}{4}} x$ is constantly decreasing.
(4) The ordered pair $(1, 0)$ belongs to both functions A and B.

Notice that the graphs of $A: y = \log_4 x$ and $B: y = \log_{\frac{1}{4}} x$ are symmetric with respect to the x-axis.

Reading in Algebra

Determine whether the graphs of the two functions described are symmetric with respect to the x-axis, the y-axis, or the line described by $y = x$. Are the two functions inverses of each other?

1. $y = \log_6 x$, $y = \log_{\frac{1}{6}} x$ **x-axis, no**
2. $y = \log_6 x$, $y = 6^x$ **y = x, yes**
3. $y = 6^x$, $y = \left(\frac{1}{6}\right)^x$ **y-axis, no**
4. $y = \log_{\frac{1}{6}} x$, $y = \left(\frac{1}{6}\right)^x$ **y = x, yes**

Written Exercises

For Ex. 5–16 and 21–22, see TE Answer Section.

For each function described, write the y-form of the equation for the inverse function.

Ⓐ 1. $y = 5^x$ **y = log₅ x**   2. $y = \log_{\frac{1}{3}} x$ **y = (⅓)ˣ**   3. $y = \log_8 x$ **y = 8ˣ**   4. $y = \left(\frac{1}{5}\right)^x$ **y = log_{\frac{1}{5}} x**

Graph each function described. Choose convenient scales for the axes. List four properties of the function.

5. $y = \log_2 x$
6. $y = \log_{\frac{1}{2}} x$
7. $y = \log_5 x$
8. $y = \log_{\frac{1}{3}} x$
9. $y = \log_6 x$
10. $y = \log_{\frac{1}{5}} x$

Ⓑ 11. $y = 4 \cdot \log_2 x$
12. $y = 2 \cdot \log_3 x$
13. $y = 3 \cdot \log_{\frac{1}{2}} x$
14. $y = 4 + \log_2 x$
15. $y = 2 + \log_3 x$
16. $y = 3 + \log_{\frac{1}{2}} x$

For each function described, write the y-form of an equation so that the graphs of both equations are symmetric with respect to the x-axis.

17. $y = \log_5 x$ **y = log_{\frac{1}{5}} x**
18. $y = \log_{\frac{1}{5}} x$ **y = log₅ x**

For each function described, write the y-form of an equation so that the graphs of both equations are symmetric with respect to the y-axis.

19. $y = 5^x$ **y = (⅕)ˣ**
20. $y = \left(\frac{1}{5}\right)^x$ **y = 5ˣ**

Ⓒ 21. Graph the function described by $y = \log_2(x - 3)$.
22. Graph the function described by $y = \log_3(x + 2)$.

LOGS OF PRODUCTS AND QUOTIENTS   15.4

Objectives **To expand the logarithm of a product or quotient**
To simplify a sum or difference of logarithms
To solve a logarithmic equation

One of the laws of logarithms concerns the logarithm of the **product** of two factors.

First law of logarithms: $\log_b (x \cdot y) = \log_b x + \log_b y$

As an example, you can show that $\log_2 (8 \cdot 32) = \log_2 8 + \log_2 32$.

$\log_2 (8 \cdot 32)$	$\log_2 8 + \log_2 32$
$\log_2 (2^3 \cdot 2^5)$	$\log_2 2^3 + \log_2 2^5$
$\log_2 2^8$	$3 + 5$
8	8

Proof Let: $\log_b x = p$ and $\log_b y = q$
Then: $x = b^p$ and $y = b^q$
$x \cdot y = b^p \cdot b^q$
$x \cdot y = b^{p+q}$
$\log_b (x \cdot y) = \log_b b^{p+q}$
$\log_b (x \cdot y) = p + q$
$\log_b (x \cdot y) = \log_b x + \log_b y$

In a similar way, you can prove another law of logarithms. It concerns the logarithm of the **quotient** of two terms.

Second law of logarithms: $\log_b \frac{x}{y} = \log_b x - \log_b y$

As an example, you can show that $\log_2 \frac{128}{8} = \log_2 128 - \log_2 8$, since

(1) $\log_2 \frac{128}{8} = \log_2 \frac{2^7}{2^3} = \log_2 2^4 = 4$ and

(2) $\log_2 128 - \log_2 8 = \log_2 2^7 - \log_2 2^3 = 7 - 3 = 4.$

Log_b of a product $\log_b (x \cdot y) = \log_b x + \log_b y \quad [x > 0, y > 0, b > 0, b \neq 1]$

Log_b of a quotient $\log_b \frac{x}{y} = \log_b x - \log_b y \quad [x > 0, y > 0, b > 0, b \neq 1]$

You can use the first two laws of logarithms to *expand* an expression like $\log_3 \frac{5c}{\pi d}$.

Example 1 Write $\log_3 \frac{5c}{\pi d}$ in expanded form.

$\log_3 \frac{5 \cdot c}{\pi \cdot d} = \log_3 5 \cdot c - \log_3 \pi \cdot d$ ◀ $\log_b \frac{x}{y} = \log_b x - \log_b y$
$= (\log_3 5 + \log_3 c) - (\log_3 \pi + \log_3 d)$ ◀ $\log_b x \cdot y = \log_b x + \log_b y$
$= \log_3 5 + \log_3 c - \log_3 \pi - \log_3 d$

Thus, $\log_3 \frac{5c}{\pi d} = \log_3 5 + \log_3 c - \log_3 \pi - \log_3 d.$

Logs of Products and Quotients

Example 2 Given the formula $L = 2\pi rh$, write $\log_b r$ in expanded form.

$$r = \frac{L}{2\pi h} \quad \blacktriangleleft \text{ Solve } L = 2\pi rh \text{ for } r.$$

$$\log_b r = \log_b \frac{L}{2\pi h} \quad \blacktriangleleft \text{ If } x = y, \text{ then } \log_b x = \log_b y.$$

$$= \log_b L - \log_b 2\pi h$$

$$= \log_b L - (\log_b 2 + \log_b \pi + \log_b h)$$

Thus, $\log_b r = \log_b L - \log_b 2 - \log_b \pi - \log_b h$, if $L = 2\pi rh$.

Reversing the steps in Examples 1 and 2 yields the logarithm of the original expression since the first two laws of logarithms are reversible. That is,

(1) $\log_b x + \log_b y = \log_b (x \cdot y)$ and (2) $\log_b x - \log_b y = \log_b \frac{x}{y}$.

Example 3 Write $\log_3 7 + \log_3 t - \log_3 4 - \log_3 v$ as the $\log_3$ of *one* expression.

$$\log_3 7 + \log_3 t - \log_3 4 - \log_3 v = \underbrace{\log_3 7 + \log_3 t}_{} - \underbrace{(\log_3 4 + \log_3 v)}_{} \quad \blacktriangleleft \text{ Rewrite.}$$

$$= \log_3 (7 \cdot t) - \log_3 (4 \cdot v) \quad \blacktriangleleft \text{ First law}$$

$$= \log_3 \frac{7t}{4v} \quad \blacktriangleleft \text{ Second law}$$

Thus, $\log_3 7 + \log_3 t - \log_3 4 - \log_3 v = \log_3 \frac{7t}{4v}$.

You can solve a logarithmic equation by using the **property of equality for logarithms:** If $\log_b x = \log_b y$, then $x = y$.

Each *apparent* solution of a logarithmic equation must be checked since $\log_b x$ is defined only for $x > 0$.

Example 4 Solve $\log_5 2t + \log_5 (t - 4) = \log_5 48 - \log_5 2$. Check.

$$\underbrace{\log_5 2t + \log_5 (t - 4)}_{} = \underbrace{\log_5 48 - \log_5 2}_{}$$

First law ▶ $\log_5 2t(t - 4) = \log_5 \frac{48}{2}$ ◀ *Second law*

$$2t(t - 4) = \frac{48}{2} \quad \blacktriangleleft \text{ If } \log_b x = \log_b y,$$

$$2t^2 - 8t = 24 \quad \text{then } x = y.$$

$$t^2 - 4t - 12 = 0$$

$$(t - 6)(t + 2) = 0$$

$$t = 6 \text{ or } t = -2$$

Check: $t = 6$. $\log_5 2t + \log_5 (t - 4) = \log_5 48 - \log_5 2$ Check: $t = -2$.
$\log_5 12 + \log_5 2 \qquad = \log_5 48 - \log_5 2$ $\log_5 2t = \log_5 (-4)$
$\log_5 24 \qquad\qquad\qquad = \log_5 24$ and $\log_5 (-4)$ is undefined
Thus, 6 is the only solution.   since $-4 < 0$.

Oral Exercises

True or false?
1. $\log_2 (3 \cdot 7) = \log_2 3 + \log_2 7$ **T**
2. $\log_5 12 = \log_5 4 + \log_5 3$ **T**
3. $\log_b 8 = \log_b 3 + \log_b 5$ **F**
4. $\log_5 2 + \log_5 3 = \log_5 6$ **T**
5. $\log_b 30 = \log_b 2 + \log_b 3 + \log_b 5$ **T**
6. $\log_b 5 + \log_b 5 + \log_b 5 = \log_b 15$ **F**
7. $\log_3 \frac{17}{4} = \log_3 17 - \log_3 4$ **T**
8. $\log_b \frac{18}{6} = \log_b 18 \div \log_b 6$ **F**
9. $\log_2 \frac{3 \cdot 5}{7} = \log_2 3 - \log_2 7 - \log_2 5$ **F**
10. $\log_2 \frac{6 \cdot 9}{5} = \log_2 6 + \log_2 9 - \log_2 5$ **T**
11. $\log_b \frac{35}{3 \cdot 8} = \log_b 35 - \log_b 3 - \log_b 8$ **T**
12. $\log_3 \frac{7}{5 \cdot 4} = \log_3 7 - \log_3 5 + \log_3 4$ **F**
13. If $\log_3 x + \log_3 y = \log_3 z$, then $x \cdot y = z$. **T**
14. If $\log_2 m - \log_2 n = \log_2 p$, then $m \div n = p$. **T**

Written Exercises

2. $\log_3 7 + \log_3 t + \log_3 w$ 6. $\log_b 8 + \log_b f - \log_b g$

Write each logarithmic expression in expanded form.

3. $\log_b 2 + \log_b m + \log_b n + \log_b p$

A
1. $\log_2 5a$ $\log_2 5 + \log_2 a$
2. $\log_3 7tw$ **above**
3. $\log_b 2mnp$ **above**
4. $\log_4 \frac{a}{3}$ $\log_4 a - \log_4 3$
5. $\log_3 \frac{4}{a}$ $\log_3 4 - \log_3 a$
6. $\log_b \frac{8f}{g}$ **above**

For each formula, write the indicated logarithm in expanded form.

7. $d = 2r$; $\log_b r$
$\log_b r = \log_b d - \log_b 2$
8. $T = \frac{1}{2}A$; $\log_5 A$
$\log_5 A = \log_5 2 + \log_5 T$
9. $3B = 2V$; $\log_5 B$
$\log_5 B = \log_5 2 + \log_5 V - \log_5 3$

Write each expression as the logarithm of one expression.

10. $\log_2 9 + \log_2 c$ $\log_2 9c$
11. $\log_3 7 - \log_3 t$ $\log_3 \frac{7}{t}$
12. $\log_b 8 + \log_b m - \log_b n$
13. $\log_5 6c^2 - \log_5 2c$ $\log_5 3c$
14. $\log_2 3t + \log_2 5t$ $\log_2 15t^2$
15. $\log_b 9c + \log_b 8c - \log_b 6c$

12. $\log_b \frac{8m}{n}$ 15. $\log_b 12c$

Solve each equation. Check.

16. $\log_2 6 + \log_2 k = \log_2 18$ **3**
17. $\log_b 12k - \log_b 4 = \log_b 15$ **5**
18. $\log_b 2t + \log_b 3t = \log_b 24$ **2**
19. $\log_5 t + \log_5 (t - 4) = \log_5 12$ **6**

B
20. $\log_3 y^2 + \log_3 y^3 = \log_3 16y$ **2**
21. $\log_b (c^2 - 4c) - \log_b c = \log_b 2$ **6**
22. $\log_5 3 + \log_5 t + \log_5 (t + 4) = \log_5 63$ **3**
23. $\log_b (y - 3) - \log_b (y - 4) = \log_b (y + 3) - \log_b y$ **6**

Write each logarithmic expression in expanded form.

24. $\log_2 \frac{3}{mn}$
25. $\log_b \frac{5c}{3d}$
26. $\log_3 \frac{1}{cd} - \log_3 c - \log_3 d$

24. $\log_2 3 - \log_2 m - \log_2 n$
25. $\log_b 5 + \log_b c - \log_b 3 - \log_b d$

Write each expression as the logarithm of one expression.

27. $\log_4 \frac{21}{mn}$ 28. $\log_b \frac{v^2}{t}$

27. $\log_4 7 - \log_4 m + \log_4 3 - \log_4 n$
28. $\log_b 2t - \log_b 3v - \log_b 6t^2 + \log_b 9v^3$

31. $\log_5 r = \log_5 4 + \log_5 A - \log_5 3 - \log_5 t$

For each formula, write the indicated logarithm in expanded form.

29. $L = 2\pi rh$; $\log_b h$
30. $V = \frac{1}{3}Bh$; $\log_2 B$
31. $A = \frac{3}{4}rt$; $\log_5 r$

29. $\log_b h = \log_b L - \log_b 2 - \log_b \pi - \log_b r$
30. $\log_2 B = \log_2 3 + \log_2 V - \log_2 h$

C
32. Prove the second law of logarithms: $\log_b \frac{x}{y} = \log_b x - \log_b y$. **See TE Answer Section.**

Logs of Products and Quotients

LOGS OF POWERS AND RADICALS 15.5

Objectives
To expand a logarithm of a power or a radical
To simplify a multiple of a logarithm
To solve a logarithmic equation

The third law of logarithms involves the logarithm of a **power**.

Third law of logarithms: $\log_b x^r = r \cdot \log_b x$

As an example, you can show that $\log_2 5^3 = 3 \cdot \log_2 5$, where $r = 3$, since

$$\log_2 5^3 = \log_2(5 \cdot 5 \cdot 5) = \log_2 5 + \log_2 5 + \log_2 5 = 3 \cdot \log_2 5.$$

Proof Let $\log_b x = p$. Then, $x = b^p$ and $x^r = (b^p)^r = b^{r \cdot p}$.
If $x^r = b^{r \cdot p}$, then $\log_b x^r = \log_b b^{r \cdot p} = r \cdot p = r \cdot \log_b x$.

Recall that $\sqrt[n]{x} = x^{\frac{1}{n}}$. Thus, the third law can be used with the logarithm of a **radical**. For example,

$$\log_2 \sqrt[3]{5} = \log_2 5^{\frac{1}{3}} = \frac{1}{3} \cdot \log_2 5, \text{ since } \log_2 5^r = r \cdot \log_2 5, \text{ where } r = \frac{1}{3}.$$

Log$_b$ of a power

$\log_b x^r = r \cdot \log_b x$ [$x > 0$, r is a real number]

$\log_b \sqrt[n]{x} = \frac{1}{n} \cdot \log_b x$ [$x > 0$, n is a positive integer].

This third law is applied in Examples 1 and 2.

Example 1 Write $\log_5(mn^3)^2$ and $\log_b \sqrt[4]{\dfrac{m^3}{n}}$ in expanded form.

$\log_5(mn^3)^2 = 2 \cdot \log_5 mn^3 = 2(\log_5 m + \log_5 n^3) = 2(\log_5 m + 3 \cdot \log_5 n)$

$\log_b \sqrt[4]{\dfrac{m^3}{n}} = \log_b \left(\dfrac{m^3}{n}\right)^{\frac{1}{4}} = \dfrac{1}{4} \cdot \log_b \dfrac{m^3}{n} = \dfrac{1}{4}(\log_b m^3 - \log_b n) = \dfrac{1}{4}(3 \cdot \log_b m - \log_b n)$

Example 2 Given $d = \dfrac{1}{2}gt^2$, write $\log_b t$ in expanded form.

Solve $d = \dfrac{1}{2}gt^2$ for t, $t > 0$.

$\dfrac{1}{2}gt^2 = d$

$gt^2 = 2d$

$t^2 = \dfrac{2d}{g}$

$t = \sqrt{\dfrac{2d}{g}}$

$\log_b t = \log_b \sqrt{\dfrac{2d}{g}} = \dfrac{1}{2} \cdot \log_b \dfrac{2d}{g}$

$\qquad = \dfrac{1}{2}(\log_b 2d - \log_b g)$

$\qquad = \dfrac{1}{2}(\log_b 2 + \log_b d - \log_b g)$

Thus, $\log_b t = \dfrac{1}{2}(\log_b 2 + \log_b d - \log_b g)$ if $d = \dfrac{1}{2}gt^2$.

You can simplify some logarithmic expressions by "reversing" the third law of logarithms.

$$r \cdot \log_b x = \log_b x^r \text{ and } \frac{1}{n} \cdot \log_b x = \log_b x^{\frac{1}{n}} = \log_b \sqrt[n]{x}$$

Example 3 **Write $5 \log_2 c + 3 \log_2 d$ as the $\log_2$ of one expression.**

$$5 \cdot \log_2 c + 3 \cdot \log_2 d = \log_2 c^5 + \log_2 d^3 = \log_2 c^5 d^3$$

Example 4 **Write $\frac{1}{3}(\log_5 t + 4 \log_5 v - \log_5 w)$ as the $\log_5$ of one expression.**

$$\frac{1}{3}(\log_5 t + 4 \cdot \log_5 v - \log_5 w) = \frac{1}{3}(\log_5 t + \log_5 v^4 - \log_5 w)$$
$$= \frac{1}{3} \cdot \log_5 \frac{tv^4}{w}$$
$$= \log_5 \sqrt[3]{\frac{tv^4}{w}}$$

Logarithmic equations with terms of the form $r \cdot \log_b x$ can be solved by using the third law of logarithms as shown below.

Example 5 **Solve $2 \log_3 t = \log_3 2 + \log_3(t + 12)$. Check.**

$$2 \cdot \log_3 t = \log_3 2 + \log_3(t + 12)$$
$$\log_3 t^2 = \log_3 2(t + 12) \quad \blacktriangleleft \text{ If } \log_b x = \log_b y,$$
$$t^2 = 2t + 24 \quad \blacktriangleleft \text{ then } x = y.$$

$$t^2 - 2t - 24 = 0$$
$$(t - 6)(t + 4) = 0$$
$$t = 6 \text{ or } t = -4$$

Check: $t = 6$

$2 \log_3 t$	$\log_3 2 + \log_3 (t + 12)$
$2 \log_3 6$	$\log_3 2 + \log_3 18$
$\log_3 6^2$	$\log_3 2 \cdot 18$
$\log_3 36$	$\log_3 36$

Check: $t = -4$

$2 \log_3 t$	$\log_3 2 + \log_3 (t + 12)$
$2 \log_3 (-4)$	$\log_3 2 + \log_3 8$

$\text{Log}_3 (-4)$ is undefined.
Thus, 6 is the solution.

Example 6 **Solve $\frac{1}{2} \log_b a + \frac{1}{2} \log_b (a - 6) = \log_b 4$. Check.**

$\frac{1}{2} \log_b x = \log_b \sqrt{x} \blacktriangleright$
$$\log_b \sqrt{a} + \log_b \sqrt{a - 6} = \log_b 4$$
$$\log_b (\sqrt{a} \cdot \sqrt{a - 6}) = \log_b 4$$
$$\sqrt{a^2 - 6a} = 4$$
$$a^2 - 6a = 16$$
$$a^2 - 6a - 16 = 0$$
$$(a - 8)(a + 2) = 0$$
$$a = 8 \text{ or } a = -2$$
Thus, 8 is the solution.

Check: $a = 8$
$\frac{1}{2} \log_b a + \frac{1}{2} \log_b (a - 6)$
$\frac{1}{2} \log_b 8 + \frac{1}{2} \log_b 2$
$\log_b \sqrt{8} + \log_b \sqrt{2}$
$\log_b \sqrt{8} \ \sqrt{2}$
$\log_b \sqrt{16}, \text{ or } \log_b 4$
Check: $a = -2$
$\text{Log}_b (-2)$ is undefined.

Logs of Powers and Radicals

Written Exercises

Write each logarithmic expression in expanded form.

(A) 1. $\log_2 a^5$ **$5 \log_2 a$**
2. $\log_3 \sqrt[5]{a}$ **$\frac{1}{5} \log_3 a$**
3. $\log_5 (mn)^2$ **$2(\log_5 m + \log_5 n)$**
4. $\log_b \sqrt{2g}$ **$\frac{1}{2}(\log_b 2 + \log_b g)$**
5. $\log_4 \left(\frac{v}{h}\right)^3$ **$3(\log_4 v - \log_4 h)$**
6. $\log_b \sqrt[3]{\frac{d}{g}}$ **$\frac{1}{3}(\log_b d - \log_b g)$**

For each formula, write the indicated logarithm in expanded form.

7. $v^2 = 2gh$; $\log_3 v$
$\log_3 v = \frac{1}{2}(\log_3 2 + \log_3 g + \log_3 h)$
8. $d = \frac{1}{2} gt^2$; $\log_b g$
$\log_b g = \log_b 2 + \log_b d - 2 \log_b t$
9. $A = \pi r^2$; $\log_4 r$
$\log_4 r = \frac{1}{2}(\log_4 A - \log_4 \pi)$

Write each expression as the logarithm of one expression.

10. $2 \log_3 x + 4 \log_3 y$ **below**
11. $\frac{1}{3}(\log_2 6 + \log_2 c)$ **below**
12. $3(\log_b 2 + \log_b t)$ **below**
13. $\frac{1}{2} \log_5 2c + \frac{1}{2} \log_5 d$ **below**
14. $4 \log_b t - 3 \log_b v$ **below**
15. $\frac{1}{4} \log_3 6y - \frac{1}{4} \log_3 5z$ **below**

10. $\log_3 x^2 y^4$ **11.** $\log_2 \sqrt[3]{6c}$ **12.** $\log_b (2t)^3$ **13.** $\log_5 \sqrt{2cd}$

Solve each equation. Check.

16. $2 \log_3 y = \log_3 4 + \log_3 (y + 8)$ **8**
17. $2 \log_b t - \log_b 2 = \log_b (2t + 6)$ **6**
18. $\frac{1}{2} \log_b a + \frac{1}{2} \log_b (a + 5) = \log_b 6$ **4**
19. $\frac{1}{3} \log_5 y + \frac{1}{3} \log_5 (y + 2) = \log_5 2$ **2**

(B) 20. $2 \log_b (2y + 2) = \log_b 16 + 2 \log_b (y - 2)$ **5**
21. $2 \log_5 (3a + 1) = \log_5 4 - 2 \log_5 (2a - 1)$ **3**
22. $\frac{1}{2} \log_3 (c + 1) + \frac{1}{2} \log_3 (c - 4) = \log_3 6$ **8**
23. $\frac{1}{2} \log_b (y - 1) - \frac{1}{2} \log_b (2y - 1) = \log_b 2 - \log_b 3$ **5**
24. $2 \log_3 (y + 5) - \log_3 (y + 2) = \log_3 (y + 10)$ **2.5**
25. $\log_5 12 - \frac{1}{2} \log_5 3 = \frac{1}{2} \log_5 2y$ **24**

14. $\log_b \frac{t^4}{v^3}$
15. $\log_3 \sqrt[4]{\frac{6y}{5z}}$
32. $\log_5 \frac{7 \sqrt{x}}{y^2}$
33. $\log_4 \frac{\sqrt[3]{10ab}}{a^2 b}$
34. $\log_b \left(\frac{mn^2}{p}\right)^3$
35. $\log_b \sqrt[4]{\frac{3x^3}{10y}}$

Write each logarithmic expression in expanded form.

26. $\log_b (t^3 v^2 w)^2$
27. $\log_5 \left(\frac{mn^2}{p^3}\right)^3$
28. $\log_b \sqrt[4]{\frac{5d}{cg^3}}$ **above**

$\frac{1}{4}(\log_b 5 + \log_b d - \log_b c - 3 \log_b g)$

26. $2(3 \log_b t + 2 \log_b v + \log_b w)$ **27.** $3(\log_5 m + 2 \log_5 n - 3 \log_5 p)$

For each formula, write the indicated logarithm in expanded form.

29. $A = P(1.09)^{12}$; $\log_{10} A$
30. $V = \frac{4}{3} \pi r^3$; $\log_b r$
31. $V = \frac{1}{3} \pi r^2 h$; $\log_2 r$ **below**

29. $\log_{10} A = \log_{10} P + 12 \log_{10} 1.09$
30. $\log_b r = \frac{1}{3}(\log_b 3 + \log_b V - \log_b 4 - \log_b \pi)$

Write each expression as the logarithm of one expression.

32. $\log_5 7 + \frac{1}{2} \log_5 x - 2 \log_5 y$ **above**
33. $\frac{1}{3} \log_4 2a + \frac{1}{3} \log_4 5b - 2 \log_4 a - \log_4 b$ **above**
34. $3(\log_b m + 2 \log_b n - \log_b p)$ **above**
35. $\frac{1}{4}(\log_b 3 + 3 \log_b x - \log_b 2 - \log_b 5y)$ **above**

Solve each equation. Check (Ex. 36–39) [*Hint*: If $\log_b x = y$, then $b^y = x$.]

(C) 36. $\log_2 (y + 4) + \log_2 y = 5$ **4**
37. $\log_4 (3y + 1) + \log_4 (y - 1) = 3$ **5**
38. $\log_5 (y^2 + 2y + 5) - \log_5 (y - 5) = 2$ **10, 13**
39. $2 \log_3 (y - 2) - \log_3 (y - 4) = 2$ **5, 8**

31. $\log_2 r = \frac{1}{2}(\log_2 3 + \log_2 V - \log_2 \pi - \log_2 h)$

For Ex. 40–41, see TE Answer Section.

40. Prove: $\log_b \sqrt[3]{y^2} - \log_b \sqrt[4]{y^3} + 4 \log_b \sqrt[3]{y} - 5 \log_b \sqrt[4]{y} = 0$.
41. Prove: $\frac{2}{5} \log_b 3 - \frac{1}{5} \log_b 2 - \frac{2}{5} \log_b 4 + \frac{1}{5} \log_b 27 = \log_b 3 - \log_b 2$.

COMMON LOGARITHMS AND ANTILOGARITHMS 15.6

Objectives **To find the base-10 logarithm of a positive rational number**
To find the base-10 antilogarithm of a rational number

The table of **common (base-10)** logarithms on page 593 can be used to find that $\log_{10} 3.26 \doteq 0.5132$, correct to four decimal places.

$\log_{10} 3.2\ 6 \dashrightarrow 0.5132$

row 32 column 6   Found in row 32 and column 6

Decimal points must be placed in "326" and "5132" so that $1 \leq 3.26 < 10$ and $0 \leq 0.5132 < 1$. If $1 \leq n < 10$, then $0 \leq \log_{10} n < 1$.

You can approximate $\log_{10} 3260$ and $\log_{10} 0.0326$ to four decimal places using $\log_{10} 3.26 = 0.5132$ and *scientific notation*. Recall that

A number is in **scientific notation** when it is written in the form $a \times 10^c$ where $1 \leq a < 10$ and c is an integer.

You can proceed as shown below.

(1) **$\log_{10} 3260$**

$3260 = 3.26 \times 10^3$
$\log_{10} 3260 = \log_{10} (3.26 \times 10^3)$
$\phantom{\log_{10} 3260} = \log_{10} 3.26 + \log_{10} 10^3$
$\phantom{\log_{10} 3260} = 0.5132 + 3$
$\phantom{\log_{10} 3260} = 3 + 0.5132,\ \text{or}\ 3.5132$
$\log_{10} 3260 = 3 + 0.5132,\ \text{or}\ 3.5132$
The integer 3 is called the **characteristic** of the log.
The decimal .5132 is called the **mantissa** of the log.

(2) **$\log_{10} 0.0326$**

$0.0326 = 3.26 \times 10^{-2}$
$\log_{10} 0.0326 = \log_{10} (3.26 \times 10^{-2})$
$\phantom{\log_{10} 0.0326} = \log_{10} 3.26 + \log_{10} 10^{-2}$
$\phantom{\log_{10} 0.0326} = 0.5132 + (-2)$
$\phantom{\log_{10} 0.0326} = -2 + 0.5132,\ \text{or}\ 8.5132 - 10$
$\log_{10} 0.0326 = -2 + 0.5132,\ \text{or}\ 8.5132 - 10$, where -2 is rewritten as $8 - 10$.
The characteristic is -2, or $8 - 10$.
The mantissa is .5132.

Definition:
Characteristic
Mantissa

If $y = a \times 10^c$ where $1 \leq a < 10$ and c is an integer, then $\log_{10} y = c + \log_{10} a$. The integer c is the *characteristic* and $\log_{10} a$ is the *mantissa* of $\log_{10} y$.

Common Logarithms and Antilogarithms

It is customary to use the *equal* sign (=), rather than the *approximately equal* sign (≐), when approximating common logarithms.

You can omit the *subscript 10* from expressions like $\log_{10} 3260$ when working with common logarithms. That is,

$\log 3260$ (without a written base) means $\log_{10} 3260$.

Example 1 **Find log 294 and log 0.00294.**

$294 = 2.94 \times 10^2$	$0.00294 = 2.94 \times 10^{-3}$
$\log 2.94 = 0.4683$	$\log 2.94 = 0.4683$
Thus, $\log 294 = 2 + 0.4683$, or 2.4683.	**Thus,** $\log 0.00294 = -3 + 0.4683$, or $7.4683 - 10$.

The base-10 logarithmic function, where $\log 1000 = \log 10^3 = 3$, can be viewed as a mapping.

$$1000, \text{ or } 10^3 \longrightarrow 3$$
$$10, \text{ or } 10^1 \longrightarrow 1$$
$$0.01, \text{ or } 10^{-2} \longrightarrow -2$$
$$10^x \longrightarrow x$$

The **inverse** of the base-10 logarithmic function is the base-10 **antilogarithmic** function.

$$3 \longrightarrow 10^3, \text{ or } 1000$$
$$1 \longrightarrow 10^1, \text{ or } 10$$
$$-2 \longrightarrow 10^{-2}, \text{ or } 0.01$$
$$x \longrightarrow 10^x$$

So, $\log 10^x = x$ and antilog $x = 10^x$ for each real number x.

Definition: Base-10 antilogarithm

antilog $x = 10^x$, for each real number x.

The table on the preceding page shows how to find that

$$\log 3.26 = \log 10^{0.5132} = 0.5132.$$

You can reverse this process to find that

$$\text{antilog } 0.5132 = 10^{0.5132} = 3.26.$$

Example 2 **Find antilog 4.9309 and antilog 7.9309 − 10.**

The mantissa is .9309. Use the table on page 593 to find that antilog $0.9309 = 10^{0.9309} = 8.53$.

	antilog 4.9309	antilog 7.9309 − 10
	antilog 4 + 0.9309	antilog −3 + 0.9309 ◀ *Rewrite 7 − 10 as −3.*
antilog $x = 10^x$ ▶	$10^{4+0.9309}$	$10^{-3+0.9309}$
$10^{m+n} = 10^m \cdot 10^n$ ▶	$10^4 \times 10^{0.9309}$	$10^{-3} \times 10^{0.9309}$
	$10^4 \times 8.53$	$10^{-3} \times 8.53$ ◀ *Scientific notation*
Ordinary notation ▶	85,300	0.00853

Thus, antilog $4.9309 = 85{,}300$ and antilog $7.9309 - 10 = 0.00853$.

Reading in Algebra

Using the equation log 346 = 2.5391, find each of the following.
1. log 346 **2.5391**
2. mantissa of log 346 **0.5391**
3. characteristic of log 346 **2**
4. log 3.46 **0.5391**
5. antilog 2.5391 **346**
6. $10^{2.5391}$ **346**
7. log 0.0346 **8.5391 − 10, or −2 + 0.5391**
8. antilog 0.5391 **3.46**
9. x, if $10^x = 34.6$ **1.5391**

Write the characteristic of each logarithm in two ways.
10. log 0.0023 = 7.3617 − 10 **7 − 10, or −3**
11. log 0.23 = −1 + 0.3617 **−1, or 9 − 10**

Express each number in scientific notation.
12. 654,000 $\mathbf{6.54 \times 10^5}$
13. 4.32 $\mathbf{4.32 \times 10^0}$
14. 0.000789 $\mathbf{7.89 \times 10^{-4}}$

Express each number in ordinary notation.
15. 3.21×10^{-1} **0.321**
16. 5.43×10^0 **5.43**
17. 8.76×10^3 **8,760**

Find each logarithm or antilogarithm.
18. log 100,000 **5**
19. log 1 **0**
20. log 0.01 **−2**
21. antilog 2 **100**
22. antilog −3 **0.001**
23. antilog 0 **1**

Find the characteristic of each logarithm.
24. log 7520 **3**
25. log 7.52 **0**
26. log 0.000752 **−4, or 6 − 10**

Written Exercises

Find each logarithm or antilogarithm.

Ⓐ
1. log 19,800 **4.2967**
2. antilog 2.5132 **326**
3. log 0.207 **9.3160 − 10**
4. antilog 7.8129 − 10 **0.00650**
5. log 0.00613 **7.7875 − 10**
6. antilog −2 + 0.2227 **0.0167**

Ⓑ
7. log 32,600,000 **7.5132**
8. antilog 6.9345 − 10 **0.000860**
9. antilog 4.9206 **83,300**
10. log $\sqrt{10}$ **$\frac{1}{2}$, or 0.5**
11. antilog $\frac{1}{3}$ **$\sqrt[3]{10}$**
12. log $\sqrt[3]{1000}$ **1**

Solve each equation for x.

Ⓒ
13. $x^2 = \log 10^{5x-6}$ **2, 3**
14. $\dfrac{x+5}{\log 100} = \dfrac{\log 1{,}000}{x}$ **−6, 1**
15. $\dfrac{\log 10^{\frac{1}{x}}}{x} = \dfrac{1}{9}$ **−3, 3**

NON-ROUTINE PROBLEMS

1. Find log (antilog (log (antilog 2.0899))). **2.0899**
2. Find $\sum\limits_{k=1}^{4}$ antilog $(k-2)$. **111.1**

CUMULATIVE REVIEW

1. Solve $\dfrac{y}{10} = \dfrac{7}{11}$. $\mathbf{6\dfrac{4}{11}}$
2. Solve $\dfrac{9}{10} = \dfrac{x}{12}$. $\mathbf{10\dfrac{4}{5}}$
3. Solve $\dfrac{y}{10} = \dfrac{11}{25}$. $\mathbf{4\dfrac{2}{5}}$

Common Logarithms and Antilogarithms

USING EXPONENTIAL EQUATIONS 15.7

Objectives **To solve an exponential equation using base-10 logarithms**
To find the base-b log of a positive number, using a table of base-10 logarithms
To solve a word problem leading to an exponential equation

An equation like $4^{2x-1} = 360$ is called an **exponential equation** since the variable appears in the exponent. You can estimate a range for x by knowing some powers of 4.

$4^4 = 256$ and $4^5 = 1{,}024$ | So, $\quad 4 < 2x - 1 < 5$
$256 < 360 < 1{,}024$ | $\quad\quad\quad 5 < 2x < 6$
$4^4 < 4^{2x-1} < 4^5$ | Thus, $2.5 < x < 3.0$.

You can find x to three significant digits using base-10 logarithms.

Example 1 **Solve $4^{2x-1} = 360$ for x to three significant digits.**

$$4^{2x-1} = 360$$
$$\log 4^{2x-1} = \log 360 \quad \blacktriangleleft \text{ If } p = q, \text{ then } \log p = \log q.$$
$$(2x - 1) \log 4 = \log 360 \quad \blacktriangleleft \log a^r = r \cdot \log a$$
$$2x - 1 = \frac{\log 360}{\log 4} \quad \blacktriangleleft \text{ Divide each side by } \log 4.$$
$$2x - 1 = \frac{2.5563}{0.6021}$$
$$2x - 1 = 4.246 \quad \blacktriangleleft \text{ Divide to 4 significant digits.}$$
$$2x = 5.246$$
$$x = 2.623$$

Thus, $x = 2.62$ to three significant digits.

In Example 1, the value of x is 2.62. Notice that this agrees with the estimate (above) that $2.5 < x < 3.0$.

A base-10 log such as log 4 is found in the table on page 593. A base-b log such as $\log_3 4$ can be found to three significant digits by solving an exponential equation using base-10 logs as shown in Example 2.

Example 2 **Find $\log_3 4$ to three significant digits.**

Let $\quad \log_3 4 = x.\quad\quad \blacktriangleleft$ Base-3 log
$\quad\quad\quad 3^x = 4 \quad\quad \blacktriangleleft$ Change the equation to exponential form.
$\quad\quad \log 3^x = \log 4 \quad \blacktriangleleft$ Base-10 logs
$\quad x \cdot \log 3 = \log 4$
$$x = \frac{\log 4}{\log 3} = \frac{0.6021}{0.4771} = 1.262$$

Thus, $\log_3 4 = 1.26$ to three significant digits.

When an investment is earning *simple* interest, the principal remains constant. If it is earning *compound* interest that is compounded *annually,* the principal is increased each year by the amount of interest earned in the previous year. If $500 is invested at a rate of 10%, compounded annually, the total amounts at the end of years 0 through 3 are as follows.

$$\$500, \$500 + \$50, \$550 + \$55, \$605 + \$60.50, \ldots$$

If the interest on an investment is compounded *semiannually,* then every 6 months the interest earned for a half year is added to the previous principal.

Recall that the formula for the total amount A of an investment earning simple interest is $A = p + prt = p(1 + rt)$. The formula for compound interest is given below.

Compound interest formula

$$A = p\left(1 + \frac{r}{n}\right)^{nt}$$

A is the total amount at the end of t years when p dollars are invested at $r\%$ per year, compounded n times each year.

Example 3 A bond that pays 12% per year, compounded quarterly (4 times each year, or every 3 mo), is bought for $6,000. How much will the bond be worth 10 yr from now? How much will the bond have earned in compound interest? Use logarithms.

$$p = 6{,}000 \quad r = 0.12 \quad t = 10 \quad n = 4$$

$$A = p\left(1 + \frac{r}{n}\right)^{nt} = 6{,}000\left(1 + \frac{0.12}{4}\right)^{40} = 6{,}000(1.03)^{40}$$

$$\begin{aligned}\log A = \log 6{,}000(1.03)^{40} &= \log 6{,}000 + 40 \cdot \log 1.03 \\ &= 3.7782 + 40(0.0128) \\ &= 4.2902\end{aligned}$$

$$A = \text{antilog } 4.2902 = 10^4 \times 1.95 = 19{,}500$$

$$A - p = 19{,}500 - 6{,}000 = 13{,}500$$

Thus, the $6,000 bond will be worth $19,500 and will have earned $13,500 in compound interest.

Some bacteria reproduce by splitting themselves into two new bacteria at the end of a growth period. This process of splitting is called *simple fission.* Suppose there are 400 bacteria at the start of an experiment and simple fission takes place every 20 min, or 3 times in each hour. The number of bacteria increases rapidly.

Growth period	0	1	2	3	4	...
Number of bacteria	400	800	1600	3200	6400	...
	400	$400 \cdot 2$	$400 \cdot 2^2$	$400 \cdot 2^3$	$400 \cdot 2^4$	...

In 6 h, there are $3 \cdot 6$, or 18, twenty-minute periods. **Thus,** at the end of 6 h the number of bacteria is $400 \cdot 2^{18}$, or more than 104 million.

Using Exponential Equations

> **Exponential growth**
>
> $$N = N_0 \cdot 2^k$$
> where N is the number of bacteria present at the end of k periods of simple fission and N_0 is the number of bacteria at the start.

Example 4   To three significant digits, find the number of bacteria produced from 500 bacteria at the end of 4 h if simple fission occurs every 12 min.

$$N = N_0 \cdot 2^k = 500 \cdot 2^{20} \quad \blacktriangleleft \text{ 20 twelve-minute periods} = 4 \text{ hours.}$$
$$\log N = \log(500 \cdot 2^{20}) = \log 500 + 20 \cdot \log 2 = 2.6990 + 20(0.3010) = 8.7190$$
$$N = \text{antilog } 8.7190 = 10^8 \times 5.24$$

Thus, there will be 524 million bacteria.

Reading in Algebra

For each equation, estimate a range for x.

(A)
1. $2^{x+1} = 70$ 2. $3^{2x} = 90$ 3. $5^{2x-1} = 200$ 4. $2^{-x} = 18$ 5. $4^{2x+1} = 200$
   5 < x < 6   **2 < x < 2.5**   **2 < x < 2.5**   **−5 < x < −4**   **1 < x < 1.5**

Written Exercises

Solve each equation for x to three significant digits.
1. $4^x = 25$ 2. $4^{2x} = 25$ 3. $4^{x-1} = 25$ 4. $7^{x+1} = 56$ 5. $8^{3x} = 72$
   2.32   **1.16**   **3.32**   **1.07**   **0.686**

Find each logarithm to three significant digits.
6. $\log_2 6$ **2.59**   7. $\log_6 2$ **0.387**   8. $\log_5 14$ **1.64**

(B)
9. $\log_3 864$ **6.15**   10. $\log_4 7250$ **6.41**   11. $\log_2 \sqrt[3]{35}$ **1.71**

Find the value of each bond and the amount of interest earned at the end of the time period for the given data. **(by logs); (by calculator)**

12. $5,000 bond paying 8%, compounded semiannually, for 6 yr. **$8,000, $3,000; $8,005, $3,005**
13. $8,000 bond paying 9%, compounded annually, for 7 yr. **$14,600, $6,600; $14,624, $6,624**
14. $7,500 bond paying 12%, compounded quarterly, for 15 yr. **$44,000, $36,500; $44,187, $36,687**

Find N, the number of bacteria present, for the given data. Use the formula $N = N_0 \cdot 2^k$. **(by logs); (by calculator)**

15. $N_0 = 750$ and simple fission occurs every 15 min for 3 h. **3,070,000; 3,072,000**
16. $N_0 = 3,000$ and simple fission occurs every 90 min for 18 h. **12,300,000; 12,288,000**
17. $N_0 = 25,000$ and simple fission occurs every 30 min for 24 h. **7.01 × 10¹⁸; 7.04 × 10¹⁸**

Solve each equation for x to three significant digits.
18. $9^{2x+1} = 624$ **0.965**   19. $7^{3x-2} = 834$ **1.82**   20. $2^{\sqrt{5}} = x$ **4.71**

(C)
21. $3^x = 4 \cdot 2^x$ **3.42**   22. $2 \cdot 5^{x+1} = 5 \cdot 2^{x+2}$ **0.756**   23. $4^{\sqrt{6}} = 2x - 1$ **15.4**

For Ex. 24–26, see TE Answer Section.

Prove that each statement is true. Each variable represents a positive number $\neq 1$.
[*Hint*: Begin each proof with "Let $\log_a y = x$."]

24. $\log_a y = \dfrac{\log y}{\log a}$ 25. $\log_a y = \dfrac{\log_b y}{\log_b a}$ 26. $\log_a y = \dfrac{1}{\log_y a}$

HIGHER-DEGREE POLYNOMIAL EQUATIONS   15.8

Objectives
To find the zeros of an integral polynomial
To factor an integral polynomial in one variable into first-degree factors
To solve an integral polynomial equation of degree greater than two

The polynomial $2x^4 + 5x^3 - 11x^2 - 20x + 12$ is an **integral polynomial** since all the coefficients are *integers*. The notation $P(x)$ is used to represent such a polynomial. The value of the polynomial $P(x)$ is represented by $P(r)$, when x is replaced by any complex number r.

If $P(x) = 2x^4 + 5x^3 - 11x^2 - 20x + 12$, then
$P(1) = 2 \cdot 1^4 + 5 \cdot 1^3 - 11 \cdot 1^2 - 20 \cdot 1 + 12$, or $P(1) = -12$.

You can factor $P(x) = 2x^4 + 5x^3 - 11x^2 - 20x + 12$ into *first-degree factors*, given that $x - 2$ and $x + 3$ are first-degree factors of $P(x)$, using *synthetic substitutions* as shown below.

$$\begin{array}{r|rrrrr} & 2 & 5 & -11 & -20 & 12 \\ \hline 2 & 2 & 9 & 7 & -6 & \boxed{0} \\ \hline -3 & 2 & 3 & -2 & \boxed{0} & \end{array}$$

◀ $P(x) = 2x^4 + 5x^3 - 11x^2 - 20x + 12$
◀ $P(x) = (x - 2)(2x^3 + 9x^2 + 7x - 6)$
◀ $P(x) = (x - 2)(x + 3)(2x^2 + 3x - 2)$

Factor $2x^2 + 3x - 2$ into $(x + 2)(2x - 1)$.

Thus, $P(x) = (x - 2)(x + 3)(x + 2)(2x - 1)$.

Notice the following for $P(x) = (x - 2)(x + 3)(x + 2)(2x - 1)$ above.
(1) $P(2) = (2 - 2)(2 + 3)(2 + 2)(2 \cdot 2 - 1) = 0$. The number 2 is called a **zero** of the polynomial $P(x)$ because $P(2) = 0$.
(2) $P(-3) = 0$, $P(-2) = 0$, and $P(\frac{1}{2}) = 0$. So, -3, -2, and $\frac{1}{2}$ are zeros of $P(x)$.

Definition: Zero of a polynomial	The number r is called a *zero* of a polynomial $P(x)$ if and only if $P(r) = 0$.

The number of "potential" integral zeros of an integral polynomial is *finite*. This can be illustrated as follows.

Let $P(x) = 2x^4 + 5x^3 - 11x^2 - 20x + 12$.
$2r^4 + 5r^3 - 11r^2 - 20r + 12 = 0$
$2r^4 + 5r^3 - 11r^2 - 20r = -12$
$r(2r^3 + 5r^2 - 11r - 20) = -12$

If r is an integral zero of $P(x)$, then $2r^3 + 5r^2 - 11r - 20$ is an integer because r is an integer. So, r is an integral factor of -12.

The "potential" integral zeros of $P(x) = 2x^4 + 5x^3 - 11x^2 - 20x + 12$ are the integral factors of the *constant term* 12: 1, -1, 2, -2, 3, -3, 4, -4, 6, -6, 12, -12.

Integral zero theorem	If an integer r is a zero of an integral polynomial $P(x)$, then r is a factor of the constant term in $P(x)$. That is, if r is an integral zero of $a_0x^n + a_1x^{n-1} + a_2x^{n-2} + \ldots + a_{n-1}x^1 + a_n$, then r is a factor of a_n.

Higher-Degree Polynomial Equations

Some of the zeros of an integral polynomial can be irrational numbers.

Example 1 Find the zeros of $P(x) = x^4 - x^3 - 8x^2 + 2x + 12$.
Factor $P(x)$ into first-degree factors.

1. List the potential integral zeros: $\pm 1, \pm 2, \pm 3, \pm 4, \pm 6, \pm 12$.
2. Test each potential zero until a *second-degree* factor is found.

$$\begin{array}{r|rrrrr}
 & 1 & -1 & -8 & 2 & 12 \\
\hline
-2 & 1 & -3 & -2 & 6 & \boxed{0} \\
3 & 1 & 0 & -2 & \boxed{0} &
\end{array}$$

◀ $P(x) = x^4 - x^3 - 8x^2 + 2x + 12$
◀ $P(x) = (x + 2)(x^3 - 3x^2 - 2x + 6)$
◀ $P(x) = (x + 2)(x - 3)(x^2 - 2)$

3. Solve $x^2 - 2 = 0$.
$x^2 = 2 \quad x = \pm\sqrt{2}$

$P(x) = (x + 2)(x - 3)(x - \sqrt{2})(x + \sqrt{2})$

Thus, the zeros of $P(x)$ are $-2, 3, \sqrt{2},$ and $-\sqrt{2}$, and $P(x) = (x + 2)(x - 3)(x - \sqrt{2})(x + \sqrt{2})$.

In Example 1, the four zeros of the *polynomial* $P(x)$ are also the solutions of the *equation* $x^4 - x^3 - 8x^2 + 2x + 12 = 0$. To solve a polynomial equation $P(x) = 0$, factor $P(x)$ into first-degree factors and find the zero of each factor.

Example 2 Solve $x^4 - 5x^2 - 36 = 0$.

Let $P(x) = x^4 - 5x^2 - 36$. Factor $P(x)$ into first-degree factors.
1. The potential integral zeros of $P(x)$ are the factors of 36: $\pm 1, \pm 2, \ldots, \pm 36$.
2. Test each potential zero, in turn, until a *second-degree* factor is found.

$$\begin{array}{r|rrrrr}
 & 1 & 0 & -5 & 0 & -36 \\
\hline
3 & 1 & 3 & 4 & 12 & \boxed{0} \\
-3 & 1 & 0 & 4 & \boxed{0} &
\end{array}$$

◀ $P(x) = x^4 + 0x^3 - 5x^2 + 0x - 36 = 0$
◀ $P(x) = (x - 3)(x^3 + 3x^2 + 4x + 12)$
◀ $P(x) = (x - 3)(x + 3)(x^2 + 4)$

3. Solve $x^2 + 4 = 0$.
$x^2 = -4 \quad x = \pm 2i$

$P(x) = (x - 3)(x + 3)(x - 2i)(x + 2i) = 0$

Thus, the solutions are $3, -3, 2i,$ and $-2i$.

The **Fundamental Theorem of Algebra** assures you that each polynomial of degree $n > 0$ has *at least one* complex zero.

Fundamental theorem of algebra If $P(x)$ is a polynomial of degree greater than 0, then there is a complex number r for which $P(r) = 0$.

Example 3 Find the zeros of $P(x) = 2x^3 - 17x^2 + 40x - 16$. Solve the equation $2x^3 - 17x^2 + 40x - 16 = 0$.

$$\begin{array}{r|rrrr}
 & 2 & -17 & 40 & -16 \\
\hline
4 & 2 & -9 & 4 & \boxed{0}
\end{array}$$

◀ $P(x) = 2x^3 - 17x^2 + 40x - 16 = 0$
◀ $P(x) = (x - 4)(2x^2 - 9x + 4)$

$2x^2 - 9x + 4$

$P(x) = (x - 4)(x - 4)(2x - 1) = 0$

Thus, 4 and $\frac{1}{2}$ are the zeros of the polynomial $P(x)$ and the solutions of the equation $P(x) = 0$.

In Example 3, notice the following three facts.
(1) The degree of $P(x)$ is 3 and $P(x)$ has 3 first-degree factors.
(2) $P(x)$ has *only* 2 distinct zeros: 4 and $\frac{1}{2}$. The zero 4 has a *multiplicity of* 2 since the factor $x - 4$ appeared *twice*. The zero $\frac{1}{2}$ has a multiplicity of 1. The sum of these two multiplicities is 3, which is the degree of $P(x)$.
(3) $P(x)$ can be factored *uniquely* into three first-degree factors and the constant factor 2, which is the coefficient of $2x^3$, as shown below.

$$P(x) = 2x^3 - 17x^2 + 40x - 16 = (x - 4)(x - 4)(2x - 1) = (x - 4)(x - 4) \cdot 2(x - \tfrac{1}{2})$$

The theorem below guarantees that a polynomial $P(x)$ of degree $n > 0$ has exactly n first-degree factors, and thus, the *sum of the multiplicities* of the zeros of $P(x)$ is equal to n.

Unique factorization theorem	Each polynomial $P(x)$ of degree $n > 0$ can be factored uniquely into n first-degree factors (not all necessarily distinct) and a constant factor which is the coefficient of the highest-degree term in $P(x)$.

Reading in Algebra

1. List the potential integral zeros of $x^4 + 23x^2 - 50$. **±1, ±2, ±5, ±10, ±25, ±50**
2. If -4 is a zero of a polynomial $P(x)$, what is one binomial factor of $P(x)$? What is one solution of the equation $P(x) = 0$? **x + 4; −4**
3. The zeros of $x^4 - 16$ are the solutions of what equation? **$x^4 − 16 = 0$**
4. Rewrite $(x + 5)(3x + 2)(2x - 5)$ to show three first-degree factors and a constant factor 6. **$6(x + 5)(x + \frac{2}{3})(x - \frac{5}{2})$**

Find the complex zeros and give the multiplicity of each zero.

5. $5x - 3$ $\frac{3}{5}$; **mult. 1**
6. $(x + 3)(2x + 6)$ **−3; mult. 2**
7. $x^2 - 8$ **$-2\sqrt{2}, 2\sqrt{2}$; mult. 1, 1**
8. $x^2 - 10x + 25$ **5; mult. 2**
9. $x^2 + 8$ **$-2i\sqrt{2}, 2i\sqrt{2}$; mult. 1, 1**
10. $(x - 6)^3(x^2 - 4)$ **6, −2, 2; mult. 3, 1, 1**

Written Exercises

1. **−1, 4, $\frac{1}{3}$; (x + 1)(x − 4)(3x − 1)**   2. **−2, $\frac{2}{3}$, −$\frac{1}{2}$; (x + 2)(3x − 2)(2x + 1)**

Find the zeros of each polynomial. Factor the polynomial into first-degree factors.

Ⓐ 1. $3x^3 - 10x^2 - 9x + 4$ **above**
2. $6x^3 + 11x^2 - 4x - 4$ **above**
3. $x^3 - 2x^2 - 7x + 14$ **below**
4. $x^3 - 3x^2 + 4x - 12$ **below**
5. $2x^3 + 3x^2 - 32x + 15$
6. $x^3 + x^2 + 8x + 8$
   3, −5, $\frac{1}{2}$; (x − 3)(x + 5)(2x − 1)   **−1, ±2i$\sqrt{2}$; (x + 1)(x + 2i$\sqrt{2}$)(x − 2i$\sqrt{2}$)**

Solve each equation.   3. **2, −$\sqrt{7}$, $\sqrt{7}$; (x − 2)(x + $\sqrt{7}$)(x − $\sqrt{7}$)**   4. **3, −2i, 2i; (x − 3)(x + 2i)(x − 2i)**

7. $2x^3 - 3x^2 - 17x + 30 = 0$ **2, −3, $\frac{5}{2}$**
8. $x^3 + 3x^2 - 5x - 15 = 0$ **−3, −$\sqrt{5}$, $\sqrt{5}$**
9. $x^3 + 5x^2 - 16x - 80 = 0$ **4, −4, −5**
10. $x^3 - 7x^2 - 5x + 75 = 0$ **−3, 5, mult. 2**

Ⓑ 11. $2x^4 + x^3 - 8x^2 - x + 6 = 0$ **1, −1, −2, $\frac{3}{2}$**
12. $3x^4 - x^3 - 37x^2 + 9x + 90 = 0$ **2, 3, −3, $\frac{-5}{3}$**
13. $x^4 - 6x^2 - 27 = 0$ **3, −3, i$\sqrt{3}$, −i$\sqrt{3}$**
14. $2x^4 + 12x + 8 = 3x^3 + 10x^2$
   2, mult. 2; −2, −$\frac{1}{2}$

Find the zeros of each polynomial.

15. $x^4 + 3x^3 - 30x^2 - 6x + 56$ **4, −7, $\sqrt{2}$, −$\sqrt{2}$**
16. $x^4 - 3x^3 - 43x^2 + 9x + 120$ **−5, 8, −$\sqrt{3}$, $\sqrt{3}$**
Ⓒ 17. $x^5 - 4x^3 - x^2 + 4$ **1, 2, −2, $\frac{-1+i\sqrt{3}}{2}$, $\frac{-1-i\sqrt{3}}{2}$**
18. $x^5 - 4x^3 - 8x^2 + 32$
19. $x^5 - 9x^3 - x^2 + 9$ **1, 3, −3, $\frac{-1+i\sqrt{3}}{2}$, $\frac{-1-i\sqrt{3}}{2}$**
20. $2x^4 + 6x^3 - 15x^2 + 15x - 50$

18. **2, mult. 2; −2, −1 + i$\sqrt{3}$, −1 − i$\sqrt{3}$**   20. **2, −5, $\frac{i\sqrt{10}}{2}$, $\frac{-i\sqrt{10}}{2}$**

Higher-Degree Polynomial Equations

RATIONAL ZERO THEOREM 15.9

Objectives
To find an upper bound and a lower bound for the real zeros of an integral polynomial
To find the zeros of an integral polynomial using the rational zero theorem
To solve an integral polynomial equation using the rational zero theorem

Let $P(x) = (2x + 1)(3x - 2)(x - \sqrt{3})(x + \sqrt{3}) = 6x^4 - x^3 - 20x^2 + 3x + 6$. The zeros of $P(x)$ are $-\frac{1}{2}, \frac{2}{3}, \sqrt{3}$, and $-\sqrt{3}$. Each of these zeros is in the interval from -2 to 2.

2 is called an **upper bound** and -2 is called a **lower bound** for the four zeros.

If you did not know the zeros of $P(x)$ above, you could find an upper bound and a lower bound for the real zeros as follows.

Upper Bounds
Find $P(x)$ for $x = 0, 1, 2, 3, \ldots$ by synthetic substitution.

x	6	-1	-20	3	6	$P(x)$
0	6	-1	-20	3	6	6
1	6	5	-15	-12	-6	-6
2	6	11	2	7	20	20

All nonnegative

3	6	17	31	96	294	294

All nonnegative

These numbers will continue to be all nonnegative, and $P(x)$ will continue to increase. So, $P(x)$ cannot be zero for $x \geq 2$.

Thus, 2 and 3 are upper bounds for the real zeros of $P(x)$.

Lower Bounds
Find $P(x)$ for $x = 0, -1, -2, -3, \ldots$ by synthetic substitution.

x	6	-1	-20	3	6	$P(x)$
0	6	-1	-20	3	6	6
-1	6	-7	-13	16	-10	-10
-2	6	-13	6	-9	24	24

Signs alternate

-3	6	-19	37	-108	330	330

Signs alternate

These signs will continue to alternate, and $P(x)$ will continue to be positive. So, $P(x)$ cannot be zero for $x \leq -2$.

Thus, -2 and -3 are lower bounds for the real zeros of $P(x)$.

Upper and lower bound theorem

Let $P(x)$ be an integral polynomial, U and L real numbers, $U \geq 0$ and $L \leq 0$. (1) Find $P(U)$ by synthetic substitution. If the last row of numbers is *all nonnegative* or *all nonpositive*, then U is an **upper bound** for the real zeros of $P(x)$. (2) Find $P(L)$ by synthetic substitution. If the last row of numbers *alternates in sign*, then L is a **lower bound** for the real zeros of $P(x)$.

Example 1 **Find an upper bound U and a lower bound L for the real zeros of $P(x) = 2x^3 - 5x^2 - 4x + 10$.**

1. For U, try $x = 0, 1, 2, \ldots$

x	2	-5	-4	10
0	2	-5	-4	10
1	2	-3	-7	3
2	2	-1	-6	-2
3	2	1	-1	7
4	2	3	8	42

All nonnegative ▶ $U = 4$

2. For L, try $x = 0, -1, -2, \ldots$

x	2	-5	-4	10
0	2	-5	-4	10
-1	2	-7	3	7
-2	2	-9	14	-18

Signs alternate ▶ $L = -2$

Thus, 4 is an upper bound and -2 is a lower bound.

An integral polynomial might not have enough integral zeros to yield its first-degree factors by using the integral zero theorem. In such a case, you can use the **rational zero theorem** to find some of the rational zeros of the polynomial.

> **Rational zero theorem**
>
> If $\dfrac{c}{d}$ (c and d relatively prime integers) is a zero of an integral polynomial $P(x)$, then c is a factor of the constant term in $P(x)$ and d is a factor of the coefficient of the highest-degree term. That is, if $\dfrac{c}{d}$ is a zero of
> $$a_0 x^n + a_1 x^{n-1} + a_2 x^{n-2} + \ldots + a_{n-1} x^1 + a_n, \text{ then}$$
> c is a factor of a_n and d is a factor of a_0.

If $P(x) = 6x^4 - x^3 - 14x^2 + 2x + 4$, you can list its "potential" rational zeros $\dfrac{c}{d}$ that are not integers, as shown below.

c is a factor of 4: 1, 2, 4
d is a factor of 6: 1, 2, 3, 6 $\Big\}$ $\dfrac{c}{d}$ can be $\pm\dfrac{1}{2}, \pm\dfrac{1}{3}, \pm\dfrac{1}{6}, \pm\dfrac{2}{3}, \pm\dfrac{4}{3}$.

Example 2 **Find the zeros of $P(x) = 6x^4 - x^3 - 14x^2 + 2x + 4$.**

1. Test the potential *integral* zeros: $\pm 1, \pm 2, \pm 4$.

	6	-1	-14	2	4
1	6	5	-9	-7	-3
2	6	11	8	18	40

All nonnegative
So, 2 is an upper bound.
Do not test 4.

	6	-1	-14	2	4
-1	6	-7	-7	9	-5
-2	6	-13	12	-22	48

Signs alternate
So, -2 is a lower bound.
Do not test -4.

So, $P(x)$ has no integral zeros.

Rational Zero Theorem

2. Test the potential *rational* zeros: $\pm\frac{1}{2}, \pm\frac{1}{3}, \pm\frac{1}{6}, \pm\frac{2}{3}, \pm\frac{4}{3}$.

$$\begin{array}{r|rrrrr} & 6 & -1 & -14 & 2 & 4 \\ \hline -\frac{1}{2} & 6 & -4 & -12 & 8 & \boxed{0} \end{array}$$ ◀ $P(x)$
◀ $P(x) = (x + \frac{1}{2})(6x^3 - 4x^2 - 12x + 8)$

$2(3x^3 - 2x^2 - 6x + 4)$ ◀ $P(x) = (x + \frac{1}{2}) \cdot 2(3x^3 - 2x^2 - 6x + 4)$

$$\begin{array}{r|rrrr} & 3 & -2 & -6 & 4 \\ \hline \frac{2}{3} & 3 & 0 & -6 & \boxed{0} \end{array}$$ ◀ $P(x) = (x + \frac{1}{2}) \cdot 2(x - \frac{2}{3})(3x^2 - 6)$

$3(x^2 + 0x - 2)$ ◀ $P(x) = (x + \frac{1}{2}) \cdot 2(x - \frac{2}{3}) \cdot 3(x^2 - 2)$

3. Solve $x^2 - 2 = 0$.   ◀ $P(x) = 6(x + \frac{1}{2})(x - \frac{2}{3})(x - \sqrt{2})(x + \sqrt{2})$
$x^2 = 2 \quad x = \pm\sqrt{2}$

Thus, the zeros of $P(x)$ are $-\frac{1}{2}, \frac{2}{3}, \sqrt{2}$, and $-\sqrt{2}$.

You can use the rational zero theorem to solve an integral polynomial equation, $P(x) = 0$, as shown in Example 3.

Example 3   **Solve $3x^4 - 5x^3 + 10x^2 - 20x - 8 = 0$.**

Let $P(x) = 3x^4 - 5x^3 + 10x^2 - 20x - 8$. Find the zeros of $P(x)$.

1. Test the potential integral zeros: $\pm 1, \pm 2, \pm 4, \pm 8$.

$$\begin{array}{r|rrrrr} & 3 & -5 & 10 & -20 & -8 \\ \hline 1 & 3 & -2 & 8 & -12 & -20 \\ 2 & 3 & 1 & 12 & 4 & \boxed{0} \end{array}$$ ◀ 2 is a zero and an upper bound.

(All nonnegative) $P(x) = (x - 2)(3x^3 + x^2 + 12x + 4)$

$$\begin{array}{r|rrrr} & 3 & 1 & 12 & 4 \\ \hline -1 & 3 & -2 & 14 & -10 \end{array}$$ ◀ -1 is a lower bound, since the signs alternate.

2. List the potential rational zeros for $Q(x) = 3x^3 + x^2 + 12x + 4$.

   c is a factor of 4: 1, 2, 4
   d is a factor of 3: 1, 3   ◀ $\frac{c}{d}$ can be $\pm\frac{1}{3}, \pm\frac{2}{3}, \frac{4}{3}$ $\left(not\ -\frac{4}{3}\right)$.

3. Test the potential rational zeros for $Q(x) = 3x^3 + x^2 + 12x + 4$.

$$\begin{array}{r|rrrr} & 3 & 1 & 12 & 4 \\ \hline -\frac{1}{3} & 3 & 0 & 12 & \boxed{0} \end{array}$$
$P(x) = (x - 2)(3x^3 + x^2 + 12x + 4)$
◀ $P(x) = (x - 2)(x + \frac{1}{3})(3x^2 + 12)$

$3(x^2 + 0x + 4)$ ◀ $P(x) = (x - 2)(x + \frac{1}{3}) \cdot 3(x^2 + 4)$

4. Solve $x^2 + 4 = 0$.   ◀ $P(x) = 3(x - 2)(x + \frac{1}{3})(x - 2i)(x + 2i)$
$x^2 = -4 \quad x = \pm 2i$

Thus, the solutions are $2, -\frac{1}{3}, 2i$, and $-2i$.

Written Exercises

Find an upper bound and a lower bound for the real zeros of each polynomial.

(A)
1. $2x^3 - 5x^2 - 8x + 18$ **4, −2**
2. $2x^3 - 7x^2 - 12x + 42$ **5, −3**
3. $6x^4 + x^3 - 8x^2 - x + 2$ **2, −2**
4. $-x^3 - 2x^2 + 10x + 20$ **4, −5**

Find the zeros of each polynomial.

5. $3x^3 - 2x^2 - 9x + 6$ **$\frac{2}{3}, \pm\sqrt{3}$**
6. $2x^3 + x^2 + 8x + 4$ **$-\frac{1}{2}, \pm 2i$**
7. $3x^4 - 5x^3 - 13x^2 + 25x - 10$ **$1, \frac{2}{3}, \pm\sqrt{5}$**
8. $4x^4 + x^3 - 27x^2 - 6x + 18$ **$-1, \frac{3}{4}, \pm\sqrt{6}$**

Solve each equation.

9. $2x^4 - 7x^3 + 8x^2 - 7x + 6 = 0$ **$2, \frac{3}{2}, \pm i$**
10. $4x^4 + x^3 + 5x^2 + 2x - 6 = 0$ **$-1, \frac{3}{4}, \pm i\sqrt{2}$**
11. $6x^4 - 5x^3 - 11x^2 + 10x - 2 = 0$ **$\frac{1}{2}, \frac{1}{3}, \pm\sqrt{2}$**
12. $6x^4 + x^3 - 31x^2 - 5x + 5 = 0$ **$-\frac{1}{2}, \frac{1}{3}, \pm\sqrt{5}$**

(B)
13. $12x^4 - 20x^3 - 11x^2 + 5x + 2 = 0$
14. $8x^4 + 20x^3 - 14x^2 - 5x + 3 = 0$
15. $8x^4 + 2x^3 + 3 = 25x^2 + 6x$
16. $6x^4 - 29x^2 = 5x^3 - 25x + 5$

13. $2, -\frac{1}{3}, \pm\frac{1}{2}$ **14.** $-3, -\frac{1}{2}, \frac{1}{2}$, mult. 2 **15.** $-\frac{1}{2}, \frac{1}{4}, \pm\sqrt{3}$ **16.** $\frac{1}{2}, \frac{1}{3}, \pm\sqrt{5}$

Find an upper bound and a lower bound for the real zeros of each polynomial.

17. $x^4 - 7x^2 + 10$ **3, −3**
18. $-x^4 + 2x^3 + 4x^2 - 10x + 5$ **4, −3**
19. $x^5 - 3x^4 - 9x^3 + 2$ **5, −2**
20. $x^6 - 13x^4 - 28x^2 - 20$ **4, −4**

Find the zeros of each polynomial.

21. $4x^4 - 4x^3 + 17x^2 - 16x + 4$ **$\frac{1}{2}$, mult. 2; $\pm 2i$**
22. $12x^4 - 5x^3 + 106x^2 - 45x - 18$ **$-\frac{1}{4}, \frac{2}{3}, \pm 3i$**
23. $36x^4 - 12x^3 - 11x^2 + 2x + 1$
24. $36x^4 - 60x^3 - 5x^2 + 25x - 6$

(C)
25. $6x^5 + x^4 + 20x^3 + 5x^2 - 16x + 4$
26. $6x^5 - 7x^4 - 19x^3 + 23x^2 + 3x - 6$

23. $\frac{1}{2}$, mult. 2; $-\frac{1}{3}$, mult. 2 **24.** $\frac{1}{2}, \frac{1}{3}, \frac{3}{2}, -\frac{2}{3}$ **25.** $-1, \frac{1}{2}, \frac{1}{3}, \pm 2i$ **26.** $1, -\frac{1}{2}, \frac{2}{3}, \pm\sqrt{3}$

CUMULATIVE REVIEW

Solve each problem.

1. In a bag of coins, there are 12 more quarters than dimes and 4 times as many nickels as quarters. The nickels and dimes together have the same value as the quarters. Find the number of each kind of coin. **quarters: 24, dimes: 12, nickels: 96**

2. Ben is twice as old as Al, and Carol is two years older than Ben. Five years ago, the sum of the two younger ages was the same as the age of the oldest. Find the present age of all three people. **Al: 7, Ben: 14, Carol: 16**

3. Some 50¢ red pens and some 30¢ blue pens are mixed to make a package of 20 pens. If the package is worth $8.40, how many red pens and how many blue pens are in the package? **red pens: 12, blue pens: 8**

4. Machine A can do a job in 15 h. If machines A and B work together, the job can be done in 10 h. How many hours would it take machine B to do the job if it works alone? **30 h**

5. Find four consecutive multiples of 4 such that the product of the second and the third numbers is 192. **−20, −16, −12, −8 or 8, 12, 16, 20**

6. How much water must be added to 4 L of a 30% iodine solution to dilute it to a 10% iodine solution? **8 L**

Rational Zero Theorem

GRAPHING POLYNOMIAL FUNCTIONS 15.10

Objective **To sketch the graph of an integral polynomial function of degree *n* with *n* distinct real zeros**

To "sketch" the graph for $y = x^3 - x^2 - 12x + 12$, you can use synthetic substitution to find ordered pairs (x, y), then plot the points and draw a "smooth" curve.

x	1	−1	−12	12	(x, y)
−4	1	−5	8	−20	(−4, −20)

Signs alternate
−4 is a lower bound.

−3	1	−4	0	12	(−3, 12)
−2	1	−3	−6	24	(−2, 24)
−1	1	−2	−10	22	(−1, 22)
0	1	−1	−12	12	(0, 12)
1	1	0	−12	0	(1, 0)
2	1	1	−10	−8	(2, −8)
3	1	2	−6	−6	(3, −6)
4	1	3	0	12	(4, 12)

All nonnegative
4 is an upper bound.

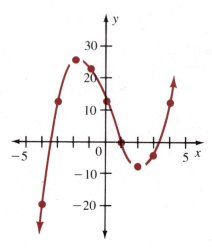

Notice that:
1. The graph of the third-degree function has a characteristic shape.
2. The "turning" points in quadrants II and IV cannot be determined at this time.
3. The polynomial has three real zeros. One is 1, one is between −4 and −3, and one is between 3 and 4.

Example 1 **Sketch the graph for $y = 8x^4 - 6x^3 - 77x^2 + 54x + 45$.**

x	8	−6	−77	54	45	(x, y)
−4	8	−38	75	−246	1029	(−4, 1029)

−4 is a lower bound.

−3	8	−30	13	15	0	(−3, 0)
−2	8	−22	−33	120	−195	(−2, −195)
−1	8	−14	−63	117	−72	(−1, −72)
0	8	−6	−77	54	45	(0, 45)
1	8	2	−75	−21	24	(1, 24)
2	8	10	−57	−60	−75	(2, −75)
3	8	18	−23	−15	0	(3, 0)
4	8	26	27	162	693	(4, 693)

4 is an upper bound.

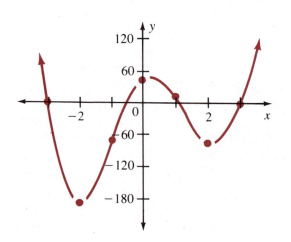

In Example 1, notice that (1) the graph of the fourth-degree function with four real zeros has a characteristic ⩗⩗ shape and (2) the turning points cannot be determined at this time.

The graph of a polynomial function of degree 2 is a *parabola*. Its turning point can be determined as follows.

If $y = ax^2 + bx + c$, $a \neq 0$, the minimum or maximum value of the function will occur when $x = \frac{-b}{2a}$.

Example 2   Sketch the graph for $y = 4x^2 - 12x - 27$.

$a = 4$, $b = -12$, $c = -27$, and $x = \frac{-b}{2a} = \frac{12}{8} = \frac{3}{2}$

If $x = \frac{3}{2}$, then $y = 4 \cdot \frac{9}{4} - 12 \cdot \frac{3}{2} - 27 = -36$, so the turning point is (1.5, −36).

x	4	−12	−27	(x, y)
−2	4	−20	13	(−2, 13)
−1	4	−16	−11	(−1, −11)
0	4	−12	−27	(0, −27)
1	4	−8	−35	(1, −35)
2	4	−4	−35	(2, −35)
3	4	0	−27	(3, −27)
4	4	4	−11	(4, −11)
5	4	8	13	(5, 13)

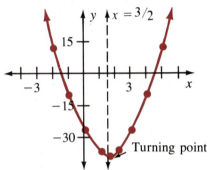

Reading in Algebra

Match the degree of a polynomial function at the left with a characteristic graph at the right.

1. degree 4   **C**
2. degree 3   **E**
3. degree 2   **A**
4. degree 1   **B**
5. degree 0   **D**

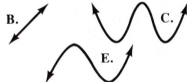

Written Exercises

Sketch the graph of each polynomial function. For each parabola, give the coordinates of the turning point.   For complete solutions, see TE Answer Section.

1. $y = 3$
2. $y = 2x - 5$
3. $y = x^2 + 4x - 5$ **(−2, −9)**
4. $y = x^2 - x - 6$ **(0.5, −6.25)**
5. $y = x^3 - 7x^2 + 4x + 12$
6. $y = x^4 - x^3 - 14x^2 + 2x + 24$
7. $y = 4x^2 - 16x + 15$ **(2, −1)**
8. $y = x^3 - x^2 - 10x + 10$
9. $y = 2x^4 - x^3 - 35x^2 + 16x + 48$
10. $y = -x^2 + 3x + 10$ **(1.5, 12.25)**
11. $y = -x^3 + x^2 + 9x - 9$
12. $y = -x^4 + 10x^2 - 9$

Graphing Polynomial Functions

APPLICATIONS: DESCARTES' RULE OF SIGNS

There are three *changes in sign* in $P(x)$, where
$$P(x) = +2x^5 \; -9x^4 \; -5x^3 \; +51x^2 \; -9x \; -54$$

There are two sign changes in $P(-x)$ as shown below.
$$P(-x) = 2(-x)^5 - 9(-x)^4 - 5(-x)^3 + 51(-x)^2 - 9(-x) - 54$$
$$= -2x^5 \; - 9x^4 \; + 5x^3 + 51x^2 \; + 9x \; - 54$$

Recall that the *sum* of the multiplicities of the zeros is 5 for a polynomial of degree 5. The *positive* real zeros of $P(x)$ are 3, with a multiplicity of *two*, and 1.5, with a multiplicity of *one*. The *sum* of these multiplicities is *three*, the same as the number of sign changes in $P(x)$. The *negative* real zeros of $P(x)$ are -1 and -2, *each* with a multiplicity of *one*. The *sum* of these multiplicities is *two*, which is the number of sign changes in $P(-x)$.

Consider $Q(x) = x^3 - 6x^2 + 10x - 8$ and $Q(-x) = (-x)^3 - 6(-x)^2 + 10(-x) - 8$
$$= -x^3 - 6x^2 - 10x - 8.$$

The zeros of $Q(x)$ are 4, $1 - i$, and $1 + i$, each of multiplicity one. Notice:
1. $Q(x)$ has *three* sign changes and *one* positive real zero.
2. $Q(-x)$ has *no* sign changes and $Q(x)$ has *no* negative real zeros.
3. $Q(x)$ has *two imaginary* zeros and they are a pair of *conjugates*.

> **Descartes' rule of signs**
>
> The sum of the multiplicities of the positive real zeros of an integral polynomial $P(x)$ is either equal to the number of sign changes in $P(x)$ or less than that number by a multiple of two.
>
> The sum of the multiplicities of the negative real zeros of $P(x)$ is either equal to the number of sign changes in $P(-x)$ or less than that number by a multiple of two.

Example 1

Find the possible combinations of the sums of the multiplicities of the positive real zeros, negative real zeros, and imaginary zeros of $P(x)$, where
$$P(x) = x^5 - 4x^3 - x^2 + 4.$$

$P(x) = +x^5 \quad -4x^3 \quad -x^2 \quad + 4$ ◀ 2 sign changes in $P(x)$

$P(-x) = (-x)^5 \quad -4(-x)^3 \quad -(-x)^2 \quad + 4$
$= -x^5 \quad +4x^3 \quad -x^2 \quad + 4$ ◀ 3 sign changes in $P(-x)$

The sum of the multiplicities of the zeros must be 5, the degree of $P(x)$. Imaginary zeros, if any, must appear in pairs.

Thus, there are four possible combinations (I–IV).

	I	II	III	IV
Positive real	2	2	0	0
Negative real	3	1	3	1
Imaginary	0	2	2	4
All zeros	5	5	5	5

◀ *Sums of the multiplicities of each type of zero.*

According to the **conjugate zero theorem**, stated below, $2 + 3i$ is a zero of $P(x) = x^4 - 5x^3 + 15x^2 - 5x - 26$ if $2 - 3i$ is a zero of $P(x)$. With this information, you can find all the zeros of $P(x)$.

Conjugate zero theorem If $a + bi$ is a zero of an integral polynomial $P(x)$, then $a - bi$ is also a zero of $P(x)$.

Example 2 Given $P(x) = x^4 - 5x^3 + 15x^2 - 5x - 26$ and $2 - 3i$ is a zero of $P(x)$, find all the zeros of $P(x)$.

$$
\begin{array}{c|ccccc}
 & 1 & -5 & 15 & -5 & -26 \\
\hline
2 - 3i & 1 & -3 - 3i & 3i & 4 + 6i & \boxed{0} \\
\hline
2 + 3i & 1 & -1 & -2 & \boxed{0} &
\end{array}
$$

$x^2 - x - 2 = (x + 1)(x - 2) = 0$
$x = -1$ or $x = 2$

Thus, the zeros of $P(x)$ are $2 - 3i$, $2 + 3i$, -1, and 2.

Written Exercises

1. pos: 2,0; neg: 1, 1; imag: 0, 2   2. pos: 0, 0; neg: 3, 1; imag: 0, 2
3. pos: 3, 1; neg: 1, 1; imag: 0, 2   4. pos: 0; neg: 0; imag: 4
5. pos: 2, 2, 0, 0; neg: 2, 0, 2, 0; imag: 0, 2, 2, 4

Find the possible combinations of the sums of the multiplicities of the positive real, negative real, and imaginary zeros of each polynomial.   6. pos: 3, 3, 1, 1; neg: 2, 0, 2, 0; imag: 0, 2, 2, 4

1. $x^3 - x^2 + 6$ **above**
2. $2x^3 + 3x^2 + 4x + 6$ **above**
3. $x^4 - 2x^3 + 2x^2 - 10$ **above**
4. $x^4 + 16$ **above**
5. $x^4 - x^3 + x^2 + x + 5$ **above**
6. $x^5 - x^3 + x^2 - x - 6$ **above**
7. $x^5 + 32$
8. $x^7 - x^6 + x^4 + x^2 + 10$
9. $2x^6 + x^5 - 3x^4 + x^3 + 8$

7. pos: 0; neg: 1; imag: 4   8. pos: 2, 0; neg: 1, 1; imag: 4, 6   9. pos: 2, 2, 0, 0; neg: 2, 0, 2, 0; imag: 2, 4, 4, 6

Use the conjugate zero theorem to find all the zeros of $P(x)$ if x_1 is a zero of $P(x)$.

10. $P(x) = x^3 - 5x^2 + 9x - 45$, $x_1 = 3i$ $3i, -3i, 5$
11. $P(x) = 2x^3 - x^2 + 50x - 25$, $x_1 = -5i$ $-5i, 5i, \frac{1}{2}$
12. $P(x) = x^4 - 6x^3 + 9x^2 + 24x - 52$, $x_1 = 3 - 2i$ $3 - 2i, 3 + 2i, 2, -2$
13. $P(x) = x^4 - 6x^3 + 29x^2 - 24x + 100$, $x_1 = 3 + 4i$ $3 + 4i, 3 - 4i, 2i, -2i$

CUMULATIVE REVIEW

1. Write $\sqrt[3]{x}$ in exponential form. $x^{\frac{1}{3}}$
2. Write $a^{\frac{1}{4}}$ in radical form. $\sqrt[4]{a}$
3. Write $\sqrt[4]{y^3}$ in exponential form. $y^{\frac{3}{4}}$
4. Write $c^{\frac{2}{3}}$ in radical form. $\sqrt[3]{c^2}$

Applications: Descartes' Rule of Signs

CHAPTER FIFTEEN REVIEW

(A) Exercises are 1–4, 6, 7, 13–17, 22–24.   (C) Exercises are starred. The rest are (B) Exercises.

Vocabulary
antilogarithm [15.6]
common logarithms [15.6]
compound interest [15.7]
exponential equation [15.7]
exponential function [15.1]
integral polynomial [15.8]
logarithmic equation [15.2]
logarithmic function [15.2]
lower bound [15.9]
multiplicity of a zero [15.8]
upper bound [15.9]
zero of a polynomial [15.8]

For complete solutions, see TE Answer Section.
Draw the graph of each function and list four of its properties. Choose convenient scales for the axes.

1. $y = \left(\dfrac{1}{6}\right)^x$ [15.1]   2. $y = \log_6 x$ [15.3]

Find each logarithm. [15.2]

3. $\log_5 625$ **4**   4. $\log_3 \dfrac{1}{9}$ **−2**   5. $\log_3 \sqrt[4]{27}$ **3/4**

Write the y-form of the equation for the inverse of each function. [15.3]

6. $y = 6^x$ **$y = \log_6 x$**   7. $y = \log_{\frac{1}{4}} x$ **$y = \left(\frac{1}{4}\right)^x$**

Write each expression in expanded form.

8. $\log_2 \dfrac{7c}{bd}$ [15.4]   9. $\log_b \sqrt[3]{\dfrac{3v^2}{k}}$ [15.5]

Write each expression as the log of one expression. [15.4]

10. $\log_b 5 + \log_b a - \log_b 3 - \log_b t$ **$\log_b \dfrac{5a}{3t}$**

11. $\dfrac{1}{3} \log_4 a + 2 \log_4 c$ [15.5] **$\log_4 c^2 \sqrt[3]{a}$**

Solve each equation. Check.

12. $\log_5 (4y + 3) - \log_5 9 = \log_5 (y - 3)$ [15.4] **6**

13. $2 \log_4 x = \log_4 3 + \log_4 (x + 6)$ [15.5] **6**

14. $\log_b 125 = 3$ **5**   15. $\log_2 y = 5$ [15.2] **32**

Find the base-10 log or antilog. (Ex. 16–17) [15.6]

16. log 3,040   17. antilog (8.2227 − 10)
    3.4829   **0.0167**

18. Solve $5^{2x-3} = 17$ for x to three significant digits. [15.7] **2.38**

19. Find $\log_3 542$ to three significant digits. [15.7] **5.73**

20. Find the value of a $6,000 bond paying 8%, compounded quarterly, for 12 yr. Use the formula $A = p\left(1 + \dfrac{r}{n}\right)^{nt}$. [15.7]

$15,500 (by logs); $15,532 (by calculator)

For each polynomial $P(x)$, (1) find an upper bound and a lower bound for the real zeros of $P(x)$, (2) factor $P(x)$ into first-degree factors, and (3) solve the equation $P(x) = 0$. (Ex. 21–23)

21. $x^4 + x^3 - 5x^2 - 3x + 6$ [15.8]
22. $4x^4 - 9x^3 + 18x^2 - 36x + 8$ [15.9]
23. $12x^3 + 8x^2 - x - 1$

24. Choose convenient scales for the axes and sketch the graph of $y = 2x^3 - x^2 - 10x + 5$. [15.10]

★25. Find $\log_t (t^3)^a$. [15.2] **3a**

★26. Find the sum of the series:
$3 \cdot \sum_{k=2}^{6} \log_{10} (0.1)^{k-5}$ [15.2] **15**

★27. Solve $\log_2 (y - 3) + \log_2 (2y + 2) = 6$. [15.5] **7**

Prove that each statement is true.

★28. $\dfrac{2}{5} \log_b 4 - \dfrac{1}{5} \log_b 2 = \dfrac{1}{5} \log_b 8$ [15.5]

★29. $\log_b 69 = \dfrac{\log_{10} 69}{\log_{10} b}$ for each positive number $b \neq 1$. [15.7]

★30. The functions defined by $y = \log_4 x$ and by $y = 4^x$ are inverse functions. [15.3]

CHAPTER FIFTEEN TEST

(A) Exercises are 1, 2, 4–7, 9, 12–14, 17, 19, 20. **(C)** Exercises are starred. The rest are **(B)** Exercises.

Draw the graph of each function and list four of its properties. Choose convenient scales for the axes.

1. $y = 2^x$
2. $y = \log_{\frac{1}{2}} x$

For Ex. 1–2, 20, 24–27, see TE Answer Section.

Solve each equation. Check.

3. $\log_5 (y + 1) - \log_5 8 = \log_5 (y - 3) - \log_5 6$ **15**
4. $2 \log_5 x = \log_5 4 + \log_5 (x + 3)$ **6**
5. $\log_b 64 = 3$ **4**

Find each logarithm. (Ex. 6–8)

6. $\log_5 125$ **3**
7. $\log_4 \frac{1}{16}$ **−2**
8. $\log_2 \sqrt[3]{4}$ **$\frac{2}{3}$**

9. Write the y-form of the equation for the inverse of $y = \log_3 x$. **$y = 3^x$**

10. Write $\log_2 \frac{a^3 \sqrt{y}}{5x}$ in expanded form.
    $3 \log_2 a + \frac{1}{2} \log_2 y - \log_2 5 - \log_2 x$

11. Write $\frac{1}{2}(\log_b 7 + \log_b c) - 2 \log_b d$ as the $\log_b$ of one expression. **$\log_b \frac{\sqrt{7c}}{d^2}$**

Find the base-10 log or antilog.

12. log 0.0374   **8.5729 − 10, or −2 + 0.5729**
13. antilog 1.7818   **60.5**
14. Solve $3^{x-1} = 42$ for x to three significant digits. **4.40**
15. Find $\log_6 724$ to three significant digits. **3.67**

16. Find the value of a $10,000 bond paying 8%, compounded semiannually, for 10 yr. Use the formula $A = p\left(1 + \frac{r}{n}\right)^{nt}$.
    $21,900 (by logs); $21,911 (by calculator)

Let $P(x) = 2x^4 - 7x^3 + 5x^2 - 7x + 3$ for Exercises 17–19.

17. Find an upper bound and a lower bound for the real zeros of $P(x)$. **4, 0**
18. Factor $P(x)$ into first-degree factors.
19. Solve the equation where $P(x) = 0$.
    19. $3, \frac{1}{2}, i, -i$ **18. $2(x-3)(x-\frac{1}{2})(x-i)(x+i)$**
20. Choose convenient scales for the axes and sketch the graph of $y = 2x^3 - x^2 - 10x + 5$.

★ 21. Find $\log_{2a} (4a^2)^3$. **6**

★ 22. Solve $\log_2 (y + 1) + \log_2 (3y - 1) = 5$. **3**

★ 23. Find the sum of the series $\sum_{j=1}^{4} \text{antilog}_{10} (3 - j)$. **111.1**

Prove that each statement is true.

★ 24. $2 \log_b xy - \log_b xy^4 = \log_b x - 2 \log_b y$

★ 25. $\frac{3}{4} \log_b 9 - \frac{1}{2} \log_b 3 = \frac{1}{3} \log_b 27$

★ 26. $\log_8 59 = \frac{\log_{10} 59}{\log_{10} 8}$

★ 27. The functions defined by $y = \log_6 x$ and by $y = 6^x$ are a pair of inverse functions.

Chapter Fifteen Test

COLLEGE PREP TEST

DIRECTIONS: Choose the one best answer to each question or problem.

1. If $x^2 = \log_2 64$ and $x > 0$, then $x =$ **B**

   (A) 2   (B) $\sqrt{6}$ (C) 6   (D) 8   (E) 16

2. $\log_b \dfrac{1}{xy} =$ **D**

   (A) $\log_b x + \log_b y$ (B) $\log_b x - \log_b y$
   (C) $1 - \log_b x - \log_b y$ (D) $-\log_b x - \log_b y$
   (E) None of these

3. If $\log_5 3x + \log_5 4x = \log_5 24$, then **C**

   (A) $x = \dfrac{24}{7}$ (B) $x = 2$
   (C) $x^2 = 2$ (D) $x^2 = 12$
   (E) None of these

4. If $P(x) = 8x^3 - 4x^2 + 6x + 1$, then $P(x)$ has **A**

   (A) no integral zeros.
   (B) 1 integral zero.
   (C) 2 integral zeros.
   (D) 3 integral zeros.
   (E) None of these

5. $4\left(\log_3 a + 2\log_3 b - \dfrac{1}{2}\log_3 c\right) =$ **D**

   (A) $\log_3 \left(\dfrac{ab}{c}\right)^4$ (B) $\log_3 \dfrac{4ab^2}{\sqrt{c}}$
   (C) $\log_3 \dfrac{ab^8}{\sqrt{c}}$ (D) $\log_3 \dfrac{a^4 b^8}{c^2}$
   (E) None of these

6. If $2^{x+1} = 50$, then **B**

   (A) $x = 24$ (B) $4 < x < 5$
   (C) $5 < x < 6$ (D) $6 < x < 7$
   (E) None of these

7. If $a = \log_b c$, then **A**

   (A) $b^a = c$ (B) $b^c = a$
   (C) $a^b = c$ (D) $a^c = b$
   (E) None of these

8. Find the number of distinct (different) real zeros of $P(x)$ if $P(x) = (x^2 + 1)(x - 2)^2$. **D**

   (A) 4   (B) 3
   (C) 2   (D) 1
   (E) None of these

9. Find the number of *potential* rational nonintegral zeros of $x^4 + x^3 - x^2 - 1$. **E**

   (A) 2   (B) 4
   (C) 6   (D) 8
   (E) None of these

10. If $\log_2 y = x$, then **B**

    (A) $y < x$ (B) $y > x$
    (C) $y = x$ (D) $y = 2x$
    (E) None of these

11. Two is a zero of the polynomial $x^5 - 4x^4 + x^3 + 10x^2 - 4x - 8$. Find the multiplicity of the zero 2. **C**

    (A) 1   (B) 2
    (C) 3   (D) 4
    (E) None of these

16 PERMUTATIONS, COMBINATIONS, AND PROBABILITY

Non-Routine Problem Solving

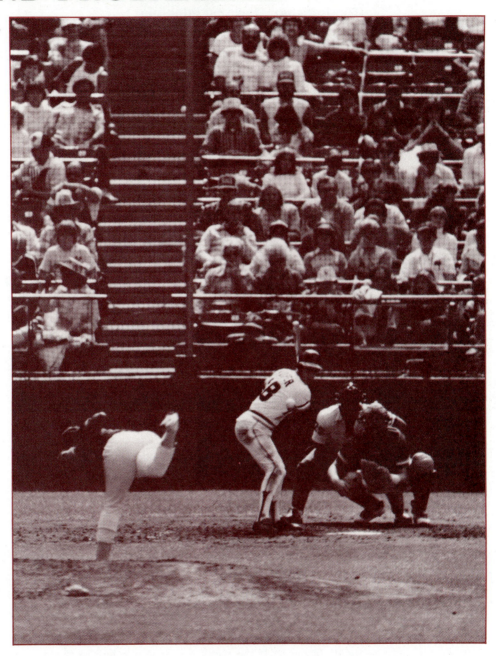

Problem 1. In how many different ways can a sports fan enter the stadium by one entrance and leave by another if the stadium has ten entrances numbered 1–10? **90**

Problem 2. In how many ways can 12 people be seated, 3 at a time, on a bench with room for only 3 people? **1,320**

FUNDAMENTAL COUNTING PRINCIPLE 16.1

Objective **To find the number of possible arrangements of objects by using the Fundamental Counting Principle**

A truck driver must drive from Miami to Orlando and then continue on to Lake City. There are 4 different routes that he can take from Miami to Orlando and 3 different routes from Orlando to Lake City.

One Miami-Orlando-Lake City route is *first* to take A and *then* take #1, or simply A1. All of the possible routes are listed below.

A1 A7 A9   C1 C7 C9   G1 G7 G9   T1 T7 T9

Notice that each route is identified by a letter and then a digit, in that order. By counting, you can find that there are 12 possible Miami-Orlando-Lake City routes. The number of possible routes can also be found without listing any of the routes.

The driver had to make 2 decisions. Mark 2 places.
First decision: Choose a Miami-Orlando route.
                  There are 4 choices: A, C, G, or T.   4
Second decision: Choose an Orlando-Lake City route.
                  There are 3 choices: 1, 7, or 9.   3

$$\underset{\begin{pmatrix}\text{number of}\\\text{letter-choices}\end{pmatrix}}{4} \times \underset{\begin{pmatrix}\text{number of}\\\text{digit-choices}\end{pmatrix}}{3} = 12 = \text{number of possible routes}$$

This suggests a principle for finding the number of possible arrangements of objects without listing all of the arrangements.

> **Fundamental counting principle**
>
> If one choice can be made in *a* ways and a second choice can be made in *b* ways, then the choices *in order* can be made in *a* × *b* different ways.

This Fundamental Counting Principle can be extended to three or more choices made *in order*. If the truck driver above had to make a round trip, Miami-Orlando-Lake City-Orlando-Miami, there would be 144 possible routes. One such route is *C79T*. The driver would have to make 4 decisions. The number of choices for each decision is shown below:

$$\underline{4} \times \underline{3} \times \underline{3} \times \underline{4} = 144$$

The Fundamental Counting Principle is used to find the number of possible arrangements of *n* objects taken *r* at a time. In Example 1, 6 digits are available and they are taken 3 at a time to form three-digit numbers.

Example 1 **How many three-digit numbers can be formed from the 6 digits, 1, 2, 6, 7, 8, 9, if no digit may be repeated in a number?**

There are three decisions to make. Choose (1) a 100s digit, then (2) a 10s digit, and then (3) a 1s digit.

$$\underline{\qquad}\ \underline{\qquad}\ \underline{\qquad}$$
$$(100s)\ (10s)\ (1s)$$

1 2 6 ⑦ 8 9 ▶ There are 6 choices for a 100s digit. $\underline{\ 6\ }\ \underline{\qquad}\ \underline{\qquad}$

1 ② 6   8 9 ▶ This leaves 5 choices for a 10s digit. $\underline{\qquad}\ \underline{\ 5\ }\ \underline{\qquad}$

1   6   ⑧ 9 ▶ Four choices remain for a 1s digit. $\underline{\qquad}\ \underline{\qquad}\ \underline{\ 4\ }$

Use the Fundamental Counting Principle. $\underline{\ 6\ } \times \underline{\ 5\ } \times \underline{\ 4\ }$

Thus, 120 three-digit numbers can be formed.

Six of the 120 three-digit numbers from Example 1 are listed below.

267   276   627   672   726   762

Each of these numbers is a **permutation** of the digits 2, 6, and 7.

Definition: Permutation	A *permutation* is an arrangement of objects in a definite order.

In Example 1, it was specified that a digit should not be repeated in a number. If such repetition is allowed, more than 120 three-digit numbers are possible. Four of these numbers, with a repeated digit, are shown below.

222   227   272   722

Example 2 **How many three-digit numbers can be formed from the digits 2, 4, 6, 8, 9, if a digit may be repeated in a number?**

$$\underline{\ 5\ } \times \underline{\ 5\ } \times \underline{\ 5\ } = 125$$

Thus, 125 three-digit numbers can be formed.

Fundamental Counting Principle

The Fundamental Counting Principle can be used to find the number of permutations of objects other than digits and letters.

Example 3 **A manufacturer makes sweaters in 6 different colors. Each sweater is available with choices of 3 fabrics, 4 kinds of collars, and with or without buttons. How many different types of sweaters does the manufacturer make?**

There are 4 decisions to be made.

$$\underset{\text{color}}{6} \times \underset{\text{fabric}}{3} \times \underset{\text{collar}}{4} \times \underset{\text{buttons}}{2} = 144$$

Thus, 144 types of sweaters are made by the manufacturer.

The number of permutations of *n* distinct objects taken *n* at a time forms a pattern. This is shown in the next example.

Example 4 **Find the number of possible batting orders (permutations) for the nine starting players on a baseball team.**

The coach has to make 9 decisions.

$$\underset{\substack{\text{Leadoff}\\\text{batter}}}{9} \times \underset{\substack{\text{Second}\\\text{batter}}}{8} \times \underset{\substack{\text{Third}\\\text{batter}}}{7} \times \underset{\substack{\text{Cleanup}\\\text{batter}}}{6} \times \underset{\substack{\text{No.}\\5}}{5} \times \underset{\substack{\text{No.}\\6}}{4} \times \underset{\substack{\text{No.}\\7}}{3} \times \underset{\substack{\text{No.}\\8}}{2} \times \underset{\substack{\text{No.}\\9}}{1} = 362{,}880$$

Thus, a baseball coach can select his batting order from among 362,880 possible orders.

In Example 4, the number of permutations, $9 \times 8 \times 7 \times \ldots \times 3 \times 2 \times 1$, can be written as 9! Recall that the expression 9! is read "9 factorial." The value of $n!$ increases rapidly as n increases.

$4! = 4 \times 3 \times 2 \times 1 = 24$
$8! = 8 \times 7 \times 6 \times 5 \times 4 \times 3 \times 2 \times 1 = 40{,}320$
$10! = 10 \times 9 \times 8 \times \ldots \times 3 \times 2 \times 1 = 3{,}628{,}800$

You should notice that the number of arrangements (permutations) of *n* objects taken *n* at a time is always *n*! for each positive integer *n*.

Oral Exercises

State each of the following expressions in factored form.

1. 6! **$6 \times 5 \times 4 \times 3 \times 2 \times 1$**
2. 2! **2×1**
3. 9! **$9 \times 8 \times 7 \times 6 \times 5 \times 4 \times 3 \times 2 \times 1$**
4. 21! **$21 \times 20 \times 19 \ldots 3 \times 2 \times 1$**

Written Exercises

A

1. How many four-digit numbers can be formed from the digits 1, 3, 5, 7, and 9, if no digit may be repeated in a number? **120**
2. If a digit may be repeated in a number, how many 4-digit numbers can be formed from the digits 2, 4, 6, 8, and 9? **625**
3. In how many ways can 6 different books be placed side by side on a shelf? **720**
4. Find the number of different ways that 8 waiters can be assigned to 8 tables if each waiter services one table. **40,320**
5. Find the number of 3-letter permutations of the letters in the word CAR if no letter is repeated in an arrangement. **6**
6. How many 4-letter "words" can be made from the letters in WEST if no letter is repeated in a "word"? **24**
7. How many different signals can be shown by arranging 3 flags in a row if 7 different flags are available? **210**
8. If a signal consists of 5 flags arranged in a row, how many signals can be made from 9 different flags? **15,120**
9. Find the number of different automobile license-plate numbers that can be formed using a letter followed by five digits. **2,600,000**
10. Find the number of 7-digit phone numbers that can be formed if the first digit of a telephone number cannot be zero. **9,000,000**

A sports stadium has eleven gates with six on the north side and five on the south side. Solve Exercises 11–14 using this information.

B

11. In how many ways can a person enter the stadium through a north gate and later leave the stadium through a south gate? **30**
12. Find the number of ways that you can enter the stadium through a south gate and then exit through a north gate. **30**
13. Find the number of ways that you can enter from the north side and then exit from the north side. **36**
14. In how many different ways can a person enter and then leave the stadium? **121**

15. A company makes trucks in 6 different sizes. Each size comes with a choice of 5 exterior colors, a choice of 3 interior colors, with or without a CB radio, and with or without an extra horn. How many different types of trucks does the company make? **360**
16. A school cafeteria offers each student 2 choices of meat, 4 choices of vegetable, 3 choices of drink, and 6 choices of fruit. How many different four-item lunch trays are available? **144**

17. Find the number of 3-letter permutations of the letters in STUDY if a letter may be repeated in a permutation. **125**
18. How many 5-letter code words can be formed from the letters in MONETARY if no letter is repeated in a code word? **6,720**

19. How many different auto license-plate numbers can be formed by 3 different letters followed by 3 digits? **15,600,000**
20. In how many different ways can 6 multiple-choice questions be answered if each question has 5 choices for the answer? **15,625**

Fundamental Counting Principle

CONDITIONAL PERMUTATIONS 16.2

Objective **To find the number of permutations of objects when conditions are attached to the arrangement**

Problems involving permutations may have specific conditions attached to the arrangement of the objects. Some examples follow:
(1) The numbers are odd numbers.
(2) The numbers contain one or more digits.
(3) The words end with the letter *t*.
(4) The numbers are less than 600.

Example 1 **How many permutations of all the letters in the word *MONEY* end with either the letter *E* or the letter *Y*?**

The first decision is to choose the 5th letter, which must be *E* or *Y*.

$$\underline{4} \times \underline{3} \times \underline{2} \times \underline{1} \times \underline{2} = 48$$

↑—— Two choices: *E* or *Y*

Thus, 48 permutations of *M*, *O*, *N*, *E*, *Y* end with *E* or *Y*.

From the digits 7, 8, 9, you can form 10 *odd* numbers containing *one or more* digits if no digit may be repeated in a number. Since the numbers are odd, there are two choices for the units digit, 7 or 9. In this case, the numbers may contain one, two, or three digits.

One-digit numbers:	7	9		
Two-digit numbers:	79	87	89	97
Three-digit numbers:	789	879	897	987

Notice that there are 2 one-digit numbers, 4 two-digit numbers, 4 three-digit numbers and that 2 + 4 + 4 = 10. This suggests that an "or" decision, like one or more digits, involves addition.

Example 2 **How many even numbers containing one or more digits can be formed from 2, 3, 4, 5, 6 if no digit may be repeated in a number?**

There are 3 choices for a units digit: 2, 4, or 6. ▶

$$3 = 3$$
$$4 \times 3 = 12$$
$$4 \times 3 \times 3 = 36$$
$$4 \times 3 \times 2 \times 3 = 72$$
$$4 \times 3 \times 2 \times 1 \times 3 = 72$$

$3 + 12 + 36 + 72 + 72 = 195$

Thus, there are 195 such even numbers.

In some situations, the total number of permutations is the *product* of two or more numbers of permutations. For example, there are 12 permutations of A, B, X, Y, Z with A, B to the left "and" X, Y, Z to the right.

| ABXYZ | ABXZY | ABYXZ | ABYZX | ABZXY | ABZYX |
| BAXYZ | BAXZY | BAYXZ | BAYZX | BAZXY | BAZYX |

Notice that (1) A, B can be arranged in 2!, or 2 ways;
(2) X, Y, Z can be arranged in 3!, or 6 ways; and
(3) A, B, X, Y, Z can be arranged in 2! × 3!, or 12 ways.

An "and" decision involves multiplication.

Example 3 **Four different algebra books and three different geometry books are to be displayed on a shelf with the algebra books together and to the left of the geometry books. How many such arrangements are possible?**

$$\underbrace{4 \times 3 \times 2 \times 1}_{\text{Algebra books (left)}} \times \underbrace{3 \times 2 \times 1}_{\text{Geometry books (right)}} = 144$$

ALG I, ALG II, ALG III, ALG IV, GEOM I, GEOM II, GEOM III

Thus, there are 4! × 3!, or 144, possible arrangements.

Written Exercises

Solve each problem.

Ⓐ 1. How many even five-digit numbers can be formed from the digits 2, 3, 4, 5, 6, if no digit may be repeated in a number? **72**

2. How many odd three-digit numbers can be formed from the digits 1, 2, 3, 4, 5, if a digit may be repeated in a number? **75**

3. Find the number of four-digit numbers that can be formed from the digits 0, 2, 4, 5, 6, 9, if a digit may be repeated in a number. (Consider 0495 as a three-digit number.) **1,080**

4. Find the number of three-digit numbers less than 500 that can be formed from the digits 0, 2, 4, 6, 8, if no digit may be repeated in a number. **24**

5. How many permutations of all the letters in the word JUNIOR begin with N? **120**

6. How many permutations of all the letters in the word NUMBERS do not end with N? **6. 4,320**

7. Six different biology books and four different chemistry books are to be placed on a shelf with the chemistry books together and to the left of the biology books. How many such arrangements are possible? **17,280**

8. A grocery store displays five brands of ground coffee and four brands of instant coffee on a shelf in a row. How many different displays of the nine brands are possible if all of the ground coffees must be shown to the right of the instant coffees? **2,880**

Conditional Permutations

9. How many three-digit or five-digit numbers can be formed from 1, 3, 5, 7, 9, if a digit may be repeated in a number? **3,250**

10. Find the number of two-digit, three-digit, or four-digit numbers that may be formed from 0, 2, 4, 6, if repetition of digits is not allowed. (Consider 024 and 24 as the same number.) **45**

(B) 11. How many numbers of one or more digits can be formed from the digits 6, 7, 8, 9, if no digit is repeated in a number? **64**

12. How many numbers of one or more digits can be formed from the digits 1, 2, 3, if the numbers are less than 400 and repetition of digits is allowed? **39**

13. Find the number of permutations of a, e, i, o, u, y that end with a, or i, or y. **360**

14. How many permutations of the letters in *UNTIL* begin with the prefix *UN-*? **6**

15. Five novels and six short stories are to be displayed. In how many ways can this be done if the novels are kept together and the short stories are kept together on a shelf? (Note: The novels can be at the left *or* the right.) **172,800**

16. A 3-volume dictionary, a 4-volume atlas, and a 6-volume collection of plays are placed on a shelf. In how many ways can this be done if volumes of the same type are kept together? **622,080**

17. How many even numbers of one or more digits can be formed from 2, 4, 6, 7, 8, 9, if repetition of digits is not allowed in a number? **1,304**

18. Find the number of odd numbers less than 1,000 that can be formed from the digits 1, 2, 3, 4, 5, 6, 7, if repetition of digits is allowed. **228**

(C) 19. In how many ways can a family of five stand together in line for tickets if the twins, Mary and Mark, are not to be separated? **48**

20. In how many ways can 6 students be seated in a row of 6 chairs if two of them are a brother and sister who do not want to sit together? **480**

21. A secretary typed 3 letters and then addressed 3 envelopes for the letters. The letters were placed in the envelopes at random with one letter per envelope. In how many different ways could this be done? **6**

22. How many even numbers of one or more digits can be formed from 0, 1, 2, 3, 4, if no digit may be repeated in a number? Note: Do not count a number like 034. [Hint: First, count the numbers ending with 0.] **163**

NON-ROUTINE PROBLEMS

In how many ways can a tennis game of mixed doubles be arranged from a group of 6 males and 4 females? **180**

DISTINGUISHABLE PERMUTATIONS 16.3

Objectives **To find the quotient of numbers given in factorial notation
To find the number of distinguishable permutations when some of the objects in an arrangement are alike**

Some permutation problems lead to expressions like $\frac{8!}{4! \times 3!}$. To find the value of such an expression, begin by dividing out common factors of the numerator and denominator.

Example 1   Find the value of $\frac{8!}{4! \times 3!}$.

One Method
$$\frac{8 \times 7 \times 6 \times 5 \times (4 \times 3 \times 2 \times 1)}{(4 \times 3 \times 2 \times 1) \times 3 \times 2 \times 1}$$
$$\frac{8 \times 7 \times \cancel{6} \times 5}{\cancel{3 \times 2 \times 1}}$$
280

Short Method
$$\frac{8 \times 7 \times \cancel{6} \times 5 \times \cancel{(4!)}}{\cancel{(4!)} \times \cancel{3 \times 2 \times 1}}$$
280

The letters in the word *Pop* are distinguishable since one of the two *p*'s is a capital letter. There are 3!, or 6, **distinguishable permutations** of *P*, *o*, *p*.

$$Pop \quad Ppo \quad oPp \quad opP \quad poP \quad pPo$$

In the word *pop*, the two *p*'s are alike and can be permuted in 2! ways. The number of distinguishable permutations of *p*, *o*, *p* is $\frac{3!}{2!}$, or 3.

$$pop \quad ppo \quad opp$$

The number of distinguishable permutations of the 5 letters in *daddy* is $\frac{5!}{3!}$ since the three *d*'s are alike and can be permuted in 3! ways. This suggests the following rule.

Number of distinguishable permutations Given n objects in which a of them are alike, the number of distinguishable permutations of the n objects is $\frac{n!}{a!}$.

You can extend this rule to find the number of distinguishable permutations of n objects in which more than one group of the objects are alike. For example, the word *pepper* contains three *p*'s which are alike and two *e*'s which are alike. The three *p*'s can be permuted in 3! ways, the two *e*'s can be permuted in 2! ways, and the number of distinguishable permutations of the 6 letters *p*, *e*, *p*, *p*, *e*, *r* is $\frac{6!}{3! \times 2!}$, or 60.

Distinguishable Permutations

Example 2 **How many distinguishable six-digit numbers can be formed from the digits of 747457?**

The 7's can be permuted in 3! ways and the 4's can be permuted in 2! ways.

$$\frac{6!}{3! \times 2!} = \frac{6 \times 5 \times 4 \times (3!)}{(3!) \times 2 \times 1} = 6 \times 5 \times 2 = 60$$

Thus, 60 such numbers can be formed.

Example 3 **How many distinguishable signals can be formed by displaying eleven flags if 3 of the flags are red, 5 are green, 2 are yellow, and 1 is white?**

$$\frac{11!}{3! \times 5! \times 2!} = \frac{11 \times 10 \times 9 \times \overset{4}{\cancel{8}} \times 7 \times \overset{1}{\cancel{6}} \times \overset{1}{\cancel{5!}}}{(\cancel{3} \times \cancel{2} \times 1) \times (2 \times 1) \times \cancel{5!}} = 27{,}720$$

Thus, 27,720 signals can be formed.

Written Exercises

Find the value of each expression.

Ⓐ 1. $\dfrac{7!}{3!}$ **840**   2. $\dfrac{10!}{7!}$ **720**   3. $\dfrac{12!}{8! \times 4!}$ **495**   4. $\dfrac{10!}{2! \times 6! \times 3!}$ **420**   5. $\dfrac{12!}{2! \times 3! \times 8!}$ **990**

How many distinguishable eight-digit numbers can be formed from the digits of each number?

6. 33553533 **56**   7. 24227242 **168**   8. 19116136 **840**   9. 88775599 **2,520**

Find the number of distinguishable permutations of all the letters in each word.

10. *root* **12**   11. *tepee* **20**   12. *puppet* **120**   13. *scissors* **1,680**

Ⓑ 14. *divided* **420**   15. *murmur* **90**   16. *nonsense* **1,680**   17. *Tennessee* **3,780**

18. How many distinguishable signals can be formed by displaying 9 flags if 3 of the flags are blue, 2 are orange, and 4 are black? **1,260**

19. In how many distinguishable ways can 4 nickels, 3 dimes, 2 quarters, and 1 penny be distributed to 10 children if each child is to receive one coin? **12,600**

NON-ROUTINE PROBLEMS

1. In how many distinguishable ways can $1.14 in pennies, nickels, dimes, quarters, and half-dollars be distributed among 12 children if each child is to receive one coin? **166,320**

Simplify.

2. $\dfrac{n!}{(n-3)!}$
   $n^3 - 3n^2 + 2n$

3. $\dfrac{(n+2)!}{n!}$
   $n^2 + 3n + 2$

4. $\dfrac{(n-4)!}{(n-6)!}$
   $n^2 - 9n + 20$

5. $\dfrac{(n+2)!}{(n-1)!}$
   $n^3 + 3n^2 + 2n$

CIRCULAR PERMUTATIONS 16.4

Objective **To find the number of possible permutations of objects in a circle**

Three objects may be arranged in a line in 3!, or 6, ways. Any one of the objects may be placed in the first position.

ABC   ACB   BAC   BCA   CAB   CBA

In a **circular permutation** of objects, there is no first position. Only the positions of the objects relative to one another are considered. In the figures below, Al, Betty, and Carl are seated in a circular arrangement with each person facing the center of the circle.

In each of the first three figures, Al has Betty to his left and Carl to his right. This is *one circular permutation* of Al, Betty, and Carl.

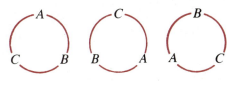

The remaining three figures each show Al with Betty to his right and Carl to his left. Again, these count as only one circular permutation of the three people.

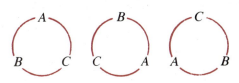

Thus, there are 2!, or 2, circular permutations of the 3 people.

To find the circular permutations of n distinct objects, begin by placing one object on the circle. Then arrange the remaining $(n - 1)$ objects in $(n - 1)!$ ways relative to the first object.

Number of circular permutations	The number of circular permutations of n distinct objects is $(n - 1)!$

Example **A married couple invites 3 other couples to an anniversary dinner. In how many different ways can all of the 8 people be seated around a circular table?**

$(n - 1)! = (8 - 1)! = 7! = 5{,}040$

Thus, there are 5,040 different ways to seat the 8 people.

Reading in Algebra

Tell which number of permutations is the greater for each pair below.
1. (a) 5 people in a line
   (b) 5 people in a circle **a**
2. (a) all the letters in *HOBO*
   (b) all the letters in *HOPE* **b**
3. (a) 6 people in a circle
   (b) all the letters in *GEESE* **a**
4. (a) all the digits in 52,466
   (b) all the digits in 63,777 **a**

Oral Exercises

Tell whether each arrangement is linear, circular, or either of these.
1. Digits in a 5-digit number **linear**
2. Keys on a key ring **circular**
3. Football players in a huddle **circular**
4. Letters in your last name **linear**
5. Books on a table **either**
6. Charms on a bracelet **circular**
7. Flags in a signal **linear**
8. Trees in a park **either**

Written Exercises

Solve each problem.

Ⓐ 1. In how many different ways can seven children be seated around a circular table at a birthday party? **720**

2. The president of a company and his seven vice-presidents are to be seated around a circular conference table. In how many ways can this be done? **5,040**

3. A football team of eleven players forms a circular huddle before each play. In how many different ways can the players be arranged? **3,628,800**

4. Find the number of different ways that a gardener may plant a dozen different bushes around a circular flower bed. **39,916,800**

Ⓑ 5. A husband and wife invite four other couples to dinner. In how many ways can they all be seated around the circular dining table? **362,880**

6. A woman and her daughter each invite three friends to lunch. Find the number of ways that all of them can be seated around one circular table. **5,040**

7. Four teachers and four students are seated in a circular discussion group. Find the number of ways this can be done if teachers and students must be seated alternately. **144**

8. In how many different ways can five girls and five boys be seated around a circular table if no two girls may be seated next to each other? **2,880**

Example **In how many different ways can 3 keys be arranged on a ring?**

A key ring has a front view and a rear view, but the two views are considered to be one arrangement.

The 3 keys can be arranged in $\frac{(3-1)!}{2}$ ways.

Front View Rear View

Thus, the 3 keys can be arranged in only 1 way.

Ⓒ 9. In how many ways can 4 keys be arranged on a ring? **3**

10. Find the number of ways that 7 charms can be arranged on a bracelet. **360**

NON-ROUTINE PROBLEMS

Simplify.
1. $(n+1) \cdot n \cdot (n-1)!$
   $(n+1)!$
2. $(n-3)(n-4)(n-5)!$
   $(n-3)!$

COMBINATIONS 16.5

Objectives
To find the number of possible selections of *n* objects taken *r* at a time without regard to order

To find the value of $\binom{n}{r}$ for nonnegative integers *n* and *r* where $n \geq r$

In a permutation, the order of the objects is important. Another type of problem involves selecting objects where the order in which they are selected or arranged does not matter. A selection of this type is called a **combination** of the objects. Some examples are listed below.

Committee: The Al-Betty-Carl committee is the same as the Carl-Al-Betty committee.

Vertices of a triangle: $\triangle DEF$ is the same as $\triangle EFD$, which is the same as $\triangle FDE$.

Value of 3 coins: A dime, a nickel, and a penny have the same total value as a nickel, a penny, and a dime.

Definition: Combination

A *combination* is a selection of objects without regard to order.

You can list all of the three-letter *combinations* that can be formed from the 5 letters *a*, *b*, *c*, *d*, and *e* and count the number of combinations.

a-b-c	a-c-d	a-d-e	b-c-d	b-d-e	c-d-e
a-b-d	a-c-e		b-c-e		
a-b-e					

Thus, there are *ten* combinations of 5 things taken 3 at a time. $\binom{5}{3}$ is the symbol for "the number of combinations of 5 things taken 3 at a time." The value of $\binom{5}{3}$ is found in the following way.

$$\binom{5}{3} = \frac{5!}{3!(5-3)!} = \frac{5!}{3! \times 2!} = \frac{5 \times 4 \times \cancel{3!}}{\cancel{3!} \times 2 \times 1} = 10$$

Number of combinations

$\binom{n}{r}$ is the number of combinations of *n* things taken *r* at a time.

$$\binom{n}{r} = \frac{n!}{r!(n-r)!}$$

Example 1 Find the value of $\binom{7}{4}$.

$$\binom{7}{4} = \frac{7!}{4!(7-4)!} = \frac{7!}{4! \times 3!} = \frac{7 \times \cancel{6} \times 5 \times \cancel{4!}}{\cancel{4!} \times \cancel{3} \times \cancel{2} \times 1} = 35$$

Combinations

Example 2   **How many different 4-member committees can be formed if 10 people are available for appointment to a committee?**

$$\binom{10}{4} = \frac{10!}{4! \times 6!} = \frac{10 \times 9 \times 8 \times 7 \times 6!}{4 \times 3 \times 2 \times 1 \times 6!} = 210$$

Thus, 210 committees can be formed.

Example 3   **Seven points, *A* through *G*, are located on a circle. How many triangles are determined by the 7 points?**

A triangle is determined by any 3 points that are not on the same line.

$$\binom{7}{3} = \frac{7!}{3! \times 4!} = \frac{7 \times 6 \times 5 \times 4!}{3 \times 2 \times 1 \times 4!} = 35$$

Thus, 35 triangles are determined.

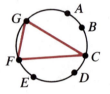

The number of combinations of 5 things taken 5 at a time is 1. Notice the consequence when the formula for $\binom{n}{r}$ is used to find $\binom{5}{5}$.

$$\binom{5}{5} = \frac{5!}{5! \times 0!} = \frac{1}{0!}, \text{ which must be equal to 1.}$$

Recall the definition of 0!

$$0! = 1$$

With this definition of 0!, the formula for $\binom{n}{r}$ can be used with any nonnegative integers *n* and *r* where $n \geq r$.

Reading in Algebra

Tell whether each collection of objects is a *permutation* or a *combination* of the objects in the collection. (Ex. 1–6)

1. A four-digit number **permutation**
2. A two-letter name of a line **combination**
3. A five-person committee **combination**
4. A six-flag signal **permutation**
5. A five-letter code word **permutation**
6. A six-card hand in a card game **combination**
7. List all of the different amounts of money that can be determined by selecting three coins from 1 penny, 1 nickel, 1 dime, and 1 quarter. **16¢, 31¢, 36¢, 40¢**
8. List a two-letter name for each line determined by the points *A*, *B*, *C*, and *D* which are located on a circle. **$\overleftrightarrow{AB}$, $\overleftrightarrow{AC}$, $\overleftrightarrow{AD}$, $\overleftrightarrow{BC}$, $\overleftrightarrow{BD}$, $\overleftrightarrow{CD}$**

Written Exercises

Find the value of each expression. (Ex. 1–6)

A
1. $\binom{7}{2}$ **21**
2. $\binom{10}{6}$ **210**
3. $\binom{18}{3}$ **816**
4. $\binom{6}{6}$ **1**
5. $\binom{63}{1}$ **63**
6. $\binom{8}{0}$ **1**

7. How many 4-letter combinations can be formed from the first half of our alphabet? **715**

8. How many different 5-member committees can be formed if 20 people are available for membership? **15,504**

9. Find the number of lines determined by 10 points, A through J, on a circle. **45**

10. Find the number of triangles determined by 15 points, A through O, on a circle. **455**

B
11. An algebra student must solve 7 out of 10 word problems. In how many different ways can the 7 problems be selected? **120**

12. In how many different ways can a team of 33 football players choose three captains for its first game of the season? **5,456**

13. During the baseball season, an 8-school athletic conference requires that each school play 3 games with each of the other schools. Find the total number of baseball games played. **84**

One bag contains 5 red marbles numbered 1 through 5 and another bag contains 7 blue marbles numbered 1 through 7. (Ex. 14–16)

14. In how many different ways can you choose 3 red marbles *and* 3 blue marbles? **350**
15. In how many different ways can you choose 3 red marbles *or* 3 blue marbles? **45**
16. In how many different ways can you choose *one or more* red marbles? **31**

17. There are 5 pellets, lettered A, B, C, D, and E, that weigh 1 g, 2 g, 4 g, 8 g, and 16 g, respectively. How many different masses can be measured by using one or more of the 5 weights on a balance scale? **31**

18. How many different amounts of money can be determined by selecting one or more coins from 1 penny, 1 nickel, 1 dime, 1 quarter, 1 half-dollar, and 1 silver dollar? **63**

For Ex. 19–20, see TE Answer Section.

You can verify that $\binom{7}{2} = \binom{7}{5}$ and that $\binom{4}{2} + \binom{4}{3} = \binom{5}{3}$.
This suggests the theorems in Exercises 19 and 20.

C
19. Prove that $\binom{n}{r} = \binom{n}{n-r}$ for all integers r and n, $0 \leq r \leq n$.

20. Prove that $\binom{n}{r} + \binom{n}{r+1} = \binom{n+1}{r+1}$ for all integers r and n where $0 \leq r < n$.

21. Find three pairs of integers (n, r) for which $\binom{n}{4} = \binom{12}{r}$. **(4, 12), (12, 8), (4, 0)**

NON-ROUTINE PROBLEMS

1. Rewrite the binomial theorem (page 382) using $\binom{n}{r}$ notation for the coefficients of the terms. $\binom{n}{0}a^n + \binom{n}{1}a^{n-1}b + \binom{n}{2}a^{n-2}b^2 + \binom{n}{3}a^{n-3}b^3 + \ldots + \binom{n}{n}b^n$

2. Find the number of different 5-card hands that may be drawn from a deck of 52 cards. **2,598,960**

Combinations

PROBABILITY: SIMPLE EVENT 16.6

Objective **To determine the probability of a simple event**

When a "fair" coin is tossed it seems reasonable to expect that the coin is just as likely to come up heads (H) as tails (T). These two outcomes, heads and tails, have equal chances of occurring. In everyday language, we say the coin has 1 chance in 2 of coming up heads and 1 chance in 2 of coming up tails. In mathematical language, we say the **probability** of obtaining a head in one flip of a coin is $\frac{1}{2}$. In symbols, $P(\text{head}) = \frac{1}{2}$.

One way of obtaining the probability of a simple event occurring is to determine the number of ways the event can occur, called *favorable outcomes*, and then divide by the *total number of all possible outcomes*.

Probability of simple event

The probability of a simple event is given by the formula:
$$P(\text{simple event}) = \frac{\text{Number of favorable outcomes}}{\text{Total number of all possible outcomes}}.$$

Example 1 **What is the probability that in a throw of a die (plural: dice) a face with 4 dots will appear? What is the probability that a face with an even number of dots will appear?**

Use $P(\text{simple event}) = \frac{\text{Number of favorable outcomes}}{\text{Total number of all possible outcomes}}.$

One face of a die contains 4 dots.	Three faces of a die contain even numbers: 2, 4, 6.
There is only 1 favorable outcome.	There are 6 faces on a die.
There are 6 faces on a die.	The number of favorable outcomes is 3.
There are 6 possible outcomes.	The number of possible outcomes is 6.
Thus, $P(\text{a face with 4 dots}) = \frac{1}{6}$.	**Thus,** $P(\text{even number of dots}) = \frac{3}{6}$, or $\frac{1}{2}$.

Example 2 **A bag contains 5 red balls and 4 black balls. One ball is drawn from the bag. What is the probability that the ball is red? What is the probability that it is black? What is the probability that it is white?**

There are 5 + 4, or 9 possible outcomes.
When looking for a red ball, there are 5 favorable outcomes.
When looking for a black ball, there are 4 favorable outcomes.
When looking for a white ball, there are 0 favorable outcomes.

Thus, $P(\text{red ball}) = \frac{5}{9}$, $P(\text{black ball}) = \frac{4}{9}$, and $P(\text{white ball}) = \frac{0}{9}$, or 0.

There are times when an event is certain to happen and times when it is impossible for an event to happen.

Probability of 1 and 0	The probability of an event that is certain to happen is 1. The probability of an event that cannot happen is 0.

As indicated, the probability of a simple event E occurring is greater than or equal to 0 and less than or equal to 1:
$$0 \leq P(E) \leq 1.$$

Example 3   **A deck of cards contains 52 cards, 13 cards of each suit. One card is drawn. What is the probability that the card is an ace? that it is the ace of hearts? that it is not an ace?**

There are 52 possible outcomes.
When looking for an ace, there are 4 favorable outcomes.
When looking for the ace of hearts, there is 1 favorable outcome.
When looking for a card that is not an ace, there are 48 favorable outcomes.

Thus, $P(\text{ace}) = \frac{4}{52}$, or $\frac{1}{13}$; $P(\text{ace of hearts}) = \frac{1}{52}$; and $P(\text{not an ace}) = \frac{48}{52}$, or $\frac{12}{13}$.

In Example 3, observe that the probability of getting an ace is $\frac{1}{13}$ and the probability of not getting an ace is $\frac{48}{52}$, or $\frac{12}{13}$.
The sum of the two probabilities is $\frac{1}{13} + \frac{12}{13} = 1$. If the probability of an event occurring is P, then the probability of the event not occurring is $1 - P$.

Definition: Odds	The *odds* that a simple event E will occur are $\frac{P(E)}{1 - P(E)}$, or $\frac{P(E)}{P(\overline{E})}$, where $P(\overline{E})$ is the probability of E not occurring.

Example 4   **A bag contains 30 balls; 18 of them are red. What are the odds that the ball drawn will be red?**

$P(E) = P(\text{red ball}) = \frac{18}{30}$, or $\frac{3}{5}$

Use $\frac{P(E)}{P(\overline{E})}$. ▶ $P(\overline{E}) = 1 - P(E)$, or $1 - \frac{3}{5} = \frac{2}{5}$

Thus, the odds of getting a red ball are $\frac{\frac{3}{5}}{\frac{2}{5}}$, or $\frac{3}{2}$, or 3 to 2.

Probability: Simple Event

Written Exercises

(A)

1. What is the probability that in a throw of a die, the 6 will appear? the 3 will appear? an odd number will appear? $\frac{1}{6}, \frac{1}{6}, \frac{1}{2}$

2. What is the probability that in a throw of a die, the 1 will appear? a prime number will appear? a number divisible by 2 will appear? $\frac{1}{6}, \frac{1}{2}, \frac{1}{2}$

3. A bag contains 8 white balls and 6 blue balls. A ball is drawn from the bag. What is the probability that the ball is white? that it is blue? $\frac{4}{7}, \frac{3}{7}$

4. A bag contains 12 red balls and 18 black balls. A ball is drawn from the bag. What is the probability that the ball is red? that it is black? $\frac{2}{5}, \frac{3}{5}$

5. A deck of cards contains 52 cards, 13 cards of each suit. One card is drawn. What is the probability that the card is a jack? that it is the jack of clubs? that it is not a jack? $\frac{1}{13}, \frac{1}{52}, \frac{12}{13}$

6. A box contains 1,000 light bulbs, 900 good ones and 100 defective ones. A bulb is drawn. What is the probability that the bulb is defective? $\frac{1}{10}$

7. A bag contains 4 red, 3 white, and 5 blue balls. A ball is drawn from the bag. What is the probability that the ball is red? that it is white? that it is blue? that it is not red? that it is not blue? $\frac{1}{3}, \frac{1}{4}, \frac{5}{12}, \frac{2}{3}, \frac{7}{12}$

8. A deck of cards contains 52 cards, 13 cards of each suit. There are 2 black suits (spades and clubs) and 2 red suits (hearts and diamonds). One card is drawn. What is the probability that the card is red? that it is black? that it is a red ace? that it is a black 7? $\frac{1}{2}, \frac{1}{2}, \frac{1}{26}, \frac{1}{26}$

(B)

9. A die is rolled 100 times. About how many times should it stop on 4? About how many times should an even number appear? About how many times should a number less than 3 appear? [Hint: First determine the probability of each event.] **16, 50, 33**

10. A die is rolled 6,000 times. About how many times should it not stop on 6? About how many times should it stop on 5? About how many times should a number greater than 2 appear? **5,000; 1,000; 4,000**

11. The probability that it will rain on any given day in December in the town of Los Altos is $\frac{8}{10}$. About how many days will it not rain in December? **6**

12. The probability of producing a defective part by a given machine is $\frac{1}{900}$. If 36,000 parts are produced in one day, about how many of the parts will be defective? **40**

13. A bag contains 16 red balls and 32 white balls. A ball is drawn from the bag. What are the odds that the ball is red? $\frac{1}{2}$, or 1 to 2

14. If the probability that it will snow on a given day is $\frac{2}{5}$, what is the probability that it will not snow? What are the odds that it will snow? that it will not snow? $\frac{3}{5}$, 2 to 3, 3 to 2

The probability of an event E occurring if the odds are m to n that it will occur is $P(E) = \dfrac{m}{m+n}$.

(C)

15. A high school senior feels her odds of getting into a college of her choice are 5 to 1. Find the probability that she will be accepted. $\frac{5}{6}$

16. The odds that the team will win are 3 to 2. Find the probability of winning. $\frac{3}{5}$

PROBABILITY: COMPOUND EVENTS 16.7

Objective **To determine the probability of an event using a sample space**

In the previous lesson, you learned how to find the probability of a simple event by determining the number of favorable outcomes and the number of all possible outcomes. In some situations, it is not easy to identify the number of outcomes. Making a list of all possible outcomes may be helpful. Such a listing is called a **sample space**.

A sample space showing all possible outcomes of tossing three coins is shown below.

1st coin ▶   Head (*H*)   or   Tail (*T*)

2nd coin ▶   *H*   or   *T*   *H*   or   *T*

3rd coin ▶   *H* or *T*   *H* or *T*   *H* or *T*   *H* or *T*

Sample space ▶

HHH	HTH	THH	TTH
HHT	HTT	THT	TTT

Example 1 **In a toss of three coins, what is the probability of obtaining two heads and a tail? What is the probability of obtaining a head, a tail, and a head, in that order?**

As shown in the sample space above, two heads and a tail can occur three times:
HHT, HTH, THH.

A head, a tail, and a head can occur in that order only once: *HTH*.

There are 8 possible outcomes. ▶   **Thus, $P(2 \text{ heads and } 1 \text{ tail}) = \frac{3}{8}$, and $P(HTH) = \frac{1}{8}$.**

Example 2 **In a toss of three coins, what is the probability of obtaining exactly two tails? What is the probability of obtaining at least one head?**

Exactly two tails can occur three times: *HTT, THT, TTH*.

At least one head can occur seven times: *HHH, HHT, HTH, HTT, THH, THT, TTH*.

Thus, $P(\text{exactly 2 tails}) = \frac{3}{8}$, and $P(\text{at least 1 head}) = \frac{7}{8}$.

Probability: Compound Events

Example 3 **In a throw of two dice, a red one and a white one, find $P(r \leq 2 \text{ or } w \leq 3)$.**

There are 36 possible outcomes.
There are 12 ways to get $r \leq 2$ and 18 ways to get $w \leq 3$, but 6 ordered pairs are common to both sets and cannot be counted twice.

		White					
		1	2	3	4	5	6
Red	1	(1, 1)	(1, 2)	(1, 3)	(1, 4)	(1, 5)	(1, 6)
	2	(2, 1)	(2, 2)	(2, 3)	(2, 4)	(2, 5)	(2, 6)
	3	(3, 1)	(3, 2)	(3, 3)			
	4	(4, 1)	(4, 2)	(4, 3)			
	5	(5, 1)	(5, 2)	(5, 3)			
	6	(6, 1)	(6, 2)	(6, 3)			

For $r \leq 2$ or $w \leq 3$, there are $12 + 18 - 6$, or 24, favorable outcomes.

Thus, $P(r \leq 2 \text{ or } w \leq 3) = \frac{24}{36}$, or $\frac{2}{3}$.

In Example 3, notice that
$P(r \leq 2) = \frac{12}{36}$, or $\frac{1}{3}$, $P(w \leq 3) = \frac{18}{36}$, or $\frac{1}{2}$, $P(r \leq 2 \text{ and } w \leq 3) = \frac{6}{36}$, or $\frac{1}{6}$ and that $\frac{1}{3} + \frac{1}{2} - \frac{1}{6} = \frac{2}{3}$, which is equal to $P(r \leq 2 \text{ or } w \leq 3)$.

This leads to the following statement about the probability of two events A or B occurring.

If two events can both occur at the same time, they are called **inclusive events**. If they cannot both occur at the same time, they are called **mutually exclusive events**.

Probability of A or B

If A and B are inclusive events, then
$$P(A \text{ or } B) = P(A) + P(B) - P(A \text{ and } B).$$
If A and B are mutually exclusive events, then
$$P(A \text{ or } B) = P(A) + P(B).$$

Example 4 **In a throw of 2 dice, what is the probability of obtaining a sum of 7? a sum of 11? a sum of either 7 or 11?**

The sum 7 occurs 6 ways: (6, 1), (5, 2), (4, 3), (3, 4), (2, 5), (1, 6).
The sum 11 occurs 2 ways: (5, 6), (6, 5).
But both sums cannot occur at the same time.

The events are mutually exclusive. Use $P(A \text{ or } B) = P(A) + P(B)$.

$P(\text{sum of 7 or sum of 11}) = \frac{6}{36} + \frac{2}{36} = \frac{8}{36}$, or $\frac{2}{9}$.

Thus, $P(\text{sum of 7}) = \frac{1}{6}$, $P(\text{sum of 11}) = \frac{1}{18}$, and $P(\text{sum of 7 or sum of 11}) = \frac{2}{9}$.

Recall that when the word "and" is used to connect two statements, the conditions of both statements must be met simultaneously if the compound statement is to be true. If you wish to find $P(A$ and $B)$, then you are interested in those ways in which event A and event B can both occur at the same time.

Example 5

In a throw of two dice, a red one and a white one, find $P(r \leq 3$ and $w \leq 2)$.

There are 36 possible outcomes.
There are 18 ways to get $r \leq 3$ and 12 ways to get $w \leq 2$, but only 6 ways to get both.

Six ordered pairs occur in both sets of ordered pairs.

Thus, $P(r \leq 3$ and $w \leq 2) = \dfrac{6}{36}$, or $\dfrac{1}{6}$.

		w				
	1	2	3	4	5	6
1	(1, 1)	(1, 2)	(1, 3)	(1, 4)	(1, 5)	(1, 6)
2	(2, 1)	(2, 2)	(2, 3)	(2, 4)	(2, 5)	(2, 6)
r 3	(3, 1)	(3, 2)	(3, 3)	(3, 4)	(3, 5)	(3, 6)
4	(4, 1)	(4, 2)				
5	(5, 1)	(5, 2)				
6	(6, 1)	(6, 2)				

In Example 5, the outcome of throwing the red die does not affect the outcome of throwing the white die. Events of this type in which neither event depends upon the other are called **independent events**.

Notice that $P(r \leq 3) = \dfrac{18}{36}$, or $\dfrac{1}{2}$, $P(w \leq 2) = \dfrac{12}{36}$, or $\dfrac{1}{3}$, and that $\dfrac{1}{2} \cdot \dfrac{1}{3} = \dfrac{1}{6}$, which is equal to $P(r \leq 3$ and $w \leq 2)$. This leads to the following statement about the probability of two events A and B occurring.

Probability of A and B

If A and B are independent events, then
$$P(A \text{ and } B) = P(A) \cdot P(B).$$

In the next example, the outcome of throwing each die is affected by the outcome of throwing the other. Events of this type in which one event depends upon the other are called **dependent events**. The sample space for dependent events is usually a reduced sample space of the original space.

Example 6

Find $P(r = 2$, given that $r + w \leq 5)$.

There are 10 ways to get $r + w \leq 5$ and so there are 10 possible outcomes. The 10 ordered pairs make up the reduced sample space.

There are 3 ways to get $r = 2$, so there are 3 favorable outcomes.

Thus, $P(r = 2$, given that $r + w \leq 5) = \dfrac{3}{10}$.

		w				
	1	2	3	4	5	6
1	(1, 1)	(1, 2)	(1, 3)	(1, 4)	(1, 5)	(1, 6)
2	(2, 1)	(2, 2)	(2, 3)	(2, 4)	(2, 5)	(2, 6)
r 3	(3, 1)	(3, 2)	(3, 3)	(3, 4)	(3, 5)	(3, 6)
4	(4, 1)	(4, 2)	(4, 3)	(4, 4)	(4, 5)	(4, 6)
5	(5, 1)	(5, 2)	(5, 3)	(5, 4)	(5, 5)	(5, 6)
6	(6, 1)	(6, 2)	(6, 3)	(6, 4)	(6, 5)	(6, 6)

> **Probability of A, given B**
>
> If A and B are dependent events, $P(A \text{ given } B) = \dfrac{P(A \text{ and } B)}{P(B)}$.

Written Exercises

18. $\frac{5}{36}$ 19. $\frac{1}{18}$ 20. $\frac{1}{36}$ 21. $\frac{1}{6}$ 22. $\frac{7}{12}$ 23. $\frac{1}{36}$

For the following exercises, use the sample space showing all possible outcomes of tossing 3 coins. Find the indicated probability.

A
1. $P(3 \text{ heads})$ $\frac{1}{8}$
2. $P(2 \text{ tails and a head})$ $\frac{3}{8}$
3. $P(1 \text{ head and 2 tails})$ $\frac{3}{8}$
4. $P(3 \text{ tails})$ $\frac{1}{8}$
5. $P(\text{TTH})$ $\frac{1}{8}$
6. $P(\text{THT})$ $\frac{1}{8}$
7. $P(\text{HHT})$ $\frac{1}{8}$
8. $P(\text{HTT})$ $\frac{1}{8}$
9. $P(\text{at least 1 tail})$ $\frac{7}{8}$
10. $P(\text{at least 2 heads})$ $\frac{4}{8}$, or $\frac{1}{2}$
11. $P(\text{at least 2 tails})$ $\frac{4}{8}$, or $\frac{1}{2}$
12. $P(\text{at least 1 head and 1 tail})$ $\frac{6}{8}$, or $\frac{3}{4}$
13. $P(\text{exactly 2 tails})$ $\frac{3}{8}$
14. $P(\text{exactly 1 head})$ $\frac{3}{8}$

For the following exercises, use the sample space showing all possible outcomes of rolling 2 dice, a red one, r, and a white one, w. Find each indicated probability.

15. $P(r = 4)$ $\frac{1}{6}$
16. $P(r \leq 3)$ $\frac{1}{2}$
17. $P(w \geq 5)$ $\frac{1}{3}$
18. $P(r + w = 8)$
19. $P(r + w = 11)$
20. $P(r + w = 2)$
21. $P(r + w \geq 10)$
22. $P(r + w \leq 7)$
23. $P(r + w < 3)$
24. $(P(r + w \leq 1)$ 0
25. $P(r \leq 3 \text{ or } w = 2)$ $\frac{7}{12}$
26. $P(r \geq 5 \text{ or } w \geq 5)$ $\frac{5}{9}$
27. $P(r \leq 2 \text{ or } w \leq 5)$ $\frac{8}{9}$
28. $P(r \leq 4 \text{ or } w \geq 4)$ $\frac{5}{6}$
29. $P(\text{sum of 6 or sum of 10})$ $\frac{2}{9}$
30. $P(\text{sum of 2 or sum of 12})$ $\frac{1}{18}$
31. $P(\text{sum of 7 or sum of 11})$ $\frac{2}{9}$
32. $P(\text{sum of 11 or sum of 12})$ $\frac{1}{12}$
33. $P(r = 2 \text{ and } w \geq 5)$ $\frac{1}{18}$
34. $P(r \geq 4 \text{ and } w = 4)$ $\frac{1}{12}$
35. $P(r \geq 5 \text{ and } w \leq 2)$ $\frac{1}{9}$
36. $P(r \leq 2 \text{ and } w \leq 4)$ $\frac{2}{9}$
37. $P(r + w = 7, \text{ given that } r = 4)$ $\frac{1}{6}$
38. $P(r = 5, \text{ given that } r + w = 9)$ $\frac{1}{4}$
39. $P(w = 4, \text{ given that } r + w = 6)$ $\frac{1}{5}$
40. $P(r + w = 9, \text{ given that } w = 6)$ $\frac{1}{6}$

A bag contains 6 red, 8 white, and 4 blue balls. A ball is drawn from the bag. Find each indicated probability for drawing a ball.

B
41. $P(\text{red or white})$ $\frac{7}{9}$
42. $P(\text{red or blue})$ $\frac{5}{9}$
43. $P(\text{white or blue})$ $\frac{2}{3}$
44. $P(\text{red, white, or blue})$ 1
45. $P(\text{blue or green})$ $\frac{2}{9}$
46. $P(\text{green or yellow})$ 0

One card is drawn from an ordinary deck. Find the indicated probability for drawing a card.

47. $P(\text{red or a queen})$ $\frac{7}{13}$
48. $P(\text{black king or a club})$ $\frac{7}{26}$
49. $P(\text{an even number or black})$ $\frac{9}{13}$
50. $P(\text{red face or a jack})$ $\frac{2}{13}$

C
51. Prepare a sample space showing all possible outcomes for boys and girls in a family with three children. **GGG, GGB, GBG, GBB, BGG, BGB, BBG, BBB**

Using the sample space in Exercise 51, find each of the following probabilities.

52. $P(3 \text{ boys})$ $\frac{1}{8}$
53. $P(2 \text{ girls and a boy})$ $\frac{3}{8}$
54. $P(2 \text{ boys and a girl})$ $\frac{3}{8}$

55. $P(\text{1st is a girl, 2nd is a boy, 3rd is a boy})$ $\frac{1}{8}$
56. $P(\text{1st is a boy, 2nd is a girl, 3rd is a boy})$ $\frac{1}{8}$

A family has 4 children. Find each indicated probability.

57. Find $P(4 \text{ boys})$ $\frac{1}{16}$
58. $P(3 \text{ girls and a boy})$ $\frac{1}{4}$
59. $P(2 \text{ girls and 2 boys})$ $\frac{3}{8}$

In a given town, there are 1,000 families, and each family has 4 children. Answer each question.

60. Approximately how many families have exactly 3 girls? exactly 3 boys? **250; 250**
61. Approximately how many families had 2 boys first and then 2 girls? **62**

Chapter Sixteen

PROBABILITY AND ARRANGEMENTS 16.8

Objective **To determine the probability of an event using the Fundamental Counting Principle, permutations, and combinations**

The Fundamental Counting Principle as well as permutations and combinations can often be used to determine the probability of a given event.

Example 1 A matching test is given. There are 5 statements in the left column and 6 possible answers in the right column. If an answer can be used only once, what is the probability that a student can guess all the answers correctly?

$$P(\text{simple event}) = \frac{\text{Number of favorable outcomes}}{\text{Total number of all possible outcomes}}$$

There is 1 way all answers can be guessed correctly, so there is 1 favorable outcome. The total number of all possible outcomes is the number of 5-answer permutations that can be formed from the 6 answers.

Thus, $P\begin{pmatrix}\text{guessing all answers using} \\ \text{each answer only once}\end{pmatrix} = \frac{1}{6 \times 5 \times 4 \times 3 \times 2} = \frac{1}{720}$.

In Example 1, to find the probability of guessing all the answers correctly if each answer can be used at least once, the Fundamental Counting Principle can be applied. The five answer choices in order can be made in $6 \times 6 \times 6 \times 6 \times 6$, or 7,776 different ways. So, $P\begin{pmatrix}\text{guessing all answers using} \\ \text{each answer at least once}\end{pmatrix} = \frac{1}{7,776}$.

The next two examples show you how to determine the probability of an event using combinations.

Example 2 A deck of cards contains 52 cards, 13 cards of each suit. Five cards are drawn. What is the probability that all 5 cards are spades?

There are $\binom{13}{5}$ ways to draw 5 spades from 13 cards.

$\binom{n}{r} = \frac{n!}{r!(n-r)!}$ ▶ $\binom{13}{5} = \frac{13!}{5!8!} = 1,287.$

There are $\binom{52}{5}$ ways to draw 5 spades from a deck of 52 cards.

$\binom{52}{5} = \frac{52!}{5!47!} = 2,598,960.$

Thus, $P\begin{pmatrix}\text{5-card hand} \\ \text{of all spades}\end{pmatrix} = \frac{1,287}{2,598,960}$, or $\frac{33}{66,640}$.

Probability and Arrangements

Example 3 A mathematics class of 25 students consists of 15 girls and 10 boys. A committee of 6 to represent the class is chosen at random. What is the probability that all 6 committee members are girls? that all 6 committee members are boys? that exactly 2 committee members are boys?

There are $\binom{15}{6}$ ways to choose 6 girls from a group of 15 girls:
$$\binom{15}{6} = \frac{15!}{6!9!} = \frac{15 \cdot 14 \cdot 13 \cdot 12 \cdot 11 \cdot 10 \cdot \cancel{9!}}{6 \cdot 5 \cdot 4 \cdot 3 \cdot 2 \cdot 1 \cdot \cancel{9!}} = 5{,}005.$$

There are $\binom{10}{6}$ ways to choose 6 boys from a group of 10 boys:
$$\binom{10}{6} = \frac{10!}{6!4!}, \text{ or } 210.$$

To choose exactly 2 boys means that 4 committee members are girls. So, 2 boys and 4 girls are chosen. By the Fundamental Counting Principle, there are $\binom{10}{2}\binom{15}{4}$ ways to choose exactly 2 boys from a group of 10 boys and 4 girls from a group of 15 girls:
$$\binom{10}{2}\binom{15}{4} = \frac{10!}{2!8!} \times \frac{15!}{4!11!} = \frac{10 \cdot 9 \cdot \cancel{8!}}{2 \cdot 1 \cdot \cancel{8!}} \times \frac{15 \cdot 14 \cdot 13 \cdot 12 \cdot \cancel{11!}}{4 \cdot 3 \cdot 2 \cdot 1 \cdot \cancel{11!}}$$
$$= 45 \quad \times 1{,}365, \text{ or } 61{,}425.$$

There are $\binom{25}{6}$ ways to choose a committee of 6 students from a group of 25 students: $\binom{25}{6} = \frac{25!}{6!19!} = \frac{25 \cdot 24 \cdot 23 \cdot 22 \cdot 21 \cdot 20 \cdot \cancel{19!}}{6 \cdot 5 \cdot 4 \cdot 3 \cdot 2 \cdot 1 \cdot \cancel{19!}} = 177{,}100.$

So, $P\binom{\text{committee of}}{\text{6 girls}} = \frac{5{,}005}{177{,}100}$, or $\frac{13}{460}$, $P\binom{\text{committee of}}{\text{6 boys}} = \frac{210}{177{,}100}$, or $\frac{3}{2{,}530}$,

and $P\binom{\text{committee with}}{\text{exactly 2 boys}} = \frac{61{,}425}{177{,}100}$, or $\frac{351}{1{,}012}$.

Example 4 A box contains 5 red marbles and 15 white marbles. What is the probability of drawing a red marble and a white marble with replacement? Without replacement? What is the probability of drawing a red marble, then a white marble with replacement? Without replacement?

$P(\text{red and white}) = P[\text{red 1st, then white or white first, then red}]$
$= P(\text{red})P(\text{white}) + P(\text{white})P(\text{red})$
$= 2P(\text{red})P(\text{white})$

Without replacement
$P(\text{red}) = \frac{5}{20}, P(\text{white}) = \frac{15}{19}$

With replacement
$P(\text{red}) = \frac{5}{20}, P(\text{white}) = \frac{15}{20}$

P(red and white) ▶ $\quad 2\left(\frac{5}{20}\right)\left(\frac{15}{19}\right) = \frac{15}{38} \qquad 2\left(\frac{5}{20}\right)\left(\frac{15}{20}\right) = \frac{15}{40}$

P(red then white) ▶ $\quad \frac{5}{20}\left(\frac{15}{19}\right) = \frac{15}{76} \qquad \frac{5}{20}\left(\frac{15}{20}\right) = \frac{15}{80}$

So, $P(\text{red and white})$ is $\frac{15}{40}$, or $\frac{3}{8}$, with replacement and $\frac{15}{38}$ without replacement.

$P(\text{red then white})$ is $\frac{15}{80}$, or $\frac{3}{16}$, with replacement and $\frac{15}{76}$ without replacement.

Written Exercises

(A)

1. An envelope contains three slips of paper with each of the numerals 1, 2, and 3 printed on exactly one slip. If one slip of paper is drawn at a time, without replacement, what is the probability that the three-digit number 231 is drawn in that order? $\frac{1}{6}$

2. A bag contains one red, one white, and one blue cube. Each cube is drawn at random exactly once. What is the probability that the order of the cubes drawn is red, blue, and white? $\frac{1}{6}$

3. A deck of cards contains 52 cards, 13 cards of each suit. Six cards are drawn. What is the probability that all 6 cards are hearts? $\frac{33}{391,510}$

4. A deck of cards contains 52 cards, 13 cards of each suit. Five cards are to be drawn. What is the probability that exactly two of the cards are diamonds? $\frac{9,139}{33,320}$

From a group of 5 boys and 3 girls, three violin students are to be selected at random to represent their school in a regional orchestra. Determine each probability.

(B)

5. The probability that all 3 students selected are girls $\frac{1}{56}$

6. The probability that all 3 students selected are boys $\frac{5}{28}$

7. The probability that exactly 1 student selected is a boy $\frac{15}{56}$

8. The probability that exactly 1 student selected is a girl $\frac{15}{28}$

9. The probability that 2 students selected are boys and 1 a girl $\frac{15}{28}$

10. The probability that 1 student selected is a boy and 2 are girls $\frac{15}{56}$

11. A box contains 6 red, 5 white, and 7 black marbles. What is the probability of drawing a black marble, then a red one, if the 1st marble is replaced before drawing the 2nd? $\frac{7}{54}$

12. A box contains 4 red, 8 white, and 10 black marbles. What is the probability of drawing a red marble, then a white marble, then a black marble, if each marble is not replaced before the next draw? $\frac{8}{231}$

Draw 2 cards *without* replacement. Find the probability that:

13. both are diamonds. $\frac{1}{17}$

14. the first is a club and the second is a heart. $\frac{13}{204}$

15. one is a club and one is a heart. $\frac{13}{102}$

16. both are red. $\frac{25}{102}$

17. the first is red and the second is black. $\frac{13}{51}$

18. one is red and one is black. $\frac{26}{51}$

Draw 2 cards, *replacing* the first before the second is drawn. Find the probability that:

19. one is a diamond and the other is a spade. $\frac{1}{8}$

20. the first is a diamond and the second is a spade. $\frac{1}{16}$

21. both are black. $\frac{1}{4}$

22. both are clubs. $\frac{1}{16}$

23. one is black and the other is red. $\frac{1}{2}$

24. *exactly one* is a diamond. $\frac{3}{8}$

(C)

25. A card is drawn from a deck of 52 cards. What is the probability that the card will be an ace, or red queen, or a ten of hearts? $\frac{7}{52}$

26. A card is drawn from a deck of 52 cards. What is the probability that the card will be an ace, king, or queen? $\frac{3}{13}$

27. A box contains 8 red marbles and 10 black marbles. Two marbles are drawn at random without replacement. What is the probability that both are the same color? That one will be red and one black? That both will be red? That both will be black? $\frac{73}{153}, \frac{80}{153}, \frac{28}{153}, \frac{5}{17}$

28. A box contains 4 red, 6 white, and 8 blue marbles. What is the probability of drawing a white marble, and then a red one, if the 1st marble is replaced before drawing the 2nd? If the 1st marble is not replaced before drawing the 2nd? $\frac{2}{27}, \frac{4}{51}$

Probability and Arrangements

Ⓐ Exercises are 1–5, 7, 9, 12, 15, 17, 19–21. Ⓒ Exercises are starred. The rest are Ⓑ Exercises.

CHAPTER SIXTEEN REVIEW

Vocabulary
circular permutation [16.4]
combination [16.5]
dependent events [16.7]
inclusive events [16.7]
independent events [16.7]
mutually exclusive events [16.7]
$n!$ [16.1]   permutation [16.1]
probability [16.6]   sample space [16.7]

Find the value of each expression.

1. $\dfrac{12!}{3! \times 8!}$ [16.3] **1,980**

2. $\binom{100}{3}$ [16.5] **161,700**

3. $\binom{9}{0}$ [16.5] **1**

4. How many four-digit numbers can be formed from the digits 4, 5, 6, 7, 8, 9, if (a) no digit may be repeated in a number and (b) a digit may be repeated in a number? [16.1] **360, 1,296**

5. How many odd four-digit numbers can be formed from the digits 4, 5, 6, 7, 8, 9, if no digit may be repeated in a number? [16.2]

6. How many even numbers of one or more digits can be formed from the digits 4, 5, 6, 7, 8, if no digit may be repeated in a number? [16.2] **195**   5. **180**

7. How many distinguishable six-digit numbers can be formed from the digits of 789778? [16.3] **60**

8. How many different auto license-plate numbers can be formed by 3 different letters followed by 4 digits? [16.1] **156,000,000**

9. Find the number of 7-letter permutations of all the letters in the word COMBINE. [16.1] **5,040**

10. Tom and Mary invite three other couples to a cookout. In how many ways can they all be seated around the fire? [16.4] **5,040**

★ 11. In how many ways can 6 keys be arranged on a ring? [16.4] **60**

12. How many different 6-member committees can be formed if 9 people are available for appointment to a committee? [16.5] **84**

13. A company makes sweaters in 7 colors. Each color is available with 5 choices of fabric and either with or without sleeves. How many different types of sweaters does the company make? [16.1] **70**

14. How many tennis games are played by 8 people if each person plays each of the others one game? [16.5] **28**

15. Three brands of canned soup and 4 brands of dried soup are to be displayed on a shelf with the dried soups together and to the right of the canned soups. How many displays are possible? [16.2] **144**

★ 16. In how many different ways can a family of seven be photographed in a line if the father and mother are not to be separated? [16.2] **1,440**

17. A deck of cards contains 52 cards, 13 cards of each suit. One card is drawn. What is the probability that the card is a king? that it is not a king? [16.6] $\frac{1}{13}, \frac{12}{13}$

18. A die is rolled 600 times. About how many times should it stop on 2? About how many times should an even number appear? [16.6] **100, 300**

For Exercises 19–21, use the sample space showing all possible outcomes of rolling 2 dice, a red one, r, and a white one, w. Find each indicated probability. [16.7]

19. $P(w \leq 4)$ $\frac{2}{3}$

20. $P(r + w = 7)$ $\frac{1}{6}$

21. $P(r + w = 6,$ given that $r = 3)$ $\frac{1}{6}$

★ 22. A family has 4 children. Find $P(3$ boys and a girl$)$. [16.7] $\frac{1}{4}$

23. A box contains 12 red, 8 white, and 10 blue cubes. What is the probability of drawing a blue cube, then a red one, if the first cube is replaced before drawing the second? If the first cube is not replaced before drawing the second? [16.8] $\frac{2}{15}, \frac{4}{29}$

Ⓐ Exercises are 1–4, 6, 8, 9, 13, 14, 16–18. Ⓒ Exercises are starred. The rest are Ⓑ Exercises.

CHAPTER SIXTEEN TEST

Find the value of each expression.

1. $\dfrac{12!}{9! \times 2!}$ **660**   2. $\binom{12}{8}$ **495**   3. $\binom{20}{20}$ **1**

4. How many three-digit numbers can be formed from the digits 2, 3, 4, 5, 6, 7, if no digit may be repeated in a number? **120**

5. A state uses 2 different letters followed by 5 digits on its auto license-plates. How many different license-plate numbers are possible in that state? **65,000,000**

6. How many distinguishable six-digit numbers can be formed from 335533? **15**

7. Find the number of 6-letter permutations of all the letters in EUCLID that end with either the letter E or the letter D. **240**

8. In how many ways can 6 bushes, all of different types, be planted around a statue? **120**

9. How many different 12-member juries can be selected from 15 qualified people? **455**

10. Find the number of distinguishable permutations of all the letters in the word BANANA. **60**

11. An amusement park has 9 gates with 5 on the north side and 4 on the south side. In how many different ways can you enter through a south gate and later exit through a north gate? **20**

12. Aluminum chips A, B, C, and D weigh 1 g, 5 g, 10 g, and 20 g, respectively. How many different masses can be measured by using one or more of the 4 weights on a balance scale? **15**

13. A box contains 15 green marbles, 12 orange marbles, and 23 purple marbles. A marble is drawn from the box. What is the probability that the marble is green? that it is not green? $\frac{3}{10}, \frac{7}{10}$

14. A deck of cards contains 52 cards, 13 cards of each suit. One card is drawn. What is the probability that the card is an ace? that it is the ace of spades? $\frac{1}{13}, \frac{1}{52}$

15. An 8-sided die is rolled 400 times. About how many times should it stop on 7? About how many times should an odd number appear? **50, 200**

For Exercises 16–18, use the sample space showing all possible outcomes of rolling 2 dice, a blue one, b, and a red one, r. Find each indicated probability.

16. $P(b = 2)$ $\frac{1}{6}$ 17. $P(b \geq 4 \text{ or } r \leq 2)$ $\frac{2}{3}$
18. $P(b \geq 2 \text{ and } r \leq 2)$ $\frac{5}{18}$

19. A bag contains 22 orange and 18 green balls. What is the probability of drawing a green ball, then an orange one, if the first ball is replaced before drawing the second? if the first ball is not replaced before drawing the second? $\frac{99}{400}, \frac{33}{130}$

★20. In how many ways can 6 cheerleaders be arranged in a line if 3 of them do not want to be separated? **144**

★21. In how many ways can 5 charms be arranged on a bracelet? **12**

★22. Solve $\binom{n}{4} = \binom{n}{1}$ for n. **5**

★23. A family has 4 children. Find $P(4 \text{ boys})$. $\frac{1}{16}$

Chapter Sixteen Test

COMPUTER ACTIVITIES

Factorials and $\binom{n}{r}$

OBJECTIVES: To evaluate factorials
To evaluate combinations

From your study of factorials, you know that $n!$ grows very large as n increases. The computer displays numbers 10^9 and larger in BASIC scientific notation. For example, 1.234E+11 is the computer display for the number 1.234×10^{11}. Values greater than approximately 10^{38} will cause the computer to display an overflow error. Run the following program, noting what value of $n!$ makes the computer switch to scientific notation, and what value of $n!$ causes an overflow error.

```
10   PRINT "WHAT IS THE NUMBER";: INPUT N
20   P = 1: FOR I = 1 TO N:P = P * I: NEXT I
30   PRINT N" FACTORIAL = "P
40   GOTO 10
```

See the Computer Section beginning on page 574 for more information.

The combination $\binom{n}{r}$ is defined to mean $\dfrac{n!}{r!(n-r)!}$.

Note that $\dfrac{n!}{(n-r)!} = n(n-1)(n-2)\ldots(n-r+1)$.

The program below uses these two facts to evaluate combinations.

```
10   PRINT "UPPER NUMBER";: INPUT U
20   PRINT "LOWER NUMBER";: INPUT L
30   P = 1: IF U = L THEN 80
40   IF L = 0 THEN 80
50   FOR I = U TO U - L + 1 STEP - 1
60   P = P * I: NEXT I
70   FOR I = 1 TO L:P = P / I: NEXT I
80   PRINT "THE VALUE OF ";U;" OVER ";L;" IS ";P
90   GOTO 10
```

EXERCISES   For Ex. 2, see TE Answer Section.

1. Test the program above by using the values from the three examples on pages 435–436.

2. Write a program to evaluate $a!/b!$ (Be careful.)

COLLEGE PREP TEST

DIRECTIONS: In each item, you are to compare a quantity in Column 1 with a quantity in Column 2. Write the letter of the correct answer from these choices:

- **A** The quantity in Column 1 is greater than the quantity in Column 2
- **B** The quantity in Column 2 is greater than the quantity in Column 1
- **C** The quantity in Column 1 is equal to the quantity in Column 2
- **D** The relationship cannot be determined from the given information

Information centered over both columns refers to one or both of the quantities to be compared.

SAMPLE ITEMS AND ANSWERS

Column 1	Column 2
S1. $3! + 2!$	$3! \cdot 2!$

The answer is B: $3! + 2! = 6 + 2 = 8$, $3! \cdot 2! = 6 \cdot 2 = 12$, and $12 > 8$.

n and r are positive integers and $n > r$.

S2. $\binom{n}{r-1}$	$\binom{n-1}{r}$

The answer is D: If $n = 3$ and $r = 1$, then $\binom{n}{r-1} < \binom{n-1}{r}$, since $\binom{3}{0} < \binom{2}{1}$.
If $n = 3$ and $r = 2$, then $\binom{n}{r-1} > \binom{n-1}{r}$, since $\binom{3}{1} > \binom{2}{2}$, or $3 > 1$.

	Column 1	Column 2	
1.	$(5-5)!$	$5! - 5!$	A
2.	$\binom{8}{3}$	$\binom{8}{5}$	C
3.	$\binom{48}{2}$	$\binom{49}{3}$	B

	Column 1	Column 2	
4.	$\binom{20}{0}$	$\binom{40}{40}$	C
5.	$\dfrac{46! \times 29!}{15!}$	$\dfrac{47! \times 28!}{16!}$	A
6.	$\binom{6}{0} + \binom{6}{1} + \binom{6}{2}$	$\binom{6}{6} + \binom{6}{5} + \binom{6}{4}$	C

n and r are positive integers and $n > r$.

	Column 1	Column 2	
7.	$\binom{n}{r}$	$\binom{n+1}{r+1}$	B
8.	$\binom{n}{r}$	$\binom{n}{r+1}$	D
9.	$\binom{n}{r-1} + \binom{n}{r}$	$\binom{n+1}{r+1}$	D

```
        B •         • C
   A •
        F •         • E
              • D
```

	Column 1	Column 2	
10.	Number of triangles determined by the points	Number of lines determined by the points	A
11.	Number of quadrilaterals determined	Number of polygons determined	B

College Prep Test

CUMULATIVE REVIEW
(Chapters 1–16)

DIRECTIONS: Choose the *one* best answer to each question or problem. (Exercises 1–15)

1. Find the solution set of $|x - 3| < 2$. **D**
   (A) $\{x | -1 < x < -5\}$ (B) $\{x | x < 3\}$
   (C) $\{x | x < 2\}$ (D) $\{x | 1 < x < 5\}$

2. Solve $3^{3x-2} = 81$ for x. **D**
   (A) $\frac{5}{3}$ (B) -2 (C) 9 (D) 2

3. Evaluate $-2xy + 3x^2y - y^2$ if $x = -2$ and $y = 3$. **C**
   (A) -33 (B) 15 (C) 39 (D) 57

4. Factor $6x^3 + 4x^2 - 2x$ completely. **C**
   (A) $2x(3x^2 + 2x - 1)$
   (B) $x(3x - 1)(2x + 2)$
   (C) $2x(3x - 1)(x + 1)$
   (D) $2x(3x + 1)(x - 1)$

5. Simplify $\dfrac{1}{n - 2} - \dfrac{3}{n^2 - 4}$. **B**
   (A) $\dfrac{-2}{n + 2}$ (B) $\dfrac{n - 1}{(n - 2)(n + 2)}$
   (C) $\dfrac{n + 5}{n - 2}$ (D) $\dfrac{-2}{(n - 2)(n + 2)}$

6. Solve $\dfrac{2a}{x - b} = \dfrac{-c}{x + b}$ for x. **B**
   (A) $\dfrac{c - 2ab}{2ac}$ (B) $\dfrac{bc - 2ab}{2a + c}$
   (C) $\dfrac{2ab - bc}{c + 2a}$ (D) $\dfrac{2ab - bc}{2ac}$

7. Simplify $\dfrac{x^{\frac{2}{3}} y^{-\frac{1}{4}}}{x^{-\frac{1}{3}} y^{-\frac{7}{4}}}$. **C**
   (A) $x^{\frac{1}{3}} y^{\frac{3}{2}}$ (B) $x^{\frac{1}{3}} y^2$
   (C) $xy^{\frac{3}{2}}$ (D) xy^2

8. Compute the discriminant and determine the nature of the solutions of $3x^2 - 7x + 6 = 0$. **D**
   (A) real and irrational (B) real and equal
   (C) real and rational (D) imaginary

9. Write an equation, in standard form, of the line that passes through $P(2, -1)$ and that is parallel to the line described by $2x - 7y = 12$. **B**
   (A) $7x + 2y = 12$ (B) $2x - 7y = 11$
   (C) $2x - 7y = -11$ (D) $7x - 2y = 16$

10. Multiply, if possible. **B**
    $\begin{bmatrix} 3 & 1 & -2 \\ -1 & 2 & 1 \end{bmatrix} \cdot \begin{bmatrix} 1 & -1 & 2 \\ 3 & 1 & -2 \\ -4 & 2 & -3 \end{bmatrix}$
    (A) $\begin{bmatrix} 15 & -1 & 1 \end{bmatrix}$ (B) $\begin{bmatrix} 14 & -6 & 10 \\ 1 & 5 & -9 \end{bmatrix}$
    (C) $\begin{bmatrix} 3 & -1 & -4 \\ 9 & 1 & -4 \\ 4 & 4 & 3 \end{bmatrix}$ (D) Not possible

11. A function f is defined by $y = -\frac{1}{2}x - 4$. Find an equation of f^{-1}. **A**
    (A) $y = -2x - 8$ (B) $y = 2x + 4$
    (C) $y = 2x + 8$ (D) $y = \frac{1}{2}x + 4$

12. Write an equation, in standard form, of an ellipse whose center is at the origin and whose x-intercepts are ± 6 and whose y-intercepts are ± 4. **A**
    (A) $\dfrac{x^2}{36} + \dfrac{y^2}{16} = 1$ (B) $\dfrac{x^2}{16} + \dfrac{y^2}{36} = 1$
    (C) $\dfrac{x^2}{36} - \dfrac{y^2}{16} = 1$ (D) $\dfrac{x^2}{16} - \dfrac{y^2}{36} = 1$

13. Convert 18 lb/ft² to ounces per square inch (oz/in.²). **C**
    (A) 24 oz/in.² (B) 0.5 oz/in.²
    (C) 2 oz/in.² (D) 1.5 oz/in.²

14. Find the value of $\binom{14}{12}$. **C**
    (A) 182 (B) $\dfrac{91}{6}$ (C) 91 (D) $\dfrac{91}{3}$

15. A deck of cards contains 52 cards, 13 cards of each suit. One card is drawn. What is the probability that the card is a red ace? **A**
    (A) $\dfrac{1}{26}$ (B) $\dfrac{2}{13}$ (C) $\dfrac{1}{13}$ (D) $\dfrac{1}{52}$

16. Solve $3x - 3(2x - 1) = -(x + 15)$. **9**

17. Simplify $\dfrac{12a^5 b^{-6}}{-2a^{-3}b^{-7}}$ and write with positive exponents. **$-6a^8 b$**

18. Factor $25a^2 - 100$ completely. **$25(a - 2)(a + 2)$**

19. The height of a triangle is 12 cm more than 4 times the length of its base. Find the length of the base and the height of the triangle if its area is 20 cm². **base: 2 cm; height: 20 cm**

20. Alberta can pick the apples in 8 days and Elias can pick them in 12 days. How long would it take them to pick the apples if they work together? **$4\frac{4}{5}$ days**

21. Simplify $\sqrt[3]{16a^5} + \sqrt[3]{54a^5}$. **$5a\sqrt[3]{2a^2}$**

22. Find the value of $81^{\frac{3}{4}}$. **27**

23. Write $3x^{\frac{2}{3}}$ in radical form. **$3\sqrt[3]{x^2}$**

24. Solve $x^2 - 8x = 5$ by completing the square. **$4 \pm \sqrt{21}$**

25. Simplify $-4 \div (3i)$ by rationalizing the denominator. **$\frac{4i}{3}$**

26. For what value(s) of k will $3x^2 + 6x - (3k - 2) = 0$ have exactly one solution? **$-\frac{1}{3}$**

27. Draw the graph of $-2x - 3y = 6$.
   For Ex. 27, see TE Answer Section.

28. Find the slope and y-intercept of the line described by $y - 3(2 - x) = 6$.
   $m = -3$, $b = 12$

29. Find the distance between $P(-3, -5)$ and $Q(5, -6)$. **$\sqrt{65}$**

30. Determine the coordinates of M, the midpoint of $\overline{AB}$, given $A(3, -1)$ and $B(7, 9)$. **(5, 4)**

31. Solve the system: $-3x + 2y + z = 8$
    $x - 3y + 2z = -5$
    $-x + y - 3z = 0$. **$(-1, 2, 1)$**

32. Let $f(x) = 2x^2 - 1$. Find $f(-3)$. **17**

33. If $f(x) = \dfrac{2x}{x^2 + 2x - 15}$, determine the domain of f.
   {all real numbers except -5 and 3}

34. Draw the graph of $y = -x^2 - 6x + 2$.
   For Ex. 34, see TE Answer Section.

35. Find an equation of a circle whose center is at $C(-3, 2)$ and which passes through the point $P(1, -7)$. **$(x + 3)^2 + (y - 2)^2 = 97$**

36. Solve the system $x^2 + y^2 \leq 36$ graphically.
    $\dfrac{x^2}{9} + \dfrac{y^2}{4} \geq 1$
   For Ex. 36, see TE Answer Section.

37. Find the sum, if it exists, of the infinite geometric series
   $1 + \frac{1}{3} + \frac{1}{9} + \ldots$ **$1\frac{1}{2}$**

38. Find the 5th term in the expansion of $(2x - 3y)^7$. Simplify. **$22{,}680 x^3 y^4$**

39. Solve $\log_3 x + \log_3(x - 2) = \log_3 15$. Check. **5**

40. How many even numbers containing one or more digits can be formed from the digits 1, 2, 3, 4 if no digit may be repeated in a number? **32**

Cumulative Review (Chapters 1–16)

17 TRIGONOMETRIC FUNCTIONS

HISTORY

Karl Wilhelm Theodor Weierstrass 1815–1897

Karl Weierstrass was born in Ostenfelde, Germany. He is famous for his work on the theory of functions and irrational numbers.

At age 19, he graduated from the Catholic Gymnasium (school) at Pearborn with honors in German, Latin, Greek, and mathematics.

At age 24, Karl enrolled at the Academy of Munster to prepare for the state teacher's examination to pursue a career as a secondary school teacher of mathematics. Two years later, at age 26, he passed the examination.

For the next 15 years, he taught in various small villages during the day and did mathematical research at night. This after-school work resulted in his being recognized as the leading mathematical analyst in Europe. Karl spent the years 1864–1897 as Professor of Mathematics at the University of Berlin.

Problem

Show that $\log_{10} 2$ is irrational.

Assume $\log_{10} 2 = \frac{a}{b}$, where a and b are integers.
Then we must have $10^a = 2^b$, which is impossible.

ANGLES OF ROTATION   17.1

Objectives   **To sketch an angle of rotation associated with a directed degree measure**
To determine a reference angle and reference triangle for any angle

An **angle of rotation** formed by rotating a ray about its end point, or vertex. Each angle formed is composed of two rays, the initial ray or side and terminal ray or side. A ray can be rotated in either a counterclockwise or clockwise direction and the terminal ray can lie in any quadrant or on any axis. Angles formed by a *counterclockwise* rotation have a *positive measure*. The figure at the right illustrates a counterclockwise rotation of a ray forming an angle with a positive measure of 150°.

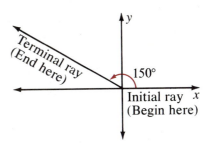

Example 1   Sketch angles of rotation with measures of 90°, 225°, and 330°.

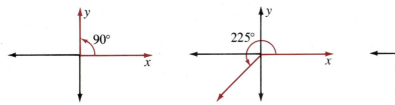

Angles formed by a *clockwise rotation* of a ray have *negative measures*. The figure at the right illustrates a clockwise rotation of a ray forming an angle with a measure of −45°.

Example 2   Sketch angles of rotation with measures of −120°, −210°, and −315°.

Angles of Rotation   455

Rays may be rotated through more than one revolution forming angles that measure more than 360°. The position of the terminal side of an angle determines the quadrant of the angle. Angles with terminal sides on an axis are called **quadrantal angles**. Angles with the same terminal sides are called **coterminal angles**.

Example 3 Sketch angles of rotation with measures of 270°, −570°, and 420°. Determine the quadrant of each angle or identify as a quadrantal angle.

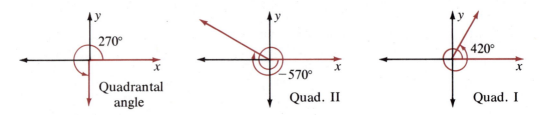

Each angle of rotation whose terminal side is not on an axis has a reference angle. A **reference angle** is the positive acute angle formed between the terminal side and the *x*-axis. The next example illustrates this idea.

Example 4 Sketch angles of rotation with measures of 135°, −120°, and 330°. Identify the reference angle and determine the quadrant of the angle of rotation.

The triangle formed by drawing a perpendicular from a point on the terminal side to the *x*-axis is called a **reference triangle**. A *reference triangle* always includes the reference angle.

Example 5 Sketch angles of rotation with measures of 60°, 200°, and −310°. Identify the reference triangle and reference angle.

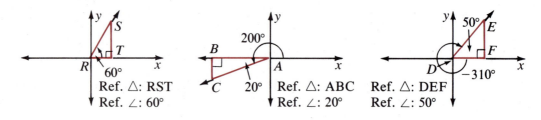

Reading in Algebra

Each item at the left can be described by one or more items at the right. Determine all items at the right that can be used to describe each item at the left. Justify.

1. 120° **B, C, F**
2. −300° **A, C, E**
3. 210° **B, D, G**
4. −30° **A, D, H**
5. 270° **B, I**

A. Clockwise rotation
B. Counterclockwise rotation
C. Ref. ∠: 60°
D. Ref. ∠: 30°
E. Quad. I
F. Quad. II
G. Quad. III
H. Quad. IV
I. Quadrantal ∠

Oral Exercises

Determine the quadrant of each angle of rotation, and state the measure of the reference angle.

1. 100° **II, 80°**
2. 330° **IV, 30°**
3. −320° **I, 40°**
4. −170° **III, 10°**

Written Exercises

Sketch angles of rotation with the given measures. Determine the quadrant of each angle of rotation or identify as a quadrantal angle (Quad. ∠).

1. 30° **I**
2. 300° **IV**
3. 210° **III**
4. 135° **II**
5. −30° **IV**
6. −150° **III**
7. −360° **Quad. ∠**
8. −225° **II**
9. −90° **Quad. ∠**
10. 225° **III**
11. 330° **IV**
12. 10° **I**
13. 515° **II**
14. −190° **II**
15. 400° **I**
16. −650° **I**
17. 100° **II**
18. 720° **Quad. ∠**

Sketch angles of rotation with the given measures. Determine the quadrant of each angle of rotation. Identify the reference triangle.

19. 75° **I**
20. 160° **II**
21. 225° **III**
22. 330° **IV**
23. −40° **IV**
24. −195° **II**
25. −245° **II**
26. −300° **I**
27. 495° **II**
28. −780° **IV**
29. −510° **III**
30. −360° **Quad. ∠**

Sketch angles of rotation with the given measures. Identify the reference triangle and reference angle of each nonquadrantal angle.

31. −50° **50°**
32. 150° **30°**
33. −135° **45°**
34. −225° **45°**
35. 315° **45°**
36. 210° **30°**
37. −405° **45°**
38. −315° **45°**
39. 780° **60°**
40. 510° **30°**
41. −90° **Quad. ∠**
42. 350° **10°**

State two more measures, one positive and one negative, for angles of rotation that have the same terminal side as the one shown. **Answers may vary.**

43. 40°
400°, −320°

44. −150°
210°, −510°

45. −200°
160°, −560°

46. 190°
550°, −170°

Angles of Rotation

SPECIAL RIGHT TRIANGLES   17.2

Objectives   **To find lengths of sides in a 30°–60° or 45°–45° right triangle**
To find points on a circle using 30°–60° or 45°–45° right triangles

Recall that in an equilateral △, an altitude bisects both the base and the vertex angle from which it is drawn. You can see that in equilateral △ABC, altitude $\overline{AD}$ bisects ∠A and base $\overline{BC}$. You can also see that in rt. △ABD, the leg opposite the 30° angle, $\overline{BD}$, has $\frac{1}{2}$ the measure of hypotenuse $\overline{AB}$.

To find AD use the Pythagorean theorem.
$$(AB)^2 = (BD)^2 + (AD)^2$$
$$10^2 = 5^2 + (AD)^2$$
$$100 = 25 + (AD)^2$$
$$(AD)^2 = 75$$
$$(AD) = \sqrt{75}, \text{ or } 5\sqrt{3}$$

| 30°–60° right triangle rule | In a 30°–60° right △:
1. the leg opposite the 30° angle has $\frac{1}{2}$ the measure of the hypotenuse
2. the leg opposite the 60° angle has $\frac{1}{2}$ the measure of the hypotenuse times $\sqrt{3}$ | |

Example 1   **In right △RST, RS = 8. Find ST and RT.**

$\overline{ST}$ is opposite the 30° angle. $\overline{RT}$ is opposite the 60° angle.

$ST = \frac{1}{2}(RS)$ $RT = \frac{1}{2}(RS)\sqrt{3}$
$\quad\ = \frac{1}{2}(8)$ $\quad\ = \frac{1}{2}(8)\sqrt{3}$
$\quad\ = 4$ $\quad\ = 4\sqrt{3}$

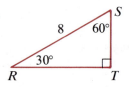

Thus, $ST = 4$ and $RT = 4\sqrt{3}$.

Example 2   **In right △XYZ, XZ = 5. Find XY and YZ.**

$\overline{XZ}$ is opposite the 30° angle. $\overline{YZ}$ is opposite the 60° angle.

$XZ = \frac{1}{2}(XY)$ $YZ = \frac{1}{2}(XY)\sqrt{3}$
$5 = \frac{1}{2}XY$ $\quad\ = \frac{1}{2}(10)\sqrt{3}$
$10 = XY$ $\quad\ = 5\sqrt{3}$

Thus, $XY = 10$ and $YZ = 5\sqrt{3}$.

Example 3   In right △ABC, BC = 6.
Find AB and AC.

$\overline{AB}$ is opposite the 30° angle.
$\overline{BC}$ is opposite the 60° angle.

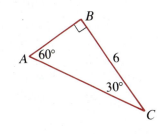

$BC = \frac{1}{2}(AC)\sqrt{3}$ $AB = \frac{1}{2}(AC)$
$6 = \frac{1}{2}(AC)\sqrt{3}$ $= \frac{1}{2}(4\sqrt{3})$
$12 = (AC)\sqrt{3}$ $= 2\sqrt{3}$
$AC = \frac{12}{\sqrt{3}}$, or $4\sqrt{3}$

Thus, $AC = 4\sqrt{3}$ and $AB = 2\sqrt{3}$.

Recall that if two angles of a triangle have the same measure, the triangle is **isosceles** and the sides opposite the two angles have the same measure. △ABC is an isosceles right triangle and AC = BC. To find AC or BC, use the Pythagorean theorem.

$(AC)^2 + (BC)^2 = (AB)^2$
$(AC)^2 + (AC)^2 = (AB)^2$
$2(AC)^2 = 8^2$
$(AC)^2 = 32$
$AC = BC = \sqrt{32}$, or $4\sqrt{2}$

45°–45° right triangle rule   In a 45°–45° right triangle, each leg has $\frac{1}{2}$ the measure of the hypotenuse times $\sqrt{2}$.

Example 4   In a 45°–45° right △, one leg measures 12.
Find the length of the hypotenuse.

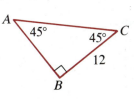

$BC = \frac{1}{2}(AC)\sqrt{2}$
$12 = \frac{1}{2}(AC)\sqrt{2}$
$24 = (AC)\sqrt{2}$
$AC = \frac{24}{\sqrt{2}}$, or $12\sqrt{2}$

Thus, the length of the hypotenuse is $12\sqrt{2}$.

Special Right Triangles

You have learned that each nonquadrantal angle of rotation has a corresponding reference angle and reference triangle. There is also a circle that is associated with an angle of rotation. The examples that follow illustrate the use of a circle with reference triangles and special triangles.

Example 5

For a circle with radius of 8 units and a reference triangle of 30°–60°–90° as shown, find the point (x, y).

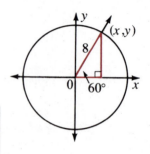

y is opposite the 60° angle.
x is opposite the 30° angle.

$|y| = \frac{1}{2}(8)\sqrt{3}$ $|x| = \frac{1}{2}(8)$
$|y| = 4\sqrt{3}$ $|x| = 4$ ◄ In Quad. I, $x > 0$, $y > 0$.
$y = 4\sqrt{3}$ $x = 4$

Thus, the point (x, y) is $(4, 4\sqrt{3})$.

Example 6

For the following circle, determine the radius r and the point (x, y) if $y = 3$.

y is opposite a 45° angle.
x is opposite a 45° angle.

$y = \frac{1}{2}r\sqrt{2}$ $|x| = |y|$
$3 = \frac{1}{2}r\sqrt{2}$ $|x| = 3$
$6 = r\sqrt{2}$ $x = -3$ ◄ In Quad. II, $x < 0$, $y > 0$.
$r = \frac{6}{\sqrt{2}}$, or $3\sqrt{2}$

Thus, $r = 3\sqrt{2}$ and the point (x, y) is $(-3, 3)$.

Oral Exercises

State the length of the indicated side.

Written Exercises

Determine the lengths of the indicated sides. Rationalize all denominators.

A

1. a=6√3, b=6
2. b=3√3, c=6
3. a=3√2, b=3√2
4. a=4, c=8
5. a=16, c=16√2
6. b=5, c=10

Determine the point (x, y) and the radius r.

7. $(-3\sqrt{3}, -3)$, r=6
8. $(4, -4)$, r=4√2
9. $(-3, 3\sqrt{3})$, r=6
10. $(5\sqrt{3}, 5)$, r=10

Complete the chart.

17. 45°, 3√2   18. 30°, 2   19. 60°, 8   20. 45°, 5√2   21. 30°, 4
22. 60°, 8   23. 60°, 10   24. 45°, 7√2   25. 30°, 8   26. 60°, 6

	Length of hypotenuse	Length of leg opposite 30° ∠	Length of leg opposite 60° ∠
B 11.	16	a 8	b 8√3
12.	c 18	9	b 9√3
13.	c 4√3	a 2√3	6

	Length of hypotenuse	Length of leg opposite 45° ∠	Length of leg opposite other 45° ∠
14.	c 3√2	3	b 3
15.	c 12√2	a 12	12
16.	14	a 7√2	b 7√2

Find the reference angle if its terminal side intersects a circle at the given point and the center of the circle is at the origin. Determine the radius of the circle.

C
17. $(-3, 3)$ 18. $(-\sqrt{3}, -1)$ 19. $(4, -4\sqrt{3})$ 20. $(5, -5)$ 21. $(-2\sqrt{3}, -2)$
22. $(4, 4\sqrt{3})$ 23. $(-5, 5\sqrt{3})$ 24. $(-7, -7)$ 25. $(4\sqrt{3}, 4)$ 26. $(3, -3\sqrt{3})$
27. $(-a, a)$ 45°, a√2   28. $(m, m\sqrt{3})$ 60°, 2m   29. $(-r, r\sqrt{3})$ 60°, 2r   30. $(n\sqrt{3}, -n)$ 30°, 2n   31. $(-r, -r)$ 45°, r√2

32. Prove the 45°–45° right triangle rule.   33. Prove the 30°–60° right triangle rule.
For Ex. 32–33, see TE Answer Section.

Special Right Triangles

SINE AND COSINE FUNCTIONS 17.3

Objectives
To find the sine and cosine of degree measures of 0°, 30°, 45°, 60°, 90° and their multiples
To determine if the sine or cosine is positive or negative for a given degree measure
To find the measure of an angle whose sine or cosine is given

Examine the four reference triangles (one in each quadrant) shown in the figures below.

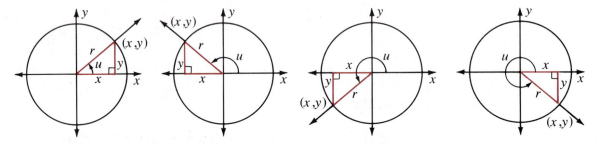

The ratios $\frac{y}{r}$ and $\frac{x}{r}$ in each of the figures determine trigonometric functions that are defined in terms of u as follows.

Definition:
Sine (sin) and cosine (cos) functions

For any angle of rotation u and any radius r, the sine of u is $\frac{y}{r}$ $\left(\sin u = \frac{y}{r}\right)$ and the cosine of u is $\frac{x}{r}$ $\left(\cos u = \frac{x}{r}\right)$. For a radius of 1, a unit circle, $\sin u = y$ and $\cos u = x$.

Example 1 **Find sin 30°, cos 30°, sin 150°, and cos 150°.**

For each angle, sketch a circle with any radius to show the angle of rotation and reference triangle.

Circle with radius 2

$|y| = \frac{1}{2}r$
$= \frac{1}{2}(2) = 1$

$|x| = \frac{1}{2}(r)\sqrt{3}$
$= \frac{1}{2}(2)\sqrt{3} = \sqrt{3}$

So, $y = 1$, $x = \sqrt{3}$,
$\frac{y}{r} = \frac{1}{2}$ and $\frac{x}{r} = \frac{\sqrt{3}}{2}$.

Circle with radius 1

$|y| = \frac{1}{2}r$
$= \frac{1}{2}(1) = \frac{1}{2}$

$|x| = \frac{1}{2}(r)\sqrt{3}$
$= \frac{1}{2}(1)\sqrt{3} = \frac{\sqrt{3}}{2}$

So, $y = \frac{1}{2}$, $x = -\frac{\sqrt{3}}{2}$

Thus, $\sin 30° = \frac{1}{2}$, $\cos 30° = \frac{\sqrt{3}}{2}$, $\sin 150° = \frac{1}{2}$, and $\cos 150° = -\frac{\sqrt{3}}{2}$.

Example 2 **Find sin 225°, cos 225°, sin 315°, and cos 315°.**

For each angle, sketch a circle with a radius of 2. Show the angle of rotation and reference triangle.

Ref. ∠: 45°

$|y| = \frac{1}{2}(r)\sqrt{2}$
$= \frac{1}{2}(2)\sqrt{2} = \sqrt{2}$

$|x| = \frac{1}{2}(r)\sqrt{2}$
$= \frac{1}{2}(2)\sqrt{2} = \sqrt{2}$

So, $y = -\sqrt{2}$, $x = -\sqrt{2}$,
$\frac{y}{r} = \frac{-\sqrt{2}}{2}$ and $\frac{x}{r} = \frac{-\sqrt{2}}{2}$.

Ref. ∠: 45°

$|y| = \frac{1}{2}(r)\sqrt{2}$
$= \frac{1}{2}(2)\sqrt{2} = \sqrt{2}$

$|x| = \frac{1}{2}(r)\sqrt{2}$
$= \frac{1}{2}(2)\sqrt{2} = \sqrt{2}$

So, $y = -\sqrt{2}$, $x = \sqrt{2}$,
$\frac{y}{r} = \frac{-\sqrt{2}}{2}$ and $\frac{x}{r} = \frac{\sqrt{2}}{2}$.

Thus, $\sin 225° = \frac{-\sqrt{2}}{2}$, $\cos 225° = \frac{-\sqrt{2}}{2}$, $\sin 315° = -\frac{\sqrt{2}}{2}$ and $\cos 315° = \frac{\sqrt{2}}{2}$.

Examples 1 and 2 suggest that the sine is positive (+) in the first and second quadrants and negative (−) in the third and fourth quadrants. They also suggest that the cosine is positive in the first and fourth quadrants and negative in the second and third quadrants.

	1st Quad.	2nd Quad.	3rd Quad.	4th Quad.
Sine	+	+	−	−
Cosine	+	−	−	+

The next example shows how to express the sine and cosine of any angle as a function of an acute angle.

Example 3 **Write each in terms of the same trigonometric function of an acute angle: sin 240° and cos 315°.**

For each angle, sketch a circle. Show the reference triangle.

Ref. ∠: 60°

sin is negative
in quadrant III.

sin 240° = −sin 60°

Ref. ∠: 45°

cos is positive
in quadrant IV.

cos 315° = cos 45°

Thus, sin 240° = −sin 60° and cos 315° = cos 45°.

Sine and Cosine Functions

Recall that angles whose terminal sides are on an axis are called *quadrantal* angles. The quadrantal angles 0°, 90°, 180°, and 270° are shown below. Their sines and cosines are summarized in the table.

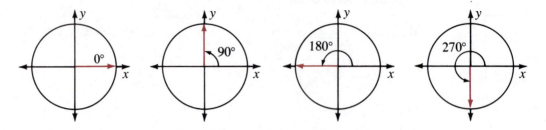

	0°	90°	180°	270°
sine: $\frac{y}{r}$	0	1	0	−1
cosine: $\frac{x}{r}$	1	0	−1	0

The following example illustrates how to solve an equation involving sin u or cos u.

Example 4 **Find two values of u between 0° and 360° for which $\sin u = -\frac{\sqrt{2}}{2}$.**

First, determine the quadrant of the angle: sine is negative in the 3rd and 4th quadrants.

Second, determine the reference angle: $\sin 45° = \frac{\sqrt{2}}{2}$.

Finally, determine the angle of rotation.

Thus, $\sin u = \frac{-\sqrt{2}}{2}$ for $u = 225°$ and $u = 315°$.

Summary   To find the measure of an angle whose sine or cosine is given:
1. Determine the quadrant from the sign (+ or −).

2. Determine the reference angle.

3. Use the reference angle to find the angle of rotation.

Reading in Algebra

Which value in each group is different from the others?

1. a) $\sin 45°$ b) $\sin 135°$ c) $\frac{\sqrt{2}}{2}$ d) $-\frac{\sqrt{2}}{2}$ **d**
2. a) $\cos 300°$ b) $\cos 60°$ c) $\cos 120°$ d) $\frac{\sqrt{3}}{2}$ **c**
3. a) $\sin 150°$ b) $\sin 210°$ c) $\sin 330°$ d) $-\frac{1}{2}$ **a**
4. a) $\frac{1}{2}$ b) $-\frac{1}{2}$ c) $\cos 120°$ d) $\cos 240°$ **a**
5. a) $\sin 0°$ b) 0 c) $\sin 180°$ d) 1 **d**

Written Exercises
For Ex. 1–36, see TE Answer Section.

A Find each of the following. Rationalize any radical denominator.
1. $\sin 60°$
2. $\cos 60°$
3. $\sin 45°$
4. $\cos 45°$
5. $\sin 30°$
6. $\cos 30°$
7. $\cos 270°$
8. $\cos 360°$
9. $\sin 180°$
10. $\cos 90°$
11. $\sin 0°$
12. $\sin 270°$
13. $\sin 135°$
14. $\cos 210°$
15. $\cos 150°$
16. $\sin 150°$
17. $\cos 180°$
18. $\cos 225°$
19. $\sin 315°$
20. $\sin 120°$
21. $\cos 330°$
22. $\sin 330°$
23. $\sin 390°$
24. $\cos 480°$

Write each in terms of the same trigonometric function of an acute angle.
25. $\cos 135°$
26. $\sin 330°$
27. $\cos 210°$
28. $\sin 225°$
29. $\sin 135°$
30. $\sin 300°$
31. $\sin 150°$
32. $\sin 210°$
33. $\sin 315°$
34. $\cos 300°$
35. $\cos 150°$
36. $\cos 330°$

B Find two values of u between $0°$ and $360°$ for which each is true.
37. $\sin u = \frac{\sqrt{2}}{2}$ **45°, 135°**
38. $\cos u = \frac{1}{2}$ **60°, 300°**
39. $\sin u = \frac{1}{2}$ **30°, 150°**
40. $\cos u = 0$ **90°, 270°**
41. $\cos u = -\frac{\sqrt{2}}{2}$ **135°, 225°**
42. $\sin u = \frac{\sqrt{3}}{2}$ **60°, 120°**
43. $\cos u = -\frac{\sqrt{3}}{2}$ **150°, 210°**
44. $\sin u = -\frac{\sqrt{3}}{2}$ **240°, 300°**

In which quadrants is each of the following true?
45. $\cos u < 0$ **II, III**
46. $\sin u > 0$ **I, II**
47. $\sin u < 0$ **III, IV**
48. $\cos u > 0$ **I, IV**

C Find two values of u between $0°$ and $720°$ for which each is true.
49. $\cos u = -\frac{1}{2}$ and $\sin u > 0$ **120°, 480°**
50. $\sin u = \frac{1}{2}$ and $\cos u > 0$ **30°, 390°**
51. $\cos u = \frac{\sqrt{3}}{2}$ and $\sin u < 0$ **330°, 690°**
52. $\sin u = -\frac{\sqrt{2}}{2}$ and $\cos u < 0$ **225°, 585°**
53. $\sin u = -\frac{\sqrt{3}}{2}$ and $\cos u > 0$ **300°, 660°**
54. $\cos u = -\frac{\sqrt{2}}{2}$ and $\sin u < 0$ **225°, 585°**

CUMULATIVE REVIEW

Complete the chart.

	0°	90°	180°	270°	360°
sin	0	1	0	−1	0
cos	1	0	−1	0	1

Sine and Cosine Functions

TANGENT AND COTANGENT FUNCTIONS   17.4

Objectives To find the tangent and cotangent of degree measures 0°, 30°, 45°, 60°, 90°, and their multiples
To determine if the tangent or cotangent is positive or negative for a given degree measure
To find the measure of an angle whose tangent or cotangent is given
To find tangents and cotangents using trigonometric relationships

Recall that for any given angle of rotation, u, the two ratios $\frac{y}{r}$ and $\frac{x}{r}$ were defined as $\sin u$ and $\cos u$, respectively. Using a reference triangle, the ratios $\frac{y}{x}$ and $\frac{x}{y}$ determine the trigonometric functions tangent and cotangent.

Definition: Tangent (tan) and cotangent (cot) functions

For any angle of rotation u and any radius r, the tangent of u is $\frac{y}{x}$ $\left(\tan u = \frac{y}{x}\right)$ and the cotangent of u is $\frac{x}{y}$ $\left(\cot u = \frac{x}{y}\right)$, $(x \neq 0, y \neq 0)$. For a unit circle, $\tan u = \frac{y}{x}$ and $\cot u = \frac{x}{y}$.

Example 1 Find tan 45°, cot 45°, tan 135°, and cot 135°.

For each angle, sketch a circle. Show the angle of rotation and reference triangle.

Circle with radius 2

$|y| = \frac{1}{2}(r)\sqrt{2}$
$= \frac{1}{2}(2)\sqrt{2} = \sqrt{2}$

$|x| = \frac{1}{2}(r)\sqrt{2}$
$= \frac{1}{2}(2)\sqrt{2} = \sqrt{2}$

So, $y = \sqrt{2}$, $x = \sqrt{2}$,
$\frac{y}{x} = \frac{\sqrt{2}}{\sqrt{2}} = 1$ and
$\frac{x}{y} = \frac{\sqrt{2}}{\sqrt{2}} = 1$.

Unit circle

$|y| = \frac{1}{2}(r)\sqrt{2}$
$= \frac{1}{2}(1)\sqrt{2} = \frac{\sqrt{2}}{2}$

$|x| = \frac{1}{2}(r)\sqrt{2}$
$= \frac{1}{2}(1)\sqrt{2} = \frac{\sqrt{2}}{2}$

So, $y = \frac{\sqrt{2}}{2}$, $x = -\frac{\sqrt{2}}{2}$,
$\frac{y}{x} = \frac{\frac{\sqrt{2}}{2}}{-\frac{\sqrt{2}}{2}} = -1$, and $\frac{x}{y} = -1$.

Thus, tan 45° = 1, cot 45° = 1, tan 135° = −1 and cot 135° = −1.

Example 2

Find tan 210°, cot 210°, tan 300°, and cot 300°.

For each angle, sketch a circle with a radius of 2. Show the angle of rotation and reference angle.

$|y| = \frac{1}{2}(r)$
$= \frac{1}{2}(2) = 1$

$|x| = \frac{1}{2}(r)\sqrt{3}$
$= \frac{1}{2}(2)\sqrt{3} = \sqrt{3}$

So, $y = -1, x = -\sqrt{3}$
$\frac{y}{x} = \frac{-1}{-\sqrt{3}} = \frac{\sqrt{3}}{3}$ and
$\frac{x}{y} = \frac{-\sqrt{3}}{-1} = \sqrt{3}.$

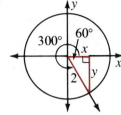

$|y| = \frac{1}{2}(r)\sqrt{3}$
$= \frac{1}{2}(2)\sqrt{3} = \sqrt{3}$

$|x| = \frac{1}{2}(r)$
$= \frac{1}{2}(2) = 1$

So, $y = -\sqrt{3}, x = 1$
$\frac{y}{x} = \frac{-\sqrt{3}}{1} = -\sqrt{3}$ and
$\frac{x}{y} = \frac{1}{-\sqrt{3}} = -\frac{\sqrt{3}}{3}.$

Thus, tan 210° = $\frac{\sqrt{3}}{3}$, cot 210° = $\sqrt{3}$, tan 300° = $-\sqrt{3}$, and cot 300° = $-\frac{\sqrt{3}}{3}$.

Examples 1 and 2 suggest that the tangent and cotangent are positive (+) in the first and third quadrants and negative (−) in the second and fourth quadrants.

	1st Quad.	2nd Quad.	3rd Quad.	4th Quad.
tangent	+	−	+	−
cotangent	+	−	+	−

The next example shows how to express the tan or cot of any angle as a function of an acute angle.

Example 3

Write each in terms of the same trigonometric function of an acute angle: tan 240° and cot 330°.

For each angle, sketch a circle. Show the reference triangle.

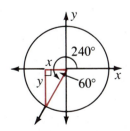

Ref. ∠: 60°

Tan is positive in quadrant III.

tan 240° = tan 60°

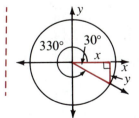

Ref. ∠: 30°

Cot is negative in quadrant IV.

cot 330° = −cot 30°.

Thus, tan 240° = tan 60° and cot 330° = −cot 30°.

Tangent and Cotangent Functions

Example 4 **Find the tan and cot of the quadrantal angles 0°, 90°, 180°, and 270°.**

$x = r, y = 0$ $\qquad$ $x = 0, y = r$ $\qquad$ $x = -r, y = 0$ $\qquad$ $x = 0, y = -r$

$\tan 0° = \dfrac{y}{x} = \dfrac{0}{r}$ $\quad$ $\tan 90° = \dfrac{y}{x} = \dfrac{r}{0}$ $\quad$ $\tan 180° = \dfrac{y}{x} = \dfrac{0}{-r}$ $\quad$ $\tan 270° = \dfrac{y}{x} = \dfrac{-r}{0}$

$\cot 0° = \dfrac{x}{y} = \dfrac{r}{0}$ $\quad$ $\cot 90° = \dfrac{x}{y} = \dfrac{0}{r}$ $\quad$ $\cot 180° = \dfrac{x}{y} = \dfrac{-r}{0}$ $\quad$ $\cot 270° = \dfrac{x}{y} = \dfrac{0}{-r}$

Thus, the tan and cot of 0°, 90°, 180°, and 270° are summarized in the table below.

	0°	90°	180°	270°
tangent	0	undefined	0	undefined
cotangent	undefined	0	undefined	0

Notice that $\tan u \cdot \cot u = \dfrac{y}{x} \cdot \dfrac{x}{y} = 1$. Since their product is 1, $\tan u$ and $\cot u$ are called *reciprocal* functions. That is, $\tan u = \dfrac{1}{\cot u}$ or $\cot u = \dfrac{1}{\tan u}$, for any *nonquadrantal* angle of rotation u.

Example 5 **If $\tan x = -\dfrac{5}{3}$, find $\cot x$.**

The reciprocal of $-\dfrac{5}{3}$ is $-\dfrac{3}{5}$. $\qquad$ **Thus,** $\cot x = -\dfrac{3}{5}$.

Example 6 **Find two values of u between 0° and 360° for which $\tan u = \dfrac{-\sqrt{3}}{3}$.**

Use the three steps in the summary on page 464.

  Tan is negative in quadrant II.

Ref. $\angle$: 30°

$u = 180° - 30°$
$ = 150°$

  Tan is negative in quadrant IV.

Ref. $\angle$: 30°

$u = 360° - 30°$
$ = 330°$

Thus, $\tan u = -\dfrac{\sqrt{3}}{3}$, for $u = 150°$ and $u = 330°$.

Notice that $\dfrac{\sin u}{\cos u} = \dfrac{\frac{y}{r}}{\frac{x}{r}} = \dfrac{y}{r} \cdot \dfrac{r}{x} = \dfrac{y}{x} = \tan u$. So, $\tan u = \dfrac{\sin u}{\cos u}$.

Similarly, $\cot u = \dfrac{\cos u}{\sin u}$.

Quotient identities	$\tan u = \dfrac{\sin u}{\cos u}$	$\cot u = \dfrac{\cos u}{\sin u}$

Example 7   If $\sin w = \dfrac{3}{5}$ and $\cos w = \dfrac{4}{5}$, find $\tan w$ and $\cot w$.

$$\tan w = \dfrac{\sin w}{\cos w} = \dfrac{\frac{3}{5}}{\frac{4}{5}} = \dfrac{3}{4} \qquad \cot w = \dfrac{\cos w}{\sin w} = \dfrac{\frac{4}{5}}{\frac{3}{5}} = \dfrac{4}{3}$$

Written Exercises

A) Find each of the following. Rationalize all radical denominators.
1. $\tan 45°$ **1**
2. $\cot 45°$ **1**
3. $\tan 60°$ $\sqrt{3}$
4. $\cot 60°$ $\dfrac{\sqrt{3}}{3}$
5. $\tan 30°$ $\dfrac{\sqrt{3}}{3}$
6. $\cot 30°$ $\sqrt{3}$
7. $\tan 90°$
8. $\cot 90°$ **0**
9. $\cot 315°$ -1
10. $\tan 330°$ $\dfrac{-\sqrt{3}}{3}$
11. $\cot 180°$
12. $\tan 180°$ **0**
13. $\tan 135°$ -1
14. $\cot 150°$
15. $\cot 300°$
16. $\cot 225°$ **1**
17. $\tan 210°$ $\dfrac{\sqrt{3}}{3}$
18. $\cot 240°$
19. $\cot 450°$ **0**
20. $\tan 120°$
21. $\tan 390°$ $\dfrac{\sqrt{3}}{3}$
22. $\tan 150°$ $\dfrac{-\sqrt{3}}{3}$
23. $\cot 360°$ undefined
24. $\tan 405°$ **1**

7. undefined   14. $-\sqrt{3}$ **20.** $-\sqrt{3}$ **11. undefined**   **15.** $\dfrac{-\sqrt{3}}{3}$ **18.** $\dfrac{\sqrt{3}}{3}$

Write each of the following in terms of the same trigonometric function of an acute angle. **For Ex. 25–36, see TE Answer Section.**
25. $\cot 110°$
26. $\tan 310°$
27. $\cot 230°$
28. $\tan 210°$
29. $\tan 340°$
30. $\cot 95°$
31. $\tan 250°$
32. $\cot 120°$
33. $\tan 150°$
34. $\cot 300°$
35. $\cot 250°$
36. $\tan 100°$

37. If $\cot w = -\dfrac{4}{3}$, find $\tan w$. $\dfrac{-3}{4}$
38. If $\tan u = \dfrac{5}{4}$, find $\cot u$. $\dfrac{4}{5}$
39. If $\tan v = -\dfrac{17}{8}$, find $\cot v$. $\dfrac{-8}{17}$
40. If $\cot u = \dfrac{12}{5}$, find $\tan u$. $\dfrac{5}{12}$
41. If $\sin m = \dfrac{4}{5}$ and $\cos m = -\dfrac{3}{5}$, find $\tan m$ and $\cot m$. $-\dfrac{4}{3}, -\dfrac{3}{4}$
42. If $\cos w = -\dfrac{12}{13}$ and $\sin w = -\dfrac{5}{13}$, find $\tan w$ and $\cot w$. $\dfrac{5}{12}, \dfrac{12}{5}$
43. If $\cos w = \dfrac{4\sqrt{41}}{41}$ and $\sin w = \dfrac{15\sqrt{41}}{41}$, find $\tan w$ and $\cot w$. $\dfrac{15}{4}, \dfrac{4}{15}$

44. 60°, 240°   45. 150°, 330°   46. 135°, 315°   47. 45°, 225°

B) Find two values of u between 0° and 360° for which each is true.
44. $\tan u = \sqrt{3}$
45. $\cot u = -\sqrt{3}$
46. $\tan u = -1$
47. $\cot u = 1$
48. $\cot u = -\dfrac{\sqrt{3}}{3}$
49. $\tan u = -\dfrac{\sqrt{3}}{3}$
50. $\cot u = -1$
51. $\tan u = 1$

120°, 300°   150°, 330°   135°, 315°   45°, 225°

C) Find two values of u between 0° and 720° for which each is true.
52. $\cot u = -\sqrt{3}$ and $\cos u < 0$ **150°, 510°**
53. $\tan u = -\sqrt{3}$ and $\sin u > 0$ **120°, 480°**
54. $\tan u = \dfrac{\sqrt{3}}{3}$ and $\sin u < 0$ **210°, 570°**
55. $\cot u = -\dfrac{\sqrt{3}}{3}$ and $\cos u > 0$ **300°, 660°**

Tangent and Cotangent Functions

SECANT AND COSECANT FUNCTIONS 17.5

Objectives **To find the secant and cosecant of degree measures of 30°, 45°, 60°, 90°, and their multiples
To determine if the secant or cosecant is positive or negative for a given degree measure
To find the measure of an angle whose secant or cosecant is given
To find secants and cosecants using trigonometric relationships**

Thus far, four trigonometric functions have been defined for any angle of rotation u. Using a reference triangle, the ratios $\frac{r}{x}$ and $\frac{r}{y}$ determine the trigonometric functions **secant** and **cosecant**. Notice that these functions are the reciprocals of the cosine and sine functions, respectively.

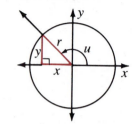

**Definition:
Secant (sec) and cosecant (csc) functions**

For any angle of rotation u and any radius r, the secant of u is $\frac{r}{x}$ $\left(\sec u = \frac{r}{x}\right)$ and the cosecant of u is $\frac{r}{y}$ $\left(\csc u = \frac{r}{y}\right)$, $(x \neq 0, y \neq 0)$.

For a unit circle, $\sec u = \frac{1}{x}$ and $\csc u = \frac{1}{y}$. Also, $\sec u = \frac{1}{\cos u}$ or $\cos u = \frac{1}{\sec u}$, and $\csc u = \frac{1}{\sin u}$ or $\sin u = \frac{1}{\csc u}$.

Example 1 **Find sec 120° and csc 120°.**

First Method: Unit circle
Use Ref. △: 30–60 rt. △.
$|y| = \frac{1}{2}(1)\sqrt{3} = \frac{\sqrt{3}}{2}$
$|x| = \frac{1}{2}(1) = \frac{1}{2}$
So, $y = \frac{\sqrt{3}}{2}$, $x = -\frac{1}{2}$
$\frac{r}{x} = \frac{1}{-\frac{1}{2}} = -2$, and
$\frac{r}{y} = \frac{1}{\frac{\sqrt{3}}{2}} = \frac{2\sqrt{3}}{3}$.

Second Method
Use reciprocal functions.

$\sec u = \dfrac{1}{\cos u}$ $\csc u = \dfrac{1}{\sin u}$

$\cos 120° = -\dfrac{1}{2}$ $\sin 120° = \dfrac{\sqrt{3}}{2}$

$\sec 120° = \dfrac{1}{-\frac{1}{2}}$ $\csc 120° = \dfrac{1}{\frac{\sqrt{3}}{2}}$

$\sec 120° = -2$ $\csc 120° = \dfrac{2}{\sqrt{3}}$ or $\dfrac{2\sqrt{3}}{3}$

Thus, $\sec 120° = -2$ and $\csc 120° = \dfrac{2\sqrt{3}}{3}$.

The sine, and therefore its reciprocal cosecant, are positive in the first and second quadrants. The cosine, and therefore its reciprocal secant, are positive in the first and fourth quadrants.

	1st Quad.	2nd Quad.	3rd Quad.	4th Quad.
secant	+	−	−	+
cosecant	+	+	−	−

The next example shows how to express the secant or cosecant of any angle as a function of an acute angle.

Example 2 **Write sec 135° and csc 240° as the same trigonometric function of an acute angle.**

For each angle, sketch a circle. Show the reference triangle.

Ref. ∠: 45°
Sec is negative in quadrant II.

sec 135° = − sec 45°

Ref. ∠: 60°
Csc is negative in quadrant III.

csc 240° = − csc 60°

Thus, sec 135° = − sec 45° and csc 240° = − csc 60°.

Since the sec and csc are reciprocals of the cos and sin respectively, the following table can be constructed.

	0°	90°	180°	270°
sec	1	undefined	−1	undefined
csc	undefined	1	undefined	−1

Example 3 **If $\sin u = -\frac{3}{4}$, find csc u. If $\cos w = -\frac{5}{7}$, find sec w.**

$$\csc u = \frac{1}{\sin u} = \frac{1}{-\frac{3}{4}}, \text{ or } -\frac{4}{3}.$$

$$\sec w = \frac{1}{\cos w} = \frac{1}{-\frac{5}{7}}, \text{ or } -\frac{7}{5}.$$

Thus, $\csc u = -\frac{4}{3}$.

Thus, $\sec w = -\frac{7}{5}$.

Example 4 **Find two values of u between 0° and 360° for which $\sec u = -\sqrt{2}$.**

Sec is negative in quadrant II.
Ref. ∠: 45°

$u = 180° - 45°$
$= 135°$

Sec is negative in quadrant III.
Ref. ∠: 45°

$u = 180° + 45°$
$= 225°$

Thus, $\sec u = -\sqrt{2}$ for $u = 135°$ and $u = 225°$.

Secant and Cosecant Functions

Oral Exercises

State whether each of the following is positive or negative.

1. sec 45° **+**
2. csc 135° **+**
3. sec 120° **−**
4. csc 300° **−**
5. sec 330° **+**
6. sec 225° **−**
7. csc 225° **−**
8. sec 150° **−**
9. csc 240° **−**
10. sec 315° **+**
11. csc 120° **+**
12. csc 330° **−**

Written Exercises

(A) Find each of the following. Rationalize any radical denominator.

1. csc 45° $\sqrt{2}$
2. sec 45° $\sqrt{2}$
3. sec 60° **2**
4. csc 60° $\frac{2\sqrt{3}}{3}$
5. sec 30° $\frac{2\sqrt{3}}{3}$
6. csc 30° **2**
7. csc 210° **−2**
8. sec 240° **−2**
9. csc 225°
10. sec 225°
11. csc 150° **2**
12. sec 150° $\frac{-2\sqrt{3}}{3}$
13. sec 300° **2**
14. csc 300°
15. sec 315° $\sqrt{2}$
16. csc 315°
17. sec 330°
18. csc 330° **−2**

9. $-\sqrt{2}$ 10. $-\sqrt{2}$ 14. $\frac{-2\sqrt{3}}{3}$ 16. $-\sqrt{2}$ 17. $\frac{2\sqrt{3}}{3}$

Write each of the following in terms of the same trigonometric function of an acute angle. For Ex. 19–30, see TE Answer Section.

19. sec 120°
20. csc 150°
21. csc 300°
22. sec 210°
23. sec 240°
24. csc 315°
25. csc 120°
26. csc 210°
27. sec 190°
28. sec 340°
29. csc 170°
30. csc 215°

31. If $\sin u = -\frac{1}{2}$, find csc u. **−2**
32. If $\cos u = -\frac{1}{3}$, find sec u. **−3**
33. If $\cos u = \frac{2}{3}$, find sec u. $\frac{3}{2}$
34. If $\sin u = \frac{4}{5}$, find csc u. $\frac{5}{4}$
35. If $\sin u = -\frac{5}{6}$, find csc u. $\frac{-6}{5}$
36. If $\cos u = -\frac{3}{7}$, find sec u. $\frac{-7}{3}$

(B) Find all values of u between and including 0° and 360° for which each is true.

37. sec $u = -1$ **180°**
38. csc $u = -1$ **270°**
39. csc $u = -\sqrt{2}$
40. sec $u = -2$
41. csc $u = \frac{2\sqrt{3}}{3}$ **60°, 120°**
42. sec $u = \frac{-2\sqrt{3}}{3}$ **150°, 210°**
43. sec $u = 2$
44. csc $u = 1$ **90°**

39. **225°, 315°** 40. **120°, 240°** 43. **60°, 300°**

(C) Find all values of u between 0° and 720° for which each is true.

45. sec $u = -1$ and sin $u = 0$ **180°, 540°**
46. sec $u = 2$ and cos $u > 0$ **60°, 300°, 420°, 660°**
47. csc $u = \frac{-2\sqrt{3}}{3}$ and tan $u > 0$ **240°, 600°**
48. csc $u = -2$ and cot $u < 0$ **330°, 690°**

NON-ROUTINE PROBLEMS

A triangle has two angles with measure 30° and 45°. The side opposite the 30° angle has length 12. Find the length of the side opposite the 45° angle. **$12\sqrt{2}$**

CUMULATIVE REVIEW

Solve each of the following using the quadratic formula.

1. $x^2 - 9x + 3 = 0$ $\frac{9 \pm \sqrt{69}}{2}$
2. $3x^2 + 5x - 4 = 0$ $\frac{-5 \pm \sqrt{73}}{6}$
3. $-2x^2 - 3x + 7 = 0$ $\frac{-3 \pm \sqrt{65}}{4}$

FUNCTIONS OF NEGATIVE ANGLES 17.6

Objective   **To find the sine, cosine, tangent, cotangent, secant, and cosecant of angles of negative measure**

The figure below illustrates two angles of rotation, u and $(-u)$, where $P(-a, b)$ is a point on the terminal side of angle u and $Q(-a, -b)$ is a point on the terminal side of the angle $(-u)$. The two reference triangles formed are congruent. Using the definitions of sin, cos, and tan of an angle, it follows that

$\sin u = \dfrac{y}{r}$ or $\dfrac{b}{r}$ and $\sin(-u) = \dfrac{y}{r}$ or $\dfrac{-b}{r}$

$\cos u = \dfrac{x}{r}$ or $\dfrac{-a}{r}$ and $\cos(-u) = \dfrac{x}{r} = \dfrac{-a}{r}$

$\tan u = \dfrac{y}{x}$ or $\dfrac{b}{-a}$ and $\tan(-u) = \dfrac{y}{x}$ or $\dfrac{-b}{-a} = \dfrac{b}{a}$.

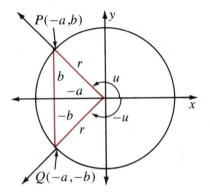

Since the cosecant, secant, and cotangent are reciprocal functions of the sine, cosine, and tangent, respectively, the following definitions are suggested.

Trigonometric functions of angles of negative measure

For any angles of rotation u and $(-u)$,

$\sin(-u) = -\sin u$ $\csc(-u) = -\csc u$
$\cos(-u) = \cos u$ $\sec(-u) = \sec u$
$\tan(-u) = -\tan u$ $\cot(-u) = -\cot u$.

Example 1   **Find $\sin(-45°)$, $\cos(-45°)$, and $\tan(-45°)$.**

Ref. ∠: 45°
$\sin 45° = \dfrac{\sqrt{2}}{2}$
$\cos 45° = \dfrac{\sqrt{2}}{2}$
$\tan 45° = 1$

$\sin(-45°) = -\sin 45°$
$= -\dfrac{\sqrt{2}}{2}$

$\cos(-45°) = \cos 45°$
$= \dfrac{\sqrt{2}}{2}$

$\tan(-45°) = -\tan 45°$
$= -1$

Example 2   **Find $\cot(-330°)$, $\sec(-330°)$, and $\csc(-330°)$.**

Ref. ∠: 30°
$\cot 30° = \sqrt{3}$
$\sec 30° = \dfrac{2\sqrt{3}}{3}$
$\csc 30° = 2$

$\cot(-330°) = -\cot 330°$
$= -(-\cot 30°)$
$= \sqrt{3}$

$\sec(-330°) = \sec 330°$
$= \sec 30°$
$= \dfrac{2\sqrt{3}}{3}$

$\csc(-330°) = -\csc 330°$
$= -(-\csc 30°)$
$= 2$

Example 3 Express sin (−315°) and cos (−315°) as functions of a positive acute angle.

$$\begin{aligned}\sin(-315°) &= -\sin 315° \\ &= -(-\sin 45°) \\ &= \sin 45°\end{aligned}$$

$$\begin{aligned}\cos(-315°) &= \cos 315° \\ &= \cos 45°\end{aligned}$$

Example 4 Express sec (−190°), tan (−300°), and csc (−210°) as functions of a positive acute angle.

$$\begin{aligned}\sec(-190°) &= \sec 190° \\ &= -\sec 10°\end{aligned} \quad \begin{aligned}\tan(-300°) &= -\tan 300° \\ &= \tan 60°\end{aligned} \quad \begin{aligned}\csc(-210°) &= -\csc 210° \\ &= -(-\csc 30°) \\ &= \csc 30°\end{aligned}$$

Oral Exercises

Express each as a function of a positive acute angle.
1. sin (−10°) **−sin 10°**   2. cos (−25°) **cos 25°**   3. tan (−50°) **−tan 50°**   4. sec (−20°) **sec 20°**
5. csc (−70°) **−csc 70°**   6. sin (−35°) **−sin 35°**   7. cos (−40°) **cos 40°**   8. tan (−5°) **−tan 5°**

Written Exercises

Ⓐ **Find each of the following. Rationalize any radical denominator.**   2. $\frac{\sqrt{3}}{2}$ 3. $\frac{-\sqrt{3}}{3}$ 7. $\frac{-\sqrt{3}}{2}$
1. sin (−30°) $-\frac{1}{2}$ 2. cos (−30°)   3. tan (−30°)   4. cot (−30°) $-\sqrt{3}$ 5. sec (−30°) $\frac{2\sqrt{3}}{3}$
6. csc (−45°) $-\sqrt{2}$ 7. sin (−60°)   8. cos (−60°) $\frac{1}{2}$ 9. sec (−60°) 2   10. csc (−60°) $\frac{-2\sqrt{3}}{3}$
11. cos (−150°)   12. tan (−150°)   13. sin (−150°) $-\frac{1}{2}$ 14. csc (−150°) −2   15. cot (−150°) $\sqrt{3}$
16. sec (−210°)   17. csc (−210°) 2   18. tan (−210°)   19. sin (−210°) $\frac{1}{2}$ 20. cos (−210°) $\frac{-\sqrt{3}}{2}$
21. cos (−315°)   22. sin (−315°)   23. cot (−315°) 1   24. tan (−315°) 1   25. sec (−315°) $\sqrt{2}$
26. **−cot 15°** 27. **−tan 10°** 28. **cos 80°** 29. **cot 85°** 30. **sin 80°** 31. **−sin 80°** 32. **−csc 15°**
Express as a function of a positive acute angle.   33. **sec 18°**   34. **csc 10°**   35. **cos 35°**
26. cot (−15°)   27. tan (−190°)   28. cos (−280°)   29. cot (−95°)   30. sin (−260°)
Ⓑ 31. sin (−100°)   32. csc (−15°)   33. sec (−18°)   34. csc (−190°)   35. cos (−35°)
36. sec (−310°)   37. cos (−350°)   38. tan (−100°)   39. sin (−350°)   40. tan (−390°)
41. csc (−195° 35′) **csc 15° 35′**   42. sin (−18° 20′) **−sin 18° 20′**   43. cos (−300° 45′) **cos 59° 15′**
36. **sec 50°** 37. **cos 10°** 38. **tan 80°** 39. **sin 10°** 40. **−tan 30°**
Ⓒ **For any angle of rotation *u* with negative measure show the following.**   11. $\frac{-\sqrt{3}}{2}$ 12. $\frac{\sqrt{3}}{3}$ 16. $\frac{-2\sqrt{3}}{3}$
44. tan (−*u*) = −tan *u*   45. cot (−*u*) = −cot *u*
46. sec (−*u*) = sec *u*   47. csc (−*u*) = −csc *u*   18. $\frac{-\sqrt{3}}{3}$ 21. $\frac{\sqrt{2}}{2}$ 22. $\frac{\sqrt{2}}{2}$
For Ex. 44–47, see TE Answer Section.

CUMULATIVE REVIEW

Solve each of the following by factoring.
1. $3x^2 - x = 0$ **0, $\frac{1}{3}$**   2. $9x^2 - 1 = 0$ **$-\frac{1}{3}, \frac{1}{3}$**   3. $6x^2 + 7x + 2 = 0$ **$-\frac{1}{2}, -\frac{2}{3}$**

VALUES OF TRIGONOMETRIC FUNCTIONS   17.7

Objective **To find the other five trigonometric functions of an angle of rotation u, given one of the six functions and the quadrant of the terminal side of u**

To find the value of five trigonometric functions of any angle of rotation u, given the value of one of them and the quadrant of the terminal side, do the following.
1. State two of the values of x, y, and r, using the definition of the given trigonometric function.
2. Find the remaining value of x, y, or r, using the Pythagorean theorem and the reference triangle.
3. Write the ratios for the other five trigonometric functions using x, y, and r.

Example 1 **Find the value of each of the other five trigonometric functions of u if $\sin u = \frac{4}{5}$ and the terminal side is in the 2nd quadrant.**

$$\sin u = \frac{4}{5} = \frac{y}{r} \text{ or } y = 4 \text{ and } r = 5$$

$x^2 + y^2 = r^2$ ◂ *Use the Pythagorean*
$x^2 + 16 = 25$ *theorem to find x.*
$x^2 = 9$
$|x| = 3$ ◂ *The terminal side*
$x = -3$ *is in the 2nd quadrant.*

Thus, $\cos u = -\frac{3}{5}$, $\tan u = -\frac{4}{3}$, $\cot u = -\frac{3}{4}$, $\sec u = -\frac{5}{3}$, and $\csc u = \frac{5}{4}$.

Example 2 **Find the value of each of the other five trigonometric functions of u if $\tan u = -\frac{5}{12}$ and $\cos u < 0$.**

Both the tan and cos are negative in the 2nd quadrant, so the terminal side lies in the 2nd quadrant.

$$\tan u = -\frac{5}{12} = \frac{y}{x} \text{ or } y = 5 \text{ and } x = -12$$

$x^2 + y^2 = r^2$ ◂ *Use the Pythagorean*
$144 + 25 = r^2$ *theorem.*
$r^2 = 169$ or $r = 13$

Thus, $\sin u = \frac{5}{13}$, $\cos u = -\frac{12}{13}$, $\cot u = -\frac{12}{5}$, $\sec u = -\frac{13}{12}$, and $\csc u = \frac{13}{5}$.

Oral Exercises

1. I, II **2.** III, IV **3.** II, III **4.** I, IV **5.** II, IV

For each of the following, tell in which quadrant(s) the terminal side of the angle may lie.

1. $\sin u = \dfrac{4}{5}$
2. $\sin u = -\dfrac{3}{4}$
3. $\cos u = -\dfrac{5}{7}$
4. $\cos u = \dfrac{7}{8}$
5. $\tan u = -\dfrac{3}{4}$
6. $\tan u = \dfrac{\sqrt{5}}{3}$
7. $\cot u = -\dfrac{5}{12}$
8. $\cot u = \dfrac{12}{5}$
9. $\sec u = \dfrac{13}{5}$
10. $\sec u = -\dfrac{13}{12}$
11. $\sin u > 0$ and $\tan u < 0$ II
12. $\sin u < 0$ and $\cos u < 0$ III

6. I, III **7.** II, IV **8.** I, III **9.** I, IV **10.** II, III

Written Exercises

For Ex. 1–30, see TE Answer section.

(A) Find the value of each of the other five trigonometric functions of u.

1. $\tan u = -\dfrac{12}{5}$ in 2nd quad.
2. $\sin u = \dfrac{12}{13}$ in 2nd quad.
3. $\cos u = -\dfrac{5}{13}$ in 3rd quad.
4. $\sin u = -\dfrac{3}{5}$ in 4th quad.
5. $\cot u = -\dfrac{3}{4}$ in 4th quad.
6. $\sec u = -\dfrac{13}{5}$ in 2nd quad.
7. $\cos u = -\dfrac{12}{13}$ in 2nd quad.
8. $\csc u = \dfrac{13}{5}$ in 1st quad.
9. $\tan u = \dfrac{4}{3}$ in 3rd quad.
10. $\sec u = \dfrac{13}{12}$ in 4th quad.
11. $\tan u = -1$ in 2nd quad.
12. $\cot u = -\sqrt{3}$ in 4th quad.

(B)
13. $\sin u = \dfrac{5}{13}$ and $\sec u > 0$
14. $\cos u = -\dfrac{3}{5}$ and $\tan u < 0$
15. $\cot u = 1$ and $\cos u > 0$
16. $\tan u = -\dfrac{3}{4}$ and $\sin u > 0$
17. $\tan u = \dfrac{\sqrt{3}}{3}$ and $\cos u > 0$
18. $\cos u = \dfrac{\sqrt{2}}{2}$ and $\csc u < 0$
19. $\sec u = -\dfrac{2\sqrt{3}}{3}$ and $\sin u > 0$
20. $\csc u = -\sqrt{2}$ and $\tan u < 0$

(C) Find the value(s) of each of the other trigonometric functions of u.

21. $\sin u = \dfrac{\sqrt{3}}{2}$ and $\tan u = -\sqrt{3}$
22. $\cos u = -\dfrac{1}{2}$ and $\cot u = \dfrac{\sqrt{3}}{3}$
23. $\cos u = \dfrac{\sqrt{2}}{2}$ and $\sin u = -\dfrac{\sqrt{2}}{2}$
24. $\tan u = -1$ and $\sec u = -\sqrt{2}$
25. $\sec u = -\dfrac{13}{5}$ and $\tan u = \dfrac{12}{5}$
26. $\csc u = \dfrac{13}{12}$ and $\sin u = \dfrac{12}{13}$
27. $\tan u = -\dfrac{5}{4}$ and $\cos u = \dfrac{4\sqrt{41}}{41}$
28. $\cot u = -\sqrt{3}$ and $\cos u = -\dfrac{\sqrt{3}}{2}$
29. $\sin u = -\dfrac{7}{25}$ and $\csc u = -\dfrac{25}{7}$
30. $\sec u = -\dfrac{5}{4}$ and $\tan u = -\dfrac{3}{4}$

CUMULATIVE REVIEW

Graph each of the following. Find the domain and the range. Determine the coordinates of the vertex and an equation of the axis of symmetry.

1. $y = x^2$
D = {all real numbers};
R = {y|y ≥ 0}; (0, 0); x = 0

2. $y = -x^2$
D = {all real numbers};
R = {y|y ≤ 0}; (0, 0); x = 0

3. $y = x^2 - 2x - 8$
D = {all real numbers};
R = {y|y ≥ -9}; (1, -9); x = 1

USING TRIGONOMETRIC TABLES   17.8

Objectives **To determine the sin, cos, tan, and cot of any angle *u* reading a trigonometric table**
To rewrite trigonometric functions using cofunctions
To find angle *u* by reading a trigonometric table when the sin *u*, cos *u*, tan *u*, or cot *u* is given

You have been finding trigonometric functions of angles with special measures like 30°, 45°, 60° and their multiples. You will now learn how to find trigonometric functions of angles with any measure by using a table. In the table, angle measure is given in multiples of 10 minutes. One degree equals sixty minutes (1° = 60').

Example 1 **Find sin 23° 30' and cot 24° 50' using the portion of a trigonometric table shown below.**

The functions at the *top* of the table must be used with degree measure in the left-hand column, (0° to 45°).

Thus, sin 23° 30' = .3987 and cot 24° 50' = 2.1609.

Example 2 **Find cos 45° 10' and tan 46° 40'.**

The functions at the *bottom* of the table must be used with degree measure in the right-hand column, (45° to 90°).

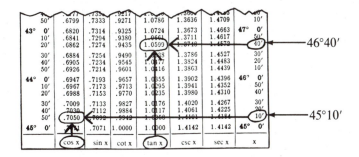

Thus, cos 45° 10' = .7050 and tan 46° 40' = 1.0599.

Using Trigonometric Tables   **477**

Since the trigonometric tables only show measures of acute angles, functions of larger angles must be written as functions of acute angles.

Example 3 **Find sin 143° 20' and cos 250° 40'.**

Write each as the same trigonometric function of an acute angle. Sketch the angle of rotation and reference triangle.

179° 60'
− 143° 20'
―――――
36° 40'

sin 143° 20' = sin 36° 40'
= .5972

250° 40'
− 180°
―――――
70° 40'

cos 250° 40' = −cos 70° 40'
= −.3311

Thus, sin 143° 20' = .5972, and cos 250° 40' = −.3311.

Example 4 **Find the value of each pair from the table.**

sin 70° and cos 20°
        ↘   ↙
        .9397

tan 33° 40' and cot 56° 20'
            ↘   ↙
            .6661

Thus, sin 70° = cos 20° = .9397, and tan 33° 40' = cot 56° 20' = .6661.

Recall that two angles are complementary if the sum of their measures is 90°. The sine and *co*sine functions are called *complementary functions,* or *cofunctions*. Other examples of cofunctions are tangent and *co*tangent as well as secant and *co*secant. Example 4 suggests the following relationships concerning cofunctions.

Cofunction relationships	$\sin u = \cos(90° - u)$ and $\cos u = \sin(90 - u)$ $\tan u = \cot(90° - u)$ and $\cot u = \tan(90 - u)$ $\sec u = \csc(90° - u)$ and $\csc u = \sec(90 - u)$

Example 5 **Express each function in terms of its cofunction.**
  cos 63°   sec 13° 40'   cot 38° 10'

Use the cofunction and the complement.
  cos 63°   sec 13° 40'   cot 38° 10'
     ↓   ↓   ↓
  sin 27°   csc 76° 20'   tan 51° 50'

Example 6 **Express each as a function of an angle measure less than 45°. Use cofunctions.**
  sin 100°   tan 290°

$\sin(90° - u) = \cos u$ ▶
  sin 100° = sin 80°   tan 290° = −tan 70°
         = cos 10°   = −cot 20°   ◀ $\tan(90° - u) = \cot u$

Thus, sin 100° = cos 10° and tan 290° = −cot 20°.

You can find u if sin u, cos u, tan u, or cot u is given.

Example 7 sin u = .5200   cos u = .8465   tan u = 1.7205
  Find u if u < 90°. Find u if u < 90°. Find u if u < 90°.

Look in the table to find the value, then find the corresponding angle u.

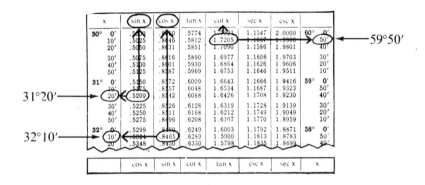

Thus, if sin u = .5200, u = 31° 20′; if cos u = .8465, u = 32° 10′; and if tan u = 1.7205, u = 59° 50′.

The next example illustrates how to solve a trigonometric equation using a table of values.

Example 8 **Find all values of u, 0° < u < 360°, for which tan u = −.6168.**

Read the table in Example 7. The measure of the ref. ∠ is 31° 40′.

The tangent is negative
in the 2nd and 4th quadrants.

Thus, u = 148° 20′ or u = 328° 20′.

Using Trigonometric Tables

Written Exercises

1. .4067 **2.** 4.7046 **3.** .9205 **4.** 1.0724 **5.** 1.0977 **6.** .9426 **7.** 2.3183 **8.** .7254

Ⓐ Use the table to find each of the following. **9.** .8718 **10.** 7.9530 **11.** .8342 **12.** .6756

1. sin 24°   **2.** tan 78°   **3.** cos 23°   **4.** cot 43°   **5.** tan 47° 40′
6. sin 70° 30′   **7.** cot 23° 20′   **8.** cos 43° 30′   **9.** sin 60° 40′   **10.** tan 82° 50′
11. tan 39° 50′   **12.** sin 42° 30′   **13.** tan 52° 50′   **14.** cos 73° 40′   **15.** cot 50° 10′
16. cos 152°   **17.** cos 138°   **18.** sin 340°   **19.** cot 155°   **20.** sin 260°
21. cot 190° 40′   **22.** tan 340° 50′   **23.** cot 140° 10′   **24.** tan 100° 40′   **25.** cos 350° 30′

13. 1.3190 **14.** .2812 **15.** .8342 **16.** −.8829 **17.** −.7431 **18.** −.3420 **19.** −2.1445 **20.** −.9848

Use the table to find u, $u < 90°$. **21.** 5.3093 **22.** −.3476 **23.** −1.1988 **24.** −5.3093 **25.** .9863

26. sin u = .9293 **68° 20′**   **27.** cos u = .9652 **15° 10′**   **28.** tan u = .5930 **30° 40′**
29. cot u = .3607 **70° 10′**   **30.** tan u = .7954 **38° 30′**   **31.** sin u = .7373 **47° 30′**
32. cos u = .7333 **42° 50′**   **33.** sin u = .3773 **22° 10′**   **34.** cot u = 6.3138 **9°**
35. tan u = 3.8667 **75° 30′**   **36.** cot u = .3249 **72°**   **37.** cos u = .2363 **76° 20′**

38. cos 55°   **39.** sin 65°   **40.** cot 22°   **41.** tan 80°   **42.** cos 42°

Express each in terms of its cofunction.
38. sin 35°   **39.** cos 25°   **40.** tan 68°   **41.** cot 10°   **42.** sin 48°
43. cos 50° 10′   **44.** tan 40° 40′   **45.** cot 55° 20′   **46.** sin 17° 40′   **47.** cos 34° 30′
  sin 39° 50′   cot 49° 20′   tan 34° 40′   cos 72° 20′   sin 55° 30′

Ⓑ Find all values of u, $0° < u < 360°$, for which each is true. **For Ex. 48–88, see TE Answer Section.**
48. cos u = .9636   **49.** tan u = −2.7725   **50.** cot u = 1.2723
51. cot u = −.2648   **52.** cos u = −.4874   **53.** sin u = −.9283
54. sin u = .8124   **55.** sin u = .1650   **56.** cos u = .7808

Express as a function of an angle measure less than 45°.
57. tan 120°   **58.** cot 200°   **59.** sin 316°   **60.** cos 110°   **61.** sin 224°
62. sin 324° 20′   **63.** cos 125° 10′   **64.** tan 212° 40′   **65.** cot 331° 50′   **66.** tan 134° 40′

Ⓒ Express each as a reciprocal function of an acute angle.
67. cos 112°   **68.** sin 140° 10′   **69.** sin 176° 40′   **70.** tan 190° 40′
71. tan 304° 20′   **72.** cot 123° 40′   **73.** cos 340° 30′   **74.** cot 129° 20′

Find each of the following.
75. sec 35° 20′   **76.** csc 150° 20′   **77.** csc 50° 40′   **78.** sec 190° 40′

Find u if $u < 90°$.
79. sec u = 1.0003   **80.** csc u = 1.0712   **81.** csc u = 1.9801   **82.** sec u = 2.1657

Find all values of u, $0° < u < 360°$, for which each is true.
83. sec u = 1.1969   **84.** csc u = −6.8998   **85.** sec u = −1.2015
86. csc u = 1.6123   **87.** sec u = 2.7904   **88.** csc u = −2.5770

CUMULATIVE REVIEW

3. log 86.45 + 3 log 425.6 − log 0.00735

Write each expression in logarithmic form. Do not solve.

1. $\dfrac{45\sqrt{13.5}}{\sqrt[3]{38}}$ **2.** $\dfrac{98^3 \sqrt{48.7}}{24(105)}$ **3.** $\dfrac{86.45(425.6)^3}{.00735}$ **above**

1. log 45 + $\dfrac{1}{2}$ log 13.5 − $\dfrac{1}{3}$ log 38   **2.** 3 log 98 + $\dfrac{1}{2}$ log 48.7 − log 24 − log 105

LINEAR TRIGONOMETRIC EQUATIONS   17.9

Objective   **To solve linear trigonometric equations**

You have learned how to determine all values of u, $0° \leq u \leq 360°$, that make equations like $\sin u = -\frac{1}{2}$ true. For $\sin u = -\frac{1}{2}$, the reference angle is 30°. The sin is negative in the 3rd and 4th quadrants. So, $u = 180° + 30°$ or 210° and $360° - 30°$ or 330°.

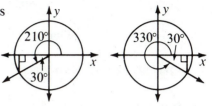

Example 1   Determine the values of u between 0° and 360° to the nearest ten minutes that make the equation $3 \sin u - 1 = 0$ true.

Solve for $\sin u$, write in decimal form, and determine the reference angle.

$3 \sin u - 1 = 0$
$3 \sin u = 1$
$\sin u = \frac{1}{3}$, or .3333   ◀ *Use the trigonometric table to find closest value.*
Ref. $\angle$: 19° 30′

$u = 19° 30′$ $u = 180° - 19° 30′$ ◀ *Sin is positive in the 1st and 2nd quadrants.*
            $= 160° 30′$

Thus, $u = 19° 30′$ or $u = 159° 30′$.

Example 2   Determine the values of θ between 0° and 360° to the nearest ten minutes that make the equations true.
$5 \tan \theta - 3 = -7$ $-3 \cos \theta - 4 = 5 \cos \theta + 2$

Solve for the trigonometric function. Determine the reference angle.

$5 \tan \theta - 3 = -7$ $-3 \cos \theta - 4 = 5 \cos \theta + 2$
$5 \tan \theta = -4$ $-6 = 8 \cos \theta$
$\tan \theta = -\frac{4}{5}$, or $-.8000$ $\cos \theta = -\frac{6}{8}$, or $-.7500$

Ref. $\angle$: 38° 40′   Ref. $\angle$: 41° 20′

$\theta = 180° - 38° 40′$ $\theta = 180° - 41° 20′$
$= 141° 20′$ (2nd quad.) $= 138° 40′$

$\theta = 360° - 38° 40′$ $\theta = 180° + 41° 20′$
$= 321° 20′$ (4th quad.) $= 221° 20′$

Thus, $\theta = 141° 20′$ or $321° 20′$.   Thus, $\theta = 138° 40′$ or $221° 20′$.

Example 3 Determine the values of t between 0° and 360° that make the equation $4 \sin t - 3 = 2 \sin t + 5$ true.

$$4 \sin t - 3 = 2 \sin t + 5$$
$$2 \sin t = 8$$
$$\sin t = 4$$

Thus, there are no solutions, since $-1 \le \sin t \le 1$.

Written Exercises

Determine the values of θ between 0° and 360° to the nearest ten minutes that make each equation true.

Ⓐ
1. $2 \sin \theta = 1$ **30°, 150°**
2. $2 \cos \theta - \sqrt{3} = 0$ **30°, 330°**
3. $3 \sin \theta - 3 = 0$ **90°**
4. $2 \cos \theta = -\sqrt{3}$ **150°, 210°**
5. $3 \sin \theta + 2 = -1$ **270°**
6. $6 \cos \theta - 2 = 0$
7. $\cot \theta = -1$ **135°, 315°**
8. $3 \tan \theta - 1 = 0$
9. $\sqrt{2} \cos \theta + 1 = 0$ **135°, 225°**
10. $2 \sin \theta + 1 = 0$ **210°, 330°**
11. $4 \cos \theta = 6$ **none**
12. $3 \tan \theta = -\sqrt{3}$ **150°, 330°**
13. $6 \cos \theta - 3 = 4 \cos \theta - 2$
14. $5 \sin \theta + 3 = 2 \sin \theta$ **270°**
15. $7 \tan \theta - 1 = 2 \tan \theta + 4$
16. $-5 \sin \theta - 6 = 5 \sin \theta - 2$
17. $-8 \tan \theta - 12 = 6 \tan \theta - 1$
18. $3 \cos \theta - 4 = 5 \cos \theta - 5$

13. 60°, **15.** 45°, 225°   **16.** 203° 30′, 336° 30′   **17.** 141° 50′, 321° 50′   **18.** 60°, 300°

Determine the values of θ between 0° and 720° to the nearest ten minutes that make each equation true. **For Ex. 19–36, see TE Answer Section. 6.** 70° 30′, 289° 30′   **8.** 18° 30′, 198° 30′

Ⓑ
19. $2 \cos \theta + \sqrt{3} = 0$
20. $\tan \theta + 1 = 0$
21. $-\cot \theta = -\sqrt{3}$
22. $5 \sin \theta - 8 = 3 \sin \theta - 9$
23. $-3 \cos \theta - 5 = -6 \cos \theta$
24. $-8 \sin \theta + 6 = -9 \sin \theta + 7$
25. $\sqrt{3} \tan \theta - 2 = 1$
26. $3 \sin \theta - 1 = -2 \sin \theta + 3$
27. $-4 \cot \theta - 7 = \cot \theta$

Ⓒ
28. $3 \sec \theta + 8 = 0$
29. $\sqrt{3} \csc \theta = 5\sqrt{3}$
30. $-\sec \theta - 12 = 4 \sec \theta$
31. $-4 \csc \theta = 4$
32. $-2 \sec \theta + 8.4 = 0$
33. $-3 \sec \theta + 1 = 0$
34. $\dfrac{\sec \theta}{4} - 3 = \dfrac{5 \sec \theta}{6}$
35. $\dfrac{1}{3} \csc \theta - \dfrac{3}{5} = 1$
36. $\dfrac{\csc \theta - 6}{3} = \dfrac{3}{4} \csc \theta + 1$

Determine the values of θ between 0° and 360° that make each equation true.
37. $\cos \theta = \sin \theta$ **45°, 225°**
38. $-\sin \theta = \cos \theta$ **135°, 315°**
39. $\tan \theta = \cot \theta$
40. $\sin \theta = \csc \theta$ **90°, 270°**
41. $\cos \theta = \sec \theta$ **0°, 180°**
42. $-\sec \theta = \cos \theta$ **none**

39. 45°, 135°, 225°, 315°

CALCULATOR ACTIVITIES

Example Determine the values of θ between 0° and 360° to the nearest ten minutes that make the equation $\sqrt{37} \tan \theta - 8 = 0$ true.

1. Solve for $\tan \theta$: $\tan \theta = \dfrac{8}{\sqrt{37}}$.

2. Use a calculator to find $\sqrt{37} \boxminus 6.0827625$.

3. Compute $\dfrac{8}{6.0828}$ to equal 1.3151837.

4. Look in the table to find $\theta = 52° 50′$ or $\theta = 232° 50′$.

INTERPOLATION WITH TRIGONOMETRY   17.10

Objectives **To use interpolation to determine functions of an angle measured to the nearest minute**
To solve an equation of the form $a \tan \theta = b$, θ to the nearest minute

You have learned how to use a table to find trigonometric values for any angle measure in multiples of 10 minutes. You will now learn how to find trigonometric values for any degree and minute measure. The process of finding values for measures not in the tables is called **interpolation.**

Example 1   **Find sin 36° 33′.**

Look in the table for known values: 36° 33′ lies between the values 36° 30′ and 36° 40′. Make a table.

	measure	sine	
$10\begin{bmatrix} 3\begin{bmatrix}\end{bmatrix} \end{bmatrix}$	36° 30′	.5948	$\Big] n \Big] .0024$
	36° 33′		
	36° 40′	.5972	

Write a proportion and solve it.

$$\frac{n}{.0024} = \frac{3}{10}$$
$$10n = .0072$$
$$n = .00072, \text{ or } n \doteq .0007$$

Add. ▶   .5948 + .0007 = .5955   **Thus,** sin 36° 33′ = .5955.

Example 2   **Find cos 311° 33′.**

The reference ∠ is 48° 27′.
Look in the table for known values: 48° 27′ lies between the values of 48° 20′ and 48° 30′. Make a table.

	measure	cosine	
$10\begin{bmatrix} 7\begin{bmatrix}\end{bmatrix} \end{bmatrix}$	48° 20′	.6648	$\Big] n \Big] -.0022$
	48° 27′		
	48° 30′	.6626	

Write a proportion and solve it.

$$\frac{n}{-.0022} = \frac{7}{10}$$
$$10n = -.0154$$
$$n = .00154, \text{ or } n \doteq -.0015$$

Add. ▶   .6648 + (−.0015) = .6633   **Thus,** cos 311° 33′ = .6633.

Interpolation with Trigonometry

The next example illustrates how to find u, if a trigonometric function of u is given.

Example 3 **If $\tan u = .3463$ ($0° < u < 90°$), find u to the nearest minute.**

Look in the table for known values: .3463 lies between .3443 and .3476.

$$10\left[n\begin{bmatrix}19° \ 0' \\ \\ 19° \ 10'\end{bmatrix}\begin{array}{|c}\text{measure} \quad\quad \text{tangent} \\ \hline \begin{matrix}.3443 \\ .3463 \\ .3476\end{matrix}\end{array}\left.\begin{matrix}\\.0020\\\end{matrix}\right].0033\right.$$

measure	tangent
19° 0'	.3443
n	.3463
19° 10'	.3476

Write a proportion and solve it.

$$\frac{n}{10} = \frac{.0020}{.0033} = \frac{20}{33}$$
$$33n = 200$$
$$n = \frac{200}{33}, \text{ or } n \doteq 6$$

Add. ▶ $19° \ 0' + 6' = 19° \ 6'$ Thus, $u = 19° \ 6'$.

Example 4 **Find all values of θ ($0° < \theta < 360°$) to the nearest minute that make the equation $-4 \tan \theta - 3 = -7 \tan \theta + 5$ true.**

Solve for $\tan \theta$. Determine the reference angle.
$$-4 \tan \theta - 3 = -7 \tan \theta + 5$$
$$3 \tan \theta = 8$$
$$\tan \theta = \frac{8}{3}, \text{ or } 2.6667$$

Use 2.6667 to find the reference angle.

measure	tangent
69° 20'	2.6511
n	2.6667
69° 30'	2.6746

Write a proportion and solve.
$$\frac{n}{10} = \frac{.0156}{.0235} = \frac{156}{235}$$
$$235n = 1560$$
$$n = \frac{1560}{235} \doteq 6$$

Add. ▶ $69° \ 20' + 6' = 69° \ 26'$

Tan is positive in the 1st and 3rd quadrants. ▶ $\theta = 69° \ 26'$
$\theta = 180° + 69° \ 26' = 249° \ 26'$

Written Exercises

(A) Find each of the following.
1. sin 43° 15′ .6852
2. cos 18° 32′ .9481
3. tan 12° 57′ .2300
4. cot 62° 38′ .5176
5. cos 72° 18′ .3040
6. tan 55° 8′ 1.4352
7. cos 33° 42′ .8320
8. cos 87° 53′ .03694
9. tan 31° 43′ .6180
10. cot 12° 24′ 4.5484
11. sin 73° 47′ .9602
12. sin 16° 18′ .2806

Find u ($0° < u < 90°$) to the nearest minute.
13. cos u = .9753 12° 46′
14. tan u = .1573 8° 56′
15. sin u = .2875 16° 43′
16. sin u = .9264 67° 53′
17. cos u = .9214 22° 52′
18. cot u = 3.5352 15° 48′
19. cot u = .3567 70° 23′
20. sin u = .7954 52° 42′
21. cos u = .4535 63° 2′

Find all values of θ ($0° < \theta < 360°$) to the nearest minute.   23. 30°, 150°   24. 38° 56′, 321° 4′
22. $2 \cos \theta + 1 = 0$ 120°, 240°
23. $5 \sin \theta - 2 = 3 \sin \theta - 1$
24. $6 \cos \theta - 4 = 3 - 3 \cos \theta$
25. $-3 \tan \theta + 5 = 4 + 3 \tan \theta$
26. $-2 \cot \theta + 8 = -5 \cot \theta$
27. $-7 \sin \theta + 5 = 7 - \sin \theta$
   9° 28′, 189° 28′   159° 27′, 339° 27′   199° 28′, 340° 32′

(B) Find each of the following.
28. cos 124° 43′ −.5695
29. sin 312° 24′ −.7384
30. tan 86° 57′ 18.7793
31. cot 204° 36′ 2.1842
32. sin 197° 22′ −.2985
33. cot 342° 53′ −3.2472
34. cos 349° 45′ .9840
35. tan 224° 27′ .9810
36. sec 134° 38′ −1.4233
37. csc 248° 17′ −1.0764
38. csc 153° 54′ 2.2731
39. sec 354° 48′ 1.0042
40. 23° 08′, 203° 08′   41. 156° 35′, 203° 25′   42. 34° 19′, 214° 19′

Find u ($0° < u < 360°$) to the nearest minute.
40. tan u = .4272
41. cos u = −.9177
42. cot u = 1.4650
43. sin u = .5987
44. sin u = −.8221
45. cos u = −.0650
   36° 47′, 143° 13′   235° 18′, 304° 42′   93° 44′, 266° 16′

(C) Find all values of θ ($0° \le \theta < 360°$).
46. $\dfrac{\sin \theta}{\cos \theta} = -1$ 135°, 315°
47. $\dfrac{\cos \theta}{\sin \theta} = -\dfrac{\sqrt{3}}{3}$ 120°, 300°
48. $\cot \theta \cdot \sin \theta = \dfrac{\sqrt{3}}{2}$ 30°, 330°
49. $\sin \theta \sec \theta = -3.2658$ 107° 2′, 287° 2′
50. $\csc \theta \cos \theta = -2.4536$ 157° 50′, 337° 50′
51. $\dfrac{\cot \theta \cdot \tan \theta}{\sec \theta} = -.7654$ 139° 57′, 220° 3′
52. $\tan \theta \cdot \csc \theta \cdot (\cos \theta)^2 = -.3857$ 112° 41′, 247° 19′
53. $\sec^2 \theta - 2 = 0$ 45°, 135°, 225°, 315°
54. $4 \csc \theta + 6 = \csc \theta$ 210°, 330°
55. $2 \csc \theta - \dfrac{4}{\sin \theta} = -8$ 194° 28′, 345° 32′
56. $\dfrac{1}{\sec \theta} + 8 \cos \theta = \sec \theta \cdot \cos \theta$ 83° 37′, 276° 23′

CUMULATIVE REVIEW

Simplify.

1. $\dfrac{1 - \dfrac{1}{3}}{4 + \dfrac{3}{5}}$ $\dfrac{10}{69}$

2. $\dfrac{\dfrac{2}{3} - \dfrac{4}{5}}{\dfrac{5}{2} + \dfrac{7}{9}}$ $\dfrac{-12}{295}$

3. $\dfrac{4 - \dfrac{5}{4x}}{7 + \dfrac{4}{3x}}$ $\dfrac{48x - 15}{84x + 16}$

Interpolation with Trigonometry

TRIGONOMETRY OF A RIGHT TRIANGLE   17.11

Objectives **To find trigonometric functions of acute angles in a right triangle**
To simplify trigonometric expressions involving special angles

Remember that for any angle of rotation there is an associated reference right triangle with six trigonometric ratios. You will now learn about trigonometry of right triangles. The following diagrams will help make comparisons.

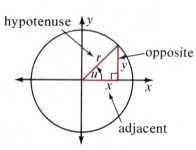

$$\sin u = \frac{y}{r} \Rightarrow \sin A = \frac{\text{length of opposite side}}{\text{length of hypotenuse}}$$

$$\cos u = \frac{x}{r} \Rightarrow \cos A = \frac{\text{length of adjacent side}}{\text{length of hypotenuse}}$$

$$\tan u = \frac{y}{x} \Rightarrow \tan A = \frac{\text{length of opposite side}}{\text{length of adjacent side}}$$

$$\cot u = \frac{x}{y} \Rightarrow \cot A = \frac{\text{length of adjacent side}}{\text{length of opposite side}}$$

Example 1 For each right triangle, find sin 30°, cos 30°, tan 30°, and cot 30°.

$\sin A = \frac{\text{opposite side}}{\text{hypotenuse}} \Rightarrow \sin 30° = \frac{1}{2}$

$\cos A = \frac{\text{adjacent side}}{\text{hypotenuse}} \Rightarrow \cos 30° = \frac{\sqrt{3}}{2}$

$\tan A = \frac{\text{opposite}}{\text{adjacent}} \Rightarrow \tan 30° = \frac{1}{\sqrt{3}}$, or $\frac{\sqrt{3}}{3}$

$\cot A = \frac{\text{adjacent}}{\text{opposite}} \Rightarrow \cot 30° = \frac{\sqrt{3}}{1}$, or $\sqrt{3}$

$\sin 30° = \frac{3}{6}$, or $\frac{1}{2}$

$\cos 30° = \frac{3\sqrt{3}}{6}$, or $\frac{\sqrt{3}}{2}$

$\tan 30° = \frac{3}{3\sqrt{3}}$, or $\frac{\sqrt{3}}{3}$

$\cot 30° = \frac{3\sqrt{3}}{3}$, or $\sqrt{3}$

Thus, $\sin 30° = \frac{1}{2}$, $\cos 30° = \frac{\sqrt{3}}{2}$, $\tan 30° = \frac{\sqrt{3}}{3}$, and $\cot 30° = \sqrt{3}$.

Example 2 For the given triangle, find sin 60°, cos 60°, tan 60°, and cot 60°.

$\sin R = \dfrac{\text{opp.}}{\text{hyp.}} \Rightarrow \sin 60° = \dfrac{\sqrt{3}}{2}$

$\cos R = \dfrac{\text{adj.}}{\text{hyp.}} \Rightarrow \cos 60° = \dfrac{1}{2}$

$\tan R = \dfrac{\text{opp.}}{\text{adj.}} \Rightarrow \tan 60° = \dfrac{\sqrt{3}}{1}$, or $\sqrt{3}$

$\cot R = \dfrac{\text{adj.}}{\text{opp.}} \Rightarrow \cot 60° = \dfrac{1}{\sqrt{3}}$, or $\dfrac{\sqrt{3}}{3}$

Thus, $\sin 60° = \dfrac{\sqrt{3}}{2}$, $\cos 60° = \dfrac{1}{2}$, $\tan 60° = \sqrt{3}$, and $\cot 60° = \dfrac{\sqrt{3}}{3}$.

From Examples 1 and 2, notice the following cofunction relationship discussed on page 478.

sin 30° = cos 60°   tan 30° = cot 60°
sin 60° = cos 30°   tan 60° = cot 30°

Example 3 Find u ($0° < u < 90°$) if cos 20° = sin u, and find u if tan 50° = cot u.

$\cos A = \sin (90° - A)$
$\cos 20° = \sin (90° - 20°)$
$ = \sin 70°$

Thus, $u = 70°$.

$\tan A = \cot (90° - A)$
$\tan 50° = \cot (90° - 50°)$
$ = \cot 40°$

Thus, $u = 40°$.

The next example illustrates how to simplify trigonometric expressions.

Example 4 Evaluate and simplify $\dfrac{\cos 30° + \tan 45°}{\sin 30°}$.

Evaluate each trigonometric function, substitute, and simplify.

$\dfrac{\cos 30° + \tan 45°}{\sin 30°} = \dfrac{\frac{\sqrt{3}}{2} + 1}{\frac{1}{2}}$ ◀ $\cos 30° = \dfrac{\sqrt{3}}{2}$, $\tan 45° = 1$

$\sin 30° = \dfrac{1}{2}$

$= \dfrac{\frac{\sqrt{3}}{2} + 1}{\frac{1}{2}} \cdot \dfrac{2}{2}$, or $\dfrac{\sqrt{3} + 2}{1}$

Thus, the value of the expression $\dfrac{\cos 30° + \tan 45°}{\sin 30°}$ is $\sqrt{3} + 2$.

Trigonometry of a Right Triangle

Oral Exercises

For each triangle, state the measure of the side adjacent to and the side opposite the indicated angle.

1. $\angle A$ 2. $\angle B$ 3. $\angle H$ 4. $\angle I$ 5. $\angle D$ 6. $\angle E$

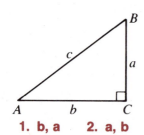

1. b, a   2. a, b   3. i, h   4. h, i   5. e, d   6. d, e

Written Exercises

For Ex. 1–6, see TE Answer Section.

(A) Find the sin, cos, tan, and cot of the measure of the indicated angle.

1. $\angle G$ 2. $\angle H$ 3. $\angle A$ 4. $\angle B$ 5. $\angle D$ 6. $\angle E$

  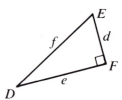

Find u, $0° < u < 90°$.

7. $\sin 34° = \cos u$ **56°**
8. $\cos 49° = \sin u$ **41°**
9. $\tan 73° = \cot u$ **17°**
10. $\cot 39° = \tan u$ **51°**
11. $\cot u = \tan 16°$ **74°**
12. $\tan u = \cot 84°$ **6°**
13. $\sin u = \cos 68°$ **22°**
14. $\cos u = \sin 12°$ **78°**

(B)
15. $\tan 12° 13' = \cot u$ **77° 47'**
16. $\sin 63° 18' = \cos u$ **26° 42'**
17. $\cos u = \sin 74° 36'$ **15° 24'**
18. $\sin 46° 45' = \cos u$ **43° 15'**
19. $\tan u = \cot 74° 18'$ **15° 42'**
20. $\cot u = \tan 33° 19'$ **56° 41'**

Evaluate and simplify. Rationalize all denominators.

21. $\dfrac{\tan 30° + \cos 60°}{\sin 30°}$ $\dfrac{2\sqrt{3} + 3}{3}$

22. $\dfrac{\sin 45° - \cos 30°}{\tan 60°}$ $\dfrac{\sqrt{6} - 3}{6}$

23. $\dfrac{\cos 45° - \sin 45°}{\tan 45°}$ **0**

24. $\dfrac{\cot 60° - \tan 45°}{\sin 60°}$ $\dfrac{2 - 2\sqrt{3}}{3}$

25. $\dfrac{\tan 60° - \tan 30°}{\tan 45°}$ $\dfrac{2\sqrt{3}}{3}$

26. $\dfrac{\sin 60° - \cot 30°}{\sin 30° + \cos 60°}$ $\dfrac{-\sqrt{3}}{2}$

(C)
27. $\sin 90° \sec 60° + \dfrac{\cot 45°}{\sin 210°} - \csc 270°$ **1**

28. $\dfrac{\cos 270°}{\sin 90°} - \dfrac{\sin 180°}{\cos 360°} - \dfrac{\csc 225°}{\sec 315°}$ **1**

29. $\dfrac{3 \cos 60° - \tan 135° + \dfrac{4 \sec 240°}{\sin 60°}}{\dfrac{1}{2} \csc 330° - \dfrac{3}{4} \cot 315°}$ $\dfrac{30 - 64\sqrt{3}}{-3}$

30. $\dfrac{\csc 30°}{\tan 135°} - \dfrac{\sec 300°}{\cot 300°} + \dfrac{3}{5} \csc 270°$ $\dfrac{-13}{5} + 2\sqrt{3}$

31. $\dfrac{\dfrac{1}{3} \sin(-60°) - \sqrt{3} \cos(-150°)}{\sec(-300°) + \csc(-315°)}$

31. $\dfrac{\sqrt{6} - 2\sqrt{3} - 9\sqrt{2} + 18}{12}$

32. $\dfrac{\sec 180° - \csc 90° + \dfrac{1}{2} \cos(-240°)}{-3 \sin(-330°)}$ $\dfrac{3}{2}$

USING TRIGONOMETRY OF A RIGHT TRIANGLE   17.12

Objectives **To find the length of a side of a right triangle given the length of one side and the measure of an acute angle**
To find the measure of an acute angle of a right triangle given the lengths of two sides
To apply right triangle trigonometry

Right triangle trigonometry is often used to find the length of one side or the measure of an acute angle of a right triangle. To do this, use a function which relates that which is given and that which is to be determined. Sometimes one function is more convenient to use than another.

Example 1 **In right triangle ABC, $\angle A$ measures $43°$, $c = 4.3$ m and m$\angle C = 90°$. Find a to the nearest meter.**

The *trigonometric* function that relates $\angle A$, a, and c is the sin function.

$$\sin A = \frac{\text{opp.}}{\text{hyp.}} \Rightarrow \sin 43° = \frac{a}{4.3}$$

Solve: $a = 4.3 \,(\sin 43°)$
$\doteq 4.3\,(.6820)$, or 2.9326

Thus, $a = 3m$ to the nearest meter.

Example 2 **Find to the nearest degree the measure of each acute angle in right triangle ABC, if $b = 5$ and $c = 13$. Angle C is a right angle.**

Draw and label the diagram.
Cos A relates $\angle A$, b, and c.

$$\cos A = \frac{\text{adj.}}{\text{hyp.}} \Rightarrow \cos A = \frac{5}{13}, \text{ or } .3846$$
$$A \doteq 67°$$

$\angle A$ and $\angle B$ are complementary.
$A + B = 90°$
$67° + B = 90$
$B \doteq 23°$

Thus, $\angle A$ measures $67°$ and $\angle B$ measures $23°$.

Using Trigonometry of a Right Triangle

In Examples 3 and 4 below, the angle of elevation or angle of depression is the angle between the horizontal and the line of sight. The angle of elevation equals the angle of depression.

Example 3 **What is the length of a shadow cast by a 10-meter pole when the angle of elevation of the sun is 54°?**

Draw and label the diagram.
The tan A or cot A relate $\angle A$, a, and b.

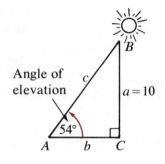

$$\tan A = \frac{a}{b}, \quad \text{or} \quad \cot A = \frac{b}{a}$$
$$\tan 54° = \frac{10}{b} \quad \quad \cot 54° = \frac{b}{10}$$
$$b = \frac{10}{\tan 54°} \quad \quad b = 10(\cot 54°)$$
$$\doteq \frac{10}{1.3764} \quad \quad \doteq 10(.7265)$$
$$\doteq 7.265 \quad \quad \doteq 7.265$$

Thus, the length of the shadow is approximately 7 m.

Example 4 **A pilot flying at an altitude of 1200 m observes that the angle of depression of an airport is 43°. Find to the nearest meter the distance to the airport from the point on the ground directly below the plane.**

Draw and label the diagram.
The tan A or cot A relate $\angle A$, a, and b.
Use cot A since the unknown will be in the numerator.

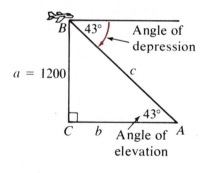

$$\cot A = \frac{b}{a}$$
$$\cot 43° = \frac{b}{1200}$$
$$b = 1200(\cot 43°)$$
$$\doteq 1200(1.0724)$$
$$\doteq 1286.88$$

Thus, the distance from C to A is 1287 m, to the nearest meter.

Oral Exercises

For Ex. 1–8, see TE Answer Section.

In right triangle *ABC*, which trigonometric function(s) relates the given parts with the unknown parts?

1. $A = 35°$, $a = 8$, $c = ?$
2. $B = 16°$, $b = 12$, $a = ?$
3. $A = 43°$, $b = 15$, $a = ?$
4. $A = 65°$, $a = 9$, $b = ?$
5. $B = 73°$, $b = 12$, $c = ?$
6. $B = 53°$, $a = 15$, $c = ?$
7. $A = 81°$, $a = 13$, $b = ?$
8. $B = 47°$, $b = 30$, $c = ?$

Written Exercises

1. $b = 8, c = 11$ 2. $a = 11, b = 15$ 3. $a = 24, c = 29$ 4. $a = 13, b = 7$ 5. $b = 61, c = 62$

(A) Find the measure of each of the other two sides of right triangle ABC, with
$m \angle C = 90$, to the nearest whole number.   6. $b = 29, c = 48$ 7. $a = 30, c = 31$ 8. $a = 10, c = 39$
1. $m \angle B = 46°, a = 7.3$ 2. $m \angle A = 36°, c = 18.3$ 3. $m \angle B = 34°, b = 16.3$
4. $m \angle A = 63°, c = 15.1$ 5. $m \angle B = 78°, a = 12.9$ 6. $m \angle A = 53°, a = 38.7$
7. $m \angle B = 18°, b = 9.6$ 8. $m \angle A = 15°, b = 37.3$ 9. $m \angle B = 82°, c = 46.5$
                                                                                   $a = 6, b = 46$

Find the measure of each acute angle of right triangle ABC, with $m \angle C = 90$, to
the nearest degree.   10. $m \angle A = 55, m \angle B = 35$ 11. $m \angle A = 37, m \angle B = 53$ 12. $m \angle A = 44, m \angle B = 46$
10. $c = 8.6 \quad b = 4.9$ 11. $b = 12.7 \quad a = 9.7$ 12. $b = 14.6 \quad c = 20.3$
13. $a = 9.1 \quad c = 12.3$ 14. $c = 18.6 \quad a = 14.1$ 15. $a = 15.1 \quad b = 7.6$
13. $m \angle A = 48, m \angle B = 42$ 14. $m \angle A = 49, m \angle B = 41$ 15. $m \angle A = 63, m \angle B = 27$

16. A ladder leans against a building. The top of the ladder reaches a point on the building that is 9 m above the ground. The foot of the ladder is 4 m from the building. Find to the nearest degree the measure of the angle that the ladder makes with the level ground. **66**

17. From the top of a lighthouse 48 m high, the angle of depression of a boat at sea measures 23° 30'. Find, to the nearest m, the distance from the boat to the foot of the lighthouse. **110 m**

18. From an airplane 925 m above sea level, the angle of depression of a ship measures 42° 20'. Find to the nearest meter the distance to the ship from the point at sea directly below the plane. **1,015 m**

19. At a point on the ground 8 m from the foot of a cliff, the angle of elevation of the top of the cliff measures 38° 40'. Find the height of the cliff to the nearest meter. **6 m**

(B) Find the measure of each of the other two sides to the nearest tenth and the measure of each acute angle of right triangle ABC, with $m \angle C = 90$, to the nearest minute. **For Ex. 20–28, see TE Answer Section.**
20. $A = 13° 23', a = 9.6$ 21. $B = 49° 18', c = 13.3$ 22. $A = 63° 38', b = 9.1$
23. $B = 74° 46', b = 12.3$ 24. $A = 11° 12', b = 7.8$ 25. $B = 16° 56', c = 11.8$
26. $A = 33° 11', c = 16.5$ 27. $B = 73° 33', a = 6.3$ 28. $A = 38° 37', a = 12.5$

(C) Determine BC to the nearest tenth.
29. $m \angle \theta = 60°, m \angle A = 45°, AD = 12$ **28.4**
30. $m \angle \theta = 45°, m \angle A = 30, AD = 10$ **13.6**
31. $m \angle \theta = 60°, m \angle A = 30°, AD = 16$ **13.9**
32. $m \angle \theta = 49° 18', m \angle A = 32° 16', AD = 14.3$ **19.7**
33. $m \angle \theta = 74° 48', m \angle A = 48° 35', AD = 28.6$ **46.8**

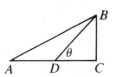

34. From point B on the ground, the angle of elevation of a kite measures 46° 18'. The kite is attached to a rope 86 m long. Find to the nearest meter the height of the kite. **62 m**

35. From an airplane flying 1250 m above sea level, the angles of depression of two ships due west measure 33° 10' and 20° 40'. Find the distance between the ships to the nearest meter. **1,401 m**

Using Trigonometry of a Right Triangle

CHAPTER SEVENTEEN REVIEW

(A) Exercises are 1–12, 17–28, 29–35, 39–42, 48–51, 56, 57, 60–62, 66.
(C) Exercises are starred. The rest are (B) Exercises.

Vocabulary
angle of rotation [17.1]   sec, csc [17.5]
cofunction [17.8, 17.11]   sin, cos [17.3]
quadrantral angle [17.3]   tan, cot [17.4]

Sketch angles of rotation with the given measures. Determine the quadrant of each nonquadrantle angle. Identify the reference angle and reference triangle. [17.1]

1. 45°   2. −150°   3. 225°   4. −240°
5. −330°   6. 450°   7. 780°   8. −420°
9. −100°   10. −270°   11. 400°   12. −720°

For Ex. 1–12, see TE Answer Section.

★ Find the reference angle if its terminal side intersects a circle at the following points and the center of the circle is at the origin. Determine the radius of the circle. [17.2]

13. $(2, 2)$ **45°, $2\sqrt{2}$**   14. $(-\sqrt{3}, 1)$ **30°, 2**
15. $(a, a\sqrt{3})$ **60°, $2a$**   16. $(a, -a)$ **45°, $a\sqrt{2}$**

Find each of the following. Rationalize any radical denominator. [17.3–17.6]

17. cos 60°   18. tan (−30°)   19. sec 150°
20. sin (−480°)   21. sin 0°   22. sec 90°
23. cot 360°   24. cos (−270°)

For Ex. 17–24, see TE Answer Section.

Write each of the following in terms of the same trigonometric function of an acute angle. [17.3, 17.6]

25. sin 130° **sin 50°**   26. cot 150° **−cot 30°**
27. cos (−160°)   28. tan (−220°)
27. −cos 20°   **28. −tan 40°**

In which quadrants is each of the following true?

29. sin < 0   30. cos > 0   31. tan < 0   32. sec > 0
29. III, IV   30. I, IV   31. II, IV   32. I, IV [17.3–17.4]

Find two values of u between 0° and 360° for which each is true. **33. 120°, 240°** [17.3]

33. $\cos u = -\dfrac{1}{2}$ 34. $\sin u = -\dfrac{\sqrt{2}}{2}$
                                 34. 225°, 315°
35. $\cot u = \dfrac{-\sqrt{3}}{3}$ **35. 120°, 300°**

For Ex. 36–38, see TE Answer Section.

★ Find the value of each of the other trigonometric functions of u. [17.7]

36. $\tan u = \dfrac{5}{12}$ and $\sin u > 0$
37. $\cos u = -\dfrac{3}{5}$ and $\cot u < 0$
38. $\sin u = \dfrac{7}{25}$ and $\sec u = -\dfrac{25}{24}$

39. **.6157**   40. **2.0965**   41. **−.7314**   [17.8]

Use the trig. table to find each of the following.

39. sin 38°   40. tan 64° 30′   41. cos 223°
42. cos 34° 17′ **.8263**   43. cot 115° 36′ **−.4792**
44. cos 215° 33′ **−.8136**   45. sin (−196° 54′) **.2907**
46. tan (−338° 28′)   47. cot (−134° 42′)
    .3946   **.9896**

Find u, $u < 90°$. [17.8, 17.10, 17.11]

48. cos u = .7790   49. sin u = .9717
50. sin 73° = cos u **17°**   51. cot u = tan 22° **68°**
48. 38° 50′   49. 76° 20′

Find all values of θ, $0° < \theta < 360°$. [17.10]

52. tan θ = −2.8258   53. cot θ = 6.9600
54. cos θ = −.7796   55. sin θ = .4562
52. 109° 29′, 289° 29′   54. 141° 13′, 218° 47′

Express in terms of the cofunction. [17.8]

56. sin 48° 25′   57. tan 64° 42′ **cot 25° 18′**
53. 8° 11′, 188° 11′   55. 27° 9′, 152° 51′   56. cos 41° 35′

Express as a function of an angle measure less than 45°. [17.8]

58. cos 320° **cos 40°**   59. cot 105° 30′
                                −tan 15° 30′

Determine the values of θ between 0° and 360° that make each equation true. [17.9]

60. $\sin \theta = -\dfrac{1}{2}$ 61. $2 \cos \theta - 1 = 5 \cos \theta + 1$
62. $5 \tan \theta - 8 = -3 \tan \theta - 5$ **60. 210°, 330°**
61. 131° 49′, 228° 11′   62. 20° 33′, 200° 33′

Evaluate and simplify. [17.11]

63. $\dfrac{\cos 45° - 3 \sin 60°}{\sec 120°}$ $\dfrac{\sqrt{2} - 3\sqrt{3}}{-4}$

64. $\dfrac{\sin 45° + \cot 45°}{\cos (-120°)}$ $-\sqrt{2} - 2$

★ 65. $\dfrac{\csc 90° + 3 \tan (-135°) - \sin 60°}{4 \cos (-315°)}$ $\dfrac{8\sqrt{2} - \sqrt{6}}{8}$

Find the measure of each of the other two sides of right triangle ABC, with m$\angle C$ = 90°, to the nearest whole number. [17.12]

66. m$\angle A$ = 46°, b = 9.2 **a = 10, c = 13**
67. m$\angle B$ = 12° 18′, a = 12.3 **b = 3, c = 13**

CHAPTER SEVENTEEN TEST

Ⓐ Exercises are 1–17, 22, 23, 26, 27, 30, 32, 35, 36, 39, 40, 42, 44. Ⓒ Exercises are starred. The rest are Ⓑ Exercises.

Sketch angles of rotation with the given measures. Determine the quadrant of each angle of rotation. Identify the reference angle and reference triangle.
1. 110°   2. −130°   3. 250°   4. −340°
For Ex. 1–4, see TE Answer Section.

Determine the lengths of the indicated sides. Rationalize all denominators.

5. $x = 6$, $y = 3\sqrt{3}$

6. $a = 8\sqrt{2}$, $b = 8\sqrt{2}$

Determine the point (x, y) and the radius r.

7. $(5, 5\sqrt{3})$, $r = 10$
8. $(-5, 5)$, $r = 5\sqrt{2}$

Find each of the following. Rationalize any radical denominator.
9. $\cos(-60°)$ $\frac{1}{2}$ 10. $\cos 180°$ 11. $\sin 420°$ $\frac{\sqrt{3}}{2}$
12. $\tan 135°$ -1 13. $\sec 180°$ 14. $\csc 420°$
10. -1 13. -1 14. $\frac{2\sqrt{3}}{3}$

Write each of the following in terms of the same trigonometric function of an acute angle.
15. $\tan 340°$ 16. $\cos 100°$ 17. $\sin 310°$
15. $-\tan 20°$ 16. $-\cos 80°$ 17. $-\sin 50°$

In which quadrants is each of the following true?
18. $\sin > 0$ 19. $\cos < 0$ 20. $\tan < 0$
21. $\cos \theta < 0$ and $\sin \theta > 0$ II
18. I and II   19. II and III   20. II and IV

Find two values of u between 0° and 360° for which each is true.
22. $\cos u = -\frac{\sqrt{3}}{2}$ 150°, 210°
23. $\sin u = \frac{1}{2}$ 30°, 150°

★ Find the values of each of the other five trigonometric functions of u.
24. $\sin u = -\frac{5}{13}$ and $\tan u < 0$
25. $\cos u = -\frac{3}{5}$ and $\sin u > 0$
For Ex. 24–25, see TE Answer Section.

Use the trig. table to find each of the following.
26. $\cos 37°$ .7986   27. $\sin 215° 40'$ −.5831
28. $\tan(-139° 46')$ .8461   29. $\cos 326° 10'$ .8307

Find u, $u < 90°$.
30. $\sin u = .5664$ 31. $\cos u = .2464$
32. $\sin 65° = \cos u$ 30. 34° 30'   31. 75° 44'
32. 25°

Find all values of θ, $0° < \theta < 360°$.
33. $\tan \theta = -.6778$ 34. $\cos \theta = -.4083$
33. 145° 52', 325° 52'   34. 114° 6', 245° 54'

Express in terms of the cofunction.
35. $\sin 34°$ cos 56°   36. $\tan 75° 26'$ cot 14° 34'

Express as a function of an angle measure less than 45°.
37. $\sin 110°$ cos 20°   38. $\cos 305° 10'$ sin 35° 10'

Determine the values of θ between 0° and 360° that make each equation true.
39. $3 \cos \theta - 1 = -2 \cos \theta$ 78° 28', 281° 32'
40. $-5 \sin \theta - 3 = 2 \sin \theta$ 205° 23', 334° 37'

Evaluate and simplify.
41. $\dfrac{\cos 60° - \sin 60°}{\cot 45°}$ $\dfrac{1 - \sqrt{3}}{2}$

Find the measure of each of the other two sides of right triangle ABC, with m$\angle C = 90$, to the nearest whole number.
42. m$\angle A = 47°$, $b = 4.7$ $a = 5$, $c = 7$
43. m$\angle B = 74° 25'$, $a = 10.3$ $b = 37$, $c = 38$

44. At a point on the ground 23 m from the foot of a building, the angle of elevation of the top of the building is 48° 37'. Find to the nearest meter the height of the building. 26 m

Chapter Seventeen Test

COMPUTER ACTIVITIES

Right Triangles

OBJECTIVE: To find the missing parts of a right triangle given two sides, or one side and an acute angle

The program below is a brief example of a menu-driven program. For each combination of known parts, the ON...GOTO statement in line 170 will branch to the appropriate portion of the program.

```
110   PRINT "ENTER THE NUMBER WHICH INDICATES WHAT YOU KNOW:"
120   PRINT    TAB( 4)"1) TWO LEGS"
130   PRINT    TAB( 4)"2) A LEG AND THE HYPOTENUSE"
140   PRINT    TAB( 4)"3) A LEG AND AN ADJACENT ANGLE"
150   PRINT    TAB( 4)"4) A LEG AND AN OPPOSITE ANGLE"
160   PRINT    TAB( 4)"5) HYPOTENUSE AND AN ACUTE ANGLE": PRINT
170   INPUT N: ON N GOTO 180,220,260,300,340
180   INPUT "ENTER THE TWO LEGS. ";A,B
190 C =  SQR (A ^ 2 + B ^ 2): PRINT "THE HYPOTENUSE = "C
200 X =  ATN (A / B) * 180 / 3.1416
210 Y =  INT (X):R =  INT ((X - Y) * 60): GOSUB 390
220   INPUT "ENTER HYPOTENUSE, THEN LEG. ";C,A
230 B =  SQR (C ^ 2 - A ^ 2): PRINT "THE OTHER LEG = ";B
240 X =  ATN (B / A) * 180 / 3.1416
250 Y =  INT (X):R =  INT ((X - Y) * 60): GOSUB 390
260   INPUT "ENTER THE LEG, THEN THE ADJACENT ANGLE ";A,B
270 B = B * 3.1416 / 180: PRINT "THE OTHER ANGLE IS "90 - B"
      DEGREES."
280 C = A /  COS (B): PRINT "THE HYPOTENUSE = "C"."
290   PRINT "OTHER LEG IS " SQR (C ^ 2 - A ^ 2)".": GOTO 380
300   INPUT "ENTER LEG, THEN OPPOSITE ANGLE. ";A,X
310   PRINT "OTHER ANGLE IS "90 - X" DEGREES.":X = X * 3.1416
      / 180
320 C = A /  SIN (X): PRINT "THE HYPOTENUSE = "C"."
330   PRINT "OTHER LEG IS " SQR (C ^ 2 - A ^ 2)".": GOTO 380
340   INPUT "ENTER HYPOTENUSE,THEN ANGLE. ";C,X
350   PRINT "THE OTHER ANGLE IS "90 - X" DEGREES. ":X = X * 3
      .1416 / 180
360 D = C *  SIN (X): PRINT "ONE LEG IS "D"."
370   PRINT "THE OTHER LEG IS " SQR (C ^ 2 - D ^ 2)"."
380   PRINT : PRINT : GOTO 110
390   PRINT "THE TWO ANGLES ARE ";Y;" DEGREES ";R;" MINUTES"
400   PRINT    TAB( 16);"AND ";89 - Y;" DEGREES ";60 - R;"
      MINUTES"
410   PRINT : PRINT : GOTO 110: RETURN
```

See the Computer Section beginning on page 574 for more information.

EXERCISES

Delete :GOSUB 390 in lines 210 and 250 and : RETURN in line 410. Renumber lines 390–410 as 212, 215, 218 and also as 252, 255, 258 and insert between lines 210–220 and lines 250–260.

1. Lines 390–410 constitute a subroutine. Rewrite the program without using a subroutine, but with the same output.

COLLEGE PREP TEST

Directions: Choose the one best answer to each question or problem.

1. The lengths of the three sides of a triangle are 5, 12, and 13. A circle passes through the vertices of the triangle. Which is the length of the radius of the circle? **A**
   (A) $\frac{13}{2}$ (B) 4 (C) 5
   (D) 6 (E) None of these

2.

   $\overline{OW}$ is a radius of the circle with center O. If $OW = 12$, which is the value of WZ? **C**
   (A) $6\sqrt{2}$ (B) 4 (C) 6
   (D) $6\sqrt{3}$ (E) None of these

3.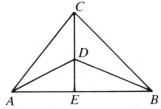

   ABC is a triangle with altitude $\overline{CE}$. $CD = 6$ and $AB = 18$. Which is the area of $ADBC$? **B**
   (A) 108 (B) 54 (C) 90
   (D) $\frac{189}{2}$ (E) None of these

4. The average of five numbers is 5.1. Four of the numbers are 4.2, 5.7, 4.9 and 5.4. Which is the fifth number? **D**
   (A) 5.06 (B) 5.14 (C) 5.1
   (D) 5.3 (E) None of these

5. A wheel rotates m times each minute. How many degrees does it rotate in t seconds? **B**
   (A) $\frac{360\,m}{t}$ (B) $6\,mt$ (C) $\frac{60\,t}{m}$
   (D) $\frac{60\,m}{t}$ (E) $\frac{360\,t}{m}$

Directions: In each question you are to *compare* a Quantity I and a Quantity II. Choose your answer in the following way. Choose:
(A) if Quantity I is greater than II,
(B) if Quantity II is greater than I,
(C) if Quantity I = Quantity II,
(D) if the relationship cannot be determined.

6.

   Quantity I $x° + y°$
   Quantity II $m° + n°$ **C**

7.

   $\overline{AB}$ is parallel to $\overline{CD}$ **C**

   Quantity I m Quantity II n

8. Quantity I   area of circle with diameter length $\frac{2\sqrt{\pi}}{\pi}$
   Quantity II area of a square whose side is 1.   **C**

College Prep Test

18 TRIGONOMETRIC GRAPHS AND IDENTITIES

CAREERS
Biological Sciences

A biologist may choose a career from among many different specialties. Two examples are entomology and microbiology.

An **entomologist** is concerned with reducing the harmful species of insects that destroy crops, buildings, and clothing or cause disease in humans or wildlife. Efforts to increase and spread the many insects that are beneficial to the "balance of nature" are also important.

A **microbiologist** studies microorganisms such as bacteria and viruses. A career in microbiology may lead to important contributions in public health, industry, medicine, or agriculture.

Project

Write a paper on the steps that an entomologist takes to protect humans and wildlife from disease.

GRAPHING y = sin x AND y = cos x   18.1

Objectives
To graph y = sin x and y = cos x
To determine the amplitude and the period of the sine and the cosine functions
To determine the quadrants in which the sin and cos functions increase or decrease

To graph the function $y = \sin x$, make a table of values by choosing special degree measures for x and finding corresponding values for y. The graph of $y = \sin x$ for $-360° \le x \le 360°$ is shown below.

x	−360°	−270°	−180°	−90°	0°	90°	180°	270°	360°
$\sin x$	0	1	0	−1	0	1	0	−1	0

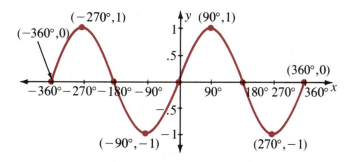

Note that the maximum value of the function is 1 and that this occurs when $x = 90°$ and when $x = -270°$. The *amplitude* of a trig function is its maximum value. The amplitude of the sine function is 1.
A *periodic function* is one that repeats itself. Note that the sine function is a periodic function with a period of 360°.

Example 1 Graph $y = \cos x$ for $-360° \le x \le 360°$ and determine its amplitude and period.

x	cos x		x	cos x
0°	1		0°	1
30°	$\frac{\sqrt{3}}{2}$ or .87		−30°	.87
90°	0		−90°	0
120°	$-\frac{1}{2}$ or −.5		−120°	−.5
180°	−1		−180°	−1
225°	$-\frac{\sqrt{2}}{2}$ or −.71		−225°	−.71
270°	0		−270°	0
330°	.87		−330°	.87
360°	1		−360°	1

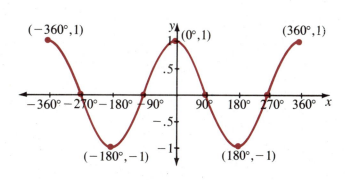

Thus, the amplitude of the cosine function is 1 and the period is 360°.

By looking at the graphs of $y = \sin x$ and $y = \cos x$, you can find where the sine increases or decreases and where the cosine increases or decreases.

1st quadrant	2nd quadrant	3rd quadrant	4th quadrant
$0° < x < 90°$	$90° < x < 180°$	$180° < x < 270°$	$270° < x < 360°$
sin increases	sin decreases	sin decreases	sin increases
cos decreases	cos decreases	cos increases	cos increases

You have learned that the values of a trig function can be found in a table. The values can also be estimated from a graph. The next example illustrates this.

Example 2   Graph $y = \cos x$ for $0° \leq x \leq 90°$. Estimate $\cos 45°$ to the nearest tenth.

The next example illustrates how to solve a system of two trigonometric equations graphically.

Example 3   Graph $y = \sin x$ and $y = \cos x$ for $-180° \leq x \leq 180°$ on the same set of axes. For which values of x does $\sin x = \cos x$?

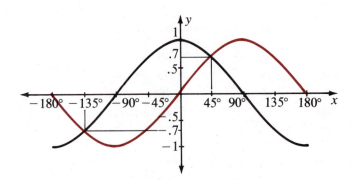

x	$y = \sin x$	$y = \cos x$
$-30°$	$-.5$	$.9$
$-90°$	-1	0
$-120°$	$-.9$	$-.5$
$-180°$	0	-1
$0°$	0	1
$30°$	$.5$	$.9$
$90°$	1	0
$120°$	$.9$	$-.5$
$180°$	0	-1

Thus, for $x = -135°$ or $x = 45°$, $\sin x = \cos x$.

Written Exercises

A) Complete the chart.

	$y = \sin x$	$y = \cos x$
1. Amplitude	1	1
2. Period	360°	360°
3. Maximum value	1	1
4. Minimum value	−1	−1
5. Quadrants where function increases	I & IV	III & IV
6. Quadrants where function decreases	II & III	I & II

Sketch $y = \sin x$ for $0° \le x \le 180°$. Estimate each of the following to the nearest tenth.

7. sin 20° **0.3**   8. sin 65° **0.9**   9. sin 75° **0.9**   10. sin 140° **0.6**   11. sin 160° **0.3**

Sketch $y = \cos x$ for $0° \le x \le 180°$. Estimate each of the following to the nearest tenth.

12. cos 40° **0.8**   13. cos 70° **0.3**   14. cos 100° **−0.2**   15. cos 130° **−0.6**   16. cos 170° **−0.9**

B)
17. Sketch $y = \sin x$ and $y = \cos x$ for $0° \le x \le 360°$ on the same set of axes. For which values of x is $\sin x = \cos x$? **45°, 225°**

18. Sketch $y = \sin x$ and $y = \cos x$ for $-360° \le x \le 180°$ on the same set of axes. For which values of x is $\sin x = \cos x$? **−135°, −315°, 45°**

19. Sketch $y = \sin x$ and $y = \cos x$ for $-180° \le x \le 360°$ on the same set of axes. For which values of x is $\sin x = \cos x$? **−135°, 45°, 225°**

20. Sketch $y = \sin x$ and $y = \cos x$ for $-360° \le x \le 0°$ on the same set of axes. For which values of x is $\sin x = \cos x$? **−315°, −135°**

C) Sketch the graph for $0° \le x \le 360°$. For Ex. 21–28, see TE Answer Section.

21. $y = -\cos x$ 22. $y = \sin(-x)$ 23. $y = -\sin x$ 24. $y = \cos(-x)$
25. $y = \sec x$ 26. $y = -\sec x$ 27. $y = \csc x$ 28. $y = \csc(-x)$

29. Sketch $y = \sin(-x)$ and $y = \cos(-x)$ for $-180° \le x \le 180°$ on the same set of axes. For which values of x is $\sin(-x) = \cos(-x)$? **−45°, 135°**

30. Sketch $y = -\sin x$ and $y = -\cos x$ for $-180° \le x \le 180°$ on the same set of axes. For which values of x is $-\sin x = -\cos x$? **−135°, 45°**

NON-ROUTINE PROBLEMS

In right triangle PRQ, the hypotenuse $\overline{QR}$ measures 20 cm and leg $\overline{PQ}$ measures 10 cm. Points S and T are on $\overline{QR}$ such that $\overline{SP}$ and $\overline{TP}$ trisect angle P. Find the length of $\overline{SP}$. **$5\sqrt{3}$ cm**

CUMULATIVE REVIEW

Find the distance between the given points. Answers should be in simplest radical form.

1. $A(3, -2)$, $B(-6, 1)$ **$3\sqrt{10}$**
2. $A(-7, 8)$, $B(7, -3)$ **$\sqrt{317}$**
3. $A(12, -8)$, $B(4, -6)$ **$2\sqrt{17}$**

DETERMINING AMPLITUDE   18.2

Objective   **To determine the amplitude of the sine and cosine functions**

You have learned that the amplitude for $y = \sin x$ and $y = \cos x$ is their maximum value, 1. You will now learn how multiplying $\sin x$ or $\cos x$ by a constant changes its amplitude.

Example 1   **Sketch $y = 2 \sin x$ and $y = \sin x$ for $-180° \leq x \leq 360°$ on the same set of axes. Determine the period and amplitude for each function.**

Make a table of values. Multiply $\sin x$ values by 2.

x	$y = \sin x$	$y = 2 \sin x$
0°	0	0
90°	1	2
180°	0	0
270°	−1	−2
360°	0	0
−90°	−1	−2
−180°	0	0

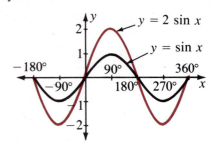

Thus, amplitude for $y = \sin x$ is 1.   Period for $y = \sin x$ is 360°.
Amplitude for $y = 2 \sin x$ is 2.   Period for $y = 2 \sin x$ is 360°.

Example 2   **Sketch $y = \frac{1}{2} \cos x$ and $y = \cos x$ for $-180° \leq x \leq 360°$ on the same set of axes. Determine the period and amplitude for each function.**

Make a table of values. Multiply $\cos x$ values by $\frac{1}{2}$.

x	$y = \cos x$	$y = \frac{1}{2} \cos x$
0°	1	$\frac{1}{2}$
90°	0	0
180°	−1	$-\frac{1}{2}$
270°	0	0
360°	1	$\frac{1}{2}$
−90°	0	0
−180°	−1	$-\frac{1}{2}$

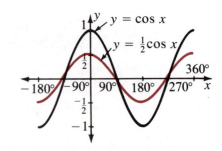

Thus, amplitude for $y = \cos x$ is 1.   Period for $y = \cos x$ is 360°.
Amplitude for $y = \frac{1}{2} \cos x$ is $\frac{1}{2}$.   Period for $y = \frac{1}{2} \cos x$ is 360°.

Example 3 Sketch $y = 3 \cos x$ and $y = \frac{1}{2} \cos x$ for $-180° \leq x \leq 360°$ on the same set of axes. Determine the period and amplitude for each function.

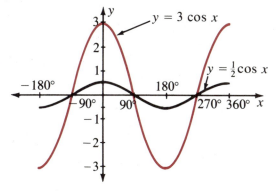

x	$y = \cos x$	$y = 3 \cos x$	$y = \frac{1}{2} \cos x$
0°	1	3	$\frac{1}{2}$
90°	0	0	0
180°	−1	−3	$-\frac{1}{2}$
270°	0	0	0
360°	1	3	$\frac{1}{2}$
−90°	0	0	0
−180°	−1	−3	$-\frac{1}{2}$

Thus, amplitude for $y = 3 \cos x$ is 3. Period for $y = 3 \cos x$ is 360°.
Amplitude for $y = \frac{1}{2} \cos x$ is $\frac{1}{2}$. Period for $y = \frac{1}{2} \cos x$ is 360°.

Amplitude and period for $y = a \sin x$ and $y = a \cos x$

If $y = a \sin x$, then the amplitude is $|a|$ and the period is 360°, for each number a.

If $y = a \cos x$, then the amplitude is $|a|$ and the period is 360°, for each number a.

Example 4 Sketch $y = -2 \sin x$ and $y = \frac{1}{2} \sin x$ for $-180° \leq x \leq 360°$ on the same set of axes. Determine the period and amplitude for each.

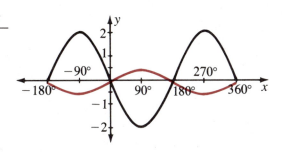

x	$y = \sin x$	$y = -2 \sin x$	$y = \frac{1}{2} \sin x$
0°	0	0	0
90°	1	−2	$\frac{1}{2}$
180°	0	0	0
270°	−1	2	$-\frac{1}{2}$
360°	0	0	0
−90°	−1	2	$-\frac{1}{2}$
−180°	0	0	0

Amplitude of $y = -2 \sin x$ is 2. Period of $y = -2 \sin x$ is 360°.
Amplitude of $y = \frac{1}{2} \sin x$ is $\frac{1}{2}$. Period of $y = \frac{1}{2} \sin x$ is 360°.

Example 5 Determine the amplitude of each trig function.

$y = 12 \sin x$ → Amplitude: 12

$y = -6 \cos x$ → Amplitude: 6

$y = -\frac{7}{2} \cos x$ → Amplitude: $\frac{7}{2}$

Oral Exercises

Determine the amplitude of each trig function.

1. $y = 5 \cos x$ **5**
2. $y = -2 \sin x$ **2**
3. $y = \frac{1}{3} \sin x$ **$\frac{1}{3}$**
4. $y = \frac{1}{5} \cos x$ **$\frac{1}{5}$**
5. $y = 15 \sin x$ **15**
6. $y = 12 \cos x$ **12**
7. $y = -\frac{4}{5} \cos x$ **$\frac{4}{5}$**
8. $y = -\frac{2}{3} \sin x$ **$\frac{2}{3}$**

Written Exercises

Ⓐ Sketch the graph for $0° \leq x \leq 360°$. Determine the period and amplitude.

1. $y = 3 \cos x$ **360°, 3**
2. $y = 2 \sin x$ **360°, 2**
3. $y = \frac{1}{2} \sin x$ **360°, $\frac{1}{2}$**
4. $y = \frac{1}{3} \sin x$ **360°, $\frac{1}{3}$**
5. $y = 4 \sin x$ **360°, 4**
6. $y = 5 \cos x$ **360°, 5**
7. $y = \frac{3}{4} \sin x$ **360°, $\frac{3}{4}$**
8. $y = \frac{4}{5} \cos x$ **360°, $\frac{4}{5}$**

9. Sketch $y = 2 \cos x$ and $y = \frac{1}{2} \cos x$ for $-360° \leq x \leq 360°$ on the same set of axes. Determine the period and the amplitude for each. **360°, 2; 360°, $\frac{1}{2}$**

10. Sketch $y = 3 \sin x$ and $y = \frac{1}{2} \sin x$ for $-360° \leq x \leq 360°$ on the same set of axes. Determine the period and the amplitude for each. **360°, 3; 360°, $\frac{1}{2}$**

Ⓑ Sketch the graph for $-360° \leq x \leq 360°$. Determine the period and amplitude.

11. $y = -\frac{1}{2} \cos x$ **360°, $\frac{1}{2}$**
12. $y = -2 \sin x$ **360°, 2**
13. $y = -\frac{1}{3} \sin x$ **360°, $\frac{1}{3}$**
14. $y = -3 \cos x$ **360°, 3**
15. $y = -3 \sin x$ **360°, 3**
16. $y = -4 \cos x$ **360°, 4**
17. $y = -\frac{2}{5} \cos x$ **360°, $\frac{2}{5}$**
18. $y = -\frac{1}{2} \sin x$ **360°, $\frac{1}{2}$**

19. Sketch $y = -\frac{1}{2} \sin x$ and $y = 2 \cos(-x)$ for $0° \leq x \leq 360°$ on the same set of axes. For how many values of x does $-\frac{1}{2} \sin x = 2 \cos(-x)$? **2**

20. Sketch $y = -2 \cos x$ and $y = 2 \sin x$ for $0° \leq x \leq 360°$ on the same set of axes. For how many values of x does $-2 \cos x = 2 \sin x$? **2**

Ⓒ Sketch the graph for $-360° \leq x \leq 360°$. Determine the period and amplitude.

21. $y = 2 \csc x$ **360°, none**
22. $y = -\frac{1}{2} \csc x$ **360°, none**
23. $y = \frac{1}{2} \sec x$ **360°, none**
24. $y = -2 \sec x$ **360°, none**

25. Sketch $y = -2 \sin x$ and $y = -\frac{1}{2} \sin x$ for $0° \leq x \leq 360°$ on the same set of axes. Estimate the values of x for which $-2 \sin x = -\frac{1}{2} \sin x$. **0°, 180°, 360°**

26. Sketch $y = -\frac{1}{2} \cos x$ and $y = -2 \cos x$ for $0° \leq x \leq 360°$ on the same set of axes. Estimate the values of x for which $-\frac{1}{2} \cos x = -2 \cos x$. **90°, 270°**

27. Sketch $y = -3 \cos x$ and $y = -2 \sin x$ for $-360° \leq x \leq 360°$ on the same set of axes. Estimate the values of x for which $-3 \cos x = -2 \sin x$. **56°, 236°, −124°, −304°**

28. Sketch $y = -2 \cos x$ and $y = \frac{1}{2} \sin x$ for $-360° \leq x \leq 360°$ on the same set of axes. Estimate the values of x for which $-2 \cos x = \frac{1}{2} \sin x$. **104°, 284°, −76°, −256°**

29. Sketch $y = \frac{1}{2} \sec x$ and $y = 2 \cos x$ for $0° \leq x \leq 360°$ on the same set of axes. Estimate the values of x for which $\frac{1}{2} \sec x = 2 \cos x$. **60°, 120°, 240°, 300°**

30. Sketch $y = \frac{1}{3} \csc x$ and $y = 2 \cos x$ for $0° \leq x \leq 360°$ on the same set of axes. Estimate the values of x for which $\frac{1}{3} \csc x = 2 \cos x$. **10°, 190°, 57°, 237°**

UNDERSEA AND UNDERGROUND EXPLORATION

Exploration below the surface of our earth, whether undersea or underground, has progressed with the invention of more and more sophisticated equipment to probe and analyze our world.

Accoustic Array, Deepstar 4000. The 18-foot-long Deepstar 4000, designed to investigate the sea bottom up to 4,000 feet below the surface, relies heavily on accoustic sensors.

Alexander the Great in glass barrel at the bottom of the sea. The first devices had no independent air supply, thus trapped air became foul and sharply limited diving time.

Lowering diver in "JIM," deep-sea diving suit designed by Sylvia Earle. Jacques Cousteau's invention of the "aqua-lung" rebreathing apparatus with a pressure regulator was the precursor to such advanced free-diving equipment as "JIM."

Cloche de Halley, diagram of old diving bell. Operating as an inverted tumbler, this early diving apparatus was supplied with air from the surface.

CNEXO, French deep submersible submarine. Small deep-sea submersibles give oceanographers a valuable vantage point for extended observations of marine life and conditions on the continental shelf.

Illustration of diving bell, Halley's Spalding. The need to move with more flexibility, without the danger of fatal capsizing, led to the design of a less unwielding diving bell.

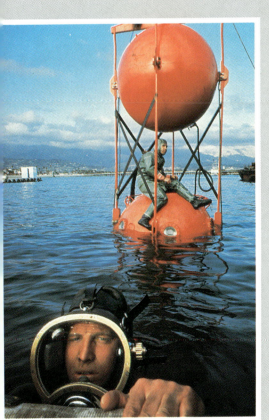

Frogmen and underwater observation chamber. During World War II frogmen were responsible for clandestine observation as well as demolition and surveying.

A map of islands and ridges plotted from deep sea soundings, Donnelly's Atlantis. In the 1860's, when the first trans-Atlantic cables were being laid, there was speculation that a high central plateau, discovered near the Azores, was the lost continent.

Earth cross-section with central core of fire. External geological disturbances such as volcanoes and geysers led early geologists to postulate that the earth's core was a ball of fire.

Landsat imagery assessment on the Gulf Coast. Satellites in orbit can scan an entire ocean basin, and when color patterns are analyzed, they provide data on marine life, currents, and pollutant levels.

Excavating the Gaillard (Culebra) cut for the Panama Canal. Five-hundred feet wide, this 8-mile-long cut was part of the 5-year effort to build the 51-mile canal, which shortened coast-to-coast voyages by 8,000 miles.

Coal pit worked by women. Early mining wells were not very wide because of the danger of collapse. This led to the use of women and children, who were smaller than men, as miners in the pits.

James Watt's steam-pumping engine at an English coal mine in the 1870's. His successive modifications such as the rotative shaft and double-action pistons made Watt's steam engine a vital implement in the progress of underground mining.

Project Famous, Interior of Alvin, July 1974. First launched in 1965, submarine Alvin is laden with instrumentation and remote controls, and is highly maneuverable at depths up to 1 mile below the surface of the sea.

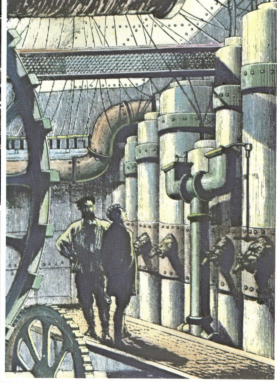

Illustration from *20,000 Leagues Under the Sea*. In 1870, when *20,000 Leagues Under the Sea* was published, Jules Verne had anticipated scientific inventions that would appear only after the latter half of the 20th century.

Putting a current meter into water. To examine underwater currents, instruments must be lowered to specific depths, and the information correlated to surface conditions such as sunlight and temperature.

Scientist estimating fish population. The life cycles of marine creatures suggest means of exploiting the abundance through fish farming and other aqua-agriculture.

Biology trainee. Woods Hole Oceanographic Institute in Massachusetts is a major research base for the study of all disciplines relating to the sea.

An ideal mine illustrating the types of support and central shaft. For safety and efficiency, rock in the roof and walls is shored by timbering, room and pillar constructions, and arches.

DETERMINING PERIOD   18.3

Objective **To determine the period for $y = \sin bx$ and $y = \cos bx$**

You have learned that the period for $y = \sin x$ and $y = \cos x$ is 360°. You will now learn how multiplying the measure of an angle by a constant changes the period.

Example 1 Sketch $y = \cos 2x$ and $y = \cos x$ for $0° \leq x \leq 360°$ on the same set of axes. Determine the period and amplitude of each function.

x	$2x$	$y = \cos 2x$
0°	0°	1
30°	60°	$\frac{1}{2}$
45°	90°	0
90°	180°	-1
135°	270°	0
180°	360°	1
225°	450°	0
270°	540°	-1
315°	630°	0
360°	720°	1

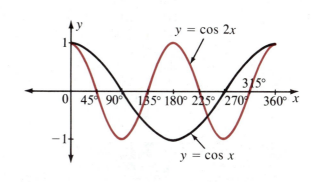

The graph of $y = \cos 2x$ completes 2 cycles from 0° to 360°. It completes one cycle every 180°. Note: $\frac{360°}{2} = 180°$.

Thus, period for $y = \cos x$ is 360°.   Amplitude for $y = \cos x$ is 1.
    Period for $y = \cos 2x$ is 180°.   Amplitude for $y = \cos 2x$ is 1.

Example 2 Sketch $y = \sin \frac{1}{2} x$ and $y = \sin x$ for $0° \leq x \leq 720°$ on the same set of axes. Determine the period and amplitude of each function.

x	$\frac{1}{2} x$	$y = \sin \frac{1}{2} x$
0°	0°	0
60°	30°	$\frac{1}{2}$
90°	45°	$\frac{\sqrt{2}}{2}$
180°	90°	1
360°	180°	0
540°	270°	-1
720°	360°	0

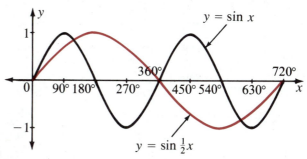

The graph of $y = \sin \frac{1}{2} x$ completes $\frac{1}{2}$ cycle from 0° to 360°. It completes one cycle every 720°. Note: $\frac{360°}{\frac{1}{2}} = 720°$.

Thus, period for $y = \sin x$ is 360°.   Amplitude for $y = \sin x$ is 1.
    Period for $y = \sin \frac{1}{2} x$ is 720°.   Amplitude for $y = \sin \frac{1}{2} x$ is 1.

Determining Period   503

Example 3 **Sketch $y = \sin 3x$ and $y = \sin x$ for $0° \leq x \leq 360°$ on the same set of axes. Determine the period and amplitude for each function.**

x	$3x$	$y = \sin 3x$
0°	0°	0
30°	90°	1
60°	180°	0
90°	270°	−1
120°	360°	0
150°	450°	1
180°	540°	0
210°	630°	−1
240°	720°	0
270°	810°	1
300°	900°	0
330°	990°	−1
360°	1080°	0

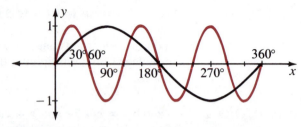

Notice that $y = \sin 3x$ completes 3 cycles from 0° to 360°. It completes one cycle every 120°.

Note: $\dfrac{360°}{3} = 120°$.

Thus, period for $y = \sin 3x$ is 120°. Amplitude for $y = \sin 3x = 1$. Period for $y = \sin x$ is 360°. Amplitude for $y = \sin x = 1$.

Period and amplitude for $y = \sin bx$ and $y = \cos bx$

If $y = \sin bx$, then the period is $\dfrac{360°}{|b|}$ and the amplitude is 1, for each *real* number $b\,(b \neq 0)$. If $y = \cos bx$, then the period is $\dfrac{360°}{|b|}$ and the amplitude is 1, for each *real* number $b\,(b \neq 0)$. Each function completes $|b|$ cycles in 360°.

Example 4 **Sketch $y = \sin(-\frac{1}{2}x)$ and $y = \sin \frac{1}{2}x$ for $0° \leq x \leq 720°$ on the same set of axes. Determine the period and amplitude for each function.**

x	$-\frac{1}{2}x$	$y = \sin(-\frac{1}{2}x)$
0°	0°	0
180°	−90°	−1
360°	−180°	0
540°	−270°	1
720°	−360°	0

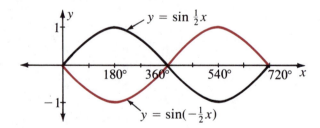

Since $\sin(-u) = -\sin u$, $\sin(-\frac{1}{2}x) = -\sin \frac{1}{2}x$. Therefore, the graph of $y = \sin(-\frac{1}{2}x)$ is symmetric to the graph of $y = \sin \frac{1}{2}x$. The x-axis is the axis of symmetry. Note: $\dfrac{360°}{\frac{1}{2}} = 720°$.

Thus, period for $y = \sin \frac{1}{2}x$ and $\sin(-\frac{1}{2}x)$ is 720°. Amplitude for $y = \sin \frac{1}{2}x$ and $\sin(-\frac{1}{2}x)$ is 1.

Example 5   **Determine the period of each trig function.**

$$y = \sin 6x$$
Period: $\dfrac{360°}{|6|} = 60°$

$$y = \cos\left(-\tfrac{1}{3}x\right)$$
Period: $\dfrac{360°}{\left|-\tfrac{1}{3}\right|} = 1080°$

$$y = \sin \tfrac{2}{3} x$$
Period: $\dfrac{360°}{\left|\tfrac{2}{3}\right|} = 540°$

Oral Exercises

Determine the period of each trig function.
1. $y = \sin x$ **360°**
2. $y = \cos x$ **360°**
3. $y = \sin \tfrac{1}{2} x$ **720°**
4. $y = \cos \tfrac{1}{2} x$ **720°**
5. $y = \cos 2x$ **180°**
6. $y = \sin \tfrac{1}{3} x$ **1,080°**
7. $y = \cos 9x$ **40°**
8. $y = \sin 12x$ **30°**

Written Exercises

Ⓐ **Sketch the graph for $0° \le x \le 360°$. Determine the period and amplitude.**
1. $y = \cos 2x$ **180°, 1**
2. $y = \cos \tfrac{1}{2} x$ **720°, 1**
3. $y = \sin 2x$ **180°, 1**
4. $y = \sin \tfrac{1}{2} x$ **720°, 1**
5. $y = \sin \tfrac{1}{3} x$ **1080°, 1**
6. $y = \sin 3x$ **120°, 1**
7. $y = \cos \tfrac{1}{3} x$ **1080°, 1**
8. $y = \cos 3x$ **120°, 1**
9. $y = \cos 4x$ **90°, 1**
10. $y = \sin \tfrac{1}{4} x$ **1440°, 1**
11. $y = \cos \tfrac{1}{4} x$ **1440°, 1**
12. $y = \sin 4x$ **90°, 1**

Ⓑ
13. $y = \sin(-2x)$ **180°, 1**
14. $y = \cos(-2x)$
15. $y = \cos\left(-\tfrac{1}{2}x\right)$
16. $y = \sin\left(-\tfrac{1}{3}x\right)$ **1080°, 1**
17. $y = \cos\left(-\tfrac{3}{2}x\right)$
18. $y = \sin\left(-\tfrac{5}{4}x\right)$ **288°, 1**
19. $y = \sin(-3x)$ **120°, 1**
20. $y = \cos(-3x)$ **120°, 1**

14. 180°, 1   15. 720°, 1   17. 240°, 1

21. Sketch $y = \cos 2x$ and $y = \sin 2x$ for $0° \le x \le 360°$ on the same set of axes. For how many values of x does $\cos 2x = \sin 2x$? **4**

22. Sketch $y = \sin \tfrac{1}{2} x$ and $y = \cos \tfrac{1}{2} x$ for $0° \le x \le 360°$ on the same set of axes. For how many values of x does $\sin \tfrac{1}{2} x = \cos \tfrac{1}{2} x$? **1**

For Exercises 25–26, see TE Answer Section.

Ⓒ **Sketch the graph for $0° \le x \le 360°$. Determine the period and amplitude.**
23. $y = \sec 2x$ **180°, none**
24. $y = \csc 3x$ **120°, none**
25. $y = \sec \tfrac{1}{2} x$
26. $y = \csc(-x)$

27. Sketch the graph of $y = \sec(-2x)$ and $y = \cos(-2x)$ in the interval $0°$ to $360°$ on the same set of axes. For how many values of x does $\sec(-2x) = \cos(-2x)$? **5**

NON-ROUTINE PROBLEMS

Solve for all real values of x and y such that:
$2 \cdot 3^x = 81y$ and $2^x = 8y$. **x = 4, y = 2**

CUMULATIVE REVIEW

1. The height of a triangle is twice the length of its base. The area is 144 cm². Find the length of the base and the height.
12 cm, 24 cm

2. The height of a triangle is 5 dm less than the length of its base. The area is 12 dm². Find the length of the base and the height.
8 dm, 3 dm

Determining Period

CHANGE OF PERIOD AND AMPLITUDE 18.4

Objectives **To sketch the functions defined by $y = a \sin bx$ and $y = a \cos bx$**
To solve a system of two trigonometric equations graphically

Recall that for the function defined by $y = 3 \sin x$, the amplitude is 3 and the period is 360°. In general, the amplitude for $y = a \sin x$ and $y = a \cos x$ is $|a|$. Also recall that for the function defined by $y = \cos \frac{1}{2}x$, the amplitude is 1 and the period is $\frac{360°}{\frac{1}{2}}$, or 720°. In general, the period of $y = \sin bx$ and $y = \cos bx$ is $\frac{360°}{|b|}$.

Example 1 Sketch $y = \frac{1}{2} \cos 3x$ for $0° \leq x \leq 360°$.

Period of $y = \frac{1}{2} \cos 3x$ is $\frac{360°}{3}$, or 120°. Amplitude of $y = \frac{1}{2} \cos 3x$ is $\frac{1}{2}$.

x	$3x$	$\cos 3x$	$y = \frac{1}{2}\cos 3x$
0°	0°	1	$\frac{1}{2}$
30°	90°	0	0
60°	180°	-1	$-\frac{1}{2}$
90°	270°	0	0
120°	360°	1	$\frac{1}{2}$
150°	450°	0	0
180°	540°	-1	$-\frac{1}{2}$
210°	630°	0	0
240°	720°	1	$\frac{1}{2}$
270°	810°	0	0
300°	900°	-1	$-\frac{1}{2}$
330°	990°	0	0
360°	1080°	1	$\frac{1}{2}$

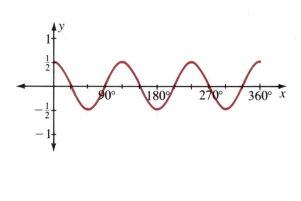

Example 2 Sketch $y = 3 \sin \frac{1}{2}x$ for $0° \leq x \leq 720°$.

Period of $y = 3 \sin \frac{1}{2}x$ is $\frac{360°}{\frac{1}{2}}$, or 720°.

Amplitude of $y = 3 \sin \frac{1}{2}x$ is 3.

x	$\frac{1}{2}x$	$\sin \frac{1}{2}x$	$y = 3 \sin \frac{1}{2}x$
0°	0°	0	0
180°	90°	1	3
360°	180°	0	0
540°	270°	-1	-3
720°	360°	0	0

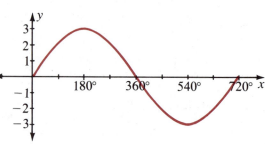

As you have learned, a system of algebraic equations may have some common solutions or no common solution. You will now study systems of two trigonometric equations. You will learn that there also may be no common solutions or many solutions. Usually, the number of points of intersection depends on the interval in which the functions are to be graphed.

Example 3 Sketch $y = 3 \sin x$ and $y = 2 \cos \frac{1}{2}x$ for $0° \leq x \leq 720°$. For how many values of x does $3 \sin x = 2 \cos \frac{1}{2}x$?

x	$3 \sin x$	$2 \cos \frac{1}{2}x$
0°	0	2
90°	3	$2(\frac{\sqrt{2}}{2})$, or 1.4
180°	0	0
270°	-3	$2(-\frac{\sqrt{2}}{2})$, or -1.4
360°	0	-2
450°	3	$2(-\frac{\sqrt{2}}{2})$, or -1.4
540°	0	0
630°	-3	$2(\frac{\sqrt{2}}{2})$, or 1.4
720°	0	2

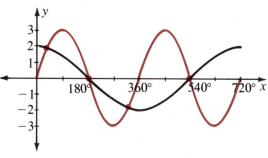

Thus, $3 \sin x = 2 \cos \frac{1}{2}x$ for 4 values of x for $0° \leq x \leq 720°$.

Example 4 Sketch $y = -\sin 2x$ and $y = 3 \cos x$ for $-90° \leq x \leq 360°$. For how many values of x does $-\sin 2x = 3 \cos x$?

x	$-\sin 2x$	$3 \cos 2x$
0°	0	3
45°	-1	0
90°	0	-3
135°	1	0
180°	0	3
225°	-1	0
270°	0	-3
315°	1	0
360°	0	3
$-45°$	1	0
$-90°$	0	-3

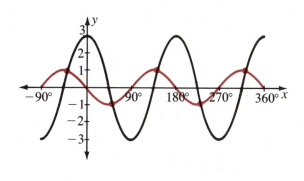

Thus, $-\sin 2x = 3 \cos x$ for 5 values of x for $-90° \leq x \leq 360°$.

Change of Period and Amplitude

Reading in Algebra

Each choice is related to the period or amplitude. Which one or ones don't belong? Defend your choices.

1. $y = 3 \sin 2x$ **c, d** a. 3 b. 180° c. 720° d. −3
2. $y = \frac{1}{2} \cos (3x)$ **b, d** a. 120° b. 1080° c. $\frac{1}{2}$ d. 180°
3. $y = -2 \sin \frac{1}{2}x$ **a, d** a. 180° b. 720° c. 2 d. −2

Written Exercises

Ⓐ Determine the period and amplitude for each of the following.

1. $y = -2 \cos \frac{1}{2}x$ **720°, 2** 2. $y = 3 \sin (-2x)$ **180°, 3** 3. $y = \frac{1}{2} \sin 2x$ **180°, $\frac{1}{2}$** 4. $y = \frac{1}{4} \cos 4x$ **90°, $\frac{1}{4}$**
5. $y = \frac{1}{3} \sin 9x$ **40°, $\frac{1}{3}$** 6. $y = -3 \cos \frac{1}{3}x$ **1080°, 3** 7. $y = -2 \sin \frac{1}{2}x$ **720°, 2** 8. $y = 12 \cos \frac{1}{12}x$ **4,320°, 12**

Sketch the graph for $0° \leq x \leq 360°$. Determine the period and amplitude.

9. $y = \frac{1}{3} \cos 3x$ **120°, $\frac{1}{3}$** 10. $y = 2 \sin 2x$ **180°, 2** 11. $y = 2 \cos 3x$ **120°, 2** 12. $y = \frac{1}{2} \sin 4x$ **90°, $\frac{1}{2}$**
13. $y = 3 \sin \frac{1}{3}x$ **1080°, 3** 14. $y = \frac{1}{2} \cos 2x$ **180°, $\frac{1}{2}$** 15. $y = 3 \sin 2x$ **180°, 3** 16. $y = 3 \cos \frac{1}{2}x$ **720°, 3**

Ⓑ 17. $y = -2 \sin \frac{1}{2}x$ **720°, 2** 18. $y = -2 \cos \frac{1}{2}x$ **720°, 2** 19. $y = -2 \sin (-\frac{1}{2}x)$ **720°, 2**
20. $y = -\frac{1}{2} \cos 2x$ **180°, $\frac{1}{2}$** 21. $y = -\frac{1}{2} \cos (-2x)$ **180°, $\frac{1}{2}$** 22. $y = -\frac{1}{2} \sin 2x$ **180°, $\frac{1}{2}$**

23. Sketch $y = \cos \frac{1}{2}x$ and $y = 3 \cos x$ for $0° \leq x \leq 360°$ on the same set of axes. For how many values of x does $\cos \frac{1}{2}x = 3 \cos x$? **2**

24. Sketch $y = \cos 2x$ and $y = \frac{1}{2} \sin x$ for $0° \leq x \leq 360°$ on the same set of axes. For how many values of x does $\cos 2x = \frac{1}{2} \sin x$? **4**

25. Sketch $y = -2 \cos x$ and $y = \sin (-\frac{1}{2}x)$ in the interval $-360°$ to $360°$ on the same set of axes. For how many values of x does $-2 \cos x = \sin (-\frac{1}{2}x)$? **4**

26. Sketch $y = \sin (-2x)$ and $y = -\frac{1}{2} \cos x$ in the interval $0°$ to $360°$ on the same set of axes. For how many values of x does $\sin (-2x) = -\frac{1}{2} \cos x$? **4**

Ⓒ Sketch the graph for $0° \leq x \leq 360°$. Determine the period and amplitude.

27. $y = \frac{1}{2} \sec 2x$ **180°, none** 28. $y = 2 \csc 3x$ **120°, none** 29. $y = -2 \sec 3x$ **120°, none**
30. $y = 2 \sec (-2x)$ **180°, none** 31. $y = 2 \sec 2x$ **180°, none** 32. $y = -2 \csc (-\frac{1}{2}x)$ **720°, none**

33. Sketch the graph of $y = -2 \sin (-\frac{1}{2}x)$ and $y = -\frac{1}{2} \cos (-2x)$ in the interval $0°$ to $360°$ on the same set of axes. For how many values of x does $-2 \sin (-\frac{1}{2}x) = -\frac{1}{2} \cos (-2x)$? **0**

34. Sketch the graph of $y = 2 \sec 2x$ and $y = \frac{1}{2} \sec \frac{1}{2}x$ for $0° \leq x \leq 360°$ on the same set of axes. For how many values of x does $2 \sec 2x = \frac{1}{2} \sec \frac{1}{2}x$? **1**

CUMULATIVE REVIEW

Express each as the same function of a positive acute angle.

1. $\sin (-68°)$ **−sin 68°** 2. $\cos 130° \, 40'$ **−cos 49° 20'** 3. $\sin 305° \, 18'$ **−sin 54° 42'**

GRAPHING y = tan x AND y = cot x   18.5

Objectives To graph $y = \tan x$ and $y = \cot x$
To determine the amplitude and period for $y = \tan bx$ and $y = \cot bx$
To determine the quadrants in which the tan increases or decreases and in which the cot increases or decreases

To graph the function $y = \tan x$, make a table of values. After graphing, observe where the function is increasing. Also observe the period and amplitude. The graph of $y = \tan x$ for $0° \leq x \leq 360°$ is shown below.

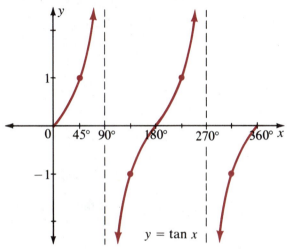

x	$y = \tan x$
0°	0
30°	.58
45°	1
60°	1.7
90°	undefined
120°	−1.7
135°	−1
150°	−.58
180°	0
210°	.58
225°	1
270°	undefined
315°	−1
360°	0

Notice that the tangent function is always increasing. It has no maximum or minimum value, hence it has no amplitude, and a period of 180°.

Example 1 Graph $y = \tan 2x$ for $0° \leq x \leq 180°$.

x	$2x$	$\tan 2x$
0°	0°	0
22.5°	45°	1
45°	90°	undefined
67.5°	135°	−1
90°	180°	0
112.5°	225°	1
135°	270°	undefined
157.5°	315°	−1
180°	360°	0

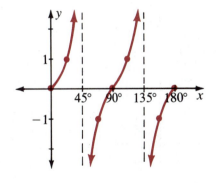

The graph of $y = \tan 2x$ completes 2 cycles from 0° to 180°. It completes one cycle every 90°. Note: $\frac{180°}{2} = 90°$.

Period for $y = \tan x$ is 180°. Period for $y = \tan 2x$ is 90°.

The function $y = \tan x$ has no amplitude.
The period of $y = \tan x$ is 180°.

If $y = \tan bx$, then the period is $\dfrac{180°}{|b|}$ for each *real* number $b (b \neq 0)$.

Example 2

Sketch $y = \cot x$ for $-180° \leq x \leq 360°$.

x	$y = \cot x$
0°	undefined
45°	1
90°	0
135°	−1
180°	undefined
225°	1
270°	0
315°	−1
360°	undefined
−45°	−1
−90°	0
−135°	1
−180°	undefined

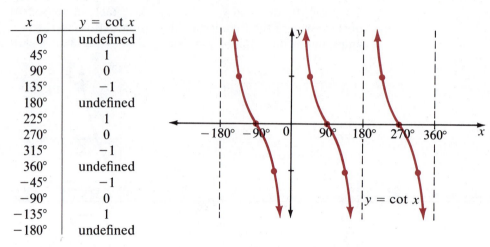

Notice that the cotangent function is always decreasing. It has no maximum or minimum value, hence it has no amplitude. Also notice that the graph repeats itself every 180°.

Example 3

Sketch $y = 2 \cot \tfrac{1}{2}x$ for $-180° \leq x \leq 360°$.

x	$\tfrac{1}{2}x$	$\cot \tfrac{1}{2}x$	$y = 2\cot \tfrac{1}{2}x$
0°	0°	undefined	undefined
90°	45°	1	2
180°	90°	0	0
270°	135°	−1	−2
360°	180°	undefined	undefined
450°	225°	1	2
540°	270°	0	0
630°	315°	−1	−2
720°	360°	undefined	undefined
−90°	−45°	−1	−2
−180°	−90°	0	0

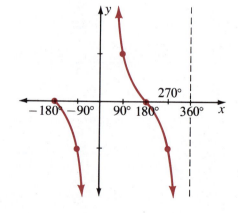

The function $y = \cot x$ has no amplitude.
The period for $y = \cot x$ is 180°.

If $y = \cot bx$, then the period is $\dfrac{180°}{|b|}$ for each real number $b (b \neq 0)$.

Written Exercises

(A) Complete the chart.

	$y = \tan x$	$y = \cot x$
1. Amplitude	None	None
2. Period	180°	180°
3. Maximum value	None	None
4. Minimum value	None	None
5. Quadrants where function increases	I, II, III, IV	None
6. Quadrants where function decreases	None	I, II, III, IV
7. Values for $-180° \leq x \leq 360°$ where the function is undefined	$-90°, 90°, 270°$	$-180°, 0°, 180°, 360°$

Determine the period for each of the following.

8. $y = 2 \tan 3x$ **60°**   9. $y = 4 \cot \frac{1}{2}x$ **360°**   10. $y = -2 \tan 3x$ **60°**   11. $y = -\frac{1}{2} \cot 3x$ **60°**
12. $y = \frac{1}{2} \tan \frac{1}{2}x$ **360°**   13. $y = -\frac{1}{2} \cot 2x$ **90°**   14. $y = 3 \tan \frac{1}{3}x$ **540°**   15. $y = -6 \cot \frac{1}{3}x$ **540°**
16. $y = -6 \cot(-2x)$ **90°**   17. $y = -\frac{1}{2} \tan(-\frac{1}{2}x)$ **360°**   18. $y = -\frac{1}{4} \cot(-\frac{1}{3}x)$ **540°**

Sketch the graph in the interval $-90° \leq x \leq 360°$. Determine the period.

19. $y = \tan \frac{1}{2}x$ **360°**   20. $y = \cot 2x$ **90°**   21. $y = \tan 3x$ **60°**   22. $y = \cot \frac{1}{2}x$ **360°**
23. $y = 2 \cot 2x$ **90°**   24. $y = \frac{1}{2} \tan 2x$ **90°**   25. $y = 2 \tan \frac{1}{2}x$ **360°**   26. $y = 4 \cot 3x$ **60°**

(B) 27. $y = -2 \tan(-2x)$ **90°**   28. $y = -\frac{1}{2} \cot(-2x)$ **90°**   29. $y = -3 \tan(-\frac{1}{2}x)$ **360°**
30. $y = -\frac{1}{2} \cot(3x)$ **60°**   31. $y = 6 \tan(-\frac{1}{3}x)$ **540°**   32. $y = -\cot(-\frac{1}{2}x)$ **360°**

33. Sketch $y = \cot \frac{1}{2}x$ and $y = \tan 2x$ for $0° \leq x \leq 360°$ on the same set of axes. For how many values of x does $\cot \frac{1}{2}x = \tan 2x$? **5**

34. Sketch $y = 2 \tan \frac{1}{2}x$ and $y = \frac{1}{2} \cot 2x$ for $0° \leq x \leq 360°$ on the same set of axes. For how many values of x does $2 \tan \frac{1}{2}x = \frac{1}{2} \cot 2x$? **4**

(C) 35. Sketch $y = -\tan(-\frac{1}{2}x)$ and $y = -2 \cot x$ for $0° \leq x \leq 360°$ on the same set of axes. For how many values of x does $-2 \cot x = -\tan(-\frac{1}{2}x)$? **2**

36. Sketch $y = -\frac{1}{2} \tan(-2x)$ and $y = \cot(-2x)$ for $0° \leq x \leq 360°$ on the same set of axes. For how many values of x does $-\frac{1}{2} \tan(-2x) = \cot(-2x)$? **0**

37. Sketch $y = \tan 2x$ and $y = \cot \frac{1}{2}x$ for $0° \leq x \leq 360°$ on the same set of axes. Estimate the values of x for which $\tan 2x = \cot \frac{1}{2}x$. **36°, 108°, 180°, 252°, 324°**

38. Sketch $y = -\tan 2x$ and $y = \sin \frac{1}{2}x$ for $0° \leq x \leq 360°$ on the same set of axes. Estimate the values of x for which $-\tan 2x = \sin \frac{1}{2}x$. **75°, 160°, 250°, 0°, 360°**

NON-ROUTINE PROBLEMS

$S_n = 1 - 2 + 3 - 4 + 5 - 6 + \ldots + (-1)^{n-1}n$ as $n = 1, 2, 3, 4, \ldots$.
Find $S_{25} + S_{36} + S_{60}$. **−35**

Graphing $y = \tan x$ and $y = \cot x$

FINDING LOGARITHMS OF TRIGONOMETRIC FUNCTIONS   18.6

Objectives **To find logarithms of the sin, cos, tan, and cot of any degree measure using a log-trig table
To find the degree measure when the log-trig value is given
To interpolate using a log-trig table**

In an earlier chapter you learned how to do computations with logarithms. Recall the following:

$$\log a \cdot b = \log a + \log b$$
$$\log \frac{a}{b} = \log a - \log b$$
$$\log a^n = n \log a$$
$$\log \sqrt[n]{a} = \frac{1}{n} \log a$$

Thus, to find $N = \dfrac{\sqrt[4]{(34.5) \times (7.3)^3}}{186.4}$ using logs, write

$\log N = \dfrac{1}{4}[(\log (34.5) + 3 \log 7.3)] - \log 186.4$ and solve.

The following examples illustrate the use of log-trig tables.

Example 1   **Find log tan 28° 10′ and log cos 27° 40′.**

The log-trig functions at the top of the table must be used with degree measures in the left-hand column. A portion of a log-trig table is shown below.

Logarithms of Trigonometric Functions

Angle	L Sin	L Tan	L Cot	L Cos	
27° 0′	9.6570	9.7072	10.2928	9.9499	63° 0′
10′	.6595	.7103	.2897	.9492	50′
20′	.6620	.7134	.2866	.9486	40′
30′	.6644	.7165	.2835	.9479	30′
40′	.6668	.7196	.2804	.9473	20′
50′	.6692	.7226	.2774	.9466	10′
28° 0′	9.6716	9.7257	10.2743	9.9459	62° 0′
10′	.6740	.7287	.2713	.9453	50′
20′	.6763	.7317	.2683	.9446	40′
30′	.6787	.7348	.2652	.9439	30′
40′	.5810	.7378	.2622	.9432	20′
50′	.6833	.7408	.2592	.9425	10′

* Subtract 10 from each entry in this table to obtain the proper logarithm of the indicated trigonometric function.

To find log tan 28° 10′ use the left-hand column and read across under L Tan (log tan) column. The log cos 27° 40′ can be found by reading across 27° 40′ under the L Cos (log cos) column. After reading the value from the table, subtract 10.

Thus, log tan 28° 10′ = 9.7287 − 10, and log cos 27° 40′ = 9.9473 − 10.

Example 2   **Find log sin 46° 40′ and log cot 45° 50′.**

Use the right column and the bottom column heads.

43° 0′	9.8338	9.9697	10.0303	9.8641	47° 0′
10′	.8351	.9722	.0278	.8629	50′
20′	.8365	.9747	.0253	.8618	40′
30′	.8378	.9772	.0228	.8606	30′
40′	.8391	.9798	.0202	.8594	20′
50′	.8405	.9823	.0177	.8582	10′
44° 0′	9.8418	9.9848	10.0152	9.8569	46° 0′
10′	.8431	.9874	.0126	.8557	50′
20′	.8444	.9899	.0101	.8545	40′
30′	.8457	.9924	.0076	.8532	30′
40′	.8469	.9949	.0051	.8520	20′
50′	.8482	.9975	.0025	.8507	10′
45° 0′	9.8495	10.0000	10.0000	9.8495	45° 0′
	L Cos	L Cot	L Tan	L Sin	Angle

Thus, log sin 46° 40′ = 9.8618 − 10 and log cot 45° 50′ = 9.9874 − 10.

Example 3   Log sin u = 9.2251 − 10   Log tan u = .7320
                   Find u, $u < 90°$.   Find u, $u < 90°$.

Angle	L Sin	L Tan	L Cot	L Cos	
9° 0′	9.1943	9.1997	10.8003	9.9946	81° 0′
10′	.2022	.2078	.7922	.9944	50′
20′	.2100	.2158	.7842	.9942	40′
30′	.2176	.2236	.7764	.9940	30′
40′	.2251	.2313	.7687	.9938	20′
50′	.2324	.2389	.7611	.9936	10′
10° 0′	9.2397	9.2463	10.7537	9.9934	80° 0′
10′	.2468	.2536	.7464	.9931	50′
20′	.2538	.2609	.7391	.9929	40′
30′	.2606	.2680	.7320	.9927	30′
40′	.2674	.2750	.7250	.9924	20′
50′	.2740	.2819	.7181	.9922	10′
	L Cos	L Cot	L Tan	L Sin	Angle

Thus, when log sin u = 9.2251 − 10, u = 9° 40′. When log tan u = .7320, u = 79° 30′.

Example 4   **Find log cos 36° 28′.**

36° 28′ lies between the values 36° 20′ and 36° 30′. Make a table.

$$10\left[\, 8\left[\begin{array}{ll}\text{Measure} & \text{log cosine} \\ 36°\ 20' & .9061 \\ 36°\ 28' & \\ 36°\ 30' & .9052\end{array}\right] n\,\right] -.0009$$

$$\frac{8}{10} = \frac{n}{-.0009} \text{ or } 10n = -.0072$$
$$n = -.00072, \text{ or } n \doteq -.0007$$
$$.9061 + (-.0007) = .9054$$

Thus, log cos 36° 28′ = 9.9054 − 10.

Finding Logarithms of Trigonometric Functions

You will now learn how to do computations with trig functions using logs.

Example 5   Find the value of $\dfrac{(39.6)^3 \sin 73°\,10'}{\sqrt[3]{0.315}}$ to four significant digits.

Use logarithms. (See pages 588–589.)
Let $s = \dfrac{(39.6)^3 \sin 73°\,10'}{\sqrt[3]{0.315}}$

$$\begin{aligned}
\log s &= 3 \log 39.6 + \log \sin 73°\,10' - \tfrac{1}{3} \log 0.315 \\
&= 3(1.5977) + 9.9810 - 10 - \tfrac{1}{3}(9.4983 - 10) \\
&= 4.7931 + 9.9810 - 10 - \tfrac{1}{3}(29.4983 - 30) \\
&= 4.7741 - (9.8328 - 10) \\
\log s &= 4.9413
\end{aligned}$$

Thus, $s = 87{,}360$.

Oral Exercises

Find the value of each of the following.
1. log sin 48°20′ **9.8733 − 10**
2. log cos 10°30′ **9.9927 − 10**
3. log tan 50°10′ **10.0788 − 10**
4. log cot 30°40′ **10.2270 − 10**
5. log tan 40°20′ **9.9289 − 10**
6. log sin 10°50′ **9.2740 − 10**

Written Exercises

For Ex. 1–21, see TE Answer Section.

(A) Find the value of each of the following.
1. log sin 53°42′
2. log cot 76°54′
3. log cos 65°26′
4. log tan 34°18′
5. log cos 18°34′
6. log sin 22°36′
7. log tan 73°44′
8. log cot 12°58′
9. log tan 27°25′
10. log cos 46°17′
11. log sin 62°33′
12. log cos 86°42′

(B) Find u, $0° < u < 90°$, for each of the following.
13. log cos u = 9.9446 − 10
14. log tan u = 9.8797 − 10
15. log sin u = 9.7727 − 10
16. log tan u = .0076
17. log cot u = .1795
18. log cos u = 9.9684 − 10
19. log sin u = 9.8111 − 10
20. log cos u = 9.4775 − 10
21. log cot u = .4057

Find the value to four significant digits. Use logarithms.

22. $\dfrac{\cos 86°\,10'}{(.437)^2}$ **0.3500**

23. $\dfrac{(31.3)^3 \sqrt{0.736}}{\tan 43°\,30'}$ **27,710**

24. $\cot 75°\,50' \sqrt{\dfrac{48.2}{5.64}}$ **0.7378**

25. $\dfrac{\sqrt[3]{46.3}\,\sin 35°\,30'}{(1.48)^2}$ **0.9520**

26. $\dfrac{\cos 23°\,36'\,(63.1)^3}{\sqrt{0.487}}$ **329,800**

27. $\dfrac{(49.3)^3 \sqrt{107.3}}{\sin 73°\,47'}$ **1,292,000**

(C)
28. $\dfrac{\sin 43°\,38'\,\cos 63°\,42'}{(\tan 17°\,14')^3}$ **10.22**

29. $\dfrac{\sqrt[3]{\cos 33°\,18'}\,\tan 16°\,47'}{\sqrt[4]{187.6}\,\sin 10°\,15'}$ **0.1182**

30. $\dfrac{(12.3)^4 (\sin 73°\,38')^3}{(\cos 15°\,36')^2 \sqrt[3]{0.1487}}$ **41,130**

31. $\dfrac{(38.6)^3 \tan 23°\,18'}{\sqrt[3]{(\sin 46°\,37')(\cos 70°\,48')}}$ **39,910**

32. $\dfrac{\sec 82°\,42'}{\sqrt{534.3}\,\tan 20°\,14'}$ **0.5608**

33. $\dfrac{(713.4)^6 \csc 12°\,18'}{\sqrt{0.493}\,\sec 46°\,37'}$ **7.303 × 10^{17}**

QUADRATIC TRIGONOMETRIC EQUATIONS

18.7

Objective **To solve quadratic trigonometric equations**

Solving trigonometric equations is similar to solving algebraic equations. Each example that follows will be preceded by an algebraic example.

To solve $4x^2 - 5 = 0$, add 5 to both sides, divide each side by 4, and find the square root.

$$4x^2 = 5$$
$$x^2 = \frac{5}{4}$$
$$x = \pm \frac{1}{2}\sqrt{5}$$

Example 1 **Determine the values of θ between 0° and 360° for which $4\sin^2\theta - 3 = 0$.**

Think of $\sin\theta$ as x. Note that $\sin^2\theta$ is written $(\sin\theta)^2$. Add 3 to both sides, divide each side by 4, and find the square root.

$$4(\sin\theta)^2 = 3$$
$$(\sin\theta)^2 = \frac{3}{4}$$
$$\sin\theta = \pm\sqrt{\frac{3}{4}}, \text{ or } \pm\frac{\sqrt{3}}{2}$$

$\sin 60° = \frac{\sqrt{3}}{2}$ ▶ $\sin\theta = \frac{\sqrt{3}}{2}$ $\sin\theta = \frac{-\sqrt{3}}{2}$

               1st quad.   2nd quad.   3rd quad.   4th quad.

Thus, $\theta = 60°, 120°, 240°,$ and $300°$.

To solve the algebraic equation $2x^2 - x = 0$, factor, set each factor equal to zero, and solve.

$$x(2x - 1) = 0$$
$$x = 0 \quad \text{or} \quad 2x - 1 = 0$$
$$x = 0 \quad \text{or} \quad x = \frac{1}{2}$$

Example 2 **Determine the values of u, $0° \leq u \leq 360°$, for which $\tan^2 u - \sqrt{3}\tan u = 0$.**

        Factor. ▶ $\tan u(\tan u - \sqrt{3}) = 0$
Set each factor ▶
equal to 0. $\tan u = 0$ or $\tan u - \sqrt{3} = 0$
                                              $\tan u = \sqrt{3}$
                      $u = 0°, u = 180°$ Ref. angle 60°
                                         1st quad. and 3rd quad.

Thus, $u = 0°, 180°, 60°,$ and $240°$.

Quadratic Trigonometric Equations

To solve the equation $3x^2 + 5x - 2 = 0$, factor, set each factor equal to zero.
$$(3x - 1)(x + 2) = 0$$
$$3x - 1 = 0 \quad \text{or} \quad x + 2 = 0$$
$$x = \frac{1}{3} \quad \text{or} \quad x = -2$$

Example 3

Determine values of u, $0° \leq u \leq 360°$, for which $2\sin^2 u + 5\sin u - 3 = 0$.

$$2\sin^2 u + 5\sin u - 3 = 0$$
$$(2\sin u - 1)(\sin u + 3) = 0$$
$$2\sin u - 1 = 0 \quad \text{or} \quad \sin u + 3 = 0$$
$$\sin u = \frac{1}{2} \qquad\qquad \sin u = -3$$

Ref. ∠: 30° 　　　　　　No solution, since $-1 \leq \sin u \leq 1$

↙　↘
1st quad.　2nd quad.

Thus, $u = 30°$ and $150°$.

To solve the equation $2x^2 - 5x + 1 = 0$, use the quadratic formula.
$$x = \frac{-b \pm \sqrt{b^2 - 4ac}}{2a} \qquad a = 2, b = -5, c = 1$$
$$x = \frac{-(-5) \pm \sqrt{(-5)^2 - 4(2)(1)}}{2(2)}$$
$$= \frac{5 + \sqrt{17}}{4} \quad \text{or} \quad \frac{5 - \sqrt{17}}{4}$$
$$= \frac{5 + 4.123}{4} \quad \text{or} \quad \frac{5 - 4.123}{4}$$
$$x = 2.281 \text{ or } .219$$

Example 4

Determine the values of u to the nearest degree between $0°$ and $360°$ for which $3\cos^2 u - 4\cos u - 2 = 0$.

Use the quadratic formula. $x = \frac{-b \pm \sqrt{b^2 - 4ac}}{2a}$, $a = 3, b = -4, c = -2$

$$\cos u = \frac{-(-4) \pm \sqrt{(-4)^2 - 4(3)(-2)}}{2(3)}$$
$$= \frac{4 \pm \sqrt{16 + 24}}{6}, \text{ or } \frac{4 \pm \sqrt{40}}{6}$$

$$\frac{4 + 6.32}{6} \qquad\qquad \frac{4 - 6.32}{6}$$

$$\cos u = \frac{10.32}{6} = 1.72 \quad \text{or} \quad \cos u = \frac{-2.32}{6} \doteq -.3867$$

↑　　　　　　　　　　　　　Ref. ∠: 67°
No solution　　　　　　　　↙　↘
　　　　　　　　　　　2nd quad.　3rd quad.

Thus, $u = 113°$ and $247°$.

Chapter Eighteen

Oral Exercises

How many values of u, $0° \le u \le 360°$, are there for each of the following?
1. $\sin u = -.7632$ **2**
2. $\cos u = 2.6543$ **0**
3. $\tan u = -6.3214$ **2**
4. $\tan u = .4132$ **2**
5. $\sin u = -3.1249$ **0**
6. $\cos u = -.2987$ **2**

Written Exercises

For Ex. 1–33, see TE Answer Section.

Ⓐ Determine all values of θ, $0° \le \theta \le 360°$, to the nearest degree for which the equations are true.
1. $2 \cos^2 \theta - 5 = -4$
2. $10 \sin \theta - 5\sqrt{3} = 0$
3. $2 \cos \theta - \sqrt{3} = 0$
4. $\sin^2 \theta - 1 = 0$
5. $\cos^2 \theta = 0$
6. $\cot^2 \theta - 3 = 0$
7. $2 \sin \theta + \sqrt{3} = 0$
8. $\cot^2 \theta + \cot \theta = 0$
9. $2 \cos^2 \theta - \cos \theta = 0$
10. $2 \sin^2 \theta + \sqrt{3} \sin \theta = 0$
11. $\sqrt{3} \tan^2 \theta - \tan \theta = 0$
12. $2 \cos^2 \theta - \cos \theta - 1 = 0$
13. $\sin^2 \theta - 7 \sin \theta + 6 = 0$
14. $2 \cos^2 \theta + 3 \cos \theta + 1 = 0$
15. $\tan^2 \theta - 2 \tan \theta + 1 = 0$

Ⓑ
16. $\tan^2 \theta - 5 \tan \theta - 2 = 0$
17. $3 \sin^2 \theta - 8 \sin \theta - 1 = 0$
18. $4 \cos^2 \theta + 5 \cos \theta - 2 = 0$
19. $5 \cos^2 \theta - 3 \cos \theta - 1 = 0$
20. $7 \cot^2 \theta + 4 \cot \theta - 3 = 0$
21. $7 \tan^2 \theta - 3 \tan \theta - 5 = 0$
22. $2 \sin^2 \theta - 5 \sin \theta = 1$
23. $\cot^2 \theta - 7 \cot \theta + 1 = 0$
24. $3 \cos^2 \theta + 6 \cos \theta + 1 = 0$

Ⓒ Determine all values of θ for which the equations are true.
25. $\sec^2 \theta - 4 = 0$
26. $4 \csc^2 \theta - 25 = 0$
27. $15 \csc^2 \theta + \csc \theta - 2 = 0$
28. $\csc^2 \theta - 7 \csc \theta + 12 = 0$
29. $2 \sec^2 \theta - \sec \theta - 15 = 0$
30. $3 \csc^2 \theta - 6 \csc \theta - 1 = 0$
31. $\sec^2 \theta - 5 \sec \theta - 3 = 0$
32. $5 \sec^2 \theta - 1 = 12$
33. $\sec^2 \theta + 7 \sec \theta - 3 = 0$

CALCULATOR ACTIVITIES

You can use a calculator to solve quadratic trigonometric equations.
Example: Solve $\sin^2 \theta - 7 \sin \theta + 5 = 0$ to the nearest ten minutes, $0 \le \theta \le 90°$.

1. Use the quadratic formula. $a = 1$, $b = -7$, $c = 5$.
2. Solve for $\sin \theta$: $\sin \theta = \dfrac{-(-7) \pm \sqrt{(-7)^2 - 4(1)(5)}}{2(1)}$
3. Compute $(-7)^2 - 4(5)$.
   Press ▶ $-7 \otimes \ominus$; $-4 \otimes 5 \ominus$; $49 \ominus 20 \ominus$
   Display ▶ 49 ; −20 ; 29
4. Use the table to find $\sqrt{29} \doteq 5.39$.
5. Compute $\dfrac{7 + 5.39}{2}$ and $\dfrac{7 - 5.39}{2}$.
6. *Press* ▶ $7 \oplus 5.39 \oslash 2 \ominus$ and $7 \ominus 5.39 \oslash 2 \ominus$
7. *Display* ▶ 6.195 and .805. Reject 6.195 since $-1 \le \sin \theta \le 1$.
8. Look in the table to find $\theta = 53° 40'$ to the nearest ten minutes.

Quadratic Trigonometric Equations

BASIC TRIGONOMETRIC IDENTITIES 18.8

Objective **To verify basic trigonometric identities using reciprocal, quotient, and Pythagorean identities**

Equations that are true for all permissible values of the variable are called identities. The equation $3 + x = x + 3$ is a statement of the commutative property for addition and is true for all real numbers. It is an example of an identity.

In this lesson you will learn how to verify certain trigonometric identities. The first two examples illustrate reciprocal identities.

Example 1 Show that $(\sin 45°)(\csc 45°) = 1$, $(\cos 60°)(\sec 60°) = 1$, and $(\tan 30°)(\cot 30°) = 1$.

$(\sin 45°)(\csc 45°)$	1	$(\cos 60°)(\sec 60°)$	1	$(\tan 30°)(\cot 30°)$	1
$\dfrac{\sqrt{2}}{2} \cdot \sqrt{2}$	1	$\dfrac{1}{2} \cdot \dfrac{2}{1}$	1	$\dfrac{\sqrt{3}}{3} \cdot \sqrt{3}$	1
$\dfrac{2}{2}$		1		$\dfrac{3}{3}$	
1				1	

Recall that $\sin A = \dfrac{y}{r}$, $\csc A = \dfrac{r}{y}$, $\cos A = \dfrac{x}{r}$, $\sec A = \dfrac{r}{x}$, $\tan A = \dfrac{y}{x}$, and $\cot A = \dfrac{x}{y}$. You will use these in Example 2.

Example 2 Show that for any degree measure A, $\sin A \cdot \csc A = 1$, $\cos A \cdot \sec A = 1$, and $\tan A \cdot \cot A = 1$.

$\sin A \cdot \csc A$	1	$\cos A \cdot \sec A$	1	$\tan A \cdot \cot A$	1
$\dfrac{y}{r} \cdot \dfrac{r}{y}$	1	$\dfrac{x}{r} \cdot \dfrac{r}{x}$	1	$\dfrac{y}{x} \cdot \dfrac{x}{y}$	1
1		1		1	

Reciprocal identities

Sin and csc, cos and sec, and tan and cot are pairs of reciprocal functions.

$$\sin A \cdot \csc A = 1 \qquad\qquad \cos A \cdot \sec A = 1$$

$$\sin A = \frac{1}{\csc A} \text{ or } \csc A = \frac{1}{\sin A} \qquad\qquad \cos A = \frac{1}{\sec A} \text{ or } \sec A = \frac{1}{\cos A}$$

$$\tan A \cdot \cot A = 1$$

$$\tan A = \frac{1}{\cot A} \text{ or } \cot A = \frac{1}{\tan A}$$

Example 3 Show that for any degree measure A, $\tan A = \dfrac{\sin A}{\cos A}$ and $\cot A = \dfrac{\cos A}{\sin A}$.

Recall that $\sin A = \dfrac{y}{r}$, $\cos A = \dfrac{x}{r}$, $\tan A = \dfrac{y}{x}$, and $\cot A = \dfrac{x}{y}$.

$\tan A$	$\dfrac{\sin A}{\cos A}$	$\cot A$	$\dfrac{\cos A}{\sin A}$
$\dfrac{y}{x}$	$\dfrac{\frac{y}{r}}{\frac{x}{r}}$	$\dfrac{x}{y}$	$\dfrac{\frac{x}{r}}{\frac{y}{r}}$
	$\dfrac{y}{r} \cdot \dfrac{r}{x} = \dfrac{y}{x}$		$\dfrac{x}{r} \cdot \dfrac{r}{y} = \dfrac{x}{y}$

Quotient identities

$$\tan A = \dfrac{\sin A}{\cos A} \qquad \cot A = \dfrac{\cos A}{\sin A}$$

Example 4 Verify the quotient identities for $A = 210°$.

$\tan 210°$	$\dfrac{\sin 210°}{\cos 210°}$	$\cot 210°$	$\dfrac{\cos 210°}{\sin 210°}$
$\dfrac{\sqrt{3}}{3}$	$\dfrac{-\frac{1}{2}}{-\frac{\sqrt{3}}{2}}$	$\sqrt{3}$	$\dfrac{-\frac{\sqrt{3}}{2}}{-\frac{1}{2}}$
	$-\dfrac{1}{2} \cdot \left(-\dfrac{2}{\sqrt{3}}\right)$		$\dfrac{-\sqrt{3}}{2}\left(-\dfrac{2}{1}\right)$
	$\dfrac{1}{\sqrt{3}}$, or $\dfrac{\sqrt{3}}{3}$		$\sqrt{3}$

Example 5 Show that for any degree measure A, $\sin^2 A + \cos^2 A = 1$, $\tan^2 A + 1 = \sec^2 A$, and $\cot^2 A + 1 = \csc^2 A$.

$\sin^2 A + \cos^2 A$	1	$\tan^2 A + 1$	$\sec^2 A$	$\cot^2 A + 1$	$\csc^2 A$
$\left(\dfrac{y}{r}\right)^2 + \left(\dfrac{x}{r}\right)^2$	1	$\left(\dfrac{y}{x}\right)^2 + 1$	$\left(\dfrac{r}{x}\right)^2$	$\left(\dfrac{x}{y}\right)^2 + 1$	$\left(\dfrac{r}{y}\right)^2$
$\dfrac{y^2 + x^2}{r^2}$		$\dfrac{y^2 + x^2}{x^2}$	$\dfrac{r^2}{x^2}$	$\dfrac{x^2 + y^2}{y^2}$	$\dfrac{r^2}{y^2}$
$x^2 + y^2 = r^2$ ▶ $\dfrac{r^2}{r^2} = 1$		$\dfrac{r^2}{x^2}$		$\dfrac{r^2}{y^2}$	

Basic Trigonometric Identities

| Pythagorean identities | $\sin^2 A + \cos^2 A = 1$, $\tan^2 A + 1 = \sec^2 A$, $\cot^2 A + 1 = \csc^2 A$ |

Example 6 **Verify the Pythagorean identities for $A = 120°$.**

The terminal side is in the 2nd quadrant. Both sin and csc are positive, all others are negative. The reference angle is $60°$.

$\sin^2 120° + \cos^2 120°$	1	$\tan^2 120° + 1$	$\sec^2 120°$	$\cot^2 120° + 1$	$\csc^2 120°$
$\left(\frac{\sqrt{3}}{2}\right)^2 + \left(-\frac{1}{2}\right)^2$	1	$(-\sqrt{3})^2 + 1$	$(-2)^2$	$\left(-\frac{1}{\sqrt{3}}\right)^2 + 1$	$\left(\frac{2}{\sqrt{3}}\right)^2$
$\frac{3}{4} + \frac{1}{4} = 1$		$3 + 1$ 4	4	$\frac{1}{3} + 1 = \frac{4}{3}$	$\frac{4}{3}$

Written Exercises   For complete solutions, see TE Answer Section.

(A) Express as a function of sin, cos, or both.

1. csc A $\frac{1}{\sin A}$
2. sec A $\frac{1}{\cos A}$
3. tan A $\frac{\sin A}{\cos A}$
4. cot A $\frac{\cos A}{\sin A}$

Verify each identity for $A = 30°$.

5. $\sin^2 A + \cos^2 A = 1$
6. $\tan^2 A + 1 = \sec^2 A$
7. $1 + \cot^2 A = \csc^2 A$
8. $\cos A = \frac{1}{\sec A}$

Verify each identity for $\theta = 240°$.

9. $\sin^2 \theta = 1 - \cos^2 \theta$
10. $\cot \theta = \frac{\cos \theta}{\sin \theta}$
11. $\csc^2 \theta - 1 = \cot^2 \theta$
12. $\sin \theta = \frac{1}{\csc \theta}$

Verify each identity for $x = 135°$.

13. $\cos^2 x = 1 - \sin^2 x$
14. $\tan x = \frac{\sin x}{\cos x}$
15. $\sec^2 x = 1 + \tan^2 x$
16. $\csc^2 x = 1 + \cot^2 x$

Verify each identity for $B = 330°$.

17. $\sin B \cdot \csc B = 1$
18. $\tan B \cdot \cot B = 1$
19. $\cos B \cdot \sec B = 1$
20. $\sin B = \frac{1}{\csc B}$

For Ex. 21–28, see TE Answer Section.
Write an equivalent expression using only sin x.
Hint: $\sin^2 x + \cos^2 x = 1$ so, $\cos x = \pm\sqrt{1 - \sin^2 x}$.

(B) 21. tan x   22. cot x   23. sec x   24. csc x

Write an equivalent expression using only cos x.

25. cot x   26. sin x   27. csc x   28. sec x

Prove each identity. Use the definitions of tan x, cot x, sec x, and csc x.

(C) 29. $\tan^2 x + 1 = \sec^2 x$ 30. $1 + \cot^2 x = \csc^2 x$

Prove each identity. Use $\sin^2 A + \cos^2 A = 1$.

31. $\tan^2 A + 1 = \sec^2 A$ 32. $1 + \cot^2 A = \csc^2 A$

PROVING TRIGONOMETRIC IDENTITIES   18.9

Objective **To prove trigonometric identities**

In the previous lesson you studied the reciprocal identities, the quotient identities, and the Pythagorean identities. You will now use these identities to prove other trigonometric identities.

Example 1 **Prove the identity:** $\tan x = \dfrac{\sec x}{\csc x}$.

Replace $\sec x$ with $\dfrac{1}{\cos x}$ and $\csc x$ with $\dfrac{1}{\sin x}$.

$$\begin{array}{c|cl} \tan x & \dfrac{\sec x}{\csc x} & \\[2ex] \tan x & \dfrac{\dfrac{1}{\cos x}}{\dfrac{1}{\sin x}} & \blacktriangleleft \text{ Substitute.} \\[3ex] & \dfrac{1}{\cos x} \cdot \dfrac{\sin x}{1}, \text{ or } \dfrac{\sin x}{\cos x} & \blacktriangleleft \text{ Divide.} \\[2ex] \tan x & \blacktriangleleft \dfrac{\sin x}{\cos x} = \tan x & \end{array}$$

Example 2 **Prove the identity:** $\csc x = \sin x + \cot x \cos x.$

$$\begin{array}{c|cl} \csc x & \sin x + \cot x \cos x & \\[1ex] \csc x & \sin x + \dfrac{\cos x}{\sin x} \cdot \cos x & \blacktriangleleft \cot x = \dfrac{\cos x}{\sin x} \\[2ex] & \dfrac{\sin^2 x + \cos^2 x}{\sin x} & \\[2ex] & \dfrac{1}{\sin x} & \blacktriangleleft \sin^2 x + \cos^2 x = 1 \\[2ex] & \csc x & \blacktriangleleft \dfrac{1}{\sin x} = \csc x \end{array}$$

Summary To prove a trigonometric identity:
(1) Work with each side separately.
(2) Substitute, using basic trigonometric identities, usually in terms of sin or cos.
(3) Simplify by adding, subtracting, multiplying, or dividing.
(4) When each side is the same expression, the identity has been proved.

Example 3 **Prove the identity:** $\csc x + \csc x \cdot \sec x = \dfrac{1 + \sec x}{\sin x}$.

$$\begin{array}{c|c}
\csc x + \csc x \sec x & \dfrac{1 + \sec x}{\sin x} \\[6pt]
\csc x (1 + \sec x) & \dfrac{1 + \sec x}{\sin x} \\[6pt]
\dfrac{1}{\sin x}(1 + \sec x) & \\[6pt]
\dfrac{1 + \sec x}{\sin x} &
\end{array}$$

Example 4 **Prove the identity:** $\dfrac{1 + \csc x}{\sec x} = \cos x + \cot x$.

$$\begin{array}{c|c}
\dfrac{1 + \csc x}{\sec x} & \cos x + \cot x \\[6pt]
\dfrac{1 + \dfrac{1}{\sin x}}{\dfrac{1}{\cos x}} & \cos x + \cot x \\[10pt]
\dfrac{\sin x + 1}{\sin x} \cdot \dfrac{\cos x}{1} & \\[6pt]
\dfrac{\sin x \cos x}{\sin x} + \dfrac{\cos x}{\sin x} & \\[6pt]
\cos x + \cot x &
\end{array}$$

Written Exercises

For Ex. 1–18, see TE Answer Section.

(A) Prove each identity.

1. $\dfrac{\cot x}{\cos x} = \csc x$
2. $\sin x \cot x = \cos x$
3. $\sec x = \csc x \cdot \tan x$
4. $\cot x + \tan x = \csc x \sec x$
5. $\sec x = \cos x + \dfrac{\tan x}{\csc x}$
6. $\sin^2 x - \cos^2 x = 2 \sin^2 x - 1$

(B)
7. $\cot x - 1 = \cos x\,(\csc x - \sec x)$
8. $\sec x = \dfrac{\cos x}{1 + \sin x} + \tan x$
9. $\sin x + \dfrac{\cot x}{\sec x} = \csc x$
10. $\dfrac{\cot x + \cos x}{1 + \sin x} = \dfrac{\cos x}{\sin x}$
11. $\tan x - \cot x = \dfrac{1 - 2 \cos^2 x}{\sin x \cos x}$
12. $\dfrac{1 - \cos^2 x}{\cos x} \cdot \csc x = \tan x$
13. $2 \csc x = \dfrac{\sin x}{1 + \cos x} + \dfrac{\sin x}{1 - \cos x}$
14. $\cos x + \sin x = \dfrac{\sec x + \csc x}{\cot x + \tan x}$

(C)
15. $\dfrac{1 - \tan^2 x}{1 + \tan^2 x} = \dfrac{2 - \sec^2 x}{\sec^2 x}$
16. $\tan(180° + \theta) + \dfrac{1}{\cot(270° - \theta)} = \dfrac{\sec(180° - \theta)}{\sin(180° + \theta)}$
17. $\dfrac{\cot^2 \theta - 1}{\cot^2 \theta + 1} = \dfrac{2 - \sec^2 \theta}{\sec^2 \theta}$
18. $\sqrt{\dfrac{\sec \theta - \tan \theta}{\sec \theta + \tan \theta}} = \dfrac{1 - \sin \theta}{\cos \theta}$

RADIAN MEASURE   18.10

Objectives **To express radian measure in degrees**
To express degree measure in radians
To evaluate expressions involving radian measure

You have learned that an angle of rotation can be measured in degrees. An angle of rotation can also be measured using real numbers called *radians*. If an angle of rotation, which is the central angle of a given circle, intercepts an arc with the same length as a radius, then the angle is said to have a measure of 1 radian. When it intercepts an arc whose length is twice the radius, the angle is said to have a measure of 2 radians. See the figures below.

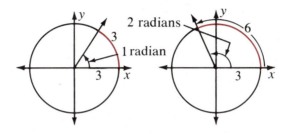

Definition of a radian	If an angle of rotation, which is the central angle of a given circle, intercepts an arc whose length is equal to the length of a radius of a circle, then the measure of the angle of rotation is 1 radian.

Example 1 **Find the measure of the angle of rotation.**

Since the length of the intercepted arc is 3 times the length of the radius, the measure of the angle of rotation, x, is 3 radians.

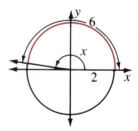

Thus, $x = 3$ radians.

Example 2 **Express 180° in radian measure.**

Circumference of a circle $= 2\pi r$
So, the length of $\widehat{AB} = \frac{1}{2}(2\pi r)$
$= \pi r$

Thus, 180° is π radians.

Radian Measure

In the preceding example you learned that 180° is π radians, or 180° = π (radians). Divide both sides by 180. The result is $1° = \frac{\pi \text{ (rad)}}{180}$. This formula will be used to change degree measure to radian measure.

Example 3   **Express each in radian measure.**

       90°   −210°

Multiply both sides by 90°. ▶

Use $1° = \frac{\pi \text{ (rad)}}{180}$.

$90° = \frac{\pi}{180}(90°)$

$= \frac{\pi}{2}$

Use $1° = \frac{\pi \text{ (rad)}}{180}$.

$-210° = \frac{\pi}{180}(-210°)$

$= -\frac{7\pi}{6}$

Thus, 90° is $\frac{\pi}{2}$ radians and −210° is $-\frac{7\pi}{6}$ radians.

Now, divide both sides of 180° = π (radians) by π. $1(\text{rad}) = \frac{180°}{\pi}$. This formula will be used to change from radian measure to degree measure.

Example 4   **Express in degree measure.**

       $\frac{3}{2}\pi$ $\frac{7}{6}\pi$

Use $1(\text{rad}) = \frac{180°}{\pi}$.

$\frac{3}{2}\pi = \frac{180°}{\pi}\left(\frac{3}{2}\pi\right)$

$= 90(3), \text{ or } 270°$

Use $1(\text{rad}) = \frac{180°}{\pi}$.

$\frac{7}{6}\pi = \frac{180°}{\pi} \cdot \left(\frac{7}{6}\pi\right)$

$= 30(7), \text{ or } 210°$

Thus, $\frac{3}{2}\pi$ radians is 270° and $\frac{7}{6}\pi$ radians is 210°.

Example 5 illustrates how to evaluate trigonometric expressions using radian measure.

Example 5   **Evaluate.**

$\sin \frac{\pi}{4}$	$\cos \frac{2\pi}{3}$	$\tan\left(-\frac{\pi}{6}\right)$	$\cot \frac{7}{4}\pi$
↓	↓	↓	↓
sin 45°	cos 120°	tan (−30°)	cot 315°
$\frac{\sqrt{2}}{2}$	−cos 60°	−tan 30°	−cot 45°
	$-\frac{1}{2}$	$-\frac{\sqrt{3}}{3}$	−1

Example 6

Evaluate $\dfrac{\sin\dfrac{5\pi}{6} - \cos\dfrac{3\pi}{4}}{\tan\left(-\dfrac{\pi}{4}\right)}$.

$$\dfrac{\sin\dfrac{5\pi}{6} - \cos\dfrac{3\pi}{4}}{\tan\left(-\dfrac{\pi}{4}\right)} = \dfrac{\sin 150° - \cos 135°}{\tan(-45°)} = \dfrac{\sin 30° - (-\cos 45°)}{-\tan 45°} = \dfrac{\dfrac{1}{2} - \left(\dfrac{-\sqrt{2}}{2}\right)}{-1} = -\dfrac{1 + \sqrt{2}}{2}$$

Written Exercises

Express each in radians.

(A)
1. 30° $\dfrac{\pi}{6}$
2. 120° $\dfrac{2\pi}{3}$
3. −200° $\dfrac{-10\pi}{9}$
4. 150° $\dfrac{5\pi}{6}$
5. 210° $\dfrac{7\pi}{6}$
6. 300° $\dfrac{5\pi}{3}$
7. −60° $\dfrac{-\pi}{3}$
8. −320° $\dfrac{-16\pi}{9}$
9. −270° $\dfrac{-3\pi}{2}$
10. 240° $\dfrac{4\pi}{3}$
11. −330° $\dfrac{-11\pi}{6}$
12. 225° $\dfrac{5\pi}{4}$
13. 450° $\dfrac{5\pi}{2}$
14. −360° -2π

Express each in degrees.

15. $\dfrac{\pi}{4}$ 45°
16. $\dfrac{3}{2}\pi$ 270°
17. $\dfrac{5\pi}{6}$ 150°
18. $\dfrac{7\pi}{6}$ 210°
19. $\dfrac{2}{3}\pi$ 120°
20. $-\pi$ −180°
21. $\dfrac{3}{4}\pi$ 135°
22. $-\dfrac{\pi}{8}$ −22.5°
23. $-\dfrac{3}{5}\pi$ −108°
24. $-\dfrac{2\pi}{9}$ −40°
25. -2π −360°
26. $\dfrac{9}{4}\pi$ 405°
27. $\dfrac{11\pi}{6}$ 330°
28. $-\dfrac{7\pi}{6}$ −210°

Evaluate each of the following.

(B)
29. $\sin\dfrac{\pi}{2}$ 1
30. $\cos\dfrac{5\pi}{2}$ 0
31. $\tan\left(-\dfrac{\pi}{4}\right)$ −1
32. $\cot\dfrac{7}{6}\pi$ $\sqrt{3}$
33. $\sin\left(-\dfrac{2}{3}\pi\right)$ $\dfrac{-\sqrt{3}}{2}$

34. $\dfrac{\tan\dfrac{5}{4}\pi - \cos\dfrac{5}{3}\pi}{\sin\dfrac{\pi}{6}}$ 1

35. $\dfrac{\cos\dfrac{5\pi}{4} + \tan\dfrac{7}{4}\pi}{\cot\left(\dfrac{5}{6}\pi\right)}$ $\dfrac{\sqrt{6} + 2\sqrt{3}}{6}$

36. $\dfrac{\sin\dfrac{\pi}{9} - \cos\dfrac{3}{2}\pi}{\tan\left(-\dfrac{3\pi}{4}\right)}$ 0.342

(C)
37. $\dfrac{\sec\dfrac{2\pi}{3} + \csc\dfrac{\pi}{6}}{\sec\dfrac{\pi}{4}}$ 0

38. $\dfrac{\csc\dfrac{2\pi}{5} - \sec\left(-\dfrac{1}{3}\pi\right) + \sin\dfrac{\pi}{3}}{\tan\left(-\dfrac{\pi}{4}\right)\cos\left(-\dfrac{\pi}{3}\right)}$ 0.17

39. $\dfrac{\cos\dfrac{11\pi}{6} - \sec\left(-\dfrac{\pi}{3}\right)}{\tan\dfrac{\pi}{6}\csc\left(-\dfrac{\pi}{4}\right)}$ $\dfrac{-3\sqrt{2} + 4\sqrt{6}}{4}$

40. $\dfrac{\sec\left(-\dfrac{\pi}{9}\right)\csc\dfrac{3\pi}{4} - \tan\left(-\dfrac{3\pi}{4}\right)}{\cos^2\dfrac{\pi}{3} + \sin^2\dfrac{\pi}{3}}$ 0.505

If x is expressed in radians, you can use the two infinite series below to approximate $\sin x$ and $\cos x$ to several decimal places.

$$\cos x = 1 - \dfrac{x^2}{2!} + \dfrac{x^4}{4!} - \dfrac{x^6}{6!} + \dfrac{x^8}{8!} - \cdots$$

$$\sin x = x - \dfrac{x^3}{3!} + \dfrac{x^5}{5!} - \dfrac{x^7}{7!} + \dfrac{x^9}{9!} - \cdots$$

Use a calculator to approximate each of the following to 5 decimal places. Use 3.14 for π.

41. $\sin\dfrac{\pi}{4}$ 0.70682
42. $\cos\dfrac{\pi}{6}$ 0.86617
43. $\sin\dfrac{5\pi}{6}$ 0.50209
44. $\cos\dfrac{\pi}{4}$ 0.70739

Radian Measure

APPLYING RADIAN MEASURE 18.11

Objectives **To sketch graphs of trigonometric functions using radian measure**
To write the solutions of trigonometric equations in radians

The table below lists some special degree-radian relationships.

Degrees	0°	30°	45°	60°	90°	120°	135°	150°	180°
Radians	0	$\frac{\pi}{6}$	$\frac{\pi}{4}$	$\frac{\pi}{3}$	$\frac{\pi}{2}$	$\frac{2\pi}{3}$	$\frac{3\pi}{4}$	$\frac{5\pi}{6}$	π

Degrees	210°	225°	240°	270°	300°	315°	330°	360°
Radians	$\frac{7\pi}{6}$	$\frac{5\pi}{4}$	$\frac{4\pi}{3}$	$\frac{3\pi}{2}$	$\frac{5\pi}{3}$	$\frac{7\pi}{4}$	$\frac{11\pi}{6}$	2π

To graph the function $y = \sin x$ using radian measure, label the horizontal axis in radians and the vertical axis in real numbers. Make a table of values by choosing special radian measures and finding the corresponding values of y. The graph of $y = \sin x$, $0 \leq x \leq 2\pi$, is shown.

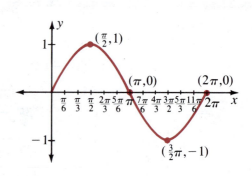

x	$y = \sin x$
0	0
$\frac{\pi}{6}$	$\frac{1}{2}$, or .5
$\frac{\pi}{3}$	$\frac{\sqrt{3}}{2}$, or .87
$\frac{\pi}{2}$	1
$\frac{2\pi}{3}$	$\frac{\sqrt{3}}{2}$, or .87
π	0
$\frac{7\pi}{6}$	$-\frac{1}{2}$, or $-.5$
$\frac{3\pi}{2}$	-1
$\frac{11\pi}{6}$	$-\frac{1}{2}$, or $-.5$
2π	0

Notice that the maximum value 1 occurs at $\frac{\pi}{2}$. The minimum value -1 occurs at $\frac{3\pi}{2}$. The period of $y = \sin x$ is 2π.

Example 1 **Graph $y = 2 \cos x$ for $0 \leq x \leq 2\pi$. Determine the period and amplitude.**

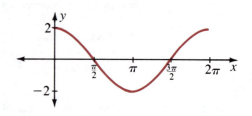

x	$\cos x$	$y = 2 \cos x$
0	1	2
$\frac{\pi}{2}$	0	0
π	-1	-2
$\frac{3\pi}{2}$	0	0
2π	1	2

Thus, period for $y = 2 \cos x$ is 2π.
   Amplitude for $y = 2 \cos x$ is 2.

Example 2

Graph $y = \sin \frac{1}{2}x$ for $0 \leq x \leq 4\pi$. Determine the period and amplitude.

x	$\frac{1}{2}x$	$y = \sin \frac{1}{2}x$
0	0	0
π	$\frac{\pi}{2}$	1
2π	π	0
3π	$\frac{3\pi}{2}$	-1
4π	2π	0

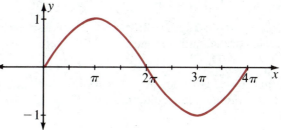

Thus, the period for $y = \sin \frac{1}{2}x$ is 4π and the amplitude is 1.

Example 3

Sketch $y = 2 \cos x$ and $y = \sin \frac{1}{2}x$ for $0 \leq x \leq 2\pi$ on the same set of axes. For how many values of x does $2 \cos x = \sin \frac{1}{2}x$?

Use Examples 1 and 2 to sketch the graphs.

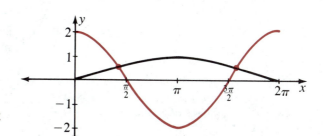

Thus, $2 \cos x = \sin \frac{1}{2}x$ for 2 values of x for $0 \leq x \leq 2\pi$.

Example 4

Solve $-3 \tan x - \sqrt{3} = 0$ if $0 \leq x \leq 2\pi$.

Solve for $\tan x$. Determine the reference angle.
$-3 \tan x = \sqrt{3}$
$\tan x = -\frac{\sqrt{3}}{3}$

Reference angle: 30°
↓   ↓
2nd quad.   4th quad.
150°   330°

◄ Tan is negative in quadrants 2 and 4.

Thus, the solutions are $\frac{5\pi}{6}$ and $\frac{11\pi}{6}$.

Example 5

Solve $2 \cos^2 x - 5 \cos x + 2 = 0$ if $0 \leq x \leq 2\pi$.

$(2 \cos x - 1)(\cos x - 2) = 0$ ◄ *Factor.*
$2 \cos x - 1 = 0$ or $\cos x - 2 = 0$ ◄ *Set each factor equal to 0.*
$\cos x = \frac{1}{2}$ $\cos x = 2$
Ref. $\angle$: 60°   No solution, since $-1 \leq \cos x \leq 1$
↓   ↓
1st quad. 4th quad.
60°   300°

Thus, the solutions are $\frac{\pi}{3}$ and $\frac{5\pi}{3}$.

Applying Radian Measure

Written Exercises

(A) Graph $y = \sin 2x$ for $-2\pi \le x \le 2\pi$ to find each value below.
1. maximum and minimum
   1, −1
2. period **π**
3. amplitude **1**

Graph $y = 3 \cos x$ for $-\pi \le x \le \pi$ to find each value below.
4. maximum and minimum
   3, −3
5. period **2π**
6. amplitude **3**

Sketch each graph for $0 \le x \le 2\pi$. Determine the period and the amplitude.
7. $y = \frac{1}{2} \sin x$ **$2\pi, \frac{1}{2}$**
8. $y = \cos 4x$ **$\frac{\pi}{2}, 1$**
9. $y = 3 \cos x$ **$2\pi, 3$**
10. $y = \sin 3x$ **$\frac{2\pi}{3}, 1$**
11. $y = 2 \cos \frac{1}{2}x$ **$4\pi, 2$**
12. $y = \frac{1}{2} \sin 2x$ **$\pi, \frac{1}{2}$**
13. $y = \frac{1}{3} \cos 3x$ **$\frac{2\pi}{3}, \frac{1}{3}$**
14. $y = 4 \sin 2x$ **$\pi, 4$**

For Ex. 15–26, see TE Answer Section.
Solve each trigonometric equation if $0 \le x \le 2\pi$.
15. $2 \sin x + 1 = 0$
16. $2 \cos x - \sqrt{3} = 0$
17. $\tan x = 1$
18. $5 \sin^2 x = 0$
19. $2 \cos^2 x - 1 = 0$
20. $2 \sin x + \sqrt{3} = 0$
21. $2 \cos^2 x + \cos x = 0$
22. $\sqrt{2} \sin^2 x - \sin x = 0$
23. $2 \cos^2 x - 7 \cos x - 4 = 0$
24. $\sin^2 x - \frac{1}{4} = 0$
25. $6 \sin^2 x - 7 \sin x - 5 = 0$
26. $4 \cos^2 - 1 = 0$

(B) 27. Sketch $y = 3 \sin x$ and $y = \cos 3x$ for $0 \le x \le 2\pi$ on the same set of axes. For how many values of x does $3 \sin x = \cos 3x$? **4**

28. Sketch $y = \cos \frac{1}{2}x$ and $y = \sin 3x$ for $0 \le x \le 2\pi$ on the same set of axes. For how many values of x does $\cos \frac{1}{2}x = \sin 3x$? **7**

Sketch each graph for $-\pi \le x \le 2\pi$. Determine the period and amplitude.
29. $y = \tan x$ **π, none**
30. $y = 3 \tan x$ **π, none**
31. $y = \tan 3x$ **$\frac{\pi}{3}$, none**
32. $y = -\cos x$ **$2\pi, 1$**
33. $y = 2 \sin (-2x)$ **$\pi, 2$**
34. $y = -\cos 2x$ **$\pi, 1$**
35. $y = -2 \sin \frac{1}{2}x$ **$4\pi, 2$**
36. $y = -3 \cos (-x)$ **$2\pi, 3$**
37. $y = -3 \sin (-2x)$ **$\pi, 3$**
38. $y = -\frac{1}{2} \sin 2x$ **$\pi, \frac{1}{2}$**
39. $y = \frac{1}{3} \cos 2x$ **$\pi, \frac{1}{3}$**
40. $y = -\frac{1}{2} \cos 3x$ **$\frac{2\pi}{3}, \frac{1}{2}$**

(C) 41. $y = \sec x$ **2π, none**
42. $y = \csc x$ **2π, none**
43. $y = \cot x$ **π, none**
44. $y = \sec 2x$ **π, none**
45. $y = -\frac{1}{2} \sec x$ **2π, none**
46. $y = \csc (-2x)$ **π, none**

Solve each trigonometric equation if $0 \le x \le 2\pi$.
47. $\sec x = -\sqrt{2}$ **$\frac{3\pi}{4}, \frac{5\pi}{4}$**
48. $\csc x = -2$ **$\frac{7\pi}{6}, \frac{11\pi}{6}$**
49. $3 \sec x - 2\sqrt{3} = 0$ **$\frac{\pi}{6}, \frac{11\pi}{6}$**
50. $2 \sec x + 1 = 0$ **No solution**
51. $2 \sec^2 x + 3 \sec x - 2 = 0$ **$\frac{2\pi}{3}, \frac{4\pi}{3}$**
52. $\csc^2 x - 4 = 0$ **$\frac{\pi}{6}, \frac{5\pi}{6}, \frac{7\pi}{6}, \frac{11\pi}{6}$**

53. Sketch $y = \sin 2x$ and $y = 2 \cos x$ for $-\pi \le x \le \pi$ on the same set of axes. For what values of x does $\sin 2x = 2 \cos x$? **$\frac{\pi}{2}, \frac{-\pi}{2}$**

54. Sketch $y = 2 \sin x$ and $y = \sin \frac{1}{2}x$ for $-\pi \le x \le 2\pi$ on the same set of axes. For what values of x does $2 \sin x = \sin \frac{1}{2}x$? **$0, \frac{5\pi}{6}$**

55. Sketch $y = -2 \sin \frac{1}{2}x$ and $y = -\frac{1}{2} \cos (-2x)$ for $0 \le x \le 2\pi$ on the same set of axes. For how many values of x does $-2 \sin \frac{1}{2}x = -\frac{1}{2} \cos (-2x)$? **2**

56. Sketch $y = \tan (-2x)$ and $y = -2 \cos \frac{1}{2}x$ for $0 \le x \le 2\pi$ on the same set of axes. For how many values of x does $\tan (-2x) = -2 \cos \frac{1}{2}x$? **5**

INVERSES OF TRIGONOMETRIC FUNCTIONS   18.12

Objectives
To determine the inverse of a trigonometric function
To graph inverses of the sin, cos, and tan functions
To apply inverse function notation

Recall that the inverse of a function is formed by interchanging the elements of each ordered pair of the function. The inverse of a function is not necessarily a function. The *inverse function* is formed when the inverse of a given function is also a function.

Example 1 **Find the inverse of $y = \cos x$.**

Interchange x and y.
$$y = \cos x$$
$$x = \cos y$$

Thus, the inverse of $y = \cos x$ is $x = \cos y$.

After finding the inverse of a function, it is customary to solve for y. $\cos y = x$ can be written in two other ways: $y = \text{arc cos } x$, or $y = \cos^{-1} x$. Each is read "y is the angle whose cos is x." In $y = \cos^{-1} x$, -1 is *not* an exponent.

Example 2 **Find u, if $u = \text{arc sin }(-\tfrac{1}{2})$, $0° \leq u \leq 360°$.**

Read $u = \text{arc sin }(-\tfrac{1}{2})$ as u is the angle whose sin is $-\tfrac{1}{2}$.
Write $u = \text{arc sin }(-\tfrac{1}{2})$ as $\sin u = -\tfrac{1}{2}$.
  Ref. ∠: 30°
  ↙   ↘
  3rd quad.   4th quad.
So, $\sin 210° = -\tfrac{1}{2}$ and $\sin 330° = -\tfrac{1}{2}$.
Thus, $u = 210°$ or $330°$.

Example 3 **Find θ, if $\theta = \text{arc cos } \dfrac{\sqrt{2}}{2}$, $0 \leq \theta \leq 2\pi$.**

Write $\theta = \text{arc cos } \dfrac{\sqrt{2}}{2}$ as $\cos \theta = \dfrac{\sqrt{2}}{2}$.
  Ref. ∠: 45°
  ↙   ↘
  1st quad.   4th quad.
So, $\cos 45° = \dfrac{\sqrt{2}}{2}$ and $\cos 315° = \dfrac{\sqrt{2}}{2}$.
Thus, $\theta = \dfrac{\pi}{4}$ or $\dfrac{7\pi}{4}$.

Inverses of Trigonometric Functions

In the next example you will learn how to graph the inverse of a trigonometric function.

Example 4 **Sketch $y = \text{arc sin } x$. Determine if the inverse is a function.**

Make a table of values for $y = \sin x$ and interchange the elements of the ordered pairs.

x	$y = \sin x$	$x = \sin y$	y
0	0	0	0
$\frac{\pi}{6}$	.5	.5	$\frac{\pi}{6}$
$\frac{\pi}{3}$	.87	.87	$\frac{\pi}{3}$
$\frac{\pi}{2}$	1	1	$\frac{\pi}{2}$
π	0	0	π
$\frac{3\pi}{2}$	-1	-1	$\frac{3\pi}{2}$
2π	0	0	2π
$-\frac{\pi}{2}$	-1	-1	$-\frac{\pi}{2}$
$-\pi$	0	0	$-\pi$
$-\frac{3\pi}{2}$	1	1	$-\frac{3\pi}{2}$
-2π	0	0	-2π

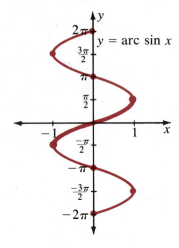

The graph does not pass the vertical line test, since there are many values of y for each x between -1 and 1.

Thus, the inverse of $y = \sin x$ is not a function.

The graphs of the inverses of $y = \cos x$ and $y = \tan x$ are shown below.

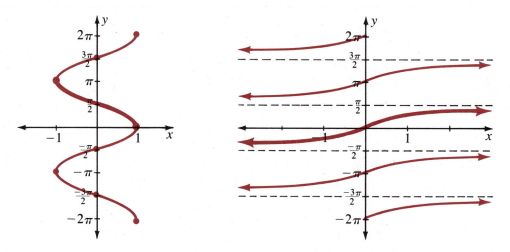

By restricting the interval of the graph of the inverse of a trig function, the inverse can be an inverse function in that interval. The shaded regions of the three graphs above are the graphs of the inverse functions.

> **Inverse trigonometric functions**
>
> To obtain inverse functions for $y = \sin x$, $y = \cos x$, and $y = \tan x$, restrict the range to values called *Principal Values*.
>
> $-\frac{\pi}{2} \leq \text{Arc sin} \leq \frac{\pi}{2}$, or $-90° \leq \text{Arc sin} \leq 90°$
>
> $0 \leq \text{Arc cos} \leq \pi$, or $0° \leq \text{Arc cos} \leq 180°$
>
> $-\frac{\pi}{2} \leq \text{Arc tan} \leq \frac{\pi}{2}$, or $-90° \leq \text{Arc tan} \leq 90°$
>
> In order to indicate principal value, capitalize the first letter of arc sin x as shown above.

The next three examples illustrate how to find a trig function given an inverse trig function.

Example 5

Find $\cos \theta$ if $\theta = \text{Arc sin}\left(-\frac{3}{5}\right)$.

Write $\theta = \text{Arc sin}\left(-\frac{3}{5}\right)$ as $\sin \theta = -\frac{3}{5}$.

The principal value of θ is

$-\frac{\pi}{2} \leq \theta \leq \frac{\pi}{2}$ or $-90° \leq \theta \leq 90°$

$\sin \theta = -\frac{3}{5}$

↙ ↘

3rd quad.   4th quad.

There is no solution in the 3rd quad. since the Principal Values for sin are in the 1st and 4th quad.

Find x.
$x^2 + y^2 = r^2$
$x^2 + 9 = 25$
$x^2 = 16$
$x = 4$

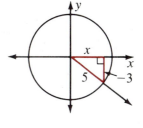

Thus, $\cos \theta = \frac{4}{5}$.

Example 6

Find $\sin \left(\text{Arc tan}\left(\frac{3}{4}\right)\right)$.

Let $A = \text{Arc tan} \frac{3}{4}$.

Write as $\tan A = \frac{3}{4}$ and $-90° \leq A \leq 90°$.

↙ ↘

1st quad.   3rd quad.
Find r.   No solution since
$x^2 + y^2 = r^2$ $-90° \leq A \leq 90°$
$9 + 16 = r^2$
$r = 5$

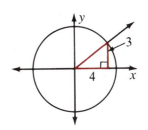

Thus, $\sin A = \frac{3}{5}$, so $\sin \left(\text{Arc tan}\left(\frac{3}{4}\right)\right) = \frac{3}{5}$.

Inverses of Trigonometric Functions

Example 7   Find $\sin^{-1}\left(\tan\frac{5\pi}{6}\right)$.

$\tan\frac{5\pi}{6} = \tan 150° = -\tan 30°$, or $-\frac{\sqrt{3}}{3}$

Let $A = \sin^{-1}\left(-\frac{\sqrt{3}}{3}\right)$.   Write as $\sin A = -\frac{\sqrt{3}}{3} = -\frac{1.732}{3}$, or $-.5773$.

Look in the table.
Ref. $\angle$: 35° to the nearest degree.

3rd quad.   4th quad.   Thus, $\sin^{-1}\left(\tan\frac{5\pi}{6}\right) = -35°$ or $-\frac{7\pi}{36}$.
No solution $-35°$

Reading in Algebra

Write A if the statement is *always* true. Write S if the statement is *sometimes* true. Write N if the statement is *never* true.

1. Inverses of trig functions are functions. **S**
2. The inverse for $y = \sin x$ is $x = \sin y$. **A**
3. The inverse for a function is formed by interchanging the elements of each ordered pair of the function. **A**
4. If $\theta = \arctan(-2)$, then $\tan \theta = -2$. **A**
5. The inverse function for $y = \sin x$ is Arc sin x. **S**
6. The Principal Values for Arc tan x are $-\frac{\pi}{2} \leq$ Arc tan $x \leq \frac{\pi}{2}$. **A**

Written Exercises

(A) Find the inverse of each function.

1. $y = \sin \frac{1}{2}x$ 2. $y = \cos x$ 3. $y = \tan x$ 4. $y = \cot x$ 5. $y = \frac{1}{2}\cos x$
   $y = 2\sin^{-1} x$ $y = \cos^{-1} x$ $y = \tan^{-1} x$ $y = \cot^{-1} x$ $y = \cos^{-1}(2x)$

Find θ for each of the following.

6. $\theta = $ Arc sin 0   **0°**
7. $\theta = $ Arc cos $\frac{1}{2}$ **60°**
8. $\theta = $ Arc sin $(-\frac{1}{2})$ **−30°**
9. $\theta = $ Arc sin $(-\frac{\sqrt{3}}{2})$ **−60°**
10. $\theta = $ Arc cos $\frac{\sqrt{2}}{2}$ **45°**
11. $\theta = $ Arc tan (-1) **−45°**
12. $\theta = $ Arc tan $(-\frac{\sqrt{3}}{3})$ **−30°**
13. $\theta = $ Arc cos (-1) **180°**

14. Find sin A if $A = $ Arc cos $(-\frac{5}{13})$. $\frac{12}{13}$
15. Find cos A if $A = $ Arc sin $\frac{3}{5}$. $\frac{4}{5}$
16. Find tan A if $A = $ Arc sin $(-\frac{12}{13})$. $-\frac{12}{5}$
17. Find sin A if $A = $ Arc tan $(-\frac{3}{4})$. $-\frac{3}{5}$

(B) Find the value of each of the following.

18. cos (Arc sin $(-\frac{\sqrt{2}}{2})$) $\frac{\sqrt{2}}{2}$
19. $\sin^{-1}(\tan(\frac{2}{3}\pi))$ **None**
20. Arc sin (cos $(-\frac{\pi}{4})$)   **45°**
21. tan (Arc cos $(\frac{\sqrt{3}}{3})$) $\sqrt{2}$
22. cos (Arc tan $(-\frac{4}{3})$) $\frac{3}{5}$
23. tan (Arc cos $(-\frac{5}{13})$) $-\frac{12}{5}$

(C) Sketch each graph. **For Ex. 24–35, see TE Answer Section.**

24. $y = $ Arc sin $2x$
25. $y = 2$ Arc cos $\frac{1}{2}x$
26. $y = -\frac{1}{2}$ Arc sin $(-2x)$
27. $y = -2$ Arc tan x
28. $y = -2$ Arc cos $(-\frac{1}{2}x)$
29. $y = $ Arc cot x
30. $y = $ Arc sin x
31. $y = $ Arc cos x
32. $y = $ Arc tan x
33. $y = 2$ Arc cot x
34. $y = $ Arc sec x
35. $y = $ Arc csc x

COMPLEX NUMBERS AND DE MOIVRE'S THEOREM   18.13

Objectives **To write a complex number in polar form**
To write the polar form of a number as a complex number
To find a power of a complex number

The relationship shown in the figure at the right is based on the Pythagorean relation $c^2 = a^2 + b^2$. The horizontal axis is the real number axis, while the vertical axis is the imaginary axis.

Absolute Value of a Complex Number

The $(a + bi)$ form for a complex number is frequently referred to as the **rectangular form** of a complex number. It is written as $r = |a + bi| = \sqrt{a^2 + b^2}$. Recall that the absolute value of a complex number like $5 - 2i$ is $r = |5 - 2i| = \sqrt{5^2 + (-2)^2}$, or $\sqrt{29}$.

According to the definition of sine and cosine, $\sin u = \dfrac{b}{\sqrt{a^2 + b^2}}$ and $\cos u = \dfrac{a}{\sqrt{a^2 + b^2}}$ since $r = \sqrt{a^2 + b^2}$, $\sin u = \dfrac{b}{r}$ or $b = r \sin u$, and $\cos u = \dfrac{a}{r}$ or $a = r \cos u$.

The rectangular form of a complex number can be written in trigonometric form or polar form as indicated below.

$$a + bi = r \cos u + (r \sin u)i \text{ or } r(\cos u + i \sin u)$$

Example 1 **Write the complex number $1 + i$ in polar form.**

$\boxed{1}_a + \boxed{1}_b i$ ▶ $a = 1$ and $b = 1$. So, $r = \sqrt{1^2 + 1^2}$, or $\sqrt{2}$.

$\sin u = \dfrac{b}{r}$ $\cos u = \dfrac{a}{r}$

$= \dfrac{1}{\sqrt{2}}$, or $\dfrac{\sqrt{2}}{2}$ $= \dfrac{1}{\sqrt{2}}$, or $\dfrac{\sqrt{2}}{2}$

Since both a and b are positive, the angle is in quadrant I. So, $u = \dfrac{\pi}{4}$, or 45°.

Substitute:

$r = \sqrt{2}, u = \dfrac{\pi}{4}$ ▶ $a + bi = r \cos u + ir \sin u$

$1 + i = \sqrt{2} \left(\cos \dfrac{\pi}{4} + i \sin \dfrac{\pi}{4} \right)$

Thus, the polar form of $1 + i$ is $\sqrt{2} \left(\cos \dfrac{\pi}{4} + i \sin \dfrac{\pi}{4} \right)$.

The polar form of a complex number can also be written in rectangular form.

Example 2   Write $2\left(\cos \frac{\pi}{3} + i \sin \frac{\pi}{3}\right)$ in rectangular form.

$\cos \frac{\pi}{3} = \frac{1}{2}$ ▶ $\quad 2\left(\cos \frac{\pi}{3} + i \sin \frac{\pi}{3}\right) = 2\left(\frac{1}{2} + i\frac{\sqrt{3}}{2}\right)$

$\sin \frac{\pi}{3} = \frac{\sqrt{3}}{2}$ ▶ $\qquad\qquad\qquad\qquad\qquad = 1 + i\sqrt{3}$

Thus, the rectangular form of $2\left(\cos \frac{\pi}{3} + i \sin \frac{\pi}{3}\right)$ is $1 + i\sqrt{3}$.

De Moivre's theorem states a method for computing powers of a complex number.

De Moivre's theorem   For all real numbers r, n, and θ, $r(\cos \theta + i \sin \theta)^n = r^n(\cos n\theta + i \sin n\theta)$.

Example 3   Compute $(-2 + 2i)^6$.

$a = -2$, $b = 2$, and $r = \sqrt{4 + 4}$, or $2\sqrt{2}$

$\sin \theta = \frac{b}{r} = \frac{2}{2\sqrt{2}}$, or $\frac{\sqrt{2}}{2}$ $\cos \theta = \frac{a}{r} = \frac{-2}{2\sqrt{2}}$, or $-\frac{\sqrt{2}}{2}$

Let $-2 + 2i = r(\cos \theta + i \sin \theta)$

$(-2 + 2i)^6 = [r(\cos \theta + i \sin \theta)]^6$

$\qquad\qquad\quad = \left[2\sqrt{2}\left(\frac{-\sqrt{2}}{2} + i\frac{\sqrt{2}}{2}\right)\right]^6$

$\cos \frac{3}{4}\pi = -\frac{\sqrt{2}}{2}$ and $\sin \frac{3}{4}\pi = \frac{\sqrt{2}}{2}$ ▶ $\left[2\sqrt{2}\left(\cos \frac{3}{4}\pi + i \sin \frac{3}{4}\pi\right)\right]^6$

$\qquad\qquad\qquad\qquad\qquad\qquad\qquad (2\sqrt{2})^6\left(\cos 6 \cdot \frac{3}{4}\pi + i \sin 6 \cdot \frac{3}{4}\pi\right)$

$\qquad\qquad\qquad\qquad\qquad\qquad\qquad 512\left(\cos \frac{9}{2}\pi + i \sin \frac{9}{2}\pi\right)$

$\cos \frac{9}{2}\pi = 0$ and $\sin \frac{9}{2}\pi = 1$ ▶ $512(0 + i(1))$, or $512i$

Thus, $(-2 + 2i)^6 = 512i$.

Written Exercises

Write each of the following in rectangular form.

Ⓐ   1. $3\left(\cos \frac{\pi}{4} + i \sin \frac{\pi}{4}\right)$ $\frac{3\sqrt{2}}{2} + \frac{3\sqrt{2}}{2}i$ 2. $\left(\cos \frac{3\pi}{4} + i \sin \frac{3\pi}{4}\right)$ $-\frac{\sqrt{2}}{2} + \frac{\sqrt{2}}{2}i$

3. $4\left(\cos \frac{\pi}{6} + i \sin \frac{\pi}{6}\right)$ $2\sqrt{3} + 2i$ 4. $5\left(\cos \frac{3\pi}{2} + i \sin \frac{3\pi}{2}\right)$ $-5i$

Compute by using De Moivre's theorem.

Ⓑ   5. $(1 - i\sqrt{3})^4$ 6. $(2 + i)^6$ 7. $(-2 + 2i)^5$ 8. $(\sqrt{3} + 2i)^3$

$-8 + 8\sqrt{3}i$ $-117 + 43.98i$ $128 - 128i$ $-5.8905\sqrt{7} + 3.7821\sqrt{7}i$

Ⓐ Exercises are 1, 3, 4, 8, 9, 14, 15, 17–19, 21–26, 31–39, 44–47, 49. Ⓒ Exercises are starred.
The rest are Ⓑ Exercises.

CHAPTER EIGHTEEN REVIEW

Vocabulary
Amplitude of a function [18.1–18.3]
Inverse of a trig function [18.12]
Periodic function [18.1–18.3]
Principal Values [18.12]
Pythagorean identities [18.8]
Quotient identities [18.8]
Radian measure [18.10]
Reciprocal identities [18.8]

Sketch each graph for $0° \leq x \leq 360°$. Determine the period and amplitude. [18.1–18.5]
1. $y = 3 \cos x$ **360°, 3**   2. $y = \sin(-2x)$
3. $y = \frac{1}{2} \sin(\frac{1}{2} x)$ 4. $y = \tan 3x$ **60°, none**
5. $y = -\frac{1}{2} \cos(-2x)$ ★6. $y = -2 \sec x$
2. 180°, 1   3. 720°, $\frac{1}{2}$ 5. 180°, $\frac{1}{2}$ 6. 360°, none

7. Sketch $y = \sin 3x$ and $y = 2 \cos x$ for $-180° \leq x \leq 180°$ on the same set of axes. For how many values of x does $\sin 3x = 2 \cos x$? **2** [18.4]

8. Find $\log \sin 28° 33'$. **9.6794 − 10** [18.6]

9. Find $\log \cot 48° 47'$. **9.9425 − 10** [18.6]

10. Find u, $0° \leq u \leq 90°$, if $\log \cos u = 9.6348 − 10$. **64° 27'**

11. Find u, $0° \leq u \leq 90°$, if $\log \tan u = 9.8732 − 10$. **36° 45'** [18.6]

Find each value to four significant digits. Use logarithms. [18.6]
12. $\frac{(\cos 43°)^4}{\sqrt{38.7}}$ **0.1454**   13. $\frac{(498)^3 \sqrt{.038}}{\tan 57° 43'}$

14. 0°, 180°   15. 0°, 180°, 360°, 45°, 135°   13. 15,210,000

Solve for θ to the nearest degree ($0° \leq \theta \leq 360°$).
14. $\cos^2 \theta - 1 = 0$ 15. $\sin^2 \theta - \frac{\sqrt{2}}{2} \sin \theta = 0$
16. $2 \sin^2 \theta + 5 \sin \theta - 6 = 0$ **62°, 118°** [18.7]

Solve for θ if $0 \leq \theta \leq 2\pi$. [18.11]
17. $\cos \theta = -\frac{\sqrt{2}}{2}$ $\frac{3\pi}{4}$ 18. $\tan^2 \theta + \tan \theta = 0$
19. $2 \sin^2 \theta - 3 \sin \theta + 1 = 0$ $\frac{\pi}{6}, \frac{5\pi}{6}, \frac{\pi}{2}$
★20. $16 \sec^2 \theta - 9 = 0$ **no solution**

Express in terms of sin, cos, or both. [18.8]
21. $\tan 23°$ $\frac{\sin 23°}{\cos 23°}$ 22. $\csc 38°$ $\frac{1}{\sin 38°}$
23. $\cot 74° 18'$ $\frac{\cos 74° 18'}{\sin 74° 18'}$ 24. $\sec(-305°)$ $\frac{1}{\cos(-305°)}$

Verify for $A = 240°$ [18.8]
25. $\sin^2 A + \cos^2 A = 1$
26. $\tan^2 A = \sec^2 A - 1$
For Ex. 25–26, 30–31, see TE Answer Section.

Write an equivalent expression using only $\cos x$. [18.8]
27. $\cos x \cdot \frac{1}{\sec x}$ $\cos^2 x$ 28. $\tan x$ $\frac{\pm\sqrt{1-\cos^2 x}}{\cos x}$
29. $\sin x \cdot \sec x$ $\frac{\pm\sqrt{1-\cos^2 x}}{\cos x}$

Prove each identity. [18.9]
30. $\frac{1}{4} \csc \theta = \frac{\sin \theta}{3 \sin^2 \theta - \cos^2 \theta + 1}$
31. $\frac{\sin \theta}{\cot \theta} + \cos \theta = \sec \theta$

Express in radians. [18.10]
32. $60°$ $\frac{\pi}{3}$ 33. $-45°$ $\frac{-\pi}{4}$ 34. $330°$ $\frac{11\pi}{6}$ 35. $-140°$ $\frac{-7\pi}{9}$

Express in degrees. [18.10]
36. $\frac{\pi}{12}$ **15°**   37. $-\frac{4}{3}\pi$ 38. 6π 39. $-\frac{3}{2}\pi$
37. −240°   38. 1080°   39. −270°

Evaluate. [18.10]
40. $\tan \frac{7\pi}{4}$ **−1**   41. $\cos\left(-\frac{\pi}{6}\right)$ $\frac{\sqrt{3}}{2}$
42. $\frac{\cos \frac{\pi}{3} - \sin \frac{\pi}{4}}{\tan\left(-\frac{\pi}{4}\right)}$ $\frac{\sqrt{2}-1}{2}$ 43. $\frac{\cot\left(-\frac{\pi}{3}\right) + \tan \frac{3\pi}{4}}{\sin \frac{11\pi}{6}}$ $\frac{2\sqrt{3}+6}{3}$

Find the inverse of each function. Is the inverse a function? **46. 150°   47. 35°** [18.12]
44. $y = \cos x$ 45. $y = 3 \sin x$
$y = \cos^{-1} x$, **no** $y = \sin^{-1} \frac{x}{3}$, **no**
Find each θ. (Ex. 46–47) [18.12]
46. $\theta = \text{Arc } \cos\left(-\frac{\sqrt{3}}{2}\right)$ 47. $\theta = \text{Arc } \tan\left(\frac{\sqrt{2}}{2}\right)$
48. Compute $(1 + i)^4$. **−4** [18.13]
49. Write $2\left(\cos \frac{\pi}{3} + i \sin \frac{\pi}{3}\right)$ in rectangular form. $1 + \sqrt{3}i$ [18.13]

18. $0, \pi, 2\pi, \frac{3\pi}{4}, \frac{7\pi}{4}$

Chapter Eighteen Review

Ⓐ Exercises are 1–4, 6, 10–16, 19–28, 38. Ⓒ Exercises are starred.
The rest are Ⓑ Exercises.

CHAPTER EIGHTEEN TEST

Sketch each graph for $0° \leq x \leq 360°$. Determine the period and amplitude.

1. $y = 2 \sin x$ **360°, 2**
2. $y = \cos\left(\frac{1}{2}x\right)$ **720°, 1**
3. $y = \frac{1}{2} \sin 2x$ **180°, $\frac{1}{2}$**
4. $y = \tan 2x$ **90°, none**
5. Sketch $y = \sin 2x$ and $y = 3 \cos x$ for $0° \leq x \leq 360°$ on the same set of axes. For how many values of x does $\sin 2x = 3 \cos x$? **2**
6. Find log cos 36° 45'. **9.9038 − 10**
7. Find u, $0° \leq u \leq 90°$, if log tan $u = 9.8685 − 10$. **36° 27'**

Find the value to four significant digits. Use logarithms.

8. $\dfrac{(\sin 38°)^3}{\sqrt{.437}}$ **0.3529**
9. $\dfrac{(137)^4 \sqrt{.0867}}{\cos 43° 36'}$ **143,240,000**

Solve for θ to the nearest degree ($0° \leq \theta \leq 360°$).

10. $2 \sin^2 \theta - 1 = 0$ **45°, 135°, 225°, 315°**
11. $2 \cos^2 \theta - \cos \theta - 1 = 0$ **120°, 240°, 0°, 360°**

Solve for θ if $0 \leq \theta \leq 2\pi$.

12. $\tan^2 \theta = \tan \theta$ **$0, \pi, 2\pi, \frac{\pi}{4}, \frac{5\pi}{4}$**
13. $\sin \theta = -\frac{\sqrt{3}}{2}$ **$\frac{4\pi}{3}, \frac{5\pi}{3}$**

Express in terms of sin, cos, or both.

14. $\tan(-230°)$ **$\frac{\sin(-230°)}{\cos(-230°)}$**
15. $\sec 43°$ **$\frac{1}{\cos 43°}$**
16. Verify for $A = 150°$. $\cos^2 A = 1 - \sin^2 A$

Write an equivalent expression using only cos x.

17. $\cot x$ **$\frac{\pm \cos x}{\sqrt{1 - \cos^2 x}}$**
18. $\tan x \cdot \csc x$ **$\frac{1}{\cos x}$**

Prove the identity.

19. $\sec \theta = \dfrac{\tan \theta}{\csc \theta} + \cos \theta$

Express in radians.

20. 45° **$\frac{\pi}{4}$**
21. −330° **$-\frac{11\pi}{6}$**

Express in degrees.

22. $-\frac{5\pi}{6}$ **−150°**
23. $\frac{5}{3}\pi$ **300°**

Evaluate.

24. $\cos\left(-\frac{2\pi}{3}\right)$ **$-\frac{1}{2}$**
25. $\dfrac{\sin \frac{\pi}{6} + \cos \frac{7}{6}\pi}{\tan\left(-\frac{\pi}{3}\right)}$ **$\frac{-\sqrt{3} + 3}{6}$**

Find the inverse of each function. Is the inverse a function?

26. $y = \tan x$ **$y = \tan^{-1} x$, no**
27. $y = -2 \sin x$ **$y = \sin^{-1}\left(-\frac{x}{2}\right)$, no**

Find θ.

28. $\theta = \text{Arc } \sin\left(-\frac{\sqrt{2}}{2}\right)$ **$\frac{-\pi}{4}$**

Find each value.

29. $\sin\left(\text{Arc } \tan \frac{5}{12}\right)$ **$\frac{5}{13}$**
30. $\tan\left(\text{Arc } \cos \frac{5}{6}\right)$ **$\frac{\sqrt{11}}{5}$**

31. Sketch $y = -\cos\left(\frac{1}{2}x\right)$ and $y = 3 \sin(2x)$ for $-\pi \leq x \leq 2\pi$ on the same set of axes. For how many values of x does $-\cos \frac{1}{2}x = 3 \sin 2x$? **6**

Sketch each graph.

★ 32. $y = \text{Arc } \cos x$
★ 33. $y = \text{Arc } \tan x$
★ 34. $y = \text{arc } \csc x$

Solve for θ if $0 \leq \theta \leq 360°$.

★ 35. $4 \csc^2 \theta - 9 = 0$ **41° 50', 138° 10', 221° 50', 318° 10'**

Evaluate.

★ 36. $\dfrac{\sec\left(-\frac{\pi}{4}\right) - \csc \frac{3}{2}\pi}{-6 \sec \frac{4\pi}{3}}$ **$\frac{1 + \sqrt{2}}{12}$**

For complete solutions, see TE Answer Section.

37. Compute $(-2 + 4i)^3$. **$39.356\sqrt{5} - 7.152\sqrt{5}i$**
38. Write $4\left(\cos \frac{3}{4}\pi + i \sin \frac{3}{4}\pi\right)$ in rectangular form. **$-2\sqrt{2} + 2\sqrt{2}i$**

COLLEGE PREP TEST

Directions: In each question you are to *compare* a Quantity I and a Quantity II. Choose your answer in the following way. Choose:
(A) if Quantity I is greater than II,
(B) if Quantity II is greater than I,
(C) if Quantity I = Quantity II,
(D) if the relationship cannot be determined

In questions 1 and 2 use the information given. Right triangle *ABC* with right angle *C*

1. Quantity I $(AC)^2 + (BC)^2$ **C**
   Quantity II $(AB)^2$

2. Quantity I $2(AC)^2$ **D**
   Quantity II $(AB)^2$

3. Information: Triangle *ABC*

   Quantity I $AC + BC$ **A**
   Quantity II AB

4. Information: Triangle *ABC* with m∠*C* < 90°. Triangle *RST* with m∠*T* > 90°.

   Quantity I AB **D**
   Quantity II RS

5. Information: Right triangle *ABC*, *C* is a right angle and m∠*B* = 30°.

   Quantity I AB **C**
   Quantity II $2AC$

6. Quantity I $\left(\dfrac{\pi}{2} + \dfrac{\pi}{6}\right)$ radians **C**
   Quantity II 30° + 90°

7. Information: Right triangle *ABC* with right angle *C*. Right triangle *RST* with right angle *T*. The measure of angle *S* equals the measure of angle *B*.

   Quantity I $(AC)(RS)$ **C**
   Quantity II $(AB)(RT)$

8. Information: Right triangle *ABC* with right angle *C* and m∠*A* = 60°. Right triangle *RST* with right angle *T* and m∠*R* = 45°.

   Quantity I *BC* **D**
   Quantity II *ST*

9. Information: m∠*A* > m∠*B*
   m∠*B* > m∠*C*

   Quantity I 2 m∠*A*
   Quantity II m∠*B* + m∠*C* **A**

10. Information:

    Quantity I $d° + c°$
    Quantity II $a° + b° + c°$ **C**

11. Information: A triangle has angles with measures of 90°, *y*°, and *z*°.
    Quantity I 45° **C**
    Quantity II *y*°

12. Information:

    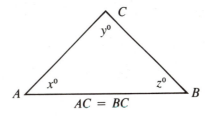

    Quantity I $x = 60°$ **C**
    Quantity II y

College Prep Test

19 TRIGONOMETRIC LAWS AND FORMULAS

Non-Routine Problem Solving

A measure, such as a measure of life expectancy of a product, is normally distributed. If many such measures are plotted, they form a **normal (bell-shaped) curve.** The normal curve reflects the mean and the variation for a given measure. For a measure that is normally distributed, approximately 68.2% of the cases will fall between ±1 standard deviations (±1 σ) of the mean, 95.4% will fall between ±2 σ of the mean, and 99.8% will fall between ±3 σ of the mean.

Percent of cases under portions of the normal curve

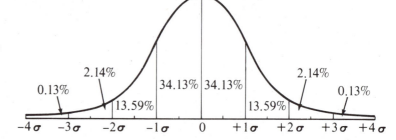

Standard deviations

Problem

The life expectancy of a certain automobile tire closely approximates the normal curve. The estimated average life is 35,000 miles, with one standard deviation of 4,000 miles. If you purchase a tire, what is the probability that it will last between 31,000 and 39,000 miles? more than 31,000 miles?

0.683; 0.864

LAW OF COSINES   19.1

Objective   **To find the length of a side of a triangle using the Law of Cosines**

If two sides and the included angle of any triangle are known, the length of the third side can be found. You will now learn how to derive formulas to do this and how to apply these formulas.

For the general triangle ABC, you can show that $a^2 = b^2 + c^2 - 2bc \cos A$.
In right $\triangle ABD$, $c^2 = h^2 + x^2$, or $h^2 = c^2 - x^2$.
In right $\triangle BDC$, $a^2 = h^2 + y^2$.

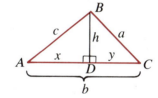

$$
\begin{aligned}
a^2 &= h^2 + y^2 \\
\text{Since } h^2 = c^2 - x^2 \quad &= (c^2 - x^2) + y^2 \\
&= (y^2 - x^2) + c^2 \\
\text{By factoring} \quad &= (y + x)(y - x) + c^2 \\
\text{Since } b = x + y,\ y = b - x \quad &= (b - x + x)(b - x - x) + c^2 \\
&= (b)(b - 2x) + c^2 \\
\text{Since } \cos A = \tfrac{x}{c},\ x = c \cos A \quad &= b^2 - 2b(c \cos A) + c^2 \\
&= b^2 - 2bc \cos A + c^2
\end{aligned}
$$

So, $a^2 = b^2 + c^2 - 2bc \cos A$.

Law of cosines

For any triangle ABC,
$a^2 = b^2 + c^2 - 2bc \cos A$
$b^2 = a^2 + c^2 - 2ac \cos B$
$c^2 = a^2 + b^2 - 2ab \cos C$

In general, the Law of Cosines states that the square of one side of a triangle equals the sum of the squares of the other two sides minus twice the product of these two sides and the cosine of the included angle.

Example 1   In $\triangle ABC$, $b = 8$, $c = 7$, and $m \angle A = 45°$. Find a to the nearest unit.

Use $a^2 = b^2 + c^2 - 2bc \cos A$.
$= 8^2 + 7^2 - 2(8)(7) \cos 45°$
$= 64 + 49 - 112\left(\dfrac{\sqrt{2}}{2}\right)$
$\doteq 64 + 49 - 56(1.414)$
$\doteq 33.8$

Thus, $a \doteq \sqrt{33.8}$, or 6.

Law of Cosines   539

Example 2 In $\triangle ABC$, $a = 4$, $b = 6$, and $m\angle c = 150°$. Find c to the nearest unit.

Use $c^2 = a^2 + b^2 - 2ab \cos C$.
$$= 4^2 + 6^2 - 2(4)(6) \cos 150°$$
$$= 16 + 36 - 48(-\cos 30°)$$
$$= 52 - 48\left(-\frac{\sqrt{3}}{2}\right)$$
$$= 52 + 24\sqrt{3}$$
$$= 52 + 24(1.732)$$
$$c^2 = 52 + 41.568$$
$$c = \sqrt{93.568} \doteq 9.673$$

Thus, $c = 10$ to the nearest unit.

For a right triangle, the Law of Cosines becomes the Pythagorean theorem. In right triangle ABC, $m\angle c = 90°$. Use the Law of Cosines to find c.

$$c^2 = a^2 + b^2 - 2ab \cos C$$
$$= a^2 + b^2 - 2ab \cos 90°$$
$$= a^2 + b^2 - 2ab(0)$$
So, $c^2 = a^2 + b^2$.

The next two examples illustrate how the Law of Cosines can be used to solve problems.

Example 3 A surveyor at point R sights two points S and T on opposite sides of a lake. Point R is 120 meters from S and 180 meters from T, and the measure of angle R is 38°. Find the distance across the lake to the nearest meter.

Use $r^2 = s^2 + t^2 - 2st \cos R$.
$$= (180)^2 + (120)^2 - 2(180)(120) \cos 38°$$
$$= 32{,}400 + 14{,}400 - 43{,}200(.7880)$$
$$= 46{,}800 - 34{,}041.6$$
$$r^2 = 12{,}758.4$$
$$r = \sqrt{12{,}758.4} \doteq 113$$

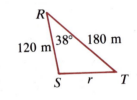

Thus, the distance across the lake $\doteq 113$ meters.

Example 4 Two forces of 30 lb and 40 lb act on a body at an angle of 60°. Find the magnitude of the resultant to the nearest pound.

$m\angle A + m\angle B = 180°$
$m\angle A + 60° = 180°$
$m\angle A = 120°$

Use $a^2 = t^2 + s^2 - 2ts \cos A$.
$$= (40)^2 + (30)^2 - 2(40)(30) \cos 120°$$
$$= 1{,}600 + 900 - 2{,}400 \left(-\tfrac{1}{2}\right)$$
$$a^2 = 3{,}700$$
$$a \doteq 60.828$$

Thus, $a = 61$, to the nearest pound.

Oral Exercises

Give the form of the Law of Cosines needed to find x.

1.
2.
3.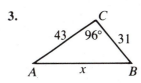

1. $b^2 = a^2 + c^2 - 2ac \cos B$
2. $a^2 = b^2 + c^2 - 2bc \cos A$
3. $c^2 = a^2 + b^2 - 2ab \cos C$

Written Exercises

A Each exercise refers to triangle ABC. Find the length of the indicated side to the nearest unit.

1. $a = 8, b = 9, m\angle C = 60°$. Find c. **9**
2. $a = 10, c = 18, m\angle B = 120°$. Find b. **25**
3. $b = 12, c = 15, m\angle A = 135°$. Find a. **25**
4. $a = 15, b = 14, m\angle C = 150°$. Find c. **28**
5. $a = 13, c = 18, m\angle B = 30°$. Find b. **9**
6. $b = 17, c = 12, m\angle A = 45°$. Find a. **12**
7. $b = 11.3, c = 13.1, m\angle A = 120°$. Find a. **21**
8. $a = 17.3, c = 19.7, m\angle B = 135°$. Find b. **34**
9. $a = 33.4, b = 22.7, m\angle C = 150°$. Find c. **54**
10. $b = 51.7, c = 73.2, m\angle A = 60°$. Find a. **65**

B
11. $b = 47.3, c = 72.9, m\angle A = 18°$. Find a. **32**
12. $a = 63.4, c = 27.6, m\angle B = 39°$. Find b. **45**
13. $a = 12.9, b = 16.3, m\angle C = 12°30'$. Find c. **5**
14. $b = 19.6, c = 35.3, m\angle A = 45°40'$. Find a. **26**
15. $b = 19.1, c = 34.6, m\angle A = 73°18'$. Find a. **34**
16. $a = 41.3, b = 16.9, m\angle C = 81°26'$. Find c. **42**
17. $a = 49.7, c = 43.1, m\angle B = 120°33'$. Find b. **81**
18. $b = 123, c = 119, m\angle A = 105°48'$. Find a. **193**

19. Two sides of a parallelogram form an angle of 72°. The lengths of two of the sides are 18 and 24 centimeters. How long is the shorter diagonal? **25**

20. A side and a base of an isosceles trapezoid form an angle of 54°. The length of the side is 14 centimeters and of the base is 34 centimeters. How long is a diagonal? **28**

21. Two ships left from the same port on paths that form an angle of 68°. Ship A traveled 400 km; ship B traveled 325 km. Find to the nearest kilometer the distance between them. **410 km**

22. Two airplanes left the same airport and formed a 65° angle in their flight paths. The first plane flew at 600 km/h, and the second flew at 800 km/h. How far apart were they after 2 hours? **1,541 km**

23. Two forces of 450 lb and 600 lb act on a body at an angle of 40°50'. Find the magnitude of the resultant to the nearest pound. **985 lb**

24. Two forces of 300 lb and 400 lb act on a body at an angle of 30°40'. Find the magnitude of the resultant to the nearest pound. **676 lb**

For Ex. 25–28, see TE Answer Section.

C
25. In $\triangle ABC$, $m\angle B < 90°$. Prove that $b^2 = a^2 + c^2 - 2ac \cos B$.

26. In $\triangle ABC$, $m\angle C < 90°$. Prove that $c^2 = a^2 + b^2 - 2ab \cos C$.

27. In $\triangle ABC$, $m\angle A > 90°$. Prove that $a^2 = b^2 + c^2 - 2bc \cos A$.

28. In $\triangle ABC$, $m\angle B > 90°$. Prove that $b^2 = a^2 + c^2 - 2ac \cos B$.

CUMULATIVE REVIEW

Use the Pythagorean relation to determine whether the following are measures of the sides of a right triangle.

1. 8, 15, 17 **Yes**
2. 5, 12, 13 **Yes**
3. 1, $2\sqrt{7}$, $3\sqrt{3}$ **Yes**

USING THE LAW OF COSINES TO FIND ANGLE MEASURES   19.2

Objective **To use the Law of Cosines to find the degree measure of an angle**

You can apply the Law of Cosines to find the measure of an angle of any triangle if the lengths of the three sides are known.

Using the Law of Cosines involving cos A, solve for cos A as follows.

$$a^2 = b^2 + c^2 - 2bc \cos A$$
$$2bc \cos A = b^2 + c^2 - a^2$$
So, $\cos A = \dfrac{b^2 + c^2 - a^2}{2bc}$.

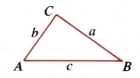

Finding an angle by using the law of cosines

For any triangle ABC,
$$\cos A = \frac{b^2 + c^2 - a^2}{2bc}$$
$$\cos B = \frac{a^2 + c^2 - b^2}{2ac}$$
$$\cos C = \frac{a^2 + b^2 - c^2}{2ab}.$$

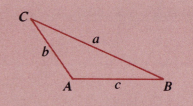

Example 1 In $\triangle ABC$, $a = 6$, $b = 7$ and $c = 4$. Find m$\angle A$ and m$\angle B$ to the nearest degree.

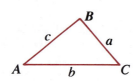

$$\cos A = \frac{b^2 + c^2 - a^2}{2bc}$$
$$= \frac{7^2 + 4^2 - 6^2}{2(7)(4)}$$
$$= \frac{49 + 16 - 36}{56}$$
$$= \frac{29}{56}$$
$\cos A \doteq .5179$

$$\cos B = \frac{a^2 + c^2 - b^2}{2ac}$$
$$= \frac{6^2 + 4^2 - 7^2}{2(6)(4)}$$
$$= \frac{36 + 16 - 49}{48}$$
$$= \frac{3}{48}$$
$\cos B \doteq .0625$

Thus, m$\angle A = 59°$ and m$\angle B = 86°$, to the nearest degree.

Example 2 In △ABC, the length of the sides are 3, 4, and 6. Find the measure of the smallest angle to the nearest degree.

The angle opposite the shortest side has the least measure.

$\cos B = \dfrac{a^2 + c^2 - b^2}{2ac}$

$= \dfrac{36 + 16 - 9}{2(6)(4)}$

$\cos B = \dfrac{43}{48}$, or .8958

$m\angle B \doteq 26°$

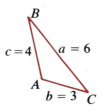

Thus, the measure of the smallest angle to the nearest degree is 26°.

Written Exercises

Find the measure of each angle to the nearest degree.

Ⓐ
1. $a = 3, b = 8, c = 7$. Find $m\angle A$. **22°**
2. $a = 9, b = 3, c = 9$. Find $m\angle B$. **19°**
3. $a = 13, b = 12, c = 5$. Find $m\angle B$. **67°**
4. $a = 2, b = 8, c = 7$. Find $m\angle C$. **54°**
5. $a = 7, b = 9, c = 3$. Find $m\angle A$. **41°**
6. $a = 6, b = 3, c = 4$. Find $m\angle C$. **36°**
7. $a = 10, b = 2, c = 9$. Find $m\angle C$. **55°**
8. $a = 10, b = 6, c = 9$. Find $m\angle A$. **81°**

Ⓑ
9. $a = 12.3, b = 9.6, c = 7.3$. Find $m\angle A$. **92°**
10. $a = 14.3, b = 19.5, c = 26.1$. Find $m\angle C$. **100°**
11. $a = 17.5, b = 16.4, c = 11.7$. Find $m\angle C$. **40°**
12. $a = 9.7, b = 23.4, c = 17.9$. Find $m\angle A$. **23°**
13. $a = 9.6, b = 17.5, c = 12.8$. Find $m\angle B$. **102°**
14. $a = 36.1, b = 44.7, c = 28.6$. Find $m\angle B$. **87°**

15. The measures of two sides of a parallelogram are 40 cm and 50 cm, and the length of one diagonal is 70 cm. Find the measures of the angles of the parallelogram. **102°, 78°**

16. Two straight roads $\overrightarrow{MN}$ and $\overrightarrow{PN}$ intersect at a town N. M is 62 km from N and P is 77 km from N. MP is 85 km. Find the angle that roads $\overrightarrow{MN}$ and $\overrightarrow{PN}$ make with each other. **75°**

In △ABC, find the measure of the smallest angle to the nearest degree.
17. $a = 6, b = 8, c = 9$ **41°**
18. $a = 12, b = 13, c = 15$ **50°**
19. $a = 9, b = 3, c = 8$ **19°**

For Ex. 20–21, see TE Answer Section.

Ⓒ
20. Prove: For any △ABC,

$\cos B = \dfrac{a^2 + c^2 - b^2}{2ac}$

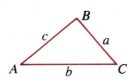

21. Prove: For any △ABC,

$\cos C = \dfrac{a^2 + b^2 - c^2}{2ab}$

Using the Law of Cosines to Find Angle Measures 543

AREA OF A TRIANGLE   19.3

Objective **To apply formulas for the area of a triangle given the measures of two sides and the included angle**

If two sides and the included angle of any triangle are known, a formula for finding the area of the triangle can be derived as follows.

Recall that the area of a triangle is one half the product of the base and the height.

Area of $\triangle ABC = k = \frac{1}{2}bh$.
But $\sin A = \frac{h}{c}$, or $h = c \sin A$.

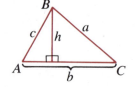

Now, substitute $c \sin A$ for h in the original formula.
$k = \frac{1}{2}b(c \sin A)$ or $\frac{1}{2} bc \sin A$

Area of any triangle

The area of a triangle is one half the product of the lengths of any two sides and the sine of the included angle measure.

For any triangle ABC,
$k = \frac{1}{2}bc \sin A$
$k = \frac{1}{2}ab \sin C$
$k = \frac{1}{2}ac \sin B$

Example 1 **In $\triangle ABC$, $b = 10$, $c = 8$ and $m\angle A = 60°$. Find the area in simplest radical form.**

$k = \frac{1}{2}bc \sin A$.
$= \frac{1}{2}(10)(8) \sin 60°$
$= 40 \cdot \frac{\sqrt{3}}{2}$, or $20\sqrt{3}$

Thus, the area of $\triangle ABC$ is $20\sqrt{3}$ square units.

Example 2 **In $\triangle ABC$, $a = 8$, $b = 12$ and $m\angle C = 140°$. Find the area to the nearest square unit.**

$k = \frac{1}{2} ab \sin C$.
$= \frac{1}{2} (8)(12) \sin 140°$
$= 48 \sin 40°$
$\doteq 48(.6428)$
$\doteq 30.8544$

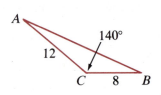

Thus, the area of $\triangle ABC$ is 31 square units to the nearest square unit.

Example 3 In △PQR, $q = 15$, $r = 60$. Find the measure of ∠P to the nearest degree if the area is 240 square units.

Use an area formula for q, r, and ∠P.

$$k = \frac{1}{2} qr \sin P$$
$$240 = \frac{1}{2}(15)(60) \sin P$$
$$\sin P = \frac{240}{450} \doteq .5333$$

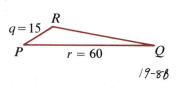

Thus, the measure of ∠P is 32° to the nearest degree.

Oral Exercises

Give the form of the area formula needed to find the area of each triangle.
1. $a = 6$, $c = 8$, m∠B = 36°
   $k = \frac{1}{2} ac \sin B$
2. $b = 9$, $c = 7$, m∠A = 73°
   $k = \frac{1}{2} bc \sin A$
3. $a = 4$, $b = 8$, m∠C = 41°
   $k = \frac{1}{2} ab \sin C$

Written Exercises

Ⓐ In △ABC, **find the area to the nearest square unit.**
1. $a = 12$, $b = 9$, m∠C = 43° **37**
2. $b = 9$, $c = 7$, m∠A = 80° **31**
3. $a = 6$, $c = 8$, m∠B = 70° **23**
4. $a = 7$, $b = 11$, m∠C = 34° **22**
5. $a = 13$, $c = 12$, m∠B = 48° **58**
6. $b = 12$, $c = 9$, m∠A = 38° **33**

In △PQR, **find the measure of ∠P to the nearest degree.**
7. $q = 8$, $r = 12$, $k = 38$ sq. units **52°**
8. $q = 14$, $r = 8$, $k = 54$ sq. units **75°**
9. $q = 16$, $r = 10$, $k = 74$ sq. units **68°**
10. $q = 6$, $r = 9$, $k = 22$ sq. units **55°**

Ⓑ In △ABC, **find the area to the nearest square unit.**
11. $a = 6.2$, $b = 4.4$, m∠C = 150° **7**
12. $a = 9.7$, $c = 4.8$, m∠B = 135° 20′ **16**
13. $a = 9.1$, $c = 7.6$, m∠B = 78° **34**
14. $b = 12.3$, $c = 13.7$, m∠A = 105° 30′ **81**

15. The lengths of two sides of a parallelogram are 24 and 36 centimeters. Their included angle measures 50°. Find the area of the parallelogram. **662 cm²**

16. The lengths of two sides of a parallelogram are 30 and 45 centimeters. Their included angle measures 130°. Find the area of the parallelogram. **1034 cm²**

Ⓒ 17. Prove: For any triangle ABC
$k = \frac{1}{2} ab \sin C$ and $k = \frac{1}{2} ac \sin B$.
For Ex. 17–18, see TE Answer Section.

18. Prove: The area of a parallelogram is equal to the product of the lengths of two adjacent sides and the sine of their included angle.

Area of a Triangle

LAW OF SINES 19.4

Objective **To apply the Law of Sines to find unknown measures in a triangle if two angles and a side of any triangle are known**

If two angles and a side of any triangle are known, the measure of the third angle and the lengths of the other two sides can be found.

Recall from the previous lesson three ways to write the area of a triangle, ABC.

$$\frac{1}{2} bc \sin A = \frac{1}{2} ab \sin C = \frac{1}{2} ac \sin B$$

Multiplying by 2. ▶ $bc \sin A = ab \sin C = ac \sin B$

Dividing by abc. ▶ $\dfrac{\cancel{bc} \sin A}{a\cancel{bc}} = \dfrac{\cancel{ab} \sin C}{\cancel{ab}c} = \dfrac{\cancel{ac} \sin B}{\cancel{a}b\cancel{c}}$

So, $\dfrac{\sin A}{a} = \dfrac{\sin C}{c} = \dfrac{\sin B}{b}$.

Law of sines

For any triangle ABC,
$$\frac{\sin A}{a} = \frac{\sin B}{b} = \frac{\sin C}{c}, \text{ or } \frac{a}{\sin A} = \frac{b}{\sin B} = \frac{c}{\sin C}.$$

In general, the Law of Sines states that the sides of a triangle are proportional to the sines of the angles opposite them.

Example 1 In $\triangle ABC$, $a = 16$, and $m\angle A = 40°$, $m\angle B = 58°$. Find b to the nearest unit.

Use $\dfrac{a}{\sin A} = \dfrac{b}{\sin B}$ since a, $m\angle A$, and $m\angle B$ are given.

$\dfrac{16}{\sin 40°} = \dfrac{b}{\sin 58°}$

$\dfrac{16 \sin 58°}{\sin 40°} = b$

$\dfrac{16(.8480)}{.6428} = b$

$21.1076 \doteq b$

Thus, $b = 21$ to the nearest unit.

Example 2   In $\triangle ABC$, $a = 80$, $b = 60$, and m$\angle A = 52°$. Find the measure of $\angle B$ to the nearest degree.

Use $\dfrac{\sin A}{a} = \dfrac{\sin B}{b}$ since a, b, and m$\angle A$ are given.

$$\dfrac{\sin 52°}{80} = \dfrac{\sin B}{60}$$

$$\sin B = \dfrac{60(\sin 52°)}{80}$$

$$= \dfrac{60(.7880)}{80}$$

$$\sin B \doteq .5910$$

Thus, the measure of $\angle B$ is 36° to the nearest degree.

Example 3   In $\triangle ABC$, $b = 18$, m$\angle A = 64°$, m$\angle C = 55°$. Find c to the nearest unit.

First, find m$\angle B$.
m$\angle B + 55° + 64° = 180°$ so m$\angle B = 61°$

Use $\dfrac{b}{\sin B} = \dfrac{c}{\sin C}$.

$$\dfrac{18}{\sin 61°} = \dfrac{c}{\sin 55°}, \text{ or } c = \dfrac{18 \sin 55°}{\sin 61°}$$

$$c \doteq \dfrac{18(.8192)}{.8746}, \text{ or } 16.86$$

Thus, $c = 17$ to the nearest unit.

You can use logarithms to solve problems involving the Law of Sines.

Example 4   In $\triangle ABC$, m$\angle A = 42° \; 10'$, m$\angle B = 73° \; 18'$, $b = 36.14$. Find a to the nearest tenth.

Use $\dfrac{a}{\sin A} = \dfrac{b}{\sin B}$.

$$\dfrac{a}{\sin 42° \; 10'} = \dfrac{36.14}{\sin 73° \; 18'}, \text{ or } a = \dfrac{36.14 \sin 42° \; 10'}{\sin 73° \; 18'}$$

Take the log of each side. ▶ $\log a = \log 36.14 + \log \sin 42° \; 10' - \log \sin 73° \; 18'$
Interpolate. ▶ $\quad\quad\quad = 1.5580 \;\; + \;\; 9.8269 - 10 \; - \; (9.9813 - 10)$
$\quad\quad\quad\log a = 1.4036$

Thus, $a = 25.3$.

Law of Sines

Oral Exercises

Give the form of the Law of Sines that is needed to find x.

1.

   1. $\dfrac{a}{\sin A} = \dfrac{b}{\sin B}$

2. (triangle with B, 11, 9, x, 65°, A, C)

   2. $\dfrac{\sin A}{a} = \dfrac{\sin C}{c}$

3.

   3. $\dfrac{b}{\sin B} = \dfrac{c}{\sin C}$

Written Exercises

Ⓐ

1. For $\triangle ABC$, $a = 28$, $b = 59$, $m\angle B = 48°$. Find $m\angle A$ to the nearest degree. **21°**

2. For $\triangle ABC$, $a = 36$, $c = 18$, $m\angle A = 57°$. Find $m\angle C$ to the nearest degree. **25°**

3. For $\triangle PQR$, $p = 17$, $m\angle P = 75°$, $m\angle Q = 38°$. Find q to the nearest unit. **11**

4. For $\triangle PQR$, $r = 29$, $m\angle P = 37°$, $m\angle R = 71°$. Find p to the nearest unit. **19**

5. For $\triangle ABC$, $a = 14$, $c = 10$, $m\angle C = 30°$. Find $m\angle A$ to the nearest degree. **44°**

6. For $\triangle PQR$, $q = 5$, $r = 7$, $m\angle Q = 40°$. Find $m\angle R$ to the nearest degree. **64°**

7. For $\triangle PQR$, $p = 22$, $\sin P = \frac{1}{5}$, $\sin Q = \frac{4}{5}$. Find q to the nearest unit. **88**

8. For $\triangle ABC$, $\sin A = .40$, $\sin B = .65$, $a = 32$. Find b to the nearest unit. **52**

Ⓑ

9. For $\triangle ABC$, $a = 16.3$, $m\angle B = 58°$, $m\angle C = 37°$. Find b, c, and $m\angle A$. **13.9, 9.8, 85°**

10. For $\triangle ABC$, $a = 12.3$, $b = 13.7$, $m\angle A = 18°\ 25'$. Find c, $m\angle B$, and $m\angle C$. **24.5, 20° 36′, 140° 59′**

11. For $\triangle ABC$, $b = 14.8$, $c = 18.9$, $m\angle C = 42°\ 18'$. Find a, $m\angle A$, and $m\angle B$. **27, 105° 54′, 31° 48′**

12. For $\triangle ABC$, $a = 15.1$, $c = 19.3$, $m\angle C = 37°\ 32'$. Find b, $m\angle A$, and $m\angle B$. **28.9, 28° 28′, 114°**

13. The distance between town B and town A is 45 kilometers. The angle formed by the roads between towns A and B and between towns A and C measures 37°, and the angle formed by $\overline{AB}$ and $\overline{BC}$ measures 110°. Find the distance from town B to town C to the nearest kilometer. **50 km**

14. Two cars P and Q are parked on the same side of the street, 50 meters apart. Car R is not on the same road. The angle formed by $\overrightarrow{PQ}$ and $\overrightarrow{PR}$ is 112° 20′, and the angle formed by $\overrightarrow{QR}$ and $\overrightarrow{PR}$ is 28° 40′. Find the distance between P and R to the nearest meter. **66 m**

Ⓒ

15. For $\triangle RST$, show that $\dfrac{\sin R}{r} = \dfrac{\sin T}{t} = \dfrac{\sin S}{s}$.

16. For $\triangle QRS$, show that $\dfrac{q}{\sin Q} = \dfrac{r}{\sin R} = \dfrac{s}{\sin S}$.

For Ex. 15–16, see TE Answer Section.

CALCULATOR ACTIVITIES

The calculator can be an aid in solving problems involving the Law of Sines.

From Example 3, $\dfrac{18}{\sin 61°} = \dfrac{c}{\sin 55°}$ or $c = \dfrac{18 \sin 55°}{\sin 61°} = \dfrac{18(.8192)}{.8746}$.

Compute ▶ 18 ⊗ .8192 ⊘ .8746 ⊜ 16.86.

THE AMBIGUOUS CASE   19.5

Objective **To determine the number of triangles that can be constructed given certain data**

In this lesson, you will compare parts of a triangle to determine the number of triangles it is possible to construct.

Example 1 **Given that $a = 18$, $b = 36$, and $m\angle A = 38°$, use the Law of Sines to find the measure of $\angle B$. How many triangles can be constructed using the given data?**

Use $\dfrac{\sin A}{a} = \dfrac{\sin B}{b}$.

$\dfrac{\sin 38°}{18} = \dfrac{\sin B}{36}$, or $\sin B = \dfrac{36 \sin 38°}{18}$

$\sin B \doteq \dfrac{36(.6157)}{18}$, or 1.2314

Thus, there is no $\angle B$ for which $\sin B = 1.2314$ since $0 \le \sin B \le 1$ and no triangle can be constructed with the given data.

In general, $\dfrac{\sin A}{a} = \dfrac{\sin B}{b}$

$\sin B = \dfrac{b \sin A}{a}$,

but $\sin B = \dfrac{h}{a}$, so $h = b \sin A$.

Compare h and a. When $a < h$, $(a < b \sin A)$, as in Example 1, no triangle can be constructed since $\sin B = \dfrac{h}{a}$ will be greater than 1. When $a = h$, $\sin B = 1$, one right $\triangle$ can be constructed.
When $a > h$, $\sin B < 1$, two $\triangle$'s can be constructed.
The following Summary is suggested.

Summary A is an acute angle and $a < b$:

1. If $a < h$, $(a < b \sin A)$,
   no triangle can be constructed.

2. If $a = h$, $(a = b \sin A)$,
   one triangle can be constructed.

3. If $a > h$, $(a > b \sin A)$,
   two triangles can be constructed.

Example 2 Given that m∠A = 34°, a = 21, b = 32, and the height, h, is 25, can a triangle be constructed?

A is an acute angle, a < b and a < h.

Thus, no triangle can be constructed.

In △ABC, ∠A is an acute angle. You will now determine how many triangles can be constructed if a = b and if a > b.

∠A is an acute angle.

 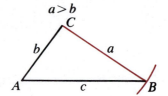

Summary ∠A is an acute angle:
1. One triangle can be constructed if a = b.
2. One triangle can be constructed if a > b.

Example 3 Given that m∠A = 130°, a = 12, and b = 13, find the measure of ∠C to the nearest degree. How many triangles can be constructed using the given data?

Use $\dfrac{\sin A}{a} = \dfrac{\sin B}{b}$.

$\dfrac{\sin 130°}{12} = \dfrac{\sin B}{13}$, or $\sin B = \dfrac{13 \sin 130°}{12}$

$\sin B \doteq \dfrac{13(.7660)}{12}$, or .8298

So, the measure of ∠B is 56° to the nearest degree.
Find m∠C. m∠A + m∠B + m∠C = 180°
130° + 56° + m∠C = 180°
186° + m∠C = 180°
m∠C = 180° − 186°, or −6°

Thus, no triangle can be constructed when angle A is obtuse and a < b.

Example 4 Given that m∠A = 130°, a = 12, b = 12, find the measure of ∠C to the nearest degree. How many triangles can be constructed?

$$\sin B \doteq \frac{12(.7660)}{12}, \text{ or } .7660 \quad \blacktriangleleft \quad \sin B = \frac{b \sin A}{a}$$

So, the measure of ∠B is 50°.
$$130° + 50° + m\angle C = 180°$$
$$m\angle C = 0°$$

Thus, no triangle can be constructed when angle A is obtuse and $a = b$.

Example 5 Given that m∠A = 130°, a = 12, b = 10, find the measure of ∠C to the nearest degree. How many triangles can be constructed?

$$\sin B \doteq \frac{10(.7660)}{12}, \text{ or } .6383$$

So, the measure of ∠B is 40° to the nearest degree.
$$130° + 40° + m\angle C = 180°$$
$$m\angle C = 10°$$

Thus, one triangle can be constructed when angle A is obtuse and $a > b$.

The last three examples suggest the following.

Summary ∠A is an obtuse angle or a right angle:
1. No triangle can be constructed if $a \leq b$.
2. One triangle can be constructed if $a > b$.

Example 6 How many triangles can be constructed if m∠A = 160°, a = 20, and b = 60?

∠A is an obtuse angle and $a < b$.

Thus, no triangle can be constructed when angle A is obtuse and $a < b$.

Written Exercises

How many triangles can be constructed?
1. m∠A = 36°, a = 12, b = 18 **2**
2. m∠A = 48°, a = 6, b = 7 **2**
3. m∠A = 40°, a = 50, b = 20 **1**
4. m∠A = 120°, a = 19, b = 21 **0**
5. m∠A = 90°, a = 96, b = 96 **0**
6. m∠A = 175°, a = 71, b = 58 **1**
7. m∠A = 15°, a = 100, b = 100 **1**
8. m∠A = 52°, a = 42, b = 38 **1**
9. m∠A = 133°, a = 16, b = 4 **1**
10. m∠A = 90°, a = 14, b = 14 **0**

SIN (A ± B) AND SIN 2A   19.6

Objective   **To apply the formulas for sin (A ± B) and sin 2A**

You can derive the formula for the sine of the sum of two angles in the following way.

Write the formulas for the areas of each of the three triangles shown at the right.
Area $\triangle ABC$ = area $\triangle ABD$ + area $\triangle BDC$

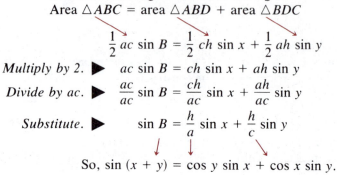

$$\frac{1}{2} ac \sin B = \frac{1}{2} ch \sin x + \frac{1}{2} ah \sin y$$

Multiply by 2. $ac \sin B = ch \sin x + ah \sin y$

Divide by ac. $\dfrac{ac}{ac} \sin B = \dfrac{ch}{ac} \sin x + \dfrac{ah}{ac} \sin y$

Substitute. $\sin B = \dfrac{h}{a} \sin x + \dfrac{h}{c} \sin y$

So, $\sin (x + y) = \cos y \sin x + \cos x \sin y$.

Sine of the sum of two angles

$$\sin (A + B) = \sin A \cos B + \cos A \sin B$$

Example 1   **Find sin 105° using special angle measures and the formula for sin (A + B). Answer may be left in simplest radical form.**

$$\sin (A + B) = \sin A \cos B + \cos A \sin B$$

$105° = 60° + 45°$ ▶ $\sin (60° + 45°) = \sin 60° \cos 45° + \cos 60° \sin 45°$

$$= \frac{\sqrt{3}}{2} \cdot \frac{\sqrt{2}}{2} + \frac{1}{2} \cdot \frac{\sqrt{2}}{2}$$

$$= \frac{\sqrt{6}}{4} + \frac{\sqrt{2}}{4}$$

Thus, $\sin 105° = \dfrac{\sqrt{6} + \sqrt{2}}{4}$.

You can derive the formula for the sine of the difference of two angle measures using the formula for the sine of the sum of two angle measures. This derivation follows on the next page.

Rewrite the difference $x - y$ as the sum $x + (-y)$.
So, $\sin(x - y) = \sin[x + (-y)]$
$= \sin x \cos(-y) + \cos x \sin(-y)$
$= \sin x \cos y + \cos x(-\sin y)$, since $\begin{cases} \cos(-y) = \cos y \\ \sin(-y) = -\sin y \end{cases}$.
$= \sin x \cos y - \cos x \sin y$.

Sine of the difference of two angles	$\sin(A - B) = \sin A \cos B - \cos A \sin B$

Example 2 Find sin 15°. Use special angle measures and the formula for $\sin(A - B)$. Answer in simplest radical form.

$\sin(A - B) = \sin A \cos B - \cos A \sin B$
$\sin(60° - 45°) = \sin 60° \cos 45° - \cos 60° \sin 45°$
$= \dfrac{\sqrt{3}}{2} \cdot \dfrac{\sqrt{2}}{2} - \dfrac{1}{2} \cdot \dfrac{\sqrt{2}}{2}$, or $\dfrac{\sqrt{6}}{4} - \dfrac{\sqrt{2}}{4}$

Thus, $\sin 15° = \dfrac{\sqrt{6} - \sqrt{2}}{4}$.

Example 3 If $\sin A = \frac{3}{5}$ and $\cos B = \frac{5}{13}$, where A and B are acute angles, find the value of $\sin(A + B)$.

$\sin(A + B) = \sin A \cos B + \cos A \sin B$
$= \dfrac{3}{5} \cdot \dfrac{5}{13} + \dfrac{4}{5} \cdot \dfrac{12}{13}$
$= \dfrac{15}{65} + \dfrac{48}{65}$
$= \dfrac{63}{65}$

Thus, $\sin(A + B) = \dfrac{63}{65}$.

Example 4 If $\sin A = \dfrac{15}{17}$ and $\angle A$ lies in the second quadrant, and if $\cos B = \dfrac{4}{5}$ and $\angle B$ lies in the first quadrant, find $\sin(A - B)$.

Using reference triangles, $\cos A = -\dfrac{8}{17}$ and $\sin B = \dfrac{3}{5}$.
$\sin(A - B) = \sin A \cos B - \cos A \sin B$
$= \dfrac{15}{17}\left(\dfrac{4}{5}\right) - \left(\dfrac{-8}{17}\right) \cdot \dfrac{3}{5}$
$= \dfrac{60}{85} + \dfrac{24}{85}$, or $\dfrac{84}{85}$

Thus, $\sin(A - B) = \dfrac{84}{85}$.

The sine of twice (double) an angle can also be derived using the formula for the sine of the sum of two angles.

Begin by rewriting $2x$ as the sum $(x + x)$.
$\sin (2x) = \sin (x + x)$
$= \sin x \cos x + \cos x \sin x$
$= \sin x \cos x + \sin x \cos x$
$= 2 \sin x \cos x.$

Sine of twice an angle	$\sin 2A = 2 \sin A \cos A$

Example 5   **Find sin 120°. Use a special angle measure and the formula for sin 2A.**

$\sin 2A = 2 \sin A \cos A$
$\sin 2(60°) = 2 \sin 60° \cos 60°$
$= 2 \left(\dfrac{\sqrt{3}}{2}\right)\left(\dfrac{1}{2}\right)$, or $\dfrac{\sqrt{3}}{2}$

Thus, $\sin 120° = \dfrac{\sqrt{3}}{2}$.

Example 6   **If $\cos A = -\dfrac{5}{13}$ and $\angle A$ lies in the third quadrant, find sin 2A.**

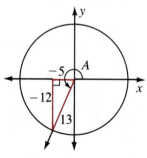

Use the reference triangle, $\sin A = -\dfrac{12}{13}$.

So, $\sin 2A = 2 \sin A \cos A$.
$= 2 \left(-\dfrac{12}{13}\right)\left(-\dfrac{5}{13}\right)$
$= \dfrac{120}{169}$

Thus, $\sin 2A = \dfrac{120}{169}$.

Oral Exercises

Write as a function of one angle.
1. $\sin 40° \cos 30° + \cos 40° \sin 30°$ **sin 70°**
2. $\sin 50° \cos 10° - \cos 50° \sin 10°$ **sin 40°**
3. $2 \sin 15° \cos 15°$ **sin 30°**
4. $\sin 125° \cos 12° + \cos 125° \sin 12°$ **sin 137°**
5. $\cos 57° \sin 33° - \cos 33° \sin 57°$ **−sin 24°**
6. $2 \cos 25° \sin 25°$ **sin 50°**
7. $\cos 70° \sin 10° + \cos 10° \sin 70°$ **sin 80°**
8. $\sin 73° \cos 62° - \cos 73° \sin 62°$ **sin 11°**

Written Exercises

Find the value of each. Use special angles and the formula for sin $(A + B)$ or sin $(A - B)$. Answers may be left in simplest radical form.

(A)
1. sin 75°   1. $\frac{\sqrt{6} + \sqrt{2}}{4}$
2. sin 135° $\frac{\sqrt{2}}{2}$
3. sin 60° $\frac{\sqrt{3}}{2}$
4. sin 195°
5. sin 285°
6. sin 165°
7. sin 150°
8. sin 255°
9. sin 435°
10. sin 375°

For Ex. 4–10, see TE Answer Section.

Find the value of each. Use special angles and the formula for sin 2A. Answers may be left in simplest radical form.

11. sin 60° $\frac{\sqrt{3}}{2}$
12. sin 90°   **1**
13. sin 300° $\frac{-\sqrt{3}}{2}$
14. sin 270°   **−1**
15. sin 450°   **1**

Find the value of sin $(A + B)$ and sin $(A - B)$.

16. $\sin A = \frac{12}{13}$ and A lies in the 2nd quad. $\cos B = \frac{3}{5}$ and B lies in the 4th quad. $\frac{56}{65}, \frac{16}{65}$

17. $\cos A = \frac{12}{13}$ and $\cos B = \frac{4}{5}$, A and B are both acute angles. $\frac{56}{65}, \frac{-16}{65}$

18. $\sin A = \frac{15}{17}$ and A lies in the 2nd quad. $\cos B = \frac{-4}{5}$ and B lies in the 3rd quad. $\frac{-36}{85}, \frac{-84}{85}$

19. $\sin A = \frac{-8}{17}$ and A lies in the 4th quad. $\cos B = \frac{12}{13}$ and B lies in the 4th quad. $\frac{-171}{221}, \frac{-21}{221}$

20. $\sin A = \frac{7}{9}$ and A lies in the 1st quad. $\cos B = \frac{-4}{11}$ and B lies in the 2nd quad. $\frac{-28 + 4\sqrt{210}}{99}, \frac{-28 - 4\sqrt{210}}{99}$

21. $\cos A = \frac{6}{13}$ and A lies in the 4th quad. $\sin B = \frac{4}{7}$ and B lies in the 1st quad. $\frac{-\sqrt{4389} + 24}{91}, \frac{-\sqrt{4389} - 24}{91}$

Find the value of sin 2A.

22. $\cos A = \frac{-4}{5}$ and A lies in the 2nd quad. $\frac{-24}{25}$
23. $\cos A = \frac{12}{13}$ and A is an acute angle. $\frac{120}{169}$
24. $\cos A = \frac{15}{17}$ and A lies in the 4th quad. $\frac{-240}{289}$
25. $\sin A = -\frac{8}{17}$ and A lies in the 3rd quad. $\frac{240}{289}$

Simplify each of the following for acute angle A.

(B)
26. $\frac{\sin^2 A}{\sin 2A}$ $\frac{\sin A}{2 \cos A}$
27. $\frac{\sin 2A}{\cos A}$ **2 sin A**
28. $\frac{\tan A}{\sin 2A}$ $\frac{1}{2 \cos^2 A}$
29. cot A sin 2A **2 cos² A**

Find the value of each. Use special angles, the formula for sin $(A + B)$, sin $(A - B)$, or sin 2A, and reciprocal relationships. Answers may be left in simplest radical form.

30. csc 75° $-\sqrt{2} + \sqrt{6}$
31. csc 135° $\sqrt{2}$
32. csc 120° $\frac{2\sqrt{3}}{3}$
33. csc 15° $\sqrt{6} + \sqrt{2}$
34. csc 90°   **1**

Prove each identity. Use the formula for sin $(A \pm B)$.

(C)
35. $\sin (\pi + \theta) = -\sin \theta$
36. $\sin (2\pi - \theta) = -\sin \theta$
37. $\sin (\pi - \theta) = \sin \theta$
38. $\sin \left(\frac{\pi}{2} - \theta\right) = \cos \theta$
39. $\sin \left(\frac{3\pi}{2} + \theta\right) = -\cos \theta$
40. $\sin \left(\frac{\pi}{2} + \theta\right) = \cos \theta$
41. $\sin (180° + \theta) = -\sin \theta$
42. $\sin (180° - \theta) = \sin \theta$
43. $\sin (90° + \theta) = \cos \theta$
44. $\sin (90° - \theta) = \cos \theta$
45. $\sin (360° - \theta) = -\sin \theta$
46. $\sin (270° + \theta) = -\cos \theta$

For Ex. 35–46, see TE Answer Section.

Sin $(A \pm B)$ and Sin 2A

COS (A ± B) AND COS 2A   19.7

Objective   **To apply the formulas for cos (A ± B) and cos 2A**

You can derive the formula for the cosine of the sum of two angles by using the formula for the sine of the difference of two angles and the cofunction relationship of sine and cosine.

In Chapter 17, you learned the cofunction relationship $\cos u = \sin(90 - u)$. If we let $u = x + y$, it follows that,

$$\cos(x + y) = \sin[90 - (x + y)]$$
$$= \sin[(90 - x) - y] \qquad \blacktriangleleft \; 90 - (x + y) = (90 - x) - y$$
$$= \sin(90 - x)\cos y - \cos(90 - x)\sin y \qquad \blacktriangleleft \; \text{using formula for } \sin(A - B), A = 90 - x, B = y$$

So, $\cos(x + y) = \cos x \cos y - \sin x \sin y$. $\qquad \blacktriangleleft \; \sin(90 - x) = \cos x$ and $\cos(90 - x) = \sin x$

Cosine of the sum of two angles

$$\cos(A + B) = \cos A \cos B - \sin A \sin B$$

Example 1   **Find cos 75°. Use special angles and the formula for cos (A + B). Answer in simplest radical form.**

$75° = 45° + 30°$ ▶

$$\cos(A + B) = \cos A \cos B - \sin A \sin B$$
$$\cos(45° + 30°) = \cos 45° \cos 30° - \sin 45° \sin 30°$$
$$= \frac{\sqrt{2}}{2} \cdot \frac{\sqrt{3}}{2} - \frac{\sqrt{2}}{2} \cdot \frac{1}{2}$$
$$= \frac{\sqrt{6}}{4} - \frac{\sqrt{2}}{4}$$

Thus, $\cos 75° = \dfrac{\sqrt{6} - \sqrt{2}}{4}$.

Example 2   **If $\cos A = -\frac{4}{5}$ and $\angle A$ lies in the second quadrant and $\cos B = \frac{12}{13}$ and $\angle B$ lies in the fourth quadrant, find $\cos(A + B)$.**

Using reference triangles, $\sin A = \frac{3}{5}$ and $\sin B = -\frac{5}{13}$.
$$\cos(A + B) = \cos A \cos B - \sin A \sin B$$
$$= -\frac{4}{5} \cdot \frac{12}{13} - \frac{3}{5} \cdot \left(-\frac{5}{13}\right)$$
$$= -\frac{48}{65} + \frac{15}{65}, \text{ or } -\frac{33}{65}$$

Thus, $\cos(A + B) = -\dfrac{33}{65}$.

Chapter Nineteen

You can derive the formula for the cos of the difference of two angles using the formula for the cosine of the sum of two angles.

Rewrite the difference $x - y$ as the sum $x + (-y)$.
$$\begin{aligned}\cos(x - y) &= \cos[x + (-y)] \\ &= \cos x \cos(-y) - \sin x \sin(-y) \\ &= \cos x \cos y - \sin x (-\sin y) \\ &= \cos x \cos y + \sin x \sin y.\end{aligned}$$
◀ $\begin{cases} \cos(-y) = \cos y \\ \sin(-y) = -\sin y \end{cases}$

Cosine of the difference of two angles

$$\cos(A - B) = \cos A \cos B + \sin A \sin B$$

Example 3 If $\sin A = -\dfrac{15}{17}$ and $\angle A$ lies in the fourth quadrant and if $\tan B = -\dfrac{4}{3}$ and $\angle B$ lies in the second quadrant, find $\cos(A - B)$.

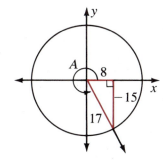

$$\begin{aligned}\cos(A - B) &= \cos A \cos B + \sin A \sin B \\ &= \left(\dfrac{8}{17}\right)\left(\dfrac{-3}{5}\right) + \left(\dfrac{-15}{17}\right)\left(\dfrac{4}{5}\right) \\ &= \dfrac{-24}{85} + \dfrac{-60}{85}, \text{ or } \dfrac{-84}{85}\end{aligned}$$

Thus, $\cos(A - B) = -\dfrac{84}{85}$.

You can derive the formula for the cosine of twice (double) an angle using the formula for the cosine of the sum of two angles.

Begin by rewriting $2x$ as the sum $(x + x)$.
So, $\begin{aligned}\cos 2x &= \cos(x + x) \\ &= \cos x \cos x - \sin x \sin x \\ &= \cos^2 x - \sin^2 x.\end{aligned}$

Cosine of twice an angle

$$\cos 2A = \cos^2 A - \sin^2 A$$

Observe that the formula for $\cos 2A$ is derived by using the formula for $\cos(A + B)$ and the formula for $\sin 2A$ is derived by using the formula for $\sin(A + B)$.

Cos $(A \pm B)$ and Cos $2A$

Example 4   **Find cos 120°. Use a special angle measure and the formula for cos 2A.**

$$\cos 2A = \cos^2 A - \sin^2 A$$
$$\cos 120° = \cos 2(60°) = \cos^2 60° - \sin^2 60°$$
$$= \left(\frac{1}{2}\right)^2 - \left(\frac{\sqrt{3}}{2}\right)^2$$
$$= \frac{1}{4} - \frac{3}{4}, \text{ or } -\frac{1}{2}$$

Thus, cos 120° = $-\frac{1}{2}$.

Example 5   **If $\cos \theta = -\frac{15}{17}$ and θ lies in the third quadrant, find cos 2θ.**

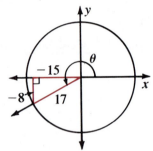

Using the reference triangle, $\cos \theta = \frac{-15}{17}$ and $\sin \theta = \frac{-8}{17}$.

$$\cos 2\theta = \cos^2 \theta - \sin^2 \theta$$
$$= \left(-\frac{15}{17}\right)^2 - \left(-\frac{8}{17}\right)^2$$
$$= \frac{225}{289} - \frac{64}{289}$$
$$= \frac{161}{289}$$

Thus, cos 2θ = $\frac{161}{289}$.

Recall from Chapter 18 that for any angle A, $\sin^2 A + \cos^2 A = 1$.
This identity can also be written as $\cos^2 A = 1 - \sin^2 A$ or $\sin^2 A = 1 - \cos^2 A$.
By using these identities, you can derive two additional formulas for cosine of twice an angle.
Begin with $\cos 2x = \cos^2 x - \sin^2 x$.

Replace $\cos^2 x$ with $1 - \sin^2 x$.
$\cos 2x = \cos^2 x - \sin^2 x$
$= (1 - \sin^2 x) - \sin^2 x$
$= 1 - 2 \sin^2 x$

Replace $\sin^2 x$ with $1 - \cos^2 x$.
$\cos 2x = \cos^2 x - \sin^2 x$
$= \cos^2 x - (1 - \cos^2 x)$
$= 2 \cos^2 x - 1$

Three formulas for cos 2A	$\cos 2A = \cos^2 A - \sin^2 A$ $\cos 2A = 1 - 2 \sin^2 A$ $\cos 2A = 2 \cos^2 A - 1$

Written Exercises

A Find the value of each. Use special angle measures and the formula for cos $(A + B)$ or cos $(A - B)$. Answers may be left in simplest radical form.

1. cos 105°
2. cos 15°
3. cos 135°
4. cos 195°
5. cos 285°
6. cos 165°
7. cos 150°
8. cos 255°
9. cos 435°
10. cos 375°

For Ex. 1–10, see TE Answer Section.

Find the value of each. Use special angles and the formula for cos $2A$. Answer in simplest radical form.

11. cos 90° **0**
12. cos 60° $\frac{1}{2}$
13. cos 240° $-\frac{1}{2}$
14. cos 450° **0**
15. cos 180° **−1**

Find the value of cos $(A + B)$ and cos $(A - B)$.

16. Cos $A = \frac{3}{5}$ and cos $B = \frac{12}{13}$. $\frac{16}{65}, \frac{56}{65}$
    A and B are both acute angles.

17. Cos $A = \frac{15}{17}$ and cos $B = \frac{5}{9}$. $\frac{75 - 16\sqrt{14}}{153}$,
    A and B are both acute angles. $\frac{75 + 16\sqrt{14}}{153}$

18. Cos $A = -\frac{5}{13}$ and A lies in the second quadrant. Cos $B = \frac{8}{17}$ and B lies in the fourth quadrant. $\frac{140}{221}, \frac{-220}{221}$

19. Cos $A = -\frac{6}{11}$ and A lies in the third quadrant. Cos $B = \frac{4}{5}$ and B lies in the first quadrant. $\frac{-24 + 3\sqrt{85}}{55}, \frac{-24 - 3\sqrt{85}}{55}$

Find the value of cos $2A$.

20. Cos $A = -\frac{4}{5}$ and A lies in the 3rd quad. $\frac{7}{25}$
21. Cos $A = \frac{12}{13}$ and A lies in the 1st quad. $\frac{119}{169}$
22. Cos $A = \frac{3}{7}$ and A lies in the 1st quad. $\frac{-31}{49}$
23. Cos $A = \frac{-7}{10}$ and A lies in the 2nd quad. $\frac{-1}{50}$

B Simplify.

24. $\dfrac{\cos 2A}{\sin^2 A - \cos^2 A}$ **−1**
25. $\dfrac{\cos^2 A - \sin^2 A}{2\cos^2 A - 1}$ **1**
26. $\dfrac{1 - 2\sin^2 A}{\cos 2A}$ **1**

C Find the value of each. Use special angle measures and the formula for cos $(A + B)$, cos $(A - B)$, or cos $2A$. Answer in simplest radical form.

27. sec 75° $\sqrt{6} + \sqrt{2}$
28. sec 135° $-\sqrt{2}$
29. sec 120° -2
30. sec 15° $\sqrt{6} - \sqrt{2}$
31. sec 90° undefined

Prove each identity. Use the formulas for cos $(A + B)$.

32. cos $(180° - \theta) = -\cos \theta$
33. cos $(360° - \theta) = \cos \theta$
34. cos $(180° + \theta) = -\cos \theta$
35. cos $(\pi + \theta) = -\cos \theta$
36. cos $(270° + \theta) = \sin \theta$
37. cos $(2\pi + \theta) = \cos \theta$
38. cos $(\frac{\pi}{2} + \theta) = -\sin \theta$
39. cos $(\pi - \theta) = -\cos \theta$
40. cos $(\frac{3}{2}\pi - \theta) = -\sin \theta$

For Ex. 32–40, see TE Answer Section.

CUMULATIVE REVIEW

1. A triangle's height and a rectangle's width are the same. The rectangle's length is three times its width. The triangle's base measures one half the rectangle's length. The rectangle's area minus the triangle's area is 9 square units. Find the height of the triangle to the nearest unit. **2**
2. A square and a rectangle have the same width. The rectangle's length is 8 cm more than three times its width. The sum of their areas is 192 square cm. Find the width of the square to the nearest centimeter. **6**

Cos $(A \pm B)$ and Cos $2A$

TAN $(A \pm B)$ AND TAN $2A$ 19.8

Objective   **To apply the formulas for tan $(A \pm B)$ and tan $2A$**

You can derive the formula for the tangent of the sum of two angles using the formulas for the sine and cosine of the sum of two angles.

Recall from Chapter 18 the identity $\tan A = \dfrac{\sin A}{\cos A}$.

$$\tan(x + y) = \frac{\sin(x+y)}{\cos(x+y)}$$

$$= \frac{\sin x \cos y + \cos x \sin y}{\cos x \cos y - \sin x \sin y} \quad \blacktriangleleft \text{ Formula for } \sin(A+B)$$
$$\quad\quad\quad\quad\quad\quad\quad\quad\quad\quad\quad\quad \blacktriangleleft \text{ Formula for } \cos(A+B)$$

Divide each term by $\cos x \cos y$ and simplify.

$$\tan(x + y) = \frac{\dfrac{\sin x \cos y}{\cos x \cos y} + \dfrac{\cos x \sin y}{\cos x \cos y}}{\dfrac{\cos x \cos y}{\cos x \cos y} - \dfrac{\sin x \sin y}{\cos x \cos y}}$$

So, $\tan(x + y) = \dfrac{\tan x + \tan y}{1 - \tan x \tan y}$.

Tangent of the sum of two angles

$$\tan(A + B) = \frac{\tan A + \tan B}{1 - \tan A \tan B}$$

Example 1   **Find tan 75°. Use special angle measures and the formula for tan $(A + B)$. Answer in simplest radical form.**

$$\tan(A + B) = \frac{\tan A + \tan B}{1 - \tan A \tan B}$$

$$\tan 75° = \tan(45° + 30°) = \frac{\tan 45° + \tan 30°}{1 - \tan 45° \tan 30°} = \frac{1 + \dfrac{\sqrt{3}}{3}}{1 - (1)\left(\dfrac{\sqrt{3}}{3}\right)}$$

$$= \frac{\dfrac{3 + \sqrt{3}}{3}}{\dfrac{3 - \sqrt{3}}{3}}, \text{ or } \frac{3 + \sqrt{3}}{3 - \sqrt{3}} \cdot \frac{3 + \sqrt{3}}{3 + \sqrt{3}} = \frac{9 + 6\sqrt{3} + 3}{9 - 3}$$

Thus, $\tan 75° = 2 + \sqrt{3}$.

You can derive the formula for the tangent of the difference of two angles using the formula for the tan of the sum of two angles.

Rewrite the difference $x - y$ as the sum $x + (-y)$.
$$\tan(x - y) = \tan[x + (-y)]$$
$$= \frac{\tan x + \tan(-y)}{1 - \tan x \tan(-y)}$$
$$= \frac{\tan x - \tan y}{1 + \tan x \tan y}, \text{ since } \tan(-y) = -\tan y.$$

Tangent of the difference of two angles

$$\tan(A - B) = \frac{\tan A - \tan B}{1 + \tan A \tan B}$$

Example 2 If $\tan A = \frac{4}{7}$ and $\tan B = \frac{13}{2}$, find $\tan(A - B)$.

$$\tan(A - B) = \frac{\tan A - \tan B}{1 + \tan A \tan B}$$
$$= \frac{\frac{4}{7} - \frac{13}{2}}{1 + \frac{4}{7}\left(\frac{13}{2}\right)}, \text{ or } \frac{-\frac{83}{14}}{\frac{66}{14}}$$

Thus, $\tan(A - B) = -\frac{83}{66}$.

You can derive the formula for the tangent of twice (double) an angle using the formula for the tan of the sum of two angles.

$$\tan 2x = \tan(x + x) = \frac{\tan x + \tan x}{1 - \tan x \tan x}, \text{ or } \frac{2 \tan x}{1 - \tan^2 x}$$

Tangent of twice an angle

$$\tan 2A = \frac{2 \tan A}{1 - \tan^2 A}$$

Example 3 Find $\tan 120°$. Use a special angle measure and the formula for $\tan 2x$. Answer in simplest radical form.

$$\tan 2A = \frac{2 \tan A}{1 - \tan^2 A}$$
$$\tan 120° = \tan 2(60°) = \frac{2 \tan 60°}{1 - \tan^2 60°} = \frac{2\sqrt{3}}{1 - (\sqrt{3})^2}, \text{ or } \frac{2\sqrt{3}}{-2}$$

Thus, $\tan 120° = -\sqrt{3}$.

Tan $(A \pm B)$ and Tan $2A$

Example 4 If $\tan \theta = -\frac{7}{6}$, find $\tan 2\theta$.

$$\tan 2\theta = \frac{2\tan\theta}{1-\tan^2\theta} = \frac{2\left(-\frac{7}{6}\right)}{1-\left(-\frac{7}{6}\right)^2}$$

$$= \frac{-\frac{14}{6}}{1-\frac{49}{36}}, \text{ or } \frac{-\frac{14}{6}}{-\frac{13}{36}}, \text{ or } \frac{-14}{6} \cdot \frac{-36}{13}$$

Thus, $\tan 2\theta = \frac{84}{13}$.

Written Exercises

(A) Find the value of each. Use special angles and the formula for $\tan(A + B)$ or $\tan(A - B)$. Answer in simplest radical form.

1. $\tan 105°$ $-2-\sqrt{3}$
2. $\tan 15°$ $2-\sqrt{3}$
3. $\tan 135°$ -1
4. $\tan 120°$ $-\sqrt{3}$
5. $\tan 375°$ $2-\sqrt{3}$

Find $\tan(A + B)$ for each of the following.

6. $\tan A = \frac{5}{3}$ and $\tan B = \frac{4}{7}$ **47**
7. $\tan A = -\sqrt{3}$ and $\tan B = \frac{\sqrt{2}}{2}$ $3\sqrt{3}-4\sqrt{2}$
8. $\tan A = -\frac{3}{4}$ and $\tan B = \frac{5}{8}$ $\frac{-4}{47}$
9. $\tan A = \frac{-5}{12}$ and $\tan B = \frac{8}{7}$ $\frac{61}{124}$

Find $\tan(A - B)$ for each of the following.

10. $\tan A = -\frac{1}{3}$ and $\tan B = -\frac{\sqrt{2}}{2}$ $\frac{-9+10\sqrt{2}}{17}$
11. $\tan A = \frac{7}{5}$ and $\tan B = -1$ -6
12. $\tan A = \sqrt{3}$ and $\tan B = \frac{\sqrt{2}}{2}$ $-3\sqrt{3}+4\sqrt{2}$
13. $\tan A = \frac{5}{12}$ and $\tan B = \frac{8}{17}$ $\frac{-11}{244}$

Find $\tan 2A$ for each of the following.

14. $\tan A = \frac{4}{5}$ $\frac{40}{9}$
15. $\tan A = \frac{15}{17}$ $\frac{255}{32}$
16. $\tan A = -\frac{5}{12}$ $\frac{-120}{119}$
17. $\tan A = -1$ **undefined**
18. $\tan A = -\frac{3}{4}$ $\frac{-24}{7}$
19. $\tan A = \frac{12}{11}$ $\frac{-264}{23}$
20. $\tan A = \frac{4}{7}$ $\frac{56}{33}$
21. $\tan A = -\frac{3}{8}$ $\frac{-48}{55}$

(B) $\tan\theta = -\frac{5}{12}$, $\cos\beta = -\frac{3}{5}$, and β lies in the second quadrant.

22. Find $\tan(\theta + \beta)$. $\frac{-63}{16}$
23. Find $\tan(\theta - \beta)$. $\frac{33}{56}$
24. Find $\tan 2\theta$. $\frac{-120}{119}$
25. Find $\tan 2\beta$. $\frac{24}{7}$

$\tan\theta = -\frac{7}{3}$, $\sin\beta = -\frac{5}{13}$, and β lies in the fourth quadrant.

26. Find $\tan(\theta + \beta)$. -99
27. Find $\tan(\theta - \beta)$. $\frac{-69}{71}$
28. Find $\tan 2\theta$. $\frac{21}{20}$
29. Find $\tan 2\beta$. $\frac{-120}{119}$

Find the value for each of the following.

30. $\frac{2\tan 15°}{1-\tan^2 15°}$ $\frac{\sqrt{3}}{3}$
31. $\frac{2\tan 22.5°}{1-\tan^2 22.5°}$ 1
32. $\frac{2\tan 67.5°}{1-\tan^2 67.5°}$ -1

(C) Prove each identity. Use the formulas for $\tan(A \pm B)$.

33. $\tan(180° + \theta) = \tan\theta$
34. $\tan(180° - \theta) = -\tan\theta$
35. $\tan(\pi + \theta) = \tan\theta$
36. $\tan(360° - \theta) = -\tan\theta$
37. $\tan(2\pi + \theta) = \tan\theta$
38. $\tan(\pi - \theta) = -\tan\theta$

For Ex. 33–38, see TE Answer Section.

HALF-ANGLE FORMULAS 19.9

Objective **To apply the formulas for $\sin \frac{A}{2}$, $\cos \frac{A}{2}$, and $\tan \frac{A}{2}$**

You can derive the formulas for the sine, cosine, and tangent of half an angle. You will also learn how to apply these formulas.
One of the formulas for $\cos 2A$ is $\cos 2A = 1 - 2 \sin^2 A$.
If we replace A with $\frac{x}{2}$, the following equation is true.

$$\cos 2\left(\frac{x}{2}\right) = 1 - 2 \sin^2\left(\frac{x}{2}\right)$$

$$\cos x = 1 - 2 \sin^2 \frac{x}{2}$$

$$2 \sin^2 \frac{x}{2} = 1 - \cos x$$

$$\sin^2 \frac{x}{2} = \frac{1 - \cos x}{2} \qquad \text{So, } \sin \frac{x}{2} = \pm \sqrt{\frac{1 - \cos x}{2}}.$$

Sine of half an angle	$\sin \frac{A}{2} = \pm \sqrt{\dfrac{1 - \cos A}{2}}$

Example 1 **Find $\sin 30°$. Use the formula for $\sin \frac{A}{2}$.**

$$\sin \frac{A}{2} = \pm \sqrt{\frac{1 - \cos A}{2}}$$

$30° = \frac{60°}{2}$ ▶ $\sin\left(\frac{60°}{2}\right) = +\sqrt{\dfrac{1 - \cos 60°}{2}}$ ◀ Sin 30° is positive.

$$= +\sqrt{\frac{1 - \frac{1}{2}}{2}}, \text{ or } +\sqrt{\frac{1}{4}}$$

Thus, $\sin 30° = \frac{1}{2}$.

Example 2 **If $\cos x = -\frac{1}{8}$ and x lies in the third quadrant, find $\sin \frac{x}{2}$.**

$$\sin \frac{x}{2} = \pm \sqrt{\frac{1 - \cos x}{2}}$$

Since x lies in the third quadrant, $180° < x < 270°$. Then $\frac{x}{2}$ lies in the second quadrant, since $90° < \frac{x}{2} < 135°$.

$\sin \frac{x}{2}$ is positive in the second quadrant. ▶ $\sin \frac{x}{2} = +\sqrt{\dfrac{1 - \cos x}{2}} = +\sqrt{\dfrac{1 - (-\frac{1}{8})}{2}}, \text{ or } +\sqrt{\dfrac{1 + \frac{1}{8}}{2}}$

$$= +\sqrt{\frac{9}{16}}$$

Thus, $\sin \frac{x}{2} = \frac{3}{4}$.

Half-Angle Formulas 563

To derive a formula for $\cos \frac{A}{2}$, use $\cos 2A = 2\cos^2 A - 1$ and replace A with $\frac{x}{2}$.

$$\cos 2A = 2\cos^2 A - 1$$

$$\cos 2\left(\frac{x}{2}\right) = 2\cos^2\left(\frac{x}{2}\right) - 1 \qquad \cos x = 2\cos^2\left(\frac{x}{2}\right) - 1$$

$$2\cos^2 \frac{x}{2} = 1 + \cos x \qquad \cos^2 \frac{x}{2} = \frac{1 + \cos x}{2}$$

So, $\cos \frac{x}{2} = \pm\sqrt{\dfrac{1 + \cos x}{2}}$

Cosine of half an angle

$$\cos \frac{A}{2} = \pm\sqrt{\frac{1 + \cos A}{2}}$$

Example 3

Find $\cos 75°$. Use the formula for $\cos \frac{A}{2}$. Answer may be left in simplest radical form.

$\cos \frac{A}{2} = \pm\sqrt{\dfrac{1 + \cos A}{2}}$. Write $75°$ as $\dfrac{150°}{2}$.

▶ $\cos 75°$ is positive. $\cos\left(\dfrac{150°}{2}\right) = +\sqrt{\dfrac{1 + \cos 150°}{2}}$

$\cos 150° = -\cos 30°$
$= -\dfrac{\sqrt{3}}{2}$.

▶ $= \sqrt{\dfrac{1 + \left(-\frac{\sqrt{3}}{2}\right)}{2}}$, or $\sqrt{\dfrac{2 - \sqrt{3}}{4}}$

Thus, $\cos 75° = \dfrac{\sqrt{2 - \sqrt{3}}}{2}$.

To derive a formula for $\tan \frac{x}{2}$, use $\tan A = \dfrac{\sin A}{\cos A}$ and replace A with $\frac{x}{2}$.

$$\tan \frac{x}{2} = \frac{\sin \frac{x}{2}}{\cos \frac{x}{2}} = \frac{\pm\sqrt{\dfrac{1 - \cos x}{2}}}{\pm\sqrt{\dfrac{1 + \cos x}{2}}} = \pm\sqrt{\frac{1 - \cos x}{1 + \cos x}}$$

Tangent of half an angle

$$\tan \frac{A}{2} = \pm\sqrt{\frac{1 - \cos A}{1 + \cos A}}$$

Example 4

If $\cos x = \frac{1}{5}$ and x lies in the 1st quadrant, find $\tan \frac{x}{2}$.

$$\tan \frac{x}{2} = \pm\sqrt{\frac{1 - \cos x}{1 + \cos x}} = +\sqrt{\frac{1 - \frac{1}{5}}{1 + \frac{1}{5}}}, \text{ or } +\sqrt{\frac{\frac{4}{5}}{\frac{6}{5}}}$$

$$= +\sqrt{\frac{4}{6}} = \frac{2}{\sqrt{6}} = \frac{2\sqrt{6}}{6}$$

Thus, $\tan \dfrac{x}{2} = \dfrac{2\sqrt{6}}{6}$ or $\dfrac{\sqrt{6}}{3}$.

Written Exercises

For Ex. 1–28, see TE Answer Section.

A Find the value of each. Use special angle measures and the formula for $\sin \frac{A}{2}$, $\cos \frac{A}{2}$, or $\tan \frac{A}{2}$. Answer in simplest radical form.

1. $\sin 15°$
2. $\cos 15°$
3. $\sin 22\frac{1}{2}°$
4. $\tan 22\frac{1}{2}°$
5. $\cos 22\frac{1}{2}°$
6. $\tan 15°$
7. $\tan 67\frac{1}{2}°$
8. $\cos 67\frac{1}{2}°$
9. $\sin 67\frac{1}{2}°$
10. $\sin 75°$

Find the value of $\sin \frac{A}{2}$ and $\cos \frac{A}{2}$ given the value of $\cos A$. Answer in simplest radical form.

11. $\cos A = \frac{3}{5}$
12. $\cos A = 0$
13. $\cos A = -\frac{5}{13}$
14. $\cos A = -\frac{1}{2}$
15. $\cos A = \frac{1}{4}$
16. $\cos A = -\frac{15}{17}$
17. $\cos A = -\frac{3}{4}$
18. $\cos A = -1$

Find the value of $\sin \frac{A}{2}$, $\cos \frac{A}{2}$, and $\tan \frac{A}{2}$. Answer in simplest radical form.

19. $\cos A = \frac{4}{5}$ and A is in the 4th quad.
20. $\cos A = \frac{-12}{13}$ and A is in the 2nd quad.
21. $\cos A = -\frac{\sqrt{3}}{2}$ and A is in the 3rd quad.
22. $\cos A = -\frac{1}{2}$ and A is in the 2nd quad.

B
23. $\cos A = \frac{1}{3}$ and A is in the 1st quad.
24. $\cos A = \frac{3}{4}$ and A is in the 4th quad.
25. $\cos A = -\frac{5}{12}$ and A is in the 2nd quad.
26. $\cos A = -\frac{4}{7}$ and A is in the 3rd quad.
27. $\cos A = a$ and A is in the 1st quad.
28. $\cos A = -b$ and A is in the 2nd quad.

C
29. Find $\tan \frac{A}{2}$ if $\sin A = -\frac{1}{2}$ and $\cos A = -\frac{\sqrt{3}}{2}$. $-\sqrt{7+4\sqrt{3}}$
30. Find $\tan \frac{\theta}{2}$ if $\cos \theta = -\frac{15}{17}$ and $90° < \theta < 180°$. **4**
31. Find $\tan \frac{\theta}{2}$ if $\sin \theta = \frac{4}{5}$ and $\cos \theta = -\frac{5}{13}$. $\frac{3}{2}$
32. Find $\tan \frac{B}{2}$ if $\cos B = -\frac{4}{5}$ and $180° < B < 270°$. **−3**
33. Prove: $\cot \frac{A}{2} = \pm \sqrt{\frac{1 + \cos A}{1 - \cos A}}$.
34. Prove: $\sec \frac{A}{2} = \pm \sqrt{\frac{2}{1 + \cos A}}$.
35. Prove: $\csc \frac{A}{2} = \pm \sqrt{\frac{2}{1 - \cos A}}$.
36. Find $\sec \frac{\theta}{2}$ If $\cos \theta = -\frac{1}{2}$ and $90° < \theta < 180°$. **2**

For Ex. 33–35, see TE Answer Section.

NON-ROUTINE PROBLEMS

Quadrilateral *ABCD* is inscribed in a circle.
Find the measure of $\angle D$ to the nearest minute.
Hint: Draw $\overline{AC}$ and use the Law of Cosines twice, both involving *AC*. **49° 27′**

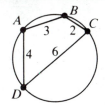

Half-Angle Formulas

DOUBLE AND HALF-ANGLE IDENTITIES 19.10

Objective **To prove identities using double and half-angle formulas**

You have derived and applied formulas for the sine, cosine and tangent of double and half angles. You will now use these formulas to prove some trigonometric identities.

Example 1 **Prove the identity $\cos 2\theta = \dfrac{\csc^2 \theta - 2}{\csc^2 \theta}$.**

Replace $\csc^2 \theta$ with $\dfrac{1}{\sin^2 \theta}$.

$\cos 2\theta$	$\dfrac{\csc^2 \theta - 2}{\csc^2 \theta}$
$\cos 2\theta$	$\dfrac{\dfrac{1}{\sin^2 \theta} - 2}{\dfrac{1}{\sin^2 \theta}}$
	$\dfrac{\dfrac{1 - 2\sin^2 \theta}{\sin^2 \theta}}{\dfrac{1}{\sin^2 \theta}}$
	$1 - 2\sin^2 \theta$
	$\cos 2\theta$

Example 2 **Prove the identity $\tan 2A = \dfrac{2 \tan A}{\sec^2 A - 2 \tan^2 A}$.**

Replace $\tan 2A$ with $\dfrac{\sin 2A}{\cos 2A}$, $\sec^2 A$ with $\dfrac{1}{\cos^2 A}$, and $\tan A$ with $\dfrac{\sin A}{\cos A}$.

$\tan 2A$	$\dfrac{2 \tan A}{\sec^2 A - 2 \tan^2 A}$
$\dfrac{\sin 2A}{\cos 2A}$	$\dfrac{2 \dfrac{\sin A}{\cos A}}{\dfrac{1}{\cos^2 A} - 2 \dfrac{\sin^2 A}{\cos^2 A}}$
$\dfrac{2 \sin A \cos A}{1 - 2 \sin^2 A}$	$\dfrac{2 \sin A}{\cos A} \cdot \dfrac{\cos^2 A}{1 - 2 \sin^2 A}$
	$\dfrac{2 \sin A \cos A}{1 - 2 \sin^2 A}$

The identity in Example 3 involves the formula for the sine of a half-angle.

Example 3   **Prove the identity** $\sin^2 \dfrac{\theta}{2} = \dfrac{\sec\theta - 1}{2\sec\theta}$.

Replace $\sec\theta$ with $\dfrac{1}{\cos\theta}$ and $\sin\dfrac{\theta}{2}$ with $\pm\sqrt{\dfrac{1-\cos\theta}{2}}$.

$$
\begin{array}{c|c}
\sin^2 \dfrac{\theta}{2} & \dfrac{\sec\theta - 1}{2\sec\theta} \\[6pt]
\left(\pm\sqrt{\dfrac{1-\cos\theta}{2}}\right)^2 & \dfrac{\dfrac{1}{\cos\theta} - 1}{2\left(\dfrac{1}{\cos\theta}\right)} \\[6pt]
\dfrac{1-\cos\theta}{2} & \dfrac{1-\cos\theta}{\cos\theta} \cdot \dfrac{\cos\theta}{2} \\[6pt]
 & \dfrac{1-\cos\theta}{2}
\end{array}
$$

Written Exercises

For Ex. 1–18, see TE Answer Section.

(A) Prove each identity.

1. $\tan\theta = \dfrac{\sin 2\theta}{1+\cos 2\theta}$

2. $\sin^2\theta = \dfrac{\tan\theta \sin 2\theta}{2}$

3. $\cot\theta = \dfrac{\sin 2\theta}{2\sin^2\theta}$

4. $\cos^2\dfrac{\theta}{2} = \dfrac{\sec\theta + 1}{2\sec\theta}$

5. $\tan\dfrac{\theta}{2} = \pm\dfrac{1-\cos\theta}{\sin\theta}$

6. $\tan\dfrac{\theta}{2} = \pm\dfrac{\sin\theta}{1+\cos\theta}$

(B)   7. $2\csc 2\theta = \dfrac{\cos\theta}{\sin\theta} + \dfrac{\sin\theta}{\cos\theta}$

8. $\sin^2\theta - 2\sin\theta\cos\theta + \cos^2\theta = 1 - 2\sin 2\theta$

9. $1 + \sin\theta = \left(\sin\dfrac{\theta}{2} + \cos\dfrac{\theta}{2}\right)^2$

10. $\cos\theta = \dfrac{\frac{1}{2}\sin^2\theta}{\sin^2\frac{1}{2}\theta} - 1$

11. $\dfrac{1-\tan^2\theta}{1+\tan^2\theta} = \cos 2\theta$

12. $\dfrac{\sin\theta + \cos\theta}{\cos\theta - \sin\theta} + \dfrac{\sin\theta - \cos\theta}{\sin\theta + \cos\theta} = 2\tan 2\theta$

(C)   13. $\tan 2\theta = \dfrac{\sin 4\theta}{1+\cos 4\theta}$

14. $\cos 3\theta = 4\cos^3\theta - 3\cos\theta$

15. $2\csc 2\theta = \sec\theta \csc\theta$

16. $\sec 2\theta = \dfrac{\sec^2\theta}{2 - \dfrac{1}{\cos^2\theta}}$

17. $\dfrac{\sec^2\theta}{2} + \tan\theta = \dfrac{1+\sin 2\theta}{1+\cos 2\theta}$

18. $\dfrac{\cos 3\theta}{\sin 3\theta} = (1+\cos 6\theta)(\csc 6\theta)$

CUMULATIVE REVIEW

Use the Pythagorean relation to determine whether the following are measures of the sides of a right triangle.

1. 6, 8, 10 **yes**
2. 0.5, 0.9, 1.2 **no**
3. $\sqrt{2}, 3\sqrt{5}, 4\sqrt{3}$ **no**

Double and Half-Angle Identities

CHAPTER NINETEEN REVIEW

(A) Exercises are 1, 2, 5, 6, 8, 11–13, 15–25, 28–32. (C) Exercises are starred.
The rest are (B) Exercises.

Vocabulary
Ambiguous case [19.5]   Law of Cosines [19.1]
Area of a triangle [19.3]   Law of Sines [19.4]

For triangle ABC, find the length of the indicated side to the nearest unit. [19.1]
1. $a = 8$, $c = 5$, m$\angle B = 42°$. Find b. **5**
2. $c = 10$, $b = 6$, m$\angle A = 130°$. Find a. **15**
3. $a = 17.3$, $b = 47.6$, m$\angle C = 67° 43'$. Find c. **44**
4. Two forces of 50 lb and 70 lb act on a body at an angle of 72°. Find the magnitude of the resultant to the nearest pound. [19.1] **98 lb**

For triangle ABC, find each angle to the nearest degree. [19.2]
5. $a = 6$, $b = 9$, $c = 4$. Find m$\angle A$. **32°**
6. $a = 4$, $b = 7$, $c = 10$. Find m$\angle B$. **33°**
7. $a = 12.7$, $b = 17.9$, $c = 23.6$. Find m$\angle C$. **100°**

For $\triangle ABC$, find the area to the nearest square unit. [19.3]
8. $a = 6$, $b = 8$, m$\angle C = 48°$. **18**
9. $b = 8.3$, $c = 9.6$, m$\angle A = 63° 20'$ **36**
10. The lengths of two sides of a parallelogram are 30 and 34 centimeters. Their included angle measures 120°. Find the area of the parallelogram. **883 cm²** [19.3]

For $\triangle PQR$, find m$\angle P$ to the nearest degree. [19.3]
11. $q = 9$, $r = 16$, area = 36 square units **30°**
12. $q = 11$, $r = 9$, area = 45 square units **65°**

13. For $\triangle ABC$, $a = 60$, m$\angle A = 63°$, m$\angle B = 47°$. Find b to the nearest unit. **49** [19.4]
14. For $\triangle ABC$, $a = 13.3$, $c = 18.5$, m$\angle A = 43° 40'$. Find m$\angle C$ to the nearest degree. **74°** [19.4]

How many triangles can be constructed? [19.5]
15. m$\angle A = 38°$, $a = 8$, $b = 12$ **2**
16. m$\angle A = 120°$, $a = 30$, $b = 10$ **1**

Find each. Use the formula for $\sin (A \pm B)$ or $\sin \frac{A}{2}$. Answer in simplest radical form. [19.6, 19.9]
17. $\sin 75°$ $\frac{\sqrt{6}+\sqrt{2}}{4}$
18. $\sin 15°$ $\frac{1}{2}\sqrt{2-\sqrt{3}}$
19. $\sin 22\frac{1}{2}°$ $\frac{1}{2}\sqrt{2-\sqrt{2}}$
20. If $\sin A = -\frac{12}{13}$ and $\angle A$ lies in the third quadrant, find $\sin 2A$. $\frac{120}{169}$ [19.6]

Find each. Use the formula for $\cos(A \pm B)$ or $\cos \frac{A}{2}$. Answer in simplest radical form. [19.7, 19.9]
21. $\cos 15°$ $\frac{1}{2}\sqrt{2+\sqrt{3}}$
22. $\cos 67\frac{1}{2}°$ $\frac{1}{2}\sqrt{2-\sqrt{2}}$
23. $\cos 22\frac{1}{2}°$ $\frac{1}{2}\sqrt{2+\sqrt{2}}$

Find $\sin (A + B)$ and $\sin (A - B)$. [19.6]
24. $\sin A = -\frac{4}{5}$ and $\angle A$ lies in the fourth quadrant. $\sin B = \frac{5}{13}$ and $\angle B$ lies in the second quadrant. $\frac{63}{65}, \frac{33}{65}$

Find $\cos (A + B)$ and $\cos (A - B)$. [19.7]
25. $\cos A = -\frac{12}{13}$ and $\angle A$ lies in the third quadrant. $\cos B = \frac{3}{5}$ and $\angle B$ lies in the fourth quadrant. $-\frac{56}{65}, -\frac{16}{65}$

26. If $\cos A = -\frac{1}{4}$ and $\angle A$ lies in the second quadrant, find $\cos \frac{A}{2}$. $\frac{\sqrt{6}}{4}$ [19.9]

Prove the identity. Use formula for $\sin (A - B)$.
★27. $\sin (\frac{3}{2} \pi - \theta) = -\cos \theta$. [19.6]
For Ex. 27, 32, and 34, see TE Answer Section.

28. Find $\tan (A + B)$ if $\tan A = -\frac{5}{12}$ and $\tan B = -\frac{3}{5}$. $-\frac{61}{45}$ [19.8]

29. Find $\tan 2A$ if $\tan A = \frac{4}{5}$. $\frac{40}{9}$ [19.8]

30. Find $\tan (A - B)$ if $\tan A = -\frac{8}{9}$ and $\tan B = \frac{4}{7}$. $-\frac{92}{31}$ [19.8]

31. Find $\tan 105°$. Use the formula for $\tan (A + B)$. $-2 - \sqrt{3}$ [19.8]

32. Prove $\cos^2 \frac{\theta}{2} = \frac{\sin \theta + \tan \theta}{2 \tan \theta}$. [19.10]

★33. Find $\tan \frac{A}{2}$ if $\sin A = \frac{\sqrt{3}}{2}$ and $\cos A = -\frac{1}{2}$. $\sqrt{3}$ [19.9]

★34. Prove: $\tan (2\pi - \theta) = -\tan \theta$. [19.8]

Ⓐ Exercises are 1, 2, 5, 6, 8, 11–13, 15–31. Ⓒ Exercises are starred.
The rest are Ⓑ Exercises.

CHAPTER NINETEEN TEST

For triangle ABC, find the length of the indicated side to the nearest unit.
1. $a = 5$, $b = 9$, $m\angle C = 50°$. Find c. **7**
2. $c = 12$, $b = 7$, $m\angle A = 140°$. Find a. **18**
3. $a = 23.4$, $c = 36.2$, $m\angle B = 73°\,18'$. Find b. **37**

4. Two forces of 14 lb and 16 lb act on a body at an angle of 48°. Find the magnitude of the resultant to the nearest pound. **27 lb**

For triangle ABC, find the angle measure to the nearest degree.
5. $a = 6$, $b = 5$, $c = 8$. Find $m\angle C$. **93°**
6. $a = 15$, $b = 7$, $c = 10$. Find $m\angle A$. **123°**
7. $a = 13.8$, $b = 24.6$, $c = 18.7$. Find $m\angle B$. **97°**

For △ABC, find the area to the nearest square unit.
8. $b = 14$, $c = 8$, $m\angle A = 73°$ **54**
9. $a = 6.3$, $b = 8.5$, $m\angle C = 30°\,40'$ **14**

10. The lengths of two sides of a parallelogram are 64 and 82 centimeters. Their included angle measures 70°. Find the area of the parallelogram. **4,932 cm²**

For △PQR, find $m\angle P$ to the nearest degree.
11. $q = 7$, $r = 8$, area = 22 sq. units **52°**
12. $q = 14$, $r = 10$, area = 60 sq. units **59°**

13. For △ABC, $a = 32$, $m\angle A = 72°$, $m\angle C = 32°$. Find c to the nearest unit. **18**

14. For △ABC, $a = 16.4$, $c = 10.8$, $m\angle A = 38°\,30'$. Find $m\angle C$ to the nearest degree. **24°**

How many triangles can be constructed?
15. $m\angle A = 28°$, $a = 7$, $b = 13$ **2**
16. $m\angle A = 130°$, $a = 16$, $b = 12$ **1**

Find each. Use the formula for sin $(A \pm B)$ or sin $\frac{A}{2}$. Answer in simplest radical form.
17. sin 15° $\frac{1}{2}\sqrt{2 - \sqrt{3}}$
18. sin 75° $\frac{\sqrt{6} + \sqrt{2}}{4}$
19. sin $22\frac{1}{2}°$. $\frac{1}{2}\sqrt{2 - \sqrt{2}}$

20. If $\sin A = -\frac{3}{5}$ and $\angle A$ lies in the third quadrant, find sin $2A$. $\frac{24}{25}$

Find each. Use the formula for cos $(A \pm B)$ or cos $\frac{A}{2}$. Answer in simplest radical form.
21. cos 135° $\frac{-\sqrt{2}}{2}$
22. cos $22\frac{1}{2}°$ $\frac{1}{2}\sqrt{2 + \sqrt{2}}$
23. cos 15° $\frac{1}{2}\sqrt{2 + \sqrt{3}}$

Find sin $(A + B)$ and sin $(A - B)$.
24. $\sin A = \frac{5}{7}$ and $\angle A$ lies in the second quadrant. $\sin B = -\frac{4}{5}$ and $\angle B$ lies in the third quadrant. $\frac{-15 + 8\sqrt{6}}{35}, \frac{-15 - 8\sqrt{6}}{35}$

Find cos $(A + B)$ and cos $(A - B)$.
25. $\cos A = \frac{12}{13}$ and $\angle A$ lies in the first quadrant. $\cos B = -\frac{4}{5}$ and $\angle B$ lies in the second quadrant. $\frac{-63}{65}, \frac{-33}{65}$

26. If $\cos A = -\frac{5}{13}$ and $\angle A$ lies in the third quadrant, find cos $\frac{A}{2}$. $\frac{-2\sqrt{13}}{13}$

27. Find tan $(A + B)$ if $\tan A = \frac{4}{5}$ and $\tan B = -\frac{2}{3}$. $\frac{2}{23}$

28. Find tan $2A$ if $\tan A = \frac{3}{5}$. $\frac{15}{8}$

29. Find tan $(A - B)$ if $\tan A = -\frac{8}{9}$ and $\tan B = -\frac{3}{7}$. $-\frac{1}{3}$

30. Find tan 75°. Use the formula for tan $(A + B)$. $2 + \sqrt{3}$

For Ex. 31–34, see TE Answer Section.

31. Prove $\sin 2A = \dfrac{2 \cot A}{\csc^2 A}$.

32. Prove $2 \csc 2\theta = \tan \theta + \cot \theta$.

Prove the identity. Use the formula for sin $(A - B)$.
★ 33. $\sin\left(\frac{\pi}{2} + \theta\right) = \cos \theta$
★ 34. Prove $\tan(\pi - \theta) = -\tan \theta$.

★ 35. Find tan $\frac{A}{2}$ if $\sin A = -\frac{1}{2}$ and $\cos A = -\frac{\sqrt{3}}{2}$. $-\sqrt{7 + 4\sqrt{3}}$

COMPUTER ACTIVITIES

Law of Cosines

OBJECTIVE: To determine the measures of angles of a triangle, given the three sides utilizing the Law of Cosines

The Law of Cosines states: $a^2 = b^2 + c^2 - 2bc \cos A$. Solving for cos A, the equation becomes $\cos A = (a^2 - b^2 - c^2)/(-2bc)$. Using this form of the Law of Cosines, you can find the measure of any angle of a triangle when you know the lengths of its three sides. The program below performs this process. First, lines 140–160 determine the cosines of all three angles of the triangle. The program then uses inverse trigonometric functions to determine the angles.

Among the inverse trigonometric functions, most forms of BASIC include only the arctangent function, symbolized by ATN. In this program, the arctangent must be expressed in terms of an angle's cosine. Line 190 does this by using the arctangent form of the following trigonometric identity:

$$\tan A = (\sqrt{1 - \cos^2 A})/\cos A.$$

All trigonometric functions in BASIC use radian measurement of angles. Lines 200, 210–230, do the necessary converting.

```
100   PRINT "WHAT ARE THE THREE SIDES OF THE TRIANGLE
      ?";: INPUT F,B,C
110   IF F + B < = B THEN 100
120   IF F + C < = B THEN 100
130   IF B + C < = F THEN 100
140 D(1) = (F ^ 2 - B ^ 2 - C ^ 2) / ( - 2 * C * B)
150 D(2) = (B ^ 2 - C ^ 2 - F ^ 2) / ( - 2 * C * F)
160 D(3) = (C ^ 2 - B ^ 2 - F ^ 2) / ( - 2 * B * F)
170   PRINT "THE ANGLES ARE "
180   FOR I = 1 TO 3
190 E =  ATN ( SQR (1 - (D(I)) ^ 2) / D(I))
200 A = E * 180 / 3.1416
210 A(I) =   INT (A)
220 S(I) =   INT ((A - A(I)) * 60)
230   PRINT  ABS (A(I));" DEGREES ";S(I);" MINUTES
240   NEXT I: PRINT : GOTO 100
```

See the Computer Section beginning on page 574 for more information.

EXERCISES   For Ex. 2, see TE Answer Section.

1. Use the program above to find the solutions to the odd-numbered problems 1 to 13 on page 543.

2. Write a program that INPUTs the measurement of an angle in degrees and outputs the values for all six trigonometric functions of that angle.

COLLEGE PREP TEST

DIRECTIONS: Choose the *one* best answer to each question or problem.

1.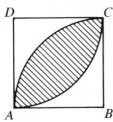

   ABCD is a square. $AB = 4$. B is the center of arc AC. D is the center of arc AC. Find the area of the shaded region.
   (A) $4(\pi - 2)$ (B) $16(\pi - 1)$ (C) $8(\pi - 1)$
   (D) $8(\pi - 2)$ (E) $16\pi - 8$ **D**

2.

   A regular hexagon is inscribed in a circle. The radius of the circle is 6 cm. Which is the area between the circle and the regular hexagon?
   (A) $36\pi - 18$ (B) $54\sqrt{3} - 36\pi$
   (C) $36\pi - 54\sqrt{3}$ (D) $54\sqrt{3} - 12\pi$
   (E) None of these **C**

3.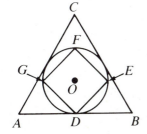

   △ABC is equilateral with side of length s. A circle is inscribed in the triangle. A square is inscribed in the circle. Which is the area of the square?
   (A) $\frac{1}{6}s^2$ (B) s^2 (C) $\frac{1}{8}s^2$
   (D) $\frac{1}{4}s^2$ (E) $\frac{1}{3}s^2$ **A**

4. For all real numbers a and b, define $a \star b = \frac{a+b}{ab}$.
   Find $(5 \star 5) \star 5$.

   (A) 2 (B) 5 (C) $\frac{2}{5}$
   (D) $\frac{27}{10}$ (E) $\frac{4}{25}$ **D**

5.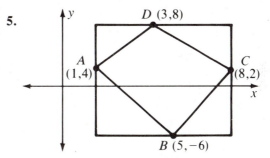

   What is the area of ABCD?

   (A) 49 (B) 98 (C) 56
   (D) 43 (E) None of these **E**

6.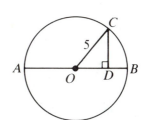

   $\overline{AB}$ is a diameter of the circle with center O and radius 5.
   If $\frac{BD}{DA} = \frac{1}{4}$, which is the value of OD?

   (A) 4 (B) 12 (C) 13
   (D) $\frac{5\sqrt{2}}{2}$ (E) 3 **E**

College Prep Test

CUMULATIVE REVIEW

(Chapters 17–19)

DIRECTIONS: Choose the *one* best answer to each question or problem. (Ex. 1–18)

1. Express sin 53° 18′ in terms of its cofunction. [17.8] **B**
   (A) csc 36° 42′ (B) cos 36° 42′
   (C) −cos 36° 42′ (D) −csc 36° 42′

2. In which quadrant is sin $u < 0$ and cot $u < 0$? [17.7] **D**
   (A) I (B) II (C) III (D) IV

3. Solve $\tan \theta = -\frac{\sqrt{3}}{3}$ for θ if $0 \leq \theta \leq 2\pi$. [18.11] **A**
   (A) $\frac{5}{6}\pi$ and $\frac{11}{6}\pi$ (B) $\frac{2}{3}\pi$ and $\frac{5}{3}\pi$
   (C) $\frac{2}{3}\pi$ and $\frac{4}{3}\pi$ (D) $\frac{5}{6}\pi$ and $\frac{7}{6}\pi$

4. Express 225° in radians. [18.10] **A**
   (A) $\frac{5}{4}\pi$ (B) $\frac{4}{3}\pi$ (C) $\frac{7}{6}\pi$ (D) $\frac{3}{4}\pi$

5. Express $\frac{2}{3}\pi$ radians in degrees. [18.10] **D**
   (A) 60° (B) 240° (C) 180° (D) 120°

6. Find the inverse of $y = -2 \sin x$. [18.12] **B**
   (A) $x = \arcsin(-\frac{y}{2})$ (B) $y = \arcsin(-\frac{x}{2})$
   (C) $x = -\arcsin(\frac{y}{2})$ (D) $y = -\arcsin(\frac{x}{2})$

7. Find θ, if $\theta = \text{Arc} \sin(-\frac{\sqrt{3}}{2})$. [18.12] **C**
   (A) 240° (B) 300° (C) −60° (D) 60°

8. Find $\cos(\text{Arc} \tan \frac{4}{3})$. [18.12] **A**
   (A) $\frac{3}{5}$ (B) $-\frac{3}{5}$
   (C) $\frac{3}{5}$ and $-\frac{3}{5}$ (D) None of these

9. Evaluate $\sec(-\frac{\pi}{4}) + \csc \frac{\pi}{2}$. [18.10] **C**
   (A) $\frac{\sqrt{2}+2}{2}$ (B) $-\frac{\sqrt{2}+2}{2}$
   (C) $\sqrt{2}+1$ (D) $-\sqrt{2}+1$

10. Write an equivalent expression for cot x using only sin x. [18.8] **D**
    (A) $\frac{1}{\sin^2 x}$ (B) $\sin^2 x$
    (C) $\pm \frac{\sin x}{\sqrt{1-\sin^2 x}}$ (D) $\frac{\pm\sqrt{1-\sin^2 x}}{\sin x}$

11. The lengths of two sides of a parallelogram are 20 and 30 centimeters. Their included angle measures 150°. Find the area of the parallelogram. [19.3] **B**
    (A) 150 cm² (B) 300 cm²
    (C) $150\sqrt{3}$ cm² (D) $150\sqrt{2}$ cm²

12. How many triangles can be constructed if $m\angle A = 46°$, $a = 16$, and $b = 14$. [19.5] **B**
    (A) 0 (B) 1 (C) 2 (D) More than 2

13. Find cos 75°. [19.6] **B**
    (A) $\frac{\sqrt{6}+\sqrt{2}}{4}$ (B) $\frac{\sqrt{2}}{4}[\sqrt{3}-1]$
    (C) $\frac{\sqrt{2}}{4}[1-\sqrt{3}]$ (D) None of these

14. If $\sin A = -\frac{5}{13}$ and $\angle A$ lies in the third quadrant, find sin 2A. [19.6] **D**
    (A) $-\frac{120}{169}$ (B) $\frac{119}{169}$ (C) $-\frac{119}{169}$ (D) $\frac{120}{169}$

15. Find sin 15°. [19.7] **C**
    (A) $\frac{\sqrt{6}+\sqrt{2}}{4}$ (B) $\frac{\sqrt{2}}{2}[\sqrt{3}-1]$
    (C) $\frac{\sqrt{6}-\sqrt{2}}{4}$ (D) $\frac{\sqrt{2}-\sqrt{6}}{4}$

16. If $\cos A = -\frac{1}{5}$ and $\angle A$ lies in the second quadrant, find $\cos \frac{A}{2}$. [19.9] **D**
    (A) $\frac{1}{5}\sqrt{15}$ (B) $-\frac{1}{5}\sqrt{15}$
    (C) $-\frac{1}{5}\sqrt{10}$ (D) $\frac{1}{5}\sqrt{10}$

17. Find $\tan(A - B)$ if $\tan A = -\frac{5}{6}$ and $\tan B = \frac{2}{3}$. [19.8] **A**
    (A) $-\frac{27}{8}$ (B) $\frac{3}{28}$ (C) $-\frac{27}{28}$ (D) $-\frac{3}{8}$

18. Find $\tan \frac{A}{2}$ if $\cos A = -\frac{1}{3}$ and $\angle A$ is in the 4th quadrant. [19.9] **B**
    (A) $\sqrt{2}$ (B) $-\sqrt{2}$
    (C) $\pm\sqrt{2}$ (D) None of these

Determine the coordinates of point P and the radius r. [17.2]

19.
(8,8) $r = 8\sqrt{2}$

20.
$(-4\sqrt{3}, -4)$ $r = 8$

Find each of the following. Rationalize any radical denominator. [17.3–17.6]

21. $\cos(-45°)$ $\frac{\sqrt{2}}{2}$
22. $\sin 150°$ $\frac{1}{2}$
23. $\tan 135°$ -1
24. $\cot(-495°)$ 1
25. $\sec 330°$ $\frac{2\sqrt{3}}{3}$
26. $\csc(-225°)$ $\sqrt{2}$
27. $\cos 180°$ -1
28. $\sin 270°$ -1
29. $\tan 90°$ undefined

Write each of the following in terms of the same trigonometric function of an acute angle. [17.3–17.6]

30. $\tan 100°$ $-\tan 80°$
31. $\sin 340°$ $-\sin 20°$
32. $\cos 210°$ $-\cos 30°$

Find two values of u between $0°$ and $360°$ for which each is true. [17.3]

33. $\sin u = -\frac{\sqrt{2}}{2}$ 225°, 315°
34. $\tan u = \frac{\sqrt{3}}{3}$ 30°, 210°

Find the value of each of the other five trigonometric functions of u. [17.7]

35. $\cos u = -\frac{12}{13}$ and $\tan u > 0$. $\sin u = -\frac{5}{13}$, $\tan u = \frac{5}{12}$, $\cot u = \frac{12}{5}$, $\sec u = -\frac{13}{12}$, $\csc u = -\frac{13}{5}$

Find all values of θ to the nearest minute ($0° < \theta < 360°$). [17.8]

36. $\tan \theta = -.1688$ 170° 25', 350° 25'

Express as a function of an angle measure less than 45°. [17.8]

37. $\cos 100°$ $-\sin 10°$
38. $\tan 322° 20'$ $-\tan 37° 40'$

Determine, to the nearest minute, the values of $0° < \theta < 360°$ that make the equation true.

39. $-6\cos\theta + 8 = 3\cos\theta + 4$. [17.9] 63° 37', 296° 23'

Evaluate and simplify. [17.11]

40. $\dfrac{8\cos 30° - \sin 45°}{-2\sec(-45°)}$ $\dfrac{1 - 4\sqrt{6}}{4}$

41. Find the measure of each of the other two sides of right triangle ABC, with m$\angle C$ = 90, to the nearest whole number. [17.12] m$\angle A$ = 40° 20', b = 3.86 $a = 3$, $c = 5$

Sketch the graph for $0° \le x \le 360°$. [18.1–18.5]

42. $y = 2\sin x$
43. $y = \cos(-2x)$
44. $y = \frac{1}{2}\sin(2x)$
45. $y = \tan 2x$

For Ex. 42–46, 51, see TE Answer Section.

46. Sketch $y = \cos 3x$ and $y = 2\sin x$ for $-180° \le x \le 180°$ on the same set of axes. For how many values of x does $\cos 3x = 2\sin x$? [18.4]

47. Find $\log \sin 34° 37'$. [18.6] 9.7544 − 10

48. Find u, $0° \le u \le 90°$, if $\log \cos u = 9.6487 - 10$. [18.6] 63° 33'

49. Find the value of the expression using logarithms. [18.6]
$\dfrac{3.95^5 \sqrt{.079}}{\tan 32° 18'}$ 427.6

50. Solve for θ to the nearest degree: $2\cos^2\theta - 3\cos\theta - 5 = 0$ ($0° \le \theta \le 360°$). [18.7] 180°

51. Prove the identity $2\cos^2\theta - 1 = \dfrac{1 - \tan^2\theta}{1 + \tan^2\theta}$. [18.9]

52. Sketch $y = \frac{1}{2}\sin 3x$ and $y = \cos \frac{1}{2}x$ for $-\pi \le x \le 2\pi$ on the same set of axes. For how many values of x does $\frac{1}{2}\sin 3x = \cos \frac{1}{2}x$? [18.11] 4

53. Sketch the graph of $y = \text{Arc}\sin x$. [18.12]

Each refers to triangle ABC. [19.1–19.2]
54. $a = 7$, $c = 6$, m$\angle B$ = 38°. Find b. 4.3
55. $a = 5$, $b = 6$, $c = 9$. Find m$\angle B$. 39°

56. For $\triangle ABC$, find the area to the nearest square unit. [19.3] $a = 10$, $b = 12$, m$\angle C$ = 52° 30' 48

For Ex. 53, 57–58, see TE Answer Section.

57. Prove the identity $\tan x = \dfrac{\sin 2x}{1 + \cos 2x}$. [19.10]

58. Prove the identity [19.10]
$\sin^2 \dfrac{x}{2} = \dfrac{\tan \frac{x}{2} \cdot \sin x}{2}$.

Cumulative Review (Chapters 17–19)

Computer Section

WRITING AND RUNNING A PROGRAM

OBJECTIVE: **To test solutions of linear equations**

For you to communicate with another human being, both of you must understand the same language. Similarly, to communicate with a computer, both you and the computer must understand the same language. Computers are designed to understand a variety of programming languages. In this text, we will use a language used by many microcomputers—BASIC (Beginner's All-purpose Symbolic Instruction Code).

The computer understands BASIC because the BASIC language has been programmed into its circuitry. In this section of the book, you will learn to use BASIC in programming your computer to solve various mathematical problems.

A computer can do many things. It can store, retrieve, process, and evaluate information and data. It performs these tasks by carrying out a computer program. A program is simply a series of numbered instructions written in a programming language the computer understands. A program can be very long and complex or as simple as the one statement PRINT 2 + 5. This instruction tells the computer to add 2 and 5 and display the answer on the monitor (screen).

PRINT is probably the most important word in the computer's vocabulary. It is used for performing mathematical computations, displaying the results, displaying messages, and performing a variety of other functions.

The computer performs mathematical calculations, using the symbols +, −, *, and / for addition, subtraction, multiplication, and division, respectively. Parentheses ( ) are used to show grouping in mathematical expressions, and the caret ∧ is used to show exponentiation (raising to a power).

When carrying out mathematical calculations, the computer follows the rules for order of operations. Type in this one-line program just as you see it.

```
10 PRINT 2 + 3 * 5     Press the RETURN key after each program line.
RUN
```

If you typed in the program exactly as shown, the computer should have displayed the result 17 on the screen. Note that the computer followed the rule for doing multiplication before addition. Now type in the following program exactly as it appears.

```
10   PRINT   3+2*8-6*4
20   PRINT   (3+2)*8-6*4
30   PRINT   3+2*(8-6)*4
40   PRINT   (3+2*8-6)*4
```

Notice that typing in the program lines does not make the computer display any results. To make the computer carry out your program and display the results, you must type in the word RUN and press the RETURN key. Before you do that, however, calculate the results yourself. Then compare your answers with the computer's. If they are not the same, there are several possible explanations.

The first explanation is that you made a mathematical error. The second explanation is that you made a typing error while entering the program. To check, type in the word LIST and press the RETURN key. The computer will display a LISTing of your program. Check it carefully. Correcting malfunctioning programs is known as "debugging."

You can program the computer to perform more useful tasks. For example, you can program the computer to check solutions of linear equations you have found.

Consider the equation $2(x + 3) = x + 11$ and its solution, 5. To check that solution, you add the value 5 into both sides of the equation. If both sides yield the same value, then the solution was correct. The following program performs this checking process for you. Before entering it, however, type in the command NEW and press the RETURN key. NEW erases the previous program stored in memory.

```
10   PRINT   2*(5+3)    Notice that an * is needed before the ( ).
20   PRINT   5+11
```

When you RUN this program, the computer will display two numbers. If they are the same, then the solution is correct.

EXERCISE:

Solve equations 2, 4, 6, and 8 on page 3. Then write and run a program that tests those solutions. [*Hint:* this program should have 8 lines, 2 for each equation.]
For Exercise, see TE Answer Section.

Computer Section

VARIABLES IN A COMPUTER PROGRAM

OBJECTIVE: **To check solutions of equations containing variables**

In BASIC, you can use variables in almost the same way you use them in algebra. The expression X + 6 means the same in both. However, 2X in algebra must be translated to 2 * X in BASIC since juxtaposition does not specify multiplication in BASIC as it does in algebra. It is necessary to use * for multiplication in BASIC.

You may assign a value to a variable by using the BASIC word LET. LET X = 4 assigns the value 4 to the variable X. Note that the variable to which the value is assigned is to the left of the =. LET 4 = X is *not* the same as LET X = 4. Note also the END statement in line 40, which halts execution of the program.

The solution of the equation 3X + 8 = 2(X − 5) is X = −18. A computer program that tests this result is:

```
10   LET X = - 18
20   PRINT  3*X+8
30   PRINT  2*(X-5)
40   END
```

If the two numbers printed out by this program are the same, then the solution is correct. If the solution is not correct, you don't have to type in both expressions again. You simply have to change line 10 by assigning another value (a new possible solution) to X.

EXERCISES:

1. Solve Exercises 3, 5, and 7 on page 3. Then write a computer program that tests your solutions.

2. Instead of solving the equation in Exercise 30 on page 3, write a program that will test each side of the equation. Use the program in this lesson as an example. Then guess a possible solution and type in a LET statement that assigns that value to the variable used in your program. Keep testing numbers until you find the correct solution. [*Hint:* The correct solution is a number between 10 and 20.]

3. Solve Exercise 28 on page 3. Then demonstrate that the solution is probably correct by running the program many times, each time using a different value for the variable.

For Ex. 1–2, see TE Answer Section.

GETTING DATA INTO A PROGRAM

OBJECTIVE: **To evaluate polynomials by using INPUT, READ, and DATA statements**

In previous programs, to change the value assigned to a variable, you had to rewrite the line that assigned the value.

There are two alternatives to this cumbersome process. The first is to use the INPUT command. INPUT instructs the computer to request a variable's value from the user when the program is run. The computer requests the value by printing a question mark and a blinking cursor.

```
10   INPUT X
20   PRINT X ^ 2 + 2 * X + 3
30   END
```

This program will ask the user for a value for X. Upon receiving it, the computer will insert the value into the polynomial $X^2 + 2X + 3$ and print the result. To evaluate the expression for a different value, you can run the program again. However, by adding the line 25 GOTO 10 the computer will return on its own to line 10. It will continue requesting new values and printing results until someone stops the program.

The second way to get values into the computer is by using the READ and DATA commands. READ X instructs the computer to look for a line that begins with the word DATA. The DATA line contains the values that are to be assigned to X. The READ statement stores those DATA values one at a time for use in the program. The DATA line may appear anywhere in a program prior to an END statement.

```
10   READ X
20   PRINT X ^ 2 + 2 * X + 3
30   GOTO 10
40   DATA  1,2,3,4,5
50   END
```

After all the values have been read, the computer will display an error message indicating there are no more numbers left. At this point, you do not need to be concerned about this type of error message.

EXERCISES:

1. Write a program using INPUT that evaluates Exercises 30–33 on page 40.
2. Replace the INPUT statement in the first program above with the READ and DATA statements. Place the numbers 1–10 in the DATA line.
   For Ex. 1–2, see TE Answer Section.

Computer Section

MULTIPLYING BINOMIALS

OBJECTIVE: **To multiply binomials by using INPUT for entering their coefficients and constants**

The PRINT statement can be used to print words, instructions, or expressions by enclosing them in quotes. We will see this illustrated in the program below in which the computer outputs the product of two binomials of the form $(AX + B)(CX + D)$.

```
10    INPUT A,B
20    INPUT C,D
30    PRINT A * C"X^2 + "A * D + C * B"X +"B * D
40    GOTO 10
50    END
```

If the binomials take the form $(AX + B)(CX - D)$, a little extra planning is needed to make their product appear correctly. See if you can pick out the strategy used to do so in the program below.

```
10    INPUT A,B
20    INPUT C,D
30    PRINT A * C"X^2";
40    IF A * D + C * D < 0 THEN 60
50    PRINT "+";
60    PRINT A * D + B * C"X"B * D
70    GOTO 10
80    END
```

In this version of the program, line 40 tests the middle term to see if it is negative. If it is, the computer goes to line 60 and prints the number with its attached minus sign. If the middle term is not negative, the computer goes to line 50 and prints a plus sign before printing the number itself in line 60. This procedure is necessary because the computer does not print signs with positive numbers. Also note the semicolons in lines 30 and 50. They are used to make the output of the following PRINT statements appear on the same line with no spaces in between.

For Ex. 1–3, see TE Answer Section.

EXERCISES:

1. Write a program similar to the one above to print the product of two binomials of the form $(AX - B)(CX - D)$.

2. Write a program similar to the previous ones that will output the product of two polynomials of the form $(AX + B)(CX^2 + DX + E)$.

3. See if you can use all the above ideas to write one program that outputs the product of two binomials regardless of the signs they contain.

CONDITIONAL BRANCHING

OBJECTIVE: **To test solutions of quadratic equations by using an IF...THEN statement**

In previous programs, you tested the solutions of an equation by evaluating each side of the equation. If the computed values of both sides were equal, then the entered solutions were correct. By using the IF...THEN command, the computer can make that determination itself.

In an IF...THEN command, you specify a comparison test in the IF portion such as IF A = B, IF A < B, IF A > B. Following THEN, you specify a line number or another command. If the comparison test following IF is true, then the computer will branch to the statement whose line number follows THEN, or will execute the command following THEN. If the test following IF is not true, then the computer executes the next line in the program.

Use the following program for testing solutions of $X^2 - 2X - 8 = 0$. Line 10 requests your solution. Line 30 tests the solution you entered. If it is correct, the computer branches to line 60, which prints the message confirming your answer. Otherwise, the computer executes line 40.

```
10   INPUT X
20   IF X = - 999 THEN 80
30   IF X ^ 2 - 2 * X - 8 = 0
     THEN 60
40   PRINT "THAT SOLUTION WAS
     NOT CORRECT."
50   GOTO 10
60   PRINT "THAT WAS A CORRECT
     SOLUTION."
70   GOTO 10
80   END
```

Line 20 contains a flag directive (see below).

Line 50 returns to line 10 for correct solution.
Line 70 returns to line 10 for the second correct solution or the flag.

The IF...THEN statement in line 20 is called a *flag directive*. It is used to end a program that otherwise would continue indefinitely. After you have entered two correct solutions, you enter -999. This number is the *flag* that line 20 specifies as the condition for branching to line 80, which terminates the program.

For Ex. 1–2, see TE Answer Section.

EXERCISES:
1. Write programs using the READ command that test solutions of the equations in Exercises 13, 34, and 35 on page 70.

2. Choose one of the equations from Exercises 55–57 on page 70 that you have not yet solved. Then write a program that tests solutions of it. After that, see if you can find both of its solutions by randomly inserting numbers, using an INPUT statement.

Computer Section

LOOPING

OBJECTIVE: **To change a fraction into a decimal using a computer program containing a FOR. . .NEXT loop and the INT function**

By using a FOR. . .NEXT loop, you can make the computer carry out a particular process any specified number of times. Such a loop begins with a FOR command, such as FOR H = 1 TO 6. The number following TO specifies the number of times the loop will be executed. The NEXT command, such as NEXT H is placed at the end of the loop. As an example, the following program prints a message 25 times.

```
10   FOR H = 1 TO 25
20   PRINT "COMPUTERS ARE FUN!"
30   NEXT H
40   END
```

You can use any numerical variable in the FOR. . .NEXT statements. The computer simply uses the specified memory storage place as a counting device. In the program on the next page, the loop is set up in line 170 and terminated in line 220. The program will convert a fraction into a decimal containing any specified number, P, of decimal places. Each execution of the loop determines one digit in the fraction's decimal conversion. The last decimal place is determined by a rounding procedure in the last portion of the program.

You will also use the BASIC function, INT, in this program. INT stands for the *greatest INTeger function*. INT(X) yields the greatest integer less than or equal to the value of X. For example, INT(2.3) = 2 and INT(−1.1) = −2. The INT function has a variety of uses. In the following program, INT is used in two different ways: 1) to determine individual digits in the quotient of a long division and 2) to round the last desired digit in such a quotient.

For example, in doing the division $3\overline{)10}$ manually, the first number you would put into the quotient is 3. When the computer divides 3 into 10, however, it finds a quotient of 3.333. . . . So, to make the computer put one digit at a time into the quotient, you instruct the computer to apply INT to the quotient, which in this example is INT(10/3) = 3. In the next program, lines 140 and 190 carry out this process.

Lines 150 and 200 determine the remainder after each division has been performed. Multiplying the remainder by 10 (lines 180 and 230) is equivalent to "bringing down" the zero in long division.

Rounding the last desired digit is accomplished in line 240 by adding .5 to the quotient before applying INT to it.

Enter and run the following program. Use it to convert several fractions into decimals of varying lengths.

```
110   PRINT "NUMERATOR";: INPUT N
120   PRINT "DENOMINATOR";: INPUT D
130   PRINT "HOW MANY DECIMAL PLACES";: INPUT P
140 A =   INT (N / D)
150 R = N - A * D
160   PRINT N"/"D" = "A".";
170   FOR I = 1 TO P - 1
180 N = 4 * 10
190 A =   INT (N / D)
200 R = N - A * D
210   PRINT A;
220   NEXT I
230 N = R * 10
240 A =   INT (N / D + .5)
250   PRINT A
260   GOTO 110
270   END
```

In lines 140, 150, 180, 190, 200, 230, and 240, note that the word LET is unnecessary in assignment statements.

See page 110 for another program that uses a FOR...NEXT loop and involves decimals and fractions.

EXERCISES:

1. The program above will diligently print out the decimal to the requested number of places even if the decimal terminates and you get a string of zeros. Make the necessary change in the program to make it stop printing as soon as the remainder is zero.

2. The variable in the loop need not be only a counter. It can be a value that is actually used in the program. For example, the following program will print out the multiples of 6 from 6 to 72 as the value of I goes from 1 to 12.

   ```
   10   FOR I = 1 TO 12
   20   PRINT I * 6
   30   NEXT I
   40   END
   ```

   Write a program using a FOR...NEXT loop that will print out the squares and cubes of the integers from 1 to 25.

3. Write a program that will ask a person how old he or she is and then wish her or him a happy birthday that many times.
   For Ex. 1–3, see TE Answer Section.

SQUARE ROOTS AND ABSOLUTE VALUE

OBJECTIVE: **To simplify radicals that utilize the BASIC functions SQR and ABS**

Two new BASIC functions that closely parallel mathematical operations are SQR( ), which calculates the square root of the number in parentheses, and ABS( ), which finds the absolute value of the number in parentheses. These two functions are utilized in the following program that simplifies radicals such as $\sqrt{8}$, which equals $2\sqrt{2}$.

First, line 130 of the program tests the quantity to see if it is a perfect square. If it is not a perfect square, the FOR...NEXT loop in lines 140 through 170 is executed. Line 150 tests the number to see if it contains any perfect squares as factors, beginning with 2^2 and continuing on with 3^2, 4^2, 5^2, etc. When a perfect square is found to be a factor, line 160 removes it from the radical. This process continues until all possible perfect squares have been tested.

Before continuing, a word of caution is in order. On occasion, the computer will output the result of a computation that is incorrect by a small amount. This is particularly likely to happen when performing calculations that involve exponents or the SQR function. These small errors are called *rounding errors*. Rounding errors result from the computer converting base-ten numerals into base-two numerals, performing computations on those base-two numerals, and converting the result back into base-ten numerals. For example, in computing the square root of 9, the computer may output a result like 2.99999999 or 3.00000002, instead of the exact answer, 3. There are techniques for compensating for the rounding errors. However, they will not be covered in this lesson. (See page 162 for more about rounding errors.)

Bearing in mind that rounding errors may occur in the computer's calculations, type in the following program and RUN it.

```
100   PRINT "WHAT IS THE RADICAND TO BE SIMPLIFIED?";
      : INPUT R
110   C = 1:S = R
120   IF  SQR (S) =  INT ( SQR (S)) THEN 190
130   FOR I = 2 TO  SQR (S)
140   IF S / (I ^ 2) <  > INT (S / (I ^ 2)) THEN 160
150   C = C * I:S = S / (I ^ 2): GOTO 140
160   NEXT I
170   PRINT "SQR ("R") = "C" * SQR ("S")"
180   GOTO 100
190   PRINT "SQR ("S") = " SQR (S)
200   GOTO 100
210   END
```

EXERCISES:

1. Change the program on page 582 so that it will simplify a cube root instead of a square root. [*Hint:* There is no BASIC function for a cube root. However, cube roots can be found by using exponentiation to raise a number to the 1/3 power.]

2. Alter the program so that it will simplify a fourth root.

3. Combine all the ideas above into one program that will simplify a radical of any index. After you have attempted to write, debug, and RUN your program, compare it with the program on page 162.

It is very convenient to use the ABS function in programs that simplify odd roots. Consider line 130 in the original program. It sets loop parameters of 2 and the square root of the number in S. In a program that simplifies cube roots, however, substituting the cube root of S for its square root in line 130 will not set up the loop correctly when S is negative. In that case, line 130 must read:

130 FOR I = −2 TO (S ∧ (1/3))

However, this line will not work for positive numbers in S. Therefore, an easy way to set up the loop for all possible cube roots is by using ABS in the following line 130:

130 FOR I = 2 TO ABS(S ∧ (1/3))

4. If you didn't input negative numbers into the programs you wrote previously, go back and rewrite your programs (if necessary), using the ABS feature noted above. Then try inputting some negative numbers.

5. Write a program that makes the computer print out a table of squares, cubes, square roots, and cube roots of the numbers from 1 to 10.

6. Write a program that uses the quadratic formula to find the solutions of a quadratic equation. Use INPUT for entering values for *a, b,* and *c.* Use the INT function to round the answers you find to the nearest hundredth.
For Ex. 1–2 and 4–6, see TE Answer Section.

ON...GOTO

OBJECTIVE: **To evaluate powers of i using an ON GOTO statement**

Thus far in this text, you have examined unconditional branching, using the GOTO statement, and conditional branching using the IF...THEN statement. In this lesson, we will introduce you to another form of conditional branching using the ON...GOTO statement. ON...GOTO gives the computer the option of going to any of several program lines depending on the appearance of any of several specified conditions.

The format for the ON...GOTO statement is as follows:

ON X GOTO 100, 200, 150, 80

In this ON...GOTO statement, the computer will branch to line 100 when X = 1. When X = 2, it will branch to line 200. When X = 3, it will branch to line 150, and when X = 4, to line 80. The ON...GOTO statement requires that values of X (or any variable following ON) be natural numbers, beginning with 1. Note that the order of the line numbers following GOTO must correspond to the branching desired as X = 1, X = 2, X = 3, etc.

We will now use ON...GOTO in writing a program that evaluates powers of i. You know that you determine the value of a power of i by dividing the exponent by 4 and checking the remainder.

Note that the program below uses an ON...GOTO statement that places the value of the remainder in R. Because ON...GOTO requires the value of R to begin with 1, a program step was added that places 4 in R when the remainder equals 0.

```
110   PRINT "WHAT IS THE POWER OF i";: INPUT P
120 N =   INT (P / 4):R = P - N * 4
130   IF R = 0 THEN R = 4
140   PRINT "i^"P" = ";
150   ON R GOTO 160,170,180,190
160   PRINT "i": GOTO 200
170   PRINT  - 1: GOTO 200
180   PRINT "-i": GOTO 200
190   PRINT 1
200   PRINT : GOTO 110
```

EXERCISE: Write a program that multiplies two complex numbers. Use Exercises 13 to 20 on page 192 as test problems.
For Exercise, see TE Answer Section.

RANDOM NUMBERS

OBJECTIVE: **To evaluate 2 × 2 determinants generated by the computer randomly**

Programming a computer to pick random integers within specific parameters is a good example of manipulating the fundamental capabilities of a computer so that it can perform a broad range of applications. For example, the BASIC function RND generates 9-digit decimal numbers between .000000000 and .999999999. These are not very useful numbers.

However, the statement X = INT (RND(1) * 10) + 1 instructs the computer to choose whole numbers from 1 to 10. This statement modifies the RND function as follows:

1. RND(1) generates a number between .000000000 and .999999999.
2. Multiplying that number by 10 yields a number between 0.00000000 and 9.99999999.
3. Applying INT to the above product yields an integer between 0 and 9.
4. Adding 1 to that generated integer yields an integer between 1 and 10.

In general, the formula X = INT (RND(1) * (B − A + 1)) + A generates random integers from A to B inclusive.

In the program below, the computer is instructed to generate integers from −5 to +5 inclusive. (Can you follow the reasoning?) Enter and RUN the following program that tests your ability to evaluate 2 × 2 determinants.

```
110   FOR I = 1 TO 4
120   A(I) =  INT ( RND (1) * 11) - 5: NEXT I
130    PRINT "WHAT IS THE VALUE OF THE DETERMINANT?"
140    PRINT  TAB( 10)A(1) TAB( 15)A(2)
150    PRINT : PRINT  TAB( 10)A(3) TAB( 15)A(4)
160   V = A(1) * A(4) - A(2) * A(3): INPUT D
170    IF V = D THEN 190
180    PRINT "NO. THE ANSWER IS ";V: GOTO 110
190    PRINT "YES. THAT WAS RIGHT."
200    FOR W = 1 TO 3000: NEXT W: GOTO 110
```

EXERCISES:

1. Rewrite the program above so that it generates integers from −9 to +9 inclusive.

2. Using the program above as a basis, write and RUN a program that evaluates 3 × 3 determinants.
   For Ex. 1–2, see TE Answer Section.

Computer Section

EXPONENTIAL EQUATIONS

COMPUTER SECTION

OBJECTIVE: **To solve exponential equations of the form $a^x = b$, using a FOR...NEXT loop**

You can solve any equation by simply inserting values into it until you find the value that works. Even when the solution of an equation is irrational, you can find a decimal solution that is accurate to as many decimal places as desired.

The computer program below finds solutions of equations of the form $a^x = b$ by inserting values into each given equation systematically. First, it begins by using a FOR...STEP...NEXT loop (lines 160–190) to insert integers from -10 to $+10$ into a given equation one at a time. When two consecutive integers, call them p and $p + 1$, are found such that the value of b lies between a^p and a^{p+1}, the program then uses the same FOR...STEP...NEXT loop to insert values from p to $p + 1$, incremented by .1 (as specified in line 210), into the given equation one at a time. When two successive numbers, call them q and $(q + .1)$, are found where $a^q \leq b \leq a^{q+.1}$, the loop plugs values from q to $(q + .1)$, incremented by .01, into the given equation one at a time. The program continues this process until it finds an exact solution, or a solution accurate to the number of decimal places you specified.

```
110   PRINT "WHAT IS A";: INPUT A
120   PRINT "WHAT IS B";: INPUT B
130   PRINT "HOW MANY DECIMAL PLACES (LIMIT 8 PLEASE)
      ";: INPUT N
140 L =  - 10:U = 10:C = 1:D = A ^ (L - C)
150   FOR M = 1 TO N + 1
160   FOR I = L TO U STEP C
170 E = A ^ I: IF B - D = 0 THEN 230
180   IF (B - D) * (B - E) < 0 THEN 200
190 D = E: NEXT I
200   PRINT I - C" , "I
210 L = I - C:U = I:C = C / 10: NEXT M
220   GOTO 110
230   PRINT "X = "I - C: GOTO 110
```

EXERCISES:

1. Write a program that uses logarithms to solve $a^x = b$. [The BASIC function for the natural logarithm is LOG( ).]

2. Use the technique in this lesson's program to "search for" an irrational solution to a quadratic equation.
   For Ex. 1–2, see TE Answer Section.
   Note: Text program will output wrong answers for X values between 10 and 11.

ARITHMETIC AND GEOMETRIC PROGRESSIONS

OBJECTIVE: **To find *n* terms and the sum of *n* terms of an A.P. or a G.P.**

Below are two programs that deal with progressions. In the first program, you enter the first three terms in a progression. Then you request the number of terms you wish to have printed out and added. In the second program, the computer generates the first three terms in a progression. Then it requests that you compute a given term in the progression, and the sum of terms through the given term. Take turns with both programs.

Program 1

```
100     INPUT "ENTER THE FIRST THREE TERMS OF THE
        PROGRESSION ";A,B,C
110     INPUT "HOW MANY TERMS DO YOU WANT ? ";N
120     IF B / A = C / B THEN 150
130     PRINT "THIS IS NOT A G.P."
140     PRINT : GOTO 100
150     PRINT A", ";
160     FOR I = 1 TO N - 1
170     PRINT A * (B / A) ^ I", ": NEXT I
180     PRINT : PRINT "THE SUM OF "N" TERMS IS ";
190     PRINT (A - A * (B / A) ^ N) / (1 - B / A)
200     PRINT : GOTO 100: END
```

Program 2

```
110 A =  INT ( RND (1) * 30) - 15
120 D =  INT ( RND (1) * 7) - 2: IF D = 0 THEN 120
130     PRINT A", "A + D", "A + D + D
140     FOR I = 1 TO 5
150 N =  INT ( RND (1) * 30) + 4
160     PRINT "WHAT IS THE "N"TH TERM?": INPUT M
170     IF M = A + (N - 1) * D THEN 190
180     PRINT "NO, THE ANSWER IS "A + (N - 1) * D".":
        GOTO 200
190     PRINT "VERY GOOD"
200     PRINT "WHAT IS THE SUM OF "N" TERMS?":
        INPUT M
210     IF M = N / 2 * (2 * A + (N - 1) * D) THEN 240
220     PRINT "NO. THE SUM OF "N" TERMS IS ";
        GOTO 250
230     PRINT N / 2 * (2 * A + (N - 1) * D)".":
240     PRINT "VERY GOOD"
250     NEXT I
```

Computations Using Logarithms

Objective — **To approximate products, quotients, powers, and radicals using common logarithms**

If x and y are positive rational numbers and n is a positive integer, you can approximate xy, $\frac{x}{y}$, x^n, and $\sqrt[n]{x}$ to three significant digits using the laws of logarithms and a table of common logarithms. The procedure is shown in the examples below.

Example 1 — Find 0.000245×3170 to three significant digits. Use logarithms.

$\log(0.000245 \times 3170)$ ◀ $\log(x \cdot y)$
$\log 0.000245 + \log 3170$ ◀ $\log x + \log y$
$6.3892 - 10 + 3.5011$
$9.8903 - 10$ ◀ *The mantissa .8903 is between .8899 and .8904 in the table.*
$9.8904 - 10$ ◀ *Use .8904, the nearer mantissa.*
$\text{Antilog}(9.8904 - 10) = \text{antilog}(-1 + .8904) = 10^{-1} \times 7.77 = 0.777$
Thus, $0.000245 \times 3170 = 0.777$ to three significant digits.

Example 2 — Find $\frac{617}{9.13}$ and $\frac{61.7}{0.0913}$ to three significant digits.

Use $\log \frac{x}{y} = \log x - \log y$.

$\log \frac{617}{9.13} = \log 617 - \log 9.13 = 2.7903 - 0.9605$ ◀ $\quad\quad 2.7903$
$\quad\quad\quad\quad\quad\quad\quad\quad\quad\quad\quad\quad\quad\quad\quad\quad\quad\quad (-) \; 0.9605$
$\quad\quad\quad\quad\quad\quad\quad\quad\quad\quad\quad = 1.8298 \quad\quad\quad\quad\quad\quad\quad\quad 1.8298$
$\quad\quad\quad\quad\quad\quad\quad\quad\quad\quad\quad \doteq 1.8299$ ◀ *Nearest mantissa is .8299.*
Antilog $1.8299 = 10^1 \times 6.76 = 67.6$

$\log \frac{61.7}{0.0913} = \log 61.7 - \log 0.0913 = 1.7903 - (8.9605 - 10)$

$\begin{array}{llll} 1.7903 - 8.9605 & \quad 1.7903 \longrightarrow & 11.7903 - 10 & \blacktriangleleft \text{ Write the} \\ \text{will not be a} \blacktriangleright & (-) \; 8.9605 - 10 & (-) \; 8.9605 - 10 & \text{characteristic} \\ \text{positive} & & \overline{\quad\;\; 2.8298} & 1 \text{ as } 11 - 10. \\ \text{mantissa}. & & & \end{array}$

Antilog $2.8298 = 10^2 \times 6.76 = 676$
Thus, $\frac{617}{9.13} = 67.6$ and $\frac{61.7}{0.0913} = 676$ to three significant digits.

Example 3 — Find $(0.0741)^3$ to three significant digits. Use logs.

$\log a^r$ ▶ $\log(0.0741)^3$
$r \cdot \log a$ ▶ $3 \cdot \log 0.0741 = 3(8.8698 - 10) = 26.6094 - 30$
$\quad\quad\quad\quad$ Antilog$(26.6094 - 30) = \text{antilog}(-4 + .6094) = 10^{-4} \times 4.07$
Thus, $(0.0741)^3 = 0.000407$ to three significant digits.

Example 4 Find $\sqrt{7{,}810}$ and $\sqrt[3]{0.0532}$ to three significant digits.

Use $\log \sqrt[n]{x} = \log x^{\frac{1}{n}} = \frac{1}{n} \cdot \log x = \frac{\log x}{n}$.

$\log \sqrt{7{,}810} = \frac{1}{2} \cdot \log 7{,}810 = \frac{3.8927}{2} = 1.94635$

Antilog $1.9465 = 10^1 \times 8.84 = 88.4$

$\log \sqrt[3]{0.0532} = \frac{1}{3} \cdot \log 0.0532 = \frac{8.7259 - 10}{3}$ ◀ *10 is not divisible by 3. Characteristic must be an integer.*

$= \frac{28.7259 - 30}{3}$ ◀ *Use $28 - 30$ for $8 - 10$. 30 is divisible by 3.*

$= 9.5753 - 10$

Antilog $(9.5753 - 10) = 10^{-1} \times 3.76 = 0.376$

Thus, $\sqrt{7{,}810} = 88.4$ **and** $\sqrt[3]{0.0532} = 0.376$.

Reading in Algebra

Given $\log x$ and $\log y$, rewrite $\log x$ so that the mantissa of $\log x - \log y$ will be positive.

1. $\quad 1.7900$
   $(-) \; 8.9600 - 10$
   $\overline{11.7900 - 10}$

2. $\quad 0.2200$
   $(-) \; 0.5100$
   $\overline{10.2200 - 10}$

3. $\quad 7.5800 - 10$
   $(-) \; 7.7400 - 10$
   $\overline{17.5800 - 20}$

4. $\quad 5.4400 - 10$
   $(-) \; 6.2200$
   $\overline{15.4400 - 20}$

Given $\log x$ and n, rewrite $\log x$ so that the characteristic of $\frac{\log x}{n}$ will be an integer.

5. $\frac{8.4400 - 10}{3}$ **28.4400 − 30**

6. $\frac{9.3200 - 10}{4}$ **39.3200 − 40**

7. $\frac{7.4800 - 10}{6}$ **57.4800 − 60**

Written Exercises

Find the value of each expression to three significant digits. Use logs.

Ⓐ 1. $39.2 \times 53{,}800$ **2,110,000**
2. 0.00286×64.3 **0.184**
3. 0.537×0.0806 **0.0433**
4. $\frac{3{,}080}{17.6}$ **175**
5. $\frac{4.52}{63.8}$ **0.0708**
6. $\frac{0.0236}{0.764}$ **0.0309**
7. $(49.4)^3$ **121,000**
8. $(0.723)^4$ **0.273**
9. $(0.0871)^2$ **0.00759**
10. $\sqrt{5{,}140}$ **71.7**
11. $\sqrt[3]{19{,}400}$ **26.9**
12. $\sqrt[4]{82.1}$ **3.01**
13. $\sqrt{0.00592}$ **0.0769**
14. $\sqrt[4]{0.0514}$ **0.476**
15. $\sqrt[3]{0.0736}$ **0.419**
Ⓑ 16. $\frac{4.25 \times 3{,}750}{543}$ **29.3**
17. $\frac{9{,}180}{2.37 \times 30.4}$ **127**
18. $\frac{0.263 \times 704}{92.2 \times 0.081}$ **24.8**
19. $(18.9 \times 4.04)^3$ **445,000**
20. $\sqrt[4]{0.263 \times 41.8}$ **1.82**
21. $\sqrt[3]{(502)^2}$ **63.2**
22. $\sqrt{\frac{735}{3{,}420}}$ **0.464**
23. $\left(\frac{0.756}{1.83}\right)^4$ **0.0291**
24. $\sqrt[3]{\frac{(0.813)^2}{283 \times 0.0461}}$ **0.370**

Computations Using Logarithms

TABLES

Squares and Square Roots

No.	Sq.	Sq. Root	No.	Sq.	Sq. Root
1	1	1.000	51	2,601	7.141
2	4	1.414	52	2,704	7.211
3	9	1.732	53	2,809	7.280
4	16	2.000	54	2,916	7.348
5	25	2.236	55	3,025	7.416
6	36	2.449	56	3,136	7.483
7	49	2.646	57	3,249	7.550
8	64	2.828	58	3,364	7.616
9	81	3.000	59	3,481	7.681
10	100	3.162	60	3,600	7.746
11	121	3.317	61	3,721	7.810
12	144	3.464	62	3,844	7.874
13	169	3.606	63	3,969	7.937
14	196	3.742	64	4,096	8.000
15	225	3.873	65	4,225	8.062
16	256	4.000	66	4,356	8.124
17	289	4.123	67	4,489	8.185
18	324	4.243	68	4,624	8.246
19	361	4.359	69	4,761	8.307
20	400	4.472	70	4,900	8.357
21	441	4.583	71	5,041	8.426
22	484	4.690	72	5,184	8.485
23	529	4.796	73	5,329	8.544
24	576	4.899	74	5,476	8.602
25	625	5.000	75	5,625	8.660
26	676	5.099	76	5,776	8.718
27	729	5.196	77	5,929	8.775
28	784	5.292	78	6,084	8.832
29	841	5.385	79	6,241	8.888
30	900	5.477	80	6,400	8.944
31	961	5.568	81	6,561	9.000
32	1,024	5.657	82	6,724	9.055
33	1,089	5.745	83	6,889	9.110
34	1,156	5.831	84	7,056	9.165
35	1,225	5.916	85	7,225	9.220
36	1,296	6.000	86	7,396	9.274
37	1,369	6.083	87	7,569	9.327
38	1,444	6.164	88	7,744	9.381
39	1,521	6.245	89	7,921	9.434
40	1,600	6.325	90	8,100	9.487
41	1,681	6.403	91	8,281	9.539
42	1,764	6.481	92	8,464	9.592
43	1,849	6.557	93	8,649	9.644
44	1,936	6.633	94	8,836	9.695
45	2,025	6.708	95	9,025	9.747
46	2,116	6.782	96	9,216	9.798
47	2,209	6.856	97	9,409	9.849
48	2,304	6.928	98	9,604	9.899
49	2,401	7.000	99	9,801	9.950
50	2,500	7.071	100	10,000	10.000

* Logarithms of Trigonometric Functions

*Subtract 10 from each entry in this table to obtain the proper logarithm of the indicated trigonometric function.

Angle	L Sin	L Tan	L Cot	L Cos	
0° 0'				10.0000	90° 0'
10'	7.4637	7.4637	12.5363	.0000	50'
20'	.7648	.7648	.2352	.0000	40'
30'	7.9408	7.9409	12.0591	.0000	30'
40'	8.0658	8.0658	11.9342	.0000	20'
50'	.1627	.1627	.8373	10.0000	10'
1° 0'	8.2419	8.2419	11.7581	9.9999	89° 0'
10'	.3088	.3089	.6911	.9999	50'
20'	.3668	.3669	.6331	.9999	40'
30'	.4179	.4181	.5819	.9999	30'
40'	.4637	.4638	.5362	.9998	20'
50'	.5050	.5053	.4947	.9998	10'
2° 0'	8.5428	8.5431	11.4569	9.9997	88° 0'
10'	.5776	.5779	.4221	.9997	50'
20'	.6097	.6101	.3899	.9996	40'
30'	.6397	.6401	.3599	.9996	30'
40'	.6677	.6682	.3318	.9995	20'
50'	.6940	.6945	.3055	.9995	10'
3° 0'	8.7188	8.7194	11.2806	9.9994	87° 0'
10'	.7423	.7429	.2571	.9993	50'
20'	.7645	.7652	.2348	.9993	40'
30'	.7857	.7865	.2135	.9992	30'
40'	.8059	.8067	.1933	.9991	20'
50'	.8251	.8261	.1739	.9990	10'
4° 0'	8.8436	8.8446	11.1554	9.9989	86° 0'
10'	.8613	.8624	.1376	.9989	50'
20'	.8783	.8795	.1205	.9988	40'
30'	.8946	.8960	.1040	.9987	30'
40'	.9104	.9118	.0882	.9986	20'
50'	.9256	.9272	.0728	.9985	10'
5° 0'	8.9403	8.9420	11.0580	9.9983	85° 0'
10'	.9545	.9563	.0437	.9982	50'
20'	.9682	.9701	.0299	.9981	40'
30'	.9816	.9836	.0164	.9980	30'
40'	.9945	.9966	11.0034	.9979	20'
50'	9.0070	9.0093	10.9907	.9977	10'
6° 0'	9.0192	9.0216	10.9784	9.9976	84° 0'
10'	.0311	.0336	.9664	.9975	50'
20'	.0426	.0453	.9547	.9973	40'
30'	.0539	.0567	.9433	.9972	30'
40'	.0648	.0678	.9322	.9971	20'
50'	.0755	.0786	.9214	.9969	10'
7° 0'	9.0859	9.0891	10.9109	9.9968	83° 0'
10'	.0961	.0995	.9005	.9966	50'
20'	.1060	.1096	.8904	.9964	40'
30'	.1157	.1194	.8806	.9963	30'
40'	.1252	.1291	.8709	.9961	20'
50'	.1345	.1385	.8615	.9959	10'
8° 0'	9.1436	9.1478	10.8522	9.9958	82° 0'
10'	.1525	.1569	.8431	.9956	50'
20'	.1612	.1658	.8342	.9954	40'
30'	.1697	.1745	.8255	.9952	30'
40'	.1781	.1831	.8169	.9950	20'
50'	.1863	.1915	.8085	.9948	10'
9° 0'	9.1943	9.1997	10.8003	9.9946	81° 0'
	L Cos	L Cot	L Tan	L Sin	Angle

Logarithms of Trigonometric Functions

Angle	L Sin	L Tan	L Cot	L Cos	
9° 0'	9.1943	9.1997	10.8003	9.9946	81° 0'
10'	.2022	.2078	.7922	.9944	50'
20'	.2100	.2158	.7842	.9942	40'
30'	.2176	.2236	.7764	.9940	30'
40'	.2251	.2313	.7687	.9938	20'
50'	.2324	.2389	.7611	.9936	10'
10° 0'	9.2397	9.2463	10.7537	9.9934	80° 0'
10'	.2468	.2536	.7464	.9931	50'
20'	.2538	.2609	.7391	.9929	40'
30'	.2606	.2680	.7320	.9927	30'
40'	.2674	.2750	.7250	.9924	20'
50'	.2740	.2819	.7181	.9922	10'
11° 0'	9.2806	9.2887	10.7113	9.9919	79° 0'
10'	.2870	.2953	.7047	.9917	50'
20'	.2934	.3020	.6980	.9914	40'
30'	.2997	.3085	.6915	.9912	30'
40'	.3058	.3149	.6851	.9909	20'
50'	.3119	.3212	.6788	.9907	10'
12° 0'	9.3179	9.3275	10.6725	9.9904	78° 0'
10'	.3238	.3336	.6664	.9901	50'
20'	.3296	.3397	.6603	.9899	40'
30'	.3353	.3458	.6542	.9896	30'
40'	.3410	.3517	.6483	.9893	20'
50'	.3466	.3576	.6424	.9890	10'
13° 0'	9.3521	9.3634	10.6366	9.9887	77° 0'
10'	.3575	.3691	.6309	.9884	50'
20'	.3629	.3748	.6252	.9881	40'
30'	.3682	.3804	.6196	.9878	30'
40'	.3734	.3859	.6141	.9875	20'
50'	.3786	.3914	.6086	.9872	10'
14° 0'	9.3837	9.3968	10.6032	9.9869	76° 0'
10'	.3887	.4021	.5979	.9866	50'
20'	.3937	.4074	.5926	.9863	40'
30'	.3986	.4127	.5873	.9859	30'
40'	.4035	.4178	.5822	.9856	20'
50'	.4083	.4230	.5770	.9853	10'
15° 0'	9.4130	9.4281	10.5719	9.9849	75° 0'
10'	.4177	.4331	.5669	.9846	50'
20'	.4223	.4381	.5619	.9843	40'
30'	.4269	.4430	.5570	.9839	30'
40'	.4314	.4479	.5521	.9836	20'
50'	.4359	.4527	.5473	.9832	10'
16° 0'	9.4403	9.4575	10.5425	9.9828	74° 0'
10'	.4447	.4622	.5378	.9825	50'
20'	.4491	.4669	.5331	.9821	40'
30'	.4533	.4716	.5284	.9817	30'
40'	.4576	.4762	.5238	.9814	20'
50'	.4618	.4808	.5192	.9810	10'
17° 0'	9.4659	9.4853	10.5147	9.9806	73° 0'
10'	.4700	.4898	.5102	.9802	50'
20'	.4741	.4943	.5057	.9798	40'
30'	.4781	.4987	.5013	.9794	30'
40'	.4821	.5031	.4969	.9790	20'
50'	.4861	.5075	.4925	.9786	10'
18° 0'	9.4900	9.5118	10.4882	9.9782	72° 0'
	L Cos	L Cot	L Tan	L Sin	Angle

Logarithms of Trigonometric Functions

Angle	L Sin	L Tan	L Cot	L Cos	
18° 0'	9.4900	9.5118	10.4882	9.9782	72° 0'
10'	.4939	.5161	.4839	.9778	50'
20'	.4977	.5203	.4797	.9774	40'
30'	.5015	.5245	.4755	.9770	30'
40'	.5052	.5287	.4713	.9765	20'
50'	.5090	.5329	.4671	.9761	10'
19° 0'	9.5126	9.5370	10.4630	9.9757	71° 0'
10'	.5163	.5411	.4589	.9752	50'
20'	.5199	.5451	.4549	.9748	40'
30'	.5235	.5491	.4509	.9743	30'
40'	.5270	.5531	.4469	.9739	20'
50'	.5306	.5571	.4429	.9734	10'
20° 0'	9.5341	9.5611	10.4389	9.9730	70° 0'
10'	.5375	.5650	.4350	.9725	50'
20'	.5409	.5689	.4311	.9721	40'
30'	.5443	.5727	.4273	.9716	30'
40'	.5477	.5766	.4234	.9711	20'
50'	.5510	.5804	.4196	.9706	10'
21° 0'	9.5543	9.5842	10.4158	9.9702	69° 0'
10'	.5576	.5879	.4121	.9697	50'
20'	.5609	.5917	.4083	.9692	40'
30'	.5641	.5954	.4046	.9687	30'
40'	.5673	.5991	.4009	.9682	20'
50'	.5704	.6028	.3972	.9677	10'
22° 0'	9.5736	9.6064	10.3936	9.9672	68° 0'
10'	.5767	.6100	.3900	.9667	50'
20'	.5798	.6136	.3864	.9661	40'
30'	.5828	.6172	.3828	.9656	30'
40'	.5859	.6208	.3792	.9651	20'
50'	.5889	.6243	.3757	.9646	10'
23° 0'	9.5919	9.6279	10.3721	9.9640	67° 0'
10'	.5948	.6314	.3686	.9635	50'
20'	.5978	.6348	.3652	.9629	40'
30'	.6007	.6383	.3617	.9624	30'
40'	.6036	.6417	.3583	.9618	20'
50'	.6065	.6452	.3548	.9613	10'
24° 0'	9.6093	9.6486	10.3514	9.9607	66° 0'
10'	.6121	.6520	.3480	.9602	50'
20'	.6149	.6553	.3447	.9596	40'
30'	.6177	.6587	.3413	.9590	30'
40'	.6205	.6620	.3380	.9584	20'
50'	.6232	.6654	.3346	.9579	10'
25° 0'	9.6259	9.6687	10.3313	9.9573	65° 0'
10'	.6286	.6720	.3280	.9567	50'
20'	.6313	.6752	.3248	.9561	40'
30'	.6340	.6785	.3215	.9555	30'
40'	.6366	.6817	.3183	.9549	20'
50'	.6392	.6850	.3150	.9543	10'
26° 0'	9.6418	9.6882	10.3118	9.9537	64° 0'
10'	.6444	.6914	.3086	.9530	50'
20'	.6470	.6946	.3054	.9524	40'
30'	.6495	.6977	.3023	.9518	30'
40'	.6521	.7009	.2991	.9512	20'
50'	.6546	.7040	.2960	.9505	10'
27° 0'	9.6570	9.7072	10.2928	9.9499	63° 0'
	L Cos	L Cot	L Tan	L Sin	Angle

TABLES

TABLES

Logarithms of Trigonometric Functions

Angle	L Sin	L Tan	L Cot	L Cos	
27° 0'	9.6570	9.7072	10.2928	9.9499	63° 0'
10'	.6595	.7103	.2897	.9492	50'
20'	.6620	.7134	.2866	.9486	40'
30'	.6644	.7165	.2835	.9479	30'
40'	.6668	.7196	.2804	.9473	20'
50'	.6692	.7226	.2774	.9466	10'
28° 0'	9.6716	9.7257	10.2743	9.9459	62° 0'
10'	.6740	.7287	.2713	.9453	50'
20'	.6763	.7317	.2683	.9446	40'
30'	.6787	.7348	.2652	.9439	30'
40'	.6810	.7378	.2622	.9432	20'
50'	.6833	.7408	.2592	.9425	10'
29° 0'	9.6856	9.7438	10.2562	9.9418	61° 0'
10'	.6878	.7467	.2533	.9411	50'
20'	.6901	.7497	.2503	.9404	40'
30'	.6923	.7526	.2474	.9397	30'
40'	.6946	.7556	.2444	.9390	20'
50'	.6968	.7585	.2415	.9383	10'
30° 0'	9.6990	9.7614	10.2386	9.9375	60° 0'
10'	.7012	.7644	.2356	.9368	50'
20'	.7033	.7673	.2327	.9361	40'
30'	.7055	.7701	.2299	.9353	30'
40'	.7076	.7730	.2270	.9346	20'
50'	.7097	.7759	.2241	.9338	10'
31° 0'	9.7118	9.7788	10.2212	9.9331	59° 0'
10'	.7139	.7816	.2184	.9323	50'
20'	.7160	.7845	.2155	.9315	40'
30'	.7181	.7873	.2127	.9308	30'
40'	.7201	.7902	.2098	.9300	20'
50'	.7222	.7930	.2070	.9292	10'
32° 0'	9.7242	9.7958	10.2042	9.9284	58° 0'
10'	.7262	.7986	.2014	.9276	50'
20'	.7282	.8014	.1986	.9268	40'
30'	.7302	.8042	.1958	.9260	30'
40'	.7322	.8070	.1930	.9252	20'
50'	.7342	.8097	.1903	.9244	10'
33° 0'	9.7361	9.8125	10.1875	9.9236	57° 0'
10'	.7380	.8153	.1847	.9228	50'
20'	.7400	.8180	.1820	.9219	40'
30'	.7419	.8208	.1792	.9211	30'
40'	.7438	.8235	.1765	.9203	20'
50'	.7457	.8263	.1737	.9194	10'
34° 0'	9.7476	9.8290	10.1710	9.9186	56° 0'
10'	.7494	.8317	.1683	.9177	50'
20'	.7513	.8344	.1656	.9169	40'
30'	.7531	.8371	.1629	.9160	30'
40'	.7550	.8398	.1602	.9151	20'
50'	.7568	.8425	.1575	.9142	10'
35° 0'	9.7586	9.8452	10.1548	9.9134	55° 0'
10'	.7604	.8479	.1521	.9125	50'
20'	.7622	.8506	.1494	.9116	40'
30'	.7640	.8533	.1467	.9107	30'
40'	.7657	.8559	.1441	.9098	20'
50'	.7675	.8586	.1414	.9089	10'
36° 0'	9.7692	9.8613	10.1387	9.9080	54° 0'
	L Cos	L Cot	L Tan	L Sin	Angle

Logarithms of Trigonometric Functions

Angle	L Sin	L Tan	L Cot	L Cos	
36° 0'	9.7692	9.8613	10.1387	9.9080	54° 0'
10'	.7710	.8639	.1361	.9070	50'
20'	.7727	.8666	.1334	.9061	40'
30'	.7744	.8692	.1308	.9052	30'
40'	.7761	.8718	.1282	.9042	20'
50'	.7778	.8745	.1255	.9033	10'
37° 0'	9.7795	9.8771	10.1229	9.9023	53° 0'
10'	.7811	.8797	.1203	.9014	50'
20'	.7828	.8824	.1176	.9004	40'
30'	.7844	.8850	.1150	.8995	30'
40'	.7861	.8876	.1124	.8985	20'
50'	.7877	.8902	.1098	.8975	10'
38° 0'	9.7893	9.8928	10.1072	9.8965	52° 0'
10'	.7910	.8954	.1046	.8955	50'
20'	.7926	.8980	.1020	.8945	40'
30'	.7941	.9006	.0994	.8935	30'
40'	.7957	.9032	.0968	.8925	20'
50'	.7973	.9058	.0942	.8915	10'
39° 0'	9.7989	9.9084	10.0916	9.8905	51° 0'
10'	.8004	.9110	.0890	.8895	50'
20'	.8020	.9135	.0865	.8884	40'
30'	.8035	.9161	.0839	.8874	30'
40'	.8050	.9187	.0813	.8864	20'
50'	.8066	.9212	.0788	.8853	10'
40° 0'	9.8081	9.9238	10.0762	9.8843	50° 0'
10'	.8096	.9264	.0736	.8832	50'
20'	.8111	.9289	.0711	.8821	40'
30'	.8125	.9315	.0685	.8810	30'
40'	.8140	.9341	.0659	.8800	20'
50'	.8155	.9366	.0634	.8789	10'
41° 0'	9.8169	9.9392	10.0608	9.8778	49° 0'
10'	.8184	.9417	.0583	.8767	50'
20'	.8198	.9443	.0557	.8756	40'
30'	.8213	.9468	.0532	.8745	30'
40'	.8227	.9494	.0506	.8733	20'
50'	.8241	.9519	.0481	.8722	10'
42° 0'	9.8255	9.9544	10.0456	9.8711	48° 0'
10'	.8269	.9570	.0430	.8699	50'
20'	.8283	.9595	.0405	.8688	40'
30'	.8297	.9621	.0379	.8676	30'
40'	.8311	.9646	.0354	.8665	20'
50'	.8324	.9671	.0329	.8653	10'
43° 0'	9.8338	9.9697	10.0303	9.8641	47° 0'
10'	.8351	.9722	.0278	.8629	50'
20'	.8365	.9747	.0253	.8618	40'
30'	.8378	.9772	.0228	.8606	30'
40'	.8391	.9798	.0202	.8594	20'
50'	.8405	.9823	.0177	.8582	10'
44° 0'	9.8418	9.9848	10.0152	9.8569	46° 0'
10'	.8431	.9874	.0126	.8557	50'
20'	.8444	.9899	.0101	.8545	40'
30'	.8457	.9924	.0076	.8532	30'
40'	.8469	.9949	.0051	.8520	20'
50'	.8482	.9975	.0025	.8507	10'
45° 0'	9.8495	10.0000	10.0000	9.8495	45° 0'
	L Cos	L Cot	L Tan	L Sin	Angle

592

TABLES

Trigonometric Functions

x		sin x	cos x	tan x	cot x	sec x	csc x	
0°	0'	.00000	1.0000	.00000		1.0000		0'
	10'	.00291	1.0000	.00291	343.77	1.0000	343.78	50'
	20'	.00582	1.0000	.00582	171.88	1.0000	171.89	40'	90°
	30'	.00873	1.0000	.00873	114.59	1.0000	114.59	30'
	40'	.01164	.9999	.01164	85.940	1.0001	85.946	20'
	50'	.01454	.9999	.01455	68.750	1.0001	68.757	10'
1°	0'	.01745	.9998	.01746	57.290	1.0002	57.299	0'
	10'	.02036	.9998	.02036	49.104	1.0002	49.114	50'	89°
	20'	.02327	.9997	.02328	42.964	1.0003	42.976	40'
	30'	.02618	.9997	.02619	38.188	1.0003	38.202	30'
	40'	.02908	.9996	.02910	34.368	1.0004	34.382	20'
	50'	.03199	.9995	.03201	31.242	1.0005	31.258	10'
2°	0'	.03490	.9994	.03492	28.636	1.0006	28.654	0'
	10'	.03781	.9993	.03783	26.4316	1.0007	26.451	50'	88°
	20'	.04071	.9992	.04075	24.5418	1.0008	24.562	40'
	30'	.04362	.9990	.04366	22.9038	1.0010	22.926	30'
	40'	.04653	.9989	.04658	21.4704	1.0011	21.494	20'
	50'	.04943	.9988	.04949	20.2056	1.0012	20.230	10'
3°	0'	.05234	.9986	.05241	19.0811	1.0014	19.107	0'
	10'	.05524	.9985	.05533	18.0750	1.0015	18.103	50'	87°
	20'	.05814	.9983	.05824	17.1693	1.0017	17.198	40'
	30'	.06105	.9981	.06116	16.3499	1.0019	16.380	30'
	40'	.06395	.9980	.06408	15.6048	1.0021	15.637	20'
	50'	.06685	.9978	.06700	14.9244	1.0022	14.958	10'
4°	0'	.06976	.9976	.06993	14.3007	1.0024	14.336	0'
	10'	.07266	.9974	.07285	13.7267	1.0027	13.763	50'	86°
	20'	.07556	.9971	.07578	13.1969	1.0029	13.235	40'
	30'	.07846	.9969	.07870	12.7062	1.0031	12.746	30'
	40'	.08136	.9967	.08163	12.2505	1.0033	12.291	20'
	50'	.08426	.9964	.08456	11.8262	1.0036	11.868	10'
5°	0'	.08716	.9962	.08749	11.4301	1.0038	11.474	0'
	10'	.09005	.9959	.09042	11.0594	1.0041	11.105	50'	85°
	20'	.09295	.9957	.09335	10.7119	1.0044	10.758	40'
	30'	.09585	.9954	.09629	10.3854	1.0046	10.433	30'
	40'	.09874	.9951	.09923	10.0780	1.0049	10.128	20'
	50'	.10164	.9948	.10216	9.7882	1.0052	9.839	10'
6°	0'	.10453	.9945	.10510	9.5144	1.0055	9.5668	0'
	10'	.10742	.9942	.10805	9.2553	1.0058	9.3092	50'	84°
	20'	.11031	.9939	.11099	9.0098	1.0061	9.0652	40'
	30'	.11320	.9936	.11394	8.7769	1.0065	8.8337	30'
	40'	.11609	.9932	.11688	8.5555	1.0068	8.6138	20'
	50'	.11898	.9929	.11983	8.3450	1.0072	8.4647	10'
7°	0'	.12187	.9925	.12278	8.1443	1.0075	8.2055	0'
	10'	.12476	.9922	.12574	7.9530	1.0079	8.0157	50'	83°
	20'	.12764	.9918	.12869	7.7704	1.0083	7.8344	40'
	30'	.13053	.9914	.13165	7.5958	1.0086	7.6613	30'
		cos x	sin x	cot x	tan x	csc x	sec x	x

Trigonometric Functions

x		sin x	cos x	tan x	cot x	sec x	csc x	
	30'	.1305	.9914	.1317	7.5958	1.0086	7.6613	30'
	40'	.1334	.9911	.1346	7.4287	1.0090	7.4957	20'
	50'	.1363	.9907	.1376	7.2687	1.0094	7.3372	10'
8°	0'	.1392	.9903	.1405	7.1154	1.0098	7.1853	0'	82°
	10'	.1421	.9899	.1435	6.9682	1.0102	7.0396	50'
	20'	.1449	.9894	.1465	6.8269	1.0107	6.8998	40'
	30'	.1478	.9890	.1495	6.6912	1.0111	6.7655	30'
	40'	.1507	.9886	.1524	6.5606	1.0116	6.6363	20'
	50'	.1536	.9881	.1554	6.4348	1.0120	6.5121	10'
9°	0'	.1564	.9877	.1584	6.3138	1.0125	6.3925	0'	81°
	10'	.1593	.9872	.1614	6.1970	1.0129	6.2772	50'
	20'	.1622	.9868	.1644	6.0844	1.0134	6.1661	40'
	30'	.1650	.9863	.1673	5.9758	1.0139	6.0589	30'
	40'	.1679	.9858	.1703	5.8708	1.0144	5.9554	20'
	50'	.1708	.9853	.1733	5.7694	1.0149	5.8554	10'
10°	0'	.1736	.9848	.1763	5.6713	1.0154	5.7588	0'	80°
	10'	.1765	.9843	.1793	5.5764	1.0160	5.6653	50'
	20'	.1794	.9838	.1823	5.4845	1.0165	5.5749	40'
	30'	.1822	.9833	.1853	5.3955	1.0170	5.4874	30'
	40'	.1851	.9827	.1883	5.3093	1.0176	5.4026	20'
	50'	.1880	.9822	.1914	5.2257	1.0182	5.3205	10'
11°	0'	.1908	.9816	.1944	5.1446	1.0187	5.2408	0'	79°
	10'	.1937	.9811	.1974	5.0658	1.0193	5.1636	50'
	20'	.1965	.9805	.2004	4.9894	1.0199	5.0886	40'
	30'	.1994	.9799	.2035	4.9152	1.0205	5.0159	30'
	40'	.2022	.9793	.2065	4.8430	1.0211	4.9452	20'
	50'	.2051	.9787	.2095	4.7729	1.0217	4.8765	10'
12°	0'	.2079	.9781	.2126	4.7046	1.0223	4.8097	0'	78°
	10'	.2108	.9775	.2156	4.6382	1.0230	4.7448	50'
	20'	.2136	.9769	.2186	4.5736	1.0236	4.6817	40'
	30'	.2164	.9763	.2217	4.5107	1.0243	4.6202	30'
	40'	.2193	.9757	.2247	4.4494	1.0249	4.5604	20'
	50'	.2221	.9750	.2278	4.3897	1.0256	4.5022	10'
13°	0'	.2250	.9744	.2309	4.3315	1.0263	4.4454	0'	77°
	10'	.2278	.9737	.2339	4.2747	1.0270	4.3901	50'
	20'	.2306	.9730	.2370	4.2193	1.0277	4.3362	40'
	30'	.2334	.9724	.2401	4.1653	1.0284	4.2837	30'
	40'	.2363	.9717	.2432	4.1126	1.0291	4.2324	20'
	50'	.2391	.9710	.2462	4.0611	1.0299	4.1824	10'
14°	0'	.2419	.9703	.2493	4.0108	1.0306	4.1336	0'	76°
	10'	.2447	.9696	.2524	3.9617	1.0314	4.0859	50'
	20'	.2476	.9689	.2555	3.9136	1.0321	4.0394	40'
	30'	.2504	.9681	.2586	3.8667	1.0329	3.9939	30'
	40'	.2532	.9674	.2617	3.8208	1.0337	3.9495	20'
	50'	.2560	.9667	.2648	3.7760	1.0345	3.9061	10'
15°	0'	.2588	.9659	.2679	3.7321	1.0353	3.8637	0'	75°
		cos x	sin x	cot x	tan x	csc x	sec x	x

Common Logarithms of Numbers

n	0	1	2	3	4	5	6	7	8	9
10	0000	0043	0086	0128	0170	0212	0253	0294	0334	0374
11	0414	0453	0492	0531	0569	0607	0645	0682	0719	0755
12	0792	0828	0864	0899	0934	0969	1004	1038	1072	1106
13	1139	1173	1206	1239	1271	1303	1335	1367	1399	1430
14	1461	1492	1523	1553	1584	1614	1644	1673	1703	1732
15	1761	1790	1818	1847	1875	1903	1931	1959	1987	2014
16	2041	2068	2095	2122	2148	2175	2201	2227	2253	2279
17	2304	2330	2355	2380	2405	2430	2455	2480	2504	2529
18	2553	2577	2601	2625	2648	2672	2695	2718	2742	2765
19	2788	2810	2833	2856	2878	2900	2923	2945	2967	2989
20	3010	3032	3054	3075	3096	3118	3139	3160	3181	3201
21	3222	3243	3263	3284	3304	3324	3345	3365	3385	3404
22	3424	3444	3464	3483	3502	3522	3541	3560	3579	3598
23	3617	3636	3655	3674	3692	3711	3729	3747	3766	3784
24	3802	3820	3838	3856	3874	3892	3909	3927	3945	3962
25	3979	3997	4014	4031	4048	4065	4082	4099	4116	4133
26	4150	4166	4183	4200	4216	4232	4249	4265	4281	4298
27	4314	4330	4346	4362	4378	4393	4409	4425	4440	4456
28	4472	4487	4502	4518	4533	4548	4564	4579	4594	4609
29	4624	4639	4654	4669	4683	4698	4713	4728	4742	4757
30	4771	4786	4800	4814	4829	4843	4857	4871	4886	4900
31	4914	4928	4942	4955	4969	4983	4997	5011	5024	5038
32	5051	5065	5079	5092	5105	5119	5132	5145	5159	5172
33	5185	5198	5211	5224	5237	5250	5263	5276	5289	5302
34	5315	5328	5340	5353	5366	5378	5391	5403	5416	5428
35	5441	5453	5465	5478	5490	5502	5514	5527	5539	5551
36	5563	5575	5587	5599	5611	5623	5635	5647	5658	5670
37	5682	5694	5705	5717	5729	5740	5752	5763	5775	5786
38	5798	5809	5821	5832	5843	5855	5866	5877	5888	5899
39	5911	5922	5933	5944	5955	5966	5977	5988	5999	6010
40	6021	6031	6042	6053	6064	6075	6085	6096	6107	6117
41	6128	6138	6149	6160	6170	6180	6191	6201	6212	6222
42	6232	6243	6253	6263	6274	6284	6294	6304	6314	6325
43	6335	6345	6355	6365	6375	6385	6395	6405	6415	6425
44	6435	6444	6454	6464	6474	6484	6493	6503	6513	6522
45	6532	6542	6551	6561	6571	6580	6590	6599	6609	6618
46	6628	6637	6646	6656	6665	6675	6684	6693	6702	6712
47	6721	6730	6739	6749	6758	6767	6776	6785	6794	6803
48	6812	6821	6830	6839	6848	6857	6866	6875	6884	6893
49	6902	6911	6920	6928	6937	6946	6955	6964	6972	6981
50	6990	6998	7007	7016	7024	7033	7042	7050	7059	7067
51	7076	7084	7093	7101	7110	7118	7126	7135	7143	7152
52	7160	7168	7177	7185	7193	7202	7210	7218	7226	7235
53	7243	7251	7259	7267	7275	7284	7292	7300	7308	7316
54	7324	7332	7340	7348	7356	7364	7372	7380	7388	7396
55	7404	7412	7419	7427	7435	7443	7451	7459	7466	7474
56	7482	7490	7497	7505	7513	7520	7528	7536	7543	7551
57	7559	7566	7574	7582	7589	7597	7604	7612	7619	7627
58	7634	7642	7649	7657	7664	7672	7679	7686	7694	7701
59	7709	7716	7723	7731	7738	7745	7752	7760	7767	7774
60	7782	7789	7796	7803	7810	7818	7825	7832	7839	7846
61	7853	7860	7868	7875	7882	7889	7896	7903	7910	7917
62	7924	7931	7938	7945	7952	7959	7966	7973	7980	7987
63	7993	8000	8007	8014	8021	8028	8035	8041	8048	8055
64	8062	8069	8075	8082	8089	8096	8102	8109	8116	8122
65	8129	8136	8142	8149	8156	8162	8169	8176	8182	8189
66	8195	8202	8209	8215	8222	8228	8235	8241	8248	8254
67	8261	8267	8274	8280	8287	8293	8299	8306	8312	8319
68	8325	8331	8338	8344	8351	8357	8363	8370	8376	8382
69	8388	8395	8401	8407	8414	8420	8426	8432	8439	8445
70	8451	8457	8463	8470	8476	8482	8488	8494	8500	8506
71	8513	8519	8525	8531	8537	8543	8549	8555	8561	8567
72	8573	8579	8585	8591	8597	8603	8609	8615	8621	8627
73	8633	8639	8645	8651	8657	8663	8669	8675	8681	8686
74	8692	8698	8704	8710	8716	8722	8727	8733	8739	8745
75	8751	8756	8762	8768	8774	8779	8785	8791	8797	8802
76	8808	8814	8820	8825	8831	8837	8842	8848	8854	8859
77	8865	8871	8876	8882	8887	8893	8899	8904	8910	8915
78	8921	8927	8932	8938	8943	8949	8954	8960	8965	8971
79	8976	8982	8987	8993	8998	9004	9009	9015	9020	9025
80	9031	9036	9042	9047	9053	9058	9063	9069	9074	9079
81	9085	9090	9096	9101	9106	9112	9117	9122	9128	9133
82	9138	9143	9149	9154	9159	9165	9170	9175	9180	9186
83	9191	9196	9201	9206	9212	9217	9222	9227	9232	9238
84	9243	9248	9253	9258	9263	9269	9274	9279	9284	9289
85	9294	9299	9304	9309	9315	9320	9325	9330	9335	9340
86	9345	9350	9355	9360	9365	9370	9375	9380	9385	9390
87	9395	9400	9405	9410	9415	9420	9425	9430	9435	9440
88	9445	9450	9455	9460	9465	9469	9474	9479	9484	9489
89	9494	9499	9504	9509	9513	9518	9523	9528	9533	9538
90	9542	9547	9552	9557	9562	9566	9571	9576	9581	9586
91	9590	9595	9600	9605	9609	9614	9619	9624	9628	9633
92	9638	9643	9647	9652	9657	9661	9666	9671	9675	9680
93	9685	9689	9694	9699	9703	9708	9713	9717	9722	9727
94	9731	9736	9741	9745	9750	9754	9759	9763	9768	9773
95	9777	9782	9786	9791	9795	9800	9805	9809	9814	9818
96	9823	9827	9832	9836	9841	9845	9850	9854	9859	9863
97	9868	9872	9877	9881	9886	9890	9894	9899	9903	9908
98	9912	9917	9921	9926	9930	9934	9939	9943	9948	9952
99	9956	9961	9965	9969	9974	9978	9983	9987	9991	9996

TABLES

Trigonometric Functions

x		sin x	cos x	tan x	cot x	sec x	csc x	
15°	0'	.2588	.9659	.2679	3.7321	1.0353	3.8637	0'
	10'	.2616	.9652	.2711	3.6891	1.0361	3.8222	50'
	20'	.2644	.9644	.2742	3.6470	1.0369	3.7817	40'
	30'	.2672	.9636	.2773	3.6059	1.0377	3.7420	30'
	40'	.2700	.9628	.2805	3.5656	1.0386	3.7032	20'
	50'	.2728	.9621	.2836	3.5261	1.0394	3.6652	10'
16°	0'	.2756	.9613	.2867	3.4874	1.0403	3.6280	0'
	10'	.2784	.9605	.2899	3.4495	1.0412	3.5915	50'
	20'	.2812	.9596	.2931	3.4124	1.0421	3.5559	40'
	30'	.2840	.9588	.2962	3.3759	1.0430	3.5209	30'
	40'	.2868	.9580	.2994	3.3402	1.0439	3.4867	20'
	50'	.2896	.9572	.3026	3.3052	1.0448	3.4532	10'
17°	0'	.2924	.9563	.3057	3.2709	1.0457	3.4203	0'
	10'	.2952	.9555	.3089	3.2371	1.0466	3.3881	50'
	20'	.2979	.9546	.3121	3.2041	1.0476	3.3565	40'
	30'	.3007	.9537	.3153	3.1716	1.0485	3.3255	30'
	40'	.3035	.9528	.3185	3.1397	1.0495	3.2951	20'
	50'	.3062	.9520	.3217	3.1084	1.0505	3.2653	10'
18°	0'	.3090	.9511	.3249	3.0777	1.0515	3.2361	0'
	10'	.3118	.9502	.3281	3.0475	1.0525	3.2074	50'
	20'	.3145	.9492	.3314	3.0178	1.0535	3.1792	40'
	30'	.3173	.9483	.3346	2.9887	1.0545	3.1516	30'
	40'	.3201	.9474	.3378	2.9600	1.0555	3.1244	20'
	50'	.3228	.9465	.3411	2.9319	1.0566	3.0977	10'
19°	0'	.3256	.9455	.3443	2.9042	1.0576	3.0716	0'
	10'	.3283	.9446	.3476	2.8770	1.0587	3.0458	50'
	20'	.3311	.9436	.3508	2.8502	1.0598	3.0206	40'
	30'	.3338	.9426	.3541	2.8239	1.0609	2.9957	30'
	40'	.3365	.9417	.3574	2.7980	1.0620	2.9714	20'
	50'	.3393	.9407	.3607	2.7725	1.0631	2.9474	10'
20°	0'	.3420	.9397	.3640	2.7475	1.0642	2.9238	0'
	10'	.3448	.9387	.3673	2.7228	1.0653	2.9006	50'
	20'	.3475	.9377	.3706	2.6985	1.0665	2.8879	40'
	30'	.3502	.9367	.3739	2.6746	1.0676	2.8555	30'
	40'	.3529	.9356	.3772	2.6511	1.0688	2.8334	20'
	50'	.3557	.9346	.3805	2.6279	1.0700	2.8118	10'
21°	0'	.3584	.9336	.3839	2.6051	1.0712	2.7904	0'
	10'	.3611	.9325	.3872	2.5826	1.0724	2.7695	50'
	20'	.3638	.9315	.3906	2.5605	1.0736	2.7488	40'
	30'	.3665	.9304	.3939	2.5386	1.0748	2.7285	30'
	40'	.3692	.9293	.3973	2.5172	1.0760	2.7085	20'
	50'	.3719	.9283	.4006	2.4960	1.0773	2.6888	10'
22°	0'	.3746	.9272	.4040	2.4751	1.0785	2.6695	0'
	10'	.3773	.9261	.4074	2.4545	1.0798	2.6504	50'
	20'	.3800	.9250	.4108	2.4342	1.0811	2.6316	40'
	30'	.3827	.9239	.4142	2.4142	1.0824	2.6131	30'
		cos x	sin x	cot x	tan x	csc x	sec x	x

Trigonometric Functions

x		sin x	cos x	tan x	cot x	sec x	csc x	
	30'	.3827	.9239	.4142	2.4142	1.0824	2.6131	30'
	40'	.3854	.9228	.4176	2.3945	1.0837	2.5949	20'
	50'	.3881	.9216	.4210	2.3750	1.0850	2.5770	10'
23°	0'	.3907	.9205	.4245	2.3559	1.0864	2.5593	67° 0'
	10'	.3934	.9194	.4279	2.3369	1.0877	2.5419	50'
	20'	.3961	.9182	.4314	2.3183	1.0891	2.5247	40'
	30'	.3987	.9171	.4348	2.2998	1.0904	2.5078	30'
	40'	.4014	.9159	.4383	2.2817	1.0918	2.4912	20'
	50'	.4041	.9147	.4417	2.2637	1.0932	2.4748	10'
24°	0'	.4067	.9135	.4452	2.2460	1.0946	2.4586	66° 0'
	10'	.4094	.9124	.4487	2.2286	1.0961	2.4426	50'
	20'	.4120	.9112	.4522	2.2113	1.0975	2.4269	40'
	30'	.4147	.9100	.4557	2.1943	1.0990	2.4114	30'
	40'	.4173	.9088	.4592	2.1775	1.1004	2.3961	20'
	50'	.4200	.9075	.4628	2.1609	1.1019	2.3811	10'
25°	0'	.4226	.9063	.4663	2.1445	1.1034	2.3662	65° 0'
	10'	.4253	.9051	.4699	2.1283	1.1049	2.3515	50'
	20'	.4279	.9038	.4734	2.1123	1.1064	2.3371	40'
	30'	.4305	.9026	.4770	2.0965	1.1079	2.3228	30'
	40'	.4331	.9013	.4806	2.0809	1.1095	2.3088	20'
	50'	.4358	.9001	.4841	2.0655	1.1110	2.2949	10'
26°	0'	.4384	.8988	.4877	2.0503	1.1126	2.2812	64° 0'
	10'	.4410	.8975	.4913	2.0353	1.1142	2.2677	50'
	20'	.4436	.8962	.4950	2.0204	1.1158	2.2543	40'
	30'	.4462	.8949	.4986	2.0057	1.1174	2.2412	30'
	40'	.4488	.8936	.5022	1.9912	1.1190	2.2282	20'
	50'	.4514	.8923	.5059	1.9768	1.1207	2.2154	10'
27°	0'	.4540	.8910	.5095	1.9626	1.1223	2.2027	63° 0'
	10'	.4566	.8897	.5132	1.9486	1.1240	2.1902	50'
	20'	.4592	.8884	.5169	1.9347	1.1257	2.1779	40'
	30'	.4617	.8870	.5206	1.9210	1.1274	2.1657	30'
	40'	.4643	.8857	.5243	1.9074	1.1291	2.1537	20'
	50'	.4669	.8843	.5280	1.8940	1.1308	2.1418	10'
28°	0'	.4695	.8829	.5317	1.8807	1.1326	2.1301	62° 0'
	10'	.4720	.8816	.5354	1.8676	1.1343	2.1185	50'
	20'	.4746	.8802	.5392	1.8546	1.1361	2.1070	40'
	30'	.4772	.8788	.5430	1.8418	1.1379	2.0957	30'
	40'	.4797	.8774	.5467	1.8291	1.1397	2.0846	20'
	50'	.4823	.8760	.5505	1.8165	1.1415	2.0736	10'
29°	0'	.4848	.8746	.5543	1.8040	1.1434	2.0627	61° 0'
	10'	.4874	.8732	.5581	1.7917	1.1452	2.0519	50'
	20'	.4899	.8718	.5619	1.7796	1.1471	2.0413	40'
	30'	.4924	.8704	.5658	1.7675	1.1490	2.0308	30'
	40'	.4950	.8689	.5696	1.7556	1.1509	2.0204	20'
	50'	.4975	.8675	.5735	1.7437	1.1528	2.0101	10'
30°	0'	.5000	.8660	.5774	1.7321	1.1547	2.0000	60° 0'
		cos x	sin x	cot x	tan x	csc x	sec x	x

TABLES

TABLES

Trigonometric Functions

x		sin x	cos x	tan x	cot x	sec x	csc x		x
38°	30'	.6088	.7934	.7673	1.3032	1.2605	1.6427	30'	
	40'	.6111	.7916	.7720	1.2954	1.2633	1.6365	20'	
	50'	.6134	.7898	.7766	1.2876	1.2662	1.6304	10'	
38°	0'	.6157	.7880	.7813	1.2799	1.2690	1.6243	0'	**52°**
	10'	.6180	.7862	.7860	1.2723	1.2719	1.6183	50'	
	20'	.6202	.7844	.7907	1.2647	1.2748	1.6123	40'	
	30'	.6225	.7826	.7954	1.2572	1.2779	1.6064	30'	
	40'	.6248	.7808	.8002	1.2497	1.2808	1.6005	20'	
	50'	.6271	.7790	.8050	1.2423	1.2837	1.5948	10'	
39°	0'	.6293	.7771	.8098	1.2349	1.2868	1.5890	0'	**51°**
	10'	.6316	.7753	.8146	1.2276	1.2898	1.5833	50'	
	20'	.6338	.7735	.8195	1.2203	1.2929	1.5777	40'	
	30'	.6361	.7716	.8243	1.2131	1.2960	1.5721	30'	
	40'	.6383	.7698	.8292	1.2059	1.2991	1.5666	20'	
	50'	.6406	.7679	.8342	1.1988	1.3022	1.5611	10'	
40°	0'	.6428	.7660	.8391	1.1918	1.3054	1.5557	0'	**50°**
	10'	.6450	.7642	.8441	1.1847	1.3086	1.5504	50'	
	20'	.6472	.7623	.8491	1.1778	1.3118	1.5450	40'	
	30'	.6494	.7604	.8541	1.1708	1.3151	1.5398	30'	
	40'	.6517	.7585	.8591	1.1640	1.3184	1.5346	20'	
	50'	.6539	.7566	.8642	1.1571	1.3217	1.5294	10'	
41°	0'	.6561	.7547	.8693	1.1504	1.3250	1.5243	0'	**49°**
	10'	.6583	.7528	.8744	1.1436	1.3284	1.5192	50'	
	20'	.6604	.7509	.8796	1.1369	1.3318	1.5142	40'	
	30'	.6626	.7490	.8847	1.1303	1.3352	1.5092	30'	
	40'	.6648	.7470	.8899	1.1237	1.3386	1.5042	20'	
	50'	.6670	.7451	.8952	1.1171	1.3421	1.4993	10'	
42°	0'	.6691	.7431	.9004	1.1106	1.3456	1.4945	0'	**48°**
	10'	.6713	.7412	.9057	1.1041	1.3492	1.4897	50'	
	20'	.6734	.7392	.9110	1.0977	1.3527	1.4849	40'	
	30'	.6756	.7373	.9163	1.0913	1.3563	1.4802	30'	
	40'	.6777	.7353	.9217	1.0850	1.3600	1.4755	20'	
	50'	.6799	.7333	.9271	1.0786	1.3636	1.4709	10'	
43°	0'	.6820	.7314	.9325	1.0724	1.3673	1.4663	0'	**47°**
	10'	.6841	.7294	.9380	1.0661	1.3711	1.4617	50'	
	20'	.6862	.7274	.9435	1.0599	1.3748	1.4572	40'	
	30'	.6884	.7254	.9490	1.0538	1.3786	1.4527	30'	
	40'	.6905	.7234	.9545	1.0477	1.3824	1.4483	20'	
	50'	.6926	.7214	.9601	1.0416	1.3863	1.4439	10'	
44°	0'	.6947	.7193	.9657	1.0355	1.3902	1.4396	0'	**46°**
	10'	.6967	.7173	.9713	1.0295	1.3941	1.4352	50'	
	20'	.6988	.7153	.9770	1.0235	1.3980	1.4310	40'	
	30'	.7009	.7133	.9827	1.0176	1.4020	1.4267	30'	
	40'	.7030	.7112	.9884	1.0117	1.4061	1.4225	20'	
	50'	.7050	.7092	.9942	1.0058	1.4101	1.4184	10'	
45°	0'	.7071	.7071	1.0000	1.0000	1.4142	1.4142	0'	**45°**
		cos x	sin x	cot x	tan x	csc x	sec x		x

Trigonometric Functions

x		sin x	cos x	tan x	cot x	sec x	csc x		x
30°	0'	.5000	.8660	.5774	1.7321	1.1547	2.0000	0'	**60°**
	10'	.5025	.8646	.5812	1.7205	1.1567	1.9900	50'	
	20'	.5050	.8631	.5851	1.7090	1.1586	1.9801	40'	
	30'	.5075	.8616	.5890	1.6977	1.1606	1.9703	30'	
	40'	.5100	.8601	.5930	1.6864	1.1626	1.9606	20'	
	50'	.5125	.8587	.5969	1.6753	1.1646	1.9511	10'	
31°	0'	.5150	.8572	.6009	1.6643	1.1666	1.9416	0'	**59°**
	10'	.5175	.8557	.6048	1.6534	1.1687	1.9323	50'	
	20'	.5200	.8542	.6088	1.6426	1.1708	1.9230	40'	
	30'	.5225	.8526	.6128	1.6319	1.1728	1.9139	30'	
	40'	.5250	.8511	.6168	1.6212	1.1749	1.9049	20'	
	50'	.5275	.8496	.6208	1.6107	1.1770	1.8959	10'	
32°	0'	.5299	.8480	.6249	1.6003	1.1792	1.8871	0'	**58°**
	10'	.5324	.8465	.6289	1.5900	1.1813	1.8783	50'	
	20'	.5348	.8450	.6330	1.5798	1.1835	1.8699	40'	
	30'	.5373	.8434	.6371	1.5697	1.1857	1.8612	30'	
	40'	.5398	.8418	.6412	1.5597	1.1879	1.8527	20'	
	50'	.5422	.8403	.6453	1.5497	1.1901	1.8444	10'	
33°	0'	.5446	.8387	.6494	1.5399	1.1924	1.8361	0'	**57°**
	10'	.5471	.8371	.6536	1.5301	1.1946	1.8279	50'	
	20'	.5495	.8355	.6577	1.5204	1.1969	1.8198	40'	
	30'	.5519	.8339	.6619	1.5108	1.1992	1.8118	30'	
	40'	.5544	.8323	.6661	1.5013	1.2015	1.8039	20'	
	50'	.5568	.8307	.6703	1.4919	1.2039	1.7960	10'	
34°	0'	.5592	.8290	.6745	1.4826	1.2062	1.7883	0'	**56°**
	10'	.5616	.8274	.6787	1.4733	1.2086	1.7806	50'	
	20'	.5640	.8258	.6830	1.4641	1.2110	1.7730	40'	
	30'	.5664	.8241	.6873	1.4550	1.2134	1.7655	30'	
	40'	.5688	.8225	.6916	1.4460	1.2158	1.7581	20'	
	50'	.5712	.8208	.6959	1.4370	1.2183	1.7507	10'	
35°	0'	.5736	.8192	.7002	1.4281	1.2208	1.7435	0'	**55°**
	10'	.5760	.8175	.7046	1.4193	1.2233	1.7362	50'	
	20'	.5783	.8158	.7089	1.4106	1.2258	1.7291	40'	
	30'	.5807	.8141	.7133	1.4019	1.2283	1.7221	30'	
	40'	.5831	.8124	.7177	1.3934	1.2309	1.7151	20'	
	50'	.5854	.8107	.7221	1.3848	1.2335	1.7082	10'	
36°	0'	.5878	.8090	.7265	1.3764	1.2361	1.7013	0'	**54°**
	10'	.5901	.8073	.7310	1.3680	1.2387	1.6945	50'	
	20'	.5925	.8056	.7355	1.3597	1.2413	1.6878	40'	
	30'	.5948	.8039	.7400	1.3514	1.2440	1.6812	30'	
	40'	.5972	.8021	.7445	1.3432	1.2467	1.6746	20'	
	50'	.5995	.8004	.7490	1.3351	1.2494	1.6681	10'	
37°	0'	.6018	.7986	.7536	1.3270	1.2521	1.6616	0'	**53°**
	10'	.6041	.7969	.7581	1.3190	1.2549	1.6553	50'	
	20'	.6065	.7951	.7627	1.3111	1.2577	1.6489	40'	
	30'	.6088	.7934	.7673	1.3032	1.2605	1.6427	30'	
		cos x	sin x	cot x	tan x	csc x	sec x		x

Glossary

The explanations given in this glossary are intended to be brief descriptions of the terms listed. They are not necessarily definitions.

absolute value (of a real number) *(p. 10)* The absolute value of the real number x is written as $|x|$. $|x| = -x$ if x is a negative number, and $|x| = x$ if x is a positive number or zero.

additive identity *(p. 26)* Zero is the additive identity and has the property $x + 0 = x$, for each number x.

additive inverse *(p. 26)* The additive inverse of x is $-x$ and $x + (-x) = 0$, for each number x.

amplitude *(p. 497)* The amplitude of a trig function is its maximum value.

angle of rotation *(p. 455)* An angle of rotation is formed by rotating a ray about its end point, or vertex.

antilogarithm (base-10) *(p. 404)* Antilog $x = 10^x$, for each real number x; antilog $2 = 10^2 = 100$.

arithmetic mean *(p. 361)* The arithmetic mean of x and y is their average, $\frac{x+y}{2}$.

arithmetic means *(p. 361)* The terms between two given terms of an arithmetic progression.

arithmetic progression (A.P.) *(p. 357)* A progression formed by adding the same number to any term to obtain the next term.

arithmetic series *(p. 364)* The indicated sum of the terms of an arithmetic progression. 3, 6, 9, … is an arithmetic progression; $3 + 6 + 9 + \ldots$ is an arithmetic series.

binomial expansion *(p. 382)* The indicated sum of the terms of the nth power of a binomial. For example, $(a + b)^3$ is $a^3 + 3a^2b + 3ab^2 + b^3$.

bounds (upper and lower) *(p. 412)* U and L are upper and lower bounds for the real zeros of a polynomial if all the real zeros are between U and L.

characteristic *(p. 403)* The integer part of a base-10 logarithm. Log $54{,}600 = 4.7372$ and log $0.0546 = 8.7372 - 10$. The characteristic of log $54{,}600$ is 4 and the characteristic of log 0.0546 is $8 - 10$, or -2.

circle *(p. 306)* The set of all points in a plane that are a given distance from a given point in the plane.

combination *(p. 435)* A selection of objects without regard to order.

combined variation *(p. 295)* A function defined by an equation of the form $y = \frac{kxz}{w}$, or $\frac{yw}{xz} = k$, where k is a nonzero constant.

complex number *(p. 187)* A number that can be written in the form $a + bi$, where a and b are real numbers and $i = \sqrt{-1}$.

conjugates *(pp. 149, 191, 419)* Two numbers of the form $x + y$ and $x - y$ are a pair of conjugates, and their product, $x^2 - y^2$, is a difference of squares. $3\sqrt{2} + 4$ and $3\sqrt{2} - 4$ are a pair of irrational conjugates, and their product, $(3\sqrt{2})^2 - 4^2$, or 2, is a rational number. $5 - 3i$ and $5 + 3i$ are a pair of imaginary conjugates, and their product, $5^2 - (3i)^2$, or 34, is a real number.

conjunction *(p. 7)* A compound sentence formed by two clauses separated by *and*. $[x > 2$ and $x < 5]$ is a conjunction.

consistent system *(p. 240)* A system of linear equations that has at least one solution.

constant function *(p. 282)* A linear function whose graph is a horizontal line, ray, or segment.

Cramer's rule *(p. 200)* Solving a system of equations using determinants.

dependent events *(p. 443)* Events in which one event depends upon the other.

dependent system *(p. 241)* A system of linear equations that has all solutions in common.

determinant *(p. 250)* A real number assigned to a matrix.

discriminant *(p. 195)* The radicand, $b^2 - 4ac$, in the quadratic formula is called the discriminant of the quadratic equation. The discriminant determines the nature of the solutions of the equation.

disjunction *(p. 7)* A compound sentence formed by two clauses separated by *or*. $[x < 2$ or $x > 5]$ is a disjunction.

domain *(p. 276)* The set of all first coordinates of the ordered pairs of a relation.

ellipse *(p. 310)* The set of all points in a plane such that for each point of the set, the sum of its distance from two fixed points is constant.

exponential equation *(pp. 33, 158, 406)* Equations such as $2^x = 16$ and $5^{4x+1} = 125$ in which the variable appears in the exponent.

factorial notation *(p. 383)* $0! = 1$ and for each positive integer k, $k! = 1 \cdot 2 \cdot 3 \cdots k$.

field *(p. 26)* A set of numbers, along with two operations, that satisfies eleven special properties.

fractional equation *(p. 116)* An equation such as $\frac{2n-9}{n-7} + \frac{n}{2} = \frac{5}{n-7}$ in which a variable appears in a denominator.

function *(p. 276)* A relation, a set of ordered pairs, in which different ordered pairs have different first coordinates.

fundamental counting principle *(p. 424)* The principle used to find the number of permutations of n objects taken r at a time.

geometric mean *(p. 372)* The geometric mean (mean proportional) of x and y is $\sqrt{xy}$ or $-\sqrt{xy}$.

geometric means *(p. 372)* The terms between two given terms of a geometric progression.

geometric progression (G.P.) *(p. 368)* A progression formed by multiplying any term by the same nonzero number to obtain the next term.

geometric series *(p. 375)* The indicated sum of the terms of a geometric progression. $3 + 6 + 12 + \ldots$ is a geometric series since $3, 6, 12, \ldots$ is a geometric progression.

hyperbola *(p. 319)* The set of all points in a plane such that for each point of the set, the difference of its distances from two fixed points is constant.

imaginary number *(p. 188)* Any complex number $a + bi$, where $b \neq 0$, such as $5 - 2i$ and $4i$.

inclusive events *(p. 442)* Events that can both occur at the same time.

inconsistent system *(p. 241)* A system of linear equations that has no solution.

independent events *(p. 443)* Events in which neither event depends upon the other.

independent system *(p. 240)* A system of linear equations that has exactly one solution.

inverse functions *(p. 285)* When both a relation and its inverse are functions, they are called inverse functions.

inverse relations *(p. 285)* The inverse of a relation is formed by reversing the order of the elements of each ordered pair.

inverse variation *(p. 291)* A function that is defined by an equation of the form $xy = k$ or $y = \frac{k}{x}$, where k is a nonzero constant.

irrational number *(p. 140)* A real number whose decimal numeral is nonterminating and nonrepeating.

joint variation *(p. 294)* A function that is defined by an equation of the form $y = kxz$ or $y = \frac{k}{x}$, where k is a nonzero constant.

law of cosines *(p. 539)* The law of cosines is used to find the length of a side of a triangle when the measures of the two other sides and included angle are known. In general, $a^2 = b^2 + c^2 - 2bc \cos A$, $b^2 = a^2 + c^2 - 2ac \cos B$, and $c^2 = a^2 + b^2 - 2ab \cos C$.

law of sines *(p. 546)* The law of sines is used to find the measure of an angle and length of a side of a triangle when the measures of two angles and a side are known. In general, $\frac{\sin A}{a} = \frac{\sin B}{b} = \frac{\sin C}{c}$.

linear function *(p. 282)* A function whose graph is a nonvertical line, ray, or segment. The equation $y = mx + b$ describes a linear function.

logarithm (base-n) *(p. 392)* For each positive number y and each positive number $n \neq 1$, if $y = n^x$, then $\log_n y = \log_n n^x = x$. For example, $125 = 5^3$ and $\log_5 5^3 = 3$.

logarithm (common; base-10) *(p. 403)* For each real number $y > 0$, the base-10 logarithm of y, or $\log y$, is defined as follows: If $y = 10^x$, then $\log y = \log 10^x = x$. Base-10 logarithms are also called common logarithms.

mantissa *(p. 403)* The decimal part of a base-10 logarithm. $\log 54{,}600 = 4.7372$, so

the mantissa of log 54,600 is .7372.
matrix *(p. 250)* An array of numbers or other elements arranged in rows and columns.
multiplicative identity *(p. 26)* One is the multiplicative identity and $x \cdot 1 = x$, for each number x.
multiplicative inverse *(p. 26)* If $x \neq 0$, the multiplicative inverse of x is $\frac{1}{x}$ and $x \cdot \frac{1}{x} = 1$, for each $x \neq 0$.
mutually exclusive events *(p. 442)* Events that cannot both occur at the same time.
parabola *(p. 327)* A set of points in a plane that are equidistant from a fixed point and a fixed line.
periodic function *(p. 497)* A function that repeats itself. If $f(x) = f(x + p)$, then f is a periodic function with a period of p.
permutation (linear) *(p. 425)* A linear arrangement of objects in a definite order.
permutation (circular) *(p. 433)* The number of circular permutations of n distinct objects is $(n - 1)!$.
principal values *(p. 531)* To obtain an inverse function of a trigonometric function, restrict the range of the inverse. The restricted values are called principal values.
probability *(p. 438)* The probability of a simple event is the number of favorable outcomes divided by the total number of all possible outcomes.
progression *(p. 357)* An ordered list of numbers formed according to some pattern; a sequence.
pure imaginary number *(p. 188)* Any complex number $a + bi$, where $a = 0$, such as $4i$.
Pythagorean relation *(p. 78)* For each right triangle ABC with right angle C, $a^2 + b^2 = c^2$. And, if $a^2 + b^2 = c^2$, then triangle ABC is a right triangle with right angle C.
quadrants *(p. 209)* The two axes in a coordinate plane separate the plane into four sections called quadrants.
quadratic equation (in one variable) *(p. 68)* The standard form of a quadratic equation in one variable is $ax^2 + bx + c = 0$, where a, b, and c are numbers and $a \neq 0$.
quadratic formula *(p. 169)* The solutions of the general quadratic equation, $ax^2 + bx + c = 0$, are given by the quadratic formula, $x = \frac{-b \pm \sqrt{b^2 - 4ac}}{2a}$.
quadratic inequality (in one variable) *(p. 71)* The standard form of a quadratic inequality in one variable is either $ax^2 + bx + c < 0$ or $ax^2 + bx + c > 0$, where a, b, and c are numbers and $a \neq 0$.
radian *(p. 523)* If an angle of rotation, which is the central angle of a given circle, cuts off an arc whose length is equal to the length of a radius of a circle, then the measure of the angle of rotation is 1 radian.
radical equation *(p. 178)* An equation like $\sqrt[3]{6x + 10} = -2$ in which a radicand contains a variable.
range *(p. 276)* The set of all second coordinates of the ordered pairs of a relation.
rational expression *(p. 91)* A rational expression is either a polynomial P or a quotient $\frac{P}{Q}$ of polynomials P and Q. $Q \neq 0$
rational number *(p. 89)* A rational number is one that can be written in the form $\frac{a}{b}$ where a and b are integers, $b \neq 0$. Examples of rational numbers are -8, 0, 7, $2\frac{3}{4}$, 9.62, $\sqrt{25}$, $\frac{22}{7}$, and $3.43434\ldots$.
real numbers *(p. 141)* All the rational numbers together with all the irrational numbers.
rectangular hyperbola *(p. 320)* A hyperbola whose equation is $xy = k$, where k is a nonzero real number.
reference angle *(p. 455)* The positive acute angle formed between the terminal side and the x-axis.
repeating decimal *(p. 89)* A nonterminating decimal in which one block of digits continues to repeat in consecutive blocks. The numeral $3.7242424\ldots$, or $3.72\overline{24}$, is a repeating decimal.
sequence *(p. 357)* An ordered list of numbers formed according to some pattern; a progression.
slope and parallel lines *(p. 228)* If two or more different nonvertical lines have the same slope, then the lines are parallel.

slope and perpendicular lines *(p. 229)* If two lines have negative reciprocal slopes, the lines are perpendicular.

standard form of a linear equation *(p. 216)* A linear equation in two variables written in the form $ax + by = c$.

synthetic division *(p. 130)* A short method of dividing a polynomial in x by a binomial of the form $x - a$, where a is a rational number.

trigonometric functions

cosecant function (p. 470) For any angle of rotation u, the cosecant (csc) of u is $\frac{r}{y}$; $\csc u = \frac{r}{y}$.

cosine function (p. 462) For any angle of rotation u, the cosine (cos) of u is $\frac{x}{r}$; $\cos u = \frac{x}{r}$.

cotangent function (p. 466) For any angle of rotation u, the cotangent (cot) of $u = \frac{x}{y}$; $\cot u = \frac{x}{y}$.

secant function (p. 470) For any angle of rotation u, the secant (sec) of u is $\frac{r}{x}$; $\sec u = \frac{r}{x}$.

sine function (p. 462) For any angle of rotation u, the sine (sin) of u is $\frac{y}{r}$; $\sin u = \frac{y}{r}$.

tangent function (p. 466) For any angle of rotation u, the tangent (tan) of $u = \frac{y}{x}$; $\tan u = \frac{y}{x}$.

trigonometric identities *(p. 518)* Trigonometric equations that are true for all permissible values of the variables.

value of a function *(p. 279)* For any ordered pair (x, y) of a function, y is the value of the function at x. For any given element of the domain of a function, there is a corresponding element in the range which is called the value of the function.

vector *(p. 50)* A directed line segment that can show the magnitude and direction of a force, displacement, or velocity.

zero (of a polynomial) *(p. 409)* A zero of a polynomial $P(x)$ is a value of x for which the value of $P(x)$ is 0. For example, $x^2 - 9$ has two zeros: 3 and -3.

Symbol List

$=$	is equal to	(x, y)	the ordered pair of numbers x, y
$<$	is less than	$\overline{AB}$	segment AB
$>$	is greater than	$\overleftrightarrow{AB}$	line AB
$\{x \mid x > 5\}$	the set of all numbers x such that x is greater than 5	AB	the distance from A to B
$\|x\|$	the absolute value of x	$\parallel$	is parallel to
$\sqrt{a}$	the principal square root of a	$\perp$	is perpendicular to
$\emptyset$	the empty set	$f(x)$	f at x (the value of function f at x)
$\doteq$	is approximately equal to	f^{-1}	the inverse of function f
$\sqrt[n]{x}$	the nth root of x	$\log x$	the base-10 logarithm of x
$a \pm \sqrt{b}$	$a + \sqrt{b}, a - \sqrt{b}$	$\log_n x$	the base-n logarithm of x
i	$\sqrt{-1}$	$\sin^{-1}$	arc sine

Annotated Answers

This section contains those answers that do not appear on the TE text page.

Page 23
34. $R(r_1 + r_2) = r_1 r_2$; $R = \dfrac{r_1 r_2}{r_1 + r_2}$; $\dfrac{1}{R} = \dfrac{r_1 + r_2}{r_1 r_2} = \dfrac{r_1}{r_1 r_2} + \dfrac{r_2}{r_1 r_2}$; $\dfrac{1}{R} = \dfrac{1}{r_2} + \dfrac{1}{r_1} = \dfrac{1}{r_1} + \dfrac{1}{r_2}$

Page 26
7.

+	0	1	2	3
0	0	1	2	3
1	1	2	3	0
2	2	3	0	1
3	3	0	1	2

·	0	1	2	3
0	0	0	0	0
1	0	1	2	3
2	0	2	0	2
3	0	3	2	1

Page 42
39. $(a + b + c)(a + b + c) = a^2 + ab + ac + ab + b^2 + bc + ac + bc + c^2 = a^2 + b^2 + c^2 + 2ab + 2ac + 2bc$

Page 52

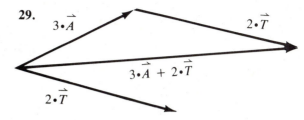

29.

Page 59
34. $(2x + 3)(5x + 4)$ **35.** $(4n - 3)(3n - 2)$
37. $(9a - 4)(2a - 3)$ **38.** $(2x - 5)(12x + 7)$
40. $(2c - 3)(10c - 9)$ **41.** $(8a + 9)(3a - 4)$
42. $(3y - 4z)(2y + 3z)$ **43.** $(5x - 6y)(x + 2y)$
44. $(9m + 5n)(4m - 3n)$ **45.** $(8c - 5d)(3c + 4d)$
46. $(5x^2 - 6)(3x^2 + 2)$ **47.** $(4x^2 - 5)(3x^2 - 2)$
49. $(3y^2 + 2)(2y^2 - 9)$ **50.** $(2n^2 - 3)(5n^2 - 8)$
51. $(3a^2 + 10)(3a^2 - 2)$ **52.** $(2x^2y^2 - 3)(x^2y^2 - 2)$
53. $(3a^2b^2 + 5)(a^2b^2 - 2)$
54. $(5m^2n^2 - 6)(m^2n^2 + 2)$
55. $(3c^2d^2 + 4)(2c^2d^2 + 5)$
56. $(5x^2 + 8y^2)(x^2 + 3y^2)$
57. $(7c^2 + 6d^2)(2c^2 - 3d^2)$
58. cannot be factored   **59.** $(x^3 - 6y^2)(x^3 - y^2)$
60. $(5m^2 + n^3)(m^2 - n^3)$

Page 67
53. Ex. 51 and 52 both contain $(x + y)(x - y)$ as a factor. $x^4 + x^2y^2 + y^4 = (x^4 + 2x^2y^2 + y^4) - x^2y^2 = (x^2 + y^2)^2 - (xy)^2 = (x^2 + y^2 + xy)(x^2 + y^2 - xy) = (x^2 + xy + y^2)(x^2 - xy + y^2)$.

Page 77 (Non-Routine Problems)
1. (O = odd; E = even) Case I: x is odd and a, b, c are odd. $ax^2 + bx + c = axx + bx + c = O \cdot O \cdot O + O \cdot O + O = O + O + O = E + O = O$; $ax^2 + bx + c$ is odd.
Case II: x is even and a, b, c are odd. $ax^2 + bx + c = axx + bx + c = O \cdot E \cdot E + O \cdot E + O = E + E + O = E + O = O$; $ax^2 + bx + c$ is odd.

Page 84
3. $(5a^2b^2 + 4)(5a^2b^2 - 3)$
4. $(2x^{n+1} - 1)(x^{n+1} - 5)$
5. $(3a + 4)(3a - 4)$
6. $(n + 4)(n^2 - 4n + 16)$
11. $(5x + y - 3)(5x - y + 3)$
12. $(x^{3m} + 5y^n)^2$
13. $16n^2 - 36$ **14.** $9a^2 - 30a + 25$
19. $3x(3x - 1)(2x - 1)$ **20.** $-1(2a - 7)(a + 2)$
21. $3(2y + 3)(2y - 3)$ **22.** $-1(y - 5)^2$
23. $(2n^2 + 3)(n + 3)(n - 3)$

T–601

24. $(2c + 3)(2c - 3)(c + 2)(c - 2)$
25. $2ab(a - 5)(a + 3)$
26. $3(x - 2y)(x^2 + 2xy + 4y^2)$ **27.** $(x^2 + 5)(x + 4)$
28. $(3c - 4d)(5a - 2b)$ **29.** $2x^{3n}(x^{2n} + 3)(x^{2n} - 3)$

Page 86
1.
```
110  A = INT ( RND (1) * 12) + 1
120  B = INT ( RND (1) * 12) + 1
130  A$ = "(X+" + STR$ (A) + ")(X+" + STR$ (B) + "
     )"
140  B$ = "(X+" + STR$ (B) + ")(X+" + STR$ (A) + ")
     "
150  PRINT "FACTOR: ";
160  PRINT "X^2 + "A + B"X +"A * B
170  INPUT C$
180  IF C$ = A$ THEN 220
190  IF C$ = B$ THEN 220
200  PRINT "THOSE FACTORS ARE NOT CORRECT."
210  GOTO 150
220  PRINT "CORRECT"
230  GOTO 110
```

2.
```
130  PRINT "FACTOR ";
140  PRINT A;"X + ";A * B
150  INPUT C$
160  IF C$ = A$ THEN 190

100  A = INT ( RND (1) * 12) + 1
110  B = INT ( RND (1) * 12) + 1
120  A$ = STR$ (A) + "(X+" + STR$ (B) + ")"

170  PRINT "THOSE FACTORS ARE NOT CORRECT."
180  GOTO 130
190  PRINT "CORRECT"
200  GOTO 100
```

Page 105

11.

Score(s)	Tally	Frequency (f)
101	||	2
100	|	1
98	||	2
97	|	1
96	||	2
83	|	1
		9

12.

93	||	2
92	|	1
73	|	1
68	|	1
58	|	1
56	|	1
54	|	1
53	|	1
		9

13.

			Sum $(f \cdot s)$
101	|	1	101
100	||	2	200
99	||	2	198
98	|||	3	294
97	|||	3	291
96	||||	4	384
94	||	2	188
93	||||	4	372
92	|||	3	276
88	||||	4	352
87	|	1	87
86	||	2	172
85	|	1	85
		32	3000

14.

20	|	1	20
19	|	1	19
18	|||	3	54
17	||	2	34
16	||||	4	64
15	|	1	15
14	||||	4	56
13	|	1	13
12	||||	4	48
11	||	2	22
10	|||	3	30
9	|||	3	27
8	|||	3	24
7	|	1	7
6	|	1	6
5	|	1	5
4	|	1	4
		36	448

Page 107

1. a, b, c, and d are integers. **1.** Def. 1
2. $\dfrac{a}{b} \cdot \dfrac{c}{d} = \dfrac{ac}{bd}$ **2.** Theorem 4
3. ac and bd are integers. **3.** Closure for mult.
4. $\dfrac{ac}{bd}$ is a rational number. **4.** Def. 1
5. $\dfrac{a}{b} \cdot \dfrac{c}{d}$ is a rational number. **5.** Statements 2 and 4

Page 129

7. $3x^2 + 8x + 7$, 30, no **8.** $3n^3 + n^2 - 2$, 0, yes
10. $3x^2 - x + 2$, -3, no **12.** $3y^3 + 2y^2 + 3y + 2$, -8, no **17.** $2n^2 - 4n + 1$, 0, yes **18.** $4x^2 + 3x + 33$, 200, no **19.** $y^2 - y - 1$, 0, yes **20.** $3a^3 - a^2 + 1$, -3, no
21. $x^{2m} + 3$, 2, no **22.** $x^{2c} - x^{c+1}$, 2, no

Page 132

33. 72, 5, 0, 9, 8, -3, 0; $x + 1$, $x - 3$ **34.** 50, 0, -6, -4, -6, 0, 50; $x + 2$, $x - 2$ **35.** 60, 0, 2, 6, 0, 20, 150; $x + 2$, $x - 1$ **36.** 285, 45, -3, -3, -3, 45, 285; none

Page 136
1.
```
10 FOR I = 15 TO 39 STEP 2
20 PRINT I;"^2 = "; INT (I ^ 2)
30 NEXT I
```

Page 162
1.
```
100 PRINT "WHAT ARE THE FIRST COEFFICIENT AND R
    ADICAND?";: INPUT A,B
110 PRINT "WHAT ARE THE SECOND COEFFICIENT AND
    RADICAND?";: INPUT C,D
120 PRINT A;" SQR(";B;") * ";C;" SQR(";D;") = ";
130 E = A * C:R = B * D
140 IF ABS ( SQR (R) - INT ( SQR (R))) < .05 THEN
    230
150 FOR I = 2 TO SQR (R)
160 P = I * I
170 IF ABS (R / P - INT (R / P)) > .001 THEN
    200
180 E = E * I:R = R / P: IF R < P THEN 210
190 GOTO 170
200 NEXT I
210 PRINT E;" SQR(";R;")"
220 GOTO 100
230 PRINT E * SQR (R)
240 GOTO 100
```

2.
```
100 PRINT "WHAT ARE THE FIRST COEFFICIENT AND R
    ADICAND?";: INPUT A,B
110 PRINT "WHAT ARE THE SECOND COEFFICIENT AND
    RADICAND?";: INPUT C,D
120 PRINT A;" CUB RT(";B;") * ";C;" CUB RT(";D;
    ") = ";
130 E = A * C:R = B * D
140 IF ABS (R ^ (1 / 3) - INT (R ^ (1 / 3))) <
    .05 THEN 230
150 FOR I = 2 TO SQR (R)
160 P = I * I * I
170 IF ABS (R / P - INT (R / P)) > .001 THEN
    200
180 E = E * I:R = R / P: IF R < P THEN 210
190 GOTO 170
200 NEXT I
210 PRINT E;" CUB RT(";R;")"
220 GOTO 100
230 PRINT E * R ^ (1 / 3)
240 GOTO 100
```

Page 177 (Non-Routine Problems)
$10\left(\dfrac{25 + 5\sqrt{37}}{2} - 20\right)\left(4 \cdot \dfrac{25 + 5\sqrt{37}}{2} - 20\right)$
$= [5(25 + 5\sqrt{37}) - 200] \cdot [2(25 + 5\sqrt{37}) - 20]$
$= (125 + 25\sqrt{37} - 200)(50 + 10\sqrt{37} - 20)$
$= (25\sqrt{37} - 75)(10\sqrt{37} + 30) = 250 \cdot 37 + 750\sqrt{37}$
$- 750\sqrt{37} - 2{,}250 = 9{,}250 - 2{,}250 = 7{,}000$

Page 184
2.
```
10 PRINT "WHAT ARE THE VALUES OF A,B, AND C?";:
   INPUT A,B,C
20 IF A = > 0 THEN 80
30 D = - B / A / 2
40 E = C - A * D ^ 2
50 PRINT "THE MAXIMUM VALUE OF ";E;" OCCURS WHE
   N X = ";D
60 PRINT : GOTO 10
80 PRINT "THERE IS NO MAXIMUM VALUE UNLESS A IS
    NEGATIVE.": PRINT : GOTO 10
```

Page 192
54. $|a + bi| \cdot |c + di| = \sqrt{a^2 + b^2} \cdot \sqrt{c^2 + d^2} = \sqrt{(a^2 + b^2)(c^2 + d^2)} = \sqrt{a^2c^2 + b^2c^2 + b^2d^2} = \sqrt{a^2c^2 + b^2d^2 + a^2d^2 + b^2c^2}$; $|(a + bi)(c + di)| = |ac + bdi^2 + adi + bci| = |(ac - bd) + (ad + bc)i| = \sqrt{(ac - bd)^2 + (ad + bc)^2} = \sqrt{a^2c^2 - 2abcd + b^2d^2 + a^2d^2 + 2abcd + b^2c^2} = \sqrt{a^2c^2 + b^2d^2 + a^2d^2 + b^2c^2}$

Page 200
37. $r + s = \dfrac{-b + \sqrt{b^2 - 4ac}}{2a} + \dfrac{-b - \sqrt{b^2 - 4ac}}{2a}$
$= \dfrac{-2b}{2a} = \dfrac{-b}{a}$; $rs = \dfrac{-b + \sqrt{b^2 - 4ac}}{2a} \cdot \dfrac{-b - \sqrt{b^2 - 4ac}}{2a} = \dfrac{(-b)^2 - (\sqrt{b^2 - 4ac})^2}{4a^2}$
$= \dfrac{b^2 - (b^2 - 4ac)}{4a^2} = \dfrac{4ac}{4a^2} = \dfrac{c}{a}$

40. If x is even, $ax^2 + bx$ is even. Therefore, if $ax^2 + bx + c$ were 0, c would have to be even. If x is odd, ax^2 and bx are odd and $ax^2 + bx$ is even. Again, if $ax^2 + bx + c$ were 0, c would have to be even. Since c is odd, x is neither an even number nor an odd number.

Page 202
Theorem 7 $(a + bi)(c + di) = (ac - bd) + (ad + bc)i$, (Def 6); $ac - bd = ac + [-(bd)]$, (Def 4); $ac + [-(bd)]$ and $ad + bc$ are real numbers (Closure for mult and add of real numbers). $(ac - bd) + (ad + bc)i$ is a complex number (Def 1). $(a + bi)(c + di)$ is a complex number (Substitution). **Theorem 8** $(a + bi)(c + di) = (ac - bd) + (ad + bc)i$, (Def 6); $= (ca - db) + (da +$

$cb)i$, (Comm for mult of real numbers); $= (ca - db) + (cb + da)i$, (Comm for add of real numbers); $= (c + di)(a + bi)$, (Def 6) **Theorem 9** $[(a + bi) + (c + di)] + (e + fi) = [(a + c) + (b + d)i] + (e + fi)$, (Def 3); $= [(a + c) + e] + [(b + d) + f]i$, (Def 3); $= [a + (c + e)] + [b + (d + f)]i$, (Assoc for add of real numbers); $= (a + bi) + [(c + di) + (e + fi)]$; $= (a + bi) + [(c + e) + (d + f)i]$, (Def 3); $= [a + (c + e)] + [b + (d + f)]i$, (Def 3); so, $[(a + bi) + (c + di)] + (e + fi) = (a + bi) + [(c + di) + (e + fi)]$, (Statements 3 and 5) **Theorem 10** $(a + bi)(x + yi) = 1 + 0i$, (given with $a + bi \neq 0 + 0i$); $(ax - by) + (ay + bx)i = 1 + 0i$, (Def 6); $ax - by = 1$ and $ay + bx = 0$, (Def 2); $ax = by + 1$ and $bx = -ay$, $x = \dfrac{by + 1}{a}$ and $x = \dfrac{-ay}{b}$, $\dfrac{by + 1}{a} = \dfrac{-ay}{b}$, $b^2y + b = -a^2y$, $a^2y + b^2y = -b$, $(a^2 + b^2)y = -b$ or $y = \dfrac{-b}{a^2 + b^2}$; then $x = \dfrac{-ay}{b} = \dfrac{-a}{b} \cdot y = \dfrac{-a}{b} \cdot \dfrac{-b}{a^2 + b^2} = \dfrac{a}{a^2 + b^2}$; thus, $x + yi = \dfrac{a}{a^2 + b^2} + \dfrac{-b}{a^2 + b^2} \cdot i$.

Page 211
For Exercises 14–47, the coordinates of two points are given. **14.** $(0, -1), (2, -7)$ **15.** $(0, 3), (3, 3)$ **16.** $(0, 2), (1, 5)$ **17.** $(0, -2), (2, 6)$ **18.** $(0, 3), (1, 7)$ **19.** $(0, 4), (1, 6)$ **20.** $(0, -3), (1, 5)$ **21.** $(0, -5), (5, 0)$ **22.** $(0, 3), (3, 0)$ **23.** $(0, -5), (1, -3)$ **24.** $(0, -5), (1, -7)$ **25.** $(0, -3)(1, -1)$ **26.** $(0, 0), (3, 3)$ **27.** $(0, 3), (1, 1)$ **28.** $(0, -5), (1, -3)$ **29.** $(0, 4), (1, 2)$ **30.** $(0, -3), (2, -2)$ **31.** $(0, 6), (4, 0)$ **32.** $(0, -5), (1, -8)$ **33.** $(0, 3), (3, -2)$ **34.** $(1, 3), (3, 4)$ **35.** $(1, 2), (3, 3)$ **36.** $(0, -3), (3, -7)$ **37.** $(2, 0), (5, 4)$ **38.** $(0, 24), (6, 0)$ **39.** $(-10, 0), \left(0, 6\dfrac{2}{3}\right)$ **40.** $(5, -3), (-5, 0)$ **41.** $(-3, 3), (-12, 4)$ **42.** $(1, 1), (4, 2)$ **43.** $(0, 4), (3, 2)$ **44.** $(0, -3), (1, 3)$ **45.** $(0, 3), (1, 2.5)$ **46.** $(10, 0), (0, -5.8)$ **47.** $(0, -1.4), (1, -2.6)$

Page 219
52. $y - y_1 = m(x - x_1)$; $y - 0 = \dfrac{b}{-a}(x - a)$; $y = \dfrac{-b}{a}(x - a)$; $ay = -bx + ab$; $bx + ay = ab$; $\dfrac{bx}{ab} + \dfrac{ay}{ab} = \dfrac{ab}{ab}$; $\dfrac{x}{a} + \dfrac{y}{b} = 1$

Page 222
For Exercises 1–20 and 33–38, the coordinates of two points are given. **1.** $(0, 3), (3, 5)$ **2.** $(0, -2), (3, -4)$ **3.** $(0, 1), (3, 2)$ **4.** $(0, 4), (2, 3)$ **5.** $(0, -1), (1, -3)$ **6.** $(0, 1), (1, -1)$ **7.** $(0, -2), (-1, 1)$ **8.** $(0, 3), (1, 8)$ **9.** $(0, 4), (1, 2)$ **10.** $(0, -1), (1, -4)$ **11.** $(0, -3), (1, 1)$ **12.** $(0, 5), (1, 4)$ **13.** $(0, 2), (2, -1)$ **14.** $(9, 0), (0, -3)$ **15.** $(0, 5), (3, 5)$ **16.** $(0, -3), (4, 0)$ **17.** $(0, 4), (1, 3)$ **18.** $(4, 0), (4, 4)$ **19.** $(-4, 0), (-4, 4)$ **20.** $(0, -3), (5, -5)$ **33.** $(1, 3), (-3, 4)$ **34.** $\left(0, \dfrac{11}{3}\right), \left(\dfrac{11}{3}, 0\right)$ **35.** $\left(0, -\dfrac{1}{2}\right), \left(1, -\dfrac{7}{6}\right)$ **36.** $(0, 1), (5, -3)$ **37.** $(2, 2), (5, -5)$ **38.** $(0, -10), (1, -7)$

Page 225
7. $2\sqrt{5}$ **8.** $\sqrt{89}$ **9.** $\sqrt{97}$ **10.** 15 **11.** $3\sqrt{13}$ **12.** $\sqrt{58}$ **13.** $\sqrt{149}$ **14.** 5 **15.** $\sqrt{41}$ **16.** $5\sqrt{5}$ **17.** $2\sqrt{5}$ **18.** $\sqrt{146}$ **19.** $\sqrt{10}$ **20.** $\sqrt{85}$ **21.** $2\sqrt{2}$ **22.** $\sqrt{433}$ **23.** $\dfrac{\sqrt{229}}{6}$ **24.** $\dfrac{7}{4}\sqrt{5}$ **25.** $\dfrac{\sqrt{221}}{6}$ **26.** $\dfrac{\sqrt{5}}{2}$ **27.** $\dfrac{17}{20}$ **28.** $\dfrac{\sqrt{2}}{3}$ **29.** $\sqrt{b^2 + 4a^2}$ **30.** $\sqrt{4x^2 + 12x + 9 + 4y^2}$ **31.** $\sqrt{9p^2 + 16s^2}$ **32.** a **33.** a **34.** $\dfrac{a - b + 2}{2}$ **35.** $\dfrac{\sqrt{b^2 + 16a^2}}{12}$ **36.** $\dfrac{\sqrt{625d^2 + 16c^2}}{30}$ **37.** $\dfrac{\sqrt{25b^2 + 16a^2}}{8}$ **53.** $c + b + \sqrt{d^2 + e^2} + \sqrt{(c - d + b)^2 + e^2}$ **55.** $\dfrac{\sqrt{6481}}{30} + \dfrac{7}{30}\sqrt{349} + \dfrac{3}{2}\sqrt{10}$ **56.** $\sqrt{(e - a)^2 + (f - b)^2} + \sqrt{(h - d)^2 + (g - c)^2}$ **57.** Let $A(0,0)$, $B(a, b)$, $C(a, 0)$ be the vertices of the triangle. Then $M\left(\dfrac{a}{2}, \dfrac{b}{2}\right) =$ the midpoint of $\overline{AB}$; $AM = \sqrt{\left(\dfrac{a}{2}\right)^2 + \left(\dfrac{b}{2}\right)^2} = \dfrac{1}{2}\sqrt{a^2 + b^2}$; $MC = \sqrt{\left(\dfrac{a}{2} - a\right)^2 + \left(\dfrac{b}{2}\right)^2} = \dfrac{1}{2}\sqrt{a^2 + b^2}$; $MB = \sqrt{\left(a - \dfrac{a}{2}\right)^2 + \left(b - \dfrac{b}{2}\right)^2} = \dfrac{1}{2}\sqrt{a^2 + b^2}$; thus, M is equidistant from the vertices of the triangle.

Page 227
41. $M(\overline{AC}) = (-1, 2)$; $M(\overline{BD}) = (-1, 2)$; $AM = MC = 5$; $BM = MD = 10$; thus, $\overline{AC}$ and $\overline{BD}$ bisect each other.

Page 233
1. $m(\overline{AB}) = \dfrac{5 - 2}{2 - 1} = 3$; $m(\overline{BC}) = \dfrac{6 - 5}{-1 - 2} = -\dfrac{1}{3}$; $\overline{AB} \perp \overline{BC}$ **2.** $m(\overline{AB}) = \dfrac{1 + 2}{5 + 1} = \dfrac{1}{2}$; $m(\overline{BC}) = \dfrac{5 - 1}{3 - 5} = -2$; $\overline{AB} \perp \overline{BC}$
3. $m(\overline{AB}) = \dfrac{3 + 5}{7 - 5} = 4$; $m(\overline{BC}) = \dfrac{5 - 3}{-1 - 7} = -\dfrac{1}{4}$; $\overline{AB} \perp \overline{BC}$ **4.** $m(\overline{AB}) = \dfrac{-1 - 3}{4 - 2} = -2$; $m(\overline{BC}) = \dfrac{1 + 1}{8 - 4} = \dfrac{1}{2}$; $\overline{AB} \perp \overline{BC}$
5. $m(\overline{AB}) = \dfrac{-8 - 2}{-1 - 1} = 5$; $m(\overline{BC}) = \dfrac{-3 + 8}{0 + 1} = 5$; $m(\overline{AB}) = m(\overline{BC})$ **6.** $m(\overline{AB}) = \dfrac{2 - 0}{1 - 0} = 2$;

$m(\overline{DC}) = \dfrac{5-3}{6-5} = 2$; $\overline{AB} \parallel \overline{DC}$; $m(\overline{AD}) = \dfrac{3-0}{5-0} = \dfrac{3}{5}$;
$m(\overline{BC}) = \dfrac{5-2}{6-1} = \dfrac{3}{5}$; $\overline{AD} \parallel \overline{BC}$

7. $AC = \sqrt{(5-1)^2 + (4-2)^2} = \sqrt{16+4} = 2\sqrt{5}$; midpoint D of $\overline{AB} = (5, 2)$; midpoint E of $\overline{BC} = (6, 4)$; $DE = \sqrt{(4-2)^2 + (6-5)^2} = \sqrt{4+1} = \sqrt{5}$; $DE = \dfrac{1}{2}AC$

8. $AB = \sqrt{(4-1)^2 + (7-2)^2} = \sqrt{9+25} = \sqrt{34}$; $BC = \sqrt{(4-7)^2 + (7-2)^2} = \sqrt{9+25} = \sqrt{34}$; $AB = BC$ **9.** $m(\overline{AB}) = 0$, $m(\overline{BC}) = \dfrac{b}{0}$, undefined, $\overline{AB} \perp \overline{BC}$; $m(\overline{AD}) = \dfrac{b}{0}$, undefined, $\overline{AB} \perp \overline{AD}$; $m(\overline{CD}) = 0$, $\overline{CD} \perp \overline{AD}$; $\overline{CD} \perp \overline{BC}$

10. Let $E\left(\dfrac{a}{2}, 0\right)$, $F\left(a, \dfrac{b}{2}\right)$, $G\left(\dfrac{a}{2}, b\right)$, and $H\left(0, \dfrac{b}{2}\right)$ be the midpoints of $\overline{AB}$, $\overline{BC}$, $\overline{CD}$, and $\overline{DA}$, respectively.
$AE = \sqrt{\left(0 - \dfrac{a}{2}\right)^2 + \left(\dfrac{b}{2} - 0\right)^2} = \dfrac{1}{2}\sqrt{a^2 + b^2}$;
similarly, $EF = FG = GH = \dfrac{1}{2}\sqrt{a^2 + b^2} = AE$; thus, $EFGH$ is a rhombus.   **11.** Let $E\left(\dfrac{a}{2}, 0\right)$, $F\left(\dfrac{a+b+d}{2}, \dfrac{c}{2}\right)$, $G\left(\dfrac{2b+d}{2}, c\right)$, and $H\left(\dfrac{b}{2}, \dfrac{c}{2}\right)$ be the midpoints of the sides of trapezoid $ABCD$. $m(\overline{ED}) = \dfrac{0 - \dfrac{c}{2}}{\dfrac{a}{2} - \dfrac{b}{2}} = \dfrac{-c}{a-b}$;
similarly, $m(\overline{GF}) = \dfrac{-c}{a-b}$, so $\overline{ED} \parallel \overline{GF}$; $m(\overline{EF})$
$= \dfrac{0 - \dfrac{c}{2}}{\dfrac{a}{2} - \dfrac{a+b+d}{2}} = \dfrac{c}{b+d}$; similarly, $m(\overline{HG}) = \dfrac{c}{b+d}$, so $\overline{EF} \parallel \overline{HG}$; thus, $EFGH$ is a parallelogram.

12. $m(\overline{SU}) = \dfrac{b}{a + \sqrt{a^2 - b^2}}$;
$m(\overline{TV}) = \dfrac{-b}{a - \sqrt{a^2 - b^2}}$; $\dfrac{b}{a + \sqrt{a^2 - b^2}} \cdot \dfrac{-b}{a - \sqrt{a^2 - b^2}} = \dfrac{-b^2}{a^2 - (\sqrt{a^2 - b^2})^2} = \dfrac{-b^2}{a^2 - (a^2 - b^2)} = \dfrac{-b^2}{b^2} = -1$; $\dfrac{b}{a + \sqrt{a^2 - b^2}}$ is the negative reciprocal of $\dfrac{-b}{a - \sqrt{a^2 - b^2}}$; $\overline{TV} \perp \overline{SU}$

13. $M(\overline{AC}) = \left(\dfrac{a+c}{2}, \dfrac{b+d}{2}\right)$;
$M(\overline{BD}) = \left(\dfrac{a+c}{2}, \dfrac{b+d}{2}\right)$

14. Let $D(a, 0)$, $E(b, c)$, and $F(a+b, c)$ be the midpoints of $\overline{AB}$, $\overline{AC}$, and $\overline{BC}$, respectively. $m(\overline{CD}) =$ $\dfrac{2c}{2b - a}$; $\overline{CD}$: $y = \dfrac{2c(x - 2b)}{2b - a} + 2c$; $m(\overline{BE}) = \dfrac{-c}{2a - b}$;
$\overline{BE}$: $y = \dfrac{-c}{2a - b}(x - b) + c$; $m(\overline{AF}) = \dfrac{c}{a+b}$;
$\overline{AF}$: $y = \dfrac{c(x - a - b)}{a + b} + c$.

Solving the system yields $x = \dfrac{2a + 2b}{3}$ and $y = \dfrac{2c}{3}$, which is the point of intersection of the medians;
thus, $T = \left(\dfrac{2a + 2b}{3}, \dfrac{2c}{3}\right)$.   **15.** Let the vertices of the square be $A(0, 0)$, $B(a, 0)$, $C(a, a)$, and $D(0, a)$. $m(\overline{AC}) = \dfrac{a - 0}{a - 0} = 1$; $m(\overline{BD}) = \dfrac{a - 0}{0 - a} = -1$; thus, $\overline{AC} \perp \overline{BD}$.

16. Let the vertices of the rectangle be $A(0, 0)$, $B(a, 0)$, $C(a, b)$, and $D(0, b)$. $AC = \sqrt{(a - c)^2 + (b - 0)^2} = \sqrt{a^2 + b^2}$; $BD = \sqrt{(b - 0)^2 + (0 - a)^2} = \sqrt{b^2 + a^2}$; thus, $AC = BD$.   **17.** Let the vertices of the quadrilateral be $A(0, 0)$, $B(a, 0)$, $C(b, c)$, and $D(0, d)$. Then $M_1(\overline{AB}) = \left(\dfrac{a}{2}, 0\right)$, $M_2(\overline{BC}) = \left(\dfrac{a+b}{2}, \dfrac{c}{2}\right)$, $M_3(\overline{CD}) = \left(\dfrac{b}{2}, \dfrac{c+d}{2}\right)$;
$M_4(\overline{CA}) = \left(0, \dfrac{d}{2}\right)$; $m(\overline{M_1 M_2}) = \dfrac{\dfrac{c}{2} - 0}{\dfrac{a+b}{2} - \dfrac{a}{2}} = \dfrac{c}{b}$;
$m(\overline{M_2 M_3}) = \dfrac{\dfrac{c+d}{2} - \dfrac{c}{2}}{\dfrac{b}{2} - \dfrac{a+b}{2}} = \dfrac{-d}{a}$; $m(\overline{M_3 M_4}) = \dfrac{\dfrac{d}{2} - \dfrac{c+d}{2}}{0 - \dfrac{b}{2}} = \dfrac{c}{b}$
$= m(\overline{M_1 M_2})$; $m(\overline{M_4 M_1}) = \dfrac{\dfrac{d}{2} - 0}{0 - \dfrac{a}{2}} = -\dfrac{d}{a} = m(\overline{M_2 M_3})$;
thus, $\overline{M_1 M_2} \parallel \overline{M_3 M_4}$ and $\overline{M_2 M_3} \parallel \overline{M_1 M_4}$.

18. Let the vertices of the equilateral triangle be $A(0,0)$, $B(a, 0)$, and $C\left(\dfrac{a}{2}, b\right)$. Then $M(\overline{AB}) = D\left(\dfrac{a}{2}, 0\right)$, $M(\overline{BC}) = E\left(\dfrac{3a}{4}, \dfrac{b}{2}\right)$, and $M(\overline{AC}) = \left(\dfrac{a}{4}, \dfrac{b}{2}\right)$;
$AE = \sqrt{\left(\dfrac{3a}{4}\right)^2 + \left(\dfrac{b}{2}\right)^2} = \dfrac{1}{4}\sqrt{9a^2 + 4b^2}$;
$BF = \sqrt{\left(a - \dfrac{a}{4}\right)^2 + \left(0 - \dfrac{b}{2}\right)^2} = \dfrac{1}{4}\sqrt{9a^2 + 4b^2}$;
$CD = \sqrt{b^2} = b$. But $\overline{AC} \cong \overline{AB}$, so $\sqrt{\left(\dfrac{a}{2}\right)^2 + b^2}$
$= a$, $\dfrac{a^2}{4} + b^2 = a^2$,
$b = \dfrac{a}{2}\sqrt{3}$, $CD = \dfrac{a}{2}\sqrt{3}$;
then $AE = BF = \dfrac{1}{4}\sqrt{9a^2 + 4\left(\dfrac{3}{4}a^2\right)}$
$= \dfrac{1}{4}\sqrt{12a^2} = \dfrac{a}{2}\sqrt{3}$; thus, $CD = AE = BF$.

19. $l_1 \perp l_3$, $l_2 \perp l_3$; $m(l_1) = -\dfrac{1}{m(l_3)}$; $m(l_2) = -\dfrac{1}{m(l_3)}$;
thus, $m(l_1) = m(l_2)$ and $l_1 \parallel l_2$.

T–605

Page 235

1.
```
10 PRINT "COORDINATES OF FIRST POINT ":: INPUT
   A,B
20 PRINT "SECOND POINT":: INPUT C,D
30 PRINT "THE SLOPE OF THE LINE THROUGH (";A;",
   ";B;") AND (";C;",";D;") = ";
40 IF A = C THEN 70
50 IF B = D THEN 90
60 PRINT D - B;"/";C - A: PRINT : GOTO 10
70 PRINT "UNDEFINED": PRINT : GOTO 10
90 PRINT "0": PRINT : GOTO 10
```

2.
```
10 PRINT "COORDINATES OF ONE ENDPOINT":: INPUT
   A,B
20 PRINT "OTHER ENDPOINT":: INPUT C,D
30 PRINT "THE LENGTH OF THE SEGMENT IS "; SQR (
   (D - B) ^ 2 + (C - A) ^ 2)
40 PRINT "THE MIDPOINT IS (";(A + C) / 2;",";(B
    + D) / 2;")"
50 PRINT : GOTO 10
```

3.
```
10 PRINT "COORDINATES OF THE ENDPOINT":: INPUT
   A,B
20 PRINT "MIDPOINT":: INPUT C,D
30 PRINT "COORDINATES OF THE OTHER ENDPOINT ARE
   "
40 PRINT "(";2 * C - A;",";2 * D - B;")"
50 PRINT : GOTO 10
```

Page 236

1. left 2, down 3   **2.** right 4, down 6   **3.** left 7
4. down 3   **5.** right 5, down 5   **6.** up 2   For Ex. 12–13, two points of each graph are given.   **12.** (0, 3), (6, 0)
13. $\left(0, \frac{3}{2}\right), \left(-\frac{3}{2}, 0\right)$ **40.** Since opposite sides have equal slopes $\left(3 \text{ and } -\frac{1}{3}\right)$, they are parallel. Therefore, $ABCD$ is a parallelogram.

$ABCD$ is a parallelogram.   **24.** Let the coordinates be $A(-a, 0)$, $B(a, 0)$, $C(0, b)$; midpoint of $\overline{AC}$ is $N\left(-\frac{a}{2}, \frac{b}{2}\right)$; midpoint of $\overline{BC}$ is $M\left(\frac{a}{2}, \frac{b}{2}\right)$; midpoint of $\overline{AB}$ is $O(0, 0)$; use the distance formula to show than $AN = NC = CM = MB = OM = ON$. $\triangle ABC$ contains 4 smaller isosceles triangles.

Page 237

5. Passes through (0, 5), (1, 3)   **6.** Passes through (0, 3), (1, 1)   **22.** Since opposite sides have equal slopes $\left(\frac{2}{3} \text{ and } \frac{-5}{3}\right)$, they are parallel. Therefore, $ABCD$ is a parallelogram.

Page 259

11. $\begin{bmatrix} 8 & -3 \\ -7 & -6 \end{bmatrix} + \begin{bmatrix} 0 & 0 \\ 0 & 0 \end{bmatrix} = \begin{bmatrix} 8 & -3 \\ -7 & -6 \end{bmatrix}$

12. $\begin{bmatrix} 8 & -3 \\ -7 & -6 \end{bmatrix} + \begin{bmatrix} 1 & -2 \\ 7 & -9 \end{bmatrix} = \begin{bmatrix} 9 & -5 \\ 0 & -15 \end{bmatrix}$;

$\begin{bmatrix} 1 & -2 \\ 7 & -9 \end{bmatrix} + \begin{bmatrix} 8 & -3 \\ -7 & -6 \end{bmatrix} = \begin{bmatrix} 9 & -5 \\ 0 & -15 \end{bmatrix}$

13. $\left[\begin{bmatrix} 8 & -3 \\ -7 & -6 \end{bmatrix} + \begin{bmatrix} 1 & -2 \\ 7 & -9 \end{bmatrix}\right] + \begin{bmatrix} 3 & 5 \\ -2 & -6 \end{bmatrix} =$

$\begin{bmatrix} 9 & -5 \\ 0 & -15 \end{bmatrix} + \begin{bmatrix} 3 & 5 \\ -2 & -6 \end{bmatrix} = \begin{bmatrix} 12 & 0 \\ -2 & -21 \end{bmatrix}$;

$\begin{bmatrix} 8 & -3 \\ -7 & -6 \end{bmatrix} + \left[\begin{bmatrix} 1 & -2 \\ 7 & -9 \end{bmatrix} + \begin{bmatrix} 3 & 5 \\ -2 & -6 \end{bmatrix}\right] =$

$\begin{bmatrix} 8 & -3 \\ -7 & -6 \end{bmatrix} + \begin{bmatrix} 4 & 3 \\ 5 & -15 \end{bmatrix} = \begin{bmatrix} 12 & 0 \\ -2 & -21 \end{bmatrix}$

14. $\begin{bmatrix} 1 & -2 \\ 7 & -9 \end{bmatrix} + \begin{bmatrix} 3 & 5 \\ -2 & -6 \end{bmatrix} = \begin{bmatrix} 4 & 3 \\ 5 & -15 \end{bmatrix}$;

$\begin{bmatrix} 3 & 5 \\ -2 & -6 \end{bmatrix} + \begin{bmatrix} 1 & -2 \\ 7 & -9 \end{bmatrix} = \begin{bmatrix} 4 & 3 \\ 5 & -15 \end{bmatrix}$

15. $\left[\begin{bmatrix} 1 & -2 \\ 7 & -9 \end{bmatrix} + \begin{bmatrix} 3 & 5 \\ -2 & -6 \end{bmatrix}\right] + \begin{bmatrix} 8 & -3 \\ -7 & -6 \end{bmatrix} =$

$\begin{bmatrix} 4 & 3 \\ 5 & -15 \end{bmatrix} + \begin{bmatrix} 8 & -3 \\ -7 & -6 \end{bmatrix} = \begin{bmatrix} 12 & 0 \\ -2 & -21 \end{bmatrix}$;

$\begin{bmatrix} 1 & -2 \\ 7 & -9 \end{bmatrix} + \left[\begin{bmatrix} 3 & 5 \\ -2 & -6 \end{bmatrix} + \begin{bmatrix} 8 & -3 \\ -7 & -6 \end{bmatrix}\right] =$

$\begin{bmatrix} 1 & -2 \\ 7 & -9 \end{bmatrix} + \begin{bmatrix} 11 & 2 \\ -9 & -12 \end{bmatrix} = \begin{bmatrix} 12 & 0 \\ -2 & -21 \end{bmatrix}$

16. $\left[\begin{bmatrix} 3 & 5 \\ -2 & -6 \end{bmatrix} + \begin{bmatrix} 1 & -2 \\ 7 & -9 \end{bmatrix}\right] + \begin{bmatrix} 8 & -3 \\ -7 & -6 \end{bmatrix} =$

$\begin{bmatrix} 4 & 3 \\ 5 & -15 \end{bmatrix} + \begin{bmatrix} 8 & -3 \\ -7 & -6 \end{bmatrix} = \begin{bmatrix} 12 & 0 \\ -2 & -21 \end{bmatrix}$;

$\begin{bmatrix} 3 & 5 \\ -2 & -6 \end{bmatrix} + \left[\begin{bmatrix} 1 & -2 \\ 7 & -9 \end{bmatrix} + \begin{bmatrix} 8 & -3 \\ -7 & -6 \end{bmatrix}\right] =$

$\begin{bmatrix} 3 & 5 \\ -2 & -6 \end{bmatrix} + \begin{bmatrix} 9 & -5 \\ 0 & -15 \end{bmatrix} = \begin{bmatrix} 12 & 0 \\ -2 & -21 \end{bmatrix}$

Page 263

5. $\begin{bmatrix} 6 & -9 & -15 \\ 18 & -30 & -18 \\ -24 & -36 & 12 \end{bmatrix}$ **6.** $\begin{bmatrix} -72 & -18 \\ 45 & -9 \\ 81 & 27 \end{bmatrix}$

7. $\begin{bmatrix} -7 & -6 & -9 \\ -8 & 1 & 5 \end{bmatrix}$ **8.** $\begin{bmatrix} -9 & 12 \\ 19 & -26 \end{bmatrix}$

9. $\begin{bmatrix} 5 & 1 & 8 \\ 11 & 20 & 0 \end{bmatrix}$ **10.** not possible   **11.** $\begin{bmatrix} 8 & -8 & 0 \\ -4 & 6 & -2 \end{bmatrix}$

12. Identity $= \begin{bmatrix} 1 & 0 \\ 0 & 1 \end{bmatrix}$; Inverse $= \begin{bmatrix} \frac{5}{2} & \frac{7}{2} \\ 2 & 3 \end{bmatrix}$

13. Identity $= \begin{bmatrix} 1 & 0 \\ 0 & 1 \end{bmatrix}$; Inverse $= \begin{bmatrix} -\frac{1}{20} & -\frac{1}{10} \\ \frac{3}{10} & -\frac{2}{5} \end{bmatrix}$

14. not possible   **15.** Identity $= \begin{bmatrix} 1 & 0 & 0 \\ 0 & 1 & 0 \\ 0 & 0 & 1 \end{bmatrix}$;

Inverse = $\begin{bmatrix} -\frac{2}{27} & -\frac{8}{27} & \frac{11}{54} \\ -\frac{19}{27} & -\frac{22}{27} & \frac{37}{54} \\ -\frac{8}{27} & -\frac{5}{27} & \frac{17}{54} \end{bmatrix}$

16. $\left[\begin{bmatrix} -3 & 1 \\ -2 & -4 \end{bmatrix} \cdot \begin{bmatrix} 1 & -6 \\ 3 & -2 \end{bmatrix}\right] \cdot \begin{bmatrix} 7 & -3 \\ 2 & 5 \end{bmatrix} =$
$\begin{bmatrix} 0 & 16 \\ -14 & 20 \end{bmatrix} \begin{bmatrix} 7 & -3 \\ 2 & 5 \end{bmatrix} = \begin{bmatrix} 32 & 80 \\ -58 & 142 \end{bmatrix}$;

$\begin{bmatrix} -3 & 1 \\ -2 & -4 \end{bmatrix} \left[\begin{bmatrix} 1 & -6 \\ 3 & -2 \end{bmatrix} \cdot \begin{bmatrix} 7 & -3 \\ 2 & 5 \end{bmatrix}\right] =$
$\begin{bmatrix} -3 & 1 \\ -2 & -4 \end{bmatrix} \begin{bmatrix} -5 & -33 \\ 17 & -19 \end{bmatrix} = \begin{bmatrix} 32 & 80 \\ -58 & 142 \end{bmatrix}$

17. $\begin{bmatrix} 1 & -6 \\ 3 & -2 \end{bmatrix} \cdot \begin{bmatrix} 7 & -3 \\ 2 & 5 \end{bmatrix} = \begin{bmatrix} -5 & -33 \\ 17 & -19 \end{bmatrix}$;

$\begin{bmatrix} 7 & -3 \\ 2 & 5 \end{bmatrix} \cdot \begin{bmatrix} 1 & -6 \\ 3 & -2 \end{bmatrix} = \begin{bmatrix} -2 & -36 \\ 17 & -22 \end{bmatrix}$;

$\therefore B \cdot C \neq C \cdot B$ **18.** $\begin{bmatrix} -3 & 1 \\ -2 & -4 \end{bmatrix} \cdot \begin{bmatrix} 1 & -6 \\ 3 & -2 \end{bmatrix} =$
$\begin{bmatrix} 0 & 16 \\ -14 & 20 \end{bmatrix}$; $\begin{bmatrix} 1 & -6 \\ 3 & -2 \end{bmatrix} \cdot \begin{bmatrix} -3 & 1 \\ -2 & -4 \end{bmatrix} =$
$\begin{bmatrix} 9 & 25 \\ -5 & 11 \end{bmatrix}$; $\therefore A \cdot B \neq B \cdot A$ **19.** $\begin{bmatrix} -3 & 1 \\ -2 & -4 \end{bmatrix} \cdot$
$\left[\begin{bmatrix} 1 & -6 \\ 3 & -2 \end{bmatrix} + \begin{bmatrix} 7 & -3 \\ 2 & 5 \end{bmatrix}\right] = \begin{bmatrix} -3 & 1 \\ -2 & -4 \end{bmatrix} \cdot$
$\begin{bmatrix} 8 & -9 \\ 5 & 3 \end{bmatrix} = \begin{bmatrix} -19 & 30 \\ -36 & 6 \end{bmatrix}$; $\begin{bmatrix} -3 & 1 \\ -2 & -4 \end{bmatrix} \cdot$
$\begin{bmatrix} 1 & -6 \\ 3 & -2 \end{bmatrix} + \begin{bmatrix} -3 & 1 \\ -2 & -4 \end{bmatrix} \cdot \begin{bmatrix} 7 & -3 \\ 2 & 5 \end{bmatrix} =$
$\begin{bmatrix} 0 & 16 \\ -14 & 20 \end{bmatrix} + \begin{bmatrix} -19 & 14 \\ -22 & -14 \end{bmatrix} = \begin{bmatrix} -19 & 30 \\ -36 & 6 \end{bmatrix}$

$\therefore A \cdot (B + C) = A \cdot B + A \cdot C$

20. $\begin{bmatrix} -3 & 1 \\ -2 & -4 \end{bmatrix} \begin{bmatrix} 1 & 0 \\ 0 & 1 \end{bmatrix} = \begin{bmatrix} -3 & 1 \\ -2 & -4 \end{bmatrix}$;

$\begin{bmatrix} 1 & 0 \\ 0 & 1 \end{bmatrix} \begin{bmatrix} -3 & 1 \\ -2 & -4 \end{bmatrix} = \begin{bmatrix} -3 & 1 \\ -2 & -4 \end{bmatrix}$; $\therefore A \cdot I = I \cdot A$

21. $\begin{bmatrix} 1 & -6 \\ 3 & -2 \end{bmatrix} \cdot \begin{bmatrix} -\frac{1}{8} & \frac{3}{8} \\ -\frac{3}{16} & \frac{1}{16} \end{bmatrix} = \begin{bmatrix} 1 & 0 \\ 0 & 1 \end{bmatrix}$

Page 267

23. Let $A^{-1} = \begin{bmatrix} x_1 & y_1 \\ x_2 & y_2 \end{bmatrix}$; then $AA^{-1} = I$ or $\begin{bmatrix} a_1 & b_1 \\ a_2 & b_2 \end{bmatrix} \cdot$
$\begin{bmatrix} x_1 & y_1 \\ x_2 & y_2 \end{bmatrix} = \begin{bmatrix} 1 & 0 \\ 0 & 1 \end{bmatrix}$; $\begin{bmatrix} a_1x_1 + b_1x_2 & a_1y_1 + b_1y_2 \\ a_2x_1 + b_2x_2 & a_2y_1 + b_2y_2 \end{bmatrix} = \begin{bmatrix} 1 & 0 \\ 0 & 1 \end{bmatrix}$,
which yields the systems $\begin{array}{l} a_1x_1 + b_1x_2 = 1 \\ a_2x_1 + b_2x_2 = 0 \end{array}$ and

$\begin{array}{l} a_1y_1 + b_1y_2 = 0 \\ a_2y_1 + b_2y_2 = 1 \end{array}$; $x_1 = \frac{b_2}{a_1b_2 - a_2b_1}$, $x_2 = \frac{-a_2}{a_1b_2 - a_2b_1}$,
$y_1 = \frac{-b_1}{a_1b_2 - a_2b_1}$, $y_2 = \frac{a_1}{a_1b_2 - a_2b_1}$; $a_1b_2 - a_2b_1 =$
$\begin{vmatrix} a_1 & b_1 \\ a_2 & b_2 \end{vmatrix}$, the determinant of A. Thus, $A^{-1} =$
$\frac{1}{\text{Determinant of } A} \begin{bmatrix} b_2 & -b_1 \\ -a_2 & a_1 \end{bmatrix}$.

Page 272

10. $x = \frac{e_1d_2 - e_2d_1}{c_1d_2 - c_2d_1}$; $y = \frac{e_1c_2 - c_1e_2}{c_2d_1 - c_1d_2}$

22. $\begin{bmatrix} 5 & 1 & 5 \\ -1 & 7 & 4 \\ 6 & -3 & 2 \end{bmatrix}$ **23.** $\begin{bmatrix} -9 & 12 & 2 \\ 11 & -12 & 13 \end{bmatrix}$

24. $\begin{bmatrix} 1 & 0 \\ 0 & 1 \end{bmatrix}$; $\begin{bmatrix} -7 & 3 \\ -2 & -1 \end{bmatrix}$; $\begin{bmatrix} \frac{1}{13} & \frac{3}{13} \\ -\frac{2}{13} & \frac{7}{13} \end{bmatrix}$

25. $\begin{bmatrix} 1 & 0 & 0 \\ 0 & 1 & 0 \\ 0 & 0 & 1 \end{bmatrix}$; $\begin{bmatrix} -3 & 1 & -2 \\ -2 & -1 & -2 \\ 3 & -2 & 1 \end{bmatrix}$;

$\begin{bmatrix} -\frac{5}{3} & 1 & -\frac{4}{3} \\ -\frac{4}{3} & 1 & -\frac{2}{3} \\ \frac{7}{3} & -1 & \frac{5}{3} \end{bmatrix}$

Page 278

1. $D = \{3, 2, -1, -2, 0\}$; $R = \{1, -3, 5, -2, 2\}$; yes
2. $D = \{8, 7, -7, -8, 5\}$; $R = \{7, 8, -2\}$; yes
3. $D = \{0, 8, -3, 4, 5\}$; $R = \{0, 7, 6, -1\}$; yes
4. $D = \{3, 4, 5, -3, -4\}$; $R = \{-3, -4, -5, 3, 4\}$; yes
5. $D = \{3, 4, -4, -3, 2\}$; $R = \{2, -2, -3\}$; yes
6. $D = \{0, 3, 1, 2\}$; $R = \{1, 4, -3\}$; No
7. $D = \{1, -1, 2, -2, 3\}$; $R = \{-2, 0, -4, 3\}$; yes
8. $D = \{9, 7, 8, -9, -8\}$; $R = \{7, 8, 9\}$; yes
9. $D = \{4, -1, -3\}$; $R = \{-1, 4, 3, -3, 0\}$; No
10. $D = \{6, -3, -6\}$; $R = \{-3, 3, 5, 6\}$; No
11. $D = \{-3, -1, 2, 3\}$; $R = \{2, -2, -1, 1\}$;
12. $D = \{-2, 0, 2, 4\}$; $R = \{2, -2, 0, 1\}$;
13. $D = \{-3, -2, -1, 0, 1, 2, 3\}$; $R = \{-3, -2, -1, 0\}$
14. $D = \{x \mid x \text{ is a real number}\}$; $R = \{y \mid 1 \leq y\}$; yes
15. $D = \{x \mid x \leq 3\}$; $R = \{y \mid y \text{ is real}\}$; no
16. $D = \{x \mid -3 \leq x \leq 3\}$; $R = \{y \mid -3 \leq y \leq 3\}$; No
17. $D = \{x \mid -3 \leq x < 2\}$; $R = \{y \mid -2 \leq y < 2\}$; yes
18. $D = \{x \mid x \text{ is real}\}$; $R = \{y \mid y \text{ is real}\}$; yes
19. $D = \{x \mid 0 \leq x \leq 4\}$; $R = \{y \mid -2 \leq y \leq 2\}$; No
20. $D = \{x \mid -4 \leq x < 4\}$; $R = \{y \mid y \text{ is an integer and } -2 \leq y \leq 2\}$; yes **21.** $D = \{x \mid 0 \leq x\}$; $R = \{y \mid y \text{ is real}\}$; No **22.** $D = \{x \mid x \text{ is any real number except } 0, -2\}$; $R = \{y \mid y \text{ is real and not equal to } f(-2) \text{ and not equal to}$

0}; yes   **23.** D = {all even integers}; R = {all integers}; yes   **24.** D = {all natural numbers > 1}; R = {squares of all natural numbers > 1}   **25.** {squares of all natural numbers > 1}; R = {all natural numbers > 1}; yes   **26.** D = {all integers}; R = {$y|y = 2x + 3$}; yes

Page 281
31. D = {all real numbers except 0}; none
32. D = {all real numbers except 0 and -2}; 1
33. D = {all real numbers except 3 and 4}; 2
34. D = {all real numbers except -1 and -3}; 0, $\frac{3}{4}$
35. D = {all real numbers except -6 and 2}; none
36. D = {all real numbers except 7 and -2}; 0

Page 284
25. darkened circle at (0, 0) and all points to the left; open circle at (0, -1) and all points to the right
26. darkened circle at (0, 0); a line from (0, 0) to the right with slope -1; a line from (0, 0) to the left with slope 1   **27.** all points right of and including $x = 2$, 2 units above x axis; all points including origin and between origin and $x = 2$, 1 unit above x axis; all points left of origin, 1 unit below x axis.   **28.** include pts: (3, 3), (2, 2), (1, 1), (0, 0), ($-1, -1$), ($-2, -2$), ($-3, -3$); move horizontally 1 unit to the right of all these points except (3, 3)   **29.** include pts: (-3, 3), (-2, 2), (-1, 1), (0, 0), (1, -1), (2, -2), (3, -3); move horizontally one unit to right of all these points except (3, -3)   **30.** include pts: (4, 6), (3, 5), (2, 4), (1, 3), (0, 2), (-1, 1), (-2, 0), (-3, -1), (-4, -2); move horizontally 1 unit to the right of all these points except (4, 6)   **31.** include points: (-4, -6), (-3, -5), (-2, -4), (-1, -3), (0, -2), (1, -1), (2, 0), (3, 1), (4, 2); move horizontally one unit to the right of all these points except (4, 2)   **33.** discontinuous x-axis, open only where irrationals would appear; discontinuous horizontal line at $y = 1$, open only where rationals would appear.

Page 287
28. The segment $\overline{PQ}$ has slope $\frac{y - x}{x - y} = -1$ and is thus $\perp$ to the line $y = x$, which has slope 1. The midpoint of $\overline{PQ}$ is $\left(\frac{x + y}{2}, \frac{x + y}{2}\right)$; this lies on the line $y = x$, so $y = x$ is the $\perp$ bisector of $\overline{PQ}$.

Non-Routine Problems
Let $(x, f(x))$, $(x + 1, f(x + 1))$, and $(x + 2, f(x + 2))$ be points on the graph of the linear function; slopes of segments joining any two of these points are the same; $\frac{f(x + 2) - f(x + 1)}{(x + 2) - (x + 1)} = \frac{f(x + 1) - f(x)}{x + 1 - x}$; $f(x + 2) - f(x + 1) = f(x + 1) - f(x)$

Page 300
1.
```
10 DEF FN A(X) = 9 * X * X * X * X + 8 * X * X
   * X + 7 * X * X + 6 * X + 5
20 DEF FN B(X) = 4 * X * X * X * X * X + 4 * X
   * X * X + 4 * X
30 DEF FN C(X) = - X * X * X * X * X - 2 * X *
   X * X * X - 3 * X * X * X - 4 * X * X - 5 *
   X - 6
40 DEF FN D(X) = - 7 * X * X * X * X * X * X -
   7 * X * X * X * X - 7 * X * X - 7
50 INPUT X
60 PRINT "FNA(";X;")= "; FN A(X)
70 PRINT "FNB(";X;")= "; FN B(X)
80 PRINT "FNC(";X;")= "; FN C(X)
90 PRINT "FND(";X;")= "; FN D(X)
95 PRINT : GOTO 50
```

Page 318
For Ex. 1–12, the directions for moving the basic graph, $y = x^2$, are given.   **1.** 1 left   **2.** 3 right   **3.** 1 right   **4.** 3 left   **5.** 1 up   **6.** 2 down   **7.** 4 down   **8.** 3 up   **9.** 2 right, 1 up   **10.** 2 left, 1 down   **11.** 1 right, 2 down   **12.** 1 right, 2 up
44. The vertex lies on the axis of symmetry, so its x-coordinate is $-\frac{b}{2a}$; substitute $-\frac{b}{2a}$ for x to find y: $y = a\left(-\frac{b}{2a}\right)^2 + b\left(-\frac{b}{2a}\right) + c$; $y = \frac{ab^2}{4a^2} - \frac{b^2}{2a} + c$; $y = -\frac{b^2}{4a} + c$; $y = \frac{4ac - b^2}{4a}$

Page 323
49. $C(1, -2)$; move the graph of $\frac{x^2}{16} - \frac{y^2}{9} = 1$ one unit right and 2 units down.   **50.** $C(-2, -3)$; move the graph of $\frac{x^2}{25} - \frac{y^2}{4} = 1$ two units left and 3 units down.   **51.** $C(2, 3)$; move the graph of $\frac{x^2}{9} - \frac{y^2}{4} = 1$ two units right and 3 units up.

Page 330
Ex. 9–16. The graphs are parabolas with vertex at the origin. Two other points are given.   **9.** (2, -1), ($-2, -1$)   **10.** (-3, 3), ($-3, -3$)   **11.** (4, 4), (4, -4)   **12.** (10, 10), (-10, 10)   **13.** (20, 20), (20, -20)   **14.** (6, 6), (-6, 6)   **15.** (3, -3), ($-3, -3$)   **16.** (-2, 2), ($-2, -2$)   **31.** The chord through any point (0, x) on the major axis has length $2y = \frac{2b\sqrt{a^2 - x^2}}{a}$. When $x = c$, $y = \frac{2b\sqrt{a^2 - c^2}}{a} = \frac{2b\sqrt{b^2}}{a} = \frac{2b^2}{a}$ **32.** $\sqrt{(x + c)^2 + y^2} - \sqrt{(x - c)^2 + y^2} = 2a$; $x^2 + 2xc + c^2 + y^2 = 4a^2$

$+ 4a\sqrt{(x-c)^2 + y^2} + x^2 - 2xc + c^2 + y^2$;
$4xc - 4a^2 = 4a\sqrt{(x-c)^2 + y^2}$; $x^2c^2 - 2xca^2 + a^4 = a^2(x^2 - 2cx + c^2 + y^2)$; $x^2c^2 - a^2x^2 - a^2y^2 = a^2c^2 - a^4$; $x^2(c^2 - a^2) - a^2y^2 = a^2(c^2 - a^2)$;
$\dfrac{x^2}{a^2} - \dfrac{y^2}{c^2 - a^2} = 1$; let $b^2 = c^2 - a^2$; $\dfrac{x^2}{a^2} - \dfrac{y^2}{b^2} = 1$

Page 331
2.
```
100 PRINT "VALUE OF A";: INPUT A
110 IF ABS (A) > 10 THEN 100
120 PRINT "VALUE OF B";: INPUT B
130 IF ABS (B) > 10 THEN 100
140 FOR Y = 12 TO -10 STEP -1
150 IF Y < > 0 THEN 170
160 FOR I = 1 TO 39: PRINT "*";: NEXT I: GOTO 3
    20
170 Y1 = B - Y: IF Y1 < 0 THEN 310
180 A1 = - A - INT ( SQR (Y1)):A2 = - A + INT
    ( SQR (Y1))
190 IF A1 < 0 THEN 210
200 C = 0:D = A1:E = A2: GOTO 240
210 IF A2 < 0 THEN 230
220 C = A1:D = 0:E = A2: GOTO 240
230 C = A1:D = A2:E = 0
240 IF C < 19 THEN 320
250 PRINT  TAB( 20 + C);"*";
260 IF D = C THEN 280
270 PRINT  TAB( 20 + D);"*";
280 IF E > 19 THEN 320
290 IF D = E THEN 320
300 PRINT  TAB( 20 + E);"*";: GOTO 320
310 PRINT  TAB( 20);"*";
320 PRINT : NEXT Y: GOTO 100
```

Page 332
1. origin **2.** y-axis **3.** x-axis **4.** none of these **5.** $(-5, -1)$; $(5, 1)$; $(5, -1)$ **6.** $(7, 2)$; $(-7, -2)$; $(-7, 2)$ **7.** $(-3, 8)$; $(3, -8)$; $(3, 8)$
8. $\left(-\dfrac{1}{7}, \dfrac{1}{8}\right); \left(\dfrac{1}{7}, -\dfrac{1}{8}\right); \left(\dfrac{1}{7}, \dfrac{1}{8}\right)$
9. (m, n); $(-m, -n)$; $(-m, n)$ **10.** $(-r, s)$; $(r, -s)$; (r, s) **13.** $x^2 + y^2 = 169$ **14.** $x^2 + y^2 = 289$
15. $x^2 + y^2 = 8$ **16.** $(x + 2)^2 + (y - 3)^2 = 25$
17. $(x - 3)^2 + (y + 2)^2 = 16$
18. $(x + 6)^2 + (y + 5)^2 = 137$
19. $(x - 1)^2 + (y + 8)^2 = 8$ **20.** $C(0, 0)$, $r = 7$
21. $C(4, -3)$, $r = \sqrt{14}$ **22.** $C(1, -2)$, $r = 4$
23. $a = \pm 6$, $b = \pm 2$; $(6, 0)$, $(-6, 0)$, $(0, 2)$, $(0, -2)$
24. $a = \pm 4$, $b = \pm 2$; $(3, -3)$, $(3, -7)$, $(7, 5)$, $(-1, 5)$
25. $a = \pm 2$, $b = \pm 6$; $(2, 0)$, $(-2, 0)$, $(0, 6)$, $(0, -6)$
28. $(0, 12)$, $(6, 12)$; $x = 3$; $V(3, 3)$ **29.** $(3, -3)$, $(-3, -3)$; $x = 0$; $V(0, 0)$ **30.** $(0, 1)$, $(2, -3)$, $(3, -2)$; move the basic graph of $y = x^2$ two units to the right and 3 units down. **31.** $(0, 5)$, $(1, 8)$, $(-1, 4)$; move the basic graph of $y = x^2$ one unit to the left and 4 units up. **36.** x-intercepts: ± 3, $C(2, -1)$; move the basic graph of $\dfrac{x^2}{9} - \dfrac{y^2}{4} = 1$ two units to the right and 1 unit down. **37.** x-intercepts: $\pm\dfrac{2}{3}\sqrt{17}$, $C(1, 2)$; move the basic graph of $\dfrac{x^2}{\frac{68}{9}} - \dfrac{y^2}{17} = 1$ one unit to the right and 2 units up.

Page 333
13. $a = \pm 4$, $b = \pm 2$, $C(-2, 1)$; move the basic graph of $\dfrac{x^2}{16} + \dfrac{y^2}{4} = 1$ two units to the left and 1 unit up.

Page 336
30. $(0, -4)$, $(2, -1)$, $(-2, -7)$ **31.** $(0, 6)$, $(4, 9)$, $(-4, 3)$ **35.** Midpt of $\overline{AB}$: $\left(\dfrac{2 + 5}{2}, \dfrac{-3 + 6}{2}\right)$; $M\left(\dfrac{7}{2}, \dfrac{3}{2}\right)$; Midpt of $\overline{BC}$: $\left(\dfrac{5 + 3}{2}, \dfrac{6 + 2}{2}\right)$; $N(4, 4)$; Midpt of $\overline{CD}$: $\left(\dfrac{3 - 2}{2}, \dfrac{2 - 1}{2}\right)$; $R\left(\dfrac{1}{2}, \dfrac{1}{2}\right)$; Midpt of $\overline{AD}$: $\left(\dfrac{2 - 2}{2}, \dfrac{-3 - 1}{2}\right)$; $S(0, -2)$; m of $\overline{MN}$: $\dfrac{4 - \frac{3}{2}}{4 - \frac{7}{2}} = \dfrac{\frac{5}{2}}{\frac{1}{2}} = 5$; m of $\overline{NR}$: $\dfrac{4 - \frac{1}{2}}{4 - \frac{1}{2}} = 1$; m of $\overline{RS}$: $\dfrac{-2 - \frac{1}{2}}{0 - \frac{1}{2}} = \dfrac{-\frac{5}{2}}{-\frac{1}{2}} = 5$; m of $\overline{MS}$: $\dfrac{-2 - \frac{3}{2}}{0 - \frac{7}{2}}$; $\dfrac{-\frac{7}{2}}{-\frac{7}{2}} = 1$; $ABCD$ is a parallelogram since the opposite sides have the same slopes. **44.** $y = 2\left(x + \dfrac{3}{2}\right)^2 - \dfrac{11}{2}$; move $y = 2x^2$ $\dfrac{3}{2}$ units left and $\dfrac{11}{2}$ down. **46.** hyperbola with $C(0, 0)$ and x-intercepts ± 3 **47.** Move $y = x^2$ two units left and 3 units down. **48.** Move ellipse $\dfrac{x^2}{9} + \dfrac{y^2}{4} = 1$ one unit right and 2 units down. **49.** Move circle $x^2 + y^2 = 9$ two units left and 2 units up.

Page 340
17. $\left(\dfrac{13 \pm 2\sqrt{46}}{3}, \dfrac{-5 \mp \sqrt{46}}{3}\right)$
24. $\left(\dfrac{16 + 12\sqrt{6}}{19}, \dfrac{-3 + 12\sqrt{6}}{19}\right)$, $\left(\dfrac{16 - 12\sqrt{6}}{19}, \dfrac{-3 - 12\sqrt{6}}{19}\right)$
26. $\left(\dfrac{-2 + \sqrt{46}}{2}, \dfrac{8 - \sqrt{46}}{2}\right)$, $\left(\dfrac{-2 - \sqrt{46}}{2}, \dfrac{8 + \sqrt{46}}{2}\right)$

Page 348

(For answers to odd-numbered exercises, see PE page 614.)
2. above, including the line containing $(0, -2)$, $(2, 1)$
4. below the line containing $(0, -5)$, $(3, 0)$
6. below the line containing $(0, -5)$, $(7, -3)$
8. above, including the line containing $(0, -6)$, $(2, -11)$
10. below the line containing $(0, -2)$, $(4, -7)$
12. below the line containing $(0, -7)$, $(1, -5)$
14. above, including the line containing $(0, 4)$, $(6, 4)$
16. left of the line containing $(5, 1)$, $(5, 5)$

(For Ex. 18–36, the coordinates of two points of each line are given. The points of a given line may or may not be part of the graph.)
18. below, including the line containing $(0, -1)$, $(1, 2)$ *and* above the line containing $(0, 4)$, $(1, 2)$
20. above, including the line containing $(0, 6)$, $(2, 1)$ *and* below, including the line containing $(0, 2)$, $(4, -1)$
22. below the line containing $(0, 5)$, $(6, 0)$ *and* above, including the line containing $(0, 3)$, $(3, -1)$
24. above, including the line containing $(0, -3)$, $(3, -5)$ *and* above, including the line containing $(0, 2)$, $(3, -2)$
26. below the line containing $(0, -8)$, $(2, -7)$ *and* below, including the line containing $(0, 2)$, $(2, 1)$
28. below, including the line containing $(0, -6)$, $(-1, -3)$ *and* above the line containing $(0, 1)$, $(1, 3)$
30. below, including the line containing $(0, 2)$, $(4, 2)$ *and* left, including the line containing $(4, 0)$, $(4, 5)$
32. above, including the line containing $(0, 5)$, $(5, 5)$ *and* right, including the line containing $(-3, 0)$, $(-3, 6)$
34. below, including the line containing $(0, -8)$, $(1, -6)$ *and* below the line containing $(0, 2)$, $(3, 1)$ *and* above, including the line containing $(0, 2)$, $(2, 1)$
36. above, including the line containing $(0, -1)$, $(1, 1)$ *and* above, including the line containing $(0, 1)$, $(1, 5)$ *and* above, including the line containing $(0, 3)$, $(1, 5)$

Page 351

(For answers to odd-numbered exercises, see PE page 615.)
2. outside, including the circle containing $(2, 0)$, $(0, 2)$
4. outside the circle containing $(4, 0)$, $(0, 4)$
6. outside the two branches of the hyperbola containing $(8, 1)$, $(-8, -1)$, asymptotes: $x = 0$, $y = 0$
8. outside, including the two branches of the hyperbola containing $(-1, 2)$, $(1, -2)$, asymptotes: $x = 0$, $y = 0$
10. outside, including the ellipse containing $(6, 0)$, $(0, 4)$
12. outside the ellipse containing $(4, 0)$, $(0, 1)$
14. inside, including the two branches of the hyperbola containing $(5, 0)$, $(-5, 0)$, asymptotes: $y = \frac{\pm 2\sqrt{2}}{5}x$
16. inside the two branches of the hyperbola containing $(5, 0)$, $(-5, 0)$, asymptotes: $y = \pm\frac{9}{5}x$
18. inside, including the ellipse containing $(3, 0)$, $(0, 1.7)$
20. inside, including the two branches of the hyperbola containing $(2, 0)$, $(-2, 0)$, asymptotes: $y = \pm\sqrt{2}x$
22. inside the two branches of the hyperbola containing $(2, 0)$, $(-2, 0)$, asymptotes:

$y = \pm\frac{5}{2}x$
24. inside the ellipse containing $(2, 0)$, $(0, 5)$

(For Ex. 26–40, the coordinates of two points of each line or curve are given. The points of a given line or curve may or may not be part of the graph.)
26. outside the circle containing $(5, 0)$, $(0, 5)$ *and* above the line containing $(0, 3)$, $(4, 6)$
28. outside the two branches of the hyperbola containing $(1, -5)$, $(-1, 5)$, asymptotes: $x = 0$, $y = 0$ *and* above, including the line containing $(0, -2)$, $(3, -4)$
30. outside, including the two branches of the hyperbola containing $(2.8, 0)$, $(-2.8, 0)$, asymptotes: $y = \pm\frac{\sqrt{2}}{2}x$ *and* below the line containing $(0, -2)$, $(1, 1)$
32. inside, including the two branches of the hyperbola containing $(5.1, 0)$, $(-5.1, 0)$, asymptotes: $y = \pm\frac{4}{3}x$ *and* below, including the line containing $(-1, 1)$, $(-5, 2)$
34. outside, including the two branches of the hyperbola containing $(3.4, 0)$, $(-3.4, 0)$, asymptotes: $y = \pm x$ *and* inside, including the circle containing $(6, 0)$, $(0, 6)$
36. outside, including the two branches of the hyperbola containing $(5.1, 0)$, $(-5.1, 0)$, asymptotes: $y = \pm\frac{1}{3}x$ *and* inside, including the ellipse containing $(2, 0)$, $(0, 5)$
38. outside, including the two branches of the hyperbola containing $(0, 3)$, $(0, -3)$, asymptotes: $y = \pm\frac{\sqrt{3}}{2}x$ *and* outside, including the ellipse containing $(5, 0)$, $(0, 2.2)$
40. inside, including the ellipse containing $(2.4, 0)$, $(0, 1.7)$, *and* inside, including the two branches of the hyperbola containing $(2.2, 0)$, $(-2.2, 0)$, asymptotes: $y = \pm\frac{\sqrt{15}}{5}x$

Page 352

(For answers to odd-numbered Ex. 13–31, see PE page 615.)
14. below the line containing $(0, 5)$, $(1, 2)$
16. above the line containing $(0, -3)$, $(5, -3)$
18. inside the two branches of the hyperbola containing $(4, 1)$, $(-4, -1)$, asymptotes: $x = 0$, $y = 0$
20. outside, including the ellipse containing $(2.4, 0)$, $(0, 1.7)$
22. outside the ellipse containing $(2, 0)$, $(0, 3)$
24. above the line containing $(0, -2)$, $(2, 3)$ *and* below, including the line containing $(1, 1)$, $(5, 5)$
26. above the line containing $\left(0, \frac{7}{3}\right)$, $\left(2, \frac{1}{3}\right)$ *and* above, including the line containing $(0, -5)$, $(2, -10)$
28. inside, including the circle containing $(5, 0)$, $(0, 5)$ *and* above the line containing $(0, 3)$, $(2, 4)$
30. outside, including the ellipse containing $(6, 0)$, $(0, 2)$ *and* below, including the line containing $(0, 4)$, $(2, 7)$
32. outside, including the ellipse containing $(3.9, 0)$, $(0, 3.4)$ *and* outside, including the two branches of the hyperbola containing $(3, 0)$, $(-3, 0)$, asymptotes: $y = \pm\frac{\sqrt{6}}{3}x$

Page 353

10. below, including the line containing $(0, -3)$, $(3, -1)$

12. inside, including the two branches of the hyperbola containing $(2.4, 0)$, $(-2.4, 0)$, asymptotes: $y = \pm x$
14. inside, including the ellipse containing $(6, 0)$, $(0, 2)$
16. above, including the line containing $(0, 2)$, $(4, 5)$ *and* below, including the line containing $(0, 2)$, $(1, 3)$
18. inside, including the two branches of the hyperbola containing $(2, 0)$, $(-2, 0)$, asymptotes: $y = \pm x$ *and* above, including the line containing $(0, -2)$, $(3, 0)$ **20.** outside, including the two branches of the hyperbola containing $(4, 1)$, $(-4, -1)$, asymptotes: $x = 0$, $y = 0$ *and* above, including the line containing $(1, -1)$, $(2, -2)$ **22.** above, including the line containing $(0, 4)$, $(1, 1)$ *and* above, including the line containing $(0, -3)$, $(2, -2)$ *and* below the line containing $(0, 3)$, $(3, 5)$

Page 354
1.
```
110 R =  INT ( RND (1) * 50) + 1
120 PRINT "THE EQUATION OF A CIRCLE IS"
130 PRINT " X^2 + Y^2 = ";R
140 PRINT
150 PRINT "INPUT THE COORDINATES OF A POINT AND I"
160 PRINT "WILL TELL YOU ITS LOCATION WITH"
170 PRINT "RESPECT TO THE CIRCLE."
180 INPUT A,B
190 PRINT "THE POINT (";A;",";B") IS ";
200 IF A ^ 2 + B ^ W = R THEN  PRINT "ON THE CIRCLE
    ."
210 IF A ^ 2 + B ^ 2 > R THEN  PRINT "OUTSIDE THE C
    IRCLE."
220 IF A ^ 2 + B ^ 2 < R THEN  PRINT "INSIDE THE CI
    RCLE."
225 FOR P = 1 TO 2500: NEXT P
230 HOME : INPUT "MORE COORDINATES? (ENTER 1 FOR YE
    S, 2 FOR NO.)";K
240 IF K = 1 THEN 110
250 END
```

2.
```
100 R =  INT ( RND (1) * 25) + 1
110 R2 =  INT ( RND (1) * 36) + R + 1: PRINT
120 PRINT "THE EQUATION OF ONE CIRCLE IS"
130 PRINT " X ^ 2 + Y ^ 2 = ":4
140 PRINT : PRINT "THE EQUATION OF ANOTHER CIRC
    LE IS"
150 PRINT "X^2 + Y^2 = ";R2
160 PRINT : PRINT "INPUT THE COORDINATES OF A P
    OINT AND I"
170 PRINT "WILL TELL YOU ITS LOCATION WITH"
180 PRINT "RESPECT TO THE TWO CIRCLES."
190 INPUT A,B
200 PRINT "THE POINT (";A;",";B") IS ";
210 IF A * A + B * B = R THEN 260
220 IF A * A + B * B = R2 THEN 270
230 IF A ^ 2 + B ^ 2 < R THEN 280
240 IF A ^ 2 + B ^ 2 > R2 THEN 290
250 PRINT "BETWEEN THE TWO CIRCLES.": GOTO 100
260 PRINT "ON THE SMALLER CIRCLE.": GOTO 100
270 PRINT "ON THE LARGER CIRCLE.": GOTO 100
280 PRINT "INSIDE THE SMALLER CIRCLE.": GOTO 10
    0
290 PRINT "OUTSIDE THE LARGER CIRCLE.": GOTO 10
    0
```

Page 360
48. A.P. with common difference d: $a, a + d, a + 2d, a + 3d, a + 4d, \ldots$; odd numbered terms: $a, a + 2d, a + 4d, \ldots$, which is an A.P. with common difference $2d$ **49.** First A.P.: $a, a + d, a + 2d, \ldots, a + (n - 3)d, a + (n - 2)d, a + (n - 1)d$; Second A.P.: $a + (n - 1)d, a + (n - 2)d, a + (n - 3)d, \ldots, a + 2d, a + d, a$, which is an A.P. with common difference $-1d$.

Page 363
32. $x, \underline{}, y$ is an A.P.; $l = a + (n - 1)d$; $y = x + 2d$; $d = \dfrac{y - x}{2}$; $x + \dfrac{y - x}{2} = \dfrac{x + y}{2}$
33. $2x - 1$ and $2x + 1$ are consecutive odd integers if x is an integer; $\dfrac{(2x - 1) + (2x + 1)}{2} = \dfrac{4x}{2} = 2x$; $2x$ is an even integer if x is an integer.

Page 367
33. $\displaystyle\sum_{k=1}^{50} 10kx = 10x + 20x + 30x + \cdots + 500x$
$= 10x(1 + 2 + 3 + \cdots + 50) = 10x \cdot \displaystyle\sum_{k=1}^{50} k$

34. $\displaystyle\sum_{j=1}^{20} (jx + jy) = (x + 2x + 3x + \cdots + 20x) + \cdots$
$+ (y + 2y + 3y + \cdots + 20y) = \displaystyle\sum_{j=1}^{20} jx + \displaystyle\sum_{j=1}^{20} jy$

35. $\displaystyle\sum_{c=1}^{30} (c + t) = (1 + t) + (2 + t) + (3 + t) + \cdots$
$+ (30 + t) = 30t + (1 + 2 + 3 + \cdots + 30)$
$= 30t + \displaystyle\sum_{c=1}^{30} c$ **36.** $(4x - 3) + (5x - 3) + (6x - 3) + (7x - 3) = 32$; $22x = 44$; 2

37. $\displaystyle\sum_{n=1}^{k} (2n - 1) = 1 + 3 + 5 + 7 + \cdots + (2k - 1)$
$= \dfrac{k}{2}[1 + (2k - 1)] = \dfrac{k}{2} \cdot 2k = k^2$;
The sum of the first k positive odd integers is k^2.

38. $\displaystyle\sum_{n=1}^{k} 2n = 2 + 4 + 6 + 8 + \cdots + 2k$
$= \dfrac{k}{2}(2 + 2k) = \dfrac{k}{2} \cdot 2(k + 1) = k(k + 1)$;
The sum of the first k positive even integers is $k(k + 1)$.

Page 374

45. $x, \_\_, y$ is a G.P.; $l = a \cdot r^{n-1}$;
$y = x \cdot r^2$; $r^2 = \dfrac{y}{x}$; $r = \pm \dfrac{\sqrt{xy}}{x}$; $x \cdot \left(\pm \dfrac{\sqrt{xy}}{x}\right)$
$= \pm\sqrt{xy}$ **46.** $a + bi, \_\_, a - bi$ is a G.P.;
$l = a \cdot r^{n-1}$; $a - bi = (a + bi) \cdot r^2$;
$r^2 = \dfrac{a - bi}{a + bi}$; $r = \pm\sqrt{\dfrac{a - bi}{a + bi}}$;
Multiply $a + bi$ by r to get $\pm\sqrt{a^2 + b^2}$

Page 378

45. $\displaystyle\sum_{k=1}^{4} ax^{k-1} = ax^0 + ax^1 + ax^2 + ax^3$
$= a \cdot 1 + a(x^1 + x^2 + x^3) = a + a \cdot \displaystyle\sum_{k=1}^{3} x^k$
46. $(x^3 + 3) + (x^4 + 3) + (x^5 + 3) = 9$;
$x^3 + x^4 + x^5 = 0$; $x^3(1 + x + x^2) = 0$;
$x^3 = 0$ or $x^2 + x + 1 = 0$; $x = 0$ or
$x = \dfrac{-1 \pm \sqrt{1 - 4}}{2}$; $0, \dfrac{-1 \pm i\sqrt{3}}{2}$

Page 383

1. $a^8 + 8a^7b + 28a^6b^2 + 56a^5b^3 + 70a^4b^4 + 56a^3b^5 + 28a^2b^6 + 8ab^7 + b^8$ **2.** $a^9 - 9a^8b + 36a^7b^2 - 84a^6b^3 + 126a^5b^4 - 126a^4b^5 + 84a^3b^6 - 36a^2b^7 + 9ab^8 - b^9$ **3.** $x^6 + 12x^5 + 60x^4 + 160x^3 + 240x^2 + 192x + 64$ **4.** $32c^5 - 240c^4 + 720c^3 - 1080c^2 + 810c - 243$ **5.** $x^{16} + 8x^{14}y + 28x^{12}y^2 + 56x^{10}y^3 + 70x^8y^4 + 56x^6y^5 + 28x^4y^6 + 8x^2y^7 + y^8$
6. $m^{10} - 10m^8n + 40m^6n^2 - 80m^4n^3 + 80m^2n^4 - 32n^5$ **7.** $16r^4 + 160r^3t + 600r^2t^2 + 1{,}000rt^3 + 625t^4$
8. $y^4 - 0.8y^3 + 0.24y^2 - 0.032y + 0.0016$
9. $64x^6 + 19.2x^5 + 2.4x^4 + 0.16x^3 + 0.006x^2 + 0.00012x + 0.000001$ **16.** $x^{12} - 8x^9y + 24x^6y^2 - 32x^3y^3 + 16y^4$ **17.** $32c^{15} + 240c^{12}d^2 + 720c^9d^4 + 1080c^6d^6 + 810c^3d^8 + 243d^{10}$ **18.** $x^6 + 6x^5\sqrt{2} + 30x^4 + 40x^3\sqrt{2} + 60x^2 + 24x\sqrt{2} + 8$
19. $16x^4 - 32x^3\sqrt{y} + 24x^2y - 8xy\sqrt{y} + y^2$
20. $16r^8 + 16r^6t + 6r^4t^2 + r^2t + \dfrac{1}{16}t^4$
21. $\dfrac{1}{x^5} + \dfrac{5}{x^4y} + \dfrac{10}{x^3y^2} + \dfrac{10}{x^2y^3} + \dfrac{5}{xy^4} + \dfrac{1}{y^5}$
22. $x^{12} + 12x^{11}y + 66x^{10}y^2 + 220x^9y^3$
23. $x^{11} - 22x^{10}y + 220x^9y^2 - 1{,}320x^8y^3$
24. $x^{20} + 30x^{18}y^3 + 405x^{16}y^6 + 3{,}240x^{14}y^9$
25. $1 + 0.07 + 0.0021 + 0.000035$ **26.** $1 + 0.16 + 0.0112 + 0.000448$ **27.** $1 + 0.18 + 0.0135 + 0.00054$
28. $\sqrt{x} + \dfrac{2\sqrt{x}}{x} - \dfrac{2\sqrt{x}}{x^2} + \dfrac{4x}{x^3}$
29. $\sqrt[3]{x^2} + \dfrac{6\sqrt[3]{x^2}}{x} - \dfrac{9\sqrt[3]{x^2}}{x^2} + \dfrac{36\sqrt[3]{x^2}}{x^3}$
30. $\dfrac{\sqrt[3]{x^2}}{x} + \dfrac{3\sqrt[3]{x^2}}{x^2} + \dfrac{18\sqrt[3]{x^2}}{x^3} + \dfrac{126\sqrt[3]{x^2}}{x^4}$

Page 386

29. $x^4 + 12x^3y + 54x^2y^2 + 108xy^3 + 81y^4$

30. $32x^{15} - 80x^{12}y + 80x^9y^2 - 40x^6y^3 + 10x^3y^4 - y^5$
31. $\dfrac{1}{m^4} + \dfrac{4}{m^3n} + \dfrac{6}{m^2n^2} + \dfrac{4}{mn^3} + \dfrac{1}{n^4}$

Page 391

(Four properties of the function are given.)

	Domain	Range	Constantly	$(x, 0)$ or $(0, y)$
1.	reals	$y > 0$	increasing	$(0, 1)$
2.	reals	$y > 0$	increasing	$(0, 1)$
3.	reals	$y > 0$	increasing	$(0, 1)$
4.	reals	$y > 0$	decreasing	$(0, 1)$
5.	reals	$y > 0$	decreasing	$(0, 1)$
6.	reals	$y > 0$	decreasing	$(0, 1)$
7.	reals	$y > 0$	increasing	$(0, 5)$
8.	reals	$y < 0$	decreasing	$(0, -1)$
9.	reals	$y > 0$	increasing	$(0, 3)$
10.	reals	$y > 0$	decreasing	$(0, 3)$
11.	reals	$y > 0$	decreasing	$(0, 2)$
12.	reals	$y < 0$	increasing	$(0, -1)$
13.	reals	$y > 4$	increasing	$(0, 5)$
14.	reals	$y > -8$	increasing	$(0, -7)$
15.	reals	$y > 6$	decreasing	$(0, 9)$
16.	$1 \le x \le 7$	$\dfrac{1}{4} \le y \le 16$	increasing	neither
17.	$-4 \le x \le 2$	$\dfrac{1}{16} \le y \le 4$	decreasing	$\left(0, \dfrac{1}{4}\right)$
18.	$0 \le x \le 4$	$1 \le y \le 81$	increasing	$(0, 1)$

Page 396

For Ex. 5–16, four properties of the function are given.

	Domain	Range	Constantly	$(x, 0)$ or $(0, y)$
5.	$x > 0$	reals	increasing	$(1, 0)$
6.	$x > 0$	reals	decreasing	$(1, 0)$
7.	$x > 0$	reals	increasing	$(1, 0)$
8.	$x > 0$	reals	decreasing	$(1, 0)$
9.	$x > 0$	reals	increasing	$(1, 0)$
10.	$x > 0$	reals	decreasing	$(1, 0)$
11.	$x > 0$	reals	increasing	$(1, 0)$
12.	$x > 0$	reals	increasing	$(1, 0)$
13.	$x > 0$	reals	decreasing	$(1, 0)$
14.	$x > 0$	reals	increasing	$\left(\dfrac{1}{16}, 0\right)$
15.	$x > 0$	reals	increasing	$\left(\dfrac{1}{9}, 0\right)$
16.	$x > 0$	reals	decreasing	$(8, 0)$

For Ex. 21–22, five points of the graph are given.

21. $(11, 3), (5, 1), (4, 0), \left(3\dfrac{1}{2}, -1\right), \left(3\dfrac{1}{4}, -2\right)$

22. $(25, 3), (7, 2), (1, 1), (-1, 0), \left(-1\dfrac{2}{3}, -1\right)$

Page 399

32. Let $\log_b x = p$ and $\log_b y = q$. Then $x = b^p$ and $y = b^q$. So, $\dfrac{x}{y} = \dfrac{b^p}{b^q} = b^{p-q}$. Thus, $\log_b \dfrac{x}{y} = \log_b b^{p-q} = p - q = \log_b x - \log_b y$.

Page 402

40. $\log_b \sqrt[3]{y^2} - \log_b \sqrt[4]{y^3} + 4\log_b \sqrt[3]{y} - 5\log_b \sqrt[4]{y}$

$= \log_b y^{\frac{2}{3}} - \log_b y^{\frac{3}{4}} + \log_b y^{\frac{4}{3}} - \log_b y^{\frac{5}{4}} = \log_b \frac{y^{\frac{2}{3}} y^{\frac{4}{3}}}{y^{\frac{3}{4}} y^{\frac{5}{4}}} =$

$\log_b \frac{y^2}{y^2} = \log_b 1 = \log_b b^0 = 0$

41. $\frac{2}{5} \log_b 3 - \frac{1}{5} \log_b 2 - \frac{2}{5} \log_b 4 + \frac{1}{5} \log_b 27 =$
$\frac{2}{5} \log_b 3 - \frac{1}{5} \log_b 2 - \frac{2}{5} \log_b 2^2 + \frac{1}{5} \log_b 3^3 =$
$\log_b 3^{\frac{2}{5}} + \log_b 3^{\frac{3}{5}} - (\log_b 2^{\frac{1}{5}} + \log_b 2^{\frac{4}{5}}) = \log_b 3^{\frac{5}{5}} - \log_b 2^{\frac{5}{5}} = \log_b 3 - \log_b 2$

Page 408

24. Let $\log_a y = x$; $a^x = y$; $\log a^x = \log y$; $x \cdot \log a = \log y$; $x = \frac{\log y}{\log a}$; $\log_a y = \frac{\log y}{\log a}$.

25. Let $\log_a y = x$; $a^x = y$; $\log_b a^x = \log_b y$; $x \cdot \log_b a = \log_b y$; $x = \frac{\log_b y}{\log_b a}$; $\log_a y = \frac{\log_b y}{\log_b a}$.

26. Let $\log_a y = x$; $a^x = y$; $\log_y a^x = \log_y y$; $x \cdot \log_y a = \log_y y^1$; $x = \frac{\log_y y^1}{\log_y a}$; $\log_a y = \frac{1}{\log_y a}$.

Page 417

1. horizontal line, y-intercept: 3 **2.** line, slope: 2, y-intercept: -5 **3.** parabola: $(-6, 7)$, $(-5, 0)$, $(-4, -5)$, $(-3, -8)$, $(-2, -9)$, $(-1, -8)$, $(0, -5)$, $(1, 0)$, $(2, 7)$ **4.** parabola: $(-3, 6)$, $(-2, 0)$, $(-1, -4)$, $(0, -6)$, $(1, -6)$, $(2, -4)$, $(3, 0)$, $(4, 6)$ **5.** S-shape: $(-2, -32)$, $(-1, 0)$, $(0, 12)$, $(1, 10)$, $(2, 0)$, $(3, -12)$, $(4, -20)$, $(5, -18)$, $(6, 0)$, $(7, 40)$ **6.** W-shape: $(-4, 112)$, $(-3, 0)$, $(-2, -12)$, $(-1, 10)$, $(0, 24)$, $(1, 12)$, $(2, -20)$, $(3, -42)$, $(4, 0)$, $(5, 184)$ **7.** parabola: $(0, 15)$, $(1, 3)$, $(2, -1)$, $(3, 3)$, $(4, 15)$, $(5, 35)$ **8.** S-shape: $(-4, -30)$, $(-3, 4)$, $(-2, 18)$, $(-1, 18)$, $(0, 10)$, $(1, 0)$, $(2, -6)$, $(3, -2)$, $(4, 18)$ **9.** W-shape: $(-5, 468)$, $(-4, 0)$, $(-3, -126)$, $(-2, -84)$, $(-1, 0)$, $(0, 48)$, $(1, 30)$, $(2, -36)$, $(3, -84)$, $(4, 0)$, $(5, 378)$ **10.** parabola: $(-3, -8)$, $(-2, 0)$, $(-1, 6)$, $(0, 10)$, $(1, 12)$, $(2, 12)$, $(3, 10)$, $(4, 6)$, $(5, 0)$, $(6, -8)$ **11.** S-shape: $(-4, 35)$, $(-3, 0)$, $(-2, -15)$, $(-1, -16)$, $(0, -9)$, $(1, 0)$, $(2, 5)$, $(3, 0)$, $(4, -21)$ **12.** W-shape: $(-4, -105)$, $(-3, 0)$, $(-2, 15)$, $(-1, 0)$, $(0, -9)$, $(1, 0)$, $(2, 15)$, $(3, 0)$, $(4, -105)$

Page 420

1. Domain: reals; range: $y > 0$; constantly decreasing; through $(0, 1)$ **2.** Domain: $x > 0$; range: reals, constantly increasing; through $(1, 0)$ **8.** $\log_2 7 + \log_2 c - \log_2 b - \log_2 d$ **9.** $\frac{1}{3} (\log_b 3 + 2 \log_b v - \log_b k)$ **21.** (1): $3, -3$; (2): $(x - 1)(x + 2)(x - \sqrt{3})(x + \sqrt{3})$; (3): $1, -2, \sqrt{3}, -\sqrt{3}$ **22.** (1): $3, 0$; (2): $4(x - 2)\left(x - \frac{1}{4}\right)(x - 2i)(x + 2i)$; (3): $2, \frac{1}{4}, 2i, -2i$ **23.** (1): $1, -1$; (2): $12\left(x - \frac{1}{3}\right)\left(x + \frac{1}{2}\right)\left(x + \frac{1}{2}\right)$; (3): $\frac{1}{3}, -\frac{1}{2}$, mult 2 **24.** S-shape: $(-3, -28)$, $(-2, 5)$, $(-1, 12)$, $(0, 5)$, $(1, -4)$, $(2, -3)$, $(3, 20)$ **28.** $\frac{2}{5} \log_b 4 - \frac{1}{5} \log_b 2 = \frac{2}{5} \log_b 2^2 - \frac{1}{5} \log_b 2 = \log_b 2^{\frac{4}{5}} - \log_b 2^{\frac{1}{5}} = \log_b \frac{2^{\frac{4}{5}}}{2^{\frac{1}{5}}} = \log_b 2^{\frac{3}{5}}$ and $\frac{1}{5} \log_b 8 = \frac{1}{5} \log_b 2^3 = \log_b 2^{\frac{3}{5}}$

29. Let $\log_b 69 = x$. Then $b^x = 69$; $\log_{10} b^x = \log_{10} 69$; $x \cdot \log_{10} b = \log_{10} 69$; $x = \frac{\log_{10} 69}{\log_{10} b}$.

30. The inverse of $y = \log_4 x$ is $x = \log_4 y$, or $y = 4^x$.

Page 421

1. Domain: reals; range: $y > 0$; constantly increasing; through $(0, 1)$ **2.** Domain: $x > 0$; range: reals; constantly decreasing; through $(1, 0)$ **20.** S-shape: $(-3, -28)$, $(-2, 5)$, $(-1, 12)$, $(0, 5)$, $(1, -4)$, $(2, -3)$, $(3, 20)$ **24.** $2 \log_b xy - \log_b xy^4 = \log_b x^2 y^2 - \log_b xy^4 = \log_b \frac{x^2 y^2}{xy^4} = \log_b \frac{x}{y^2}$ and

$\log_b x - 2 \log_b y = \log_b x - \log_b y^2 = \log_b \frac{x}{y^2}$

27. The inverse of $y = \log_6 x$ is $x = \log_6 y$, or $y = 6^x$.

Page 437

19. $\binom{n}{r} = \frac{n!}{r!(n-r)!} = \frac{n!}{(n-r)!\,r!}$; and
$\binom{n}{n-r} = \frac{n!}{(n-r)!\,[n-(n-r)]!} = \frac{n!}{(n-r)!\,r!}$.
So, $\binom{n}{r} = \binom{n}{n-r}$.

20. $\binom{n}{r} + \binom{n}{r+1} = \frac{n!}{r!(n-r)!} + \frac{n!}{(r+1)!(n-r-1)!} =$
$\frac{(r+1)\,n!}{(r+1)\,r!(n-r)!} + \frac{n!(n-r)}{(r+1)!(n-r)(n-r-1)!} =$
$\frac{n!(r+1+n-r)}{(r+1)!(n-r)!} = \frac{(n+1)!}{(r+1)!(n-r)!}$; and
$\binom{n+1}{r+1} = \frac{(n+1)!}{(r+1)!(n+1-r-1)!}$
$= \frac{(n+1)!}{(r+1)!(n-r)!}$. So, $\binom{n}{r} + \binom{n}{r+1} = \binom{n+1}{r+1}$.

Page 450

2.
```
100 PRINT "WHAT IS A?";: INPUT A
110 PRINT "WHAT IS B?";: INPUT B
120 P = 1
130 PRINT A;"!/";B;"! = ";
140 IF A = B THEN 190
150 IF A > B THEN 200
```

```
160  PRINT "1 / ";
170  FOR I = A + 1 TO B:P = P * I: NEXT I
180  PRINT P: GOTO 220
190  PRINT 1: GOTO 220
200  FOR I = B + 1 TO A:P = P * I: NEXT I
210  PRINT P
220  PRINT : GOTO 100
```

Page 453
27. $(0, -2), (-3, 0), (3, -4)$
34. axis of symmetry: $x = -3$; turning point: $(-3, 11)$; $(0, 2), (-6, 2), (-3, 11)$
36. the circle $(x^2 + y^2 = 36)$ and all its interior points with the exception of the interior points of the ellipse $\left(\dfrac{x^2}{9} + \dfrac{y^2}{4} = 1\right)$

Page 461
32. $c^2 = a^2 + b^2$; $c^2 = x^2 + x^2$; $c^2 = 2x^2$; $c = x\sqrt{2}$
33. Assume $\triangle ABC$ is an equilateral $\triangle$, with $m\angle A = m\angle B = m\angle C = 60$.
Draw altitude $\overline{AD}$; $\overline{AD}$ bisects $\angle A$; $m\angle BAD = 30$;
$\overline{BD} \cong \overline{DC}$; $\boxed{BD = \dfrac{1}{2}AB \text{ or } AB = 2BD}$
$(AB)^2 = (BD)^2 + (AD)^2 = \left(\dfrac{AB}{2}\right)^2 + (AD)^2$;
$(AB)^2 = \dfrac{(AB)^2}{4} + (AD)^2$; $4(AB)^2 - (AB)^2 = 4(AD)^2$;
$3(AB)^2 = 4(AD)^2$; $(AD)^2 = \dfrac{3}{4}(AB)^2$; $AD = \dfrac{1}{2}\sqrt{3}(AB)$

Page 465
1. $\dfrac{\sqrt{3}}{2}$ **2.** $\dfrac{1}{2}$ **3.** $\dfrac{\sqrt{2}}{2}$ **4.** $\dfrac{\sqrt{2}}{2}$ **5.** $\dfrac{1}{2}$ **6.** $\dfrac{\sqrt{3}}{2}$
7. 0 **8.** 1 **9.** 0 **10.** 0 **11.** 0 **12.** -1 **13.** $\dfrac{\sqrt{2}}{2}$
14. $-\dfrac{\sqrt{3}}{2}$ **15.** $\dfrac{-\sqrt{3}}{2}$ **16.** $\dfrac{1}{2}$ **17.** -1 **18.** $\dfrac{-\sqrt{2}}{2}$
19. $\dfrac{-\sqrt{2}}{2}$ **20.** $\dfrac{\sqrt{3}}{2}$ **21.** $\dfrac{\sqrt{3}}{2}$ **22.** $-\dfrac{1}{2}$
23. $\dfrac{1}{2}$ **24.** $-\dfrac{1}{2}$ **25.** $-\cos 45°$ **26.** $-\sin 30°$
27. $-\cos 30°$ **28.** $-\sin 45°$ **29.** $\sin 45°$
30. $-\sin 60°$ **31.** $\sin 30°$ **32.** $-\sin 30°$ **33.** $-\sin 45°$
34. $\cos 60°$ **35.** $-\cos 30°$ **36.** $\cos 30°$

Page 469
25. $-\cot 70°$ **26.** $-\tan 50°$ **27.** $\cot 50°$ **28.** $\tan 30°$
29. $-\tan 20°$ **30.** $-\cot 85°$ **31.** $\tan 70°$
32. $-\cot 60°$ **33.** $-\tan 30°$ **34.** $-\cot 60°$
35. $\cot 70°$ **36.** $-\tan 80°$

Page 472
19. $-\sec 60°$ **20.** $\csc 30°$ **21.** $-\csc 60°$
22. $-\sec 30°$ **23.** $-\sec 60°$ **24.** $-\csc 45°$
25. $\csc 60°$ **26.** $-\csc 30°$ **27.** $-\sec 10°$
28. $\sec 20°$ **29.** $\csc 10°$ **30.** $-\csc 35°$

Page 474
Use the diagram on page 473 of the text.
44. $\tan u = \dfrac{b}{-a} = -\dfrac{b}{a}$; $\tan(-u) = \dfrac{b}{a}$; $\tan(-u) = -\tan u$
45. $\cot u = \dfrac{-a}{b} = -\dfrac{a}{b}$; $\cot(-u) = \dfrac{a}{b}$; $\cot(-u) = -\cot u$
46. $\sec u = \dfrac{r}{-a} = -\dfrac{r}{a}$; $\sec(-u) = \dfrac{r}{-a} = -\dfrac{r}{a}$;
$\sec(-u) = \sec u$ **47.** $\csc u = \dfrac{r}{b}$; $\csc(-u) = \dfrac{r}{-b} =$
$-\dfrac{r}{b}$; $\csc(-u) = -\csc u$

Page 476
1. $x = -5, y = 12, r = 13$; $\sin u = \dfrac{12}{13}$, $\cos u = -\dfrac{5}{13}$, $\cot u = -\dfrac{5}{12}$, $\sec u = -\dfrac{13}{5}$, $\csc u = \dfrac{13}{12}$ **2.** $x = -5$, $y = 12, r = 13$; $\cos u = -\dfrac{5}{13}$, $\tan u = -\dfrac{12}{5}$, $\cot u = -\dfrac{5}{12}$, $\sec u = -\dfrac{13}{5}$; $\csc u = \dfrac{13}{12}$ **3.** $x = -5, y = -12$, $r = 13$; $\sin u = -\dfrac{12}{13}$, $\tan u = \dfrac{12}{5}$, $\cot u = \dfrac{5}{12}$, $\sec u = -\dfrac{13}{5}$, $\csc u = -\dfrac{13}{12}$ **4.** $x = 4, y = -3, r = 5$; $\cos u = \dfrac{4}{5}$, $\tan u = -\dfrac{3}{4}$, $\cot u = -\dfrac{4}{3}$, $\sec u = \dfrac{5}{4}$, $\csc u = -\dfrac{5}{3}$ **5.** $x = 3, y = -4, r = 5$; $\sin u = -\dfrac{4}{5}$, $\cos u = \dfrac{3}{5}$, $\tan u = -\dfrac{4}{3}$, $\sec u = \dfrac{5}{3}$, $\csc u = -\dfrac{5}{4}$
6. $x = -5, y = 12, r = 13$; $\sin u = \dfrac{12}{13}$, $\cos u = -\dfrac{5}{13}$, $\tan u = -\dfrac{12}{5}$, $\cot u = -\dfrac{5}{12}$, $\csc u = \dfrac{13}{12}$ **7.** $x = -12$, $y = 5, r = 13$; $\sin u = \dfrac{5}{13}$, $\tan u = -\dfrac{5}{12}$; $\cot u = -\dfrac{12}{5}$, $\sec u = -\dfrac{13}{12}$, $\csc u = \dfrac{13}{5}$ **8.** $x = 12, y = 5, r = 13$; $\sin u = \dfrac{5}{13}$, $\cos u = \dfrac{12}{13}$, $\tan u = \dfrac{5}{12}$, $\cot u = \dfrac{12}{5}$, $\sec u = \dfrac{13}{12}$ **9.** $x = -3, y = -4, r = 5$; $\sin u = -\dfrac{4}{5}$, $\cos u = -\dfrac{3}{5}$, $\cot u = \dfrac{3}{4}$, $\sec u = -\dfrac{5}{3}$, $\csc u = -\dfrac{5}{4}$
10. $x = 12, y = -5, r = 13$; $\sin u = -\dfrac{5}{13}$, $\cos u = \dfrac{12}{13}$, $\tan u = -\dfrac{5}{12}$, $\cot u = -\dfrac{12}{5}$, $\csc u = -\dfrac{13}{5}$
11. $x = -1, y = 1, r = \sqrt{2}$; $\sin u = \dfrac{\sqrt{2}}{2}$, $\cos u = -\dfrac{\sqrt{2}}{2}$, $\cot u = -1$, $\sec u = -\sqrt{2}$, $\csc u = \sqrt{2}$
12. $x = \sqrt{3}, y = -1, r = 2$; $\sin u = -\dfrac{1}{2}$, $\cos u = \dfrac{\sqrt{3}}{2}$, $\tan u = -\dfrac{\sqrt{3}}{3}$, $\sec u = \dfrac{2\sqrt{3}}{3}$, $\csc u = -2$
13. $x = 12, y = 5, r = 13$; $\cos u = \dfrac{12}{13}$, $\tan u = \dfrac{5}{12}$, $\cot u = \dfrac{12}{5}$, $\sec u = \dfrac{13}{12}$, $\csc u = \dfrac{13}{5}$ **14.** $x = -3$,

$y = 4$, $r = 5$; $\sin u = \frac{4}{5}$, $\tan u = -\frac{4}{3}$, $\cot u = -\frac{3}{4}$, $\sec u = -\frac{5}{3}$, $\csc u = \frac{5}{4}$ **15.** $x = 1$, $y = 1$, $r = \sqrt{2}$; $\sin u = \frac{\sqrt{2}}{2}$, $\cos u = \frac{\sqrt{2}}{2}$, $\tan u = 1$, $\sec u = \sqrt{2}$, $\csc u = \sqrt{2}$ **16.** $x = -4$, $y = 3$, $r = 5$; $\sin u = \frac{3}{5}$, $\cos u = -\frac{4}{5}$, $\cot u = -\frac{4}{3}$, $\sec u = -\frac{5}{4}$, $\csc u = \frac{5}{3}$ **17.** $x = \sqrt{3}$, $y = 1$, $r = 2$; $\sin u = \frac{1}{2}$, $\cos u = \frac{\sqrt{3}}{2}$, $\cot u = \sqrt{3}$, $\sec u = \frac{2\sqrt{3}}{3}$, $\csc u = 2$ **18.** $x = 1$, $y = -1$, $r = \sqrt{2}$; $\sin u = -\frac{\sqrt{2}}{2}$, $\tan u = -1$, $\cot u = -1$, $\sec u = \sqrt{2}$, $\csc u = -\sqrt{2}$ **19.** $x = -\sqrt{3}$, $y = 1$, $r = 2$; $\sin u = \frac{1}{2}$, $\cos u = -\frac{\sqrt{3}}{2}$, $\tan u = -\frac{\sqrt{3}}{3}$, $\cot u = -\sqrt{3}$, $\csc u = 2$ **20.** $x = 1$, $y = -1$, $r = \sqrt{2}$; $\sin u = -\frac{\sqrt{2}}{2}$, $\cos u = \frac{\sqrt{2}}{2}$, $\tan u = -1$, $\cot u = -1$, $\sec u = \sqrt{2}$ **21.** $\cos u = -\frac{1}{2}$, $\cot u = -\frac{\sqrt{3}}{3}$, $\sec u = -2$, $\csc u = \frac{2\sqrt{3}}{3}$ **22.** $\sin u = -\frac{\sqrt{3}}{2}$, $\tan u = \sqrt{3}$, $\sec u = -2$, $\csc u = -\frac{2\sqrt{3}}{3}$ **23.** $\tan u = -1$, $\cot u = -1$, $\sec u = \sqrt{2}$, $\csc u = -\sqrt{2}$ **24.** $\sin u = \frac{\sqrt{2}}{2}$, $\cos u = -\frac{\sqrt{2}}{2}$, $\cot u = -1$, $\csc u = \sqrt{2}$ **25.** $\sin u = -\frac{12}{13}$, $\cos u = -\frac{5}{13}$, $\cot u = \frac{5}{12}$, $\csc u = -\frac{13}{12}$ **26.** 1st quadrant: $\cos u = \frac{5}{13}$, $\tan u = \frac{12}{5}$, $\cot u = \frac{5}{12}$, $\sec u = \frac{13}{5}$; 2nd quadrant: $\cos u = -\frac{5}{13}$, $\tan u = -\frac{12}{5}$, $\cot u = -\frac{5}{12}$, $\sec u = -\frac{13}{5}$ **27.** $\sin u = -\frac{5\sqrt{41}}{41}$, $\cot u = -\frac{4}{5}$, $\sec u = \frac{\sqrt{41}}{4}$, $\csc u = -\frac{\sqrt{41}}{5}$ **28.** $\sin u = \frac{1}{2}$, $\tan u = -\frac{\sqrt{3}}{3}$, $\sec u = -\frac{2\sqrt{3}}{3}$, $\csc u = 2$ **29.** 3rd quadrant: $\cos u = -\frac{24}{25}$, $\tan u = \frac{7}{24}$, $\cot u = \frac{24}{7}$, $\sec u = -\frac{25}{24}$; 4th quadrant: $\cos u = \frac{24}{25}$, $\tan u = -\frac{7}{24}$, $\cot u = -\frac{24}{7}$, $\sec u = \frac{25}{24}$ **30.** $\sin u = \frac{3}{5}$, $\cos u = -\frac{4}{5}$, $\cot u = -\frac{4}{3}$, $\csc u = \frac{5}{3}$

Page 480
48. $15° 30'$, $344° 30'$ **49.** $109° 50'$, $289° 50'$ **50.** $38° 10'$, $218° 10'$ **51.** $104° 50'$, $284° 50'$ **52.** $119° 10'$, $240° 50'$ **53.** $248° 10'$, $291° 50'$ **54.** $54° 20'$, $125° 40'$ **55.** $9° 30'$, $170° 30'$ **56.** $38° 40'$, $321° 20'$ **57.** $-\tan 60° = -\cot 30°$ **58.** $\cot 20°$ **59.** $-\sin 44°$ **60.** $-\cos 70° = -\sin 20°$ **61.** $-\sin 44°$ **62.** $-\sin 35° 40'$ **63.** $-\cos 54° 50' = -\sin 35° 10'$ **64.** $\tan 32° 40'$ **65.** $-\cot 28° 10'$ **66.** $-\tan 45° 20' = -\cot 44° 40'$ **67.** $-\cos 68° = -\sin 22° = -\frac{1}{\csc 22°}$ **68.** $\frac{1}{\csc 39° 50'}$ **69.** $\sin 3° 20' = \frac{1}{\csc 3° 20'}$ **70.** $\tan 10° 40' = \frac{1}{\cot 10° 40'}$ **71.** $-\tan 55° 40' = -\cot 34° 20' = -\frac{1}{\tan 34° 20'}$ **72.** $-\cot 56° 20' = -\tan 33° 40' = -\frac{1}{\cot 33° 40'}$ **73.** $\cos 19° 30' = \frac{1}{\sec 19° 30'}$ **74.** $-\cot 50° 40' = -\tan 39° 20' = -\frac{1}{\cot 39° 20'}$ **75.** 1.2258 **76.** 2.0204 **77.** 1.2929 **78.** -1.0176 **79.** $1° 20'$ **80.** $69°$ **81.** $30° 20'$ **82.** $62° 30'$ **83.** $33° 20'$, $326° 40'$ **84.** $188° 20'$, $351° 40'$ **85.** $146° 20'$, $213° 40'$ **86.** $38° 20'$, $141° 40'$ **87.** $69°$, $291°$ **88.** $202° 50'$, $337° 10'$

Page 482
19. $\cos \theta = -\frac{\sqrt{3}}{2}$; $\theta = 150°, 210°, 510°, 570°$ **20.** $\tan \theta = -1$; $\theta = 135°, 315°, 495°, 675°$ **21.** $\cot \theta = \sqrt{3}$; $\theta = 30°, 210°, 390°, 570°$ **22.** $2 \sin \theta = -1$; $\sin \theta = -\frac{1}{2}$; $\theta = 210°, 330°, 570°, 690°$ **23.** $-3 \cos \theta - 5 = -6 \cos \theta$; $3 \cos \theta = 5$; $\cos \theta = \frac{5}{3}$; no solution **24.** $\sin \theta = 1$; $\theta = 90°, 450°$ **25.** $\tan \theta = \frac{3}{\sqrt{3}} = \sqrt{3}$; $\theta = 60°, 240°, 420°, 600°$ **26.** $5 \sin \theta = 4$; $\sin \theta = \frac{4}{5} = 0.8000$; $\theta = 53° 10', 126° 50', 413° 10', 486° 50'$ **27.** $5 \cot \theta = -7$; $\cot \theta = -\frac{7}{5} = -1.4000$; $\theta = 144° 30', 324° 30', 504° 30', 684° 30'$ **28.** $\sec \theta = -\frac{8}{3} = -2.6667$; $\theta = 112°, 248°, 472°, 608°$ **29.** $\csc \theta = 5$; $\theta = 11° 30', 168° 30', 371° 30', 528° 30'$ **30.** $5 \sec \theta = -12$; $\sec \theta = -\frac{12}{5} = -2.4000$; $\theta = 114° 40', 245° 20', 474° 40', 605° 20'$ **31.** $\csc \theta = -1$; $\theta = 270°, 630°$ **32.** $\sec \theta = \frac{8.4}{2} = 4.2000$; $\theta = 76° 10', 283° 50', 436° 10', 643° 50'$ **33.** $\sec \theta = \frac{1}{3} = 0.3333$; no values **34.** $3 \sec \theta - 36 = 10 \sec \theta$; $7 \sec \theta = -36$; $\sec \theta = -\frac{36}{7} = -5.1429$; $\theta = 101° 10', 258° 50', 461° 10', 618° 50'$ **35.** $\frac{1}{3} \csc \theta = \frac{8}{5}$; $\csc \theta = \frac{24}{5}$

$= 4.8000$; $\theta = 12°, 168°, 372°, 528°$ **36.** $4 \csc \theta - 24$
$= 9 \csc \theta + 12$; $5 \csc \theta = -36$; $\csc \theta = -\dfrac{36}{5}$
$= -7.2000$; $\theta = 188°, 352°, 548°, 712°$

Page 488
1. $\sin G = \dfrac{g}{i}$; $\cos G = \dfrac{h}{i}$; $\tan G = \dfrac{g}{h}$; $\cot G = \dfrac{h}{g}$
2. $\sin H = \dfrac{h}{i}$; $\cos H = \dfrac{g}{i}$; $\tan H = \dfrac{h}{g}$; $\cot H = \dfrac{g}{h}$
3. $\sin A = \dfrac{a}{c}$; $\cos A = \dfrac{b}{c}$; $\tan A = \dfrac{a}{b}$; $\cot A = \dfrac{b}{a}$
4. $\sin B = \dfrac{b}{c}$; $\cos B = \dfrac{a}{c}$; $\tan B = \dfrac{b}{a}$; $\cot B = \dfrac{a}{b}$
5. $\sin D = \dfrac{d}{f}$; $\cos D = \dfrac{e}{f}$; $\tan D = \dfrac{d}{e}$; $\cot D = \dfrac{e}{d}$
6. $\sin E = \dfrac{e}{f}$; $\cos E = \dfrac{d}{f}$; $\tan E = \dfrac{e}{d}$; $\cot E = \dfrac{d}{e}$

Page 490
1. sin **2.** tan **3.** tan **4.** cot **5.** sin **6.** cos
7. cot **8.** sin [Also, reciprocal fcns]

Page 491
20. $\tan 13° 23' = \dfrac{9.6}{b}$; $b = \dfrac{9.6}{.2379} = 40.4$; $\sin 13° 23' = \dfrac{9.6}{c}$; $c = \dfrac{9.6}{.2314} = 41.5$; $B = 90° - 13° 23' = 76° 37'$
21. $\sin 49° 18' = \dfrac{b}{13.3}$; $b = 13.3(.7581) = 10.1$;
$\cos 49° 18' = \dfrac{a}{13.3}$; $a = 13.3(.6521) = 8.7$;
$A = 90° - 49° 18' = 40° 42'$ **22.** $\tan 63° 38' =$
$\dfrac{a}{9.1}$; $a = 9.1(2.0174) = 18.4$; $\cos 63° 38' = \dfrac{9.1}{c}$;
$c = \dfrac{9.1}{.4441} = 20.5$; $B = 90° - 63° 38' = 26° 22'$
23. $\sin 74° 46' = \dfrac{12.3}{c}$; $c = \dfrac{12.3}{.9649} = 12.7$, $\tan 74° 46'$
$= \dfrac{12.3}{a}$; $a = \dfrac{12.3}{3.6723} = 3.3$, $A = 90° - 74° 46' =$
$15° 14'$ **24.** $\tan 11° 12' = \dfrac{a}{7.8}$; $a = 7.8(.1980) = 1.5$;
$\cos 11° 12' = \dfrac{7.8}{c}$; $c = \dfrac{7.8}{.9810} = 7.95$ or 8.0;
$B = 90° - 11° 12' = 78° 48'$ **25.** $\cos 16° 56' =$
$\dfrac{a}{11.8}$; $a = 11.8(.9567) = 11.3$; $\sin 16° 56' = \dfrac{b}{11.8}$;
$b = 11.8(.2913) = 3.4$; $A = 90° - 16° 56' = 73° 4'$
26. $\sin 33° 11' = \dfrac{a}{16.5}$; $a = 16.5(.5473) = 9.0$;
$\cos 33° 11' = \dfrac{b}{16.5}$; $b = 16.5(.8369) = 13.8$;
$B = 90° - 33° 11' = 56° 49'$ **27.** $\cos 73° 33' =$
$\dfrac{6.3}{c}$; $c = \dfrac{6.3}{.2832} = 22.2$; $\tan 73° 33' = \dfrac{b}{6.3}$;
$b = 6.3(3.3770) = 21.3$; $A = 90° - 73° 33' = 16° 27'$
28. $\sin 38° 37' = \dfrac{12.5}{c}$; $c = 20.0$; $b = 15.6$; $B = 51° 23'$

Page 492
1. I, 45° **2.** III, 30° **3.** III, 45° **4.** II, 60°
5. I, 30° **6.** 90° **7.** I, 60° **8.** IV, 60° **9.** III, 80°
10. 90° **11.** I, 40° **12.** 0° **17.** $\dfrac{1}{2}$ **18.** $-\dfrac{\sqrt{3}}{3}$
19. $-\dfrac{2\sqrt{3}}{3}$ **20.** $-\dfrac{\sqrt{3}}{2}$ **21.** 0 **22.** undefined
23. undefined **24.** 0 **36.** $\sin u = \dfrac{5}{13}$,
$\cos u = \dfrac{12}{13}$, $\cot u = \dfrac{12}{5}$, $\sec u = \dfrac{13}{12}$,
$\csc u = \dfrac{13}{5}$ **37.** $\sin u = \dfrac{4}{5}$, $\tan u = -\dfrac{4}{3}$,
$\cot u = -\dfrac{3}{4}$, $\sec u = -\dfrac{5}{3}$, $\csc u = \dfrac{5}{4}$
38. $\cos u = -\dfrac{24}{25}$, $\tan u = -\dfrac{7}{24}$, $\cot u = -\dfrac{24}{7}$,
$\csc u = \dfrac{25}{7}$

Page 493
1. II, 70° **2.** III, 50° **3.** III, 70° **4.** I, 20°
24. 4th quadrant: $\cos u = \dfrac{12}{13}$, $\tan u = -\dfrac{5}{12}$,
$\cot u = -\dfrac{12}{5}$, $\sec u = \dfrac{13}{12}$, $\csc u = -\dfrac{13}{5}$
25. 2nd quadrant; $\sin u = \dfrac{4}{5}$, $\tan u = -\dfrac{4}{3}$,
$\cot u = -\dfrac{3}{4}$, $\sec u = -\dfrac{5}{3}$, $\csc u = \dfrac{5}{4}$

Page 499
(For Ex. 21–28, the coordinates of four points of each curve are given.) **21.** (0°, −1), (90°, 0), (180°, 1), (270°, 0)
22. (0°, 0), (90°, −1), (180°, 0), (270°, 1) **23.** (0°, 0), (90°, −1), (180°, 0), (270°, 1) **24.** (0°, 1), (90°, 0), (180°, −1), (270°, 0) **25.** (0°, 1), (90°, undefined), (180°, −1), (270°, undefined) **26.** (0, −1), (90°, undefined), (180°, 1), (270°, undefined)
27. (0, undefined), (90°, 1), (180°, undefined), (270°, −1)
28. (0, undefined), (90°, −1), (180°, undefined), (270°, 1)

Page 505
25. 720°, none **26.** 360°, none

Page 514
1. $9.9063 - 10$ **2.** $9.3668 - 10$ **3.** $9.6188 - 10$
4. $9.8339 - 10$ **5.** $9.9768 - 10$ **6.** $9.5847 - 10$
7. $10.5350 - 10$ **8.** $10.6378 - 10$ **9.** $9.7150 - 10$
10. $9.8395 - 10$ **11.** $9.9481 - 10$ **12.** $8.7601 - 10$
13. 28° 20' **14.** 37° 10' **15.** 36° 20' **16.** 45° 30'
17. 33° 29' **18.** 21° 36' **19.** 40° 20' **20.** 72° 32'
21. 21° 27'

Page 517
1. 45°, 135°, 225°, 315° **2.** 60°, 120° **3.** 30°, 330°
4. 90°, 270° **5.** 90°, 270° **6.** 30°, 150°, 210°, 330°
7. 240°, 300° **8.** 90°, 270°, 135°, 315° **9.** 90°, 270°, 60°, 300° **10.** 0°, 180°, 240°, 300° **11.** 0°, 180°, 30°, 210° **12.** 120°, 240°, 0°, 360° **13.** 90°

14. 120°, 240°, 180°   15. 45°, 225°   16. 79°, 259°, 160°, 340°   17. 187°, 353°   18. 71°, 289°   19. 33°, 327°, 104°, 256°   20. 67°, 247°, 135°, 315°   21. 47°, 227°, 147°, 327°   22. 191°, 349°   23. 8°, 188°, 82°, 262°   24. 101°, 259°   25. 60°, 120°, 240°, 300°   26. 24°, 156°, 204°, 336°   27. no solution   28. 14°, 166°, 19°, 161°   29. 114°, 246°, 71°, 289°   30. 28°, 152°   31. 80°, 280°   32. 52°, 128°, 232°, 308°   33. 98°, 262°

Page 520

5. $\left(\frac{1}{2}\right)^2 + \left(\frac{\sqrt{3}}{2}\right)^2 = 1; \frac{1}{4} + \frac{3}{4} = 1; 1 = 1$

6. $\left(\frac{1}{\sqrt{3}}\right)^2 + 1 = \left(\frac{2}{\sqrt{3}}\right)^2; \frac{1}{3} + 1 = \frac{4}{3}; \frac{4}{3} = \frac{4}{3}$

7. $1 + (\sqrt{3})^2 = 2^2; 1 + 3 = 4; 4 = 4$ 8. $\frac{\sqrt{3}}{2} = \frac{\frac{1}{2}}{\frac{1}{\sqrt{3}}}; \frac{\sqrt{3}}{2} = \frac{\sqrt{3}}{2}$ 9. $\left(\frac{-\sqrt{3}}{2}\right)^2 = 1 - \left(-\frac{1}{2}\right)^2; \frac{3}{4} = 1 - \frac{1}{4}; \frac{3}{4} = \frac{3}{4}$ 10. $\frac{1}{\sqrt{3}} = \frac{\frac{-1}{2}}{\frac{-\sqrt{3}}{2}}; \frac{1}{\sqrt{3}} = \frac{1}{\sqrt{3}}$

11. $\left(\frac{-2}{\sqrt{3}}\right)^2 - 1 = \left(\frac{1}{\sqrt{3}}\right)^2; \frac{4}{3} - 1 = \frac{1}{3}; \frac{1}{3} = \frac{1}{3}$

12. $-\frac{\sqrt{3}}{2} = \frac{\frac{-2}{\sqrt{3}}}{\frac{-2}{\sqrt{3}}}; -\frac{\sqrt{3}}{2} = -\frac{\sqrt{3}}{2}$

13. $\left(-\frac{1}{\sqrt{2}}\right)^2 = 1 - \left(\frac{1}{\sqrt{2}}\right)^2; \frac{1}{2} = 1 - \frac{1}{2}; \frac{1}{2} = \frac{1}{2}$

14. $-1 = \frac{\frac{1}{\sqrt{2}}}{\frac{-1}{\sqrt{2}}}; -1 = -1$ 15. $(-\sqrt{2})^2 = 1 + (-1)^2; 2 = 1 + 1; 2 = 2$ 16. $(\sqrt{2})^2 = 1 + (-1)^2; 2 = 2$ 17. $\left(-\frac{1}{2}\right)(-2) = 1; 1 = 1$

18. $\left(-\frac{1}{\sqrt{3}}\right)\left(-\frac{\sqrt{3}}{1}\right) = 1; 1 = 1$ 19. $\frac{\sqrt{3}}{2} \cdot \frac{2}{\sqrt{3}} = 1 = 1 = 1$ 20. $-\frac{1}{2} = \frac{1}{-2}; -\frac{1}{2} = -\frac{1}{2}$

21. $\frac{\sin x}{\cos x} = \frac{\sin x}{\pm\sqrt{1 - \sin^2 x}}$ 22. $\cos x = \frac{\pm\sqrt{1 - \sin^2 x}}{\sin x}$

23. $\frac{1}{\cos x} = \frac{1}{\pm\sqrt{1 - \sin^2 x}}$ 24. $\frac{1}{\sin x}$

25. $\frac{\cos x}{\sin x}; \frac{\cos x}{\pm\sqrt{1 - \cos^2 x}}$ 26. $\pm\sqrt{1 - \cos^2 x}$

27. $\frac{1}{\sin x}; \frac{1}{\pm\sqrt{1 - \cos^2 x}}$ 28. $\frac{1}{\cos x}$

29. $\tan^2 x + 1 = \sec^2 x; \left(\frac{\sin x}{\cos x}\right)^2 + 1 = \sec^2 x;$ $\frac{\sin^2 x}{\cos^2 x} + 1 = \sec^2 x; \frac{\sin^2 x + \cos^2 x}{\cos^2 x} = \sec^2 x; \frac{1}{\cos^2 x} = \sec^2 x; \sec^2 x = \sec^2 x$ 30. $1 + \cot^2 x = \csc^2 x;$ $1 + \left(\frac{\cos x}{\sin x}\right)^2 = \csc^2 x; \frac{\sin^2 x}{\sin^2 x} + \frac{\cos^2 x}{\sin^2 x} = \csc^2 x;$ $\frac{\sin^2 x + \cos^2 x}{\sin^2 x} = \csc^2 x; \frac{1}{\sin^2 x} = \csc^2 x; \csc^2 x = \csc^2 x$

31. $\tan^2 A + 1 = \sec^2 A; \left(\frac{\sin A}{\cos A}\right)^2 + 1 = \sec^2 A; \frac{\sin^2 A}{\cos^2 A} + 1, = \sec^2 A; \frac{\sin^2 A}{\cos^2 A} + \frac{\cos^2 A}{\cos^2 A} = \sec^2 A; \frac{\sin^2 A + \cos^2 A}{\cos^2 A} = \sec^2 A; \frac{1}{\cos^2 A} = \sec^2 A; \sec^2 A = \sec^2 A$

32. $1 + \cot^2 A = \csc^2 A; 1 + \frac{\cos^2 A}{\sin^2 A} = \csc^2 A; \frac{\sin^2 A}{\sin^2 A} + \frac{\cos^2 A}{\sin^2 A} = \csc^2 A; \frac{\sin^2 A + \cos^2 A}{\sin^2 A} = \csc^2 A; \frac{1}{\sin^2 A} = \csc^2 A; \csc^2 A = \csc^2 A$

Page 522

1. $\frac{\cot x}{\cos x} = \csc x; \frac{\cos x}{\sin x} \div \cos x = \csc x; \frac{\cos x}{\sin x \cos x} = \csc x; \frac{1}{\sin x} = \csc x; \csc x = \csc x$

2. $\sin x \cdot \cot x = \cos x; \sin x \cdot \frac{\cos x}{\sin x} = \cos x; \cos x = \cos x$ 3. $\sec x = \csc x \cdot \tan x; \sec x = \frac{1}{\sin x} \cdot \frac{\sin x}{\cos x}; \sec x = \frac{1}{\cos x}; \sec x = \sec x$ 4. $\cot x + \tan x = \csc x \cdot \sec x; \frac{\cos x}{\sin x} + \frac{\sin x}{\cos x} = \csc x \cdot \sec x;$ $\frac{\sin^2 x + \cos^2 x}{\sin x \cdot \cos x} = \csc x \cdot \sec x; \frac{1}{\sin x \cdot \cos x} = \csc x \cdot \sec x; \frac{1}{\sin x} \cdot \frac{1}{\cos x} = \csc x \cdot \sec x; \csc x \cdot \sec x = \csc x \cdot \sec x$ 5. $\sec x = \cos x + \frac{\tan x}{\csc x};$ $\sec x = \cos x + \frac{\sin x}{\cos x} \div \frac{1}{\sin x}; \sec x = \cos x + \frac{\sin^2 x}{\cos x};$ $\sec x = \frac{\cos^2 x}{\cos x} + \frac{\sin^2 x}{\cos x}; \sec x = \frac{\cos^2 x + \sin^2 x}{\cos x};$ $\sec x = \frac{1}{\cos x}; \sec x = \sec x$ 6. $\sin^2 x - \cos^2 x = 2\sin^2 x - 1; \sin^2 x - \cos^2 x = 2\sin^2 x - 1;$ $2\sin^2 x - (\sin^2 x + \cos^2 x); 2\sin^2 x - \sin^2 x - \cos^2 x;$ $\sin^2 x - \cos^2 x$ 7. $\cot x - 1 = \cos x(\csc x - \sec x); \cot x - 1 = \cos x\left(\frac{1}{\sin x} - \frac{1}{\cos x}\right); \cot x - 1 = \frac{\cos x}{\sin x} - \frac{\cos x}{\cos x}; \cot x - 1 = \cot x - 1$ 8. $\sec x = \frac{\cos x}{1 + \sin x} + \tan x; \sec x = \frac{\cos x}{1 + \sin x} + \frac{\sin x}{\cos x};$ $\sec x = \frac{\cos^2 x + \sin x + \sin^2 x}{\cos x(1 + \sin x)}; \sec x = \frac{1 + \sin x}{\cos x(1 + \sin x)};$ $\sec x = \frac{1}{\cos x}; \sec x = \sec x$ 9. $\sin x + \frac{\cot x}{\sec x} = \csc x; \sin x + \frac{\cos x}{\sin x} \div \frac{1}{\cos x} = \csc x; \sin x + \frac{\cos^2 x}{\sin x} = \csc x; \frac{\sin^2 x + \cos^2 x}{\sin x} = \csc x; \frac{1}{\sin x} = \csc x;$

$\csc x = \csc x$ **10.** $\dfrac{\cot x + \cos x}{1 + \sin x} = \dfrac{\frac{\cos x}{\sin x} + \cos x}{1 + \sin x}$;

$\dfrac{\frac{\cos x + \sin x \cos x}{\sin x}}{1 + \sin x}$; $\dfrac{\cos x(1 + \sin x)}{\sin x} \cdot \dfrac{1}{1 + \sin x}$; $\dfrac{\cos x}{\sin x}$

11. $\tan x - \cot x = \dfrac{1 - 2\cos^2 x}{\sin x \cos x}$; $\dfrac{1}{\sin x \cos x} - \dfrac{2\cos^2 x}{\sin x \cos x}$;

$\dfrac{\sin^2 x + \cos^2 x}{\sin x \cos x} - \dfrac{2\cos^2 x}{\sin x \cos x}$; $\dfrac{\sin^2 x + \cos^2 x - 2\cos^2 x}{\sin x \cos x}$;

$\dfrac{\sin^2 x - \cos^2 x}{\sin x \cos x}$; $\dfrac{\sin^2 x}{\sin x \cos x} - \dfrac{\cos^2 x}{\sin x \cos x}$; $\dfrac{\sin x}{\cos x} - \dfrac{\cos x}{\sin x}$;

$\tan x - \cot x$ **12.** $\dfrac{1 - \cos^2 x}{\cos x} \cdot \csc x = \tan x$; $\dfrac{\sin^2 x}{\cos x}$

$\cdot \dfrac{1}{\sin x} = \tan x$; $\dfrac{\sin x}{\cos x} = \tan x$; $\tan x = \tan x$

13. $2 \csc x = \dfrac{\sin x}{1 + \cos x} + \dfrac{\sin x}{1 - \cos x}$; $2 \csc x =$

$2 \csc x = \dfrac{\sin x - \sin x \cos x + \sin x + \sin x \cos x}{1 - \cos^2 x}$;

$2 \csc x = \dfrac{2 \sin x}{\sin^2 x}$; $2\left(\dfrac{1}{\sin x}\right)$; $2 \csc x =$

$2 \csc x$ **14.** $\cos x + \sin x = \dfrac{\sec x + \csc x}{\cot x + \tan x}$;

$\cos x + \sin x = \left(\dfrac{1}{\cos x} + \dfrac{1}{\sin x}\right) \div \left(\dfrac{\cos x}{\sin x} + \dfrac{\sin x}{\cos x}\right)$;

$\cos x + \sin x = \left(\dfrac{\sin x + \cos x}{\cos x \sin x}\right) \div \left(\dfrac{\cos^2 x + \sin^2 x}{\sin x \cos x}\right)$;

$\cos x + \sin x = \dfrac{\sin x + \cos x}{\cos^2 x + \sin^2 x}$;

$\cos x + \sin x = \sin x + \cos x$ **15.** $\dfrac{1 - \tan^2 x}{1 + \tan^2 x}$

$= \dfrac{2 - \sec^2 x}{\sec^2 x}$; $\dfrac{2 - (1 + \tan^2 x)}{1 + \tan^2 x}$; $\dfrac{1 - \tan^2 x}{1 + \tan^2 x}$

16. $\tan(180 + \theta) + \dfrac{1}{\cot(270 - \theta)}$; $\tan \theta + \tan(270 - \theta)$;

$\tan \theta + \tan(90 - \theta)$; $\tan \theta + \cot \theta$; $\dfrac{\sin \theta}{\cos \theta} + \dfrac{\cos \theta}{\sin \theta}$;

$\dfrac{\sin^2 \theta + \cos^2 \theta}{\cos \theta \sin \theta}$; $\dfrac{1}{\cos \theta \sin \theta}$; $\dfrac{1}{\cos \theta} \cdot \dfrac{1}{\sin \theta}$; $\dfrac{\sec \theta}{\sin \theta}$;

$\dfrac{-\sec \theta}{-\sin \theta}$; $\dfrac{\sec(180 - \theta)}{\sin(180 + \theta)}$ **17.** $\dfrac{\cot^2 \theta - 1}{\cot^2 \theta + 1} = \dfrac{2 - \sec^2 \theta}{\sec^2 \theta}$;

$\dfrac{2 - (1 + \tan^2 \theta)}{1 + \tan^2 \theta}$; $\dfrac{1 - \tan^2 \theta}{1 + \tan^2 \theta}$; $\dfrac{\frac{1 - \tan^2 \theta}{\tan^2 \theta}}{\frac{1 + \tan^2 \theta}{\tan^2 \theta}}$; $\dfrac{\frac{1}{\tan^2 \theta} - 1}{\frac{1}{\tan^2 \theta} + 1}$;

$\dfrac{\cot^2 \theta - 1}{\cot^2 \theta + 1}$ **18.** $\sqrt{\dfrac{\sec \theta - \tan \theta}{\sec \theta + \tan \theta}} = \dfrac{1 - \sin \theta}{\cos \theta}$;

$\sqrt{\left(\dfrac{1}{\cos \theta} - \dfrac{\sin \theta}{\cos \theta}\right) \div \left(\dfrac{1}{\cos \theta} + \dfrac{\sin \theta}{\cos \theta}\right)} = \dfrac{1 - \sin \theta}{\cos \theta}$;

$\sqrt{\left(\dfrac{1 - \sin \theta}{\cos \theta}\right) \div \left(\dfrac{1 + \sin \theta}{\cos \theta}\right)} = \dfrac{1 - \sin \theta}{\cos \theta}$;

$\sqrt{\dfrac{1 - \sin \theta}{1 + \sin \theta} \cdot \dfrac{1 - \sin \theta}{1 - \sin \theta}} = \dfrac{1 - \sin \theta}{\cos \theta}$;

$\sqrt{\dfrac{(1 - \sin \theta)^2}{1 - \sin^2 \theta}} = \dfrac{1 - \sin \theta}{\cos \theta}$; $\sqrt{\dfrac{(1 - \sin \theta)^2}{\cos^2 \theta}}$

$= \dfrac{1 - \sin \theta}{\cos \theta}$; $\dfrac{1 - \sin \theta}{\cos \theta} = \dfrac{1 - \sin \theta}{\cos \theta}$

Page 528

15. $\sin x = -\dfrac{1}{2}$; $x = \dfrac{7\pi}{6}, \dfrac{11\pi}{6}$ **16.** $\cos x = \dfrac{\sqrt{3}}{2}$;

$x = \dfrac{\pi}{6}, \dfrac{11\pi}{6}$ **17.** $\dfrac{\pi}{4}, \dfrac{5\pi}{4}$ **18.** $\sin x = 0$; $x = 0$,

$\pi, 2\pi$ **19.** $\cos^2 x = \dfrac{1}{2}$; $\cos x = \pm \dfrac{1}{\sqrt{2}} = \pm \dfrac{\sqrt{2}}{2}$;

$x = \dfrac{\pi}{4}, \dfrac{7\pi}{4}, \dfrac{3\pi}{4}, \dfrac{5\pi}{4}$ **20.** $2 \sin x = -\sqrt{3}$;

$\sin x = \dfrac{-\sqrt{3}}{2}$; $x = \dfrac{4\pi}{3}, \dfrac{5\pi}{3}$ **21.** $\cos x(2 \cos x + 1)$

$= 0$; $\cos x = 0$; $\cos x = -\dfrac{1}{2}$; $x = \dfrac{\pi}{2}, \dfrac{3\pi}{2}, \dfrac{2\pi}{3}, \dfrac{4\pi}{3}$

22. $\sin x(\sqrt{2} \sin x - 1) = 0$; $\sin x = 0$; $\sin x = \dfrac{\sqrt{2}}{2}$;

$x = 0, \pi, 2\pi, \dfrac{\pi}{4}, \dfrac{3\pi}{4}$ **23.** $(2 \cos x + 1)(\cos x - 4)$

$= 0$; $\cos x = -\dfrac{1}{2}$; $\cos x = 4$ (reject); $x = \dfrac{2\pi}{3}, \dfrac{4\pi}{3}$

24. $\sin^2 x = \dfrac{1}{4}$; $\sin x = \pm \dfrac{1}{2}$; $x = \dfrac{\pi}{6}, \dfrac{5\pi}{6}, \dfrac{7\pi}{6}, \dfrac{11\pi}{6}$

25. $(3 \sin x - 5)(2 \sin x + 1) = 0$; $\sin x = \dfrac{5}{3}$

(reject); $\sin x = -\dfrac{1}{2}$; $x = \dfrac{7\pi}{6}, \dfrac{11\pi}{6}$

26. $4 \cos^2 x = 1$; $\cos^2 x = \dfrac{1}{4}$; $\cos x = \pm \dfrac{1}{2}$;

$x = \dfrac{\pi}{3}, \dfrac{2\pi}{3}, \dfrac{4\pi}{3}, \dfrac{5\pi}{3}$

Page 532

For Exercises 24–35, points on the curve are given.

24. $(0, 0)$, $\left(\dfrac{\sqrt{2}}{2}, \dfrac{\pi}{2}\right)$, $(1, \pi)$, $(0, 2\pi)$

25. $(2, 0)$, $\left(1, \dfrac{\pi}{6}\right)$, $\left(0, \dfrac{\pi}{4}\right)$, $\left(-1, \dfrac{\pi}{3}\right)$

26. $\left(\dfrac{1}{2}, \pi\right)$, $\left(\dfrac{\sqrt{3}}{4}, \dfrac{2\pi}{3}\right)$, $\left(\dfrac{\sqrt{2}}{4}, \dfrac{\pi}{2}\right)$, $(0, 0)$

27. $\left(2\sqrt{3}, -\dfrac{\pi}{3}\right)$, $\left(2, -\dfrac{\pi}{4}\right)$, $(0, 0)$, $\left(-2, \dfrac{\pi}{4}\right)$

28. $(-2, 0)$, $\left(0, -\dfrac{\pi}{4}\right)$, $\left(1, -\dfrac{\pi}{3}\right)$, $\left(2, -\dfrac{\pi}{2}\right)$

29. $(0, 0)$, $\left(1, \dfrac{\pi}{4}\right)$, $\left(-\dfrac{\sqrt{3}}{3}, -\dfrac{\pi}{3}\right)$, $\left(-1, -\dfrac{\pi}{4}\right)$

30. $\left(-1, -\dfrac{\pi}{2}\right)$, $\left(-\dfrac{\sqrt{3}}{2}, -\dfrac{\pi}{3}\right)$, $(0, 0)$, $\left(1, \dfrac{\pi}{2}\right)$

31. $(1, 0)$, $\left(\dfrac{\sqrt{2}}{2}, 4\right)$, $\left(\dfrac{1}{2}, \dfrac{\pi}{3}\right)$, $\left(0, \dfrac{\pi}{2}\right)$

32. $(0, 0)$, $\left(-1, -\dfrac{\pi}{4}\right)$, $\left(1, \dfrac{\pi}{4}\right)$, $\left(\sqrt{3}, \dfrac{\pi}{3}\right)$

33. $(0, 0)$, $\left(1, \dfrac{\pi}{2}\right)$, $\left(-1, -\dfrac{\pi}{2}\right)$, $\left(\dfrac{\sqrt{3}}{3}, \dfrac{2\pi}{3}\right)$

34. $(1, 0)$, $\left(\sqrt{2}, \dfrac{\pi}{4}\right)$, $\left(2, \dfrac{\pi}{3}\right)$, $(-1, \pi)$

35. $\left(-1, -\dfrac{\pi}{2}\right)$, $\left(-\sqrt{2}, -\dfrac{\pi}{4}\right)$, $(0, 0)$, $\left(1, \dfrac{\pi}{2}\right)$

Page 535

25. $\sin^2 240° + \cos^2 240°$; $\left(\frac{-\sqrt{3}}{2}\right)^2 + \left(-\frac{1}{2}\right)^2$
$= \frac{3}{4} + \frac{1}{4} = 1$ **26.** $(\sqrt{3})^2 = (-2)^2 - 1$;
$3 = 4 - 1$; $3 = 3$ **30.** $\frac{\sin\theta}{3\sin^2\theta + (1 - \cos^2\theta)}$
$= \frac{1}{4}\csc\theta$; $\frac{\sin\theta}{3\sin^2\theta + \sin^2\theta} = \frac{1}{4}\csc\theta$; $\frac{\sin\theta}{4\sin^2\theta}$
$= \frac{1}{4}\csc\theta$; $\frac{1}{4\sin\theta} = \frac{1}{4}\csc\theta$; $\frac{1}{4}\csc\theta = \frac{1}{4}\csc\theta$

31. $\sin\theta \div \frac{\cos\theta}{\sin\theta} + \cos\theta = \sec\theta$; $\frac{\sin^2\theta}{\cos\theta} + \frac{\cos\theta}{1}$
$= \sec\theta$; $\frac{\sin^2\theta + \cos^2\theta}{\cos\theta} = \sec\theta$; $\frac{1}{\cos\theta} = \sec\theta$;
$\sec\theta = \sec\theta$

Page 536

16. $\cos^2 150° = 1 - \sin^2 150°$; $\left(-\frac{\sqrt{3}}{2}\right)^2$
$= 1 - \left(\frac{1}{2}\right)^2$; $\frac{3}{4} = 1 - \frac{1}{4}$; $\frac{3}{4} = \frac{3}{4}$

19. $\sec\theta = \left(\frac{\sin\theta}{\cos\theta} \div \frac{1}{\sin\theta}\right) + \cos\theta$; $\sec\theta =$
$\frac{\sin^2\theta}{\cos\theta} + \frac{\cos\theta}{1}$; $\sec\theta = \frac{\sin^2\theta + \cos^2\theta}{\cos\theta}$;
$\sec\theta = \frac{1}{\cos\theta}$; $\sec\theta = \sec\theta$

32. $(1, 0)$, $\left(\frac{\sqrt{2}}{2}, \frac{\pi}{4}\right)$, $\left(0, \frac{\pi}{2}\right)$, $(-1, \pi)$

33. $\left(-1, -\frac{\pi}{4}\right)$, $(0, 0)$, $\left(1, \frac{\pi}{4}\right)$, $\left(\text{undef.}, \frac{\pi}{2}\right)$

34. $\left(-1, -\frac{\pi}{2}\right)$, $\left(-\sqrt{2}, -\frac{\pi}{4}\right)$, $(\text{undef.}, 0)$, $\left(1, \frac{\pi}{2}\right)$

Page 541

25. $b^2 = h^2 + y^2$, but $h^2 = c^2 - x^2$
$b^2 = c^2 - x^2 + y^2$
$b^2 = (y^2 - x^2) + c^2$
$b^2 = (y + x)(y - x) + c^2$
$b^2 = (a - x + x)(a - x - x) + c^2$
$b^2 = a(a - 2x) + c^2$
$b^2 = a^2 - 2ax + c^2$ $\begin{cases} a = x + y \\ y = a - x \end{cases}$
$b^2 = a^2 - 2a(c\cos B) + c^2$ $\begin{cases} \cos B = \frac{x}{c} \\ x = c\cos B \end{cases}$
$b^2 = a^2 + c^2 - 2ac\cos B$

26. $c^2 = h^2 + y^2$, but $h^2 = b^2 - x^2$
$c^2 = b^2 - x^2 + y^2$
$c^2 = (y^2 - x^2) + b^2$
$c^2 = (y + x)(y - x) + b^2$
$c^2 = (a - x + x)(a - x - x) + b^2$ $\begin{cases} a = x + y \\ y = a - x \end{cases}$
$c^2 = a(a - 2x) + b^2$
$c^2 = a^2 - 2ax + b^2$
$c^2 = a^2 + b^2 - 2a(b\cos C)$ $\begin{cases} \cos C = \frac{x}{b} \\ x = b\cos C \end{cases}$
$c^2 = a^2 + b^2 - 2ab\cos C$

27. $c^2 = h^2 + x^2$ or $h^2 = c^2 - x^2$
$a^2 = (b + x)^2 + h^2$
$a^2 = (b + x)^2 + (c^2 - x^2)$
$a^2 = b^2 + 2bx + x^2 + c^2 - x^2$
$a^2 = b^2 + c^2 + 2bx$
$a^2 = b^2 + c^2 - 2bc\cos A$ $\begin{cases} \cos A = -\frac{x}{c} \\ x = -c\cos A \end{cases}$

28. $c^2 = h^2 + x^2$ or $h^2 = c^2 - x^2$
$b^2 = h^2 + (a + x)^2$
$b^2 = c^2 - x^2 + a^2 + 2ax + x^2$
$b^2 = a^2 + c^2 + 2ax$
$b^2 = a^2 + c^2 - 2ac\cos A$ $\begin{cases} \cos B = -\frac{x}{c} \\ x = -c\cos B \end{cases}$

Page 543

20. For any $\triangle ABC$, $b^2 = a^2 + c^2 - 2ac\cos B$;
$2ac\cos B + b^2 = a^2 + c^2$; $2ac\cos B = a^2 + c^2 - b^2$;
$\cos B = \frac{a^2 + c^2 - b^2}{2ac}$ **21.** For any $\triangle ABC$,
$c^2 = a^2 + b^2 - 2ab\cos C$; $2ab\cos C = 2ab\cos C$;
$c^2 + 2ab\cos C = a^2 + b^2$; $2ab\cos C = a^2 + b^2 - c^2$;
$\cos C = \frac{a^2 + b^2 - c^2}{2ab}$

Page 545

17. $\sin C = \frac{h}{a}$; $h = a \cdot \sin C$; but $k = \frac{1}{2}bh$;
$k = \frac{1}{2}b(a \cdot \sin C)$; $k = \frac{1}{2}ab\sin C$

18. Area of $\triangle ABD = \frac{1}{2}m \cdot n \cdot \sin A$; Area of
parallelogram $= 2 \cdot$ Area of $\triangle ABD =$
$2\left(\frac{1}{2}m \cdot n \cdot \sin A\right)$; Area of parallelogram $= m \cdot n \cdot \sin A$

Page 548

15. $\frac{1}{2}st\sin R = \frac{1}{2}tr\sin S = \frac{1}{2}sr\sin T$;
$\frac{st\sin R}{str} = \frac{tr\sin S}{str} = \frac{sr\sin T}{str}$; $\frac{\sin R}{r} = \frac{\sin S}{s} = \frac{\sin T}{t}$;

16. $\frac{1}{2}rq\sin S = \frac{1}{2}rs\sin Q = \frac{1}{2}qs\sin R$;
$\frac{rq\sin S}{rqs} = \frac{rs\sin Q}{rsq} = \frac{qs\sin R}{rsq}$; $\frac{\sin S}{s} = \frac{\sin Q}{q} = \frac{\sin R}{r}$;
$\frac{s}{\sin S} = \frac{q}{\sin Q} = \frac{r}{\sin R}$

Page 555

4. $\sin 195° = \sin(135° + 60°)$; $\sin(135° + 60°) =$
$\sin 135°\cos 60° + \cos 135°\sin 60° = \frac{\sqrt{2}}{2} \cdot \frac{1}{2} + \left(\frac{-\sqrt{2}}{2}\right) \cdot$
$\frac{\sqrt{3}}{2} = \frac{\sqrt{2}}{4} - \frac{\sqrt{6}}{4} = \frac{\sqrt{2} - \sqrt{6}}{4}$ **5.** $\sin 285° =$
$\sin(225° + 60°) = \sin 225°\cos 60° + \cos 225°\sin 60° =$
$\left(\frac{-\sqrt{2}}{2}\right)\left(\frac{1}{2}\right) + \left(\frac{-\sqrt{2}}{2}\right)\left(\frac{\sqrt{3}}{2}\right) = \frac{-\sqrt{2}}{4} - \frac{\sqrt{6}}{4} =$
$-\left(\frac{\sqrt{2} + \sqrt{6}}{4}\right)$ **6.** $\sin 165° = \sin(135° + 30°) =$
$\sin 135°\cos 30° + \cos 135°\sin 30° =$
$\frac{\sqrt{2}}{2} \cdot \frac{\sqrt{3}}{2} + \left(\frac{-\sqrt{2}}{2}\right) \cdot \frac{1}{2} = \frac{\sqrt{6}}{4} - \frac{\sqrt{2}}{4} = \frac{\sqrt{6} - \sqrt{2}}{4}$

7. $\sin 150° = \sin(90° + 60°) = \sin 90° \cos 60° + \cos 90° \sin 60° = 1 \cdot \frac{1}{2} + 0 = \frac{1}{2}$ **8.** $\sin 255° = \sin(225° + 30°) = \sin 225° \cos 30° + \cos 225° \sin 30° = \left(\frac{-\sqrt{2}}{2}\right)\left(\frac{\sqrt{3}}{2}\right) + \left(\frac{-\sqrt{2}}{2}\right)\left(\frac{1}{2}\right) = \frac{-\sqrt{6}}{4} - \frac{\sqrt{2}}{4} = -\left(\frac{\sqrt{6} + \sqrt{2}}{4}\right)$ **9.** $\sin 435° = \sin(405° + 30°) = \sin 405° \cos 30° + \cos 405° \sin 30° = \frac{\sqrt{2}}{2} \cdot \frac{\sqrt{3}}{2} + \frac{\sqrt{2}}{2} \cdot \frac{1}{2} = \frac{\sqrt{6}}{4} + \frac{\sqrt{2}}{4} = \frac{\sqrt{6} + \sqrt{2}}{4}$ **10.** $\sin 375° = \sin(315° + 60°) = \sin 315° \cos 60° + \cos 315° \sin 60° = -\left(\frac{\sqrt{2}}{2}\right)\left(\frac{1}{2}\right) + \frac{\sqrt{2}}{2} \cdot \frac{\sqrt{3}}{2} = \frac{-\sqrt{2}}{4} + \frac{\sqrt{6}}{4} = \frac{\sqrt{6} - \sqrt{2}}{4}$ **35.** $\sin(\pi + \theta) = -\sin \theta$; $\sin \pi \cos \theta + \cos \pi \sin \theta = -\sin \theta$; $0 + (-1)\sin \theta = -\sin \theta$; $-\sin \theta = -\sin \theta$ **36.** $\sin(2\pi - \theta) = -\sin \theta$; $\sin 2\pi \cos \theta - \cos 2\pi \sin \theta = -\sin \theta$; $0 - \sin \theta = -\sin \theta$ **37.** $\sin(\pi - \theta) = \sin \theta$; $\sin \pi \cos \theta - \cos \pi \sin \theta = \sin \theta$; $0 - (-1)\sin \theta = \sin \theta$; $\sin \theta = \sin \theta$ **38.** $\sin\left(\frac{\pi}{2} - \theta\right) = \cos \theta$; $\sin \frac{\pi}{2} \cos \theta - \cos \frac{\pi}{2} \sin \theta = \cos \theta$; $1 \cdot \cos \theta - 0 = \cos \theta$; $\cos \theta = \cos \theta$ **39.** $\sin\left(\frac{3\pi}{2} + \theta\right) = -\cos \theta$; $\sin \frac{3\pi}{2} \cos \theta + \cos \frac{3\pi}{2} \sin \theta = -\cos \theta$; $(-1)\cos \theta + 0 = -\cos \theta$; $-\cos \theta = -\cos \theta$ **40.** $\sin\left(\frac{\pi}{2} + \theta\right) = \cos \theta$; $\sin \frac{\pi}{2} \cos \theta + \cos \frac{\pi}{2} \sin \theta = \cos \theta$; $1 \cdot \cos \theta + 0 = \cos \theta$; $\cos \theta = \cos \theta$ **41.** $\sin(180° + \theta) = -\sin \theta$; $\sin 180° \cos \theta + \cos 180° \sin \theta = -\sin \theta$; $0 + (-1)\sin \theta = -\sin \theta$; $-\sin \theta = -\sin \theta$ **42.** $\sin(180° - \theta) = \sin \theta$; $\sin 180° \cos \theta - \cos 180° \sin \theta = \sin \theta$; $0 - (-1)\sin \theta = \sin \theta$; $\sin \theta = \sin \theta$ **43.** $\sin(90° + \theta) = \cos \theta$; $\sin 90° \cos \theta + \cos 90° \sin \theta = \cos \theta$; $1 \cdot \cos \theta + 0 = \cos \theta$; $\cos \theta = \cos \theta$ **44.** $\sin(90° - \theta) = \cos \theta$; $\sin 90° \cos \theta - \cos 90° \sin \theta = \cos \theta$; $\cos \theta - 0 = \cos \theta$; $\cos \theta = \cos \theta$ **45.** $\sin(360° - \theta) = -\sin \theta$; $\sin 360° \cos \theta - \cos 360° \sin \theta = -\sin \theta$; $0 - (1)\sin \theta = -\sin \theta$; $-\sin \theta = -\sin \theta$ **46.** $\sin(270° + \theta) = -\cos \theta$; $\sin 270° \cos \theta + \cos 270° \sin \theta = -\cos \theta$; $-1 \cdot \cos \theta + 0 = -\cos \theta$; $-\cos \theta = -\cos \theta$

Page 559

1. $\frac{\sqrt{2} - \sqrt{6}}{4}$ **2.** $\frac{\sqrt{2} + \sqrt{6}}{4}$ **3.** $\frac{-\sqrt{2}}{2}$ **4.** $\frac{-\sqrt{2} - \sqrt{6}}{4}$ **5.** $\frac{-\sqrt{2} + \sqrt{6}}{4}$ **6.** $\frac{-\sqrt{2} - \sqrt{6}}{4}$ **7.** $\frac{-\sqrt{3}}{2}$ **8.** $\frac{-\sqrt{2} - \sqrt{6}}{4}$ **9.** $\frac{-\sqrt{2} + \sqrt{6}}{4}$ **10.** $\frac{\sqrt{2} + \sqrt{6}}{4}$ **32.** $\cos(180° - \theta) = -\cos \theta$; $\cos 180° \cos \theta + \sin 180° \sin \theta = -\cos \theta$; $-1 \cdot \cos \theta + 0 = -\cos \theta$; $-\cos \theta = -\cos \theta$ **33.** $\cos(360° - \theta) = \cos \theta$; $\cos 360° \cos \theta + \sin 360° \sin \theta = \cos \theta$; $1 \cdot \cos \theta + 0 = \cos \theta$; $\cos \theta = \cos \theta$ **34.** $\cos(180° + \theta) = -\cos \theta$; $\cos 180° \cos \theta - \sin 180° \sin \theta = -\cos \theta$; $-1 \cdot \cos \theta - 0 = -\cos \theta$; $-\cos \theta = -\cos \theta$ **35.** $\cos(\pi + \theta) = -\cos \theta$; $\cos \pi \cos \theta - \sin \pi \sin \theta = -\cos \theta$; $-1 \cdot \cos \theta - 0 = -\cos \theta$; $-\cos \theta = -\cos \theta$ **36.** $\cos(270° + \theta) = \sin \theta$; $\cos 270° \cos \theta - \sin 270° \sin \theta = \sin \theta$; $0 - (-1)(\sin \theta) = \sin \theta$; $\sin \theta = \sin \theta$ **37.** $\cos(2\pi + \theta) = \cos \theta$; $\cos 2\pi \cos \theta - \sin 2\pi \sin \theta = \cos \theta$; $1 \cdot \cos \theta - 0 = \cos \theta$; $\cos \theta = \cos \theta$ **38.** $\cos\left(\frac{\pi}{2} + \theta\right) = -\sin \theta$; $\cos \frac{\pi}{2} \cos \theta - \sin \frac{\pi}{2} \sin \theta = -\sin \theta$; $0 - 1 \cdot \sin \theta = -\sin \theta$; $-\sin \theta = -\sin \theta$ **39.** $\cos(\pi - \theta) = -\cos \theta$; $\cos \pi \cos \theta + \sin \pi \sin \theta = -\cos \theta$; $-1 \cdot \cos \theta + 0 = -\cos \theta$; $-\cos \theta = -\cos \theta$ **40.** $\cos\left(\frac{3\pi}{2} - \theta\right) = -\sin \theta$; $\cos \frac{3\pi}{2} \cos \theta + \sin \frac{3\pi}{2} \sin \theta = -\sin \theta$; $0 + (-1)\sin \theta = -\sin \theta$; $-\sin \theta = -\sin \theta$

Page 562

33. $\tan(180° + \theta) = \frac{\tan 180° + \tan \theta}{1 - \tan 180° \tan \theta} = \frac{0 + \tan \theta}{1 - 0} = \tan \theta$ **34.** $\tan(180° - \theta) = \frac{\tan 180° - \tan \theta}{1 + \tan 180° \tan \theta} = \frac{0 - \tan \theta}{1 + 0} = -\tan \theta$ **35.** $\tan(\pi + \theta) = \frac{\tan \pi + \tan \theta}{1 - \tan \pi \tan \theta} = \frac{0 + \tan \theta}{1 - 0} = \tan \theta$ **36.** $\tan(360° - \theta) = \frac{\tan 360° - \tan \theta}{1 + \tan 360° \tan \theta} = \frac{0 - \tan \theta}{1 + 0} = -\tan \theta$ **37.** $\tan(2\pi + \theta) = \frac{\tan 2\pi + \tan \theta}{1 - \tan 2\pi \tan \theta} = \frac{0 + \tan \theta}{1 - 0} = \tan \theta$ **38.** $\tan(\pi - \theta) = \frac{\tan \pi - \tan \theta}{1 + \tan \pi \tan \theta} = \frac{0 - \tan \theta}{1 + 0} = -\tan \theta$

Page 565

1. $\frac{1}{2}\sqrt{2 - \sqrt{3}}$ **2.** $\frac{1}{2}\sqrt{2 + \sqrt{3}}$ **3.** $\frac{1}{2}\sqrt{2 - \sqrt{2}}$ **4.** $\sqrt{3 - 2\sqrt{2}}$ **5.** $\frac{1}{2}\sqrt{2 + \sqrt{2}}$ **6.** $\sqrt{7 - 4\sqrt{3}}$ **7.** $\sqrt{3 + 2\sqrt{2}}$ **8.** $\frac{1}{2}\sqrt{2 - \sqrt{2}}$ **9.** $\frac{1}{2}\sqrt{2 + \sqrt{2}}$ **10.** $\frac{1}{2}\sqrt{2 + \sqrt{3}}$ **11.** $\pm\frac{1}{5}\sqrt{5}, \pm\frac{2}{5}\sqrt{5}$ **12.** $\pm\frac{1}{2}\sqrt{2}, \pm\frac{1}{2}\sqrt{2}$ **13.** $\pm\frac{3}{13}\sqrt{13}, \pm\frac{2}{13}\sqrt{13}$ **14.** $\pm\frac{1}{2}\sqrt{3}, \pm\frac{1}{2}$ **15.** $\pm\frac{1}{4}\sqrt{6}, \pm\frac{1}{4}\sqrt{10}$ **16.** $\pm\frac{4}{17}\sqrt{17}, \pm\frac{\sqrt{17}}{17}$ **17.** $\pm\frac{1}{4}\sqrt{14}, \pm\frac{1}{4}\sqrt{2}$ **18.** $\pm 1, 0$ **19.** $\frac{1}{10}\sqrt{10}, -\frac{3}{10}\sqrt{10}, -\frac{1}{3}$ **20.** $\frac{5}{26}\sqrt{26}, \frac{1}{26}\sqrt{26}, 5$ **21.** $\frac{1}{2}\sqrt{2 + \sqrt{3}}, -\frac{1}{2}\sqrt{2 - \sqrt{3}}, -\sqrt{7 + 4\sqrt{3}}$ **22.** $\frac{1}{2}\sqrt{3}, \frac{1}{2}, \sqrt{3}$ **23.** $\frac{1}{3}\sqrt{3}, \frac{1}{3}\sqrt{6}, \frac{1}{2}\sqrt{2}$ **24.** $\frac{1}{4}\sqrt{2}, -\frac{1}{4}\sqrt{14},$

$-\frac{1}{7}\sqrt{7}$ **25.** $\frac{1}{12}\sqrt{102}, \frac{1}{12}\sqrt{42}, \frac{1}{7}\sqrt{119}$
26. $\frac{1}{14}\sqrt{154}, -\frac{1}{14}\sqrt{142}, -\frac{1}{3}\sqrt{33}$ **27.** $\frac{1}{2}\sqrt{2(1-a)}$,
$\frac{1}{2}\sqrt{2(1+a)}, \frac{1}{1+a}\sqrt{1-a^2}$ **28.** $\frac{1}{2}\sqrt{2+2b}$,
$\frac{1}{2}\sqrt{2-2b}, \frac{\sqrt{1-b^2}}{1-b}$ **33.** $\cot\frac{A}{2} = \left(\cos\frac{A}{2}\right) \div$
$\left(\sin\frac{A}{2}\right) = \left(\pm\sqrt{\frac{1+\cos A}{2}}\right) \div \left(\pm\sqrt{\frac{1-\cos A}{2}}\right)$
$= \pm\sqrt{\frac{1+\cos A}{1-\cos A}}$ **34.** $\sec\frac{A}{2} = \frac{1}{\cos\frac{A}{2}}$
$= 1 \div \pm\sqrt{\frac{1+\cos A}{2}} = \pm\sqrt{\frac{2}{1+\cos A}}$ **35.** $\csc\frac{A}{2}$
$= \frac{1}{\sin\frac{A}{2}} = 1 \div \pm\sqrt{\frac{1-\cos A}{2}} = \pm\sqrt{\frac{2}{1-\cos A}}$

Page 567

1. $\tan\theta = \frac{2\sin\theta\cos\theta}{1+2\cos^2\theta-1}$; $\tan\theta = \frac{2\sin\theta\cos\theta}{2\cos^2\theta}$;
$\tan\theta = \frac{\sin\theta}{\cos\theta}$; $\tan\theta = \tan\theta$ **2.** $\sin^2\theta = \frac{\sin\theta}{\cos\theta} \cdot$
$\frac{2\sin\theta\cos\theta}{2}$; $\sin^2\theta = \frac{2\sin^2\theta}{2}$; $\sin^2\theta = \sin^2\theta$
3. $\cot\theta = \frac{2\sin\theta\cos\theta}{2\sin^2\theta}$; $\cot\theta = \frac{\cos\theta}{\sin\theta}$; $\cot\theta = \cot\theta$
4. $\cos^2\frac{\theta}{2} = \left(\frac{1}{\cos\theta}+1\right) \div \left(2 \cdot \frac{1}{\cos\theta}\right)$;
$\cos^2\frac{\theta}{2} = \frac{1+\cos\theta}{\cos\theta} \cdot \frac{\cos\theta}{2}$; $\cos^2\frac{\theta}{2} = \frac{1+\cos\theta}{2}$;
$\cos^2\frac{\theta}{2} = \cos^2\frac{\theta}{2}$ **5.** $\tan\frac{\theta}{2} = \pm\frac{1-\cos\theta}{\sin\theta}$;
$\pm\sqrt{\frac{1-\cos\theta}{1+\cos\theta} \cdot \frac{1-\cos\theta}{1-\cos\theta}} = \frac{1-\cos\theta}{\sin\theta}$;
$\pm\sqrt{\frac{(1-\cos\theta)^2}{1-\cos^2\theta}} = \frac{1-\cos\theta}{\sin\theta}$; $\pm\sqrt{\frac{(1-\cos\theta)^2}{\sin^2\theta}} =$
$\frac{1-\cos\theta}{\sin\theta}$; $\pm\frac{1-\cos\theta}{\sin\theta} = \pm\frac{1-\cos\theta}{\sin\theta}$ **6.** $\tan\frac{\theta}{2} =$
$\pm\frac{\sin\theta}{1+\cos\theta}$; $\pm\sqrt{\frac{1-\cos\theta}{1+\cos\theta} \cdot \frac{1+\cos\theta}{1+\cos\theta}} =$
$\pm\frac{\sin\theta}{1+\cos\theta}$; $\pm\sqrt{\frac{1-\cos^2\theta}{(1+\cos\theta)^2}} = \pm\frac{\sin\theta}{1+\cos\theta}$;
$\pm\sqrt{\frac{\sin^2\theta}{(1+\cos\theta)^2}} = \pm\frac{\sin\theta}{1+\cos\theta}$; $\pm\frac{\sin\theta}{1+\cos\theta} =$
$\pm\frac{\sin\theta}{1+\cos\theta}$ **7.** $2\csc 2\theta = \frac{\cos^2\theta+\sin^2\theta}{\sin\theta\cos\theta}$;
$2\csc 2\theta = \frac{1}{\sin\theta\cos\theta}$; $2\csc 2\theta = 1 \div \frac{\sin 2\theta}{2}$;
$2\csc 2\theta = \frac{2}{\sin 2\theta}$; $2\csc 2\theta = 2\csc 2\theta$ **8.** $\sin^2\theta -$
$2\sin\theta\cos\theta + \cos^2\theta = 1 - 2\sin 2\theta$;
$\sin^2\theta + \cos^2\theta - 2\sin\theta\cos\theta = 1 - 2\sin 2\theta$;
$1 - 2\sin\theta\cos\theta = 1 - 2\sin 2\theta$; $1 - \sin 2\theta =$
$1 - \sin 2\theta$ **9.** $1 + \sin\theta = \sin^2\frac{\theta}{2} + 2\sin\frac{\theta}{2}$
$\cos\frac{\theta}{2} + \cos^2\frac{\theta}{2}$; $1 + \sin\theta = 1 + 2\sin\frac{\theta}{2}\cos\frac{\theta}{2}$;

$1 + \sin\theta = 1 + 2\sqrt{\frac{1-\cos\theta}{2}} \cdot \sqrt{\frac{1+\cos\theta}{2}}$;
$1 + \sin\theta = 1 + \frac{2}{2}\sqrt{1-\cos^2\theta}$; $1 + \sin\theta =$
$1 + \sqrt{\sin^2\theta}$; $1 + \sin\theta = 1 + \sin\theta$ **10.** $\cos\theta =$
$\left(\frac{1}{2}\sin^2\theta - \sin^2\frac{1}{2}\theta\right) \div \sin^2\frac{1}{2}\theta$; $\cos\theta =$
$\left(\frac{1}{2}\sin^2\theta - \left(\frac{1-\cos\theta}{2}\right)\right) \div \frac{1-\cos\theta}{2}$; $\cos\theta =$
$\frac{\sin^2\theta - 1 + \cos\theta}{1-\cos\theta}$; $\cos\theta = \frac{1-\cos^2\theta - 1 + \cos\theta}{1-\cos\theta}$;
$\cos\theta = \frac{\cos\theta(1-\cos\theta)}{1-\cos\theta}$; $\cos\theta = \cos\theta$
11. $\left(1 - \frac{\sin^2\theta}{\cos^2\theta}\right) \div \left(1 + \frac{\sin^2\theta}{\cos^2\theta}\right) = \cos 2\theta$;
$\left(\frac{\cos^2\theta - \sin^2\theta}{\cos^2\theta}\right) \div \left(\frac{\cos^2\theta + \sin^2\theta}{\cos^2\theta}\right) = \cos 2\theta$;
$\cos^2\theta - \sin^2\theta = \cos 2\theta$; $\cos 2\theta = \cos 2\theta$
12. $\frac{-(\sin\theta+\cos\theta)^2 + (\sin\theta-\cos\theta)^2}{\sin^2\theta - \cos^2\theta} = 2\tan 2\theta$;
$\frac{-\sin^2\theta - 4\sin\theta\cos\theta - \cos^2\theta + \sin^2\theta + \cos^2\theta}{\sin^2\theta - \cos^2\theta} =$
$2\left(\frac{2\tan\theta}{1-\tan^2\theta}\right)$; $\frac{-4\sin\theta\cos\theta}{\sin^2\theta - \cos^2\theta} = \left(4 \cdot \frac{\sin\theta}{\cos\theta}\right) \div$
$\left(1 - \frac{\sin^2\theta}{\cos^2\theta}\right)$; $\frac{4\sin\theta\cos\theta}{\cos^2\theta - \sin^2\theta} = \left(4 \cdot \frac{\sin\theta}{\cos\theta}\right) \div$
$\left(\frac{\cos^2\theta - \sin^2\theta}{\cos^2\theta}\right)$; $\frac{4\sin\theta\cos\theta}{\cos^2\theta - \sin^2\theta} =$
$\frac{4\sin\theta\cos\theta}{\cos^2\theta - \sin^2\theta}$ **13.** $\tan 2\theta = \frac{\sin 2(2\theta)}{1+\cos 2(2\theta)}$;
$\tan 2\theta = \frac{2\sin 2\theta\cos 2\theta}{1+2\cos^2 2\theta - 1}$; $\tan 2\theta =$
$\frac{2\sin(2\theta)\cos(2\theta)}{2\cos^2(2\theta)}$; $\tan 2\theta = \frac{\sin 2\theta}{\cos 2\theta}$; $\tan 2\theta = \tan 2\theta$
14. $\cos 3\theta = 4\cos^2\theta - 3\cos\theta$; $\cos(\theta + 2\theta) =$
$4\cos^2\theta - 3\cos\theta$; $\cos\theta\cos 2\theta - \sin\theta\sin 2\theta =$
$4\cos^2\theta - 3\cos\theta$; $\cos\theta[2\cos^2\theta - 1] - \sin\theta$
$[2\sin\theta\cos\theta] = 4\cos^2\theta - 3\cos\theta$; $2\cos^3\theta -$
$\cos\theta - 2\sin^2\theta\cos\theta = 4\cos^2\theta - 3\cos\theta$;
$2\cos^2\theta - \cos\theta - 2[1-\cos^2\theta]\cos\theta = 4\cos^2\theta -$
$3\cos\theta$; $2\cos^2\theta - \cos\theta - 2\cos\theta + 2\cos^2\theta =$
$4\cos^2\theta - 3\cos\theta$; $4\cos^2\theta - 3\cos\theta = 4\cos^2\theta -$
$3\cos\theta$ **15.** $\frac{2}{\sin 2\theta} = \sec\theta\csc\theta$; $\frac{2}{2\sin\theta\cos\theta} =$
$\sec\theta\csc\theta$; $\frac{1}{\sin\theta\cos\theta} = \sec\theta\csc\theta$; $\frac{1}{\cos\theta} \cdot \frac{1}{\sin\theta} =$
$\sec\theta\csc\theta$; $\sec\theta\csc\theta = \sec\theta\csc\theta$ **16.** $\sec 2\theta =$
$\left(\frac{1}{\cos^2\theta}\right) \div \left(\frac{2\cos^2\theta - 1}{\cos^2\theta}\right)$; $\sec 2\theta = \frac{1}{\cos^2\theta} \cdot$
$\frac{\cos^2\theta}{2\cos^2\theta - 1}$; $\sec 2\theta = \frac{1}{2\cos^2\theta - 1}$; $\sec 2\theta = \sec 2\theta$
17. $\frac{1}{2\cos^2\theta} + \frac{\sin\theta}{\cos\theta} = \frac{1+\sin 2\theta}{1+\cos 2\theta}$; $\frac{1+2\sin\theta\cos\theta}{2\cos^2\theta}$
$= \frac{1+\sin 2\theta}{1+\cos 2\theta}$; $\frac{1+\sin 2\theta}{\cos 2\theta + 1} = \frac{1+\sin 2\theta}{1+\cos 2\theta}$; $\frac{1+\sin 2\theta}{1+\cos 2\theta} =$

$\dfrac{1+\sin 2\theta}{1+\cos 2\theta}$ **18.** $\dfrac{\cos 3\theta}{\sin 3\theta} = \dfrac{1+\cos 6\theta}{\sin 6\theta} \cdot \dfrac{\cos 3\theta}{\sin 3\theta} =$
$\dfrac{1+\cos 2(3\theta)}{\sin 2(3\theta)} \cdot \dfrac{\cos 3\theta}{\sin 3\theta} = \dfrac{1+(2\cos^2 3\theta - 1)}{2 \sin 3\theta \cos 3\theta} \cdot \dfrac{\cos 3\theta}{\sin 3\theta} =$
$\dfrac{1 + 2\cos^2 3\theta - 1}{2\cos 3\theta \sin 3\theta} \cdot \dfrac{\cos 3\theta}{\sin 3\theta} = \dfrac{2 \cos^2 3\theta}{2 \sin 3\theta \cos 3\theta} \cdot \dfrac{\cos 3\theta}{\sin 3\theta} =$
$\dfrac{\cos 3\theta}{\sin 3\theta}$

Page 568

27. $\sin \dfrac{3\pi}{2} \cos\theta - \sin\theta \cos \dfrac{3\pi}{2} = -1 \cos\theta -$
$\sin\theta (0) = -\cos\theta$ **32.** $\cos^2 \dfrac{\theta}{2} = \left(\sin\theta + \dfrac{\sin\theta}{\cos\theta}\right) \div$
$\dfrac{2\sin\theta}{\cos\theta}$; $\cos^2 \dfrac{\theta}{2} = \dfrac{\sin\theta \cos\theta + \sin\theta}{2 \cos\theta} \cdot \dfrac{\cos\theta}{\sin\theta}$; $\cos^2 \dfrac{\theta}{2} =$
$\dfrac{(\cos\theta + 1)}{2}$; $\cos^2 \dfrac{\theta}{2} = \dfrac{1 + \cos\theta}{2}$; $\cos^2 \dfrac{\theta}{2} = \cos^2 \dfrac{\theta}{2}$
34. $\tan (2\pi - \theta) = \dfrac{\tan 2\pi - \tan\theta}{1 + \tan 2\pi \tan\theta} = \dfrac{0 - \tan\theta}{1 - 0(\tan\theta)} =$
$-\tan\theta$

Page 569

31. $\sin 2A = \dfrac{2\cos A}{\sin A} \div \dfrac{1}{\sin^2 A}$; $\sin 2A = \dfrac{2\cos A}{\sin A} \cdot$
$\dfrac{\sin^2 A}{1}$; $\sin 2A = 2 \cos A \sin A$; $\sin 2A = 2 \sin A \cos A$;
$\sin 2A = \sin 2A$ **32.** $2 \csc 2\theta = \tan\theta + \cot\theta$;
$2 \csc 2\theta = \dfrac{\sin\theta}{\cos\theta} + \dfrac{\cos\theta}{\sin\theta}$; $2 \csc 2\theta = \dfrac{\sin^2\theta + \cos^2\theta}{\cos\theta \sin\theta}$;
$2 \csc 2\theta = \dfrac{1}{\sin\theta \cos\theta}$; $2 \csc 2\theta = \dfrac{2}{2 \sin\theta \cos\theta}$;
$2 \csc 2\theta = \dfrac{2}{\sin 2\theta}$; $2 \csc 2\theta = 2 \csc 2\theta$
33. $\sin \dfrac{\pi}{2} \cos\theta + \cos \dfrac{\pi}{2} \sin\theta = 1 \cos\theta + 0 \sin\theta =$
$\cos\theta$ **34.** $\tan (\pi - \theta) = \dfrac{\tan\pi - \tan\theta}{1 + \tan\pi \tan\theta} =$
$\dfrac{0 - \tan\theta}{1 + 0 \cdot \tan\theta} = -\tan\theta$

Page 570

2.
```
10 PRINT "ANGLE IN DEGREES ":: INPUT D
20 R = 3.1416 * D / 180
30 PRINT "SIN(";D;") = "; SIN (R)
40 PRINT "COS(";D;") = "; COS (R)
50 PRINT "TAN(";D;") = "; TAN (R)
55 IF TAN (R) = 0 THEN 65
60 PRINT "CTN(";D;") = ";1 / TAN (R)
65 IF COS (R) = 0 THEN 75
70 PRINT "SEC(";D;") = ";1 / COS (R)
75 IF SIN (R) = 0 THEN 90
80 PRINT "CSC(";D;") = ";1 / SIN (R)
90 PRINT : GOTO 10
```

Page 573

42. $(0, 0)$, $(45°, \sqrt{2})$, $(90, 2)$, $(360°, 0)$ **43.** $(0°, 1)$,
$(45°, 0)$, $(90, -1)$, $(135°, 0)$ **44.** $(0, 0)$, $\left(45°, \dfrac{1}{2}\right)$,
$(90°, 0)$, $(180°, 0)$ **45.** $(0°, 0)$, $\left(22\dfrac{1}{2}°, 1\right)$, $\left(67\dfrac{1}{2}°, -1\right)$,
$(90°, 0)$ **46.** 2 values   **51.** $2 \cos^2\theta - 1 = \dfrac{1 - \tan^2\theta}{1 + \tan^2\theta}$;
$2 \cos^2\theta - 1 = \left(1 - \dfrac{\sin^2\theta}{\cos^2\theta}\right) \div \left(1 + \dfrac{\sin^2\theta}{\cos^2\theta}\right)$; $2 \cos^2\theta - 1$
$= \dfrac{\cos^2\theta - \sin^2\theta}{\cos^2\theta} \div \dfrac{\cos^2\theta + \sin^2\theta}{\cos^2\theta}$; $2 \cos^2\theta - 1 =$
$\cos^2\theta - \sin^2\theta$; $2 \cos^2\theta = \cos^2\theta - (1 - \cos^2\theta)$;
$2 \cos^2\theta - 1 = 2 \cos^2\theta - 1$ **53.** $(0, 0)$, $\left(-1, -\dfrac{\pi}{2}\right)$,
$\left(-\dfrac{\sqrt{3}}{2}, -\dfrac{\pi}{3}\right)$, $\left(1, \dfrac{\pi}{2}\right)$ **57.** $\tan x = \dfrac{2 \sin x \cos x}{1 + 2 \cos^2 x - 1}$;
$\tan x = \dfrac{2 \sin x \cos x}{2 \cos^2 x}$; $\tan x = \dfrac{\sin x}{\cos x}$; $\tan x = \tan x$
58. $\pm \sin^2 \dfrac{x}{2} = \left(\pm \sqrt{\dfrac{1 - \cos x}{1 + \cos x}} \cdot \sin x\right) \div 2$;
$\pm \left(\dfrac{1 - \cos x}{2}\right) = \left(\pm \sqrt{\dfrac{(1 - \cos x)^2}{1 - \cos^2 x}} \cdot \sin x\right) \div 2$;
$\pm \left(\dfrac{1 - \cos x}{2}\right) = \pm \dfrac{(1 - \cos x)}{2 \sin x} \cdot \sin x$; $\pm \left(\dfrac{1 - \cos x}{2}\right) =$
$\pm \left(\dfrac{1 - \cos x}{2}\right)$

Index

Abscissa, 209
Absolute value
  of a complex number, 189, 192, 533
  equation property for, 10
  equations involving, 10, 70
  inequalities involving, 11, 73
  inequality properties for, 11
  of a real number, 10
Addition
  of complex numbers, 188, 201
  of logarithms, 397
  of matrices, 258–259
  of polynomials, 38
  properties of matrices, 258–259
  of radicals, 145, 153
  of rational expressions, 96
  in solving systems of two or three linear equations, 244–249
  of vectors, 50–51
Additive identity, 26, 201, 258
Additive identity matrix, 258
Additive inverse (opposite), 26, 96–97, 188, 202, 258
  of complex numbers, 188, 202
  of matrices, 258
Ambiguous case, 549–551
Amplitude, 497, 503–504, 509–510
Angle(s)
  acute, 463
  cosecant of, 470–471
  cosine of, 462–464
  cosine of sum or difference of two, 556–557
  cotangent of, 466–469
  coterminal, 456
  of depression, 490
  double, 554, 557, 561, 566
  of elevation, 490
  half, 563–564, 566
  initial side of, 455
  and law of cosines, 542–543
  measure of, 455–456
  negative functions of, 473–474
  of negative measure, 455
  obtuse, 551
  of positive measure, 455
  quadrantal, 456
  reference, 456
  right, 551
  of rotation, 455–456, 475
  secant of, 470–471
  sine of, 462–464
  sine of sum or difference of two, 552–553
  tangent of, 466–469
  tangent of sum or difference of two, 560–561
  terminal side of, 455
Antilogarithm, 404, 407, 588–589
Applications, 26, 83, 106–107, 181, 201–202, 234, 297, 344–345, 418–419
Approximately equal ($\doteq$), 140
Arc cosine, 529–531
Arc sine, 529–531
Arc tangent, 531
Arithmetic mean(s), 361–362
Arithmetic progressions, 357–358
Arithmetic sequences, 357–358
Arithmetic series, 364–366
Arrangements, probability and, 445–446
Associative properties, 26, 106, 202, 259, 262
Asymptotes, 320–322
Augmented matrix, 265–266
Averages, 102–105, 361
  mean, 102–105
  median, 103
  mode, 102
Axes, 209
Axis of symmetry, 303, 314–317

Base(s)
  of exponents, 31, 35, 390
  of natural logarithms, 385
Base-b logarithms, 392, 406
Base-10 antilogarithms, logarithms, 403–404, 406, 588–589
Binomial(s), 37, 57–58, 60, 128, 130–131, 146
Binomial expansions, 382–383
Binomial theorem, 382, 437
Biography, 186, 356, 454
Bounds, upper and lower, 412–414

Calculator activities, 82, 90, 127, 133, 174, 177, 242, 360, 371, 378, 482, 517, 548
Career(s), 138, 208, 275, 496
Chapter Test, see Tests
Characteristic, 403
Circle, 306–309, 324–325, 354
Circular permutations, 433–434
Clockwise rotation, 455
Closure properties, 26, 107, 201–202
Coefficients, 38, 115, 129, 195, 382, 409, 413
Cofunctions, 478
College Prep Test, see Tests
Combinations, 435–436, 445
Combined variation, 295
Common binomial factor, 66
Common difference, 357, 364
Common monomial factor, 65–66
Common ratio, 368, 375, 379
Commutative properties, 26, 201–202, 209
Complementary functions, 478
Completing the square, 165–167
Complex fraction, 99
Complex numbers, 187–188, 190–191, 195, 201–202, 409–410, 419, 533–534
Complex rational expressions, 99
  simplifying, 99–100
Computations using logarithms, 588–589
Computer activities, 86, 110, 136, 162, 184, 235, 300, 331, 354, 450, 494, 570
Computer section, 574–587
Conditional permutations, 428–429
Conic section(s), 302–333
  circle, 306–309
  ellipse, 310–313
  hyperbola, 319–323
  parabola, 314–318
Conjugate zero theorem, 419
Conjugates, property of, 149, 191, 199, 419
Conjunctions, 7–9, 11, 71
Consecutive integers, 15, 74
Consistent systems, 240
Constant functions, 282–284
Constant of proportionality, 289
Constant term, 409, 413
Constant of variation, 288, 290, 291, 293
Converse of the Pythagorean relation, 224
Conversion factor, 94
Coordinate geometry, 209–237

623

Coordinate plane, 209, 303–304
Coordinates of a point, 209
Cosecant, 470–471
Cosine(s), 462–464, 556–558, 564
  as cofunction of sine, 478
  as complementary to sine, 478
  as infinite series, 384, 525
  law of, 539–540
Cotangent, 466–469
Coterminal angle, 456
Counterclockwise rotation, 455
Cramer's rule, 250, 254
Cube roots, 152, 193–194
Cubes, 61, 193
  difference of two, 61, 159
  factoring difference of two, 63, 154, 159
  factoring sum of two, 63, 154, 159, 194
  sum of two, 61, 159
Cumulative Reviews, see Reviews

Decimal
  as irrational number, 140
  nonrepeating, 140
  nonterminating, 140
  part of logarithm, 403
  rational number as, 89–90
  repeating, 89, 140, 380
  terminating, 89, 380
Degree of a polynomial, 37, 409–411
Degrees, radians, 523–524
De Moivre's theorem, 534
Denominator(s), 96–97, 99–100, 115–117
  of a complex rational expression, 99–100
  least common multiple (LCM) of, 97
  rationalizing, 148–150, 153–154, 191
Dependent events, 443
Dependent systems, 241
Descartes' (rule of signs), 418
Descending order of exponents, 39, 129
Determinants, 250–256
  Cramer's rule, 250, 254
  minor's, 255
Differences, 96, 100
  of complex numbers, 187–188
  of radicals, 145
  of two cubes, 61, 63, 154, 159
  of two squares, 60, 62, 146
Digits, 268–269
Dimensional analysis, 94–95

Direct variation, 288–289
Directrix of a parabola, 327
Discriminant, 195–196
Disjunction, 7–9, 11, 71
Distance, 47–48, 223–224
Distance formula(s), 47–48, 122–123, 224
Distinguishable permutations, 431–432
Distributive property, 2, 26, 41
Dividend, 128, 130
Division
  of complex numbers, 191
  of imaginary numbers, 191
  of polynomials, 128, 131
  of positive numbers using logarithms, 397–398
  of powers, 31, 35
  of radicals, 148, 153, 159
  of rational expressions, 92
  synthetic, 130–132
Divisor, 128, 130
Domain
  of an exponential function, 390–391
  of a function, 279
  of a logarithmic function, 392, 394–396
  of a rational function, 280
  of a relation, 276
Double angle identities, 566
Double angles, 554, 557–558, 561
Dry mixtures, 43–44

e, e^x, 385
Ellipse, 310–313, 324, 328–330
  equation of, 328–330
  identifying of, 324–325
  properties of, 329
Equality for
  complex numbers, 201
  exponentials, 33, 158
  logarithms, 398
  matrices, 257
Equation(s)
  absolute value and, 10, 70
  of the asymptotes of a hyperbola, 320–322
  of axis of symmetry of a parabola, 315–317
  of a circle, 306–308, 324
  derived, 116
  of an ellipse, 310–312, 324, 328–330
  exponential, 33, 158, 390, 393, 406
  fourth-degree, 69, 193, 410

  fractional, 115–117
  general form of conic section, 324
  of a higher degree, 409–411
  of a hyperbola, 319–321, 324
  of a hyperbola with x-intercepts, 319–320
  of a hyperbola with y-intercepts, 321
  with imaginary number solutions, 193
  of a line, see Linear equation(s)
  linear, see Linear equation(s)
  of linear-quadratic systems, 338–339
  linear trigonometric, 481–482
  literal, 125–126
  logarithmic, 393, 398, 401
  of a parabola, 314–315, 317, 324
  polynomial, 410, 414
  properties, 1
  quadratic, see Quadratic equations
  of quadratic systems, 341–342
  quadratic trigonometric, 515–516
  radical, 178–179
  of a rectangular hyperbola, 321
Equation property for absolute value, 10
Event(s)
  compound, 441–444
  dependent, 443
  exclusive, 442
  inclusive, 442
  independent, 443
  mutually exclusive, 442
  simple, 438–439
Exponent(s), 31, 35, 143
  irrational number as, 390–391, 408
  rational number as, 155–156
  zero as, 34
Exponential equations, 33, 158, 390, 393, 406
Exponential expression(s)
  in radical form, 143, 155–156
  simplifying, 31–32, 34–35, 157–159
Exponential form, 155
Expression(s)
  exponential, 31–32, 34–35, 155–157
  logarithmic, 392, 397–398, 400–401
  radical, 142–143, 145–146, 148–150, 152–156
Extraneous solution(s), 116, 178, 398, 401
Extremes of a proportion, 117

Factor(s), 57, 62–63, 65–66
  greatest common, 65–66
  perfect square, 142–143
  of a perfect square trinomial, 62
  of a trinomial, 57–58, 86
Factor theorem, 128
Factorials, 383–384, 426, 431–433, 435–436, 450
Factoring, 57–58, 65–66
  completely, 65–66
  a difference of two squares, 62
  a perfect square trinomial, 62
  a polynomial into its first-degree factors, 409
  and quadratic equations, 68–69
  and quadratic inequalities, 71–72
  a sum or difference of two cubes, 63
Field, 26, 106, 201
Foci
  of an ellipse, 328–329
  of a hyperbola, 319
  of a parabola, 327
FOIL method, 41, 146
Formula(s)
  for area of a rectangle, 79
  for area of a triangle, 79, 544–545
  for the arithmetic mean of two numbers, 362
  bacteria, for counting, 408
  binomial theorem, 382
  for circular permutations, 433
  compound interest, 407
  for cos x as infinite series, 384, 525
  for cosine of the difference of two angles, 557
  for cosine of the sum of two angles, 556
  for cosine of twice an angle, 557
  distance, 47–48, 122–123, 224
  for distinguishable permutations, 431
  for e as infinite series, 385
  for e^x as infinite series, 385
  for expected value, 83
  fundamental counting principle, 424
  for geometric mean of two numbers, 373
  $h = vt - 5t^2$, 23, 173
  for half-angles, 563–564
  as a logarithmic equation, 398, 400
  for $\log_b x \cdot y$ and $\log_b \frac{x}{y}$, 397
  for $\log_b x^r$ and $\log_b \sqrt[n]{x}$, 400
  midpoint, 226, 232
  for number of combinations, $\binom{n}{r}$; 435
  for nth term of A.P., 358
  for nth term of G.P., 369
  for perimeter of a rectangle, 18, 125
  Pythagorean, 78, 115
  quadratic, 169–170, 172–173, 175
  rth term of $(a + b)^n$, 383
  for simple interest, 22
  simple pendulum, 181
  for sin x as infinite series, 384, 525
  for sine of the difference of two angles, 553
  for sine of the sum of two angles, 552
  for sine of twice an angle, 554
  slope, 231–232
  special product, 60–61
  for the sum of finite geometric series, 375, 376
  for the sum of the first n terms of arithmetic series, 364–366
  for the sum of infinite geometric series, 379
  for tangent of the difference of two angles, 561
  for tangent of the sum of two angles, 560
  for tangent of twice an angle, 561
  for total amount, 22
  for total amount A
    including compound interest, 407
    including simple interest, 22
  for total resistance, 23
  volume, 176, 181
Fourth-degree equations, 69, 193, 410
Fraction(s), 99, 115
Frequency table, 103
Function(s), 276–296
  amplitude of, 497, 500–501, 503–504, 506–507, 509–510
  composite, 280, 300
  constant, 282–283
  cosecant, 470–471
  cosine, 462–464
  cotangent, 466–469
  determining whether the inverse is a, 286
  exponential, 390
  greatest integer, 284
  inverse, 285–286, 392, 394–395
  linear, 282
  logarithmic, 392
  of negative angles, 473–474
  periodic, 497
  polynomial, 409–414, 416–417
  reciprocal, 468
  as a relation, 276
  secant, 470–471
  sine, 462–464
  tangent, 466–469
  trigonometric, 462–475, 529–532
  values of, 279–281
  values of trigonometric, 475
Fundamental counting principle, 424–426, 445–446
Fundamental theorem of algebra, 410

Geometric mean, 372–373
Geometric progressions, 368–369
Geometric series, 375–377, 379–380
Geometry, coordinate, 209–237
Graph(s), 277, 279, 282–283, 286
  absolute value inequality, 11
  compound linear inequalities, 7–8
  conjunctions, 7–8
  disjunctions, 7–8
  of the equation of a circle, 306–308
  of the equation of an ellipse, 310–312
  of the equation of a hyperbola, 319–322
  of the equation of a parabola, 314–318
  of exponential functions, 390–391
  of inverses of trigonometric functions, 529–530
  of a linear equation in two variables, 210
  of linear inequalities, 4–5, 7–8
  of linear inequality systems, 346–347
  of linear-quadratic inequality systems, 349–350
  of linear-quadratic systems, 338
  of logarithmic functions, 394–396
  of an ordered pair in a coordinate plane, 209
  of a quadratic function (parabola), 314–318, 328
  quadratic inequality in one variable, 71–72
  of quadratic systems, 341
  of a relation, 276–277
  slope-intercept method for, 220–221
  of systems of two linear equations, 240–241
  of systems of two trigonometric

equations, 507
of trigonometric functions, 497–498, 509–510
of trigonometric functions applying radian measure, 526–527
Greatest common factor, 65–66

Half-angle formula, 563–564
Half-angle identities, 567
Hexagon, 19
Horizontal line(s), 214
Hyperbola, 319–322
Hypotenuse, 78, 175

Identity (Identities)
additive, 26, 201
double angle, 566
half-angle, 567
multiplicative, 26, 202
Pythagorean, 520
quotient, 519
reciprocal, 518
trigonometric, 518–522, 566–567
Imaginary numbers, 188, 193–195, 201–202, 418, 419
Inconsistent systems, 241
Independent events, 443
Independent systems, 240
Index, 152
Inequality (inequalities)
absolute value and, 11, 73
compound linear, 7–8
linear, see Linear inequalities
polynomial, 72
properties of, 4
properties for absolute value, 11
quadratic, 71–72
solutions of, see Solution set(s)
systems of linear, 346–347
Infinite geometric series, 379–380
Infinite series, special, 384–385
Integer(s), 15, 26, 74, 89
Integral zero theorem, 409
Interpolation, 483–484, 512–513
Inverse
additive, 26, 96, 188, 201
of exponential function, 394
of log function, 394–395
multiplicative, 26, 92, 106, 202
Inverse functions, 285, 392, 394–395, 529–532
Inverse relations, 285
Inverse variation, 291–292
Irrational numbers, 140–141, 175
as exponents, 390–391, 408

Joint variation, 294–295

Kovalevsky, Sonya, 186

Law of cosines, 539–540, 542–543, 570
Law of sines, 546–547
Laws of logarithms, 397–398, 400–401
Least common denominator (LCD), 97, 99, 115
Least common multiple (LCM), 97
Like radicals, 145
Like terms, 38, 145
Line(s)
determining whether a point is on, 220–221
equation of, 216–218
horizontal, 214
negative slope of, 213–214
parallel, 228
perpendicular, 229–230
positive slope of, 213–214
slope of, 212–214
undefined slope of, 214
vertical, 214
y-intercept of, 217
zero slope of, 214
Linear equation(s)
consistent system of, 240
dependent system of, 241
inconsistent system of, 241
independent system of, 240
in one variable, 1–2
in two variables, 210, 216
point-slope form of, 216–217
properties of, 1
slope-intercept form of, 217–218, 235
solving problems using systems of, 268–270
solving a system of three, 247–248
solving systems of three using matrix transformations, 265–266
solving by using Cramer's rule, 250, 254
solving by using determinants, 250–251, 253–254
solving by using minors, 255
system of two, 240–241
system of two solved by addition, 244–246
system of two solved by inverse of a matrix, 264–265
system of two solved by substitution, 243
Linear functions, 282
Linear inequalities, 4–5, 7–8

Linear inequality systems, 346–347
Linear polynomials, 37
Linear-quadratic systems, 338–339
Linear-quadratic inequality systems, 349–350
Linear trigonometric equations, 481–482
Literal equations, 125–126
Logarithm(s), 392–405, 588–589
base-b, 392, 406
base of natural, 385
base-10, 403–404
characteristic of, 403
common, 403
computations using, 588–589
decimal part of, 403
integer part of, 403
mantissa of, 403
of a power, 400–401, 588–589
of a product, 397–398
of a quotient, 397–398
of a radical, 400–401
of trigonometric functions, 512–514
Logarithmic equations, 397–398, 400–401
Logarithmic expressions, 399, 401–402
Logarithmic function, inverse of, 394–395
Log-trig table, 590–592
Lower bound, 412

Major axis, of an ellipse, 329
Mantissa, 403
Matrix (matrices), 257–267
addition properties of, 258–259
additive identity, 258
additive inverse, 258
associative addition property of, 259
augmented, 265–266
commutative addition property of, 259
equality of, 257
multiplication properties of, 260–262
multiplicative identity, 261
multiplicative inverse, 261–262
product of scalar and, 260
product of two, 260–261
properties of, 257–262
property of sum of two, 258
square, 257
transformations, 265–266
Maximum value, 167–168
Mean (average), 102–105
Mean(s)

arithmetic, 361–362
average, 102–105
geometric, 372–373
of a proportion, 117
proportional, 372–373
Measure
of central tendency, 102
degree, 523–524
radian, 523–527
Median, 102–103
Midpoint formula, 226, 232
Minor axis, of an ellipse, 329
Minors, 255
Mixture(s), 43–46
dry, 43–44
wet, 44–45
Mode, 102
Monomial(s), 37
Motion problems, 47–48, 122–123
Multiplication
of complex numbers, 190–191, 202
of exponentials, 31, 35, 157
FOIL method, 41, 146
of polynomials, 41–42, 60–61
properties of matrices, 260–262
of rational expressions, 92
of a scalar and a vector, 52
zero product, property of, 68–69
Multiplicative identity, 26, 202
Multiplicative identity matrix, 261
Multiplicative inverse, 26, 92, 106, 202
Multiplicative inverse matrix, 261
Multiplicity, 411, 418
Mutually exclusive events, 442

Natural logarithms, 385
Nature of solutions of quadratic equations, 195
Negative angles, functions of, 473–474
Negative exponents, 34–35
Negative measure of angle of rotation, 455
Negative slope of a line, 213–214
Nonrepeating decimal, 140
Non-routine problems, 3, 42, 46, 77, 82, 88, 101, 114, 118, 144, 151, 164, 177, 197, 222, 239, 287, 302, 323, 337, 340, 389, 405, 423, 430, 432, 434, 437, 472, 499, 505, 511, 538, 565
Nonterminating decimal, 140
nth root, 152–154
nth term, 358, 369
Number(s)
averages of, 102–105

complex, 187–188, 190–192, 195, 201–202, 410, 419, 533–534
conjugate complex, 191, 419
conjugate irrational, 149, 199
finding in word problems, 13–15, 74–75, 172
imaginary, 188, 190–194, 201–202
irrational, 140–141, 175–176, 195
line, 4, 7, 10–11, 71–72
pure imaginary, 188, 190
rational, 26, 89–90, 106–107, 115, 140–141, 195
real, 141, 188, 195, 390
Number line, 4, 7, 10–11, 71–72

Obtuse angle, 551
Opposites, 26, 96, 188, 201
Ordered pair, 209, 416
Ordered triple, 247, 254
Ordinate, 209
Origin, 10, 209, 306, 310, 319

Parabola, 314–318, 324–325, 327–328, 331, 417
axis of symmetry, 314–317
Parallel lines, 228
Pascal, Blaise, 356
Pentagon, 19
Perfect cube, 193
Perfect square, 139
Perfect square factor, 142–143
Perfect square trinomial, 60–61, 165
Perimeter, 18, 125
Period of a function, 500–501, 503–507, 509–510
Period of a simple pendulum, 181
Periodic function, 497
Permutation(s), 425–426, 428–429, 431–433, 445
Perpendicular lines, 229–230
Plane, coordinate, 209, 223–224
Point(s)
circle as set of, 306
collinear, 231
coordinates of, 209–210
determining whether on a line, 220–221
distance between two, 223–224
ellipse as set of, 310, 328
hyperbola as set of, 319
parabola as set of, 327
symmetric, 303–304
turning, 314, 416–417
Point-slope form of an equation of a line, 216–217
Polynomial(s), 37–39, 409–419
binomial factor of, 128–131

degree of, 37, 410, 416
division of, 128–131
equation, 409
factoring completely, 65–66, 409–411, 413–414
function, 416
integral, 409
multiplication of, 41, 60–61
quadratic, 37
subtraction of, 38–39
zero of, 409
Positive measure of angle of rotation, 455
Positive slope of a line, 213–214
Power(s)
of a complex number, 534
logs of, 400–401, 588–589
of a power, 32, 157
of a product, 32, 157
product of, 31–32, 157
of a quotient, 32, 157
quotient of, 31–32, 157
Power series, 385
Principal value, 531
Probability, 438–447
Problem solving
age, 20
applying formulas, 22–25
area, 78–80
compound interest, 407–408
digit, 268–271
electrical circuit, 23
geometric, 78–81, 175–176
height of object sent upward, 23, 25, 167, 173
maximum income, area, height, 167–168
mixture, 43–45
motion, 47–49, 122–123
non-routine problems, 3, 42, 46, 77, 82, 88, 101, 114, 118, 144, 151, 164, 177, 197, 222, 239, 287, 302, 323, 337, 340, 389, 405, 423, 430, 432, 434, 437, 472, 499, 505, 511, 538, 565
number, 13–17, 74–76, 172
perimeter, 18
quadratic equations, 74–75, 78–80, 172–173, 175–176
simple interest, 22
solving by using combined variation, 295
solving by using direct variation, 289
solving by using inverse variation, 292
solving by using joint variation, 295

627

solving by using the quadratic formula, 172–173, 175–176
solving by using a system of linear equations, 268–271
steps in solving word, 15, 172
work, 119–121
writing word phrases in algebraic terms, 13, 15
Problem-solving method
formulating a situation, 56
four steps, x
planning how to solve, 30
Product(s)
of complex numbers, 190–191, 201–202
formula for special, 60–61
logarithms of, 397–398, 588–589
of the means and extremes of a proportion, 117
nth root of, 142–143, 152–153
of a number and a logarithm, 401
of powers, 31–32, 157
in proportions, 117
of pure imaginary numbers, 190
of radicals, 145, 153, 159
of rational expressions, 92
of a scalar and a matrix, 260
of a scalar and a vector, 52
of solutions of a quadratic equation, 198–200
square root of, 142
of square roots, 145–146
of the sum and difference of two terms, 60
of two binomials, 41
of two matrices, 260–261
Progression(s), 357–358, 368–369
Proof(s), 106–107, 201–202
Property (properties)
absolute value, 10–11
additive identity, 26, 201, 258
additive inverse, 26, 96–97, 188–202, 258
associative, 26, 106, 202, 259, 262
binomial expansions, 382–383
closure, 26, 107, 201–202
commutative, 26, 201–202, 259
complex numbers, 187–192, 201–202
of conjugates, 149, 191, 199, 419
conjunctions, 7–9
discriminant, 195
disjunctions, 7–9
distributive, 2, 26, 41
of an ellipse, 329
equality for,
complex numbers, 201

exponentials, 33, 158
logarithms, 398
matrices, 257
equation for absolute value, 10
equation, linear, 1
exponential functions, 390–391
of exponents, 31–32, 34, 155–156
field, 26
i and $-i$, 187
inequality
of absolute value, 11
compound linear, 7–8
polynomial, 72
quadratic, 71
logarithmic functions, 395–396
of matrices, 257–262, 266
of matrix addition, 257–259
of matrix multiplication, 260–262
multiplicative identity, 26, 148, 202, 261
multiplicative inverse, 1, 26, 92, 202, 261
negative product of two factors, 71
nth root of a product, 142–143, 152
nth root of a quotient, 150, 153
positive product of two factors, 71
power
of a power, 32
of a product, 32
of a quotient, 32
powers
of i, 187, 191–192
of a binomial, 382–383
product
of powers, 31
of radicals, 145, 152
of scalar and matrix, 260
of scalar and vector, 52
of two matrices, 260
proof(s), 106–107, 201–202
proportion, 117
quotient of powers, 31
quotient of radicals, 148, 153
rational numbers, 89, 106–107
sum and product of solutions of quadratic equation, 198
of sum of two matrices, 258
vector, 50–52
zero-product, 68
Proportion, 117, 126
Pure imaginary numbers, 188, 190
Pythagorean identities, 520
Pythagorean relation, 78, 175, 223–224

Quadrant(s), 209
Quadrantal angle, 456
Quadratic equations, 68–70
coefficients of, 195–196
discriminant of, 195–196
factoring, 68–69
with imaginary solutions, 187–188, 193–194
nature of solutions, 195–196
product of solutions of, 198–200
solving by completing the square, 165–166
solving by using the quadratic formula, 169–170, 184
solving by using square root, 139–140
standard form of, 68
sum of solutions of, 198–200
and word problems, 74–75, 78–80, 172–173, 175–176
Quadratic formula, 169–170, 172–173, 175, 193, 195
Quadratic inequality (inequalities)
factoring, 71–72
graphing the solutions of, 71–72
solution set of, 71–72
Quadratic polynomial, 37
Quadratic systems, 341–342
Quadratic trigonometric equations, 515–516
Quadratic trinomial, 57
Quotient(s)
of complex numbers, 190–191
identities, 519
logarithm of, 397–398, 588–589
nth root of, 153
of numbers in factorial notation, 383–384, 431–432, 435–436
of polynomials, 128–131
power of, 32
of powers, 31, 157
of radicals, 148–150, 153–154
of rational expressions, 92

Radian measure, 523–527
Radical(s), 139, 152–153
combining, 145
cube roots, 152, 193–194
exponents as, 390–391, 408
fifth roots, 152
fourth roots, 152
logarithms of, 400–401, 588–589
nth root, 152
products of, 145–146, 153
quotients of, 148–150, 153–154
simplifying, 142–143, 145–146, 148–150, 152–154, 162
Radical equations, 178–179, 181, 373

Radical expression(s), 142–143, 145–146, 148–150, 152–156, 159
Radical form, 155
Radical sign, 139, 152
Radicand, 139, 152, 195
Radius, 306
Range
 of an exponential function, 390–391
 of a function, 279
 of a logarithmic function, 395–396
 of a relation, 276
 of a set of data, 104
Rational expression(s)
 addition of, 96–97
 complex, 99–100
 definition of, 91
 division of, 92
 evaluating, 91
 multiplication of, 92
 polynomial as, 91
 products of, 92
 quotients of, 92
 reciprocals of, 92
 simplifying, 91
 subtraction of, 96–97
 undefined, 91
 zero as denominator of, 91
Rational function, 280
Rational number(s), 26, 89–90, 106–107, 140–141, 195
Rational number exponents, 155–158
Rational zero theorem, 413
Rationalizing the denominator, 148–150, 153–154, 191
Ray, 455
Reading in Algebra, 6, 9, 12, 15, 20, 33, 36, 39, 64, 67, 70, 73, 93, 118, 123, 132, 141, 143–144, 150, 171, 188, 192, 197, 211, 215, 219, 241, 249, 284, 287, 348, 359, 370, 377, 396, 405, 408, 411, 417, 433, 436, 457, 465, 508, 532
Real number(s), 141, 188
Reciprocal(s), 1, 92, 106
 function(s), 468, 470
 identities, 518
 of n-factorial, 384
Rectangular hyperbola, 321
Reference angle, 456
Reference triangle, 456
Relation(s), 276–277
 and combined variation, 295
 and direct variation, 288–289
 domain of, 276
 graph of, 276–277
 inverse of, 285

 and inverse variation, 291–292
 and joint variation, 294
 range of, 276
Remainder, 128–131
Remainder theorem, 132
Repeating decimal, 89, 110, 140, 380
Resultant of vectors, 50–51, 540–541, 568–569
Reviews
 Chapter, 27, 53, 84, 108, 134, 160, 182, 203, 236, 272, 298, 332, 352, 386, 420, 448, 492, 535, 568
 Cumulative, 112–113, 206–207, 335–336, 452–453, 572–573
Right triangle, 78, 175, 223–224, 458–460
 hypotenuse of, 78, 175
 law of cosines and, 539–540
 Pythagorean relation and, 78, 175, 223–224
 trigonometry of, 486–487, 489–490
Root(s)
 cube, 152, 194
 of an equation, 1, 68
 fifth, 152
 fourth, 152
 logarithm of, 400–401, 588–589
 of negative numbers, 187–188
 nth, 152
 square, 139–140, 142–143, 145–146, 148–150, 187–188
Rotation, 455–456, 475
rth term of $(a + b)^n$, 383
Rule of signs, Descartes', 418–419

Sample space, 441–443
Scalar, 260
 and a matrix, 259
 and a vector, 52
Scientific notation, 35, 403
Secant, 470–471
Segment(s)
 endpoints of, 212, 226
 length of vertical or horizontal, 212–214
 midpoint of, 226
Series
 arithmetic, 364–366
 geometric, 375–377
 infinite, special, 384–385
 infinite geometric, 379–380
 infinite power, 385
Sigma (Σ, σ), 104–105, 365–366, 377
Simplest radical form, 142, 150

Sine(s), 462–464
 as cofunction of cosine, 478
 as complementary to cosine, 478
 of the difference of two angles, 553
 graph of function, 497–498
 of half an angle, 563
 as infinite series, 384
 law of, 546–547
 negative, 463
 positive, 463
 of the sum of two angles, 552
 of twice an angle, 554
Slope
 of a line, 212–214
 negative, 213–214
 of parallel lines, 228
 of perpendicular lines, 229–230
 positive, 213–214
 undefined, 214
 zero, 214
Slope-intercept form of an equation of a line, 217
Solution(s)
 of exponential equations, 33, 158, 390, 393, 406
 extraneous, 116, 178, 398, 401
 of fractional equations, 115–117
 imaginary numbers as, 193–194
 of linear equations, 1–2
 of linear inequalities by graphing, 4–5, 7–8
 of linear inequality systems, 346–347
 of linear-quadratic inequality systems, 349–350
 of linear-quadratic systems of equations, 338–339
 product of, 198–200
 of quadratic equations, 68–69, 139–140, 165–166, 169–170, 188, 193–194
 of quadratic systems, 341–342
 of radical equations, 178–179
 sum of, 198–200
 of two trigonometric equations by graphing, 507, 527
Solution set(s)
 of a compound linear inequality, 7–8
 of an inequality involving absolute value, 11
 of a linear inequality, 4–5
 of a quadratic inequality, 71–72
Special Topics
 careers, 138, 208, 275, 496
 normal (bell-shaped) curve, 538
 Pascal's Triangle, 356

standard deviation, 104–105
Square(s)
  of a binomial, 60
  completing, 165–168
  difference of, 62
  perfect, 139–140, 142
  perfect square trinomial, 60, 62, 165
Square matrix, 250, 261
Square root(s), 139–140
  approximating to the nearest hundredth, 140
  of an exponential expression, 143
  of a negative number, 187–188
  of a product, 142
  product of, 145–146
Standard deviation, 104–105
Statistics, 102–105
Substitution method for solving linear system of equations, 243
Sum(s)
  of complex numbers, 188
  of infinite geometric series, 379–380
  involving a common denominator, 96
  involving different denominators, 96–97
  of logarithmic terms, 397–398, 401
  of n terms of arithmetic series, 364–366
  of radicals, 145
  of solutions, 198–200
  of the terms of finite geometric series, 375–377
  of two cubes, 61, 63, 154, 159
  of two matrices, 258
Symmetric points, 303–304
Symmetry, 391, 395–396
Synthetic division, 130–131, 136
Synthetic substitution, 38, 131, 409, 412, 416

Tangent(s), 466–469
  amplitude of, 509–510
  of the difference of two angles, 561
  graph of, 509–510
  of half an angle, 564
  inverse of, 529–532
  negative, 467
  positive, 467
  of the sum of two angles, 560
  of twice an angle, 561–562
Terminating decimal, 89, 110, 140
Tests
  Chapter, 28, 54, 85, 109, 135, 161, 183, 204, 237, 273, 299, 333, 353, 387, 421, 449, 493, 535, 569
  College Prep, 29, 55, 87, 111, 137, 163, 185, 205, 238, 274, 301, 334, 355, 388, 422, 451, 495, 537, 571
Transformations, matrix, 265–266
Translation(s), 311, 322
Triangle(s)
  ambiguous case, 549–551
  area of, 79, 544–545
  isosceles, 459
  reference, 456
  right, 78, 175, 223–224, 458–460, 486–487, 489–490, 494
Triangular form, 265
Trigonometric equation(s)
  linear, 481–482
  quadratic, 515–516
Trigonometric expressions, evaluating using radian measure, 524–525
Trigonometric function(s), 462–475, 529–532
  amplitude of, 500–501, 503–504, 506–507, 509–510
  of angles of negative measure, 473–474
  graphs of, 497–498, 500–501, 503–504, 506–507, 509–510, 526–527
  interpolation with, 483–484
  inverses of, 529–532
  logarithms of, 512–513
  period of, 500–501, 503–507, 509–510
  values of, 475
Trigonometric identities, 518–522, 566–567
Trigonometric tables, 590–592, 594–596
Trigonometry
  interpolation with, 483–484
  of a right triangle, 486–487, 489–490
Trinomial(s), 37, 57–58, 60, 165
Turning point, 314, 416–417

Undefined rational expressions, 91
Upper bound, 412

Value(s)
  of functions, 279–280
  principal, 531
  of trigonometric functions, 475
Variance, 104
Variation(s)
  combined, 295
  direct, 288–289
  inverse, 291–292
  joint, 294
Vectors, 50–52
  collinear, 51
  difference, 51
  direction, 50
  equivalent, 50
  initial point, 50
  magnitude, 50
  opposite, 50
  scalar multiple, 52
  sum, 50
  terminal point, 50
Vertex (vertices)
  of an ellipse, 329
  of a parabola, 314
Vertical line(s), 214

Weierstrass, Karl, 454
Wet mixtures, 44–45
Word phrases in algebraic terms, 13–15
Word problems, see Problems

x-axis, 209
  and symmetric points, 303
x-coordinate, 209
x-intercept,
  of an ellipse, 310, 329
  of a hyperbola, 319–321

y-axis, 209
  and symmetric points, 304
y-coordinate, 209
y-intercept, 217
  of an ellipse, 310, 329
  of a hyperbola, 319–322

Zero, 26, 34, 91, 167, 201, 214
Zero factorial, 383, 436
Zero of a polynomial, 409, 412–413, 416, 418–419
Zero-product property, 68–69
Zero(s) of a rational function, 280